AF376187

Infinite Science
Publishing

Student Conference Proceedings 2022

11th   Conference on Medical Engineering Science
7th   Conference on Medical Informatics
5th   Conference on Biomedical Engineering
4th   Conference on Auditory Technology
2nd   Conference on Biophysics
2nd   Conference on Robotics and Autonomous Systems

Lübeck, March 9-11, 2022

Editors in Chief

T. M. Buzug, H. Handels, C. Hübner, S. Klein, A. Mertins, P. Rostalski

Associate Editors

C. Debbeler, K. Gräfe, J.-H. Wrage, K. Ehlers, R. Pallenberg, Y.-H. Song, S. Venker

Editors

H. Abbas, S. Abdul-Karim, M. Ahlborg, E. Barth, P. Bartmann, M. Berekovic, R. Birngruber, H. Botterweck, R. Brinkmann,
H. Busch, T. M. Buzug, J. Ehrhardt, F. Ernst, S. Fischer, T. Friedrich, H. Gehring, J. Ghofrani, M. Gräser, M. Grzegorzek,
T. Gutsmann, H. Hamann, H. Handels, M. Heinrich, H. Hellbrück, M. Henke, R. Huber, C. Hübner, H. Husstedt, G. Hüttmann,
J. Ingenerf, T. Jürgens, S. Karpf, S. Klein, F.-L. Lau, M. Leucker, N. Linz, K. Lüdtke-Buzug, T. Martinetz, A. Mertins,
S. Müller, N. T. Nguyen, H. Paulsen, M. Rafecas, M. Rahlves, P. Rostalski, E. Rückert, G. Schildbach, J. Schröter,
A. Schweikard, Y.-H. Song, F. Spitzenberger, M. Urban, J. Weil, R. Wendlandt, M. Zille

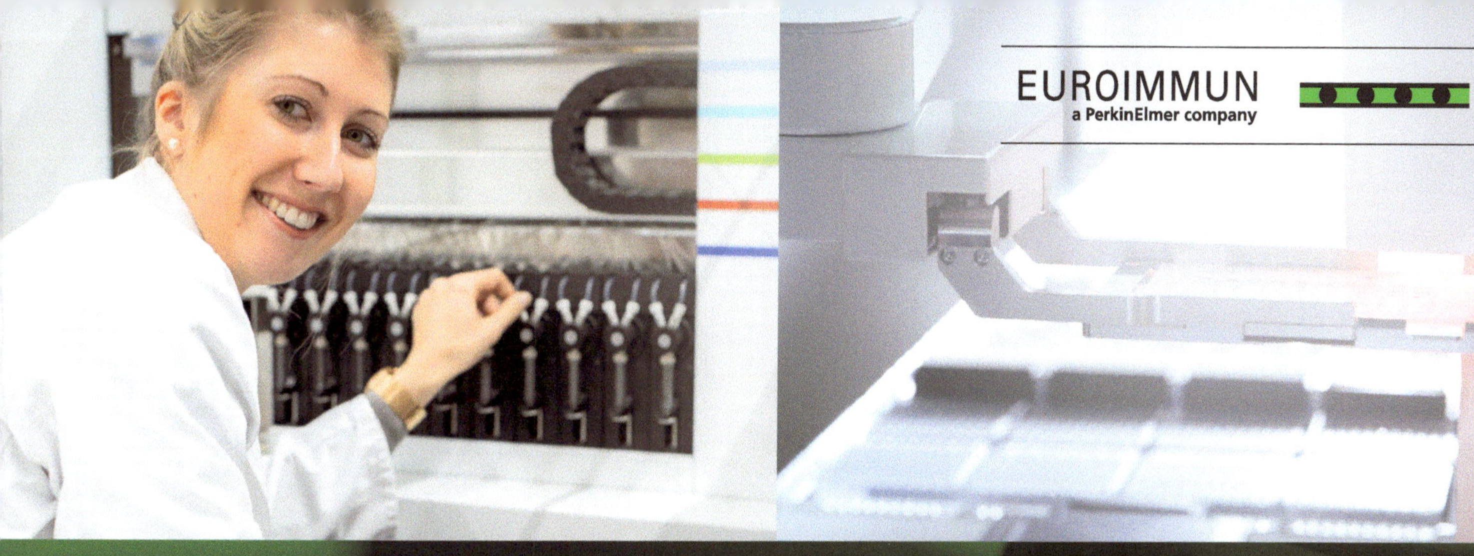

LABORAUTOMATEN & SOFTWARE FÜR DIE MODERNE DIAGNOSTIK

EUROIMMUN gilt als einer der führenden Hersteller von Testsystemen, Geräten und Software für die medizinische Labordiagnostik. Unsere Produkte dienen der Gesundheit des Menschen. Im Vordergrund steht die Entwicklung und Produktion von Testsystemen, mit denen im Serum von Patienten verschiedenste Antikörper bestimmt und dadurch Autoimmun- und Infektionskrankheiten sowie Allergien diagnostiziert werden können.

Ergänzend dazu entwickelt EUROIMMUN Laborgeräte und Softwarelösungen zur Automatisierung und Digitalisierung der Laborprozesse. Dazu gehören zum Beispiel Geräte zur vollautomatisierten Abarbeitung der EUROIMMUN-Testpalette sowie Liquidhandling-Systeme. Wir stützen uns dabei auf modernste, zum Teil weltweite patentierte Produktionsverfahren und Mikroanalysetechniken.

Weltweit arbeiten mehr als 6.000 Labore in über 150 Ländern mit unseren Produkten. So leisten wir täglich einen wertvollen Beitrag zur globalen Gesundheit und behalten stets unser Ziel vor Augen, den Menschen mit unserer Arbeit zu helfen.

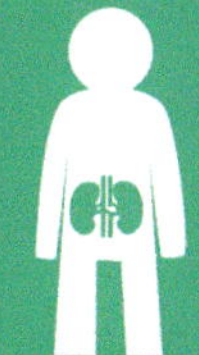

B. BRAUN AVITUM
AT A GLANCE

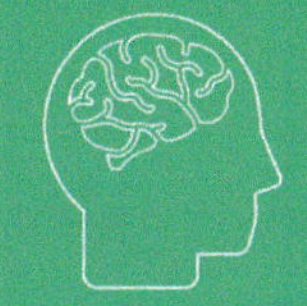

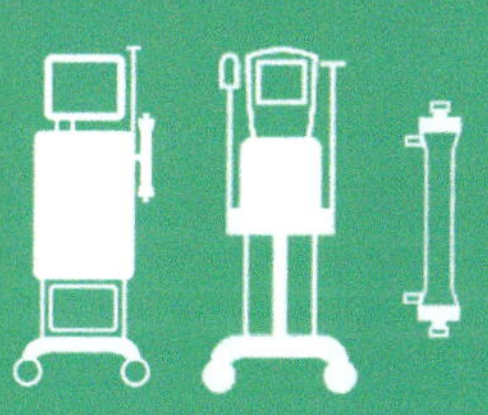

PHILIPS
Careers
Don't just
make a living
Make a
difference
philips.to/campus_de

OP Technologie &
Medizinrobotik

Bildgebung & Visualisierung

Elektronik & Instrumentierung

Modellierung & Digital Twins

Künstliche Intelligenz

Additive Fertigung

Studien & Regulatory Affairs

Algorithmendesign &
Modularisierung

Zelltechnik, Bio- und
Lebensmitteltechnologie

Fraunhofer
IMTE

Fraunhofer-Einrichtung für
Individualisierte und Zellbasierte
Medizintechnik IMTE

Wir bieten Abschlussarbeiten, Praktika oder
einen Direkteinstieg in die angewandte Forschung.
Bewirb dich bei uns am
Fraunhofer IMTE!

jobs@imte.fraunhofer.de

Student Conference Proceedings 2022

Conference Chairs

Thorsten M. Buzug	Fraunhofer IMTE and Institute of Medical Engineering, Universität zu Lübeck
Heinz Handels	Institute of Medical Informatics, Universität zu Lübeck
Christian Hübner	Institute of Physics, Universität zu Lübeck
Stephan Klein	Medical Sensors and Devices Laboratory, Lübeck University of Applied Sciences
Alfred Mertins	Institute for Signal Processing, Universität zu Lübeck
Philipp Rostalski	Fraunhofer IMTE and Institute for Electrical Engineering in Medicine, Universität zu Lübeck
Hartmut Gehring	Department of Anesthesiology, UKSH, Lübeck

Local Coordination

Christina Debbeler	Institute of Medical Engineering, Universität zu Lübeck
Ksenija Gräfe	Institute of Medical Engineering, Universität zu Lübeck
Jan-Hinrich Wrage	Institute of Medical Informatics, Universität zu Lübeck
Kristian Ehlers	Institute of Computer Engineering, Universität zu Lübeck
René Pallenberg	Institute for Signal Processing, Universität zu Lübeck
Young-Hwa Song	Institute of Physics, Universität zu Lübeck
Silke Venker	Medical Sensors and Devices Laboratory, Lübeck University of Applied Sciences

Scientific Program Committee

Abbas, Dr. Hossam	Institute for Electrical Engineering in Medicine, Universität zu Lübeck
Abdul-Karim, Saif	Medical Sensors and Devices Laboratory, Lübeck University of Applied Sciences
Ahlborg, Dr.-Ing. Mandy	Fraunhofer IMTE, Lübeck
Barth, Prof. Dr. Erhardt	Institute for Neuro- and Bioinformatics, Universität zu Lübeck
Bartmann, Peter	Department of Electrical Engineering and Computer Science, Lübeck University of Applied Sciences
Berekovic, Prof. Dr.-Ing. Mladen	Institute of Computer Engineering, Universität zu Lübeck
Birngruber, Prof. Dr. Reginald	Institute of Biomedical Optics, Universität zu Lübeck
Botterweck, Prof. Dr. Henrik	Department of Applied Natural Sciences, Lübeck University of Applied Sciences
Brinkmann, Dr. Ralf	Medizinisches Laserzentrum Lübeck GmbH and Institute of Biomedical Optics, Universität zu Lübeck
Busch, Prof. Dr. Hauke	Lübeck Institute of Experimental Dermatology, Universität zu Lübeck
Buzug, Prof. Dr. Thorsten M.	Fraunhofer IMTE and Institute of Medical Engineering, Universität zu Lübeck
Ehrhardt, Dr. Jan	Institute of Medical Informatics, Universität zu Lübeck
Ernst, Prof. Dr. Floris	Institute for Robotics and Cognitive Systems, Universität zu Lübeck
Fischer, Prof. Dr. Stefan	Institute of Telematics, Universität zu Lübeck
Friedrich, Dr. Thomas	Fraunhofer IMTE, Lübeck
Gehring, Prof. Dr. Hartmut	Department of Anesthesiology, University Medical Center Schleswig-Holstein
Ghofrani, Dr. Javad	Institute of Computer Engineering, Universität zu Lübeck
Gräser, Prof. Dr. Matthias	Fraunhofer IMTE and Institute of Medical Engineering, Universität zu Lübeck
Grzegorzek, Prof. Dr. Marcin	Institute of Medical Informatics, Universität zu Lübeck
Gutsmann, Prof. Dr. Thomas	Research Center Borstel, Leibniz Lungcenter
Hamann, Prof. Dr.-Ing. Heiko	Institute of Computer Engineering, Universität zu Lübeck
Handels, Prof. Dr. Heinz	Institute of Medical Informatics, Universität zu Lübeck
Heinrich, Prof. Dr. Mattias	Institute of Medical Informatics, Universität zu Lübeck
Hellbrück, Prof. Dr. Horst	Department of Electrical Engineering and Computer Science, Lübeck University of Applied Sciences and Institute of Telematics, Universität zu Lübeck

Scientific Program Committee - Cont'd

Henke, Dr. Maria	Institute for Robotics and Cognitive Systems, Universität zu Lübeck
Huber, Prof. Dr. Robert	Institute of Biomedical Optics, Universität zu Lübeck
Hübner, Prof. Dr. Christian	Institute of Physics, Universität zu Lübeck
Husstedt, Dr. Hendrik	German Institute of Hearing Aids, Lübeck
Hüttmann, Prof. Dr. Gereon	Institute of Biomedical Optics, Universität zu Lübeck
Ingenerf, Prof. Dr. Josef	Institute of Medical Informatics, IT Center for Clinical Research, Universität zu Lübeck
Jürgens, Prof. Dr. Tim	Department of Applied Natural Sciences, Lübeck University of Applied Sciences
Karpf, Prof. Dr. Sebastian	Institute of Biomedical Optics, Universität zu Lübeck
Klein, Prof. Dr. Stephan	Medical Sensors and Devices Laboratory, Lübeck University of Applied Sciences
Lau, Dr. Florian-Lennert	Institute of Telematics, Universität zu Lübeck
Leucker, Prof. Dr. Martin	Institute for Software Engineering and Programming Languages, Universität zu Lübeck
Linz, Dr. Norbert	Institute of Biomedical Optics, Universität zu Lübeck
Lüdtke-Buzug, Dr. Kerstin	Institute of Medical Engineering, Universität zu Lübeck
Martinetz, Prof. Dr. Thomas	Institute for Neuro- and Bioinformatics, Universität zu Lübeck
Mertins, Prof. Dr. Alfred	Institute for Signal Processing, Universität zu Lübeck
Müller, Prof. Dr. Stefan	Medical Sensors and Devices Laboratory, Lübeck University of Applied Sciences
Nguyen, Ph.D. Ngoc Thinh	Institute for Robotics and Cognitive Systems, Universität zu Lübeck
Paulsen, PD Dr. Hauke	Institute of Physics, Universität zu Lübeck
Rafecas, Prof. Dr. Magdalena	Institute of Medical Engineering, Universität zu Lübeck
Rahlves, Prof. Dr. Maik	Institute of Biomedical Optics, Universität zu Lübeck
Rostalski, Prof. Dr. Philipp	Fraunhofer IMTE and Institute for Electrical Engineering in Medicine, Universität zu Lübeck
Rückert, Prof. Dr. Elmar	Institute for Robotics and Cognitive Systems, Universität zu Lübeck
Schildbach, Prof. Dr. Georg	Institute for Electrical Engineering in Medicine, Universität zu Lübeck
Schröter, Jörg	Medical Sensors and Devices Laboratory, Lübeck University of Applied Sciences
Schweikard, Prof. Dr.-Ing. Achim	Institute for Robotics and Cognitive Systems, Universität zu Lübeck
Song, Dr. Young-Hwa	Institute of Physics, Universität zu Lübeck
Spitzenberger, Prof. Dr. Folker	Fraunhofer IMTE and Centre for Regulatory Affairs in Biomedical Sciences, Lübeck University of Applied Sciences
Urban, Prof. Dr. Max	Medical Sensors and Devices Laboratory, Lübeck University of Applied Sciences
Weil, Prof. Dr. Joachim	Medical Clinic II - Cardiology and Angiology of the Sana Clinics Lübeck
Wendlandt, Dr. Robert	Clinic for Orthopedic and Trauma Surgery, Biomechanics Laboratory, University Medical Center Schleswig-Holstein, Lübeck
Zille, Dr. Marietta	Institute for Experimental and Clinical Pharmacology and Toxicology, Universität zu Lübeck

Preface and Acknowledgements

After the great success of the previous meetings from 2012 to 2021, the Student Conference 2022 shows continuing growth both in quality and quantity of scientific contributions. In this year, the 11th Student Conference on Medical Engineering Science is held together with the 7th Student Conference on Medical Informatics, the 5th Student Conference on Biomedical Engineering, the 4th Student Conference on Auditory Technology, the 2nd Student Conference on Biophysics and the 2nd Student Conference on Robotics and Autonomous Systems.

The organization team has worked to provide an excellent conference, where master students of the campus present their recent research results to a broad public of academics and industry.

The contributions show how new approaches and methods in medical engineering and medical informatics can advance medicine, health, and health care. Moreover, this conference offers a good opportunity for both students and companies to get in touch at the Recruiticon, i.e. a satellite recruiting fair with industrial exhibition.

Students from the Life Sciences programs present their results from projects carried out at the laboratories, clinics, and institutes of Lübeck's Universities, in international research facilities, or research-oriented industrial companies. The conference focus has been placed on topics ranging from medical engineering to medical informatics. The interdisciplinary field of medical engineering has been established at the Lübeck University of Applied Sciences for decades, and Medical Engineering Science (Medizinische Ingenieurwissenschaft – MIW) is an important bachelor and master program at the Universität zu Lübeck as well. Both universities jointly offer the international master degree programs Biomedical Engineering (BME) and Auditory Technology (Hörakustik und Audiologische Technik – HAT). Furthermore, in the master program Medical Informatics (Medizinische Informatik – MI) the Student Conference on Medical Informatics is integrated as an important element where project results in the emerging field of digital medicine are presented by the students. These subject areas are widened by the master's programs Robotics and Autonomous Systems (Robotik und Autonome Systeme – RAS) and Biophysics (Biophysik – BP).

As Conference Chairs, we want to thank everybody who worked with enthusiasm and dedication to make the conference a successful event. We want to thank the companies who support the meeting. Moreover, our thanks go to Infinite Science GmbH for producing these proceedings and organizing the Recruiticon meeting supporting the Student Conference.

Personally and on behalf of all colleagues of the Student Conference Committee, we especially want to thank Christina Debbeler and Ksenija Gräfe, they have been the central contact points for all questions of students and the program committee members, as well as Jan-Hinrich Wrage for editing the proceedings and furthermore Silke Venker, René Pallenberg, Kristian Ehlers and Young-Hwa Song. Their in-depth overview of all details of this event is the key to the success of the conference.

.

Lübeck, March 9-11, 2022

Prof. Dr. Thorsten M. Buzug
Chair of the 11th Student Conference on Medical Engineering Science

Prof. Dr. Heinz Handels
Chair of the 7th Student Conference on Medical Informatics

Prof. Dr.-Ing. Stephan Klein
Chair of the 5th Student Conference on Biomedical Engineering

Prof. Dr.-Ing. Alfred Mertins
Chair of the 4th Student Conference on Auditory Technology

Prof. Dr. Christian Hübner
Chair of the 2nd Student Conference on Biophysics

Prof. Dr.-Ing. Philipp Rostalski
Chair of the 2nd Student Conference on Robotics and Autonomous

Contents

Sensor Data Analysis

Regulatory Affairs

Biochemical Physics

Safety and Quality

Machine Learning / AI

E-Health

Image Processing

Biomedical Engineering

Biomechanics

Biomedical Optics

Medical Imaging

Medical Electronics

Auditory Technology

1

Sensor Data Analysis

Feature Augmentation and Functional Improvement of ROS-Mobile

Nico Studt [1], Nils Rottmann [2]
[1] Robotics and Autonomous Systems, University of Luebeck, nico.studt@student.uni-luebeck.de
[2] Institute for Robotics and Cognitive Systems, University of Luebeck, nils.rottmann@rob.uni-luebeck.de

Abstract

Monitoring and controlling is an increasingly requested capability among users, developers and researchers of robotic systems. Even working with the open source Robot Operating System framework, especially in conjunction with mobile robot control in wide-area environments, still can be a challenging task. Therefore, ROS-Mobile an Android application with a continuously growing number of users was introduced in 2020. This work aims to extend the functionality of the app and present a more structured system to implement additional tools namely a battery indicator, a dynamic time plot and a 2D layer visualization to meet the requirements and requests of the app users. Future extensions can be added more easily with the proposed interface structure.

1 Introduction

The Robot Operating System (ROS) was developed as a robotic middleware system which simplified hardware abstraction, message-passing between individual processes, package management and other tasks [1]. Since then, ROS has become the open source alternative to other frameworks and the quasi-standard in the robotics community. Many robotic systems, and particularly mobile robots during development and testing phases, use the ROS framework for navigation and control. Often additional controllers and devices are required to monitor these robots, especially while dealing with operation in outdoor environments. As a solution to missing support of ROS applications for mobile devices, such as smartphones or tablets, *ROS-Mobile* [2] was introduced, which is an open source Android application with an intuitive user interface for controlling and monitoring robotic systems utilizing the *rosjava* [3] framework. ROS-Mobile is constantly developed and fits the needs for a large variety of tasks with different kinds of robots. With a growing number of users of the app, including researchers and robotics developers, functionality requirements and demand for advanced tools are increasing. One of the most requested features is the ability to visualize multiple, transformed data in a combined tool comparable to *Rviz* [4] as an addition to ROS-Mobile, but the proposed widget structure of the app complicates the implementation of more sophisticated tool chains, e.g. receiving and displaying multiple different ROS message types. Therefore this work presents a reworked tool structure and implementation interfaces to simplify the creation and integration of new widgets in ROS-Mobile. Section 2 describes how tools are categorized and advanced features can be encapsulated in a more organized tree based structure using the introduced architecture. The presented tools in section 3, show the appliance and capabilities of the new structure.

2 Tool Categorization

In order to comply with the requirements of the app functionality and to ensure easy extensibility, it was necessary to adapt the tool structure. The widget structure of the previous app version prevented the creation of more complex functionalities such as a multi-layer display without implementing a separate dedicated widget for every layer. Furthermore, besides simple publisher/subscriber widgets, adding widgets not connected to the ROS network, e.g. a label, was not possible directly, only with great effort and unnecessary overhead. Existing as well as future tools are now divided into categories based on their dependency and are each organized in a tree structure. Widgets are standalone tools, whereas layers belong to a widget but contain an independent function, e.g. Touch Goal Layer (Section 3.3, Item 5).

2.1 Interaction Types

Depending on the interaction with the ROS system, widgets and layers (referred as entities hereafter) are additionally divided into the following subcategories:

1. Publisher entities send data in the form of ROS messages to a topic selected by the app user. One example usage is the control of a mobile robot with velocity commands.

2. Subscriber entities receive ROS messages, process them and/or present them visually in their entirety or

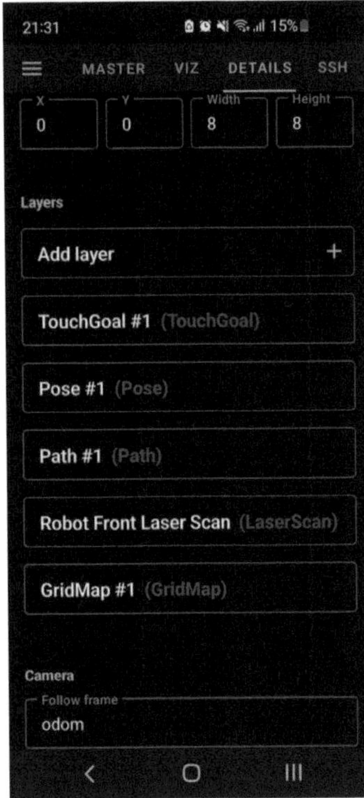 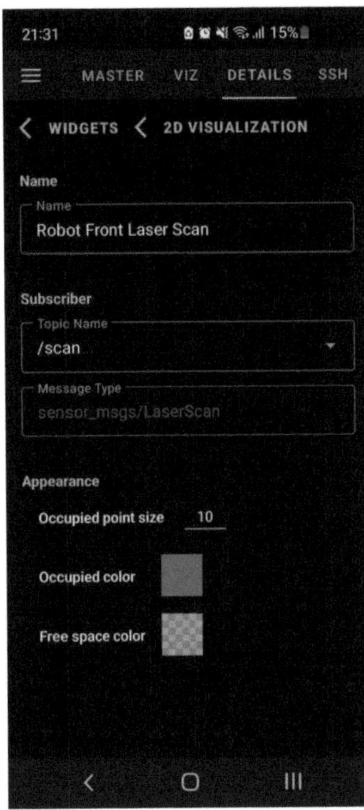 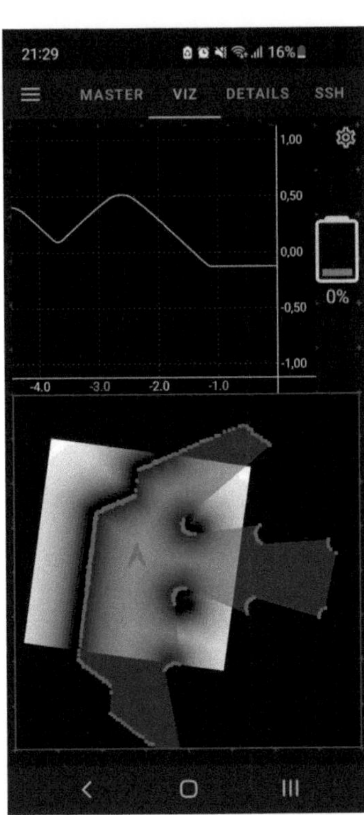

Figure 1: Illustrative example using ROS-Mobile with the advanced entity system and new tools for control and monitoring of a mobile Robot. From left to right: (a) List of visible layers, added to a 2D visualization widget, (b) detail page of a laser scan layer with its configurable settings and (c) the visualization tab showing a live view during an experiment with multiple widgets (here Rqt-plot, battery indicator and stacked 2D layers).

in partial form, such as the dynamic Rqt-plot in section 3.2.

3. Silent/Anonymous entities represent the tools not interacting with the ROS system. They can be used, for instance, to improve the clarity of the user interface or to provide additional information like a text label or a counter.

4. Group entities serve as containers that pass user inputs as well as messages to child entities, but like the 2D layer visualization in section 3.3, can also function as standalone widgets.

2.2 Required Interfaces

Each implementation of a new entity must follow a specific interface structure introduced in version 2.0 of the app, which serves as additional guidance for contributors to ensure reliable functionality and maintainability of the app. An entity model defines the interaction type in conjunction with the entity type, all possible settings and default values. User interaction with the options is handled in an entity detail view holder based on the input types (as seen in Fig. 1b) of the according setting tab in the app. Finally the visual appearance and gesture controls are implemented in an entity view class. For a more detailed information of the actual implementation, see the guide and examples at the ROS-Mobile wiki page [5].

3 Additional Entities

In response to feedback and improvement requests from ROS Mobile users, either as issues on the Github page or as comments on the app's page in the Google Play Store, the existing widgets have been revised and additional ones added. The new entities, which are described in more detail in the following subsections, include a battery status readout tool, a dynamic scaling time plot, and the ability to display and interact with multiple two-dimensional sensor or state data sets.

3.1 Battery Indicator

Figure 2: Visual representation of the battery indicator and its various statuses in the app.

The battery widget provides the app user with the possibility to visually check the battery status of e.g. a robot or a device by subscribing to a BatteryState [6] ROS topic. An icon indicates the current state of charge divided in five partitions from 0% to 100% or whether the battery is currently charging as seen in Fig. 2. In addition, a text display optionally shows either the charge percentage or the voltage of the battery.

3.2 Dynamic Rqt-Plot

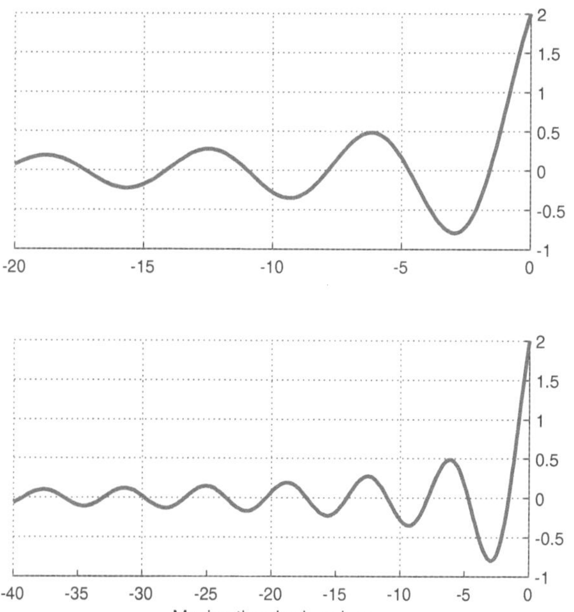

Figure 3: Different configurations of the Rqt-plot with dummy data.

Visualizing raw data points in a time plot is crucial for a realtime debugging app. The new Rqt-plot widget allows the user to select a numeric subfield of any subscribable ROS topic to display its data in a moving plot. The updating process may be stopped to be able to analyse the latest data. Via pinch and stretch gestures, the maximum time horizon can be controlled so that the data can be displayed in more or less detail (Fig. 3). The upper and lower limits of the Y axis are calculated regularly based on the currently stored data and rounded to the most visually appealing numbers e.g. multiples of $0.1, 1, 10, \dots$.

3.3 2D Layer Visualization

Similar to the visual presentation of ROS messages in Rviz [4] this work introduce a tool, which is capable of transforming data from its coordinate frame into a global reference frame of a two dimensional coordinate system using the transform library [7]. Drag and zoom gestures are supported to change the camera frame and perspective. Multiple layer entities with different interaction types can be added and configured to match the user needs as in Fig. 4. Listed below are the layers created as part of this work:

1. The *Grid Map Layer* transforms received messages of type *OccupancyGrid* [8] into rectangular map patterns. Different color coding of the map cells may be added later, currently displayed in grayscale colors. This

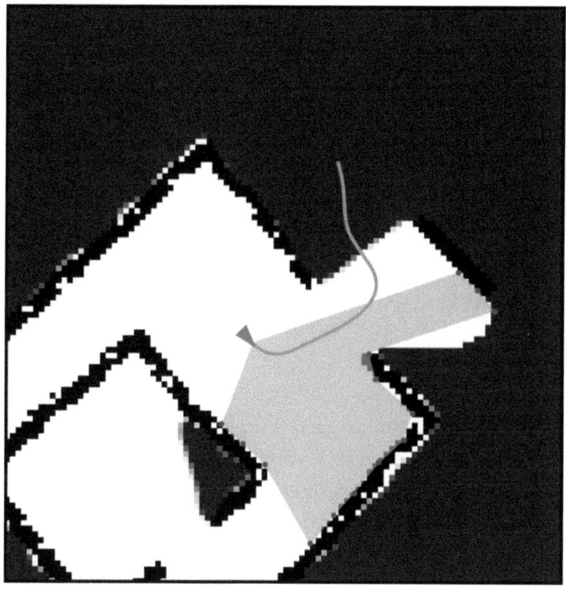

Figure 4: Example view of the 2D visualization widget with multiple visible layers which shows an occupancy grid map, robot pose, planned global trajectory and current laser scan.

layer replaces the standalone *Grid Map Widget* of the previous app version.

2. Laser scans from *Laserscan* [9] messages can be displayed either as occupied area or as measured points in the respective coordinate frames.

3. *Pose* [10] messages are forwarded to a *Pose Layer*, which draws a triangle shaped marker with its orientation given by the pose data. More shapes and optional color schemes are planned.

4. The *Path Layer* subscribes to a *Path* [11] topic and displays multiple waypoints as connected lines which may be used e.g. to visualize the planned trajectory of a robot.

5. The app user should not only be able to visualize data from a robot, but also interact with it. Therefore, a function to send orientation data via touch gesture was added to the 2D layer widget. Activated by a double tap and drag gesture, a *PoseStamped* [10] message is published with the touched position relative to a selected transformation frame and oriented in the dragged direction.

4   Conclusion

The reworked entity system and several new tools were presented in this paper to fulfill feature requests and requirements of the users of ROS-Mobile and to simplify future tool implementation for main developers and other contributors by introducing a guide in form of several needed interfaces. The additional widgets and layers presented in this work enhance the functionality by providing new visual debug tools and robot control options. In the course of the ongoing development of the app, further improvements are planned, such as more configurable entity settings and integration of the latest ROS version 2 communication interface. With the restructured entity system various new tools like a three dimensional visualization, diagnostic display and support of hardware and visual sensors of mobile devices can easily be added.

Acknowledgement

This project was supervised by the Institute for Robotics and Cognitive Systems, University of Luebeck and has received funding from the Deutsche Forschungsgemeinschaft (DFG, German Research Foundation) No #430054590 (TRAIN) and from the Bundesministerium für Bildung und Forschung (BMBF, Federal Ministry of Education and Research) No #13N14862 (A-DRZ).

Author's Statement

We wish to confirm that there are no known conflicts of interest associated with this publication and there has been no significant financial support for this work that could have influenced its outcome.

5 References

[1] M. Quigley, K. Conley, B. Gerkey, J. Faust, T. Foote, J. Leibs, R. Wheeler, A. Y. Ng *et al.*, "Ros: an open-source robot operating system," in *ICRA workshop on open source software*, vol. 3, no. 3.2. Kobe, Japan, 2009, p. 5.

[2] N. Rottmann, N. Studt, F. Ernst, and E. Rueckert, "Ros-mobile: An android application for the robot operating system," *arXiv preprint arXiv:2011.02781*, 2020.

[3] D. Kohler and K. Conley, "rosjava–an implementation of ros in pure java with android support," 2011.

[4] H. R. Kam, S.-H. Lee, T. Park, and C.-H. Kim, "Rviz: a toolkit for real domain data visualization," *Telecommunication Systems*, vol. 60, no. 2, pp. 337–345, 2015.

[5] N. Studt, "Ros-mobile guide," (Date last accessed 22-Jan-2022). [Online]. Available: https://github.com/ROS-Mobile/ROS-Mobile-Android/wiki/How-to-contribute%3F#create-a-new-widget

[6] Autogenerated, "Batterystate message," (Date last accessed 20-Jan-2022). [Online]. Available: http://docs.ros.org/en/melodic/api/sensor_msgs/html/msg/BatteryState.html

[7] T. Foote, "tf: The transform library," in *Technologies for Practical Robot Applications (TePRA), 2013 IEEE International Conference on*, ser. Open-Source Software workshop, April 2013, pp. 1–6.

[8] Autogenerated, "Occupancygrid message," (Date last accessed 20-Jan-2022). [Online]. Available: http://docs.ros.org/en/melodic/api/nav_msgs/html/msg/OccupancyGrid.html

[9] ——, "Laserscan message," (Date last accessed 20-Jan-2022). [Online]. Available: https://docs.ros.org/en/melodic/api/sensor_msgs/html/msg/LaserScan.html

[10] ——, "Posestamped message," (Date last accessed 20-Jan-2022). [Online]. Available: http://docs.ros.org/en/melodic/api/geometry_msgs/html/msg/PoseStamped.html

[11] ——, "Path message," (Date last accessed 20-Jan-2022). [Online]. Available: http://docs.ros.org/en/melodic/api/nav_msgs/html/msg/Path.html

Combination of two binary sensors to determine the inspiratory oxygen concentration in ventilation devices
– A theoretical analysis –

Mike Gugel[1], Annika Polert[2], Christian Neuhaus[3], Matthias Graeser[4,5]
[1] Medical Engineering Science, University of Lübeck, mike.gugel@student.uni-luebeck.de
[2] Weinmann Emergency Medical Technology GmbH + Co. KG, A.Polert@weinmann-emt.de
[2] Weinmann Emergency Medical Technology GmbH + Co. KG, C.Neuhaus@weinmann-emt.de
[4] Institude of Medical Engineering, University of Lübeck, matthias.graeser@imte.fraunhofer.de
[5] Fraunhofer Research Institution for Individualized and Cell-based Medical Engineering

Abstract

The monitoring of the inspiratory oxygen FiO2 is not standard in ventilation devices for emergency use. Current solutions with oxygen sensors have disadvantages in terms of cost or durability or are not suitable for portable devices. In particular, the FiO2 measurement is difficult when the gas mixture consists of more than two components and contains humidity. In this work, a new concept for the determination of FiO2 is developed. Two binary sensors for gas measurement are combined together with a temperature and a humidity sensor to a sensor-fusion model. The sensor-fusion is implemented in a theoretical model, while the influence of humidity is compensated using a correction term. Propagation of uncertainty is calculated for the validation of the sensor-fusion model. Based on an error calculation of the sensor-fusion model, the requirements for the sensors are established to fulfill the request for oxygen measurements in medical devices

1 Introduction

Currently, ventilators in clinical medicine determine the inspiratory oxygen FiO2 for better monitoring. In DIN EN ISO 80601-2-55 [1] the requirements for respiratory gas measurement in medical devices are defined. It is not standard practice for emergency ventilators to measure FiO2 due to the additional price, dimension, ruggedness and energy consumption requirements. This work developes and analyzes a promising concept to determine the FiO2 value, fulfilling the requirements, which defines the maximum deviation of the oxygen measurement. This sensor-fusion concept combines two non-selective sensors based on different physical principles. Several possible sensors are compared in terms of commercial availability, price, dimension, required performance, and durability. The desired gas compositions are mixed together in the ventilator by two adjustable valves of atmospheric air and medical oxygen. To achieve several mixtures, atmospheric air is used, consisting of nitrogen and oxygen, while the humidity was controlled using a water vaper.

1.1 Fundaments of the sensor principles

The thermal conductivity λ is specific for every gas (Table 1). In a static thermal condition, the energy transfer per time is

$$\dot{Q} = \lambda \frac{A}{s}(T_1 - T_2). \qquad (1)$$

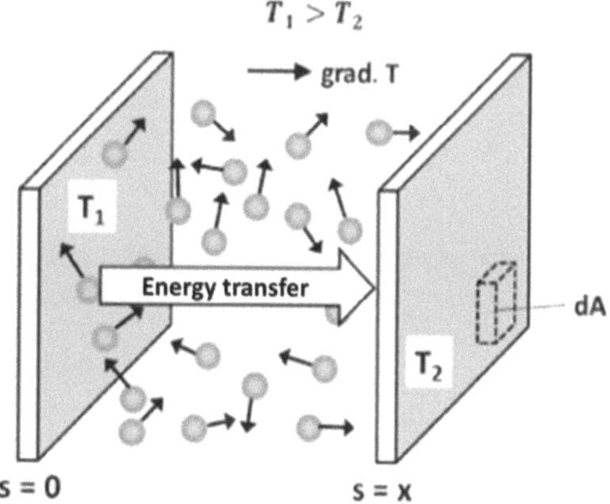

Figure 1: Heat transfer by thermal conduction. Figure taken from [2].

As shown in Figure 1, A is the area of the sensor plate, s is the distance between the plates and $\Delta T = \|T_1 - T_2\|$ is the temperature difference between the plates. A higher thermal conductivity results in a higher $\dot{Q}$. The energy transfer per time is measured with the power needed to keep the heat plate at a constant temperature T_1. The entire energy transfer per time is additionally influenced due to the thermal radiation, the convection, and the electrical connection wires. These effects for the energy transfer per time can be assumed to be constant and independent of the gas [2]. Table 1 shows the thermal conductivities, the molar mass, the

Table 1: Specifications of relevant gases [3] (at a temperature of 300° C)

Gas	λ	M	C_P	η
	$m\mathrm{W}/(\mathrm{mK})$	$\mathrm{g\,mol}^{-1}$	$\mathrm{J\,K}^{-1}$	$\mu\mathrm{Pa\,s}$
Oxygen	45.81	32.00	0.92	35.79
Nitrogen	42.72	28.01	1.04	34.74
Argon	48.58	39.95	0.52	39.82
Water vapor	40.57	18.02	1.87	20.57

heat capacity C_p and the dynamic viscosity η for different gases [3]. The thermal conductivity of serveral mixture of gases is well determined in the literature and is described as

$$\Lambda = \sum_i \frac{x_i \lambda_i}{\sum_j x_i F_{ij}} \qquad (2)$$

where

$$F_{ij} = \frac{\left[1 + (\eta_i/\eta_j)^{1/2}(M_j/M_i)^{1/4}\right]}{\sqrt{8(1 + M_i/M_j)}} \qquad (3)$$

with the partial gas concentrations x_i and the particular values for the thermal conductivity λ, the molar mass M, and the dynamic viscosity η [2].

To gain more information about the mixture the speed of sound within the measurement chamber is determined by measuring the phase shift in an acoustic medium. The sound is transmitted into a measurement tube at a transmission frequency and is received after a certain distance [2]. The speed of sound c for a gas mixture is calculated for an ideal gas, and low-frequency by the equation

$$c = \sqrt{\frac{RT}{\sum_i x_i M_i} \frac{\sum_i x_i C_{Pi}}{\sum_i (x_i C_{Pi} - R)}}, \qquad (4)$$

with the universal gas constant R, the temperature T and the heat capacity C_P. Furthermore, the temperature and the humidity are determined [4].

Based on a theoretical physical model [5], the absolute humidity H is defined depending on the relative humidity ϕ as

$$H = \frac{\rho_\mathrm{W}}{\rho_\mathrm{A} + \rho_\mathrm{W}} \qquad (5)$$

where

$$\rho_\mathrm{W} = \phi \cdot \frac{611.2\,\mathrm{hPa} \cdot \exp\left(\frac{22.46 \cdot (T - 273.15\,\mathrm{K})}{T - 30.03\,\mathrm{K}}\right)}{461.51\,(\mathrm{J/kg\,K}) \cdot T},$$

$$\rho_\mathrm{A} = \frac{1013.2\,\mathrm{hPa} - 611.2\,\mathrm{hPa} \cdot \exp\left(\frac{22.46 \cdot (T - 273.15\,\mathrm{K})}{T - 30.03\,\mathrm{K}}\right)}{R_\mathrm{A} \cdot T}.$$

ρ_W is the density of water vapor and ρ_A is the density of dry air. The molar gas constant for dry air R_A depends on the composition of the gas mixture and is calculated as a linear combination of the components. The absolute humidity specifies the concentration of water vapor in the gas mixture. The water vapor is assumed to be another proportion of the gas mixture to compensate for the humidity.

In addition, propagation of uncertainty is calculated to validate the developed sensor-fusion model for the requirements on medical oxygen sensors. Finally, accuracies for the sensors are proposed, with which the requirements in the theoretical test are fulfilled.

2 Material and Methods

The sensor-fusion concept is implemented in a theoretical model. The required accuracies of the sensors are determined to fulfill the requirements defined in DIN EN ISO 80601-2-55. The uncertainty of the oxygen measurement must be less than $\Delta E = \pm 2.5\,\mathrm{Vol.\text{-}\%} \pm 2.5\,\mathrm{Vol.\text{-}\%} \cdot x_{O_2}$, where x_{O_2} is the oxygen concentration. This requirement applies for all conditions and must be validated for oxygen concentrations of $O_2 = \{15\,\%,\ 21\,\%,\ 40\,\%,\ 60\,\%,\ 100\,\%\}$. An ideal gas mixture of oxygen rest nitrogen is assumed for the observed gas. The temperature is assumed to be $T = 23°\mathrm{C}$ and unsaturated vapor with a humidity of $\phi = 0\,\%$. Furthermore, the influence of pressure and flow on the sensors is not considered. After the required accuracies of the sensors are determined in this work, other cases that are more relevant for the application are evaluated. The model is evaluated for different temperatures, humidities, and gas mixtures.

2.1 Implementation of the sensor-fusion model

First of all, the Equations 2, 4 and 5 are implemented to calculate the speed of sound, the thermal conductivity and the absolute humidity for a mixture of gases. The concentration of the gas mixture is determined by solving a system of equations using the Scipy[1] $fsolve$ function. The system of the equations consists of three equations

$$\Lambda([x_{O_2}, x_{N_2}, x_{Ar}]) - \lambda = 0,$$
$$C([x_{O_2}, x_{N_2}, x_{Ar}], T) - c = 0,$$
$$x_{O_2} + x_{N_2} + x_{Ar} + H([x_{O_2}, x_{N_2}, x_{Ar}], T, \phi) - 1 = 0. \qquad (6)$$

The parameters x_{O_2}, x_{N_2}, x_{Ar} are the unknown concentrations of oxygen, nitrogen and argon in the gas mixture. A function $S(\lambda, c, T, \phi)$ is defined that solves the system of equations for the measured values and determines the oxygen concentration. A diagram of the sensor-fusion model is shown in Figure 2.

2.2 Sensor requirements

A suitable sensor to measure the temperature and the humidity is the SHT35[2]. The uncertainties for the SHT35 sensor given in the datasheet are $\delta T = 0.3\,\mathrm{K}$ and $\delta H = 2\,\%$. The value $\delta c = 0.56\,\mathrm{m\,s}^{-1}$ for a speed-of-sound sensor is taken from [6], where the uncertainty is calculated with

---

[1]https://docs.scipy.org/doc/scipy/index.html
[2]SENSIRION https://sensirion.com/de

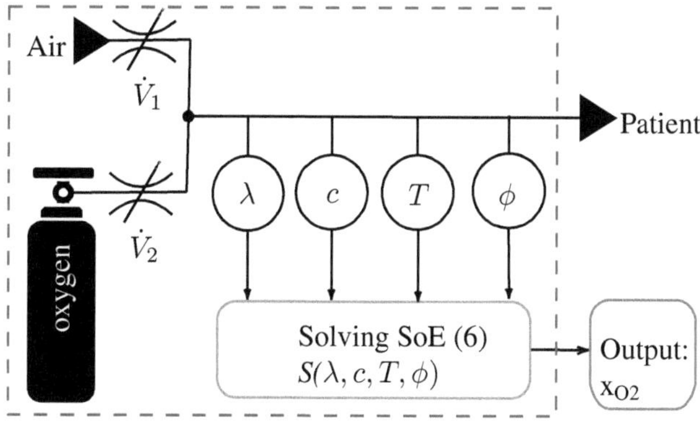

Figure 2: The gas composition is mixed together from atmospheric air and medical oxygen. Two adjustable valves are control the flows $\dot{V}_1$ and $\dot{V}_2$. Four sensors are measure the thermal conductivity λ, the speed of sound c, the temperature T and the relative humidity ϕ. Subsequently, the sensor fusion model determiens the inspiratory oxygen x_{O2} for the measured values.

a general error propagation analysis [6]. The uncertainty δc can also be written as $0.16\,\%$ of the measured value $C = 343.2\,\mathrm{m/s}$ (in dry air at $20°$ C). The uncertainty for the thermal conductivity sensor $\delta\lambda$ is determined depending on the requirement for the accuracy of the sensor-fusion model.

The propagation of uncertainty for the sensor-fusion model $S()$ is calculated to determine the permittet uncertainty $\delta\lambda$. The partial derivative is determined numerically with the difference quotient because of the solving algorithm in the sensor-fusion model $S()$. The partial derivations are determined for the oxygen concentrations specified above. The unknown inaccuracies $\delta\lambda$ are chosen according to the maximum error ΔR for the particular O_2 concentration. For this purpose, the formula for the propagation of uncertainty [7] is converted to

$$\delta\lambda^2 = \left(\Delta R^2 - (\partial S/\partial T \cdot 0.3\,\mathrm{K})^2 - (\partial S/\partial H \cdot 2\,\%)^2 \right.$$
$$\left. - (\partial S/\partial c \cdot 0.56\,\mathrm{m\,s^{-1}})^2 \right) (\partial S/\partial c)^{-2} .$$

The maximum uncertainty for the thermal conductivity sensor corresponds to the smallest value of $\delta\lambda$, which is used in the following calculations.

2.3 Consideration of practical use cases

For the further procedure the determined uncertainties are used. The error propagation is calculated for temperatures between 0 and $50°$ C and humidities between 0 and $100\,\%$. Finally, the result of the error propagation for different temperatures and humidities is considered and evaluated. It is assumed, that the medical oxygen is extracted with an oxygen concentrator. Table 2 shows the nitrogen and argon concentrations for different possible oxygen concentrations. For these values, the error propagation is calculated. The temperature is $23°$ C, and the humidity is $0\,\%$.

Table 2: Gas mixture composition

$O_2/Vol. - \%$	$N_2/Vol. - \%$	$Ar/Vol. - \%$
21	78	1
40	57.4	2.6
60	35.7	4.3
93	0	7

3 Results and Discussion

3.1 Sensor requirements

Table 3 applies the result of the propagation of uncertainty for the different concentrations of oxygen. The proportions of oxygen O_2 and the maximum error ΔR were selected according to the requirements of DIN EN ISO 80601-2-55. The uncertainty for the thermal conductivity sensor $\delta\lambda$ is calculated for each oxygen concentration. The smallest cal-

Table 3: Error propagation comparison

O_2	ΔR	$\delta\lambda$
Vol.-%	Vol.-%	W/(mK)
15	2.9	0.7×10^{-4}
21	3.0	0.9×10^{-4}
40	3.5	1.3×10^{-4}
60	4.0	1.8×10^{-4}
100	5.0	2.7×10^{-4}

culated uncertainty for the thermal conductivity sensor is $0.7 \times 10^{-4}\,\mathrm{W/(mK)}$ or $0.15\,\%$ of the measured value of the thermal conductivity. The main errors (cross-sensitivities and the thermostatization sensor block) can be almost completely compensated with a reference sensor with a known gas concentration in the measurement chamber [2]. The reference sensor needs to have the same geometry as the thermal conductivity sensor. If the sensor block temperature changes, the reference sensor and the measuring sensor change in the same way. The sensor is assembled together with the reference sensor and two resistors in a Wheatstone bridge, to get a higher sensitivity. A maximum accuracy of a thermal conductivity sensor is possible with sufficiently electronics and well calibrated sensors. A suitable sensor is the thermal conductivity pellistor VQ6MB[3]. The sensor has small dimensions and low power consumption. Furthermore, this sensor is cheaper than the galvanic sensors usually used for the FiO2 measurement and has longer durability.

The proportion of the propagation of uncertainty is calculated for each sensor. Table 4 shows the proportional errors for the different concentrations of oxygen rest nitrogen. The speed of sound measurement causes the most significant proportional error. The most significant influence of the speed of sound error is the measurement chamber's spreading due to the temperature [6]. A sensor that reduces the inaccuracy of the speed-of-sound sensor with temperature compensation would best improve the sensor-fusion

[3]SGX Sensortech https://www.sgxsensortech.com/

Table 4: Proportional errors
(oxygen rest nitrogen with 23° C and 0 % humidity)

O_2 Vol.-%	ΔT Vol.-%	ΔH Vol.-%	$\Delta\lambda$ Vol.-%	Δc Vol.-%
15	0.51	0.06	0.60	1.69
21	0.54	0.06	0.60	1.74
40	0.54	0.05	0.60	1.80
60	0.57	0.05	0.60	1.91
100	0.63	0.06	0.58	2.08

model. An optical length measurement to compensate this error could improve the accuracy of the model also. Humidity has the slightest influence on the error. Nevertheless, the humidity measurement is essential to increase the accuracy of the model. Especially in a practical implementation, humidity can have a more significant influence on the thermal conductivity than assumed in this work.

3.2 Consideration of practical use cases

For different temperatures and humidities, the propagation of uncertainty ΔS is determined and is visualized in Figure 3. The concentration of the observed gas is 21Vol.-% oxygen in nitrogen.

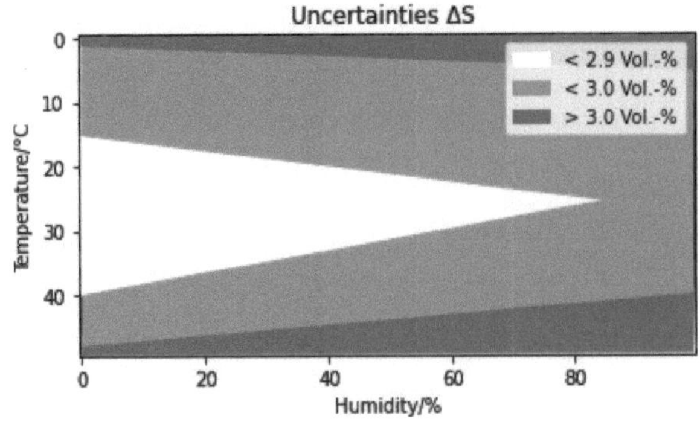

Figure 3: The propagation of uncertainty ΔS for different temperatures and humidities. The gas composition is 21Vol.-% oxygen in nitrogen. The data is classified in three areas by a linear classifier.

The dark gray area in the Figure shows the conditions where the requirements for the oxygen sensor are not fulfilled ($\Delta E_{21Vol.-\%} > 3$Vol.-%, see Table 3). The temperatures at which the sensor-fusion model does not fulfill the requirements are above 40°C and below 5°C for a maximum humidity of 100 %. For the temperatures and humidities in the white and light grey area, the sensor-fusion model fulfills the requirements with the described specification and the assumed sensor accuracies. These results need to be evaluated in further practical studies.

The propagation of uncertainty for the gas concentrations in Table 2 differs only slightly from the errors in Table 3. Consequently, the additional argon content in the gas mixture has no critical influence on the error. As a result, the developed sensor-fusion model fulfills the requirements for the assumptions regarding the accuracy of the thermal conductivity sensor.

4   Conclusion

A sensor-fusion model has been theoretically implemented to determine oxygen concentration in emergency ventilators. The sensor-fusion model combines a thermal conductivity sensor, a speed-of-sound sensor, a humidity sensor and a temperature sensor. An error calculation has determined the requirements for the sensors to fulfill the request for clinical ventilators. The results of the theoretical model are promising for further practical investigation.

Acknowledgement

The work has been carried out at Weinmann Emergency Medical Technology GmbH + Co. KG and supervised by Prof. Dr. Matthias Gräser, Universität zu Lübeck

Author's Statement

Conflict of interest: Authors state no conflict of interest.

5   References

[1] DIN (DIN EN ISO 80601-2-55:2019-03): Medical electrical equipment - Part 2-55: Particular requirements for the basic safety and essential performance of respiratory gas monitors, Beuth-Verlag, Berlin, 2019.

[2] G. Wiegleb, *Gasmesstechnik in Theorie und Praxis: Messgeräte, Sensoren, Anwendungen.*, Springer Verlag, Wiesbaden, 2016.

[3] VDI-Gesellschaft Verfahrenstechnik und Chemieingenieurwesen, *VDI-Wärmeatlas. Mit 320 Tabellen.*, Springer Verlag, 11th ed. Berlin, Heidelberg, 2013.

[4] N. Elsner: *L. Zipser: Akustische Analyse von Gasgemischen und Aerosolen.*, Berlin, Akademie-Verlag, 1973.

[5] H. Häckel: *Meteorologie*, TB; 8th ed., 2016.

[6] T. Lofqvist and J. Delsing: *Speed of sound measurements in gas-mixtures at varying composition using an ultrasonic gas flow meter with silicon based transducers*, FLOMEKO 2003 - 11th IMEKO TC9 Conference on Flow Measuremen, Luleå, 2003.

[7] G. Hermann: *Dokumentation in der Mess- und Prüftechnik: Messen - Auswerten - Darstellen Protokolle - Berichte - Präsentationen*, Wiesbaden, Springer Fachmedien Wiesbaden, 2014.

Evaluation of a Segmentation Algorithm for Drone Localization using Fused Lidar and Vision Data

Laslo Guenther [1], Hiren Patel [2], and Jonas Gruner [3]

[1] Robotics and Autonomous Systems, Universität zu Lübeck, laslo.guenther@student.uni-luebeck.de
[2] Institute for Electrical Engineering in Medicine, Universität zu Lübeck, hiren.patel@uni-luebeck.de
[3] Institute for Electrical Engineering in Medicine, Universität zu Lübeck, j.gruner@uni-luebeck.de

Abstract

Transmitted data from an incident site to the hospital can decrease the mortality rate by helping the hospital to prepare for the care of potential patients. This data transmission requires a good internet connection. In occluded environments, this is hard to obtain. A quadrocopter based mobile network relay can address this problem. To operate safely, the quadrocopter needs to act autonomously to avoid obstacles. Data from the environment can be acquired by a lidar sensor and a camera. Segmentation of the lidar data for precise localization of the quadrocopter in the environment is required to accomplish autonomy. This paper proposes a segmentation algorithm composed of ready-to-use software components. The feasibility of the algorithm is demonstrated with simulated data. Further findings reveal the need for more specialized datasets for the visual detection part.

1 Introduction

This work is part of the public-funded MOMENTUM project which explores new possibilities to enhance medical care with new technologies like the 5G technology. The new 5G communication standard can improve communication with novel types of transmitted data such as video or real-time information about the patient [1]. This kind of communication technology requires a stable connection to cell towers. A network relay can gain better connectivity, for occluded locations, where signal coverage is low. A tethered quadrocopter is proposed to realize such a network relay since it can adjust its altitude and may always find the optimal altitude for signal coverage [2]. The quadrocopter needs to operate autonomously to avoid obstacles. Therefore, it is crucial to map the environment and localize the quadrocopter accurately in 3D space.

There are already a wide variety of solutions on how to achieve obstacle avoidance on a drone. One common solution is to use onboard sensors and computers [3]. This configuration is advantageous because of its compactness and freedom of movement. The major drawback is the limited power supply [4]. Since emergencies on the incident site can vary in duration, such a solution is not favorable [2]. A ground-based obstacle avoidance system with a tethered quadrocopter is used to adress these problems.

This paper discusses a potential solution to localize a tethered quadrocopter based on environment segmentation. The segmentation algorithm uses a ground-based lidar sensor and a camera. The algorithm is evaluated with simulated data.

2 Material and Methods

2.1 System Overview

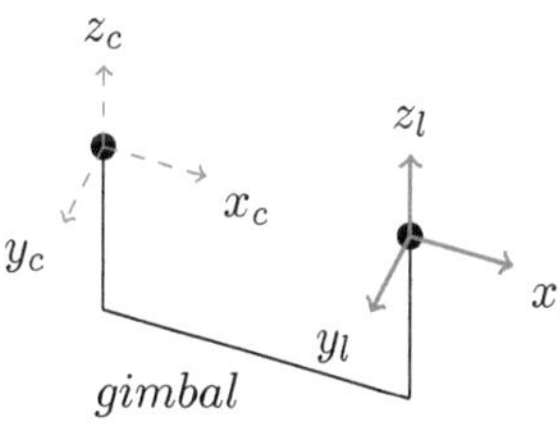

Figure 1: Coordinate frame for camera and lidar. Index c defines the camera frame (line) and index l defines the lidar coordinate frame (dashed). The sensors are illustrated as black dots. Both sensors are mounted parallel and facing up (in the z-direction).

The Livox Avia lidar sensor[1] is used for environment mapping. Lidar means light detection and ranging and is comparable to radar. Furthermore, a camera is used to detect the quadrocopter. Both sensors are mounted on a gimbal. Camera data is captured in the camera frame and lidar data is represented in the lidar frame as shown in Fig. 1. The data of the sensors is then processed on an Nvidia Xavier AGX computer[2].

The cable which connects the quadrocopter and the ground system is anchored close to the sensors. Data (e.g., quadro-

---

[1] Avia LiDAR Sensor https://www.livoxtech.com/de/avia
[2] Jetson AGX Xavier Developer Kit
https://developer.nvidia.com/embedded/jetson-agx-xavier-developer-kit

copter position relative to obstacles) and power get transmitted through the cable to the quadrocopter to increase flight time and to achieve an interference free data transmission [2]. However, flight space is reduced compared to a non-tethered system due to physical (e.g., cable length) and systematic (e.g., quadrocopter needs to be in the view area of the sensors) limitations. The quadrocopter mostly varies its altitude z_c and has no significant offsets in x_c or y_c.

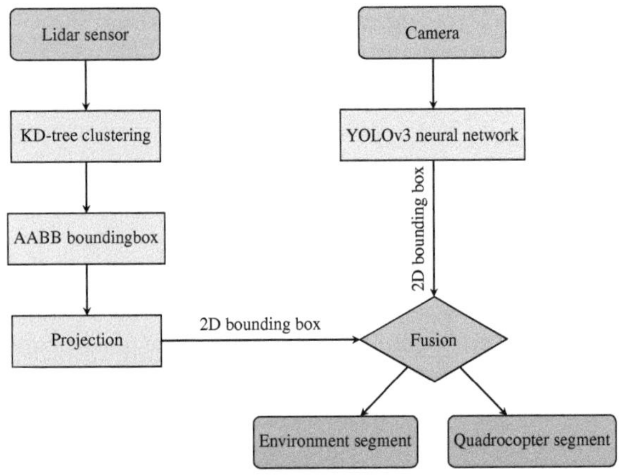

Figure 2: Fusion of the lidar detection (left strain) and the visual detection (right strain) to get a segmentation of the environment and the quadrocopter.

2.2 Lidar Data

The Livox Avia lidar sensor has a maximum range of 450 m, a Field of view (FOV) of 70.4° x 77.2°, and precision of two centimeters.

The sensor data is used in a ROS message format[3]. The data is first received with the Livox ROS driver[4] and gets published as a point cloud message.

A ROS node that subscribes to the Livox driver segregates the data into clusters of separate objects. The data flow of the lidar data can be seen on the left strain in Fig. 2. A Kd-tree clustering algorithm is used for clustering the data [12]. After cluster detection, Axis-Aligned Bounding Boxes (AABB) are applied for each cluster [10]. The clustered data is then published as a custom ROS message. The custom message contains the bounding boxes and the corresponding lidar points of each object. The segmentation step later uses this to segment the point cloud.

2.3 Visual Detection of the Quadrocopter

Visual data is suitable for segmenting an environment [11]. There are already ready-to-use object detection algorithms like the YOLOv3 algorithm [5]. It shows suitable detection and real-time performance for drone tracking [6]. The YOLOv3 algorithm uses convolutional neural nets to predict bounding boxes of various object classes in an image.

A derivative of the YOLOv3 algorithm called YOLOv3-tiny was used for retraining. This neural net is a smaller version of the YOLOv3 model to suit embedded systems and can run on the Xavier AGX in real-time. To detect drones, the YOLOv3-tiny algorithm is extended by a drone class. Training took place on an Nvidia Quadro P5000. The neural net was trained for 100 epochs with a batch size of 8. The dataset used for training and validation contained 2600 different annotated drone images.

The training/validation sets were split at an 80/20 ratio. This neural network estimates the quadrocopter's x_c and y_c position. The neural network was implemented in PyTorch[5] using Python 3.6[6]. Since ROS melodic was build to work with Python 2.7, the neural network runs outside the ROS melodic environment and communicates via sockets with a ROS node. This ROS node publishes the data (x_c and y_c position of the quadrocopter) to the ROS melodic environment.

2.4 Lidar and Vision Fusion

Segmentation of lidar data is possible but requires 3D samples for retraining segmentation models which is tedious to obtain [7][9]. In contrast, datasets of drone images for detection are widely available. Fusion of vision and lidar data for segmentation shows a suitable performance which is even better than direct segmentation from lidar data [8].

A ROS node is used to fuse the data from the vision and lidar ROS nodes. The segmented lidar data can then be used for quadrocopter localization.

The lidar 3D bounding boxes are projected into the 2D camera image, to fuse the data. The 3D bounding boxes needed a translation transformation into the camera frame and can then be projected.

The visual bounding box predicted by the YOLOv3-tiny algorithm is also projected into the camera image.

Intersection over Union (IoU) is calculated between lidar and vision 2D bounding boxes. The best two matching bounding boxes result in the highest IoU value.

Therefore the 3D lidar bounding box whose 2D representation is the best matching is the quadrocopter and can be separated from the lidar data. A segmented 3D data set is obtained.

2.5 Simulated Data

For a first evaluation, it is easier to use simulated data instead of getting data from real flight tests. Multiple test cases can be generated quickly and failures during testing do not lead to loss of hardware. The data was generated with Blender[7] and Blensor[8] to simulate the lidar sensor.

For testing, the segmentation algorithm shown in Fig. 2 was realized inside a Python module instead of the later used ROS melodic environment. The Python module received

---

[3]ROS https://www.ros.org/

[4]Livox ROS Driver https://github.com/Livox-SDK/livox_ros_driver

[5]PyTorch https://pytorch.org/

[6]Python https://www.python.org/

[7]Blender https://www.blender.org/

[8]Blensor https://www.blensor.org

rendered images from blender as camera input and binary data from Blensor as lidar input. Sensor data is set to be noiseless in simulation. The module generates bounding boxes for each lidar cluster on top of the camera image. The bounding boxes predicted by the YOLOv3-tiny algorithm are dashed. The whole process is static and does not run in real-time.

In the simulation, a translation between camera and lidar points does not need to be considered since both sensors are mounted at the same location. Blensor, unfortunately, does not offer the Livox Avia sensor. The Velodyne HDL-64E rotary lidar sensor[9] is used in the simulation. The rotation got limited to -35° and 35° to get a horizontal FOV of 70°, so it is comparable to the Livox Avia sensor. The vertical FOV is fixed to +2.0° to -24.9°. The HDL-64E sensor is rotated around the x_l-axis by $\phi = 11.45$ to get the FOV symmetrical aligned with the camera's z_c-axis as in Fig. 3. For testing, this is sufficient because tests do not take place outside a FOV of $\theta = 26.9$.

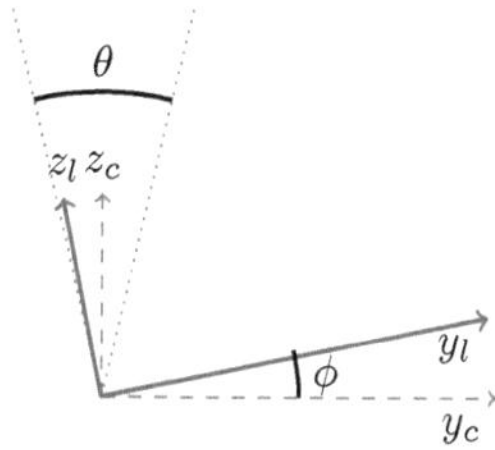

Figure 3: Alignment of lidar (line) and camera (beneath dotted line) FOV. Lidar frame is a line and camera frame is dashed. (line)

3 Results and Discussion

In simulation four different environment settings were used to get a better understanding of the limits of the proposed algorithm. For each setting three different altitudes of the quadrocopter (10 m, 20 m, 30 m) were tested. The quadrocopter hovers centered above the ground sensors.

Setting 1 is used as a baseline where only the quadrocopter is visible and the sky is bright. In setting 2 the sky is darker than in setting 1 but otherwise identical to setting 1. An environment object (cube) was added to setting 3. The cube was positioned $x_c = -3$ m and $z_c = 16$ m. The setting is otherwise identical to setting 1.

A tethered quadrocopter was also tested in setting 4. The cable's base position is distant $x_c = 1.5$ m to the sensors. The setting is otherwise identical to setting 1.

The results in Fig. 4 only show the detection and projection part of the segmentation algorithm. The segmentation was successful in all cases where visual detection was successful too.

In setting 1 a high contrast is achieved because of the bright sky. The lidar detection is a little bit off due to resolution

<hr>

[9]HDL-64E Lidar Sensor https://velodynelidar.com/blog/hdl-64e-lidar-sensor-retires/

errors. Those resolution issues get worse at higher altitudes of the quadrocopter. The visual detection returns confidence of > 0.9 for 10 and 20 m. For 30 m detection is not possible because of low resolution and lack of training data at these altitudes. The visual bounding box has quite a consistent size across 10 and 20 m where it should be smaller at 20 m. False segmentations are possible if the discrepancy between the visual and lidar bounding box is too high and objects are close to the quadrocopter. Since the quadrocopter is not considered to operate close to other objects this is not a problem.

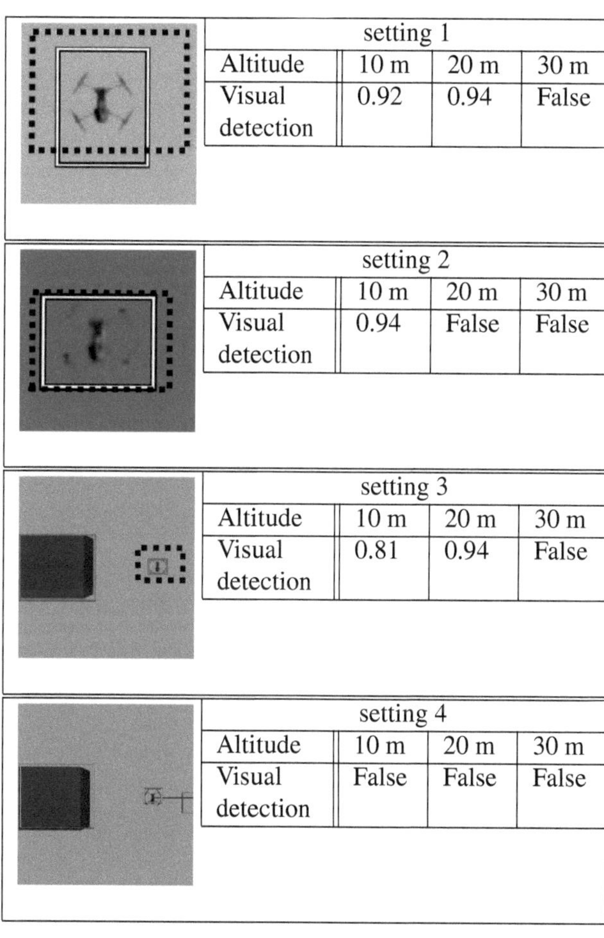

setting 1			
Altitude	10 m	20 m	30 m
Visual detection	0.92	0.94	False

setting 2			
Altitude	10 m	20 m	30 m
Visual detection	0.94	False	False

setting 3			
Altitude	10 m	20 m	30 m
Visual detection	0.81	0.94	False

setting 4			
Altitude	10 m	20 m	30 m
Visual detection	False	False	False

Figure 4: Simulated test settings. Visual detection state (dashed bounding box) either as a confidence value between 0 and 1 or a boolean. Lidar detection (line bounding box) was in all settings successful.

Setting 2 has a lower contrast due to the dark sky. Visual detection seems to be the same at 10 m as in setting 1. At higher altitudes, the visual detection fails. Higher contrast between quadrocopter and environment is therefore beneficial. More training data at these settings may also help.

The object in setting 3 reduced the confidence in visual detection at 10 m to 11 % but regains the same performance as in setting 1 at 20 m. Close objects to the quadrocopter seem to reduce visual detection performance. This is not a problem because flights of the quadrocopter are considered to take place in a free area without obstacles.

The later use case does not consider close objects to the quadrocopter, as earlier mentioned.

In setting 4 the visual detection fails at all altitudes. The cable may be too close to the quadrocopter. Setting 3 already shows that close objects affect performance. More training data with tethered drones is required. Lidar detection of the cable is not precisely aligned due to resolution errors but mainly to close detection artifacts. A cutoff to close lidar points would be beneficial. The cable should also not be mounted too close to the sensors.

4   Conclusion

A segmentation algorithm for a quadrocopter based on a camera, a lidar sensor, and ready-to-use algorithms was proposed. The algorithm was evaluated with simulated data, acquired with Blender and Blensor. Test results show that the algorithm works as long as the training data of the visual detection is similar to deployment scenarios. Most training data did not account for dark skies and tethered quadrocopters. More training data with these contents needs to be added to gain better results. Another visual detection algorithm may be also beneficial for detection performance like Faster RCNN or DETR [6]. Visual detection seems to be the bottleneck in this approach. Lidar detection is only troublesome with close detections. The next step is retraining of the visual detection model, such that hardware tests can be performed in the future.

Acknowledgement

The work has been carried out at the Institute for Electrical Engineering in Medicine, Universität zu Lübeck. I would like to thank Prof. Dr. Philipp Rostalski for the opportunity to work on this project. Special thanks to Jonas Gruner and Hiren Patel for guidance and advise.

Author's Statement

Research funding: The work was funded by the German Federal Ministry of Education and Research under grantno. 16KIS1027. Conflict of interest: Authors state no conflict of interest.

5   References

[1] M. Rockstroh et al., *Towards an Integrated Emergency Medical Care using 5G Networks*. Current Directions in Biomedical Engineering, Vol. 6 (Issue 3), pp. 5-8, doi: 10.1515/cdbme-2020-3002, 2020.

[2] J. Gruner, A. Bloße, M. Rockstroh, T. Neumuth and P. Rostalski, *Requirement Analysis for an Aerial Relay in Emergency Response Missions*. Current Directions in Biomedical Engineering, Vol. 6 (Issue 3), pp. 16-19, doi: 10.1515/cdbme-2020-3005, 2020.

[3] N. Gageik, P. Benz and S. Montenegro, *Obstacle Detection and Collision Avoidance for a UAV with Complementary Low-Cost Sensors*. In: IEEE Access, vol. 3, pp. 599-609, doi: 10.1109/ACCESS.2015.2432455, 2015.

[4] B. Boroujerdian, H. Genc, S. Krishnan, A. Faust, V. J. Reddi, *Why Compute Matters for UAV Energy Efficiency?* Edge Computing Lab, Harvard University, 2018.

[5] J. Redmon, S. K. Divvala, R. B. Girshick, A. Farhadi, *You Only Look Once: Unified, Real-Time Object Detection*. CoRR, abs/1506.02640, 2015.

[6] B. K. S. Isaac-Medina, M. Poyser, D. Organisciak, C. G. Willcocks, T. P. Breckon and H. P. H. Shum, *Unmanned Aerial Vehicle Visual Detection and Tracking using Deep Neural Networks: A Performance Benchmark*. CoRR, abs/2103.13933, 2021.

[7] C. R. Qi, H. Su, K. Mo and L. J. Guibas, *PointNet: Deep Learning on Point Sets for 3D Classification and Segmentation*. CoRR, abs/1612.00593, 2016.

[8] G. P. Meyer, J. Charland, D. Hegde, A. L. and C. Vallespi-Gonzalez, *Sensor Fusion for Joint 3D Object Detection and Semantic Segmentation*. CoRR, abs/1904.11466, 2019.

[9] G. P. Meyer, A. Laddha, E. Kee, C. Vallespi-Gonzalez and C. K. Wellington, *LaserNet: An Efficient Probabilistic 3D Object Detector for Autonomous Driving*. CoRR, abs/1903.08701, 2019.

[10] E. Hastings and J. Mesit, *Optimization of Large-Scale, Real-Time Simulations by Spatial Hashing*. In: Proc. Summer Computer Simulation Conference 2005, Vol. 37, No. 4, pp. 9-17, 2005.

[11] L. Porzi, et al., *Seamless Scene Segmentation*. CoRR, abs/1905.01220, 2019.

[12] A. Moore, W., *Very Fast EM-Based Mixture Model Clustering Using Multiresolution Kd-Trees*. NIPS , 1998.

The LNSeismograph:
A LoRaWAN-Capable Seismograph For Historic Preservation Of Monuments

Melanie Badura,
Robotics and Autonomous Systems, Universität zu Lübeck, melanie.badura@student.uni-luebeck.de

Abstract

A considerable aspect for historic preservation is the monitoring of ground movement. Damage can occur to monuments if tremors are too high. A seismograph can help, find sources of ground movement. However, seismographs are relatively expensive and do not generate data that is accessible over distributed systems. We introduce the LNSeismograph that operates with LoRaWAN. Evaluating the data generated in historic monuments can be done from any system. We test if the recorded velocity corresponds to inflicted vibrations. The recorded data shows the ground moving at the expected time. We transmit only data over a certain threshold over LoRaWAN. So it is not possible to evaluate earthquakes, but it suffices to provide data for monument protection.

1   Introduction

The world has many monuments that need protection from environmental influences. One such influence is earth vibration. They can not only be caused by earthquakes but also by traffic and humans in general. There are different specifications to evaluate tremors. For example, the Richter Scale demonstrates the magnitude of an earthquake [1]. Another parameter is the velocity, which can be calculated by how fast the ground is shaking. In this project, a seismograph is built, which monitors ground movement by sending critical velocities over *Long Range Wide Area Network* (LoRaWAN) to an *Internet of Things* (IoT) platform.

IoT describes a world where everything connects by sensors and actuators. These sensors can not only sense but can also communicate. What is notable about this is that some of these physical information systems work without human intervention [2]. LoRaWAN is a protocol for *Long Range* (LoRa), which uses chirp pulses to send data over radio waves to gateways. The gateways forward the data over Ethernet or *Global System for Mobile Communications* (GSM) to a network server [3]. The *LoRaWAN seismograph* (LNSeismograph) measures seismic events from 10 to 100 Hz and not only sends them over LoRaWAN to a gateway but also over WiFi to a local webserver. It also evaluates the seismic events. Historic monuments are at risk if they are above a certain threshold. Ultimately the prototype works as expected and shows that an inexpensive ground surveilling system with LoRaWAN is possible.

2   Related Works

This project relates to all the seismographs that have been developed since. Specifically, two were of great interest for the LNSeismograph. Firstly, the RaspberryShake was developed as a low-cost alternative for hobby seismologists. It generates data, that is sent through an ethernet cable to a community platform, as well as saved on a miniSD card [4]. It uses a geophone as a seismogram, whereas the QuakeSense gets their data through a *micro-electro-mechanical system* (MEMS), which instead of a mass and a coil, uses capacitors to find ground movement [5]. The QuakeSense also uses LoRaWAN to forward its data.

The LNSeismograph combines the best abilities of both seismographs while staying inexpensive.

3   Material and Methods

We use LoRaWAN to let the LNSeismograpgh communicate with the gateways. The hardware of the LNSeismograpgh consists of two microcontrollers, Teensy 4.0 and Wireless Stick Lite, and a self-build *Printed Circuit Board* (PCB). The PCB includes a Geophone and a circuit that amplifies and offsets the signal of the geophone. We also implement software to sample the *Analog-to-digital converter* (ADC) readings and afterward send the recordings to an IoT Platt-form. It is important to note that seismic waves have a frequency range from 1 Hz to 100 Hz.

3.1   LoRaWAN

LoRaWAN is a radio technology developed for the IoT sector. It defines the communication protocol as well as the network system architecture. Nodes, gateways, a network server, and an application server form the network. The nodes broadcast data via LoRa, and the gateways receive that data. Then they can acknowledge the receivement of the data, should that be requested from the node. Afterward, they forward the data to the network server via a cellular connection or WiFi. The application server then can get the data from the network server over *Message Queuing Telemetry Transport* (MQTT) or Webhook.

3.2 Hardware

The hardware consists of a geophone, two microcontrollers, and a circuit.

Geophone Geophones convert ground vibrations into analog voltage signals. They consist of a spring, which couples a coil and a permanent magnet. The motion of the geophone causes motion between the magnet and the coil, and this movement induces a voltage in the coil [6]. Since seismic waves have a frequency range from 1 Hz to 100 Hz, we want a geophone that covers that range. The best, inexpensive geophone we found is the SM-24. It has a bandwidth from 10 Hz to 240 Hz and the sensitivity is $28.8\frac{Vs}{m}$.

Teensy 4.0 The Teensy 4.0 is a microcontroller with an integrated ADC. We choose the microcontroller for their converter, which has a sampling rate of 44 kHz. Since the Teensy has an *Advanced RISC Machine* (ARM) processor, no additional buffer is needed, as would have been the case if we only used an ADC.

Wireless Stick Lite The Wireless Stick Lite integrates an ESP32-Pico. It also has a LoRa node chip SX1276. The SX1276 uses 868 MHz and 915 MHz frequencies. Those are the LoRa frequencies in Europe and North America. Additionally, the board has a WiFi and Bluetooth chip.

Circuit The geophone produces a small analog signal, which converts to a digital signal by an ADC. We must amplify the incoming analog signal of the geophone to prevent it from aliasing. This also ensures that the signal is picked up well by the converter. We insert an offset since the ADC only works in the positive range. Fig. 1 shows this circuit.

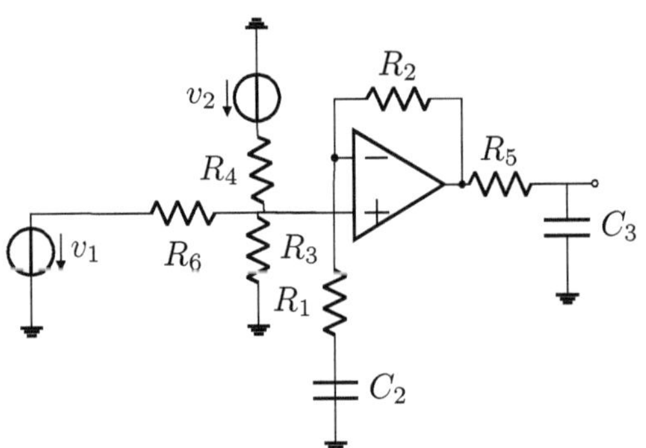

Figure 1: Offset and non-inverting Amplifier, where v_1 is the signal of the geophone and v_2 the voltage source.

Calculations One can calculate the amplification V_u with:

$$V_u = \frac{R_1 + R_2}{R_1} = 1 + \frac{R_2}{R_1} = 1 + \frac{10k\Omega}{1k\Omega} = 11 \quad (1)$$

with resistor (R).
The offset is implemented by:

$$\text{If } R_3 = R_4 => \text{Offset} = \frac{v_2}{2} \quad (2)$$

DC failures while amplifing are ignored by a high-pass filter. To add a high-pass filter another capacitor C_2 is needed. Now we have a high-pass filter and a low-pass filter in series. First, the *time constants of the high-pass filter* (τ_{HP}) and the *time constants of the low-pass filter* (τ_{LP}) are calculated. With those we calculate the *frequency of the high-pass filter* (f_{HP}) and *frequency of the high-pass filter* (f_{LP}). Since we were not interested in the low-pass filter, C_2 was chosen such that the frequency response for the low pass filter is high. We use the high-pass filter only for DC failures, which indicates that this filter should be small. To choose a value for C_2, we do the following calculations.

$$C_2 = \frac{\tau_{HP}}{(R_1 + R_2)} \quad (3)$$

$$C_2 = \frac{\tau_{LP}}{(R_1)} \quad (4)$$

$$\tau_{HP} = \frac{1}{2\pi \cdot f_{HP}} \quad (5)$$

$$\tau_{LP} = \frac{1}{2\pi \cdot f_{LP}} \quad (6)$$

We want a high-pass filter, which does not exceed 10 Hz, because our interest lies in the low frequencies. We choose f_{HP} aproximately as 10Hz, which means, that τ_{HP} equals 16.5 ms. The next step is to calculate C_2 from τ_{HP}. By using Eq. 6, we get a result of 150 nF for the capacitor. Forthwith, it is necessary to calculate the cut-off frequency of the low-pass filter. We achieve this by applying Eq. 5 and Eq. 7. The result for f_{LP} equals to roughly 106 Hz.

After the signal was offset and amplified, it runs through a low pass filter. To calculate C_3 and R_5, the *cut-off frequency* f_c has to be definied.

$$f_c = \frac{1}{2 \cdot \pi \cdot R_5 \cdot C_3} \quad (7)$$

$$<=> R_5 \cdot C_3 = \frac{1}{2 \cdot \pi \cdot f_c} \quad (8)$$

Our cut-off frequency f_c would preferably be 240 Hz in accordance with the datasheet of the geophone. Unfortunately, resistors and capacitors come in specific units. Thus we could not achieve our ideal cut-off frequency. We choose 10 kΩ for R_5 and 47 nF for C_3 and with Eq. 9 f_c equals to 338 Hz, which is in an adequate range. Lastly, there is R_6 in the circuit, which has a value of 1 kΩ. In compliance with the datasheet this damps the response curve of the geophone .

3.3 Software

Since we are using two different microcontrollers, the software splits into two parts (see Fig. 2 and Fig. 3).
Firstly, the Teensy gets the analog sampling of the geophone, and the ADC converts the samples into digital ones, as can be seen in Fig. 2. The sample value can then be calculated into a voltage with:

$$\text{voltage} = \frac{(\frac{\text{sampleValue} \cdot 3.3\,V \cdot 1000}{\text{bitTotal}}) - 1650\,mV)}{11} \quad (9)$$

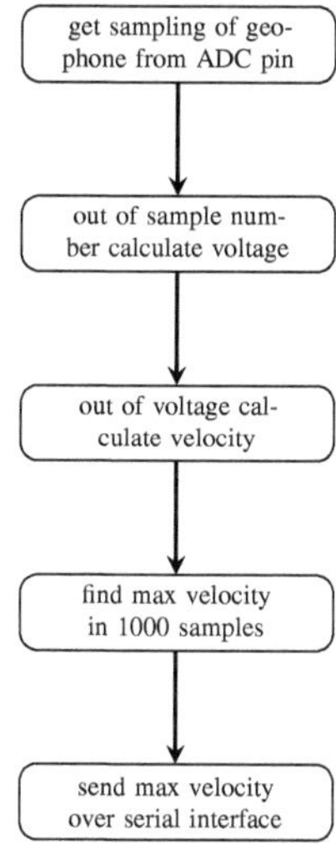

Figure 2: The Teensy Flow shows the sampling of the geophone. Afterward the Teensy converts the samples into the velocity.

with bitTotal being the maximum amount of different numbers the ADC can represent. In Eq. 9, we also subtract the offset and divide by the amplification so that we get a realistic voltage output. Now we calculate the velocity out of the voltage by using Eq. 10.

$$\text{velocity} = \frac{\text{voltage}}{\text{sensitivityGeophone}} = \frac{\text{voltage}}{28.8 \frac{\text{mV}}{\frac{\text{mm}}{\text{s}}}} \quad (10)$$

Afterward, the Teensy sends the velocity to the Wireless Stick Lite over *Universal asynchronous receiver-transmitter* (UART).

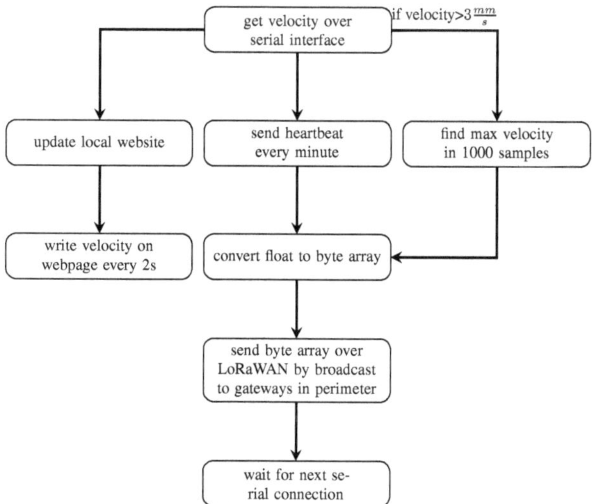

Figure 3: The Wireless Stick Lite Flow shows the three different tasks it has to perform. It sends a velocity via LoRaWAN, while also updating the values on the website. As the third task it sends a heartbeat via LoRaWAN.

Fig. 3 shows the further signal processing. The microcontroller differentiates between three events. If the received velocity is over $3\frac{\text{mm}}{\text{s}}$, it sends the converted data over LoRaWan to a gateway. Every minute it sends a heartbeat,

including the velocity, to a gateway. Furthermore, every two seconds, a local webserver shows the velocity graphically. The gateway sends the data to a middleware. There it is further processed.

4 Results and Discussion

We conduct an experiment to verify the performance of the seismograph.

Test Setup The LNSeismograph was placed on a table and connected to a power supply. A test subject is sitting in front of the table. The person performs four motions to simulate the ground movement. First, the test subject got up without touching the table. Afterward, they banged on the table. Third, the person sits down while gripping the table. Lastly, they stomp on the ground with their feet while still sitting at the table.

Fig. 4a shows the resulting velocities plotted on a serial plotter.

4.1 Discussion

We can observe from Fig. 4a that our experiment worked. The 100 sample mark shows our test subject getting up. Afterward, at the 180th sample, the second motion takes place. After a few samples, the velocity of the third and right afterward, the fourth motion. This shows that the LNSeismograpgh can measure ground movement. We can also see the offset implementation in the circuit working.

Fig. 4b shows us 20 seconds of data on the website. Since the data points arrive roughly every two seconds, we cannot see the tremors. This also applies to the IoT Plattform. However, the data is presented on the website, which indicates that the LNSeismograph works in this regard. The data larger than $3\frac{\text{mm}}{\text{s}}$ also arrives on the Plattform.

5 Conclusion

We build a functioning seismograph within the scope of this internship. The seismograph sends its data without human intervention. This data comes from the selected geophone and is converted into digital velocities by the ADC in the Teensy. The Wireless Stick Lite ensures the transmission via LoRaWAN to a middleware. To avoid losing negative values, we shift the analog signal beforehand. We also scale the data by a factor of eleven to enable better differentiation between the data. Both modifications were applied in hardware. In the software, we convert those changes again. We can install the LNSeismograph in any building that has electricity. Comparing the LNSeismograph to the QuakeSense [5] a few advantages are seen, but also many aspects that were not implemented. At the moment, we only measure the vibrations in one direction. Ideally, all three directions of the coordinate system would be covered. The QuakeSense has the advantage of using a MEMS with a 3-axis gyroscope. An advantage of the LNSeismograpgh is the small size. Additionally, we can retrieve the velocity values

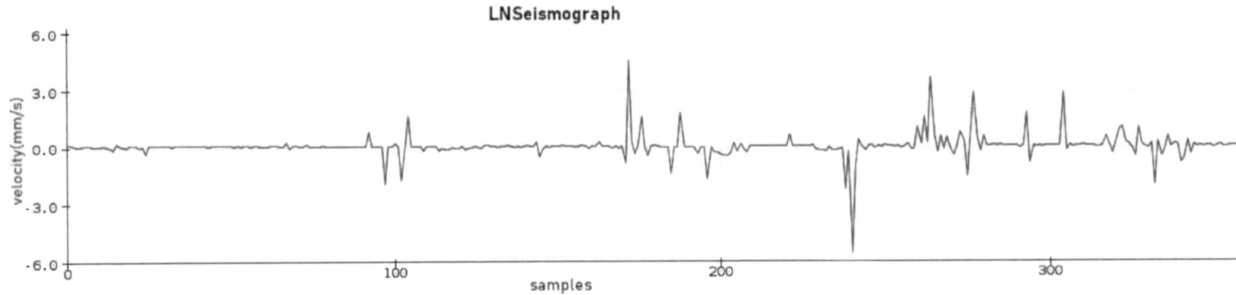

(a) Results of the experiment shown on a serial plotter. Around the 100th sample shows the movement of a person getting up. Between the 180th and the 200th samples, the velocities of the banging on the table can be seen. The next big outliers show the person sitting down while gripping the table, and the many outliers afterward show stomping of feet on the ground.

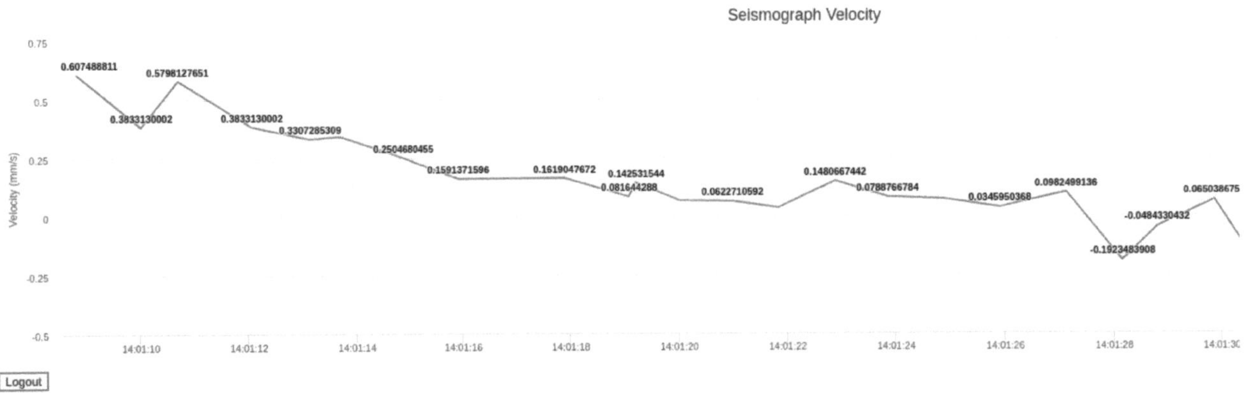

(b) Excerpt of velocities showing up on the website. The velocities were send every two seconds by the LNSeismograpgh. It can be seen, that even small movements are recorded.

Figure 4: Presentation of ground movement, the LNSeismograpgh transmitted. The two figures show the seismograph working as expected.

via a website, which is not possible with the QuakeSense since no WiFi chip is included. However, the QuakeSense is battery-powered, while we can only theoretically power the LNSeismograph with batteries. But, we did not consider the energy consumption, so this should be looked at beforehand. Since we did not have another seismograph to verify our results, more tests are to be done. With additional experiments, the project can be used as a basis for further seismographs of this kind.

Acknowledgement

The work has been carried out at TraveKom, Lübeck and supervised by Dr. Florian-Lennart Lau at the Institute of Telematik, Universität zu Lübeck.

Author's Statement

Conflict of interest: Authors state no conflict of interest.

6   References

[1] U.S. Department of the Interior | U.S. Geological Survey, "Usgs-measuring earthquakes," https://pubs.usgs.gov/gip/earthq1/measure.html, accessed: 2021-09-28.

[2] S. Madakam, V. Lake, V. Lake, V. Lake *et al.*, "Internet of things (iot): A literature review," *Journal of Computer and Communications*, vol. 3, no. 05, p. 164, 2015.

[3] LoRa Allience, "What is lorawan," https://lora-alliance.org/wp-content/uploads/2020/11/what-is-lorawan.pdf, accessed: 2021-09-24.

[4] S. Derouin, "Desktop seismology: How a maker-inspired device is changing seismic monitoring," https://www.earthmagazine.org/article/desktop-seismology-how-maker-inspired-device-changing-seismic-monitoring/, accessed: 2021-09-12.

[5] P. Boccadoro, B. Montaruli, and L. A. Grieco, "Quakesense, a lora-compliant earthquake monitoring open system." IEEE, 2019.

[6] J. M. Reynolds, *An introduction to applied and environmental geophysics.* John Wiley & Sons, 2011.

Evaluation of 3D sensors for autonomous vehicles and forklifts in warehouse environments

Katharina Spitzley [1]

[1] Robotics and Autonomous Systems, Universität zu Lübeck, katharina.spitzley@student.uni-luebeck.de

Abstract

Autonomous vehicles are becoming more widespread in the warehouse environment and are now used for autonomous load handling (pick, drop, etc.). Objects specific to these environments need to be detected reliably, e.g. pallets and shelves. This paper evaluates three Time-of-Flight sensors regarding their general technical specifications, like noise and stability against varying lightning conditions as well as their detection abilities. The sensors were tested on four different types of pallets, verifying their detection performance and position, dimension and orientation accuracy. All sensors have similar properties such as resolution and opening angle with the sensor dimension and date of development as their main difference. They all perform good in the tests but show specific strengths and weaknesses related to the use case.

1 Introduction

The development of 3D sensors, especially Time-of-Flight and laser imaging, detection, and ranging (Lidar) sensors changed rapidly in the last years [2] due to the increasing demand of industrial applications as well as the increasing importance of autonomous vehicles [3]. This paper is about the use of Time-of-Flight sensors in the use case of shelf and pallet detection in order to store or pick up loads in warehouses. Existing forklifts and other vehicles (see Fig. 1) are automated by adding more sensors and a robotics system. The Time-of-Flight sensor will be placed between the forks. The robotic system will generate measurements of the load and shelf to autonomously perform picks and drops of loads. The vehicles also drive autonomously to the shelves and have to position themselves in front of the pallet, so the required operating range of the sensor is about 1 m to 5 m.

This paper evaluates three sensors in relation to several different load types and surroundings. The sensor names have been pseudonymized as requested by the Still GmbH.

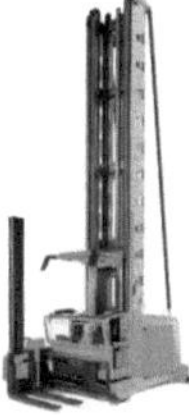

Figure 1: The STILL Very narrow aisle truck MX-X [1] can be upgraded to an autonomous vehicle

Table 1: Sensor properties according to the data sheet

	Sensor A	Sensor B	Sensor C
Size [mm]	162x93x78	80x70x77	100x81x64
Weight [g]	1400	520	690
Resolution [px]	640x480	512x424	640x480
opening angle (hor) [°]	70	70	67
opening angle (vert) [°]	56	60	51
range [m]	0.5-60	< 16	<10

2 Sensors

The three sensors tested here are Time-of-Flight depth cameras. These emit light and measure the phase shift of the reflected light. This allows an estimation of the light's travel time and thus of the distance to the object that reflected the light [4]. The sensors are precise, fast and cost efficient. However, problems with direct sunlight and reflections can occur. Table 1 shows the specifications of the sensors. Sensor A is the largest and heaviest. Sensors B and C are both about half the size and weight. The opening angle is nearly the same for all three sensors, but Sensor B has the lowest resolution. According to the data sheets the range differs much between the sensors, therefore, one of the tests verifies the actual range.

Sensors often use built-in filters to reduce noise. These filters can also cause other effects, e.g. clipping at a long range. We mainly used the standard parameters in the evaluation and turned off all filters where possible to get comparable results. That was not possible in all cases, especially sensor C has filters that could not be deactivated. The data was accessed through the tools provided by the manufacturers. These point cloud data are provided in a .pcd-file which stores the coordinates and the rgb or intensity value of every data point.

3 Tests

First, basic tests like minimum and maximum range, the noise and the influence of reflectors were executed to verify the general performance of the sensors. Then the outdoor compatibility and the influence of sun- and daylight were tested. Finally, the performance of the sensors was tested in the use case of pallet detection.

The tests were performed in the same setting for each sensor. For each situation and each sensor at least five measurements of the point cloud were recorded. The point clouds could be examined visually to get the noise of the data and the sharpness of objects. They are also used to run and evaluate the existing object detection algorithms. This detector was used to recognize the holes in pallets and the position, orientation, size and probability of the rectangular holes. These values and the detection reliability are used to evaluate the quality of the sensor output.

These tests were chosen because they are concerned with the performance of sensors to detect pallets in different situations and lighting conditions.

3.1 General Tests

These tests verify the claims of the data sheet: how the sensors performs at minimum and maximum range, the noise when measuring a plain wall and the influence of highly reflective objects in the measurement area. Reflectors are mounted in warehouses to provide information for the navigation and localization of autonomously driving vehicles and should not influence the performance of the sensors. The minimum distance was tested by placing a brown box in 40 cm and 60 cm distance to the sensor. To verify both the maximum distance and the detection performance on low remissive objects in large distances, a black and a brown box were used. All tests were evaluated visually.

3.2 Outdoor Tests

Because all other tests were executed in a large hall without much daylight, these tests verify the outdoor compatibility and influence of extraneous light, especially sunlight. Sunlight consists partially of light in the near-infrared spectrum, which is also used in the Time-of-Flight sensors [4]. Thus, we verified whether this influences the results. First the sensors were placed in a doorway to receive intermediate amounts of daylight. Then, we placed the sensors outside and took measurements of a forklift standing in the sunlight (see Fig. 4).

3.3 Pallet Detection

For the use case of pallet detection we tested four different pallets: The common europallet made out of wood, a black

Figure 2: Different pallet types: europallet (upper left), black pallet (middle left), grid pallet (lower left), grid box wrapped (right)

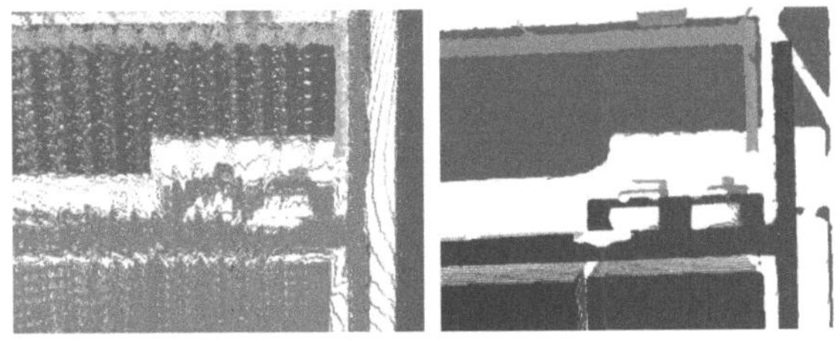

Figure 3: Point cloud of an europallet. Recorded with Sensor A (left) and Sensor B (right)

plastic pallet, a gird pallet and a grid box (see Fig. 2). There are two main difficulties: Different colors and smaller bars have to be detected and pallets often are wrapped in a thin transparent sheet to protect the load. This sheet reflects the light and can disturb the sensor results.

The sensors were placed in a distance of two metres from the pallets. After the recording of at least 8 point clouds the data was evaluated visually and an object detection algorithm was executed which detects the holes of the pallets. This algorithm returns the position, dimension, orientation and probability of the detected objects. These values were compared for every pallet and every sensor.

4 Results

4.1 General Tests

When testing the minimal distance Sensor A was not able to detect the outlines of the box properly. Sensor B detected a difference between the box and the background, but without clear and distinct edges. Sensor C was able to detect the outlines at both distances clearly. Sensor A detected the brown box at a maximum distance of approximately 6 m and the black box at 5.3 m. Sensor B has a larger maximum distance with 11 m to 12 m detection of both boxes. Sensor C was able to detect the brown box at a distance of 10 m (as written in the data sheet) and the black box up to a distance of 7.5 m.

All three sensors had a good detection of a straight wall without much noise. At a distance of 6 m the reflectors were visible with all sensors and had no further influence to the rest of the sensor image. At 9 m distance the sensors project phantom reflectors near to the sensor (Sensor C) or behind

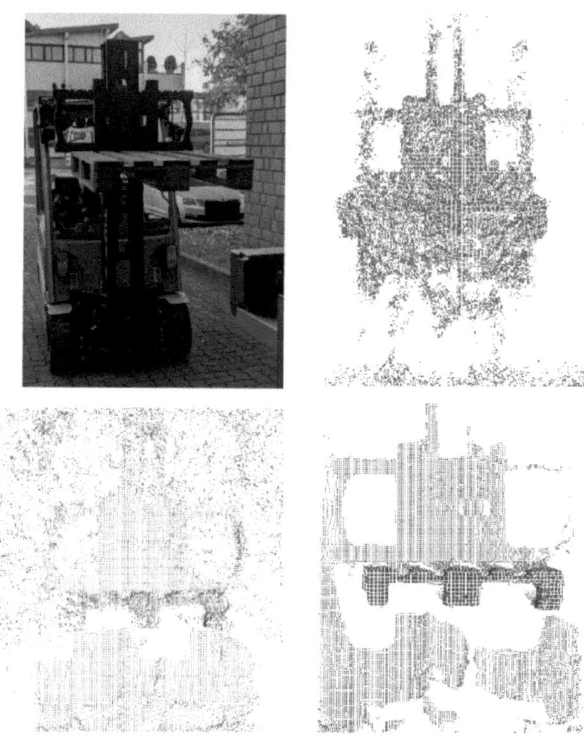

Figure 4: Point cloud of an forklift loaded with an euro-pallet. Ground truth (upper left), Sensor A (upper right), Sensor B (lower left), Sensor C (lower right)

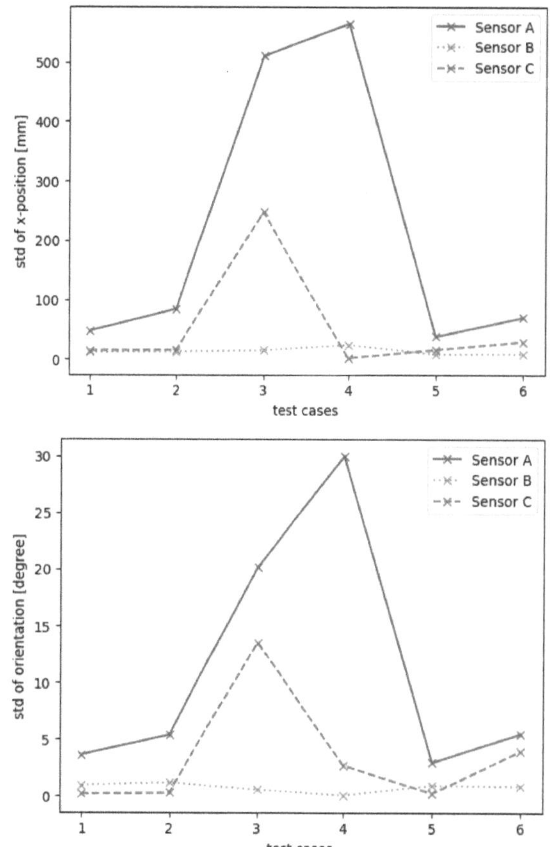

Figure 5: Standard deviation of the x-position (top) and orientation changes (bottom) in different test situations for all sensors: euro-pallet (1), euro-pallet with influence of reflector (2), black pallet (3), black pallet with wrapping (4), grid pallet (5), grid box with wrapping (6)

the wall (Sensor B) which can easily be filtered because of their position outside of the region of interest.

4.2 Outdoor Tests

First the sensors were placed in a doorway to receive intermediate amounts of daylight. Because this did not influence the performance of the sensors, the forklift was then placed in the direct sunlight. Sensor A detected the outlines of the forklift but could not distinguish between the forklift and the pallet. Sensor B had much noise but was able to differentiate between pallet and forklift. Sensor C had no noise and the pallet was detected with straight edges. This was possibly a result of the filters used in this sensor.

4.3 Pallet Detection

First, the noise of the data was evaluated visually. Fig. 3 shows the difference between a point cloud from Sensor A (left) and Sensor B (right). The right one has much less noise and clearer edges than the left. Sensor C performs as good as Sensor B in this case. Afterwards, the pallet detection algorithm was executed and the properties of the detected holes were evaluated.

Fig. 5 (top) shows that for nearly every test situation the measured distance between sensor and pallet varies in a range of less than 100 mm. Only the values from Sensor A differ more. The same behaviour is seen when looking at the orientation (see Fig. 5 (bottom)). Sensor B and C differ less than five degree from the mean (with one exception),

Table 2: Mean object detection probabilities

	europallet	black pallet
Sensor A	0.894	0.777
Sensor B	0.795	0.775
Sensor C	0.955	0.852

but Sensor A has much higher values. Both figures show that Sensor C and Sensor A have problems with the black pallets, because the data varies more than with other pallet types.

In Fig. 6 and 7 the dimensions of the detected objects are shown. These show that sensors B and C are better at measuring the correct size of the pallet holes than Sensor A. Additionally, Sensor C performs worse when detecting the black pallet. The probability of the object detector also supports this (see Table 2).

4.4 Discussion

The older Sensor A is not as good as the other two sensors, the results were noisier, with less precision in pallet detection and a lower operating range (see Table 3). Sensor B has problems with noise, especially under the influence of

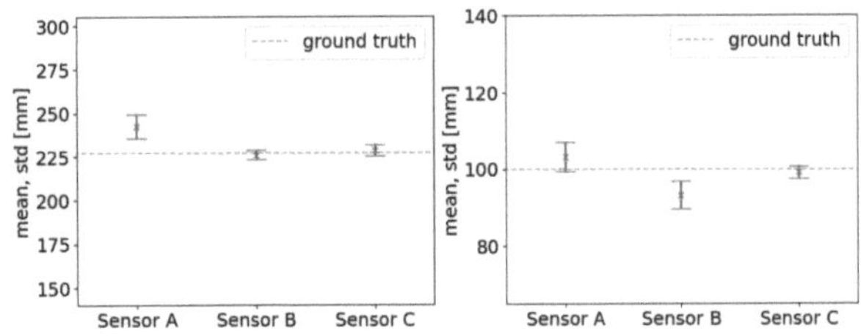

Figure 6: Dimension of detected holes in an europallet: horizontal (left), vertical (right)

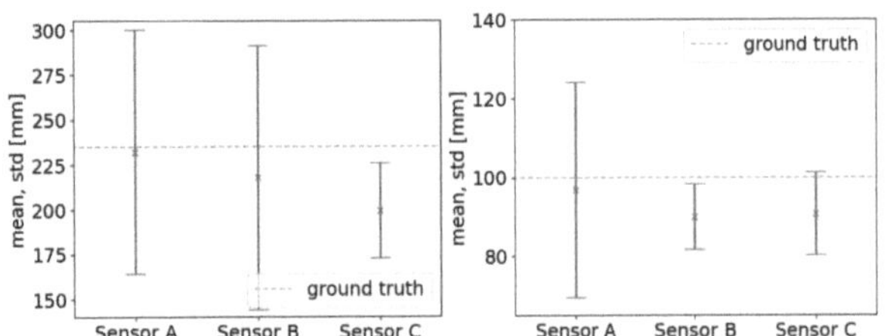

Figure 7: Dimension of detected holes in an black pallet: horizontal (left), vertical (right)

day- and sunlight. Sensor C has filters that could not be deactivated, so the distance is limited to ten meters. On the one hand the filter probably results in a worse performance with black objects, on the other hand it leads to the least noisy results. In pallet detection both sensors B and C deliver good or very good output, depending on the type of the pallet. It is important to say that the results are probably not comparable because of the filter used.

Table 3: Results of the evaluation

	Sensor A	Sensor B	Sensor C
Palette detection	o	++	++
Outdoor	-	o	+
Max. Distance	- (6m)	+ (11.5m)	+ (10m)
Distance < 1m	- -	-	o
Influence of black objects	-	+	-
Noise	-	+	++
Influence of reflectors	o	o	o

5 Conclusion

This paper evaluates three Time-of-Flight depth sensors which are used in autonomous vehicles to handle loads autonomously. Tests were executed to verify the minimum and maximum range, the noise of the sensors in different situations, the influence of day- and sunlight and the pallet detection performance. The evaluation shows that it depends on the use case which sensor is better, but sensors B and C clearly outperform Sensor A.

Acknowledgement

The work has been carried out at Still GmbH, Department Mobile Automation - Intelligent Autonomous Software Development, Berzeliusstraße 10, 22331 Hamburg and supervised by the Institute of Electrical Engineering in Medicine, Universität zu Lübeck. The author would like to thank Matthias Haase from Still GmbH for helping with the tests and the evaluation of the results.

Author's Statement

Conflict of interest: Authors state no conflict of interest. The sensors are named under a pseudonym as requested by the Still GmbH.

6 References

[1] STILL Very narrow aisle truck. Available: https://www.still.de/fahrzeuge/gabelstapler-und-lagertechnik/schmalgangstapler/mx-x.html [last accessed on 2022-01-23]

[2] Time of Flight (ToF) Sensors Bring Autonomous Applications to Market. https://www.automate.org/industry-insights/time-of-flight-tof-sensors-bring-autonomous-applications-to-market [last accessed on 2022-01-23]

[3] Automation in logistics: Big opportunity, bigger uncertainty. https://www.mckinsey.com/industries/travel-logistics-and-infrastructure/our-insights/automation-in-logistics-big-opportunity-bigger-uncertainty [last accessed on 2022-01-23]

[4] V. Castaneda, N. Navab, *Time-of-flight and kinect imaging.* In: Kinect Programming for Computer Vision, Citeseer, 2011.

Inspection of Wind Turbines with Autonomous Drones

Jasper Pflughaupt [1]

[1] Robotics and Autonomous Systems, Universität zu Lübeck, jasper.pflughaupt@student.uni-luebeck.de

Abstract

With the expansion of renewable energies the number of operational wind turbines is increasing steadily, both onshore and offshore. These turbines need to be inspected and maintained to avoid serious accidents, especially in the harsher offshore environments where damage repair is complex and cost intensive. To cope with an increasing demand for such inspections in an economically efficient manner with low down times, new methods are required that allow fast and early detection of possible damage. A prototype drone system for autonomous inspection of offshore wind turbine rotor blades is being developed in the Autonomous Offshore Drone (AOD) project. It carries both thermographic and conventional cameras that allow the detection of anomalies up to 18 cm deep into the structure. Close proximity to the rotor blade, accurate positioning of the drone, and resistance to turbulent winds are required to ensure data quality, necessitating a fast and reliable control approach.

1 Introduction

Wind power is one of the corner stones of sustainable power generation. With an increasing number of operational wind turbines, the demand for inspection and maintenance rises. Traditionally, these inspections are done by industry climbers or from helicopters, which is cost-intensive, time consuming, and dangerous. As the wind turbines need to be halted during the inspections, additional opportunity costs arise. Using drones for the inspection missions instead can reduce costs, required time, and danger to humans.

A patented method developed by ROLAWIND GmbH allows detection of anomalies up to 18 cm deep into the structure of the wind turbine's rotor blades by capturing thermographic images [1]. In the Autonomous Offshore Drone (AOD) project, a prototype drone system for the autonomous inspection of offshore wind turbine rotor blades is being developed [2]. It utilizes both thermographic and conventional imaging to gather more data than traditional approaches and aims to reduce the inspection time via automation.

At the Universität zu Lübeck, the control software for the prototype is developed. In this paper, methods for sensor data processing as well as planning and execution of the inspection flight trajectory are presented. This includes in particular the ascend and approach to the wind turbine, the detection of the turbine's orientation and the control of distance and angle relative to the rotor blade.

1.1 Inspection Flight Profile

Each of the three rotor blades of a wind turbine is inspected individually, due to the limited flight time of the drone. The wind turbine is halted such that the blade that is to be in-spected points straight up. The inspection flight consists of a takeoff maneuver from a ship that transported the drone to the wind turbine, the ascent to the nacelle height, approaching the nacelle up to a distance of about 10 m, the ascent and descent along all four sides of the rotor blade while keeping a predefined distance and angle to the blade and capturing both thermographic and conventional images, and finally the landing maneuver on the ship.

The distance to the wind turbine varies drastically over the course of the mission, preventing the use of some sensors during some phases. Thus, specialized sensors and control approaches for each phase are required. Due to the close proximity to the rotor blade when capturing the images and possibly strong and gusty wind conditions, fast and accurate position control is needed.

The goal of the AOD project is to develop a prototype that is able to carry out the entire mission without human input. At the same time, due to both safety and legal reasons, a pilot is able to intervene at any point.

2 Hard- and Software

To reduce the development time of the prototype, a commercially available drone is used as a basis. It is then outfitted with a computer to run the control software and additional sensors that provide the necessary data. Fig. 1 shows the prototype in the current configuration.

2.1 Drone Platform

The DJI Matrice 300 RTK (M300) was chosen as the platform to build the prototype upon. The M300 is a 3.6 kg (6.3 kg with batteries) quad-rotor drone with a maximum

Figure 1: DJI M300 with Zenmuse H20T on bottom gimbal, control computer and Hokuyo UST-20LX on top.

Figure 2: AirSim simulation of a drone in Unreal environment with several wind turbines. The picture shows the view from a simulated FPV camera with 145° FOV.

payload weight of 2.7 kg that is designed for commercial use in inspection, mapping, or search and rescue operations. Three gimbal mounts allow the installation of multiple sensor solutions. It can withstand winds of up to 15 m/s and is equipped with redundant sensor and flight control systems, allowing operation in offshore conditions. A control computer running Ubuntu 20.04 with ROS Noetic is mounted on top of the drone to handle sensor data processing, flight planning and position control. It interfaces with the built in flight control system to exchange sensor data and issue control commands.

2.2 Sensors

The M300 has stereo cameras and infrared sensors for obstacle detection in all directions and a first-person-view (FPV) camera with 145° field-of-view (FOV). To accurately measure the distance and angle to the wind turbine's rotor blade a Hokuyo UST-20LX light detection and ranging (LIDAR) sensor with a range of 20 m is mounted on top of the M300. Thermal images are recorded with a DJI Zenmuse H20T hybrid camera module on the bottom gimbal position. Using both the gimbal and electronic image stabilization (EIS) it provides smooth image data that is not disturbed by pitch and roll maneuvers of the drone due to wind or acceleration.

2.3 Simulation

To test control algorithms, the AirSim [3] simulator is used. It features physics-based simulations in photorealistic Unreal Engine environments. This allows the creation of complex scenarios in which most aspects of the control algorithms can be tested. Fig. 2 shows the drone in a simulated environment.

3 Control Algorithms

In different phases of the inspection mission, the distance between drone and wind turbine varies drastically. During the initial ascend, a safety distance larger than 50 m is likely, so the turbine is out of LIDAR range. When inspecting the blade, a distance of 8 to 10 m must be maintained, requiring fast and accurate position control to avoid collisions under windy conditions. Different wind turbine models vary in shape and size, requiring adaptive control. Thus, specialized algorithms for sensor data processing and control are necessary for each part of the mission. In the following sections, a selection of these algorithms is presented.

3.1 Surface Angle Measurement

For the thermographic images to contain the wanted data and in order to prevent reflection issues, the drone must keep its distance and angle to the rotor blade within certain intervals. Hence, a way to measure said angle is required. This is non-trivial, because the surface of the blade is not flat but curved. Calculating an angle from distance measurements requires at least two points. Which points are chosen and the distance between them has an influence on the measured angle, if the surface is not flat between them. The algorithm presented in this section describes a method of estimating the angle of a surface to a LIDAR ray at the point of contact. This is done by using multiple pairs of measurements around this point of contact and averaging the angles obtained from them.

Fig. 3 illustrates the geometry of these measurements. $\overline{SP_1}$ and $\overline{SP_2}$ are one measurement pair. $\overline{SP}$ coincides with the central LIDAR ray, however, their length is only equal if the surface is flat between P_1 and P_2. When the green and blue lines coincide, i.e. when d_1 and d_2 are equal and α is zero, the object's surface is perpendicular to $\overline{SP}$.

Let without loss of generality d_1 be larger than d_2. $\overline{SP}$ is an angle bisector, thus d can be calculated as:

$$d = \frac{2 \cdot d_1 \cdot d_2 \cdot \cos(\delta)}{d_1 + d_2}. \tag{1}$$

Per the law of cosines l_1 and l_2 are:

$$l_{1/2} = \sqrt{d^2 + d_{1/2}^2 - 2 \cdot d \cdot d_{1/2} \cdot \cos(\delta)}. \tag{2}$$

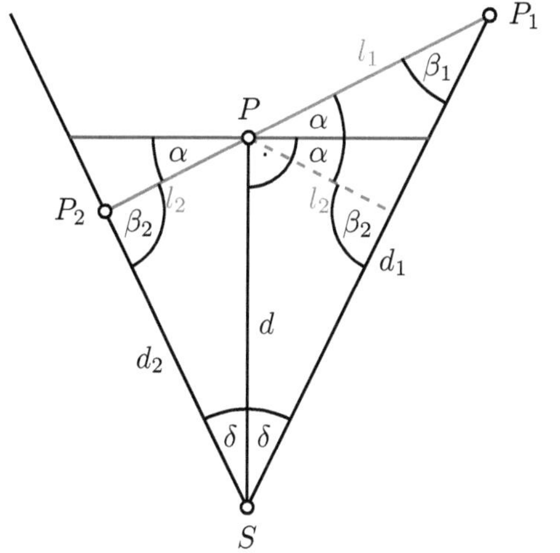

Figure 3: Geometric relations between LIDAR rays and object surface. S is the location of the LIDAR sensor, d_1 and d_2 are distances measured by the LIDAR to the points P_1 and P_2 respectively. δ is the opening angle between the measurements. The green line indicates the surface of the object (assumed to be flat between P_1 and P_2). α is the sought-after angle.

The law of cosines is used again to calculate β_1 and β_2:

$$d^2 = d_{1/2}^2 + l_{1/2}^2 - 2 \cdot d_{1/2} \cdot l_{1/2} \cdot \cos(\beta_{1/2}),$$
$$\beta_{1/2} = \cos^{-1}\left(\frac{d_{1/2}^2 + l_{1/2}^2 - d^2}{2 \cdot d_{1/2} \cdot l_{1/2}}\right). \tag{3}$$

Finally, the surface angle α follows from the sum of angles in the upper right triangle spanned by l_1 and the dashed l_2:

$$180 = 2 \cdot \alpha + \beta_1 + (180 - \beta_2),$$
$$\alpha = 0.5 \cdot (\beta_2 - \beta_1). \tag{4}$$

If d_1 is larger than d_2, α is positive, otherwise it is negative. It describes how far the surface is from being perpendicular to the central LIDAR ray ($\overline{SP}$ in Fig. 3).

Depending on the required accuracy and stability of the result, multiple measurements with increasing opening angle δ can be averaged. This also helps to reduce noise from the LIDAR data. By selecting the increment between opening angle steps, the total opening angle can be controlled.

3.2 Computer Vision

During the initial phases of the mission, when ascending and approaching the wind turbine, the distance to it will likely be larger than 20 m. Thus, it is out of LIDAR range and other methods to find and approach it are required. It cannot be assumed that the drone will be facing in the direction of the turbine after takeoff, so simply moving forward until the distance is less than 20 m is not an option. While asking a user to specify the GPS coordinates of the turbine is possible, it is not a convenient solution. Other methods of detecting the turbine at range would make operating the drone more versatile and flexible.

Figure 4: Edges detected in FPV camera images are drawn as red lines. On the left, the towers of two wind turbines in the AirSim simulation are detected. On the right, the edges of a 5 m high dummy are detected by the M300.

The wind turbine's color and distinct shape make it possible to design image processing algorithms specifically for detecting it. The drone can process images from the FPV or stereo cameras to detect the turbine and then turn and move towards it until it is within LIDAR detection range. It is sufficient to acquire a rough estimate of the direction in which the turbine is located: starting 100 m away from the turbine, the drone would still come within the 20 m LIDAR detection range with an error in the estimated direction of up to $\tan^{-1}\left(\frac{20\mathrm{m}}{100\mathrm{m}}\right) \approx 11.31°$.

The comparatively more challenging task is differentiating the turbine that is to be inspected from the others in the wind park.

3.2.1 Edge Detection

Edge detection is one of the most basic computer vision approaches. The wind turbine's tower offers two prominent edges which can be detected and used to determine in which direction the turbine is located.

Convolving the camera image with an edge detection kernel and applying the probabilistic Hough line transformation [4] yields the locations of prominent edges in the image. Filtering for vertical edges excludes those detected on the ground, the horizon, and at the rotor blades. This results in a robust detection of the wind turbine's tower. Fig. 4 shows two example results, where all detected vertical edges are drawn as red lines into the images.

3.2.2 Depth from Motion

The edge detection algorithm described above will often also detect turbine towers in the background of the image. As this is a likely scenario in an offshore wind farm, methods to work around this limitation are required. A depth from motion approach [5] can be used to acquire depth information for the images. This allows selection of those lines belonging to the closest turbine.

When capturing two images of a scene while moving the camera sideways, objects close to the camera will appear at different places in the resulting images, while those further away will not move as much. Calculating the relative distance each pixel moved between the two images by correlating features in both images results in an estimate of the optical flow, which describes the apparent motion of objects in the images. As all turbine towers in the scene are static and the drone can simply be moved sideways to capture images from different perspectives, those towers with higher apparent motion in the images must be closer to the drone. While this does not provide an absolute depth measurement, it allows selection of those lines that correspond to the closest wind turbine. This is enough to detect the direction in which the turbine is located.

Traditional feature tracking algorithms like the Lucas-Kanade method [6] track pixels with high information, such as corners. Experiments showed that this is not suitable for the features provided by the wind turbines, a more specialized feature tracker is required, e.g. one that directly tracks edges.

3.3 Determining Nacelle Orientation and Dimensions

Knowledge of the orientation of the wind turbine's nacelle can be used to infer the rotor blade position and orientation or for positioning the drone at specific locations relative to it. This section describes a maneuver for measuring the nacelle's orientation, independent of its exact shape and the approach direction of the drone.

Initially, the drone approaches the nacelle up to a distance of 10 m. Distance to and direction of the nacelle are measured with the LIDAR. The drone then moves laterally while holding the distance and facing the nacelle. This results in an elliptic trajectory around the nacelle. During this motion the distance between the outermost points detected by the LIDAR is calculated, its maximum value is stored. It serves as a rough size measurement of the nacelle side currently facing the drone.

After a quarter turn the drone has measured both the longer and the shorter side of the nacelle. It now rotates back around the nacelle, until the current size measurement roughly matches the maximum observed previously. At that point, the drone is located next to the wide side of the nacelle. It now centers itself in front of this wide side by moving laterally until the distance measured to both nacelle edges is the same. The nacelle orientation is now equal to the drone's orientation plus or minus $90°$, depending on whether the drone is on the left or right side of the nacelle. This can be determined by ascending a few meters above the nacelle. The rotor blade facing up is now the only object detected by the LIDAR. It is either to the left or to the right, which determines the nacelle's orientation.

4   Conclusion and Outlook

In the AOD project a prototype drone system for the autonomous inspection of wind turbine rotor blades is being developed. The hardware is completed and the software architecture upon which the control logic is build is in place. Simple flight tasks have been demonstrated both in simulation and with the M300 at a model airfield. This includes avoiding objects detected by LIDAR, holding a fixed distance to a moving target, circling objects at various distances and speeds, and detection of dummy wind turbines with the FPV camera.

In simulation more complex maneuvers have been successfully implemented. This includes:

1. detecting the wind turbine in FPV camera images
2. the maneuver for determining the nacelle's orientation
3. ascent and descent along all four sides of the blade.

These have been combined to a near complete inspection mission.

Until now, tests with the M300 could only be conducted with dummy models like the one shown in Fig. 4. To validate the algorithms tested in simulation, test flights at an onshore turbine are planned for the near future. This will allow execution of the complex maneuvers on an installation with the correct scale. At the same time the weather conditions will not be as harsh as offshore, resulting in a safer test environment. Once these tests are successful, flights at offshore turbines will be scheduled.

Acknowledgement

The work has been carried out at the Institute for Electrical Engineering in Medicine, Universität zu Lübeck and supervised by M. Sc. Carlos Castelar.

Author's Statement

Conflict of interest: Authors state no conflict of interest.

5   References

[1] "ROLAWIND GmbH, thermographic inspections," https://www.rolawind.de/thermografie/, accessed: 14.12.2021.

[2] "Autonomous Offshore Drone (AOD) funding project," https://www.vde.com/renewables/newsroom/autonomous-offshore-drone, accessed: 14.12.2021.

[3] S. Shah, D. Dey, C. Lovett, and A. Kapoor, "AirSim: High-Fidelity Visual and Physical Simulation for Autonomous Vehicles," in *Field and Service Robotics*, 2017.

[4] J. Matas, C. Galambos, and J. Kittler, "Robust detection of lines using the progressive probabilistic hough transform," *Computer Vision and Image Understanding*, vol. 78, no. 1, pp. 119–137, 2000.

[5] G. Sperling and B. A. Dosher, "Depth from motion," *Early vision and beyond*, pp. 133–142, 1994.

[6] B. Lucas and T. Kanade, "An iterative image registration technique with an application to stereo vision (IJCAI)," vol. 81, 04 1981.

Evaluation and Error Correction of Position Tracking Systems of Autonomous Vehicles for Indoor Localization

Robin Angelstein [1], Sven Ole Schmidt[2], Fabian John[2], and Horst Hellbrück [2]

[1] Applied Information Technology, Technische Hochschule Lübeck, robin.angelstein@alumni.msoe.edu
[2] Department of Electrical Engineering and Computer Science, Technische Hochschule Lübeck, {sven.ole.schmidt, fabian.john, horst.hellbrueck}@th-luebeck.de

Abstract

The number of autonomous vehicles on the market increased during the last few years. For those vehicles, a correct working position tracking system (PTS) with input coordinates and target coordinates is most important. PTS enables locating the movements of objects. The paper covers the analysis of the PTS of an autonomous vehicle which is not verified. This means that it is not known if the target coordinates follow the input coordinates. If the target coordinates of an autonomous vehicle are unknown, it is not valid ground truth. Therefore, the system is not applicable as an autonomous segway. The focus of this paper is to develop and implement an error correction for the PTS with a following evaluation of the corrected system. The evaluation compares movement patterns to an already validated PTS. The system is evaluated with real measurements. The measurement results show that the implemented error correction algorithm improves the results for basic movements but more iterations are needed to run the system as a fully autonomous vehicle.

1 Introduction

Autonomous driving devices become more and more common nowadays. When a self-driving system moves without any human interaction it must be reliable. Talking in terms of the under-laying coordinate system, input coordinates must match the target coordinates. It does not matter how good the artificial intelligence is if the calculated coordinates do not match the target coordinates. This paper focuses on the evaluation of a not validated PTS of an autonomous vehicle. We implemented different movement patterns to test the PTS. We programmed a graphical User Interface (GUI) application to navigate the autonomous vehicle. To evaluate the system we compared the results to a validated PTS. By analyzing the results we derived an error-correcting algorithm. We choose very general methods and a generic measurement setup to apply to other evaluations for autonomous vehicles.

2 Analysed autonomous vehicle

This section gives an overview of the analyzed autonomous vehicle. At first, it provides a general overview, afterward, the section describes the coordinate system, followed by a description of two navigation modes. Lastly, it explains the defined movement patterns for the evaluation and the error correction.

2.1 Description of autonomous vehicle

The analyzed autonomous vehicle is the so-called Loomo Segway Robot from the company Segway. [1] In the following, it will be called Loomo. The Loomo combines a self-balanced vehicle (SBV) and an autonomous robot. In robot mode, the Loomo operates as an autonomous vehicle and receives input coordinates and processes them. The Loomo runs on Android and is programmed in Java. The company Segway supplies a Software Development Kit (SDK) with all needed functionality to control the Loomo. [2]

The under-laying coordinate system of the Loomo consists of an X, Y, and Z plane. The system shown in Figure 1 is a Cartesian coordinate system.

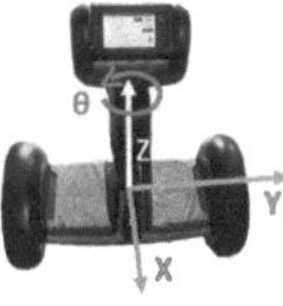

Figure 1: Loomo coordinate system

To reach a desired coordinate it takes X and Y coordinates and the angle θ (theta) as an input. By setting a new checkpoint to X, Y, and a defined angle θ, Loomo will arrive at the point in the X-Y plane and afterward, it will rotate itself by the angle θ. The value of θ ranges from $-\pi$ to π. Increasing a coordinate by one leads to a movement of approximately one meter. The problem is that it is not ensured that increasing a coordinate by one leads to a movement of exactly one meter.

The Loomo internally has the two navigation modes odometry and Visual Localization System (VLS). Section 3 compares them against each other. As default, Loomo operates with relative odometry to determine its position. With odometry, it counts the wheel revolutions by incremental encoders. This mode is considered inaccurate because it is based on the assumption that wheel revolutions directly translate into the covered distance. Following J. Borenstein and L. Feng, a heavy location offset occurs by wheel slippage. [3]

Because of this, the manufacturer of the Loomo claims accuracy for odometry in a range of 0.25 meters. [1] VLS takes depth images to calculate the coordinate data. [4] According to the company Segway this has higher accuracy.

2.2 Implemented Movement Pattern

Section 3 evaluates the PTS of the Loomo. Therefore it is part of this paper to define movement patterns. We defined the following criteria for movement pattern:
1. Include uni-directional and multi-directional movements
2. Take each measurement for different distances
3. Take each measurement for odometry and VLS mode
4. Compare rotation by using angle θ vs. rotation by using X and Y coordinates at the same time

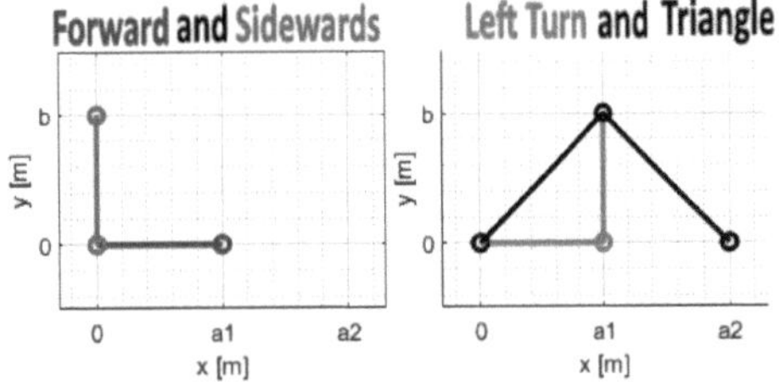

Figure 2: Movement Pattern

As shown in Figure 2 two movement patterns are defined as uni-directional. Point P(0;0) is the starting point for all measurements. The left plot shows in blue a forward movement and a sidewards movement in red. Each movement applies for individual distances a1 and b. These two patterns are basic movements. In addition, these two patterns fulfill criteria 1,2 and 3.

Figure 2 also shows two multi-directional movements in the right plot. The two patterns are called left turn and triangle. The left turn pattern shown in green moves forward until a1 is reached. The Loomo performs a 90 ° rotation using the angular rotation functionality. Another forward movement follows after the rotation. In comparison, the triangle pattern shown in black performs a rotation by applying the X and Y coordinate at the same time. This addresses criteria 4. The next section utilizes all four movement patterns.

3 Error Correction of the Position Tracking System

To solve the issue of not validated PTS measurements we perform measurements in order to evaluate the system. Therefore this section is divided into the measurement procedure and the error correction of the PTS.

3.1 Data Analysis

To detect if the Loomo coordinates match a ground truth this section tests the previously defined pattern. Figure 3 shows the measurement setup as a MQTT Publish/-Subscribe Model. Message Queuing Telemetry Transport (MQTT) is a network messaging protocol based on a publish-subscribe communication. [5]

The Technische Hochschule Lübeck owns a validated indoor Ultra-Wideband (UWB) localization system. [6] Multiple anchors communicate with a sensor tag.

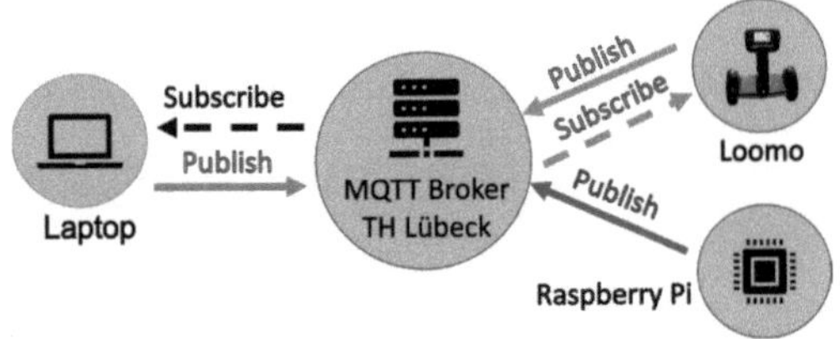

Figure 3: Measurement Setup

In mean the maximum deviation is 20 cm. This localization system operates as the ground truth (GT) in this paper. The tag of the GT is connected serially to a Raspberry Pi and mounted on the Loomo. The microcontroller sends GT coordinates via MQTT. To implement the movement patterns part of this paper is to write a Java application. In a Graphical User Interface (GUI) the user selects the movement pattern and sets the distance for X and Y coordinates. A user controls the application via direct touch input at the Loomo display or via MQTT. The Loomo MQTT Client follows the Paho Android Service MQTT Client. [7] A laptop saves and visualizes the data and sends and receives commands.

We start a measurement by sending a remote command to not influence the setup by any human interaction. This paper defines the actuator size as set value. Having up to three different distances for each movement pattern for each navigation mode leads to the desired number of 22 measurements. Each measurement is recorded two times. The evaluation uses the mean values. The goal of the overall measurement is to find an algorithm to equal the Loomo coordinates, the GT, and the set coordinates. Because the Loomo and the GT are using coordinate systems with different origin, the absolute value |XY| of the X and Y coordinate is taken in order to compensate this difference.

Table 1: Measurement Results

Movement Pattern	Mode	Loomo | XY | [m]	GT | XY | [m]	Set value | XY | [m]
Forward 1 m	Odo	1.391	1.291	1.000
Forward 2 m	Odo	2.010	2.380	2.000
Forward 3 m	Odo	3.192	3.376	3.000
Forward 1 m	VLS	1.182	1.307	1.000
Forward 2 m	VLS	2.260	2.364	2.000
Forward 3 m	VLS	3.202	3.352	3.000
Sidewards 1 m	Odo	1.080	0.897	1.000
Sidewards 2 m	Odo	2.119	1.943	2.000
Sidewards 3 m	Odo	2.911	3.090	3.000
Sidewards 1 m	VLS	0.971	0.819	1.000
Sidewards 2 m	VLS	1.919	1.900	2.000
Sidewards No. 2 m	VLS	1.977	1.973	2.000

Movement Pattern	Mode	Loomo angle θ [°]	GT angle θ [°]	Set angle θ [°]
Left Turn 1 m	Odo	126.993	79.922	90.000
Left Turn 2 m	Odo	80.508	84.774	90.000
Left Turn 3 m	Odo	50.928	87.314	90.000
Left Turn 1 m	VLS	74.458	72.157	90.000
Left Turn 2 m	VLS	84.483	74.255	90.000
Left Turn No. 2 m	VLS	82.474	80.201	90.000
Triangle 1 m	Odo	55.587	94.510	90.000
Triangle 2 m	Odo	42.902	78.119	90.000
Triangle 1 m	VLS	107.495	100.371	90.000
Triangle 2 m	VLS	41.479	73.815	90.000

By only applying either X or Y direction optimizations a time during the movement, the absolute value $|XY|$ is directly applicable as the X or Y coordinate although they are absolute.

$$|XY| = \sqrt{X^2 + Y^2} \qquad (1)$$

Table 1 shows the most important measurement results. It shows the odometry and VLS navigation mode for the distances between 1 m, 2 m, or 3 m. For the forward measurements, there is no correlation for the odometry mode. Table 1 shows also the results of the sidewards movement pattern. The bottom of Table 1 shows the multi-directional movements. For left turn patterns VLS delivers the best results. This is valid for the distance, as well as for the angular rotation. The most interesting about this measurement is the angular rotation in comparison to the rotation by using X and Y coordinates. The triangle movement pattern covers the multi-directional movement by using X and Y coordinates instead of the angular rotation. The results for triangle movements are way worse compared to the angular rotation. Therefore this paper focuses on angular rotation rather than rotation by using X and Y coordinates.

Table 2: Correction Offsets

Movement Pattern	Loomo offset [cm]	Set value offset [cm]	Total corrected offset [cm]
Forward	+13	+35	5
Sidewards	-6	-10	10

Movement Pattern	Loomo angle offset θ [°]	Set angle offset θ [°]	Total corrected offset θ [°]
Left Turn	-5	-14.5	6

Because the GT is the measurement reference, the goal is to correct the set value and the Loomo coordinate. Table 2 shows the derived offset optimizations from the different measurements. In most cases, we detect a pattern only for VLS mode. For this reason, we decide to continue with VLS mode for the error correction algorithm. It is not possible to combine the two movement modes. For the forward movements calculating the mean over the offsets for all three measurements show, that the absolute Loomo coordinates must be increased by a constant factor of 13 cm and the set coordinate must be increased by additional 35 cm. Applying this optimization would lead to a maximum offset of 5 cm between all three absolute coordinates. The other results are interpretable in the same manner. For the sidewards patterns, the correction is tested for a left sidewards movement and a right sidewards movement and is valid in both cases. Verifying these offsets in the evaluation would be desirable because the results will be theoretically lower than the claimed 25 cm offset from the manufacturer.

3.2 Error Correction

For error correction, the GT is used as a reference. The goal is to match all three coordinate sources. Therefore it is tried to find an algorithm that corrects the coordinates after each movement. The measurements from the previous subsection show that the VLS mode performed best. Another important conclusion is obtained from the two multi-directional movement patterns. It is analyzed that the angular rotation performs better than using X and Y coordinates

at the same time to do a 90 ° rotation. Applying the Y coordinate is only performed when no X coordinate is involved in the movement. The reason for this is that the angle movement works better for rotations.

Algorithm 1: Error correction algorithm

  Result: Correct Loomo and set coordinates to match the GT
  Data: $|XY|$ describes absolute value of X and Y
1 **if** *movement into X direction in VLS mode* **then**
2   Loomo: X = $|XY|$ + 13 cm;
3   Set: X = $|XY|$ + 35 cm;
4 **else if** *movement into X and Y direction in VLS mode* **then**
5   handle it as forward movement followed by a 90° rotation, followed by another forward movement ;
6   Loomo: $\theta = \theta$ - 5°;
7   Set: $\theta = \theta$ - 14.5°;
8   For the X coordinate apply the X direction offset;
9 **else if** *movement into Y direction in VLS mode* **then**
10   Loomo: Y = $|XY|$ - 6 cm;
11   Set: Y = $|XY|$ - 10 cm;

Algorithm 1 summarizes the previous derived rules. For the previous measurements, the correction algorithm looks promising.

4 Evaluation

To show that the error correction algorithm works as wanted, an evaluation with additional measurements is performed. The measurement procedure is equal to section 3. The Loomo contains a software version including the error correction algorithm. The algorithm is evaluated with eight movement patterns. All results show the average over three repetitions. The section starts testing five basic patterns again.

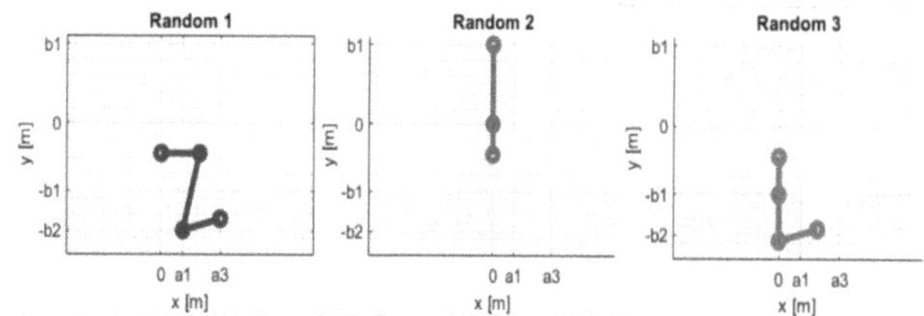

Figure 4: New random Movement Pattern

In addition, the evaluation introduces three random movement patterns. Those three shown in Figure 4, evaluate if the new algorithm is also applicable for more complex patterns.

Table 3: Comparison of corrected measurements

Movement Pattern	Mode	Loomo $|XY|$ [m]	GT $|XY|$ [m]	Set value $|XY|$ [m]
Forward	VLS	2.420	2.444	2.350
SidewardsLeft	VLS	0.930	0.839	0.900
SidewardsRight	VLS	-1.100	-1.040	-1.100
Random 2	VLS	3.480	3.51	3.250

Movement Pattern	Mode	Loomo angle θ [°]	GT angle θ [°]	Set angle θ [°]
Left Turn	VLS	82.800	79.900	75.500
Right Turn	VLS	76.000	75.800	75.500
Random 1 1.Angle	VLS	80.200	85.139	75.500
Random 1 2.Angle	VLS	82.474	80.201	75.500
Random 3 1.Angle	VLS	82.58	75.69	75.500

The results from the first row in Table 3 show that the average error for a forward measurement between the absolute Loomo coordinate and the GT coordinate is 2 cm.

In addition, the average error between the absolute GT coordinate and the set Coordinate is 9 cm. This shows that the algorithm works as desired for the forward movement pattern. The second and third row of Table 3 shows the average result of both sidewards movements. These results look almost as good as for the forward movement pattern. On average for both cases, the Loomo absolute coordinate differs from the GT absolute coordinate approximately 6 cm, and the GT and the absolute set coordinate differ also about 6 cm. These results are slightly better than the ones without the error correction. The next row shows the Random 2 movement pattern. It is listed here because it also consists of a straight movement containing a left sidewards movement, followed by two forward movements. Here the worst difference between all three absolute coordinates is 23 cm. The offset is still within the 25 cm accuracy of the manufacturer, but it can be seen for the first time that combining different patterns leads to a higher error.

Table 3 compares the angles between the three coordinate sources for all measurements containing angular movements. For the average of the left turn and the right turn movement patterns, all three angles match very well. For both the average error between the Loomo angle and the GT angle is 1.55 ° and the average error between the GT angle and the set angle is 2.35 °. These results are way better than without the error correction algorithm. This shows that the algorithm improves the results also for these two movement patterns.

Table 3 also shows the angular results for the last two random functions. Random 1 movement pattern contains a right turn, followed by a left turn. Random 3 movement pattern consists of a right sidewards movement followed by a left turn. As shown in the table the angular results are very similar to the ones from the basic movement turns.

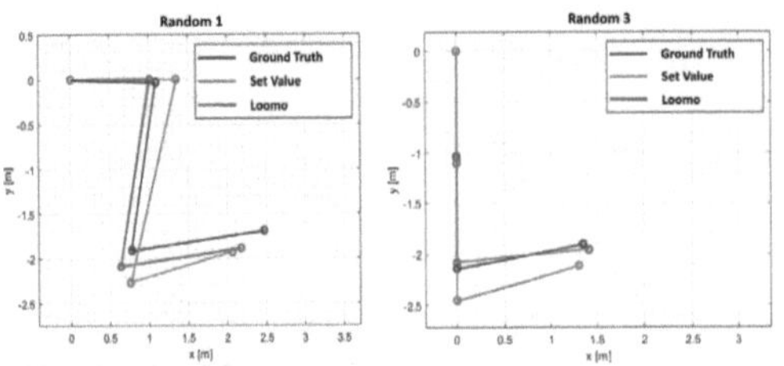

Figure 5: Results of Random 1 and Random 3

A combination of two turning movements like in the Random 1 movement pattern increases the error. This is also illustrated in Figure 5. In total, the target coordinates of all three sources are within 50 cm. The final GT coordinate is about 30 cm away from the Loomo and the set coordinate. Here the error correction algorithm has to be improved. The same is valid for the Random 3 movement pattern also shown in Figure 5. The measurement show there is still a small offset in the angular rotation. The figure points out that for a sidewards movement followed by a left turn, the set coordinates have to be optimized, especially P(1.35;0) and P(0;-2,45) have a huge offset compared to the GT and Loomo coordinate. The evaluation shows that for basic movement patterns, the error correction algorithm worked out well, but for many combined patterns with rota-

tions, the algorithm has to be optimized in another iteration to obtain better results.

5 Conclusion and Future Work

This paper shows one generic possibility to evaluate a PTS. It is stated that the coordinates of the Loomo are not trustworthy for an autonomous vehicle. In addition, an error correction algorithm is derived and implemented by using an external GT. Using this algorithm improves the coordinates of the Loomo. The evaluation section shows that for basic movement patterns, the error correction algorithm work out very well. For combined movement patterns with many rotations, the algorithm has to be optimized in another iteration to obtain even better results. By just requesting basic movements and applying the derived error correction algorithm the Loomo is usable as an autonomous vehicle. Because of these constraints the usage as a fully autonomous vehicle some future investigations are recommended. One possible way is taking all the output of the evaluation measurements from section 4 and expanding the error correction algorithm in further iterations. In addition, tests with larger movement distances lead to further evaluation of the error propagation. Another future study of interest is implementing a Kalman filter to linearize the mean target coordinate and the covariance [8]. Implementing this filter for both navigation modes would be out of scope for this paper but the output might improve the results.

Acknowledgement

This publication results from the research of the Center of Excellence CoSA at the Technische Hochschule Lübeck. Horst Hellbrück is an adjunct professor at the Institute of Telematics of University of Lübeck.

Author's Statement

Conflict of interest: Authors state no conflict of interest.

6 References

[1] SegwayRobotics, "Loomo sdk," 2019. [Online]. Available: https://developer.segwayrobotics.com/developer/documents/segway-robots-sdk.html

[2] ——, "Sdk robot," 2019. [Online]. Available: http://developer.segwayrobotics.com/developer/documents/java-doc.html?section=ConnectivitySDK-Robot

[3] J. Borenstein and L. Feng, "Measurement and correction of systematic odometry errors in mobile robots," *IEEE Transactions on Robotics and Automation*, vol. 12, no. 6, pp. 869–880, Dec. 1996.

[4] SegwayRobotics, "Loomo overview," 2019. [Online]. Available: https://developer.segwayrobotics.com/developer/documents/segway-robot-overview.html

[5] C. Bernstein, K. Bush, and A. Gillis, "Mqtt (mq telemetry transport)," 2021. [Online]. Available: https://internetofthingsagenda.techtarget.com/definition/MQTT-MQ-Telemetry-Transport

[6] S. Leugner and H. Hellbrueck, "Lessons learned: Indoor ultra-wideband localization systems for an industrial iot application," *University of Luebeck, Germany*, 2018.

[7] S. Kock, "Paho android service - mqtt client library encyclopedia," 2015. [Online]. Available: https://www.hivemq.com/blog/mqtt-client-library-enyclopedia-paho-android-service/

[8] O. Laureano Casanova, A. Fragaria, and F. Machaca, "Robot position tracking using kalman filter," vol. 2171, 07 2008.

Underwater Ultrasonic Multipath Diffraction Model for Short Range Communication and Sensing Applications with Reflector

Lukas Paul Schön [1], Fabian John [2], Horst Hellbrück [2]
[1] Applied Information Technology, Technische Hochschule Lübeck, lukas.paul.schoen@stud.th-luebeck.de
[2] Technische Hochschule Lübeck - University of Applied Sciences, Germany , Department of Electrical Engineering and Computer Science, Center of Excellence CoSA fabian.john@th-luebeck.de, horst.hellbrueck@th-luebeck.de

Abstract

In ultrasonic transmission, the line of sight influences the quality and the surroundings too. Due to reflections and other factors, the communications environment matters a lot. In this report, it gets investigated how the transmission of ultrasonic waves behaves. There will be a metal rod gradually distancing itself perpendicular to the line of sight and a reflector on the opposing side. The additional reflection was added parallel to the line of sight and fully reflects the signal. The simulation handles this by adding the send signal with a time delay and reducing it by the water's factor. The measurement does it by reflecting off of the transition of the water to the glass, together with the change of the glass to the air, both of which have a high reflection factor.

1 Introduction

Short-range wireless communication is a very well-documented and researched field. But the vast majority of research is made for signals in the air. There is far less research done in underwater environments, specifically ultrasonic waves. There are ways for communication to Work underwater [2, 7], even some involving acoustic signals [1], but very few with ultrasonic waves. It is well researched how ultrasonic soundwaves travel in the air [6], but the combination of water and ultrasonic waves is open for investigation. In [3] a model gets developed that simulates the properties of ultrasonic waves underwater and adapts to an obstructing metal rod between the sender and receiver. The publication showcases the effects of the metal rod obstructing the line of sight between the sender and receiver. It also documents a Matlab script to simulate ultrasonic waves in an underwater situation. It explains the results of different angles with the line of sight of the obstructing object. In this work, this gets supplemented by an additional reflector on the opposing side of the moving metal rod. This includes:

- An equation to optimize the result to the expected, measured result

- Expansion of the original document by the reflection mentioned above

- A specific Matlab script to evaluate the results of the simulation.

2 Material and Methods

2.1 Measurement setup

We set up a water tank with a sufficient framework to measure the ultrasonic sound waves. The water tank in use is equipped with a setup that enables the sender and receiver to dangle from the top into the water [4]. It is also equipped with a framework that holds the obstructing object. Either the object or the transmitter/receiver setup is moved with motors equipped in the framework. The motor moves with a precision of around $0.2\,\mathrm{mm}$ [5]. On the other side of the moving object is the wall of the tank. This composition is shown in Figure 1. Since the transition between water and glass and then between glass and air has a high reflection factor for ultrasonic signals, it will work well as the additional reflecting wall. The measurement setup is shown in Figure 1

To measure how the ultrasonic waves perform, the object gets placed between the sender and the receiver and gets repositioned away from them between each measurement.

2.2 Equipment for the Simulation

This simulation heavily builds upon the simulation [3]. This simulation works in a Matlab environment and calculates how the ultrasonic signal should behave given a specific distance of the object in the middle. Given no reflective side opposite of the moving object, it calculates the ultrasonic waves accurately within a particular range of the moving object. It does so by calculating the different factors that go into the transmission of ultrasonic signals. The major components are the diffraction of the object, LOS transmission

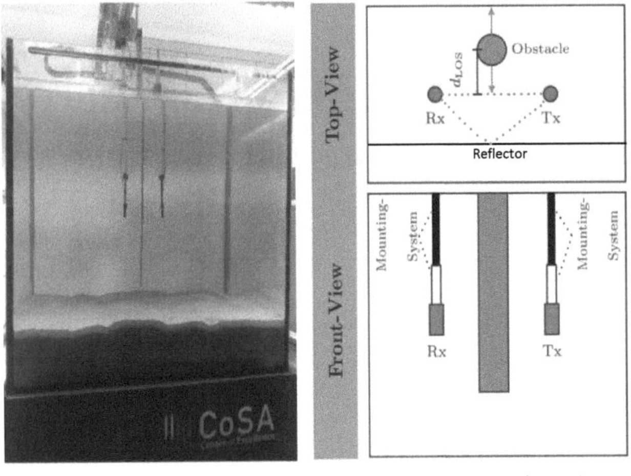

(a) Front view.

(b) Schematic top and front view.

Figure 1: The measurement setup used. ©2021 IEEE, adapted from [3].

between the sender and the receiver, and the reflection of an object. To make the best use of the simulation model, we applied the specific variables that determine the simulation outcome. These include, but are not limited to:

- obstacle diameter

- LOS distance of transducers

- reflection factors

- frequency band.

This script gets modified to include a wall on the other side of the moving object. For this, the distance of the reflected signals has to be calculated. Since the distance from the object, reflection, and transmitters is known, this is calculated with trigonometric functions. Those are implemented in the simulation. In the simulation, they will cause a delay for the transmission signal to arrive. A second variable that is important for the simulation is the distance of the line of sight of the object, which is also derived from simple trigonometric functions.

2.3 Procedure

To achieve the results, there have to be a few steps, those are:

- adapting the simulation

- measurement of actual data

- comparison of real and simulated data.

The simulation used in this report is the same that is used in [3]. This got adapted by additional formulas to optimize the results.

The measurement is made in a saltwater tank, which is shown in Figure 1. The setup itself is dangled from the top of the water into it, with a metal rod going perpendicular away from the line of sight of the transmitters. The added

reflection is on the opposing side of the obstructing metal rod.

The comparison with the measurement is made with the original Matlab plot, which shows both the original measurement and the simulation. This visualizes the differences of the simulation and measurement.

The comparison is made with an additional Matlab plot that uses an algorithm to compare the simulation of a specific distance of the metal rod with the measurements to determine which is most similar. This gets repeated for all simulation results, which results in a graph. Then those are done with and without the simulation to show the improvement the additional reflector had on the transmission.

3 Results and Discussion

3.1 Adaptation of the previous model

Since the previous Matlab script did not include a total reflection in the calculation, it had to be adapted. The most efficient way to do this was to add another object to the calculation that reflects all of the energy received and is a certain distance away. This is adapted in Matlab by adding an array with the coefficients into the variables list instead of just a single integer. Further, it is implemented that the first added variable in the array moves further away as it did in [3], while the second one is stationary.

3.2 Result adjustment of simulation

Adjusting the simulation results with the real measurements is a crucial part of correcting the outcome of the performance of the simulation. Similar to [3] the first and last few millimeters are not very accurate. This will not be fixed, but instead just be evaluated in the range between 40 and 80 mm.

The unmodified simulation is way too low, which is shown in Figure 2. The local minima and maxima are missing in the unmodified comparison, but the tendency still derives. Suppose a modification in the form of an adjustment to the frequency, the distance of the obstructing object, and a direct component gets added. In that case, these local minima and maxima get visible. With the modifications in place, it gets clear that the simulation is close to the actual measurements. This is shown in Figure 3.

3.3 Reformatting the results for the evaluation

An efficient way to compare two Signals is to subtract them and integrate the result. The lower the result of this integration, the more the signals look like each other. This method is only partially applicable to this problem, for three reasons:

- A digital signal is not continuous, but discreet

- A significant part of the signal has no relevant information

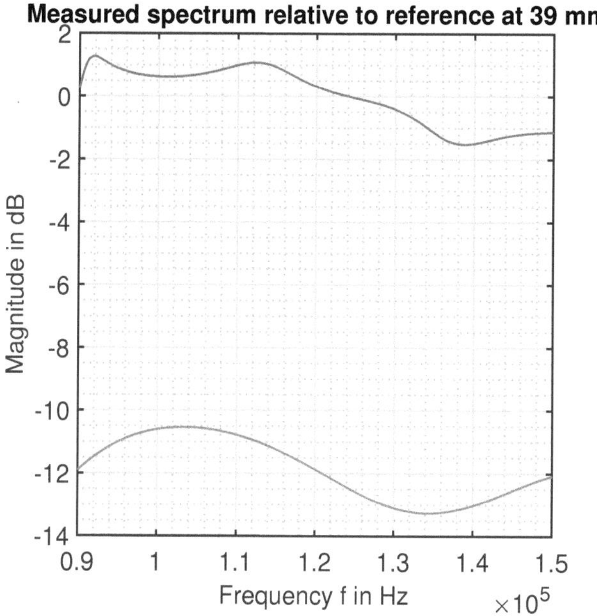

Figure 2: Unmodified comparison of the measured and simulated signal. Blue is the measurement, red is the simulation.

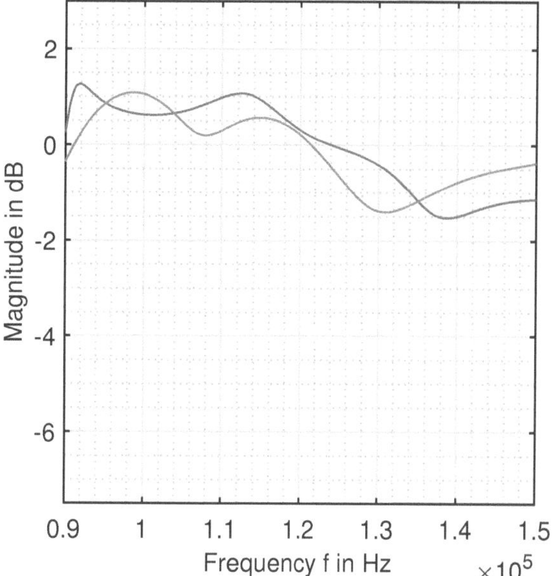

Figure 3: Modified comparison of the measured and simulated signal. Blue is the measurement, red is the simulation.

- The Signals include different direct components.

Different approaches solve these problems. The discrete values from this measurement are summed up instead, which yields the same result. For this approach to work, you will have to line up the frequencies perfectly, which gets achieved through interpolation. This loses a bit of accuracy, but since the frequencies lie very close to each other, this will not be significant.

Since the signal spectrum includes every frequency, not just the ones that have to be compared, this leads to wrong interpretations because the part of the signal with no relevant information is throwing off the algorithm. This is fixed by cutting the irrelevant parts of the signal off. This also fixes the issue of having such different direct components. This is because now the Matlab function for precisely this purpose is used.

3.4 Evaluating the simulation

The comparing algorithm tries to prove a working simulation by comparing it to the measurement and reassigning it the most fitting one. This gets done by eliminating the direct component, subtracting one signal from the other, and taking the absolute value of the result. After adding those results for every relevant frequency, you get a factor representing how much the signals differ from each other. The lowest value for each position the object traveled gets saved into an array. This array is plotted in Figure 4. There we show the different guesses of the algorithm applied with an x and the diameter of the metal rod applied with grey dotted lines.

In Figure 4 we show the graph for the signal with an additional reflection. The optimal outcome would be a function whose values never differ from the function $f(x) = x$. The current result shows that the simulation has clearly defined weaknesses and strengths. The simulation does not produce accurate results for lower distances of the obstructing object. This tendency gets better at around 25 mm, which is a turning point and has good results up to approximately 80 mm with few values that vary significantly from the function $f(x) = x$. Since the algorithm could recognize the distance of the obstructing object very accurately in the specified range, the simulation model is evaluated to determine the object position in presence of an additional reflector.

4   Conclusion

This experiment proves that the simulation from [3] is also applicable with an added reflection and still be accurate, regarding the transmission of ultrasonic waves. This is proven with a Matlab script that calculates the distance of the object in the line of sight of the sender and receiver. This is an improvement to evaluate the simulation to more realistic scenarios. Future Work on this is the inclusion of the form of the object to be any shape instead of just a round metal rod. Furthermore, suppose you wanted to improve the result of this paper itself. In that case, you could run an optimization algorithm on the adjustment of the simulation to enhance the quality of the simulation even further.

Acknowledgement

This publication results from the research of the Center of Excellence CoSA at the Technische Hochschule Lübeck

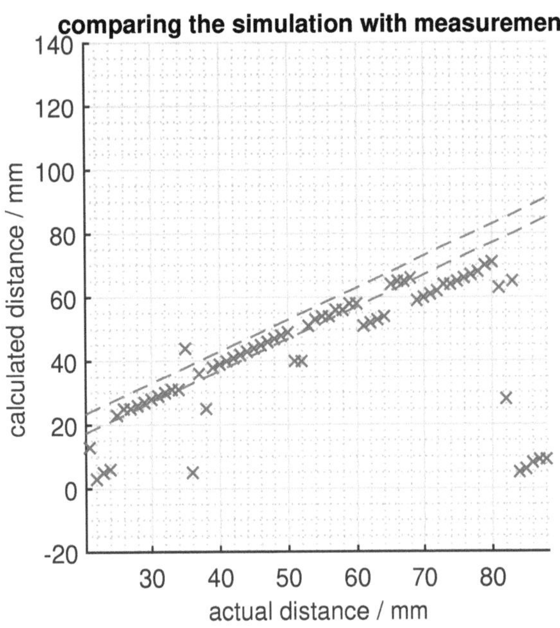

Figure 4: Result after comparing the simulation with the measurement.

and is funded by the Federal Ministry of Economic Affairs and Energy of the Federal Republic of Germany (Id 03SX467B, Project EXTENSE, Project Management Agency: Jülich PTJ). The Product of the Module "Scientific Project" is part of the Applied Information Technology Curriculum. Horst Hellbrück is an adjunct professor at the Institute of Telematics of University of Lübeck.

Author's Statement

Conflict of interest: Authors state no conflict of interest.

5 References

[1] Gunilla Burrowes and Jamil Y Khan. Short-range underwater acoustic communication networks. *Autonomous Underwater Vehicles*, pages 173–198, 2011.

[2] J. Joe and S. H. Toh. Digital underwater communication using electric current method. In *OCEANS 2007 - Europe*, pages 1–4, 2007.

[3] Fabian John, Marco Cimdins, and Horst Hellbrück. Underwater ultrasonic multipath diffraction model for short range communication and sensing applications. *IEEE Sensors Journal*, 21(20):22934–22943, Oct 2021.

[4] Fabian John, Sven Ole Schmidt, and Horst Hellbrück. Flexible arbitrary signal generation and acquisition system for compact underwater measurement systems and data fusion. In *Global Oceans 2021: San Diego - Porto*, pages 1–6, Oct 2021.

[5] Fabian John, Sven Ole Schmidt, and Horst Hellbrück. High precision open laboratory 3d positioning system for automated underwater measurements. In *Global Oceans 2021: San Diego - Porto*, pages 1–5, Oct 2021.

[6] Chuan Li, David A. Hutchins, and Roger J. Green. Short-range ultrasonic digital communications in air. *IEEE Transactions on Ultrasonics, Ferroelectrics, and Frequency Control*, 55(4):908–918, 2008.

[7] Paul A Van Walree. Propagation and scattering effects in underwater acoustic communication channels. *IEEE Journal of Oceanic Engineering*, 38(4):614–631, 2013.

Integration of Arrowhead Framework on Fischertechnik for Building a Digital Twin

Gowtham Reddy Eeda[1] and Javad Ghofrani[2]
[1] Robotics and Autonomous Systems, Universität zu Lübeck, gowtham.eeda@student.uni-luebeck.de
[2] Javad Ghofrani, Institute of Computer Engineering, Universität zu Lübeck, ghofrani@iti.uni-luebeck.de

Abstract

The fourth industrial revolution introduces interoperability and interconnectivity between automated systems using disruptive technologies like Artificial Intelligence, Big-Data Analytics, Robotics and Digital Twins. Problems in vertical integration in industrial setting can be resolved by building Digital Twins. They are virtual representation of the real-time digital counterparts of physical objects or processes. It consumes real world data about physical object and returns an output as in the form of predictions or simulations. With the help of the Eclipse Arrowhead Framework, the digitization of production systems leads to the interoperability between systems and system of systems (SoS). In this paper we report our experience on utilization of Eclipse Arrowhead Framework and Eclipse Ditto Framework to build a Digital Twin for a model production system. We introduce the basic concepts as a model to create a Digital Twins for educational purpose in industrial automation.

1 Introduction

As technology advances day-to-day, so does the way humans produce products. Fig. 1 illustrates different phases in the industrial revolution. In industry 1.0, water and steam powered machines were developed to help workers in the mass production of goods. In the second industrial revolution, the main contributor was the development of machines that uses electrical energy where as in industry 3.0, the speed of production increased considerably through Programmable Logic Controllers (PLC) and computer systems. At present, we are in fourth industrial revolution where automation and networking technologies (information and communication) are being used [1]. The main objectives of industry 4.0 are more efficiency, less capital costs and interoperability [2]. Industry 4.0 delivers continuous improvements, highest safety standards and solutions to new business models making itself even more competitive in the market. Emerging disruptive technologies such as M2M (Machine to Machine), IoT (Internet of Things), Big-Data, Cyber-Physical Systems (CPS), Digital Twins, Augmented Reality, Additive Manufacturing, Cybersecurity, Edge Computing and Cloud Computing are applied to achieve the objectives of industry 4.0 [3].

Digital Twins as one of the enabling technologies for industry 4.0 which arose interests in industrial and academic communities. Digital Twins are virtual dynamic representations of a physical system, which is connected to it over the entire life cycle for bidirectional data exchange [4]. Elements that constitute for building Digital Twins are sensors, communication networks (communication protocols like OPC-UA, MQTT, HTTP and AMQP) and digital plat-

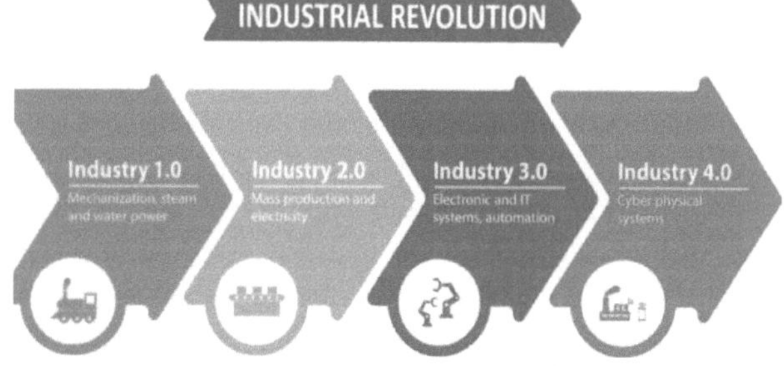

Figure 1: Different phases in industrial revolution [Source: MI Translations].

forms such as Ditto Framework [5].

Implementing the Digital Twins contains lots of technical and technological complexities which makes it less attractive for industrial use case. The objectives of Eclipse Ditto Framework is to facilitate the creation of digital Twins from industrial IoT systems. In this paper, we report our experience with Eclipse Ditto which we are integrating with Arrowhead Framework to build a Digital Twin from a model factory made by Fischertechnik. Our contribution can be used as a basic models and starting concept by researcher and practitioners who are involved in the digitization of industrial processes with application of Digital Twins.

1.1 Background and Related Work

The importance of Digital Twins in the industry 4.0 revolution is a necessary step for improving the self-adaptive behavior of interconnected cyber-physical systems (CPS).

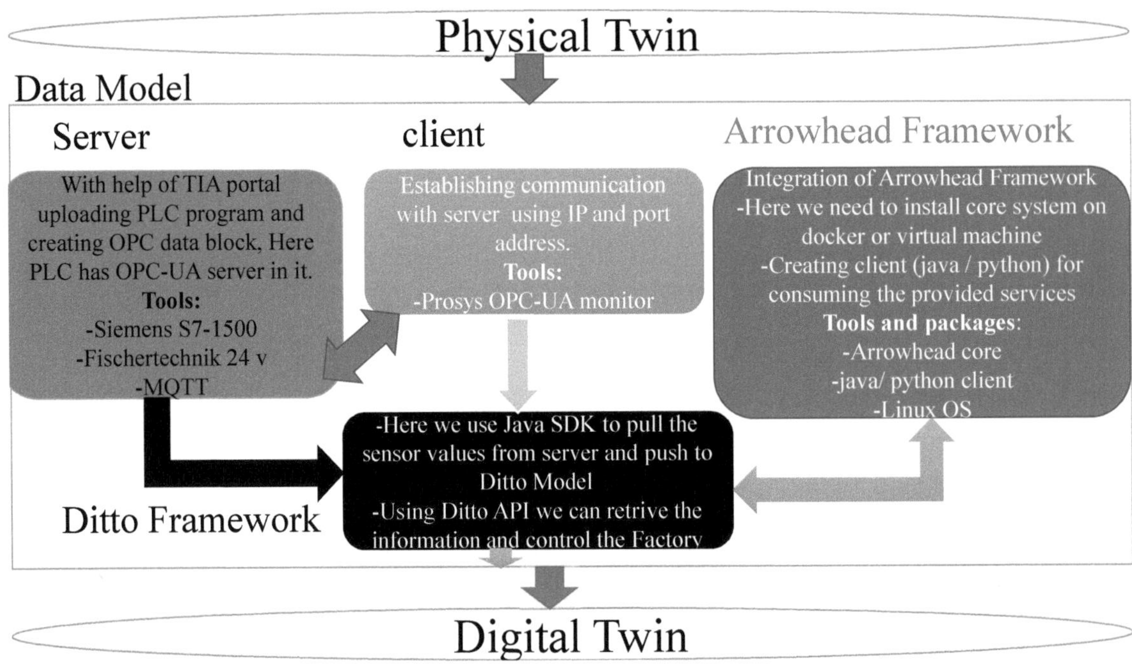

Figure 2: Overview of Digital Twin for Fischertechnik factory.
The above blocks illustrates PLC, prosys OPC-UA monitor, Arrowhead Framework and Ditto model are represented by red, green, blue and black colours respectively and arrows indicates the information flow.

In the context of manufacturing, the use of Digital Twins allows the practical integration of simulation like discrete-event simulation and system dynamics approach. Digital Twins are not only limited to manufacturing but are also widely used in applications like construction, aviation, automotive, healthcare and smart cities [6]. The focus of this paper is on smart manufacturing. It is a technology driven approach that utilizes internet-connected machinery to monitor the production process. Generally, in manufacturing, we mainly focus on the Product LifeCycle Management (PLM). It consists of stages such as design, production and service [7]. The primary goal of any production system is to improve competitiveness, productivity and efficiency. Based on different categorization methods, Digital Twins can be classified in terms of level of integration, technology, applications and platforms. In the context of industry 4.0, Digital Twins are applied for mainly predictive maintenance, process planning, optimization, product design and virtual prototyping. The rest of this paper is structured as follows. In Section 2 we will discuss the implementation of Arrowhead Framework on Fischertechnik factory for building industrial automation and implement Ditto Framework for building Digital Twin. Furthermore, in Section 3 we will discuss results and challenges.

2 Material Methods and Implementation

We used a smart factory model produced by Fischertechnik, and we built a Digital Twin based on its components. Fig. 2 illustrates the overall concept of our project. In order to provide a systematic method for building Digital Twins, we proposed a six layered model as shown in Fig. 5. This model was inspired by automation pyramid of ISA-95 (International Society of Automation).

2.1 Fischertechnik 4.0 Training Factory

Fischertechnik 4.0 training as shown in Fig. 3 is a physical simulation model used in educational settings and businesses for research, training and learning. It integrates multiple industry 4.0 features used in real world industrial plants. The production plant is completely automated and digitized with the help of sensors and actuators. It is composed of different sections, or modules, which are interconnected and controlled by a Programmable Logic Controller (PLC). These modules contain sensors, motors, conveyor belts, pneumatic cylinders, ejectors, cameras, NFC readers and lights. The factory layout has different parts containing High-Bay Warehouses (HBW), Vacuum Gripper Station (VGR), Multi-Processing Station (MPO), Sorting Line (SLD), Environmental Sensors (SSC) and the Delivery and Pickp Station (DPS) as shown in Fig. 3.

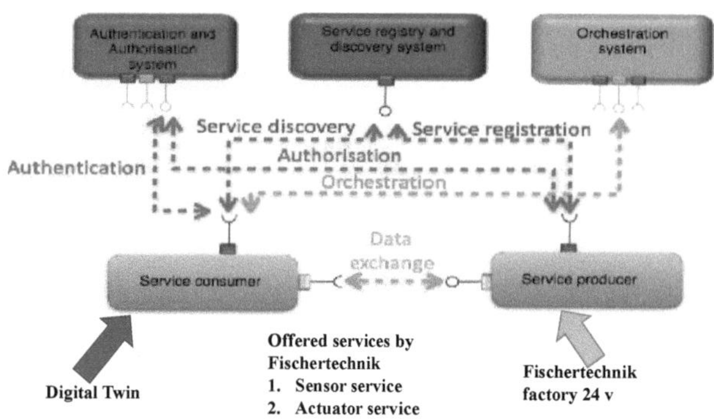

Figure 4: Arrowhead local cloud [8].

Figure 3: Factory layout (Top view) [Source: Fischertechnik.de].

2.2 Arrowhead Framework

The Arrowhead Framework acts as a middleware between operation technology (OT) and information technology (IT) to build IoT based automation. It supports the collaborations of Cyber-Physical Systems (CPS) based on principle of Service Oriented Architecture (SOA) and makes it possible to implement the concepts of industry 4.0 through digitization [8]. The main idea of the Arrowhead Framework is based on considering the IoT devices as a service provider or service consumer. It enables IoT devices to collaborate and to communicate, which leads to interoperability between and within IoT based systems. The Arrowhead Framework consists of main component, called mandatory systems namely service registry, orchestration and authorization as shown in the Fig. 4. The Arrowhead Framework introduces the concept of local clouds, which support the implementation of system of systems (SoS). Local clouds enable IoT services to get connected to Arrowhead core system without security concerns. Furthermore, IoT services from different local clouds collaborate with each other. In the service registry, the registration of services takes place, which allows them to be discovered by other systems in the cloud. The orchestration establishes guidelines that specify the systems need to consume which services are offered by the system. For security reasons, some systems might not be allowed to communicate, they might need to prove that they are allowed to consume certain services. The authorization system keeps a list of what services a system can consume and generates tokens that can validate that a system is allowed to consume other services.

2.3 Digital Twin Model

To generate a Digital Twins for complex systems, the new requirements aspects are application, technology, modelling object and modelling method aspect. According to [9], to solve new requirements in the system, a five Dimensional (5D) Digital Twin contains Physical Entity (PE), Virtual Entity (VE), Services for Physical and Virtual Entities (SS), Digital Twin Data (DD) and Connection between different parts (CN). With considering these aspects, we conceptualize a model for building digital twins from physical facilities. We use the Fischertechnik smart factory model as an example to show the application of our model. Our model consists of six layers which are illustrated in Fig. 5. Layer 1 includes the physical model where devices, sensors and actuators exist. Layer 2 consists of Siemens PLC and an OPC-UA server. Layers 1 and 2 are connected With the help of adaptor boards. The PLC connects to the IoT gateway (layer 3) using the OPC-UA communication protocol. Here, MQTT acts as an IoT gateway to broadcast the information to the above layers. Layer 4 acts a Fischertechnik TxT controller which provides an interface for connecting cameras to the factory. In Layer 5, Arrowhead local cloud is responsible for storing information about the services provided by sensors of physical model. Final layer is the Digital Twin which represents the data model built from the physical model, i.e. Fischertechnik factory model in our case. Fig. 4 illustrates how Arrowhead local cloud can be used in layer 5. The sensors and actuators in the physical model are considered as service provider, while the data model in layer 6 is considered as service consumer. The IoT devices from physical model should be registered in service registry of Arrowhead local cloud. However, defining all details of IoT devices is error-prone and time-consuming. Eclipse Ditto Framework makes it possible to define the IoT devices as things and specify their features. In our case, things are Fischertechnik factory assets and features are all data providing elements. Furthermore, Eclipse Ditto stores defined things within a cloud. Interaction with the things in the cloud is possible through Eclipse Ditto client SDK (Software Development Toolkit). Which is available in programming languages like Java this way, registered service providers in Arrowhead local cloud can send the information to their corresponding virtual representations in Eclipse Ditto cloud.

3   Results and Discussion

We used our proposed model to implement the Digital Twin of the introduced Fischertechnik factory model, According to the proposed automation stack model, our implementation consists of a Siemens PLC S7-1500 series in the sec-

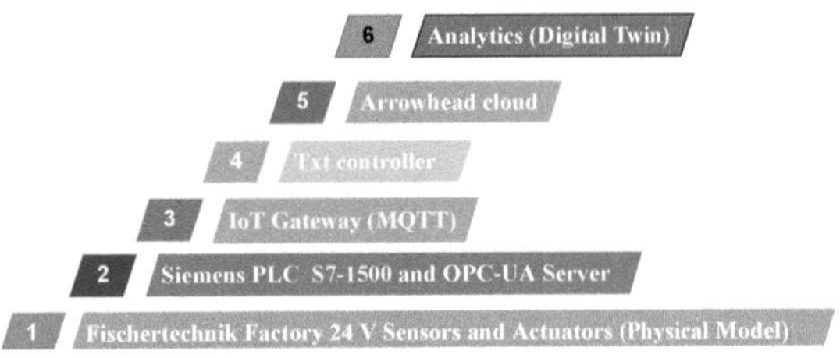

Figure 5: Automation stack model.

ond layer. We create a PLC program as per the standards of IEC 61131-3.

In order to establish communication between PLC and prosys OPC-UA monitor referred as client by using IP and port address. Furthermore, we created a OPC data model to store all the sensor and actuator values obtained from the Fischertechnik factory, in order to build IoT based automation and digitization for this factory we make use of the concept local cloud offered by the Eclipse Arrowhead Framework. Our implementation demonstrates two services and these services are registered in service registry of the local cloud. In order to build our Digital Twin, we use the Eclipse Ditto Framework. This Framework connected to Arrowhead service providers pull the sensor and actuator values from the factory and push them to the Ditto model using a Java SDK. To control the factory we use Ditto API to retrieve and modify the state of assets. However, to achieve digitization few challenges have been faced such as, we require a true time sensor data integration and data fusion which is highly challenging to emulate accurately. Furthermore, implementing feedback control on two-way transmission and real time analysis remains challenging.

4   Conclusion

This paper endeavors to test open source Frameworks (Eclipse Arrowhead and Eclipse Ditto) within the settings of industry 4.0 by accomplishing the interoperability between the systems and system of systems (SoS). We introduced a systematic model and showed how to build a Digital Twin based on this model. We have shown the utilization of the model on a simple example from Fischertechnik training equipment. However, more use cases, case studies and examples are required to show the feasibility of using these Frameworks in bigger projects. Beside the complexities in IT and OT systems, these Frameworks are still in their early stages. The main advantage of using these Frameworks are achieving real time operation, security and security controlled collaborations. Digital Twins are systems that not only consume data, but also provide services to other systems. It allows the manufacturer to constantly monitor machines and collect data on the actual performance of a device and its expected work performance.

Acknowledgement

The work has been carried out at Institute of Computer Engineering, supervised by Dr. Javad Ghofrani, Universität zu Lübeck.

Author's Statement

Conflict of interest: Authors state no conflict of interest.

5   References

[1] S. Teker and T. Koc, "Industrial revolutions and its effects on quality of life," *Pressacademia*, vol. 9, pp. 304–311, 07 2019.

[2] A. Albers, B. Gladysz, T. Pinner, V. Butenko, and T. Stürmlinger, "Procedure for defining the system of objectives in the initial phase of an industry 4.0 project focusing on intelligent quality control systems," *Procedia Cirp*, vol. 52, pp. 262–267, 2016.

[3] S. Wang, J. Wan, D. Zhang, D. Li, and C. Zhang, "Towards smart factory for industry 4.0: a self-organized multi-agent system with big data based feedback and coordination," *Comput. Networks*, vol. 101, pp. 158–168, 2016.

[4] J. Trauer, S. Schweigert-Recksiek, C. Engel, K. Spreitzer, and M. Zimmermann, "What is a digital twin? – definitions and insights from an industrial case study in technical product development," *Proceedings of the Design Society: DESIGN Conference*, vol. 1, pp. 757–766, 05 2020.

[5] V. Damjanovic-Behrendt and W. Behrendt, "An open source approach to the design and implementation of digital twins for smart manufacturing," *International Journal of Computer Integrated Manufacturing*, vol. 32, pp. 1–19, 04 2019.

[6] A. Fuller, Z. Fan, C. Day, and C. Barlow, "Digital twin: Enabling technologies, challenges and open research," *IEEE Access*, vol. 8, pp. 108952–108971, 2020.

[7] D. Kozma, G. Soos, and P. Varga, "Supporting digital production, product lifecycle and supply chain management in industry 4.0 by the arrowhead framework - a survey," 07 2019.

[8] P. Varga, F. Blomstedt, L. Ferreira, J. Eliasson, M. Johansson, J. Delsing, and I. Martínez de Soria, "Making system of systems interoperable – the core components of the arrowhead framework," *Journal of Network and Computer Applications*, vol. 81, pp. 85–95, 08 2016.

[9] F. Tao, M. Zhang, and A. Nee, *Five-Dimension Digital Twin Modeling and Its Key Technologies*, pp. 63–81. 01 2019.

A High-accuracy, Low-budget Sensor Glove

Robin Denz [1], Rabia Demirci [2], Elmar Rueckert [3] and Nils Rottmann [4]

[1] Robotics and Autonomous Systems, Universität zu Lübeck, robin.denz@student.uni-luebeck.de
[2] Robotics and Autonomous Systems, Universität zu Lübeck, rabia.demirci@student.uni-luebeck.de
[3] Chair of Cyber-Physical-Systems, Montanuniversität Leoben, rueckert@unileoben.ac.at
[4] Institute for Robotics and Cognitive Systems, Universität zu Lübeck, rottmann1990@gmx.de

Abstract

Sensor gloves are gaining importance in tracking hand and finger movements in virtual reality applications as well as in scientific research. They introduce an unrestricted way of capturing motion without the dependence on direct line of sight as for visual tracking systems. With such sensor gloves, data of complex motion tasks can be recorded and used for modeling probabilistic trajectories or teleoperation of robotic arms. While a multitude of sensor glove designs relying on different functional principles exist, these approaches require either sensitive calibration and sensor fusion methods or complex manufacturing processes. In this paper, we propose a low-budget, yet accurate sensor glove system that uses flex sensors for fast and efficient motion tracking. We furthermore evaluate the performance of our sensor glove, such as accuracy and latency.

1   Introduction

This article is a shorter version of [1], that has been previously published at the International Conference on Advanced Robotics (ICAR) 2021. The following chapters therefore give insight into the full work from [1].

In times of virtual reality (VR) and augmented reality (AR) applications, motion tracking of human body parts gains more importance, especially in the field of robotics, the precise tracking of hand and finger movements is of high interest. The tracked data can be used for trajectory model learning of complex tasks or for real world applications with robotic arms like human-robot collaboration.

The tracking and display of the user's field of view is a mature technique and available for the mainstream consumer. However, the availability of tracking extremities like hands is lagging far behind and is mostly exclusive for high-budget products. For the former, VR goggles use internal sensors and can therefore be used anywhere and without external setup. Meanwhile, the tracking of extremities is restricted to external optical sensor setups, e.g. Valve Index [2], Optitrack [3]. These setups rely on optical markers that are worn by the user, in combination with multiple, precisely calibrated cameras to capture the relative poses of the trackers in the setup space. Thus, those systems are not very versatile, since they require a dedicated setup of the working space as well as equipping the user with external markers. Furthermore, these setups incur a high expense, especially when a high accuracy and the tracking of smaller body parts, like fingers, is required.

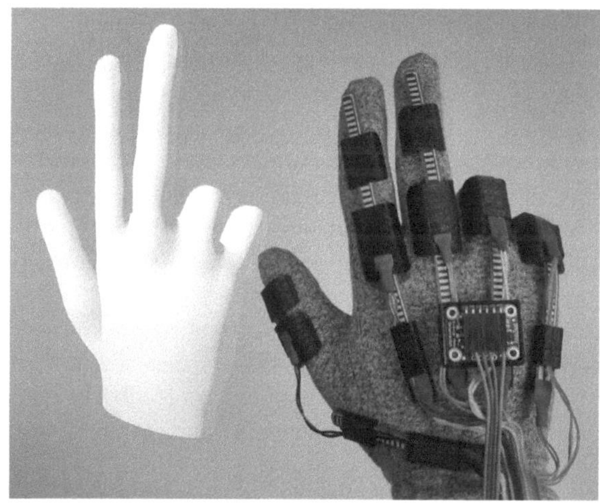

Figure 1: Example of the here developed sensor glove together with an illustration using Unity.

Other approaches for tracking finger motions are for example realized by a camera attached to the VR goggles, which points down towards the hands of the user. Devices such as Leap Motion [4] abstract the position of the bones in each finger to display the three dimensional model of the hand including the finger joints. However, this still restricts the working area of the sensor to its line of sight.

The goal of this project is to design a versatile, while low-budget sensor glove for modeling and comparing different hand movements.

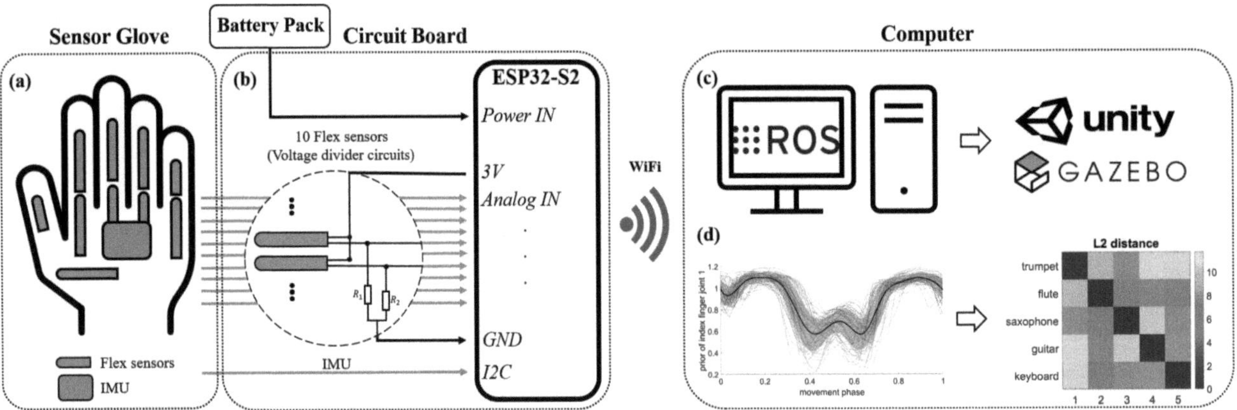

Figure 2: Simplified schematic diagram of the system architecture for our sensor glove design: (a) Glove layout with sensor placements, the orange fields denote the flex sensors, while the IMU is marked as a green rectangle, (b) Circuit board which is wired with the sensor glove, has 10 voltage dividers for reading each flex sensor connected to ADC pins of the microcontoller ESP32-S2 and the IMU is connected to I2C pins, (c) The ESP32-S2 sends the raw data via WiFi as ROS messages to the computer, which allows a real-time visualization in Unity or Gazebo, (d) Post-processing of the recorded data, e.g. learning probabilistic movement models and searching for similarities

1.1 Contribution & Paper Organization

Our aim is to construct a simple, budget-friendly, yet accurate sensor glove with low latency, suited for recording hand motion data to learn trajectory models. Most sensor glove implementations require elevated techniques for construction and are not suited for easy and low-cost replication. Our contribution here is threefold: (1) an open-source realization of an intuitive sensor glove which keeps up with the state-of-the-art hand motion capture techniques, in which we payed attention to the user experience by avoiding bulkiness and movement confining parts in the glove design, since the sensor is ultimately developed for transfer learning and teleoperation, (2) a ROS (Robot Operating System [5]) interface for realizing wireless and straight forward access to the sensory data and complex visualization in Unity [6] for direct visual feedback, and (3) an evaluation of the performance of our proposed sensor glove.

We start with Section 2 by giving insights into the hardware design and software realization of the sensor glove. We proceed in Section 3 by evaluating the accuracy and latency of our implementation. In Section 4 we conclude our findings and give an outlook on future work regarding our sensor glove.

2 Material and Methods

In this section, we present the methods we used to design and build our sensor glove seen in Fig. 1, as well as the underlying software implementation.

2.1 Sensor glove design

For easy donning and doffing, the sensors and further components are mounted on a fabric glove. To achieve a highly versatile glove that fits different users and does not disturb the users by being too tight or by having excess material especially on the finger tips, we chose to use special soft fabric gloves made for osteoarthritis patients. These gloves are highly stretchable and therefore adjust to many hand types and sizes, while being very pleasant to wear.

We used 10 flex sensors in total [7], [8], where two of them are smaller to fit the thumb and pinkie finger (see Fig. 3). The reading of each sensor requires a voltage divider circuit, illustrated in Fig. 2(b). Since we use two different types of flex sensors, two different resistors are required. We chose 47kΩ resistors for the long flex sensors and 10kΩ for the smaller ones to ensure the voltage change lies between the range of the input pins. As shown in Fig. 3, our sensor glove measures the flexion of the proximal interphalangeal (PIP) and metacarpophalangeal (MCP) joints for each finger in addition to the interphalangeal (IP) and carpometacarpal (CMC) joints of the thumb. To mount the sensors on the glove, we designed and printed custom 3D pieces shown in Fig. 4, out of slightly flexible filament. First, the sensors were fixed on these pieces and afterwards glued onto the glove. The flex sensors were secured in place through this construction, while allowing the fingers to move without constraints. This enables a consistent bending of the sensors to maintain the precision of our measurements.

As for the controller of our sensor glove, we chose ESP32-S2 [9] microcontroller which provides a built-in WiFi module for connectivity, a 10-channel 13-bit analog to digital converter (ADC) for the connection of the 10 bend sensors and an I2C-bus for the IMU, that will be used. This controller holds just the right amount of analog input channels for our sensor layout while providing a higher resolution than most competitive microcontrollers. It is

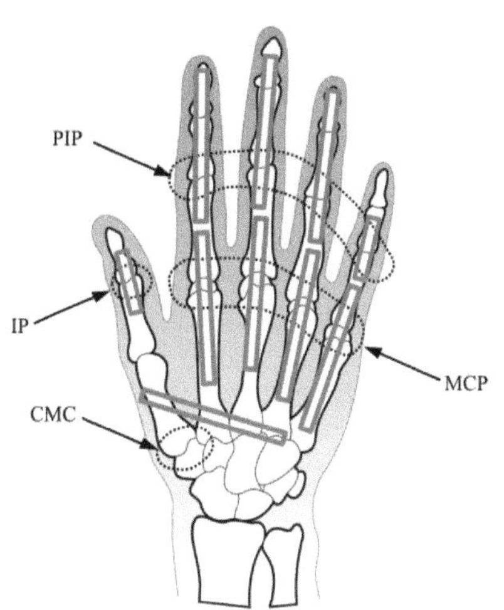

Figure 3: Placement of the flex sensors to measure human hand articulations: Proximal interphalangeal (PIP), metacarpophalangeal (MCP), interphalangeal (IP) and carpometacarpal (CMC) joints.

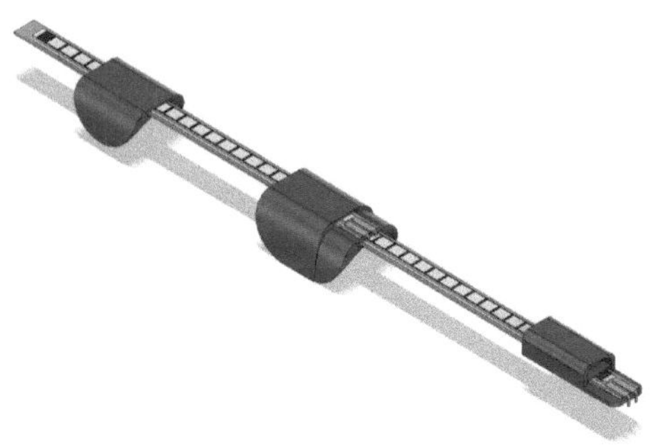

Figure 4: Two flex sensors mounted onto the 3D printed guiding shapes. This represents the arrangement on one finger.

also programmable through the widely-used Arduino IDE [10] that allows the use of many pre-build libraries.

For tracking rotation and hand movement, we attached an Adafruit BNO055 [11] Inertial Measurement Unit (IMU) to the center of the back of the glove, which allows for reasonably good orientation tracking of the hand posture. This breakout board provides a 9-degree-of-freedom IMU coupled with a controller to calculate the absolute orientation, then making it accessible through the I2C-bus that is hooked up to our microcontroller.

Furthermore, a WiFi connection offers more flexibility by being able to broadcast information to every other device in one network, while a Bluetooth connection is restricted to peer-to-peer communication with one other device. The resulting availability of the sensor data is furthermore supported by our usage of a ROS interface. In summary, we benefit from the intuitive, inexpensive and accurate way of realizing a sensor glove by using flex sensors. Additionaly, we are elevating our system by incorporating a WiFi connection, offering a ROS interface and providing a real-time visualization.

2.2 Software implementation

As mentioned before, the microcontroller used in our setup is programmable through the widely-used Arduino IDE [10]. We therefore can make use of many prebuild libraries that allow the implementation of a ROS interface on the glove, publish the sampled sensor data in ROS messages and make it available to any other machine running ROS in the WiFi network. We furthermore implemented a complex simulation environment in Unity, shown in Fig. 1, that can display the published data in real time, allowing the user a good understanding of the raw sensor data generated by the glove. Since this is only available on Windows platform, we also implemented a simpler simulation in Gazebo. For better visualization, we also simulated the flexion of the interphalangeal joints of each finger, that has no flex sensor attached to it, by presuming it to be half as flexed as the proximal interphalangeal joints of that finger.

Moreover, as proposed in [12], we also used hyper sampling for collecting smooth sensor data, which means for each flex sensor 50 measurements are taken, 10% of the highest and lowest values were discarded. Afterwards, the average of the remaining 80% of measurements are taken. This minimizes jittering in the sensor data and drastically smoothes the output.

3   Results and Discussion

In this section, we evaluate the performance of our sensor glove.

3.1   Performance

The performance of our sensor glove is evaluated by three criteria, namely, the *accuracy* of the flex sensors in our design, the *sampling rate*, in which the sensor data is collected and also the *latency* of the whole system connected to the simulation.

The measurement of the accuracy is, in our terms, described with the repeatability of the sensors when performing the same finger motion. By performing the same bending positions for different finger joints with the assistance of 3D-printed guiding shapes to secure the finger, we evaluated the accuracy of our flex sensors and therefore of our sensor glove to be below $2°$.

The second important aspect is the sampling rate of the whole sensor glove. This is highly dependent on the used microcontroller, or precisely the analog-to-digital converter to read the sensor values. While the overall sampling rate of the integrated ADC is sufficient to reach high amounts of measurements, the switching between the several channels is the main cause for longer sampling duration. Therefore, the amount of samples per sensor used for the hyper sampling and smoothing of the data is not very influential on the whole sampling rate. Due to these properties of the ADC, we reached a sampling rate of 20 Hz. To achieve a higher sampling rate, one should include external ADC chips with a high sample rate for each flex sensor and then hook them up to a serial peripheral interface (SPI) bus. We did not take this approach, as it would raise the costs of our design significantly while also complicating the building and replicating process. Furthermore, our achieved sampling rate of 20 Hz is sufficient for our purpose of transfer learning and teleoperation.

Especially for the teleoperation of robotic arms, it is important to have a real-time system with very low latency. In our setup with a normal WiFi router, we observed none to little latency of 1-10 milliseconds. This is not a noticeable delay for the human eye, so online teleoperation can be done with fast reactions to external influences. However, it should be considered, that this can vary and is highly dependent on the available network, but since we are working with very small data packages, the latency should always stay in an acceptable range for real-time applications.

In summary, the performance of our sensor glove has some minor drawbacks in the sampling rate while keeping a high measurement accuracy and also providing a real-time latency. But it fully satisfies the constraints of being low-budget (below 200€), highly accurate and very versatile, which makes it an appropriate fit for transfer learning of movement trajectories and also for teleoperating robot arms.

4   Conclusion

In in this paper, we proposed a low-cost sensor glove design that satisfies the prerequisites for trajectory learning with dynamic movement primitives. Our sensor glove reaches a high accuracy and repeatability as well as very low latency, which renders it suitable for other real-time application, for example teleoperation of robotic manipulators. Furthermore, our approach has a high flexibility due to the integrated WiFi connection and the implemented integration to ROS in combination with a real time simulation in Unity. It is also easily replicable, due to its low-cost design, i.e., only requires some 3D-printed parts. Our whole implementation is open source and available on GitHub (`https://github.com/ai-lab-science/SensorGloves`). The main drawback of this simple approach is in the low sampling rate of the whole setup, which might be counteracted by extending the design with external analog-to-digital converters to get faster measurements of the multiple flex sensors.

Acknowledgement

This project has been supervised by the Institute for Robotics and Cognitive Systems at the Universität zu Lübeck and has received funding from the Deutsche Forschungsgemeinschaft (DFG, German Research Foundation) No #430054590 (TRAIN).

Author's Statement

Parts of this article were previously published at ICAR 2021 [1].

5   References

[1] R. Denz, R. Demirci, M. E. Cansev, A. Bliek, P. Beckerle, E. Rueckert, and N. Rottmann, "A high-accuracy, low-budget sensor glove for trajectory model learning," in *2021 20th International Conference on Advanced Robotics (ICAR)*.   IEEE, 2021, pp. 1109–1115.

[2] V. Corporation. Valve index vr system. [Online]. Available: https://www.valvesoftware.com/de/index

[3] Optitrack. Optitrack - motion capture systems. [Online]. Available: https://optitrack.com/

[4] Ultraleap. Leap motion controller. [Online]. Available: https://www.ultraleap.com/product/leap-motion-controller/

[5] M. Quigley, K. Conley, B. Gerkey, J. Faust, T. Foote, J. Leibs, R. Wheeler, A. Y. Ng *et al.*, "Ros: an open-source robot operating system," in *ICRA workshop on open source software*, vol. 3, no. 3.2.   Kobe, Japan, 2009, p. 5.

[6] U. Technologies. Unity. [Online]. Available: https://unity.com/

[7] . F. S. Systems. Flexpoint bend sensor. [Online]. Available: https://www.flexpoint.com

[8] S. Electronics. Sparkfun bend sensor. [Online]. Available: https://www.sparkfun.com/

[9] E. Systems. Esp32s2. [Online]. Available: https://www.espressif.com/en/products/socs/esp32-s2

[10] Arduino. Arduino ide. [Online]. Available: https://www.arduino.cc/en/software

[11] Adafruit. Adafruit bno055. [Online]. Available: https://www.adafruit.com/product/2472

[12] O. Nisar, M. A. Imtiaz, S. Hussain, and O. Saleem, "Performance optimization of a flex sensor based glove for hand gestures recognition and translation," *International Journal of Engineering Research & Technology*, vol. 3, no. 5, pp. 1565–1570, 2014.

Simple Setup for Generation and Spectrometry of Vapors

Joop Dölemeyer [1,2] and Oliver Wille [2]

[1] Biomedical Engineering , University of Applied Sciences Lübeck, joop.doele@googlemail.com
[2] Optical Sensor Development , Dräger Safety AG & Co. KGaA

Abstract

Explosive gases can be detected by measuring their characteristic spectral absorption in the infrared wavelength region. The knowledge of the respective absorption spectrum of a specific compound allows for the exact prediction, how a sensor will detect a given concentration. Most explosive compounds appear as vapors. We present a simple laboratory setup for the generation and spectrometry of vapors of desired concentrations. Furthermore, we present a numerical correction algorithm for measured spectra and discuss errors and their propagation throughout the system.

1 Introduction

On industrial sites toxic and explosive gases are often products or side products of petrochemical processes. Many explosive compounds appear as vapors. Vapors are liquid substances in their gaseous phase with a temperature lower than their critical temperature[1]. Gas detection systems increase the safety of the employees and installations and allow for an early detection of hazardous concentrations in industrial environments. One possible way of measuring gas concentrations are optical detection systems. A typical measurement setup of an industrial infrared gas sensor is presented in Fig. 1.

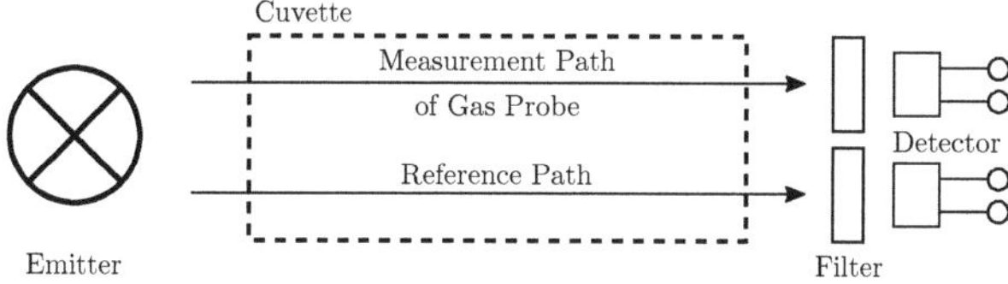

Figure 1: Typical setup of an industrial IR gas detection device.

Optical gas detection devices are based on the principal of absorption spectrometry. Spectrometry is an experimental method of splitting radiation into single frequencies or wavelengths. The resulting intensity distributions are called spectra [2]. In this work we use a FT-IR spectrometer [3] that measures all wavelengths at the same time. A transmission spectrum indicates how much radiation of a specific wavelength transmits through a - in this case - gaseous substance. Every gas has a characteristic transmission spectrum. This fact can be used in industrial optical gas sensors to detect them. In Fig. 1 a light source emits an electromagnetic spectrum which passes through the gas cuvette and the optical filters until it reaches the detecting elements. The reference detector measures the transmission signal in a spectral region where no or very few light is absorbed by the gas while the measurement detector acquires the transmission signal in a region where a significant fraction of light can be observed. The ratio of the two measured absorption amplitudes is a characteristic function of the concentration of the gas. The usage of the ratio makes the signal independent from the emitter's intensity [4]. Since all gases have different absorption spectra one can, by selection of filter parameters for both detectors, measure various hazardous substances. For the development of future gas sensors at Dräger Safety AG & Co. KGaA the availability of high-resolution spectra for the compounds of interest is crucial. The Pacific Northwest National Laboratory (PNNL) and the National Institute of Standards and Technology (NIST) both offer spectrum based libraries [5] [6], but the PNNL ceased work on their library, it is no longer available, while the NIST's quantitative database contains just a few substances and the IR spectra from NIST's substance database do not comply with Dräger's requirements. Our work aims for the setup of an extended spectra database for hazardous substances. So far, we have measured the spectra of Acetone, Ethanol, Heptane, Hexane, Methanol, Nonane, Pentane and Toluene. To demonstrate our methodology we here focus on the two substances Heptane and Toluene. For both, the spectra from the PNNL database are available [5].

The preparation of vapors used for the spectrometry is not trivial, especially when precise concentrations are required. We present a compact and simple setup of a vaporizer for the generation of vapors with given concentrations in a laboratory environment.

2 Material and Methods

The general experimental setup consist of three main parts, first is the vaporizer that turns a liquid into a vapor and fills the cuvette of the spectrometer. Second is a calibration setup with two IR detectors that enhance the measurements done in the spectrometer, which is the third part.

2.1 Vaporizer

The carrier gas in the system is Nitrogen, which is heated in a water bath to 50 °C. A Flow controller (Bronkhorst F-201EV [7]) reduces the flow to 3 $\frac{1}{min}$. The Nitrogen is then guided into the vaporizer flask (see Fig. 2), that is heated with a resistor heating element. The substances are dosed with the Metrohm Dosino 800 [8].

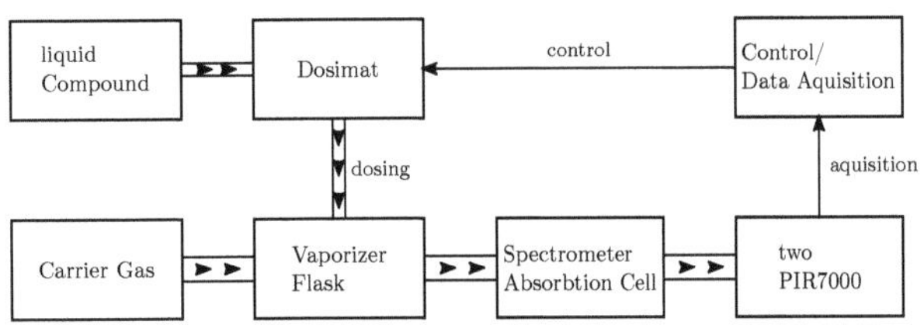

Figure 2: Experimental setup of the vaporizer with indicated gas flow, liquid flow and data lines

The dosimat has a substance cuvette with a volume of 5 ml and can dose the liquid with 10.000 Steps at a minimum flow rate of 0.025 $\frac{ml}{min}$.

$$(P + \frac{a}{V_m^2}) \cdot (V_m - b) = R \cdot T \tag{1}$$

$$\dot{V}_{dose} \approx \frac{M_m \cdot \dot{V}_{N_2} \cdot c}{\rho \cdot V_m} \tag{2}$$

The calculation of the molar volume V_m is determined numerically from the van der Waals Equation for real gasses [1] (see (1)). Afterwards, the dosing volume $\dot{V}_{dose}$ is calculated by means of (2). The van der Waals equation is not the only equation of state for real gases. It is widely used due to its simple analytic approach but higher accuracy can be reached with the use of more complex equations like the Soave-Redlich-Kwong Equation of State [9]. But this is beyond the scope of our current work. To achieve constant evaporation the system temperature is kept constant at 50 °C. Comparing the different vapor pressures for different compounds between 0 and 100 %-LEL (lower explosion limit, minimum concentration where an ignition of the gas can be reached) shows, that a temperature of 50 °C is sufficient for the desired concentration range. E.g. Nonane, our target substance with the lowest partial pressure can be vaporized with 2.4 Vol.-% at ambient conditions of 1013 hPa and 50 °C. This corresponds to 342 %-LEL [10]. Over a valve system the output can switch between carrier gas and the produced vapor. The output flow can be adjusted with two needle valves. The vapor is filled into the absorption cell of the spectrometer (see Fig. 2). The cell is designed as a single pass cell with calcium fluoride windows with a usable wavelength range from 0.2 to 10.5 μm. The vapor passes through the absorption cell and is then guided into two Dräger PIR 7000 [11] IR detectors for a calibration measurement as seen in Fig. 2. The PIR 7000 is a standard industrial gas sensor (see Fig. 1) for explosive gases with

filters for the reference detector @ 3.09 μm and measurement detectors @ 3.3 or 3.4 μm.

2.2 Calibration

One experimental problem with the spectrometer is, that it needs to be purged with Nitrogen to suppress the influence of ambient water and carbon dioxide absorption. The purging takes up to 25 min, during this period a concentration reduction of the vapor is observed in the absorption cell. So when measuring multiple spectra for averaging every single spectrum has a different concentration level. To recalibrate the spectra, two Dräger PIR 7000 are integrated into the vapor output of the vaporizer. Both are using different filter combinations for the measurement and reference path. The filters have certain center wavelengths and filter widths. The Dräger PIR 7000 are measuring the absorption as an integration of the wavelength over the product of the vapor spectrum $V(\lambda)$, the emitter spectrum $L_{BB}(\lambda)$ and the filter spectrum $F(\lambda)$. The emitter is a filament lamp and its spectrum is assumed as a black body spectrum with a temperature of 1900 K. The ratio (3) of this expression to the integral of just the product of the emitter spectrum $L_{BB}(\lambda)$ and the filter spectrum $F(\lambda)$ results in the relative amplitude which is measured by the vaporizer setup.

$$A_{rel} = \frac{\int_{\lambda_1}^{\lambda_2} L_{BB}(\lambda) \cdot V(\lambda) \cdot F(\lambda)}{\int_{\lambda_1}^{\lambda_2} L_{BB}(\lambda) \cdot F(\lambda)} \tag{3}$$

An auto calibration is used in the post processing of the spectra measurement. Due to concentration loss all measured spectra for a given compound have different concentration levels. This calibration uses Beer's law [12] to rescale the measured spectra to the PIR 7000's relative amplitudes. The ratio of a spectral intensity $I(\lambda)$ to a spectral intensity $I_0(\lambda)$ of two different gas concentrations c and c_0 is based on the ideal gas equation [1] (4).

$$\frac{I(\lambda)}{I_0(\lambda)} = exp^{\frac{c}{c_0}} \tag{4}$$

The correction factor of the resulting spectrum is calculated using a classical, linear, Gauss-Newton fit. Additionally the routine does a zero-line calibration to compensate an additional offset that arises from small deviations in positioning the spectroscopy cell between obtaining the measurement and the reference spectra. The exponent of (4) is numerically optimized, to minimize the deviations of the calculated amplitude values from the PIR 7000.

2.3 Spectrometer

The spectrometer that is used for measuring absorption spectra is a Bruker Vertex 70 FT-IR spectrometer [3]. The measurements have a spectral range of 8000 cm^{-1} to 625 cm^{-1} with a maximum spectral resolution of 0.5 cm^{-1}. The before mentioned purging flow is set to 3.5 $\frac{1}{min}$. Spectra are obtained every 5 min.

3   Results and Discussion

The resulting Heptane and Toluene transmission spectra are shown in Fig. 3. Six single Toluene spectra and eleven single Heptane spectra were averaged before the auto calibration step. Therefore all single spectra were rescaled to one spectrum of the same concentration level and then averaged, before calibrating the result to the PIR 7000 measurements.

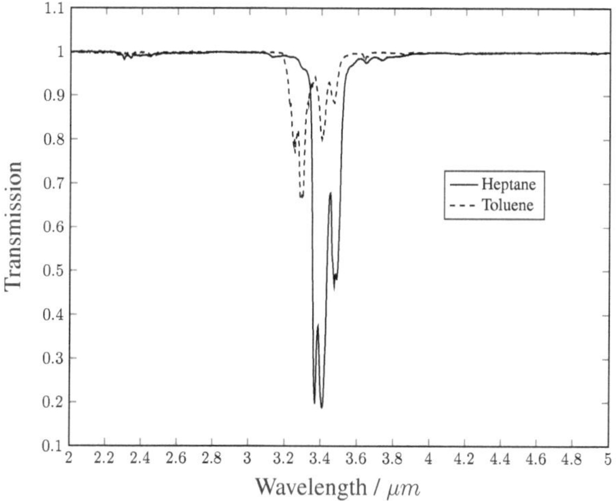

Figure 3: Transmission spectrum of Heptane and Toluene with a vapor concentration of 0.4 Vol.-%.

All substances were dosed with a prepared program and the Dosino unit. This leads to uncertainties regarding the dosing quantity, as it is not clear how accurately the system doses the substances. Therefore the vapor concentration of Toluene was increased step by step and the relative amplitudes were measured with the *PIR 7000* devices. Fig. 4 shows a linear progression of the logarithm of the amplitudes with increasing concentration for both sensors. This indicates a constant dosing ability of the Dosino unit.

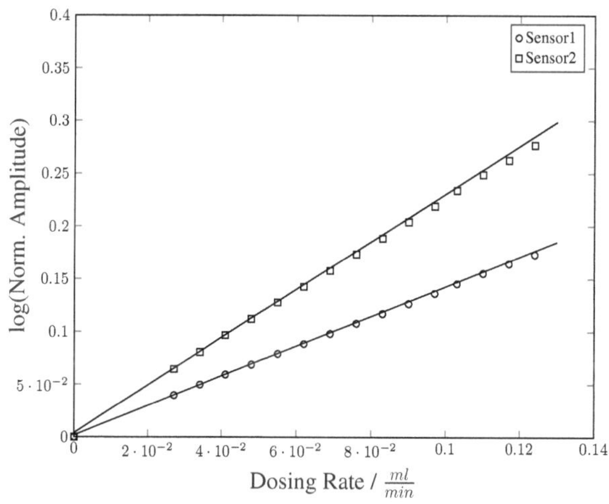

Figure 4: Step wise increase of the vapor concentration from 20 %-LEL to 90 %-LEL in steps of 5 %-LEL with Toluene. Data points of both IR detectors and linear regression.

Fig. 5 shows the resulting standard deviation after post processing with multiple spectra averaging and the auto calibration. The standard deviation has higher uncertainties in the range from 2 to 2.4 μm and in other regions where lower absorption levels are observed. But overall the values for the standard deviation are exceptionally small with values in the range of 10^{-5} and below. For a future usage of the obtained spectra a precise knowledge of the ambient conditions, the concentrations and the path length are needed to rescale the spectrum with (4) to any desired concentration or ambient condition. The relative errors of the measured relative amplitudes were in the range of 1.5 to 2 % for both sensors and both substances. The error propagation through the weighted Gauss-Newton fit yields a relative statistical error of the concentration level of 0.7 % for Heptane and 3 % for Toluene. Comparing the resulting averaged and calibrated spectra with the PNNL spectra one recognizes that the overall shape of the spectra matches almost perfectly.

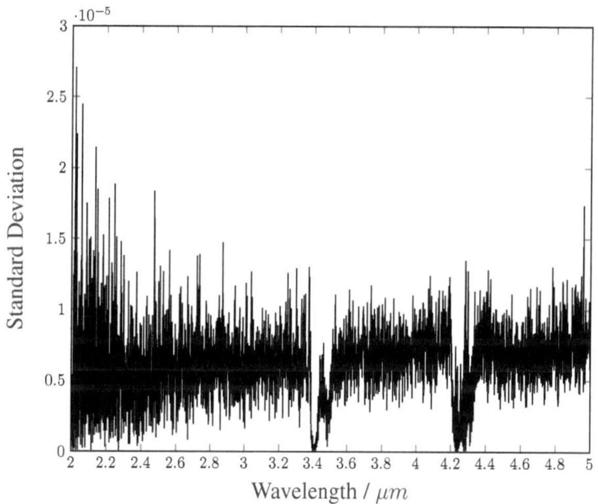

Figure 5: Standard Deviation of Fig. 3, calculated from the averaging step for Heptane measurements.

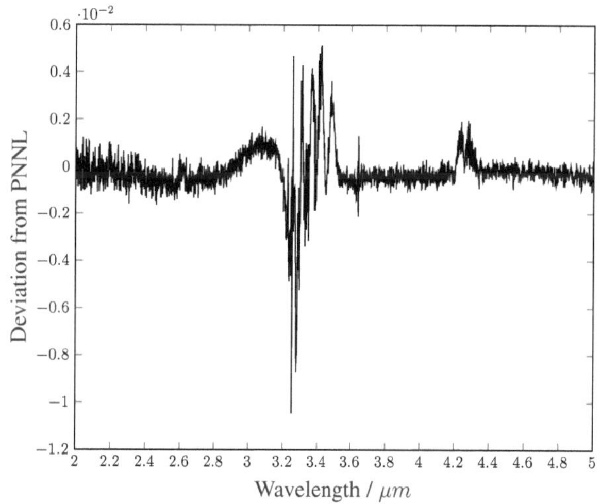

Figure 6: Deviation of measured Toluene from PNNL spectrum [5], concentration levels adjusted.

Fig. 6 shows the difference between our post-processed spectrum and the PNNL spectrum for Toluene after the concentration levels have been adjusted. Unfortunately the concentration levels show a surprising deviation of 9.2 % for Toluene and 14 % for Heptane, a completely unexpected deviation of presumably systematic origin! An extensive analysis of all the possible error sources shows that the dosing itself seems to be very reproducible, the flow controller has a very low specified error and the purity of the substances is very high (99.5 %). The PNNL experimental method of generating the vapors is much more sophisticated compared to our method. They measure the dosed quantity of the liquid accurately, evaporate into an evacuated cell and directly measure the resulting partial pressure in the low-pressure regime, while we are calculating the needed quantity of the liquid based on the van der Waals equation (1). Further investigation is needed to determine, what the corresponding concentration to each of our measured spectra is.

4 Conclusion

Precise knowledge of all the contributing properties and parts is crucial to understand the behavior of a new infrared gas sensor device that shall exactly detect gases in industrial environments. One key factor for that is the availability of good vapor spectra. Looking at the resulting spectra of Heptane and Toluene (Fig. 3), we have demonstrated that we can measure such spectra with acceptable experimental effort and small deviations from spectra in public databases (Fig. 6). A calibration routine compensates the offset and concentration levels of the spectra for experimental imperfections of the setup. The vaporizer has a reproducible dosing ability (Fig. 4). By averaging multiple spectra we can achieve small standard deviations in the 10^{-5} range for the spectra (Fig. 5). The resulting relative statistical error of the concentration level of the spectra arises from small fluctuations in the vapor generation and is in the range of 1-3 %. Further improvements in the calculation of the real gas molar volume can reduce the systematic deviations of the measured spectra from published spectra. Additionally, we could calibrate our setup by using calibration gases of specified concentration. Furthermore, a correction of the wavelength axis should be performed and the dosing ability of concentrations should be verified for different gases. So far, in addition to Heptane and Toluene, other compound's spectra were measured like Acetone, Hexane, Methanol, Nonane and Pentane. Our goal is to increase the number of substances to build up a large database that can be used for future products at Dräger Safety AG & Co. KGaA.

Acknowledgement

The work has been carried out at Dräger Safety AG & Co. KGaA, Gas Detection Sensors and supervised by Prof. Dr. Stefan Müller, Institute for medical Sensors, University of applied Sciences Lübeck.

Author's Statement

Conflict of interest: Dräger Safety AG & Co. KGaA covered the costs of the work and both authors are employed at the company.

5 References

[1] P. W. Atkins, J. de Paula, *Kurzlehrbuch Physikalische Chemie*. Wiley-VCH Verlag, 2020.

[2] D. Meschede, *Optik, Licht und Laser*. Vieweg und Teubner, 3rd, 2008.

[3] Bruker Corporation, *Bruker Vertex 70*. Available: https://www.bruker.com [last accessed on 2022-01-14]

[4] W. Jessel, *Gase-Dämpfe-Gasmesstechnik*. Dräger Safety AG & Co. KGaA, Lübeck, 2001.

[5] S. W. Sharpe, T. J. Johnson, R. L. Sams, P. M. Chu, G. C. Rhoderick and P. A. Johnson, *Gas-Phase Databases for Quantitative Infrared Spectroscopy*. Society for Applied Spectroscopy, Vol. 58, No. 12,2004.

[6] P. M. Chu, F. R. Guenther, G. C. Rhoderick and W. J. Lafferty, *NIST Quantitative Infrared Database*. Available: https://webbook.nist.gov/chemistry/quant-ir/ [last accessed on 2021-11-12]

[7] Bronkhorst High-Tech B.V., *Low Pressure Drop Mass Flow Meters/Controllers*. Available: https://www.bronkhorst.com [last accessed on 2022-01-07]

[8] Metrohm GmbH & Co. KG, *846 Dosing Interface: Automatic Liquid Handling*. Available: https://www.metrohm.com [last accessed on 2022-01-08]

[9] B. E. Poling, J. M. Prausnitz, J. P. O'Connel, *The Properties of Gases and Liquids*. McGraw-Hill, 2001.

[10] Gefahrenstoffinformationssystem IFA, *GESTIS-Stoffdatenbank*. Available: https://gestis.dguv.de [last accessed on 2022-01-5]

[11] Dräger Safety AG & Co. KGaA, *Dräger PIR 7000*. Available: https://www.draeger.com [last accessed on 2022-01-18]

[12] A. Beer, *Bestimmung des rothen Lichts in farbigen Flussigkeiten*. Ann. Physik 162, 78-88, 1852.

Effect of Object Diameter Specific Reflections to Underwater Ultrasonic Multipath Diffraction Measurements

Hauke Petersen [1], Fabian John [2], Horst Hellbrück [2]
[1] Applied Information Technology, Technische Hochschule Lübeck, h.petersen@stud.th-luebeck.de
[2] Technische Hochschule Lübeck - University of Applied Sciences, Germany , Department of Electrical Engineering and Computer Science, Center of Excellence CoSA fabian.john@th-luebeck.de, horst.hellbrueck@th-luebeck.de

Abstract

In ultrasonic-based object detection and distance determination, reflections have a not inconsiderable influence on the measurement result. In this paper, the effects of additional reflection on objects to the measured spectrum are evaluated for a given model. We focus on the investigation of the position- and diameter-dependent multipath reflection components. Measurements from natural environments are compared with measurements from simulations to ensure that both provide the same results. To work with position depending reflections of an object, the code of the original model is adapted. Finally, we evaluate the performance of our modifications with the comparison of the localization results between the original and modified model. The high accuracy of the simulation values to the real measured values is confirmed.

1   Introduction

The ultrasonic multipath scanning was continuously improved over the last decades, and with the improvement, the field of application also grew. Some of the main features are that this technology is used in environments where humans have no easy access or in areas where no physical damage to the environment is allowed due to the scanning technology. An extraordinary use case where both features are combined is detecting high voltage power cables in the seafloor. For humans, it is hazardous to operate in high depth, and also the marine biosphere is very sensitive to most physical methods like digging. In this particular field of application, the COSA of the Technische Hochschule Lübeck is in charge of a project that targets this use case with a newly modified ultrasonic multipath solution. The project is based on an ultrasonic transceiver with wideband pulses and advanced spectral signal processing, which achieves up to 97% accuracy [2].

The main question this paper has to answer is how the influence of different diameters of an object in terms of multipath reflections on the signal path is investigated by the system. The formation of reflection is directly connected to the diameter of an object. The thicker an object is, the larger is the surface where reflections occur. So the most significant impact should be expected by larger object diameters.

To get valid data we are later able to compare and evaluate, we are going through the following points:

1. Measurement :

At first, we have done multiple measurements with different obstacles diameters and obstacle distances to the transmitting and receiving units to generate a valid first data set representing the natural behavior.

2. Simulation:

To compare the measurement and simulation results correctly, we integrated the same obstacles in our Matlab- Program and simulated them to generate the second data set.

3. Evaluation:

This paper aims to optimize the simulation so that both data sets from the measurement and simulation have similar results, and all differences between the real model and simulation model are optimized to a minimum.

In the following pages, we are going to describe our idea more detailed. Furthermore, we also take a closer look at our measurement model and the simulation model to explain the basics of our working principle. The last section is the evaluation part, where we take a closer look at our results and how they were achieved. Finally, we give an outlook into the future and possible extensions and application areas.

2   Related Work

In the context of the Extense Project, there have been no observations in the direction of this paper so far. Therefore, a reference to previous works is not possible. However, in the general environment around the topic :"handling the effects of reflections in ultrasonic measurements", some studies have already been done. There are also different approaches to the studies. The effects of reflections on the measurement itself are investigated, but there are also approaches in which the reflection is used as an additional source of information or even only the reflection is used.

The first approach, "Ultrasonic reflection of a bounded beam at Rayleigh and critical angles for a plane liquid-solid interface ", [5] also considers effects of reflections on the measurement. It is about the surface scanning of the seabed from the aircraft. Different materials reflect the measuring beams at different angles, and this leads to measurement errors.

The second work, "Ground Reflection Elimination Algorithms for Enhanced Distance Measurement to the Curbs Using Ultrasonic Sensors" [6], is investigate algorithms that detects and eliminates the reflections. The field of application is slightly different, but the purpose is the same. Autonomous vehicles should calculate the distance to the curb to avoid possible collisions. However, field tests showed that reflections caused by the curb itself negatively affected the measurement result and therefore had to be eliminated by an algorithm.

Finally, one more work is to be mentioned: "Ground material distinction method using reflection intensities obtained by ultrasonic sensor" [1], which uses the reflections to conclude the type of material in the ground. The reflected signal energy is measured, and thus the scanned materials are uniquely assigned.

In the following, this paper will also deal with the core topic of the three previous papers: What influence do reflections have on our system?

3 Measurements

The measurements will be performed in the laboratory setup of the Extense project at the COSA of the Lübeck University of Applied Sciences [2] (Figure 2). The laboratory setup consists of a water tank filled with saltwater and a bottom covered with sediments. The unit in the water tank, consisting of the transmitter and receiver, is mounted on a holding frame. The object that is to cause the reflection is mounted on a carriage system that is moved over the entire surface of the water tank [4].

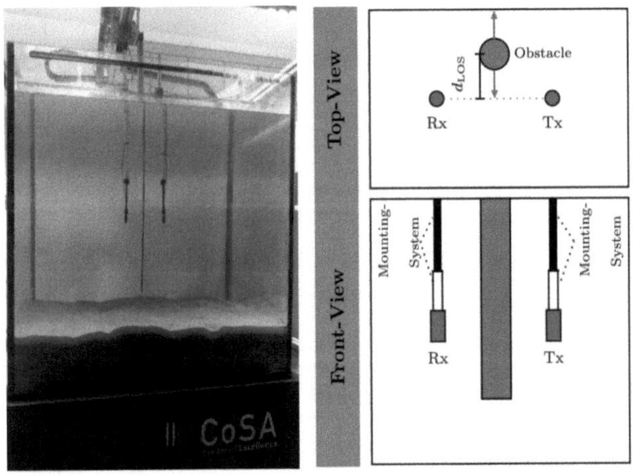

Figure 1: Watertank for measurements with the described hardware setup in the COSA-laboratory ©2019 IEEE [2].

3.1 Setup measurement process

In the laboratory water tank, the two measuring units (Tx, Rx) are placed exactly in the center of the water tank to delay reflections from the sidewalls, the bottom, and the water surface as much as possible by maximizing the distance to them. The measurement object is then clamped exactly centered between the two transmitting and receiving units, with a slight offset from the centerline between the Tx and Rx Unit. The measurement setup is then started via a MATLAB script. The ultrasonic frequency in the measurement and simulation is between $90\ to\ 150kHz$.

During the measurement, the positioning system [4] continuously moves the measuring object out of the measuring range in a defined increment of 1mm along the orthogonal center axis. This results in a separate measured value for the reflection for each positioning.

3.2 Used Hardware

The following components were used during the measurements:

Table 1: Measurement Equipment.

Type	Device	Description
Transdurcer	Teledyne TC4013	Ultrasonic transmitter and receiver
Amplifier	PCB board with OP-Amps	Self designed printed circuit board for analog signal amplification [3]
Data acquisition	RedPitaya 125-14	Embedded 2 channel oscilloscope Signal and 2 channel signal generator
Signal Generator	RedPitaya 125-14	Embedded 2 channel oscilloscope Signal and 2 channel signal generator
PC	Windows PC with Intel i7	Control and automation unit
Velocity measurement	CTD60Mc probe (Sea and Sun Technology GmbH)	Measurement of the sound velocity during measurements (c = 1503 m=s)
Water tank	Saltwater filled tank 125-14	60 cm x 80 cm x 50 cm (length,width, and water height)

4 Model

The simulations are performed with the Matlab Simulation of the EXTENSE Project[1]. The Matlab simulation is designed to be able to get an idea of the system's behavior without implementing the measurement setup in the real physical world or the water tank in the laboratory. Therefore, this simulation is very helpful to gain fast and easy first measurement impressions. It also allows the testing of

[1] https://git.mylab.th-luebeck.de/fabian.john/us_multipath_model

some proper design features before implementing them into the physical measurement setup.

The previously created measurements had to be executed the same in the simulation program to be able to compare them afterward. For this purpose, the program was supplemented and adapted in some places. In the original model, it was already possible to set reflections, but these were not dynamic but only adjustable with constant position and constant reflection coefficient. With the conversion, the position had to be changed over the entire measurement, and also the influence of the reflection coefficient is in constant change over the measurement.

4.1   Calculation of the Position

The position-accurate determination of the reflection length is therefore essential to accurately determine the signal energy of the reflection. The signal power emanating from the reflection depends on how far it is away from the transmitter unit of the measurement setup. We consider the transmitter's output signal as a spherical signal, which has its center in the transmitter. Figure 3

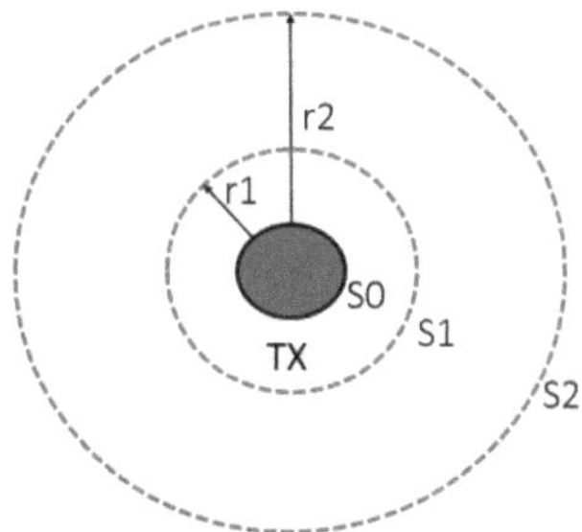

Figure 2: Spherical signal arround the Transmitter.

The total signal power $S0$ of the transmitter is distributed over the enveloping surface of the sphere signal. The enveloping surface of the beads grows quadratically with the diameter.

$$A = (2 * r)^2 * \pi \tag{1}$$

Thus, the incoming signal power at the reflection object and the distance from the signal source to the reflection object are in a direct relationship.

The length of the signal path to the reflection object is calculated using the Pythagorean theorem. Since we always place the measurement object centrally between the transmitter and the receiver (4.1). To calculate the total signal path from the transmitter via the reflection object and to the receiver, the simple signal path d_{refl} is multiplied by a factor of 2 by simplification:

$$d_{Object} = \sqrt{d1^2 + (d_{los} - \frac{b}{2})^2} \tag{2}$$

$$d_{refl} = 2 * d_{Object} \tag{3}$$

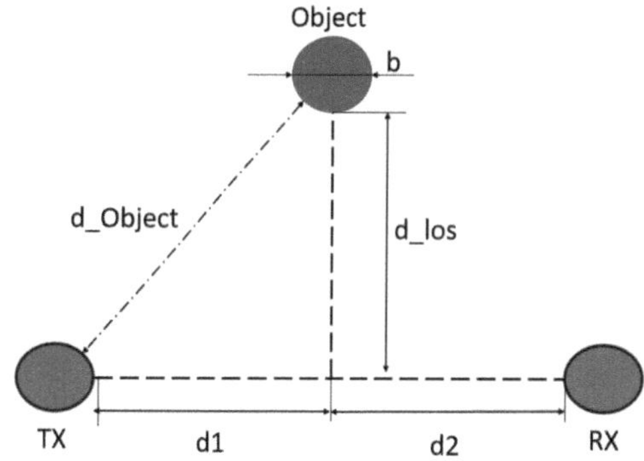

Figure 3: Topview on measurementsetup.

4.2   Calculation of Position-dependent reflection coefficient

The refection coefficient also changes due to the distance-dependent signal power that arrives at the reflection object. Again, the further away the reflective object is from the transmitter, the lower the refection coefficient becomes. Thereby the punctual signal power on the enveloping surface then decreases with a cubic factor.

$$S(r) = R_0 * d_{refl}(r = 0)^2 / d_{refl}(r)^2 \tag{4}$$

Through these two main adjustments and other smaller ones, it was possible to carry out the previously performed real measurements in the program with the simulation. The following section will show how congruent the two methods are.

5   Evaluation

A Matlab script was used to compare the measurements and simulations, which was also developed within the EXTENSE project. The script adjusts the measurement and model data to cover the same frequency spectrum and then compares the acquired measurement data with the simulation data. The previously mentioned model for the simulation of the reflection objects is embedded in this script and calculates the simulation values at runtime. For this reason, the previous adjustments in the model must also be implemented in this script. As a reflection object, a round steel bar made of solid material with a diameter of $b = 35\,mm$ was chosen in order to create a reflective surface as large as possible to gain a strong multipath reflection.

The object was then evaluated in the laboratory measurement and the simulation on the distance of $(d_{LOS} = [3.5; 12]\,cm)$. We perform a fingerprinting to calculate the object position $d_{calc}(d_{LOS})$ for each measured spectrum $Y_{ms}(d_{LOS}, f)$ compared with all model calculated spectra $Y_{mdl}(\overline{d}_{LOS}, f)$ with Equation 5 and 6.

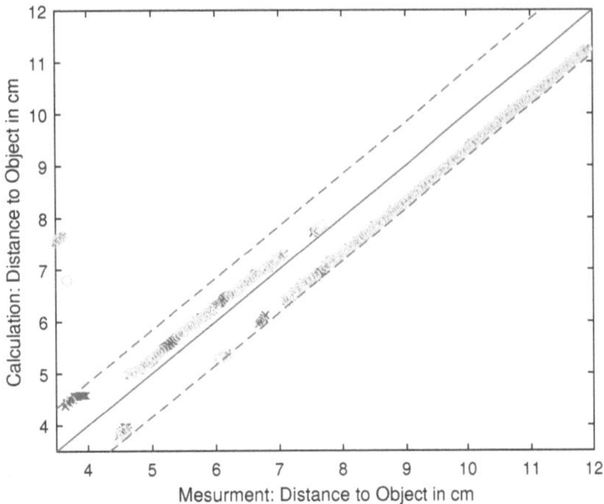

Figure 4: Evaluation plot for object position estimation.

$$\Delta(d_{\text{LOS}}, \overline{d}_{\text{LOS}}) = \int |Y_{\text{ms}}(d_{\text{LOS}}, f) - Y_{\text{mdl}}(\overline{d}_{\text{LOS}}, f)| \, df \quad (5)$$

$$d_{\text{calc}}(d_{\text{LOS}}) = \min\{\Delta(d_{\text{LOS}}, \overline{d}_{\text{LOS}})\} \quad (6)$$

In the plot in Figure 4, it is shown that the calculated position (blue cross) corresponds to the expected position given by the measurement. In contrast, the yellow circles in the plot show the expected position of the nonmodified model. The solid red line represents the expected position of the object, and the blue crosses the calculated position. In addition, the dashed lines indicate the diameter of the measured object. We observe the calculated positions of the object with high precision that the deviation is in the range of the object diameter. In the beginning, there are few positions estimated with a more significant deviation.

6 Conclusion and Future Work

Figure 5 clearly shows that the adjustment in the model has brought a significant improvement. The blue data points are, to a large extent, still within the object diameter. In contrast, the yellow data circles also show a very good calculation of the actual position, but we can see that the modified model has a more accurate calculation. The estimation overall positions was calculated to prove this first conclusion. Therefore the average deviation from the calculated to the actual position of the object was calculated. By the modified model, there was an average deviation of $2.06 cm$, and by the unmodified model, it was at $2,32 cm$. This difference shows that the adjusted model is working more precisely than the older one.

In a possible future adaptation of the model the reflections could serve as an additional source of information and thereby refine or extend the measurement results. A possible application of the additional data would be an environmental scan around the object to be measured. Due to the transit time shifts of the two received signals for the

measurement object and the reflection object could be easily distinguished from each other and thus be recorded and processed differently. Another possibility to use the measurement data is to draw conclusions about the condition of the cable by different reflection strengths and angles.

Acknowledgement

This publication results from the research of the Center of Excellence CoSA at the Technische Hochschule Lübeck and is funded by the Federal Ministry of Economic Affairs and Energy of the Federal Republic of Germany (Id 03SX467B, Project EXTENSE, Project Management Agency: Jülich PTJ). The Product of the Module "Scientific Project" is part of the Applied Information Technology Curriculum. Horst Hellbrück is an adjunct professor at the Institute of Telematics of University of Lübeck.

Author's Statement

Conflict of interest: Authors state no conflict of interest.

7 References

[1] Daisuke Iwasaki, Kazuo Haruyama, Shenglin Mu, Huimin Lu, Kanya Tanaka, Yuhki Kitazono, Yuji Wakasa, Seiichi Serikawa, and Shota Nakashima. Ground material distinction method using reflection intensities obtained by ultrasonic sensor. *IEEE/SICE International Symposium on System Integration (SII)*, 2012.

[2] Fabian John, Marco Cimdins, and Horst Hellbrück. Underwater ultrasonic multipath diffraction model for short range communication and sensing applications. *IEEE Sensors Journal*, 21(20):22934–22943, 2021.

[3] Fabian John, Roman Kusche, Felix Adam, and Horst Hellbrück. Differential ultrasonic detection of small objects for underwater applications. In *Global Oceans 2020: Singapore–US Gulf Coast*, pages 1–7. IEEE, 2020.

[4] Fabian John, Sven Ole Schmidt, and Horst Hellbrück. High precision open laboratory 3d positioning system for automated underwater measurements. In *Global Oceans 2021: San Diego - Porto*, pages 1–5, 2021.

[5] Werner G. Neubauer. Ultrasonic reflection of a bounded beam at rayleigh and critical angles for a plane liquid-solid interface. In *Journal of Applied Physics*, pages 44–48, 1973.

[6] Joon Hyo Rhee and Jiwon Seo. Ground reflection elimination algorithms for enhanced distance measurement to the curbs using ultrasonic sensors. In *Proceedings of the 2018 International Technical Meeting of The Institute of Navigation*, pages 224 – 231, January 29 - 1, 2018.

2

Regulatory Affairs

Regulatory strategy for the introduction of a laboratory tracking system into the United States and United Kingdom markets

César Alexis Ruiz Rodríguez [1,2], Yannick Timo Böge[2]

[1] Biomedical Engineering, Luebeck University of Applied Sciences, cesar.alexis.ruiz.rodriguez@stud.th-luebeck.de
[2] Science department, Smart 4 Diagnostics GmbH, München, timo.boege@s4dx.com

Abstract

This paper aims to explain the technical-regulatory assessment performed to the Smart 4 Diagnostic (S4DX) system to develop a regulatory authorization procedure for its introduction into the U.S. and U.K. markets. The main function of the S4DX system is to monitor and evaluate relevant pre-analytical data of the clinical laboratory in real-time. The assessment was performed by consulting the FDA, FCC regulations, and the U.K. statutory instruments. It was confirmed that the S4DX system does not fall under the scope of FDA regulations. Nevertheless, FCC regulations shall be considered for the hardware. Regarding U.K, the product can enter freely until 2023 due to its CE Mark. This paper concludes that the S4DX system is classified as software that integrates and analyzes laboratory quality control data, identifying increased random errors, including radiofrequency emission hardware that must fulfill the applicable radiofrequency regulations.

1   Introduction

Smart 4 Diagnostic (S4DX) has developed a system that offers a digital market-leading solution for physicians and laboratories to support, monitor, and digitally document pre-analytical processes in medical laboratory diagnostics testing. The pre-analytical phase of the total testing process in in-vitro diagnostic testing is much more prone to errors than all other steps in the highly fragmented process, and is responsible for up to 70% of all laboratory errors, [1], [2].

The pre-analytical steps, which are usually performed at the sample collection site, include patient preparation, laboratory order placing, sample collection, storage, and transportation procedures before samples arrive in the medical laboratory. Thus, these steps are by no means controlled by the clinical medical laboratory, where standardized sample processing steps are performed under the control of experienced laboratory personnel nor are they in an automated manner on laboratory analyzers.

The cloud-based data platform of the S4DX system monitors and evaluates relevant pre-analytical data in real-time. The data platform combined with the operating system front-end solution, guarantees a verifiable pre-analytical sample process for complete quality assurance and supply chain monitoring of all human blood samples. Data is integrated via an interface into the laboratory/hospital information management system. The S4DX solution is composed of the Smart Tube©, Gateway, Satellite/Courier App, and Web Service, which functions are described in Fig.1 The S4DX system is implemented in several German laboratories, and pilots are running in further European countries.

The S4DX system is not classified as a medical device

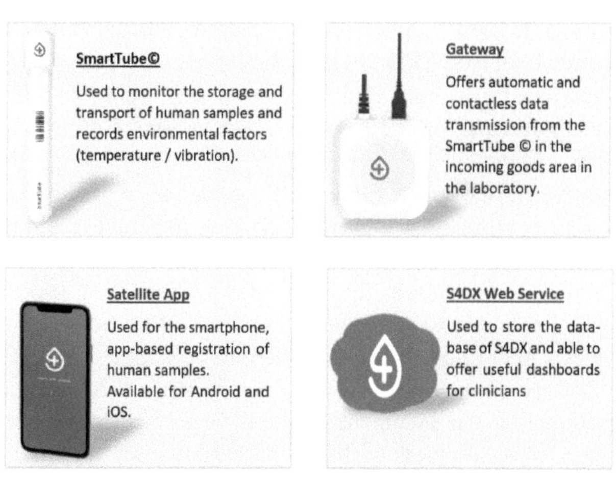

Figure 1: Elements that integrates the S4DX system

and does not fall under the scope of the In-Vitro Diagnostic medical device Regulations (IVDR), the Medical Device Regulation (MDR), and MDCG guidelines, because according to the manufacturer intended use of each component, and the whole system per se, does not fit to the definition of a medical device (the system does not diagnose or cure [3]), in vitro diagnostic medical device (the system does not examine specimens like blood or tissues from human body), specimen receptacle (Smart Tube does not contain or preserve any specimen from human body[4]), and/or software as a medical device(the stored data is not used to perform any diagnosis[5]).

The S4DX company aims to expand its market scope by introducing the product into the U.S. and U.K market, therefore, a regulatory evaluation for both markets shall be performed to know if the S4DX system can be considered as a medical device or not in this country, and then follow the due requirements according the applicable regulations.

2   Material and Methods

A regulatory evaluation was performed at the Regulatory Affairs and Quality Management department of the company Smart 4 Diagnostics. To perform it, an analysis of the S4DX system was done. On the one hand, the existing technical documentation of the hardware (Smart Tube © and Gateway devices), were consulted, also the specific functions, type of states, electronics features and connection modes of the devices were defined. On the other hand, the software section was tested following the "Test Suite" functionality and Graphic User Interface (GUI) templates, these documents mention all functions and all gathered data by the software, being useful for identifying the system functions. For the U.S., the Code of Federal Regulations (CFR) was consulted. This compendium of laws is divided into 50 titles, where the 21st title contains the Food and Drug Administration (FDA) regulations, [6], and title 47th contains the Federal Communications Commission (FCC) regulations, [7], which regulates telecommunication devices. The FDA and the International Medical Device Regulators Forum (IMDRF) guidelines regarding the classification of medical devices, IVD devices, Software as a Medical Device (SaMD), Mobile Medical App (MMA), were consulted to demonstrate the status of the S4DX system as a medical device in the U.S, [8], [9].

Moreover, title 47th CFR and FCC guidelines regarding integration of modular transmitter of year 2020 and labelling guideline of 2018 were consulted, [11], to determine the devices classification, the applicable authorization procedure, laboratory tests and labels for the S4DX hardware devices. According to the 47 CFR part 2 subpart J and the SDoC guideline of 2019, [11], specific labels, compliance information and documentation shall be performed and presented at the moment of the product exportation.

For the U.K, the statutory instruments legislation was consulted, where the Medical Devices Regulation from 2002 defines the characteristics of a medical device. The guidance "placing manufactured goods on the market in Great Britain" of the U.K. government was consulted to clarify the U.K. status in the E.U. after the Brexit, [10]. Finally, A regulatory evaluation report which contains tests results, and authorization procedure of the hardware devices was performed to demonstrate the due requirements to introduce the product in the U.S. ans U.K. market.

3   Results and Discussion

The S4DX system elements do not fall under the FDA and the UKMDR regulations as can be seen in Table 1. This decision had been performed, because according to the intended use of the S4DX system, which is to monitor and record relevant sample workflow data from time-point and environmental parameters of the blood sample collection room until its laboratory arrival, it is not aimed to make diagnose, prevent or cure any disease, hence, it does not fit in the FDA definition of medical device. Furthermore, the United Kingdom Medical Device Regulation (UKMDR) has insignificant differences between the EU Medical Device Directive (MDD), where the S4DX components are also not considered as a medical device or similar. However, it can be seen in Table 1 that Smart Tube © and the Gateway fall under the Radio Equipment Regulation (RER) and FCC regulations.

Table 1: Applicable regulation for the S4DX system

	FDA	FCC	UKMDR	RER
Smart Tube ©	X	✓	X	✓
Gateway	X	✓	X	✓
Satellite App	X	X	X	X
Courier App	X	X	X	X
S4DX Web Service	X	X	X	X

3.1   U.S. Market

Table 2 shows different types of medical devices to which the S4DX system was compared, to identify the correct FDA classification. The SaMD definition is software intended to be used for one or more medical purposes that perform these purposes without being part of a hardware medical device. To be classified as a SaMD, a device or software should be interfaced with other medical devices, either hardware medical device or other SaMD; this is not the case of the S4DX system.

A MMA is a software that meets the definition of a medical device and it is deployed on a mobile platform. The FDA issued the guidance "Policy for Device Software Functions and Mobile Medical Applications", [9], where in its appendix A "Examples of Software Functions that are not Medical Devices", in the point 21 appears the example of apps that are intended to transfer, store, convert formats or display clinical laboratory tests and results. The S4DX apps gather general data from blood samples in the pre-analyticial phase, in order to present data in dashboards, hence, the gathered data is not processed or analyzed to do a diagnose. Then, the courier app and satellite app are not classified as a MMA.

Table 2: S4DX system's classification according the FDA

	SaMD	MMA	BECS	MD	SQC
S4DX	X	X	X	X	✓

Blood Establishment Computer Software (BECS) regulations were analyzed, due to the S4DX system is related indirectly in a patient-blood environment. The Table 2 shows that the BECS regulations are not applicable for the S4DX system, because it is not focused on blood transfusions and identification of blood receiver patients.

The IMDRF guideline: Frameworks for risk categorization, in its section 10 "Appendix: Software that are not SaMD" mentions that a software that integrates and analyzes laboratory System Quality Control (SQC) data to identify increased random errors or trends in calibration on IVDs is not a SaMD, [8]. The S4DX system has exactly the intended purpose of the IMDRF guideline, because it collects, sorts and uploads the gathered pre-analytical information from the clinical laboratory to the cloud via web services for the creation of dashboards, aiming to reduce random errors in pre-analytical phase, hence, the S4DX system is not a medical device.

3.1.1 Device classification

Nevertheless, the Smart Tube[©] and Gateway do fall under the FCC regulations, due to their telecommunication functions. They are digital devices that emits radiofrequency to send or receive data. Table 3 shows the FCC classification of the S4DX devices, which determines the applicable regulation and authorization procedure to follow, both devices have the same classification, where CIR is for Component Intentional Radiator.

Table 3: FCC device classification

	Gateway	Smart Tube[©]
UR with integrated CIR	✓	✓
Class B digital device	✓	✓
Mobile device	✓	✓

The PCB board and passive elements of the Gateway and Smart Tube[©] emit unintentional radiations, [7], however, both devices contain intended radiator modules, either Bluetooth or WIFI module. For these types of devices, the applicable guideline is the "Module integration guide V02", which mentions the CFR points shown in Table 4, which shall be fulfilled to introduce this type of telecommunication digital device into the U.S. market.

The Gateway and Smart tube shall be classified as class B digital device because they are aimed for use in a residential environment like medical offices and clinical laboratories. Since the Smart Tube[©] and Gateway are intended to be used mostly in a higher distance than 20 centimeters between any human, they can be classified as a mobile device.

Table 4: S4DX applicable CFR points

	Gateway	Smart Tube[©]
47 CFR §15.109(a)	✓	✓
47 CFR §15.107	✓	X
47 CFR §15.209(a)	✓	✓
47 CFR §15.247	✓	✓
47 CFR §15.407	✓	X

According to the "Module Integration Guide v02", Gateway and Smart Tube[©] shall fulfill the points shown in Table 4. On the one hand, the Gateway FCC tests performed by the raspberry pi inc. company can be used to fulfill the points of Table 4, this is possible because the S4DX company did not modify any element in the hardware of the raspberry, therefore, these test reports can be used for our purposes

with the permission of the Raspberry Inc, due to it is an open source hardware. On the other hand, in the case of the Smart Tube[©], the unintentional radiation test(§15.109) and spurious emissions test (§15.209, §15.247) shall be done and approved by an FCC certified laboratory, also known as Telecommunication Certification Bodies (TCB). The objective of these tests, is to check for emissions that may occur due to the intermixing of emission with the digital circuitry.

3.1.2 Supplier Declaration of conformity

The supplier declaration of conformity authorization procedure was selected over the certification process, due to its faster application and its lower price, according to the quotes received by different TCB. To follow the SDoC procedure, the requirements shown in Table 5, shall be performed in the Smart Tube[©] and Gateway, the processes are the same for both devices.

Table 5: SDoC requirements

	Gateway	Smart Tube[©]
FCC Unique identifier	✓	✓
47 CFR §2.1077	✓	✓
"Contains FCC ID" legend	✓	✓

On the one hand, in Fig. 2 is shown how the final product of Smart Tube[©] shall be labeled. On the left side of the same figure, are located the FCC unique identifier and the "Contains FCC ID" legend, which are label requirements. On the other hand, the Gateway should have on the bottom part of the device analogous label requirements.

Figure 2: Applicable FCC labels on the Smart Tube[©]

Regarding the 47 CFR §2.1077 "Compliance information", the instructions manual of the Smart Tube[©] and Gateway should have a specific section, available for the user, with the data of the responsible party, in this case the S4DX company, also its address and contact information, and the FCC ID of the certified module, which in the case of the Smart Tube[©] is "QOQ13" and for the Gateway is "2ABCB-RPI4B". Moreover, the instructions manual shall contain the FCC logo and the statement mentioned in the 47 CFR §15.19(a)(3), picturized in Fig. 3. Finally, in the regulatory evaluation report shall be attached the respective certified test reports and the grant of certification of the certified module.

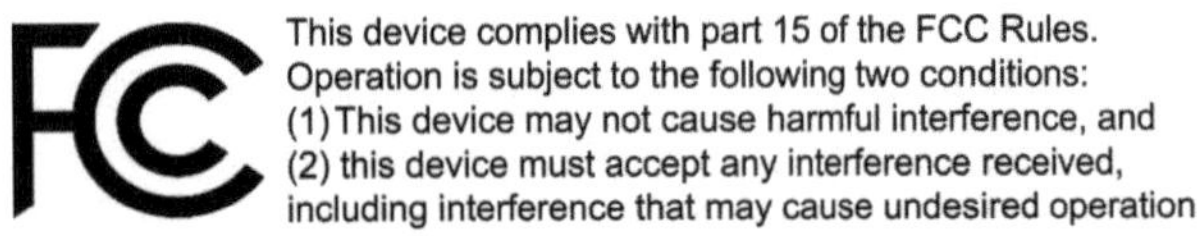

Figure 3: FCC logo and compliance information §15.19(a)(3)

3.2 U.K. Market

Due to the U.K. has exited of the EU, and according to the "Guidance-placing manufactured on the market in Great Britain", all goods which previously required the CE marking will now require the United Kingdom Conformity Assessed (UKCA) marking from 1st of January 2023. The Smart Tube© and Gateway fall under this statement, thus, before this due date, both can enter freely into the market without restrictions during all 2022 because of their CE Mark. If the product is introduced in the market in 2023 or after, further considerations regarding the UKCA Mark and DoC shall be treated, however, barely changes in documentation would be performed because the radiofrequency equipment regulations between the UK and EU are analogous. Company's plan is to introduce the product in 2022, hence, no deeper research was necessary for the objective of this paper.

4 Conclusion

The S4DX system is considered a highly innovative product which cannot be associated to a particular regulatory product category of the FDA restrictions of the U.S. or the UKMDR. The S4DX system hardware must fulfill the applicable FCC regulatory requirements, due to its telecommunication functions to be marketed in the U.S.

The Smart Tube © and Gateway are unintentional radiators that integrate a certified module that works as an intentional radiator; therefore, both devices shall follow the SDoC process (tests, labels and technical documentation) and attach in the regulatory evaluation report the respective certified test reports and the grant of certification of the certified module.

The Smart Tube © needs to fulfill sections §15.109, §15.247 and §15.209, therefore, radiated emission test for unintentional radiator and radiated spurious emission test shall be done. Regarding the Gateway, no additional tests are required, due to the certified Raspberry PI 4B has not been modified. Tests done by the Raspberry inc. can be used for the regulatory evaluation report.

Both devices shall follow the SDoC authorization procedure, present the compliance information of the FCC part 15 B and the FCC logo in their instruction manuals. The devices shall present the corresponding FCC ID on a visible part of the device surface as well.

The product can enter freely in the U.K. market until 2023, it will require the UKCA mark and a DoC with the UK regulations. Following this regulatory strategy, the S4DX system will be able to enter the U.S. and U.K. markets.

Acknowledgement

The work has been carried out at the start-up company Smart 4 Diagnostics GmbH, München.
Furthermore, I would like to thank CONACYT and DAAD for the scholarship support during my master program.

Author's Statement

Conflict of interest: Authors state no conflict of interest.

5 References

[1] A.S. Sakyi, A.F. Laing, R.K. Ephraim, O.F. Asibey, O.K. Sadique, *Evaluation of Analytical Errors in a Clinical Chemistry Laboratory: A 3 Year Experience.* Annals of Medical and Health Sciences Research, University of Science and Technology, Kumasi, Ghana, January 2015.

[2] M. Zaniotto, A. Tasinato, A. Padoan, G Vecchiato, A. Pinato, L. Sciacovelli, M. Plebani, *Effects of sample transportation on commonly requested laboratory tests.* Clinical Chemistry and Laboratory Medicine, October 2012.

[3] European Union, *point 1 of article 2 regulation(EU) 2017/745*, Official Journal of European Union, 2017.

[4] European Union, *point 2 and 3 of article 2 of regulation(EU) 2017/746.* Official Journal of European Union, 2017.

[5] Medical Device Coordination Group, *Decision step 3 of point 3.3 of MDCG 2019-11*, 2019

[6] Code of Federal Regulations, *Title 21"Food and Drugs"| sub-chapter H "Medical Devices".* November 2021.

[7] Code of Federal Regulations, *Title 47 "Telecommunication" part 15 B and 15 C.* November 2021.

[8] IMDRF working group, *"Software as a Medical Device": Possible Framework for Risk Categorization and Corresponding Considerations*, IMDRF, 2014.

[9] Food and Drugs Administration, *Policy for Device Software Functions and Mobile Medical Applications*, FDA Division of digital health, September 2019.

[10] Department for Business, Energy Industrial Strategy, *Placing manufactured goods on the market in Great Britain*, Available: https:www.gov.uk/guidance, 2021.

[11] Office of Engineering and Technology Laboratory Division, *FCC General guidelines.* Federal Communications Commission, 2020.

Establishing and implementing a risk management process for a software as a medical device

Seyma Inceli [1]

[1] Biomedical Engineering, Luebeck University of Applied Sciences, seyma.inceli@stud.th-luebeck.de

Abstract

The European Union Regulation (EU) 2017/745 also called Medical Device Regulation (EU MDR) and U.S. Food and Drug Administration (FDA) regulations require the medical device manufacturers to establish a continuous risk management process through the life cycle of the medical devices including software as a medical device. The goal of this project is to establish and implement a risk management process for a software as a medical device (SaMD). The results of the risk management process such as risk analysis, risk evaluation, and risk control strongly depend on the characteristics and intended use of the medical device. It is crucial to identify and control risks related to performance, user failure, assets, threats, and vulnerabilities associated with the SaMD for developing an effective process.

1 Introduction

Medical device manufacturers are obligated to provide, implement, document, and maintain a risk management process by EU MDR and FDA regulations. A risk management process is established to ensure that medical devices that are intended to be put into the market do not adversely affect the health of patients as well as users or any other person when the device is used in accordance with the intended use. The risk management process is required to follow the elements of International Organization for Standardization (ISO) 14971:2019 [1]. The term "Software as a Medical Device (SaMD)" means that a software is intended to be used for medical purposes such as diagnosis, prevention, monitoring, treatment, or alleviation of disease or injury [2] and it can achieve its medical purpose independently of a hardware. The risk management process is obligated to be implemented throughout the total product life cycle of the software including design plan, design development, design verification, design validation, release, and post-production activities. While establishing a risk management process, hazards related to cybersecurity, usability, and performance depending on the characterization of SaMD are required to be identified, evaluated, controlled, and monitored. Safety management by implementing a cybersecurity risk management process is necessary for protecting sensitive data such as Protected Health Information (PHI) including demographic information, intellectual property, and test and laboratory results [3]. Moreover, interruption in the delivery of patient care due to cyberattacks is prevented by implementing cybersecurity measures. Hence, the device performance is maintained. The goal of this project is to establish and implement a risk management process for SaMD that transforms preacquired 2D medical images into 3D models and visualizes 3D imaging holograms of the patient. There are different viewing principles of this software platform such as on a head-mounted display via head-mounted display application or PC monitor via web application. The device utilizes mixed reality (MR) technology for displaying medical images and visualizing 3D imaging holograms. This technique not only allows projecting virtual objects on real-world surfaces but also allows users to interact with the digital content by using gesture methods.

2 Material and Methods

The elements of ISO 14791:2019 and some guidance on the application of ISO 14971 were followed to define good practices for implementing each step of the risk management process. According to ISO 14971:2019, a risk management process includes the steps of risk analysis, risk evaluation, risk control, evaluation of overall residual risk, risk management review, production, and post-production activities [4].

2.1 Risk Analysis

According to ISO 14971:2019, the first step of risk analysis requires identifying the intended use and characteristics that could have an impact on the safety of the medical device. One way of doing this was to ask a series of questions present in Annex A of EN ISO 24971:2020 concerning the manufacture, intended users, intended use, reasonably foreseeable misuse, and security from the point of view of all the individuals involved (e.g., users, patients, etc.) [5]. The next step of the risk analysis was the identification of hazards and hazardous situations associated with the

Table 1: The risk matrix with respect to probability and severity levels [5], and estimation of risk levels/classes (R1, R2, R3, R4, R5, R6) by multiplying the given values of severity levels of harm and probability levels of harm

	Negligible(1)	Minor(2)	Serious(3)	Critical(4)	Catastrophic(5)
Frequent(5)	R1 (5x1=5)	R3(5x2=10)	R5(5x3=15)	R6(5x4=20)	R6(5x5=25)
Probable(4)	R1 (4x1=4)	R3 (4x2=8)	R4(4x3=12)	R5(4x4=16)	R6(4x5=20)
Occasional(3)	R1(3x1=3)	R2(3x2=6)	R3(3x3=9)	R4(3x4=12)	R5(3x5=15)
Remote(2)	R1(2x1=2)	R1(2x2=4)	R2(2x3=6)	R3(2x4=8)	R3(2x5=10)
Improbable(1)	R1(1x1=1)	R1(1x2=2)	R1(1x3=3)	R1(1x4=4)	R1(1x5=5)

medical device. Hazard is described as a potential source of harm based on the intended use, safety characteristics, and foreseeable misuse, while hazardous situations, which might lead to harm, arise from a sequence or combination of events of random faults, systematic faults, and security vulnerabilities. Since contact with software cannot contribute to an injury, harm can be considered as a threat to property or information stored in the medical device itself. After completing the identification of hazards and hazardous situations, the last step of the risk analysis was estimating the risks that is defined as a combination of the occurrence probability of harm and the severity of that harm. Several approaches can be used for risk estimation. In this study, 5x5 risk matrix was used as presented in Table 1 [5] in which the probability levels of the harm are rows while the severity levels of the harm are columns. The acceptance levels of risks are not specified by ISO 14971:2019. Therefore, medical device manufacturers are responsible for deciding how many probability levels and severity levels are appropriate and providing the definitions for each level [5]. The risk levels (R1, R2, R3, R4, R5, R6) was calculated by multiplying the given values of severity levels of harm and probability levels of harm. The values are presented in Table 1.

2.2 Risk Evaluation

The second step of the risk management process, risk evaluation, is performed by comparing estimated risks with the risk acceptability criteria and determining whether the criteria is met. If the evaluation, which is multiplying the given values of severity levels of harm and probability levels of harm, is from 1 to 5 by means R1 region risk, very low, the risk was judged acceptable. The rest of the estimated risks (R2, R3, R4 and R5) were not judged acceptable. Therefore, they had to be reduced by implementing risk control measures.

2.3 Risk Control

If the risk is not judged acceptable, then the risk control measures are implemented. Risk control measures include three methods: Inherent safe design and manufacture, protective measures in the medical device itself or the manufacturing process, and information for safety such as providing user manual and training to users, where applicable [4]. The next step after the implementation of risk control measures is residual risk evaluation. Residual risk evalua-

tions are performed by the same method used for risk acceptability of initial risks such as comparing residual risks with the risk acceptability criteria and determining whether these criteria are met. As the result of the residual risk evaluation step, if a residual risk is not judged acceptable, the manufacturer is obligated to either implement further risk control measures or perform a benefit-risk analysis that is used to determine whether the benefits outweigh the residual risk. For instance, although X-rays are harmful to a patient body, the benefits of diagnostic imaging outweigh the risks under controlled conditions [5]. As the last step of the risk control process, it is significant to ensure that implementing a risk control measure does not introduce new risks or increase the risk ratio of other risks including those previously evaluated to be acceptable.

2.4 Evaluation of Overall Residual Risk

After implementation and verification of all the risk control measures, the overall residual risk, which is the contribution of all remaining residual risks, must be evaluated from a broad perspective to determine whether the medical device, as a whole, meets the risk acceptance criteria

2.5 Risk Management Review

The risk management review is performed prior to the commercial release of the medical device in order to verify that the risk management plan and activities are implemented, overall residual risk is acceptable.

2.6 Production and Post-Production Activities

The production and post-production process need to be monitored to collect and review information about the medical device in terms of safety.

3 Results and Discussion

The result of identification of the device characteristics showed that the medical device can not directly harm the patients since it is a software and there is no direct connection between the device and individuals. Moreover, the device is used only for informational purposes as complementary device to the conventional medical imaging systems such

Table 2: Some of the hazards, hazardous situations, harm, risk control measures and risk classification before and after implementing risk control measures associated with the device

Hazard	Hazardous Situations	Harm	Risk Class	Risk Control Measures	Risk Class
Hazards due to mistakes and judgment errors	The user selects the wrong patient's data to support a decision.	User does not have accurate information to support a diagnostic decision and might cause indirect harm to patient due to wrong information/decision.	R5	Training and IFU on how to create patient ID without entering any PHI to recognize patients are provided. Data in 3D shall be compared with data in 2D	R1
Hazards due to misuse of the device	The device is used for diagnostic purposes	User might decide a wrong therapy due to wrong diagnosis resulting deterioration of the patient health indirectly	R6	By training and IFU, the intended use is highlighted. The device is only can used together with the conventional imaging devices.	R3
Hazards due to mistakes and judgment errors	The user, by using drawing tools such as pen to mark a region, might miss marked/not marked some important image structures	User does not have clear information to support a diagnostic decision.	R2	By training and IFU the information on how to draw and how to delete drawing is provided	R1
Hazards due to inadequate information provided to user	The user enter PHI of patients when saving the case	Unauthorized access to private health information in case of cyberattacks	R5	By training and IFU, the importance of not entering patient private health information to system is provided.	R3
Hazards due to uncontrolled distribution of passwords	Unauthorized access to users' accounts	Unauthorized access to patient data (medical scans)	R3	The user needs a Microsoft Account Access to the device requires a Microsoft Account. Microsoft has security verification that is sent when, for e.g., there is a login from a new device/location. Moreover, since medical scans are anonymized within the system, confidentiality is preserved.	R1
Hazards due to lack of anonymization of uploaded DICOM data	Cyber attacks	PHI is revealed	R4	DICOM data is anonymized automatically by DICOM processor. Inherently safe by design.	R1
Hazards due to incorrect coloring of given tissues	Colors of given tissues do not represent the reality	User does not have clear information to support a diagnostic decision might cause indirect harm to patient due to wrong information/decision.	R3	Automatic coloring of tissues is implemented by Natural Rendering feature. Before each release, the results of each type of renderings are compared with previous results to ensure that the Natural Rendering Feature works properly.	R1

as CT and MRI, and the device is not intended for diagnostic use. In total, 86 risks were detected related to user failure, performance and cybersecurity of the device. Initially, 38 of the total risks were judged acceptable since the risk estimation value for each one them were between 1 and 5. The rest of the not acceptable risks, which are R2, R3, R4 and R5 were reduced as far as possible. Some examples of the risks, which were not judged acceptable prior to implementing risk control measures are given in Table 2. When the risks related to cybersecurity of the device were estimated, the harm was considered as a threat to property or information stored in the medical device itself as presented in Table 2. While defining the harm, the worst-case scenario was determined in case of the hazardous situation. Hence, the maximum risk classification was obtained by using the severity and probability levels. The risk control measures were defined internally together with the regulatory affairs team and software developers. All risks related to user failure were reduced by providing the user training and accompanying document which is the instructions for use of the device (IFU). The risk control measures and risk classification before/after implementing the risk control measures are presented in Table 2. As it can be seen from the Table 2, the residual risks due to user error were still not judged acceptable. In this case, the manufacturer is obligated to either implement further risk control measures or perform a benefit-risk analysis that is used to determine whether the benefits outweigh the residual risk [4]. In this project, all the residual risks due to user error were decreased as far as possible. To do so, the intended use was defined as that the device transforms preacquired 2D medical images into 3D models and visualizes 3D imaging holograms of the patient for pre-operative planning. The images are viewed are for informational purposes only and not intended for diagnostic use. Moreover, the training and accompanying documents are provided to increase the awareness of the users. The remaining risks are included in the instructions for use. Furthermore, a literature research for benefit risk analysis was conducted on the scientific papers published in between 2011 and 2021 by using keywords "Benefit and risk analysis", "augmented reality", "mixed reality" and "user failure" via Google Scholar. As the result of the research, nothing specific for the risks related to the users could be found. Therefore, the benefit-risk analysis was done internally based on the manufacturer security policy and the result showed that the risks were reduced as far as possible. Under controlled conditions (e.g. using the device only for supporting decisions made by using conventional medical imaging systems), the benefits outweigh the residual risks. Initially, 86 risks were detected related to user failure, performance and cybersecurity of the device and 38 of the total risks were judged acceptable which had the risk value between 1 and 5. The rests, 48 of the total risks, were reduced as far as possible by implementing the risk control measures. In the end, 20 of the residual risks, which were mostly belonged to R2 class, were not judged acceptable. The benefit risk analysis was provided as described above. After implementation and verification of all

risk control measures, the overall residual risk was evaluated by reviewing adverse events of similar medical devices via a database called Manufacturer and User Facility Device Experience (MAUDE). Since the devices with brand name SurgicalAR and OpenSight were defined predicate devices of the subject device due to same intended use, the search was done based on these devices. No records were found with brand name SurgicalAR and OpenSight. Hence, the overall residual risk analysis was done internally and it was found at the acceptable level. Due to time considerations and limitations, the results post-production process will be obtained via post-market surveillance (PMS) system in the future

4   Conclusion

In this study, a risk management process was established and implemented for a software as a medical device in accordance with ISO 14971:2019. A view on ISO 14971:2019 is essential for improving the compliance with EU MDR and Food and Drug Administration (FDA) regulations. However, the additional requirements set out by the regulations must be addressed for ensuring full compliance.

Acknowledgement

The work has been carried out at apoQlar GmbH and supervised by the Prof. Dr. Folker Spitzenberger, Technische Hochschule Luebeck. I would also like to thank Liliana Duarte, apoQlar GmbH, for her continuous guidance.

Author's Statement

Author states no conflict of interest.

References

[1] International Electrotechnical Commission, IEC 62304: 2006 + AMD1:2015 Medical device software - Software life cycle processes. IEC, 2015

[2] International Medical Device Regulators Forum, IMDRF/SaMD WG/N10 FINAL:2013 Software as a Medical Device (SaMD): Key Definitions. IMDRF, 2013

[3] International Medical Device Regulators Forum, IMDRF/CYBER WG/N60FINAL:2020 Principles and Practices for Medical Device Cybersecurity, IMDRF, 2020

[4] International Organization for Standardization, ISO 14971:2019 Medical Devices-Application of risk management to medical devices. ISO, 2019

[5] International Organization for Standardization, ISO/TR 24971:2020 Medical devices - Guidance on the application of ISO 14971. ISO, 2020

Design verification for the Heart-Lung-Machine

Arpan Pokharel [1]

[1] Biomedical Engineering,Luebeck University of Applied Sciences, arpan.pokharel@stud.th-luebeck.de

Abstract

The heart-lung machine is a medical device that functions as heart and lung substitutes during surgery, leading to a complex medical device with many subsystems. During the manufacturing process, the developing team has to perform the verification of all the subsystems of the heart-lung machine according to EN ISO 13485: 2016 and 21 CFR parts 820.30 to comply with EU and FDA requirements respectively.Since there are no prespecified methods and techniques to implement the verification tests for a subsystem of the heart-lung machine ,this research aims to understand how the involvement of a risk management system (EN ISO 14971), the implementation of EN/IEC standards, the Gage R and R study are essential for the design of a suitable verification test(for example:mechanical impact test on display unit of heart lung machine) which can be validated.

1   Introduction

Effective verification procedures are essential in the medical device development lifecycle and consume almost half of the project development time [1]. In addition, it plays a crucial role in the regulatory approval process. The heart-lung machine is a class II medical device whose primary function is to draw the venous blood out of the heart, oxygenate it and pump it back to the arterial system, which leads to a complex machine system with roller pumps, a centrifugal pump, filters, sensors, display systems, etc. Each subsystem of the heart-lung machine has to pass verification procedure. The verification procedure for the heart-lung machine consists of the following steps:

• Planning of the verification

• Creating the verification protocols

• Executing the verification protocols and documenting the verification results

Each of these steps becomes part of the so-called design history file. The verification planning includes:

• Determining the requirements to be verified;

• Select the type of verification test method for each requirement;

• Selecting the sample size for each verification test method. Subsequently,a test protocol is developed that documents the overall necessary information to execute the test, such as sample size justification, acceptance criteria, the template for the data collection, and test method validation. Before running a test, the test methods are validated. For example, for the test that includes parametric measurement, the measuring instruments used have to be calibrated and verified by performing a Gage R and R study. The electrical components that are part of the test are qualified by performing a test according to the manufacturer's guidance to ensure that their specification are met. Once the test protocol is ready with valid information, an engineering test is executed to ensure every test detail.After the succession of an engineering test, a verification test is carried out whose overall activities are recorded and documented.Finally, the verification test report is written, including the findings or failures, the deviations, and its justification.

The standard EN ISO 13485 includes risk-related activities (ENISO 14971) in the section 7.3.2 "Design and development inputs" [2]. For the subsystem of heart-lung machines for example to design a mechanical impact test on the display unit, risk-related activities and risk control measures are implemented when developing design and developments inputs. Furthermore, they are also used for determining the sample size for a given verification test.

This research aims to understand how the different verification tests like mechanical, electrical, demonstration and analysis are performed for the subsystem of the heart-lung machine within the framework of EN: ISO 13485:2016 and 21 CFR parts 820.30.

2   Material and Methods

2.1   Development and Validation of a test method

For the development of the test process, standardized test methods are used if they comply with the design. However, in the case of a complex design that demands various kinds of tests and does not have the standard procedure, verification expertise creates and applies suitable test techniques to verify the particular design feature based on commonly accepted techniques. These test techniques are designed with a high level of repeatability and precision not to cause any changes in the device's performance. The developed tests should be validated as it is a regulatory requirement and an essential aspect of quality control. In test method validation, a performance characteristic, criteria to the method and recognition of influences that may affect these characteristics are established. It ensures that the test method used on the various components of the heart-lung machine is of the highest reliability and provides valid and compliant results. In addition, in test method validation, the equipment qualification is done. After determining all the necessary factors for test method validation, a test protocol is developed to ensure the test process is consistent between varying sets of testing and different operators.

2.2 Design Inputs

Design inputs are significant aspects of medical device product development. The involvement of the risk management process as a part of design input reduces chances of unacceptable risk being discovered at the end phase of the design process. This allows the design process to be more efficient. The same strategy is applied for deriving the design inputs for the verification of the heart-lung machine. The process of estimating risks involves following steps:
• Set out intended use
• List hazards ($P1$= probability of a hazardous situation occurring)
• Describe hazardous situations leading to Harm ($P2$= Probability of a hazardous situation leading to harm)
• Estimate risk (Probability of occurrence of harm ($P=P1*P2$) Severity of harm)
Example: One of the failure modes for the heart-lung machine is the breakage of the display unit. The effects of the breakage of the display unit on the operator, the machine, the surgeon, and the patient are determined. Next, the possible causes that lead to breakage are identified, and the occurrence rate is calculated. Finally, a design input that addresses the intended use, the safety, and performance is derived or estimated based on these risk estimations.

2.3 Sample Plan

The valid sample plans are developed and documented for carrying out verification tests. It is designed in a manner so that it conforms to requirements and admits all questions in a test case to be answered. The selection of the sample size is based on statistically valid rationale and risk management [2].

2.3.1 Risk-based strategy to develop suitable sample sizes for verification tests

With the help of factors as the sampling class, the U.S. FDA medical device classification, required conformance, and confidence level, the sample size required for a verification test is determined. It ensures the verification test will give a valid result based on risk acceptance criteria set by regulatory requirements.
Example: For creating a verification test related to the display system. The appropriate sample size that justifies the requirements needs to be selected. These can be done

within the risk-based approach, where in a first step, the severity (catastrophic, critical, serious, minor, negligible, no harm) of all the failure modes related to display mode is identified. In the second step, the probability of a hazardous situation leading to harm is evaluated. There are two approaches for estimating the probability of occurrence of harm (quantitative or qualitative).After estimation of the level of probability of occurrence. The residual occurrence level is derived. With the help of residual occurrence level and severity, the residual risk level is determined using the risk level Assessment matrix. After determining the residual risk level of all the failure modes, the final sampling class (medium, low, and high) is assigned as an average of all residual risk levels. Finally, the sampling class with selected conformance, Confidence level, and data type are used to obtain the Sample size that complies with the design input needed to be tested.

2.4 IEC Standards

EN and IEC standards contain the basis for conformity testing of electrical and electronic testing. They are significant for activities related to quality and risk management. They contain definitions and instructions used to test, maintain, and repair electrical and electronic equipment and systems. Many electronic and electrical tests for the subsystem of the heart-lung machine are created based on the IEC standards. For example: IEC 62262, IEC 60068-2-75, EN 62353, EN 62366 etc.
For test methods that include measurements and measuring equipment, different types of variation are observed. The familiar sources of variations are equipment/instrument, operators,complex calculation, test methods, lack of Standard Operating Procedures (SOP). All these sources of the variations are identified, and the following aspects are considered to resolve these variations for test method validation:
• Accuracy,
• Precision,
• Gage R and R

3 Results and Discussion

The design of mechanical impact protects test of the display unit of the heart-lung machine with the implementation of risk management and IEC standards.

3.1 Design Requirements

The four failure modes of the display unit (Details cannot be disclosed in these reports) of the heart lung machine are identified, and their effects, possible causes, occurrence rate, and occurrence rate justification are determined. The test requirement, which can cover best all the identified failures, is derived and identified. The derived test requirement is: The display unit is designed to meet IEC 62262, class IK06. The display unit can resist the mechanical energy as per IEC 62262, class IK06 standard.

The IEC 62262 standards explain how external mechanical impact protection tests on the enclosure are done and classify the degrees of protection provided by the enclosure for electrical equipment [3].

3.2 Determination of total sample size

For determining the total display units needed so that the objectivity of the Mechanical impact test is met, the following steps are followed. At first, The four failure modes of the display unit (Details cannot be disclosed)are mitigated through risk control measures. The risk control measures results indicate the risk level as low. Furthermore, the most appropriate FDA class level of the heart-lung machine according to FDA regulation as per risk class is II. These results in the level of evidence applied (minimum conformance (P=80 percentage) and confidence level (90 percentage). Next, the Mechanical impact test type is a demonstration test, and the data type is an attribute. This qualitative analysis leads to a sample size of 15 display units needed to comply with the design input of the mechanical impact test. Several factors are considered when calculating the appropriate sample size, and one of them is determining the risk level associated with the device. The risk level associated with device helps to figure out the confidence and conformance level needed to perform the given test. Prior to the execution of verification protocol, the FDA requires the sampling plan, including sampling points, number of samples, and the sampling frequency for each unit operation and attribute to be clearly described[4].

3.3 The Device under Test and Configuration

The device under test is the display unit of the heart-lung machine. It must be in a clean and new condition, complete with all its parts in place except the following parts:
• Knob
• Encoder
• Quick-release

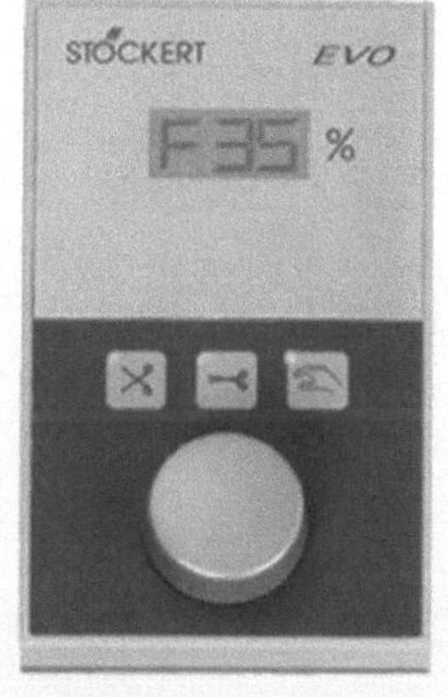

Figure 1: Display unit[5].

3.4 Test Setup and Execution

The Equipment and Accessories needed for the test are:
• Drop Ball
• Rope
• Fixation point
• Device under test Fixation plane
The temperature range of the room where the test is carried out must be in the range of 15°C to 35 °C. The air pressure in the room must be in the range of 86kPa to 106kPa (860 mbar to 1060 mbar).
The test setup is a simple pendulum composed of a drop ball, rope, and a fixed point. The Device under the test plane is kept with a rigid mount surface to avoid displacement once the ball hits the display unit. These allow impact energy to hit on exact target. In addition, the ball must be stable (only rope and gravity forces are applied to the ball). The Fig. 2 represents the test setup described in IEC 60068-2-75.

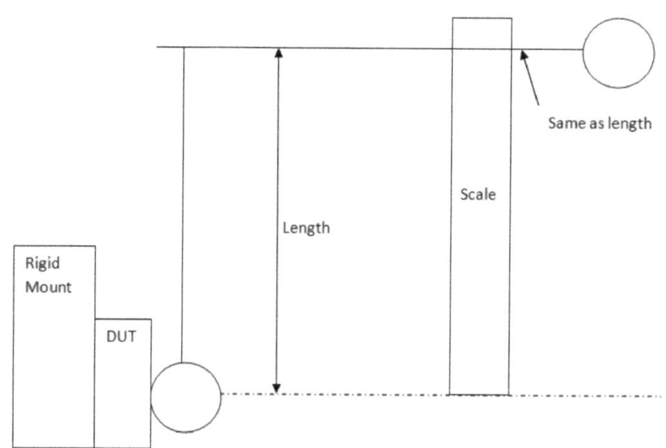

Figure 2: Test setup.

After the test setup, the drop ball is released from the defined initial position to hit the display unit with an impact energy of 1 joule (IK06).
Based on this impact energy, the initial position or defined height is calculated:
E: energy (j)
M: mass of a drop ball (Kg)
g: the gravitational field (N/Kg)
L: rope length (m)
h: Height (m)
the measuring tape is used to measure the height and the rope length:

$$E = mgh \tag{1}$$

$$h = E/mg \tag{2}$$

The adjusted Height is:

$$h' = h + measuringtapeaccuracy \tag{3}$$

To hit the target, the rope length must be:

$$L = h' - balldiameter/2 \qquad (4)$$

The five different areas of display unit are striked with the impact energy of 1 joule.

These test is repeated on 15 different sample of display unit. After test, all the samples are checked to see if any damage is seen. There was no damage seen on any of the display units. Hence the test is passed confirming that the display unit interface of the heart-lung machine is designed to meet IK06 according to IEC 62262.

4 Conclusion

With regards to the use of standards such as EN ISO 13485, ISO 14971, the 21 CFR 820.30 requirements, the verification test for the different subsystems of the heart-lung machine is designed. When no specific test methods are available, the risk-based approach, Use of IEC standards, the Gage R and R study, and the identification and analysis of different variations are implemented to design and validate the verification procedure. The mechanical impact test on the display system demonstrates how the factors mentioned above play an influential role in developing a verification test method that meets a suitable degree of objectivity.

Acknowledgement

I want to express my sincere gratitude to Mr. Gerhard Lang and my Internship Supervisor, Mr. Khaled Jbir, for providing me an opportunity to do my internship and project work in "LivaNova Deutschland Gmbh." I also appreciate Mr. Abdelbasset Eloufrhani for his guidance and encouragement to execute verification activities. Finally, I would like to thank Prof .Dr. Folker Spitzenberger for guiding and supervising my Internship.

Author's Statement

Conflict of interest: Authors state no conflict of interest.

5 References

[1] S. S. Altayyar, "The essential principles of safety and effectiveness for medical devices and the role of standards," *Medical Devices (Auckland, NZ)*, vol. 13, p. 49, 2020.

[2] M. Donawa, "Effectively incorporating risk management into quality systems." *Medical device technology*, vol. 17, no. 5, pp. 28–30, 2006.

[3] S.-E. 62262, "Degrees of protection provided by enclosures for electrical equipment against external mechanical impacts (ik code)," 2008.

[4] M. Durivage, "Sample sizes: How many do i need?" *Quality Magazine. http://www. quality-mag. com/articles/91991-sample-sizes-how-many-do-i-need*, 2014.

[5] K. Kim, C. Ball, P. Grady, and S. Mick, "Use of del nido cardioplegia for adult cardiac surgery at the cleveland clinic: perfusion implications," *The journal of extra-corporeal technology*, vol. 46, no. 4, p. 317, 2014.

Comparative regulatory analysis of technical dossiers for different kinds of health and consumer products

Magzhan Ospanbay [1]

[1] Biomedical Engineering, Luebeck University of Applied Sciences, magzhan.ospanbay@stud.th-luebeck.de

Abstract

The paper analyses a variety of formats, procedures, and aspects of regulatory activities in a comparative perspective with health and consumer products by defining the operating principle of the formats when creating technical dossiers. The different formats, procedures, and legislation used for different products in the European Union. It is therefore important to describe how they are applied, and which medicinal product procedures are more effective. Besides, the legislation paper presents information and provides comparative analysis on the Common Technical Documentation format used for the dossier compilation of medicinal products, Annex II/III of Medical Device Regulation, and Table of Content form. Furthermore, the different types of formats have been defined that are used for the submission of medical device technical documentation. Finally, various formats and legislation directly influence the dossier's content and have been developed in connection with the necessity for the European Union to regulate different products.

1   Introduction

In addition to a significant number of product types, there are also a variety of legislation regulating products in the European Union (EU). The aim is to provide a comparative regulatory analysis of technical dossiers to obtain regulatory approval to market medicinal products (MP), medical devices (MD), and food supplements (FS). There are different types of procedures for placing MP on the market in the EU, which are classified as different routes to obtain authorisation. Furthermore, the paper thoroughly explains what MPs are and how they differ from MD and FS. The requirements for obtaining an approval for FS may not be different within the law of one state because FS do not fall into different risk classes, like MD, whose classification is based on the use and complexity of the device and the risk, ranging from Class I (surgical instruments) to Class IIa, Class IIb, and Class III which are the most invasive devices, such as pacemakers. Therefore, the MD requirements for obtaining CE - marking for placing a product on the market differ depending on the risk class specified in the Medical Device Regulation (MDR). In the studies, through analysis of the EU directives and regulations, it is possible to obtain the more detailed information needed for the pharmaceutical industry to establish regulatory conformity for different products. Today, there are various dossiers depending on the type of product that can be created and documented according to the Common Technical Documentation (CTD), Annex II/III of MDR, and Table of Content (ToC) formats, each with its own form of content. Finally, the studies in this paper analyze how legislation varies significantly in relation to various health, consumer products and the format for submission of technical documentation.

2   How the European Union regulates medicinal products, medical devices and food supplements

To protect the population's health and ensure that high-quality, safe, and effective MPs are available to European citizens, all MPs must be authorised for use prior to being placed on the EU market. The different authorization routes can be described as follows: *National Procedure (NP), Mutual Recognition Procedure (MRP), Decentralised Procedure (DCP), and Centralised Procedure (CP)* working with the European Medicines Agency (EMA). MRP and DCP are national applications designed to be authorised in more than one country. Most MPs available in the EU are authorised through NP and marketed in one member state. Alternatively, it can be done through CP. The big advantage of CP for EU citizens is that the MP is approved for all EU citizens at the same time. However, it requires a significant amount of time and is costly compared to other procedures [1]. Transformation of the European directive on MP for human use into German national law is illustrated in Fig. 1:

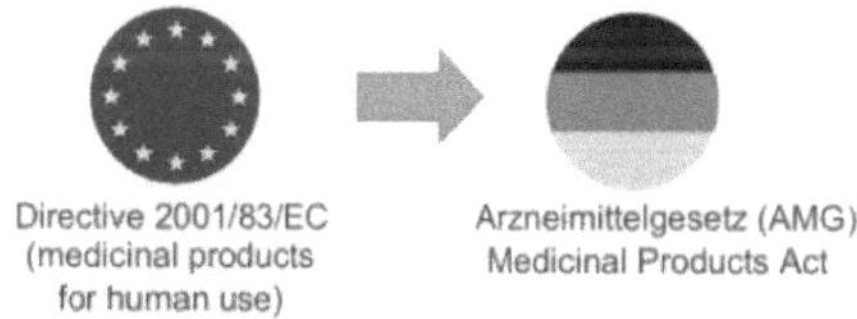

Figure 1: National implementation of European directive into German Medicinal Products Act.

MP for human use means any substance or combination of

substances represented as having properties for treatment or preventing disease in humans; or any substance or combination of substances which may be used in or administered to humans either to restore, correct, or modifying physiological functions by exerting a pharmacological, immunological or metabolic action [2].

Whereas, FSs are defined as foodstuffs intended to supplement the normal diet and are concentrated sources of nutrients or other substances with a nutritional or physiological effect [3]. They cannot be counted as MD and as such cannot exert a pharmacological or metabolic action. The list of vitamins and minerals substances as well as food additives used by manufacturers is prescribed by EU legislation on FS - Directive 2002/46/EC and Regulation (EC) No 1333/2008 respectively. In addition, FS has another compulsory Regulation (EU) No 1169/2011 that covers mandatory information on foodstuffs: name, ingredients, mandatory nutrition declaration, instruction for use, storage conditions, country of origin, and minimum shelf life, all of which must be indicated on the product packaging [4].

Concerning the new legislation on MD, the EU medical device sector has faced considerable changes as a result of the new MDR EU 2017/745, which came into *force on May 26, 2017 and is fully applicable since May 26, 2021*. As the name implies, this is a regulation, not a directive, and all companies trading MD in the EU must directly comply with this new Regulation. MD itself means any instrument, apparatus, software, material, reagent, or other article intended by the manufacturer to be used, alone or in combination for human beings for the following specific medical purposes: diagnosis, prevention, monitoring, prediction, treatment, or alleviation of disease. Additionally, the main effect is *not due to* pharmacological, immunological or metabolic means of action [5], as the MD shown in Fig 2.

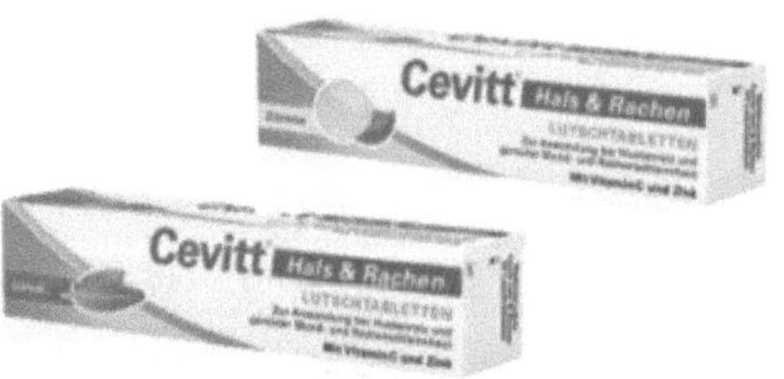

Figure 2: The Cevitt Hals and Rachen is a MD for the treatment of irritated mucous membranes of mouth and throat.

For instance, Fig. 2 shows a Hermes product that was classified as a Class I MD in accordance with the rules of Medical Device Directive (MDD) 93/42/EEC. In addition, the MD is reclassified to Class IIa under the MDR and has an additional grace period for placing on the market until May 26, 2024. It must also be re-certified in accordance with the MDR after the expiration date. The MDR was designed to improve the safety of MD. It has replaced the MDD and Fig. 3 shows the result of the conversions.

After all, all manufacturers of MD must be prepared to com-

Figure 3: New regulations of medical devices.

ply with the new requirements. It cannot be assumed that the new regulation is entirely different from the old directive, but it is more revised to ensure a high level of public health protection. Some key changes can be highlighted from the new regulation: the definition of MD covered by the MDR has been considerably extended to include MD which does not have intended medical purposes; systematic clinical evaluation of Class IIa and Class IIb medical devices; the manufacturers of products must now identify at least one person in their organisation who is ultimately responsible for all aspects of compliance with the new regulation; the MDR prescribes the use of Unique Device Identification (UDI) mechanisms. This requirement is expected to improve the ability of manufacturers to trace specific devices in the supply chain, and to facilitate the rapid and effective recall of medical devices that have been found as representing a safety risk. Finally, the MDR requires manufacturers to reclassify MD according to risk, contact duration, and invasiveness that were lawfully placed on the market before the 26 May 2017 in accordance with old directives, with a transition period until 25 May 2021 [6].

3 The way of providing a common format for technical documentation of health and customer products

Depending on the food classification and product type, the requirements for compliance and registration of foods and FS may be different. The registration is referred to as a simple notification or approvable application. Any registration process requires a dossier, technical file, or set of technical documents collected in so-called Product Master File (PMF) which, for example for FS, includes several or a combination of the following: Composition and Specification of the finished product, Manufacturing Process, Packaging Material, Allergen and GMO Statements.

In the case of MP, it can only be placed on the market with an approval. The approval means evidence of quality, effectiveness, and harmlessness provided and documented in accordance with state law. The approval can only be granted if the benefit-risk-assessment is positive and the benefits exceed the risks. The basis to perform the benefit-risk-assessment by the authority is a dossier documentation in the so-called electronic Common Technical Documentation (eCTD) format.

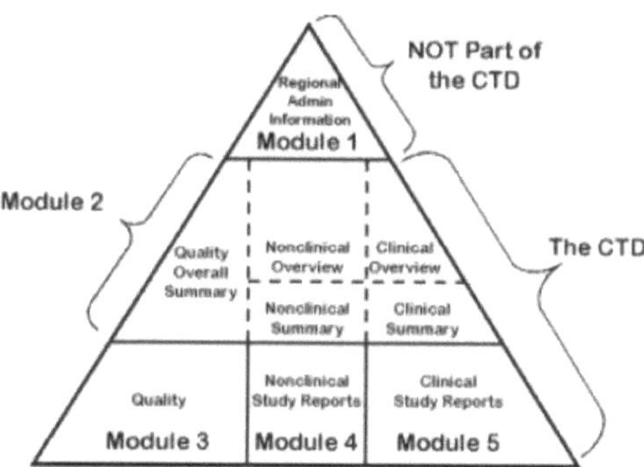

Figure 4: The CTD triangle is organised into five modules.

The CTD itself has a modular structure and is divided into five main modules, as shown in Fig. 4; Module 1: Administration and Prescribing Information which is not strictly included in the CTD because it contains documents specific to each region. Module 2: Overviews and Summaries. The information presented in Module 2 is based on the basic material presented in Module 3, 4 and 5: for quality information. Module 3 of the CTD includes quality data relevant to MP registration: active ingredients, ingredient review, drug descriptions and formulation developments, packaging material, stability studies; Module 4: nonclinical information; Module 5: clinical information [7].

Two technical documentation formats can be distinguished for MD, which are different from CTD. Firstly, the European technical documentation is given in Annex II and III of MDR. The MDR not only defines the general safety and performance requirements for the device that have been written in Annex I, but also defines the requirements for the documentation itself in Annex II, which must include device description and specification; information supplied by the manufacturer; design and manufacturing information; general safety and performance requirement; benefit – risk analysis and risk management; product verification and validation. Annex III: demonstrates the list of technical documentation on Post – Market Surveillance (PMS), including PMS Plan, PMS Report, and Periodic Safety Update Report (PSUR). In addition to the technical documentation in Annex II/III, the MDR contains other Annexes: based on a quality management system and assessment of the technical documentation, EU - type examination, product conformity verification, clinical evaluation, and investigations the requirements to be complied with depend on the risk class of the medical devices [5].

Secondly, the International Medical Device Regulation Forum (IMDRF) has developed a single format ToC for all MD submissions which differs in content from the Annex II and III. This guidance is intended to help manufacturers and regulators understand the structure, content and require-

ments for various MD products and facilitates application for different jurisdictions. The ToC for Non-In Vitro Diagnostic (nIVD) device consists of 7 sections which is illustrated in Fig 5. The ToC for In Vitro Diagnostic (IVD) device also consists of 7 sections, with one difference: Chapter 3 – Analytical Performance and other Evidence [8].

| 1. Regional Administrative. |
| 2. Submission context. |
| 3. Non-Clinical Evidence. |
| 4. Clinical Evidence. |
| 5. Labelling and Promotional Material. |
| 6A. Quality Management System Procedures. |
| 6B. Quality Management System Device Specific Information. |

Figure 5: The ToC of nIVD developed by IMDRF.

4 Results and Discussion

The paper demonstrates that regulatory affairs has developed from the task of health authorities to protect individuals and public health by controlling safety, efficacy, and quality of products. As shown in the paper, the key to differentiating between MP and MD lies in interpreting the basic terms that define them: therapeutic effect and mechanism of action. MP and MD have in common the therapeutic effect, while they differ in the mechanism of action by which they achieve their effect. MPs have a pharmacological, immunological or metabolic mechanism of action, while MD does not have such mechanisms.

It was revealed from the paper that FS refers to more European regulations and directives compared to MD and MP. In addition, the required documentation and the format, content of the technical documentation differ for all these three types of products mentioned in the above sections.

For MD, in addition to the technical documentation set out in Annexes II and III of the MDR, obtaining approval to place a product on the market becomes more stringent, depending on the higher risk class. In this case, the manufacturer must ensure compliance with the other requirements laid down in the MDR. For example, to obtain CE - marking for a MD in risk class IIb it is necessary to obtain *EU - type examination* certification from a notified body as part of the Annex X procedure for that product. In addition, the manufacturer must comply with all product quality assurance requirements in accordance with Annex XI - part A and perform clinical evaluation. As can be seen, the ways of obtaining the CE - marking depends on the risk classes of the MD.

Once the type of product has been clarified as prescribed in European regulations and directives, an appropriate product regulation format can be selected for further documentation procedures, respecting the format framework of the CTD, ToC, Annex II/III and the requirements of the authorities for obtaining an approval. The paper shows that CTD, ToC, and Annex II/III work as a guide for RA managers and also as a framework for health authorities not to go beyond the boundaries of these formats and request certifications that are not specified in them.

5   Conclusion

Finally, MD, FS and MP are regulated in Europe by separate legislation with different requirements and classifications. For MP, companies can go through a centralised procedure at European level, working with the EMA, or a decentralised procedure, working with the competent authority of the member state to obtain approval. In the meantime, FS are considered to be foodstuffs, and the responsibility for them lies with the food business operator. Moreover, the legislation on FS is completely different from the other two products and there is no analogue of the CTD or ToC formats for FS. For MD, companies must work with a notified body approved by the EU, as there is no equivalent of the EMA. The classification of MD is based on the use and complexity of the device and the degree of risk.

In conclusion, the CTD, ToC, Annex II/III of MDR affect the content of the dossier by forcing a consistent order of different technical documentation. The significant advantages include the following points: harmonisation, standardised review, communication, time management. As demonstrated, the CTD is intended entirely for the regulation of MP as well as the ToC and Annex II/III are for the regulation of MD. The difference is that the ToC was developed by an international group of volunteers from worldwide to create a strong format for global harmonisation of MD regulation, whereas Annex II/III of MDR are mandatory regulation in the EU. Finally, the future development of the ToC format in MD may explore the possibility of registering devices in a more flexible and time-saving way worldwide.

Acknowledgement

The work has been carried out at Hermes Arzneimittel GmbH, Regulatory Affairs International Department and supervised by Dr. C. Kuhn and by Prof. Dr. sc. hum. F. Spitzenberger, Fachbereich Angewandte Naturwissenschaften, Technische Hochschule Lübeck.

Author's Statement

Conflict of interest: Authors state no conflict of interest.

6   References

[1] M. Nieto-Gutierrez, *Marketing Authorization Routes in the EU* , London, 2011.

[2] "Directive 2001/83/EC of the European Parliament and of the Council of 6 November 2011 on the Community code relating to medicinal products for human use". Available: https://eur-lex.europa.eu/legal-content/EN/TXT/?uri=CELEX[last accessed on 2021-12-07].

[3] "Directive 2002/46/EC of the European Parliament and of the Council of 10 June 2002 on the approximation of the laws Member States relating to food supplements". Available: https://eur-lex.europa.eu/legal-content/EN/TXT/?uri=CELEX[last accessed on 2022-01-25].

[4] "Regulation (EU) No 1169/2011 of the European Parliament and of the Council of 25 October 2011 on the provision of food information to consumer". Available: https://eur-lex.europa.eu/legal-content/EN/TXT/?uri=CELEX[last accessed on 2022-01-27].

[5] "Regulation (EU) No 2017/745 of the European Parliament and of the Council of 5 April 2017 on medical devices and repealing Council Directives 90/38/EEC and 93/42/EEC". Available: https://eur-lex.europa.eu/legal-content/EN/TXT/?uri=CELEX[last accessed on 2021-12-07].

[6] TÜV SÜD, *Significant Changes ahead for medical device manufacturers*, Available: https://www.tuvsud.com/en/industries/healthcare-and-medical-devices/medical-devices-and-ivd/medical-device-market-approval-and-certification/medical-device-regulation [last accessed on 2021-12-07].

[7] D. Jordan, *An overview of the Common Technical Document (CTD) regulatory dossier* , Medical Writing, vol. 23, no. 5, 2014.

[8] A. Oldfield, *IMDRF ToC Health Canada's new submission format* , Medical Devices Bureau, Health Canada, Canada.

Development of a digital solution to optimize reports related to performance and post-market surveillance data required by the EU IVDR

Maria Magdalena Rodriguez Ruiz [1], and David Bensaude [2]

[1] Biomedical Engineering, University of Applied Sciences of Lübeck, maria.magdalena.ruiz@stud.th-luebeck.de
[2] Roche Diagnostics GmbH, Mannheim, david.bensaude@roche.com

Abstract

The In Vitro Diagnostic Regulation (IVDR) is the legal basis for placing In vitro diagnostic medical devices on the market of the European Union. Compared to the pre-existing In Vitro Diagnostic Directive (IVDD), IVDR introduces new obligations aiming at improving health. One of these new obligations is the periodic reporting on post-market surveillance. The effective and efficient execution of this new obligation requires good findability, accessibility, interoperability, and reusability of product performance and post-market surveillance data. Thus, an algorithm, which optimizes the data provision for planning, creating, and maintaining periodic reports, has been developed by analyzing the EU regulatory framework and by applying an information extraction technique. The digital solution created depicts a best-practice approach to acquire and reuse relevant data compared to the existing manual method. However, this paper suggests a fundamental change, to move from document management systems to content management systems.

1   Introduction

In vitro diagnostic medical devices (IVDs) are medical devices, such as reagent, reagent product, calibrator, control material, kit, instrument, apparatus, piece of equipment, software or system, intended by the manufacturer to use *in vitro* for the examination of specimens solely or principally to provide information for diagnostic, monitoring or compatibility purposes [1]. In the European Economic Area, IVDs will be regulated by the In-Vitro Diagnostic Regulation (EU) 2017/746 or IVDR, which will fully replace the more than 20-year-old Directive 98/79/EC or IVDD on May 26th, 2022. IVDR is intended to better protect the patient and user health and to close gaps that technological advances have brought by introducing meticulous requirements [3]. For instance, manufacturers must follow up on the performance of their devices during the on-market phase and periodically report on post-market surveillance. For large manufacturers, the effective and efficient execution of these new obligations requires good findability, accessibility, interoperability, and reusability of data. On the other hand, large manufacturers who still store performance or post-market surveillance data to documents such as text documents face challenges [4]. Indeed, data or data assessment results are harder to find, access, and re-use if they are stored solely to documents than if they are stored in structured databases. A Python-based solution has been designed to extract information stored in hundreds of Performance Evaluation Reports (PERs) to a table structured around device identifiers. This spreadsheet can certainly be reutilized by other software systems such as these ones to plan and create Post–Market Surveillance (PMS) activities.

2   Material and Methods

This study presents a pragmatic approach to develop a digital solution that optimizes the planning, creation, and maintenance of Post Market Surveillance Report (PMSR) and Periodic Safety Update Report (PSUR) by breaking down and analyzing the EU regulatory framework. As a result, the project was segmented into two main parts. The first one, which was the examination and collection of requirements by reviewing regulatory documents following a hierarchy model, depicted in Fig. 1. And the second one was the development of an algorithm for practical information extraction based on a regex pattern technique. The regular expression module is embedded inside Python and is accessible through the *re module* [5]. This module provides regular expression matching operations that will help to match or find indispensable information. In addition, Python was selected because it gives a general approach to data wrangling.

2.1   Analysis of regulatory requirements applicable to IVDs

- The Regulation (EU) 2017/746 of the European Parliament and of the Council of 5 April 2017 on in vitro

Figure 1: Hierarchy model of regulatory requirements.

diagnostic medical devices was consulted as a primary legislation source. Two annexes were scrutinized. The first one was Annex XIII as it includes the specific components to plan and report the performance evaluation of IVDs, and the post-market performance follow-up demands. The second one was Annex III, as it contains the requirements to plan and report the Post-Market Surveillance activities.

- A guidance drafted by the Medical Device Coordination Group (MDCG) for Medical Devices was reviewed and used as a secondary source. MDCG is a group established by Article 103 of the Medical Device Regulation (EU) 2017/745 (MDR) and shall ensure a harmonized implementation of the MDR and the IVDR. This source was exploited to identify dependency between post-market surveillance data and pre-market IVDR elements.

- The third source of regulatory requirements used was the Quality Management System of Roche Diagnostics (QMS). Work habits of employees following the QMS to execute their tasks were also considered, to compare and evaluate the manual method of data collection with the automated alternative proposed in this project.

2.2 Automated extraction of content from hundreds of PERs by applying a regular expression approach

An algorithm was developed to extract and transfer content from the PER, such as the unique device identifier(s), also called "material number(s)", the justification to carry out or not a post-market performance follow-up (PMPF) plan, and the filename of the PER, to an intermediate spreadsheet structured around the device identifier. To achieve this aim, patterns at the word level were used to extract fine information from text. Fig. 2 depicts an overview of the model built to extract data points from docx files. The information extraction block of the model consists of four major stages: filtering, reading and converting to raw text, processing and extracting, and data resulting formatting.

2.2.1 Filtering

Roche records performance evaluation of its devices to hundreds of PERs. Most of them are already several years old and have been updated multiple times. Indeed, Article 56 of the IVDR mandates to update PERs for class C and D devices at least once a year. An example is the PER for the analyte Troponin I, with four versions. Only the approved version should be used by the PSURs creators. Therefore, the algorithm identify the most recent version and extract content from this version only. However, this can be a real challenge in large enterprises if version naming is not standardized and diverse. This impacts the filtering logic as it becomes complex to obtain the most recent version. Hence, the algorithm uses the created date property of each file to identify the most recent one. Created date correlated with the approval date. The modified date property was not considered a viable property since it is more likely that a user adjusts a document, but that does not entail that it is the latest version. After, identifying the latest version of each file, the next step was converting its content into raw data that could be processed. This filtering step has replaced the manual effort of searching documents stored in electronic document management systems, saving Roche employees time.

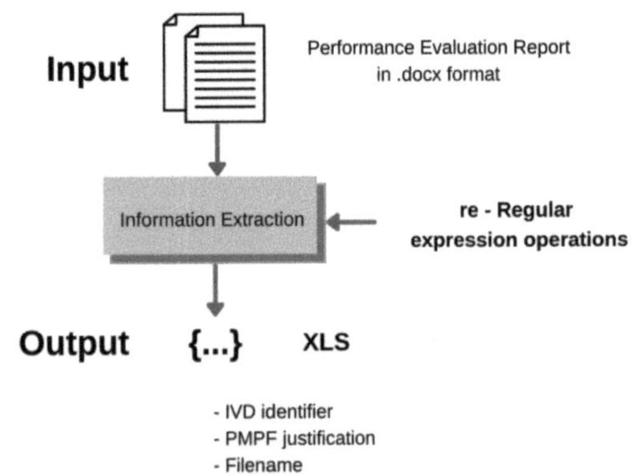

Figure 2: Content extraction model following a regular expression approach.

2.2.2 Read and convert to raw text

Using the python library *docx2txt*, a pure python-based utility to extract text, it was possible to read files and convert them to raw text. This raw data is returned as an array of lines. All text lines can be later parsed to obtain the required data.

2.2.3 Processing and extraction

Regular expression operations were applied to segments of the raw text to find and extract the required information from each PER, such as the material numbers depicted in Fig. 3. First, it was necessary to read the files by blocks, where the header of each section has played a boundary role. This mechanism reduces the possibility of extracting irrelevant information. Instead, the algorithm focuses on locating the essential information. After that, the raw text is scanned for matching patterns, and finally, the relevant information is extracted. In semi-structured data, such as the content of PER, the required text can be extracted by using regular expression operators because the relevant information is expected to be labeled to a reference keyword. Or the data value can be matched following a specific search pattern in the text [5]. For instance, the device identifiers are always in the vicinity of the *Material* word, which is used in this case as the reference keyword, in the PER. In total, three elements were extracted out of the PERs, which have been mentioned at the beginning of section 2.2. The next step was to organize the data.

Roche

Performance Evaluation Report

Version: 1.0
Product name: Micral-Test
Material numbers: 11544039005, 11544039171, 11544039172

Author:

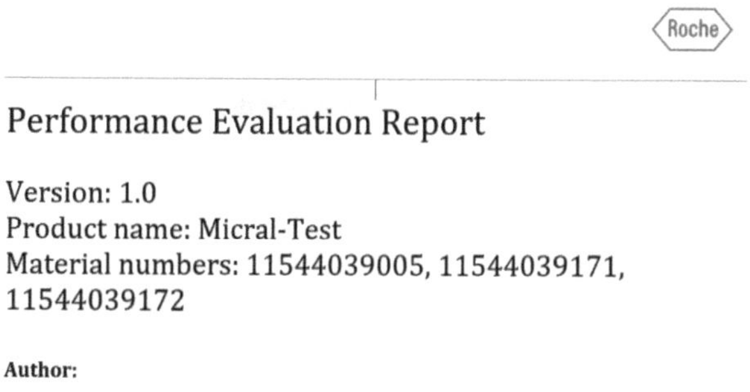

Figure 3: Section of the Performance Evaluation Report template.

2.2.4 Format resulting data

Once all required data was extracted from PERs, the information was organized using a table method (rows and columns). In this case, the *pandas library* was used, specifically pandas DataFrame() constructor was implemented to achieve this proposal. Each row of the table was identified by a material number. In the first iteration, the expected number of rows was unknown. However, if data was missing, it was easy to identify since the list of material numbers or IVDs (reagents, calibrator-, controls materials, kits), which must be certified, are always known. Finally, the data was exported in a spreadsheet that software systems can easily re-used.

2.3 Evaluation of data availability, validity, and reusability

Firstly, data availability was assessed by the following assumption: all IVDs *(class A, B, C and D)* shall have a PER according to Article 56 of the IVDR. To sum up, for each device identifier, a PER is expected. Secondly, the validity of the extracted justified decisions to perform the PMPF activities or not was assessed by comparing them to requirements of IVDR Annex XIII, Part B. In this section, IVDR establishes that "if PMPF is not deemed appropriate for a specific device, then a justification shall be provided and documented within the PER" [1]. In essence, there should be either a PMPF plan or a justification in the PER why a PMPF is not deemed appropriate for each device. If the two conditions above are fulfilled, then the PMPF data can be re-used in the PSUR or PMSR, which shall contain either the main findings of the PMPF or a justification why no PMPF is performed [3]. Over and above that, 31 PERs were randomly selected and divided into three groups: small, medium, and large, to assess how well a digital tool could replace the manual effort of migrating information from one source to another. For demonstration purposes, the information gathering process of the manual method starts with having already identified and accessed the information. The findability and accessibility of data in large enterprises can take up to three working days [4].

3 Results and Discussion

3.1 Interdependence between PMPF and other IVDR elements

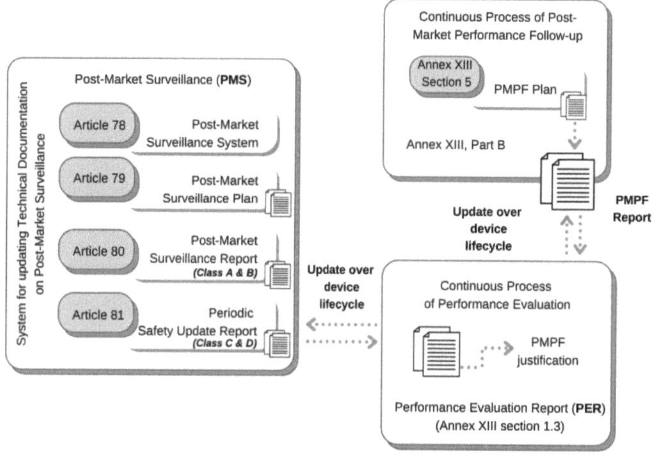

Figure 4: Dependencies between PMPF and other IVDR elements.

The PMPF is a continuous process for following up the performance of on-market IVDs, according to IVDR Annex XIII, Part B. This new concept builds a bridge between the evidence collected at the pre-market and post-market stages. Fig.4 was created to illustrate the relationship between the PMPF process and other IVDR processes based on a PMPF diagram suggested by MedTech Europe [3]. The PMPF concept appears in the Performance Evaluation Plan for the first time, and the PMS process consumes it. Therefore, the new findings of the PMPF report could trigger a direct update of the PER. Similarly, it is expected that some PMS activities could trigger an update of the PER because in some cases, PMS activities are enough to evaluate the performance of IVDs. For instance, the emergence of new mutations will likely trigger PMPF, which, in turn, will generate a reassessment of the clinical evidence of the device [3].

3.2 Manual vs. Automated Extraction

A total of 391 docx files (PERs) were processed. However, from this list, only 327 PERs were leaked as valid PERs, based on the most recent created date attribute. The other 64 files were identified as old versions. From these 327 documents were extracted their filename, 1099 device identifiers, and corresponding PMPF statement. Table 1 depicts the data points extracted from a PER.

Table 1: Data points extracted from a PER

Filename	Material number(s)	PMPF statement
F_PER_MICRAL_TEST V1.DOCX	11544039005 11544039171 11544039172	Post market performance follow up for standardized and established product is covered by other Post Market Surveillance activities as outlined in the Post Market Surveillance plan.

The time of content collection was improved, as shown in Table 2. The table depicts the comparison between the time it takes to collect and migrate the set of values using a manual extraction (ME) approach against an automatic extraction (AE) approach. The total automated time extraction of the 327 documents was 12 seconds, which is even less than the data migration of a single document, which is around 1 min. Therefore, information extraction using a regular expression approach is a superior method to acquire and reuse relevant data compared to the existing manual procedure.

Table 2: Manual Vs Atomated Extraction

Sample	Number of PERs	ME Time	AE Time
Small	1	00:01:07	00:00:02
Medium	10	00:11:05	00:00:04
Large	20	00:26:35	00:00:08

The manual method only provides the PMPF statement if a PMPF plan needs to be carried out, leading to a lack of information because all PMPF justifications are required according to Annex XIII part B. The digital tool solves this drawback by extracting all PMPF justifications from PERs. Last but not least, the digital tool has contributed to identify the lack of PERs, which leads to a lack of PMPF statements. The lack of this information directly affects the creation of PMSRs and PSURs. This was demonstrated by comparing the list of IVDs, which require certification, with all extracted device identifiers contained in the PERs.

4 Conclusion

The IVDR has developed different mechanisms to ensure an effective and efficient performance evaluation of the IVDs, such as the development of principles and rules to implement the performance evaluation plan, the PMS plan, and the PMPF plan. These three processes are tightly connected to each other by the PMPF element. Thus, the planning, creation, and periodic update of PMS reports and PSURs represent a Herculean task for manufacturers with an extensive product catalogue and for manufacturers that still have not arranged their data to a pre-set data model. A digital solution that can make data stored in files available for re-use is temporarily adequate. This solution allows to focus more time and effort to answer questions that really matter, such as identifying which products require a PMPF plan or when a PER requires an update. Even though this digital solution depicts a best-practice approach to acquire and re-use relevant data compared to the manual method, this method has demonstrated a dependence on the expected information, which means that the format of this document has been set and cannot be suddenly modified. However, practice has shown that content creators constantly ignore or forget the format conditions, which directly affects the correct extraction of information. Hence, this paper suggests a fundamental change to move from document management systems to content management systems. The latter simplifies the creation and editing of digital content.

Acknowledgement

The work has been carried out at Roche Diagnostics GmbH in Mannheim and supervised by Prof. Dr. Folker Spitzenberger, Centre for Regulatory Affairs in Biomedical Sciences (CRABS), University of Applied Sciences of Lübeck. Especially gratitude to Mr. Bensaude for his guidance.

Author's Statement

Conflict of interest: Authors state no conflict of interest.

5 References

[1] European Union, *Regulation (EU) 2017/746 of the European Parliament and of the Council of 5th April, 2017 on in vitro diagnostic medical devices and repealing Directive 98/79/EC and Commission Decision 2010/227/EU*. Official Journal of the European Union, 2017.

[2] Medical Device Coordination Group, *Draft Guidance on Periodic Safety Update Report (PSUR) according to Regulation 2017/745*, 2021.

[3] MedTech members, *Clinical Evidence Requirements for CE certification under the in-vitro Diagnostic Regulation in the European Union*. MedTech Europe, First Edition, pp. 70-82, 2020.

[4] V. Hoffmann, *Evaluation of challenges for large in vitro diagnostic manufacturers in the context of IVDR and Implementation of a solution approach for content-based documentation for Roche Diagnostics*. Master's thesis, Ruprecht-Karls-Universität Heidelberg, Heidelberg, 2021.

[5] Aggarwal, A., Garhwal, S., and Kumar, *HEDEA: a Python tool for extracting and analysing semi-structured information from medical records*. Healthcare informatics research, vol 24, no. 2, pp. 148-153, 2018.

3

Biochemical Physics

Investigation of Surface-Active Therapeutics on Lipid Monolayers

Carla M. Neitzke [1], Thomas Gutsmann [2], and Christian Nehls [2]
[1] Biophysics, University of Lübeck, carla.neitzke@student.uni-luebeck.de
[2] Division of Biophysics, Research Center Borstel, {tgutsmann, cnehls}@fz-borstel.de

Abstract

Membranes of human cells usually consist of phospholipid bilayers. One of the few exceptions is the monolayer lining the lungs, the pulmonary surfactant. Aerosolized active substances such as drugs interact with these membranes. Therefore, this study is used to investigate the interaction between a 1,2-Dipalmitoyl-sn-glycero-3-phosphocholine (DPPC) monolayer prepared on a Langmuir-Blodgett film balance that serves as a model system for pulmonary surfactant monolayers and an inhalable drug (Tyloxapol). Drugs can be applied to a monolayer on the film balance either by nebulization or by injection into the medium. In this work, both techniques were compared. For this, the area change of the lipid monolayers due to drug delivery was measured at a constant lateral pressure of 20 mN/m using a Langmuir-Blodgett (LB) film balance. It could be shown that with nebulization already a lower drug concentration leads to a larger area of the lipid monolayer than after injection. Furthermore, the system gets saturated when nebulization is used. The topography of the lipid monolayer was investigated using atomic force microscopy (AFM).

1 Introduction

The bronchial system of the lungs branch out more finely the deeper the lungs are looked into. The smallest instance are the alveoli. These are lined with a phospholipid monolayer, the pulmonary surfactant. The word "surfactant" is made up of the words "surface active agent". As the word already expresses, the surfactant is surface-active and prevents the collaps of the alveoli during the expiration with its surface pressure reducing property. The surfactant consists of phospholipids (90 %) and sufactant specific proteins (10 %). [1] [2]

The Langmuir-Blodgett film balance was used to examine the interaction between a drug and a monolayer at the water-air interface . To simulate the conditions in the lungs, the subphase consists of distilled water and the lipid monolayer of dipalmitoylphosphatidylcholine (DPPC), which is the main lipid component of the pulmonary surfactant. This creates a simplified model of the lung. [3]

So far, the injection of the surface-active agent under the lipid into the subphase is the standard technique for investigations regarding the interaction of a lipid monolayer and an surface-active agent using the film balance. But that is not the way the drug would interact with the pulmonary surfactant in the lungs, because the active ingredient interacts with the lipid monolayer from the water phase. However, for example asthma medication is inhaled. Thus, it interacts with the membrane from the air phase. In this project a new technique, the nebulization of the active ingredient onto the lipid monolayer, was used for experiments and compared with the injection technique, where the active ingredient is injected under the lipid monolayer. The results of the injection technique were generated in an earlier work.

2 Material and Methods

2.1 Material

The used lipid monolayer consisted of 1,2-Dipalmitoyl-sn-glycero-3-phosphocholine, short DPPC, (Avanti Polar Lipids, Alabama, USA) and the surface-active therapeutic was Tyloxapol (Sigma-Aldrich, Darmstadt, Germany). The subphase in the film balance trough was distilled water. The Wilhelmy plate was cut of chromatography paper (GE Healthcare, Chicago, USA) with the dimensions of 10 mm x 25 mm. For the AFM imaging, the lipid monolayer must be transferred to a solid substrate, in this case a mica plate (20 mm x 16 mm) (Electron Microscopy Sciences, Hatfield, USA).

2.2 Methods

The measurement setup to investigate two different drug delivery methods on lipid monolayers consisted of the KSV Nima KN2001 Langmuir Blodgett trough (Biolin Scientific, Gothenburg, Sweden) and the corresponding components described in Fig. 1. The trough and barriers were cleaned thoroughly using ethanol and distilled water prior to each measurement. A thermostat kept the temperature constant at 21 ± 1 °C by a circulating water system.

The subphase, which consisted of 60 mL distilled water, was filled into the teflon trough. DPPC solved in chloroform (0.5 mg/ml) was spread dropwise onto the subphase using a Hamilton syringe. The applied volume depended on whether the nebulization or the injection technique was used. For the nebulization technique it was between 30 µL and 35 µL and for the injection technique it was 15 µL. Since the chloroform evaporated shortly after the spreading, a monolayer of DPPC was formed. Tyloxapol diluted with distilled water at various concentrations was either nebulized or injected. Tyloxapol was chosen because it reduces the surface tension and it is often a component in asthma medication [4]. Injection and nebulization of distilled water were used as a control experiment.

For subsequent analysis with the atomic force microscope (AFM) (Asylum Research, Santa Barbara, USA), the lipid monolayer was transferred onto the mica subtrate 50 min after the nebulization or the injection respectively. For this transfer, the mica substrate was pulled from the trough at a rate of 0.1 mm/min

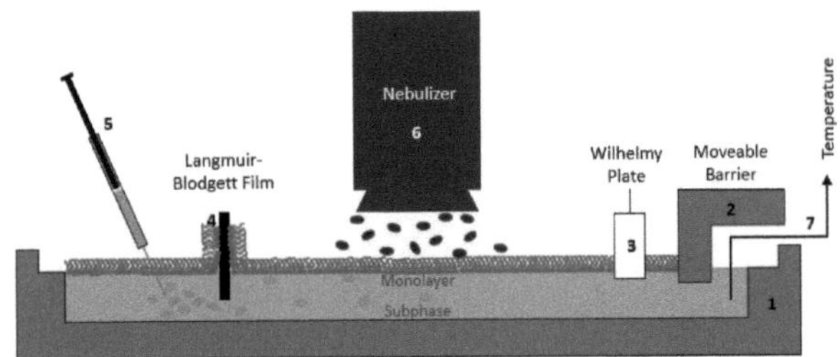

Figure 1: Schematic illustration of the film balance. It is composed of a teflon trough (1) with moveable barriers (2) with which a lipid monolayer on top of a subphase can be compressed or decompressed. During such a compression the Wilhelmy plate (3) measures the lateral pressure. With a Langmuir-Blodgett transfer (4) the lipid film can be transferred for investigation with an AFM. With the Hamilton syringe (5) or the nebulizer (6) an active compound is brought into the system. The temperature of the system (7) can be controlled with a circulating water system. Modified after [5].

2.2.1   Measurement of Isotherms

For the measurements of isotherms, distilled water was filled inside the teflon trough and the DPPC solution was applied. After 5 min the solvent completely evaporated so the compression of the monolayer was started by closing the barriers at a rate of 10 mm/min. This method provides information on the phase behavior of the lipid used. In addition, using this technique, the optimal volume of the dropped DPPC solution for the nebulization and injection experiments was determined.

2.2.2   Nebulization Method

For the nebulization technique one of the moveable barriers was replaced by a 3D printed funnel guiding the aerosols

onto the DPPC monolayer connected to a nebulizer (beurer, Ulm, Germany). The teflon trough was filled with distilled water and the DPPC solution was spread. Once the solvent has evaporated a DPPC monolayer has formed. This monolayer was then compressed to a lateral pressure of 20 mN/m which is equivalent to the pressure in biological membranes [6]. The lateral pressure was measured with the Wilhelmy plate. After an equilibration of 20 min, the nebulizer was started for 5 min. Tyloxapol dissolved in water in various concentrations was nebulized at a rate of 0.4 mL/min leading to the final concentrations of 0.5 µg/mL, 1 µg/mL, 1.5 µg/mL, 2 µg/mL, 3 mg/mL, 5 µg/mL, 7 µg/mL, and 10 µg/mL in the trough assuming that the whole drug amount reached the film balance. Following the nebulization, the increase in area was recorded at constant lateral pressure.

2.2.3   Injection Method

For the injection technique the DPPC solution wass spread onto distilled water forming a monolayer. Again the monolayer was compressed to a lateral pressure of 20 mN/m. Then 40 µL Tyloxapol diluted in water was injected into the subphase using a Hamilton syringe with various concentrations leading to the final concentrations of 125 µg/mL, 100 µg/mL, 70 µg/mL, 50 µg/mL, 20 µg/mL, 10 µg/mL, and 5 µg/mL in the trough. In accordance with the nebulization technique, the area change was recorded at constant lateral pressure after the injection.

2.2.4   Atomic Force Microscopy

Atomic force microscopy (AFM) was used to investigate the effects of Tyloxapol on the DPPC monolayer topography. In the experiments described here, the tapping mode in air was used. The cantilever is set in vibration and moved over the sample in a grid. The interaction forces with the sample surface modulate the resonant frequency of the vibration, resulting in a different amplitude and phase, making the surface structure visible. This deflection of the cantilever is measured using a laser aimed at the top of the cantilever tip. The path of the reflected light beam ends at a photodetector. [7]

An OMCL-AC160TS-R3 (Olympus Micro Cantilever, Hamburg, Germany) cantilever was used.

3   Results and Discussion

3.1   Problems with the Nebulizer

After the first measurments with the nebulizer it could be observed that Tyloxapol foamed inside the nebulizer. Sometimes even bigger bubbles were blocking the attached tube so no aerosol could get to the lipid monolayer. Also occasionally not the whole 2 mL of Tyloxapol could interact with the lipid monlayer since the foam can not be nebulized. These problems lead to imprecise Tyloxapol concentrations which can be seen in Fig. 2. When carrying out these five experiments, everything was made the same, but the graphs differ. Hence, a criterion is needed to decide which experiments worked the way they should have. For this the pressure vs. time diagrams were used. When the pressure is increasing/steady for the 5 min of nebulization as can be seen in the curves for the experiments CMN063, CMN064, and CMN066 in Fig. 2 the criterion is met.

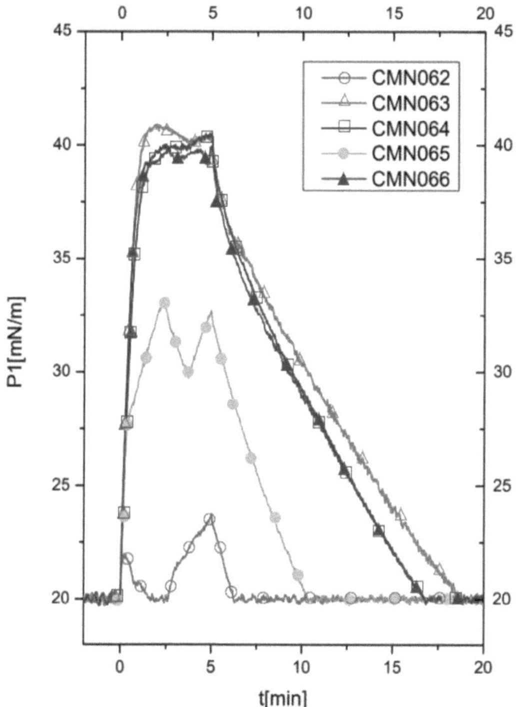

Figure 2: Pressure vs. time diagram. It displays measurements where Tyloxapol was nebulized with the final concentration of 1.5 µg/mL in the trough. The pressure should be increasing for the 5 min of nebulization. This is the case for the CMN063, the CMN064, and the CMN066 graph but not for the CMN062 and the CMN065 one. The reason for this is the foaming of Tyloxapol in the nebulizer.

3.2 Comparison of Injection and Nebulization Method

When the nebulization method was used it could be observed that the Tyloxapol concentration in the monolayer saturates (Fig. 3). So, even with more Tyloxapol the area of the lipid film did not expand over 220 %. The area of 100 % is equivalent to the area when the monolayer is compressed to 20 mN/m and before any interactions have taken place. Aside from the saturation, less Tyloxapol is needed to interact with the lipid film for it to expand extensivly compared to the injection technique. For example, a Tyloxapol concentration between 0.5 µg/mL and 1 µg/mL in the trough lead to a lipid area of around 160 % after 40 min.

With the injection method one can see an approximatly linear correlation between the Tyloxapol concentration and the lipid area (Fig. 4). So, the lowest Tyloxapol conentration of 5 µg/mL lead to no area change after 40 min but the high-

est concentration of 120 µg/mL lead to an area of approximatly 250 %. Furthermore, the lipid area of 160 % which is reached with a Tyloxapol concentration of around 1 µg/mL with the nebulization method is reached with a Tyloxapol concentration of 70 µg/mL using the injection method.

In summary, it can be said that the nebulization method produces larger changes even at lower concentrations, but the injection method leads to higher maximal changes at high concentrations.

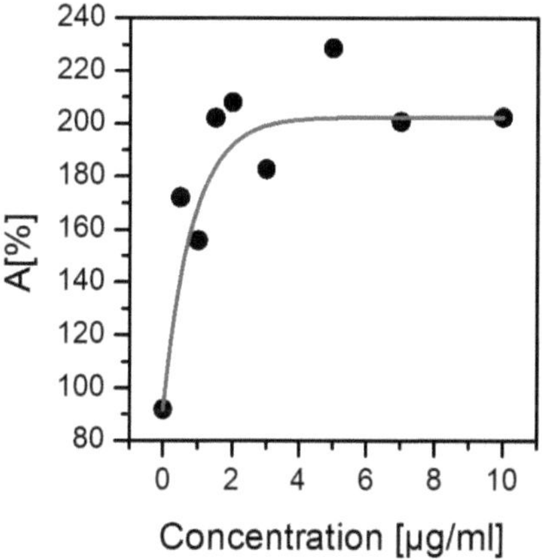

Figure 3: Schwarzplot of nebulization measurements. It displays the area in percent 40 min after the nebulization of every Tyloxapol concentration mentioned above. With higher Tyloxapol concentrations the area does not get larger anymore, so a saturation curve is displayed. The used measurements were selected according to the pressure criterion.

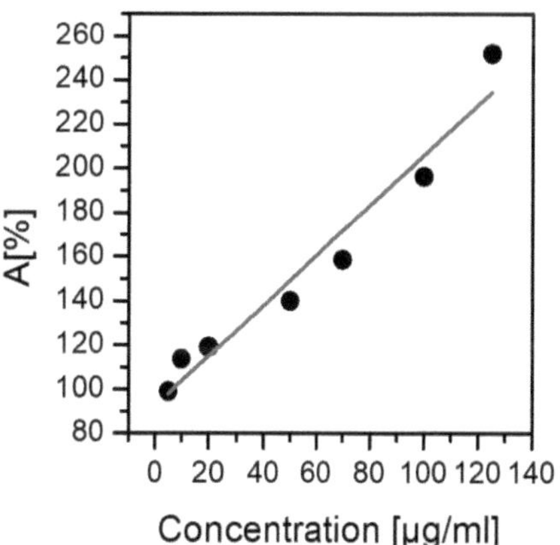

Figure 4: Schwarzplot of injection measurements. It displays the area in percent 40 min after the injection of every Tyloxapol concentration mentioned above. A linear correlation between the size of the area and the Tyloxapol concentration injected is displayed.

The fact that the nebulization method leads to larger lipid surfaces at lower Tyloxapol concentrations is probably due to the fact that more of the Tyloxapol introduced into the system interacts with the lipid monolayer. This could either be due to the nebulizer, which brings the aerosols to the lipid surface with a certain pressure. Or it could be due to the fact that the Tyloxapol that is nebulized onto the lipid surface has no other option than to interact with the lipid. In contrast, Tyloxapol that was injected into the subphase can either linger in the subphase or interact with the lipid layer, which is probably the reason why higher concentrations are needed for a larger increase in the lipid area when using the injection method.

3.2.1 AFM Imaging

For the AFM imaging Langmuir-Blodgett-Transfers with various combinations of Tyloxapol concentrations, DPPC films, and method used were made. The three pictures in Fig. 5 were chosen exemplaryly. The light areas show hills, the darker areas show pits.

The AFM image of the DPPC film which was compressed up to 20 mN/m and was not treated with Tyloxapol shows a flat level which is a continuous lipid monolayer (Fig. 5, c). When Tyloxapol was nebulized onto the compressed DPPC film round/oval lipid domains have formed (Fig. 5, a). On the other hand, when Tyloxapol was injected under the compressed DPPC film lipid domains with edges have formed (Fig. 5, b). It can be concluded that the interaction of Tyloxapol with the lipid depends on the side of the lipid that the Tyloxapol encounters.

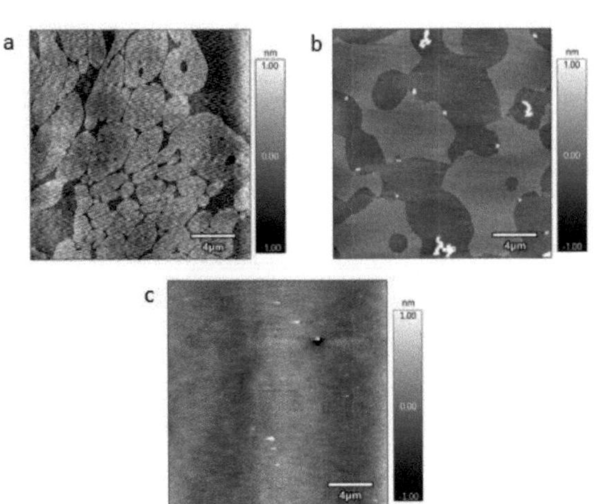

Figure 5: AFM images. The lipid was treated with Tyloxapol via nebulization (a) or via injection (b). The control lipid was not treated with Tyloxapol just compressed (c). The scales show the hight differences up to 2 nm of the surface. The light areas are peaks, the dark areas are valleys. With the injection technique (b) angular lipid domains has been formed. With the nebulization technique (a) round lipid domains has been formed. Without treatment (c) no domains has been formed.

4   Conclusion

In this paper two methods for the film balance were compared: the injection method - the surface active drug Tyloxapol was injected under a DPPC monolayer - and the nebulization method, where Tyloxapol was nebulized on top of a DPPC monolayer. It could be observed that with the nebulization method the system gets saturated and the area of the monolayer does not expand further. The injection method displayed a linear correlation - the higher the Tyloxapol concentration the larger the area. In addition, with the nebulizer method the area increase was larger at lower Tyloxapol concentrations. Also, with the AFM imaging it could be seen that the two different methods induce different lipid domains.

In summary, it can be said that Tyloxapol is interacting differently with DPPC depending on whether the active compound is delivered from the water side or the air side.

Since the nebulizer made Tyloxapol foam a number of further experiments should be executed with a different nebulizer which is based on ultrasound for example. Since the nebulization would be based on a different technique, Tyloxapol might not foam.

Acknowledgement

The work has been carried out at the division of biophysics, Research Center Borstel, supervised by Thomas Gutsmann, head of the division of biophysics and Christian Nehls, postdoc at the division of biophysics.

Author's Statement

Conflict of interest: Authors state no conflict of interest.

5   References

[1] J. G. Betts et al., *Anatomy and Physiology*. OpenStax, 2013.

[2] K. M. Heschl, *SFTA2 - Charakterisierung eines neuen lungenspezifischen Proteins*. Dissertation, 2018.

[3] B. Piknova and V. Schram and S. B. Hall, *Pulmonary Surfactant: Phase Behavior and Function*. Current Opinion in Structural Biology, vol. 12, no. 4, pp. 487–494, 2002.

[4] bene-Arzneimittel GmbH, *Tyloxapol*. Available: https://www.tacholiquin.de/tacholiquin/beratung/tyloxapol [last accessed on: 2022-02-04].

[5] A. David, *Biophysikalische Charakterisierung der Wechselwirkung des antimikrobiellen humanen Kathelizidins hCAP18 mit Modellmembranen*. Dissertation, 2004.

[6] S. Baoukina and L. Monticelli and S. J. Marrink and D. P. Tieleman, *Pressure-Area Isotherm of a Lipid Monolayer from Molecular Dynamics Simulations*. Langmuir, vol. 23, no. 25, pp. 12617-12623, 2007.

[7] Oxford Instruments Company, *Asylum Research Application Guide*. User Guide 3, Version 16, Revision: A-2053, 28.09.2018.

Investigation of the interaction of Polymyxin B
with small unilamellar vesicles of DOPC and PE:PG
with fluorescence intensity measurements

Tom Handrianz [1], Christian Nehls [2] and Thomas Gutsmann [3]

[1] Biophysics, Universität zu Lübeck, Tom.Handrianz@student.uni-luebeck.de
[2] Research Center Borstel, Leibniz Lung Center, Research Group Biophysics, cnehls@fz-borstel.de
[3] Research Center Borstel, Leibniz Lung Center, Research Group Biophysics, tgutsmann@fz-borstel.de

Abstract

Membranes of mammal cells differ from bacterial membranes and therefore, it is obvious that antibiotics interact with these membranes in different ways. This study investigates whether the antimicrobial peptide Polymyxin B (PMB) interacts differently with parts of the major components of mammalian cell membranes (for example as 1,2-dioleoyl-sn-glycero-3-phosphocholine (DOPC)) and bacterial membranes (1,2-dipalmitoyl-sn-glycero-3-phosphoethanolamine (PE) and 1,2-dipalmitoyl-sn-glycero-3-phospho-(1'-rac-glycerol (PG)). For this purpose, small unilamellar vesicles (SUVs) were prepared from DOPC or PE:PG and labelled with the fluorophores 1,2-dipalmitoyl-sn-glycero-3-phosphoethanolamine-N-(7-nitro-2-1,3-benzoxadiazol-4-yl) (NBD-PE) and Rhodamine-PE. Changes in fluorescence intensity have shown that PMB interacts with both SUV types, with the interaction being more intense for PE:PG SUVs and dependent on PMB concentration. Experiments on Förster resonance energy transfer (FRET) have confirmed this. The stronger mode of action of PMB on the PE:PG SUVs suggests that PE and PG, as two of the major components of bacterial membranes, may play a role in the antimicrobial effect of PMB.

1 Introduction

The different lipid composition of mammalian cell membranes and bacterial membranes gives rise to the question of what role certain lipids and lipid compositions play for antimicrobial substances. Two of the main lipid components of bacterial membranes are 1,2-dipalmitoyl-sn-glycero-3-phosphoethanolamine (PE) and 1,2-dipalmitoyl-sn-glycero-3-phospho-(1'-rac-glycerol) (PG) [1], while mammalian cell membranes are dominated by Phosphatidylcholine (PC) (for example as 1,2-dioleoyl-sn-glycero-3-phosphocholine (DOPC) and 1-palmitoyl-2-oleoyl-sn-glycero-3-phosphocholine (POPC)) [2].

This work investigates how differently the antimicrobial peptide polymyxin B (PMB) [3] interacts with the lipids DOPC and PE:PG and what role the main lipid components thus play in the antibiotic effect of PMB. As a membrane model small unilamellar vesicles (SUVs) are prepared and fluorescently labelled. The fluorescence intensity of the samples is then determined for different PMB concentrations using a 96-well plate reader from TECAN[©]. Depending on how the fluorescence intensity is changed by an interaction with PMB (for example by quenching), conclusions can be drawn about the interaction between PMB and the vesicles. Furthermore, investigations were carried out on the Förster resonance energy transfer (FRET) [7]. A central property of the energy transfer efficiency is a dependence on the distance between donor and acceptor. Thus, changes in the efficiency of energy transfer may indicate interactions between PMB and the SUVs.

2 Material and Methods

2.1 Preperation of small unilamellar vesicles

For the preparation of the SUVs, lipid stock solutions of DOPC and PE:PG (ratio 1:1) with a concentration of 10 mM and stock solutions of the fluorophores NBD-PE and Rhodamine-PE with a concentration of 1 mM were used. All lipids and fluorophores were from Avanti Polar Lipids[®] (Alabaster, USA). Fluorophores were added to the lipids to give a fluorescent label of 1 mol%. Samples for fluorescence intensity measurements were labelled with NBD-PE or Rhodamine-PE and samples for Förster resonance energy measurements were labelled with both fluorophores. The lipids were then evaporated under a gentle stream of nitrogen and rehydrated with a buffer solution to give a lipid concentration of 100 mM in buffer [4]. The buffer solution used was a Hepes-NaCl buffer, which consisted of 20 mM Hepes and 10 mM NaCl and was adjusted to a physiological pH of 7.4.

To generate the SUVs, the sample solution was sonicated

with an ultrasonic probe (HTU-SONI 130, G. Heinemann, Schwäbisch Gmünd, Germany) for 2 min with a power of 39W (30% of the maximum intensity). Finally, the SUVs were put into a temperature cycle consisting of 3 cycles of 30 min at 60 °C (heating block) and 30 min at 4 °C (refrigerator). The quality of the SUVs produced were checked with a ZetaSizer (Nano-ZS90 from Malvern Instruments, Malvern, United Kingdom) via dynamic light scattering (DLS) by analysing the size and size distribution.

2.2 Fluorescence intensity measurements and Förster resonance energy transfer

For the measurement of fluorescence intensities a 96-well plate reader from TECAN© is used (infinite M200 Pro, TECAN, Männedorf, Switzerland). Various measurement procedures can be created for this device, in which certain measurement parameters can be specified. Liposomes labelled only with NBD-PE were excited at 460 nm and detected at 535 nm with a gain of 50 % (Amplification of the signal with 50 % of the maximum possible amplification). Liposomes labelled only with Rhodamine-PE were excited at 550 nm and detected at 590 nm with a gain of 25 %. Liposomes labelled with both NBD-PE and Rhodamine-PE are suitable for FRET measurements because the emission spectrum of NBD-PE and the absorption spectrum of Rhodamine-PE overlap spectrally. This is a prerequisite for FRET. The efficiency of the energy transfer can be calculated as follows [6], [7]:

$$E = I_A/(I_A + I_D). \tag{1}$$

where I_D stands for the intensity of the donor and I_A for the intensity of the acceptor, which was excited via the donor (and thus via FRET). To determine the donor fluorescence, the sample was excited at 460 nm and the emission was detected at 535 nm. For the determination of the acceptor fluorescence, NBD-PE (donor) was excited at 460 nm and the acceptor (Rhodamine-PE) was detected at 590 nm. Both measurements were performed at a gain of 50 %.

3 Results and Discussion

3.1 Interaction of PMB with DOPC-SUVs

In the first experiment in this section, DOPC SUVs were labelled with either NBD-PE or Rhodamine-PE (Fig. 1). The fluorescence intensity of the SUVs labelled with NBD-PE was slightly lower when PMB was present in the sample. In general, this PMB effect does not seem to be concentration-dependent because the fluorescence intensity fluctuated around a nearly constant level across the PMB concentrations. The SUVs labelled only with Rhodamine-PE showed a stronger reduction of fluorescence intensity by PMB than the SUVs labelled with NBD-PE. Furthermore, these Rhodamine-SUVs showed a reduction in fluorescence intensity with increasing PMB concentration. In both cases, a reduction of the fluorescence intensity by PMB could

be detected (for the Rhodamine-PE SUVs more strongly than for the NBD-PE SUVs) and for the Rhodamine-PE samples even a concentration dependence. This all points to an interaction of the PMB with the liposomes, whereby the fluorophores are quenched directly by the PMB or indirectly via the membrane by self-quenching [5].

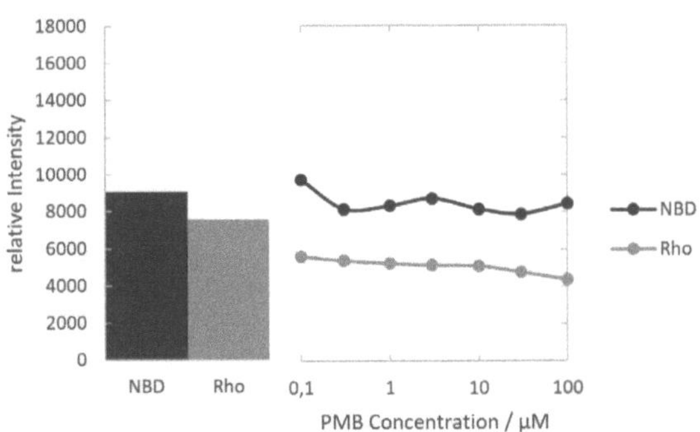

Figure 1: The fluorescence intensity of DOPC SUVs labelled with NBD-PE ("NBD", dark) or Rhodamine-PE ("Rho", light). The bars on the left show the fluorescence intensity of the SUVs without PMB addition, while the curves on the right show the fluorescence intensity of the SUVs to which PMB of different concentration was added. Importantly, the SUVs labelled with Rhodamine-PE were detected with a different gain (25%) than the SUVs labelled with NBD-PE (50%), which accordingly affects the intensities (see chapter 2 Materials and Methods).

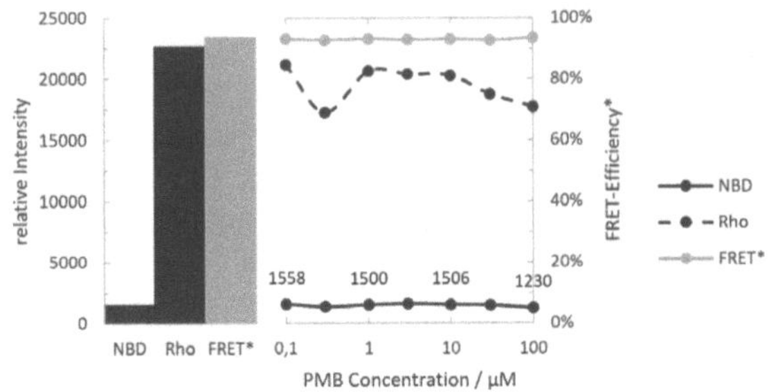

Figure 2: The fluorescence intensity of DOPC SUVs labelled with NBD-PE ("NBD", donor) and Rhodamine-PE ("Rho", acceptor) (left axes) and the resulting FRET efficiency (right axis). The bars on the left side show the fluorescence intensity and the FRET efficiency of a sample to which no PMB was added. The curves on the right show the development of the fluorescence intensity and the FRET efficiency of the sample with increasing PMB concentration. (Bars and curves referring to FRET efficiency are marked with an asterisk).

Next, double fluorescence-labelled DOPC SUVs were investigated with regard to their FRET efficiency (Fig. 2). The FRET efficiency was calculated as described in formular (1). The influence of PMB on the fluorescence intensity of NBD-PE is similar as in the previous experiment. The only difference is the lower intensity level, which is due to FRET. The PMB influence on Rhodamine-PE here excited by the donor is also similar to the previous experiment; the changed intensity level is also due to FRET.

A change in FRET efficiency due to PMB could not be observed, neither in comparison to the PMB free sample nor with increasing PMB concentration. On the one hand, this is due to the low NBD-PE intensity level (donor) and the comparatively low decrease in Rhodamine intensity (acceptor) at a higher intensity level. On the other hand, the intensity of the NBD-PE also decreases slightly with increasing PMB concentration.

The fluorescence intensities of the dyes show again that PMB influences at least the fluorophores, possibly also via an interaction with the liposomes. The constant FRET efficiency indicates that the PMB interaction is not related to an insertion of the peptide into the membrane. If this were the case, the surface area of the liposomes would increase, which would reduce the strongly distance-dependent FRET efficiency.

3.2 Interaction of PMB with PE:PG-SUVs

In order to compare the influence of PMB on the two lipid variants DOPC and PE:PG, the same experiments were repeated for PE:PG as for DOPC in chapter 3.1.

First, liposomes labelled either with NBD-PE or with Rhodamine-PE were examined (Fig. 3). For the SUVs labelled with NBD-PE, it was observed that the fluorescence intensity increased with increasing PMB concentration. An exception is the sample containing 0.1 µM PMB. It is possible that the amount of PMB is too low to have a measurable influence on the fluorescence intensity. Furthermore, it can be seen that the fluorescence intensity increases more slowly at higher PMB concentrations and even decreases slightly at 100 µM PMB. This shows that PMB influences the PE:PG SUVs and that the interaction here is more complex than with the DOPC SUVs.

The PE:PG SUVs labelled with Rhodamine-PE show a reduction in fluorescence intensity up to 0.3 µM under PMB influence. Subsequently, a further increase of the PMB concentration no longer has any influence on the intensity of the Rhodamine-PE. Thus, these SUVs also show that there is an interaction with PMB, and that this interaction is more complex than with DOPC.

In the second part, PE:PG SUVs labelled with both NBD-PE and Rhodamine-PE were investigated for their FRET efficiency and fluorescence intensity (Fig. 4). The FRET efficiency was again calculated according to equation (1).

The fluorescence intensities show that PMB increases the Rhodamine-PE intensity and decreases the NBD-PE intensity (at higher PMB concentration the intensities stabilise at a constant level). Accordingly, PMB provides an increase in FRET efficiency. Thus, the effect of PMB on the fluorescence intensities is exactly the opposite of the experiments with the single-labelled SUVs (Fig. 3). Unexpected as well is the increase in FRET efficiency caused by PMB, as this

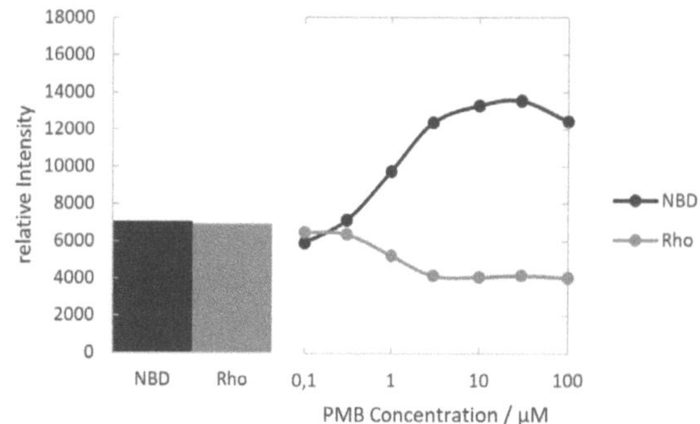

Figure 3: The fluorescence intensity of PE:PG (1:1) SUVs labelled with NBD-PE ("NBD", dark) or Rhodamine-PE ("Rho", light). The bars on the left show the fluorescence intensity of the SUVs without PMB addition, while the curves on the right show the fluorescence intensity of the SUVs to which PMB of different concentration was added. Importantly, the SUVs labelled with Rhodamine-PE were detected with a different gain (25%) than the SUVs labelled with NBD-PE (50%), which accordingly affects the intensities (see chapter 2 Materials and Methods).

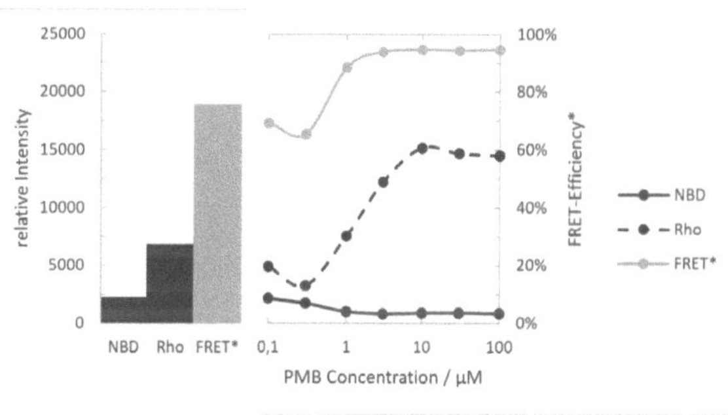

Figure 4: The fluorescence intensity of PE:PG SUVs labelled with NBD-PE ("NBD", donor) and Rhodamine-PE ("Rho", acceptor) (left axes) and the resulting FRET efficiency (right axis). The bars on the left side show the fluorescence intensity and the FRET efficiency of a sample to which no PMB was added. The curves on the right show the development of the fluorescence intensity and the FRET efficiency of the sample with increasing PMB concentration. (Bars and curves referring to FRET efficiency are marked with an asterisk).

means that PMB reduces the distance between the donor and acceptor.

The results in this part show again that the interaction of PMB with the PE:PG SUVs seems to be complex and can vary greatly depending on which fluorophores were used. This raises the question of whether there is an interaction between the fluorophores and the PMB. In general, it seems necessary to obtain more information about the interactions between the individual compartments for a more precise interpretation of the data, which is beyond the capabilities of this method. Nevertheless, the results show that the influence of the PMB on the PE:PG SUVs is significantly stronger than on the DOPC SUVs.

4   Conclusion

The results show that PMB affects the fluorophores in the PE:PG SUVs more than it does in the DOPC SUVs, which is expected due to the bacterial background of the PE:PG lipids. However, the interaction especially between PMB and the PE:PG liposomes seems to be so complex that a comparison with the DOPC SUVs leaves room for interpretation. Now, for a more precise interpretation of the data, it is necessary to obtain more information about the exact interactions between the lipids, peptides and fluorophores. An interaction between the PMB and the PE of the fluorophores cannot be excluded. Own experiments (not shown) on this have mainly shown that the preparation of PE SUVs with the method described here is difficult and yields liposomes with very strong size variation.

Acknowledgement

This work was carried out at the Research Centre Borstel, Leibnitz Lung Centre, Research Group Biophysics. The work was supervised by Prof. Dr Thomas Gutsmann (head of the research group) and Dr Christian Nehls and Sabrina Groth (both members of the research group).

Author's Statement

Conflict of interest: Authors state no conflict of interest

5   References

[1] K. Murzyn, T. Róg and M. Pasenkiewicz-Gierula, *Phosphatidylethanolamine-Phosphatidylglycerol Bilayer as a Model of the Inner Bacterial Membrane.* Biophysical Journal, vol. 88, pp. 1091–1103, February 2005.

[2] G. Shahane, W. Ding, M. Palaiokostas and M. Orsi, *Physical properties of model biological lipid bilayers: insights from all-atom molecular dynamics simulations.* Journal of Molecular Modeling, vol. 25, articel number 76, 2019.

[3] A. Goode, V. Yeh and B. B. Bonev, *Interactions of polymyxin B with lipopolysaccharide-containing membranes.* Faraday Discussions, vol. 232, 2021.

[4] T. Gutsmann, C. Nehls, S. Groth, *internal protocoll for the generation of SUVs.* Research Centre Borstel, Leibnitz Lung Centre, Research Group Biophysics

[5] R. S. Brown, J. D. Brennan and U. J. Krull, *Selfquenching of nitrobenzoxadiazole labeled phospholipids in lipid membranes.* The Journal of Chemical Physics, vol. 100, 1994.

[6] M. Hoefling, N. Lima, D. Haenni, C. A. M. Seidel, B. Schuler and H. Grubmüller, *Structural Heterogeneity and Quantitative FRET Efficiency Distributions of Polyprolines through a Hybrid Atomistic Simulation and Monte Carlo Approach.* PLOS ONE, 24-05-2011.

[7] T. Förster *Zwischenmolekulare Energiewanderung und Fluoreszens.* Annalen der Physik 2: 55–75.

Functional Characterization of Bacterial Galactosaminidases

Irina Walter [1], Cristian Roth [2], Christian Hübner [3]
[1] Biophysics, Universität zu Lübeck, irina.walter@student.uni-luebeck.de
[2] Max Planck Institute of Colloids and Interfaces, Carbohydrates:Structure and Function, christian.roth@mpikg.mpg.de
[3] Institut für Physik, Universität zu Lübeck, huebner@physik.uni-luebeck.de

Abstract

Many microorganisms form colonies on attaching surfaces to form biofilms. In mature biofilms they are integrated in an extracellular matrix, which also contains e.g. polysaccharides and proteins. One way to disrupt a biofilm could be the hydrolysis of polysaccharides. In this study, the hydrolase mechanism for members of the GH114 family, especially Tm114, is investigated. Tetra-galactosamine can be hydrolysed by Tm114 as well as samples from purified *A. Oryzae* and *A. Nidulans*. To further understand the hydrolase mechanism, the activation sites for hydrolysis are investigated. Therefor, two amino acids, which are thought to be important for the mechanism, Asp225 and Glu156, are replaced by Gln and Ala. Activity assays should reveal the role of these respective proteins in the hydrolysis mechanism. Both investigated mutants show similar results. For succesful investigation of the hydrolyse mechanism it is important to find a more suitable substrate.

1 Introduction

Biofilm formation is an important virulence factor for pathogenic fungi. Investigating the formation process of fungi may help preventing human infections in the future. In many fungal biofilms Galactosaminogalactan (GAG) is an integral component of the extracellular matrix (ECM) to increase adherence between cells and prevent cellular attacks [1,2]. A group of enzymes are responsible for the hydrolysis of GAG such as Sph3 and Ega3 [3-5]. Depletion of fungal biofilms may prevent fungal colonisation and reduce established fungal biofilms. Here, enzymes of the GH114 family are investigated, which hydrolyse in particular α-1,4- polygalactosamine, an exopolysaccharide found in fungal biofilms.

2 Material and Methods

2.1 Cell Lysis

Harvested cells, previously prepared from *E. Coli* NEB5α gold strain (Fisherscientific), were resuspended in 15ml 50mM Tris buffer (Sigma Aldrich) and lysed two times using french press (Avestin). Afterwards the sample was centrifuged at 50000xg and 4°C for 45 minutes. The supernatant was collected for IMAC analysis.

2.2 Immobilized Metal Ion Affinity Chromatography (IMAC)

The sample was loaded onto a His-Trap FF column using a Biorad NGC FPLC system. Immobilized nickel binds protein via an affinity tag like a His-tag. Afterwards the column was washed with 50mM Tris buffer and bound protein was eluted with 50mM Tris buffer containing 500mM imidazole (Carl Roth).

2.3 Sodium Dodecyl Sulfate Polyacrylamide Gel Electropheresis (SDS-PAGE)

IMAC fractions were mixed with 0.25mM Tris buffer containing 1% SDS (Thermo Fisher Scientific), heated at 95°C for 5 minutes and transferred to a gel column. The loaded gel was transferred into the electrode assembly. After running the gel was washed with water, boiled and stained with Coomassie Brilliant Blue R-250 Dye (Thermo Fisher Scientific).

2.4 Gel Filtration Chromatography (GFC)

For separation based on molecule sizes, a Sephadex S200 16/600 (Bio-Rad) column containing dextran polymers equilibrated with 20mM Tris buffer was used. Fractions of 1.5ml volume were collected and examined concerning containing protein via SDS-PAGE. Fractions containing TM114 were pooled and concentrated.

2.5 Purification of Exopolysaccharides

Exopolysaccharides were prepared from *A. Nidulans*. Culture medium was suspended in ethanol (Carl Roth) (1:2) and centrifuged at 5500rpm for 20 minutes. Supernatant was discarded and pellet was dissolved in water, centrifuged at 5500rpm for 20 minutes and supernatant was frozen and dried by lyophilizer (Christ).

2.6   Thin Layer Chromatography (TLC)

Samples containing $85\mu l$ 50mM Hepes buffer (Carl Roth), $10\mu l$ substrate and $5\mu l$ enzyme were incubated at 50°C and 800rpm for 2 hours. $2\mu l$ were pipetted on a TLC plate. Afterwards, the plate was transferred to a chamber with isopropanol, dried in air and dipped into 10% sulfuric acid in ethanol and heated for 10 minutes under 150°C.

2.7   Fluorescein Isothiocyanate (FITC) labelled TLC

FITC was dissolved in anhydrous DMSO (Serva) to 22mM and added to TLC samples in a 1:10 ratio. The mixture was incubated in darkness for 2 hours. $2\mu l$ were pipetted on a TLC plate and run in a closed chamber in chloroform/isopropanol and dried in air. The result was observed under UV light.

3   Results and Discussion

3.1   Protein purification

Protein synthesis procedure was performed for two mutants of Tm114, Tm114-D156A and Tm114-E225Q. After cell lysis the crude product was subjected several purification steps.

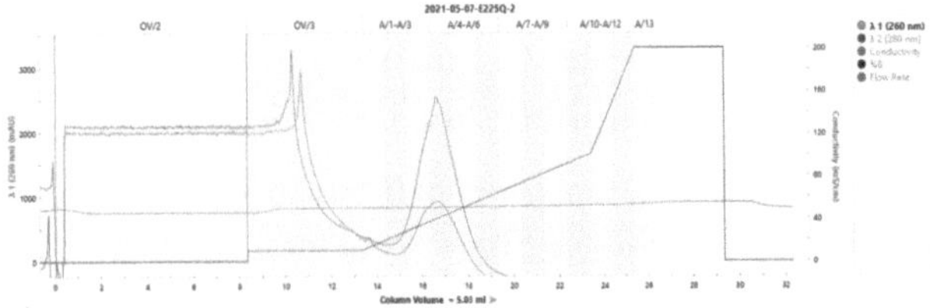

Figure 1: **IMAC result of Tm114-E225Q** Sample was applied to Ni-NTA column. The lower line represents absorbance at 260nm and the higher line shows absorbance at 280nm.

First, IMAC was carried out as shown in figure 1. The protein with a His6-tag was trapped on Ni-NTA column through chelation before elution. The peak of absorbance at 280nm indicated capturing of Tm114-E225Q. Columns containing the respective protein were analyzed by SDS-PAGE as seen in figure 2. All fractions show protein with a molecular weight of 37 kDa. The respective fractions were pooled and concentrated for GFC.

The next step in protein purification is gel filtration chromatography. The sample was injected on a HiLoad 16/60 Superdex 200 prep grade column for purification and buffer exchange. Figure 3 shows results of GFC. Absorbance at 280nm showed a peak which indicated protein detection. Fractions containing protein were analyzed by SDS-PAGE

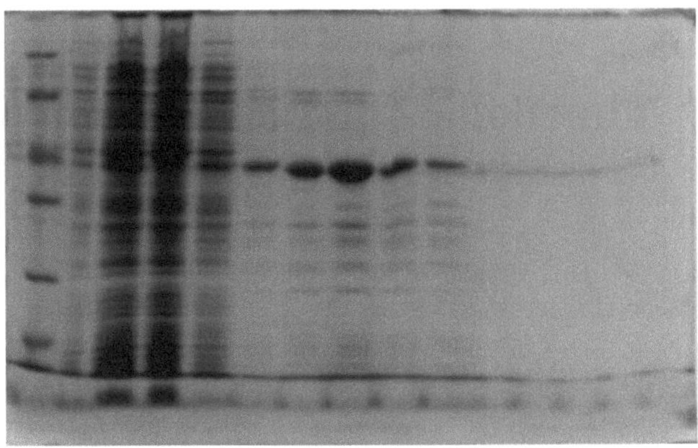

Figure 2: **SDS-PAGE result of Tm114-E225Q** Columns contain protein marker, crude extract, wash, flowthrough and the fractions containing respective protein collected by IMAC

as displayed in figure 4. All respective fractions contained Tm114-E225Q. They were pooled and concentrated.

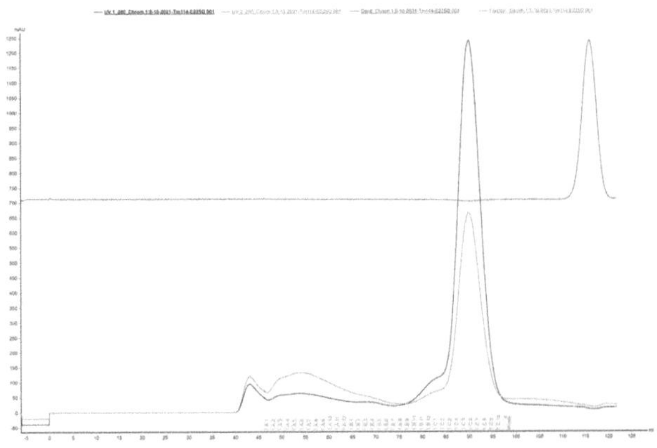

Figure 3: **SEC result of Tm114-E225Q** Lower line is absorbance at 260nm, higher line indicated absorbance at 280nm

The same protocol was carried out for another mutant, Tm114-D156A.

3.2   Hydrolysis of Polysaccharides

For functional characterization of the enzymes a suitable substrate that is hydrolized by the Tm114 enzyme has to be identified. Therefor, different polysaccharides that were obtained from *Aspergillus Oryzae*, *A. Nidulans* and *A. Niger* cultures were anaylzed using TLC.

The polysaccharides were previously purified from respective cultures and stored completely dry at -20°C. Before use they were hydrated with MQ water.

For the hydolase product samples were prepared as described in chapter 2.6. On the way up, the isopropanol

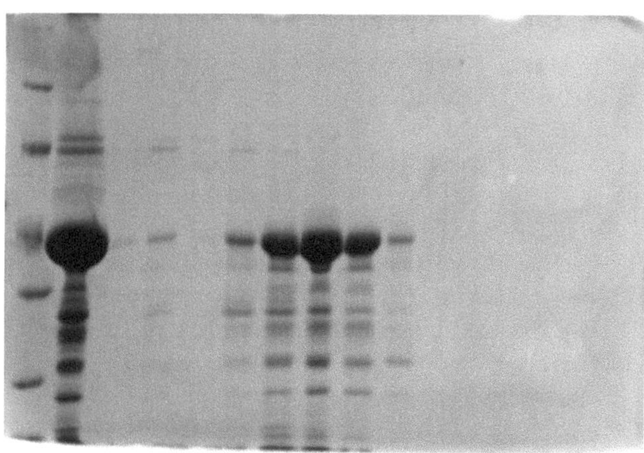

Figure 4: **SDS-PAGE result of Tm114-E225Q** Columns contain protein marker, crude extract and the fractions containing respective protein collected by GFC

would interact with the sample spots. The results of the TLC analysis revealed that substrates derived from *A. Nidulans* showed visible differences between the pure substrate and the respective hydrolase product. These results indicated that the Tm114 enzyme interacts with the polysaccharides of the substrate and hydrolyses them. So one of these substrates might be suitable for further investigation of the Tm114 mutants. It was decided to furthermore work with polysaccharides obtained from *A. Nidulans*.

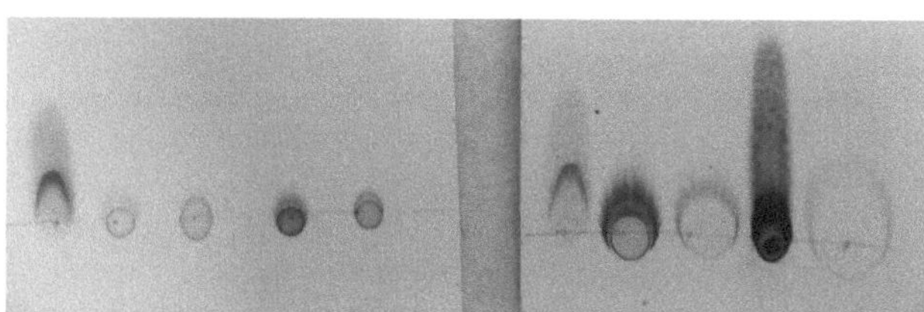

Figure 5: **TLC of hydrolysis of different polysaccharides** Spots from left to right show pure Galactosamine, *A. Niger*, *A. Niger*+Tm114, *A. Oryzae*, *A. Oryzae*+Tm114, Glactosamine, *A. Nidulans*, *A.Nidulans*+Tm114, *A. Nidulans*, *A. Nidulans*+Tm114

3.3 Purification of Polysaccharides from *A. Nidulans*

For futher experiments, new polaysaccharides obtained from *A. Nidulans* culture had to be purified. Therefore, previously expressed *A. Nidulans* DSM 820 supernatant which was stored at -20° had to be thawed. For purification the protocol as described in section 2.5 was carried out.

For further experiments the dried pellet was weighted and afterwards dissolved in ultrapure water to 10mg/ml. The

solution was stored at -20°C until further use.

3.4 Hydrolysis of *A. Nidulans* polysaccharides

After finding and purifying a supposably suitable substrate for hydrolase experiments with the two mutants and wildtype Tm115, activity assays with TLC were carried out. Therefore, samples were prepared containing the following mixture: $85\mu l$ buffer F, $15\mu l$ substrate (polysaccharides derived from *A. Nidulans*) and $5\,\mu l$ enzyme (Tm114 wildtype, Tm114-E225Q or Tm114-D156A). The samples were incubated overnight on a shaker at 50°C and 800rpm.

3.4.1 TLC without staining

Results for the hydrolysis of the purified substrate from *A. Nidulans* with Tm114 wildtype and two mutants are shown in figure 6. From left to right the spots represent the pure substrate, the substrate with Tm114 wildtype, the substrate with Tm114-D156A and the substrate with Tm114-E225Q.

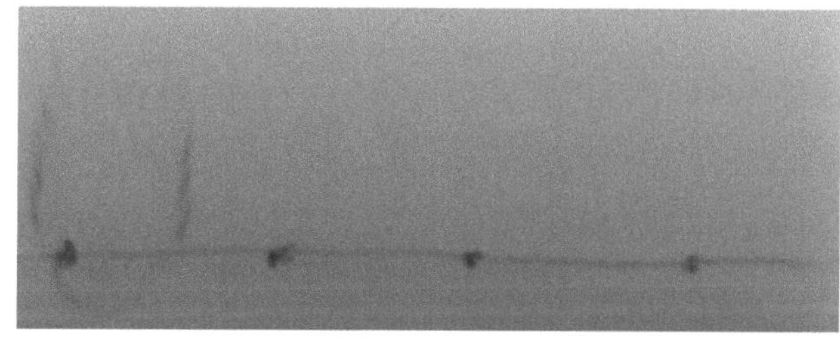

Figure 6: **TLC of hydrolase products** Spots from left to right: substrate, substrate+Tm114 wildtyoe, substrate+Tm114-D156A, substrate+Tm114-E225Q

The TLC result in figure 6 show a subtle difference between the spots representing the pure substrate (left) and the hydrolase products (right), since the spot on the left appears a lot darker and bigger than the others. That indicates that reaction between the substrate and the enzymes did happen. To further investigate differences between the reactions of the different enzymes with the substrate, many attempts were made to make the spots better visible by for example varying the amount of sample pipetted on the plate, trying different incubation times or varying the heating temperature and time. Unfortunately none of these attempts were successful. That is why FITC labelling of the samples was tried.

3.4.2 FITC-labelled TLC

As the TLC results without any further staining were not clear to interpret, a FITC labelling was used to make differences in substrates and hydrolase products better visible. Staining with FITC labelling was carried out as described in chapter 2.7.

Before the FITC labelling was successful, it needed many attempts to find a proper ratio of hydrolase product to FITC solution. In the first tries, the sample would be overstained leading to blurred results. Also the samples would not run in a straight line to the top of the plate because of a not well chloroform-saturated chamber. After several attempts a suitable amount of FITC and running solution was found. A volume of $19,5\mu l$ of hydrolase product was taken from each sample and stained with $0,5\mu l$ 22mM FITC.

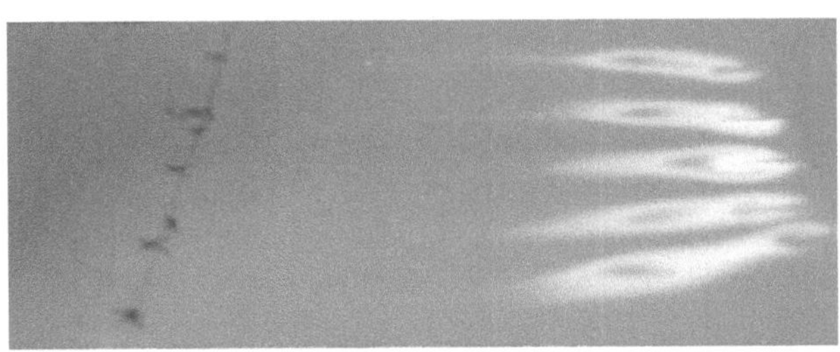

Figure 7: **FITC labelled TLC** Spots from top to bottom show FITC labelled substrate, substrate+buffer F, hydrolase product with wildtype Tm114, hydrolase product with Tm114-D145A, hydrolase product with Tm114-E225Q

A result of FITC-labelled TLC is shown in figure 7. Even here after many attempts to optimize the staining and running it is still visible that the samples are slightly overstained, which leads to blurring of the single spots, and the running lines are not straight.

Also, when comparing these results with the TLC results without staining as shown in figure 6 the results came out quite different as the spots visible without staining are mainly near the bottom of the film and the ones with FITC labelling seemed to be run until the top edge of the film. Even after many repetitions the two different approaches (with and without staining) did not seem to be comparable. That led to the assumption that the samples did not really react with the FITC label and what is detected during the experiments is only the FITC interacting with the running solution and running to the top of the film. After these findings, FITC-labelling was not further used to analyse the hydrolase products.

4   Conclusion

Concluding, expression and purification of two mutants of Tm114, Tm114-D156A and Tm115-E225Q including different purification steps, could successfully be performed. This was also confirmed by analysis of the respective sequencing after plasmid DNA isolation. Nevertheless, the purification of substrates for the hydrolase experiment,

namely polysaccharides from *A. Nidulans* culture is thought not to be successful. For that reason activity assays perfomed with TLC experiments can not be considered as proof for the activity of the expressed enzymes. For further experiments, a better suitable substrate for hydrolase activity assays, e.g. exppolysaccharides prepared from *A. Fumigatus*, has to be identified. Also, before starting to perform activity tests, it is found to be helpful to examine the substrate regarding the content e.g. by mass spectronomy. This way, succesful purification of the desired exopolysaccharides can be proofen.

Acknowledgement

The work has been carried out at Max Planck Institute of Colloids and Interfaces, Carbohydrates:Structure and Function and supervised by the Institute of Physics, Universität zu Lübeck. Furthermore, we would like to thank Michael Krummhaar, Ruben Ananian, Aylin Mizmizlioglu and Dominik Bierbaum for their kind assistance and advices.

Author's Statement

Conflict of interest: Authors state no conflict of interest. Informed consent: Informed consent has been obtained from all individuals included in this study.

References

[1] Bamford, N. C. et al. Sph3 is a glycoside hydrolase required for the biosynthesis of galactosaminogalactan in aspergillus fumigatus. *J. Biol. Chem.* **290**, 27438–27450 (2015)

[2] Fontaine, T. *et al.* Galactosaminogalactan, a new immunosuppressive polysaccharide of Aspergillus fumigatus. *PLoS Pathog.* **7**, 1002372 (2011)

[3] Bamford, N. C. *et al.* Ega3 from the fungal pathogen Aspergillus fumigatus is an endo-α-1,4-galactosaminidase that disrupts microbial 47 biofilms. *Journal of Biological Chemistry* **294**, 13833–13849 (2019)

[4] Colvin, K. M. *et al.* PelA deacetylase activity is required for Pel polysaccharide synthesis in Pseudomonas aeruginosa. *J. Bacteriol.* **195**, 2329–39 (2013)

[5] Baker, P. *et al.* Exopolysaccharide biosynthetic glycoside hydrolases can be utilized to disrupt and prevent *Pseudomonas aeruginosa* biofilms. *Sci. Adv.* **2**, e1501632 (2016)

Biochemical Characterization of the Interaction between CNPase and an Ultrahigh-affinity Camelid Nanobody

Marcel Schröder [1,2], Arne Raasakka [2], Sigurbjörn Markússon [2], Aleksi Sutinen [3], Ommolbanin Asadpour [4], Felipe Opazo [4], and Petri Kursula [2,3]

[1] Biophysics, Universität zu Lübeck, Lübeck, Germany, Marcel.Schroeder@student.uni-luebeck.de

[2] Department of Biomedicine, University of Bergen, Bergen, Norway, Arne.Raasakka,petri.kurusla@uib.de, sigurbjorn.markusson@student.uib.no

[3] Faculty of Biochemistry and Molecular Medicine & Biocenter Oulu, University of Oulu, Oulu, Finland, Aleksi.Sutinen@oulu.fi

[4] Center for Biostructural Imaging of Neurodegeneration, Göttingen, Germany, asadpour.ommolbanin@med.uni-goettingen.de, fopazo@gwdg.de

Abstract

Nanobodies (Nbs) are single-domain antibody fragments from camelids that can be used for structural and functional studies. The nanobody NbCNP-8D was characterized in complex with two $2',3'$-cyclic nucleotide $3'$-phosphodiesterase (CNPase) constructs. CNPase, the specific target antigen of NbCNP-8D, is essential for the function of the central nervous system and linked to neurological and psychiatric disorders. Using size-exclusion chromatography, enzyme-Nb complexes were obtained, and biochemical, biophysical, and structural methods were employed to observe the impact of the Nb. The investigation revealed a high binding affinity ($K_D < 3$ nM) coupled with a stabilizing and slightly activating effect, with no large alterations in protein conformation. The results indicate a very high affinity interaction, which can be utilized in further structural and functional studies on CNPase. In general, other disciplines, like high-resolution imaging or functional *in vitro/vivo* studies, could benefit from the small molecular weight and the high binding affinity of Nbs.

1 Introduction

$2',3'$-cyclic nucleotide $3'$-phosphodiesterase (CNPase) is a peripheral membrane enzyme of the nervous system myelin sheath, which accelerates nerve impulses by insulating axons. CNPase is found in oligodendrocytes and in smaller quantities in Schwann cells [1]. Its function is not limited to non-compact myelin formation and stabilization, but helps in maintaining the axon after myelination [2]. CNPase plays a crucial role in the $2',3'$-cyclic AMP-adenosine and $2',3'$-cyclic AMP-guanosine pathways, protecting neurons from an accumulation of toxic $2',3'$-cyclic AMP [3]. The absence of CNPase leads to severe deficiencies in both myelin structure and myelin-axon interfaces, eventually leading to swelling and degradation of axons [4]. Alterations in CNPase expression are linked to various neurological and psychiatric disorders, like Alzheimer's disease and Down syndrome [1], [5].

The variable heavy chain domain of the unique heavy chain antibodies from the serum of camelids can provide a soluble monomeric antibody known as a nanobody (Nb). Nbs contain three hypervariable regions (complementarity-determining regions, CDRs) linked by four conserved framework regions, resulting in a similar overall structure to traditional variable heavy chain domains [6]. Nbs offer a variety of advantages for research, starting with easy expression in simple prokaryotic systems, tolerance towards severe chemical and thermal conditions, and preservation of binding activity under high pH, temperature, and salt conditions, including reversible denaturation. Other interesting properties include a molecular weight of ~15 kDa, a binding affinity in the nM to pM range, no immune response in human and the use as crystallization chaperones [6], [7].

In this study, two CNPase constructs were complexed with a Nb, and various biochemical and structural methods were applied to characterize the CNPase-Nb interaction.

2 Material and Methods

2.1 Materials

Chemicals, unless stated otherwise, were obtained from Sigma Aldrich-Merck (MA, US) and materials from Thermo Fisher Scientific (MA, US). The CNPase substrate, $2',3'$-cyclic NADP$^+$, was purchased from BioLog Life Science Institute (Bremen, Germany). Protein chromatography was performed using an Äkta Pure chromatography system (equipped with an F9-C fraction collector) (GE healthcare, IL, USA) at 4-10 °C. Reagents for protein crystallization were from Molecular Dimensions (Sheffield, UK).

2.2 Protein Expression and Purification

Mouse CNPase constructs 4-2, the full-length enzyme composed of the PNK-like domain, the catalytic CPDase domain and the C-terminal tail, and 6-1, the catalytic CPDase domain only, were expressed and purified as described [8]. In brief, a Ni-NTA purification was performed, followed by an overnight TEV protease digestion and a second Ni-NTA step to remove His_6-tags and TEV protease. The eluted proteins were subjected to size-exclusion chromatography (SEC) using a HiLoad Superdex 75 and 200 16/600 pg SEC columns for CNPase 6-1 and 4-2, respectively. Purified CNPase 6-1 was stored on ice at 4 °C and CNPase 4-2 was snap frozen with liquid N_2 and stored at -80 °C.

2.3 Anti-CNPase Nanobody and Complex Formation

The anti-CNPase Nb-8D (NbCNP-8D) was subjected to SEC using a HiLoad Superdex 75 16/600 pg column to improve the purity.

For all experiments except for small angle X-ray scattering (SAXS), synchrotron radiation circular dichroism (SRCD) and isothermal titration calorimetry (ITC), a pre-prepared complex was used. For this, the two components were mixed with a 1.2-fold molar excess of NbCNP-8D, kept on ice for 15 min and subjected to SEC using a Superdex 200 Increase 10/300 GL column. The SEC buffer contained 20 mM HEPES, 200 mM NaCl, 0.5 mM TCEP, pH 7.5. The concentrated complexes were stored on ice.

2.4 Protein Quality Control

The purity and monodispersity of the proteins were checked. Samples from each purification step and the SEC fractions were analyzed by sodium dodecyl sulfate-polyacrylamide gel electrophoresis (SDS-PAGE) using Mini-PROTEAN TGX 4-20% precast 15-well gradient gels (BioRad Laboratries, CA, USA). Gels were stained using InstantBlue Coomassie stain (Abcam, Cambridge, UK) for 15 min, destained overnight and imaged using a Chemi-Doc XRS+ (BioRad, CA, US). Dynamic light scattering was performed using a Zetasizer Nano ZS DLS instrument (Malvern Panalytical) on 2 mg/ml samples in a 3-mm quartz cuvette (Hellma analytical) to check for monodispersity before experiments. SEC multi-angle light scattering (SEC-MALS) was performed to check the quality of the protein samples and as an analytical technique, which provides the mass and stoichiometry of the protein complex. For NbCNP-8D, a Superdex 75 Increase 10/300 GL SEC column, and for the CNPase constructs and complexes, a Superdex 200 Increase 10/300 GL column was used. The injection mass for NbCNP-8D was 200 µg and for CNPase constructs and complexes 100 µg. The experiment was performed with a Shimadzu Prominence-I LC-2030C HPLC system coupled to Wyatt miniDAWN TREOS MALS and an ERC RefractoMAX 520 RI detector. The columns were at room temperature, and the sample chamber of the HPLC was thermostatted at 4 °C. The system was calibrated using bovine serum albumin.

2.5 Interaction of NbCNP-8D with CNPase

To obtain information about the impact of NbCNP-8D on CNPase, different experiments were performed.

Differential scanning fluorimetry (DSF) provides information on the thermal stabilization due to Nb binding. The instrument was a Prometheus NT.48 (NanoTemper Technologies, Munich, Germany). DSF samples were at 2 mg/ml in SEC buffer.

To check if NbCNP-8D affects CNPase catalysis, activity assays were performed. The $3'$-phosphodiester bond hydrolysis of $2',3'$-cyclic $NADP^+$ was coupled to the oxidation of D-glucose-6-phosphate (G6P) using G6P dehydrogenase, as described [9]. The activity of $5 - 10$ ng CNPase was measured in 100 mM Bis-Tris pH 6.0, 10 mM $MgCl_2$ in the presence of 1 U G6P dehydrogenase, 10 mM G6P and 0-2 mM $2',3'$-cyclic $NADP^+$ at 25 °C in 96-well plates in a Tecan Infinite M Nano. Initial velocities were obtained by fitting the linear portions of the measured activity curves, plotted as a function of substrate concentration. Kinetic parameters were obtained by fitting the data directly to the Michaelis-Menten function using non-linear regression in GraphPad Prism 7.

ITC was performed on a MicroCal iTC200 instrument (Malvern Panalytical, Malvern, UK) to gain information about binding thermodynamics and affinity. The experiments were carried out in SEC-buffer with $8 - 10 \, \mu M$ CNPase 6-1 in the cell and $80 - 95 \, \mu M$ Nb-8D in the syringe at 25 °C, with a reference power of 5 µcal/s.

2.6 Structural Characterization

SRCD data were obtained from 0.5 mg/ml samples in 20 mM Tris-HCl (pH 7.5) on the AU-CD beamline at ASTRID2 (ISA, University of Aarhus, Denmark). Spectra were recorded from 170 nm to 280 nm at 25 °C in 100-µm closed circular cells (Hellma analytics, Müllheim, Germany). Background spectra were subtracted and CD units converted to $\Delta\epsilon$ ($M^{-1} \, cm^{-1}$).

To study the solution structure of CNPase in complex with NbCNP-8D, SAXS experiments were performed on the SWING beamline at the SOLEIL synchrotron (Gif-sur-Yvette, France) [10]. For the data collection of the complex, NbCNP-8D was mixed in a 1.2-fold molar excess with the CNPase constructs. A SEC-SAXS setup was used, with SAXS images collected during the elution of the protein in SEC-buffer from an Agilent Bio SEC-3, 300Å, 4.6 x 300 mm SEC column. SAXS data processing and analysis were carried out with ATSAS 3.0.4 [11]: for frame selection and buffer subtraction, CHROMIXS was used, primary analysis was carried out with PRIMUS and distance distribution function (P(r)) analysis in GNOM. *Ab initio* models and multiphase modelling were done using GASBOR and MONSA, respectively.

3 Results and Discussion

3.1 Recombinant CNPase Purification

For the investigation of proteins, it is indispensable to use pure protein samples. The purity, dispersity and the molar mass for each protein and complex were determined using SEC-MALS (Fig. 1).

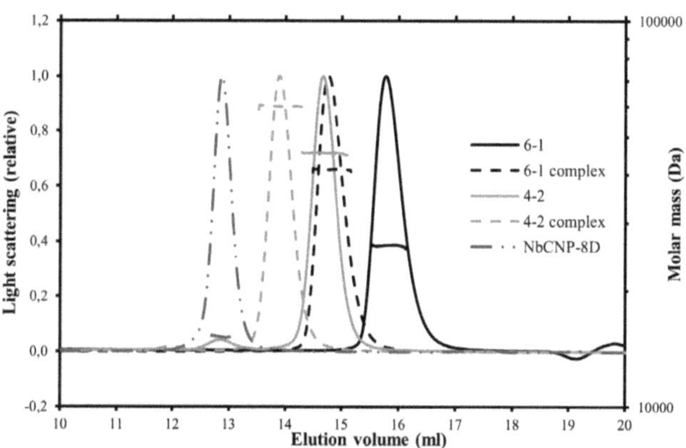

Figure 1: SEC-MALS spectra of all proteins. The horizontal line indicates the corresponding molecular weight of each peak. Note that different SEC columns were used for the NbCNP-8D and the CNPase constructs.

CNPase, NbCNP-8D and their complex were monodisperse. The error of the measured mass of the complexes (obtained through SEC-MALS) compared to the summed protein masses was approximately 6% and 1% for the 6-1 and 4-2 complexes, respectively. The SEC-MALS peaks showed that the samples are monomeric, or 1:1 in case of the complexes.

3.2 Impact of NbCNP-8D Binding on CNPase

The influence of NbCNP-8D binding on CNPase was investigated using DSF, SRCD, activity assays and ITC (Fig. 2, Table 1 & 2, ITC not shown). These experiments provide insights into Nb-CNPase complex formation and the subsequent effects on CNPase activity, stability and structure.

The 1:1 complex showed a tight binding affinity (ITC only for 6-1 construct: $K_D < 3$ nM). Additionally, DSF measurements revealed a stabilizing effect by the Nb (increase of T_m by 16-17 °C, Table 1). In case of CNPase 6-1, the binding did not significantly affect the enzymatic activity; the Nb does not act as an inhibitor (Table 2). For the full-length enzyme (CNPase 4-2), activity increased, which suggests that the Nb somehow improves the activity or stabilizes an active state. One explanation for the increased activity would be that inhibition by the C-terminal tail is affected due to the binding of the Nb (unpublished data). Similar K_M values were observed earlier. There were no major differences in the SRCD spectra between complexes and CNPase constructs, which means there is no alteration in the overall secondary structure content (Fig. 2). The result does not rule out local conformational changes that can be expected for a Nb-antigen interaction.

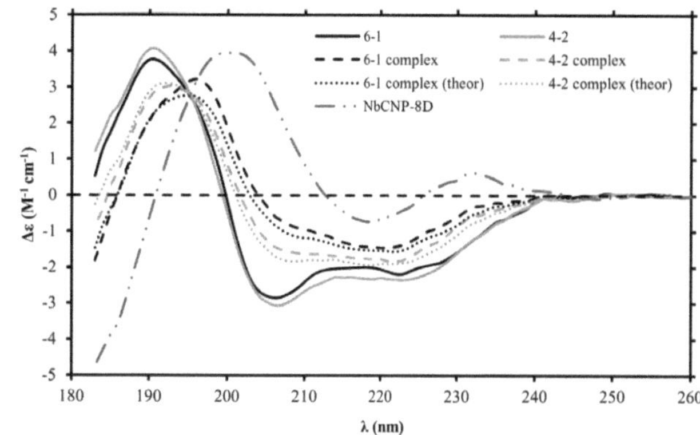

Figure 2: SRCD spectra of CNPase 6-1, 4-2, NbCNP-8D and the formed complexes. Additionally shown are spectra of theoretical complexes (theor), by summing the individual spectra after multiplying with their mass fraction in the complex.

Table 1: Melting temperatures of CNPase 6-1, 4-2, NbCNP-8D and the complexes. All measurements were performed in triplicate.

Protein	T_m (°C)	Nb-complex T_m (°C)	ΔT_m (°C)
CNPase 6-1	60.4 ± 0.01	76.5 ± 0.05	$+ 16.1$
CNPase 4-2	59.7 ± 0.03	77.4 ± 0.12	$+ 17.7$
NbCNP-8D	56.4 ± 1.31	-	-

Table 2: Kinetic parameters of CNPase 6-1 & 4-2 with and without bound Nb. All measurements were performed in triplicate.

Protein	K_M (μM)	k_{cat} (s^{-1})	k_{cat}/K_M (μM^{-1}s^{-1})
CNPase 6-1	468.7 ± 21.3	398.7 ± 6.7	0.9 ± 0.3
6-1 complex	328.2 ± 22.2	389.6 ± 8.7	1.2 ± 0.4
CNPase 4-2	531.7 ± 77.5	35.2 ± 2.0	0.1 ± 0.03
4-2 complex	374.5 ± 54.9	74.8 ± 3.8	0.2 ± 0.07

3.3 Low-resolution Structures of CNPase Constructs, NbCNP-8D and the Complexes

SAXS was used to obtain information about molecular size and shape in solution, providing a low-resolution structure of the proteins and their complexes (Fig. 3).

The modelling based on the SAXS data shows that all components remain globular, whether in complex or not. MONSA & GASBOR reveal a Nb binding to the end of the CPDase domain of CNPase.

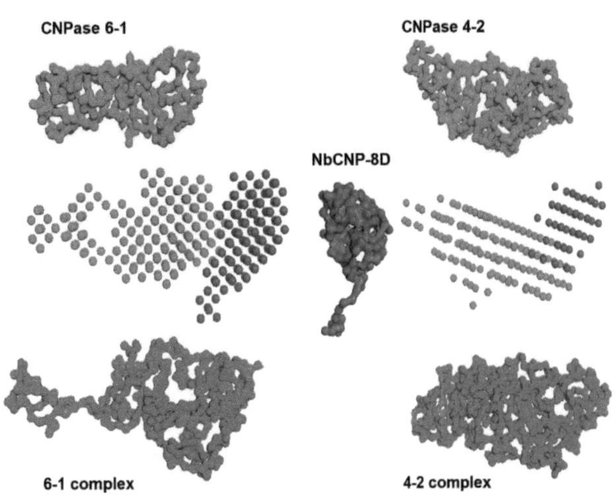

Figure 3: GASBOR & MONSA modelling of 6-1, 4-2, NbCNP-8D and the complexes. The software MONSA created the multiphase models (center, spherical map), (estimating possible binding sites considering the scattering curves). The outer models show the low-resolution GASBOR models.

4 Conclusion

It is necessary to analyze the interaction of each Nb with the target protein to understand the influence of the binding and to develop applications. Nbs mostly have a stabilizing impact on the bound protein, as in this study, but destabilizing effects could occur as well. In general, Nb binding could have an enhancing or inhibiting effect on the bound protein. For future drug design or functional studies in cells and *in vivo*, Nbs could be used to alter the function of the protein by *e.g.* blocking binding sites for ligands or stabilizing specific conformations. Common fluorescence microscopy as well as high-resolution imaging techniques, like STED microscopy, likewise benefit from the use of Nbs because of the smaller size compared to full-size antibodies. Structural biologists are already utilizing Nbs as crystallization chaperones to provide protein crystals that do not normally form. The crystallization of the CNPase constructs has been started, and upcoming work includes X-ray diffraction data collection of the CNPase 6-1/NbCNP-8D. Additionally, NbCNP-8D could be used to analyze other CNPase constructs or the full-length CNPase in combination with other Nbs to obtain the crystal structure.

Acknowledgement

We express our gratitude towards Biophysics, Structural Biology and Screening (BiSS) facilities at University of Bergen and the Biocenter Oulu Protein Analysis Core Facility. We gratefully acknowledge the synchrotron radiation facilities and the beamline support at ASTRID2 and SOLEIL. The work has been carried out at the University of Bergen, Norway and supervised by Dr. Young-Hwa Song, Institute of Physics, Universität zu Lübeck.

Author's Statement

The authors declare no conflict of interest.

5 References

[1] A. Raasakka and P. Kursula, *The myelin membrane-associated enzyme 2′,3′-cyclic nucleotide 3′-phosphodiesterase: on a highway to structure and function.* Neuroscience Bulletin 30, 956–966, 2014.

[2] E. K. Jackson, *Handbook of Experimental Pharmacology.* Springer New York LLC: 2017; Vol.238, pp 229-252

[3] E. K. Jackson, Z. Mi, K. Janesko-Feldman, T. C. Jackson and P. M. Kochanek, *2′,3′-cGMP exists in vivo and comprises a 2′,3′-cGMP-guanosine pathway.* American Journal of Physiology-Regulatory, Integrative and Comparative Physiology, vol. 316, no. 6, pp. R783–R790, Jun. 2019.

[4] J. M. Edgar et al., *Early ultrastructural defects of axons and axon–glia junctions in mice lacking expression of Cnp1.* Glia, vol. 57, no. 16, pp. 1815–1824, 2009.

[5] R. Vlkolinský, N. Cairns, M. Fountoulakis and G. Lubec, *Decreased brain levels of 2′,3′-cyclic nucleotide-3′-phosphodiesterase in Down syndrome and Alzheimer's disease.* Neurobiology of Aging, vol. 22, no. 4, pp. 547–553, Jul. 2001.

[6] S. Muyldermans, *Nanobodies: natural single-domain antibodies.* Annual Review of Biochemistry, vol. 82, 775-797, 2013.

[7] J. R. Ingram, F. I. Schmidt and H. L. Ploegh, *Exploiting Nanobodies' Singular Traits.* Annual Review of Immunology, vol. 36, no. 1, pp. 695–715, 2018.

[8] M. Myllykoski and P. Kursula, *Expression, purification, and initial characterization of different domains of recombinant mouse 2′,3′-cyclic nucleotide 3′-phosphodiesterase, an enigmatic enzyme from the myelin sheath.* BMC Research Notes, vol. 3, Jan. 2010.

[9] M. Myllykoski et al., *Crystallographic analysis of the reaction cycle of 2′,3′-cyclic nucleotide 3′-phosphodiesterase, a unique member of the 2H phosphoesterase family.* Journal of Molecular Biology, vol. 425, no. 22, 4307-4322, Nov. 2013.

[10] A. Thureau, P. Roblin and J. Pérez *BioSAXS on the SWING beamline at Synchrotron SOLEIL.* Journal of Applied Crystallography, 54, 1698-1710, 2021.

[11] K. Manalastas-Cantos et al., *ATSAS 3.0: expanded functionality and new tools for small-angle scattering data analysis.* Journal of Applied Crystallography, vol. 54, no. 1, pp. 343–355, Feb. 2021.

Biophysical characterization of the interaction between the antimicrobial peptide polymyxin B and living bacteria of the genus *Salmonella*

Helena S. Fabritz [1], Christian Nehls [2], and Thomas Gutsmann [3]
[1] Biophysics, Universität zu Lübeck, helena.fabritz@student.uni-luebeck.de
[2] Division of Biophysics, Research Center Borstel, Leibniz Lung Center, cnehls@fz-borstel.de
[3] Division of Biophysics, Research Center Borstel, Leibniz Lung Center, tgutsmann@fz-borstel.de

Abstract

Drug resistances among bacteria cause worldwide health threats and rises the urgent need for new compounds. Antimicrobial peptides (AMPs) are a promising alternative to conventional antibiotics for the treatment of bacterial infections. Research into the mode of action of AMPs against living bacteria represents an essential aspect for the development of new therapeutics against infectious diseases. In this report, the interaction between the AMP polymyxin B (PMB) and living bacteria of the genus *Salmonella* was investigated. Dynamic light scattering (DLS) experiments and zeta potential measurements were carried out studying the effect of PMB on bacterial growth behavior and surface potential. PMB was found to exert a significant influence on time-dependent growth at various concentrations represented by a decrease in bacterial particles and average size. Regarding the bacterial surface (zeta) potential, the PMB addition resulted in a decrease in this parameter. These results provide clues to elucidate the mode of action of AMPs.

1 Introduction

The envelope of bacterial cells has a very complex structure and shows differences in its composition between Gram-positive and Gram-negative bacteria. In contrast to Gram-positive bacteria, Gram-negative bacteria are surrounded by two membranes: a plasma membrane and an outer membrane (OM) [1]. The outer leaflet of the OM contains negatively charged glycolipids, known as lipopolysaccharide (LPS) or endotoxin, which are capable of virulence factors [2]. Due to their increasing antibiotic resistance, Gram-negative bacteria are of particular (bio)medical interest [1]. This increased frequency of resistance to conventional antibiotics has led to a lack of effective treatment options. Antimicrobial peptides (AMPs) are being developed in order to overcome this problem, due to the low emergence of microbial resistance to this class of molecules [2] [3].

AMPs are found in the human body and are part of the innate immune system [3]. These membrane-active defense molecules are highly effective in the permeabilization of bacterial lipid layers. The antibiotic polymyxin B (PMB) is an AMP isolated from *Bacillus polymyxa* with significant bactericidal activity, especially against Gram-negative bacteria [2] [3] [4]. This selective activity is based on the high affinity for LPS [3]. Regarding the mechanism of membrane disruption, complex formation by PMB with LPS is expected to be the first step. As a result, PMB changes the packing order of LPS and increases the permeability of the OM [5].

In this work, the effect of PMB on different parameters of living Gram-negative bacteria of the genus *Salmonella* was investigated. The analyses revealed that PMB interacts with the bacteria resulting in biophysical changes that reduce the number of bacterial particles, the average particle size, and the surface potential. The effect of the AMP was studied as a function of different PMB and growth media concentrations and was already observed at values below the minimal inhibitory concentration (MIC).

2 Materials and Methods

2.1 Materials

In this work, the R60 mutant of *Salmonella enterica* serovar Minnesota, or short *S.* Minnesota, was used. Polymyxin B was sourced from Sigma-Aldrich. The used buffer solution was a 20 mM HEPES (4-(2-hydroxyethyl)-1-piperazine-ethanesulfonic acid), 10 mM NaCl (sodium chloride) buffer adjusted to pH 7.4. It was prepared with demineralized water and sterile filtrated. LB-broth (lysogeny broth) was used as growth medium and was also prepared in demineralized water. After preparation the medium was autoclaved at 121 °C for 15 minutes. Trifluoroacetic acid (TFA) was used in a concentration of 0.01 % to dissolve the peptide.

2.2 Bacterial growth of pre-cultures

Pre-cultures were prepared by inoculating 2 ml of the growth medium with one cryo culture pearl obtained from the Research Center Borstel. These cultures were grown overnight at 37 °C and 150 revolutions per minute (rpm). For further use the overnight culture was diluted in LB-broth (1:50) and incubated for another 2 hours (37 °C, 150 rpm) to reach the exponential growth phase. Subsequently, the optical density (OD) of the bacteria suspension was determined using a cell density meter (WPA Biowave CO8000) at a wavelength of 600 nm. By comparing the OD to a reference curve the cell density was adjusted at 1×10^7 colony forming units per milliliter (cfu/ml) and 1×10^6 cfu/ml for the MIC test respectively.

2.3   MIC test

The MIC is defined as the lowest concentration of antimicrobial agents which prevents visible growth of microorganisms. The MIC of PMB against *S.* Minnesota represents a crucial parameter for most of the experiments performed in this work and was determined as a function of the medium concentration. The corresponding values served as a guidance for the peptide concentrations used in the subsequent experiments.

For MIC determination a 96-well flat-bottom-microtiter plate was prepared with six different LB concentrations and serial peptide dilutions (90 µl). Subsequently, to each well a defined volume of the bacteria suspension (10 µl, 1×10^6 cfu/ml) was added, except for the negative control. The inoculated wells were incubated overnight in a wet chamber at 37 °C and 150 rpm. After 24 hours of inoculation the plates were read out using a microplate reader (Tecan Sunrise) at a wavelength of 620 nm.

2.4   Dynamic light scattering (DLS)

The technique of dynamic light scattering (DLS) is used to determine the size distribution of small particles in aqueous solution. This determination is based on the relation between Brownian motion and particle size [6].

DLS-experiments were performed using Malvern's Zetasizer Nano-ZS90 to study the influence of PMB on the time-dependent growth behavior of *S.* Minnesota in terms of the average bacterial particle size and the number of particles as a function of the medium concentration. The laser of the zetasizer emits light at a wavelength of 532 nm. The light intensity scattered on the particles was analyzed at a scattering angle of 90 °. All measurements were performed at 37 °C. The growth behavior was observed over a period of four hours with a total of five acquisitions at intervals of one hour each.

The growth medium used was diluted in buffer to the six concentrations of 10, 20, 40, 60, 80 and 100 % (pure medium). The bacteria were prepared as described in section 2.2 with the cell density adjusted to 1×10^7 cfu/ml in the various LB dilutions. Regarding the added peptide concentrations, variations of the previously determined MIC values were used (0.5 x MIC, 2 x MIC). DLS-experiments were carried out by filling the zetasizer cuvettes (67.754; Sarstedt) with 1 ml of the sample solution. Samples were stored at 37 °C until usage.

2.5   Zeta potential measurements

The zeta potential is used to characterize the surface potential of charged particles in solution relative to the surrounding aqueous phase. This potential can be deduced from the velocity of the particles in an electric field [6].

Zeta potential measurements were performed to study the influence of bacterial growth on the surface potential of *S.* Minnesota and to analyze the effect of PMB on this parameter. The experiments were also carried out using Malvern's Zetasizer at 37 °C. While the zeta potential of *S.* Minnesota without PMB was observed over a period of four hours with a total of five acquisitions at intervals of one hour each, the bacterial potential under the influence of an addition of PMB (2 x MIC) was investigated over three hours with a total of four measurements. Finally, the bacterial potential was studied as a function of different variations of the MIC value.

The high ionic strength of the pure medium would cause Joule heating and therefore affect the mobility of the dispersed particles [6]. Thus, LB-broth was diluted in demineralized water to a concentration of 10 %. The bacteria were prepared as described before and the cell density (1×10^7 cfu/ml) was adjusted in the diluted (1:10) medium. The cuvettes (DTS1070; Malvern Panalytical) were filled with 700 µl of the sample solution. The samples were stored at 37 °C until usage.

3   Results and Discussion

3.1   MIC values of PMB against *S.* Minnesota at various growth media concentrations

Since most of the experiments were carried out using the six LB-broth concentrations mentioned previously, the MIC of PMB against *S.* Minnesota was determined as a function of these concentrations. Table 1 shows the results of the MIC test.

Table 1: MIC values of PMB against *S.* Minnesota at various growth media concentrations

LB-broth concentration	MIC of PMB against *S.* Minnesota
10 %	4 µg/ml
20 %	2 µg/ml
40 %	2 µg/ml
60 %	2 µg/ml
80 %	2 µg/ml
100 %	2 µg/ml

The test revealed different values for the minimal inhibitory concentration of PMB. At a LB concentration of 20 to 100 %, a MIC of 2 µg/ml was obtained, whereas a higher value of 4 µg/ml was observed at a medium dilution of 1:10. These results suggest a possible influence of the LB concentration on the concentration of PMB preventing visible growth of *S.* Minnesota.

3.2   Investigation of the influence of PMB on the growth behavior of *S.* Minnesota

First, the growth behavior of *S.* Minnesota without peptide was measured to obtain information on the relation between bacterial growth and the number of particles and the bacterial particle size, respectively. Measurements were performed using various LB dilutions. The corresponding results are shown in Fig. 1.

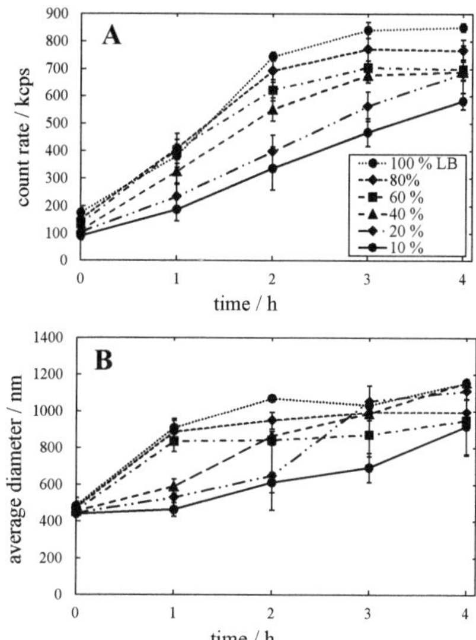

Figure 1: Evolution of the number of particles (A), referred to as count rate, and the average diameter (B) of *S.* Minnesota as a function of growth medium (LB) concentration over time. LB was diluted in a 20 mM HEPES, 10 mM NaCl buffer (pH 7.4). Measurements were performed at 37 °C with 1×10^7 cfu/ml. The results shown are averages from three subsequent acquisition cycles.

It can be seen that the number of detected bacterial particles, referred to as count rate, and the particle size are dependent on the LB concentration. In undiluted growth medium the count rate takes the highest value at almost every point. At LB concentrations in the range of 40 to 100 %, the parameter appears to reach a saturation value after about three hours, whereas no flattening of the curves can be observed at 10 and 20 % LB-broth. In terms of the size distribution of *S.* Minnesota, the largest values for the average diameter of the particles were usually measured using the undiluted growth medium. Although an overall increase in bacterial size is observed over time, the lines seem to follow a rather random course.

Measurements using two different MIC values (0.5 and 2 x MIC) of PMB were performed to study the bacterial growth behavior under the influence of peptide. The evolution of the count rate and the average particle size of *S.* Minnesota over time using PMB with a concentration of 0.5 x MIC is shown in Fig. 2. These experiments were carried out with a single peptide addition immediately after the first measurement. After addition, a significant influence of PMB on the culture growth was observed. Regarding the count rate there is almost no increase in the number of particles at all LB concentrations. Instead, a decay in the count rates can be observed after three hours. This suggests that no bacterial growth occurs due to the addition of PMB. The average size of the bacteria

generally reaches its maximum after one hour at any medium concentration and remains almost constant in the subsequent period. The same observations on bacterial growth behavior could be made at a higher PMB concentration of 2 x MIC (data not shown).

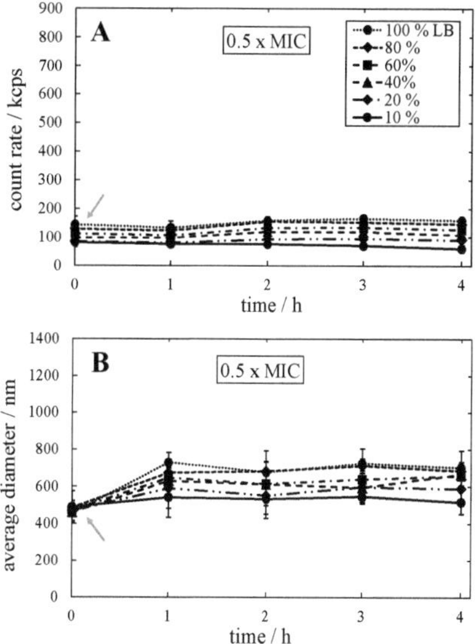

Figure 2: Count rate (A) and average diameter (B) of *S.* Minnesota over time under the influence of PMB. The same experimental conditions were chosen as for the measurement shown in Fig. 1. The single PMB addition with a concentration of 0.5 x MIC is indicated by an arrow.

3.3 Investigation of the influence of PMB on the zeta potential of *S.* Minnesota

In order to evaluate the effect of PMB on the surface potential of *S.* Minnesota, the zeta potential without peptide was studied first. These measurements were carried out with a medium concentration of 10 % over a period of four hours to investigate whether changes in the zeta potential occur due to bacterial growth. The corresponding results are shown in Fig. 3 (●). It can be seen that a non-linear dependence of the zeta potential on the time-dependent bacterial growth exists. This statement is based on the observation that the initial potential takes a lower value at the end of the measurement period. After two hours, the zeta potential reaches an almost constant value of about -13 mV.

Two different experiments were carried out to investigate the influence of PMB on the zeta potential of *S.* Minnesota. The first measurement was performed with a single addition of peptide to inhibit bacterial growth. In the second one the peptide concentration was increased by successive additions.

The time of the single peptide addition (2 x MIC; 8 µg/ml) was chosen as the time at which the zeta potential had already almost reached its minimum value. Therefore, the PMB was added immediately after the second data acquisition and the zeta potential was detected over a total period of three hours. The addition of PMB results in a further decrease of the zeta potential, as seen in Fig 3 (■).

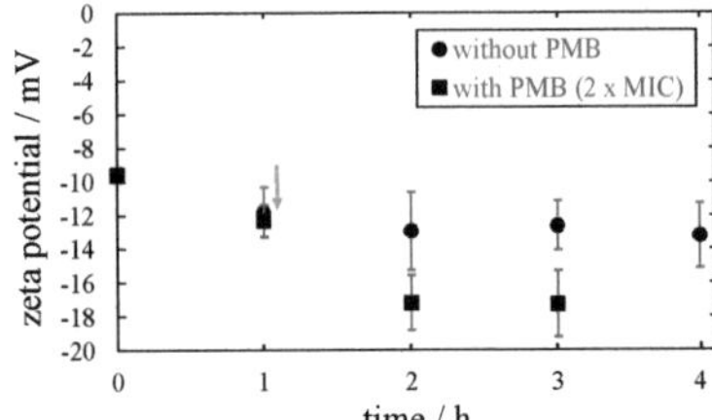

Figure 3: Zeta potential of *S.* Minnesota at 10 % LB without (●) and with (■) PMB over time. LB was diluted in demineralized water. Measurements were performed at 37 °C with 1×10^7 cfu/ml. The PMB addition (2 x MIC; 8 µg/ml) is indicated by an arrow.

Four peptide additions were performed to investigate the effect of successively increasing PMB concentration on the bacterial zeta potential. Thus, the PMB concentration was gradually increased from 0.25 x MIC after the first addition to 2 x MIC after the final one. Each addition was followed by an incubation period of one hour. As seen in Fig. 4, the successive addition of PMB results in an almost exponential decay in the zeta potential. At a peptide concentration of 0.5 x MIC the potential reaches a value of about -15 mV. This value remained almost constant even when higher PMB concentrations were used.

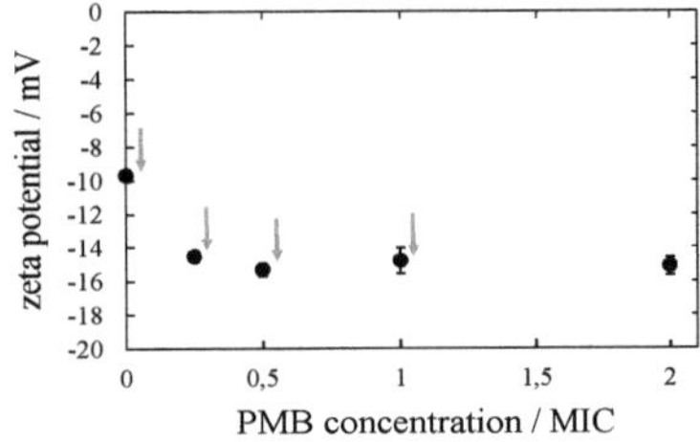

Figure 4: Zeta potential of *S.* Minnesota as a function of the PMB concentration (MIC). The same experimental conditions were chosen as for the measurement shown in Fig. 3. Data acquisitions were carried out at intervals of one hour each.

3.4 Discussion

The observation that bacterial cultures in high LB concentrations reach the stationary phase faster allows the assumption that higher concentrated LB-broth favors bacterial growth. In contrast to the count rate, the increase in the particle size over time cannot be explained directly by the growth of the bacteria and is therefore still unclear. To clarify this observation, the bacterial growth over the measured period could be investigated more closely using suitable techniques such as atomic force microscopy.

For both MIC values (0.5 and 2 x MIC) the results of the measurements for studying the influence of PMB on the growth behavior of *S.* Minnesota were almost identical. Since no considerable difference in terms of the count rate and the particle size could be observed for all six LB concentrations, it can be concluded that the concentration of the medium has no discernable influence on the bacterial growth at both MIC values when PMB is used.

The observed reduction in the zeta potential could be due to PMB-induced modifications of membrane components,

such as LPS [7]. This assumption is supported by the fact that even at the highest PMB concentration of 2 x MIC there is no neutralization of the zeta potential which would result from a total collapse of the bacterial membrane. It is more likely that only changes in the surface structure of the bacterial membrane occur.

4 Conclusion

The experiments performed in this work may provide initial clues to elucidate the mechanisms of action of PMB against Gram-negative bacteria of the genus *Salmonella*. In future work it is important to investigate the effect of further peptide concentrations on the growth behavior of *Salmonella* and other strains as well as the structural changes of the bacterial membrane induced by AMPs. The findings can be used for the development of new therapeutics against bacterial infections.

Acknowledgement

The work has been carried out at the Research Center Borstel, Leibniz Lung Center and supervised by T. Gutsmann, Division of Biophysics.

Author's Statement

Conflict of interest: Authors state no conflict of interest.

5 References

[1] L. A. Clifton et al., *Asymmetric phospholipid: lipopolysaccharide bilayers; a Gram-negative bacterial outer membrane mimic.* Journal of the Royal Society Interface, pp. 1-11, 2013

[2] A. B. Schromm et al., *Cathelicidin and PMB neutralize endotoxins by multifunctional mechanisms including LPS interaction and targeting of host cell membranes.* PNAS, vol. 118, no. 27, pp. 1-12, 2021

[3] M. M. Domingues et al., *Biophysical characterization of polymyxin B interaction with LPS aggregates and membrane model systems.* Wiley Online Library, Peptide Science, vol. 98, no. 4, pp. 338-344, 2012

[4] V. H. Tam et al., *Pharmacodynamics of polymyxin B against Pseudomonas aeruginosa.* Antimicrobial Agents and Chemotherapy, vol. 40, no. 9, pp. 3624-3630

[5] R. Daugelavičius, E. Bakienė and D. H. Bamford, *Stages of polymyxin B interaction with Escherichia coli cell envelope.* Antimicrobial Agents and Chemotherapy, vol. 44, no. 11, pp. 2969-2978, 2000

[6] J. Wernecke, *Biophysical characterisation of the fungal peptide toxin Ece1-III and its interaction with lipid membranes.* Dissertation, Lübeck, Universität zu Lübeck, 2016

[7] A. Saathoff, *Biophysical investigations into the interactions between antimicrobial peptides of the epithelial defense and microbial cell envelopes.* Dissertation, Lübeck, Universität zu Lübeck, 2013

4

Safety and Quality

Improvement approaches for the complaint handling and the supplier evaluation in a medical device company

Vüsal Bayramov [1]

[1] Medical Engineering Science, Universität zu Lübeck, vuesal.bayramov@student.uni-luebeck.de

Abstract

For a medical device company the suppliers are of great importance, that the medical products can be build, but these products have to comply with the regulatory requirements. For this, the suppliers have to be evaluated with regard to the quality of their products regularly, which is in general done by the quality management system of the medical device company. This complex process however often suffers from an unprecise evaluation of the quality of the suppliers. In this work we will analyze different external suppliers of the Business Unit Hospital Consumables & Accessories at the Drägerwerk AG & Co. KGaA, a large medical device manufacturer in Lübeck. Therefore the regulatory standards and different tools and strategies of the Dräger internal processing of complaints and of the supplier evaluation were the main research basis. The developed approach is implemented and shows promising results.

1   Introduction

The main tasks of the quality management system of a medical device company are the complaint handling and the supplier evaluation. Because of that, every medical device company has it's own quality management system, which is different from the quality management system of other companies. The aim of every quality management system of a medical device company is to check the quality of their products regularly and to guarantee the customer satisfaction. In addition, it is important to ensure, that the quantity of yearly customer complaints is as low as possible.

The Business Unit Hospital Consumables & Accessories (BU HCA) at Dräger in Lübeck, founded in 2020, consists of various departments that are responsible for the quality management, purchasing and the development of medical products, for example ventilation and anesthesia equipment. These medical products are constructed from consumables and accessories, such as ventilation tubes, which are produced by external suppliers. The department of Quality & Regulatory Affairs (QRA) is responsible for the quality management in the BU HCA, mainly for the processing of complaints. The department of purchasing is responsible for the supplier evaluation. The products made from consumables and accessories represent independent medical products in the sense of the Medical Device Regulation (MDR) and must comply with the requirements of the International Standardization Organization (ISO) [1].

A legal manufacturer of these products, a medical device company is obliged to check the suppliers and to evaluate them regularly with regard to the quality of their products.

2   Material and Methods

The complaint handling and the supplier evaluation with the Delivery & Quality Performance Evaluation Tool and the more accurate Supplier Evaluation Tool, which are used by Dräger internally and their advantages and disadvantages will be explained. The Delivery & Quality Performance Evaluation Tool is used two times a year and the Supplier Evaluation Tool is used once a year.

The analysis of the complaint handling and of the supplier evaluation is based on different suppliers of the BU HCA. The developed methods can be used for the evaluation of all external suppliers of the BU HCA. The aim is to make the complaint handling as easy as possible and to reach a more qualitative and reliable evaluation of suppliers.

A complaint is generally any written, electronic or oral communication that alleges deficiencies related to the identity, quality, durability, reliability, safety, effectiveness or performance of a device after it is released for distribution [2]. The first step for the complaint handling is to check whether the complaint or the Failure on Arrival (FOA) are caused by a supplier of the BU HCA. Complaints and FOAs are field complaints (FCs) and they are corresponding to customer complaints. FOAs describe processes, in which a customer finds one or more defects of the medical product by the first application. The differentiation between FCs and supplier caused FCs (SCFCs) is an important step, because SCFCs are processed individually: A 8D-report will always be sent to the supplier, to check and analyze the reason of the error and also to describe the correction process, if the FC is caused by the supplier. A 8D-report includes eight different questions to the viewed supplier, which have to be answered by the supplier. First of all, the supplier has to name the members of the team, who are responsible to solve the problem. In the second step, the supplier has to describe the failure and to explain the immediate containment actions. In addition, the supplier has to describe the root cause of the problem, the planned corrective actions, the implemented corrective actions and the preventive actions. Finally, the supplier has to confirm, that the problem has been solved completely. In this case the quality notification is evaluated as a SCFC and registered in the SAP-system. The SAP-system is the Enterprise-Resource-Planning (ERP)-system of the BU HCA. In addition, the Dräger internal quality database is a part of the SAP-system. In dependence of the differentiation, there will be registered a different amount of SCFCs in the SAP-system. This is important for the supplier evaluation, because the quality of an evaluation is based on the amount of SCFCs, which were taken into account for the calculation of the quality performance in percentage [4]. Because of that, the second relevant step is to include the SCFCs, these quality notifications, into the calculation of the quality performance of a supplier. This is realized by an extension of the Supplier Evaluation Tool.

Only if a FC is evaluated as a SCFC, the quality notification can be registered in the SAP-system. Afterwards different processes of the complaint handling can be performed, based on the recommended procedure, which is illustrated in Fig. 1.

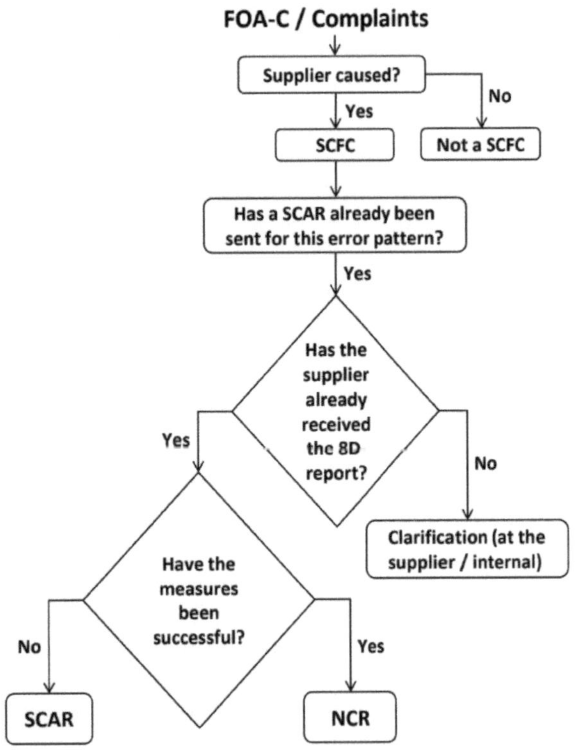

Figure 1: Recommended strategy for the processing of a supplier caused field complaint in the BU HCA

One quality criteria is, that only FOAs from the type C (FOA-C) can be evaluated as a SCFC. This class of FOAs includes errors in the arrival of products, which have a high negative impact on the user, the customer. A FOA-C is for instance a failure in function or a damage in the delivery of a product to the customer. In contrast to that, a FOA-A has a low negative impact on the user. FOAs from the type B (FOA-B) have the second strongest negative impact on the user and are for example deviations in the quantity of delivered medical products. Complaints and FOAs will not be evaluated as a SCFC, if they are for example caused by label mistakes or design errors. These simple failure events in the delivery of medical products to the customers will be processed by Dräger independently and these quality notifications, non-SCFCs, will not be registered in the SAP-system. Other criteria for a SCFC are partial or complete loss of function, missing parts of the product or defect parts. In general, the reason for a SCFC is a large, systematic, product-specific defect with a high negative consequence for the user (therefore only FOAs of category C). In contrast to that, a non-SCFC is mostly caused by a design error, label error, simple, inferior, non-product specific, possibly also random and non-systematic error with minor negative consequences for the user.

The quantity of total registered SCFCs in the SAP-system is very low. Because of that, a more precise differentiation between supplier caused and non-supplier caused FCs is important and necessary for the supplier evaluation: SCFCs are the most important type of complaints for a more reliable supplier evaluation, which includes the calculated quality, the quality performance of a supplier in percentage by considering the SCFCs. For the supplier evaluation and also the complaint handling it is important, to register all SCFCs in the SAP-system.

Fig. 1 illustrates the recommended procedure for the complaint handling in the BU HCA. It shows the strategy, how to process different FCs, complaints and FOAs, in dependence whether they are caused by the viewed supplier. In the first step, there is the process of checking, whether the quality notification is a SCFC. In dependence of this, the complaint handling is continued in different steps. In the case, that the FCs are caused by the suppliers and SCFCs are existing, different aspects have to be considered. If the quality notification of a SCFC has the same root cause as in the past or an 8D-report has already been sent to the same supplier because of the same failure and it has not been corrected until today, an 8D-report must be sent to the supplier again. The supplier has to confirm, that the problem has been solved and must fill out the 8D-report correctly and completely. The 8D-report is helpful for the BU HCA to analyze whether the viewed supplier implemented corrective actions for the failures. If the BU HCA accepts the 8D-report, the quality notification will be registered in the SAP-system as completed. If not, the supplier must fill out the 8D-report correctly until it is accepted by the BU HCA, which will be checked by the Supplier Quality Engineer. In both cases, the quality notification is registered as a SCFC in the SAP-system. The 8D-report is one of many possible tools to analyze and search for the reason of the complaints and only one type of a Supplier Corrective Action Response (SCAR) [6]. In this case a statement of the supplier is required. If an 8D-report is not necessary, the quality notification in the SAP-system gets marked as a Non-Conformity Report (NCR) and an 8D-report will not be sent to the supplier [6]. In this case, a statement of the supplier is not required.

For the further process of the complaint handling and particularly of the supplier evaluation, an extension of the Supplier Evaluation Tool has been developed, because this tool does not consider SCFCs in the calculation of the quality performance of a supplier. The extension is done by the development of the tool Supplier Quality Performance Evaluation (SQPE) based on Microsoft-Excel. The quality notifications including the SCFCs can be exported from the SAP-system in form of a Microsoft-Excel list to the tool SQPE, in which the quality performance of the viewed supplier can be calculated in percentage. Equation (1) describes, how the quality of a supplier can be calculated in dependence of the amount of SCFCs. 0% is the worst, 100% the best achievable quality. Important values for the calculation of the quality are the amount of SCFCs and the amount of all FCs of the viewed supplier of a year, which include FOAs and complaints. The quality can be calculated then by

$$Q = 100 - \frac{N_{\text{SCFCs}}}{N_{\text{FCs}}} \cdot 100. \qquad (1)$$

The calculated quality Q in % is defined by N_{SCFCs}, the quantity of SCFCs and N_{FCs}, which is the quantity of FCs of the viewed supplier.

This formula for the calculation of the quality is an easy and fast method, which can be applied for all external suppliers of the BU HCA and can also be used in the same way for the calculation of the delivery performance of a supplier. For the calculation of the quality performance of one supplier, only the list of quality notifications in the SAP-system must be filtered regarding one year and the viewed supplier. This can be done by the input of the supplier number in the SAP-system, which is individual for every supplier of the BU HCA.

Another improvement approach is to make the process of the complaint handling easier. The supplier master data are documented in the SAP-system, which includes their locations and also orderings of consumables from the BU HCA to them. Therefore, a strategy to differentiate between different production locations and the correct order addresses has been developed within this work. The developed strategy enables an efficient and more flexible administration of the supplier master data. This is an important strategy particularly for the bigger suppliers of the BU HCA, which have many different locations with different functions worldwide and their documentation is required. The reasons for the documentation of complaints in form of quality notifications and of the supplier master data in the SAP-system are the regulatory requirements. These must be considered in the process of the complaint handling in the BU HCA [3], [5].

Table 1 shows the regulatory requirements for the complaint handling and the supplier evaluation in the BU HCA.

The documentation is an efficient and flexible method to en-

Regulatory requirement	Detailed description
MDR	Medical Device Regulation
MPG	Medizinproduktegesetz
ISO 13485:2016	Medical Devices
ISO 9001:2015	Quality Management System
21 CFR 820	Quality System Regulation

Table 1: Regulatory requirements for the complaint handling and the supplier evaluation in the BU HCA

sure a correct working quality management in the BU HCA, which enables a fast backtracking and correction process of complaints, particularly of customer complaints.

3 Results and Discussion

3.1 Supplier evaluation with the developed tool SQPE

After following the recommended procedure for the handling with SCFCs based on the differentiation between FCs and SCFCs, the quality notifications of SCFCs will be registered in the SAP-system. Now a more precise evaluation of the quality of suppliers can be performed based on the quality notifications of SCFCs in the SAP-system.

The quality of an external supplier of the BU HCA, including the SCFCs from the SAP-system, is calculated with the tool SQPE based on (1). This has been proposed in this article. For the viewed supplier (supplier number: 112419) for the year 2021, the tool SQPE calculates a quality, the quality performance of 85%. Because of the result of the year 2021, the viewed supplier is one of the suppliers of the BU HCA, which has a good quality performance.

The results of this work represent an extension of the tools, which are used by the BU HCA for the supplier evaluation.

3.2 Method for the administration of supplier master data based on the SAP-system

To explain the developed strategy for the administration of supplier master data based on the SAP-system, the connections between the production locations of an external supplier (supplier number: 112419) of the BU HCA are illustrated in Fig. 2. In addition, Fig. 2 shows, how the different locations of one supplier can be correctly assigned, so that orderings to different production locations are easy to understand in the SAP-system. In this context the Corporate Supplier-number (CS-number) and the supplier number, which is referred to as creditor, are used.

The viewed supplier is one of many suppliers of the BU HCA, which has different locations in the world. It is one of the bigger suppliers of the BU HCA, which has a more complex infrastructure. This supplier is a good example to describe, how the supplier master data can be documented in the SAP-system and at the same time guaranteed, that the data are usable with a high flexibility. Moreover the supplier

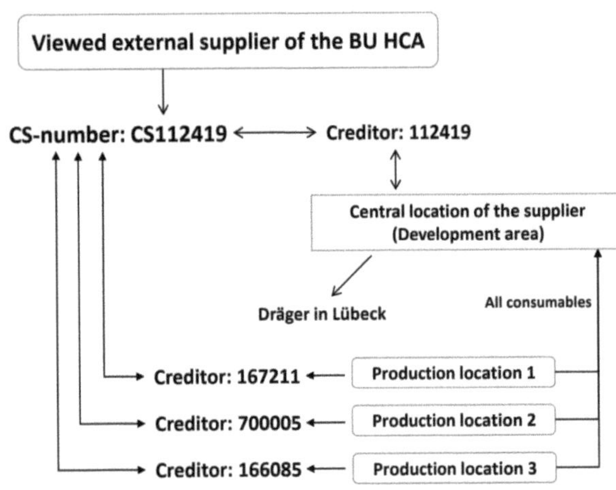

Figure 2: Developed strategy for the administration of supplier master data using the example of an external supplier (supplier number: 112419) of the BU HCA

master data can be used in a compact way, to ensure a stable and flawless complaint handling in the BU HCA.

The complete information about this supplier with all it´s production locations and development areas is coded with the CS-number. The creditor includes only the information about one specific location, for example about one production location of this supplier. Because of that, the creditor is an important number to encode the information, the address and also the orderings to a specific location of a supplier. Every supplier has different legal locations, which are characterized, for example by different names. The locations are not always production locations.

Every creditor encodes the order address of one location of a supplier. For example, one creditor can encode the headquarter and the others the other locations of this supplier. The headquarter can be, for example the central place of administration and development and the other locations can be logistic areas, for example warehouses of the supplier. Every produced consumable from the different production locations of the viewed supplier, which have different creditors, is brought to the headquarter, which is also coded by an own creditor. From there the consumables will be brought to the medical technology company. In this example the consumables will be brought to Dräger in Lübeck. Furthermore, every creditor encodes the information about the purchased goods from one location of a supplier. In addition, every different creditor of a supplier is assigned to the same CS-number of the same supplier. The CS-number includes the information about all creditors of one supplier in the SAP-system. This results in a lower complexity and a higher flexibility of the administration of supplier master data and also of the complaint handling in the BU HCA. In addition, orderings to one location of a supplier can easily be illustrated by the input of the creditor in the SAP-system. This is helpful for example in the case, that a customer complaint exists and the complete process, how the complaint has arisen, must be analyzed.

4 Conclusion

Within this work improvement approaches for a more qualitative complaint handling in the BU HCA have been developed. In addition, the developed, recommended strategy for the complaint handling includes an extension of the Supplier Evaluation Tool, which can be used by the BU HCA. In the future, the developed strategies and improvement approaches for the complaint handling will be used by the BU HCA, because the methods, which were explained, are important for Dräger as a medical technology company. The reason for that is the influence of the quality on medical products. An example for that is the recommended strategy for the handling with SCFCs and the 8D-report. It is already planned at the BU HCA, that this workflow is going to be used in the complaint handling in future.

Acknowledgement

The work has been carried out in the BU HCA at the Drägerwerk AG & Co. KGaA in Lübeck and supervised by Prof. Rostalski, Institute for Electrical Engineering in Medicine, Universität zu Lübeck.

Author's Statement

Conflict of interest: Authors state no conflict of interest.

5 References

[1] Health Sciences Authority, *Medical Device Guidance - Guidance on Complaint Handling of Medical Devices (Revision 2.2)*. Helios, Singapore, 2017.

[2] U.S. Food and Drug Administration, *CFR - Code of Federal Regulations Title 21*. Complaint Files, Silver Spring, 2021.

[3] M. Jagtap and S. Teli, *Effect of Supplier Quality Management (SQM) on performance measures of manufacturing organization*. 11th ISDSI International Conference, Trichy, India, 2017.

[4] Sparta Systems, *Turning Medical Device Complaints Into Quality Management Intelligence*. New Jersey, 2013.

[5] N. Ahmad and Z. M. Zaidi, *Total Quality Management (TQM) Practices and Operational Performance in Manufacturing Company*. University Tun Hussein Onn, Johor, Malaysia, 2020.

[6] A. Padilla, *Supplier Handbook*. Available: https://literature.thomsonlinear.com/ [last accessed on 2022-01-17]. Thomson Linear Motion Systems, 2019.

Development of a roll-out concept for intelligent substations according to line lengths

Lennart Joachim Heins [1], Gerrit Green [2] and J.-Christian Töbermann [3]

[1] Applied Information Technology, Luebeck University of Applied Sciences, lennart.heins@stud.th-luebeck.de

[2] Wissenschaftszentrum für intelligente Energienutzung (WiE), Luebeck University of Applied Sciences, gerrit.green@th-luebeck.de

[3] Wissenschaftszentrum für intelligente Energienutzung (WiE), Luebeck University of Applied Sciences, christian.toebermann@th-luebeck.de

Abstract

Intelligent substations can simplify the supply in medium voltage (MV) grids and allows to restore supply more quickly in the event of faults. The reduction of the re-supply time can be achieved by remote control of circuit breakers in the intelligent substations. Since converting a substation to a intelligent substation is cost-intensive, not all substations in the grid can be converted to an intelligent substation instantaneously. Ensure a high re-supply capability of the grid at all times, a roll-out concept for the stepwise integration of intelligent substations has to be developed. With the help of the open source power systems analysis tool "pandapower" the modeling and simulation of grids is possible. The algorithm determine the longest line length without an intelligent substation to find the next substation to be converted. Furthermore, a comparison between this algorithm of the concept and an adapted algorithm with a beforehand selection of headend stations is presented.

1 Introduction

Due to requirements of the regulatory authorities and increasing complexity due to decentralized power generation and sector coupling, there is a demand for optimization in medium-voltage grids. This optimization can be implemented by means of intelligent substations, taking into account the first paragraph of the Energy Industry Act. This describes the most secure, affordable, consumer-friendly, efficient and environmentally compatible grid-based supply of electricity possible to the general public [1]. This is because, in addition to the technical implementation of a smart grid, regulatory and economic aspects must also be taken into account for its realization [2].

Smart grids detect and analyze faults in the grid independently and can restore the supply autonomously [3]. These faults are mostly ground faults, which can be detected by ground fault location procedures [4]. These procedures indicate the cable in which direction the fault is located. With several locating devices, the defective cable can thus be narrowed down to specific line sections. The faults can already be narrowed down from the control center. The topology of the grid can then be changed by remotely controlling circuit breakers in the substations. This is done on the basis of measured values from intelligent substations. It is therefore important to have as many measuring points as possible that are well distributed in the grid. Thus, in the event of a fault, each additional intelligent substation can reduce the amount

of energy not supplied [5]. Consequently, intelligent substations can increase the reliability of the grid and speed up the resupply after the fault through automated switches [6].

Before a grid becomes a complete smart grid with only intelligent substations, the existing substations must be replaced. Compared to a conventional substation, an intelligent substation consists of several additional components. These components include telecontrol technology, remotely readable measuring devices and remotely controllable load-break switches. The conversion of substations should be implemented as economically and strategically as possible. Therefore, a roll-out concept has to be generated in order to integrate intelligent substations into the existing grid step by step [7]. This can be done according to different criteria and depends on many technical, geographical and economic aspects. In grids that cover longer distances, it is particularly costly to search for faults in lines. Therefore, it is advisable to distribute the intelligent substations geographically well.

2 Material and Methods

By using intelligent substations, troubleshooting can be performed quickly. The control center can limit the fault to as few consumers as possible by remotely setting load break switches [4]. In the best case, troubleshooting can already be performed completely from the control center and thus all consumers can be supplied with electricity again. If, on the other hand, no or only a few load-break switches can be remotely controlled, the fault must be located and rectified by maintenance personnel on site, which takes a great deal of time. Since substations have been in operation for several decades, a large proportion of substations cannot be remotely controlled to date. Therefore, it is in the interest of the grid operator to realize the optimal integration of intelligent substations into the grid with given economic means. This should result in the shortest possible downtimes in the event of a fault and keep the SAIDI value (System Average Interruption Duration Index) low.

Since the conversion of all substations is not possible for economic reasons, a concept must be created according to which the grid operator replaces the local substations. It would be possible to replace the stations according to their age, but this would not lead to the best possible division of the grid area. Another possibility would be to halve the number of substations. This is shown in Figure 1. Especially in rural areas with longer lines and larger distances between substations, having fewer intelligent substations could lead to longer travel times for the maintenance personal. This would lead to a longer downtime. Therefore, splitting by line length can be a time-saving alternative. This is also shown in Figure 1. In Figure 1, the open separation point is also drawn, which creates an open ring topology. The substation where the sectioning point is located is called the headend station. The sectioning point could also be located on the right side of substation No. 5 in Figure 1, and would have the same effect on the grid. If the disconnection point were moved to substation No. 5, there would be no load shifting in the grid.

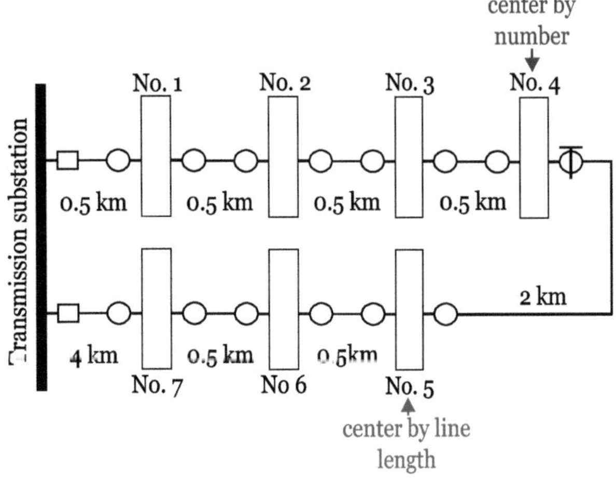

Figure 1: MV grid in an open ring operation mode with no intelligent substation. By halving the number of stations, substation No. 4 would be the first intelligent station. By considering the line length, on the other hand, station No. 5 would be the first substation to be rebuilt.

The given grid is based on a real grid. It consists of two open rings. Both rings are connected to the same Transmission substation. In addition, there are connections to other grid areas on both rings, but they are open in normal operation. In the given grid area there is only one intelligent substation, which is located at the transition to another grid area. All other substations are not remote controllable. This means that the so-called headend stations cannot be controlled remotely either. Headend stations are substations with sectioning points.

This example grid was modelled in the open source power systems analysis tool pandapower (University of Kassel and Fraunhofer Institute for Energy Economics and Energy System Technology)[8]. This tool is an easy to use grid calculation program, which is composed of the toolboxes PYPOWER and pandas. Through this a detailed model of the grid can be created. For this purpose, all necessary components of the grid were first created in the software. Afterwards, each element of the grid can be examined in a table data structure. In addition, various results, such as load flows and others can be taken from the table data structure. In the tables new properties can be added to the elements of the grid. In pandapower it is not possible by default to declare a substation or a switch as remote controllable. Therefore, this had to be stored in the tables of these elements.

Using pandapower, a query was made to determine the longest line distance between each substation and the intelligent substations. At the beginning, the intelligent stations only included the one substation and the circuit breakers at the substation. After the longest line distance has been determined, the station closest to the center of this line is identified. This substation should therefore be the first to be converted to a intelligent substation. This station is added to a list and then set as remote controllable in pandapower. This is done within a loop, which runs through each substation and saves the one with the longest line. The next loop can then be started, which again determines the longest line. Thus, the next substation to be rebuilt. The number of runs and thus the number of substations to be rebuilt can be set at the beginning of the script. The list with the new intelligent substations can be output at the end of the script.

Since the script takes into account the open switches at the headend stations, the headend stations are probably not among the first substations to be rebuilt according to the algorithm. The remote controllability of the open disconnection points should be prioritized first, since otherwise the disconnection point would always have to be closed manually in the event of a fault. This would therefore always mean a longer outage if the headend station cannot be remotely controlled. Furthermore, another intelligent substation, which is operated with closed switches, must always be integrated in the half-ring to isolate the fault. This substation could then switch off the faulty area and leave the fault-free grid area in operation.

In a second algorithm, the headend substations have been changed to intelligent substations at the beginning. Thus, even before the first search for the longest line, there are already four intelligent substations in the grid area. This is three substations more than in the original grid. These two different algorithms with the previously converted headend stations and with the original grid are compared with each other in the following.

3   Results and Discussion

In this section, the obtained results are presented and then compared and discussed. First, the results are presented and explained without the prior consideration of the head-end stations. The results after eight runs are shown in Table 1. Table 1 shows the respective substation numbers, which should be replaced in order from one to eight. This order can also be seen in Figure 2, which is represented by the numbers on the substations (blue dots). The basic structure of the grid is shown in Figure 2, but the real line lengths are not shown for protection reasons. Figure 2 shows the substation with the yellow square in the center. The other yellow squares are the connections to the other grid areas. Open switches are represented by white squares in Figure 2. The only intelligent substation to date is located in the upper left corner of Figure 2 and can be identified by the green circle around the substation. Table 1 further shows the longest line distance before the respective substation modification. Thus, without another intelligent substation, the longest continuous line distance is 9.1km long. The distribution of the intelligent substations is shown in Figure 2, where it can be seen that four of the first five substations to be replaced are located in the upper left half-ring. This can be explained by the geographical location of the substations, as they are distributed over several villages. This is not the case to the same extent in the other half rings. Furthermore, it is noticeable that in the longest line lengths, the line length is increased again after the reconstruction of the sixth substation. This can be explained by the modification of the substation in question, since no remote-controllable switch was installed beforehand and the algorithm therefore does not include the open line in the calculation. This open line is represented by the unfilled square in Figure 2 below substation No. 6. This error could not be corrected during the course of the project. However, it does not occur when the headend stations are converted prematurely. This is explained in the following.

Table 1: Ranking of the substations to be rebuilt without taking into account the head-stations.

Ranking	Station Index	longest line before
1	24	9.1 km
2	27	4.8 km
3	23	4.3 km
4	8	2.9 km
5	25	2.3 km
6	21	2.2 km
7	16	2.7 km
8	4	1.7 km

In the roll-out concept with the prior conversion of the head-end stations, the three missing headend stations must first be converted in the given grid. These are marked in Figure 3 by orange circles around the substation. Then, the longest line length in the grid area is 7.9km, which can be taken from Table 2. Table 2 again shows the first eight substations to be rebuilt. The order for the rebuild can also be seen in Fig-

ure 3. In this figure, it is again clear that the half-ring at the top left has the longest line length, since the first four substations to be rebuilt are located in this half-ring.

Table 2: Ranking of the substations to be rebuilt with taking into account the head-stations.

Ranking	Station Index	longest line before
1	25	7.9 km
2	24	4.4 km
3	27	3.5 km
4	23	3.1 km
5	16	2.7 km
6	6	2.2 km
7	10	1.9 km
8	4	1.6 km

When comparing the two roll-out concepts from Figures 2 and 3, it is noticeable that the same substations are often selected for conversion. In the left-hand part of the grid, the same substations should be rebuilt, only the order changes slightly. In the right part of the grid, the potential substations are shifted to the neighboring station or an additional substation is to be rebuilt. Due to the early Reconstruction of the headend stations in the second rollout concept, not only the first eight substations but the first eleven remote controllable substations are shown in Figure 3 after the concept.

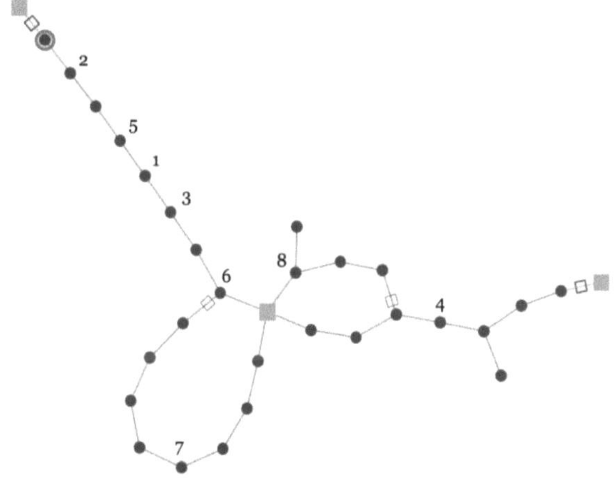

Figure 2: Grid with four outgoing lines from the transmission substation (yellow square in the center). White squares are open switches and the two yellow squares at the left and right are a external grid. The local grid stations in blue cannot be switched remotely. Local grid stations with an additional green circle can already be switched remotely. The numbers on the local grid stations are the order in which the respective station should be replaced.

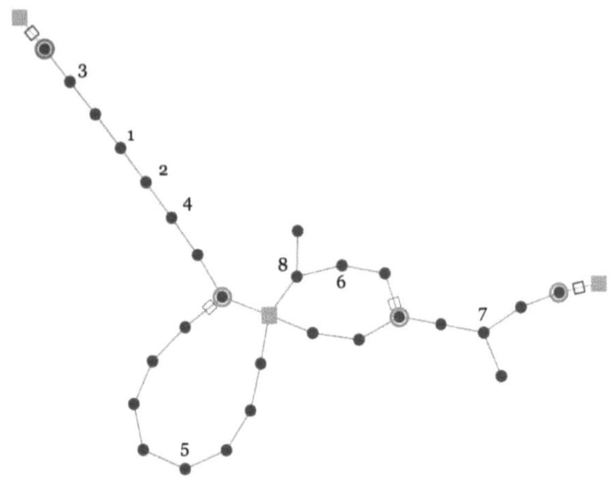

Figure 3: Same grid as in figure 2. Here, the orange circles mark in addition the headend stations, which are replaced first. The numbers on the local grid stations are the order in which the respective station should be replaced.

4   Conclusion and Outlook

In summary, it can thus be said that if intelligent substations are already integrated in the grid area and in the half rings, the next intelligent substation can be determined quickly and easily according to the criterion of line length.

Thus, the substation in the lower left area of the grid in figure 2 and 3 should not be rebuilt only at the fifth or seventh position. This circumstance is not taken into account in the algorithm. This must be considered by the user. This circumstance thus indicates that the algorithm is not yet mature. Furthermore, the division according to other criteria such as the number of substations cannot be made. The division according to line length is probably only recommended for longer line lengths and longer travel times of the maintenance personal. In addition, no other criteria of the grid are taken into account, such as the age of the substation, the cost of rebuilding this particular substation, planned grid conversions or inaccessible substations. This has to be further investigated and implemented in the further course of the project. The next step in this process is to implement a concept according to the number of non-intelligent substations, which is already shown in Figure 1. Subsequently, the age of the substation should be taken into account in the algorithm, as it makes more economic sense for some stations than for others. By implementing the other criteria, the use of heuristic optimization methods seems reasonable. The use of heuristic methods is also recommended for larger grids. So far, all possible substations are investigated and checked this will take more computing time for larger grids. However, the computation time is not too critical for this type of optimization.

Acknowledgement

This thesis was written as part of the Scientific Project module of the Applied Information Technology master's degree program. This work was carried out at the Wissenschaftszentrum für intelligente Energienutzung (WiE), Lübeck, and supervised by the Department of Electrical Engineering and Computer Science at the University of Applied Sciences Lübeck.

Author's Statement

Conflict of interest: Authors state no conflict of interest.

5   References

[1] Bundesministerium der Justiz *Gesetz über die Elektrizitäts- und Gasversorgung (Energiewirtschaftsgesetz - EnWG) § 1 Zweck und Ziele des Gesetzes.* Berlin, 2005.

[2] C. Etezadzadeh, *Smart City – Made in Germany.* Springer Vieweg, Stuttgart, 2020.

[3] J. Scheffler, *Verteilnetze auf dem Weg zum Flächenkraftwerk.* Springer Vieweg, Merseburg, 2016.

[4] C. Tengg, K. Schoaß, R Scharanz, M. Marketz and G. Druml, *Practical evaluation of new directional earthfault detection methods.* ETG-Fachbericht 129, Erfurt, 2011.

[5] T. Werth, *Investitionsstrategien für Mittelspannungskabel - Zuverlässigkeit und Wirtschaftlichkeit von Investitionen und Netzautomatisierung.* Springer Vieweg, Hungen, 2014.

[6] C. Czajkowski, T. Kneiske, D. Lohmeier, C. Spalthoff, L. Thurner and J. Kupka, *ANaPlan - Automatisierte Netzausbauplanung im Verteilnetz.* Frauenhofer IEE, Kassel, 2019.

[7] D. Heuberger, *Roadmap for Smart Grids: Four Steps to an intelligent electrical distribution grid.* International ETG Congress, Bonn, 2015.

[8] L. Thurner, A. Scheidler, F. Schäfer et al, *pandapower - an Open Source Python Tool for Convenient Modeling, Analysis and Optimization of Electric Power Systems,* IEEE Transactions on Power Systems, pp. 6510-6521, 2018.

Application of the Drone Development Platform RflySim

Phillip Overlöper [1], Carlos Castelar [2], Georg Schildbach [2]

[1] Robotics and Autonomous Systems, Universität zu Lübeck, phillip.overloeper@student.uni-luebeck.de
[2] Institut für Medizinische Elektrotechnik, Universität zu Lübeck, {carlos.castelar,georg.schildbach}@uni-luebeck.de

Abstract

This project deals with the application of a development platform (toolchain) for drones which is based upon the PX4 hardware and software standard and MATLAB/Simulink. The toolchain used is called *RflySim* and developed and published by BUAA Reliable Flight Control Group. It enables the developer to simplify the design and programming of low level controllers and high level applications in Matlab/Simulink and the subsequent software-in-the-loop and hardware-in-the-loop simulation and testing. The toolchain is demonstrated by using the example of the testing of a hold position controller.

1 Introduction

The use and technology of unmanned aerial vehicles (UAV) has been growing rapidly in recent years giving more importance to the idea of a proper development process of the software of these UAVs. For example, a simple controller design consists of three distinct phases. Firstly, the software-in-the-loop (SIL) phase, in which the controller is tested independently of the UAV hardware in a simulation. Secondly, the hardware-in-the-loop (HIL) phase, in which the controller is tested on the actual hardware of the controller, but still within a simulation environment. And lastly, the testing of the controller during the flight of the UAV.

While each one of these steps is self-contained, it can be an arduous task to switch between the different steps, since they involve different simulation programs or programming languages.

The *RflySim* development toolchain [4][5], developed and published by BUAA Reliable Flight Control Group, promises to standardise and simplify tasks like the design of a controller. It is based upon MATLAB/Simulink [6] and its PX4PS package [3] aswell as the open-source UAV hardware standard PX4 [2]. It enables the user to program and to test (SIL and HIL) without needing to worry about switching between the different phases. In the following this toolchain shall be illustrated using the example of the testing of a hold position controller.

2 RflySim Architecture

The entire process of developing and testing a controller with the *RflySim* toolchain can be seen in Fig. 1. The application (e.g., a PID controller) is designed/programmed in Simulink. The *RflySim* toolchain provides a simulation

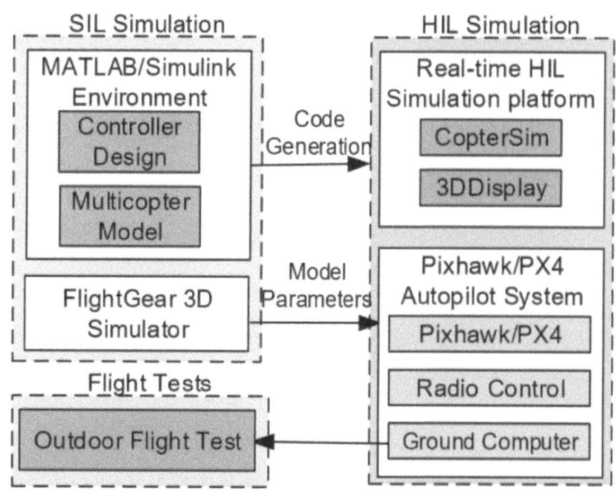

Figure 1: The architecture of *RflySim*, taken from [5]. The design and SIL test of the controller happens in Simulink in combination with the *FlightGear 3D* simulation program. A compiler generates code and loads it onto the Pixhawk hardware. The HIL test is performed with the simulation software *3DDisplay* as well as the control software *Copter-Sim*. Afterwards the controller is tested in flight.

environment to conduct the SIL test (in Fig. 1 under "SIL Simulation"). Once the application is tested successfully, it is loaded onto the Pixhawk hardware. PX4 provides an open-source hardware standard for drone development. The specific hardware used in this internship is the Pixhawk 4 mini developed in collaboration with Holybro and Auste-rion [1].

Since the PX4 firmware is programmed in C/C++, the *Rfly-Sim* toolchain includes a compiler that automatically generates C/C++ code from a Simulink model. So after the controller has been successfully tested in the SIL phase, the

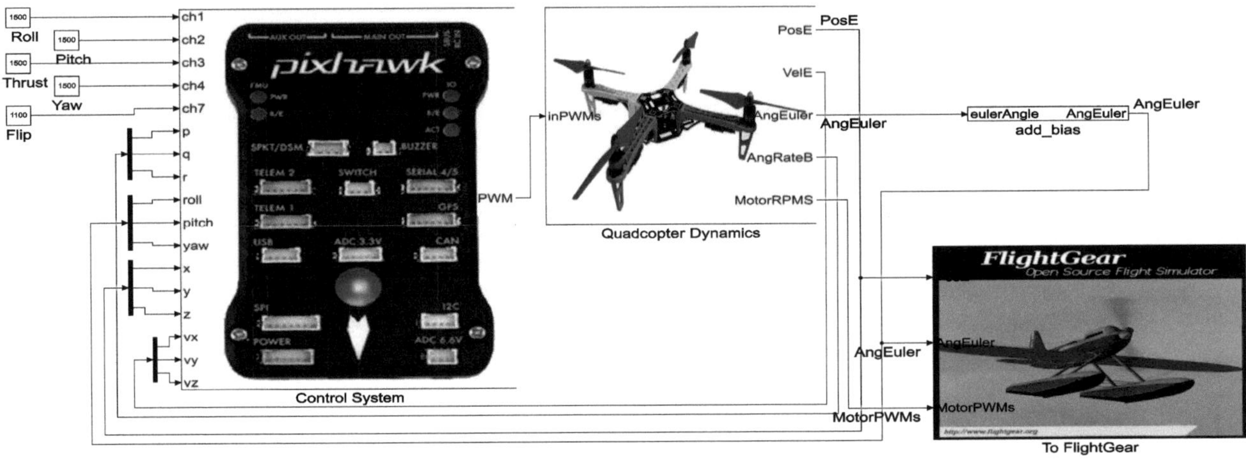

Figure 2: The Simulink model of the setup in the SIL phase. Since no hardware is involved, the motor controls from the controller are sent to a drone model which then computes the changes of the velocity, position etc. The simulation "Flight Gear" provides visual feedback.

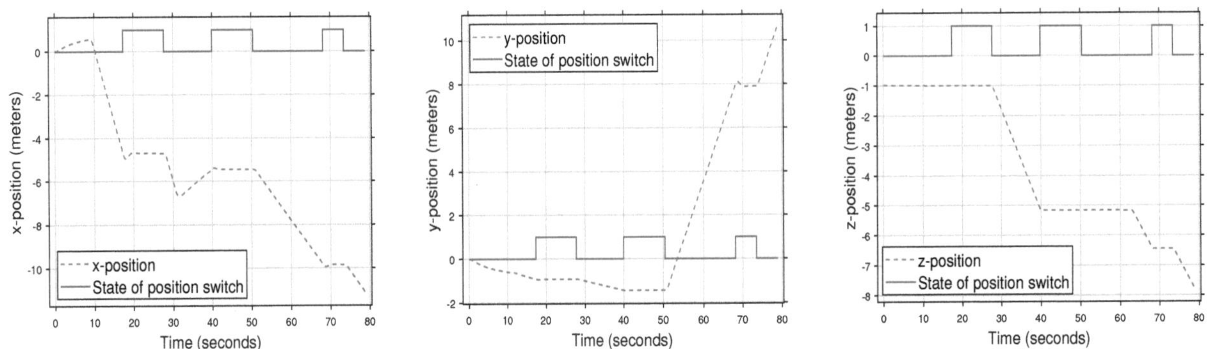

Figure 3: Three plots of the position of the simulated drone over time (dashed lines during the SIL phase). From left to right: x-position, y-position and the z-position. The solid line indicates the state of the hold position switch ("1" if activated).

model is built and uploaded onto the Pixhawk. When the upload is completed, HIL testing is performed with a simulation environment provided by *RflySim*. During this phase, *RflySim* also blocks the actuator outputs for the actual drone to which the hardware is connected (in Fig. 1 under "HIL Simulation").

After a satisfactory completion of the HIL test, the last step consists of testing the application on the drone in actual flight. For that the application is again uploaded to the hardware, but without blocking the actuator outputs this time.

3   Experiment and Results

In the following, this entire toolchain shall be illustrated with an example. A position controller is tested during the SIL and HIL phases. The position controller used here is based on a simple flight stabilisation PID controller and expands it with a position module that is activated once a switch on the remote control is flipped. In normal flight, the manual inputs from the remote control are forwarded directly to the PID controller, however, once a specific switch

on the remote control is flipped, the position module steps in. It computes the roll and pitch angles, as well as the thrust, which are necessary to hold the current position (or fly back to it if required). These are then forwarded to the PID controller. As long as the switch is flipped, all other manual input from the remote control is overwritten.

3.1   Software in the loop

In Fig. 2 an example of how the SIL simulation is conducted in Simulink is shown. On the left the controller is depicted (it is masked by a Pixhawk 1, however, it does not influence which Pixhawk versions are compatible or not). Its inputs are (from top to bottom): the simulated input from the remote control (can be modified during simulation), the current angular velocities, the current rotation expressed in Euler angles, the current position and the current linear velocities. Based on these inputs, the controller forwards the appropriate motor controls to the model of the drone (in the middle). The parameters of the model, e.g., weight or rotations per minute (RPM), can be freely configured to match those of the drone used later for real flight. Once the data

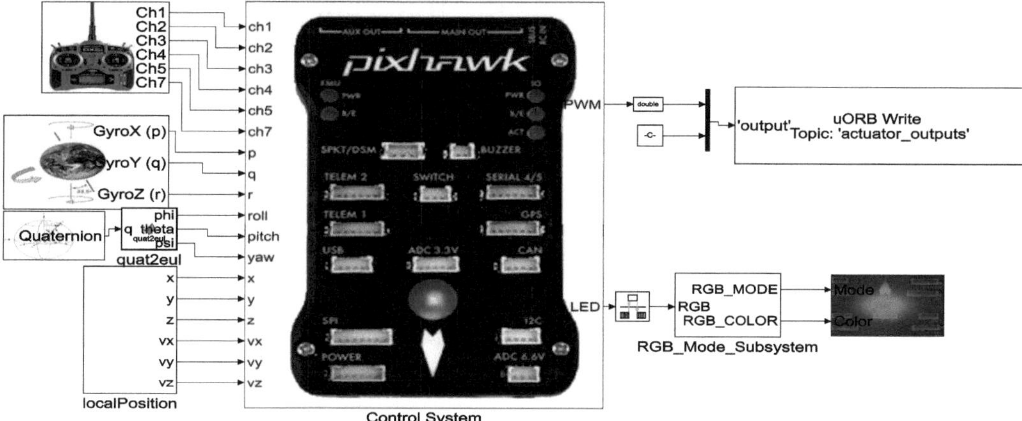

Figure 4: The Simulink model of the setup in the HIL phase. The controller is connected to read and write modules which read data directly from the Pixhawk hardware and send them back to it. The simulation is done outside of Simulink once the controller has been uploaded to the Pixhawk hardware.

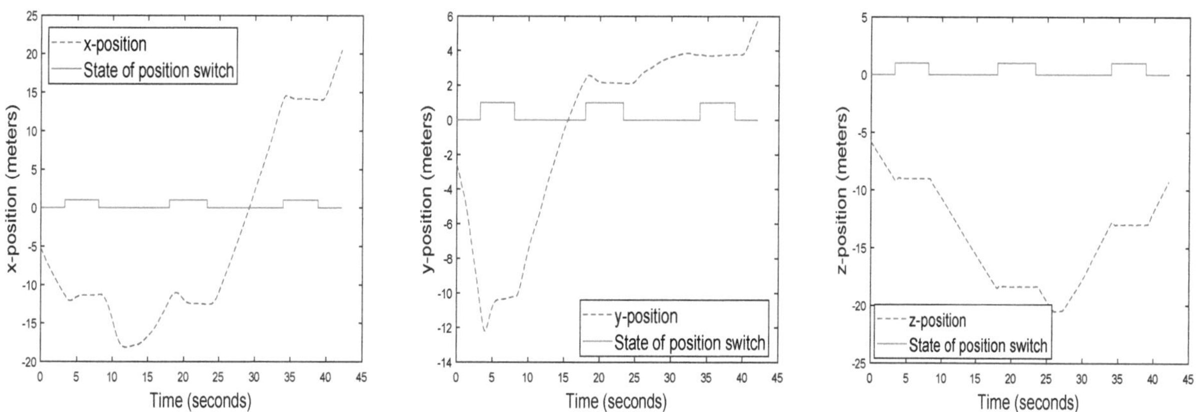

Figure 5: Three plots of the position of the simulated drone over time (dashed lines) during the HIL phase. From left to right: the x-position, y- position and z-position. The solid line indicates the state of the hold position switch ("1" if activated).

are received, the model then computes the change in position, velocity and orientation and loops these data back to the controller and the simulation program *Flight Gear* (on the right). This setup makes it possible to control the drone in the simulation and to test whether the designed controller works as intended.

Fig 3 shows the three plots of the drone's x-, y- and z-positions within the simulation over time (dashed line) as well as the state of the position hold switch (solid line, "1" if activated). It is observable that the drone holds its current position in all three dimensions, whenever the position hold switch is flipped. It can even be observed that, due to high velocities, the drone overshoots at some instances but flies back to the position that was active when the switch was flipped. It can be concluded that the designed controller works, i.e., the SIL phase has been successfully concluded and it is possible to commence the HIL phase.

3.2 Hardware in the loop

In Fig. 4 one can see an example of how HIL simulation is conducted in Simulink. In the middle, there is the controller. Connected to it are several *read* and *write* modules. These modules can receive from or send data to the Pixhawk hardware, such as the current velocity or the commands for the motors. On the left, there are the *read* modules (from top to bottom) for the remote control input, the current angular velocities, the current rotation (received in quaternions but converted to Euler angles) and the current position and rotation. All of these data are forwarded to the controller. The controller computes the appropriate response and sends it to the *write* modules on the right. In this case, they are the status LED, e.g., for signaling whether the drone is armed or whether the hold position switch has been flipped. Since it is still the HIL phase, the motor commands are not sent directly to the actual motors but rather to the simulation software. Once this configuration is completed, the Simulink model is built and uploaded to the Pixhawk hardware. The

HIL simulation is now performed on the Pixhawk hardware with a real remote control.

Fig. 5 shows the three plots of the x-, y- and z-position of the drone within the simulation over time (dashed line) as well as the state of the hold position switch (solid line, "1" if activated). As can be seen in each plot, the drone manages to hold its current position in all three dimensions, once the switch is flipped. It can also be observed that the drone tends to overshoot due to high velocities, but then flies back to the position the drone had, when the switch was flipped. It can be concluded that the designed controller works, i.e., the HIL phase has also been successfully concluded and it is now possible to test the controller in real flight.

3.3 Real flight

When the HIL phase has been successfully completed, the designed controller would be tested on the real hardware in flight under real conditions. The only difference to the Simulink model of the HIL phase (Fig. 4) is that this time the motor commands are sent directly to the motors and not to the simulation.

4 Conclusion

In conclusion, it can be said that the *RflySim* development platform applied in this paper greatly facilitates the programming of software for drones and the subsequent testing of it in different stages. It is not necessary to know the underlying structure of the PX4 hardware and how to program in the C/C++ languages (in which the Pixhawk firmware is written). Instead "programming" is conducted in the graphical environment Simulink, making it easy even for beginners to design applications for drones. However, the real simplification lies in the transitions between the different testing stages. The converting for and simulation during the SIL and HIL phases is taken over by *RflySim*, leaving the user with more time to concentrate on the designed application itself. A next step for the presented example could be to try out the designed controller during real flight. If that works, it could be made more robust, for example by adding noise into the simulation.

Acknowledgement

The work has been carried out and supervised by Prof. Dr. Georg Schildbach and Carlos Castelar, Institute for Electrical Engineering in Medicine, Universität zu Lübeck. Special thanks go to Jonas Gruner.

Author's Statement

Conflict of interest: Author states no conflict of interest.

5 References

[1] Pixhawk/PX4, *Pixhawk 4 mini*. Available: https://docs.px4.io/master/en/flight_controller/pixhawk4_mini.html [last accessed on 2022-01-16].

[2] Pixhawk/PX4, *Pixhawk*. Available: https://pixhawk.org/ [last accessed on 2022-01-16].

[3] Mathworks Pilot Engineering Group, *Pixhawk Pilot Support Package User Guide*. Available: https://ww2.mathworks.cn/content/dam/mathworks/mathworks-dot-com/hardware-support/files/Simulink_Pixhawk_Support_v2.0.pdf [last accessed on 2022-01-16], 2016

[4] Dai, Xunhua and Ke, Chenxu and Quan, Quan and Cai, Kai-Yuan, *RFlySim: Automatic test platform for UAV autopilot systems with FPGA-based hardware-in-the-loop simulations*, Aerospace Science and Technology, Elsevier, 2021

[5] Wang, Shuai and Dai, Xunhua and Ke, Chenxu and Quan, Quan, *RflySim: A Rapid Multicopter Development Platform for Education and Research Based on Pixhawk and MATLAB*, 2021 International Conference on Unmanned Aircraft Systems (ICUAS), IEEE, 2021

[6] MATLAB, *MATLAB R2021a*, The Mathworks Inc., 2021

[7] Quan Quan, Yunhua Dai, Shuai Wang, *Multicopter Design and Control Practice: A Series Experiments based on MATLAB and Pixhawk*, Springer, 2020

Early design reliability prediction of mechatronic subsystems using reliability block diagrams

Melina Beyerlein [1]
[1] Biomedical Engineering, Luebeck University of Applied Sciences, melina.beyerlein@stud.th-luebeck.de

Abstract

To ensure high product reliability, failure prediction must be applied. Especially in the healthcare sector, reliability is crucial for patient safety and to ensure lowest overall life cycle costs. This paper presents the reliability block diagram method and applies it to a medical technology product to predict the reliability of a subsystem composed of several independent components. The Weibull distribution is used to model stochastic time-to-failure phenomena of a component and to calculate the failure rate per year in percentage. The results show that, based on specific reliability data, electronic components are the most critical parts with constant failure rates. Reliability data typically describes the worst-case scenario, which does not apply to this project. This results in overestimated failure rates. Nevertheless, the method serves as an orientation for possible design changes and allows a prediction of expected down-times and spare parts based on the quality of the input data.

1 Introduction

Failures, especially in healthcare, can have severe consequences. That is why engineering products should provide a high quality over time, also called reliability. The reliability of an item is the "ability to perform a required function under given conditions for a given time interval." [1]. With the ever increasing complexity of products, the need for reliability engineering rises as well. The aim of reliability engineering is to identify causes of failures and developing techniques to reduce and prevent them. Thus, reliability techniques are defined, planed and executed to avoid unsatisfactory product reliability due to early life failures, constant product failure rates during stable operation and any failures before end of life due to wear out effects [2]. A failure occurs when one or more intended functions of a product are no longer fulfilling the customer's satisfaction. Reliability is essential for customer perceived quality, patient safety and low overall life cycle costs [3]. The lower the reliability the more interruptions will occur in the process and lower availability leads to customer inconvenience. At the same time a low reliability leads to more service calls and corrective maintenance costs which lead to higher costs for the manufacturer and the customer [2].

Reliability addresses the entire life cycle. Nevertheless, the earlier a reliability deficiency can be identified, the less costly is a design fix. An error during the design concept can be easily solved and has low impact on the costs. Whereas it costs a multiple when an error is detected during the manufacturing. That is why particularly in the development phase, the reliability analysis is primarily used for early detection and elimination of weak points [2], [3].

The word ability of the original definition of reliability can be extended by probability. Using the concept of probability distributions to describe reliability data, takes into account that failures have stochastically distributed causes. A stochastic model can be especially used for reliability prediction of a component. One of the most commonly used distributions in reliability to model time-to-failure phenomena and predict the reliability is the Weibull distribution [2]. To evaluate the reliability of a system composed of several independent components, reliability block diagrams (RBD) can be used. Furthermore, reliability assessment is based on the analysis of reliability data which presupposes the availability of reliability data of a component or system such as failure rates [1], [2].

This paper applies the method of reliability block diagrams for reliability calculations as a model to predict the system's design reliability in order to estimate the number of necessary spare parts and life cycle costs as well as early detection and elimination of weak points. Therefore, the RBD is applied to a medical technology product and its related components such that the failure rate per year in percentage can be calculated. Based on this, the parts most prone to failure can be figured out to predict the number of spare parts, maintenance and thus life time costs.

2 Methods

2.1 Reliability block diagrams (RBD)

A reliability block diagram is a logical, graphical representation of the combination of subsystems that lead to a successful functioning of the system. The goal of an RBD is to ensure the reliability of all subsystems so the overall system is reliable. Therefore, the subsystems are represented by blocks and the function flow describes the logical connections in such a way that the overall system can operate successfully [2], [4]. The blocks can represent two states, either success or failure. In case the blocks behave independently from the order of failure occurrence and each other, the blocks can be repaired or non-repaired components and probabilistic calculations can be time dependent. If this is not the case, i.e. the order of failures matters or for dynamic systems, other modelling techniques (e.g. Markov model) may be more suitable [4].

The blocks of an RBD can be in series, parallel or a combination of both. By application of probabilistic models, the reliability of each block can be calculated such that the RBD can be used to determine the overall reliability of the system [4].

In case reliability-critical subsystems exist, the failure behavior of the components is to be figured out [2].

2.2 Calculation of RBD

As mentioned in section 2.1, blocks can be in series or parallel. In case all independent blocks are required for the successful operation of the system, all blocks of the RBD will be connected in series [4]. Thus, the total reliability of a system R_s is the product of the individual reliabilities of block R_n as in (1) [2].

$$R_s = R_1 \cdot R_2 \cdot ... \cdot R_n \tag{1}$$

In general, with time dependent probabilities and n blocks this leads to (2):

$$R_s(t) = \prod_{i=1}^{n} R_i(t) \tag{2}$$

In this case, the whole system fails if one of the components fails.

If two parts exist which assure the same functionality, they are modeled as parallel blocks. As it is not applied in this project, the mathematics is no further explained.

2.3 Weibull distribution

To calculate the reliability of a component or a system at different points in time, the Weibull-Model is used as it can be adjusted to fit many life distributions. The model uses a density function with the shape parameter b. For $b < 1$ the failure rates decrease with increasing service life such that it is used to describe early failures. With a shape parameter $b = 1$ the failure rate is constant, and the density function equals the exponential distribution. This is why it is used to describe random failures. For shape parameters $b > 1$, the density function always starts at $f(t) = 0$, then reaches a maximum with increasing service life and finally drops which represents the life cycle of a wear out part [2]. The two-parameter Weibull distribution always describes failures from the point in time $t = 0$. The probability of survival R is calculated by the two-parameter Weibull distribution specified in (3).

$$R(t) = e^{-(\frac{t}{T})^b} \tag{3}$$

with T the characteristic life time . Respectively, the probability of failure F is calculated by (4).

$$F(t) = 1 - R(t) = 1 - e^{-(\frac{t}{T})^b} \tag{4}$$

The probability of failure is the number of failures at a specific time t. In other words, the reliability corresponds to the time-dependent probability $R(t)$ for no failure [5].

The failure rate λ, so the number of failures at time t compared to the sum of the units that are still intact, is calculated by (5) [2].

$$\lambda(t) = \frac{f(t)}{R(t)} = \frac{b}{T} \cdot \left(\frac{t}{T}\right)^{b-1} \tag{5}$$

It can be interpreted as the probability of a part failing, assuming that it has not failed up to this point and thus can be used to predict the number of fails in the next time interval [2]. This means, in case 100 subsystems are installed, a failure rate of 5% per year of a component means that five components need maintenance or to be exchanged.

2.4 Reliability metrics

To measure the reliability of a system, firstly reliability specifications in terms of how long ("life" in hours, years, cycles) a product should perform at a certain level should be stated as well as its intended use. For non-repairable systems the term Mean-Time-To-Failure (MTTF) describes the failure-free time of a system . All are used in case of constant (random) failure rates [2]. For random errors ($b = 1$) the MTTF is calculated by (6):

$$MTTF = \frac{1}{\lambda} = T \tag{6}$$

The lifetime B_x of a system indicates the point in time at which a specific percentage of x% of all parts have failed. I.e. a reliability target of $B_{10} = 5$ years means after 5 years 10% systems are defect [2].

3 Application of RBD to project

The aforementioned methods shall be demonstrated by applying the RBD method to an automated solution of a medical technology product which is in development. The aim of the reliability assessment is to analyse how the failure rate changes over time, to detect weak points and to predict the number of necessary spare parts and unplanned service calls.

For the calculations, the following subassemblies were identified to be relevant for the reliability: power supply, embedded control, power over ethernet (PoE) switch, programmable logic controller (PLC), smart camera, pneumatic gripper, two wear resistant tubes, energy chain gripper, energy chain z-drive, motorized x-, y- and z-axis cycles, motorized x-, y- and z-axis distance, LED pole, motor controller x-, y-, motor controller z-axis.

As all functions are necessary to ensure a correct working of the subsystem, the reliability block diagram results in a series structure. For the motorized axes, the distance and the cycles need to be considered separately as either reaching the maximum number of cycles or the maximum distance can lead to failure. Respectively, the two failure modi are represented by two independent series blocks in the RBD.

For electrical parts the Weibull shape parameter was set to be $b = 1$ as unpredictable random errors occur, such that the failure rate is constant. For parts which suffer from wear-out effects, the Weibull shape parameter was chosen to be $b = 2$. The characteristic life time of a component was determined based on the supplier data, data from the reliability plan or calculated by (6). The target product life is at least ten years. Thus, the corresponding failure rates for each component over ten years have been calculated and displayed in a matrix such that the annual failure rate of the whole system was computed as the sum. Finally, the failure rate in percentage for each component was plotted over the next ten years in one graph.

4 Results

To predict the expected number of service calls for an automated solution of a medical technology product, the failure rates of the components have been calculated by (5) in a first step. Next, the annual failure rate of the total subsystem was calculated by summing the failure rates of all subsystems. The expected annual failure rates of the subsystem for the life time of ten years are shown in Table 1.

Table 1: Total annual failure rate of the subsystem in percentage for ten years

Year	Annual failure rate in %
1	20.30
2	20.66
3	21.02
4	21.38
5	21.74
6	22.10
7	22.47
8	22.83
9	23.19
10	23.55

The total annual failure rate of the first year is 20.30%. Subsequently, the total failure rate increases linearly over the years and reaches a total annual failure rate of 23.55% in the tenth year.

The failure rate per year in percentage for each component is shown in Fig. 1. The respective graphs for the calculated points were linearly approximated. As the graphs of the three motor controllers for the three different axes were identical, they are represented by only one line; explicitly it is stated that it is not the sum of the failure rates. The same is applicable for the two tubes as well as the cycles of the motorized x- and y-axis.

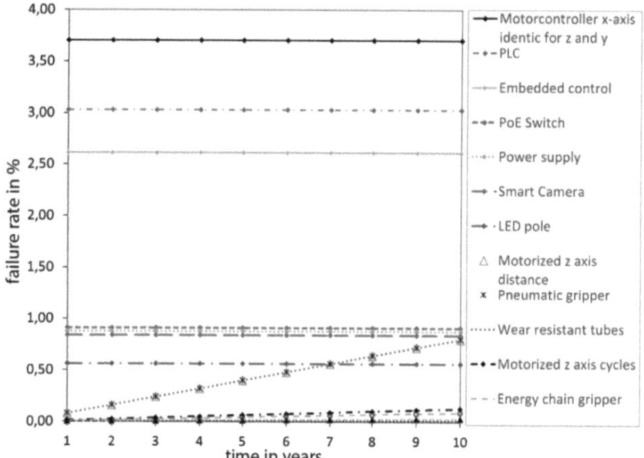

Figure 1: Failure rate in percentage for each component of the subsystem for a life time of ten years.

Most components show a constant failure rate. With 3.7% each of the motor controllers shows the highest failure rate per year. With 3.03% the PLC and with 2.61% the embedded control are following the motor controllers. All other components show a significantly lower failure rate with smaller than 1%. The smallest failure rate was calculated for the case that the motorized y-axis reaches the maximum distance within the lifetime. Within the first ten years it is approaching a failure rate of 0.01%.

5 Discussion

To predict the reliability of a medical technology subsystem, i.e. the number of maintenance and spare parts, the RBD method was applied. For the project only blocks in series exist, as there is no redundancy (which is quite typical for a lot of systems), such that the failure rates could be summed up. Blocks in series lead to the effect that if one component fails, the complete subsystem fails. By means of the resulting RBD, supplier reliability data and the Weibull distribution with different shape parameters, the component failure rates and total annual failure rate were calculated for a life time of ten years. For the first year, a failure rate of 20% was calculated. Typically, datasheet reliability values refer to high-use scenarios (e.g. 100% duty cycle) which do not necessarily apply to this project. As this project is still under development, field data is not available yet. It was not part of the scope of this work to adapt these values regarding this project. A value adjustment of the high-use scenario would increase the data quality and therefore provide a more precise interpretation of the results. The total annual failure rate increases linearly over time as the failure rates of the single components increase as well.

Attention needs to be paid when two reliability parameters may independently lead to a failure of one component. I.e for the motorized axes, the distance and the cycles needed to be considered separately as either reaching the maximum number of cycles or the maximum distance can lead to failure. Considering only one of the reliability parameters would have a negative impact on the reliability prediction. Most components show a constant failure rate which means that the probability of occurrence of a failure is the same at every point in time. The majority of components are electrical parts which show a random failure behavior while non-electrical parts fail due to wear-out. Respectively, parts as the pneumatic gripper and the tubes show a linear increasing failure rate in Fig. 1 as they suffer from wear-out. Hence, the wear-out parts become more critical over the years, resulting in increasing importance for reliability considerations, but with reaching a failure rate of 0.8% after ten years they are not reaching the failure rate of the motor controllers. In Fig. 1 it can be seen that electronic components are the most prone to failure. Especially the motor controllers are the components most prone to failure with a failure rate of 3.7%. This failure rate means for 100 installed subsystems, a failure of a motor controller can be expected every four month which leads to three failed motor controllers per year (with a duty cycle of 100%). As the motor controllers have the highest impact on the overall reliability of the subsystem, the data needs to be analyzed regarding the quality of the stated MTTF to determine whether a design change is needed to decrease the overall failure rate.

Even though the failure rates were successfully calculated by applying the RBD method, some limitations of the method need to be taken into consideration. Firstly, subsystems can only have two states, functioning or failing. Another limitation of static statistical modeling is that repairability of systems, dynamic states or load-dependent failure rates cannot be represented [2]. RBD is a simple model and for complex systems of larger size it might be prone for errors. With increasing complexity of a system, assumptions and simplifications may be made which lowers the reliability. Systems which can be modeled by series blocks have no redundancy and thus are more critical systems. Contrarily, RBDs are often used as they are easy to understand, no matter if talking to an engineer or manager. Furthermore, RBDs can be easily built as the blocks follow the functional logic of the system. This is why it is a commonly used model for prediction of failure rates as a proactive handling of faults [6].

In general, a failure refers to misbehavior that can be observed by the user [3]. If a component is built of an assembly, one subassembly inside the component may fail, which may not be visible and would not result in a failure of the overall component. Thus, it can be said, that as long as errors occur that do not have an impact on the output, there is no failure [6]. With the applied method explained in this paper, these partial failures are included within the failure rates. This is why the failure rates are overestimated for these cases. Thus, it needs to be verified, whether the failure rates stated in data sheets are applicable to the current project. Nevertheless, the calculation of failure rates can be used to estimate the number of spare parts and life time costs especially for electronic components which are not further tested.

6 Conclusion

This paper investigates the use of RBD to analyse the reliability of mechatronic subsystems. In general, series and parallel blocks exist for RBD. The component reliability is calculated by means of the Weibull distribution. Besides the probability of survival, the failure rate is an important factor to describe reliability. The RBD method was applied to a project to examine the method. The results show a successful application of RBD but the predicted number of failures are overestimated. Nevertheless, the values serve as an orientation on how much maintenance will be expected as well as detection of weak points. Thus, service calls as well as the resulting overall life cycle costs can be estimated.

Acknowledgement

The work has been carried out at a medical technology company. I want to thank the company for enabling me to realize my research internship in the R&D department. Additionally, Prof. Dr.-Ing. Stephan Klein, the dean of applied sciences of the Luebeck University of Applied Sciences, supervised my work.

Author's Statement

The author states no conflict of interest.

7 References

[1] Verein Deutscher Ingenieure, *VDI 4002-2: Zuverlässigkeitsingenieur/ Zuverlässigkeitsingenieurin Anforderungen an die Qualifizierung*. VDI-Handbuch Zuverlässigkeit, 2011.

[2] B. Bertsche and G. Lechner, *Zuverlässigkeit im Fahrzeug- und Maschinenbau: Ermittlung von Bauteil- und System-Zuverlässigkeiten*. 3., überarb. und erw. Aufl. Berlin [u.a.]: Springer (VDI), Kapitel 1-3, 2004.

[3] G. S. Wasserman, *Reliability Verification, Testing, and Analysis in Engineering Design*. Marcel Dekker, Inc., 2003.

[4] IEC, *IEC 61078 - Reliability block diagrams*. Geneva, Switzerland, 2016.

[5] P. D. T. O'Connor, *Practical Reliability Engineering*. 4th ed., John Wiley Sons, UK, 2010, pp.143-147.

[6] F. Afsharnia, "Failure Rate Analysis", *Failure Analysis and Prevention*, IntechOpen, 2017, doi: 10.5772/intechopen.71849, pp. 99-115.

Improving the Usability of a User Interface for the Creation of PCR Test Specifications

Christian Seitzer [1], Artur Piet[2], Fiona Schilling [3], and Marcin Grzegorzek [2]

[1] Medical Informatics, Universität zu Lübeck, christian.seitzer@student.uni-luebeck.de
[2] Institute of Medical Informatics, Universität zu Lübeck, {artur.piet,marcin.grzegorzek}@uni-luebeck.de
[3] EUROIMMUN Medizinische Labordiagnostika AG, f.schilling@euroimmun.de

Abstract

As one of their products the company EUROIMMUN Medizinische Labordiagnostika AG develops and distributes kits that enable laboratory technicians to perform real-time polymerase chain reaction (PCR) tests. To guide through the process of performing PCR tests, EUROIMMUN provides software detailing the steps that are necessary to perform the tests. EUROIMMUN specialists utilize an in-house tool to specify the PCR tests. The user interface of this tool should be improved to save time and avoid costly errors. In this work, several changes were made to the user interface by using a heuristic approach and fixing problems that were pointed out by the users.. These modifications were then evaluated through a survey of 6 users with varying levels of familiarity with the tool. The evaluation confirmed that the usability of the in-house tool has been improved by those changes, which is expected to result in time and cost savings in the future.

1   Introduction

The importance of polymerase chain reaction (PCR) assays has never been more evident than in the period from late 2019, when a new coronavirus, SARS-CoV-2, emerged [1]. Through the current pandemic of SARS-CoV-2, PCR tests have been very important to curb the spread of the virus, due to their higher sensitivity [2]. However, PCR tests can also help to diagnose other diseases in individual patients. The company EUROIMMUN distributes test kits with the necessary chemicals and instructions for performing real-time PCR tests for various pathogens: SARS-CoV-2, SARS-CoV-2/Influenza A/B, Zika Virus, Herpes Simplex Virus (HSV-1/-2), and Mycobacterium Tuberculosis. To support laboratories in performing these tests, EUROIMMUN offers the EURORealTime Analysis Software (ERTA) [3]. This software provides instructions for mixing the chemicals, temperature profiles for the PCR cyclers, and includes visualizations and evaluations of the test results. This allows ERTA to guide laboratory technicians through each phase of the test.

In order to enable EUROIMMUN to create the specifications for these tests, an internal tool is used: the Master Data Editor (MDE). With the MDE, EUROIMMUN staff can create test specifications for the development of new tests, and they can export these test specifications so that EUROIMMUN customers can perform the tests with ease.

While developing for this project most attention was paid to the UI for the main ERTA program, while the MDE component was neglected. Internal tools are often not developed with the same care for the UI as software products intended for customer use. Markus estimates that in the United States "bad Intranet Web Design will cost $ 50-100 billion per year in lost employee productivity in 2001" [4].

In our approach, we made changes to the MDE UI according to Jakob Nielsen's 10 general principles for interaction design [5] and our own user feedback. After implementing the changes to improve the UI, we conducted usability tests with our users and had them fill out questionnaires to evaluate our changes.

This paper is divided into Material and Methods where our approach is explained, followed by Results and Discussion and finally our Conclusion

2   Material and Methods

Our main goal was to improve the usability of the MDE. To this end, we first evaluated the ticket system in which users of the MDE already proposed some improvements.

Many of them amounted to simple bug fixes and improvements in error messages when trying to save data, that were inconsistent with the database model.

In the original MDE, many error messages were just the information intended for software developers detailing on which location in the code an error occurred.

These were very hard to understand for the user and contained a lot of unnecessary information.

With the presented changes, these hard to understand error messages were caught beforehand and replaced with explanations why the state of the data is inconsistent and how to best return to a consistent state.

Another frequently requested feature was the option to duplicate test specifications. The use case is that users are copying test specifications and changing only individual parameters of the copied test, running these updated specifications, and then checking for improvements to develop a better test. Copying a test specification involved a lot of work and was very error-prone. Creating a way to duplicate tests could simplify this process immensely.

2.1 Ease Further Software Development

During development, it became apparent that there is another stakeholder for the MDE: the software developer.

However, software development on the MDE was complex and relied on a custom framework for plugin management and UI element design. Which was created to reuse resources across EUROIMMUN but became obsolete. Usage of this framework complicated development, because tools in the development environment, like autocompletion, would not work and changing the design of UI elements needed approval from a different department.

By replacing this framework with the widely used industry standard Prism [6], we were able to reduce complexity for software developers working on this project in the future.

However, by changing the framework, part of the design was lost and had to be reimplemented.

2.2 Improved Usability for New Users

Another focus of this work was improving the usability for inexperienced users.

During first explorations it became apparent that it was very confusing for new users to operate the MDE. One of the reasons for this was that several of the parameters for the tests were shown in a table view. The rows in this Table could be expanded to show more options. But there was no difference between rows that could be expanded and rows that could not be expanded, which confused multiple new users. The old design of a row can be seen in Figure 1. To remedy this situation we marked rows that could be expanded with a downward pointing arrow which turned to an upward pointing arrow when the row was expanded as can be seen in Figure 2. By doing this, we were able to improve the consistency of the MDE and follow Jakob Nielsen's 4th rule ("Users should not have to wonder whether different words, situations, or actions mean the same thing. Follow platform and industry conventions."). In our new version, the rows with different behaviour also differ in how they look. Rows were also only selected when an element in the row was clicked and clicking on empty space did not select a row. This behaviour was also changed in our new version of the MDE. In the original version, rows were not fully expanded on selection but a second button had to be pressed to expand a row. In our new version, rows fully expand as soon as they are selected.

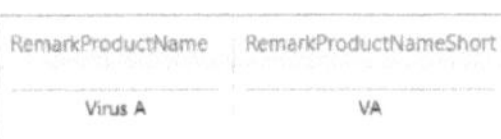

Figure 1: The old row design, without any indication, showing that this row can be expanded.

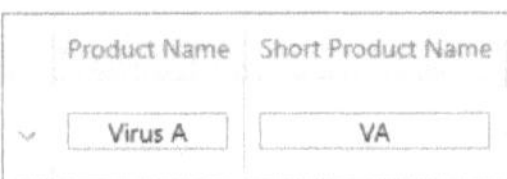

Figure 2: The new row design, showing that it can be expanded by a downward pointing arrow.

2.3 UI Evaluation

To evaluate the changes in a realistic context we recruited 6 test participants to perform certain tasks. During these tasks, we observed everything that our subjects did. To allow recruitment of subjects outside EUROIMMUN, all real data for test specifications were replaced by fabricated specifications.

2.3.1 User Testing

To compare the performance of the new user interface with the old state of the MDE, two versions of the system were created: one with the old version of the MDE before we implemented our changes and one with our changes. The tasks were performed both with the old version of the MDE and with the new version. The order of the tasks was kept the same for every participant, but some participants started on the old version of the MDE and then the new version and others started on the new version and then the old version. This was done to balance against performance improvements due to participants learning the tasks.

Participants were then asked to perform the following 6 tasks:

Task 1: Change Temperature: Change the temperature of "Temperature Profile 1" in Step 1 of the Cycle to 60°C.

Task 2: Copy Test Virus A: Copy test specification of "Virus A" and name it "Virus A-2".

Task 3: Delete Test Virus A-2: Delete test specification "Virus A-2".

Task 4: Update Cycler Version: Update supported version of "Cycler 1" to 1.2.

Task 5: Find Temperature Profile: Find out which temperature profile is selected for test specification "Bacterium C" for "Cycler 1".

Task 6: Find Volume Mix Value : What is the Volume Mix Value for "Reaction B" for the test specification "Bacterium C".

The results of the user testing are shown in Section 3.2.

2.3.2 Questionnaire

After the participants had completed these tasks, further data were collected with the help of a questionnaire. The original statements where written in German, but have been translated here. The following statements were presented to the participants, and they were then asked to rate them with the following options: Strongly Agree, Agree, Neutral, Disagree, Strongly Disagree.:

Statement 1: Error Messages: For an internal tool it is not important what error messages look like.

Statement 2: New Design: Many elements of the updated Master Data Editor have a new design.

Statement 3: Test Specification Save: The restrictions that test specifications can only be copied when data were saved is bothersome.

Statement 4: Translation: Translating the Master Data Editor into German would make work easier.

Statement 5: Easier Row Selection: Rows are easier to select in the new Master Data Editor.

Statement 6: External Tool Standards: The same standards should be applied to internal tools and software for the end user.

Statement 7: Easier Usage: The new Master Data Editor is easier to use.

Statement 8: Identifiable Row Expansion : It is easy to see which rows can be expanded in the new Master Data Editor.

Statement 9: Automatic Row Expansion: The implemented automatic expansion of rows is bothersome.

3 Results and Discussion

3.1 Questionnaire Results

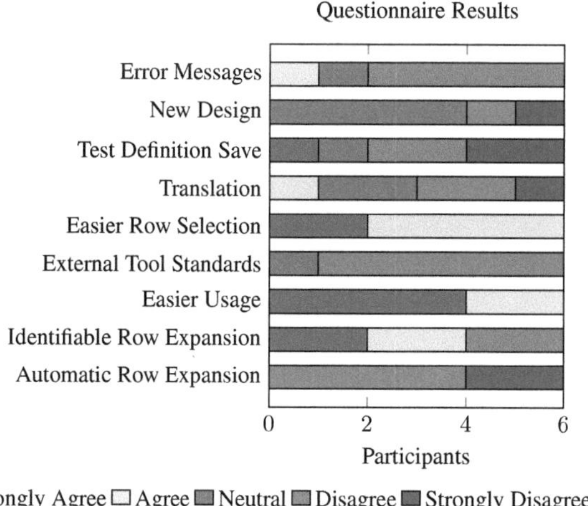

Figure 3: The results of the questionnaire. The full text of the statements can be found in Section 2.3.2 Questionnaire.

The results for our questionnaire in Fig. 3 show that it was correct to improve the error messages, since a majority of our participants disagreed that error messages are not important for internal tools.

Seeing that none of our participants agreed that the MDE has a new design it seems, that we were successful in reimplementing the design that was lost due to replacing the internal framework of the plugin manager.

Several statements of the questionnaire were aimed at finding out if our improvements of the selection and expandability of rows had the desired effect of improving the usability. All our participants agreed that rows were easier to select and that the automatic expansion of rows on selection was not obstructive. Regarding the statement, if expandable rows are easier identifiable, there is a less clear picture. Two of our participants disagreed with the statement that expandable rows are identifiable. Making expandable rows more identifiable by giving them a different design could be another change for the future to improve usability even further.

All except one of our participants disagreed with the statement, that internal tools should have the same standards applied as external tools, created for the customer. There seems to be an acceptance that highly specialised internal tools may have a UI that is less polished than UI intended for a public audience.

All users agreed or strongly agreed that the MDE is easier to use in the new version, which is a solid indicator that our approach is successful.

The answers to our questionnaire indicates that our changes improved usability. To ensure that this was not just subjective bias and that our changes actually transform into time savings, we also checked how long participants needed for certain tasks. The results for these tests can be found in the Section 3.2.

3.2 User Testing Results

As can be seen in Fig. 4, the number of clicks for the new version is less than the number of clicks needed in the old version across all tasks. This could be an indicator for better

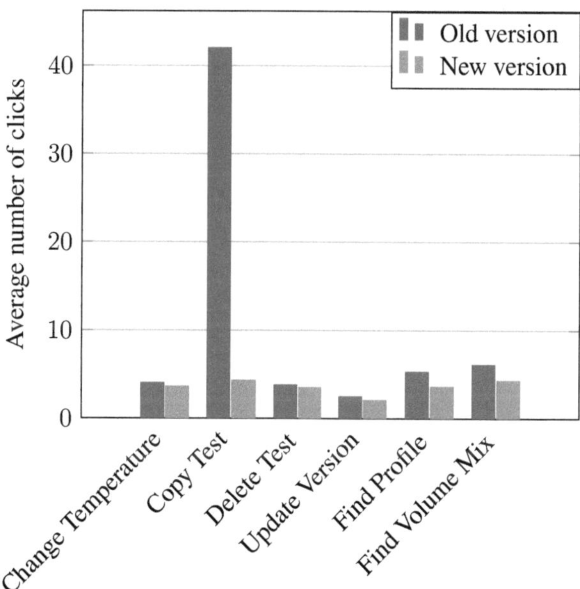

Figure 4: Results of the UI evaluation for counting clicks needed to finish the task. Descriptions for the tasks can be found in 2.3.1 User Testing

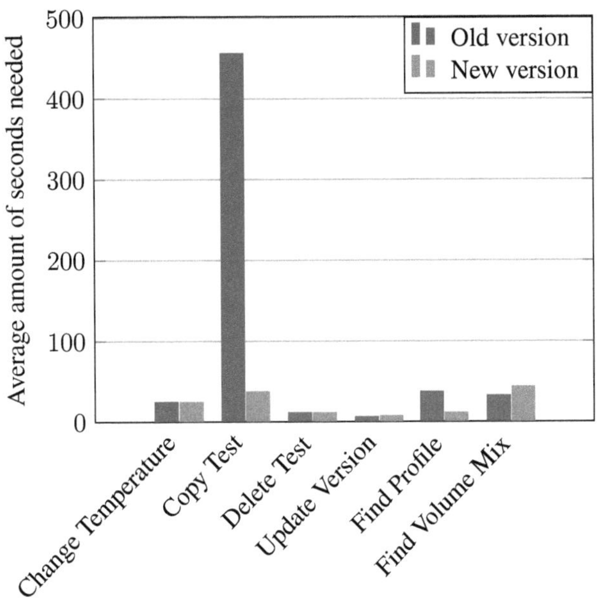

Figure 5: Results of the UI evaluation for measuring time needed to finish the task. Descriptions for the tasks can be found in 2.3.1 User Testing

usability. But to ensure that this is actually the case, we also measured the time needed for these tasks.

As can be seen in Fig. 5, the time needed for our specified tasks decreased or stayed the same for the new version. This difference can especially been seen in the task "Copy Test". Participants also had to be corrected multiple times when copying a test manually, because several parameters were hidden deeper in the UI structure. So replacing the manual steps needed for copying a test with the press of a button should lead to a reduction of mistakes as well as a reduction of time needed.

Unexpectedly participants performed worse for the new version in the task "Find Volume Mix". Possible reasons could be that experienced users were confused by the fabricated specifications and the different behaviour of the UI, that did not correspond to their learned experience and led to slower performance with the new version. It should be noted, that every change in UI behaviour can lead to users being confused and having to relearn how to do certain tasks. This could also explain our results in the "Find volume mix" test case. Another possible reason may be that we tested only a small number of subjects, but due to time restraints and not wanting to keep too many of the employees of EUROIM-MUN from their work, it would have been difficult to acquire more test participants.

Apart from this one case our collected data show that our changes are moving the usability of the MDE in the right direction.

4   Conclusion

After evaluating the results of our questionnaire and the time needed to perform our use cases we come to the con-clusion that our approach was successful. Especially the introduction of a button to duplicate test specifications will save users of the MDE a lot of time in creating new tests. This enables test specification developers to create new PCR tests in a faster manner and reduce time-to-market for new products, as well as create updates to existing products more quickly. Of course, there is still a lot that can be improved for the MDE in the future to achieve further gains.

Acknowledgement

The work has been carried out at EUROIMMUN Medizinische Labordiagnostika AG and supervised by the Universität zu Lübeck.

Author's Statement

The authors report no conflicts of interest. The authors alone are responsible for the content and writing of the paper.

References

[1] P. Zhou, X.-L. Yang, X.-G. Wang, B. Hu, L. Zhang, W. Zhang, H.-R. Si, Y. Zhu, B. Li, C.-L. Huang *et al.*, "A pneumonia outbreak associated with a new coronavirus of probable bat origin," *nature*, vol. 579, no. 7798, pp. 270–273, 2020.

[2] J. Lee, J.-U. Song, and S. R. Shim, "Comparing the diagnostic accuracy of rapid antigen detection tests to real time polymerase chain reaction in the diagnosis of sars-cov-2 infection: A systematic review and meta-analysis," *Journal of Clinical Virology*, p. 104985, 2021.

[3] EUROIMMUN, "Eurorealtime analysis product page," 2021, accessed 29-November-2021. [Online]. Available: https://www.euroimmun.com/products/ automation/software-solutions/eurorealtime-analysis/

[4] A. Marcus, "2 chapter - user interface design's return on investment: Examples and statistics," in *Cost-Justifying Usability (Second Edition)*, second edition ed., ser. Interactive Technologies, R. G. Bias and D. J. Mayhew, Eds.   San Francisco: Morgan Kaufmann, 2005, pp. 17–39. [Online]. Available: https://www.sciencedirect. com/science/article/pii/B978012095811550002X

[5] J. Nielsen, "Ten usability heuristics," 2005. [Online]. Available: https://pdfs.semanticscholar.org/5f03/ b251093aee730ab9772db2e1a8a7eb8522cb.pdf

[6] The Prism Team, "Prism library," 2021, accessed 09-February-2022. [Online]. Available: https: //prismlibrary.com

Development and Documentation of a Use Case for intra-hospital Transport of a ventilated Patient

Larissa Hamann[1], Uwe Becker[2], and Maria Henke[3]

[1] Medical Engineering Science, Universität zu Lübeck, larissa.hamann@student.uni-luebeck.de
[2] Drägerwerk AG Co. KGaA, Lübeck, uwe.becker@draeger.com
[3] Institute for Robotics and Cognitive Systems, Universität zu Lübeck, henke@rob.uni-luebeck.de

Abstract

The motivation of this paper is, to present the development of a use case within the framework of requirements engineering based on the SysML modelling language for a ventilator for intra-hospital transport. SysML allows documentation of stakeholder requirements in a structured way. The requirements were identified in the determination step through a literature research and an exchange with experts. These are subsequently implemented into the SysML program MagicDraw, which is the documentation step. It is shown that use case modelling leads to a structured graphical representation with all requirements. Through this, a precise model of the process can be created. At a later development phase of the ventilator, the requirements must be reviewed and agreed. Further it is necessary to keep the requirements up to date, which is the managing step. As the project is in preliminary development, this paper only discusses the determination and documentation of the use case.

1   Introduction

For the development of a new product, requirements and system engineering are used in order to be able to determine the exact requirements of various stakeholders with the help of use cases.

A use case describes a typical interaction of a user with a system or a product that leads to an event desired by the user [1]. The creation of use cases is an important task in research and development, because they answer both, technical and regulatory requirements. The regulatory requirements are described in the General Safety and Performance Requirements [2] and with regards to Usability Engineering specified more detailed in IEC- 62366-1. The latter answers essential questions about the acceptability of the residual risk in terms of usability[3].

In addition, step-by-step modelling can also reveal problems as well as possible customer requirements.

According to IEC-62366, the use specification contains [3]:

- intended medical indication,

- intended patient population,

- intended part of the body or type of tissue applied to or interacted with,

- intended USER PROFILE,

- USE ENVIRONMENT,

- operating principle.

There are four main activities of requirements engineering [4], which includes the use case creation:

- determining,

- documenting,

- checking and agreeing and

- managing requirements.

In this paper, the first two steps are covered.

The use case considered in this paper is the intraclinical transport of ventilated patients.

2   Material and Methods

Requirements engineering is almost indispensable for meeting customer requirements and budget and time schedules at the same time. To achieve this, requirements engineering documents customer requirements as completely as possible in good quality to identify and correct errors at an early stage. The four main activities of requirements engineering as well as the involvement of the primary stakeholders are the basis of the system development process [4]. Stakeholder and operator have, as all components of the product, safe operating conditions, expected error rates and predictable performance. The range for associated metrics is variable. Nevertheless, as with any other component of a large system, analysing the capabilities of the actor helps to create reliable and effective systems. This reduces risks from human-machine system errors.

This is part of user-centered design which shapes the interface according to the capabilities and needs of the operators. Rather than displaying information that focuses on the technologies, user-centred design integrates this information to meet the goals, tasks and needs of the users. The goal is optimal functioning of the entire human-machine system [5]. First, the system context is defined, which has an influence on the system to be developed and thus determines requirements for it [4]. In this context, intra-clinical transport is described as follows: *The ventilated patient must be transported in bed within the hospital with all therapy and monitoring procedures because of a medical examination. The standard of care in the intensive care unit (ICU) must be guaranteed during transport (basic care and monitoring without interruption). Preparation, transport route, length of stay at the destination and postprocessing must be considered at patient transport [6].*

2.1 Determination

With this delimitation, the first step — determination — can be started. In summary, there are three sources of requirements used: Documents (literature, norms/standards), stakeholders (users, system operators, developers, customers) and systems in operation [4].

This determination step took place through a small literature research. During this research, suitable sources on intraclinical transport were sought. The database of PubMed was used to find clinical trials on the topic, as well as the Arbeitsgemeinschaft der Wissenschaftlichen Medizinischen Fachgesellschaften (AWMF) guidelines search to include guidelines from the Deutsche Interdisziplinäre Vereinigung für Intensiv- und Notfallmedizin (DIVI) in the requirements. In addition, further information was obtained through an exchange of experts with various departments. This was done through interviews with employees, who know customer requirements for example through contact with various stakeholders, their own professional experience or from previous projects of the company.

2.2 Documentation

With this information the second step — documentation — is started. In requirements engineering, the various information must be documented in order to appropriately capture the requirements for the system to be developed [4]. First, the information is structured in a table as seen in Table 1. This is subsequently transferred in a system use case diagram, which is created with the program MagicDraw, based on the System Modeling Language (SysML) [7], which is a tool of model based system engineering (MBSE).

MBSE is a subtype of systems engineering that uses models instead of documents for representation. MBSE is the formalized application of modeling to support representing system requirements, analysis, and design. Furthermore it supports verification and validation tasks that begin in the conceptual design phase. It is also used in development and later lifecycle phases [8].

Thus, the task is to implement the requirements gathered about intra-hospital transport in a structured way in MagicDraw. Starting point is creating the use case diagrams (Fig. 1). Activity diagrams can be added to these one level deeper. These represent the sequence's steps exactly in the form of a flow chart (Fig. 2).

SysML activity diagrams consider the flow of control between activities or actions of the system. In a sequential order, a subsequent activity is executed when the previous action/activity terminates. Activity diagrams have two activity nodes with defined semantics, the initial node and the final node, which represent the beginnig of execution and termination. The representation of alternative control flows is enabled by decision nodes. Conditions for choosing an alternative control flow are annotated [4]. Furthermore, there are synchronization bars that allow to display concurrent or parallel control flows. The object flow is a special form of control flow. It expresses in contrast to the control flow not only a pure control dependency, but an additional data dependency [9]. Also to be explained is the accept event that waits for a certain event to occur. When this event occurs, the action is not finished, but waits for further events [10].

3 Results and Discussion

This section presents the results obtained. The procedure and thus the requirements for intra-hospital transport are explained. This results in the structure of the use case.

Table 1 shows the results on the respective points that need to be adressed when identifying a use case.

As it can be seen, the essential steps are divided into transport preparation, transport and transport follow-up. The intention of the system environment and the reaction of the system were described for each step, but cannot be listed here due to lack of space. There are 27 steps in total. A more specific example will be considered. This concerns the step "Announcement at treating station", which is part of the preparation for transport. The reaction of the system is the confirmation that the transport can take place when there is enough capacity for examination.

In terms of time behavior, it should also be noted that transport preparations begin 30 - 45 minutes before the planned start of the examination. The duration of the transport must be taken into account. This is on average one hour. The transport time is to be minimized for patient safety. There must also be a reserve capacity of 30 minutes on the ventilator. Various transport ventilators therefore have a battery life of four to six hours. With this information, the use case diagram (Fig. 1) and activity diagram (Fig. 2) can be created.

The use case diagram contains various use cases, with ventilation itself at the center. As can be seen in Fig. 1, the use cases are shown as ovals that contain the name. Intra-hospital transport is linked to this by means of extend. This means that the sequence of interactions from intra-hospital transport contains the sequence of interactions of ventilation.

Table 1: Use case table showing the results for intra-hospital transport. (Template according [1])

Name	Intra-hospital transport of a ventilated patient		
Actor	ICU nurse, doctor		
Trigger	Medical examination, operation, transfer to different supply level		
Brief description	Ventilated patient must be transported within the hospital with standard care of ICU guaranteed		
Precondition	Stable vital parameters, benefit > risk		
Essential steps	**Intention of the system environment**		**System response**
	Transport preparation		
	...		...
	Transport		
	...		...
	Transport follow-up		
	...		...
Exceptional cases	Seriously injured, severe oxygenation disorder, other therapies, perhaps MRI		
Postcondition	Check patients vital parameters		
Time behavior	Preparation, route (outward and return), length of stay, follow-up		
Availability	Battery back-up: 4-6 h		
Questions, comments	- The entire monitoring is considered to show possible lack of space - Special medicines are required for transport (interesting for ventilation: against spasms, circulatory stabilisers) - For the acceptance of the product, especially the caregivers are interesting as actors		

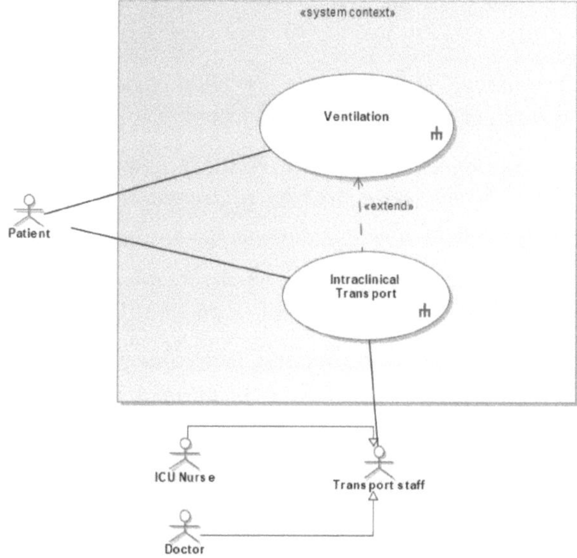

Figure 1: Use case diagram

The system boundaries separate the parts of the diagram that belong to the system from parts that are outside the system (actors). The actors are related to the use case with which communication takes place during the execution [4]. They are summerised with the command generalise, which means that ICU nurse and doctor are combined into the actor transport staff, as their tasks during intra-hospital transport are the same.

The corresponding activity diagram in Fig. 2 is located within the use case for in-hospital transport. This structures the information that has already been collected in Table 1. The patient himself and his monitoring data are listed as "accept event action" and are included in the diagram as object flow. The data dependency is detected with the square pins on the activity steps.

The steps of identifying the patient and measuring his vital signs lead to the decision whether the transport has more benefit than risk and thus should take place. If this is the case, the sequence continues. Otherwise, return to the first step to illustrate that transport only takes place when vital parameters are stable.

In the scope of this paper, only the first steps of the sequence are considered. The rest of the sequence is modelled with the same elements and includes the transport and its follow-up.

4 Conclusion

The diagrams clearly show the advantages of the MBSE. The systematic approach especially brings advantages with interaction of different specialist areas. The graphic representation brings better understanding due to its structured form and the technical exchange works more efficiently. Since some words are used by different departments, but with different meanings, misunderstandings can arise. These are remedied through the clarity of MBSE. A use case is a clear strucutred representation and contains further information in the activity diagrams contained in it. By avoiding misunderstandings and handling the structured documents, time is saved, which in turn can lead to cost savings. There is no loss of information due to migration of these from one software to another. In the next step the developed use case will be checked and agreed. The solution will be discussed with various stakeholders in a later step to verify that the requirements and ideas of the customers are adequately represented. There are various techniques for this, which are selected depending on the current project circumstances and objectives. There are different forms of reviews of requirements such as statements, inspections or walkthroughs.

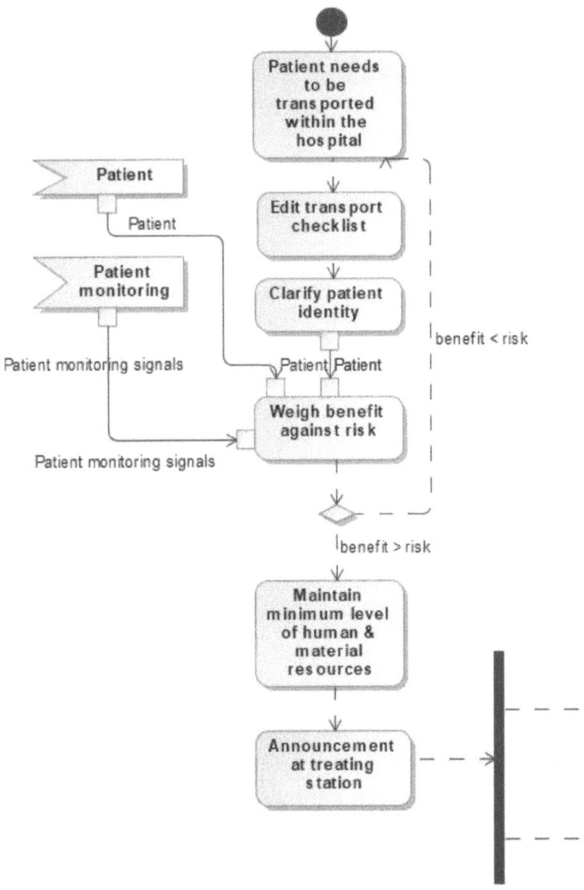

Figure 2: First steps of the activity diagram of intra-clinical transport

The use of prototypes and checklists is also helpful. In addition for coordination it is necessary to identify conflicts between stakeholders, analyze them and resolve them appropriately [4]. An example of this for a ventilator for in-hospital transport is the size of the screen. From the point of view of the nursing staff, it should not be too small in order to be able to see the vital parameters clearly. The display should also not be too large due to increasing battery power.

The last step is managing of requirements which is one of the main activities in requirements engineering. The goal is to make the documented requirements and other relevant information persistently available over the entire life cycle of the product or system, to structure them in a meaningful way, and to ensure selective access to this information [4]. The last two steps are not considered in this paper, as they can only be started at a later stage and development status of the product. Therefore, they are only given as an outlook on what needs to be done next in the requirements engineering process.

Acknowledgement

The work has been carried out at Drägerwerk AG & Co. KGaA. It is a listed company in Lübeck with around 15.800 employees (first quarterly report 2021). The company develops, produces and sells devices and systems in the fields of medical and safety technology.
In the Medtech sector, Dräger specializes in anesthesia, ventilation and heat therapy.
The work was supervised by the Institute for Robotics and Cognitive Systems, University of Luebeck.

Author's Statement

This paper contains a generalized use case, as not all information can be published due to trade secrets. The results in section 4 were taken from an internal unpublished source. It includes testing of MagicDraw as the only program used in various projects for requirements engineering in the company.

5   References

[1] C. Rupp & die SOPHISTen, *Requierements-Engineering and -Management: aus der Praxis von klassisch bis agil.*, Hanser, vol. 6, pp.172-189, 2014.

[2] *Verordnung (EU) 2017/45 des Europäischen Parlaments und des Rates.* 2017. Anhang I, Kapitel 5.1.

[3] *International Standard IEC 62366-1 - Application of usability engineering to medical devices.* IEC, Geneva, Switzerland, 2020.

[4] K. Pohl and C. Rupp, *Basiswissen Requierements Engineering.*, dpunkt.verlag, vol. 3, pp.19-151, 2011.

[5] J.D. Oury, F.E. Ritter, *Building Better Interfaces for Remote Autonomous Systems - An Introduction for Systems Engineers.* Springer, pp.21-22, 2021.

[6] Deutsche Interdisziplinäre Vereinigung für Intensiv- und Notfallmedizin, *Empfehlung der DIVI zum innerklinischen Transport kritisch kranker, erwachsener Patienten*, pp.1-3, 2004.

[7] MagicDraw, *No Magic, Inc., a Dassault Systèmes company*, Software Development, Stable release June 29,2020.

[8] A. Mahboob, *Modelling and use of SysML behaviour models for achieving dynamic use cases of technical products in different VR-systems.* Institut für Maschinen- und Gerätekonstruktion, TU Ilmenau, vol. 38, pp.31-34, 2021.

[9] Business Informatics Group, *Ojektorientiere Modellierung, Aktivitätsdiagramm.* Institue of Software Technology and Interactive Systems, Vienna University of Technology, pp. 4-7, 2012.

[10] *https://docs.nomagic.com/display/MD183/Accept+Event+Action.* MagicDraw 18.3 Documentation, October 15, 2015. Access on January 4, 2022.

Implementation of film dosimetry
for quality assurance of flattening filter free radiation

Birte Michael [1], Karsten Gerull [2], Florian Cremers [3],

[1] Medical Engineering Science, Universität zu Lübeck, birte.michael@student.uni-luebeck.de
[2] Klinik für Strahlentherapie, Universitätsklinikum Schleswig-Holstein, Karsten.Gerull@uksh.de
[3] Klinik für Strahlentherapie, Universitätsklinikum Schleswig-Holstein, Florian.Cremers@uksh.de

Abstract

Due to the further development of radiochromic films, film dosimetry has become an important tool for quality assurance of dose distributions. The aime was to implement film dosimetry as a standardized method for validation of treatment plans with flattening filter free radiation. 3D-conformal, stereotactic fields were radiated, using coplanar and uniplanar beam set-ups. The dose distributions of the irradiated films (FilmAnalyze) and the treatment planning system were compared with VeriSoft^{TM}. The comparisons were performed with the gamma analysis using a 10% cut-off level. Gamma analysis (conditions: 1 *mm distance-to-agreement and 3% dose difference with reference to maximum dose of measured slice*) gives a high compliance for the different orientations of the film towards the beam. Film dosimetry is a very accurrate way to verify dose distributions. However, it is quite elaborate and time-consuming, thus limiting its use in clinical routine.

1 Introduction

Over the past years, film dosimetry has become a useful tool for treatment verification and quality assurance (QA) of irradiation plans. A huge advantage of using films is the high, two-dimensional spatial resolution which qualifies the film to resolve small fields with a steep dose gradient as it is typical for stereotactic irradiation,which is used for small target volumes, such as brain metastases. The further development and improvement of the GAFchromic^{TM} EBT Film generation (Ashland^{TM}, USA) with the latest version EBT3, simplifies the working process. Contrary to the predecessor, the EBT3 film has less problems with artefacts, including polarisation, Newton rings or energy dependence. In general, QA of the dose distribution of a treatment plan is performed by using the electronic portal imaging device (EPID) of the linear accelerator. In the case of a stereotactic irradiation, it is typical to use flattening filter free (FFF) radiation to achieve a higher maximum dose and a steeper dose gradient. Verifying FFF radiation with EPID is not possible because, according to the manufacturer, the EPID is not designed for high dose peaks. At the moment, the verification is carried out with a point dose measurement. Due to the small fields and the steep dose gradient, this kind of verification is problematic. Even a very small displacement of the ionisation chamber can result in a huge dose deviation, if the ionisation chamber is placed in the gradient. Additionally, this method verifies only one point of the three-dimensional dose distribution. QA of FFF treatment plans via film dosimetry could offer increased safety. Aim of these measurements is to improve film dosime-

try as procedure of QA for three-dimensional, stereotactic irradiation plans with 6 MV FFF radiation. Therefore, the workflow using the evaluation software FilmAnalyze (PTW, Freiburg), as well as the calibration and resolving power, must be determined.

2 Material and Methods

2.1 GAFchromic^{TM} EBT3 film

The GAFchromic^{TM} EBT3 radiochromic film (Ashland^{TM} USA) is subdivided into three layers. The middle layer, or active layer, is about $27\,\mu$m thick and is coated by two $120\,\mu$m thick polyester layers. The active component is lithium pentacosa-10,12-diyonate (LiPCDA) in crystalline form. This monomer is sensitive towards heat, photon and electron irradiation and UV light. Thatswhy working with the films may only be done in artificial light to avoid a pre-irradiation by UV light. If the monomer is exposed to dose, there is a configuration change from the monomer to a polymer [4]. This polymerisation results in a change of color from yellow to blue in the irradiated part of the film. Adiitionally, the polymerisation is energy independent related to the type of radiation [5]. This is important for measuring FFF radiation as it has a wide energy spectrum. After the irradiation, the film must be stored in a dark environment during the darkening process. The self-development of the radiochromic EBT3 film is finished after about $18 - 24\,$h [5], [6]. An advantage of the further development of the EBT3 film is the crosslamination of the encasing polyester layers, which means that the layers are twisted 90 degrees to each other. This reduces the polarisation effects and the polyester also decreases artefacts like Newton rings [1].

2.2 Scanner

To digitalize the radiated films, an EPSON Expression 11000XL in reflective mode was used. The scan area of the flatbed scanner is able to capture films up to a maximum size of DIN A3. Additionally, the integrated software EpsonScan offers the possibilities to switch off any kind of color correction and select between a large range of settings, which were set to *positive film, no colour correction, 48-bit RGB and 72 dpi resolution.* As light source, the scanner is equipped with a xenon fluorescent lamp, which needs a warm-up period of 30 min and five warm-up scans [2]. After the scan procedure the resulting scans are 48 bit RGB-colored images saved in a TIFF format. This format offers the opportunity to separate the color channels to red, green and blue, which is important for the succeeding work with the software FilmAnalyze. To prevent scanning artefacts like the lateral scan effect, dust, damage of the film or the polarisation effect, the film and the scanning process must be handled with care. Especially the orientation (landscape) and the position on the scanner bed must be the same for every scan. Therefore, a template was used. Ideally, the film is placed in the sensitive area of the scanner to avoid artefacts originated from the inhomogenities of the scanner. The measurement of the nearly homogenous part of the scanner was performed following the protocol of Paelinck *et al.* [2]. For this, 70 measurement points were defined on the scanner bed and subsequently set in relation to the middle point of the scanner bed, which is defined as optimum according to illumination. For deviations between +/- 2%, the area was determined to be homogenous.

2.3 Software

As mentioned in the previous section, the scanning process was performed with the software EpsonScan. To calibrate and analyse the scans, the modules Film Calibration (Film-Cal) and Film and Image Analysis (FilmAnalyse) of the software package FilmAnalyze were used. For the investigation of scans, this software only uses the red channel from the RGB channels because the monomer inside the active layer has an absorption maximum in the wavelength range of red light. This procedure is called single channel dosimetry [4]. The program automatically filters the scan so there is no extra image processing step needed. To create the calibration curve, the program FilmCal uses a polynominal interpolation between every data point. The calibration curve gives the connection between the irradiated dose and the optical density of the respective film. The following gamma analysis between the calibrated film and the dose distribution calculated by the treatment planning system was performed by VeriSoft™ (PTW, Freiburg).

2.4 Calibration curve

The set-up of the calibration curve was executed after a given protocol [3] - 50 mm RW3 slab phantoms (polysterol with admixture of titanium oxide), 20 mm RW3 slab phantoms with integraded ion chamber mount for the $0.3\,cm^3$ Rigid Stem Chamber Type 30016 (effective measurement point 63 mm), 43 mm RW3 slab phantoms, source-surface distance (SSD) 95 cm - First 100 MU (monitor units) were mesaured three times. The average dose was taken to calculate the needed MU for the doses of 0.5 Gy, 1 Gy, 1.5 Gy, 2 Gy, 2.5 Gy, 3 Gy, 3.5 Gy and 4 Gy (see Table 1). The calculated MUs were irradiated and the actual doses determined by the ionisation chamber. For the irradiation of the film, the set-up was modificated to 63 mm RW3 slab phantoms, the film and on top 50 mm RW3 slab phantoms. With a SSD of 95 cm, there are the same conditions as in the first measurement. The film was cut into ten $10 \times 2\,cm^2$ pieces. The irradiation was performed separately for each dose with a $8 \times 8\,cm^2$ field. The film was irradiated perpendicular to the central beam axis with the same MUs as calculated and measured in the first calibration step. After irradiation, the film pieces were stored in a dark envelope for 18 h. Before starting the scanning procedure, it was switched on and warmed up for 30 min. Five warm-up scans were also performed [2]. The pieces were located in the middle of the scan area along the central axis of the scanner, perpendicular to the scan direction (landscape). Five scans per irradiated film were taken with the settings: no color correction, professional mode, reflective mode, 48-bit Color, 72 dpi. Afterwards, the TIFF files were opened with Film-Cal. The dose values of the measurement with the ionisation chamber were entered in the look up table. The scanner values were determined by the software FilmCal by means of 15×15 pixel areas in the middle of the scans. The measurement of the scanner values was performed for each scan of each dose. The final scanner values of one irradiated film is the average of the five measurements. The resulting calibration curve is shown in Fig. 1.

2.5 Verification of dose distributions

To imitate an environment that emulates a patient the RUBY phantom (PTW, Freiburg) was used. The field reference point was placed in the film plane and coincides with the isocenter of the linear accelerator. The dose distribution is comprised of five 3D-conformal fields ($0\,°$, $45\,°$, $90\,°$, $270\,°$, $315\,°$) of 6 MV flattening filter free (FFF) photon radiation. The jaws where set to the smallest usually used formation of $1.6 \times 1.6\,cm^2$. To modulate the field the multi-leaf-collimator (MLC) was used so that the 80% isodose encloses the target volume completely. To double check the dose distribution a second measurement was done using the EPID and a third by a two-dimensional array of ionisation chambers. The calculated dose distributions were exported with a size of $9 \times 17\,cm^2$ and 512×512 pixel. This gives a spatial resolution of 0.2 mm. To verify the dose distribution in all planes the RUBY phantom was irradiated three times in different positions (sagittal, coronal, transversal). After the irradiation all film pieces were stored dark for 18 h for self-development. The calibration was performed with help of the software Film and Image Analysis. The absolute dose distribution in the film plane was evaluated and compared with VeriSoft^{TM}.

2.6   Resolving Power

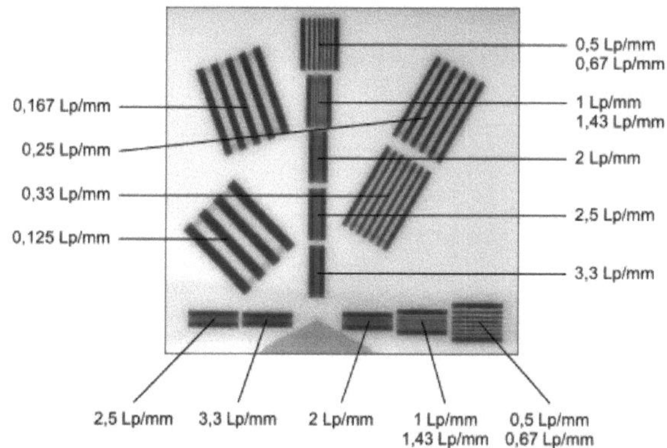

Figure 1: High contrast structures of the EPID Phantom. Resolvingpower is indicated in linepaires per millimeter as marked at the margin.

To determine the resolution of the radiochromic film, the EPID Phantom (PTW, Freiburg) was used. The resolving-power is indicated in Lp/mm (linepaires per millimeter). Fig. 1 shows the high resolution contrast structures. The horizontal and vertical high contrast test patterns in Fig. 1 are highly absorbent, so only the transmitted radiation forms the phantom equivalent pattern on the film. The measurement set-up was built by five 1 mm RW3 slab phantoms to eliminate backscattering of the patient table. The EBT3 film and EPID phantom were placed on top of these. The film was located in the isocenter of the linear accelerator and irradiated with 200 MU. This way, the high contrast structure is depicted on the film.

3   Results and Discussion

3.1   Calibration curve

Table 1 shows the calculated and irradiated MU and the corresponding measured doses. The doses were assigned to the scanner values of the films irradiated with equivalent MU number. The resulting diagram is shown in Fig. 2. To obtain a complete calibration curve, a polynomial interpolation was carried out between the pair of points.

Table 1: Measurement dose values of the calibration curve.

irradiated MU	0	53	107	160	213	267	320	373	427
measured dose in Gy	0	0.498	1.003	1.499	1.996	2.503	3.001	3.499	4.006

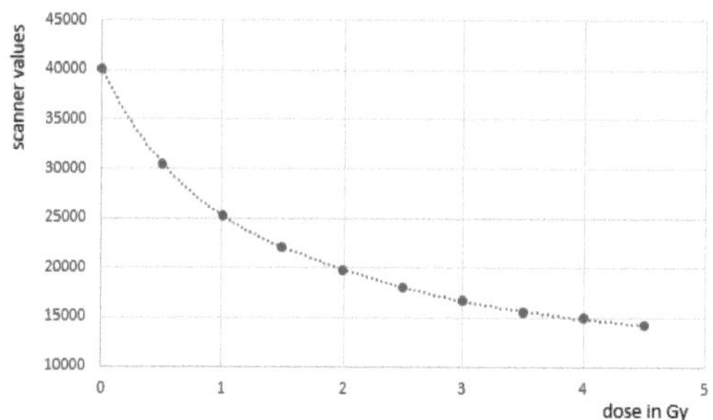

Figure 2: Calibration curve. Scanner values in dependence of the irradiated dose. Curve adapted by polynominal interpolation between each pair of points.

3.2   Verification of dose distributions

The control measurements via EPID and 2D-Array both showed a compliance with the dose distribution of the treatment planning system of over 98%. The comparison of the dose distributions, given by the treatment planning system and the irradiated film, shows a high conformity of 97.8% up to 98.9% for the different planes. Fig. 3 presents the failing points where the dose is either too high or too low. The choosen criterium for the comparison was the 2D gamma criterium under the conditions 1 *mm Distance-To-Agreement and 3% Dose Difference with ref. to the Max.dose of measured slice*. This conditions were chosen because the producer of the linear accelerator guaranteed a maximum dose difference of 3% and stereotactic irradiation has to be extremely accurate, so the distance-to-agreement was determined to 1 mm. The measurements also show that it is possible to measure the dose distribution in different planes in the room with multiple orientations between film and gantry. Contrary to the previously used point dose, the film dose distribution enables a 2D and accurate evaluation and verification of the irradiation plan in different planes.

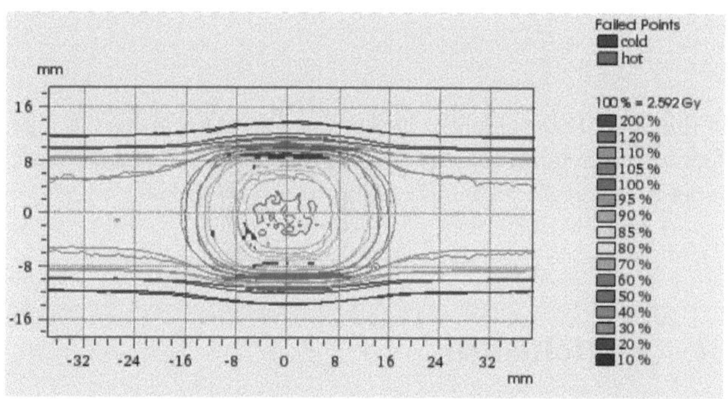

Figure 3: Comparison of the dose distributions given by the treatment planning system and the irradiated, calibrated film with failing points (blue: dose too low, red: dose too high).

3.3   Resolving power

To evaluate the high contrast structures Film and Image Analysis was used. Fig. 4 (A) shows the expected course of the high contrast stripes as a dotted line and the course of the scanner values (points) of the structure of 0.5-0.67 Lp/mm resolution. Thereby, one Lp corresponds to two maxima/minima. As can be seen in Fig. 4 the maxima and minima of the measured points and the high contrast stripes correspond to each other. In contrast to this is Fig. 4 (B). For a resolution of 1.43 Lp/mm the measured values and the expected striped pattern do not fit together anymore. The reason for this is the limited density of measurement points with a minimum distance of 0.4 mm. To resolve one pair of lines, at least three values (e.g. max, min, max) are needed. Because of this, the most accurate resolution is 1 Lp per 0.8 mm or normalized to 1 mm it is 1.25 Lp/mm. The fine part of the structure (Fig. 4 (B), x:3,5-5,5) has a resolution of 1.43 Lp/mm and that's why no longer resolvable for the software.

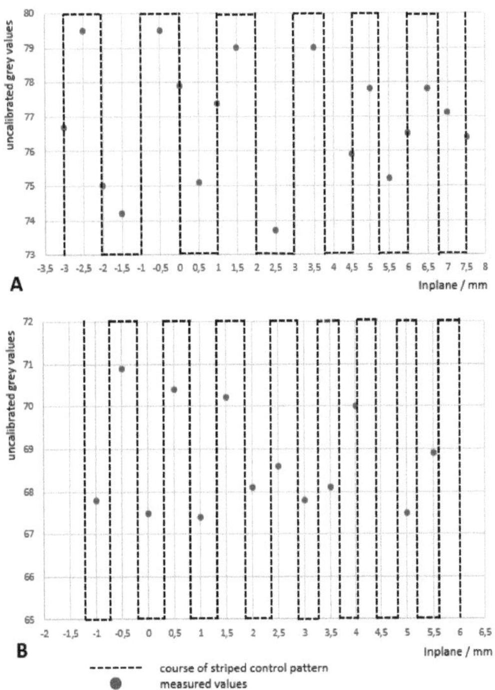

Figure 4: Points: course of the scanner values along the high contrast structures. Dotted line: expected course of the striped control pattern. A. 0.5-0.67 Lp/mm. B. 1-1.43 Lp/mm.

4 Conclusion

Verifying dose distributions of 6 MV FFF radiation with film dosimetry is an effective tool. The high two-dimensional resolution in all planes offers the opportunity to accurately analyse the dose distribution. However, the film has a long development time. It takes approximately 20 h to verify a dose distribution, which is likely too long to successfully implement it in a clinical routine. For control measurements, which usually take place every half a year, film dosimetry is very useful and accurate. The resolution power is limited by the distance between two measurement points. The software is not able to generate enough values

for an independent evaluation process. An alternative is to evaluate the pixels of the scanned film with MATLAB. The analysis of the pixels results in a higher number of measuring points and so probably for enough values for an evaluation process.

Acknowledgement

The work has been carried out and supervised at *Universitätsklinikum Schleswig-Holstein, Klinik für Strahlentherapie*. Special thanks to all the team members of the clinic for radiotherapy including the radiation oncologists, the therapists and particularly medical physicists.

Author's Statement

Preceding planning of filmdosimetry verification measurements was performed by Karsten Gerull. Planning and implementation of the experiment was done by Karsten Gerull and Birte Michael. Birte Michael evaluated and analysed the data and wrote the paper with input of all authors. Conflict of interest: Authors state no conflict of interest.

5 References

[1] N. Farah, Z. Francis, M. Abboud, *Analysis of the EBT3 Gafchromic film irradiated with 6 MV photons and 6 MeV electrons using reflective mode scanners*. Physica Medica, vol. 30, pp. 708–712, 2014.

[2] L. Paelinck, W. De Neve, C. De Wagter, *Precautions and strategies in using a commercial flatbed scanner for radiochromic film dosimetry*. Physics in Medicine and Biology, vol. 52, pp. 231–242, 2006.

[3] M. Grehn, *Evaluierung des radiochromen GafChromic^{TM} EBT3 Filmes für die Dosimetrie in zwei Dimensionen*. In: Student Conference on Medical Engineering Sciense 2016, Univerität zu Lübeck, pp. 1–38, 2016.

[4] T. Pócza, Z. Zongor, B. Melles-Bencsik, D. Zita Tatai-Szabó, T. Major, C. Pesznyák, *Comparison of three film analysis softwares using EBT2 and EBT3 films in radiotherapy*. Sciendo, vol. 54, no. 4, pp. 505–512, 2020.

[5] V. Casanova Borca, M. Pasquino, G. Russo, P. Grosso, D. Cante, P. Sciacero, G. Girelli, M. Rosa La Porta, S. Tofani, *Dosimetric characterization and use of GAFCHROMIC EBT3 film for IMRT dose verification*. Journal of applied clinical medical physics, vol. 14, no. 2, pp. 158–171, 2013.

[6] A.L Palmer, D. Bradley, A. Nisbet, *Evaluation and implementation of triple-channel radiochromic film dosimetry in brachytherapy*. Journal of applied clinical medical physics, vol. 15, no. 4, pp. 280–296, 2014.

Safety evaluation of AI-based prediction models for optical blood analysis

Nils Alexander Heywinkel [1,2], Benjamin Kern [2], Reza Behroozian [2] and Stefan Müller [2]

[1] Biomedical Engineering, Luebeck University of Applied Sciences, nils.alexander.heywinkel@stud.th-luebeck.de

[2] Medical Sensors and Devices Laboratory, Luebeck University of Applied Sciences{benjamin.kern, reza.behroozian, stefan.mueller}@th-luebeck.de

Abstract

This paper explores possible test scenarios of AI and focuses on the safety assessment of AI-based prediction models for mobile optical blood analysis. Input transmission spectra of blood samples in the visible spectral range were used as input of AI models, which predict clinical relevant blood parameters. To evaluate the AI models, typical fault cases for mobile devices were analyzed and tested using black-box testing, more specifically back-to-back and metamorphic testing. Noise and shift of the spectrum are the fault cases tested in this paper. Testing was done by modifying input transmission spectra and evaluating the outputs of the disturbed AI models. The test of the AI-based prediction models showed a low bias regarding the examined fault cases and a low deviation to their reference values, acquired with a lab device. It was proven that the trained models can reliably withstand the typical faults for mobile applications.

1 Introduction

In recent years more and more medical and in-vitro diagnostic devices use AI to improve their functionality, for example to increase accuracy and robustness. This results out of the recent advances in the field of AI. With respect to existing regulatory requirements, AI must be tested as part of the embedded software. This is especially important in biomedical applications since faults can result in harm to the patients. With respect to existing regulatory requirements, AI must be tested as part of the embedded software. This is especially important in biomedical applications since faults can result in harm to the patients.[1]

AI-based predictive models being developed for mobile optical blood analysis have been investigated for their insensitivity to perturbations. In the context of this work, the main focus was on spectral drift and noise on the optical spectra as these disturbances are among the most significant. One of the main challenges to properly test the models is the regulatory framework regarding in-vitro diagnostics devices which demands a high level of safety. Furthermore, the limited number of test data was another challenge, because the aim is to test as extensive as possible. Therefore different types of black-box tests were used to evaluate the models.[2]-[5]

2 Material and Methods

The implementation of AI-based prediction models in devices for in-vitro diagnostics and also the use of mobile low power devices pose special challenges. These challenges will be explained in regard to the regulatory requirements as well as the generation of test cases.

2.1 AI-based models and input

Five different models for five different blood parameters were given to test. The parameters to be determined by the models were the concentration of total hemoglobin (ctHb) , the fraction of oxyhemoglobin (O_2Hb), the fraction of carboxyhaemoglobin (COHb), the fraction of deoxyhemoglobin (HHb) and the fraction of methemoglobin (MetHb). Input to the models were transmission spectra from blood samples irradiated from 450 to 700 nm at 1 nm spacing. The data were normalized to the incident light intensity without blood sample. Models of ctHb and O_2Hb were selected to show the test process exemplary. Output ranges for the two models varied between 30 - 230 g/l for ctHb and 0 - 100 % for O_2Hb.

The models were given as TensorFlow files. They are considered by the German Notified Bodies Alliance as static black-box AI, an "AI model which does not explain how the results are derived" [3]. This influences the test methods chosen in this work.

2.2 Regulatory framework

Since AI is relatively new in medical or in-vitro diagnostic devices the regulatory framework is not well defined for implementing AI. Currently AI models according to the In-vitro-Diagnostic Device Regulation (IVDR) have to be tested as software and not specifically as AI. Over the next years this will change due to regulations that will be released regarding the use of AI.[2]-[4],[6]

The IVDR specifies that the safety, efficiency as well as technical equivalence to devices with the same use case have to be proven as is stated in the IVDR ANNEX I. In order to obtain the CE marking, evidence of the safety of the device, part or software is required. Because regulations do not give explicit information on how to test the models, best practice guides were used as a point of reference. The "Curriculum of the Korean Software Testing Qualification Board" gives information on the possible types of tests divided in black and white box test methods.[1]-[3],[5]

Black-box tests were considered as the best testing strategy. In black-box testing the inner construction of the test subject is not known and only input-output relations are examined. Two black-box testing methods were chosen for the project. They were chosen due to the models being static black-box AI, which means that the inner construction of the model is not sufficiently explainable.[7]

2.3 Test generation

Tests were done by using two different black-box testing methods stated in the best practice guide. Both methods deal with the so called test oracle problem of AI, which means that the expected results for an input is unknown and analytically not derivable. This is the case because the possible combinations of inputs are high. [7]

2.3.1 Back-to-back tests

Back-to-back tests deal with the oracle test problem in a way where an existing device with certification is used on the same samples as the AI models. Testing was done with 45 blood samples resulting in spectra and reference values provided by a laboratory device. The spectra were used without modifications.[7],[8]

2.3.2 Metamorphic tests

The main idea of metamorphic testing is to generate new test cases from successful test cases. In terms of the metamorphic tests a successful test case means that no error is found. This method was used to successfully test AI models in industrial settings.[7],[9]

As successful test cases the back-to-back tests are used, which means that the test data as well as the reference and predictions are used as a starting point. To generate further test cases so-called metamorphic relations were defined. Metamorphic relations are the expected change of the output resulting from an input change of a successful test case. This can be done in a way where for a changed input an expected change is known but cannot be quantified as an exact value.[7],[9]

For metamorphic tests, it was defined that the test should be used to evaluate the result of typical faults. It was analysed which typical faults could occur and which must be handled by the AI-based predictive models. This was done because many faults can be detected or eliminated before the results are predicted and used for further decision making. Two of the analysed fault cases were selected and tested:

First, the noise on the signal by the electric parts as well as the behaviour of the spectrometer depending on signal intensity was investigated. Second, the wavelength shift of the transmission spectra, which might result from a temperature change was tested.

These two fault cases were chosen because they are among the most significant. Furthermore, the models were already trained with these faults considered in the generation, which made it possible to define metamorphic relations. [10],[11]

Noise

Noise modelling was done by multiplying the transmission spectrum with random values for every wavelength to mimic real occurring noise. The signal-noise-ratio was used to model random noise in the resulting spectrum due to low reference intensity of the light source. The lowest intensity was subtracted from the spectrum of the light source to eliminate the offset and then divided by the highest value to normalize the spectrum. Afterwards, the spectrum was inverted.[11]

This modulation was multiplied with a normal distributed random function. A normal distribution was chosen because it represents the fault resulting out of the detection with the spectrometer and is the most probable. To account for the randomness of noise, each sample was disturbed 100 times by randomized noise. For the normal distribution a mean value of zero and a standard deviation of 0.4 were chosen. The random vectors were then multiplied by a factor of 0.01, which represents 1% of the maximum spectral intensity. This noise value was observed during measurements. These final modulated noises were lastly added on the spectrum.[11]

As a metamorphic relation, it was defined that a deviation to the reference compared to the back-to-back test results should change. The deviation should change low in terms of a bias of the results. The reason was that this was the aim of the models.

Spectral shift

The shift of the transmission spectrum is a common error that can be observed due to temperature changes. To simulate this perturbation, spectra of all 45 blood samples were shifted by -2 nm to +2 nm in 1 nm steps. A shift of 2 nm was chosen because it slightly exceeds the change expected due to temperature changes for mobile application of +/-40 °C. Ordinary grating spectrometers have a spectral shift of maximum 0.05 nm/°C.[10]

As a metamorphic relation it was defined that a deviation to the reference compared to the back-to-back test results should change. The deviation should show a small change in terms of a bias but show higher variation of the results. This is reasoned by the training for the model to compensate for this shift.

Statistical evaluation

For the results the focus was on the deviation of the reference value. This means that for the calculation the absolute deviation from the reference value was used. Along these deviations the mean deviation and standard deviation (SD) were calculated. Furthermore, the mean reference value for each parameter out of the reference values was calculated. Plots showing the predicted values versus their reference values served as a second evaluation tool. In these plots the identity line from the lower left side to the upper right side represents the perfect match between reference and prediction.

3 Results and Discussion

3.1 Results

The results were analysed for the two different models separately. The results were plotted in the prediction plots, the x-axis shows the reference values and the y-axis shows the corresponding values of the tests. Results of the back-to-back tests are plotted as "O", the noise test as "×" and shift test with a "△". Results for the noise and shift test display the biggest deviation found in the tests.

3.1.1 Total hemoglobin (ctHb)

The results for ctHb are displayed in Fig. 1 with the values given in g/l. It can be seen for the back-to-back test lower deviation for the lower values and an increased deviations for the higher values. The mean absolute deviation was -0.8 g/l with a standard deviation of 2.41 g/l (Table 1). Furthermore, an error range from -5.72 g/l to 3.04 g/l is given. As a mean value for the reference data 112.87 g/l was calculated.

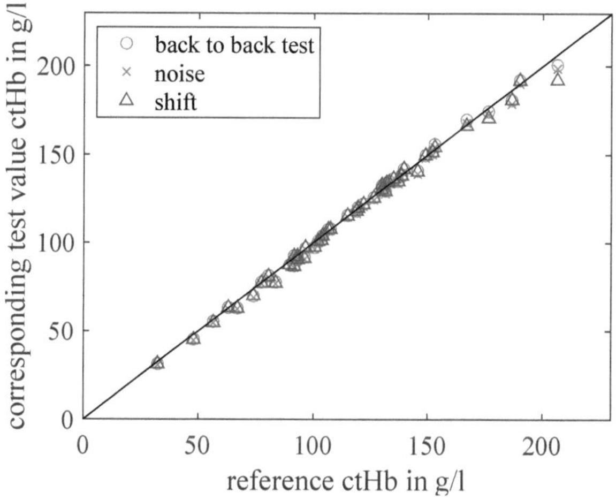

Figure 1: Total hemoglobin (ctHb) test results

Compared to the results of the back-to-back test the noise test shows a notable change in deviation only for the higher values. For the the noise test the mean deviation changed to -0.87 g/l and the standard deviation increased to 2.46 g/l see

Table 1. Further, the error range increased to a range with the boundaries -7.87 g/l and 4.54 g/l.

Shifting the spectrum showed a change in deviation similar to the noise test. For high values the change in sample prediction error changed in some cases more for noise than for spectral shift and vice versa. The model for ctHb shows a mean deviation for spectral shift of -0.65 g/l and standard deviation of 2.60 g/l (Table 1). The error increased to a range from -14.35 g/l to 4.43 g/l.

3.1.2 Oxygenated hemoglobin (O$_2$Hb)

In Fig. 2 the results are shown for the O$_2$Hb model. The given unit is % as these are fractions of the total hemoglobin. Compared to the ctHb model the results for the back-to-back test deviate the most between 30 % and 60 %. The O$_2$Hb model showed a mean deviation of 0.99 % to the reference results and a standard deviation of 2.82 % (Table 1). Further, an error range from -6.14 % to 6.91 % was observed. As a mean value for the reference values 61.14 % was calculated.

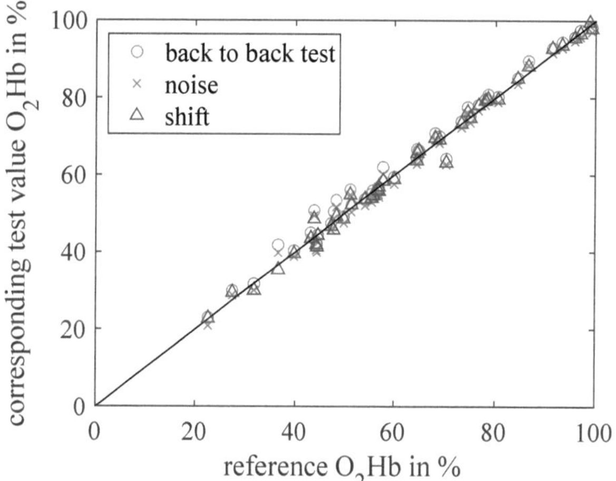

Figure 2: Oxygenated hemoglobin (O$_2$Hb) test results

The noise test had a low change compared to the back-to-back test results. Values with an already high deviation of the reference showed the biggest change to the back-to-back test results. The mean deviation for the noise test for O$_2$Hb changed to 1.05 % and the standard deviation increased to 2.95 % (Table 1). An increase of the error range to -7.67 % up to 7.84 % was observed.

Table 1: Mean and standard deviation for all models and tests

Parameter	ctHb in g/l	O$_2$Hb in %
mean	-0.80	0.99
SD	2.41	2.82
mean noise	-0.87	1.05
SD noise	2.46	2.95
mean shift	-0.65	0.90
SD shift	2.60	2.81

The spectral shift had comparable low difference in the deviation to the noise test. Some test spectra show more deviation for noise some less. For the O_2Hb models the mean deviation changed to 0.90 % with an increased standard deviation of 2.81 % (Table 1).The error range increased with a lower value of -7.13 % and higher value of 9.41 %.

3.2 Discussion

In general, the results for the AI models showed a low mean deviation lower then -0.9 g/l for ctHb and 1.1 % for O_2Hb (Table 1). These mean deviations did not change by more than 0.2 g/l for ctHb and 0.1 % for O_2Hb for the noise and shift tests, which showed only low bias added to the prediction. Furthermore, this could be observed by the low increase for the standard deviation. The prediction plots show the same low bias in the difference between the tests and the reference. All predicted values were close to the identity line, which represents a perfect prediction.

The bigger change in the prediction results from the tests can be seen in the variance of the results. This is shown in the error range compared to the back-to-back tests and the Fig. 1 and 2 where the results for tests deviate further. A comparison between these individual models is not possible due to the difference in prediction scales as seen in the mean reference values and the value ranges (subsection 2.1). These test results showed that the models achieved the goal of a low bias due to perturbation.

4 Conclusion

Overall, the tests demonstrated that the models perform as expected. The models showed a low bias and low variance despite the interference from noise and spectral shift. As it was shown by the small change in standard deviation in the different tests as well as the small changes seen in Fig. 1 and 2. The biggest deviations were found in the higher values where clinical consequences are negligible as the errors here would not lead to a changed treatment decision. Furthermore, it was shown how metamorphic testing can be used to further evaluate AI models, by increasing the number of test cases as well as checking deviating behaviour.

To achieve better results, it is likely that noise and spectral shift can be further reduced by using improved preprocessing methods in AI model generation. These were not considered in this project and will be investigated in further studies. Furthermore, the training can be continued with other fault cases in the training data, which would increase the resistance against these fault cases. For the resulting models the tests could then be repeated and metamorphic testing can be used to test other fault cases, as shown by the project.

Acknowledgement

This work has been carried out at the Medical Sensors and Devices Laboratory at Lübeck University of Applied Sciences.

Author's Statement

Conflict of interest: Authors state no conflict of interest.

5 References

[1] U. J. Muehlematter, P. Daniore and K. N. Vokinger *Approval of artificial intelligence and machine learning based medical devices in the USA and Europe (2015–20):a comparative analysis.* Lancet Digit Health 2021, vol. 3, pp. e195–203, 2021.

[2] Prof. Dr. C. Johner, *Regulatorische Anforderungen an Medizinprodukte mit Machine Learning.* Available: https://www.johner-institut.de/blog/regulatory-affairs/regulatorische-anforderungen-an-medizinprodukte-mit-machine-learning/ [last accessed on 2022-01-17].

[3] *Fragenkatalog „Künstliche Intelligenz bei Medizinprodukten".* Interessengemeinschaft der Benannten Stellen für Medizinprodukte in Deutschland, 2020.

[4] *Perspectives and Good Practices for AI and Continuous Learning Systems in Healthcare.* Xavier University, Cincinnati Ohio, 2018.

[5] *Regulation (EU) 2017/746 of the European Parliament and of the Council of 5 April 2017 on in vitro diagnostic medical devices and repealing Directive 98/79/EC and Commission Decision 2010/227/EU.* OJ L 117, pp.176–332, 2017.

[6] *Proposal for a REGULATION OF THE EUROPEAN PARLIAMENT AND OF THE COUNCIL LAYING DOWN HARMONISED RULES ON ARTIFICIAL INTELLIGENCE (ARTIFICIAL INTELLIGENCE ACT) AND AMENDING CERTAIN UNION LEGISLATIVE ACTS.* COM/2021/206 final, 2021.

[7] *Certified Tester AI Testing – Testing AI-Based Systems (AIT – TAI) Foundation Level Syllabus.* Korean Software Testing Qualification Board, 2019.

[8] M. A. Vouk, *"On back-to-back testing"* Computer Assurance, 1988. COMPASS '88, pp. 84-91, 1988

[9] Chen, T. Y., Kuo, F. C., Liu, H., Poon, P. L., Towey, D., Tse, T. H. and Zhou, Z. Q, *Metamorphic testing: A review of challenges and opportunities.* ACM Computing Surveys (CSUR), vol. 51, no. 1, pp. 1-27, 2018.

[10] S. Salim, N. Fox, E. Theocharous, T. Sun, and K. Grattan *Temperature and nonlinearity corrections for a photodiode array spectrometer used in the field.* Lancet Appl. Opt., vol. 50, pp. 866-875, 2011.

[11] W. Liu and W. Lin, *Additive White Gaussian Noise Level Estimation in SVD Domain for Images.* IEEE Transactions on Image Processing,vol. 22, no. 3, pp. 872-883 , 2013,.

Model predictive control for trajectory tracking of a Ballbot Robot

Mohamed Sawah [1]
[1] Robotics and Autonomous Systems, Universität zu Lübeck, mohamed.sawah@student.uni-luebeck.de

Abstract

Controlling the movement of robots using their wheels requires complex planning of each driven wheel, especially in the case of rotational motions, as they cannot rotate instantly at any point in every direction. An agile mobile called the ballbot robot can move in every direction while being installed on a sphere ball. This paper presents a procedure for controlling and balancing the robot's motion above the ball in addition to implementing trajectory tracking using model predictive control (MPC). As a result of applying the proposed approach in a simulated procedure, the robot appears to be balanced on the ball during its motion and perform a tracking path. However, achieving the stability and reference track on the actual ballbot is still under examination.

1 Introduction

Currently, Building wheeled robots faces several challenges. One of the main fundamental challenges is movement and navigation in cluttered spaces. Since wheels take a large room during rotation and translation to change the robot's direction. Commonly, wheeled robots are considered stable mobile machines, while afterward, more groups are curious about designing dynamically stable wheeled robots that actively balance [1]. The presented work during this paper pursues the development of dynamically stable mobile robots that are dynamically agile, human-sized, and sufficiently slim to simply move in crowded environments. The ballbot is one of the developed omnidirectional mobile systems that utilize wheels, however, they are used to balance the robot's body above a ball that carries and drives the robot [2]. The first ballbot was developed in 2006 [2], It was composed of human-like dimensions that allowed interaction with a person at eye level, and, accordingly, the ballbot can efficiently communicate with humans and transport and rotate in a crowded space and open environment made for humans as well.

The given ballbot, shown in Fig. 1, which is constructed at the IME (Institute for Electrical Engineering in Medicine) at the Universität zu Lübeck, consists of three parts: a body placed over the ball containing the sensors and the microcontrollers, motors attached to the body and the ball. Similar to an inverted pendulum, the body is sustained upward around its unstable equilibrium point by controlling the ball [2].

In this paper, an implementation of a ballbot balanced on a ball using a Proportional Derivative (PD) controller established on three decoupled 2D planes model is presented, in addition to the execution of a tracking trajectory using model predictive control (MPC) technique [3], which based

on two decoupled 2D planes. In the first step, the structure of the ballbot system will be introduced, then applying the system identification approach to determine the parameters that define the dynamics of the ballbot system. Afterward, an Implementation of a control architecture for the stability of the robot and reference tracking will be performed.

Figure 1: The illustration of the ballbot structure

2 Material and Methods

In this section, the description of the ballbot, the identification of the ballbot dynamic model, in addition to the controller design for the system will be presented.

2.1 Ballbot System

The considered ballbot, illustrated in Fig. 1, is represented by three different parts, the body, the propulsion system, and the ball. The body is placed at the top where it contains the battery, microcontroller, (inertial measurement Unit) IMU, power electronics, and Raspberry Pi [4]. Under the body, a propulsive system consisting of three omni wheels operated by the same number of motors with gear heads is installed. Each omni wheel is placed at 120° from the other and orthogonal to the ball's plane.

An encoder is coupled to each motor for measuring the motor's position and speed. The ball is attached to a holder from the bottom. The IMU sensor measures the particular gravity and angular rate of the attached object. The states of the body's rotation and speed are provided to the microcontroller from the IMU. A Raspberry Pi receives that information to implement an algorithm allowing for a particular motion of the wheels to ensure the body's stability on the ball. The process of receiving the states of the body and transmitting feedback information to the wheels to balance the body above the ball is repeated iteratively to secure system stability. The structure of the ballbot allows ease of motion in tight spaces containing dynamic obstructions as the robot maintains its orientation during its movement.

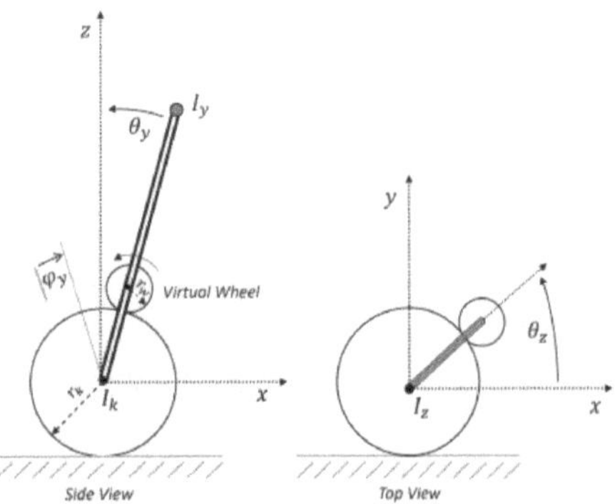

Figure 2: Sketch of the side and top views of the ballbot, xz plane (left) and xy plane (right) [5].

2.2 Dynamic model

The system dynamics can be represented by three decoupled 2D motion planes xz, yz, and xy without taking into account the coupling impact between each plane. Each plane is defined by the angle of the body θ_y, θ_x and θ_z, and the angular position of the ball is ϕ_y, ϕ_x and ϕ_z for planes xy, yz, and xz respectively. By applying the Euler-Lagrange formulation to solve the dynamic equations in each plane, the dynamic model of the xz plane is given by:

$$M(q)\ddot{q} + C(q,\dot{q}) + D(\dot{q}) + G(q) + \varepsilon(q,\dot{q}) = \tilde{B}\tau_y, \quad (1)$$

where $q = \begin{bmatrix} \phi_y & \theta_y \end{bmatrix}^\top$ represents the state of the angular position of both the ball and the body, the combined torque acting on the ball delivered from the three motors in the xz plane is represented by τ_y . The terms M, C, D, G, B represent the mass/inertia matrix, the vector of Coriolis and centrifugal forces, the frictional torque vector, the vector of gravitational forces, and the input vector respectively, given in (2):

$$M = \begin{bmatrix} b_1 & -b_2 + \ell r_b \cos(\theta_y) \\ -b_2 + \ell r_b \cos(\theta_y) & b_3 \end{bmatrix},$$

$$C = \begin{bmatrix} -\ell r_b \sin(\theta_y)\dot{\theta}_y^2 \\ 0 \end{bmatrix}, G = \begin{bmatrix} 0 \\ -\ell g \sin(\theta_y) \end{bmatrix}, \quad (2)$$

$$\tilde{B} = \begin{bmatrix} b_4\dot{\phi}_y \\ 0 \end{bmatrix}, D = \begin{bmatrix} \frac{r_b}{r_w} \\ -\frac{r_b}{r_w} \end{bmatrix}.$$

Additionally, r_b is the radius of the ball, r_w is the omni wheel radius, g is the gravitational acceleration and b_1, b_2, b_3, b_4, ℓ represents the dynamic and kinematic physical parameters of the system that need to be identified to describe the dynamic model. Information on the physical parameters of the system can be found in [5].

Since both of the xz and yz planes is symmetric, the dynamic equation which contains the angular positions of the ball ϕ_x and ϕ_y, the body θ_x, θ_y, and also the torque of the τ_x and τ_y are set to be identical in both planes. Unlike the opposite two planes, the xy plane is only represented through the position of the body θ because the ball only moves within the xy plane.

The control input of the system is identified by the virtual torques of the model τ_x, τ_y, τ_z. These torques are different from the particular torques engaged on the wheels, as they use a virtual wheel to balance the robot. Since the virtual torques are computed by the controller within the closed-loop, they have to be transformed to the particular torques on the omni wheels using the following kinematic relation:

$$\begin{bmatrix} \tau_1 \\ \tau_2 \\ \tau_3 \end{bmatrix} = \begin{bmatrix} \frac{2}{3\cos\alpha} & 0 & \frac{1}{3\sin\alpha} \\ -\frac{1}{\sin\alpha} & \frac{\sqrt{3}}{3\cos\alpha} & \frac{1}{3\sin\alpha} \\ -\frac{1}{\sin\alpha} & -\frac{\sqrt{3}}{3\cos\alpha} & \frac{1}{3\sin\alpha} \end{bmatrix} \begin{bmatrix} \tau_x \\ \tau_y \\ \tau_z \end{bmatrix}, \quad (3)$$

where τ_1, τ_2, τ_3 are the torques applied on the wheels in the 3D plane and they depend on the zenith angle α, which is the angle between the vertical axes and the inclination of the omni wheel.

The dynamic equation of the model in the xz plane can be linearized around the origin using the jacobian method for linearization to determine (4), where the state vector is $\begin{bmatrix} q^\top & \dot{q}^\top \end{bmatrix}^\top$. Then it can be transformed as the state-space model in (5).

$$\begin{bmatrix} \dot{q} \\ \ddot{q} \end{bmatrix} = \begin{bmatrix} \dot{q} \\ M^{-1}(-C - D - G - \varepsilon) + M^{-1}\tilde{B}\tau_y \end{bmatrix} \quad (4)$$

$$\dot{x} = Ax + Bu \quad (5)$$

where $x = \begin{bmatrix} \phi_y & \theta_y & \dot{\phi}_y & \dot{\theta}_y \end{bmatrix}^\top$ is the state vector, $u = \tau_y$ is the control input and

$$A = \begin{bmatrix} 0 & 0 & 1 & 0 \\ 0 & 0 & 0 & 1 \\ 0 & p_1 & p_2 & 0 \\ 0 & p_4 & p_5 & 0 \end{bmatrix}, B = \begin{bmatrix} 0 \\ 0 \\ p_3 \\ p_6 \end{bmatrix}, \quad (6)$$

with $p_1, p_2, \cdots, p_6$ being constant parameters of the linearized model which are related to the parameters of the nonlinear model in (1), i.e. b_1, b_2, b_3, b_4, ℓ.

2.3 Parameters Identification

The main target in this section is to review briefly the identification of the parameters of the linearized model $p_1, p_2, \cdots, p_6$ implemented (6). Since the system is unstable, identifying $p_1, p_2, \cdots, p_6$ in (6) requires an implementation of a closed-loop system identification. For this purpose, a simple PD controller has been utilized for both of the symmetric planes xz-and yz-plane with the coefficients K_p and K_d.

The closed-loop system identification using the system identification toolbox of Matlab as detailed in [5] enables determining a transfer function of the closed-loop, subsequently, identifying the system parameters using controller Information.

Subsequently, a cross-validation method has been utilized to compare the experimental data of the system with the identified model to specify its accuracy as shown in Fig. 3.

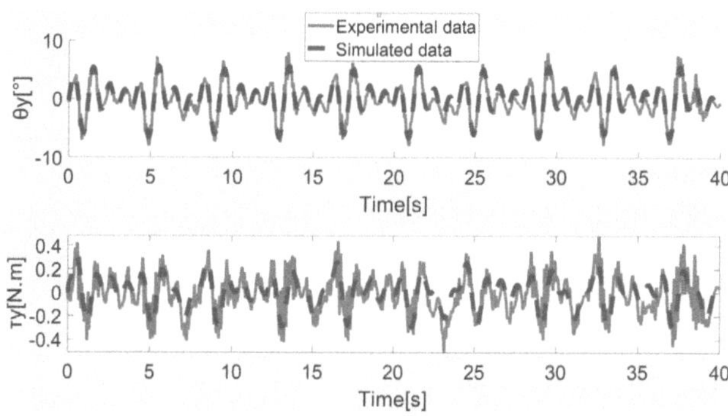

Figure 3: Comparison between the experimental and simulated output data, experimental (red), simulated data with linear model (blue-dashed).

Therefore, the linearized constant parameters of the system $p_1, p_2, \cdots, p_6$ in (6) have been obtained, as illustrated in Table 1.

Table 1: Linear and non linear parameters [5]

$p_1 - 342.603$	$p_2 - 52.8301$	$p_3 - 1425.9$
$p_4 - 36.073$	$p_5 - 9.147$	$p_6 - 251.847$
$b_1 - 251.847$	$b_2\ 0.0236$	$b_3\ 0.1431$
$b_4\ 0.084$	$\ell\ 0.297$	$r_\mathrm{b}\ 0.12$
$r_\mathrm{w}\ 0.05$	$g\ 9.81$	

2.4 Controller Configuration for Trajectory Tracking

As prementioned, the principal objective of the given project is to implement a control system aiming at balancing the body of the ballbot on the ball and enabling tracking of a reference trajectory. The PD controller is utilized to achieve the stability of the ballbot using its identified model of the xz-plane or the yz plane and considering associated tunning coefficients $K_p = 6.75$ and $K_d = 0.007$.

The implementation of the trajectory tracking using MPC is carried out as an optimization problem of the inner loop of the balanced model, as demonstrated in Fig. 4, where the system in the inner loop represents the model of the ballbot of the xz and yz planes, which is controlled by the PD controller discussed in section 2.3. The system for trajectory tracking controls the angular position of the ball based on the referenced path of ϕ_r where the MPC computes the required angular position of the body θ for tracking the desired trajectory. For the MPC implementation, its required to consider a discrete-time model.

$$x_{k+1} = A_\mathrm{d} x_k + B_\mathrm{d} u_k, \tag{7}$$

with $x = \begin{bmatrix} \phi_y & \theta_y & \dot{\phi}_y & \dot{\theta}_y & \phi_x & \theta_x & \dot{\phi}_x & \dot{\theta}_x \end{bmatrix}^\top$ is the state vector, $u = \begin{bmatrix} \theta_y^r & \theta_x^r \end{bmatrix}^\top$ is the control input, and k represents the sampling instant and A_d, B_d are the discrete-time system matrices with the proper size as illustrated in (7).

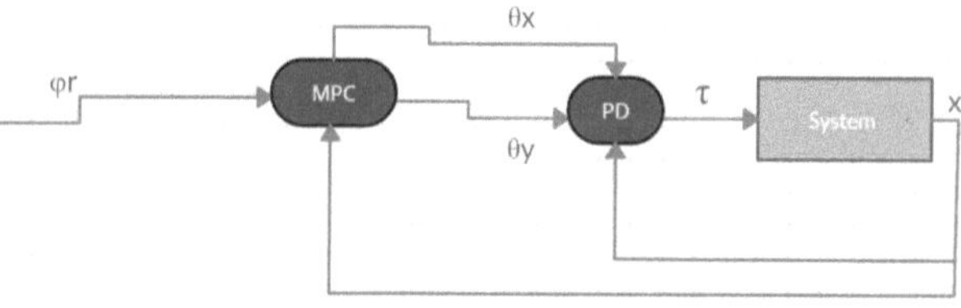

Figure 4: Closed-loop System for reference tracking

The implementation of the reference tracking (MPC) is applied on the xz plane and yz plane as there is no translational motion for the ball in the xy plane. Thus, the state vector is composed of the angular position of the ball ϕ, the angular position of the body θ, the angular velocity of the ball $\dot{\phi}$, and the angular velocity of the body $\dot{\theta}$ of both planes xz and yz. Likewise, the control input is defined as a reference for the angular position of the body θ at the balancing instance of the robot in the aforementioned planes as shown in (7). For the implementation of the MPC, a discrete state-space model is implemented in MatLab with a sampling time of 5ms. Besides using the Yalmip interface to solve the following optimization problem of the MPC trajectory tracking [6].

$$\min_{u_{0|k}, \cdots u_{N-1|k}} (Cx_{k|N} - \phi_{k|N}^r)^\top P(Cx_{k|N} - \phi_{k|N}^r)$$

$$+ \sum_{i=0}^{N-1} (Cx_{k|N} - \phi_{k|N}^r)^\top Q(Cx_{k|N} - \phi_{k|N}^r) + u_{i|N}^\top R u_{i|k}$$

$$\text{s.t. } x_{i+1|k} = A_\mathrm{d} x_{i|k} + B_\mathrm{d} u_{i|k} \tag{8}$$

$$x_\mathrm{min} \leq x_{i|k} \leq x_\mathrm{max}, i = 1, \cdots, N,$$

$$u_\mathrm{min} \leq u_{i|k} \leq u_\mathrm{max}, i = 0, \cdots, N-1,$$

where $C = \begin{pmatrix} 1 & 0 & 0 & 0 & 0 & 0 & 0 & 0 \\ 0 & 0 & 0 & 0 & 1 & 0 & 0 & 0 \end{pmatrix}$, the prediction horizon is N, the trajectory tracking is $\phi^r = \begin{bmatrix} \phi_y^r & \phi_x^r \end{bmatrix}$, the prediction steps $i \mid k$ indicated by step i of N at instance

k, $P = P^\top \succ 0$ represents the weight of the terminal cost $(Cx_{k|N} - \phi^r_{k|N})^\top P (Cx_{k|N} - \phi^r_{k|N})$, that bounds the tail of the infinite horizon cost from N to infinity. Q and R are set to be as follow: $Q = \mathrm{diag}(1,1)$, $R = \mathrm{diag}(0.01, 0.01)$. P is computed from Q and R using the Riccati equation [7], Moreover, the $x_{\min}$, $x_{\max}$, $u_{\min}$, and $x_{\max}$ are defined below in (9).

$$x_{\max} = \begin{bmatrix} 1000 & \frac{\pi}{3} & 4\pi & 4\pi & 1000 & \frac{\pi}{3} & 4\pi & 4\pi \end{bmatrix}^\top,$$

$$x_{\min} = -x_{\max}, \qquad (9)$$

$$u_{\max} = \begin{bmatrix} \frac{\pi}{3} & \frac{\pi}{3} \end{bmatrix}^\top,$$

$$u_{\min} = -u_{\max},$$

3 Results and Discussion

The presented PD controller has been executed in simulation for the two symmetrical planes xz and yz. As illustrated in Fig. 5, the tilt angle θ_x and the input torque τ_x of the yz plane generated by the PD controller appears to be stable, as their values go to zero along the time axis. The same model was used also for θ_y and τ_y in the xz plane.

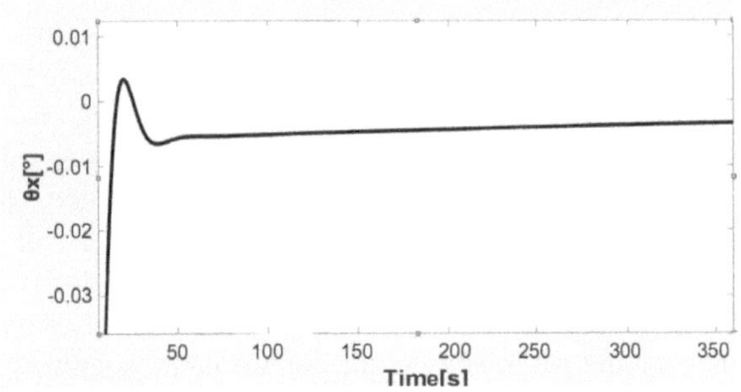

Figure 5: balancing control results of the tilt angle (θ_x) for yz plane by using PD Controller.

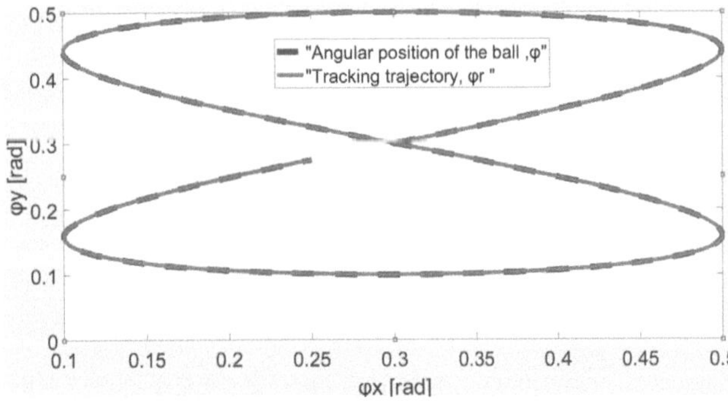

Figure 6: Trajectory tracking of the ball's angular position (ϕ_y) and (ϕ_x) in 2D by using MPC.

The MPC reference tracking of the ballbot model is implemented in simulation on a two-dimensional plane, where the angular position of the ball, represented by ϕ, tracks a curved trajectory, defined as ϕ_r. As illustrated in Fig. 6, the displacement of the ballbot is very close to the tracking path. Therefore, successful tracking has been implemented.

4 Conclusion

The purpose of this paper is to implement an appropriate control strategy for balancing and trajectory tracking of a ballbot. First, the identification of a linear model of the ballbot by linearizing the dynamic non-linear model of the system. Subsequently, The obtained model is utilized for tuning a PD controller for balancing the robot on the ball to achieve a stable motion of the tilt angle of the body that results in the stability of the ballbot. The last phase corresponds to applying an MPC with proper constraints, which control the angular position of the ball at the current state to track the trajectory. These implementations have been carried out in Matlab simulations and resulted in satisfying results. In future work, the proposed control approach will be applied to the real robot to determine the validity of the simulated data in the real world and accomplish a stabilized control system for the ballbot, which is capable of tracking a given trajectory path balancedly.

Acknowledgement

The work has been carried out at the Institute for Electrical Engineering in Medicine (IME) at Universität zu Lübeck and supervised by Dr.Hossam Abbas.

Author's Statement

I am Mohamed Sawah, a master's student in Robotics and Autonomous Systems.

5 References

[1] Umashankar Nagarajan, George Kantor, Ralph Hollis, *The ballbot: An omnidirectional balancing mobile robot*, International Journal of Robotics Research (IJRR), 2013.

[2] P. Frankhauser and C. Gwerder, *Modeling and control of a ballbot*, ETH Zürich, Swiss Federal Institute of Technology, Zürich, 2010.

[3] Jespersen, Kølbæk, Ahdab, Mohamad, Méndez, Flores, Damgaard, Rørmose, Hansen, Damkjær, Pedersen, Rasmus, Bak, Thomas, *Path-Following Model Predictive Control of Ballbots*, Aalborg University, Denmark, 2020.

[4] Warren Gay, Raspberry Pi Hardware Reference (1st. ed.), Apress, 2014.

[5] Max Studt, Ievgen Zhavzharov, and Hossam S. Abbas, *Parameter Identification and LQR/MPC Balancing Control of a Ballbot*. IME Department, Uni Luebeck, 2021.

[6] J. Lofberg, *YALMIP : a toolbox for modeling and optimization in MATLAB*, IEEE International Conference on Robotics and Automation, 2004.

[7] T. Chen and B. A. Francis, *Discrete-Time H2-Optimal Control*. London: Springer London, 1995.

Extended Ventilation Test Framework to Compare Different Ventilation Prototypes

Anja Chlebusch [1], Marcin Grzegorzek[2]

[1] Medizinische Informatik, Universität zu Lübeck, anja.chlebusch@student.uni-luebeck.de
[2] Institut for Medical Informatics, Universität zu Lübeck, marcin.grzegorzek@uni-luebeck.de

Abstract

Ventilation systems require a high level of safety and reliability. Therefore, a comprehensive review of requirements is necessary to ensure the quality and efficiency of the system. An automated test framework can help to check the quality and safety of a system more quickly and thus increase the efficiency of product development. In addition to different ventilation systems, replacing different components, such as sensors, in a particular ventilation system also can help to find best fitting resources in context of performance. This paper is about extending an automated test framework for competitive ventilation systems to one comparing different components of the same ventilation system. For this a graphical user interface (GUI) extended by suitable functionality in this paper, enables the user to add configurations and store it in an Influx database (InfluxDB). In this way, different ventilation prototypes with different components could be compared and evaluated prospectively.

1 Introduction

The main function of the respiratory system of the human is to ensure adequate tissue oxygenation and carbon dioxide removal. In a two phase cycle of inspiration and exspiration the volumes of alveolar gas are transported from the ambient air to the alveoli and vice versa [1]. When a human system cannot ensure this function a mechanical ventilation system has to step in. Besides different types of mechanical ventilation taking over the function of the human respiratory system, there are a variety parameters that need to be controlled. In worst case, defects in ventilation systems can lead to death of the patient. In development phase, an automated test framework can help to evaluate different ventilation systems and improve their performance and safety. In this context, the automation can assists with verifying faster and increasing the efficiency. Due to lack of virtually representing value of testing ventilation systems with all nuances as fluctuation in the actuator and measurements, a physically test setup is a necessary part of evaluating ventilator systems [2]. Different ventilation systems have different applications and in order to optimize them or compare their performance an automated test framework with a physical test setup seems to be a good method. However, during development phase of a ventilation system it can also be necessary to evaluate not only the entire system but also individual components/items of the system. This can help the developer to identify weak points and to change hardware settings or replace non-running construction units. Additionally, changing different components such

as valves or sensors can allow the systems builder to compare them in performance and cost-effectiveness.

This paper presents the extension of an automated test framework to a framework which compares different components. In order to create an option to also change components physically, which is not part of this work, the possibility to exchange the components is supported by functions of an extended user interface. This user interface includes data storage in a connected database also. In this manner, it propose a method to compare the same ventilation system with different components virtually. Later on, this can be used to find a visual configuration to improve the performance of the system.

2 Material and Methods

2.1 Experimental Setup

As mentioned in the introduction, an optimal test requires a physical test setup combined with an automated test software. Based on this, the test setup is shown in Fig. 1. The components are working together as part of the automated test framework. More specifically, the ASL 5000 is an active breathing simulator which can simulate the entire range of patients, from neonates to adults. It helps saving time, speeds up development, and provides better control over testing protocols [3]. Furthermore, ASL customization requires that the computer, including the graphical user interface, can be used as a suitable (configurable) control device by adding and editing breath test settings. The ASL is connected with the specific ventilation system to simulate the procedure of ventilation. An Speedgoat computer can be integrated optionally to test faster. The Speedgoat would automatically change ventilation settings for a specific test scenario such as PEEP (Positive End Expiratory Pressure) or PinsP (Inspiration pressure). Normally this does the user on the ventilator. This becomes necessary due to different test scenarios for one test sessions. Parameters such as PEEP need to be changed on the ventilation system to represent the requirements of a specifically simulated patient. A MATLAB simulink program on the Speedgoat connected with the ventilator takes on this task. But this Speedgoat has to be implemented for the specific ventilation system to fit in the specific interface. All in all this setup helps to simulate a typical ventilation situation on humans in the hospital and furthermore to evaluate and to improve ventilation systems.

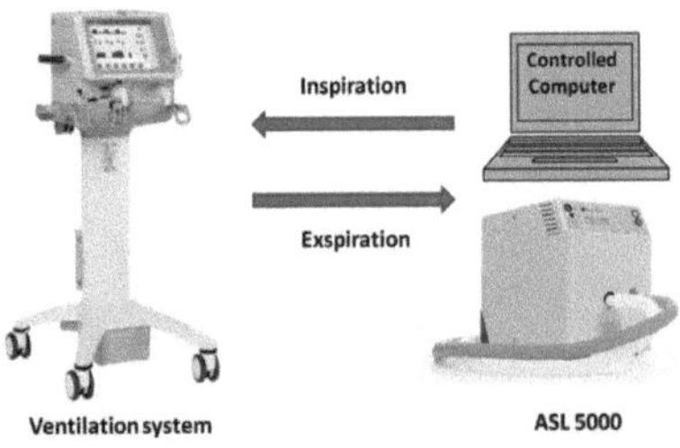

Figure 1: Experimental setup: ASL (active breathing simulator) controlled by a computer to simulate exspiration and inspiration is connected with the specific ventilation system Savina [4] [3].

2.2 Graphical User Interface

To allow the engineer to instruct the software which components and devices are integrated into the physical test setup, a good GUI makes it easy, practical and efficient [5]. In context of this work, both the GUI as a front-end as well as the underlying back-end are written in the Python programming language. To create a GUI window quickly and easily, this paper uses PyQT5 Designer, a tool from QT, which can be used to create a GUI by easy dragging and dropping objects on a blank form. With a graphical interface of this designer, it is possible to create independent main or sub windows as well as input dialogues. This leads to the advantage of editing or adding windows by different developers. In the specific case of testing different ventilation systems, several evaluating criteria must be taken into account in the GUI. These are already defined in the existing GUI. A GUI should contain options to configure different test scenarios with various parameters as PEEP or PinsP in a sequence of a test session as well as device information. These test scenarios are previously generated by an engineer and represent requirements in context of a specific patient (as COVID-19 patient) to a ventilation system. The test sequences can then be applied and compared either to different ventilation systems or several times to the same ventilation system. Fig. 2 shows the main window of test framework to which this paper refers. The buttons of removing, deleting and editing

pass to sub windows which can be used to create, edit and delete the specific tests related to the given requirements to the ventilation system.

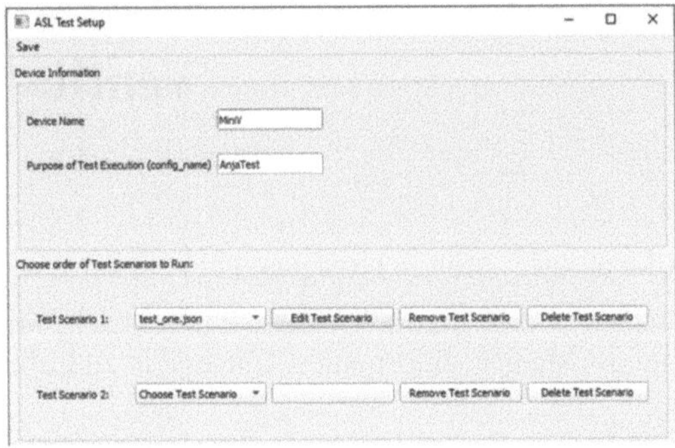

Figure 2: GUI main window of the test framework: The upper part contains information about the ventilation system (name, purpose of execution), the test scenarios are defined in the lower part.

In order to extend the system to compare different ventilation prototypes of the same ventilation system, certain requirements must be defined. Based on the already existing framework and listing to the engineers needs the requirements are defined. These are described in the use case diagrams Fig. 3 and Fig. 4.

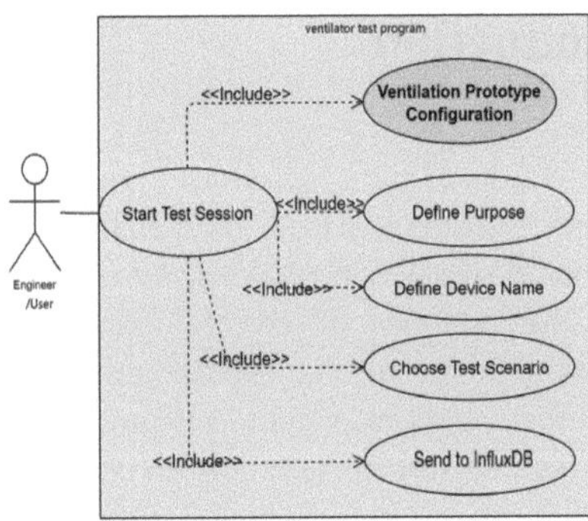

Figure 3: Use case diagram of main process of the starting test session (extension printed in bold).

The main process of starting a test on a ventilation system is shown in Fig. 3. When a user (systems engineer) starts a test session, he has to specify the purpose, the device name and test scenario as shown in Fig. 2. "Ventilation Prototype Configuration" symbolizes the extension of the framework to compare different prototypes.
Therefore Fig. 4 shows the requirements for this activity in detail respectively with his sub-activities. If a systems engineer wants to choose a device configuration, he can do so from the inventory or delete or edit/create one which is optional. Especially the creation is necessary when a new prototype or item is physically developed. Furthermore, if the user decides to edit or create a new device configuration, he must choose configuration items. This includes to choose a category and to define the name or optional to delete the item. These use case diagrams are used in the next chapter to create the extension of the GUI.

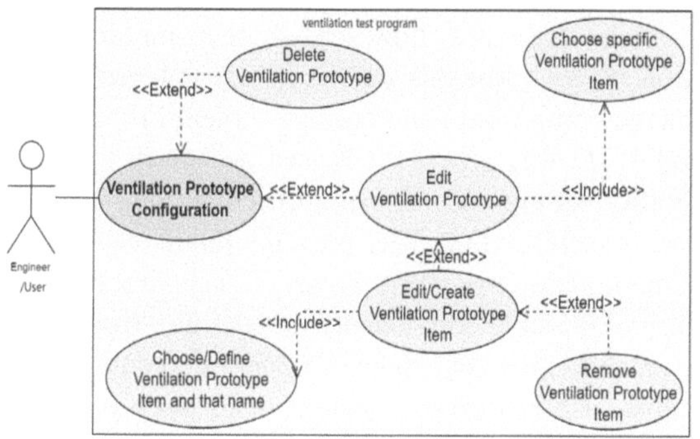

Figure 4: Use case diagram of the extended test session.

2.3 InfluxDB

As shown in Fig. 3, the test data of the ventilation systems must be stored in a database in order to be able to compare them. In the context of ventilation system test scenarios, time is an important factor. Each time a test is executed, the ASL generates a time stamp. In total, this results in a huge amount of data which needs to be processed. InfluxDB is chosen to store the data. InfluxDB is an schemaless open-source database which is optimized to handle times series data and has optional closed-source components. The most important concept of every InfluxDB database is time and with a discrete timestamps in a time column the associated data are stored [6].

Using a Python API with a client, it is possible to send data from ASL to InfluxDB.

```
#import InfluxDB client
import influxdb_client
from influxdb_client.client.write_api import SYNCHRONOUS

bucket=""
org=""
token=""

#Store the URL of your influxDB instance
url="https://us-west-2-1.aws.cloud2.influxdata.com"

client= influxdb_client.InfluxDBClient(
url=url,
token=token,
org=org
)
write_api= client.write_api(write_options=SYNCHRONOUS)

#create point sending to InfluxDB
p= influxdb_client.Point("mv_measurement").tag("location
    ","Prague").field("temperature",25.3)
```

As shown in the computer code listing [7], the database is accessed by importing the InfluxDB client via the API. To do this, the existing **bucket**, the **organization**, and the InfluxDB authorization **token** within the organization must be named to define the client.

Afterwards, a **point** p ("mv _ measurement") with **tag** ("location","Prague") and **field** ("temperature",25.3) can be send to the database. In this context the **measurement**, the field and the tag are key concepts of InfluxDB. The measurement explains our field of concept while tag and field are key-value pairs. The difference between fields and tags

is that tags are indexed and due to this queries on them are faster [6].

With an InfluxDB user interface and dashboard, both the user and other users can see the data as it is sent to the database and monitor it with tools such as Grafana (time series analytics open-source software) or Telegraf (product of the InfluxData ecosystem). While Telegraf collects and stores the data, Grafana analyses and visualizes them [6].

3 Results and Discussion

3.1 Extended Graphical User Interface

The GUI is the interface for the engineer to configure the virtual test setup. The aspects defined in the GUI determine which settings are stored with the ASL data in the InfluxDB. This gives the possibility to compare and distinguish test settings and sessions. Nevertheless, it is the engineer's requirement to extend the GUI to compare the ventilation systems with changing components. The requirements for this have already been shown in section 2.2 with Fig. 3 and Fig. 4.

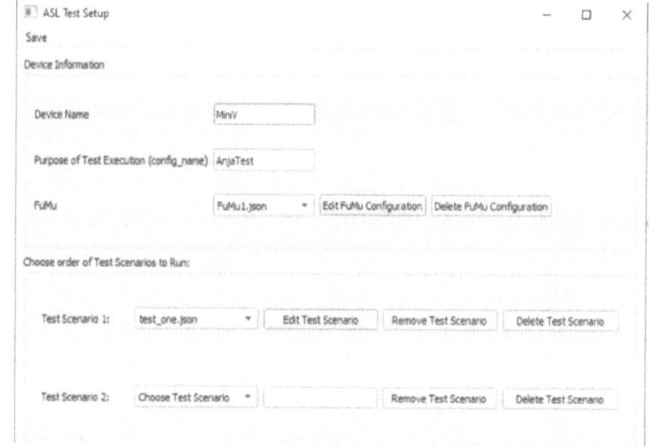

Figure 5: GUI main window of the test framework with extension.

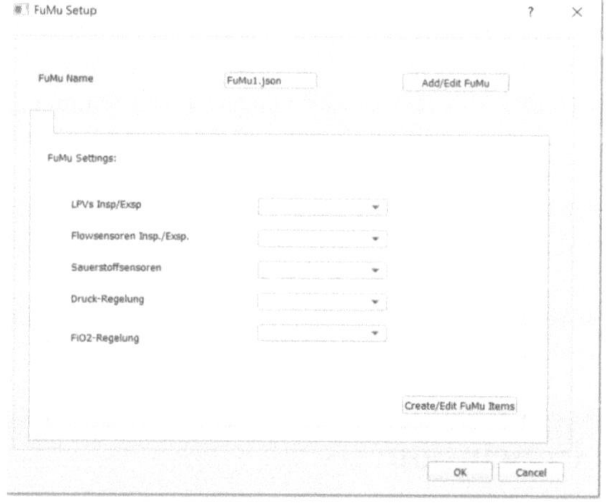

Figure 6: GUI sub window to configure the ventilation setup.

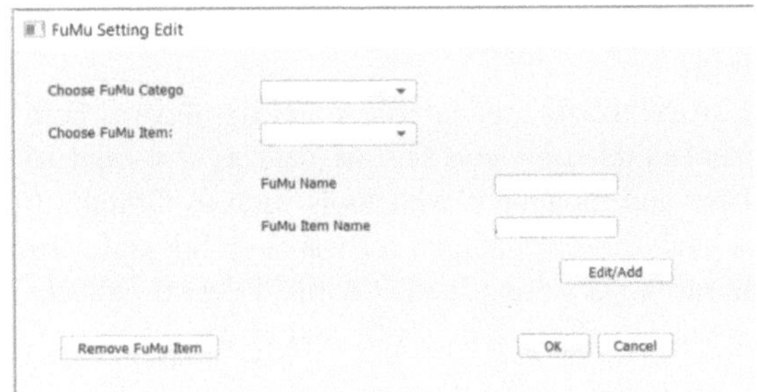

Figure 7: GUI sub window to choose/configure items.

The Fig. 5, Fig. 6 and Fig. 7 show the implemented use cases of Fig. 3 and Fig. 4. Before starting with the specific test case, a prototype of the ventilation system (here: 'FuMu') must be chosen. In this context, as shown in Fig. 5, it is optionally to define a new prototype or edit or delete an old one. By editing or defining a new prototype a new window will open. In this window (Fig. 6) the user has the possibility to define the different items as valves or sensors of the already existing prototype or open a new window (Fig. 7) to configure new valves or sensors.

3.2 Storing extended GUI Information in In-fluxDB

In the test setup for the ventilation, as shown in section 3.1, the engineer is able to send test sessions to InfluxDB using the API from section 2.3. With the new extension, the database now includes not only test sessions with test scenarios, but also buckets with an inventory of the ventilation system configuration and an inventory of specific items.

3.3 Discussion

Planing and building a good GUI is not an easy task but can be optimize the communication of the user to the system. The extension of the framework gives an engineer the possibility to change components in a ventilation system virtually and to care an inventory of different components. Later on this might be helpful to change components of the actual state of the physical system. It would be possible to compare different valves or sensors and evaluate their advantages or to verify their performance. These requirements are based on the existing framework as well as listening to the engineer.

But due to the fact that physically changing the components is not realized yet the framework advantages could not be evaluated. Only by evaluating the the system physically it can be measured if enhancing the performance of a ventilation system is satisfied.

4 Conclusion

Before the extension of the testing framework, it was indeed possible to compare the performance and use of different ventilator systems but not different prototypes with different items such as flow sensors or inspiration and expiration valves. For this reason, it has not been possible to change components and configure them in the GUI and InfluxDB. In this process it is also necessary to have the possibility adding and choosing different components and storing it in InfluxDB. This prevents redundancies due to different notation and helps the user to keep track of the inventory of possible components. In addition, replacing elements also has the advantage of selecting the most suitable elements such as different sensors or identify non-running elements. But due to the fact that changing components is not physically tested yet this can not be evaluated. This will be the next step in corporation with the system engineers. All in all, prospectively the new test framework could help to test if changing components of the same ventilation system improve the performance and efficiency of ventilation systems. It will be possible to build systems with changing components such as valves or sensors. If this enhance performance and efficiency of the system may then be proven.

Acknowledgement

The work has been carried out at Dräger AG Co. KGaA, Lübeck and supervised by Prof. M. Grzegorzek, Institute for Medical Informatics, Universität zu Lübeck.

Author's Statement

Conflict of interest: Authors state no conflict of interest

References

[1] D. Shelledy and J. Peters, *Mechanical Ventilation*. Jones and Bartlett Learning, 2019, ISBN: 9781284125931. [Online]. Available: `https : / / books . google . de / books ? id = wmKLDwAAQBAJ`.

[2] J. van Amerongen and P. Breedveld, "Modelling of physical systems for design and control mechatronic systems.," pp. 87–117, 2003.

[3] I. Medical, last accessed on 2021-12-19. [Online]. Available: `https : / / www . ingmarmed . com / product / asl – 5000 – breathing – simulator/`.

[4] D. A. /. C. KGaA, "Dräger savina® 300 select intensivbeatmung und lungenmonitoring," 2022.

[5] B. J. Jansen, "The graphical user interface," *ACM SIGCHI Bulletin*, vol. 30, no. 2, pp. 22–26, 1998.

[6] S. N. Z. Naqvi, S. Yfantidou, and E. Zimányi, "Time series databases and influxdb," *Studienarbeit, Université Libre de Bruxelles*, vol. 12, pp. 8–11, 2017.

[7] I. Influxdata, last accessed on 2021-12-19, 2022. [Online]. Available: `https://docs.influxdata . com / influxdb / cloud / api – guide / client-libraries/python/`.

Automatic Generation of Synthetic Data for Domain Randomization and Adaptation

Abhishek Dinkar Jagtap[1], Marian Himstedt [2] Mattias Heinrich [2]
[1] Robotics and Autonomous Systems, Universität zu Lübeck, abhishek.djagtap@student.uni-luebeck.de
[2] Institute for Medical Informatics, University of Lübeck, {marian.himstedt, mattias.heinrich}@uni-luebeck.de

Abstract

Most of the AI and machine learning algorithms are data-driven and to achieve the state of the art performance it highly relies on a large amount of high quality real world data. The collection of this real world data is not only challenging but also raises issues on privacy infringement. Generating synthetic data that can capture all the relevant information of the real world event can help overcome these underlying problems. Furthermore, the deployment of deep neural networks for medical diagnosis often suffers from domain shifts due to varying patient populations and clinical settings. The concept of domain randomization and adaptation are widely used to encounter the entailed domain gap. We focus our work on generating synthetic data for endoscopy and extend it to bronchoscopy while simultaneously introduce a concept to automate the process of data collection and provide a basis for autonomous navigation. This synthetic data can then help to serve as the base for implementing domain randomization and adaptation techniques.

1 Introduction

In general, the most common problem faced in artificial intelligence is the lack of sufficient data. Most of the data available are redundant in nature, for the data to be useful it has to be meticulously cleaned, sorted and labeled which becomes an exhaustive work and also collection of data belonging to various sub-fields of the same domain becomes impractical. As artificial intelligence is integrated into different aspects of the industry, classic anonymization of the data simply fails, raising the issues regarding privacy. In some domains such as healthcare, real data is not only valuable but also sensitive[1]. This reason exactly makes synthetic data a perfect solution to overwhelming problems. Synthetic data is the data generated from computer simulations that can mimic real world characteristics. One of the recently published work in this area is Unity Perception: Generate Synthetic Data for Computer Vision [2]. It introduces many toolkits that can be extended to *Unity Editor* to achieve synthetic data generation in an simulated environment. Our paper attempts to shed a light on the usage of this tools and toolkits for generating synthetic data for endoscopy and bronchoscopy. The fact that an infinite amount of data can be generated for the training of deep neural networks, makes it perfect for the application of AI algorithms. However, many problems arise during deployment and the prominent one among them is *domain transfer*. Thus techniques such as *Domain Randomization* and *Domain Adaptation* play a major role in the deployment of the models for real world applications [3].

The main purpose of domain randomization is to provide substantially variable simulated data such that the model can generalize to real world data. It is of paramount importance to realize that domain randomization is practically applicable to only simulated data as some of the parameters such as textures, materials, occlusions and coat masks are controllable in a simulated environment. In domain adaptation the primary focus is on adapting the model across the target domain by dealing with data distribution across domains [4]. Instead of gathering data and annotating it for the target domain which would be highly time consuming and inefficient, the idea is to create large synthetic data with all the required variations and diversity. By doing so, we can train the neural network on the synthetic dataset and test it on the targeted dataset. A visual representation of domain adaptation is shown in Fig. 1.

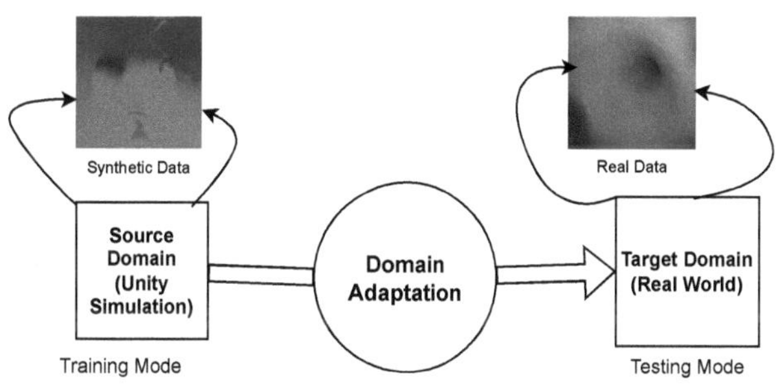

Figure 1: Illustration of Domain Adaptation.
Training a deep learning model on large synthetic data with maximum variations in order to get good accuracy on test holdout with real world data

2 Material and Methods

The process of generating synthetic data for endoscopy and bronchoscopy is carried out using *Unity Virtual Engine* as it provides high definition render pipeline for our simulation environment that can produce a realistic rendering of images. This virtual engine is commonly used for game development along with that it can be used for AI research as it provides a powerful graphical simulation platform not only to generate synthetic data but also to develop and test algorithms for various computer vision tasks. The whole process of artificially creating data is firstly achieved on endoscopy based on the open source project VR-Caps: A Virtual Environment for Capsule Endoscopy [5]. Furthermore, a bronchoscopic environment is created from scratch based on a simple 3D model of lung airways based on CT scans. The experimental setup to convert the data into realistic images follows the usage of *Blender* for modeling, sculpting, UV-editing and shading, Unity simulation environment to give realistic patterns to the datasets with varied textures.

2.1 Setting up a Simulation Environment

We initiate an endoscopic environment with an artificial light source and incorporate toolkits such as *Unity ML-Agents Toolkit* [6] that can provide simulations to serve as an environment for training agents. The open source project [5] comes with a built-in clinical scene that can be used as a starting point for synthetic data generation. Unity comes with a built-in scripting concept supported by C-sharp language widely used for game development which allows us to treat the environment and game objects as behavioral components. A script can give access to a virtual engine that can later help us in writing algorithms to facilitate the autonomous data collection process.

2.2 Creating Synthetic Data

The main aim is to generate synthetic data which is closer to real data, an 3D CT scan of a colon is exported to the blender for UV editing. Generally, a mesh is created surrounding the organ that can be edited along the vertices of the object. This mesh allows the object for UV editing. *UV mapping* is a technique for projecting a 3D model surface onto a 2D plane for texture mapping. A UV editing tool built-in blender offers the possibility of unwrapping the 3D object onto a 2D plane where textures can be applied seamlessly throughout the region of the colon. This texture gives a realistic pattern to the object. Standard shaders in blender offer the possibility of changing material properties corresponding to colon such as surface IOR, secular tint and anisotropy to further enhance the realism. The 3D object is then exported to unity where parameters such as lighting, materials, occlusions, transparency, coat mask and many other attributes are altered to give it a more realistic look. These attributes are carefully curated by trial and error so that they can mimic real world data. A 3D model of the capsule with pre-defined attributes of a camera is placed inside the colon which can be used for data col-

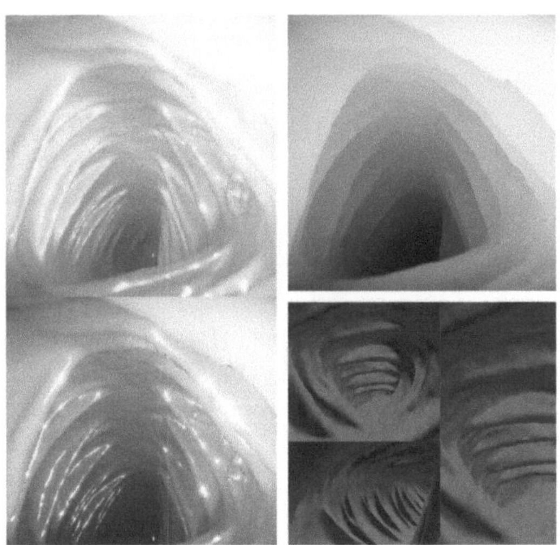

Figure 2: Synthetic Data Generated for Colon
Substantially variable simulated data with additional textures for endoscopy, suitable for domain randomization

lection in further processes. Adjusting these parameters is crucial for both mimicking real endoscopy and augmenting the data. The table below shows the camera parameters and post-processing effects required to achieve a fully synthetic model of the colon. Fig. 2 shows artificially generated data for colon with varying textures.

Attributes	Values
Surface Metallic	0.3
Surface Smoothness	0.7
Lens Intensity	0.1
Chromic Abberation	0.5
Coat Mask	0.435
Camera's Field of View	91.375
Focal length	159.45
ISO	200
Aperture	16
Anisotropy	1

Table 1: Camera parameters and Post-processing Effect

2.3 Autonomous Image Rendering

Manually collecting data for endoscopy or bronchoscopy becomes highly ineffective when creating synthetic datasets consisting of all the required variation and diversity. For example, in domain randomization we need to collect sequences of images each time with different textures and materials which would be highly time consuming. Thus an approach to automate the process of data collection is introduced, which allows us to collect samples from every region with different simulating conditions. Once the synthetic model of the endoscopy and bronchoscopy are completed we make use of the *scripting API* offered by unity which gives access to the simulation environment and game components via executable scripts. Firstly, a capsule with a camera attached is introduced to the colon and lung airways

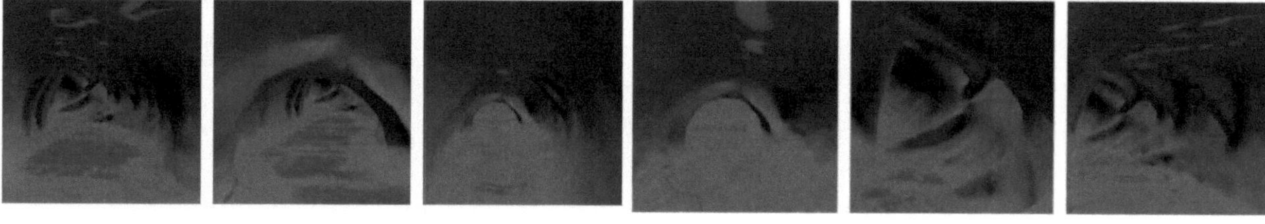

Figure 3: Image Sequences Rendered Autonomously for Endoscopy
Sequences of images rendered at an time interval of 2 seconds at 30 FPS in endoscopic environment

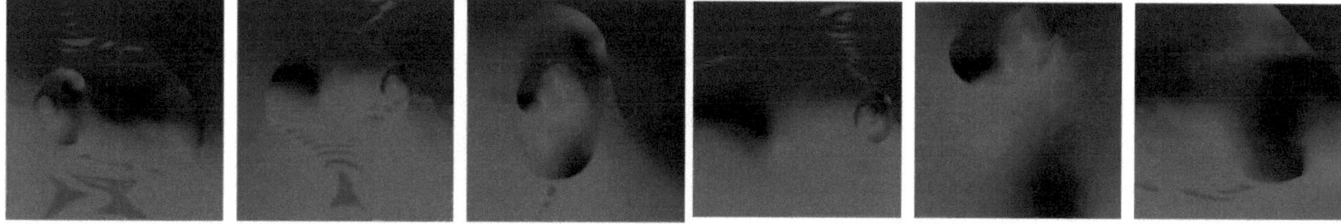

Figure 4: Image Sequences Rendered Autonomously for Bronchoscopy
Sequences of images rendered at an time interval of 2 seconds at 30 FPS in bronchoscopic environment

which can travel along a path and collect image sequences. The general idea is to create a waypoint system as shown in Fig. 5, consisting of real time 3D coordinates. The aim is to create a pathway for the capsule to travel without touching the walls or tissue segments or else it can raise unforeseen complications such as punctured chest wall or hemothorax.

2.3.1   Centerline Extraction

To generate a pathway along with the airway segmentation a centerline extraction process such as Tube Segmentation Framework (TSF) method is generally used [7]. The scope of using it is out of focus of this paper, but rest assured TSF offers automatic segmentation of airways and extraction of the centerline. Similarly, in endoscopy centerline extraction can be achieved based on the paper Automatic Centerline Extraction for Virtual Colonoscopy [8].

2.3.2   Scripting for Rendering Image Sequences

Once the information pertaining to the centerline coordinates is made available, our focus is to develop a path following algorithm using the scripting API in unity. The path following script consist of two parts the first one being the *Update* function where components behavior such as movement, triggering actions are hardcoded and secondly a *start* function which initializes the script before running the simulation. This instance of the path following script is then attached to the game objects such as a capsule along with it a recorder script is created which records the image sequences viewed from the perspective of the camera which is directly attached to the capsule. The camera can be controlled and set at any angle with the help of a control script. As a result, once the gameplay of the simulation begins the capsule starts moving along the path and a camera records the image sequences rendered at a targeted (FPS) frame rate. Thus automating the whole process of data genera-

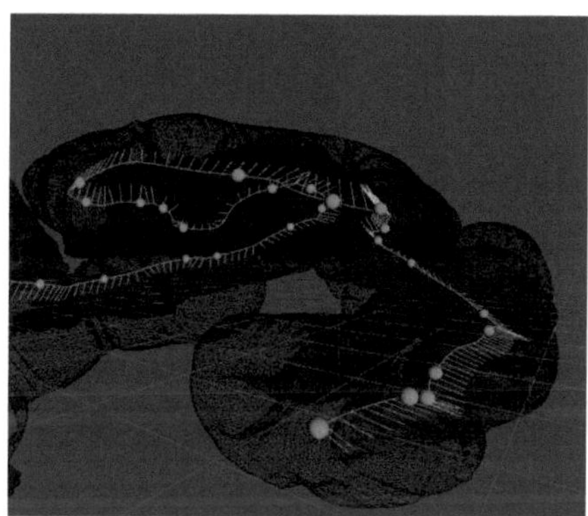

Figure 5: Waypoint System
Random path generated for testing the path following algorithm, The green line shows the path traced and circles represents the key-frame coordinates

tion. This procedure is incorporated in both endoscopic and bronchoscopic environments and a sample of the respective images rendered are shown in Fig. 3 and Fig. 4 respectively. As the whole process of automatic generation of synthetic data is achieved, the same concept can further be utilized for autonomous navigation in bronchoscopy and endoscopy as a navigational guidance system becomes an area of particular interest. For example, navigating to a desired location of interest to find lesions but a prior knowledge on the location of the lesion would be needed.

3   Results and Discussion

We focused our work on generating synthetic data that can be actually used for training deep learning models, Fig. 6 shows the comparison of real and virtual data for endoscopy and Fig. 7 shows the comparison between real and virtual data for bronchoscopy. The obtained synthetic model can

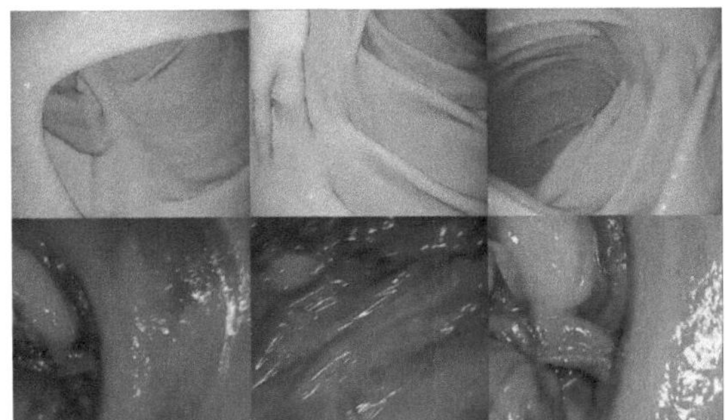

Figure 6: Top; Virtual Data; Bottom; Real Data
Comparison of synthetic and real data for endoscopy

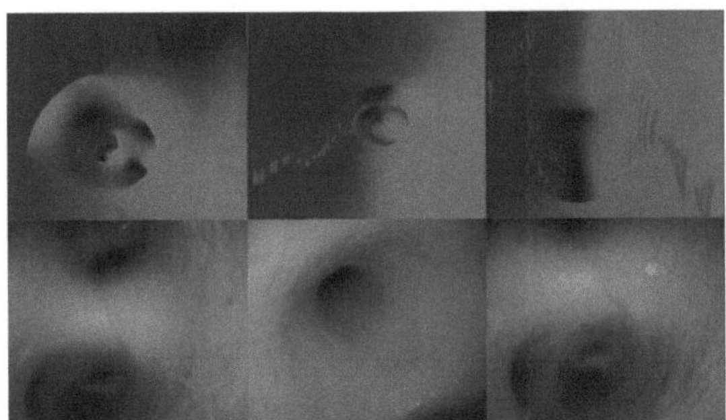

Figure 7: Top; Virtual Data; Bottom; Real Data
Comparison of synthetic data and real data for bronchoscopy

be reproduced by following the steps listed and using the camera parameters and post-processing effects listed in table 1. These attributes are carefully curated on a trial and error basis to achieve a working model of the data. Furthermore, we achieved the automatic rendering of images using the concept of waypoint system which is widely used in game development. As we can observe from Fig. 6 and Fig. 7, synthetic data tries to mimic the real world data but it still lacks the photo realistic patterns to achieve the realism required. However, synthetic data with need for actual realism is still an active area of discussion [1].

4   Conclusion

Artificially created data has promising advantages over real world data and above everything its completely privacy friendly. This work can be extended further to compare the results between domain randomization and adaptation for bronchoscopy navigation by conducting experiments of shuffling synthetic data in training and testing modes [9]. This paper aims at providing data that serves as a base in the future for implementing and testing techniques such as domain randomization and adaptation.

Acknowledgement

The work has been carried out at the Institute of Medical Informatics, and supervised by Dr. Marian Himstedt, Instituteof Medical Informatics, Universität zu Lübeck

Author's Statement

Conflict of interest: Authors state that they have no conflict of interest.
Informed consent: This article does not contain patient data

5   References

[1] S. I. Nikolenko, "Synthetic data for deep learning," 2021.

[2] S. Borkman, A. Crespi, S. Dhakad, S. Ganguly, J. Hogins, Y.-C. Jhang, M. Kamalzadeh, B. Li, S. Leal, P. Parisi et al., "Unity perception: Generate synthetic data for computer vision," arXiv preprint arXiv:2107.04259, 2021.

[3] J. Tremblay, A. Prakash, D. Acuna, M. Brophy, V. Jampani, C. Anil, T. To, E. Cameracci, S. Boochoon, and S. Birchfield, "Training deep networks with synthetic data: Bridging the reality gap by domain randomization," in Proceedings of the IEEE conference on computer vision and pattern recognition workshops, 2018, pp. 969–977.

[4] A. Choudhary, L. Tong, Y. Zhu, and M. D. Wang, "Advancing medical imaging informatics by deep learning-based domain adaptation," Yearbook of medical informatics, vol. 29, no. 01, pp. 129–138, 2020.

[5] K. Incetan, I. O. Celik, A. Obeid, G. I. Gokceler, K. B. Ozyoruk, Y. Almalioglu, R. J. Chen, F. Mahmood, H. Gilbert, N. J. Durr, and M. Turan, "Vr-caps: A virtual environment for capsule endoscopy," 2020.

[6] A. Juliani, V.-P. Berges, E. Teng, A. Cohen, J. Harper, C. Elion, C. Goy, Y. Gao, H. Henry, M. Mattar et al., "Unity: A general platform for intelligent agents," arXiv preprint arXiv:1809.02627, 2018.

[7] P. J. Reynisson, M. Scali, F. Smistad, E. F. Hofstad, H. O. Leira, F. Lindseth, T. A. Nagelhus Hernes, T. Amundsen, H. Sorger, and T. Langø, "Airway segmentation and centerline extraction from thoracic ct–comparison of a new method to state of the art commercialized methods," PloS one, vol. 10, no. 12, p. e0144282, 2015.

[8] M. Wan, Z. Liang, Q. Ke, L. Hong, I. Bitter, and A. Kaufman, "Automatic centerline extraction for virtual colonoscopy," IEEE transactions on medical imaging, vol. 21, no. 12, pp. 1450–1460, 2002.

[9] V. Seib, B. Lange, and S. Wirtz, "Mixing real and synthetic data to enhance neural network training–a review of current approaches," arXiv preprint arXiv:2007.08781, 2020.

5

Machine Learning / AI

Behaviour-based Approach for Solving a Sparse Swarm Environment Exploration Task

Till Aust [1], Lars Schilling [2], and Heiko Hamann [1]

[1] Institute of Computer Engineering, Universität zu Lübeck, till.aust@student.uni-luebeck.de
[2] Institute for Robotics and Cognitive Systems, Universität zu Lübeck, lars.schilling@dfki.de

Abstract

Precision Farming is increasingly coming into focus as a way to solve today's agricultural challenges. As a foundation for further research in this area, we are developing a framework for solving a sparse swarm environment exploration task using a behaviour-based approach. In addition to the exploration of the environment, we introduce further constraints, such as the safety of the individual robots and energy-saving navigation. We implement these conditions by combining several sub-behaviours in a probabilistic state machine to model a more complex overall behaviour. Furthermore, we introduce the parameter α, which represents the risk-taking of the robot, regarding connectivity to its swarm. We simulate the complex overall robot behaviour in 8 experiments using the Gazebo environment. We conclude, besides the feasibility of the system, that α should be chosen conservatively in order to keep connectivity between robots.

1   Introduction

Although Precision Farming has been around since the 1980s [1], it becomes increasingly important due to food and sustainability needs [2]. The main goal of Precision Farming is to improve efficiency while reducing the impact on environment [3]. To this end, new technologies are used, such as multi-robot systems, also called robot swarms. This technology is used in an agricultural context due to its advantage over single robots in terms of effectiveness, efficiency, flexibility and fault tolerance [4]. We propose a novel approach to solve a robot swarm environment exploration task via a behaviour-based approach, to address the challenges of complexity and needed infrastructure for Precision Farming by using only decentralized swarm approaches.

Exploration is a fundamental (sub-)task for swarm robotics [5] and the first step in the direction of developing Precision Farming from a behaviour-based view.

For solving this task, algorithms like the Hungarian Algorithm [6], Leader Follower Exploration Algorithm [6], Role-Based Exploration Algorithm [7] and Frontier-Based Exploration Algorithm [8] have been developed. All these algorithms have in common, that they are developed and tested for indoor environments. Further, they only consider optimal exploration, means they try to maximize the explored area over time. These are two fundamental differences from our work. First, we assume an outdoor environment. Second, we also consider safety of the individual robots. Safety is defined as connectivity to the swarm, because if a robot has an accident (e.g., gets stuck in a muddy field), it can be perceived by other robots and they can forward this problem and get help or at least save the data of the stuck robot. Another issue, which is only considered in some approaches, is the limited capability of communication (e.g., limited communication range) between robots. We overcome this issue by following the sparse-swarm concept [9].

Further, we are interested in energy efficient navigation. This is ensured by trying to always navigate at the same altitude (assuming driving uphill costs more energy). In contrast, we assume, that higher altitudes are more desirable, because it is a more energy efficient state (driving down costs less) and other beneficial features, such as a wider communication range.

2   Material and Methods

For the conducted simulation experiments, we use Gazebo 9.16.0 as the simulation environment and ROS Melodic Morenia as middleware to interface the robot. We use the robot platform Jackal UGV by Clearpath. We extend the Jackal with an Ouster OSI-16 lidar sensor to be able to sense the environment and create topographic maps (grid map, which tracks the altitude of each cell), which we later use for path planning. We implemented a communication module, based on a client-server architecture, to mimic local communication between individual robots. Allowing peer-to-peer communication up to 20 m.

2.1   Robot Behaviour

We implemented the behaviour of the robots as a probabilistic state machine, as shown in Fig. 1. Focus on that probabilistic state machine allows to be extended easily with

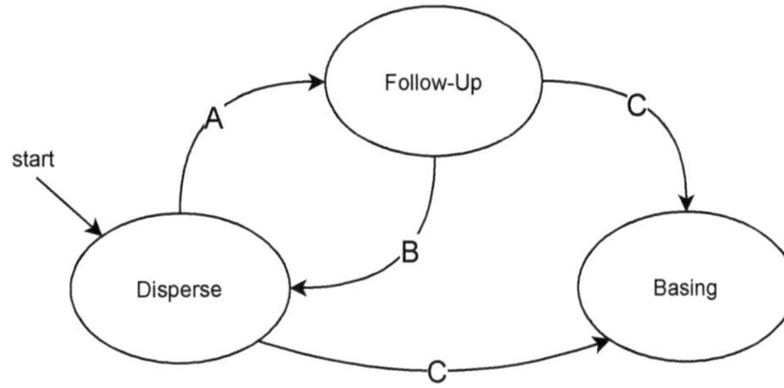

Figure 1: Probabilistic state machine of the behaviour of a robot.

new behavioural states, which makes it a good fundamental framework for further research. The robots send a ping message, with their position, every 5 s.

2.1.1 Sub-behaviours

Disperse is the basic exploration behaviour. It is build up on frontier exploration and composed of three steps for computing the next position of interest for the robot. The algorithm can be seen in Fig. 2. First, the robot swarm center is

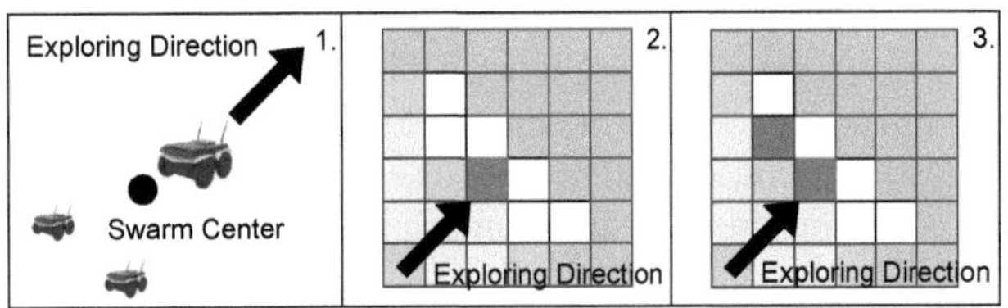

Figure 2: Procedure of the Disperse-sub-behaviour.

calculated, with locally available information (GPS and position information, gathered from peers through ping messages). Next, the robot calculates the exploring direction by subtracting the swarm center from its current position. Second, following along the exploration direction, starting from the robot's position until the end of its known world (orange, yellow and light green tiles; grey tiles represent the unknown areas), the last cell (red tile) marks the center of the frontier (white tiles), that the robot should explore. In the final step, the robot calculates which cell (considering all frontier cells) is the next position of interest (green tile). Therefore, a metric is used, which depends on the altitude of the cell (assuming higher altitude is a more advantageous position) and the distance to the robot (less distance leads to faster exploration of unknown territory).

Follow-Up is responsible for keeping up connectivity (robots stay in communication range with each other) throughout the robot swarm, which is required to ensure fault tolerance. It is implemented as follows: First, the robot searches for the closest robot in its vicinity. Afterwards, it takes the current position of this neighbour and its current target position (position the neighbour drives to) and draws a line between them. Depending on the distance to that neighbour, the robot chooses a point on this line (closer to the neighbour, the robot chooses the point closer to the

neighbour's current position). This is the robot's new way point. If the robot loses connection to the swarm (i.e., no neighbour in communication range), the robot drives back to its initial position, i.e. the base.

Basing This behaviour is chosen when the robot is low on battery. The robot returns to the base in the most energy efficient way.

2.1.2 State Transitions

The robot has the opportunity to change its current state every 10 s. This ensures that the robot does not fluctuate between two states, when, e.g., the robot is on the edge of the communication range. Thus, the transition probabilities need to be calculated in the same frequency.

Before calculating the transition probabilities, the robot first recomputes a risk-taking variable α, which indicates how venturesome the robot is. This variable is in range $[0, 1]$, where $\alpha = 1$ means maximum risk. This parameter configures the exploration-exploitation trade-off. There are three transition probabilities we need to calculate (A, B, C in Fig. 1). The first transition probability is given by

$$p_A = \begin{cases} 0 & \text{, if active neighbours} > 1 \\ 1 - \alpha & \text{, else} \end{cases} . \quad (1)$$

It describes the transition from *Disperse* to *Follow-Up*. It ensures exploration, when the robot is in communication range with other robots.

The next transition probability calculates as

$$p_B = \begin{cases} 1 & \text{, if active neighbours} > 1 \\ \alpha & \text{, else} \end{cases} . \quad (2)$$

Describing the transition from *Follow-Up* to *Disperse*, this equation ensures to keep a connectivity throughout the robot swarm.

The last transition probability is given as

$$p_C = \begin{cases} 0 & \text{, if } b > 40 \\ (1 - \alpha)(-\frac{1}{30}b + \frac{4}{3}) & \text{, if } 10 < b \leq 40 \\ 1 & \text{, else} \end{cases} , \quad (3)$$

where b is the battery level of the robot. This function is used as a threshold to ensure that the robots reach the base at the end of the experiment.

2.2 Path Planner

The path planner is implemented to ensure an energy efficient navigation of the robot. It is based on the A*-algorithm. The underlying graph is generated from the topographic map, which every robot generates individually, while driving around. First, all cells, which contain obstacles, are removed. Second, the cost map, with the remaining cells, based on the height difference and the height change (from robot to target) is calculated:

$$m_{i,j} = 10 \left(|m_{x_i, y_i} - m_{i,j}| + \frac{\delta m}{\delta x} + \frac{\delta m}{\delta y} \right) + 2, \quad (4)$$

where m is the cost map. The +2 is needed, for including cost for unknown territory. The height difference is represented by the linear term of this equation and the height change (height difference from cell to cell) is taken into account by the first derivative. The factor 10 is a scale that allows to consider height differences down to 10 cm.

2.3 Experimental Set-up

We conduct our experiments on three different outdoor maps. The first map (*Flat Obstacle*) is an area of 40 m $\times$ 40 m, bounded by four walls. In addition, this map has a wall in the middle, which divides the map into two spaces. This map is mainly used for proving obstacle and wall detection. The second map (*Four Peaks*) is 200 m $\times$ 200 m and contains four 20 m high hills. It is used to test the value approximation for way points and to ensure that the weighting for path planning is correct. The last map (*Elevation Obstacle*) is also 200 m $\times$ 200 m. It combines the individual features of the other two maps, so that we have obstacles and different altitudes. This map allows testing the overall behaviour. The resolution of a cell of the topographic map is 1 m $\times$ 1 m for all the experiments.

We tested different dynamic approaches for calculating α. First, we chose α purely depending on the number of neighbours in communication range (*Neighbours*),

$$\alpha = \begin{cases} 0.9 & \text{, if neighbours} > 1 \\ 0.7 & \text{, if neighbours} > 0 \\ 0.2 & \text{, else} \end{cases} \quad (5)$$

Second, we chose α randomly from a uniform distribution (*Random*), to check if doing a complex calculation of α is beneficial or if it does not impact the overall behaviour at all. Last, we chose α as a weighted sum

$$\alpha = \sum_{i=1}^{4} x_i k_i, \quad (6)$$

based on distance to the base (x_1), number of neighbours (x_2), distance to neighbours (x_3) and battery level (x_4) (*Weighted Sum*). The weights $k_1, ..., k_4$ are chosen, such that $||\mathbf{k}||_1 = 1$ and they are all set to 0.25. A summary of the experimental set-up is given in Table 1. For experiment EX1 and EX2 three, and for EX3 to EX8 five independent runs have been conducted.

Table 1: Overview experimental set-up

Name	Map	Robots	α-Approach
EX1	Flat Obstacle	1	Neighbours
EX2	Flat Obstacle	4	Neighbours
EX3	Four Peaks	4	Random
EX4	Four Peaks	4	Neighbours
EX5	Four Peaks	4	Weighted Sum
EX6	Elevation Obstacle	4	Random
EX7	Elevation Obstacle	4	Neighbours
EX8	Elevation Obstacle	4	Weighted Sum

3 Results and Discussion

Besides showing the feasibility of a behaviour-based robot control framework for outdoor environments and providing a tool chain to create experiments, we also targeted other challenges, such as the exploration-exploitation trade-off and energy-efficient navigation. To confirm the feasibility of our novel approach, we have conducted several experiments in simulation.

3.1 Results

The first experiment (EX1) proves that the robot is able to explore the given environment in average 15 min, while avoiding walls and other obstacles. Also, it showed that the

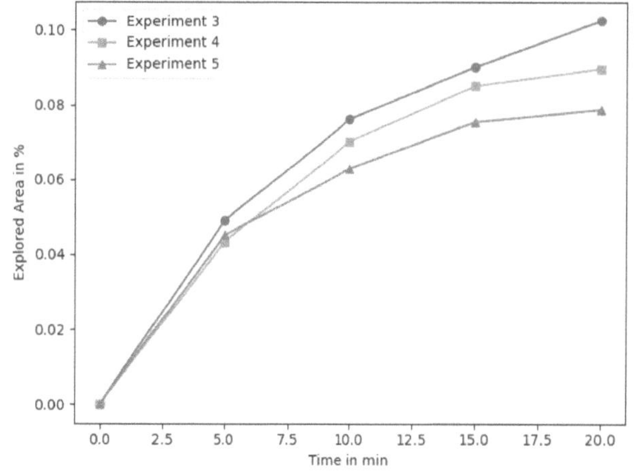

Figure 3: Explored area over time from experiments EX3, EX4 and EX5. It is averaged over 5 runs for each experiment.

Basing-sub-behaviour works properly. The second experiment (EX2) adds to these findings, that the *Follow-up*-sub-behaviour works. It took the swarm in average five min to explore the whole map. The next three experiments (EX3, EX4 and EX5) prove that the exploration in topographic maps work. In Fig. 3, the explored area over time for the three different approaches of α is plotted. Table 2 presents an overview of measured parameters of the experiments. In Fig. 4, we plotted the area coverage over time for experiments EX6, EX7 and EX8.

Table 2: Overview Measured Metrics: **Number of average Neighbours**, number of robots, which are in the communication range of a given robot. **Inter-Swarm-Distance**, average distance between each robot with each other - excluding the closest neighbour. **Closest Neighbour**, robot, which is closest to the given robot. All metrics are averaged over time and individual runs.

Experiment	Number of average Neighbours	Inter-Swarm-Distance (m)	Closest Neighbour (m)
EX1	n.a	n.a	n.a
EX2	0.556	127.49	18.23
EX3	0.172	39.62	21.97
EX4	0.174	35.24	16.68
EX5	0.374	27.58	15.19
EX6	0.201	27.45	17.63
EX7	0.302	26.62	16.43
EX8	0.284	25.35	15.41

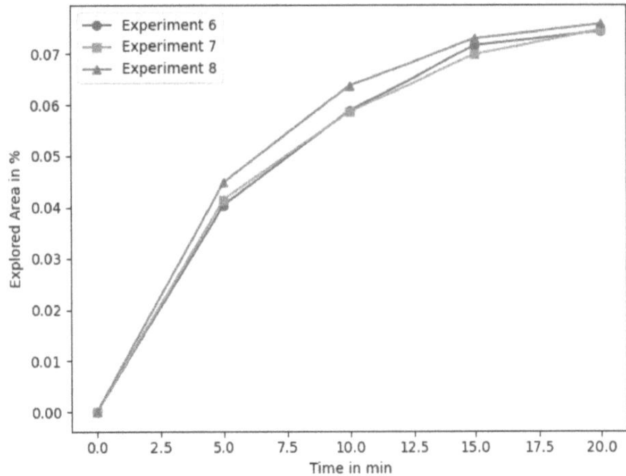

Figure 4: Explored area over time from experiments EX6, EX7 and EX8, averaged over 5 runs for each experiment.

3.2 Discussion

Experiment EX1 and EX2 prove a speed up in exploration, using a robot swarm. This shows the benefit of using robot swarms over individual robots.

From Fig. 3 and Table 2 we can conclude that the *Random*-approach for α explores the most at the expense of connectivity, meaning the robots are more likely to move alone. Putting this in context with the choice of α we observe, that in average, when we take α from a uniform distribution we get $\alpha = 0.5$, which is quite high compared to the more conservative calculation strategies *Neighbours* and *Weighted Sum*. For the experiments EX6, EX7 and EX8, the exploration performance is approximately the same. The *Inter-Swarm-Distance* was also similar in all the experiments, see Table 2. Here, the *Weighted Sum* approach of α was able to keep robot pairs closer together. This indicates that a more complex calculation of α improves the performance of safety by keeping equally good exploration performance. Further, we found in all approaches that the robots start teaming up in pairs, to exploit the given metrics, which is an interesting and desirable behaviour.

4 Conclusion

We were able to show that the proposed approach of using a probabilistic state machine to model behaviour-based exploration, with α used to trade-off exploration and exploitation, is applicable to outdoor environments and achieves promising results. Experiments on how to select α indicated that choosing it small works best in terms of exploration but loosen the swarm connectivity. Thus, we propose to choose it more conservative, to meet potential safety constraints, needed for outdoor scenarios. Interesting behaviour emerged from the given metrics. Further research, includes fine tuning calculation strategies for α and transition probabilities.

Acknowledgement

The work has been carried out at the Institute of Computer Engineering, Universität zu Lübeck. Supplementary Materials are available online at `https://doi.org/10.5281/zenodo.5997532`.

Author's Statement

Conflict of interest: Authors state no conflict of interest.

5 References

[1] P. C. Robert, *Precision agriculture: a challenge for crop nutrition management*. In: Progress in Plant Nutrition: Plenary Lectures of the XIV International Plant Nutrition Colloquium, Springer Netherlands, Dordrecht, pp. 143–149, 2002.

[2] M. Ofori and O. El-Gayar, *Drivers and challenges of precision agriculture: a social media perspective*. In: Precision Agriculture, Springer, 2021.

[3] S. Blackmore, *Precision Farming: An Introduction*. In: Outlook on Agriculture, vol. 23, no. 4, pp. 275–280, 1994.

[4] J. J. Roldán, J. del Cerro, D. Garzón-Ramos P. Garcia-Aunon, M. Garzón, J. de León and A. Barrientos, *Robots in Agriculture: State of Art and Practical Experiences*. In: Service Robots, IntechOpen, Rijeka, 2018.

[5] K. M. Wurm, C. Stachniss and W. Burgard, *Coordinated multi-robot exploration using a segmentation of the environment*. In: 2008 IEEE/RSJ International Conference on Intelligent Robots and Systems, IEEE, pp. 1160–1165, 2008.

[6] A. E. Shenawy, K. Mohammed and H. M. Harb, *Exploration Strategies of Coordinated Multi-Robot System: A Comparative Study*. In: IAES International Journal of Robotics and Automation, vol. 7, no. 1, pp. 48–58, 2018.

[7] J. de Hoog, S. Cameron and A. Visser, *Role-Based Autonomous Multi-robot Exploration*. In: 2009 Computation World: Future Computing, Service Computation, Cognitive, Adaptive, Content, Patterns, pp. 482–487, 2009.

[8] J. Faigl and M. Kulich, *On determination of goal candidates in frontier-based multi-robot exploration*. In: 2013 European Conference on Mobile Robots, pp. 210–215, 2013.

[9] D. Tarapore, R. Gross and K. P. Zauner, *Sparse robot swarms: Moving swarms to real world applications*. In: Frontiers in Robotics and AI, vol.7, 2020.

MuZero: Model-Based Reinforcement Learning

Harsh Yadav [1], Honghu Xue [2], Ngoc Thinh Nguyen [2]
[1] Robotics and Autonomous Systems, Universität zu Lübeck, harsh.yadav@student.uni-luebeck.de
[2] Institute for Robotics and Cognitive Systems, Universität zu Lübeck, {xue, nguyen}@rob.uni-luebeck.de

Abstract

The model-based Reinforcement Learning (RL) algorithm MuZero has recently become the first algorithm to achieve superhuman performance in strategic games like go, chess and shogi and other visually complex games like Atari 2600. This brings the deployment of reinforcement learning algorithms in real world a step closer since a model-based approach is more sample-efficient than model-free approaches can potentially reduce the long training cycle of model-free RL. Our study aims to build a working prototype of this algorithm and study how it can be deployed in practical environments. The results show that MuZero outperforms model-free RL algorithm DQN, however, it is still computationally very expensive. Some key insights shall be presented on what causes the long training time and how could it be mitigated.

1 Introduction

In recent years deep learning has enabled many advancements in the field of perception and inference. However when it comes to decision making, classical methods such as search trees, model predict control etc. are still heavily used, which are sub-optimal as compared to deep learning in utilising a large amount of data efficiently. The use of neural networks to make decisions falls under the umbrella of deep reinforcement learning (RL). RL along with supervised and unsupervised learning constitute machine learning. A typical RL algorithm contains an agent interacting with an environment, perceiving its states, and taking actions to maximise its long-term reward. Usually, RL algorithms use value function to predict the expected reward achievable when the agent starts from a particular state and simultaneously policy function is deployed to predict an action at the current state. Such algorithms are known as model-free RL. If an algorithm also models the dynamics and immediate reward functions of the environment, then it is considered as model-based RL.

The earliest breakthrough in the model-free deep RL was Deep Q-Network (DQN) [1]. The algorithm was able to achieve superhuman performance on Atari 2600 [2] games. While AlphaZero [3], the state of the art in model-based deep RL, can beat the world champion in the game of Go, chess and shogi. The recent model-based RL algorithm Muzero [4] can perform at superhuman level in the game of Go, chess, shogi as well as Atari 2600. This signifies that MuZero can work with complex visual tasks and plan long term strategies. Real-world applications demand the agent to work with both complex visual tasks as well as long term strategic planning and this makes MuZero a good choice for practical applications.

2 Materials and Methods

In this section we shall discuss the background knowledge of RL and MuZero.

2.1 Abstract Markov Decision Process

The mathematical formulation of any RL algorithm is done with Markov Decision Process (MDP). Once an MDP is constructed, Monte Carlo Tree Search (MCTS) [4] can be used to do the planning over it. However, this approach has not worked in the visually complex problem e.g. Atari, due to compounding error in modelling. This problem can be mitigated with Abstract Markov Decision Process (AMDP) [5], which focuses on end-to-end learning of the value function. The idea is to construct an AMDP model such that planning on this model is equivalent to planning in the real environment in terms of receiving the same total reward, given that planning starts from the same initial state. AMDP facilitates the flexibility that the states of its transition model need not match the states in the real environment, rather the MDP model is seen as a deep neural network that can be unrolled to match its values with the cumulative Temporal Difference (TD) reward of the played games.

2.2 MuZero

This section will cover the design of the AMDP model and how MCTS is used to plan over this model. We shall also discuss, how the selected actions are applied on the environment by the agent. Finally, light will be shed on the training methodology of the AMDP model.

2.2.1 AMDP Model

The model (neural network) in MuZero is μ_θ, with θ as the model parameters. During the interaction with environment, predictions are made at every time step t by unrolling the model for further $k = 0, 1, ..., K$ steps. The model is conditioned on the previous observations $o_1, o_2, ..., o_t$ and the future actions $a_{t+1}, a_{t+2}, ..., a_{t+K}$ for $k > 0$ and predicts three quantities: (1) policy: $p_t^k \approx \pi(a_{t+k+1}|o_1, ..., o_t, a_{t+1}, ..., a_{t+K})$, (2) value function: $v_t^k \approx E[u_{t+k+1} + \gamma u_{t+k+2} + ...|o_1, ..., o_t, a_{t+1}, ..., a_{t+k}]$ and (3) immediate reward: $r_t^k \approx u_{t+k}$, where π is the policy which is used to select actions, γ is the discount factor and u is the reward observed in played games. The model consists of three functions: dynamics function g_θ, prediction function f_θ and representation function h_θ. For a generalised representation of these functions subscript t is dropped. The dynamics function (g_θ) is a recurrent process $r^k, s^k = g_\theta(s^{k-1}, a^k)$, which takes the previous hidden state and current action as input and outputs immediate reward and the next hidden state. Here the hidden states are just the embodied representation of true states and have no semantic meaning, except to accurately predict policy, value and immediate reward. The prediction function f_θ takes the current hidden state as input and predicts policy and value, $p^k, v^k = f_\theta(s^k)$. The representation function h_θ builds a hidden state representation based on the previous observation and actions, $s^0 = h_\theta(o_1, ..., o_t, a_1, ..., a_t)$.

2.2.2 Planning

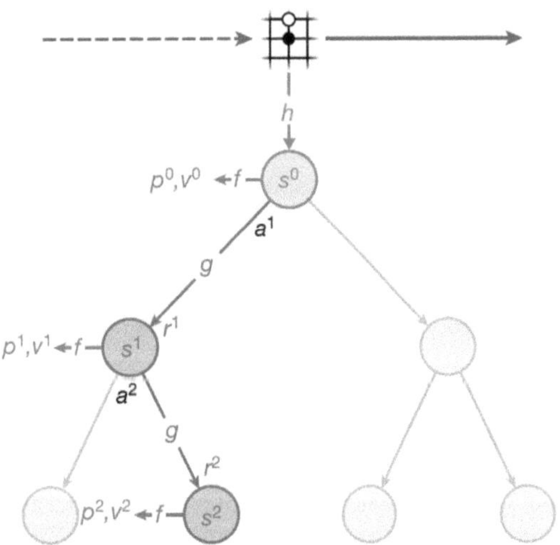

Figure 1: Planning on AMDP, Source [4]

At every time step t, planning is done with MCTS as shown in Fig. 1. First, the stacked observations and actions $(o_1, ..., o_t, a_1, ..., a_t)$ are passed to representation function h_θ, which outputs the hidden state representation s^0 of it. From s^0, an MCTS is unrolled for fixed number of simulations. In these simulations, the state s^0 goes as input to prediction function f_θ, which outputs policy p^0 and value v^0. Based on the policy p^0 action a^1 is selected. The dynamics function takes a^1 and s^0 as inputs and outputs the

next state s^1 and the immediate reward r^1. Similarly, for a state s^k, also referred to as a node, the model computes the following: value v^k, policy p^k and immediate reward r^{k+1}. For efficient exploration in MCTS, at the start of each simulation, prior exploration noise, modelled with Dirichlet distribution, is added to all the children of the root node.

Search statistics. For every node s^{k-1}, corresponding to every possible action a^k, which can be taken from that node, the tree stores the following statistics, containing $\{N(s^{k-1}, a^k), Q(s^{k-1}, a^k), P(s^{k-1}, a^k) = p^{k-1}(a^k), R(s^{k-1}, a^k) = r^k, S(s^{k-1}, a^k) = s^k\}$, where N is number of visits, Q is value, P is policy, R is immediate reward, S is immediate next sate.

Action selection. The action a^k is selected based on the stored statistics of node s^{k-1}, by maximising the Probabilistic Upper Confidence Tree (PUCT) bound, shown in Equation 1.

$$a^k = \arg\max_a \left\{ \bar{Q}(s,a) + P(s,a) \frac{\sqrt{\sum_b N(s,b)}}{1 + N(s,a)} \left[c_1 + \log\left(\frac{\sum_b N(s,b) + c_2 + 1}{c_2} \right) \right] \right\}, \tag{1}$$

$$\bar{Q}(s,a) = \frac{Q(s,a) - \min_{s,a \in \text{Tree}} Q(s,a)}{\max_{s,a \in \text{Tree}} Q(s,a) - \min_{s,a \in \text{Tree}} Q(s,a)}$$

and c_1 and c_2 are constants. $\bar{Q}\left(s^{k-1}, a^k\right)$ is the normalised $Q\left(s^{k-1}, a^k\right)$, which is used to mitigate the problem of unbounded return from the environment.

Leaf node expansion. In every simulation, the algorithm starts from s^0 and finishes by adding a new leaf node s^l to the tree, by expanding the node s^{l-1} along the action a^l, which has never been explored before this simulation. The statistics of s^l are initialised along all possible actions a^{l+1} as $\{N(s^l, a^{l+1}) = 0, Q(s^l, a^{l+1}) = 0, P(s^l, a^{l+1}) = p^l(a^{l+1})\}$, while $\{R(s^{l-1}, a^l) = r^l, S(s^{l-1}, a^l) = s^l\}$ are added to the statistics of node s^{l-1} along action a^l.

Value backup. At the end of the simulation, the statistics $N(s^{k-1}, a^k)$ and $Q(s^{k-1}, a^k)$ are updated along all the nodes that have been visited during that simulation, shown in Equation 2.

$$Q\left(s^{k-1},\ a^k\right) := \frac{N\left(s^{k-1}, a^k\right) \times Q\left(s^{k-1}, a^k\right) + G^k}{N\left(s^{k-1}, a^k\right) + 1},$$

$$N\left(s^{k-1}, a^k\right) := N\left(s^{k-1}, a^k\right) + 1, \tag{2}$$

$$\text{where } G^k = \sum_{\tau=0}^{l-1-k} \gamma^\tau r_{k+1+\tau} + \gamma^{l-k} v^l$$

and v^l is the value of the leaf node.

2.2.3 Acting

Once MCTS is completed, the search policy $\pi_t = P[a_{t+1} \mid o_1, \ldots, o_t]$ is constructed for the root node, which

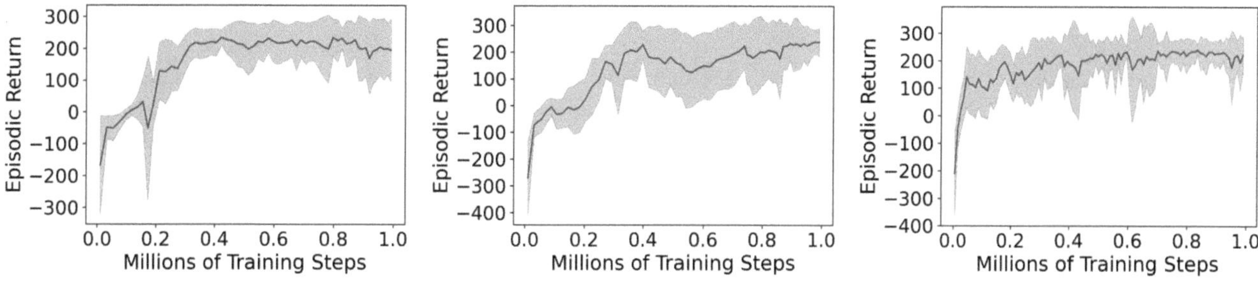

Figure 2: Evaluation of MuZero throughout training on lunar lander

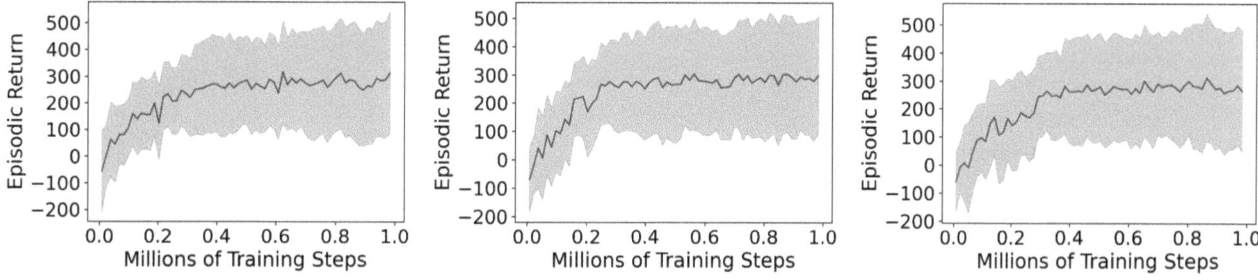

Figure 3: Evaluation of DQN throughout training on lunar lander

is proportional to the number of visits to immediate children of the root nodes, shown in Equation 3.

$$\pi(a \mid s) = \frac{N(s,a)^{1/T}}{\sum_b N(s,b)^{1/T}}, \tag{3}$$

where T, temperature parameter, controls the degree of exploration vs exploitation. The agent acts with the selected action a_{t+1} and receives the immediate reward u_{t+1} and next observation o_{t+1}. The recursive process of using MCTS to construct a better policy for selecting an action runs until either the game finises or a maximum number of game steps are played. The finished games are then stored in a replay buffer.

2.2.4 Training

At every iteration of training, a k-step sequence of a played game is selected from replay buffer, starting at time step t. The samples are drawn as per the priority replay buffer [4]. The model is then trained to match predicted policy p_t^k with search policy π_{t+k}, predicted root node value v_t^k with n step TD-return z_{t+k} and predicted immediate reward r_t^k with immediate reward from game u_{t+k}. Here $z_t = u_{t+1} + \gamma u_{t+2} + \ldots + \gamma^{n-1} u_{t+n} + \gamma^n v_{t+n}$, is the n-steps TD return. The loss function is given in Equation 4.

$$l_t(\theta) = \sum_{k=0}^{K} l^{\mathrm{p}} \left(\pi_{t+k}, p_t^k \right) + \sum_{k=0}^{K} l^{\mathrm{v}} \left(z_{t+k}, v_t^k \right)$$
$$+ \sum_{k=1}^{K} l^{\mathrm{r}} \left(u_{t+k}, r_t^k \right) + c\|\theta\|^2, \tag{4}$$

where c is the regularisation parameter.

3 Results and Discussion

In this section, we present the results as well as other important setup problems such as the choice of hyperparameters and neural network architecture.

3.1 Results

In this study, Muzero [4] was evaluated on the lunar lander problem in OpenAI Gym [6] and the corresponding results are shown in Fig. 2. The algorithm was able to solve the lunar lander problem only after 0.4 million training steps. The results were compared with DQN [1], shown in Fig. 3, which also achieves the same level of performance after a similar number of training steps. It is also worth noticing that the standard deviation in episodic return in case of MuZero is smaller than that in case of DQN. Furthermore, it is important to mention the corresponding training times for 1 million training steps. MuZero took around 15 hours while DQN took approximately 1 hour. There are two reasons behind the longer training time in MuZero. First, the number of decision steps were 2.9 million in MuZero as compared to 1 million in DQN. Second, while there is no planning in DQN, in MuZero MCTS planning incurs cost at every decision step. For more details, the training time of MuZero on Atari was even larger and while the algorithm did show initial signs of learning after 7 days of training, a complete evaluation was not feasible due to lack to computational resources. Also as reported by the inventors of MuZero, it surpasses the standard RL algorithms already in early training phase but to get better results they ran it for 20 billion decision steps.

Table 1: Fully connected neural network architecture for lunar lander

Network	Input	Hidden	Output
Representation	Stacked observations + action broadcast (8*33 + 8*32)	NA	Hidden state(10)
Dynamic state	Hidden state + one hot action (10 + 4)	64	Hidden state(10)
Dynamic reward	Hidden state(10)	64	Full support size (21)
Prediction policy	Hidden state(10)	64	Action space (4)
Prediction value	Hidden state(10)	64	Full support size(21)

3.2 Hyperparameters

The hyperparameters used to evaluate MuZero on lunar lander are shown in Table 2.

Table 2: Hyperparameters for lunar lander

Hyperparameter	Value
Stacked observations	32
MCTS simulations	50
Maximum episodic length	700
MCTS prior dirichlet noise, α	0.25
PUCT parameter c_1	1.25
PUCT parameter c_2	19652
Batch size (games)	64
Optimizer	Adam
Weight decay	10^{-4}
Momentum	0.9
Constant learning rate	$5 * 10^{-4}$
Replay buffer (games)	5000
TD-steps for n-step return	30
Length of games for training (k)	10

3.3 Network Architecture

The model only consist of fully connected layers as inputs were not visual in nature. It consisted of 5 individual networks shown in Table 1. The environment is modified to only take discrete value of actions and hence action space becomes $\{0, 1, 2, 3\}$, corresponding to thrust, left, right and nothing. The observation space of the lunar lander is a vector of length 8. The input to the representation network is recent observation stacked with the latest 32 observations and corresponding actions. The stacked actions are broadcasted into a vector of length same as observations, e.g. if thrust action was selected in a particular time step, then the action broadcast will be $\{0, 0, 0, 0, 0, 0, 0, 0\}$. The hidden states are vectors of length 10 in all the networks. The one hot action in the input to dynamic state network would be $\{1, 0, 0, 0\}$ corresponding to thrust as action. Before computing the loss, n-step TD return z^{t+k} and reward u^{t+k} are scaled using an invertible transformation $h(x) = \text{sign}(x)(\sqrt{|x| + 1} - 1) + 0.001x$ followed by a transforming the scalar value and reward to categorical representation. The categorical representation is of length 'Full support size' from -10 to +10. Under this transformation each scalar is represented as the linear combination of the adjacent supports and the original value can be recovered by the function, $x = x_{\text{low}} \times p_{\text{low}} + x_{\text{high}} \times p_{\text{high}}$.

4 Conclusion

Even though MuZero has been able to outperform other model-free RL algorithms, the training time remains an issue for any practical application. Two possible hurdles to overcome are sample inefficiency i.e. large number of samples are needed for a certain level of learning and slow planning with MCTS.

Acknowledgement

The work was carried out and supervised by the Institute for Robotics and Cognitive Systems, Universität zu Lübeck.

Author's Statement

Conflict of interest: Authors state no conflict of interest.

5 References

[1] V. Mnih, K. Kavukcuoglu, D. Silver, A. Graves, I. Antonoglou, D. Wierstra, and M. Riedmiller, "Playing atari with deep reinforcement learning," 2013.

[2] M. G. Bellemare, Y. Naddaf, J. Veness, and M. Bowling, "The arcade learning environment: An evaluation platform for general agents," *Journal of Artificial Intelligence Research*, vol. 47, pp. 253–279, 2013.

[3] D. Silver, T. Hubert, J. Schrittwieser, I. Antonoglou, M. Lai, A. Guez, M. Lanctot, L. Sifre, D. Kumaran, T. Graepel, T. Lillicrap, K. Simonyan, and D. Hassabis, "Mastering chess and shogi by self-play with a general reinforcement learning algorithm," 2017.

[4] J. Schrittwieser, I. Antonoglou, T. Hubert, K. Simonyan, L. Sifre, S. Schmitt, A. Guez, E. Lockhart, D. Hassabis, T. Graepel, T. Lillicrap, and D. Silver, "Mastering atari, go, chess and shogi by planning with a learned model," *Nature*, vol. 588, p. 604–609, Dec 2020.

[5] D. Silver, H. van Hasselt, M. Hessel, T. Schaul, A. Guez, T. Harley, G. Dulac-Arnold, D. Reichert, N. Rabinowitz, A. Barreto, and T. Degris, "The predictron: End-to-end learning and planning," 2017.

[6] "Lunarlander-v2." https://gym.openai.com/envs/LunarLander-v2/. Accessed: 14/01/2022.

Development of a Federated Learning based classification of handwritten digits for a mobile application

Philipp Goldbach [1]

[1] Medical Informatics, Universität zu Lübeck, philipp.goldbach@student.uni-luebeck.de

Abstract

A major problem in machine learning is the lack of data to sufficiently train a model [1]. In many use cases, required data exists at competitors, partners, or academic institutions that cannot be shared for privacy reasons. With the federated learning method (FL), models can be generated and trained on external data in order to subsequently merge the learned knowledge of many distributed models without having to send the training data. It was investigated to what extent the classification performance of the FL model of models differs from the "conventional learning method", Centralized Learning (CL), when trained with the same set of image data. As a result, it is found that the generated FL model doesn't achieve the same classification performance as the CL model. Further optimization techniques are investigated to counteract this. Nevertheless, the FL is a suitable method to train models confidentially over distributed datasets while respecting privacy.

1 Introduction

Considering at potential applications in the field of medicine where AI systems can assist in the diagnosis of diseases, a lot of medical images are needed to train a classifier. The creation of these required medical images is a standard procedure in the treatment of patients. In about 80% of full inpatient treatments, at least one medical image of the patient is taken. This corresponds to 13.2 million medical images in 16.4 million patients in 2020 [2]. Data, similar to this scenario, often exists in distributed locations and could be used collectively to train a model. The distributed data is often subject to privacy protections, such as being personal data, as is the case with medical images, or cannot be shared for other privacy reasons.

This is where federated learning comes in, as it can be used to train a model over distributed data sets without having to exchange the data [3]. Models are generated at the data set, trained locally at the respective site, and then the learned distributed knowledge is merged at a central instance by sending only the abstract learning parameters and no training data. Thus, the learned knowledge of many models is combined to produce a stronger classifier at a central location.

In this work, the centralized learning (CL) method was compared with the federated learning (FL) method in terms of classification performance. For this purpose, a mobile application for handwritten digit recognition was developed, which was used on multiple devices and successively generated training data by user input. On the one hand, a model with CL was trained with all images from all participating end devices. On the FL model, the locally trained models were aggregated. We investigated how the classification performance of the two methods differs. In addition, it was investigated to what extent the number of nodes involved in FL influences the classification performance during aggregation. Furthermore, it was investigated whether the exclusion of weak nodes in FL can be used to positively influence the learning process or to prevent a deterioration.

2 Methods

In order to make the FL method comparable with CL, a distributed application was first implemented. The centralized FL approach was taken into account, in which a central instance coordinates the learning process with several participating nodes [4]. In this section, we will describe which tasks are performed by the central instance and which by the participating nodes. In this context, we will take a closer look at how the individual training is implemented on the participating nodes and how the distributed knowledge is reunited at the central instance. Finally, it will be described how the centralized approach of the FL is compared to the CL approach in order to make their classification performance comparable given the same training data.

2.1 System architecture

The tasks of the central instance at the FL are the generation of a base model, the provision of the current model, which can be obtained from nodes and further trained individually, and the merging of the models adapted by the nodes. The latter means that the learning parameters of the further trained models are aggregated in such a way that an updated model is created, which combines the accumulated knowledge of the nodes and, in the optimal case, leads to a stronger classifier as described in subsection 2.4.

On the user side, a mobile application is provided that generates further training data through user input, which is used by the terminal for local training. This results in a total of N individually adapted models for N end devices, which are then returned to the central instance for the purpose of merging.

Communication between the central instance and nodes is based on HTTP requests. The central instance provides a REST API via which the current model can be obtained from the server via HTTP GET request (Figure 1 a). An individually trained model of a node can be sent to the central instance via HTTP PUT request (Figure 1 b). Thus, by feeding back the further trained models from generation to generation, a new model is created at the central instance, generated by merging the different models of the nodes (Figure 1 c).

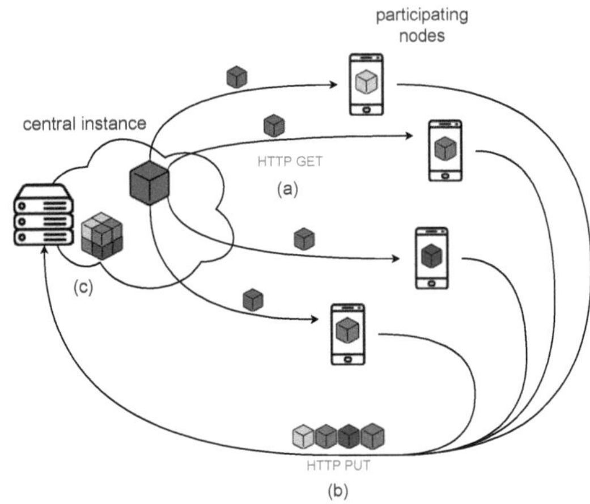

Figure 1: (a) Using a REST API, nodes can obtain a model. (b) Updated models can be returned to the server. (c) The aggregation of the models happens at the central instance.

2.2 Creation and Provision of the Base Model

At the central instance, a model is first generated, which is then to be obtained from participating nodes and trained further. This base model consists of two convolutional layers for feature extraction and a fully-meshed layer for classification of the digit in the input image based on extracted features.

A dataset of 100 images in total is used to train this base model, which is insufficient to produce a strong classifier. In order to simulate a real use case as well as possible, the base model should not yet achieve a strong classification performance (>90%) in recognizing handwritten digits, because a common problem in the implementation of AI projects is the lack of available data [1]. Second, improvements in classification performance should be measurable and detectable by applying the FL method, which would be challenging should the classifier already be very strong. By training with this self-generated dataset, the model achieved

an accuracy of 50% ± 2% when evaluated with a test dataset of also 100 images.

After training, the base model in the first generation of the FL can be obtained from nodes and further trained within the mobile applications using their own local training data. In general, for the FL, nodes in all subsequent generations always obtain the last updated / aggregated model from the central instance.

2.3 Further training on nodes

Nodes obtain the current model of the central instance when the mobile application is started. In the first generation, this means that the pre-trained base model is obtained. In subsequent generations, the last aggregated model. The sourced model is used during the session in the mobile application to classify entered handwritten digits. To do this, the user enters a digit using a drawing field (Figure 2 (a)).

Then, the input is classified by the referred model. After that, the user can rate the prediction as correct or incorrect (Figure 2 (b)). If the prediction is correct, the input image with the prediction is saved locally as a label and kept for the later training process. If the prediction is incorrect, the user can now correct the prediction (Figure 2 (c), Figure 2 (d)). The input image is then also saved locally with the corrected label. In this way, new training data is successively created on the end device.

This process is repeated until a total of more than 30 training images are available locally and the data set is balanced, i.e. there is exactly the same number of training images for each class. This is to prevent the model from specializing in a particular class.

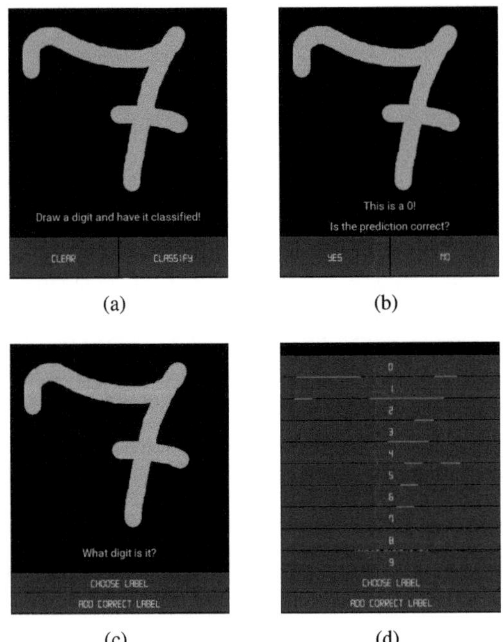

Figure 2: (a) User enters handwritten digit. (b) Prediction to users input; user can rate it as correct or incorrect. (c) User can set the correct label. (d) Listing of possible labels.

2.4 Aggregation of the models

Nodes can continue to train a model of the same generation any number of times. A model further trained by a node is sent back to the central instance after the local training process on the end device. The transmission is done via HTTP PUT request, which ensures that only the last further trained model of a node is stored in each generation [5].

The central instance starts aggregating the returned models after a certain number of nodes have returned a model. This threshold parameter is customizable in the application, which allows to better investigate a correlation between increasing number of nodes and increase of classification performance of the aggregated model (Figure 3).

The aggregation of the model is done according to the FedAvg method [3] by forming the arithmetic mean for each weight w over all node models K.

An addition to this approach, which resulted from an observation in section 3, is the filtering of weak node models. Only node models whose classification performance exceeds that of the last aggregated model by a certain threshold are included in the aggregation. This threshold is also adjustable in the application of the central instance. This procedure is to avoid that weak node models affect the new aggregated model during the aggregation in order to obtain an improved model in the result.

2.5 Generation of the reference model

The reference model is trained with CL and is intended to create a comparison between the FL and the CL. The reference model therefore requires the images generated on the nodes. Contrary to the principle of the FL process, the mobile application has been extended in this regard. After each local training process on a node or terminal, the images used for training are sent to the central instance in order to train the reference model at a later time. This should provide the reference model with the same image data for training that was implicitly included in the aggregated model when the learning parameters were merged.

The basis for the reference model of a new generation is always the base model delivered at the beginning or rather the model of the first generation. This is further trained with the help of all image data uploaded up to this point. Like the aggregated model, the reference model is trained as soon as sufficient nodes have sent their image data or learning parameters for aggregation to the central instance. Thus, the two resulting models of one generation will become direct opponents, which can be compared in the evaluation.

3 Results and Discussion

Finally, the CL was compared with the FL by generating a model for each of the two methods with the same training data. By using the mobile app by test persons, training data could be successively generated on the end devices. On the one hand, these were used to train models on participating nodes, which then flowed into the aggregation of a FL model. On the other hand, the training data was sent to the central instance to generate a CL model.

This was done generation by generation, i.e. as soon as a sufficient number of nodes sent a model back to the central instance (here: 3, 5 or 7 nodes) the node models were aggregated. At the same time, the image data collected on the terminals up to this exact same point in time was collected at the central instance to train a CL model with it. Thus, the CL model and the FL model of the same generation became direct opponents, since they were implicitly further trained with the same training data. The described procedure was applied in a total of six different scenarios.

The scenarios differed in the number of nodes or end devices used and in the activated or deactivated filtering of weak nodes as described in subsection 3.2.

3.1 Increasing the number of nodes

Figure 3 (a-c) shows the classification performance of the CL model and the FL model at 3, 5 and 7 nodes without filtering weak models; Figure 3 (d-f) shows the same development with filtering weak models. For both methods, it is noticeable in the FL model that the more nodes involved in the aggregation, the steeper the classification performance curve becomes. At three nodes, an outstanding increase in performance occurs in the fifth generation compared to previous generations. For five nodes, such an exceptional increase already exists in the third generation and for seven nodes in the second generation.

3.2 Filtering of weak clients

Figure 3 (d-f) shows the evolution of the classification performance with three, five and seven nodes, where weak models (see subsection 2.4) are excluded during aggregation in each case. Here it can be seen that filtering weak models results in devastating drops in classification performance being greatly reduced, but performance drops cannot be completely avoided. This effect is particularly noticeable in the scenario with five nodes involved Figure 3 (e). Filtering weak models prevents a strong drop of the classification performance from the fourth to the fifth generation (compare Figure 3 (b) & (e)). This effect can also be seen in the scenario with seven nodes in the further development of the second to the third generation. Here, the weak models impair the performance in the scenario with seven nodes without filtering (Figure 3 (c)) to such an extent that the strong increase, which can be seen in the third generation with the same number of nodes and activated filtering (Figure 3 (f)), completely fails to appear.

3.3 Discussion

Overall, it was found that as the number of nodes increases, a noticeable increase in performance occurs sooner. The more nodes that participate, the higher the overall performance of the FL process appears to be. However, this would have to be investigated in more detail by measurements with

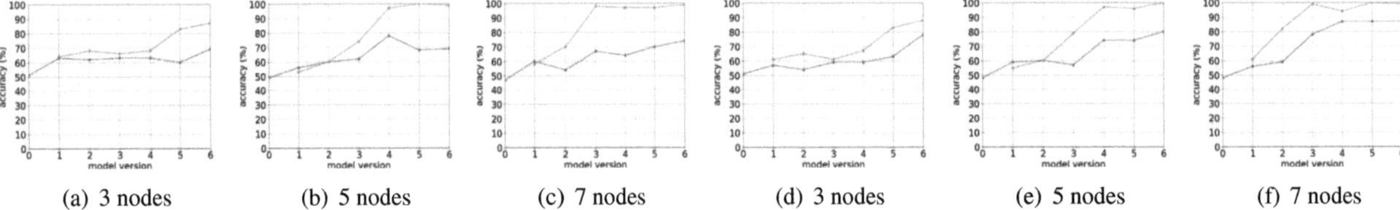

| (a) 3 nodes | (b) 5 nodes | (c) 7 nodes | (d) 3 nodes | (e) 5 nodes | (f) 7 nodes |

Figure 3: Comparison of classification performance of Federated Learning models (blue line) with the Centralized Learning models (orange line) over a total of 6 generations. (a-c) show the evolution of each classification performance when weak nodes are not filtered. (d-f) show the evolution of the classification performance when weak nodes are filtered.

more nodes over more generations. This could not be realized here due to the small number of participants.

Filtering the models with low classification performance could greatly reduce strong performance drops, but not completely prevent them. This could also be used to prevent manipulation of classification performance, where users intentionally attach false labels to data to degrade models.

However, it is also noticeable that the classification performance of the aggregated models generated using the FL principle were still inferior to the CL models in terms of classification performance. Further measures could be investigated to counteract this effect. For example, a higher weighting of strong models in the aggregation would be conceivable or the investigation of alternative aggregation algorithms instead of the FedAvg algorithm [6].

4   Conclusion

The result is a web application that, through the FL method, provides a way to train a classifier without exchanging image data but only the learning parameters of the node models. Thus, compared to CL, where image data would be exchanged, it offers the advantage of increased privacy, since confidential data does not have to be sent over an insecure channel. The FL thus provides a way to bring together the knowledge of many distributed entities in a confidential manner, and can thereby foster cooperation among many companies or academic institutions in generating a common classifier that they would not be able to create individually due to insufficient data. It was shown that such a merging of learning parameters can lead to a stronger model with higher classification performance. Here it could be investigated that by increasing the number of involved nodes during aggregation leads to an earlier increase of the classification performance. It was also shown that filtering weak models can prevent severe losses in terms of classification performance and can be suitable as a method to prevent deliberate manipulation of the aggregated model. The developed application is so adaptable that the underlying model is interchangeable. Likewise, other aggregation algorithms can be used and specifically adapted to the use case. Thus, it does not form a final product but an adaptable system. A next step in the further development of the system is therefore the transfer to a medical context. In this context, the application is to be used in the context of a project in the

field of AI-supported diagnostics for the detection of eye diseases (PASBADIA at Technische Hochschule Lübeck).

Acknowledgement

The work has been carried out at Technische Hochschule Lübeck, CoSA and supervised by Prof. H. Hellbrück, Institute of Telematics, Universität zu Lübeck.

Author's Statement

Conflict of interest: Authors state no conflict of interest.

5   References

[1] T. Hagendorff and K. Wetzel, *15 challenges for AI: or what AI (currently) can't do.* AI & Society 35, 355–365, 2020.

[2] Statistisches Bundesamt, *Operationen und Prozeduren der vollstationären Patientinnen und Patienten in Krankenhäusern (4-Steller) 2020.* Available: https://www.statistischebibliothek.de/mir/receive/ DEHeft_mods_00136766 [last accessed on: 2022-02-08]

[3] McMahan, H. Brendan; Moore, Eider; Ramage, Daniel; Hampson, Seth; Arcas, Blaise Agüera y., *Communication Efficient Learning of Deep Networks from Decentralized Data.* AISTATS. 2016.

[4] Kairouz, Peter; McMahan, H. Brendan; Avent, Brendan; Bellet, Aurélien; et. al, *Advances and Open Problems in Federated Learning.* Foundations and Trends in Machine Learning Vol 4 Issue 1. 2019.

[5] IETF, *Hypertext Transfer Protocol.* Available: https://datatracker.ietf.org/doc/html/rfc2616 [last accessed on 2021-01-04].

[6] Ek, Sannara; Portet, François; Lalanda, Philippe; Vega, German, *A Federated Learning Aggregation Algorithm for Pervasive Computing: Evaluation and Comparison,* PerCom 2021, 2021.

Perspective Transformation and Reconstruction using Neural Networks in the Context of Automotive Systems

Daniel Henning [1], Julian Petzold [2], and Heiko Hamann [2]

[1] Robotics and Autonomous Systems, Universität zu Lübeck, daniel.henning@student.uni-luebeck.de
[2] Institute of Computer Engineering, Universität zu Lübeck, {petzold, hamann}@iti.uni-luebeck.de

Abstract

Autonomous driving systems are expected to react quickly in safety-critical situations. To achieve this, it is essential to predict the behavior of other road users. The question arises as to whether their perspective offers more insight into future states and actions compared to that of the vehicle alone. The approach presented here transforms the perspective of the vehicle to a nearby pedestrian using only local sensor information, i.e. an RGBD-camera. Using simulation data, we show that the pedestrian's perspective can be reconstructed to a certain degree, which is limited by the vehicle's perception. Missing information is filled in using semantic inpainting, producing results that are close to the actual perspective. However, large uncertainty regions can lead to image artifacts and significant deviations.

1 Introduction

Advanced driver assistance systems (ADAS) such as adaptive cruise control (ACC), lane-keeping systems (LKS) or collision avoidance systems (CAS) [1], are already common features that can be found in many modern cars today. They not only support the driver in terms of a more relaxed driving experience but also contribute to road safety [2]. With this, autonomous driving is increasingly moving into the realm of reality, as many real-world tests are currently being carried out on the roads [2]. Ensuring that these systems are reliable and safe for all road users is paramount.

Pedestrians are particularly at risk here, as they are very vulnerable in the event of an accident. While today's systems typically only use people's recent trajectories to predict their future states and actions, humans have a far more developed and intuitive understanding of how other road users might behave [3]. Not only their recent trajectories, but also viewing direction, gestures, body postures and even facial expressions are taken into account to predict future actions [3]. This is where current systems lack intuition and interaction, such as eye contact, which cannot be easily replicated with autonomous systems. Crosswalks for example are a common challenge that requires the ability to determine whether a pedestrian wants to cross the road or not. Understanding where the pedestrian is looking and what he or she is seeing might be a good clue.

This is the main concern of this internship, which is part of the NEUPA project [4]. It aims to demonstrate the explainability of artificial neural networks (ANN) with the development of a driver assistance system (DAS) as a use case to enable certification of these systems in the medium term. The particular project task that this internship is dedicated to is reconstructing the perspective of the pedestrian, using local information of the vehicle only. It is like sitting in the vehicle and imagining what the pedestrian might be seeing from their point of view.

2 Related Work

Along with recent developments of advanced driver assistance systems (ADAS), predicting dangerous behavior of pedestrians has become a relevant research task [3]. After all, the sooner a hazard is detected, the sooner the vehicle can react. Whether a pedestrian is about to behave dangerously can be predicted by understanding what the pedestrian is seeing. Here, this is done by geometrically transforming the vehicle's perspective.

Perspective transformation is a common term in image processing that is referred to as deforming the pixel grid of an image to change its perspective [5]. This is mainly done to geometrically align an image of a planar surface, for example to obtain the bird's eye view of the road surface for lane detection [6]. Here we refer to it as changing the vehicle's perspective to that of a pedestrian. That means using local information only, obtained by various vehicle sensors, reconstructing the image that the pedestrian perceives from their position and orientation. This can be done geometrically on the basis of a 3D world model.

The resulting image might contain gaps, i.e., missing information. These can be filled using semantic inpainting, which is the learning-based approach to image inpainting [7]. Semantic inpainting approaches [8] are able to produce results based on the context of the image. This can be trained using convolutional neural networks or generative adversarial networks (GANs) [9].

3   Perspective Transformation

In the context of this paper, perspective transformation refers to reconstructing an image from the perspective of a pedestrian, using only local information from the vehicle. For the purpose of this project, traffic situations are simulated using CARLA [10]. CARLA is an open-source API based on Unreal Engine 4 that is used to develop, train and validate autonomous driving systems. Here it was used to collect vehicle sensor data like RGB images and depth information (Fig. 1), as well as the ground truth perspective of the pedestrian. For this purpose, the vehicles were equipped with depth cameras and left to drive around autonomously to generate a dataset. To model the pedestrian perspective, images with randomly varied transformations were taken in front of the vehicle. Thus eliminating the need to capture actual perspectives of pedestrians in the simulation. This simplifies the process without compromising the result because, after all, a pedestrian cannot see him or herself. In addition to the image of this perspective, the transformation is also saved in order to be able to reconstruct it in the following step.

(a) rgb image   (b) depth image

Figure 1: Sensor data captured by the vehicle within the CARLA [10] simulation.

3.1   Geometric Transformation

As a basis for the perspective transformation, a geometric approach is used. It is the first step, in which both the RGB and the depth image from the vehicle are combined to form RGB-D image, such that each RGB pixel is mapped to its corresponding position (x, y, z) in space. Let

$$x = \frac{z(u - c_x)}{f_x}, \quad y = \frac{z(v - c_y)}{f_y}, \quad z = \frac{d}{d_{scale}}, \quad (1)$$

where u and v define the position of any pixel on the sensor with respect to the offset from the image center c_x and c_y. The focal length in each direction is given by

$$f_x = w \left(2 \tan \left(\alpha_x \pi / 360\right)\right)$$
$$f_y = h \left(2 \tan \left(\alpha_y \pi / 360\right)\right), \quad (2)$$

where the field of view (FOV) in degrees is denoted by α_x, α_y and the height h and width w represent the resolution of image.
This can be visualized as an RGB point cloud (Fig. 2), allowing the perspective to be transformed to any other position and orientation in space. The pedestrians transformation is assumed to be known and was saved along with the ground truth perspective as extrinsic matrix. It is defined as

$$M_{\text{ext}} = \begin{bmatrix} R_{3\times3} & T_{3\times1} \\ 0_{1\times3} & 1 \end{bmatrix}_{4\times4} \quad (3)$$

with $R_{3\times3}$ being the rotation and $T_{3\times1}$ the translation matrix. Here, the camera is transformed within the three-dimensional representation of the point cloud, resulting in an image of the pedestrians perspective as captured by the vehicle.

Figure 2: The RGB point cloud obtained by combining the rgb and depth image.

3.2   Semantic Inpainting

The geometrically transformed image might be missing information caused by occlusions, a limited field of view or the limited resolution of the sensory data. This can be filled using semantic inpainting, for which a generative adversarial network (GAN) [9] based approach was used. The general idea of a GAN is to train two networks simultaneously. Firstly, a generator generates images based on a certain input and secondly, a discriminator tries to distinguish the generated image from the real one, leading to a competition between the two networks to outperform the other.
Here, *Hyperrealistic Image Inpainting with Hypergraphs* [11] was used, capable of learning complex relationships within the data using hypergraph convolution on spatial features. Conventional convolutional neural networks cannot capture global features of the image since they are limited by the size of their kernel [11]. Relationships between these local features can be learned and represented using hypergraph layers, creating high-level feature maps. Compared to simple graph structures, hypergraphs can connect more than two nodes with one edge. It is a two-stage network, which initially fills the missing information of the input image only roughly and then refines it. That means, the rough prediction of the coarse network is used as the input for the refined network. The loss function is defined as the weighted sum of L1, adversarial, perceptual and edge-preserving loss, which compares the prediction to the ground truth perspective of the pedestrian. It was trained for 75 epochs and a batch size of 8 using 26 000 images of

| | | | |
| (a) vehicle image | (b) perspective transformation | (c) predicted perspective | (d) true perspective |

Figure 3: Perspective transformation steps from the image the vehicle perceives (a), transformed to the perspective of a pedestrian (b), to the resulting through semantic inpainting (trained using CARLA) supplemented image (c), in comparison to the ground truth image (d).

the geometrically transformed perspective, described in the section 3.1. This took 96 hours using an Nvidia GeForce RTX 3090.

4 Results and Discussion

For evaluation, a distinction between the results from the geometric transformation and semantic inpainting must be made.

The geometric transformation determines the projection of the RGB point cloud from the perspective of a pedestrian, for which the transformation, i.e., position and orientation, is assumed to be known. As shown in Fig. 3b, the perspective can be represented well using this three-dimensional data structure. However, some pixel that would be hidden by objects that are not completely captured by the sensory data are now visible. For example, the street behind the rear part of the car in the lower figure of 3b should not be visible. This influences the result since given pixels are not replaced in the inpainting step. There are approaches that remove hidden points [12] under the assumption that the captured object is roughly uniform. This is not possible here without knowing the shape of the object.

As for image inpainting, this can be evaluated both quantitatively and qualitatively.

Quantitative Results: The reconstruction error is generally not a good measure for analyzing the performance of inpainting approaches given that there can be multiple reasonable results for an input image [11]. But since we want to achieve results that approximate and reconstruct the pedestrian's actual perspective as closely as possible, it certainly makes sense here. Table 1 shows the mean ℓ_1, ℓ_2 loss and the peak signal-to-noise ratio (PSNR) for a test set of

3 000 images created from the perspective transformation (see Fig. 3b).

dataset	ℓ_1 loss	ℓ_2 loss	PSNR
unreconstructed image	42,26%	25,54%	6,24 dB
Places2 256x256	18,03%	5,27%	13,01 dB
CARLA	10,56%	2,63%	16,28 dB

Table 1: Quantative results comparing models that were trained on different datasets in terms of mean ℓ_1, ℓ_2 loss and the peak signal-to-noise ratio (PSNR).

The ℓ_1 and ℓ_2 losses, defined as

$$\ell_1 = \sum_{i=1}^{n} \left| y_{\text{true}} - y_{\text{predicted}} \right|, \tag{4a}$$

$$\ell_2 = \sum_{i=1}^{n} \left(y_{\text{true}} - y_{\text{predicted}} \right)^2, \tag{4b}$$

are a measure of how close the prediction is to the ground truth, where smaller is better. The peak signal to noise ratio is given by

$$\text{PSNR} = 10 \cdot \log_{10} \left(\frac{\text{MAX}_I^2}{\ell_2} \right). \tag{5}$$

In addition to the mean squared error (ℓ_2 loss), the PSNR contains the maximum possible pixel value MAX_I of the image to make results with different color depths comparable. Here, higher is better. These are common metrics that are used to compare inpainting approaches.

Qualitative Results: Figure 3c shows three examples of inpainted images. The network was able to fill in missing parts. It has no problems with small regions and even larger uniform regions like the sky or the road. Outlines of new buildings were created and sidewalks were expanded. However, more complex relationships within the image are not captured by this inpainting approach resulting in crosswalks and other road lines being cut off. Some objects like pedestrians and cars are partially or completely missing. As already mentioned, this is also due to some inaccuracy in the input image.

5   Conclusion

The approach presented here demonstrates the ability to transform the perspective well using a simulation. This was due to the two step process of first creating a three dimensional data representation, i.e., the point cloud, in which the perspective was changed to a desired one, namely the pedestrian's, and inpainting any missing information in the second step.

Upcoming tasks for this project include training on larger datasets with more variety and a higher resolution, as well as using other inpainting architectures and examining different types of inputs (e.g. a larger FOV or semantic segmentation). There is also the idea of using the raw point cloud as input for a neural network to mitigate the problem of having visible hidden points as the input for inpainting. Translating this approach into reality poses new challenges, namely, determining the position and viewing direction of the pedestrian, as well as examining whether the model trained on simulation data needs to be retrained and if so, how to obtain real world data. Once this method has been proven to be reliable, it can be used to predict dangerous behavior of pedestrians.

Acknowledgement

The work has been carried out at the Institute of Computer Engineering, Universität zu Lübeck.

Author's Statement

NEUPA is funded by the *Federal Ministry of Education and Research* (BMBF), grant agreement no. 01 IS 19078.

6   References

[1] R. Rajamani, *Vehicle Dynamics and Control*. 2nd ed., Mechanical Engineering Series, Springer, 2012.

[2] T. Winkle, *Safety Benefits of Automated Vehicles: Extended Findings from Accident Research for Development, Validation and Testing*. In: Autonomous Driving, Springer, pp. 335–364, 2016.

[3] A. Rudenko, L. Palmieri, M. Herman, K. M. Kitani, D-M. Gavrila and K. O. Arras, *Human Motion Trajectory Prediction: A Survey*. In: The International Journal of Robotics Research, vol. 39, no. 8, Sage Publications Sage UK, London, pp. 895–935, 2020.

[4] H. Hamann, *NEUPA*. `https://www.neupa.net`, Institute of Computer Engineering, Universität zu Lübeck.

[5] G. Bradski and A. Kaehler, *Learning OpenCV: Computer Vision with the OpenCV Library*. O'Reilly Media, Inc., 2008.

[6] M. Bertozzi and A. Broggi, *GOLD: A Parallel Real-Time Stereo Vision System for Generic Obstacle and Lane Detection*. IEEE Transactions on Image Processing, vol. 7, no. 1, IEEE, pp. 62–81, 1998.

[7] C. Guillemot and O. Le Meur, *Image Inpainting: Overview and Recent Advances*. IEEE signal processing magazine, vol. 31, no. 1, IEEE, pp. 127–144, 2013.

[8] D. Pathak, P. Krähenbühl, J. Donahue, T. Darrell and A. Efros, *Context Encoders: Feature Learning by Inpainting*. In: Proceedings of the IEEE Conference on Computer Vision and Pattern Recognition, 2016.

[9] I. Goodfellow et al., *Generative Adversarial Nets*. Advances in Neural Information Processing Systems, vol. 27, 2014.

[10] A. Dosovitskiy, G. Ros, F. Codevilla, A. Lopez and V. Koltun, *CARLA: An Open Urban Driving Simulator*. In: Proceedings of the 1st Annual Conference on Robot Learning, pp. 1–16, 2017.

[11] G. Wadhwa, A. Dhall, S. Murala and U. Tariq, *Hyper-realistic Image Inpainting with Hypergraphs*. In: Proceedings of the IEEE/CVF Winter Conference on Applications of Computer Vision, pp. 3912–3921, 2021.

[12] S. Katz, A. Ayellet and R. Basri, *Direct Visibility of Point Sets*. In: ACM SIGGRAPH 2007 papers, pp. 24–es, 2007.

Semantic Segmentation on Embedded Systems in the Context of Automotive Applications

Johann Engster[1], Julian Petzold[2], and Heiko Hamann[2]

[1] Robotics and Autonomous Systems, Universität zu Lübeck, Johann.Engster@student.uni-luebeck.de
[2] Institute of Computer Engineering, Universität zu Lübeck, {Petzold, Hamann}@iti.uni-luebeck.de

Abstract

Semantic Segmentation is a machine learning task that can be helpful for automotive driving applications. While classification only offers one class label of an object per image and object detection only offers an approximative bounding box of an object, semantic segmentation can give pixelwise classification for different objects in the picture. Thus, the segmented image can be used to track traffic components for safety. We present a two-task split problem. Firstly, semantic segmentation is implemented on embedded hardware and is thoroughly evaluated for different resolutions and ResNet architectures. Secondly, the combination of both simulated and real-world data is tested. The first task shows that the hardware meets real-world performance requirements. The second task shows that the already shown performance can be improved by combining the simulated and real-world data. Further research of combined training should be done.

1 Introduction

During the last two years of the coronavirus pandemic, the automotive industry was heavily hit by microchip shortages [1]. These microchips are especially important for *Advanced Driver Assistance Systems* (ADAS). Thus, maximising the efficient usage and performance of such microchips is an important task. The purpose of this internship is further research of semantic segmentation for automotive applications. It is part of the project *new explainability supported by world models for predictions in autonomous driving* (NEUPA) [3]. NEUPA's main focus is to make AI explainable and that way enable certifications of AI systems for use as ADAS. We want to prepare a *System-on-Chip* (SoC) for semantic segmentation in real-time. For data generation we use the CARLA simulator, which focuses on support, development and validation of autonomous driving systems [2]. We use this data to train neural networks (NN), which we then port onto the SoC. Afterwards, we evaluate if learning from simulated data is sufficient and to what degree combined training with simulated and real world data can increase the performance.

2 Material and Methods

2.1 Datasets

Dataset selection is a critical task, especially if humans can get harmed due to ineffective or wrong generalisation of the learned NN. To prevent this, two datasets have been acquired, combining simulated day cycle and weather data together with real-world data. Both datasets have the same six labelled classes (unlabeled, pedestrian, road, sidewalk, vehicle and crosswalk) for training.

2.1.1 CARLA Simulator

To obtain simulated data we use the CARLA simulator, which is based on the Unreal Engine [2]. It allows the user to spawn different actors such as pedestrians and cars in different environments such as a big city center, or a rural town, in order to simulate traffic scenarios. CARLA has built in support for RGB cameras and for semantic cameras. Thus, data for training can be generated with the help of a Python script. By changing the daytime, weather and maps, variance can be generated, which is needed for generalisation of the NN.

2.1.2 Audi Autonomous Driving Dataset

In addition to simulated data, we use real-world data. For this, RGB images as well as corresponding labeled masks from the Audi Autonomous Driving Dataset (A2D2) [4] are taken. The A2D2 images were recorded in different cities and with different camera perspectives.

2.2 Nvidia Jetson Nano

As an SoC, the NVIDIA Jetson Nano Developer Kit [5] is chosen. Combined with a quad-core ARM processor, 4 GB of RAM and a Maxwell GPU with 128 CUDA cores, the Jetson offers a mobile platform for machine learning (ML) applications. Aside from the hardware, the NVIDIA Jetson also offers software running on Ubuntu. We use the Jetson Inference library [6]. The library provides prebuild packages to run ML tasks like classification, object detection, and semantic segmentation models using SegNet [8] on the Jetson. This library uses NVIDIA TensorRT, which accelerates the performance of learned models on NVIDIA GPUs. The NVIDIA Jetson Nano is shown in Fig. 1.

Figure 1: The NVIDIA Jetson Nano.

2.3 Pytorch and ResNet

For training the NN, PyTorch [7] 1.7.1+cu110, which enables CUDA support, is used. With help of a NVIDIA RTX 3090, 1-2 days of training time per NN can be reached. Afterwards, the model can be exported onto the Jetson. As a backbone encoder ResNet [9] is chosen, which offers a variety of different and complex architectures. The advantage of ResNets lies in the use of so-called residual layers, shown in Fig. 2. With these residual layers the diffusion of gradients is averted by using skip connections which skip over the layers in the middle. This allows better results and speeds up the learning process for deeper NNs.

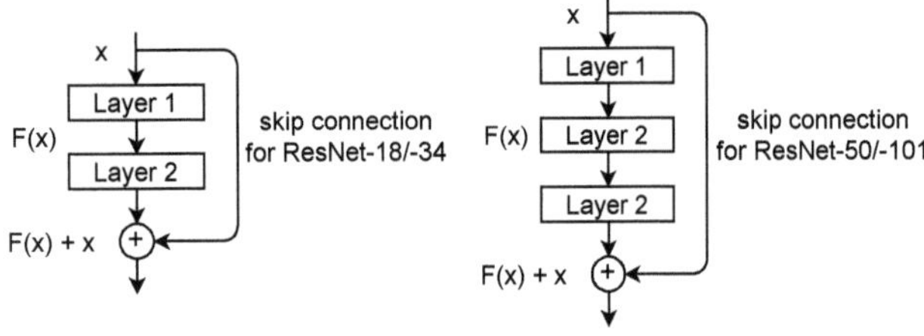

Figure 2: Residual layers that use a so-called skip connection which skips over the layers in the middle. The left side shows the skip connection for ResNet-18/-34 and the right side shows the skip connection for ResNet-50/-101 [9].

Table 1 shows the architectures of the used ResNet-18, ResNet-34, ResNet-50 and ResNet-101. Skip connections are added to each pair of filters, shown in brackets. As explained in Fig. 2, ResNet-18/-34 and ResNet-50/-101 skip over different amounts of layers inbetween. Thus, ResNet-18/-34 are very similar and only the amount of filter pairs changes, which is also the same for ResNet-50/-101. However, aside from the different amount of filter pairs, the ResNet-18/-34 and ResNet-50/-101 use different convolutions inside these filter pairs and thus offer different output resolutions on the Jetson running with Jetson Inference. For the ResNet-18/-34 the classification grid will always be 32 times smaller than the output resolution, but for ResNet-50/-101 it will be only 8 times smaller due to the higher convolution sizes. The reason is that ResNet-50/-101 output convolutions are all 4 times larger. The downscaling inside Jetson Inference is implemented because the deconvolution operation is computationally demanding and would not allow real-time computation.

layer	out size	ResNet-18/-34 layer	ResNet-50/-101 layer
conv1	112×112	7×7, 64, stride 2	
conv2_x	56×56	3×3 max pool, stride 2	
		$\begin{bmatrix} 3 \times 3, 64 \\ 3 \times 3, 64 \end{bmatrix} \times 2/3$	$\begin{matrix} 1 \times 1, 64 \\ 3 \times 3, 64 \\ 1 \times 1, 256 \end{matrix} \times 3/3$
conv3_x	28×28	$\begin{bmatrix} 3 \times 3, 128 \\ 3 \times 3, 128 \end{bmatrix} \times 2/4$	$\begin{matrix} 1 \times 1, 128 \\ 3 \times 3, 128 \\ 1 \times 1, 512 \end{matrix} \times 4/4$
conv4_x	14×14	$\begin{bmatrix} 3 \times 3, 256 \\ 3 \times 3, 256 \end{bmatrix} \times 2/6$	$\begin{matrix} 1 \times 1, 256 \\ 3 \times 3, 256 \\ 1 \times 1, 1024 \end{matrix} \times 6/23$
conv5_x	7×7	$\begin{bmatrix} 3 \times 3, 512 \\ 3 \times 3, 512 \end{bmatrix} \times 2/3$	$\begin{matrix} 1 \times 1, 512 \\ 3 \times 3, 512 \\ 1 \times 1, 2048 \end{matrix} \times 3/3$
	1×1	average pool, 1000-d fc, softmax	
FLOPs		$1.8/3.6 \times 10^9$	$3.8/7.6 \times 10^9$

Table 1: The architectures of the used ResNet-18, ResNet-34, ResNet-50 and ResNet-101 [9]. Skip connections are added to each pair of filters, shown in brackets. The "X/Y" after each filter denotes the amount of filter pairs for the Resnet-18/-34 or Resnet-50/-101 respectively, where X stands for the number of filter pairs for ResNet-18/-50 and Y for ResNet-34/-101. ResNet-18/-34 are very similar and only the amount of filter pairs changes, which is the same for ResNet-50/-101. In addition the amount of floating point operations per second (FLOPs) is shown.

3 Results and Discussion

In preparation the data is split as follows: The 41.277 A2D2 real world image and mask pairs are split into 28.000 pairs for training, 7.000 pairs for validation and the remaining 6.277 pairs are saved for final testing. An equivalent amount of simulated (sim) pairs has been created, with the same amount of training and validation pairs. However, because data generation is only a matter of computation time, 10.000 sim pairs have been generated for final testing.

3.1 SoC Performance

Training for the SoC is only conducted with sim data. This way the performance of the different architectures is easier to evaluate. In total, four different architectures are evaluated on five different resolutions. Afterwards, the trained models are ported to the Jetson. This way, the performance can be measured in frames per second (FPS) on the limited hardware. Around 6 FPS are sufficient for cyclist tracking tasks and for pedestrian tracking tasks the number of necessary FPS is even lower, with 1.2 FPS while still staying at 80% precision [10]. Combining all information, 5 FPS are set as a real-time capable performance bound. This corresponds to 200 ms to segment a full image. The results combine three different performance metrics. Firstly, there is the mean Intersection over Union (mean IOU), which is the average over all class IOUs. The class IOU gives the percentage of correctly classified pixels over all images for that class. Then, there is the FPS on the Jetson running with Jetson Inference and finally, there is the output mask resolution size given in pixels. As explained in the above section, the output mask resolution is different for the ResNet-18/-34 and ResNet-50/-101.

Table 2 shows the mean IOU, FPS and output resolution using Jetson Inference for different architecture and input resolution combinations.

mean IOU				
resolution	ResNet-18	ResNet-34	ResNet-50	ResNet-101
1280×800	86.4	87.3	92.4	92.6
960×576	84.4	85.1	91.5	91.6
640×384	80.8	81.7	89.7	90.0
480×288	-	-	88.0	88.1
320×192	-	-	84.7	84.9
frames per second				
resolution	ResNet-18	ResNet-34	ResNet-50	ResNet-101
1280×800	5.45	3.03	0.41	0.25
960×576	9.82	5.46	0.75	0.46
640×384	20.51	11.67	1.69	1.03
480×288	-	-	2.89	1.78
320×192	-	-	6.09	3.77
output resolution with Jetson Inference				
resolution	ResNet-18	ResNet-34	ResNet-50	ResNet-101
1280×800	40×25		160×100	
960×576	30×18		120×72	
640×384	20×12		80×48	
480×288	-		60×36	
320×192	-		40×24	

Table 2: The mean IOU, FPS and output resolution for different architecture and input resolution combinations.

In general, a higher input resolution or a more complex ResNet architecture increase the mean IOU and output resolution, but come with a trade-off in lower FPS. The mean IOU of the different ResNet-18/-34 and ResNet-50/-101 pairs are similar, however the FPS vary. ResNet-18 is nearly 2 times faster than ResNet-34, while only having a 1-2 percent points worse mean IOU. These speed differences are similar for the performance of ResNet-50 compared with ResNet-101 respectively. Thus, ResNet-18 or ResNet-50 architectures should be preferred for higher FPS.

Considering the output resolution, a ResNet-18/-34 on the highest input resolution 1280×800 is nearly the same as a ResNet-50/-101 on lowest input resolution 320×192. Higher output resolutions are hard to achive with ResNet-18/-34 without training on extremely high input resolutions. Thus, if a high resolution is preferred, ResNet-50/-101 architectures should be used.

The images in Fig. 3 bring the actual output mask resolutions into perspective by showing four different mask size output resolutions using Jetson Inference (3(c), 3(d), 3(e), 3(f)) on the same input image. Additionally the input image 3(a) and the true mask 3(b) are shown.

The focus will mainly lie on the crosswalks (bright grey pixels) in the images. With lower output resolution, information can and will be lost. On 3(c) (20×12) the output is nearly unusable in comparison to the actual true mask 3(b). The crosswalks in the back can not be recognised and the crosswalk in the foreground is labeled in a too imprecise way. While the foreground crosswalk is labelled visibly on 3(d) (40×25), the background crosswalks lying

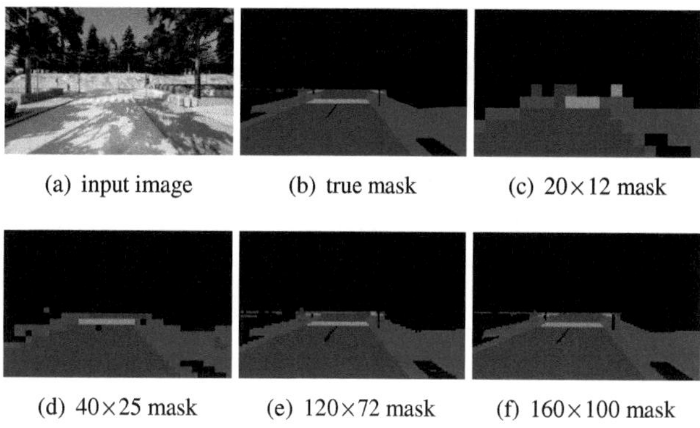

(a) input image   (b) true mask   (c) 20×12 mask

(d) 40×25 mask   (e) 120×72 mask   (f) 160×100 mask

Figure 3: Four different mask size output resolutions using Jetson Inference (3(c), 3(d), 3(e), 3(f)) on the same input image. Additionally the input image 3(a) and the true mask 3(b) are shown.

ahead are still not visible. This changes on 3(e) (120×72) and all crosswalks are labelled visibly. With an increase to 3(f) (160×100), the crosswalks become a bit clearer. Even higher resolutions would not be able to increase the visibility much further.

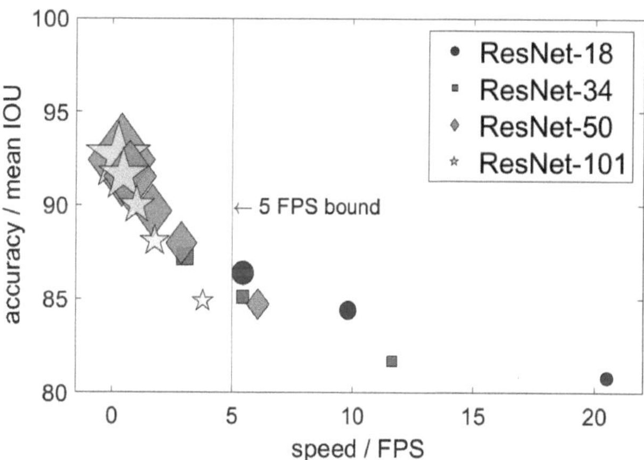

Figure 4: The speed vs accuracy trade-off for different architecture and input resolution combinations. A bigger marker size corresponds to a higher output resolution.

The observations from Fig. 4 support the results from Table 2 above. Either high speed (FPS) or high accuracy (mean IOU / output resolution) is achievable for the Jetson. However, as also stated above, some ResNets are better at this trade-off than others (ResNet-18 and ResNet-50 are nearly twice as fast than the ResNet-34 or ResNet-101 without losing much mean IOU points). The ResNet-18 on 1200×800 has the highest mean IOU above 5 FPS, which has been the performance criterion that is defined above. Most importantly, the accuracy over speed does not scale linearly. With better hardware, a better trade-off could be achieved. A ResNet-50/-101 at high resolution might be too slow on the limited Jetson Nano, but this paper should only function as a proof of concept. Real-time capable framerates are possible in general.

3.2 Combining Both Datasets

For this task, the performance on the Jetson is not a criterion anymore and only the mean IOU performance is important. Thus, the ResNet-18 has been chosen combined with a high resolution of 1200×800. The table 3 shows the mean IOU for NNs trained with different amounts of real & sim data.

mean IOU for different real & sim data combinations			
real amount	sim amount	mean IOU on real	mean IOU on sim
35 000	0	74.8	40.6
35 000	17 500	75.7	84.4
35 000	35 000	76.6	85.0
17 500	17 500	75.2	85.6
17 500	35 000	77.1	86.5
0	35 000	37.7	86.4

Table 3: The mean IOU for NNs trained with different amounts of real & sim data.

Unsurprisingly, the NN trained on real data only performs poorly on the sim test data. The real data only contains daytime and moderate weather changes in its images and therefore the NN can not generalize to the different daytimes and weather cycles in the simulation dataset. Similarly, the NN that has been trained on only sim data performs poorly on the real test data. Firstly, the real-world is different from the rather perfect sim data. This is also the reason why the mean IOU in general is higher on sim data than on real data. Secondly, the real data uses five different camera perspectives, while the sim data only uses one. The NNs trained on both real and sim data boost the performance on the real test set, while the performance on the sim test becomes worse. One explanation for the increase might be the bigger amount of data. Therefore, a NN with 17 500 real & 17 500 sim images is trained. Thus, it uses the same number of images as the NN trained on only real / only sim images. While it performs worse than the NN trained with 35 000 real & 35 000 sim images, it still performs better on the real test set than the NN trained on only real images. In addition, the NN trained on 17 500 real & 35 000 sim images uses less images than the NN trained on 35 000 real & 35 000 sim images and still performs better on both test sets. So, this improvement can not be explained by higher data set size. The results clearly show that combining real and sim data can boost the real-world performance. Because the system has to function in the real world, the lower sim performance is less relevant, but still a good measure for transferability.

4 Conclusion

Semantic segmentation for automotive applications is an important field that may lead to safer and better autonomous driving. Usually, segmentation is processed on top-of-the-line hardware, but for this paper the hardware has been limited to the Jetson Nano. In this work four different ResNet architectures are trained on five different resolutions. All trained NNs are evaluated and especially the speed vs. accuracy trade-off is examined. As a proof of concept, the technology is ready for real-world performance, despite the hardware limitations of the Jetson Nano. Based on this, next

level ADAS could be deployed soon. In addition, the benefits of combining real and simulated data are tested. The performance increases, but further research must be done. The next goal could be reducing the functional differences between real and simulated data. This way combined training can be the answer to create more robust and reliable performance for automotive-focused semantic segmentation.

Acknowledgement

The work has been carried out at the Institute of Computer Engineering, Universität zu Lübeck.

Author's Statement

NEUPA is funded by the *Federal Ministry of Education and Research* (BMBF), grant agreement no. 01 IS 19078.

5 References

[1] L. Kelion, *Why is there a Chip Shortage for Computers and Cars?*. https://www.bbc.com/news/technology-55936011, 2021 [last accessed on 2022-02-07].

[2] A. Dosovitskiy, G. Ros, F. Codevilla, A. López and V. Koltun, *CARLA: An Open Urban Driving Simulator*. CoRR, 2017.

[3] H. Hamann, *NEUPA*. https://www.neupa.net/, Institute of Computer Engineering of the University of Lübeck. 2020.

[4] J. Geyer et al., *A2D2: Audi Autonomous Driving Dataset*. 2020.

[5] NVIDIA, *Jetson Nano Developer Kit*. https://developer.nvidia.com/embedded/jetson-nano-developer-kit [last accessed on 2022-02-07].

[6] Dustin Franklin, *Jetson Inference*, GitHub repository, https://github.com/dusty-nv/jetson-inference [last accessed on 2022-02-07].

[7] A. Paszke et al., *PyTorch: An Imperative Style, High-Performance Deep Learning Library*. Advances in Neural Information Processing Systems 32, pp. 8024-8035, 2019.

[8] V. Badrinarayanan, A. Kendall and R. Cipolla, *SegNet: A Deep Convolutional Encoder-Decoder Architecture for Image Segmentation*. 2016.

[9] K. He, X. Zhang, S. Ren and J. Sun, *Deep Residual Learning for Image Recognition*. 2015.

[10] A. Mohan et al., *Determining the Necessary Frame Rate of Video Data for Object Tracking under Accuracy Constraints*. 2018 IEEE Conference on Multimedia Information Processing and Retrieval (MIPR), pp. 368-371, 2018.

Pupil and Iris Segmentation for Pupil Light Reflex Evaluation on Mobile Devices

Katharina Gercke [1], Paul Kaftan [2], and Andreas Trabhardt [3]

[1] Medical Engineering Science, Universität zu Lübeck, katharina.gercke@student.uni-luebeck.de
[2] Medical Informatics, Universität zu Lübeck, paul.kaftan@student.uni-luebeck.de
[3] Drägerwerk AG & Co. KGaA, Lübeck, andreas.trabhardt.contractor@draeger.com

Abstract

The pupillary light reflex test is a quick, non-invasive medical diagnostic procedure that indicates brain and optic nerve injury or alcohol and drug influence. To quickly detect these cases, we are trying to develop a smartphone application that will make this test possible at any time and for any person without prior knowledge. Therefore, the pupil size must be recorded over time based on images. By first determining the segmentation of the pupil using a convolutional neural network, the pupil radius can be determined. To quickly obtain a large training dataset, artificially generated eye images were combined with a dataset of real eye images. However, it was found that these were not sufficient as training data. A segmentation network is presented that already shows promising results for the determination of the pupil radius.

1 Introduction

The pupillary light reflex (PLR) is the adaption of pupil size to light conditions to regulate the amount of light in the eye. High light intensity on the retina leads to a constriction of the pupil, i.e. a reduction of the pupil radius. Since especially brain or optic nerve injuries as well as the intake of certain drugs change the PLR, it is an important medical diagnostic procedure [1]. Especially in life threatening situations, it is particularly important that a measurement of the PLR can be performed quickly at any time, in any place and under all possible external conditions. In clinics, there are reliable, standardized devices, such as the pupillometer. Outside clinics, the measurement is performed using a simple flashlight test. However, this requires specific training and the assessment is an individual judgement of the observer. Using an application for smartphones, this test could be performed precisely by any person and at the same time be well documented.

To perform PLR analysis on smartphones, tracking of pupil radius over time is required while the eye is stimulated with a light pulse. For an image-based determination of the pupil radius, the segmentation of the pupil is required first. As a change in distance to the eye during exposure results in a change in pupil size in the image without its actual size having changed, a reference size is essential. Since the size of the human iris does not change, the ratio of pupil size to iris size can be used. Therefore, the iris radius must also be determined, so segmentation of the pupil as well as the iris is necessary.

To ensure that the application works under all environmental conditions, a convolutional neural network (CNN) is used for segmentation. A particular difficulty is that training CNNs requires a large, diverse dataset. Mariakakis et al. [2] therefore recorded videos of the PLR in 42 healthy subjects and used these as training data. To quickly obtain a large dataset, we tested a combination of artificially generated images and an existing dataset of eye images from 46 different subjects as training dataset.

2 Material and Methods

To obtain a large set of labeled training images, a tool called UnityEyes [3] was used. It was introduced to generate synthetic training data for research in the field of eye tracking. The advantage of UnityEyes is the ability to generate a huge labeled eye dataset in short time. The tool randomly changes the appearance of the eye region and adjacent skin in terms of color, texture and age. It also changes the viewing direction, camera position, lighting, and reflections. The pupil size is also randomly generated, but the tool only creates left eyes. Fig. 1 shows an example of an UnityEyes image and the automatically generated landmarks. 32 of the landmarks outline the iris, these can be used to determine the ground truth labels for pupil and iris segmentation. The label for the pupil can therefore be determined by a simple thresholding procedure within the iris landmarks. The iris label then includes all remaining pixel within the landmarks. To use it as training data, the image itself is cropped to fit the eye, converted to a gray-scale image, and resized to 256×192 pixel (px) (see Fig. 2).

In order to additionally train with real images, the Multimedia University Iris Database 2 (MMU2) [4], published for research in iris recognition, is used. The ground truth

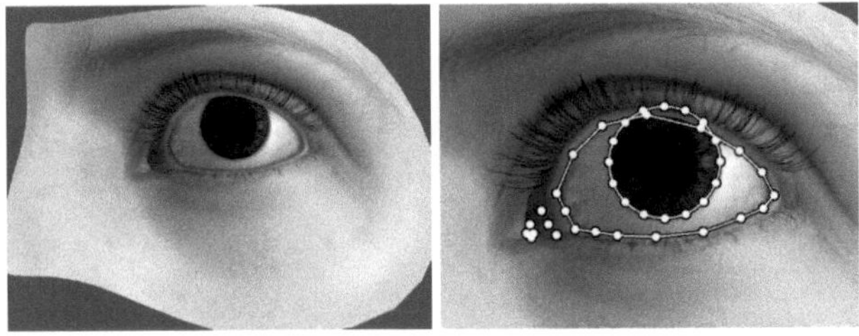

Figure 1: Raw data generated by the UnityEyes tool. On the left, the UnityEyes image and on the right overlaid with the landmarks provided by the tool for eye-tracking research.

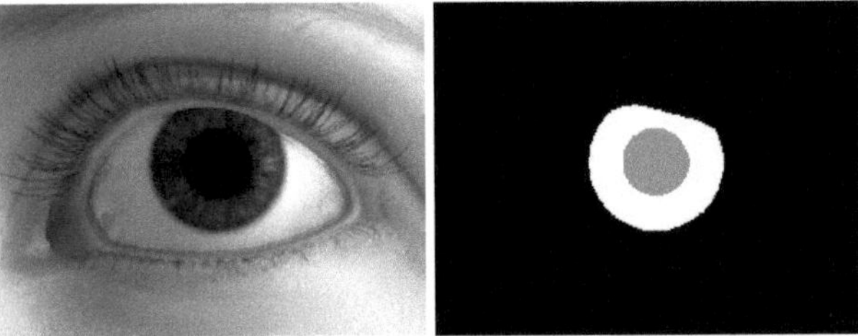

Figure 2: UnityEyes image preprocessed for training (left) and generated ground truth label map (right).

labels had to be determined manually. The dataset consists of five left and five right eye images of 100 persons and was published for research in the field of iris recognition. Unlike the UnityEyes, this dataset has the disadvantage that the images were all taken under bright illumination and thus contain only small pupils. Furthermore, since not all images were suitable, only 262 of this dataset are used. The images were also cropped to fit the eye and resized to 256×192 px (see Fig. 3). To avoid overfitting on UnityEyes, only 262 images of this dataset were additionally used. Randomly selected 20% of the 262 MMU2 and the 262 UnityEyes images were combined as a validation dataset. The remaining 80% of each were combined as the training dataset.

To obtain a realistic test dataset, images of 16 different people in different lighting conditions were taken with the camera of a One Plus 8 Pro smartphone and manually annotated (see Fig. 4). The images cover particularly difficult cases, such as reflections in the eye, lower contrast between iris and pupil, and reflections through eyeglasses. The test dataset also consists of equal numbers of left and right eyes, so that the resulting evaluation is meaningful for both eyes. In order to transfer the segmentation of the retina and the iris to a mobile device, a Mobile-Unet [5] was used.

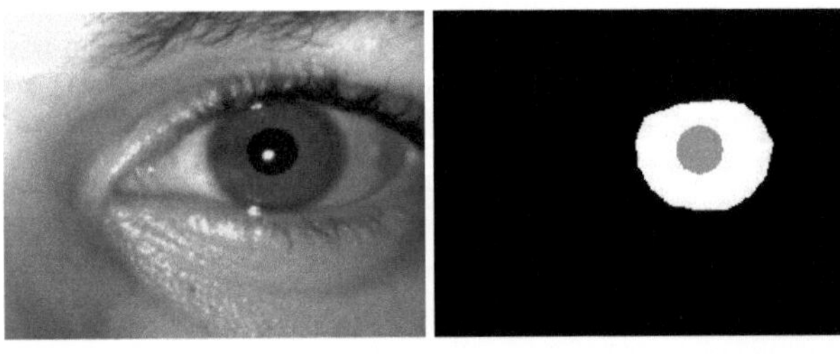

Figure 3: MMU2 image preprocessed for training (left) and manually generated ground truth label map (right).

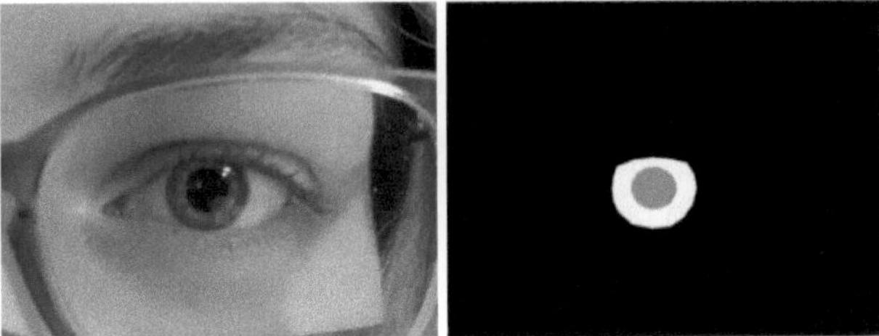

Figure 4: Test dataset image preprocessed (left) and manually generated ground truth label map (right).

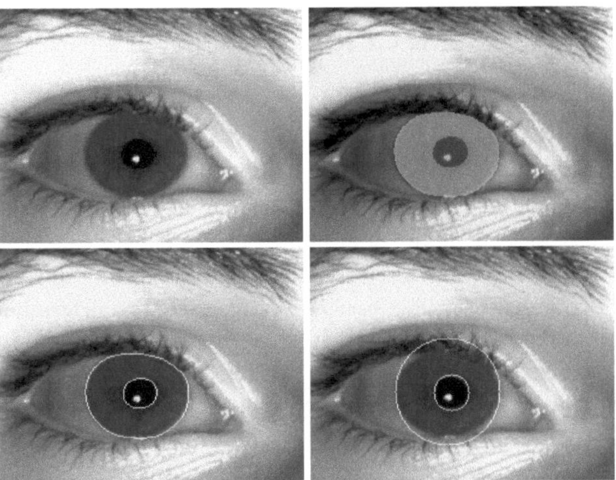

Figure 5: Determination of the pupil and iris radius of an MMU2 validation image. 1. Segmentation result overlaid on the input image (top right). 2. Outer contour of corresponding segmentation (bottom left). 3. Circle enclosing the corresponding contour whose radius is the searched pupil or iris radius (bottom right).

3 Experiments

The Mobile-Unet was trained on 256×192 px image crops with an initial learning rate of 1e-3. After one and two thirds of the training process, the learning rate was decreased by a factor of ten. For the training over 150 epochs, a batch size of four was used. In addition, the Adam optimizer [6] and cross entropy loss were used. A $256 \times 192 \times 3$ image was generated as output, since three labels, background, pupil, and iris, are to be predicted. For each training, the epoch, accuracy, recall, and Dice score were considered for the model of the best epoch on validation data. Here the best epoch was determined by the best Dice score averaged over pupil and iris predictions on the validation dataset. To estimate how dependent the network's performance is on the initialization, the algorithm was made deterministic by seeding. The network was then evaluated for seeds 1, 11, 42, 420. All following experiments were performed with the seed that yielded the best results. To improve the network, a learning rate optimization was done. For this the initial learning rates 1e-1, 1e-2, 1e-3, and 1e-4 were chosen. The model of the best validation epoch trained with the best seed and learning rate was then run on the test dataset. To achieve better generalization, data augmenta-

Table 1: Best validation results per seed.

Seed	Epoch	Accuracy	Recall	Dice
1	102	0.9958	0.9480	0.9505
11	104	**0.9960**	0.9501	**0.9524**
42	119	0.9959	**0.9512**	0.9523
420	146	0.9957	0.9472	0.9504

Table 2: Best validation results per learning rate (LR).

LR	Epoch	Accuracy	Recall	Dice
1e-1	15	0.9769	0.8609	0.7471
1e-2	104	**0.9963**	**0.9537**	**0.9561**
1e-3	104	0.9960	0.9501	0.9524
1e-4	145	0.9914	0.8771	0.8917

Table 3: Best model results for different augmentations on the validation and test dataset.

Augm	Ep	Data	Accuracy	Recall	Dice
None	104	Val	0.9963	0.9537	0.9561
		Test	0.9930	0.6808	0.7187
Flip	104	Val	**0.9965**	0.9545	0.9577
		Test	**0.9941**	0.7402	**0.7673**
Noise	105	Val	0.9964	0.9543	0.9575
		Test	0.9934	0.7547	0.7599
Blur	111	Val	0.9960	0.9490	0.9539
		Test	0.9933	0.7154	0.7278
Flip & Noise	128	Val	0.9962	0.9532	0.9564
		Test	0.9931	**0.7618**	0.7547
Flip & Blur	136	Val	**0.9965**	**0.9574**	**0.9583**
		Test	0.9937	0.7380	0.7500
Noise & Blur	109	Val	0.9962	0.9495	0.9551
		Test	0.9932	0.7453	0.7529
All	148	Val	0.9964	0.9540	0.9577
		Test	0.9938	0.7542	0.7634

tion experiments were performed. First, a horizontal flip was performed with a probability of 50%. Further, Gaussian noise with a variance range of 5 to 20 and a probability of 50% was applied. Last, the images were blurred with a maximum kernel size of 7 and a probability of 30%. These three augmentations were used individually, in pairs, and all together. Since the test data is very different from the training and validation data, the performance on the test dataset was also considered. Afterwards, the segmentation results on validation and test dataset were assessed visually. Based on the segmentation, the calculation of the radii of the pupil and iris is required for the detection of a PLR. To calculate these, the outer contour of the corresponding segmentation is determined and a circle is fitted around it. The radius of this circle is then the sought pupil or iris radius. Fig. 5 visualizes these steps on an MMU2 image of the validation dataset. The pupil radius results were also evaluated and the median absolute error was calculated.

4   Results and Discussion

Table 1 shows the results of the trainings with the four different seeds, 1, 11, 42, 420, the epoch, accuracy, recall and Dice score of the best validation epoch. All metrics were rounded to four decimal places, as they differ only in the third or fourth decimal place between the different seeds. The best value per metric is highlighted in the table. The recall has a mean of 0.9491 and a standard deviation of 0.0016 and the Dice score has a mean of 0.9514 and a standard deviation of 0.0010. Since these are very small deviations, the performance of the network is probably independent of the initialization. All following runs were performed with a seed of 11, as this yielded the best values for accuracy and Dice score. As shown in Table 2, the learning rate search revealed that training with the learning rate 1e-2 can improve the accuracy by 0.0003, the recall by 0.0036 and the Dice score by 0.0037 compared to the previously used learning rate of 1e-3. As the learning rates 1e-1 and 1e-4 degrade the metrics, the best results were obtained with a learning rate of 1e-2. Further training would likely be to no avail as all best epochs are well within the amount of total epochs. Therefore, this learning rate is used for the following augmentation experiments along with a seed of 11.

When applying the best model to the validation and test data, as seen in Table 3, all metrics yield lower results during testing. The accuracy differs only by 0.0033, but the recall by 0.2729 and the Dice score by 0.2374. The different augmentations improve performance for both datasets. Applying a random horizontal flip yields better results during testing, likely since this can compensate for an unequal distribution of left and right eyes in the training dataset. Particularly as the test dataset is completely balanced in terms of the number of left and right eyes. For the validation dataset, the recall improves by up to 0.0037 and the Dice score by up to 0.0022 when horizontal flip and motion blur are performed as augmentations. On the test dataset, there is an improvement of 0.081 for the recall using the combination of horizontal flip and Gaussian noise and 0.0486 for the Dice score using only horizontal flip. Because the results for the test dataset are always lower than for the validation dataset, the segmentations were qualitatively examined. For this the segmentations generated by the model trained with only the horizontal flip augmentation were considered, as this yielded the best results during testing. When looking at the test segmentation results overlaid on the input images, it becomes apparent that the network has particular problems with reflections or poor lighting. Fig. 6 shows an image of the test dataset where a reflection in the lower right corner of the pupil seems to cause this area to be segmented as iris. Given that the validation data always achieves higher metrics than the test data, partly due to mirroring, it can be assumed that the validation data are too similar to the training data and might not adequately represent reality.

In Fig. 5, the lower right image shows that a real iris is not a perfect circle, but oval, which can cause miscalculations of the radii. The UnityEyes images instead have very round irides (see Fig. 7), which therefore do not sufficiently simulate real eyes. On both datasets, MMU2 and UnityEyes,

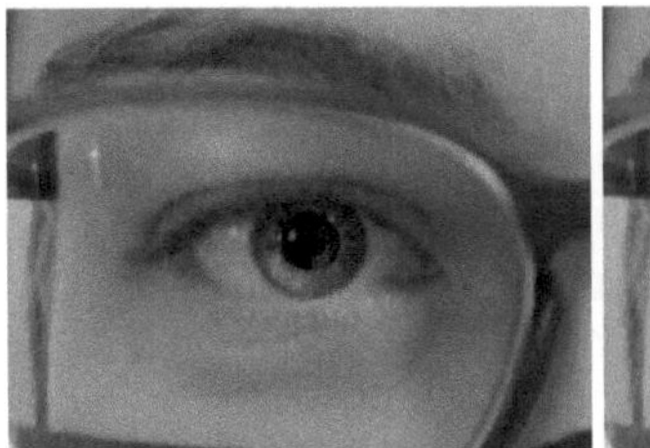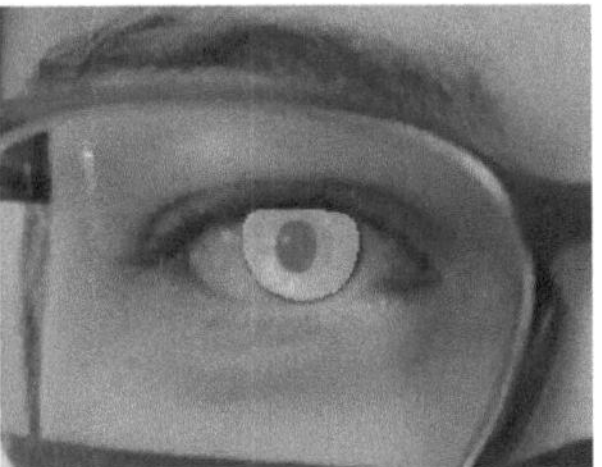

Figure 6: Segmentation result on test image. Due to a reflection in the lower left of the pupil, this area is incorrectly labeled as iris pixel.

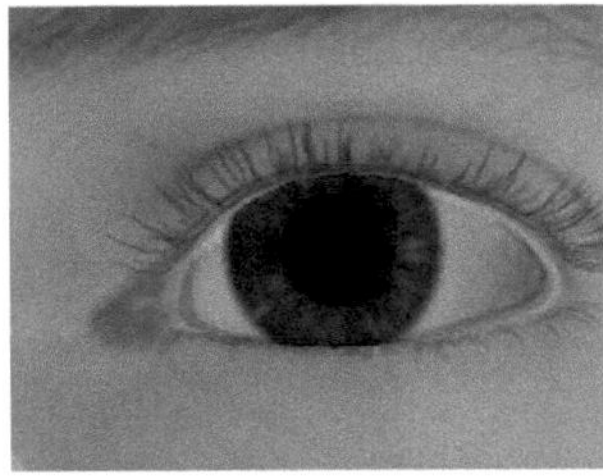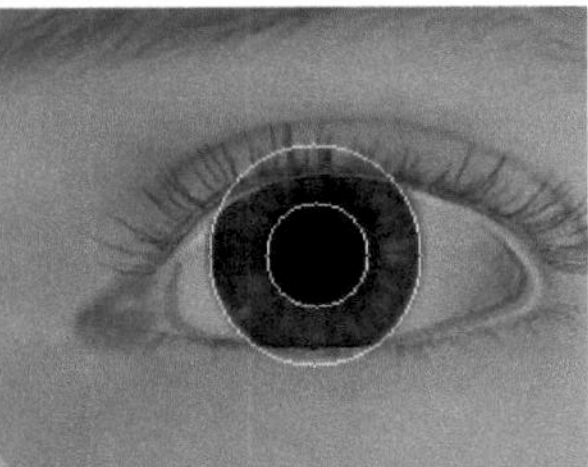

Figure 7: UnityEye validation image with circles visualizing the calculated radii for pupil and iris. The resulting circle fits exactly on the round iris and pupil of the Unity-Eye.

it becomes clear that due to the eye curvature the radius is distorted when the subject looks to the left or right of the camera (see Fig. 8). This optical distortion also leads to calculated radii that no longer correspond to the real radii. Therefore, it is important that the camera is aimed frontally at the pupil during the recording. For the evaluation of the radii, the median absolute error was calculated on the validation and test data. Here, the deviation between the radius, calculated using predicted segmentation as described above, and the radius, calculated using ground truth segmentation, was considered. Errors due to distortions are not detected here, since they also occur when calculating the radii using ground truth segmentation. These must be prevented in advance by only taking frontal image of the pupil. For the validation data, the median absolute error for the predicted iris radius is 0.5440 px and for the pupil radius 0.3919 px. For the test dataset, which provides lower segmentation metrics, the median absolute error for the iris radius nevertheless has a similar value of 0.5025 px. For the pupil radius, the value here is 1.5802 px. This reflects the previous analysis that for the test dataset, the network has particular problems with reflections in the pupil.

5   Conclusion

The trained network has a very good performance for the validation data. However, it performs significantly worse for the test dataset. The evaluations revealed that especially reflections and different exposures seem to be responsible for this. Since the test dataset consists of images captured with a smartphone, it best represents the final application for PLR analysis. Whereas, combining synthetic UnityEyes

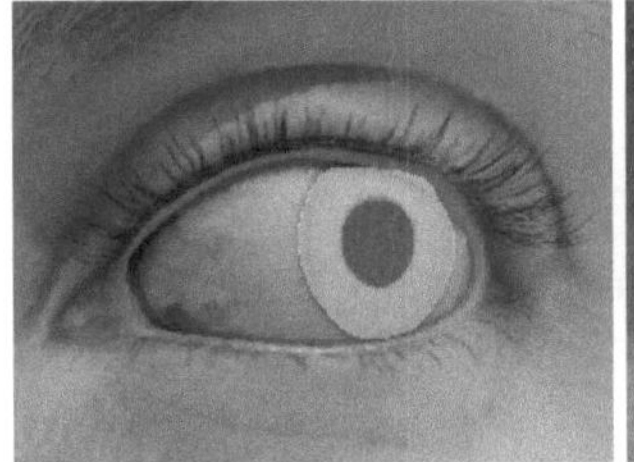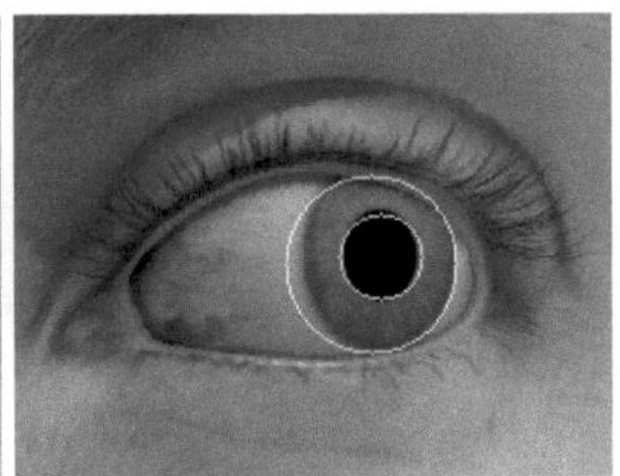

Figure 8: UnityEye validation image with correct segmentation (left), but wrong radii for pupil and iris (right), due to optical distortions caused by a gaze direction to the side.

images with the MMU2 dataset does not seem to contain all of such realistic challenges. Thus, more real training data is needed to further improve generalization. Nevertheless, the approach is very promising. Despite these data limitations and the small number of images, it was already possible to calculate accurate radii for pupil and iris for many images based on the current segmentation performance.

Acknowledgement

The work has been carried out at Drägerwerk AG & Co. KGaA, Lübeck and supervised by Prof. Erhardt Barth, Institute of Neuro- and Bioinformatics, Universität zu Lübeck.

Author's Statement

Conflict of interest: Authors state no conflict of interest.

6   References

[1] A. P. Belliveau, A. N. Somani, and R. H. Dossani, "Pupillary light reflex," 2019.

[2] A. Mariakakis *et al.*, "Pupilscreen: using smartphones to assess traumatic brain injury," *Proceedings of the ACM on Interactive, Mobile, Wearable and Ubiquitous Technologies*, vol. 1, no. 3, pp. 1–27, 2017.

[3] E. Wood, T. Baltrušaitis, L.-P. Morency, P. Robinson, and A. Bulling, "Learning an appearance-based gaze estimator from one million synthesised images," in *Proceedings of the Ninth Biennial ACM Symposium on Eye Tracking Research & Applications*, 2016, pp. 131–138.

[4] Multimedia University. (2006) Multimedia University iris database 2 (MMU2). Accessed: 2021-01-10. [Online]. Available: https://andyzeng.github.io/irisrecognition

[5] J. Jing, Z. Wang, M. Rätsch, and H. Zhang, "Mobile-unet: An efficient convolutional neural network for fabric defect detection," *Textile Research Journal*, 2020.

[6] D. P. Kingma and J. Ba, "Adam: a method for stochastic optimization," 2017.

AI-Showcase
– How to Convey and Visualize Artificial Intelligence –

Lennart Landt [1] and Tim Suthau [2]

[1] Medical Engineering Science, Universität zu Lübeck, lennart.landt@student.uni-luebeck.de

[2] UniTransferKlinik Lübeck GmbH, suthau@unitransferklinik.de

Abstract

We present the planned buildup for a laboratory designed for collaboration on and demonstration of Artificial Intelligence (AI). The target is to create a space for AI to be conveyed. Since AI became an important topic in the public debate and new possibilities of AI rapidly arised in the past few years, it became a necessity to make AI understandable and to demonstrate its abilities. Therefore, by gathering information from papers, scientific talks and interviews with experts a plan for the setup of the laboratory for collaboration and demonstration of AI was constructed, and partially realized. The purpose of the laboratory is to bring comprehension for AI to the broad masses and thereby boosting the potential of new AI innovations.

1 Introduction

In the last few years, AI became a highly discussed topic and reached the public debate. Questions about the trustworthiness of AI and its impact on society lead to a growing controversy over AI [3]. This could rise to a general reservation on adapting AI. On the other hand, AI will have a strong influence on the economics and will change the way business is made [4]. Therefore, it is crucial to give people a connection to AI, teach them about AI and show them how to adapt AI. There are already existing approaches to break down the complexity of AI as in [5] or [6], that try to show how small enterprises could use AI to their advantage, without needing to be an AI expert themselves. However, for a true comprehension and willingness to adapt AI, it is crucial to get to use and experience AI for yourself. Therefore, to strengthen the development of new AI solutions, in this paper we describe how to set up a laboratory for collaboration on and demonstration of AI. The two main questions addressed in this work are: How to convey AI to the broad masses? And interlaced with that: How to demonstrate AI?

2 Material and Methods

The first step was to gather information about the fundamentals of AI. Thus, paper and lectures regarding the topic AI got reviewed with the target to break down the groundwork someone needs to understand AI. Furthermore, AI-centred conferences, namely the *University:Future Festival* (UFF) and the *KI-SIGS Summit* were attended. Additionally, experts in the field of AI were consulted.

2.1 University:Future Festival 2021

The University:Future Festival 2021 [1] was a three-day-long conference organized by the *Hochschulforum Digitalisierung*. Its main aspect was the digital transformation of the teaching at schools and university. Therefore, one of the major topics was AI, with many talks and workshops around the topic AI. Here, a great variety of workgroups presented their projects and models for the integration of AI. One topic, for example, was *Natural Language Processing* (NLP), which plays a major part in how chatbots work. The UFF was used to get in touch with different workgroups and get an impression on the various approaches regarding the possibilities of AI.

2.2 KI-SIGS

KI-SIGS (KI-Space für intelligente Gesundheitssysteme) [2] is a cooperation of various AI-institutes from northern Germany in Bremen, Hamburg, Schleswig-Holstein, and medical devices developers with the help of university hospitals. Their mission is the implementation of an AI-Space for smart health systems. This includes adaptive medical systems as well as learning robotic assistance systems and smart-living home-assistants.

On the first day of their conference "KI-SIGS Summit - The Northern AI-Booster" the different workgroups of KI-SIGS presented their projects and discussed their progress with the other consortium partners. Whereas on the second day, external partners were invited to give insights into the use and the difficulties of AI in economy driven businesses.

This gave a good overview of various different AI projects and the issues they get confronted with, like the accessibility of sufficient data necessary to train an AI adequately.

2.3 Demonstration of applied AI

Age related macular degeneration (AMD) is a degenerative illness of the macular, interlaced with a degradation of the central field of view. The best therapy option is the periodic injection of so-called VEGF-inhibitors. Since the timing of this injection is critical for the success of the therapy, the company Visotec built an OCT system, that patient get to use on their own at home, the *HomeOCT* [7]. With the help of deep learning networks and medical image processing, an AI shall evaluate the OCT images and highlight those biomarkers typical for an acute AMD. Fig. 1 shows how this could look like.

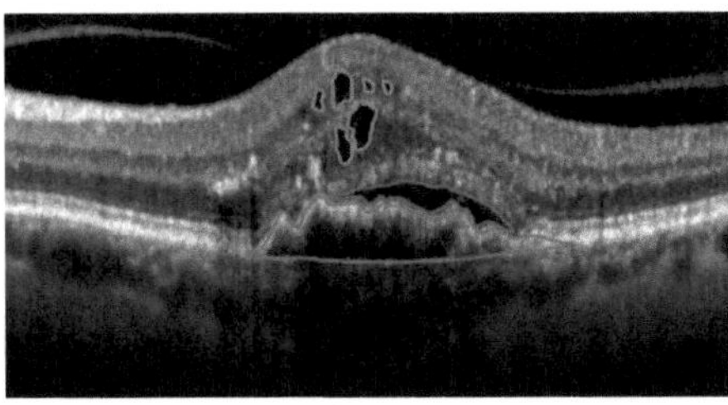

Figure 1: OCT image of a human retina altered by AMD, with highlighted AMD-biomarkers – from the top intraretinal fluid, pigment epithelium detachment and subretinal fluid [7].

Since the HomeOCT should be used multiple times a week by the patient, a great amount of data is generated. To analyse all the data by hand would cost a lot of time, and additionally, the lower resolution of the HomeOCT makes it harder to detect AMD biomarkers by eye. This increases the time it would cost to analyse the data by hand even further. Fortunately, an AI can help with the analysis separating the biomarkers, so a doctor just needs to inspect those images highlighted by the AI [7].

As a part of this work, the HomeOCT was demonstrated to an audience at the KI-SIGS Summit. Thus, giving the possibility to evaluate the use of the HomeOCT for the demonstration of AI. The most common questions like those regarding the accuracy of the used AI or the way the AI was trained were taken into account by the preparation of the HomeOCT-demonstrator for the laboratory.

2.4 AI Interviews

To gain insights about the currently relevant topics in the field of AI, experts were interviewed about the issues they are presently working on. The first interview was with Prof. Dr. Martin Leucker, consortium leader of KI-SIGS and head of the Institute for Software Engineering and Programming Languages at the University to Lübeck (UzL). He talked about Machine Learning (ML) as an alternative to classic programming, how AI and classic programming differentiate from each other and the task to create *Trustworthy AI*. Furthermore, he argued that ML is still at its beginning and where AI still lacks certain abilities. Additionally, he talked about the difficulty and necessity to certify AI systems. The second interview was with Prof. Dr. Esfandiar

Mohammadi, Professor at the Institute for IT-Security, UzL. His topic is the protection of data. Therefore, he explained that someone could unravel the ML model, that was used in an AI application, to get information about someone whose data was used to train the model. For this reason, he works on methods to protect personal data, for example methods from the field of *Federal Learning*.

The key aspects Trustworthy AI, the certification of AI and data security when working with AI shall be included in the laboratory. At this stage, Trustworthy AI is one of the elements the laboratory setup is focussed on, while certification and data security will be considered more proficiently in later iterations of the laboratory.

3 Results and Discussion

Combining the gathered input, it was possible to construct a plan for the laboratory. The main components are: An AI-assisted OCT-System specialized for the therapy control of AMD namely the Home-OCT by the company Visotec, a visual preparation of the AI fundamentals, and some methods to visualize how an AI works.

3.1 HomeOCT

For the laboratory setup, one instance of this OCT system is used to demonstrate one application for AI. This illuminates the possibilities of AI since on the one hand it is shown that AI can help to process huge amounts of data, that otherwise would have had to be examined by a human, thus saving resources while simultaneously leading to better results. And on the other hand, this presents the possibilities for small businesses and startups to establish themselves by AI. Furthermore, by using the OCT system themselves and having their own retina analysed, AI can be experienced first-hand.

3.2 AI Fundamentals

To gain access to the complex topic of AI, one part of the laboratory is designed for the purpose to teach the fundamentals of AI. First, the terminology often used in context of AI, terms like bias, overfitting and underfitting, strong/ weak AI, explainable AI, Internet of Things (IoT), machine learning etc. are explained with the help of digital index cards, displayed on a large-scaled touchscreen. When one card gets flipped by getting touched, it reveals an explanation of the term that was on the top side of the card. Sometimes the explanation is supplemented by further visual keys. For example when the overfitting-/ underfitting-card is flipped, the basic problem of over- and underfitting AI is explained, that if you train your AI with too few data, the AI is not able to generate a fitting model, thus making mistakes – underfitting, and if you train your AI with too many data, the AI is not able to adequately generalize, thus making mistakes as well – overfitting. Such explanation would then be aided by a graph or a more unmathematical representation, as in Fig. 2.

Figure 2: Common example for underfitting: The AI gets the task to find motorcycles in images. But in addition to the motorcycles, the AI also marks the bicycle – probably because it only learned to recognize motorcycles by the wheels.

Next, typical use cases are presented, two of those being chatbots and object detecting. To demonstrate the first one, chatbots, one instance is shown on the screen with the possibility to interact with it. For this, an openly accessible chatbot, which is still to be determined, will be used. Hereby, one can test this new form of machine-human-communication first hand. Moreover, to show the capabilities of AI-object-detecting, a video is played showing an object detection algorithm detecting different objects in real time.

The *KI-Campus* is a project, funded by the Federal Ministry of Education and Research, that was published in 2020. It was one of the organizer of the UFF. The main aspect is the building of a digital platform for teaching AI. Currently, multiple AI-institutes are developing this platform. All learning opportunities are free of charge. Thereby, the KI-Campus includes a wide range of AI related topics, basics as well as interdisciplinary questions and specializations in certain domains, such as machine learning. Furthermore, specific teachings for particular work fields, like AI in medicine, are included as well. The different AI topics are prepared in form of classic lectures, short films or podcasts in German and English and the platform is constantly extended. Therefore, some contents of the KI-Campus will be chosen for our laboratory.

3.3 Visualization of AI

To understand AI it is helpful to know the mechanisms working in the background of AI, unfortunately it is not an easy task to know exactly what an AI is doing in the hidden layers of its *Neural Network*. This is not just a problem for beginners of AI as there is a need, generally, to be able to explain the decisions an AI is making – key word: *Trustworthy AI*. Thereby, by visualizing the Neural Network of an AI, people are able to gain trust in AI more easily on the one hand and on the other hand it helps with making an AI better by helping to find biases in the model and adjusting the training data. This could also be useful

for the certification of AI systems.

Therefore, three works, chosen for their interactive features, will be implemented, that try to visualize how an AI is making decisions. For the implementation, data provided by those projects will be used.

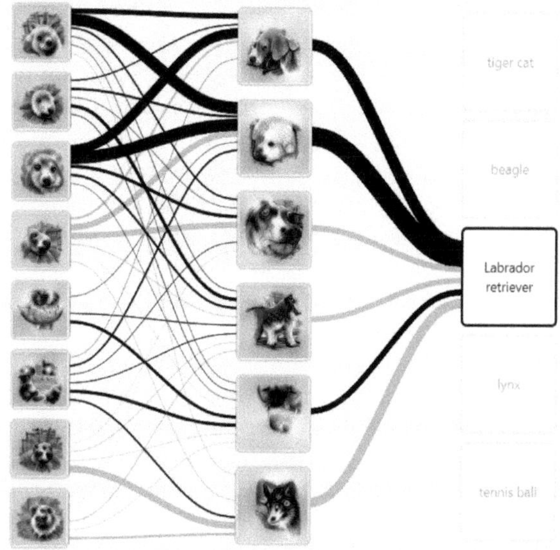

Figure 3: Visualization of how an AI classifies a Labrador. The darker lines represent a positive influence, while the brighter lines indicate a negative influence. For example, you can see that floppy ears seem to be important for identifying Labradors, while pointy ears are a negative influence [8].

The Building Blocks of Interpretability is a work by Chris Olah and others [8], where they explore different possibilities of designing an interface that makes you retrace how an AI comes to a classification layer by layer. By combining feature visualization and attribution or in other words by adding what a neuron is looking for to how the neuron affects the output, you get an interface for exploring how the AI decided between different classes. The interface shown in Fig. 3 makes you see which features the model used most to identify one class or the other. Furthermore, by hovering over the visualized features, it shows you where in the image the AI found this feature. Moreover, it makes you observe how the understanding of the AI evolves over the different layers, from detecting edges in the early layers, to more sophisticated shapes and recognizable object parts in the late layers. This gives a more comprehensible insight into AI. For example, the interface shows, that in one layer the AI connects a part of the lower snout of the dog with the idea of a tennis ball. Maybe because of the green background or maybe because, in the training data for the model, dogs got frequently presented with a tennis ball in their snout [8].

A different approach of visualizing the work of a Neural Network can be found in *Experiments in Handwriting with a Neural Network* by Shan Carter et al. [9], where they used a generative model of handwriting and visualized it in different ways. This model is trained with handwritten text and therefore is able to produce handwritten text by itself or to predict the next stroke of someone's writing. The here implemented interface lets you write something and the AI will try to predict your next stroke, in real time. Shan Carter et al. used a technique to visualize what the model would have done at each point if it had taken over for the person at that moment, where strokes got coloured green where the model would have veered rightward from the chosen stroke and orange where it would have veered leftward. Sophisticatedly, this produced a diagram, that shows the activation of the cells (neurons) over time, where each column in a heatmap represents one line segment of the writing and each row represents one cell coloured by its activations on that part of the stroke. Thus making it possible to find patterns in the way certain neurons activate for certain strokes [9].

The last tool to visualize AI presented in our laboratory setup is the *Neural Network Playground* created by Daniel Smilkov and Shan Carter [10]. This web interface lets its user tinker with a Neural Network for classification. The user chooses between one of four different datasets, that consist of different arrangements of clusters of blue and orange dots, representing the values -1 and 1 respectively. Then you can pick between different features for the first layer of the network, like a horizontal separation, or a combination of those features. In the next step, the amount of hidden layers and neurons is chosen. Here, the output of every neuron can be inspected by hovering of the neuron. At the end, the output, the classification based on the previous made choices of the network is shown. Thereby giving a nice invitation to play with the network on your own and finding the most fitting setup for the best classification. This creates a great way of learning how to build a Neural Network and comprehending the theoretical fundamentals practically without needing to be able to write code [10].

4   Conclusion

By gathering information on AI from various sources, the concept of a laboratory to teach, experience and develop AI was created. The main parts of this laboratory are the AI assisted HomeOCT to emphasize how AI could be used in a complex scenario, a dashboard for teaching AI fundamentals in an interactive way, including AI basics, a chatbot and a presentation of object detection, and some methods to visualize AI to counter the black-box problem AI often encounters. This laboratory should help people to comprehend AI, on the one hand, so they accept, trust and use AI and on the other hand to get more people to work on AI. The next step will be the realisation of this concept, building the laboratory. Then the usability of the laboratory will be evaluated. This is just the starting concept for the laboratory, over time it will be extended with more projects and ideas. Not only on the field of AI but also on topics like *Smart Systems* and *Digital Twin*, which is why the integration of

a 3-D endoscope combined with different kinds of displaying the 3-D image (3-D monitor, Virtual Reality (VR) and Augmented Reality (AR) lenses) and the construction of a Digital Twin of the laboratory is planned, as well.

Acknowledgement

The work has been carried out at the UniTransferKlinik Lübeck GmbH and supervised by the Institute of Software Engineering and Programming Languages, Universität zu Lübeck.

Author's Statement

Conflict of interest: Authors state no conflict of interest.

5   References

[1] University:Future Festival 2021. Available: https://festival.hfd.digital/de/ [last accessed on 2022-02-04].

[2] KI-SIGS. Available: https://www.ki-sigs.de/ [last accessed on 2022-02-04].

[3] J.-M. Chauvet, *The 30-year cycle in the AI debate.* arXiv preprint arXiv:1810.04053, 2018.

[4] M. Trajtenberg, *AI as the next GPT: a Political-Economy Perspective.* National Bureau of Economic Research, 2018.

[5] S. Stowasser, O. Suchy, N. Müller, N and others, *Einführung von KI-Systemen in Unternehmen.* Gestaltungsansätze für das Change-Management. Whitepaper aus der Plattform Lernende Systeme, München, 2020.

[6] BSP Business School Berlin 2021 Berlin, *KI-Kochbuch.* Available: https://www.mittelstand-digital.de/MD/Redaktion/DE/Publikationen/zentrum-kommunikation-ki-kochbuch.html [last accessed on 2021-12-21].

[7] C. v.d.Burchard, M. Moltmann, J. Tode, C. Ehlken, H. Sudkamp, G. Hüttmann, et al. *Self-examination low-cost full-field OCT (SELFF-OCT) for patients with various macular diseases.* In: Graefe's Archive for Clinical and Experimental Ophthalmology, vol. 259, pp. 1503-1511, 2020.

[8] C. Olah, A. Satyanarayan, I. Johnson, S. Carter, L. Schubert, K. Ye, A. Mordvintsev, *The Building Blocks of Interpretability*, In: Distill, 2018.

[9] S. Carter, D. Ha, I. Johnson, C. Olah, *Experiments in Handwriting with a Neural Network*, In: Distill, 2016.

[10] S. Carter and D. Smilkov, *Playground Tensorflow*, Available: http://playground.tensorflow.org/ [last accessed on 2022-01-04].

Integrating phase-contrast and fluorescence microscopy data for the analysis of axonal degeneration in a model of intracerebral hemorrhage using deep learning

Yamil Maluje [1,2], and Marietta Zille [2,3]

[1] Biomedical Engineering, Luebeck University of Applied Sciences, yamil.maluje@stud.th-luebeck.de
[2] Fraunhofer Research Institution for Individualized and Cell-Based Medical Engineering (IMTE), Luebeck, Germany.
[3] University of Vienna, Department of Pharmaceutical Sciences, Division of Pharmacology and Toxicology, Vienna, Austria, marietta.zille@univie.ac.at

Abstract

Axonal degeneration (AxD) occurs in many neurological diseases and describes a process, in which axons disintegrate, leading to axonal swellings and axonal fragmentation. However, the quantification of these morphological hallmarks of AxD remains challenging, in particular with respect to automatic high-throughput analyses. We present a hybrid convolutional neural network (CNN) model to analyze automatically the vitality and morphology of axons using phase-contrast (PhC) and fluorescence images. We adapted a u-net with ResNet-50 encoder and used an ensemble of eight different CNNs models. We demonstrate that the CNN model trained with PhC images allowed better analysis of axonal swellings and fragments, whereas the fluorescence CNN model was more suitable to analyze axonal areas in experimental hemorrhagic stroke. This CNN hybrid model opens new possibilities for the analysis of AxD. It can be used to further understand and explore different mechanisms of AxD, which will help develop novel treatments for neurological diseases.

1 Introduction

AxD can occur during nervous system aging or development as well as in neurodegenerative diseases [1], [2]. AxD is characterized by two morphological hallmarks 1) Axonal fragments are created when the axon disintegrates into individual separated parts, whereas 2) axonal swellings develop when axonal transport is interrupted and the cytoskeleton and organelles are disorganized [3].

The quantification of the AxD hallmarks is a challenge with respect to high-throughput analyses. Available tools to quantify axons such as NeuronJ, NeurphologyJ, or AxonQuant [3] have several disadvantages: 1) They rely on manual thresholding or binarization to recognize structures, 2) the regions of interest need to be manually defined, 3) the quantification of axons in low-density areas is difficult, 4) and time-consuming, and 5) they do not differentiate between the different morphological hallmarks of AxD.

CNNs can be used to extract information effectively and have been demonstrated to outperform other methods in medical imaging analysis [4]. We present a hybrid CNN model that can analyze the morphology and vitality of axons using PhC and fluorescence images automatically. Fluorescently labeled antibodies or markers can be used to detect specific target antigens or molecules with the advantage that fluorescence imaging provides a better target-to-noise ratio than PhC microscopy. We used calcein acetoxymethyl ester (calcein AM) to visualize the vitality of the neurons at a specific time point. This marker enters living cells through the plasma membrane and binds to calcium ions leading to green fluorescence [5]. However, fluorescence microscopy entails certain disadvantages, such as photobleaching during repeated imaging, or in some cases the need for cell fixation which allows only to observe a single time point. On the other hand, PhC microscopy exploits differences in the refractive index of different materials to distinguish structures. This allows to visualize the morphology of the axons without cell fixation and thus to investigate the morphological hallmarks of AxD over time.

Here, we hypothesized that an integrated analysis of PhC and fluorescence images would lead to an improved analysis of the different morphological hallmarks of AxD. We used previously published datasets of axons exposed to the hemolysis product hemin to mimic hemorrhagic stroke [3], [6] and assessed whether the potential drug candidate deferoxamine (DFO) is able to reverse hemin-induced AxD.

2 Material and Methods

2.1 Datasets

AxD was induced in an *in vitro* model of hemorrhagic stroke as described previously [3]. Primary cortical neurons were isolated from mice at embryonic day 14 in accordance with the German Animal Welfare Act and the local ethics committee (Ministerium für Landwirtschaft, Umwelt und ländliche Räume, Kiel, Germany, license number 2017-07-06 Zille). The neurons were placed into a microfluidic device that uses culture medium volume differences to drive axonal outgrowth, which spatially separates the axons from their cell bodies. After seven days, the axons were treated and recorded using time-lapse PhC microscopy (Olympus IX81) at baseline (1 hour, start of recording) and 24 hours (24 h). Then, the axons were stained with calcein AM and recorded by fluorescence microscopy (Zeiss Axiovert 200). For the first dataset (previously published [3], [6]), axons were treated with different hemin concentrations (0, 50, 100, and 200 µM). The images of the second dataset were provided by Alessa Pabst (Fraunhofer EMB). Axons were exposed to 0 or 100 µM hemin or 100 µM hemin and DFO (0.01, 0.1, 1, and 10 µM). We used DFO as a potential drug candidate that has already been described to provide protection in brain hemorrhage [7].

2.2 Pre-processing of microscopic images

An ImageJ script was created to batch-process the datasets. The script converts the images from 16- to 8-bit to make them compatible with the CNN models. It creates a 2-image stack according to the time point at which the image was taken (baseline and 24 h after hemin treatment). The baseline images were PhC whereas at 24 h, they were either PhC or fluorescence (with calcein AM). Thereby, two types of image stacks were created, i.e., pre-processed PhC and pre-processed calcein stack. As the images may have shifted during the recording, they were aligned using the ImageJ plug-in "Linear Stack Alignment with Swift", with parameters previously described in [3]. Finally, all images were cropped to eliminate black edges, the microgrooves, and other undesirable objects.

2.3 Hybrid CNN model

We developed a new algorithm that integrates the two already trained and validated CNN models [3], [6] into one model to segment the new image stacks. The architectures of both CNN models are adaptations of a standard u-net with ResNet-50 encoder [8]. Also, each CNN model is an ensemble-based CNN, which combines the output predictions of eight models to a final output prediction as this was shown to be superior to individual CNNs [9]. This model (Fig. 1) is formed by three main stages: (1) Input data of microscopic images, (2) PhC CNN model and fluorescence CNN model (fluo CNN model), (3) output segmentation with respective quantification measurements. The data goes sequentially through each stage to calculate the segmentation of each image stack. Depending on the type of the input (pre-processed PhC or pre-processed calcein), the image stack is analyzed by the PhC or fluo CNN model. Here, the model calculates the output segmentation which is the prediction of the different classes, i.e., ax-

ons, axonal swellings, and axonal fragments using different gray values. The darker gray value corresponds to the class 'axon', the middle range gray value to 'axonal swelling' and the lighter gray value to 'axonal fragment'.

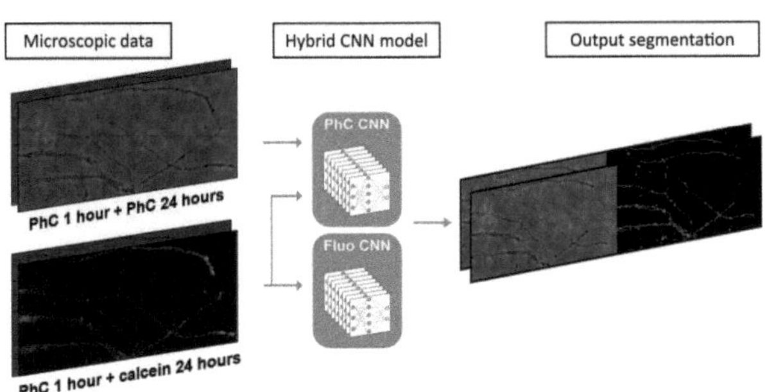

Figure 1: Schematic of the hybrid CNN model. Depending on the input, the 2-image stack goes through the PhC or fluo CNN model to calculate the respective segmentation output.

For quantification, the normalized area of all pixels per class on all images for each hemin concentration per time point and experimental day were summed (baseline and 24 h). To determine the changes of the classes *axon*, *axonal swelling*, and *axonal fragment* over time, we calculated the percentage of pixels of the corresponding class over the sum of the pixels of all three classes at baseline since axonal outgrowth in each device may vary for each experimental day.

$$normalized\ 'class'\ area\ (t_{24}) = \frac{'Class'_{t_{24}} \times 100}{Axon_{t_1} + Axonal\ swelling_{t_1} + Axonal\ fragment_{t_1}} \quad (1)$$

3 Results

3.1 Analysis of CNN segmentation output in the PhC dataset

First, we compared the 2-image PhC stacks to the previously published time-lapse PhC stacks [3] to investigate whether the new model can achieve similar results. We observed that the overall changes were similar with our previous results (Fig. 2). The axon area decreased after 24 h (with respect to baseline) for all hemin concentrations with the largest difference at 200 µM hemin, whereas the axons continued to grow in the vehicle condition (0 µM). Both axonal swelling area and axonal fragment area increased at 24 h for all three hemin concentrations compared to the baseline with the largest difference at 50 µM.

3.2 Integration of the fluorescence dataset into the hybrid CNN model

We next integrated the fluorescence images into the hybrid CNN model to determine the best method to quantify axons, axonal swellings, and axonal fragments (Fig. 2). The axonal area at 24 h was similar in both CNN models with decreasing relative area while increasing hemin concentration. The fluorescence results present a more lin-

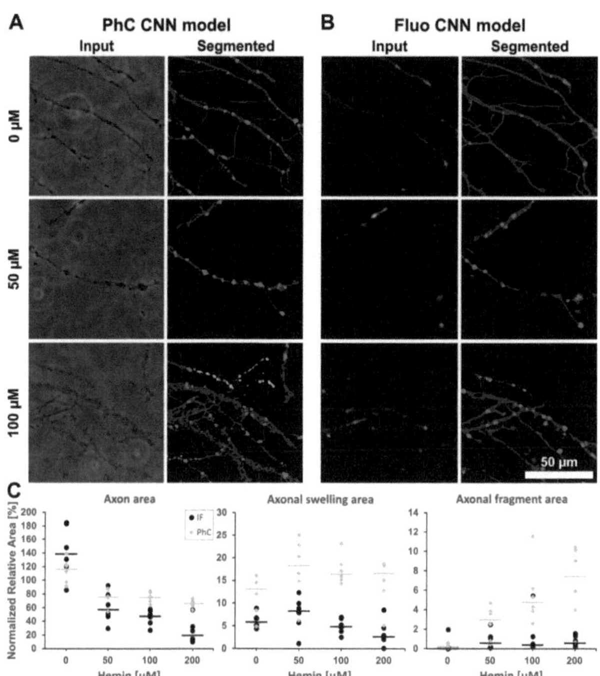

Figure 2: PhC and fluo CNN model segmentation and quantification at 24 h after hemin exposure. Input PhC image and the segmented output of the PhC CNN model (A) and input fluorescence image and the segmented output of the fluo CNN model (B) are shown. (C) Normalized area of axons, axonal swellings, and axonal fragments. Medians given for n=6 independent experiments.

ear response and bigger fold change for recognizing axons across all hemin concentrations than the PhC results. At 0 µM hemin, both models are comparable in the segmented output as the axons are healthy. At 100 µM hemin, fewer axons were detected with calcein AM compared to PhC, indicating that these axons degenerated. The axonal swelling area was similar in both CNN models with an increasing relative area with increasing hemin concentration in the PhC dataset. The PhC results present a bigger fold change for recognizing axonal swellings. The axonal fragment area was also different in both CNN models with an increasing relative area with increasing hemin concentration in the PhC dataset, whereas no noticeable change was observed in the fluorescence dataset. Taken together, the fluorescence CNN model was more suitable to detect changes in the axonal area, whereas the PhC CNN model more sensitively detected axonal swellings and axonal fragments.

3.3 Application of the CNN hybrid model to investigate AxD in hemorrhagic stroke

It has been demonstrated that neuronal cell death upon exposure to hemolysis products can be abrogated by DFO [7]. Thus, we applied our hybrid CNN model to investigate the effect of DFO on axons exposed to hemin (Fig. 3).
Similar to the previous data, there was a decrease in the axonal area after exposure to 100 µM hemin at 24 h compared with 0 µM hemin. When the axons were treated with DFO,

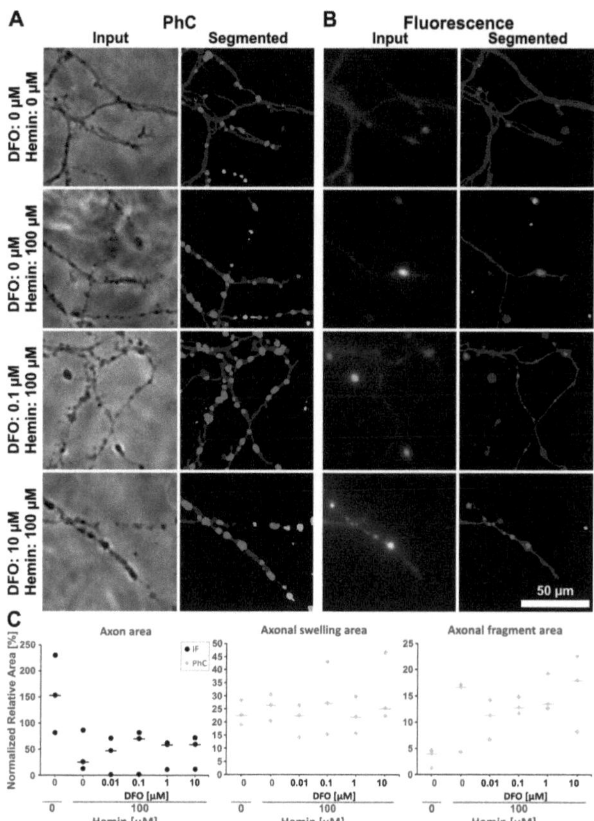

Figure 3: PhC and fluo CNN model segmentation and quantification at 24 h after hemin and DFO exposure. Input PhC image and the segmented output of the PhC CNN model (A) and input fluorescence image and the segmented output of the fluo CNN model (B) are shown. (C) Normalized area of axons, axonal swellings, and axonal fragments. Medians given for n=3 independent experiments.

there was a slight increase in axonal area at 0.01 µM DFO, which further increased at 0.1 µM DFO and then remained stable across the higher DFO concentrations. Compared to the previous data, the axonal swelling area was increased at 0 µM hemin, which may be the reason why only a small increase in axonal swelling area was detected after exposure to 100 µM hemin. The effect of DFO treatment is inconclusive as the variance was very high across the three replicates. Similarly, the axonal fragment area at 0 µM hemin was also larger compared to the previous data, but further increased at 100 µM hemin. DFO treatment may have reduced the axonal fragment area between 0.01 and 1 µM. However, the variance across the three replicates was again high, precluding firm conclusions.

4 Discussion

We successfully developed a hybrid CNN model which can quantify the different morphological hallmarks of AxD automatically by segmenting PhC and fluorescence microscopy images conjointly. We observed that the fluo CNN model was more suitable to quantify the axon, whereas the

PhC CNN model better detected axonal swellings and axonal fragments. Finally, we were able to apply the hybrid model to a new dataset assessing a potential therapeutic candidate, DFO.

The currently available tools cannot automatically differentiate the hallmarks of AxD on both PhC and fluorescence images [3], [6], which is desirable as each imaging modality seems to be differently sensitive to the morphological hallmarks. For instance, the fluo CNN model was more suitable to quantify the axon likely because of its wider dynamic range, whereas the PhC CNN model was more suitable to quantify axonal swellings and axonal fragments as calcein AM is not cleaved by esterases in degenerated axons. Hence, the calcein AM that is inside the cell is released, resulting in a less visible axon or invisible axonal fragments in fluorescence images whereas they are still present in the PhC images. Future studies should employ a 3-image stack of baseline and 24 h PhC images as well as a 24 h fluorescence image. Importantly, normalization to baseline allows to account for variations in axonal outgrowth at different experimental days, which was previously not possible for fluorescently stained images that can only be taken at a single time point. However, combining PhC and fluorescence microscopy can overcome this limitation.

We then successfully applied the hybrid CNN model to a new dataset using the potential therapeutic candidate DFO that was previously shown to prevent neuronal cell death after brain hemorrhage. However, the data was highly variable and no clear DFO effect on AxD was observed. Hence, the sample size needs to be increased in future studies.

5   Limitations and outlook

As current results are based on baseline and 24 h, only a specific stage of AxD can be analyzed and further information about the development of AxD, i.e. different AxD patterns [3], were not investigated.

We used calcein AM as a marker to visualize the vitality of the neurons. However, the model should be further validated using different fluorescently labeled antibodies that detect specific target antigens.

6   Conclusion

We successfully developed a hybrid CNN model which can segment PhC and fluorescence microscopical images for the analysis of AxD. This opens new possibilities of data analysis in a higher throughput manner and to further explore different mechanisms of AxD to develop novel therapeutic approaches for neurological diseases.

7   Acknowledgement

The work has been carried out at Fraunhofer (IMTE), in Luebeck, Germany. It was supported by a Fraunhofer MEF grant (project 900199) and by the Joachim Herz Stiftung (grant 850022) to Marietta Zille. We would like to thank Luisa Bartram and Philipp Grüning for computational assistance as well as Alex Palumbo and Alessa Pabst for providing the microscopic datasets.

8   Author's Statement

Conflict of interest: Authors state no conflict of interest.

9   References

[1] L. Luo and D. D. O'Leary, "Axon retraction and degeneration in development and disease," *Annual Review of Neuroscience*, vol. 28, 7 2005.

[2] N. Salvadores, M. Sanhueza, P. Manque, and F. A. Court, "Axonal degeneration during aging and its functional role in neurodegenerative disorders," *Frontiers in Neuroscience*, vol. 11, 9 2017.

[3] A. Palumbo, P. Grüning, S. K. Landt, L. E. Heckmann, L. Bartram, A. Pabst, C. Flory, M. Ikhsan, S. Pietsch, R. Schulz, C. Kren, N. Koop, J. Boltze, A. M. Mamlouk, and M. Zille, "Deep learning to decipher the progression and morphology of axonal degeneration," *Cells*, vol. 10, p. 2539, 9 2021. [Online]. Available: https://www.mdpi.com/2073-4409/10/10/2539

[4] S. M. Anwar, M. Majid, A. Qayyum, M. Awais, M. Alnowami, and M. K. Khan, "Medical image analysis using convolutional neural networks: A review," *Journal of Medical Systems*, vol. 42, 11 2018.

[5] F. Miles, J. Lynch, and R. Sikes, "Cell-based assays using calcein acetoxymethyl ester show variation in fluorescence with treatment conditions," *Journal of Biological Methods*, vol. 2, 10 2015.

[6] N. Menon, A. Palumbo, P. Gruening, S. Landt, L. Heckmann, L. Bartram, A. M. Mamlouk, and M. Zille, *Segmentation of fluorescently labelled axons in a model of brain hemorrhage-induced axonal degeneration using convolutional neural networks.* Infinite Science Publishing, 12 2020.

[7] M. Zille, S. S. Karuppagounder, Y. Chen, P. J. Gough, J. Bertin, J. Finger, T. A. Milner, E. A. Jonas, and R. R. Ratan, "Neuronal death after hemorrhagic stroke in vitro and in vivo shares features of ferroptosis and necroptosis," *Stroke*, vol. 48, 2017.

[8] O. Ronneberger, P. Fischer, and T. Brox, *U-Net: Convolutional Networks for Biomedical Image Segmentation.* Springer, Cham, 2015.

[9] A. O. Vuola, S. U. Akram, and J. Kannala, "Mask-rcnn and u-net ensembled for nuclei segmentation." IEEE, 4 2019, pp. 208–212.

Simulating Vulnerable Road Users in CARLA using Neural Networks

Rui Song[1]
[1] Robotics and Autonomous Systems, Universität zu Lübeck, rui.song@student.uni-luebeck.de

Abstract

One of the most important factors that must be taken into consideration about autonomous vehicles is road safety. In other words, it is necessary for autonomous vehicles to possess the ability to detect obstacles, other vehicles, and other vulnerable road users such as pedestrians, cyclists, and e-scooters, especially in urban environments. This work developed a neural-network-controlled pedestrian model and a neural-network-controlled cyclist model in a realistic simulator for the later development of a driver assistant system. The models of vulnerable road users are trained using two approaches, the traditional supervised learning with expert demonstrations and the minimize surprise approach in a street-crossing scenario.

1   Introduction

For newly developed intelligent transportation systems, such as autonomous vehicles, road traffic safety is particularly important. The first guideline for connected and autonomous vehicles according to the European Commission is to 'reduce physical harm to persons' [1]. Therefore, it is necessary for autonomous vehicles to possess the capabilities to detect other road users and interact with them to avoid traffic accidents.

Among all kinds of traffic accidents, those involving non-motorized road users are the most dangerous [2]. The non-motorized road users including pedestrians, cyclists, motorcyclists, and 'persons with disabilities or reduced mobility and orientation' are defined as vulnerable road users by the European Commission in the ITS (Intelligent transport systems) directive [3]. Vulnerable road users (VRUs) are vulnerable in the transportation system due to their characteristics. Unlike vehicles, most VRUs are not protected by metal frames. This means that VRUs, who are involved in a traffic accident, are more likely to be seriously injured or even killed. As the European Commission reported in 2019, "vulnerable road-users account for 70% of road deaths in urban areas" [2]. Among them, 40% of fatalities are pedestrians. As is asserted by Sweden's Vision Zero initiative, "it can never be ethically acceptable that people are killed or seriously injured when moving within the road system" [4]. Moreover, VRUs are usually small and flexible, which makes them hard to detect targets, that can't be tracked easily by most sensors. It's even harder to recognize their movements or gestures like eye contact and body language which could be used to predict their movements and behavior. Additionally, most VRUs possess the ability to make sudden motions [5]. As reported by the European Commission in 2018, mistakes by the human opera-

tor and the other road users are responsible for most accidents involving autonomous vehicles [6]. As a result, detecting and predicting the movement of VRUs can be seen as a crucial ability for autonomous vehicles to avoid accidents. Simulated urban traffic environments were developed since intelligent transportation products should not be trained in real traffic conditions in preliminary experiments, especially when VRUs are involved. Additionally, there is a need for realistic VRU models to be developed as well.

This work focused on the simulation of the two most common VRUs, pedestrian and cyclist, in a street-crossing situation. The full simulated VRU agent consists of three parts, a 3D model with communication motions, a motion scheduler, and action-generating neural networks. The VRU agent behaves based on a first-person visual observation. Both the pedestrian model and the cyclist model are trained using two approaches, the traditional supervised learning with expert demonstrations [7] and the minimize surprise approach [8] with behavioral penalties.

2   Material and Methods

2.1   CARLA Simulator

In this work, the simulation of the VRU agent behavior, as well as the background urban environment, is done in CARLA 0.9.10. CARLA is an open-source simulator developed by Alexey et al. based on the Unreal Engine 4 for researches in the field of autonomous driving [9]. CARLA was chosen as the simulator because it can simulate a realistic urban environment with various road conditions and static and dynamic actors including VRUs and vehicles. Furthermore, CARLA also provides various realistic sensors including an RGB camera, advanced sensors such as the semantic segmentation camera, and other sensors such as a collision detector. Additionally, 3D-related digital assets are free to use as well, which means that the layout of maps, the appearance of actors, and the motion animation can all be customized to fit the research objective.

2.2 JAAD Dataset

It is important to understand VRUs' behavior in reality for both VRUs motion design and model training. The Joint Attention in Autonomous Driving dataset (JAAD) proposed by Amir et al. [10] was used to analyze VRUs' behavior. The JAAD dataset contains more than 300 videos that cover different seasons, weather, illumination, and various road structures using a naturalistic approach, in order to reduce the influence of the collector on the pedestrians. [11] Furthermore, the dataset does not only focus on the behaviors of the pedestrians but also the reactions of drivers in a road-crossing situation.

2.3 Pedestrian and Cyclist 3D Model Motion

As pointed out in research proposed by Amir et al., non-verbal communication, such as gazing, hand gestures, or nodding, with other road users plays an important role at the point of crossing. According to their findings, pedestrians show their intention of crossing using non-verbal communication in over 90% of the circumstances [11]. However, there is no non-verbal communicational motion provided for VRU's 3D models in the CARLA simulator. Therefore, a set of common motions such as walking and communicational motions such as waving hands and a supporting motion scheduler are developed for the pedestrian using a skeleton controller provided by CARLA. However, there is no similar controller provided for the cyclist motion. Therefore, the communicational motions for the cyclist model are directly added to the animation asset.

2.4 Expert Demonstration

The first approach was using the traditional supervised learning with predefined expert demonstrations. The idea comes from the learning from demonstrations (LfD) research proposed by Beomjoon et al. [7]. They combined LfD with reinforcement learning to avoid insufficient or inaccurate demonstrations. Unfortunately, this approach was so computational demanding that the reinforcement learning algorithm could not be implemented. Instead, this work followed traditional supervised learning to train the network. The action generation problem is treated as an image classification task. Different from the network used in the minimize surprise approach, the network in this approach only generates discrete actions according to the current observation. For training the networks, a total of 146 predefined scenarios consisting of observation sequences and corresponding action commands was recorded for the VRU model in the CARLA simulator using data from the JAAD dataset. The predefined scenarios contain mainly two kinds

of behaviours of the VRUs, crossing the street using a cross-walk, and walking (riding) on the sidewalk (non-motorized path). Additionally, there are also some scenarios that contain other road users such as vehicles. In this case, the pedestrian and cyclist models will signal other road users when crossing or stop and wait for other road users to pass. Since the predefined scenarios do not include the accident situation such as when the VRUs deviated from the correct path, models which are trained using these scenarios may be less robust. However, as opposed to the situations of following a correct path which can only appear on the side-walk (non-motorized path) or crosswalk, an accident situation can appear anywhere. Therefore, it is difficult to cover the accident situations in a limited number of scenarios.

2.5 Minimize Surprise

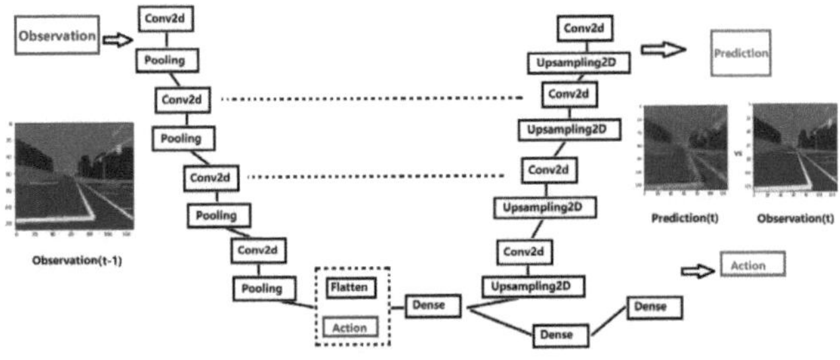

Figure 1: A U-net-like architecture generates the prediction and action commands based on the current observation and actions.

In this work, the NN-controlled pedestrian model is trained using the minimize surprise approach proposed by Hamann as an important part of the training loss [8]. The main idea of the minimize surprise comes from the free energy principle proposed by Friston et al. [12]. A self-maintaining system has to minimize the surprise (prediction error) of the states that it visits to stay statable [12]. In Hamann's research, the behaviors of agents are trained by evaluating the prediction error. Each agent is controlled by two independent networks, the action network, and the prediction network. The action network takes sensor values and the agent's last action as inputs and generates the next action of the agent as output. The prediction network uses the same input as the action network and generates the next time step's predicted sensor values as outputs, minimizing the difference between the predicted sensor values and actual sensor values to train both networks. The network which controls the VRU model was trained in a similar process in this work. The sensor values are the first-person observation of the VRU agent, and the actions are direction, speed, and the motion command. Instead of using two independent networks, the VRU model is equipped with one U-net-like network, as shown in Fig. 1, that uses current observation and action command as input and gives both the next action and prediction as output. Moreover, the network uses a hybrid loss to encourage the network to generate the complex behavior of the VRU agent. A collision penalty was added to encourage the agent to avoid collisions. Furthermore, a distance penalty was added to encourage the agent to go further from its start position. The network was trained with a simple evolutionary algorithm. The algorithm initializes 20 networks for 20 agents at the beginning and runs them generation by generation. Each agent starts at a random location on the sidewalk (for pedestrian models) or the non-motorized path (for cyclist models) and runs for 200 frames as one episode. After each run, the better half remains and replaces the other half with mutated copies of the better half. Each parameter of their network has a 0.5 to 0.2 chance to be changed using a normal distribution with a 0.01 variance.

3 Results and Discussion

3.1 Expert Demonstration

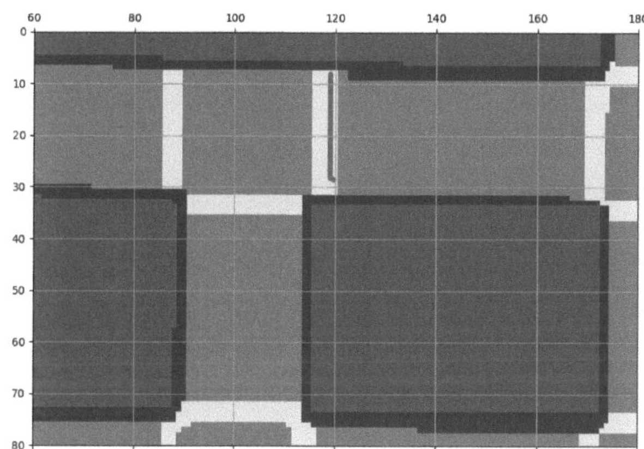

Figure 2: An example trajectory of the trained pedestrian model

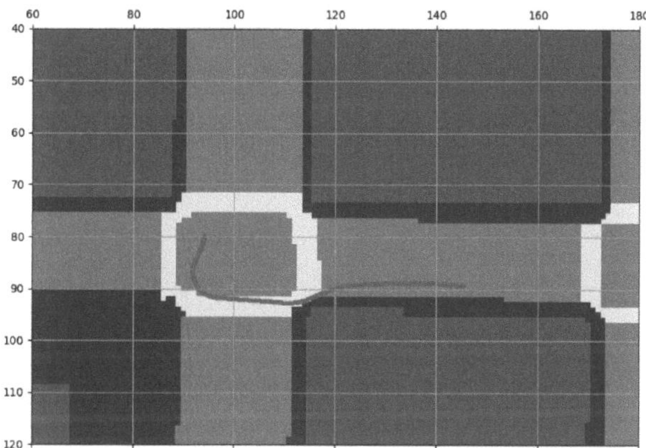

Figure 3: An example trajectory of the trained cyclist model

Fig. 2 and Fig. 3 show a top-view of a street crossing. As can be seen from these two Figures, the VRU models are able to recognize the sidewalk/bicycle path and crosswalk to some extent. We did not find a suitable quantified method to evaluate the generated action. However, it can be seen that the VRU models still cannot follow the 'safe path' perfectly. This can be caused by three reasons. One of the reasons is, as pointed out in the research of Beomjoon et al., [7] the predefined demonstrations themselves can be inaccurate. Furthermore, the insufficient number of demonstrations can also reduce the generalizability of the network. The other critical reason is that using only one frame as input lacks temporal information and can lead to an incorrect action selection.

3.2 Minimize Surprise

Figure 4: Actual observation of the next frame and prediction of the next frame made by the pedestrian model

Figure 5: Actual observation of the next frame and prediction of the next frame made by the cyclist model

Since there is no behavioural restriction that encourages the VRU models to find a safe path, the models trained using this approach are evaluated by the accuracy of the prediction instead of their trajectories. In Fig. 4 and Fig. 5, the left image is the actual current observation and the right one is the prediction made using the observation of the last frame. As can be seen from these Figures, the models can predict one frame (0.05 s) ahead to some extent. The network was trained without any behavioural restriction but only the mean squared error between the prediction and actual next observation. The agent then learned a behaviour that runs into a wall as soon as possible to reduce the prediction difficulty. A collision penalty was added to the loss to avoid this case. Furthermore, a distance penalty was added as well to encourage the agent to travel further. However, these two extra losses do not only increase the difficulty of the task but dilute the effect of the prediction loss as well. Models that trained with hybrid loss learn better behaviours but a worse ability to predict.

4   Conclusion

The models trained with traditional supervised learning with predefined expert demonstrations are able to learn relatively realistic behaviour, but there are still some limitations of predefined scenarios that exist which lead to bad performance. On the other hand, the models trained with minimizing surprise can predict the next frame observation to some extent, but they require suitable behavioural restrictions to learn the desired behaviour.

There are some future works that can be done to improve the performance. The first improvement that can be made is to customise motions directly using the animation asset instead of using the bone controller and the motion scheduler. Designing and simulating new motions using the bone controller and the motion scheduler is very complex and resource-consuming. On the other hand, editing the CARLA animation asset could be used to add new motions easier, but it requires extra knowledge of Unreal Engine. The other improvement can be behaviour quantification. An annotated 2D map can easily help evaluate the learned behaviour according to the trajectories. Additionally, the map can also be used for a reinforcement learning approach to avoid the limitation of LfD. Last but not least, for both approaches, increasing temporal information such as using more previous frames as the input of the network might lead to a performance improvement.

Acknowledgement

The work has been carried out at Universität zu Lübeck and supervised by Julian Petzold and Heiko Hamann, Institute of Computer Engineering, Universität zu Lübeck.

Author's Statement

Conflict of interest: Authors state no conflict of interest.

5   References

[1] J. F. Bonnefon, D. Černy, J. Danaher, N. Devillier, V. Johansson, T. Kovacikova, ... & K. Zawieska, *Ethics of connected and automated vehicles: recommendations on road safety, privacy, fairness, explainability and responsibility.* (2020).

[2] European Commission, *2019 road safety statistics: what is behind the figures?.* Available: `https://ec.europa.eu/commission/presscorner/detail/en/qanda_20_1004` [last accessed on 2021-08-11].

[3] European Commission, *Intelligent transport systems ITS and vulnerable road users.* Available: `https://ec.europa.eu/transport/themes/its/road/action_plan/its_and_vulnerable_road_users_en` [last accessed on 2021-08-11].

[4] C. Tingvall, & N. Haworth, *Vision Zero-An ethical approach to safety and mobility.* 6th ITE International Conference Road Safety & Traffic Enforcement: Beyond 2000 (1999).

[5] M. Bieshaar, G. Reitberger, S. Zernetsch, B. Sick, E. Fuchs, & K. Doll, *Detecting intentions of vulnerable road users based on collective intelligence.* arXiv preprint arXiv:1809.03916 (2018).

[6] European Commission, *Autonomous vehicles and traffic safety summary 2018.* Available: `https://ec.europa.eu/transport/road_safety/sites/roadsafety/files/pdf/ersosynthesis2018-autonomoussafety-summary.pdf` [last accessed on 2021-08-11].

[7] B. Kim, A.M. Farahmand, J. Pineau, & D. Precup, *Learning from limited demonstrations..* In:NIPS, pp.2859–2867 (2013).

[8] H. Hamann, *Evolution of collective behaviors by minimizing surprise.* In:Artificial Life Conference Proceedings 14, pp.344–351 (2014).

[9] A. Dosovitskiy, G. Ros, F. Codevilla, A. Lopez, & V. Koltun, *CARLA: An open urban driving simulator.* In:Conference on robot learning, pp.1–16 (2017).

[10] A. Rasouli, I. Kotseruba, & J. K. Tsotsos, *Understanding pedestrian behavior in complex traffic scenes.* In:IEEE Transactions on Intelligent Vehicles, vol. 3, no. 1, pp.61–70 (2017).

[11] A. Rasouli, I. Kotseruba, & J. K. Tsotsos, , *Agreeing to cross: How drivers and pedestrians communicate.* In:Conference on robot learning, pp.264–269 (2017).

[12] K. Friston, *The free-energy principle: a rough guide to the brain?.* In:Trends in cognitive sciences, vol. 13, no. 7, pp.293–301 (2009).

The simulation of pedestrian and e-scooter rider behaviors with CARLA

Liwei Tao [1]

[1] Robotics and Autonomous Systems, Universität zu Lübeck, liwei.tao@student.uni-luebeck.de

Abstract

The target of this research project is to realize the simulation of pedestrian and e-scooter rider behavior in traffic, which is an important part of the prediction model for autonomous driving. After training with the simulation, the prediction model of the vehicle is supposed to predict the actions of real pedestrians and e-scooter riders, and then take corresponding actions to avoid them. The simulation was based on an artificial neural network (ANN) and a hand-coded controller. It was realized in Python and was tested with the simulator CARLA. In this project, the models of a pedestrian and an e-scooter rider were created. Several ANNs, such as ResNet20, Resnet50 and networks evolved with NEAT were tried. They have been trained to show the predicted vision and give an action command to the model, which enables the model to behave like a real pedestrian or e-scooter rider. show the predicted vision.

1   Introduction

Autonomous driving is a relatively popular technology under development in recent years. This technology aims at enabling vehicles to move independently with the input of sensors instead of a human driver. The target of this internship is to realize the simulation of pedestrian and e-scooter rider behavior, which is an important part of the prediction model for autonomous driving. In this paper, the models of pedestrian and e-scooter rider behavior were realized in Python and were tested with the simulator CARLA [3]. CARLA is an open-source simulator for autonomous driving research, which enables users to test the prediction model in a digital city with pedestrians and e-scooter riders. It provides an API for the simulation and the various sensors of the vehicle to get detailed inputs. With the help of CARLA, the model of the pedestrian can be trained and tested to predict the complex behavior of pedestrians. The final prediction output will be used to warn the driver as a driver assistance system (DAS).

The tasks will focus on the simulation of pedestrians and e-scooter riders in the CARLA simulator. Taking a pedestrian as an example, the simulation will be based on a corresponding neural network and a controller of a pedestrian. The neural network will generate the motion control commands which are appropriate to a pedestrian in the real world. The controller maps their input commands to the corresponding action of a pedestrian in the simulator. In brief, the behavior of the simulation will be based on the development of the neural network and the controller. The development of models will start with the pedestrian, and then go on to the e-scooter rider.

NEUPA [1] aims at building a driver assistance system to demonstrate explainability in artificial neural networks (ANN) to enable their certification in the medium term. The DAS will predict the behavior of pedestrians and e-scooter riders and will warn the driver if a dangerous traffic situation might arise. For this project, the goal is to train an ANN prediction model of a pedestrian and an e-scooter rider for this DAS.

2   Material and Methods

2.1   Model of Pedestrian and E-Scooter Rider

CARLA contains 3D-models of pedestrians, which can be directly used. For the model of the e-scooter, it was prepared and then added to CARLA as a blueprint. A free model of the e-scooter [7] was downloaded as the base model, and it was modified with the 3D software Maya to connect with the skeleton provided by CARLA. To enable the model to move like a vehicle, the blueprint of an animation was created and the model of the driver was also modified to match the model of the e-scooter.

2.2   Design of Motion

To know which kind of motion is necessary to be realized in CARLA for a pedestrian, the video data of JAAD [5, 6] was used as reference. These videos are about pedestrian and driver behaviors at the time of crossing the street and factors that influence them. To this end, the JAAD dataset provides a richly annotated collection of 346 short video clips (5-10 sec long) extracted from over 240 hours of driving footage. We selected some typical videos as examples and summarized the behavior in these videos. This way, the desired behavior of the pedestrian in simulation was determined, which will help the DAS to distinguish which pedestrian may cross the street.

For the e-scooter rider, the generated behavior was inspired by the behavior of the pedestrian. The walking motion of the pedestrian was replaced with the motion of riding.

2.3 Realization of Motion

Hwang et al. [4] use feature descriptors for recognizing human actions from three-dimensional (3D) motion capture video sequences. With 3D geometric relational features, their model can represent the geometric relations among body points of a pose and classify the human action with a Markov model.

We use these approaches to realize the different actions of the pedestrian. Different motions were realized through the Python-API "bone transformation" provided by CARLA. With the API in CARLA, we can use simple Python code to send bone transformation data to CARLA and realize the action of the pedestrian. CARLA contains some blueprints of pedestrians, and one of them was chosen as the template of the simulation. A blueprint of the pedestrian contains all elements to create a dynamic pedestrian model in simulation. It contains many important bones of a pedestrian, which are able to be rotated and moved (Fig. 1). The combination of rotations of different bones enables the pedestrian to move different body parts, and the rotations of different bones at every time step form a complete action. In addition, the motions of the pedestrian are also based on the changing of the pedestrian's position at every time step. The position of the pedestrian is calculated according to the last position and the current motion. In brief, the motion of the pedestrian is the combination of the bone transformation and the change of position.

Figure 1: Pedestrian blueprint in CARLA.

The motions are separated into three kinds: *endless*, *once* and *once-keep*. The *endless* actions contain walking, fast walking, slow walking and standing, which would be executed until another *endless* action is given. The *once* actions contain looking to different directions and corresponding actions of looking back, and *once* means the pedestrian would return to the previous behavior after executing the *once* action. The *once-keep* actions contain waving arms and the return action, which means that the pedestrian would keep waving until the command of the return

action is given. Different actions need the rotation of different bones and therefore the priority of actions and the occupation of bones had been ensured to prevent the situation that different actions move the same bone, and the pedestrian behaves strangely. For example, if the pedestrian needs to wave their arms when walking, the rotation of the legs will be kept, but the rotation of the upper body will change from the action sequence of walking to the one of waving arms. You can see from the image below that the pedestrian keeps walking with the lower part of the body and waves their arms with the upper body. The priority of the *once* and *once-keep* actions is higher than the priority of *endless* actions, and therefore, these two actions will take place when the *endless* action is executed. For example, the arm and some relative parts of the body of the pedestrian will execute the *once-keep* action of waving the arm, when the pedestrian still does the *endless* action of walking with other parts of their body. For actions of the same kind, the next action will wait for the current action to finish. Fig. 2 shows a pedestrian walking while raising his arm.

Figure 2: Pedestrian walking and raising arm.

For the e-scooter rider, we tried another way to realize the rider behavior. It uses the animation blueprint in the Unreal Engine to set the motions of the rider. In the blueprint, the motions are set as states, and when certain conditions are satisfied, the state of the rider will change. When the transition of the state takes place, the angle of the bone of the rider will be adjusted to the preset angle of current state and the model of the e-scooter rider will be changed simultaneously. For example, the steering angle set in the code will be adjusted in different situations. The driver will raise the arm when the steering angle reaches a certain value, and the driver will turn right or left when the steering angle still increases and reaches another threshold.

2.4 Design of Scenarios

This part is about the work of designing scenarios, so that the pedestrian and the e-scooter rider move in the right places and execute their predetermined actions. For pedestrians in these scenarios, they would cross streets or walk along the sidewalk. For example, in the scenarios of crossing the street, the pedestrian will look around first, raise their arm if there is a vehicle and walk fast. These scenarios are the behavior that the pedestrian should be able to do after the training.

The e-scooter model in the scenarios should show behavior like a normal scooter in reality. The driver would drive along the road, and they would wave arms and look in the same direction for some time and then turn. One thing that should be noted in the design is that the driver should drive the e-scooter legally, which means going in a direction which is not allowed by traffic regulations was avoided during scenario design.

In these scenarios, the positions of the pedestrians and vehicles, the type, duration and starting time of different actions, and the moving angles of the pedestrian were set and used as the input data with the saved vision data in these scenarios. The vision data is the first-person vision of the pedestrian and the rider, which is recorded with the semantic camera provided by CARLA. The left part of Fig. 3 and Fig. 4 are examples. The semantic camera can distinguish objects, which is useful for the neural network. In this project, the size of the vision image is fixed to 128×128.

2.5 Neural Network

The data is dealt with in a neural network. The purpose of this network is to make the pedestrian and the e-scooter act like the behavior in the scenarios. During the simulation of the scenarios, the data which was used to train the neural network was recorded in each time step. It was saved in code. The action data and vision data were used as input of the neural network and some existing neural networks like ResNet20 and ResNet50 were tried. The output is the action made by pedestrian or rider.

In addition, neural networks evolved with minimizing surprise, which has been proposed and tested by Hamann [2], were also tried. The evolutionary algorithms evolve the weights of the neural network during training. An existing method called NEAT [8] was tried. It is an evolutionary algorithm that creates artificial neural networks and evolves them, automatically changing the network topology. If the input and the required output of the vision data is too big, NEAT will use many weights and will take a long time for training. Therefore, the input data of the vision data is encoded with a regular convolutional autoencoder [9] to compress the network's input data size, and the output prediction data is also decoded with the autoencoder to compare with the input vision data. The autoencoder compresses the input data by a factor of 32. It contains four convolutional layers with eight filters each, and four maxpooling layers. The fitness function is set as the difference between the current vision and the predicted vision of the last time step by using the mean squared error. The compression is done by convolution to accelerate the training when retaining the features. A normal network is also designed and used for the evolution. In this project, a normal ANN was tried and evolved this way. The networks use the vision data and the

action as input and output the next action and a predicted vision image. The fitness function is the same as the one in NEAT. The current vision is a 128x128 image, which is the first-person vision of the pedestrian observed by the semantic camera, and the prediction image is also a 128x128 image. The fitness function calculates the mean squared error of these two visions and adds it up for all time steps, and the final value is the score of the neural network. The weights randomly change in a small interval of the original value during the evolution, and the weights are saved if they lead to a higher fitness than other changed weights. In this project, ten models were trained together, which means each generation five are kept and five are replaced by the better ones, and then all of them change their weights randomly. Through this way the weights of neural network are able to be evolved.

3 Results and Discussion

The first tried neural networks are ResNet20 and ResNet50. The action data and vision data were used as the input of the neural networks. The output is the action made by pedestrian or rider in the next time step. Then the trained models were used in the simulation, but the behaviors of the pedestrian were not as expected. The pedestrian would not walk along the road and not always wave arms to the vehicles, which were expected to be done like the pedestrian in the scenarios. The e-scooter just kept circling, which may be caused by the lack of the scenarios, or the structures of the neural networks were not complex enough. In brief, the actions output by the neural network were not corresponding to the actions in the original scenarios.

The second one used is the NEAT. This time, the action data and vision data were used as the input of the neural networks, and the output is the action and vision image of the pedestrian and the rider in the next time step. It evolves with minimizing surprise by setting the fitness function as the difference between the current vision and the predicted vision of the last time step with mean squared error. However, with the help of an autoencoder, the training still took a long time and the result was not as expected. The training will take nearly one day with a GTX 1070 GPU in the situation that the vision data is encoded, and the result of the prediction data is blurry compared to the next neural network(Fig. 3).

Finally, a regular neural network is designed and used for the evolution. The networks use the vision data and the action as input and output the next action and predicted vision image. It contains four convolutional layers, four maxpooling layers and two dense layers to predict the next action, and it also contains four convolutional layers and four upsampling2D layers to predict the vision in the next time step. The fitness function calculates the mean squared error of these two visions and adds it up for all time steps, and the final value would be the score of the neural network. After the evolution, the model can get a quite good fitness score. The fitness score of the model was over 25 before training. After training, it decreased to nearly one,

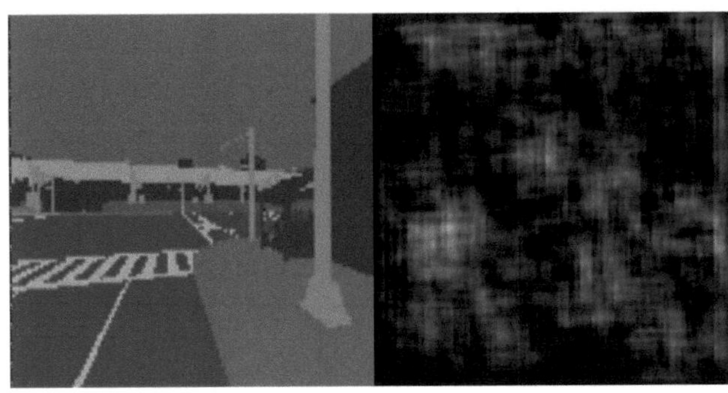

Figure 3: Predicted vision of the neural network evolved with NEAT.

which means that the output prediction image is similar to the next image gotten in the simulation. Fig. 4 shows the the similarity between the output prediction image and the next image gotten in the simulation.

Figure 4: The left part is the image recorded with the semantic camera, and the right part is the predicted image.

4   Conclusion

In this research project, the simulation of pedestrian and e-scooter rider behavior in traffic was realized. Different actions were designed and realized, and then used to created scenarios of pedestrians and e-scooter riders for the training of the neural networks. Three kinds of networks which were used to give action commands were created and trained. After the training, they were not able to make a good choice for the action the pedestrian and the e-scooter should take. This may be caused by the training time, the complexity of the neural network, or the lack of diversity in the scenarios. However, the last neural network evolved with minimizing surprise made a good prediction of the vision image in the next time step. It may show that the neural network has the potential to enable the pedestrian and the e-scooter rider to behave like a real one and may potentially be used for autonomous driving.

Acknowledgement

The work has been carried out at Universität zu Lübeck, and supervised by Julian Petzold and Heiko Hamann, Institute of Computer Engineering, Universität zu Lübeck.

Author's Statement

Conflict of interest: Authors state no conflict of interest.

5   References

[1] Hamann, H., NEUPA, Institute of Computer Engineering, Universität zu Lübeck. URL https://www.neupa.net/

[2] Hamann, H., *Evolution of Collective Behaviors by Minimizing Surprise*. In: Artificial Life Conference Proceedings 14, pp. 344–351. MIT Press (2014)

[3] Dosovitskiy, A., Ros, G., Codevilla, F., Lopez, A., Koltun, V.: *CARLA: An Open Urban Driving Simulator*. In: Proceedings of the 1st Annual Conference on Robot Learning, pp. 1–16 (2017). URL https://CARLA.org/

[4] Hwang, J.N., Thuc, H.L.U., Ke, S.R., Tuan, P.V., Chau, T.N.:*An Effective 3d Geometric Relational Feature Descriptor for Human Action Recognition*. In: 2012 IEEE RIVF International Conference 5 on Computing and Communication Technologies, Research, Innovation, and Vision for the Future, pp. 1–6. IEEE (2012)

[5] Rasouli, A., Kotseruba, I., Tsotsos, J.K.: *Are They Going to Cross? A Benchmark Dataset and Baseline for Pedestrian Crosswalk Behavior*. In: Proceedings of the IEEE Conference on Computer Vision Workshops, pp. 206–213 (2017)

[6] Rasouli, A., Kotseruba, I., Tsotsos, J.K.: *Agreeing to Cross: How Drivers and Pedestrians Communicate*. In: IEEE Intelligent Vehicles Symposium (IV) , pp. 264–269 (2017)

[7] Gesta2: "Electric Scooter" (https://skfb.ly/6RBZw) Licensed under Creative Commons Attribution (http://creativecommons.org/licenses/by/4.0/).

[8] Stanley, K.O., Miikkulainen, R.: *Evolving Neural Networks through Augmenting Topologies*. Evolutionary Computation 10(2), 99–127 (2002)

[9] Bank, D., Koenigstein, N., Giryes, R.: *Autoencoders*. arXiv preprint arXiv:2003.05991, 2020.

Interactive Segmentation of 2D Femoral Artery Ultrasound Data

Hanju Chen [1], Niclas Bockelmann [2]
[1] Robotics and Autonomous Systems, Universität zu Lübeck, hanju.chen@student.uni-luebeck.de
[2] Institute for Robotics and Cognitive Systems, Universität zu Lübeck, bockelmann@rob.uni-luebeck.de

Abstract

Ultrasound images are often used in clinical medicine due to their advantages such as being radiation-free and fast which can be beneficial in endovascular treatments. However, ground truth data is difficult and costly to obtain, leading to limited amount of data that can impede training with deep learning methods. Interactive segmentation introduces user interaction, allowing users to click on the objects they want to segment or by other means to obtain better segmentation results even with limited data. In this work, we investigate the feasibility of interactive segmentation methods for 2D femoral artery ultrasound data with two different algorithms: 1. Click-based interactive segmentation algorithm; 2. U-Net algorithm. Afterwards, post-processing methods are used to further improve the segmentation results. The final difference in evaluation coefficients between the two networks is close, which illustrates the feasibility of interactive segmentation methods for image segmentation of ultrasound data.

1 Introduction

Endovascular treatments, due to the use of X-ray imaging, expose patients and doctors to constant X-ray radiation during diagnostic and surgical procedures, which is harmful to living cells and can lead to the development of life-threatening diseases [1]. For this reason ultrasound imaging is an alternative that can be used to guide endovascular procedures as a radiation-free, low-cost and real-time low-risk method. To further improve surgical outcomes, automated analysis of ultrasound images is required, which relies on accurate automatic segmentation. However, the basis for such a step is a large amount of segmented training data, which is difficult and costly to obtain, especially in medical applications.

In order to train with limited data and acquire image segmentations in fast and reliable fashion, interactive segmentation methods need to be introduced. The usual method is to allow the user to manually annotate to get feedback and manually intervene and optimise the segmentation results, a semi-automatic segmentation method. Semi-automatic interactive segmentation methods can be divided into three categories: (1) Bounding boxes [2]: users needed to supply bounding boxes around the objects of interest to guide segmentation; (2) Extreme points [3], [4]: user needs to mark the four extreme points of the object: leftmost, rightmost, top and bottom, to guide the image segmentation; (3) Click-based [5], [6]: user clicks on the object to be segmented and determines the object to be segmented based on the result of the click. These three types of studies have demonstrated that semi-automatic methods based on deep learning can provide more accurate results while providing users with ways to interact and potentially get a better insight of the network's predictions. Compared to bounding boxes and

extreme points, click-based is the easiest and least time-consuming way to interact. The user only needs to click on the object he or she wants to segment, whereas the other two methods require a more detailed selection of the object's border position. This work is therefore based on an interactive click-based segmentation method.

2 Material and Methods

2.1 Data sources

The data used in this work is consistent with the work in [7]. This data consisted of five probands, for each proband three data acquisition were conducted using 2D ultrasound imaging along the femoral artery. Those recordings are referred to as run 1, 2, or 3. We use run 1 from all probands for training purposes and run 2 for validation and run 3 for testing. There are 3623 ultrasound images and mask images in the training data, 3524 in the validation data and 1916 in the test data respectively. All ultrasound images with a size of 512×323 pixels and an intensity range of $(0,255)$. The mask is a binary ground truth mask consisting of two classes: background and femoral artery $(0,1)$. As two different network structures were used, each with different requirements for the data, the data was not pre-processed at the very first step, but rather as the process encountered requirements, such as normalisation of the images, resizing of the images, etc.

2.2 ICAN

The click-based segmentation network consists of a VGG-19 network [8] and a Context Aggregation Network (CAN)

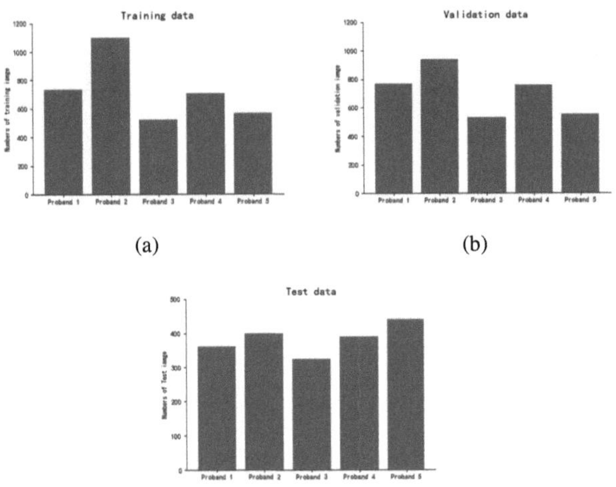

(a)   (b)

Figure 1: Training, validation and test data composition, both algorithms use this dataset for network training

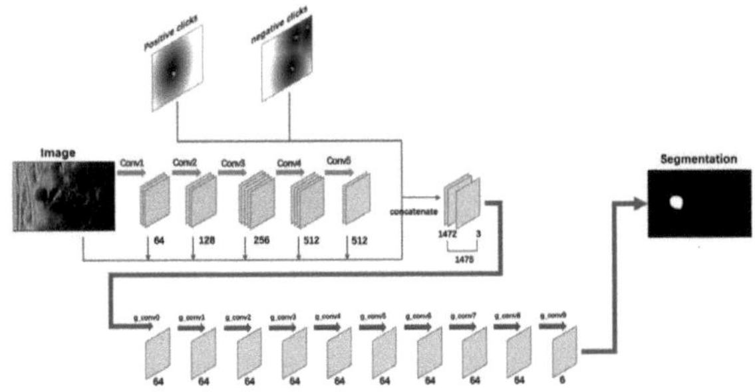

Figure 2: ICAN network structure with input images and user clicks on the input. The algorithm generates two separate channels for positive and negative clicks. The input image is passed through VGG-19 to form an augmented input. The channel generated by the click map is then concatenated to enhance the input image channel, passing through the segmentation network. The final output is a click-selected segmentation.

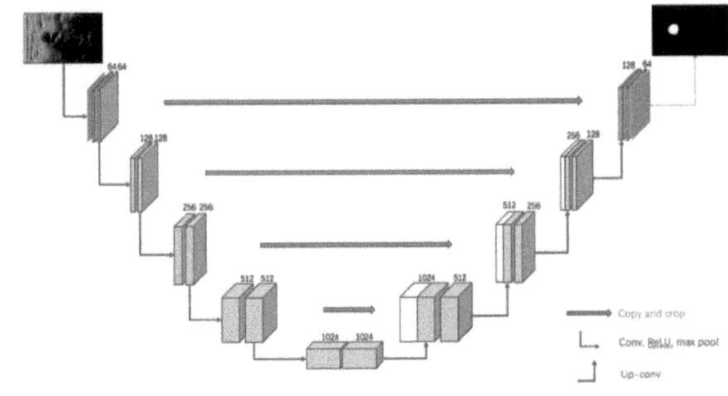

Figure 3: U-Net architecture. The traditional U-Net network, which performs well in most cases, is used to compare the data with the ICAN network results.

[9]. The overall network structure is referred to as ICAN and is shown in Fig. 2. The input to the ICAN network in this work is a 512×323 ultrasound image and the output is the predicted segmentation of the femoral artery.

The input image is first passed through the VGG-19 network to extract the feature maps for each layer, which improves model generalisation. The feature maps are bilinearly upsampled to a resolution of 512×323 and are appended to the input tensor. User interaction positive and negative clicks are converted into a Euclidean distance map, which is appended to the input tensor as an additional channel. Different users do not select the same object at the same click position, so random sampling is used to automatically generate positive and negative click pairs. Euclidean distance images by clicks are concatenated with input tensor to form an input with a channel number of 1475 into the CAN. We use a CAN network based on dilation convolution, which is divided into eleven layers: each layer consists of numbers of kernels, kernel size, dilation rate and LeakyReLU. The previous 1475 channels of input are reduced in dimension to 64 channels, then passed through middle layers of expanding receptive fields. The final output layer has six channels, each of which generated a segmentation mask. The reason for this is that according to the paper [5] illustrate that most of the datasets tend to perform smoothly and reach a fairly high IoU performance when the number of clicks reach six, we therefore also experiment with the same number. Six channels will generate six masks. The best performing mask is selected for output by selecting the network.

The input data we use are images training, validation and test datasets with the size of 512×323. Typically, the input images for image segmentation are RGB three-channel images. However, we use a single-channel image in grayscale for this work. Therefore, the number of input channels to the network is adjusted to one channel.

The ICAN workflow is divided into three parts: 1) the training data is 100 epochs, while the validation data is continuously updated with parameters and weights. Validation data is used for generalisation estimation; 2) input the test data into the trained network to obtain the segmented image; 3) compare ICAN generated segmentation mask with ground truth mask to evaluate network performance.

2.3   U-Net

The structure of the U-Net network [10] is shown in Fig. 3. Using the same input image as the ICAN network, the output is a segmented image of the same size.

U-Net first downsamples the image five times through the convolutional and pooling layers, shrinks the image and extracts the feature values; then deconvolutes for upsampling, cropping the previous lower layer feature map for fusion; then upsamples the data again. This process is repeated leading to the characteristic U shape architecture, until the size of the output feature map is restored to 512×323 , after which the segmentation result map is obtained by softmax. Finally the U-Net generated mask is compared with the ground truth mask to evaluate the results of the U-Net.

2.4 Experiments

In all subsequent experiments, we will use the Dice coefficients, to evaluate the similarity of mask and ground truth, to assess the merit of the network segmentation results. The formula for the Dice coefficients is defined in the following:

$$Dice = \frac{2|A \cap B|}{|A| + |B|} \qquad (1)$$

$A \cap B$ is the intersection between A and B. The sub-tables |A| and |B| denote the number of elements of A and B, where the factor of 2 in the numerator is due to the existence of double counting of the common elements between A and B in the denominator. Dice coefficients take values in the range [0,1], with 0 indicating no overlap and 1 indicating complete overlap.

We trained the networks using Tensorflow (ICAN) and Py-Torch (U-Net) respectively, and utilized a GPU for training accelaration. Using Adam as the optimisation method, a maximum epoch number of 100 and a learning rate of 0.0001.

We conducted a total of two experiments. In the first experiment, we investigated the segmentation effect of two methods without any post-processing of the prediction results. The Dice coefficients of 0.8014 and 0.7824 were obtained for the ICAN and U-Net, respectively. In the second experiment, we investigated different post-processing methods to improve the segmentation results for the networks used (mainly for ICAN). We post-processed the experimental results because certain ICANs predicted significantly different segmentation results from the ground truth. For example, in Fig. 4, one example segmentation results is depicted. As can be seen, the ICAN segments mulitple fermoral artery areas which is not possible in this particular image.

In order to make the ICAN predictions more similar to the ground truth, we decided to post-process the segmentation results. The main post-processing methods are the morphological operation: Closing and Opening, and finding the largest connected components (Lcc). Opening is an erosion followed by a dilation that tends to remove pixels or open gaps in the image. Closing is a dilation followed by an erosion that tends to close gaps in the image. Finding largest connected components is a way to make the predicted mask more consistent with the ground truth mask by removing the smaller parts of the mask and keeping the larger ones.

3 Results and Discussion

3.1 Model performance

All training results are presented in Table 1. Results of 80.1% and 78.2% were obtained after the test dataset. The ICAN we used achieved an accuracy of 80.1%, even a little better than the results of U-Net. But there are still some shortcomings in segmentation. Our prediction target is only the femoral artery, so there should be only one femoral artery area in prediction result. However, a few of the segmentation images obtained multiple false positve areas.

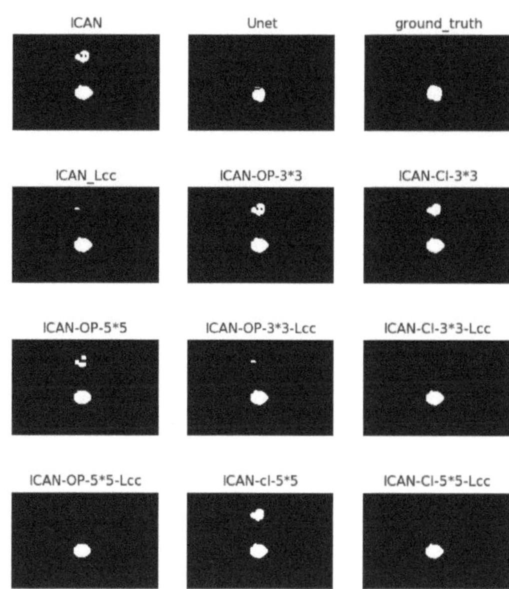

Figure 4: Post-processing methods for segmentation

This is probably related to the randomly generated click images we use. When randomly generating click pictures for images with actual objects, it is possible to obtain positive and negative click images efficiently. However, when generating interaction maps for femoral artery ultrasound images, a valid pair of click image is not always obtained, the main problem being that a valid positive click images cannot be generated. This is probably due to the fact that the ultrasound data images are not as clear as the layers of a normal image. This resulted in only a small percentage of the click maps making it into the network during training. This may further lead to a reduction in Dice score, as the click maps do not always point to the location of the femoral artery, so some of the segmentation maps produce multiple artery areas. The U-Net does not rely on click maps, so sometimes the segmented images of the U-Net look closer to the ground truth.

3.2 Post-processing

After applying different post-processing methods to the ICAN segmentation results, we obtained the results shown in Table 1. As can be seen, ICAN's Dice coefficient decreases as the size of the kernel increases after doing the Opening; the situation is similar for Lcc, where the Dice coefficient decreases when used alone; only the Dice coefficient for Closing increases. However, as we can see in Fig. 4, the segmentation results are still not sufficient and the incorrectly segmented areas still exists. So we tried to use morphological operations and Lcc at the same time to get better segmented images. We can clearly observe that the incorrectly segmented masks are removed, especially Closing + Lcc, which almost completely removes the incorrect masks, and the Dice coefficients are also improved.

Overall, the post-processing method of Closing+Lcc on

Table 1: The Dice coefficients obtained by different methods. Dice Std: Dice Standard deviation ; OP: Opening; Cl: Closing; 3×3, 5×5: kernel size; Lcc: largest connected components

Networks	Post-processing	Dice Mean	Dice Std
ICAN	-	0.8014	0.1908
U-Net	-	0.7824	0.2568
ICAN	OP-3×3	0.7986	0.2058
ICAN	Cl-3×3	0.8025	0.1902
ICAN	OP-5×5	0.7672	0.2736
ICAN	Cl-5×5	0.8050	0.1888
ICAN	Lcc	0.7987	0.2115
ICAN	OP-3×3-Lcc	0.7942	0.2243
ICAN	Cl-3×3-Lcc	0.8026	0.2060
ICAN	OP-5×5-Lcc	0.7631	0.2821
ICAN	Cl-5×5-Lcc	0.8059	0.2036

segmented images can effectively remove false positives, increase the Dice score of ICAN network segmentation results and improve network performance, which is an essential step in the whole experiment.

4   Conclusion

In conclusion, we have successfully implemented image segmentation of ultrasound images using ICAN networks and obtained good results. Compared to the traditional and commonly used U-Net network segmentation results, ICAN shows superior performance. This demonstrates the feasibility of an interactive segmentation approach for ultrasound images, where user interaction can be used to improve and correct the training results in a limited data set. The post-processing method, as a complement to the ICAN network, improves the accuracy of the generated segmentation masks and brings them closer to the ground truth masks. In future work, we will investigate and develop interactive graphical user interfaces to allow real-time user interactions. Additionally, shape regularizations as well as the expansion from 2D to 3D will be investigated to further improve the segmentation results on ultrasound-based femoral artery data.

Acknowledgement

This work was carried out and supervised by the Institute for Robotics and Cognitive Systems. The data used in this work was acquired within the project Nav EVAR (German Federal Ministry of Education and Research grant number: 13GW022).

Author's Statement

Conflict of interest: Authors state no conflict of interest.

5   References

[1] Sultan A Meo, Abdul Majeed Al Drees, Syed Za Zadi, Salah AL Damgh, and Ali S Al-Tuwaijri. Hazards of x-ray radiation on the quantitative and phagocytic functions of polymorphonuclear neutrophils in x-ray technicians. *Journal of occupational health*, 48(2):88–92, 2006.

[2] David Acuna, Huan Ling, Amlan Kar, and Sanja Fidler. Efficient interactive annotation of segmentation datasets with polygon-rnn++. In *Proceedings of the IEEE conference on Computer Vision and Pattern Recognition*, pages 859–868, 2018.

[3] Eirikur Agustsson, Jasper RR Uijlings, and Vittorio Ferrari. Interactive full image segmentation by considering all regions jointly. In *Proceedings of the IEEE/CVF Conference on Computer Vision and Pattern Recognition*, pages 11622–11631, 2019.

[4] Kevis-Kokitsi Maninis, Sergi Caelles, Jordi Pont-Tuset, and Luc Van Gool. Deep extreme cut: From extreme points to object segmentation. In *Proceedings of the IEEE Conference on Computer Vision and Pattern Recognition*, pages 616–625, 2018.

[5] Zhuwen Li, Qifeng Chen, and Vladlen Koltun. Interactive image segmentation with latent diversity. In *Proceedings of the IEEE Conference on Computer Vision and Pattern Recognition*, pages 577–585, 2018.

[6] Ning Xu, Brian Price, Scott Cohen, Jimei Yang, and Thomas S Huang. Deep interactive object selection. In *Proceedings of the IEEE Conference on Computer Vision and Pattern Recognition*, pages 373–381, 2016.

[7] Ana Estrada Lugo, Niclas Bockelmann, and Felix von Haxthausen. Sequential u-net architecture for automatic femoral artery segmentation in ultrasound images. *Current Directions in Biomedical Engineering*, 7(1):158–161, 2021.

[8] Md Foysal Haque, Hye-Youn Lim, and Dae-Seong Kang. Object detection based on vgg with resnet network. In *2019 International Conference on Electronics, Information, and Communication (ICEIC)*, pages 1–3. IEEE, 2019.

[9] Peng Gao, Jiasen Lu, Hongsheng Li, Roozbeh Mottaghi, and Aniruddha Kembhavi. Container: Context aggregation network. *arXiv preprint arXiv:2106.01401*, 2021.

[10] Olaf Ronneberger, Philipp Fischer, and Thomas Brox. U-net: Convolutional networks for biomedical image segmentation. In *International Conference on Medical image computing and computer-assisted intervention*, pages 234–241. Springer, 2015.

Automatic Nightmare Recognition Using Convolutional Neural Network

Marvin Raasch [1], Marcin Grzegorzek [2]

[1] Medical Informatics, Universität zu Lübeck, marvin.raasch@student.uni-luebeck.de

[2] Institut für Medizinische Informatik, Universität zu Lübeck, marcin.grzegorzek@uni-luebeck.de

Abstract

At the current point in time, human sleep still raises unresolved questions. Many of these analyses depend strongly on manual work of experts. In this work, a convolutional neural network (CNN) for the detection of nightmares from Electroencephalography (EEG) sleep data is presented. To train and test this CNN, a sleep lab provided EEG data of patients from the current EDOREM2 sleep study. Each patient was recorded over a minimum of two nights with at least one normal night and one night with a nightmare. The proposed model was able to detect the occurrence of a nightmare with an accuracy of 76%. It can be used to automate the annotation process of sleeping data and therefore facilitate the evaluation of sleep quality. Since sleep quality can be correlated with overall health, this work may help in monitoring the health of patients.

1 Introduction

Technologically, the field of sleep medicine has evolved rapidly and increasingly integrates techniques from many fields, such as data science and machine learning. This can be seen in the work of Sajad Mousavi with SleepEEGNet [1], Ye Yuan et al. with HybridAtt [2] or Xinyu Huang et. al with a deconvolutional CNN [3]. Each of those projects introduced machine learning algorithms to automatically recognize sleep patterns or other specific events, like sleep spindles. This trend can be explained by the time, effort, and knowledge needed to do such analysis manually. As seen in the study of Luyster et al. [4] a big factor of mental and physical health is healthy sleep. There are many factors that cause unhealthy sleep, but for patients with Post-traumatic stress disorder (PTSD) or borderline personality disorder, nightmares are one of the most prominent symptoms. While monitoring the sleep of such patients, a lot can be learned about the progress of their therapy. Currently, the most widely used technique to detect nightmares is a simple conversation with the patient. In most cases this is sufficient, but some patients may forget they had a nightmare or give false statements. Therefore, empirical data should be included in this process to obtain an objective view. One option could be to study the recorded EEG data and discuss it. This work presents how a neuronal network can be used to classify a patient which had a nightmare or a night of normal sleep.

2 Material and Methods

2.1 Sleep Stages

While sleeping the brain cycles through four stages of sleep. The occurrence of dreams and nightmares is inherently linked to the last two stages of sleep, primarily the last stage. The first three stages are considered non-rapid eye movement (NREM) sleep. NREM stage 1 (N1) is the first stage and is defined as a transition period between wakefulness and sleep. During the NREM stage 2 (N2) body temperature, awareness and motor functions are reduced. While a person is in NREM stage 3 (N3) noises and surrounding activity may be insufficient to wake them up since they are in deep sleep. Stage four is called rapid eye movement (REM) sleep. Here are the voluntary movements repressed and dreams can occur. It can be recognized by the eponymous rapid movement of the eyes behind closed eyelids [5].

2.2 EDOREM2 Dataset

The data was recorded during the EDOREM2 study, in which the effects of imagery rehearsal therapy (IRT) are currently being researched. In IRT patients discuss their nightmares with a therapist and try to find non-frightening endings for specific dream scenarios or how to alter negative dreams. During the study, the sleep of healthy subjects and patients with borderline personality disorder were recorded over two to four nights. For this work the two groups were viewed as one because no information about the patients diagnosis could be disclosed. Between recording sessions the patients received IRT. The study is jet to be published.

The following table gives an overview of the data provided.

Table 1: Data proportion in the EDOREM2 dataset

	Recorded nights
Total	108
Missing	33
Usable Total	75
Scored Nights	29
Nightmares	14
Normal Nights	15
Unscored Nights	46
Nightmares	21
Normal Nights	25

In total 108 nights were recorded and analysed. The total is reduced by the nights where data is not available. This is due to the fact that not every patient was recorded over four days and some recordings were faulty. The dataset is furthermore divided into scored and unscored data. The scored data has already been annotated so every part of the recording was matched with the corresponding sleep stage.

Each participant was questioned after a recorded night of sleep if they had a nightmare or a night of normal sleep. From this question the ground truth for the CNN was derived. The data was collected with a sampling rate of 256 Hz and a sampling interval of 3.90625 ms. Figure 1 shows the arrangement of 6 EEG electrode leads and two reference electrodes following the 10-20 system, which were used to collect the sleep data.

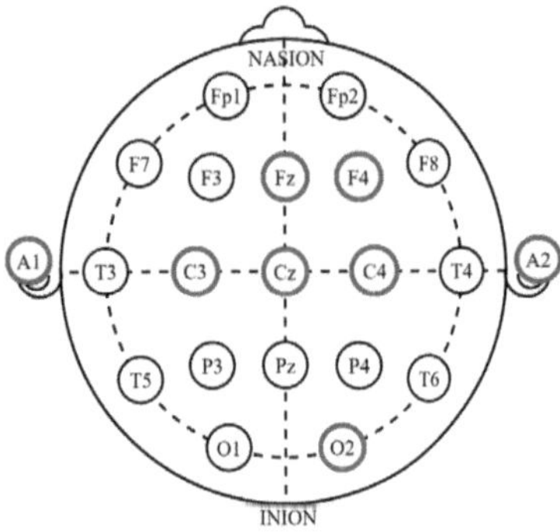

Figure 1: Arrangement of the electrode leads of EEG in the 10 - 20 system

As seen above, the electrodes were placed on Fz, F4, C3, C4, Cz, O2, A1, and A2. In order to monitor the movement of the eyes, two more electrodes are attached near the eyes. One Electromyography (EMG) channel on the arm and one Electrocardiography (ECG) channel were documented. To record all this data a device from the company Somnomedics was used [7]. Some nights have already been annotated. In those cases every patient has an additional text file with the corresponding sleep stage for every 30 second window of the night. Since the patients did not sleep exactly the same amount of time, the number of scored frames vary. On average each patient has around 1000 30-second frames per night. The importance of using scored data will be explained in the discussion.

2.3 Convolutional Neural Network

In order to train the system to automatically classify a nightmare in the patient data, a Convolutional Neural Network (CNN) was implemented. In a basic neural network, three types of layers are used. First, the data is given to the input layer. CNNs were originally developed for two-dimensional image data but time series data has only one dimension. Note that with multiple channels the time series data becomes two-dimensional. The input of the network is the recorded voltage of each electrode. From this point, the network is compiled of hidden layers. In these layers the system can learn by creating or deleting new connections between data points or by assigning connections higher or lower relevance. Lastly, there is an output layer, which defines what result is shown when information is passed through. In this case, only two outputs are needed, because the output is binary, either a nightmare or no nightmare. After building such a network it has to be trained. In order to achieve this, data with the correct label must be given to the network. Every patient was given one label, either nightmare or normal sleep. The network reads the data and checks if it got the correct result and changes accordingly. The advantage of a network with convolution is that the context of the data is not disregarded. Instead of reading in each data point independently, the network reads in a group of data. This concept is shown in Figure 2.

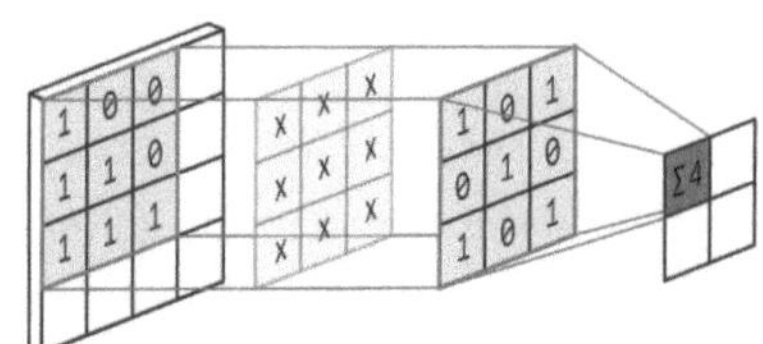

Figure 2: Schematic diagram of convolutional layer [8]

Using a CNN is imperative because a frame of the voltage over time can give more information than a single voltage value. In related works, the usage of a CNN has shown high versatility and sufficient results. For example, the SleepEEGNet [1], HybridAtt [2], and the network described in the paper of Ozal Yildirim [9] all use forms of convolutional neural networks. This may be due to the fact that the convolution enables deeper learning with fewer parameters. Since the quantity of data is a point of contention the importance of this advantage is further amplified.

For this project, an artificial neural network with deep learning capability seems the best choice. A CNN was used to deal with different sensor modalities and different sensor channels. With this, the network does not need complete supervision and is capable of a better learning process. Another advantage is the fact that deep learning networks are able to learn their own features. This may enable the network to find determining factors of nightmare detection, which are currently unknown to current detection standards.

2.4 Implementation details of the CNN

Table 2 shows the architecture of the implemented convolutional neural network.

Table 2: Architecture of CNN

Layer number	Layer type	Activation function	Kernel
0-1	2D convolutional	Relu	32
1-2	batch normalization	Relu	-
2-3	2D convolutional	Relu	32
3-4	max-pooling	-	(3, 2)
4-5	2D convolutional	Relu	64
5-6	max-pooling	-	(2, 1)
6-7	2D convolutional	Relu	64
7-8	max-pooling	-	(2, 1)
8-9	fully-connected	Relu	192
9-10	dropout	-	0.5
10-11	sigmoid	sigmoid	2

For preprocessing the data was normalized. Different channels have their own value ranges and normalization reduces the mean and unit variance to zero. To input the time series data into the 2D convolutional layer the time series was split into chunks. Then the chunks were placed below each other to generate a second dimension. The size of a chunk depends on the length of the time series data. The CNN was used with a binary cross-entropy loss function to predict the probability of a nightmare or normal night. In order to test the A batch normalization layer can be applied after the first convolutional layer. Following are multiple convolutional layers with max-pooling layers in-between. The max-pooling layer is implemented to condense and reduce the feature dimension in the previous layer, which improves the learning efficiency. The sizes of the used kernels are 32, 64, and 128. A batch size of 64 was used in this model and the network was trained over 250 epochs. In order to test the network the data of 20 patients from Table 1 was randomly selected. This data was not used in training to avoid overfitting. Training and testing was repeated multiple times with a different randomly chosen testing dataset.

3 Results and Discussion

3.1 Experiment Results

The following shows the best accomplished performance using CNN based on the accuracy. Accuracy is defined by the following formula:

$$Accuracy = \frac{TP + TN}{TP + TN + FP + FN} \quad (1)$$

Where TP equals True Positives, TN equals True Negatives, FP equals False Positives, and FN equals False Negatives. Firstly the network, with the complete dataset as input, achieved an accuracy of 76%. When using only the EEG data the accuracy was reduced to 68%. In another case only the remaining ECG, EOG and EMG data was used as input to the CNN. This resulted in an accuracy of 55%.

In Table 3 the performance of sleep nightmare detection using the CNN model based on various EEG channels is displayed.

Table 3: Performance comparison of CNN sleep nightmare classification based on different EEG channels

EEG Channel	Accuracy
Fz	52%
F4	55%
C3	57%
C4	54%
Cz	52%
A1	50%
A2	51%
O2	52%

Since a small amount of data can influence the accuracy of a DNN (Deep Neural Network), some EEG channels were grouped together and used for the training of a CNN model. The results are shown in the following Table 4.

Table 4: Performance comparison of CNN sleep nightmare classification based on different EEG channel combinations

EEG Channel	Accuracy
C3 & C4 & Cz	66%
F4 & Fz	61%
A1 & A2	58%
O2	52%

The input data was separated by sleep stage to observe their individual relevance to distinguish a nightmare from a normal night. Afterwards, the CNN was trained and tested with the data from only one sleep stage. The first three sleep stages (Awake, N1, N2) only reached an accuracy from around 50% to a maximum of 53%. When training and testing with the N3 sleep stage an accuracy of 60% could be reached. The highest accuracy of 69% could be achieved with the REM sleep stage.

3.2 Discussion

For a fully independent analysis, without manual influence, the accuracy should be increased further. Nevertheless, this algorithm can be used to indicate the sleep quality of patients. Additionally, this could greatly reduce the time needed for a manual evaluation of EEG data by highlighting possible patients with nightmares. The proposed approach is planned to be made usable for sleep researchers. For this a simple graphical user interface is currently being developed. From the accuracy resulting from the exclusion of specific data, a few conclusions can be derived. As shown in Table 4, the central "C" electrodes show the highest amount of relevant information for the learning process. Therefore exist two possible reasons: Firstly, the central electrodes occupy the most channels and thus have the biggest amount of data. With more data, the CNN can fully learn relevant features for each sleep class and avoid shallow learning. Shallow learning is defined as learning the sleep features based on the empirical knowledge of the expert. Secondly, the areas of the brain where the "C" electrode is attached are commonly used to represent EEG activity and determine stages of sleep. Since nightmares are closely related to sleep stages, this information density of the central electrode is to be expected. Furthermore, the results shown when training and testing with separate sleep stages confirm the current understanding of the correlation between sleep stages and the detection of dreams. The REM stage is the most relevant sleep stage because it is the sleep stage in which dreams take place. It should be noted that stage N3 is not unimportant, as the 60% accuracy indicates. This supports the current theory, that dreams can begin to develop at the end of stage N3. The first three sleep stages (Awake, N1, N2) were expected to not contribute much relevant information to classify nightmares because nightmares can not occur in these stages. As expected, the accuracy for any of these stages has not gone above 53%. In a binary classification this equates to randomly guessing or just picking one classification for every test.

On average, the EEG data was collected over a time frame of about eight hours and the resulting file has a size of around three to four gigabytes. Computing with this amount of data is very time-consuming and requires adequate computing power. Since nightmares occur almost exclusively in the REM (rapid eye movement) sleep phase, the data can be reduced. By filtering out all other stages, namely Awake, N1, N2 and N3, the data can be reduced by around 75%. This leads to a reduction of the time frame of an average file to two hours and a size of close to one gigabyte, which reduces the needed computing time drastically. It is currently not proven if the non-REM sleep stages give reliable information about the appearance of nightmares. A leading theory is that nightmares correlate with the three major symptoms. The first two are an increased amount of high alpha waves (10 – 14.5 Hz) in REM sleep and low alpha waves (7.75 – 9 Hz) in non-REM sleep. Lastly, persistently high alpha waves were observed through the night [10]. This correlates with the results of this work. The accuracy was reduced from 76% to 69% when only the REM sleep stage was used. It is evident that the reduction of the data has advantages, but it comes with the disadvantage of loss of accuracy. In conclusion, this reduction is an option, if computing time and power are problematic. In other cases, the data should be used as a whole to get the best accuracy and preserve the context of the data.

4   Conclusion

This work has shown that the usage of a CNN for the detection of nightmares is possible and can be beneficial for sleep research. It enables a fast and low manual effort way of evaluating sleep quality. Like most modern usages of machine learning in medicine, the proposed model is not intended to replace skilled medical personnel but to assist them by preselecting cases. It was also shown that the REM sleep stage is not the only relevant stage for the occurrence of nightmares. In addition, this work ranked the information value of each biopotential with their influence on nightmare detection. In the future, this model will be tested in a real medical environment, as soon as the GUI is implemented. With this, the practical usage of this model in a real-life environment can be determined.

Acknowledgement

The work has been carried out at the Center of Brain, Behavior and Metabolism, Department of Psychiatry and Psychotheray with the Sleep and Mental Health Team and supervised by Prof. Dr.-Ing. habil. Marcin Grzegorzek, Institut für Medizinische Informatik, Universität zu Lübeck.

Author's Statement

Author state no conflict of interest.

5   References

[1] S. Mousavi et al., *SleepEEGNet: Automated sleep stage scoringwith sequence to sequence deep learningapproach.* PLOS ONE

[2] Yuan, Y., Jia, K., Ma, F. et al., *A hybrid self-attention deep learning framework for multivariate sleep stage classification.* BMC Bioinformatics 20, 586 (2019).

[3] X. Huang, M. Grzegorzek et al., *Sleep stage classification for child patients using DeConvolutional Neural Network.* Artificial Intelligence in Medicine, 2020

[4] Faith S. Luyster, et al., *Sleep: A Health Imperative.* Sleep, Volume 35, Issue 6, 1 June 2012, Pages 727–734, https://doi.org/10.5665/sleep.1846

[5] E. Suni, Dr. N. Vyas *Stages of Sleep.* Sleep Foundation. Available: https://www.sleepfoundation.org/stages-of-sleep

[6] V. Valiulis, *The effect of transcranial magnetic stimulation on brain bioelectric activity.* ResearchGate. (2014).

[7] Somnomedics, *Produkte Polysomnographie.* Available: https://somnomedics.de/de/produkte/schlafdiagnostik/ (2020).

[8] Peltarion, *2D Convolution.* Available: https://peltarion.com/static/2d_convolution_pa3.png (2018).

[9] O. Yildirim et al., *A Deep Learning Model for Automated Sleep Stages Classification Using PSG Signals.* Int. J. Environ. Res. Public Health, EISSN 1660-4601, Published by MDPI (2019).

[10] P. Simor et al., *Fluctuations between sleep and wakefulness.* In: Biological Psychology, Volume 94, Issue 3, 2013.

Decentralized Multi-Robot Movement
for Warehouses with high Scalability Goals

Kai Pfister [1], Heiko Hamann [2]

[1] Robotics and Autonomous Systems, Universität zu Lübeck, kai.pfister@student.uni-luebeck.de
[2] Institute of Computer Engineering, Universität zu Lübeck, heiko.hamann@iti.uni-luebeck.de

Abstract

Innovative and efficient solutions concerning warehouse handling are vital to the logistic branch. Pocket sorters combine beneficial advantages such as modularity and variability. We developed a hybrid approach of centralized path planning and decentralized reactive robot control for the system by EMHS. This system extends these pocket sorter advantages by autonomous bags. The goal for our approach is high scalability and a reliable logistics system. It is based on swarm robotic features such as reactive control, local sensing and local communication to avoid collisions. In addition we developed a simulator to evaluate the advantages of the system and test our approach in multi-robot scenarios. Our experiments show scalability in the number of robots and the complex trade-off between speed, storage space and costs.

1 Introduction

Logistics is a strong growing field and continuously strives for improvement. Delivering goods as fast as possible to the customer is a crucial task. For the year 2021 a total growth of the logistics branch by +4,4 % (nom.) respectively +3,1 % (real) was estimated [1]. In recent years online trade has seen significant increases in sales. Especially in sectors of stationary trade, e.g. fast moving consumer goods or drugs, online trading is becoming more and more relevant [1]. However, with increasing online trade the number of returns is growing as well as the costs [2]. Therefore, tasks such as sorting, transportation, storing and cost optimization demand innovative solutions.

Pocket sorters hold a great potential with respect to variability, speed and sorting [3]. Pocket sorters are capable of picking hanging, lying or any other goods that fit into a grocery bag [3].Each bag in the system serves as single shelf location. They are moved by a chain in a guide rail and are mostly made of polyester fabric. The pocket sorter has the unique ability to pick hanging textile goods by hooking it directly to the rolling adapter connected to the dragging chain. To identify goods in bags, RFID-tags or barcodes are used [3]. The high modularity of the system, offers efficient use of space [3]. Pocket sorters are mainly used for picking goods, storage and buffer systems and handle the continuously growing occurrence of returns.

AFLE - a German abbreviation for *Autonomous Driving Storage Unit* is an innovative low-cost robot used for a stand-alone storage system. Currently under development by the EMHS GmbH it serves as the foundation for the featured work in this paper. Extending the technique of pocket sorters, where all bags are connected to one dragging chain, EMHS uses a motorized robot, the AFLE, for each bag. This prevents the system from loosing speed to the slowest unit. Each agent is powered by the tracks on which they run. While a pocket sorter can reach maximal speeds of $2\ m/s$ the AFLE can move with a speed of $4\ m/s$. Its unique ability is to independently handle individual goods at higher speeds. Thus, throughput is increased. Focusing on the urgent issue of returns [4], the AFLEs are capable to take the returns back to the storage and prevent possible destruction of the goods. To further improve the system we pursue a hybrid approach of decentralized swarm robotics [5] and centralized path planning. The robots communicate only with local neighbors. In addition the robot controllers are limited to behavior-based, reactive control after they received their route from the server. This promises high scalability and results in a robust, high-throughput system.

1.1 Related Work

Automating warehouses by using multiple logistic robots is an innovative topic nowadays that finds great interest. Currently there are only a few systems such as Amazon Robotics and the System of the Ocado Group, to name two of them. Amazon Robotics, former Kiva Systems, uses several robots to move entire shelves to the picking stations. The robots are designed for this particular task with a height of $400\ mm$ and a width of $1200\ mm$ [6]. They are build rather flat enabling them to position themselves directly under the shelves, to move them in a grid pattern through the warehouse. The Ocado Group utilizes a robot system of 3000 units in the context of online grocery shopping. The robots pack the ordered grocery bags while being controlled by an artificial intelligence robot controller [7]. The cube-shaped robots move along a grid of rails positioned above boxes that are stacked on top of each other. In order to move these boxes to a picking station or a new location the robots grab them with claws and pull them into their interior. With the use of several robots it is important to address upcoming topics like collision avoidance and dead lock avoidance. Zhou et al. [8] examined a distributed method to avoid higher-order deadlocks in multi-robot systems. Therefore each robot checks future states to decide whether its move can cause higher-order deadlocks. Deadlocks are crucial problems that can lead to a stoppage of the system and or degrade its performance.

1.2 Approach

We present a hybrid approach of decentralized swarm robotics and centralized path planning. The AFLE-System is still under development with only a small number of robots available. Therefore we developed a simulator to examine their reactive behaviour and to evaluate the advantages of the system with our approach in multi-robot scenarios. The simulator aims to verify estimated and calculated data of predefined use cases. It is necessary for the simulator to be as close as possible to real scenarios to derive comparable results. Visualized as a top-down view of a warehouse the simulation needs a 2D-collision handling. The collision handling is crucial to prevent agents from driving through each other and detect possible crashes. The simulated agents are capable of local sensing and can only communicate with neighbors in their close vicinity. Furthermore, local agent-to-agent communication is required for a collision avoidance protocol. At the beginning of a run the simulated AFLEs receive their route, which is centralized planned on the server, via server-to-agent communication. To evaluate the effects of different layouts, distinct warehouse environments for given parameters have to be built. To achieve reliable data handling a log file is created afterwards. This file contains essential parameters and results of the respective simulation.

2 Methods and Simulation

The simulation models the real environment with relevant physical constraints. This includes the dimensions of the robot 750 mm × 100 mm, its maximal speed of 4 m/s, its acceleration of 1 m/s^2 and its enforced communication range of 5 m via Bluetooth. In real-life the fixed communication range is achieved by pre-filtering the Bluetooth-signal according to its strength. Each AFLE has a time of flight sensor at the front for collision detection. The prototype of the AFLE has no path planning and collision avoidance yet. Therefore suitable methods, developed and tested with the simulator, will be applied in the future system. The simulated warehouses contain three kinds of rails with a distance of 750 mm in between. First, storage rails, where robots without an order are parked and serve as shelf locations. These rails can be driven in both directions. Second,

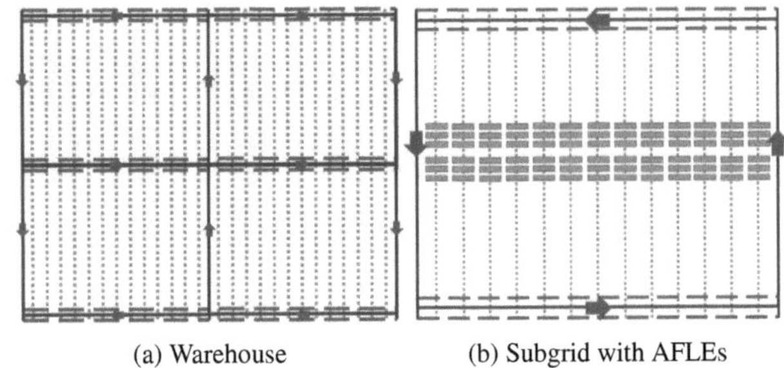

(a) Warehouse
(b) Subgrid with AFLEs

Figure 1: (a) Warehouse consisting of 2 × 2 subgrids with alternating one-way rails. There must be an odd number of subgrids in height an width otherwise unreachable dead ends appear like in the top right corner. (b) Subgrid in which the pointed lines represent the storage rails, the dashed lines are buffer rails and the solid lines depict one-way rails (direction marked by arrows). AFLEs are equally distributed from the middle of the subgrid to its borders.

buffer rails, which serve as evasive rails to free the way for blocked robots. They have no fixed direction as well. Third, one way rails, which build the main connections to the goal positions. They are used as highways enabling the actual movement within the warehouse. The warehouse is divided in so-called subgrids shown in Fig. 1 consisting of these rail types. The simulator then constructs a warehouse with an optimal fit for a given subgrid size.

2.1 Challenges

Within the simulation both, collision detection and communication, have to be modeled sufficiently. This ensures that a robot can detect robots in front of it (simulating the time-of-flight sensor and detect "physical" crashes) and prevent crashes in general (especially at intersections). To navigate the robots through the rail system a path finding algorithm is an essential component. These three topics presented challenges for the simulation.

Euclidean vs. Grid-based Collision Handling: We implemented a Euclidean collision handling but it lacked scalability. This is due to the high computational cost of the Euclidean distance for a large number of agents. Discarding the Euclidean method we started testing grid-based collision handling leading to satisfactory results. Therefore, the environment of the robots is divided in a grid. The resolution of the grid (size of the cells) is thereby the minimal detectable distance, in our case 0.1 m. During a run each agent becomes member of the respective grid cells at its actual position. Knowing the resolution we can then compute the distance between two robots and detect crashes once a grid cell inhabits more than one member.

Grid-based vs. Graph-based Path Finding: Path finding is one of the key features of the simulator. In previous tests we ran the simulation with a grid-based A* algorithm. We implemented a cost map which reflected the rail system of the warehouse, representing one-way rails in the wrong direction as an obstacle. Later this was discarded because the cost map was not updated and freed paths remained an obstacle. We solved that problem by exchanging the A* algorithm with a weighted directed graph for path planning. Fig. 2 shows an example of such a graph.

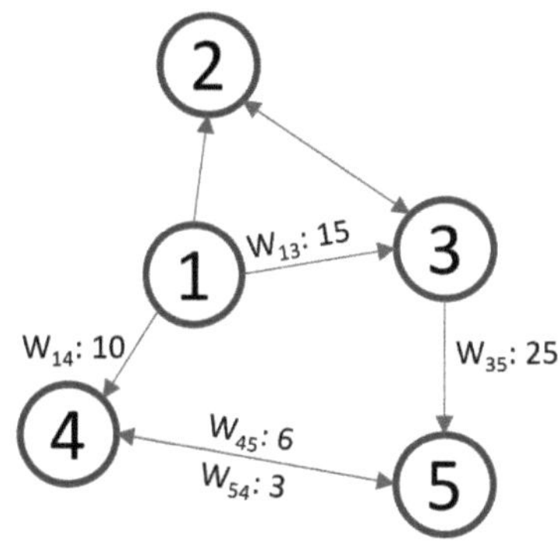

Figure 2: Example of a weighted directed Graph with 5 nodes (nodes represent intersections). Route from node 1 to node 5 via node 4 is cheaper than via node 3.

The weighted directed graph is suitable for the one-way rail problem with its directed edges. In the graph each node represents an intersection in the simulated warehouse. For path planning we use the Dijkstra-algorithm to find the shortest path to the goal node within the directed graph. All nodes in the graph, excluding the start node, are initialized as the set of unvisited nodes. Tentative weights are applied - 0 for the starting point and infinity for unvisited nodes. The weight to each unvisited node is then updated accordingly to the summed up shortest distance beginning from the starting node. Once we have the distance weight of a node it is marked as visited and removed from the unvisited set. Once the goal node is reached the path with the shortest distance is selected. For a warehouse of 90 m × 60 m, the Dijkstra-algorithm is applied on a graph with 1488 nodes. The mean computational time for the path planning at this size is at around 0.21 s per robot. With the present small number of robots in the simulation and in real scenarios (AFLE is still under development) this is not a big problem. However, in future scenarios with up to one million agents requiring a path, this has to be optimized.

Communication and Collision Avoidance: Communication within a swarm of agents is an essential feature. In order to implement it in the simulation we pursue a similar approach to the former described collision handling. Using a communication grid each robot is a member of a particular grid cell. It can communicate with robots within range depending on the grid's resolution. In the case that an agent is selected for an order it receives a random priority from 1 - low priority to 4 - high priority. In a real-life scenario this is used to control the urgency of different orders. Without an active order their priority is 0 - still in the storage. To avoid collisions the robots have to check if their position for the next movement is blocked by other robots. The robots

have to verify if the following track section (actual position to next intersection) intersects with robot paths in their vicinity, to detect possible collision points (PCP). Robots only communicate with the nearest most relevant neighbor. Neighbors that are either behind this robot on the same rail, already waiting for it, past the PCP or that have priority 0 are excluded. To achieve collision avoidance based on the communication between the agents, we set up different rules to avoid deadlocks. Deadlocks are a certain constellation of agents and their set of rules that lead to congestion, ultimately resulting in a vicious circle of waiting. The defined rules specify which robot has to wait based on priorities (low priority has to wait) and the distances to the PCP.

3   Results

We evaluate three scenarios regarding the warehouse size, the subgrid size and the number of agents influencing the order time. The order time defines the time a robot needs to move from its starting position to its goal position. The different scenarios are conducted with the same procedure: The robots are equally distributed in the middle of the subgrids. Each robot is assigned a random goal (uniform distribution) in one of the four corners of the warehouse. After each robot receives the route to its goal, the robots start their task simultaneously. The scenario in Fig. 3a shows three warehouse sizes consisting of subgrids with a distinct size, resulting in a warehouse with one, 3 × 3 and 5 × 5 subgrids respectively. The experiment shows the mean order times of the robot patches over the different warehouse sizes. With increasing size of the warehouse a seemingly exponential increase in time is visible.

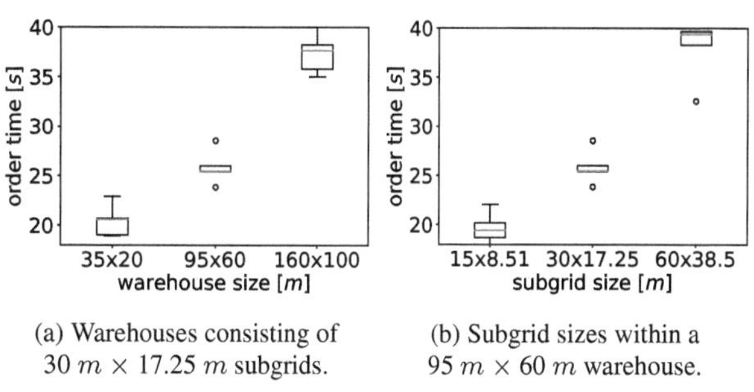

(a) Warehouses consisting of 30 m × 17.25 m subgrids.

(b) Subgrid sizes within a 95 m × 60 m warehouse.

Figure 3: Investigation of the mean order time over different warehouse and subgrid sizes with 10 robots. For 5 runs the mean of all robots was calculated. The boxplot represents these 5 mean values.

At first glance the warehouse with only one subgrid seems to be the best solution, comparing it to the warehouse with many subgrids. Nevertheless, if we examine these results in contrast to the experiment in Fig. 3b we notice a contrary outcome. In this experiment we examine a fixed warehouse size filled with different subgrid sizes. The first plot shows 25 small subgrids, the next one holds 9 subgrids and the last one has only one big subgrid. Here, the lowest mean order times are derived from the warehouse with many small subgrids, while the warehouse with one subgrid results in the longest mean order times. The influence of the warehouse size on the order time in Fig. 3a can be explained by general path lengths in warehouses. Considering a fixed warehouse size as in Fig. 3b we observe that the size of the subgrids influences the mean order times as well. With many small subgrids the robots can reach various goals more easily (more one-way rails and intersections). In addition, smaller subgrids encounter less passive movement over all (robots that necessarily have to be moved to free a path for a requested robot), resulting in lower electricity costs. However, a lot of storage space is lost due to intersections and one-way rails, the former are also the most expensive rail-parts. This results in a complex trade-off between speed, storage space and costs.

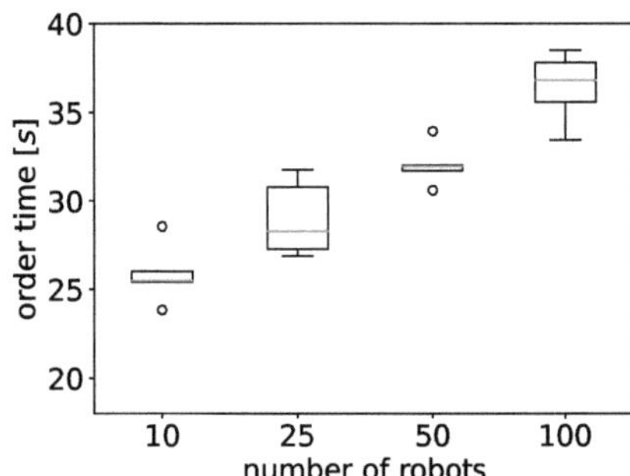

Figure 4: Influence of on the mean order time. Each boxplot consists of 5 runs and the respective means of all robots, tested within a $95\ m \times 60\ m$ warehouse with $30\ m \times 17.25\ m$ subgrids.

Next, we study how the number of robots influences the scalability. For the trial we analyse different sizes of robot patches in the same warehouse. Noteworthy is that the spatial distribution changes with each added robot respectively (the robots are distributed equally in the subgrids beginning in the top-left subgrid). However, this should not have a big impact on the test runs. In Fig. 4 we observe an almost linear increase of the mean order times with an increasing number of agents. This is an indicator for scalable behaviour. Deviations from a linear course, especially for 100 robots may occur due to non-optimized waiting times of the robots at intersections. The inefficient waiting times are due to the state of development of the simulator and will be improved. The variance differences of the various runs can be led back to inconveniently located goal positions, as it cannot be tracked back to the number of robots (see Fig. 4 variances of 25 and 100 robots).

4   Conclusion

We successfully implemented a simulator to evaluate real scenarios for the AFLE robot system. In our trial we have shown that in smaller sized warehouses the robots reach their destinations faster. In real scenarios the size of a warehouse is mostly not an adjustable factor. To reduce the mean order time many small subgrids can be used. However, small subgrids are more expensive due to many intersections. It also entails a reduced storage space because of shorter or fewer storage rails, leading to a complex trade-off between speed, storage space and costs. Even though the size of the warehouse and subgrid seems to increase the mean order time exponentially, the amount of robots shows a linear course regarding the mean order times. This leads to the conclusion that the featured approach has scalable properties resulting in an innovative and promising storage solution. For future work the trade-off between speed, storage space and costs can be further analysed. Furthermore, it would be useful to run trials with greater amounts of robots to verify the scalable behaviour of the system.

Acknowledgement

The work has been carried out at EMHS GmbH, efficient material handling solutions and supervised by H. Hamann, Institute of Computer Engineering, Universität zu Lübeck.

Author's Statement

Conflict of interest: Authors state no conflict of interest.

5   References

[1] *Logistik 2021 - Stabilitätsfaktor in der Kriese und Stütze des Aufschwungs*, 2021.

[2] S. Weinfurtner, G. Zellner, and S. Münch, "Auswirkungen der Digitalisierung im Handel am Beispiel des Retourenprozesses," vol. 53, 2016, p. 98–108.

[3] K.-H. Wehking, *Technisches Handbuch Logistik 1 - Fördertechnik, Materialfluss, Intralogistik.* Springer, 2020, no. 1.

[4] N. Kampffmeyer and C.-O. Gensch, "Nachhaltiger Konsum durch Digitalisierung?" Tech. Rep., 2019. [Online]. Available: https://www.oeko.de/fileadmin/o ekodoc/WP-Konsum-Digitalisierung.pdf

[5] H. Hamann, *Swarm Robotics: A Formal Approach.* Springer, 2018.

[6] J.-t. Li and H.-j. Liu, "Design optimization of Amazon robotics," *Automation, Control and Intelligent Systems*, vol. 4, no. 2, p. 48–52, 2016.

[7] M. Davis. (2021) 3,000 robots working in Ocado's automated warehouse for faster online grocery. [Online]. Available: https://www.sciencetimes.com/articles/30 881/20210427/fleet-3-000-robots-working-ocados-aut omated-warehouse-making-online.htm

[8] Y. Zhou, H. Hu, Y. Liu, S.-W. Lin, and Z. Ding, "A distributed method to avoid higher-order deadlocks in multi-robot systems," *Automatica*, vol. 112, p. 108706, 2020.

Investigation and Development of Class-Incremental Learning Methods

Anna Larsen [1] and Pauline Lux [2]

[1] Robotics and Autonomous Systems, Universität zu Lübeck, anna.larsen@student.uni-luebeck.de
[2] R&D Innovation - Research, Basler AG, Ahrensburg, pauline.lux@baslerweb.com

Abstract

Incremental Learning (IL) is an important concept to ensure the transferability of Deep Learning (DL) methods to real-world applications with continually new emerging data or challenges. This work aims to mitigate the appearing Catastrophic Forgetting of previously learned knowledge while performing a classification task over time. Therefore, an IL environment is built up and the impact of different methods, such as the use of Exemplars or additional retraining of the classification layer, is examined. Further, we provide the Class Separation Loss (CSL) as an additional term in the objective function of our model to generate an improved representation of the data in the feature space. By combining the proposed methods the performance of the incrementally trained model increases by 25.4% in average classification accuracy and therefore contributes to the usability of DL systems in real-world applications.

1 Introduction

In recent years, Deep Neural Networks (DNNs) have become a widely used tool in a variety of Computer Vision (CV) applications. In particular, object recognition and image classification are some of the most fundamental image processing tasks in this area. DNNs solve these tasks quite reliably in an offline manner, assuming a fully predefined task as well as the availability of a sufficiently large amount of training data. However, this static approach of learning encounters serious difficulties in real-world applications caused by non-stationary environmental conditions. If new data or even new tasks appear over time, the standard approach of learning requires full retraining of the network. This fact is not only a weakness in terms of efficiency and computational cost, but also leads to the necessity of storing all data at once, which may be impractical or even impossible due to privacy restrictions or limited storage space.

Thus, the aim is to develop algorithms that are capable of adapting to new data while preserving the previously learned knowledge as well. This kind of method is also subject of current research and referred to as Incremental Learning (IL).

In comparison to human brains, standard DNNs have inherently no memory or any capability of transferring knowledge from previously learned tasks to later emerging challenges. Hence, preceding works on IL encountered the same main problem: *Catastrophic Forgetting* (CF). This term denotes a huge performance drop on previous data by retraining the network only on new emerging data samples. A more detailed investigation into the problem of CF was done in [1].

This work focuses on IL in the context of image classification tasks. In particular, we investigate the successive insertion of new classes to the network's knowledge over time, which is referred to as *Class-Incremental Learning* (CIL) in the literature. Our following experiments demonstrate, on the one hand, the severe influence of CF on a given IL problem. On the other hand, they provide a chance to evaluate already existing methods, that were constructed to inhibit the loss of classification accuracy during retraining. Finally, we aim to present a new loss function that turned out to be beneficial for incremental training, as it enforces a large class separability by finding a suitable feature representation.

2 Material and Methods

The implementation was done with the machine learning framework PyTorch and is based on FACIL (Framework for Analysis of Class-Incremental Learning) [2]. All experiments were carried out on a single GPU of type NVIDIA RTX 2080 TI.

2.1 Class-Incremental Learning Setting

In the following, the standard setting of the CIL approach is presented, which serves as a basis for all subsequent experiments. It is inspired by the often reused experimental setup from [3] and is illustrated in Fig. 1. The used dataset was recorded beforehand and is not affected by any bias caused by changing environmental conditions over time. The training is divided into T timesteps. Each time step corresponds to a different classification task, that is learned one after

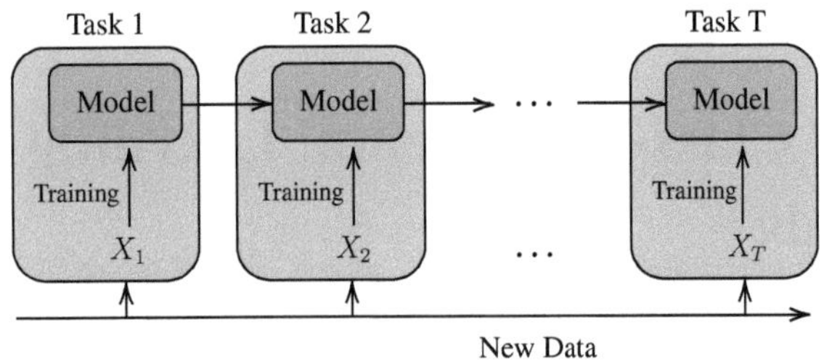

Figure 1: Experimental setup of the incremental training for a number of T different classification tasks.

another. Therefore, the dataset X is split up into disjoint subsets $\{X_1, \ldots, X_T\}$ of the same size, with

$$\bigcap_{i=1}^{T} X_i = \varnothing. \tag{1}$$

Each of the subsets correspond to one of the associated subtasks $1 - T$. Moreover, each subset X_i is composed of image data from N different classes $c_1^{(i)}, \ldots, c_N^{(i)}$. The classes are also disjoint, it holds:

$$\forall k, l \in \{1, \ldots, N\}, i \neq j: \quad c_k^{(i)} \neq c_l^{(j)}. \tag{2}$$

The model is trained at each timestep t with the corresponding dataset X_t and the resulting network weights are stored for the case of inference. Afterwards, the training proceeds at the following time step $t + 1$ with the current configuration of weights. This procedure continues until the data of all subtasks has been shown to the network.

2.2   Dataset and Model

To ensure reproducibility and comparability of the experiments, the network structure ResNet32 [4] was used along with the dataset CIFAR100 [5]. The set contains 100 different categories, each with 600 training images. A fixed subset of 100 images was held back as a test set and another 10% per class was split off as a validation set. The training was carried out with the remaining 450 images per category. Except for additional class outputs in the classification layer, the network size was retained over the entire duration of an experiment, which limits the computational cost as well as the required amount of storage space for the model during inference to a maximum value.
According to the procedure from section 2.1, the dataset was split in $T = 10$ classification tasks, each containing the data of $N = 10$ classes. The order of classes is arbitrary but fixed and obtained from [3].

2.3   Experiments

The network was initialized randomly and trained with Stochastic Gradient Descent (SGD) for 200 epochs per task, based on the cross-entropy loss function [6]. The initial learning rate (lr) was set to 0.1 and divided by 3 every time the validation loss has not decreased for more than 10

epochs. Moreover, early stopping was implemented with a stopping criterion of lr $\leq 10^{-5}$. Since this work does not focus on hyperparameter tuning, these parameters were kept for all following experiments.
Initially, for the purpose of evaluation, an *upper* and *lower bound* for the network performance were created.
The upper bound corresponds to the conventional training of neural networks without any incremental steps and is referred to as joint training. The measured accuracy in any task t refers to the classification accuracy of a randomly initialized network, jointly trained on the subset of all previously encountered datasets $\{X_1, \ldots, X_t\}$.
The lower bound, on the other hand, refers to the training procedure introduced in section 2.1, without any additional extensions. It is referred to as finetuning in the following.

2.3.1   Exemplars

A common tool for CIL is the usage of *exemplars*, first introduced by [3]. Thereby, a certain number M of representative example images are stored from all previously seen classes and added to the training dataset of the current task to mitigate the forgetting. Several different methods can be used for the selection of exemplars, such as herding [3] or entropy-based selection [7]. For simplicity, the exemplars in this work were chosen randomly only. Nevertheless, the required capacity to store the training data increases through the use of exemplars linearly over time, proportional to the number M. After examining the tradeoff between classification accuracy and memory requirements, the number of $M = 20$ exemplars per class turned out to be an appropriate choice. This fraction corresponds to a proportion of less than 5% of the training data.

2.3.2   Classifier Retraining

The second method aims to reduce the Catastrophic Forgetting of the previous classes with additional training of the classification layer. Based on the assumption that the network is influenced by class imbalance, caused by the use of exemplars, the weights of the classifier were adjusted again after a new feature representation was found. To do so, the network was trained as usual with 20 exemplars per class. Subsequently, all weights of the feature extractor were frozen and the network was trained for another 50 epochs to adjust the classification layer accordingly. This

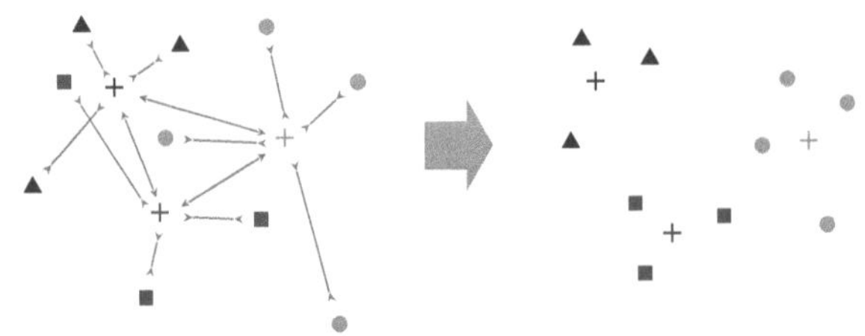

Figure 2: Basic principle of the Class Separation Loss presented for three exemplary classes in a 2-dimensional feature space.

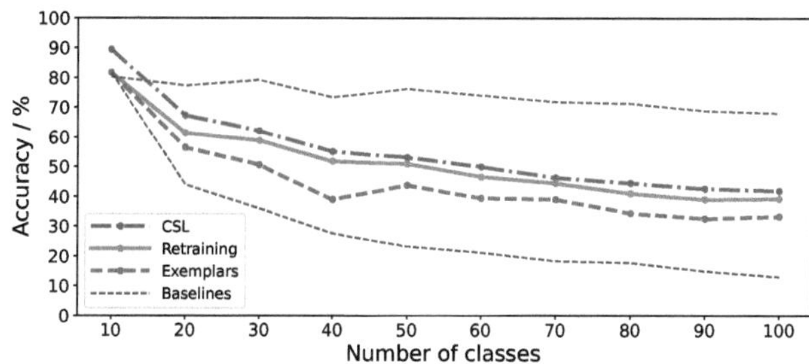

Figure 3: Resulting accuracy of all presented methods. The lower baseline corresponds to the finetuning approach and the upper baseline contains the results of the joint training.

2.3.3 Class Separation Loss

The last experiment incorporated a different loss function that acts directly in the feature space. Fig. 2 illustrates the basic idea of the *Class Separation Loss* (CSL). In order to localize particular classes, each of them is assigned to a class center (+) in the form of a learnable parameter. The CSL penalizes large distances from the class samples to their corresponding center as well as small distances between the centers themselves. Distances are calculated using the euclidean norm, resulting in formula

$$CSL(X, C, \lambda) = \lambda \cdot \sqrt{\frac{\sum_{i} \|x_i - c_{y_i}\|}{\sum_{\substack{k,l \\ k \neq l}} \|c^k - c^l\|}}, \quad (3)$$

where $X = \{x_1, \ldots, x_b\}$ denotes the dataset with batchsize b. The corresponding label of sample x_i is referred to as y_i and $C = \{c_1, \ldots, c_{N \cdot t}\}$ represents all classes that have appeared until the current time step t. Moreover, λ acts as a scaling factor, that needs some adjustment and incorporates i.a. the batchsize or total number of classes.

To envolve the CSL to the standard cross entropy (CE) loss function, the training criterion is constructed using

$$\begin{aligned} Loss(X, Y, C) = {}&(1-\alpha) \cdot CE(X, Y) \\ &+ \quad \alpha \cdot CSL(X, C, \lambda) \end{aligned} \quad (4)$$

with the hyperparameter $\alpha \in [0, 1]$ to regulate the influence of the CSL across different time steps. An appropriate choice of the parameter α was obtained by gridsearch.

3   Results and Discussion

To evaluate the performance of the trained network, the classification accuracy was determined after every task by using a test dataset containing all previously seen classes. The resulting accuracy of the presented methods is shown

Table 1: Resulting average accuracies of the five conducted experiments. Each column represents the accuracies obtained by averaging for each method over the accuracies that were measured after every training step performed so far.

	Average accuracy up to task / %									
	1	**2**	**3**	**4**	**5**	**6**	**7**	**8**	**9**	**10**
Finetuning	81.7	62.8	53.8	47.3	42.5	38.9	36.0	33.7	31.6	29.8
Joint	80.3	78.8	78.9	77.5	77.3	76.7	76.0	75.5	74.7	74.1
Exemplars	81.7	69.0	62.9	56.9	54.3	51.8	50.0	48.0	46.3	45.0
Retraining	81.7	71.5	67.2	63.4	60.9	58.5	56.5	54.6	52.8	51.5
CSL	89.4	78.3	72.9	68.4	65.4	62.8	60.4	58.4	56.7	55.2

in Fig. 3 for all ten tasks. Additionally, Table 1 supplements as a second measure the average accuracy, which is obtained by averaging the resulting accuracies from previous tasks including the current one.

The first experiment (finetuning), that is described in Section 2.3 and corresponds to the lower baseline, shows clearly the catastrophic forgetting. Already after the second task, the accuracy drops from initial 81.7% to only 43.9% (visualized in Fig. 3). Together with the second non-incrementally trained baseline (joint training) a huge performance gap becomes apparent that is caused by the network's lack of memory. The joint training achieves an average accuracy of at least 74.1% after the tenth task, whereas finetuning ends up with only 29.8% average accuracy. Thus, the emerging gap between both training methods, which is attempted to be closed by CIL methods, extends up to a difference of 44.3%.

Still, the assumption taken from the literature, that the use of exemplars can alleviate the forgetting, was confirmed. The training with 20 exemplars per class amounts in an overall accuracy of at least 45.0% in contrast to the previous 29.8%, which can be considered as a success compared to the effort.

However, a closer look at the network performance on the

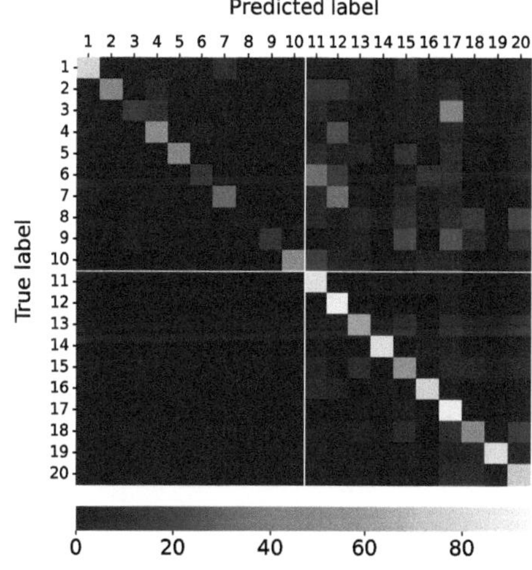

Figure 4: Confusion matrix after training task 2 with the use of 20 exemplars per class.

---

additional training of the classifier was carried out with the dataset of exemplars exclusively, to mitigate the introduced bias towards classes belonging to the present time step.

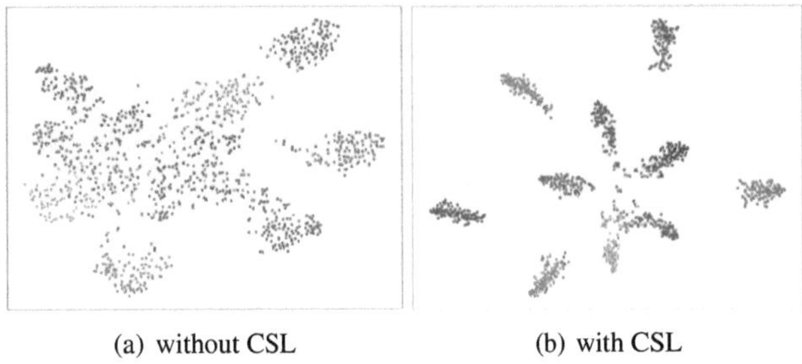

(a) without CSL   (b) with CSL

Figure 5: Resulting class distributions in a 2D-feature space after learning the first ten classes in task 1. The left-hand side pictures the representation after training without CSL, whereas the right-hand side illustrates the class distribution with the use of the CSL. The dimensionality reduction is achieved by t-SNE.

classes of each individual task reveals some large deviations from the average task accuracy. An example that illustrates this point particularly well is the confusion matrix in Fig. 4. It visualizes a comparison of predicted class labels and actual true labels of the classes $1 - 20$ after training task 2. It seems that the network remembers the previously learned classes $11 - 20$ from task 2 quite well, whereas samples from old classes are often incorrectly assigned to one of the classes $11 - 20$ (upper right quadrant). This leads to the assumption that the network prefers the classes of the current task due to the uneven class distribution in the training data, evoked by the small number of 20 exemplars per class.

Retraining the classification layer alleviates the problem partially and increases the accuracy on testing samples from previous tasks. Therefore, the average accuracy improves in every task from 2 up to 10 by $2 - 6\%$, as it is demonstrated in Table 1. It already reaches an average accuracy of more than 50% after the last task.

Finally, we consider the results using the CSL. It turns out that it achieves also an increase in the classification accuracy over all tasks. Especially after training the first task, it is already increased by 7.7% compared to the method only using the cross-entropy as loss function, which suggests that the CSL is also beneficial for non-incremental classification problems. To visualize the effects of the CSL on the location of classes in the feature space, Fig. 5 shows the feature space as a 2D-plane by using the dimensionality reduction method t-SNE [8]. On the left-hand side, the locations of classes $1 - 10$ are shown after the training without CSL. The right-hand side presents the spatial distribution of classes with the use of the CSL function and indicates a better separation of classes in the feature space.

4   Conclusion

Within this work, the presence of Catastrophic Forgetting in incrementally trained classification tasks has been confirmed but not entirely prevented. Nonetheless, our results show, through an increase in classification accuracy, a positive impact on the forgetting when using an appropriate

selection of the training data from previous classes as well. Additionally, it was shown that a subsequent training of the classification layer with an even class distribution is able to further reduce the forgetting.

The main contribution of our work is an enhanced separability of the individual classes in the feature space. For this purpose, we provide the Class Separation Loss as an additional term in the objective function of the model. It turned out to improve the classification results of both incremental and non-incremental training methods.

Acknowledgement

The work has been carried out at Basler AG, Ahrensburg and supervised by Prof. Dr. Mattias Heinrich, Institute of Medical Informatics, Universität zu Lübeck. We thank Basler AG for its financial and scientific support.

Author's Statement

Conflict of interest: Authors state no conflict of interest.

5   References

[1] I. J. Goodfellow, M. Mirza, X. Da, A. C. Courville, and Y. Bengio, "An empirical investigation of catastrophic forgetting in gradient-based neural networks," *CoRR*, vol. abs/1312.6211, 2014.

[2] M. Masana, X. Liu, B. Twardowski, M. Menta, A. D. Bagdanov, and J. van de Weijer, "Class-incremental learning: survey and performance evaluation," *ArXiv*, vol. abs/2010.15277, 2020.

[3] S.-A. Rebuffi, A. Kolesnikov, G. Sperl, and C. H. Lampert, "icarl: Incremental classifier and representation learning," *2017 IEEE CVPR*, pp. 5533–5542, 2017.

[4] K. He, X. Zhang, S. Ren, and J. Sun, "Deep residual learning for image recognition," *2016 IEEE CVPR*, pp. 770–778, 2016.

[5] A. Krizhevsky and G. Hinton, "Learning multiple layers of features from tiny images," University of Toronto, Ontario, Tech. Rep. 0, 2009.

[6] P. Golik, P. Doetsch, and H. Ney, "Cross-entropy vs. squared error training: a theoretical and experimental comparison," *INTERSPEECH*, pp. 1756–1760, 08 2013.

[7] A. Chaudhry, P. K. Dokania, T. Ajanthan, and P. H. S. Torr, "Riemannian walk for incremental learning: Understanding forgetting and intransigence," *ArXiv*, vol. abs/1801.10112, 2018.

[8] L. van der Maaten and G. Hinton, "Visualizing data using t-sne," *Journal of Machine Learning Research*, vol. 9, no. 86, pp. 2579–2605, 2008.

Analytics on battery electric vehicle data

Dominic Janczyk [1], Kerstin Hadler [2], Christoph Schuler [3] and Alfred Mertins [4]

[1] Robotics and Autonomous Systems, Universität zu Lübeck, dominic.janczyk@student.uni-luebeck.de
[2] Mercedes-Benz AG, kerstin.hadler@daimler.com
[3] Mercedes-Benz AG, christoph.schuler@daimler.com
[4] Institute for Signal Processing, Universität zu Lübeck, alfred.mertins@uni-luebeck.de

Abstract

This work aims to preprocess and cluster data of battery electric vehicles for further analysis on the capacity loss of batteries. The raw data is transformed to a synchronous and faultless state, before the clustering can be applied. The goal of clustering, is to discover dissimilar user group which show different aging processes of the battery. For the clustering task, 4 different types of clustering techniques are used, each approaching another way to build cluster. All clustering methods used are unsupervised and builds cluster groups out of feature similarities. As measure of quality and to tune hyperparameters, quality indices are needed. Three of these quality indices are used, silhouette coefficient, Calinski-Harabasz index and S_Dbw. These are calculated to an cumulative index to evaluate the result. Based on these build groups with various aging processes and therefore capacity losses, further analysis can be done. These aim to discover vehicle signals with high impact to the capacity loss, to counteract a rapid loss of battery lifetime. This would not only build a better guideline for a customer to drive more sustainable, but also cause a positive impact in pollution reduction.

1   Introduction

Electric vehicles exist since the development of the automobile, but were less relevant until lithium based batteries were adapted to them. From 2015 to 2019 the worldwide supply of electric vehicle rose on average by 54 % per year, resulting in a total of 7,89 million vehicles [1]. The "IEA" (International Energy Agency) estimates, that by 2050 two third of worlds car sales will account to electric driven vehicles [1]. To meet this oncoming demand, it is crucial to not only develop more performant batteries, but to enhance existing battery performances and their lifetime. This will not only benefit the user satisfaction, it will also reduce the pollution caused by the production and the waste disposal of batteries. To meet this goal, vehicle users need to be grouped by their usage, to identify crucial factors for a low battery lifespan. Clustering seems to be a favorable method for this demand.

2   Material and Methods

With a telemetry device more than 40 signals were tracked from around 140 vehicles over the period of one year. These signals were recorded with different sampling rates. Lower rates for less dynamic signals like the drive position (drive, park, reverse) and high sampling rates for more fluctuating signals like the battery current. Not all signals were recorded without errors, some signals consisted on implausible outliner or did not record for a certain time. Therefore the signals had to be evaluated and filtered to clean these

values. After the data cleaning, the remaining data had to be synchronised. To get a synchronous signal path, some signals had to be upsampled to either fill missing values or fit the highest sampling rate. Thereafter the data was processed and could be analysed. To extend the lifespan of a battery, it is necessary to understand the causes of the loss of capacity. The battery lifespan is dependent on many internal and external properties like environment temperature, current or SoC of a battery. A huge indicator for the aging state is the SoH (State of Health) [2]. To set the capacity loss in relation to their performance, there are some possibilities. The ETP (energy throughput) suits this demand well [3]. It is often used to declare a maximum throughput a battery can handle, before its capacity loss becomes too high for the battery and is declared as unusable for automotive applications. Also the odometer can be an indicator for the battery performance, such as the charging behavior. The odometer indicates the total driven distance of the vehicle. There are several influences of the charging process linked to the capacity loss of a battery and therefore to the SoH. High temperatures, high SoC, overcharging, high cycling rates, high DoD (Depth of discharge) and geometric imperfections are all causes for a capacity loss during the charging process. Due to no additional power source of the telemetry device, the recording of the signals was linked to the ignition of the vehicle, therefore only a fraction of this influences could be considered [4]. To break down all causes of the loss of the capacity, it is necessary to consider the effect of more potential causes and use all tracked signals. But it would cost too much time to analyse all the data-sets of every car over the period of one year and much more to understand the data and recognize some patterns by hand. In order to obtain groups with similar characteristics, it would be advantageous to use machine learning. Machine learning can be applied supervised, which means the data is labeled (the result is known) or unsupervised, which is used with unlabeled data. The used data is unlabeled, because the similarities are unknown. A favorable method to find these similarities, is clustering. This technique can be separated into 4 different approaches to cluster data: partitioning clustering, hierarchical clustering, density-based clustering and distribution-based clustering. Partitioning clustering techniques, also called centroid based, subdivide the data by finding the center of certain cluster. These centers are represented by central vectors and data points are assigned by proximity, therefore points are allocated to the closest central vector. There exist several distance metrics for this proximity measurement, like manhatten, euclidean or minkowski distance, but here euclidean distance was used. The used methods of this techniques are the k-Means and the mean shift algorithm. With hierarchical clustering method, no particular number of clusters has to be determined at the beginning, by which the data is partitioned. This is a huge advantage against other clustering methods. Instead this technique consists on a series of cluster operations, starting from n clusters, each contain only one data point, up to only one cluster containing all n data points. This process can be divided into two principal techniques, the **agglomerative** methods and the **divisive** methods. While agglomerative methods starts with n cluster and successively fuses the data points into bigger clusters, the divisive methods begins the other way around. They build only one cluster at the beginning and subdivide the clusters into finer and finer groups, to the point of n clusters. But independent of the method, once a cluster is fused or divided, the operation stays permanent and can not be undone by further steps. To cluster two data points, the proximity matrix is the most important part. This matrix displays the similarity between points and is symmetrical. It is usually calculated by the euclidean distance from one data point to every other data points. To evaluate the matrix, there are several linkage methods like average, single, complete/maximum or ward. The used method was HAC (Hierarchical Agglomerative Clustering) with a single linkage. The previous methods were based on distance matrices like similarity and proximity, while the density-based approach takes the density into account. In most clustering methods, the data is assumed to be loose from noise and is purely geometrically shaped. This makes these methods prone to noise, because outliners can not be ignored and they are limited to certain attribute shape, mostly elliptical or circular. With this approach on the other hand, only densities are considered, which makes it possible to ignore data points inside arbitrary shapes and exclude outliners from the clustering. One of these methods is DBSCAN (Density-Based Spatial Cluster of Applications with Noise) [5]. Density in this case, can be described as number of data points per unit volume of the feature space. A huge advantage here is that contrary to k-Means there is no need for the determination of the number of clusters to look for. The number of clusters are solely established through the data, but there are still some hyperparameter to define and to fine tune, like the $/epsilon$ and N_{min} which are needed to define a cluster. The distribution based clustering techniques are gaussian or normal based, meaning they assume the data is gaussian or normal distributed and cluster after this schema. Some clustering algorithms are assuming circular or elliptical distributions and are therefore limited to this expectation. A GMM (Gaussian Mixture Model) is such a probabilistic model. One can think of it as k-Means algorithm with information of the covariance structure of the data, which extends the search space to an additional dimension. GMM assumes K gaussian distributions with unknown parameters, which consist on a mean $\vec{\mu_k}$ and a covariance matrix Σ_k. The GMM parameter are estimated using either the EM (Expectation Maximization) algorithm of MAP (Maximum A Posteriori) estimation (which is not considered further here). EM is often used, when data is incomplete. This "missing" data can also be considered as the labels in unsupervised learning or the number of clusters and are called latent variables. To get a quality value for this process, there are a number of supervised and also unsupervised methods, but here only unsupervised methods could be used. There are several factors, which can influence the result of a validation method. These are noisy data, skewed distributions, various density, close subclusters or monotonicity. To include as much factors as possible, 3 validation indices were used. The silhouette coefficient SC evaluates the compactness inside a cluster and how separated it is to its nearest cluster. The quality value lies between 1 (optimal solution) and -1 (worst solution) with 0 indicating that a point lies between two clusters [6]. With the Calinski-Harabasz index CH, also variance ratio criterion, the cohesion inside one cluster compared to the separation to other cluster is measured. The cohesion c is based on the proximity of cluster points to the cluster center and the separation s is based on the distance from cluster center to global center. This is calculated by the variance of summed squared distances from data point to cluster center is divided by the squared sum of the distance from cluster center to cluster mean. The S_Dbw validation method, which is the combination of Scatt and Dens_bw evaluates the inter-cluster density (Scatt) and the intra-cluster variance (Dens_bw) and builds a trade-off between these factors. The inter-cluster density $Dens_{bw}$ calculates the ratio of the cluster density of cluster to the average cluster density.

3   Results and Discussion

To get a initial impression of the data, the clustering was done firstly with only 2 features, the ETP and SoH. The ETP (energy throughput) represents here some kind of stress for the batteries of the vehicle. It represents the total amount of energy, transmitted by the battery and therefore is an indicator for its performance. The SoH on the other hand represents the loss of capacity, compared to the initial capacity. Therefore the SoH is a suitable indicator for the usage of the battery. Afterwards the clustering could be continued with all available features, which consisted on ETP, Odo (total odometer), SoH (State of Health), "dCharged" (mean charged current per loading process) and "dOdo" (mean odometer difference per vehicle usage).

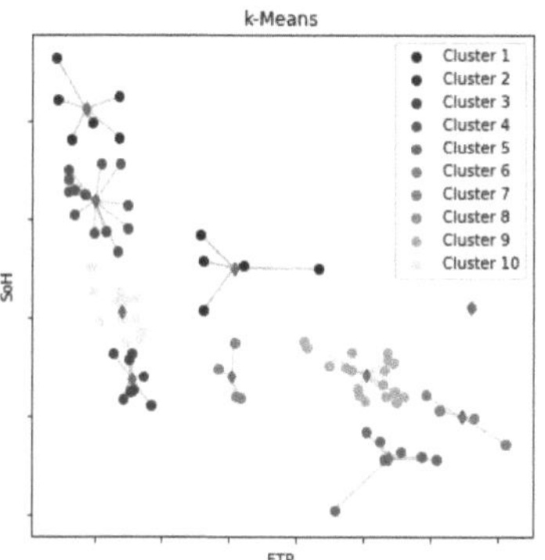

Figure 1: k-Means clustering with $K = 10$ cluster centers

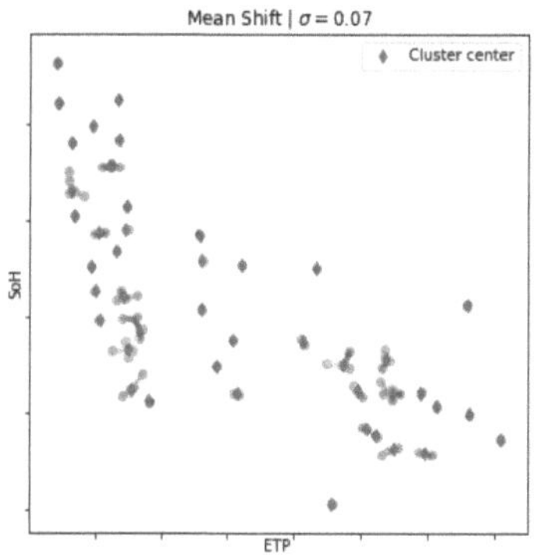

Figure 2: Mean Shift clustering with $\sigma = 0.07$

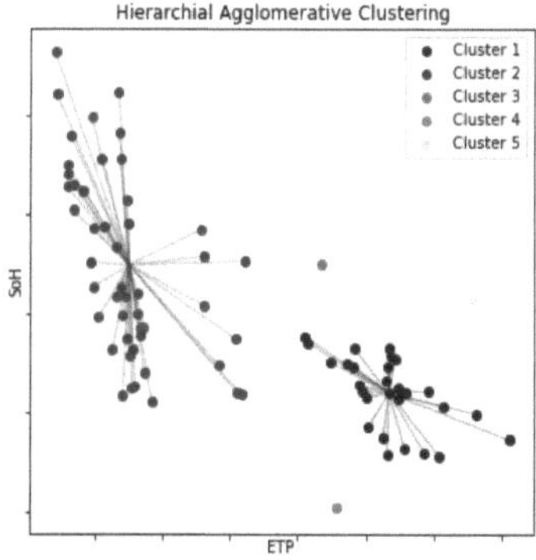

Figure 3: HAC clustering with $K = 5$

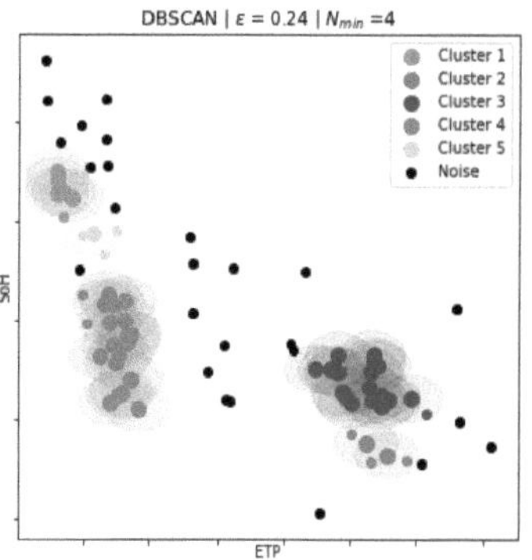

Figure 4: DBSCAN clustering with $\epsilon = 0.24$ and $N_{min} = 4$

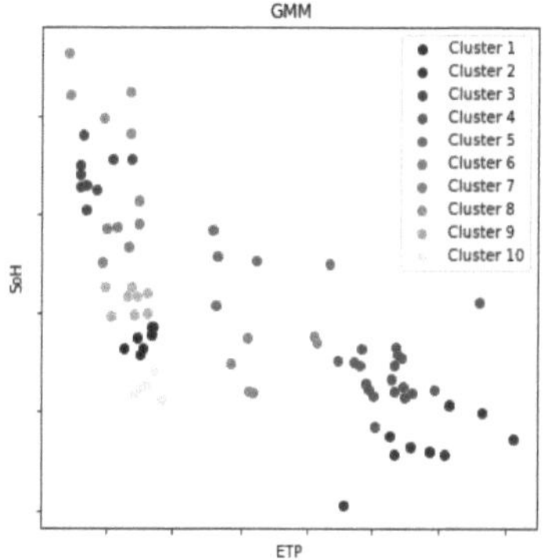

Figure 5: GMM clustering with $K = 10$

Method	Parameter	SC	CH	S_Dbw
kMeans	K=10	0.4181	163.83	0.1996
MeanShift	σ=0.07	0.2966	276.92	0.0227
HAC	K=5	0.2665	34.69	0.1254
DBSCAN	ϵ=0.24	0.1577	18.17	0.2991
GMM	K=10	0.4150	151.04	0.2083

Table 1: Quality indices for the examined methods best cluster results with 2 features

With 2 features, the k-Means, the DBSCAN and the GMM showed advantageous clustering results (see Figure 1 4, 5), because the could separate vertically quite good, which was the expected separation between cars with the same usage (ETP) and different stress (SoH). The Mean Shift method was evaluated well considering the results of the quality indices, despite its displayed clustering results (see table 1 and Figure 2). This might correlate with the visual expectation bound to this clustering. Also the HAC algorithm was evaluated quite well, even though it did not cluster the data in a plausible form (see Figure 3). This results could easily be identified with only 2 feature. But with more than 3 features, it would be quite impossible to evaluate these clusters without the quality indices. To prevent replicated results of high quality indices paired with disadvantageous clustering, these methods were not considered anymore with added features. So in further analyzes, only the k-Means, DBSCAN and GMM algorithms were used. It is to see, that the k-Means algorithm overall delivered worse quality indices with 5 features, than with only 2 features. Even though it still resulted in $K = 10$, which was also the result with only 2 features (see table 2 and Figure 1). It is also to mention, that the k-Means algorithm overall resulted in the best results over the other used methods. DBSCAN also resulted in worse quality indices in every category with 5 features and even revealed negative silhouette coefficients with ϵ values below 0,62, which indicates a cluster result of low quality. The build clusters showed a beneficial vertical distribution of clusters, but a lot of data points were considered as noise (see Figure 4). The GMM algorithm was the only method, to result in a better quality index, with the additional features. Here the silhouette coefficient delivered better, but also not outstanding results. The remaining quality indices are still lower with 5 features, than with 2 features (see table 1, 2).

Method	Parameter	SC	CH	S_Dbw
kMeans	K=10	0.3560	40.47	0.2587
DBSCAN	ϵ=1.04	0.3319	24.96	0.5094
GMM	K=10	0.3616	41.63	0.3067

Table 2: Quality indices for the examined methods best cluster result with 5 features

4 Conclusion

The clustering with 2 features delivered a crucial insight to the data and also showed interesting clustering results. With this insight into the clustering behaviour, the Mean Shift and the HAC clustering algorithms, could be excluded for further analysis. This is because with the used quality indices (silhouette coefficient, Calinski-Harabasz index and S_Dbw), no clustering parameter could be found to build reasonable cluster. With the extension of 3 additional features, the quality indices were significantly worse for all used clustering algorithms. This implies, that the clustering with 2 features delivered better results. This could indicate, that the additionally used features were not optimally chosen and should be more accurately inspected in further analyzes. One method would be to try different combinations without using all features at once. Due to temporary malfunction of the telemetry devices of singular vehicle, some signals could not be recorded or were in such poor quality, that the number of features were reduced. Additionally some analyzes could not be made for some vehicles and this resulted in less data points than expected. It is highly recommended to repeat this clustering with more features available and also examine the features more accurately before the usage. Nevertheless, significant information could be gathered through the clustering and further analysis on the found clusters with only 2 features is recommended. Especially the k-Means and the GMM delivered satisfying

results. The next step would include detailed investigations inside the build clusters and the comparison with the distinct clusters.

Acknowledgements

The work has been carried out at the Mercedes-Benz AG and supervised by the Institute of Signal Processing, Universität zu Lübeck.

Author's Statement

Kerstin Hadler designed the study. Christoph Schuler and Prof. Dr.-Ing. Alfred Mertins supervised the study. Dominic Janczyk preprocessed the data. Kerstin Hadler and Dominic Janczyk analysed the data. Dominic Janczyk wrote the paper with input from all authors.

5 References

[1] I. E. Agency, *Electric and Plug-in Hybrid Electric Vehicles*, 2009. [Online]. Available: https://www.oecd-ilibrary.org/content/publication/9789264088177-en

[2] N. Noura, L. Boulon, and S. Jemeï, "A review of battery state of health estimation methods: Hybrid electric vehicle challenges," *World Electric Vehicle Journal*, 2020.

[3] B. De Beer, A. J. Rix, and A. J. Rix, "Influences of Energy Throughput on the Life of Various Battery Technologies," *6th South African Solar Energy Conference*, no. October 2016, 2016. [Online]. Available: https://www.researchgate.net/publication/313249361

[4] M. De Gennaro, E. Paffumi, G. Martini, A. Giallonardo, S. Pedroso, and A. Loiselle-Lapointe, "A case study to predict the capacity fade of the battery of electrified vehicles in real-world use conditions," *Case Studies on Transport Policy*, vol. 8, no. 2, pp. 517–534, 2020. [Online]. Available: https://doi.org/10.1016/j.cstp.2019.11.005

[5] M. Ester, H. peter Kriegel, J. Sander, and X. Xu, "A density-based algorithm for discovering clusters in large spatial databases with noise." AAAI Press, 1996, pp. 226–231.

[6] J. Han, M. Kamber, and J. Pei, "Cluster analysis," *Data Mining*, pp. 443–495, 2012. [Online]. Available: https://linkinghub.elsevier.com/retrieve/pii/B9780123814791000101

Detecting behavioral changes of an individual using a Random Forest classifier on floor sensor data

Jasmin Walter [1], Laura Liebenow [2], Raoul Hoffmann [3], and Marcin Grzegorzek [4]

[1] Medical Informatics, Universität zu Lübeck, j.walter@student.uni-luebeck.de
[2] Institute of Medical Informatics, Universität zu Lübeck, laura.liebenow@student.uni-luebeck.de
[3] SensProtect GmbH, Altlaufstraße 35, 85635 Höhenkirchen-Siegertsbrunn, raoul.hoffmann@sensprotect.com
[4] Institute of Medical Informatics, Universität zu Lübeck, marcin.grzegorzek@uni-luebeck.de

Abstract

As the world's population continues to age, many elderly people end up in nursing homes and require special care. Since medical staff in nursing homes is understaffed, it is important to find other ways to monitor residents' behavior and detect changes early. One easy-to-use and unobtrusive way to collect data about residents is to use a sensor floor. In this paper, we present a transformation of the collected sensor floor data, which is then used in a Random Forest classifier to distinguish a resident's behavior before and after a hospitalization. To do this, the raw sensor data was transformed into a set of features that are then used in the classifier. The classifier is evaluated with cross-validation and various metrics. The results show that it is possible to use the sensor data to detect behavioral changes in individuals and thus can be used to monitor people in need of care.

1 Introduction

Medical and nursing staff are often understaffed and lack the necessary resources to monitor their patients. Especially in nursing homes, where elderly residents need around-the-clock care, it can be easy for understaffed nursing staff to miss changes in residents' behavior. To assist nursing staff, residents' sensor data can be used to monitor their behavior. It is assumed that human behavior changes in response to health status and that these changes can be detected [1]. One device for collecting such data is a capacitive sensor floor, which is installed under the normal floor and detects people's movement behavior. It can collect data completely unobtrusive, which makes the collected behavioral data more natural [2]. The sensor floor is a relatively new device and initial analyses of gait patterns have recently been performed [2], but its potential applications are not clear at this time. Our goal for this work is to determine if it is possible to analyze human behavior, particularly its changes, using only sensor floor data. To do this, we use collected data from a sensor floor in a nursing home to distinguish the behavior of a nursing home resident before and after a hospitalization, assuming that the behavior changes after such an event [5]. We are specifically concerned with the behavior of an individual resident. From the data collected under natural living conditions, we create a set of features that can then be used for classification. We use a Random Forest for classification and evaluate the results with a K-fold cross-validation. Thus, our contribution consists of two steps: (1) extracting a set of features from the sensor floor data, and (2) using a Random Forest classifier on these features to distinguish the resident's behavior before and after a hospitalization.

2 Material and Methods

The following chapter describes the material and methods used in this work.

2.1 Sensor Floor

The sensor floor used for data acquisition in this work is the *SensFloor®* by the company Future-Shape GmbH [4]. It is a contact-less system that detects changes in electric capacitances on a grid of sensor fields and measures through a flooring layer on top of it. Currently, this system is used in elderly care facilities for fall detection and for assisted living at home. It is an excellent solution for these purposes because it is easy to install and blends in well with its surroundings, as it is hidden under the normal flooring. It also allows around-the-clock observation without affecting a person's behavior [2].

The *SensFloor®* consist of a three layer composite with a thin aluminum foil at the bottom, a polyester fleece of 3mm in the middle and a thin polyester fleece coated in metal at the top to make it electrically conductive. The system is organized as a grid consisting of independent modules. Each module has a microcontroller board in its center, which is connected to eight triangular sensor shapes, that measure the electric capacities, and to the power supply lines. It is possible to cut the modules in the right shape so that they fit in the corners of a room, as long as the microcontroller circuit remains undamaged. The whole installation can be powered by a single power supply unit of 12V, that can be placed at any position at the edge of the sensor floor [2].

There are three standard types of *SensFloor*® with low resolution (1m x 0,5m modules, 16 sensors/m^2), high resolution (0,5m x 0,5m / 32 sensors/m^2) and gait resolution (0,38m x 0,38m / 55 sensors/m^2) [2]. In the nursing home, a low resolution *SensFloor*® is used for the data acquisition. The rooms there have a size of 20.5/m^2 with an additional 4.7/m^2 bathroom.

The modules of the sensor floor send messages to a receiver. These stored messages can be displayed in a text file. The messages are stored in two different types: module capacity information and trajectory information [2]. We use the trajectory information files for our approach, because they provide more information about the movement of people and objects. From these raw data we create the features that are then used in the classifier. The *SensFloor*® stores two types of data messages in the trajectory information: the Diff messages, which describe the location changes of various objects in the room, and the State messages, which are collected every 10 seconds and describe the current state of the objects in the room. We only use Diff messages in our approach, since we are interested in the location changes of the resident.

2.2 Feature Extraction

The data is collected around-the-clock, but stored in a separate file for each day so that the data samples start and end at midnight. Therefore we transformed the raw data into trajectories separately for each day. In the data files, all data points belonging to the same object, and thus to the same trajectory, are identified by the same object ID. When a trajectory ends, the ID is released. We filtered the raw data by ID and stored all data points with the same object ID in a list of dictionaries, where each dictionary contains the information for one data point. Thus, one complete trajectory $\mathbf{T}$ is given by a list of data points $\mathbf{v}_i$ with variable length n that make up the trajectory. Each data point contains the information timestamp t_i, x-coordinate x_i and y-coordinate y_i. A trajectory can thus be described as follows:

$$\mathbf{T} = <\mathbf{v}_1, ..., \mathbf{v}_n>$$
$$\mathbf{v}_i = (t_i, x_i, y_i)^\mathsf{T}, i = 1, ..., n. \tag{1}$$

We deleted all trajectories that were not completed at the end of the day. These could be easily identified because they still contain their ID. After completing the list of daily trajectories, we had to filter it again to get only the trajectories that most likely belong to the resident of the room and not to visitors or caregivers. For this we made some assumptions:

1. We assume that the resident is alone in the room at the beginning of each data file, which is at midnight of each day.

2. If the resident is in the room and another trajectory starts in the area of the door, it belongs to another person.

3. If a person other than the resident is in the room and the trajectory ends in the area of the door, the person leaves the room.

4. If the resident is alone in the room and the trajectory ends in the area of the door, the resident leaves the room.

5. If there is no one in the room, the first trajectory, which starts in the area of the door, belongs to the resident.

With this assumptions, we deleted most trajectories belonging to persons other than the resident. An example trajectory placed in the resident's room can be seen in Fig. 1.

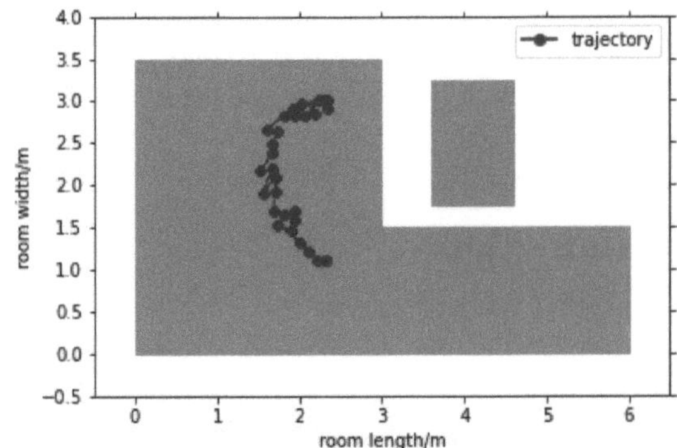

Figure 1: A plan of the resident's room in the nursing home. The axes describe the size of the room in meters. An example trajectory is plotted for visualization.

From the remaining trajectories, we did not use the trajectory data directly for the classifier, but extracted additional information:

- f_d the distance traveled, or length of the trajectory, given in meters

- $f_{\Delta t}$ the duration from start to end of the trajectory, given in milliseconds

- f_r the maximum radius of the circle containing all trajectory points, given in meters

For the distance we used the Euclidean distance to first calculate the distances between each two points $\mathbf{p}_i$ of a trajectory, which are then summed up:

$$f_d = \sum_{i=1}^{n-1} \sqrt{(\mathbf{p}_{i+1} - \mathbf{p}_i)^2}, \quad \mathbf{p}_i = \begin{pmatrix} x_i \\ y_i \end{pmatrix}, \tag{2}$$

where x_i, y_i are from $\mathbf{v}_i$.

For the duration we used the stored Unix timestamps t_i of the data points and calculated the time difference between the start time t_1 and the end time t_n of the trajectory with the following equation:

$$f_{\Delta t} = t_n - t_1, \quad \text{where } t_i \text{ is from } \mathbf{v}_i. \tag{3}$$

For the radius we decided to calculate the maximum radius, which encloses all data points of the trajectory. To do this, we first calculated the mean vector $\overline{\mathbf{p}}$ for each trajectory and then calculated the distance between the mean vector $\overline{\mathbf{p}}$ and each point $\mathbf{p}_i$ of the trajectory to save the maximum distance d_{max}. This was done with the following equation:

$$f_r = d_{max}(\mathbf{p}_i, \overline{\mathbf{p}}) = \max_{i \in [1,...,n]} \sqrt{(\overline{\mathbf{p}} - \mathbf{p}_i)^2}. \tag{4}$$

Then, we filtered the trajectories and deleted those with a radius or distance less than 0.5, assuming that they were noise from different objects in the room.

To better format the features that are used for the Random Forest classifier, we decided to divide all trajectories of a day into the 24 hours of the day. To do this, we summarized the distances, durations and radii of all trajectories that started within an hour and stored them with the number of trajectories passed in an hour. Thus, in the end, we have a 3-dimensional vector, with the following axes:

1. Number of days from which we have used the trajectory information

2. The 24 hours of each day.

3. The trajectory information for each hour, which includes the number of trajectories f_c, the total distance, the total duration and the total radius of the trajectories.

The first axis yielded 108 days of trajectory data that we used. We used data from a single resident who was hospitalized with an infection for a specific period of time. Since our goal is to distinguish the resident's behavior before and after hospitalization, we used 54 days before and 54 days after hospitalization to obtain a balanced set of training data. For training, the data was labeled 0 for pre-hospitalization data and 1 for post-hospitalization data.

The final feature vector with size $(108, 24, 4)$ had to be converted to a 2-dimensional size of $(108, 96)$ for the classifier we wanted to use. We also normalized the features to get more balanced information. The final set of features contains one feature vector $\mathbf{f}$ for each day that describes the behavior of a single day and looks like this:

$$\begin{aligned} \mathbf{f} = (&f_{c,1}, \ldots, f_{c,24}, f_{d,1}, \ldots, f_{d,24}, \\ &f_{\Delta t,1}, \ldots, f_{\Delta t,24}, f_{r,1}, \ldots, f_{r,24})^{\mathsf{T}}, \end{aligned} \tag{5}$$

with every $f_{x,h}$ being the sum of all the feature values that were calculated for every trajectory that took place in one hour of the day, for $h \in [1, \ldots, 24]$ and for all feature types $x \in \{c, d, \Delta t, r\}$. The complete data set, that was used as input for the classifier, consists of 108 of these vectors for the 108 days considered.

We selected the described features from the trajectories and the split in 24 hours to obtain a simple but meaningful representation of a person's daily behavior. It is assumed that the total number of trajectories traveled by the resident and the values of the extracted features provide a good description of the resident's activity and thus health status. Therefore, it should be possible to use these values and also their distribution over the 24 hours of a day to identify a daily behavioral routine of the resident and thus detect changes.

2.3 Random Forest Classifier

For classification, we chose a Random Decision Forest as classifier, because it achieves very good classification results [3]. Random Forests are a combination of individual decision trees that work as predictors. A decision tree classifies the given data based on various features of the data. With these features, the decision tree tries to separate the classes as best as possible. In a Random Forest, a large number of individual decision trees operate as an ensemble, which is used to vote for the most popular class, with each tree casting a unit vote. The class with the most votes becomes the prediction of the model. Random Forests are good because a large set of individual predictors outperforms any individual model, but for this the correlation between the individual decision trees must be minimal. This is achieved by using randomly selected inputs or combinations of inputs at each node of a tree to grow the tree [3].

2.4 Evaluation

To evaluate our classification results, we performed a K-fold cross-validation with five folds to split our data into a training data set and a testing data set. We used this method because we only have a small amount of data. For the evaluation, the metrics accuracy, F1 score, precision and recall were calculated independently for each fold.

3 Results and Discussion

In this section, we look at and discuss the results of the Random Forest classifier. The results for accuracy, F1 score, precision and recall of the classifier are shown in Table 1. It shows the results for the five different folds of the K-fold cross-validation, as well as the mean and standard deviation of these five folds.

The results vary between the different folds. Overall, the precision has the highest mean value and even reaches 100% in fold 4. However, it also has the largest standard deviation, which means that precision varies the most between folds. The results for accuracy and F1 score are slightly lower than precision, but they also vary less between folds. Recall has the lowest mean of all metrics and a high standard deviation of 0.10, making it the worst of all the metrics. It reaches the worst value in fold 3 with only 55%. In Fig. 2 the confusion matrix of the results is plotted, showing the

Table 1: Results with Cross-Validation in 5 folds

K Fold	Accuracy	F1 Score	Precision	Recall
Fold 1	0.7727	0.7684	0.8333	0.7692
Fold 2	0.7272	0.7250	0.6364	0.7777
Fold 3	0.7272	0.7054	0.7143	0.5555
Fold 4	0.9048	0.9045	1.0	0.8182
Fold 5	0.8095	0.8056	0.8333	0.8333
Mean	**0.79**	**0.78**	**0.80**	**0.75**
Std	0.07	0.07	0.12	0.10

accuracy results for both classes. The plot shows that the classifier predicts the correct class in most cases, namely 81% for class 0 and 76% for class 1.

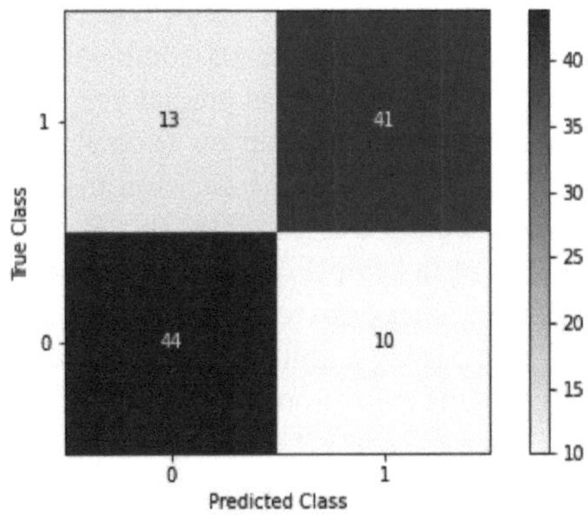

Figure 2: Confusion matrix of the results of the classifier. It shows the accuracy for both classes, with 0 for pre-hospitalization and 1 for post-hospitalization.

Overall, with an average mean of 78% for all metrics, the Random Forest classifier provides great and meaningful results. We only use a few features from the trajectories extracted from the raw *SensFloor*® data, but the results are good enough to work with. Expanding the list of features from the sensor data and using more detailed features could improve the results even more.

Also, we train the classifier with data collected from only one person, so the results are good for that particular person. We cannot say anything about general behavior patterns because each person is individual and the effects before and after hospitalization can be very different depending on the individual. Since the person we are referring to is an elderly person, we can assume that the difference between the resident's behavior before and after hospitalization is relatively large and easier to detect [5]. Therefore, the results of the classifier could be different for data from other individuals and also for general behavioral approaches.

4 Conclusion

In this paper, a new approach for human behavior analysis using sensor floor data was defined. First, the raw data from a nursing home resident's *SensFloor*® was used to create trajectories and define features from them. Then, the new features were used in a Random Forest classifier to distinguish between pre- and post-hospitalization data. K-fold cross-validation was used to evaluate the Random Forest classifier. The results show that the classifier can use the given features to distinguish the resident's behavior before and after hospitalization. Therefore, it is possible to detect changes in human behavior using only sensor floor data.

Future work includes extending the features used for the Random Forest classifier since the extracted features used in this work are very simple. An extension of the features could lead to even better results. In addition, this work only used data from a single individual. Future work could include data from other individuals or even a general approach to analyzing people's behavior.

Acknowledgement

This work has been carried out at the Institute of Medical Informatics, Universität zu Lübeck and was supervised by Prof. Dr.-Ing. habil. Marcin Grzegorzek of the Institute of Medical Informatics, Universität zu Lübeck.

Author's Statement

Conflict of interest: Authors state no conflict of interest. Ethical approval: The research related to human use complies with all the relevant national regulations, institutional policies and was performed in accordance with the tenets of the Helsinki Declaration, and has been approved by the authors' institutional review board or equivalent committee.

5 References

[1] M. Grzegorzek, X. Huang, T. Irshad, C. Lauterbach, A. Steinhage and R. Hoffmann, *Ethikantrag zur Studie: Sensorbasierte Bewegungsanalyse von Pflegeheimbewohnern im Zusammenhang mit Medikamenteneinnahmen*, Universität zu Lübeck, 2020.

[2] R. Hoffmann, H. Brodowski, A. Steinhage and M. Grzegorzek, *Detecting Walking Challenges in Gait Patterns Using a Capacitive Sensor Floor and Recurrent Neural Networks*, Sensors, vol. 21, no. 4, 2021. https://doi.org/10.3390/s21041086

[3] L. Breiman, *Random Forests*, Machine Learning, no. 45, pp. 5–32, 2001. https://doi.org/10.1023/A:1010933404324

[4] Future-Shape GmbH, Available: https://future-shape.com, 2018 [last accessed on 2022-01-12].

[5] D. Cook, J. Thompson, S. Prinsen, J. Dearani and C. Deschamps, *Functional Recovery in the Elderly After Major Surgery: Assessment of Mobility Recovery Using Wireless Technology*, The Annals of Thoracic Surgery, vol. 96, no. 3, pp. 1057–1061, 2013. https://doi.org/10.1016/j.athoracsur.2013.05.092

Comparison of Data Augmentation Techniques for Phenotype Data of Parkinson's Patients

Christian Trense[1], Artur Piet[2], Xinyu Huang[2], and Marcin Grzegorzek[2]

[1] Medical Informatics, Universität zu Lübeck, christian.trense@student.uni-luebeck.de

[2] Institute of Medical Informatics, Universität zu Lübeck, {artur.piet, x.huang, marcin.grzegorzek}@uni-luebeck.de

Abstract

Parkinson's disease is a degenerative disease spreading at an alarming rate. The treatment needs to know which genetic defect is the cause. Therefore, a genetic test is necessary. Unfortunately, this test is not available for all Parkinson's patients, especially in poor regions. A machine learning algorithm can counteract this by determining the genotype (i.e. the genetic Parkinson's subtype) based on the patient's phenotype. However, the problem arises that the frequency of some genotypes is much lower than others. Classification algorithms recognise these rare genotypes less well. The expansion of the genotype classes counteracts the misclassification. We use machine learning methods to generate synthetic data samples in this work. The methods used were CTGAN, SMOTE and RandomOverSample. The latter method achieved the best results with an F1 score of 76.5% and an accuracy of 87.5%.

1   Introduction

Nowadays, there are more and more fields of application that use machine classification. In the meantime, methods have also been developed to generate new images for under-represented classes for image data, for example. However, the problem with underrepresented classes is not only found in image data. Underrepresented classes should also be classified in table data, such as the MDSGene dataset. The Movement Disorder Society Genetic Mutation Database (MDSGene) data set deals with Parkinson's patients. It considers which gene expression has led to Parkinson's and which phenotype the patient has. However, the database has two problems: first, the inequality mentioned earlier of the classes and secondly, much data is missing. The database is significant because Parkinson's disease is becoming more widespread. The ageing of society is driving Parkinson's disease, which is a degenerative disease[6]. In treatment, it is shown that the genetic cause influences the success of therapy. However, the determination of this cause is not possible everywhere. In order to enable optimal treatment for all patients, the genotype should be determined based on the phenotype. However, due to the unequal distribution of the classes, the larger classes are preferred. We attempted to generate new data with algorithms to counteract this. For this purpose, different methods are tested and compared in this thesis. No attention is paid to data classification in this paper because, in another paper, this was already done [3].

2   Material and Methods

In the following Chapter 2.1, the disease and the database are explained. After that, in Chapter 2.2 the methods are explained, which are used for the classification of the data samples and also for the generation of new samples.

2.1   Disease & Database

The disease that is under research in this work is Parkinson's disease. This disease is a degenerative disease that was reported first by James Parkinson in 1817 [2]. Since then, much work has been done in the direction of Parkinson's disease. This work showed that the disease is mainly active in the central nervous system. The most impact is localised in the motor capabilities, but there are also symptoms in the cognitive condition that can be monitored. Misfolded proteins clog the cells, leading to the death of many cells, which trigger these symptoms. These cell deaths are primarily located in the midbrain, which is the cause for broad problems Parkinson patients experience. In addition, society over the world is ageing. Therefore it is essential to improve the treatment options.

This work uses the dataset MDSGene [5] that was created by the Movement Disorder Society. In this dataset, the scientist recorded 2222 Parkinson patients and measured for every patient 37 features. The features consist of demographic and clinical values. A total of 6 different Parkinson's disease genotypes are included in the dataset. These are subdivided according to genetic types. The amount of patients is not the same in the different classes. How significant the differences are can be seen in table 2.1.

DJ1	Parkin	LRRK2	VPS35	PINK1	SNCA
33	1003	822	67	151	146

Table 1: This table shows the size differences between the classes.

Another challenge is that in the dataset, many records are missing. Around 72% of the entries are empty. In addition, we excluded two features (parkinsonism and motor signs) because, in 99% of the data samples, the entry was the same, which does not benefit the classifier.

2.2   Methods

With the data mentioned in 2.1, we tried to predict the genetic outcome automatically from the features. Therefore a random forest was used. We chose this method because, in previous work, this method proved to be the best among other methods such as SVM, CNN, or MLP [7]. Simple decision trees are the basis of this technique. However, to obtain the data of a patient that needs to be classified, we use a group of decision trees. Then every decision tree takes an individual decision in which class the data fit the most. Finally, we added up the different results and chose the most votes to get the final decision. Additionally, every tree receives a different subset of the given features to prevent relying on the same features.

Due to the significant data imbalance, different methods were used to generate artificial data for better random forest training.

The simplest method that we used is the RandomOverSample. This method picks random samples of the original data and modifies it slightly. The classes with fewer samples provide the drawn samples. Then the algorithm modifies these random samples at a random amount of randomly chosen features. Logically, this approach generates now more data, but there is no protection against unrealistic data. The Synthetic Minority Over-sampling Technique (SMOTE)[1] generates more realistic data than the RandomOverSample. This data augmentation technique is repeated for every synthetic sample. The process uses two samples from the original samples to calculate the new sample. The calculation starts with picking a random data sample. After that, the method chose the nearest neighbour. These two samples are then subtracted from each other, which shows the simple point-by-point difference between them. This difference is then added to the first drawn sample, but the new sample would reach the second original point without a modification. Therefore, a random variable between 0 and 1 is multiplied to the difference to prevent that. With this addition, the generated sample is located between both original samples. This work used a variation of this technique because the data contain many categorical values, which also requires special handling when creating new samples.

2.2.1   GAN

The last and most complex method is the Conditional Tabular Generative Adversarial Net (CTGAN). This CTGAN is a variant of a Generative Adversarial Network (GAN)[4] as in a normal GAN, two deep learning networks are working against each other to create new data. In this composition, seen in Figure 1, one of the networks is dedicated to generating new data (generator). The other is designed to detect if the presented sample belongs to the original data or the synthetic data samples (discriminator). At the beginning of the process, the generator processes random data samples, which the discriminator can easily detect. Nevertheless, as the generator improves, it gets more complicated for the discriminator to decide. As the generator improves, the discriminator also learns, so the generator is taught to produce even better data. This training will be continued until the discriminator can only guess if it is synthetic or not.

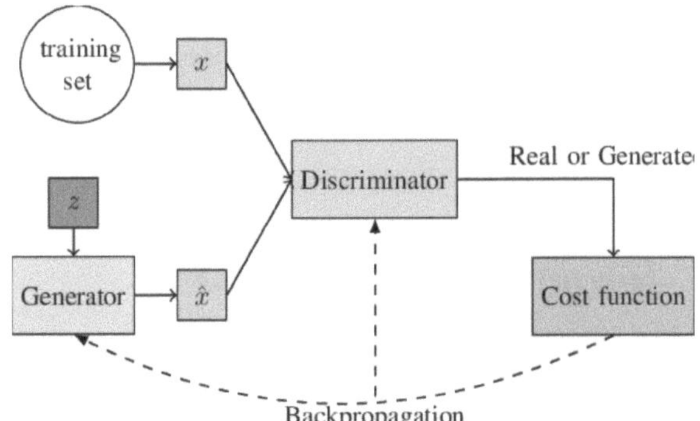

Figure 1: This figure illustrates the process of the training of a GAN.

The Tabular Generative Adversarial Net (TGAN) has the advantage that it is designed to use tabular data and also to output this data type. Furthermore, due to the representation of the MDSGene dataset in a large table, the TGAN seems to be a suitable option.

Through the restriction that the MDSGene dataset contains many columns with categorical features, another characteristic had to be added to the TGAN. The result is the CTGAN. Based on the additional condition, it can be determined which columns consist of categorical and continuous characteristics. This characteristic ensures that the CTGAN generates new data that fits into the columns' existing value ranges. Another benefit is that the values that need to be tested in the columns with categorical features are not as many. Therefore, the time the algorithm needs is reduced significantly. This reduction enables the system to do more training iterations.

2.2.2   Evaluation Metrics

During the evaluation of the methods, we divided the data randomly. This division can cause a split of the data favouring the classification method. The cross-validation schema can be seen in Figure 2 to prevent getting a false understanding of the performance. This method splits the data into k parts instead of splitting the data into only two parts of test and training data. We used one part for the testing and the rest to train the algorithm. After that, we execute the process with this split. After this, the process is repeated but with another combination of the data splits. These repetitions will be performed until every data split is used as testing data. The results from the individual repetitions are now averaged, and this average is then the final result.

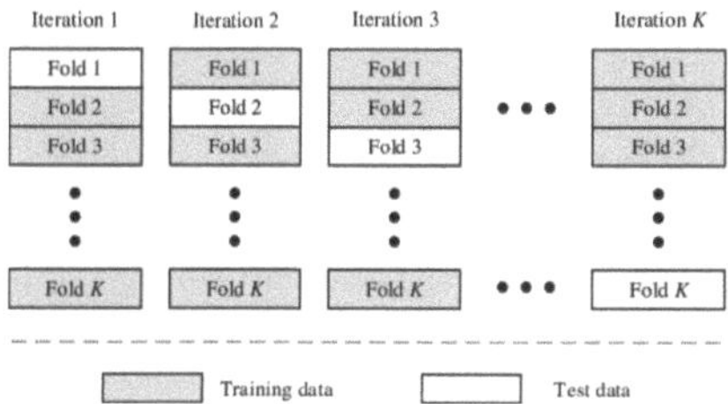

Figure 2: This figure illustrates the process of a k-fold cross-validation. In this work k is chosen as five.

We measured the results with four metrics: average F1-Score, accuracy, precision and recall. The accuracy is a standard metric to measure classifier. Because this metric is one of the most basic. It is calculated by

$$accuracy = \frac{|correct\ classified\ samples|}{|samples|} \tag{1}$$

This metric allows a general overview of the performance of the algorithm. The following two metrics are essential because the F1 score is based on them. The precision indicates how many percent of the data samples returned as correct should also be classified as such. This can determine whether the algorithm only returns the correct samples or classifies other classes incorrectly. The calculation looks as follows:

$$precision = \frac{|a \cap b|}{|a|} \tag{2}$$

$$recall = \frac{|a \cap b|}{|b|} \tag{3}$$

a = {correct classified data samples}
b = {all correct data samples}

The recall is similar to the precision but aims at the ratio of the correct data found by the algorithm. Therefore the recall is calculated like in equation 3.

The most important is the average F1-Score because this score rates the test accuracy and considers that the classes are not equally significant. The F1-Score is computed with the precision and recall as seen in equation 4, and this is done for every class, and then the average is taken.

$$F1 = 2 * \frac{precision * recall}{precision + recall} \tag{4}$$

That the approach is considering every class is outstanding because it is also possible that small classes get recognised with this.

3 Results and Discussion

We used our RF classification without any data augmentation. This classification enabled us to compare the later results with this baseline. The following metrics: average F1-Score, accuracy, precision and recall, allow to compare the different approaches and visualise possible improvements. This test and also every following test that was carried out was performed with a random forest with 150 decision trees and cross-validation with five folders. The result with only the original data was an F1-Score of 70,3% and an accuracy of 85,9%. The confusion matrix of this result is shown in Figure 3 and displays the accuracy for the classification of the different classes.

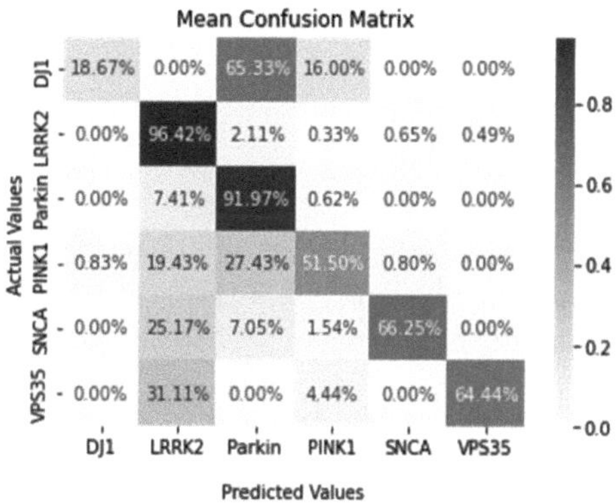

Figure 3: In this figure you can see that the two big classes "LRRK2" and "Parkin" are mostly correctly classified. Only with the smaller classes are many data samples classified incorrectly.

The metrics results show that the algorithm generally works well, but we can see that the class DJ1 is largely misclassified. Also, the nearly perfect classes belong to the most significant classes. That shows that the classification performance correlates with the number of available samples. Afterwards, we carried out the other tests with the augmentation methods. The best performing method was the RandomOverSample method with an F1-Score of 76,5% and an accuracy of 87,5%. Shortly after that, the SMOTE algorithm ranked second, and the CTGAN achieved the lowest scores. Remarkable is that the confusion matrix of the test with the RandomOverSample shows in Figure 4 that the smallest class DJ1 has a significant increase of the correct classified samples, and the other classes lose only a few percent.

In addition to these tests, the synthetic data also was used alone to test how realistic these data samples are. Unfortunately, the performance was terrible with an F1-Score of 30 to 40 %, and the other metrics also had a drop of around 40 % compared to the combined approach. This performance leads to the assumption that the generated data is not good enough to perform alone but has achieved a significant impact in combination with the original data. That shows that the information saved in the original data is mostly enough to classify the data correctly. However, the small classes do have not enough samples to be relevant to the classifier.

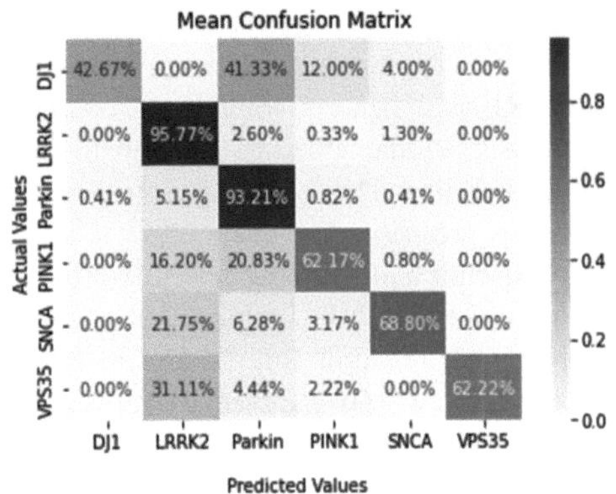

Figure 4: Compared to Figure 3, the smaller classes were clearly better classified. In addition, the values of the two large classes have only decreased minimally.

Method	F1-Score	Accuracy
No Augmentation	70,3%	85,9%
GAN	66,1%	93,7%
SMOTE	72,4%	86,2%
RandomOverSample	76,5%	87,5%
Only Synthetic	41,7%	43,8%

Table 2: This table shows the results that were achieved. In direct comparison, it can be seen that the RandomOver-Sample works best.

The table 2 summarises the results for an overview. Once again, it is clear that the accuracy does not differ significantly, but this is the case with the F1 score.

Another point that can be seen in this work is that the few data samples from the "DJ1" are not enough for the CTGAN to learn the proper representation, and therefore it cannot realistically reproduce new samples.

A possible explanation for the excellent performance of the RandomOverSample method is that it was not possible to restrict the categorical columns with this method. This freedom could lead to values that were impossible for the other methods. Of course, this would be unrealistic data, but the classifier still uses it, and perhaps these abnormal values help it in training.

4   Conclusion

The executed tests showed that the distinction between the classes works very well for the big classes and not so well for the small ones. The data augmentation significantly improved the allocation performance. However, it also showed that the complex methods did not perform very well compared to the simple approaches, implying either that GANs cannot represent this type of data or that the amount of training data was too small. In conclusion, we can say that the data augmentation methods achieve significant improvements, but this is not yet sufficient for use in the real world. More data needs to be collected, or the methods used must be extended and refined. In the future, other methods can

be tested that generate data. In addition, combinations of the existing methods could be tested and the use of under-sampling methods.

Acknowledgement

The work has been carried out at Universität zu Lübeck and was also supervised by the Institute of Medical Informatics, Universität zu Lübeck.

Author's Statement

Conflict of interest: Authors state no conflict of interest.

5   References

[1] Nitesh V. Chawla Kevin W. Bowyer, Lawrence O. Hall / Kegelmeyer, W. Philip *SMOTE: Synthetic Minority Over-sampling Technique* 2002-02, Journal of Artificial Intelligence Research 16 (2002) 321–357 , Vol. 16, p. 321-357

[2] Parkinson, James An Essay on the Shaking Palsy 1817 London: Sherwood Neely and Jones

[3] Piet, Artur, *Erkennung des Genotyps anhand vom Phänotyp bei Parkinson-Patienten mittels Deep Learning*, 2021-08 Universität Siegen

[4] Goodfellow, Ian / Pouget-Abadie, Jean / Mirza, Mehdi / Xu, Bing / Warde-Farley, David / Ozair, Sherjil / Courville, Aaron / Bengio, Yoshua *Generative Adversarial Nets*, 2014, Ghahramani, Z. / Welling, M. / Cortes, C. / Lawrence, N. / Weinberger, K. Q. (Eds.), Advances in Neural Information Processing Systems , Vol. 27 Curran Associates, Inc.

[5] University of Lübeck. MDSGene Movement Disorder Society Genetic Mutation Database. https://mdsgene.org. Retrieved 17.01.2022.

[6] GBD 2016 Parkinson's Disease Collaborators. *Global, regional, and national burden of Parkinson's disease, 1990-2016: a systematic analysis for the Global Burden of Disease Study 2016.* Lancet Neurol. 2018 Nov;17(11):939-953. doi: 10.1016/S1474-4422(18)30295-3. Epub 2018

[7] N.S. Kuhnke. Bachelor Thesis: *Applying Machine Learning Methods for Predicting the Genetic Forms of Parkinson's Disease Based on Clinical Presentation.* Nov 2020.

Building an autonomous Apple Collector

Tim-Henrik Traving [1] and Floris Ernst [2]

[1] Robotics and Autonomous Systems, University of Lübeck, timhenrik.traving@student.uni-luebeck.de
[2] Institute for Robotics and Cognitive Systems, University of Lübeck, ernst@rob.uni-luebeck.de

Abstract

Tending a garden can be very fulfilling, but it can also be quite laborious. To eliminate the lengthy task of collecting fallen fruit before lawn mowing, an affordable robotic platform is designed and constructed. First, different collection mechanisms are presented, followed by the design of the overall robot. Both take the environmental conditions of the test scenario into consideration. After that, the execution and construction are described. This includes a framework for classifying fallen apples by decomposition stage is proposed, as well as a detailed description of the pick-up and drop-off sequence. In the end, an outlook is given, showing potential uses of the platform besides collecting fallen fruit.

1   Introduction

Tending a garden becomes more laborious when fall approaches. In order to mow the lawn, it has to be freed from fallen leaves, twigs and fruits. This paper describes an extendable robotic platform that is capable of collecting fallen fruits. The platform is not self-propelled, but was designed such that it may be towed through a garden. It is affordable and suitable for non-commercial home use. An image classification system allows to collect only specific fruits. This is an advantage compared to other, more low-tech solutions, which collect every fruit (and possibly more) that is lying on the ground, and which are usually hand-operated.

2   Material and Methods

Working with fallen fruit meant that at least a part of the project had to take place during late summer and fall. Simulation of the different decomposition states by buying fruits from a store and letting them rot during another time of the year would not yield reliable results. On the one hand, insects (such as e.g. wasps) that play a major role in the decomposition may not be active at the particular time of the year. On the other hand, fruit from a store might have been subject to special treatment, such as waxing or application of fungicides, which alters the decomposition.

The project was thus split into two parts: Planning and execution. During the planning phase, the environmental conditions were surveyed and multiple different build-options were drawn out, the best of which was then implemented in the execution phase. However, this does not mean that everything went to according to plan. The execution phase still consisted of a lot of experimentation.

2.1   Environmental Conditions

The proving grounds for the robot was a garden in northern Germany with five apple-, one pear-, and one plumtree. The apple trees grow up to 1000 kg of apples each year, many of them ending up as fallen fruit. While edible, most of them are rather small, between four and seven centimeters in diameter, weighing between 100 and 300 grams, which makes the collection for consumption of individual apples unattractive. Furthermore, fallen apples quickly fall pray to wasps, which feed on them, or are damaged otherwise, leaving marks and fostering rot.

The garden itself measures approximately 1500 square meters, most of it covered with lawn and moss. Flowerbeds, paveways and other zones where the robot should not collect apples are not always convex, but still well defined. However, some tree branches hang down to a few centimeters above ground level, making for dynamically changing, with regard to ultrasonic and electromagnetic sensors semi-transparent obstacles. There are no major hills or slopes, but local variation in ground height of up to 20 cm may occur.

2.2   Robot Setup and further Proceedings

The robotic platform used for the project was based on an XY-portal robot architecture, as used e.g. for CNC-milling and laser-engraving. The XY-portal is used to position a pneumatic piston, which in turn uses one of different interchangeable, custom built manipulators to pick up soft objects, like e.g. apples and pears. A routine for picking up and dropping such objects was developed. The whole construction was tested in the proving grounds. For testing purposes, a loose framework to categorize apples based on their state of decomposition was developed.

3   Results and Discussion

A lot of time was spent on planning and thinking of different ways to achieve the task of picking up apples. During this process, complexity was further and further reduced, until a feasible concept was found. Having only a general concept meant that still a lot of experimentation was necessary in the execution phase, but it allowed to move on quickly and also gave the execution phase the necessary flexibility to deal with problems that arose.

3.1   Planning

The environmental conditions played a major role in the design process. In order to reduce complexity, the design was split into different, mostly independent parts. A sketch of the overall construction can be seen in Fig. 1.

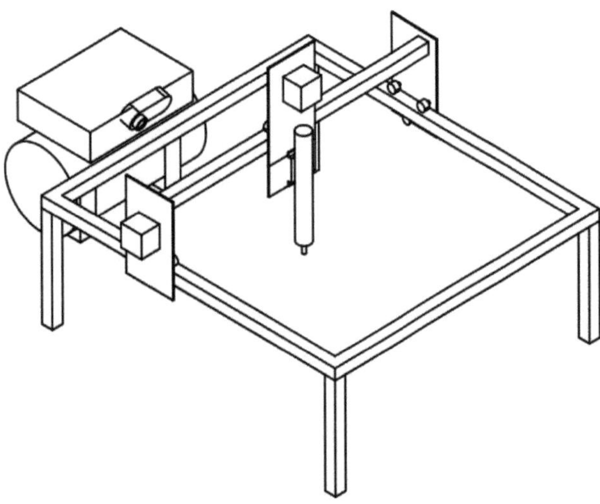

Figure 1: An overview of the portal construction with a tank for pressurized air to actuate the piston, the piston itself and a camera for object detection.

3.1.1   Portal Construction

The first draft of the construction was a self-propelled portal with a piston that used a dedicated shove-off mechanism to drop collected fruits onto a conveyor belt, which in turn moved the apples into a container atop the construction for intermediate storage. It became quickly apparent that building such a robot from scratch would be beyond the scope of the project. The next option was to adapt and build an existing robot. One possibility would have been the *Acron Precision Farming Robot* by *Twisted Fields* [1]. However, this would have exceeded the project's budget.

Nonetheless, the projects scope had to be further reduced to a non-self-propelled platform. However, the platform should provide the opportunity for future expansion, especially with regard to a propulsion system. Therefore the propulsion of the platform still played a major role in the robot's design. In the end, the choice fell on a portal construction that would be towed by a *Jackal Unmanned*

Ground Vehicle (UGV) from *Clearpath Robotics* [2]. This provided numerous advantages: First, the separation of concerns into a collection and a propulsion system made the project feasible at all, given the circumstances. Second, the Jackal robots were readily available at the *Institute for Robotics and Cognitive Systems* (ROB) of the University of Lübeck. Third, the majority of the construction could be done with common-off-the-shelf (cots) parts, reducing uncertainty. And fourth, the Jackal's resources, such as a *NVIDIA Jetson AGX Xavier* [5] for image recognition or an onboard power supply, could later be used for the platform.

3.1.2   Piston

A pneumatic two-way piston should be used to shoot the collection mechanism into an apple and then lift it up. This means that an onboard compressed air reservoir is necessary for the piston actuation. The reservoir can be filled beforehand by an external air compressor, so there is no need for an air compressor on board or on site. The airflow is controlled by a two way solenoid valve and a bleed-off system. The piston's tip should have some kind of flexible mounting system, so that the collection mechanism could be changed easily, if necessary.

3.1.3   Collection Mechanism

The goal of the project was to free the lawn from apples in order to mow the lawn. Collecting apples one by one is just a means to an end[1]. This implies that damaging apples, potentially leading to their complete destruction, is acceptable. Working with this implication not only opened new possibilities, but was also necessary for the feasibility of the project, as collecting apples e.g. by the use of a robotic hand/claw, or sucking them up like an enormous vacuum cleaner would have exceeded the scope of the project.

Three different collection mechanisms were designed, all of them to be shot from directly above down into the apple, and then lifting it up against the force of gravity. From most complicated to least complicated, these mechanisms were:

1. An arrow tip with ex-/retractable wings: The arrow would have been shot into the apple. Then the wings would have been extracted through a system of wires in the hollow shaft of the tip's construction, cutting its way through the apple and ensuring that the apple would not fall off when picked up. To drop the apple, the wings would have been retracted and the apple could have been shoved off from the piston.

2. A bolt with a dedicated shove-off-mechanism: The bolt would have been shot into the apple, expecting the apple to cling onto the bolt when lifted up. At the drop-off position, a small metal ring would have been moved along the bolt, shoving anything off that is larger than its diameter (such as apples). The ring would have been guided by a metal rail, which is actuated towards the tip of the bolt by a cord, running

---

[1]Other means could e.g. be cutting down the apple trees (unacceptable), or introducing apple-eating animals to the garden (not feasible).

around a diverter pulley, and retracted back to the base of the bolt by a rubber band.

3. A blank bolt with a rotatable piston: The bolt gets shot into the apple, the apple clings onto the bolt during lift up. The piston rotates sideways and shoots the apple off to one side.

One of these mechanisms was to be mounted at the end of a pneumatic piston, which would shoot the mechanism downwards at considerable speed, driving it deep into the apple below. It would then retract and be positioned at a drop-off zone, where the apple would be removed.

3.1.4 Software

Even though the software for an autonomous apple collector consist of many complex parts, only three of them were mainly interesting for this project: Portal-actuation for positioning the piston, piston actuation for picking up and dropping off apples, and image recognition for detecting apples. For the image recognition an artificial neural network (NN), pretrained on the *Common Objects in Context* (COCO) dataset [4] was to be used for transfer learning, in order to specialize it in detecting and classifying different kinds of apples. It locates the center of an apple and forwards the coordinates to software running on an *Arduino Uno* [6], which converts the coordinates from image-space (where an apple is in the image) to portal-space (where an apple is w.r.t. the portal). These portal-coordinates are handed over to a Grbl-controller, which is the heart of the portal-actuation. Grbl is a parallel-port-based motion control system, used e.g. for CNC-milling and 3D-printing [3]. It provides an easy to use interface, where basically just the coordinates of a point need to be given, and the controller deals with moving an actuator to the desired point. This removes the necessity of controlling the hardware (mostly stepper-motors) directly. The software for the piston is much simpler. It has only two states: retracted and extruded. When actuated, a valve is opened and pressurized air moves the piston. The piston can also be rotated, which is achieved by a servo that is controlled via pulse wave modulation (PWM). The piston actuation and the servo control is done by the Arduino.

3.2 Execution

Compared to the design phase, the execution phase was rather short, but very productive. Construction took place at the author's parents' workshop, where all the necessary tools were available and the proving grounds were right at hand, which allowed for quick tests and short iterations.

3.2.1 Data Collection and NN Training

Even before the first parts arrived, the collection of training data for the NN began and continued throughout the execution phase. In order to achieve a high and stable quality of the training data, a small camera-jig was improvised, that held a camera at the same height above and the same angle

towards ground for every picture and could be operated one-handedly. This made data-collection very easy and allowed to map all of the proving grounds in less than an hour. Apples are classified by decomposition stage. Please refer to Fig. 2 for an example of each stage.

(a) Pristine. $\sim$0% of the visible surface damaged. These apples could be consumed directly.

(b) Processable. < 10% of the visible surface damaged. These apples could be used for further processing, e.g. making juice.

(c) Damaged. < 50% of the visible surface damaged. These apples show major damages and signs of rot.

(d) Severely Damaged. < 100% of the visible surface damaged. These apples were typically pecked thoroughly by birds, but have not turned foul yet.

(e) Rotten. $\sim$100% of the visible surface damaged. These apples are completely brown or even black.

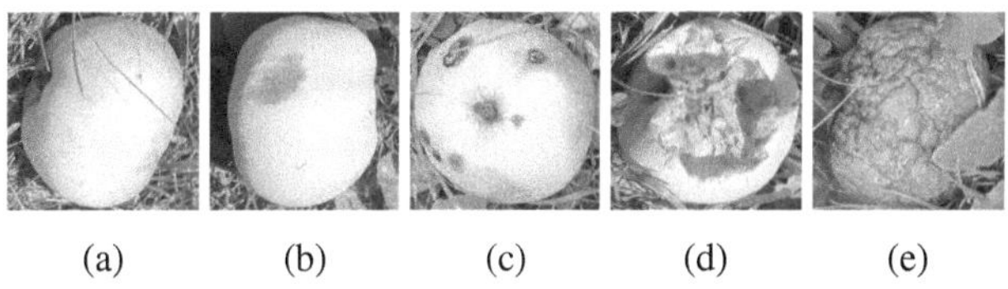

Figure 2: Different stages of an apple's decomposition. (a) Pristine, (b) Processable, (c) Damaged, (d) Severely Damaged, (e) Rotten

The current setup is only capable of perceiving a maximum of 50% of an apples surface, as the other half is facing the ground. However, this is not a problem for this application. In the end, it is sufficient if the robot recognizes the presence of an apple. The differentiation between different decomposition-stages is not necessary when the robot is supposed to collect *all* apples.

3.2.2 Piston and Pick-Up Mechanism

The first hardware that was constructed was the piston, together with the pressurized air system. The piston can extrude 20 cm and has a threading at its tip, so that different collection mechanisms can be exchanged quickly. The operating pressure is 8 bars, dropping with each actuation. The pressurized air tank has a volume of 3 liters, which results into 24 liters of compressed air, enough for actuating the piston approximately 120 times. In order to keep the effort as low as possible, the fist collection-mechanism that was built was the least complicated one, the blank bolt with a rotating piston. At first, the construction did not work, because the piston retracted as fast as it extruded, ripping the bolt back out of just pierced up apples. This was solved by reducing the diameter of the air-outlet, limiting airflow and therefore slowing the drop of the air pressure, which resulted in a much slower piston-retraction. The relative number off successful pick-ups immediately increased from nearly 0% to almost 100%. These results eliminated the need for a more complex pick-up system. Dropping the apples is achieved by actuating the pistons while facing sideways. The apple is accelerated to such a high velocity, that it is not only ejected off the bolt, but shot 50 to 200 cm through the air. Pristine apples fly farther than rotten ones.

3.2.3 XY-Portal

The use of cots-parts made the construction of the XY-portal tremendously easy. A cheap, yet durable kit, originally intended for laser-engraving, was used as a base. The construction was as easy as following the manual and leaving away the last steps of the laser assembly. Instead, a servo was mounted instead of the laser, using a custom made, 3D-printed adapter. The servo is capable of rotating and holding a weight of more than 500 grams on the extruded piston, which is an effective lever of more than 20 cm. The apples found in the proving grounds are between 100 and 300 grams and are rotated while the piston is retracted, resulting in an effective lever of approximately 5 cm. For prototyping purposes, the servo does not hold the piston directly, but a sheet of perforated metal. This makes swapping the piston for other tools that might be developed easier. One problem that occurred was that the actuation of the piston's valve was an electromagnetic event of such magnitude, that the interference-sensitive parallel-port connection between the Arduino and the Grbl-controller was lost. The problem was solved by adding a freewheeling diode to the valve's power supply, in order to block high voltage-peaks that occur when disabling the valve. The stepper motors driving the portal allow for a precision in the scale of one tenth of a millimeter. This precision is useless however, if the actuator's position is changed by external forces, as the real-world position does not align with the internally calculated position anymore. In such a case, the actuator's position needs to be reset to a known position, so that the real-world and internally calculated positions align again, which is achieved through a procedure called *homing*. This is necessary after shooting an apple away to one side, as the stepper motors do not have enough power to prevent the piston from moving in the opposite direction.

3.2.4 Pick-Up and Drop-Off Sequence

First, the apple needs to be detected by the NN. The apple's picture coordinates are transformed into XY-portal coordinates, where the piston is then positioned. Actuation of the piston shoots the bolt into the apple, which clings onto the bolt when the piston retracts. Then the whole piston rotates about 135 degrees to one side. Once positioned, the piston is actuated, ejecting the apple away. Finally, the piston is rotated back downwards and auto-homing begins.

4   Conclusion

The robot is a base to further build upon. The next step is the use of a propulsion-system, so that the platform can autonomously move around. But having the functional systems separated, the robot can be used in all kinds of different scenarios. It may operate stationary, processing apples that were collected by other means. The NN can be used in further scenarios such as sorting apples for different applications (e.g. human consumption, processing, animal consumption, composting, ...). Pristine apples could be spared in order to be hand-collected later. Or the network can be used for another robot, which thoroughly inspects collected apples from all sides. Transfer-Learning can be used to train the NN on other kinds of fruit in order to easily adapt it to new tasks. Furthermore, the piston can be exchanged for other actuators, opening a broad range of applications, from high precision fertilizing to targeted weed control.

Acknowledgement

The work has been carried out at the author's parents house and garden and has been supervised by Prof. Dr. rer. nat. habil. Floris Ernst, Institute for Robotics and Cognitive Systems, University of Lübeck. Furthermore, the author would like to thank M. Sc. Lars Schilling for his supervision of the planning phase and Dipl. Geol.-Paläont. Martina Galler for her supervision of the patenting-process.

Author's Statement

Conflict of interest: Authors state no conflict of interest.

5   References

[1] Twisted Fields, *Solar-powered precision farming Rover*. Available: https://www.twistedfields.com/research [last accessed on 2022-01-22].

[2] Clearpath Robotics, *Jackal UGV - small weartherproof Robot*. Available: https://clearpathrobotics.com/jackal-small-unmanned-ground-vehicle/ [last accessed on 2022-01-22].

[3] Grbl Contributors, *Grbl*. Available: https://github.com/grbl/grbl [last accessed on 2022-01-22].

[4] T.-Y. Lin et al., *Microsoft COCO: common Objects in Context*. CoRR, vol. abs/1405.0312, 2014. Available: http://arxiv.org/abs/1405.0312 [last accessed on 2022-01-22].

[5] NVIDIA, *KI-gestützte autonome Maschinen im richtigen Umfang - NVIDIA Jetson AGX Xavier*. Available: https://www.nvidia.com/de-de/autonomous-machines/embedded-systems/jetson-agx-xavier/ [last accessed on 2022-01-22].

[6] Arduino, *Arduino UNO R3*. Available: https://docs.arduino.cc/hardware/uno-rev3 [last accessed on 2022-01-22].

Developing a Basis for Learning from Demonstrations on a Robotic Arm

Rabia Demirci [1], Robin Denz [2], Elmar Rueckert [3], and Nils Rottmann [4]

[1] Robotics and Autonomous Systems, Universität zu Lübeck, rabia.demirci@student.uni-luebeck.de
[2] Robotics and Autonomous Systems, Universität zu Lübeck, robin.denz@student.uni-luebeck.de
[3] Chair of Cyber-Physical-Systems, Montanuniversität Leoben, rueckert@unileoben.ac.at
[4] Institute for Robotics and Cognitive Systems, Universität zu Lübeck, rottmann1990@gmx.com

Abstract

One of the most important goals in robotics is to generalize robot control, especially in dynamic environments. In this regard, the approach, learning by demonstration, offers very good results. This is realized by human instructors who teach robots a new task. Then, parameterized models are trained with the data obtained from the demonstrations, so that the learned skill can be executed and updated under different conditions. In this paper, we establish a basis for such demonstration learning processes. Therefore, we implement a calibration method for a robotic arm and an optitrack system. We furthermore make use of a sensor glove to record movement data which is used to learn representative motion models as a proof of concept. As a result, our work provides a solid basis for the realization of more complex tasks for a robotic arm that can be performed by learning from demonstration using a sensor glove.

1   Introduction

This article contains parts of [1], that has been previously published at the International Conference on Advanced Robotics (ICAR) 2021. Parts of the following chapters therefore give insight into the work from [1].

Nowadays the demand for robots in co-working scenarios increases, especially considering the developments in Industry 4.0 practices [2]. The application field can be divided into robot interaction with a passive environment (e.g. objects or surfaces) or the interaction with another operator as a co-manipulator (e.g. a robot or a human). These applications can require the adaptation of the robot behavior to dynamic environments and more intuitive human-like movements for co-working. In [3] the robotic arm has to adapt to changing surfaces while executing a wiping task. While in [4] a collaborative human–robot sawing is realized.

One of the most efficient methods to implement complex, more human-like robot motion, without requiring a heavy amount of data or a long learning period, is done with Learning from Demonstration (LfD) [5]. This approach is a combination of data acquisition and reinforcement learning. The trajectory learning can be done using movement primitives. Thus, data acquired from human demonstrations can be used to learn weights of the movement primitives.

In this paper we give a base for implementing co-working scenarios using low-cost methods. We realize the teleoperation of a robotic arm with hand movement using an optitrack system and implement a calibration method. This

Figure 1: Example of remotely controlling the Panda robot arm only using markers held in the users hand after realizing the hand-eye calibration method (left to right, top to bottom). The shown gripping task was done manually and is not part of this experiment.

allows us an intuitive robot control, which then can be used to generate movement trajectories of the robot arm as a base for LfD. Moreover, we demonstrate the ability to derive movement primitives from data of a low-cost sensor glove [1]. To successfully realize LfD, generated trajectories are used to learn movement primitives. In this work we focus on hand movements, which can later be used to learn a robotic gripper. As a result, our work establishes a foundation for learning from demonstrations on a robotic arm using hand movements.

2 Material and Methods

First we introduce the robotic arm and sensor glove that have been used in this study and highlight additional requirements of our setup.

Our aim is an intuitive robot control, where the robotic arm should move forward when the operator's hand moves forward. Thus, the coordinate frames of the robotic arm and the hand have to be matched. Therefore, we demonstrate in the next section a hand-eye calibration, which allows the teleoperation of the robotic arm using hand movements.

Afterwards, this setup should be used for trajectory learning. In this paper we demonstrate the learning strategy by recording only finger movements using a sensor glove to generate movement primitives.

2.1 Experimental Setup

We work with the robotic arm Panda by Franka Emika with seven degrees of freedom, which allows us to realize more complex behaviors. Additionally, Franka Emika provides a control interface *franka_ros* for easy robot control. In the first part of this study we focus on teleoperating the robotic arm. This is done by using an optitrack system and markers on the operators hand. Our setup includes four cameras. For the detection of the hand, an asymmetrical layout of the markers is required. As a result, we are able to track the position and orientation of the hand in the workspace.

In the second part of this project, we realize a trajectory model learning algorithm. For our proof-of-concept, we concentrate on learning movement primitives from finger movements. To obtain movement trajectories for each joint, we use a low-cost sensor glove [1], which we developed in a previous work. This sensor glove uses two flex sensors for each finger to sufficiently measure its joint movement. Furthermore, an internal measurement unit (IMU) on the back of the hand provides us with information about the orientation of the hand. An integrated Wifi module allows this setup to be operated remotely.

The sensor glove and the optitrack system as well as the robotic arm are operated using the Robot Operating System (ROS) [6]. ROS enables the communication between each client in the network, so that measured sensor data can be easily received by some listeners and message commands can be sent to actuators for execution.

2.2 Hand-Eye Calibration

Since our goal includes an intuitive remote control of the robotic arm employing the user's hand, the sensor data acquired from the optitrack system needs to be transformed into a matching endeffector position for the robotic arm. Considering the different coordinate frames of the robot and the user's hand, a direct transfer of the measured hand position as the desired endeffector position would yield a different result. Therefore, we implemented a hand-eye calibration method called *QR24 calibration algorithm* as proposed in [7].

In this setup, a marker is put on the robot's endeffector and the optitrack system is used to locate it. The general principle of tool/flange and robot/world calibration can be described using the following relation, where $\mathbf{M}$ denotes the transformation matrix from the robot base to its endeffector, $\mathbf{X}$ the transformation matrix from the endeffector to the marker, $\mathbf{Y}$ the transformation matrix from the robot base to the tracker and $\mathbf{N}$ the transformation matrix from the tracker to the marker.

$$\mathbf{M}\mathbf{X} = \mathbf{Y}\mathbf{N} \tag{1}$$

Here, the fixed matrices $\mathbf{X}$ and $\mathbf{Y}$ are unknown. Once these transformation matrices are found, the measured position and orientation of a users hand can be transformed into robot base coordinate frame. Therefore, n measurements of different robot poses were taken to obtain n equations. After rearranging Equation (1) and using n measurements it yields to,

$$\mathbf{M}_i\mathbf{X} - \mathbf{Y}\mathbf{N}_i = 0, \text{ with } i = 1, ..., n \tag{2}$$

Afterwards, the elements of $\mathbf{X}$ and $\mathbf{Y}$ are written as components of a vector $w \in \mathbb{R}^{24}$ and Equation (2) is combined into a system of linear equations

$$\mathbf{A}w = b, \tag{3}$$

where $\mathbf{A} \in \mathbb{R}^{12n \times 24}$ and $b \in \mathbb{R}^{12n}$. Each matrix $\mathbf{A}_i \in \mathbb{R}^{12 \times 24}$ and vector $b_i \subset \mathbb{R}^{12}$ is based on Equation (4).

$$\mathbf{A_i} = \begin{bmatrix} \mathcal{R}[\mathbf{M}_i](\mathbf{N}_i)_{1,1} & \mathcal{R}[\mathbf{M}_i](\mathbf{N}_i)_{2,1} & \mathcal{R}[\mathbf{M}_i](\mathbf{N}_i)_{3,1} & \mathbf{Z}_{3\times3} & \\ \mathcal{R}[\mathbf{M}_i](\mathbf{N}_i)_{1,2} & \mathcal{R}[\mathbf{M}_i](\mathbf{N}_i)_{2,2} & \mathcal{R}[\mathbf{M}_i](\mathbf{N}_i)_{3,2} & \mathbf{Z}_{3\times3} & \\ \mathcal{R}[\mathbf{M}_i](\mathbf{N}_i)_{1,3} & \mathcal{R}[\mathbf{M}_i](\mathbf{N}_i)_{2,3} & \mathcal{R}[\mathbf{M}_i](\mathbf{N}_i)_{3,3} & \mathbf{Z}_{3\times3} & -\mathbf{E}_{12} \\ \mathcal{R}[\mathbf{M}_i](\mathbf{N}_i)_{1,4} & \mathcal{R}[\mathbf{M}_i](\mathbf{N}_i)_{2,4} & \mathcal{R}[\mathbf{M}_i](\mathbf{N}_i)_{3,4} & \mathcal{R}[\mathbf{M}_i] & \end{bmatrix}$$

$$\text{and } b_i = \begin{bmatrix} \mathbf{Z}_{9\times1} \\ \mathcal{T}[\mathbf{M}_i] \end{bmatrix} \tag{4}$$

The elements consist of the rotational part of $\mathbf{M}_i$ denoted as $\mathcal{R}[\mathbf{M}_i \in \mathbb{R}^{3\times3}]$, the translational part of $\mathbf{M}_i$ as $\mathcal{T}[\mathbf{M}_i] \in \mathbb{R}^3$, $\mathbf{Z}_{m\times n}$ is a $m \times n$ zero matrix and $\mathbf{E}_k$ is an $k \times k$ identity matrix. The system of linear equation can then be solved as a least squares problem using QR-factorization and resulting vector w contains the elements of unknown transformation matrices $\mathbf{X}$ and $\mathbf{Y}$. For further insight on the derivation of this solution refer to [7].

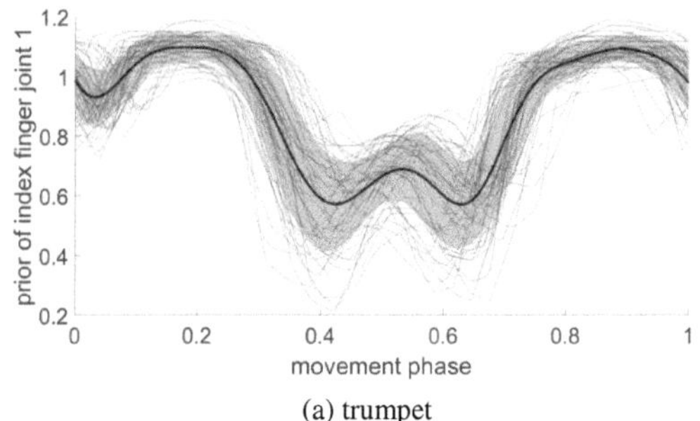

(a) trumpet

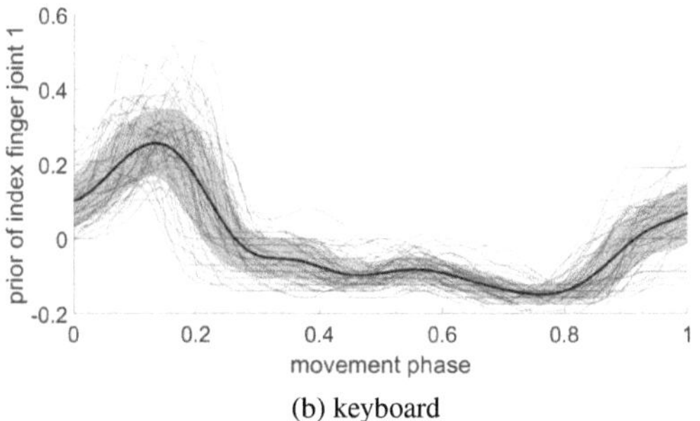

(b) keyboard

Figure 2: Exemplary movement primitives in unit time derived from trajectories of the index finger MCP joint for different music instruments: (a) trumpet and (b) keyboard. The data-set of trajectories is illustrated as the thin blue lines, while the mean of the given trajectories is shown as a solid black line. The shaded area illustrates the 1-sigma confidence level.

2.3 Trajectory Learning with Movement Primitives

This section was previously published in [1] at ICAR 2021.

We assume that measurements are given as trajectory, i.e. a sequence of multi-dimensional sensor values denoted by the matrix $\tau = [y_1, y_2, \ldots, y_T]$. Here, τ is modeled through a basis function approximation using the parameter vector w,

$$p(\tau|w) = \prod_{t=1}^{T} \mathcal{N}(y_t|\Phi_t w, \Sigma_y) = \mathcal{N}(y_{1:T}|\Phi_{1:T} w, \Sigma_y),$$
$$(5)$$

where Φ_t denote the basis function matrix defined in [8]. Σ_y denotes the measurement noise which is often specified manually, i.e. we used $\Sigma_y = 10^{-12}I$, where I is the identity matrix with convenient dimension.

To compute the movement model $p(\tau)$, the parameter vector w is integrated out, i.e.

$$p(\tau) = \int p(\tau|w)p(w)dw$$
$$= \int \mathcal{N}(y_{1:T}|\Phi_{1:T} w, \Sigma_y)\mathcal{N}(w|\mu_w, \Sigma_w)dw \quad (6)$$
$$= \mathcal{N}(y_{1:T}|\Phi_{1:T} w, \Phi_{1:T}\Sigma_w\Phi_{1:T}^T + \Sigma_y).$$

This marginalization process can be computed in closed form and results in a single Gaussian distribution. The core of the movement model is the Gaussian prior $\mathcal{N}(w|\mu_w, \Sigma_w)$, which can be computed from the sensor measurements (i.e. the training data) through maximum likelihood [9] or in the simplest case through ridge regression,

$$w^{[i]} = (\Phi_{1:T}^{\mathsf{T}}\Phi_{1:T} + \lambda I)^{-1}\Phi_{1:T}^{\mathsf{T}}\tau^{[i]}. \quad (7)$$

The regularization term λ is set to $\lambda = 10^{-6}$ in our experiments.

Note that the correlation of the multi-dimensional data is captured through the covariance matrix Σ_w. This covariance can be used to predict individual dimensions.

In the following, we summarize how to compute model similarities. In a training phase, measurements of elementary movement primitives like the playing of a guitar were collected. This training data is used, as discussed above, to compute model priors, like $\mathcal{N}(w_{\text{guitar}}|\mu_{\text{guitar}}, \Sigma_{\text{guitar}})$, $\mathcal{N}(w_{\text{sax}}|\mu_{\text{sax}}, \Sigma_{\text{sax}})$, etc. To evaluate the similarities of different movement primitives, we computed the Kullback-Leibler divergence of the two Gaussian prior distributions,

$$\mathrm{KL}(\mathcal{N}_1||\mathcal{N}_2) = \frac{1}{2}\log\frac{|\Sigma_2|}{|\Sigma_1|} - n + \mathrm{tr}(\Sigma_2^{-1}\Sigma_1)$$
$$+ (\mu_2 - \mu_1)^{\mathsf{T}}\Sigma_2^{-1}(\mu_2 - \mu_1), \quad (8)$$

where $\mathcal{N}_1$ denotes a Gaussian distribution with the n-dimensional mean μ_1 and the covariance Σ_1. The symbol tr denotes the matrix trace. This probabilistic measure in the parameter space is also compared to a deterministic measure computing Euclidean distances in the trajectory space. The evaluation results are shown in Subsection 3.2 in the experiments.

3 Results and Discussion

In this section, we briefly sum up our results regarding the teleoperation of the robotic arm and discuss our results on trajectory model learning.

3.1 Teleoperation

To execute the hand-eye calibration we manually set four poses to obtain measurements, which are then used to solve the system of linear equations mentioned in 2.2. Consequently, we were able to steer the robotic arm using the optitrack system, as shown in Figure 1. The user has markers in his hand, which are detected by the cameras. The position and orientation of the markers, are then transformed into poses in the base coordinate frame of the robotic arm using the transformation matrices obtained from hand-eye calibration. These poses are then used to control the robotic arm by setting the actual pose of the markers as the desired pose of the endeffector.

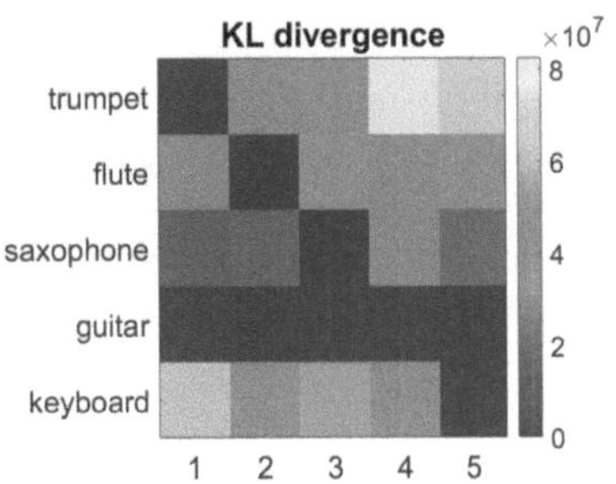

Figure 3: Kullback-Leibler divergence of model similarities between each instrument (the numbers on the x-axis correspond to the instruments from trumpet to keyboard). A low distance value denotes a high similarity and vice versa.

3.2 Movement Evaluation

This section was previously published in [1] at ICAR 2021.

To analyze the performance of our sensor glove with respect to motion model learning scenarios, we recorded the hand motion while playing music instruments: a trumpet, a flute, a saxophone, a guitar and a keyboard. For every instrument, a short melody is repeatedly performed. This results in a data-set of 60 trajectories of each joint for each instrument respectively. In our experiment, we only captured the right hand finger motion. This is sufficient for our proof-of-concept. However, for future work, the additional data from the left hand may increase the detection of differences between the hand motions while playing these instruments. We were able to generate movement primitives which allow a good representation of the given motion of each joint. Two exemplary generated trajectories are shown in Figure 2. Furthermore, we calculated the learned model similarities between each instrument (see Figure 3). It shows that the sensor glove is able to capture accurate hand motion to learn a motion model and also to show differences between them.

4 Conclusion

In this paper we developed a basis for learning from demonstration applications on a robotic arm using a sensor glove and an optitrack system. We successfully realized the hand-eye calibration method QR24, which allows the user a remote and intuitive control of the robotic arm. Furthermore, we realized a proof-of-concept for modeling trajectories generated by the sensor glove. In combination, this could be used for a complete transfer learning system from human motions to a robotic arm. For this, the finger movement obtained from the sensor glove can be applied to the robot's endeffector in a future work.

Acknowledgement

This project has been supervised by the Institute for Robotics and Cognitive Systems at the University to Luebeck and has received funding from the Deutsche Forschungsgemeinschaft (DFG, German Research Foundation) No #430054590 (TRAIN).

Author's Statement

Parts of this article were previously published at ICAR 2021 [1].

5 References

[1] R. Denz, R. Demirci, M. E. Cansev, A. Bliek, P. Beckerle, E. Rueckert, and N. Rottmann, "A high-accuracy, low-budget sensor glove for trajectory model learning," in *2021 20th International Conference on Advanced Robotics (ICAR)*. IEEE, 2021, pp. 1109–1115.

[2] I. Karabegović, "The role of industrial and service robots in the 4th industrial revolution–" industry 4.0"," *Acta Technica Corviniensis-Bulletin of Engineering*, vol. 11, no. 2, pp. 11–16, 2018.

[3] A. Kramberger, E. Shahriari, A. Gams, B. Nemec, A. Ude, and S. Haddadin, "Passivity based iterative learning of admittance-coupled dynamic movement primitives for interaction with changing environments," in *2018 IEEE/RSJ International Conference on Intelligent Robots and Systems (IROS)*. IEEE, 2018, pp. 6023–6028.

[4] L. Peternel, N. Tsagarakis, D. Caldwell, and A. Ajoudani, "Robot adaptation to human physical fatigue in human–robot co-manipulation," *Autonomous Robots*, vol. 42, no. 5, pp. 1011–1021, 2018.

[5] M. E. Cansev, H. Xue, N. Rottmann, A. Bliek, L. E. Miller, E. Rueckert, and P. Beckerle, "Interactive human–robot skill transfer: A review of learning methods and user experience," *Advanced Intelligent Systems*, p. 2000247, 2021.

[6] M. Quigley, K. Conley, B. Gerkey, J. Faust, T. Foote, J. Leibs, R. Wheeler, A. Y. Ng *et al.*, "Ros: an open-source robot operating system," in *ICRA workshop on open source software*, vol. 3, no. 3.2. Kobe, Japan, 2009, p. 5.

[7] F. Ernst, L. Richter, L. Matthäus, V. Martens, R. Bruder, A. Schlaefer, and A. Schweikard, "Non-orthogonal tool/flange and robot/world calibration," *The International Journal of Medical Robotics and Computer Assisted Surgery*, vol. 8, no. 4, pp. 407–420, 2012.

[8] A. Paraschos, C. Daniel, J. Peters, G. Neumann *et al.*, "Probabilistic movement primitives," *Advances in neural information processing systems*, 2013.

[9] E. Rueckert, J. Mundo, A. Paraschos, J. Peters, and G. Neumann, "Extracting low-dimensional control variables for movement primitives," in *2015 IEEE International Conference on Robotics and Automation (ICRA)*. IEEE, 2015, pp. 1511–1518.

Segmentation of the dermal-epidermal junction in OCT image data using deep learning

Caroline Florack [1], Timo Kepp [2], Felix Hilge [3], Michael Wang-Evers [3], Dieter Manstein [3] and Heinz Handels [2]

[1] Medical Engineering Science student, Universität zu Lübeck, caroline.florack@student.uni-luebeck.de
[2] Institute of Medical Informatics, Universität zu Lübeck, {timo.kepp, heinz.handels}@uni-luebeck.de
[3] Cutaneous Biology Research Center, Massachusetts General Hospital, Boston

Abstract

The topic of this paper is the segmentation of the dermal-epidermal junction (DEJ) in optical coherence tomography (OCT) volume data. The DEJ is a complex structure primarily responsible for epidermis to dermis attachment. It has been shown that intrinsic and extrinsic aging are associated with important DEJ structural modifications. The goal is to design a deep learning-based segmentation that automatically segments the DEJ consistently for each 2D image for evaluation of aging processes. This paper shows that the use of a neural network is useful for segmenting the DEJ.

1 Introduction

Due to the aging society, a better understanding of the cell aging process is necessary. In this work, the focus is placed on the skin, as it is the largest organ in the human body and the first barrier between body and environment [1]. The organ consists of three layers: dermis, epidermis and subcutis [2].

The top two layers are responsible for protection from the environment and their transition is described as dermal-epidermal junction (DEJ) [3]. The dermal epidermal junction is a complex structure primarily responsible for epidermis to dermis attachment. The DEJ thus warrants cohesion and mechanical resistance of the skin [1]. Despite the current broad knowledge about the skin, the function and the topology of the DEJ is not yet sufficiently studied and understood.

It has been shown that intrinsic and extrinsic aging are associated with important DEJ structural modifications including flattening, duplication and thinning [4]. Therefore it could be important and useful to analyse these intrinsic and extrinsic effects on the dermal-epidermal junction, which requires a three-dimensional imaging technique like optical coherence tomography (OCT).

OCT is a non-invasive imaging technique based on interference of light. It uses low coherence light to acquire two or three dimensional images of the optical scattering medium. The technique has a resolution in the range of micrometers. Among other things, it allows the measurement of living organic tissue non invasievly. Due to the absorbing properties of the tissue OCT is used in medicine mainly in the field of ophthalmology and dermatology [5].

Optical coherence tomography enables visual detection of DEJ in vivo. But due to a low contrast between dermis and epidermis, a skin boundary detection method is required for DEJ segmentation [2]. Yuwei Liu et al. [6] already show that tissue segmentation in OT images using NN is useful. Therefore, we want to find out whether it is also suitable for DEJ segmentation.

In this paper the U-net architecture [7] is used to generate a segmentation of the dermal-epidermal junction in OCT B-scan data. The goal is to design a Deep learning (DL)-based segmentation that automatically segments the DEJ consistently for each 2D image.

2 Material and Methods

For the ssegmentation of the DDEJ in OCT data a U-Net is used [7].

The following section first describes the problem definition, which includes the tasks and difficulties of segmentation. Next, the data that have been provided are described. A short overview of the preprocessing of the data is given in section 2.3. Afterwards the used network, a U-net, is described.

2.1 Problem definition

The problem discussed in this paper is shown in Fig. 1. At the top the OCT B-scan gray scale image is shown and at the bottom the gray scale image is pictured, which has been overlaid with a hand segmented ground truth. The aim is to find the top edge of the junction. The segmentation of the dermal epidermal junction is an classification problem with two classes. The first class is the background and the second class is below the DEJ, so the edge of the two classes maps the topology of the DEJ. The task is to generate a segmentation of the dermal-epidermal junction in OCT B-scan data with an U-net architecture.

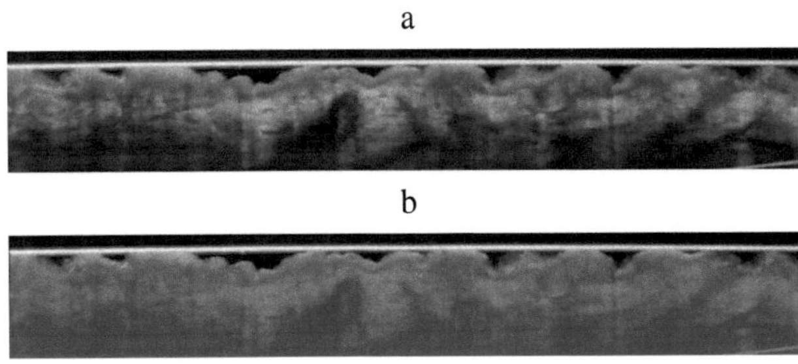

Figure 1: This Figure shows the problem that is discussed in this paper. The dermal-epidermal junction is to be segmented. At the top (a) the OCT B-Scan gray scale image, which should be segmented, is shown and at the bottom (b) is the gray scale image, which has been overlaid with a hand segmented ground truth (here red).

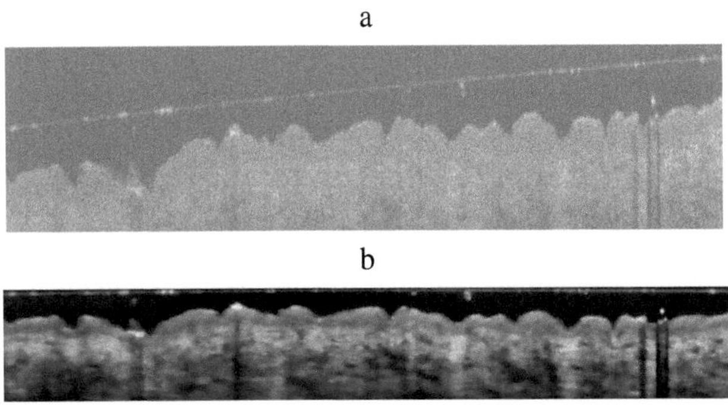

Figure 2: This Figure shows (a) the Original OCT B-scan image of the patients before preprocessing and (b) the OCT B-scan after preprocessing.

2.2 Dataset

The dataset which is used is provided by the Cutaneous Biology Research Center, Mass General Research Institute. The data are acquired with a Thorlabs Telesto III. The scanner used is the OCTG13 and the lens is the OCT-LK3 with a lateral resolution of 13 µm and a pixel spacing of 6.5 µm. The field of view has a size of $6x6$ mm. The dataset contains recordings of skin layers of the forearms of 75 patients. Ten images have been taken per patient resulting in a dataset which consists of 750 images.

The ground truth is manually segmented by experts using ITK-snap. Since this tool does not support contour segmentation [8], the segmentation is started at the upper edge of the DEJ and applied to the rest of the image. In postprocessing, further steps are taken to extract the upper edge from the segmented data.

The data used in this paper are OCT gray scale images of size $923x256$.

2.3 Preprocessing

First the original images of the patients, shown in Fig. 2, are flattened and cropped. For this purpose, the RANSAC algorithm [9] is used to find the vertical high intensity line in the

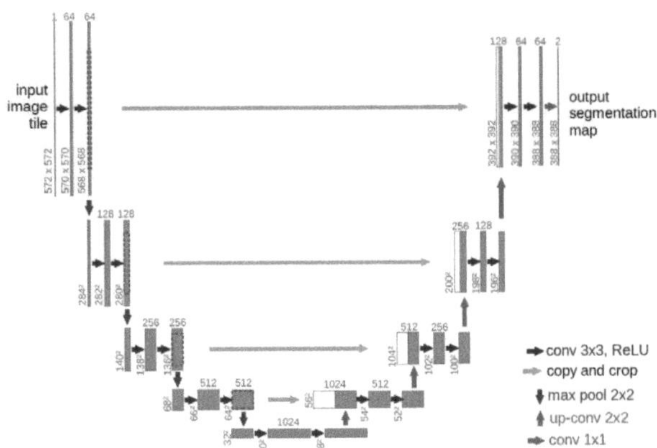

Figure 3: Figure taken from „U-net by O. Ronneberger et al.". Used U-net architecture [7]. Each blue box corresponds to a multi-channel feature map. The number of channels is denoted on top of the box. The $x - y$-size is provided at the lower left edge of the box. White boxes represent copied feature maps. The arrows denote the different operations.

images for each column. And then the image is transported so that the line is on a row. After that total variation denoising is applied with a weight of $w = 0.05$ to the transposed image to reduce noise whilst simultaneously preserving the edges. Furthermore histogram equalization is used to adjust the contrast of the images.

Subsequently, the images are divided patient-wise into train and test data, with a distribution of 80 to 20 percent.

2.4 Network architecture

The network used is based on the network described in the paper by Ronneberger et al. [7]. The design of the U-net is shown in Fig. 3. It consists of three contracting encoding steps, three expanding decoding steps and the bottleneck. The input and output layer of the U-net have a dimension of $256x512$, where the output is a pixel-by-pixel labeled image.

In the extracting path multiscale features are extracted from the input image. The larger the receptive field or the more stages this path has, the more complex are the features that can be learned by the network. Each encoding step consists of two convolutional layers with a kernel size of $3x3$, an activation function Rectified Linear Unit (ReLU) after the convolutional step and a maxpooling layer with a kernel size of $2x2$. The skip-connections and featured maps are the output of each encoding step.

The bottleneck consists of two consecutive convolutional layers with the activation function at the end.

The expanding path combines the location information of the skipped connections with the learned features of the feature maps and creates a pixel-by-pixel labeled image. The decoding step consists on two convolutional layers, also with a kernel size of $3x3$ and an upconvolutional layer, with a kernel size of $2x2$. The skip-connections and the featured maps are the input data for each decoding step. The expanding decoding step generates spatially resolved prediction of individual pixels for segmentation.

Also, batch normalization is applied after each convolutional layer.

2.5 Training

During training data augmentation methods were applied on the fly to the training data set. This method results in a larger amount of data and variance in the data. The augmentation methods used are: horizontal flip, non linear intensity shift, scaling by 15 % and vertical shift. Other augmentation techniques as vertical shift will be evaluated in future work. In addition, random patches of size $256x512$ are cropped and only on these the segmentation for the training was performed.

The generated Dice Loss [10]

$$GDL = 1 - 2\frac{\sum_{l=1}^{2} w_l \sum_n r_{ln}p_{ln} + 1}{\sum_{l=1}^{2} w_l \sum_n r_{ln} + p_{ln} + 1} \qquad (1)$$

is widely used loss function in the computer vision community and is selected here as the loss function. The generated Dice Loss is well suited for images with a lot of background, as it looks at the intersection of the Prediction p with the ground truth r [10]. w_l is used to provide invariance to different label set properties. To ensure that the function is not undefined a one is added in numerator and denominator.

For the initial test runs, a starting learning rate of 0.001, a batch size of seven, and an epoch count of 300 are chosen. The learning rate is adjusted after each epoch and ADAM-Optimizer is selected as optimizer [11].

3 Results and Discussion

In order to evaluate the segmentation first a visual evaluation has been done. Fig. 5 shows an example segmentation after the first test runs of the training. This segmentation resembles the ground truth. However, some small errors are still visible. One of these errors are visible gaps under the upper edge of the segmentation as can be seen in Fig. 5 (b) on closer examination of the left-hand side. Furthermore only one edge should be visible in each column of the segmentation for a functioning image post-processing. However this is not the case for every segmentation.

As can be seen in Fig. 5 at the top, there are some interfering properties, such as low intensity horizontal lines and high intensity vertical lines. These artifacts could be randomly distributed in the sense of a data augmentation method. In this way the network could learn to deal better with these artifacts.

In Fig. 4 the generalized Dice Loss is shown as a function of the number of epochs. At the beginning, the graph drops exponentially during both training and testing. Towards the end, the training values converges to 0.001 and the testing values at 0.0015.

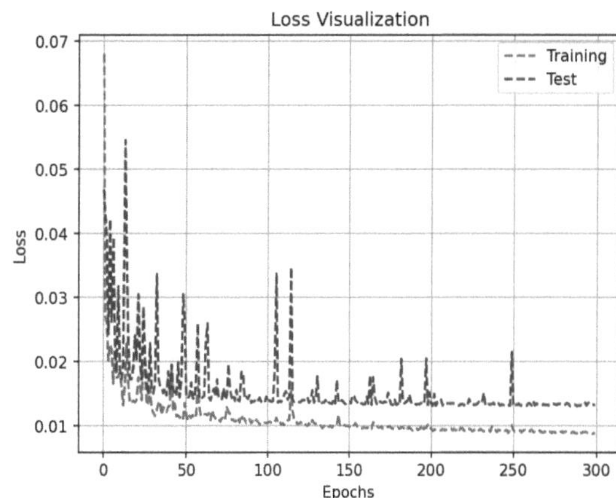

Figure 4: Visualization of the training and test loss of the training procedure

4 Conclusion

The network used is basically suitable for the segmentation of OCT grayscale images. Even though adaptations, such as further data enhancement methods, are still needed to improve the segmentation. The goal of segmenting the topology of the DEJ in 2D has not yet been achieved. First of all, the segmentation in 2D has to be improved.

In addition, post-processing is necessary and must be implemented. For this purpose, the top edge of the segmentation is extracted to obtain a line representing the topology of the DEJ after post-processing.

However, the results obtained suggest that the method is suitable for segmenting the DEJ.

Acknowledgement

I would like to take this opportunity to thank the stuff of the Cutaneous Biology Research Center, who provided the data. The work has been carried out and supervised by the Institute of Medical Informatics, Universität zu Lübeck.

Author's Statement

Conflict of interest: Authors state no conflict of interest.

5 References

[1] W. C. J. Briggaman RA, "The epidermal-dermal junction," *J Invest Dermatol*, vol. 65, pp. 71–84, 07 1975.

[2] A. Taghavikhalilbad, S. Adabi, A. Clayton, H. Soltanizadeh, D. Mehregan, and M. R. Avanaki, "Semi-automated localization of dermal epidermal junction in optical coherence tomography images of skin," *Applied optics*, vol. 56, no. 11, pp. 3116–3121, 2017.

a

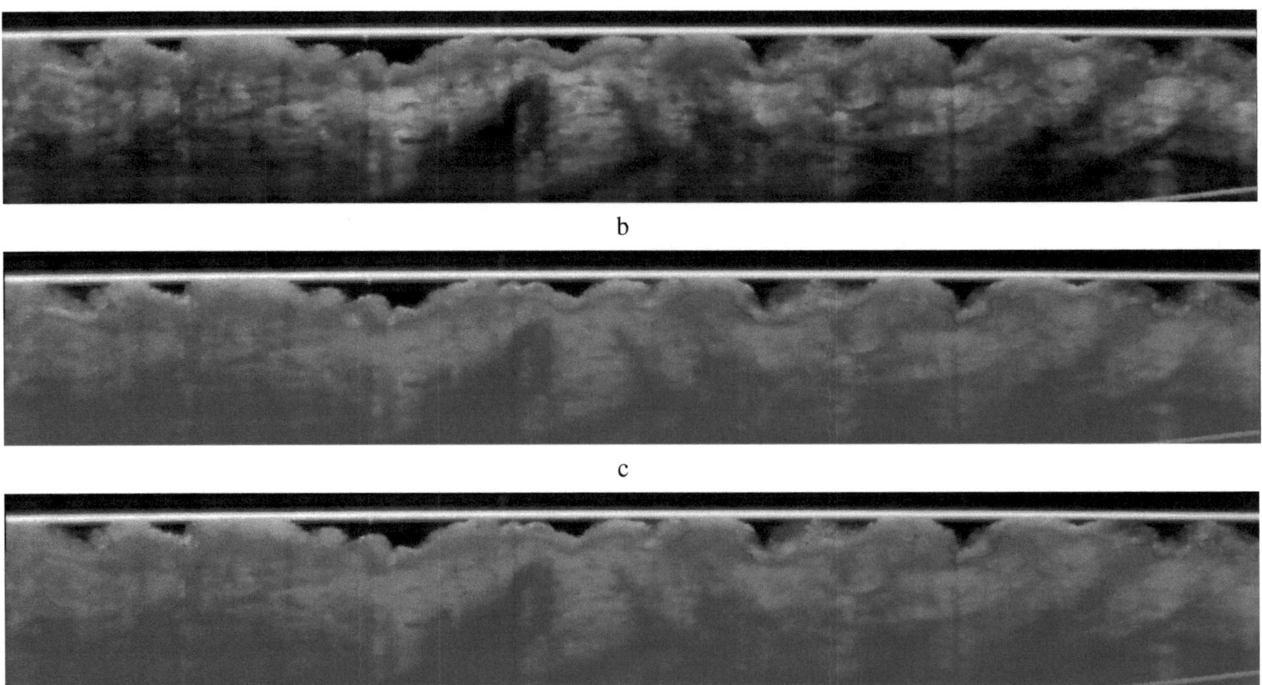

b

c

Figure 5: Segmentation result for the dermal-epidermal junction after the first test runs of the training. At the top (a) the OCT B-Scan gray scale image, which should be segmented, is shown and in the middle (b) is the gray scale image, which has been overlaid with the generated prediction (here red). The ground truth overlaid with the gray scale Image is shown in subfigure (c).

[3] M. Aleemardani, M. Z. Z. Trikić, N. H. H. Green, and F. Claeyssens, "The importance of mimicking dermal-epidermal junction for skin tissue engineering: A review," *Bioengineering*, vol. 8, no. 11, p. 148, 2021.

[4] S. G. Lagarrigue, J. George, E. Questel, C. Lauze, N. Meyer, J.-M. Lagarde, M. Simon, A.-M. Schmitt, G. Serre, and C. Paul, "In vivo quantification of epidermis pigmentation and dermis papilla density with reflectance confocal microscopy: variations with age and skin phototype," *Experimental dermatology*, vol. 21, no. 4, pp. 281–286, 2012.

[5] J. G. Fujimoto, C. Pitris, S. A. Boppart, and M. E. Brezinski, "Optical coherence tomography: an emerging technology for biomedical imaging and optical biopsy," *Neoplasia*, vol. 2, no. 1-2, pp. 9–25, 2000.

[6] Y. Liu, R. Adamson, M. Galan, B. Hubbi, and X. Liu, "Quantitative characterization of human breast tissue based on deep learning segmentation of 3d optical coherence tomography images," *Biomedical Optics Express*, vol. 12, no. 5, pp. 2647–2660, 2021.

[7] O. Ronneberger, P. Fischer, and T. Brox, "U-net: Convolutional networks for biomedical image segmentation," 2015.

[8] P. A. Yushkevich, Y. Gao, and G. Gerig, "Itk-snap: An interactive tool for semi-automatic segmentation of multi-modality biomedical images," in *2016 38th Annual International Conference of the IEEE Engineering in Medicine and Biology Society (EMBC)*. IEEE, 2016, pp. 3342–3345.

[9] M. A. Fischler and R. C. Bolles, "Random sample consensus: a paradigm for model fitting with applications to image analysis and automated cartography," *Communications of the ACM*, vol. 24, no. 6, pp. 381–395, 1981.

[10] S. Jadon, "A survey of loss functions for semantic segmentation," in *2020 IEEE Conference on Computational Intelligence in Bioinformatics and Computational Biology (CIBCB)*. IEEE, 2020, pp. 1–7.

[11] D. P. Kingma and J. Ba, "Adam: A method for stochastic optimization," *arXiv preprint arXiv:1412.6980*, 2014.

Detecting empty Pallet Spaces with Sensor Fusion in a Warehouse on the Inventairy XL Drone System

Janik Klingert [1] and Mike Becker [2]

[1] Robotics and Autonomous Systems, Universität zu Lübeck, janik.klingert@student.uni-luebeck.de
[2] doks.innovation GmbH, mike.becker@doks-innovation.com

Abstract

For an automated stocktaking process, it is interesting to not only find existing goods, but also empty spaces in a warehouse. This helps warehouse managers to keep an overview over their storage capabilities. The doks.inventairy system scans pallet spaces in a warehouse with a combination of an Autonomous Guided Vehicle (AGV) and an attached drone. Until recently, it was only possible to detect full pallet spaces with barcode scanners attached to the drone. In this paper, a method of sensor fusion is evaluated to detect empty pallet spaces. The system uses an ultrasound sensor, barcode scanners and a neural network to process drone images. It was implemented in the Robot Operating System (ROS) in C++. Overall, the methods performed well on the given task, but live neural network predictions were shown to be processing intensive and had to be queued and processed in a delayed manner.

1   Introduction

At doks.innovation GmbH (doks), Kassel, Germany, a robotic system is developed to automate stocktakes in large pallet warehouses. Traditionally, this was a dangerous task, involving manual labor in a cage on a forklift high up in the racks to scan the barcodes of the stored pallets. The doks.inventairy XL system consists of two main components: an AGV and a drone which is located on top of the AGV and connected via a cable on a winch supplying powers. Fig 1 shows the system in operation.

The workflow of an automated stocktake consists of multiple steps: first, the warehouse manager starts the stocktake for a certain amount of racks via the web interface, the AGV starts driving from its charging station at the warehouse to the specified rack. Satelite based navigation, such as GPS is not feasible for indoor navigation due to the design of warehouses and weak signal strength. A traditional Simultaneous Localization and Mapping (SLAM) algorithm based on a LIDAR scanner [1] is used on the AGV to navigate. When the AGV arrives at a specified starting point in front of the rack, it aligns itself to the rack and starts the stocktake: the drone begins to fly up at the center of the first pallets. When the top is reached, the AGV and the drone simultaneously drive to the next pallet position and the drone flies down. At the bottom, the vertical zigzag continues until the end of the rack is reached. Then, the drone lands on top of the AGV and it drives back to the charging station or to the next rack. The system can operate on its own for about five hours until the battery runs out. The long flight time of the drone is only possible because it flies tethered to the AGV through its power cable. The power cable is rolled up on a winch

Figure 1: Doks.inventairy XL system: AGV and flying tethered drone above. Picture: Doks.innovation GmbH.

in the AGV. It feeds out just enough cable for the drone to reach its assigned altitude. During the stocktake, the AGV tracks the position of the drone from below using a camera. It assigns incoming barcode scans from the drone to the position at which the drone is currently at. The drone is equipped with sensors in the front: an ultrasound sensor, three barcode scanners, and a high resolution camera with a LED to capture images of the pallets. Up to now, empty pallet spaces are not detected. In this paper, a method of sensor fusion is evaluated to detect empty pallet spaces with an ultrasound sensor, multiple barcode scanners and pallet detection in images with a neural network.

2 Material and Methods

As stated above, the drone has multiple sensors to observe its surroundings. The barcode detection is relatively simple, as readily available barcode scanners work well if the pallet labels are readable, reasonably sized and within the barcode sensors range. Multiple barcode scanners are deployed to simultaneously detect small barcodes across the whole width of a pallet. For an empty pallet space, a barcode scanner alone is not sufficient, since a missing label does not imply an empty pallet space. However, it can be a useful input, as described further below.

A more relevant sensor for this task is the attached ultrasound sensor. When polled, it returns the distance of the next object based on the runtime of an emitted ultrasound beam and the detection of the reflected sound with a microphone. Since sound waves do not travel in perfect lines, the sensor range of an ultrasound sensor is cone shaped rather than a point measurement. Its perception area varies with the distance of the object to the drone. As an alternative, a laser distance measure could be used, but ultrasound sensors tend to be lighter and more cost effective [2].

Thirdly, the drone's camera is used in this task. The images of the pallet spaces are captured during the flight approximately every 3-4 seconds and transferred to the AGV in realtime. To detect, whether an image shows a pallet or an empty space, the pallet itself is chosen as an indicator. A dataset is created and the pallets are labeled with bounding boxes. A pretrained PyTorch [3] ResNet-50 [4] is trained on the dataset. The resulting bounding box predictions can be used to detect a pallet, or to detect a missing pallet - an empty space. Due to the construction of the large warehouse racks as shown in Fig. 1, it can not be expected that the image contains only a single pallet or no pallet. To filter out the background and neighbouring pallets, a predicted bounding box is discarded if it is not in the center of the image or too small. Crucial for this approach is, that the drone always flies in the vertical center of the pallets.

The robotic foundation for the doks.inventiary XL system is built in ROS nodes, which communicate with each other via ROS-topics. Interactions between nodes are done with service calls to provided service functions. Here, the communication speed has to be high, to not start a chain reaction of nodes waiting for service call results due to blocked calls. To evaluate the data of the sensors for the pallet detection, three ROS nodes are implemented in C++. The ultrasound sensor and the barcode scanner are evaluated at the same time in a single node, since this is an online task and cannot be delayed. The image detection is done in a separate node due to the heavy processing required. A third node is used to trigger the data collection of the different sensors at the same time, fuse the input together, queue the processing and publish a result when the evaluation is done.

During an inventory stocktake flight, at the start of each pallet space the fuse node gets triggered by the system, passing on a pallet ID to identify the result. The flowchart shows the processing of a single pallet space (Fig. 2). The fuse node triggers the start of the ultrasound and barcode measurement, and starts to store the incoming images from the drone. The ultrasound node measures the distance at a sampling rate of 20 Hz [5]. The ultrasound sensor itself operates at a device specific maximum rate of 10 Hz. Since the

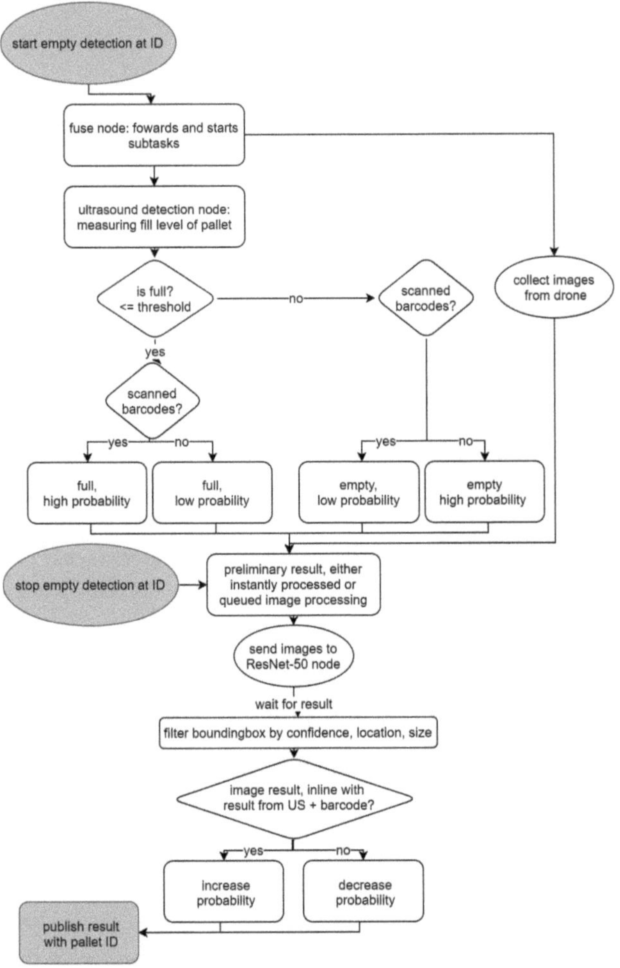

Figure 2: Flowchart of nodes processing and measuring one pallet position

planned distance between drone and rack is known, measured ultrasound distances above a certain threshold count as empty data points, distances below the threshold count as full pallet space data point. A hysteresis is applied to even out the noisy ultrasound measurements. The measurement ends with a service call to the fuse node, which it forwards to the ultrasound node. In the ultrasound node, when a certain user defined percentage threshold of empty vs. full data points is reached, the pallet space is counted as empty or full according to the ultrasound sensor. Additionally, an uncertainty measure is influenced by the barcode scanners: when a barcode is detected during the measurement but the space is empty the uncertainty measure is increased, otherwise decreased. This data is returned and stored together with the recorded images in the fuse node.

Two methods of data fusion analysis are implemented. Their main difference is the timing of data processing which is described in the following section.

In the first method, the fuse node instantly triggers the image object detection neural network on a single image of the pallet space. Depending on whether the drone is flying up or down, the first or the last image in the series is selected. As soon as the image object detection is done, the final fused result is published to a ROS topic by the fuse node. Again, when the image object detection contradicts the expected outcome of the ultrasound sensor, the uncertainty measure is increased whilst otherwise it is decreased.

In the second method, data gathered by the fuse node is not processed live during a stocktake but saved in a queue. The queue consists of a full set of data for each pallet space: the position ID, the ultrasound measurement result, barcode likelihood, images captured during the stocktake of that position, a boolean indicating the down or upwards flight. At the end of a stocktake flight, the system triggers the processing of the queue while the drone has landed and the AGV is either driving or charging. The fuse result is generated by the same logic as in the instant processing method.

As a result, the system gives a boolean and a uncertainty measure between 0 and 1, to show the user where the pallet spaces are empty. The uncertainty measure serves the user as an error indication which is also shown in the user interface.

3   Results and Discussion

The ultrasound sensor is capable of measuring in the sub-cm range. Therefore, the depth information from this sensor is accurate enough to detect if an object was at the expected distance (indicating a full pallet space) of usually around one meter or if the ultrasound was hitting the backside of the next pallet or a wall (empty space) at around 2.5 meters. The measurement cone's surface area is greater than expected. Every measurement needs to be started slightly delayed, because otherwise the horizontal beam of the rack would still be taken into account of the measurement. The stopping of the measurement also leads to a similar effect and has to be done with a little distance to the next beam. This leads to complications with narrow pallet spaces, as only an even shorter vertical distance can be used for the ultrasound measurement. The exact measurement of the drones current flying height turns out to be necessary. A falsely measured and programmed height of the horizontal rack beams would disturb the ultrasound measurement and leads to errors. During development of this method, the team at doks found a bug in the drone tracking software component which lead to an erroneous offset of 5cm per 1m of altitude of the drone compared to its actual height over ground. After the bug was fixed, the ultrasound measurements reflect the fill height of a pallet.

The barcode scanners give a good additional indication for the presence of a pallet, but sometimes also scan the neighbouring pallet space or additional barcodes. A more precise result is achieved if the barcode type of the pallet is specified and filtered. Additionally, it might be useful to filter out barcodes that are not coming from the centered barcode scanner or at least decrease their influence on the probability.

The processing response time of the fuse node is in the 20-30ms range as all the heavy processing is done in different nodes or background threads. This is an acceptable response time. It is responsive enough that the drone can simply continue and run the measurements along the way.

The bounding box prediction of the neural network node is accurate as shown in Table 1 in the ResNet-50 variant and at a resolution of 960x720 pixels after 10 epochs. On

Table 1:   Results from ResNet-50 evaluation after 10 epochs. AP = Average Precision, AR = Average Recall, max Dets = maximum number of detections

AP/AR	IoU	Area	max Dets	Result
AP	0.50:0.95	all	100	0.582
AR	0.50:0.95	all	10	0.654

lower resolutions, the detection rate was reduced. The results show that the average precision of 0.58 on the Intersections over Union (IoU) is good with good recall. Meaning: if a pallet is detected the probability of it being correct is high, but some might not be detected correctly.

The instant image detection method without a queue is not successful. A simulated image prediction on the AGV shows processing times of around 1-2s at a 100% usage of up to 8/16 CPU cores if only a bare-bone linux operating system is running. This changes as soon as the other ROS systems required for the drone flight are running. In idle state, the CPU is occupied by ROS at around 60% and in flight around 75%, which does not leave enough capacity for fast processing of the images: A simulated image prediction in this state results in 6-8s at 100% CPU usage of all 16 cores for a single image of the ResNet-50 model. Lower resolution images do not change the processing times considerably whilst reducing the bounding box prediction to an intolerable state. At standard flight speed, the drone passes a pallet space in 2-4 seconds, depending on the height of the pallet space. The image processing of the instant image object detection is not able to keep up with the processing of the fuse node. Before implementation, it is estimated that the transition time between two pallet stacks or two vertical flights might be enough to make up for the timing of the processing. This turns out to be false. The instant processing overloads the CPU in combination to the other ROS nodes, to a state where the drone control gets instable. The queued processing yielded more promising results, as the system uses a second independent computer in the AGV for navigating within the warehouse. Since the drone is not in a flying state after the stocktake or on the charging pad, it is safe to use the remaining CPU power for image processing. In a real world test, 108 out of 238 drone images are correctly detected as containing a pallet. Upon manual verification, 94 images have a traditional wooden pallet on them (see Fig. 3) and 48 show a warehouse specific custom metal pallet (see Fig. 4) which are not in the training data set. As a result, only some of these specific pallets are detected correctly at the high confidence greater than 0.95. Pallets bounding box filtering is tested successfully but might need

Figure 3: Sample of detected pallet, green bounding box

Figure 4: Sample of a detected but rejected pallet due to bounding box size and centering constraints. Left: an undetected special pallet.

further tuning to filter out pallets in the vertically next or previous spaces (also shown in Fig. 4). In a newer robot firmware, the image frequency was increased from 0.25 Hz to 1-3 Hz, which could increase the accuracy of the system in the future. Processing time for a sample of 57 positions is around 7 minutes, but will be further reduced in future revisions of the system. The reason for sensor fusion in this setting is to fill the reality gaps and limitations of the different sensors with other sensory inputs. A barcode scanner alone would not have sufficed as stated above. The ultrasound sensor might lead to a rough estimation of the pallet space. The neural network image detection however would overload the system if it was solely used, as probably more than one image should be analyzed per pallet space in this setting. The fusion of these different sensors allows for a fast and responsive system, which improves its results once the image processing is done while still not overloading the CPU. At the same time, contradictory sensor values help to create an uncertainty measure. This can be used as an indication which pallet spaces might have been a measurement error thus suggesting the necessity of a manual check of the pallet space. This can either be done manually, or by verifying the position images captured by the drone. Further optimizations on the CPU load may be achievable with oneshot neural networks, and/or by using the ultrasound and barcode measurements contradictions to only analyze images of questionable sensor values.

4 Conclusion

The results show, that a combination of sensors as used in a sensor fusion system is advantageous in certain scenarios. An empty or mostly empty pallet space might not be solely detected by the ultrasound sensor as an occupied space, but by using additional information such as readings from barcode scanners and pallet detections from a neural network, empty pallet spaces can be detected reliably. Additionaly, sensor fusion allows to create a measure of uncertainty adapted to the tasks needs. In the cases shown above, this information helps to identify questionable results and enables the user to only manually check the questionable pallets. To enable the inventairy XL system to perform heavier image processing it might be interesting to utilize an onboard GPU of the computer or offload heavy processing to a battery independent processing unit at the charging station or in the cloud. A next step could be to feedback the user's manually verified result into the system and refine the uncertainty measure or improve the image detection. It could be of interest to omit the manual balancing act in the fuse node and feed all the sensors datastreams into a neural network to let the system learn this behavior on a predetermined dataset and then autonomously improve in an online manner on the user's manual checks.

Acknowledgement

The work has been carried out at doks.innovation GmbH, Kassel and supervised by the Institute of Robotics and Cognitive Systems, Universität zu Lübeck.

Conflict of Interest

Authors are employees at doks.innovation GmbH.

5 References

[1] W. Hess, D. Kohler, H. Rapp, and D. Andor, "Real-time loop closure in 2D LIDAR SLAM," in *2016 IEEE International Conference on Robotics and Automation (ICRA)*. IEEE, pp. 1271–1278. [Online]. Available: http://ieeexplore.ieee.org/document/7487258/

[2] B. Szlachetko and M. Lower, "A Surrounding World Knowledge Acquiring by Using a Low-cost Ultrasound Sensors," vol. 35, pp. 93–100. [Online]. Available: https://linkinghub.elsevier.com/retrieve/pii/S1877050914010539

[3] A. Paszke, S. Gross, S. Chintala, G. Chanan, E. Yang, Z. DeVito, Z. Lin, A. Desmaison, L. Antiga, and A. Lerer, "Automatic differentiation in PyTorch." [Online]. Available: https://openreview.net/forum?id=BJJsrmfCZ

[4] K. He, X. Zhang, S. Ren, and J. Sun. Deep Residual Learning for Image Recognition. [Online]. Available: http://arxiv.org/abs/1512.03385

[5] C. E. Shannon, "A Mathematical Theory of Communication," vol. 27, p. 55.

Machine learning powered design framework for healthcare applications

Shaarang Mishra[1], Ashwani Harkara[2]
1 Robotics and Autonomous Systems, Universität zu Lübeck,shaarang.mishra@student.uni-luebeck.de
2 Volmo Ltd, aharkara@volmopl.com

Abstract

Nowadays, image segmentation is one of the most used methods in computational medical image analysis. These semantic segmentations should be accurate and precise on Computer Tomography and Magnetic Resonance Imaging volume images which helps in clinical decision making for patient treatment. Every deep learning architecture performs differently based on the shape of the organ or the type of the image, and many other deep learning methods are performing better than other state of the art architecture. In this paper, we review hip joint image segmentation using U-Net architecture and how to improve the performance by altering the hyperparameters.

1 Introduction

Recently, in the field of medical imaging there is a lot of research going on and is becoming increasingly important. The increase of studies in segmentation has also helped to extend different segmentation architectures. Image segmentation is the specific image into a plurality of areas with unique properties and make the object of interest and technical process.[1] The idea of image segmentation is to categorise 2D and 3D medical images into various regions and then focus on the regions of interest as illustrated in Fig.1. Through this information we can analyse special tissues and can also diagnose. Apart from that other applications are location of diseased tissue, treatment planning and computer guided surgery. Many studies showed[3] that there are still many issues in medical image segmentation that need a solution as medical images are more complicated than natural images. Another point of concern is that natural images are much easier to obtain as compared to medical images and it is very difficult to obtain large scale medical data to train our model. Secondly, natural images have good contrast and resolution and moreover it is easy to identify the visual features. However, medical images have usually signal strength value and mostly have low signal to noise ratio.

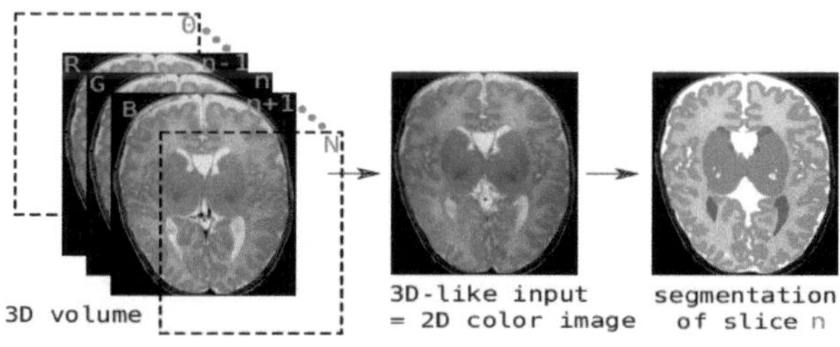

Figure 1: Segmentation of a 3D medical image performed on a set of 2D color image [2]

Furthermore, medical image segmentation can be distinguished into three methods, manual segmentation, semi automatic segmentation and fully automatic segmentation.In manual segmentation an experienced person has to draw the boundaries of tissues or bones on original image on a computer through an image editor. This method shows good accuracy but on the down side it demands time and it is laborious. The semi automatic segmentation is the combination of data processing power and analysis through algorithm to complete image segmentation via human computer interaction. In automatic segmentation, as the name suggests, the segmentation is carried out completely automatic without any help of manual hand of a human.

After the introduction, this paper is structured as follows; in section 2 we will discuss the development of state of the art architectures in deep learning and the segmentation of deep learning in medical images.Then, in section 3 we will take a look into the methods of the experiment. Whereas, in section 4 will discuss results and what changes were made to get the better segmentation followed by discussion, challenges and conclusion.

2 Related Work

2.1 Deep Learning

Deep learning evolved on artificial neural networks and in initial period implementation was made on model perceptrons, where a single neuron can be trained. But a single layer perceptron as shown in Fig.2 was only useful for linear problems. Then later an backpropagation algorithm came and it solved the non linear classification problems. But unaccessibilty of computer hardware restricted the development of the algorithms. Then the concept of deep learning was introduced and that got a huge attention as it was able to solve the issue of vanishing gradients and also enhanced running speed of GPUs. Now it is being used in for many applications like natural language processing, face recognition, speech recognition.

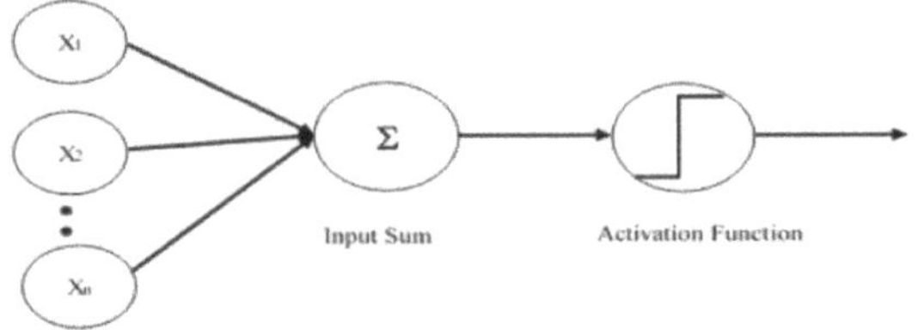

Figure 2: Single layer perceptron [4]

2.2 Medical Segmentation

Usually, for training a model a large number of dataset is required. But through deep learning training a model on a small dataset is possible, which was a problem as research and private organizations usually do not give access to huge amount of data. So a CNN model can be used for medical image semantic segmentation problem. And due to the advancements in deep learning it is possible to process the whole image at a time with the help of fully convolutional network.

2.3 Segmentation Based on U-net

U-Net is one of the architectures which is used for medical image segmentation and that can segment images using less amount of data. These qualities have made U-Net the primary tool for segmenting medical images. Now usually every medical image type like MRI, CT scans and X-rays can be used in U-Net architecture. Still its potential is increasing which has lead to more development in its architecture.

3 Methods

3.1 U-Net Module

Basically, the U-Net architecture consist of two sections. The First section is known as "contracting path" which uses a CNN architecture. In Fig.3 every block in this section consist of two successive 3x3 convolutions and after these convolutions there is a ReLU activation function and a layer of max pool.
The other section is known as "expansive path" in which at every stage the feature map will be upsampled by using 2x2 up convolution. After this the feature map in the corresponding layer in the contracting path will be cropped and also concatenated into the upsampled feature map. This will be followed by two successive 3x3 convolutions and a ReLU activation unit. At last one extra 1x1 convolution will be applied so that it can reduce the feature map to the required number of channels and get the segmented image. In this architecture cropping is essential as pixel features

in the edges have the less amount of information and thus need to be excluded. So this will resemble a U shape which propagates the information alongside the model which allows it to segment objects in an area using context from a larger overlapping area.

The energy function of the model is described by the following equation:

$$E = \sum w(x) log(p_{k(x)}(x)) \tag{1}$$

where p_k is a pixel wise SoftMax function.

$$p_k(x) = \frac{e^{a_k(x)}}{\sum_{k'=1}^{k} e^{a_k(x)'}} \tag{2}$$

Where $a_k(x)$ is the activation in channel k.

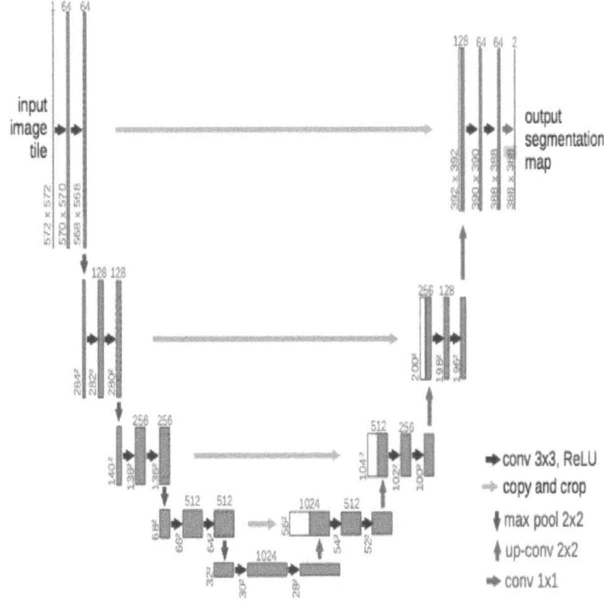

Figure 3: U-Net architecture. The arrows defines the various operations, the blocks on left side represents feature maps and the unshaded blocks on right side represent the cropped feature maps from the contracting path [5]

3.2 3D U-Net

It is a variant of U-Net in which the inputs are 3D images. The structure is same as discussed above which also contains 2 sections "contracting and expansive path". But there is a slight change all the 2D operations are replaced with 3D operations, 3D convolutions, 3D max pooling and 3D up convolutions which will result in 3D segmented image. 3D U-Net segments images with less labeled data. That is due to the reason that 3D images have plenty of repeating structures and shapes which lead to a faster training process. Its application has extensive use in volumetric CT and MR image segmentation which includes bones, tumors, multi organ segmentation.

3.3 Training

The input and the segmentation maps are used to train the network with the stochastic descent implementation. The output image is smaller then input image by a constant width due to the unpadded convolutions. So to minimize the overhead we have to make maximum use of the GPU and will consider large input tiles rather than large batch size and therefore reduce the batch to a single image. The energy function can be computed by a pixel wise soft max for the output feature map and also with the cross entropy loss function.

3.4 Data Augmentation

Data augmentation is important to train, if there are not much training data. If training samples are microscopical images then we need shift and rotation variance to gray value variations and deformations[8]. Especially when we have very little annotated images than random elastic deformations of the training samples can work to train a segmentation model.

3.5 Image modalities

3.5.1 Magnetic resonance imaging(MRI)

MRI is a well known radiology imaging technique which is used to capture pictures of tissues inside the human body. Especially for U-Net, MRI is the most popular image modality for segmentation. It is also an useful tool for diagnosis of various tumors such as white matter tissue in brain and fetal brain development.

Cancer is one of the major cause of deaths all around the world and there has been an U-Net implementation applied on cardiovascular MR images [6] to segment structures of the human heart and MR is one of the most trustable methods for identifying of various types of cancers.

3.5.2 Computed tomography(CT)

CT scans are one of the other medical analysis tools for organs and tissues. CT scans helped in creating better surgery, diagnosis and improve treatment of injuries and cancer. CT scans have many advantages over other image modalities, it is easily available and it can be performed in less time which allows the professionals to analyse the issue rapidly[9].

4 Results

In the experiment, we segmented medical image on U-Net. The segmentation was performed on hip joint datasets as shown in Fig.4,5 which are segmented at different slices. The experiment shows that U-Net architecture performs decently well on hip joint datasets in which there were 200 images for medical segmentation tasks.

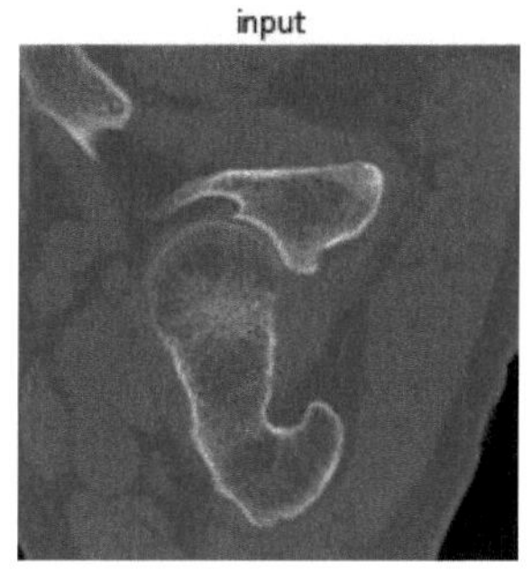
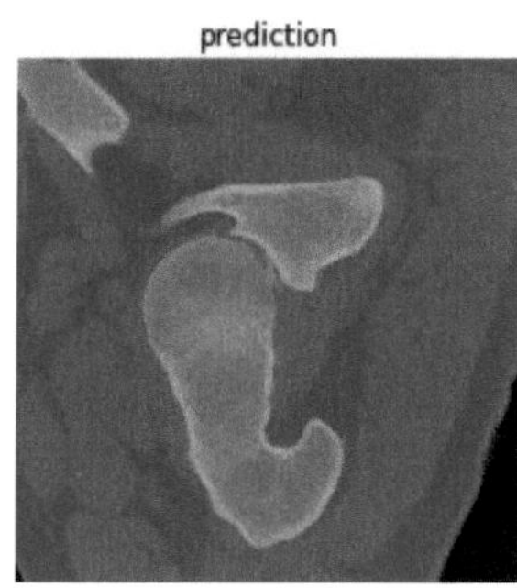

Figure 4: Highlighted area in prediction is the segmented part of hip joint

But in prediction we can see that the segmentation was not completely accurate, there are some instances where the model does not predict some spots in the bone as shown in Fig.5.

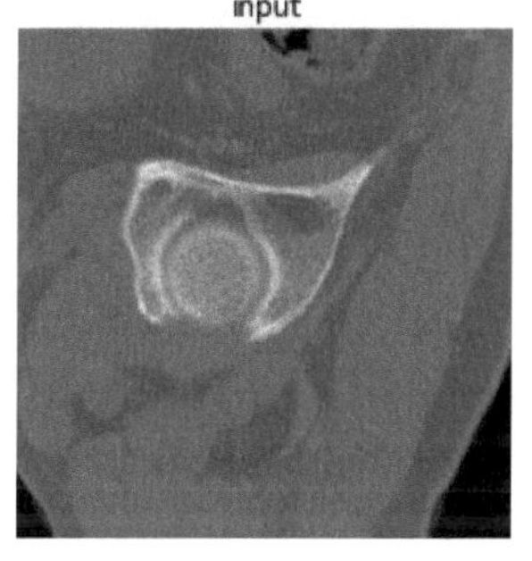
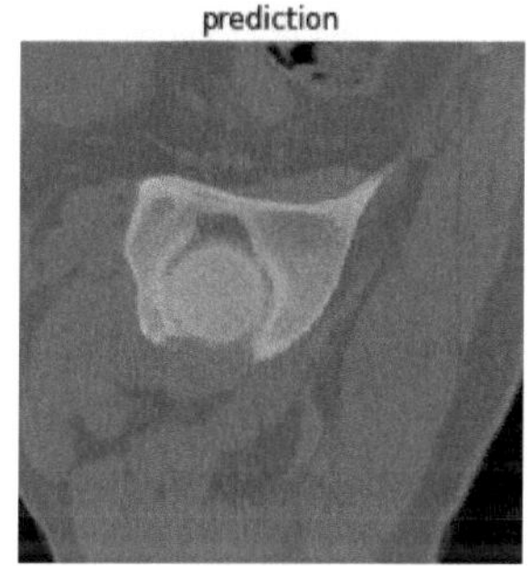

Figure 5: Highlighted area in prediction is the segmented part of hip joint. In this particular slice upper left corner of the bone is not predicted by model.

Fig.6,7 represent some evaluation metric for the model that we worked on. The model was trained for 150 epochs at a learning rate of 0.0001. Also we focused on altering the hyperparameters of the model just to get the better results. Hyperparameters are crucial for deep learning algorithms as they can impact the trend of training algorithms and thus can enhance the performance of the model. But to alter the hyperparameters to get the best results, some professional knowledge and experience required, and usually the other way is just to try brute force search. So in our model when we increased the number of epochs to 300 and added more data to the training dataset, we got the accuracy of 0.71 and previously we got accuracy of 0.53.

We got the above results on 2D U-Net. We have also implemented the segmentation model on 3D U-Net and if we compare the performance of both variants of U-Net we can conclude that 2D U-Net gave better results than 3D U-Net. We used same number of data of hip joint to train 3D U-Net model and got the accuracy of 0.56, which is significantly less than accuracy of 2D U-Net with accuracy of 0.71. Both the models were tuned at the same set of hy-

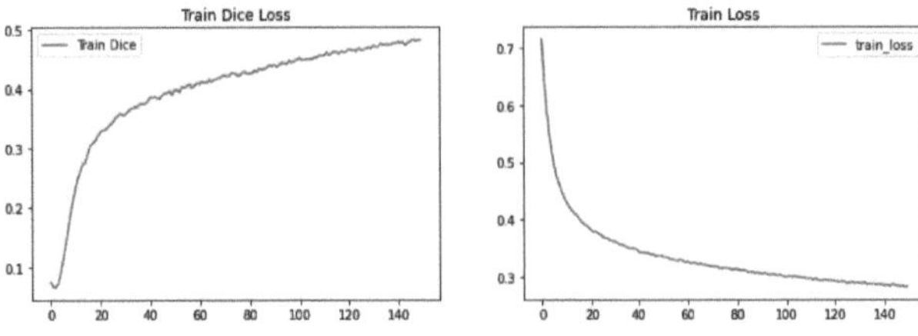

Figure 6: Graphs of training dice and training loss

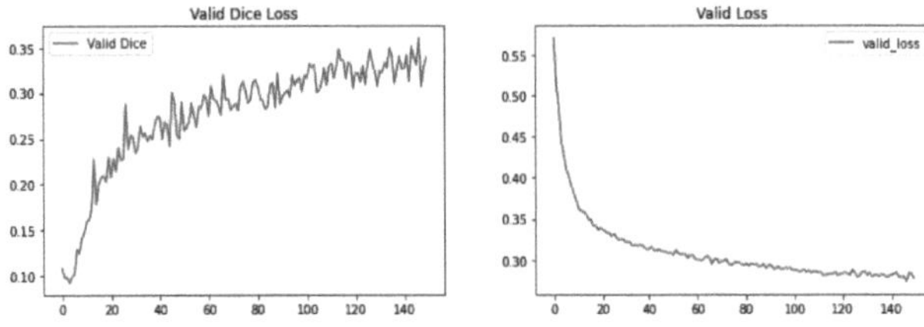

Figure 7: Graphs of validation dice and validation loss

perparametes. We suppose, 3D U-Nets can overperform 2D U-Nets but because of deeper layering structure a 3D U-Net needs much more annotated training data than 2D U-Nets and thus needed more GPU computation power that can reduce the memory effort in training the model.

5   Discussion

Deep learning algorithms like U-Net have seen immense use in medical image analysis. Through image processing, variety of tasks can be executed such as localization, detection, classification and many more. In medical imaging, segmentation tasks are the most demanding topic. In some surveys[7] it has been found that base U-Net can be proved to be a prominent method. U-Net can provide many possible applications since it can be tuned highly and build network models quickly. Still it has a lot of potential for improvement since its architecture allows it to make advancements by new ideas.

6   Challenges

For advancements in medical diagnosis it is necessary for deep learning to get better. As of now medical analysis demands algorithms to get results with minimal error. One of the major limitation in solving this problem is computational power. For powerful algorithms require more time to train and therefore are less feasible. Another major problem that we have mentioned previously also is the scarcity of annotated data for training.

7   Conclusion

It is concluded that U-Net architecture can get ground breaking results and is valuable in medical image analysis. However, this method is not completely perfect and it has some drawbacks: during the training the model memory footprint was too large and the results can get easily influenced by the number of training sets.

Acknowledgement

The work has been carried out at Volmo Ltd under supervision of Ashwani Harkara and supervised by Florist Ernst of the Institute for Robotics and Congintive Systems, Universität zu Lübeck.

Author's Statement

Conflict of interest: Authors state no conflict of interest.

8   References

[1] Haralick, R.M., Shapiro, L.G.: Image segmentation techniques. Comput. Vis. Graph. Image Process. 29(1), 100–132 (1985)

[2] Puybareau, Elodie Zhao, Zhou Khoudli, Younes Carlinet, Edwin Xu, Yongchao Lacotte, Jerome Géraud, Thierry. (2019). Left Atrial Segmentation in a Few Seconds Using Fully Convolutional Network and Transfer Learning: 9th International Workshop, STACOM 2018, Held in Conjunction with MICCAI 2018, Granada, Spain, September 16, 2018, Revised Selected Papers. 10.1007/978-3-030-12029-037.

[3] Sharma, N., Aggarwal, L.M.: Automated medical image segmentation techniques. J. Med.Phys./Assoc. Med. Phys. India 35(1), 3 (2010)

[4] https://www.tutorialspoint.com/tensorflow/tensorflow single layer perceptron.htm

[5] Ronneberger, Olaf, Philipp Fischer, and Thomas Brox. *"U-net: Convolutional networks for biomedical image segmentation."* International Conference on Medical image computing and computer-assisted intervention. Springer, Cham, 2015.

[6] Q. Tong, M. Ning, W. Si, X. Liao, and J. Qin, *"3D deeply-supervised U-net based whole heart segmentation,"* 2017, pp. 224–232.

[7] S. Minaee, Y. Boykov, F. Porikli, A. Plaza, N. Kehtarnavaz, and D. Terzopoulos, *"Image Segmentation Using Deep Learning: A Survey,"* ArXiv200105566 Cs, Apr. 2020.

[8] Shorten, C., Khoshgoftaar, T.M. A survey on Image Data Augmentation for Deep Learning. J Big Data 6, 60 (2019).

[9] Power, Stephen P et al. *"Computed tomography and patient risk: Facts, perceptions and uncertainties."* World journal of radiology vol. 8,12 (2016)

Advancements in Path Planning and Coordination of Autonomous Mobile Robot Fleets

Fabian Domberg [1], Carlos Castelar Wembers [2], Georg Schildbach [2], Jannik Abbenseth [3] and Stefan Dörr [3]

[1] Robotics and Autonomous Systems, Universität zu Lübeck, fabian.domberg@student.uni-luebeck.de

[2] Institute for Electrical Engineering in Medicine, Universität zu Lübeck, {carlos.castelar, georg.schildbach}@uni-luebeck.de

[3] NODE Robotics GmbH, Stuttgart, {jannik.abbenseth, stefan.doerr}@node-robotics.de

Abstract

Warehouse robots have seen an immense population growth around the world in recent years. Navigation and coordination for such robots, especially heterogeneous fleets of them, is a highly non-trivial task. In cooperation with a young robotics startup, this work tries to solve real world implementation issues. The tasks include improving memory efficiency for path planning, path execution optimization through reduction of waiting times and trajectory modeling and optimization for overall smoother paths. The algorithms are implemented in C++ and interact directly with the robots through ROS. Besides unit testing, the implemented behaviors are also verified directly on the robot hardware at the company's testing site. The work achieves to implement the above mentioned improvements and therefore takes part in advancing the company's navigation software to a production-ready state. As the software is run on more and more customers' robots, further improvements will have to be made to maintain smooth operation.

1 Introduction

Robots are indispensable in today's factories. Whether they are picking and placing objects on a conveyor belt, welding together parts of new cars or even performing autonomous surveillance of dangerous chemical processes they make up an important part of the workforce, especially with the advent of Industry 4.0. Getting the components of a product to the assembly location safely and storing away the finished goods quickly is not an easy task, as it requires high precision and good coordination with other participants on the factory floor. Research into autonomous mobile robots (AMRs) has enabled big leaps in the automation of factory logistics. These robots must be able to plan and execute an optimal trajectory, usually while operating in a large, potentially unknown, dynamically changing environment with other robots and moving obstacles around them. One of the state-of-the art algorithms to solve these tasks is A* [4], along with its many derivatives, which is the basis for most of the path planning and optimization done in this work.

2 Material and Methods

2.1 Autonomous Mobile Robots

Forklifts, and their analog counterpart the pallet jack, have been used to haul goods within factories for decades. Because they are simple in design and their environments are rather structured, research into their automation is reasonably advanced and most big manufacturing sites use them

in one way or another. However, these robots are not only used in industrial production settings, but also in various other domains, such as hospitals or retail.

Figure 1: The company's testing fleet made of 5 AMRs.

These AMRs are usually build in a similar fashion. Most of them are differential-drive with two driven wheels, others are omnidirectional and can thus move in any direction at any time, see Fig. 1. They are equipped with LiDAR sensors front and rear, while some additionally carry cameras. Due to the cost of such sensors, these robots often cost tens of thousands of dollars. Depending on their use case and manufacturer, they can reach comparatively high top speeds and carry up to multiple tons of load. Most of them come with their own basic navigation software, which often does not scale well or allow for cross-vendor communication.

2.2   ROS

The Robot Operating System (ROS) is a software suite consisting of libraries, tools and frameworks for the operation, programming and control of robots [1]. It is an open-source project lead by Open Robotics. ROS contains an extensive set of fully implemented algorithms for common robotics use-cases such as planning, computer vision and control. If a desired algorithm is not included with the default installation, one can easily search the vast community contributions or even write and publish their own package for others to use. ROS supports multiple programming languages natively, such as Python and C++, but also has bindings and bridges to all sorts of additional languages, e.g. Javascript. Due to the size and nature of the underlying industrial application of this work, most of the code is written in C++. Due to its lower level, its efficiency and therefore scalability far outperforms that of Python [2].

2.3   NODE Robotics

NODE Robotics GmbH is an early stage startup, founded in 2020 by researchers of the Fraunhofer Institute for Manufacturing Engineering and Automation (IPA) [3]. The company hosts a testing site in the *ARENA 2036* research facility, seen in Fig. 2, where a number of different robots are available for testing, see Fig. 1. NODE's goal is to enable autonomous intra-logistics for warehouses and factory floors. Being a software company, it strives to make its products hardware independent, such that customers can run multi-vendor heterogeneous robot fleets, seamlessly. The different software products include everything required to operate a robot fleet: from individual robot control and sensor evaluation, fleet management to cloud-based planning and remote monitoring. NODE is currently seeking to broaden its customer base and bring new investors on-board in its seed-round. The individual teams work mostly sovereign and with practically non-existent hierarchies among members. Within teams, the product development framework *SCRUM* is used and agile software development methods and practices are employed.

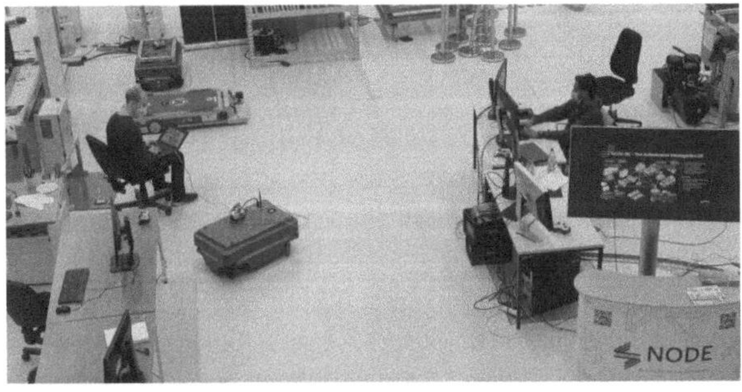

Figure 2: NODE Robotics' testing site in the *ARENA 2036* research facility in Stuttgart.

2.4   Robot Path Planning

The focus of this research is path planning and optimization for bespoke AMRs. Once a robot receives an order to go somewhere, the first thing it has to do is figure out how to get there. Path finding algorithms like $A*$ have been around for decades and are still readily used today, although with some additional improvements [4]. Finding a path is then fairly straight forward. However, the path planning has to adhere to certain constraints such as not getting too close to walls, being able to react to dynamic obstacles, or follow factory-internal traffic laws. Solving these non-trivial problems often comes down to mathematical optimization.

2.5   Path Downsampling

A simple, yet highly valuable contribution is the downsampling of a planned path to reduce factory-internal bandwith use. Since at large production sites there can be multiple hundreds of robots working simultaneously, communication bandwidth limits have to be considered.

One of NODE's customers found that their Wifi was commonly under heavy load due to each robot publishing its currently followed plan. One such plan consists of starting and end pose, as well as waypoints spaced evenly along the trajectory, e.g. every 50 cm. For large factory floors, this adds up to a sizable chunk of data, which has to be reported to a central server every so often. A possible solution to this problem is sparsing the plan, i.e. removing redundant waypoints. The resulting set of waypoints is no longer evenly spaced, but for example straight sections just consist of two points, as can be seen in Fig. 3. To achieve this, a modified version of the *Ramer-Douglas-Peuker* algorithm is used [6].

2.6   Zone Leasing

To communicate the factory's traffic laws to the robot, different kind of areas within its operating environment can be defined - so called zones. A zone can have a variety of rules, e.g. a one-way-zone or a limited-capacity-zone where only a certain amount of robots can be in at the same time. When executing a planned path, the robot will drive along its desired route until it comes to such a zone, where it will request a lease to go through. A central server provides this lease, if and only if it is currently not being held by any other robot. After gaining the lease, the robot will continue on its path and eventually return the lease upon leaving the zone. This whole system works similar to a Mutex in traditional Computer Science [5].

Since there had been some issues with this system in the past, like robots not returning leases or waiting indefinitely, some optimizations had to be done. There where also some inefficiencies like having to stop right in front of a zone, requesting a lease and immediately getting it and starting to go full speed again. The underlying code was rewritten such that now a predictive look-ahead-controller -like behavior is used. It requests leases upon coming in the vicinity of a zone, such that when they are granted it is able to continue driving seamlessly.

After implementing it, the new system had to be rigorously tested both in simulation and on real robots. For this, multiple testing zones were defined both in Gazebo and at NODE's private testing area, see Fig. 2. A variety of different error-prone customer scenarios where identified and recreated. Every feature was thoroughly tested before finally releasing the improved version to the customers.

2.7 NURBS Fitting

Non-Uniform Rational B-Splines (NURBS) and their relatives, the Bézier-Curve and B-Splines, are used extensively, for example in computer graphics. They are often used for their continuous smoothness. In this work, however, they are used to create a smooth path from a given set of waypoints. Connecting these waypoints with straight lines would have the robot drive towards a point, stop at it and orient itself towards the next one before continuing to drive towards it. To create a smoother and more continuous motion, a NURBS is fit to the waypoints. This is done by solving the system of linear equations

$$A \times \vec{c} = \vec{p} \tag{1}$$

for the control points $\vec{c} \in R^{(n+1)}$ of the NURBS curve. Here, $\vec{p} \in R^{(n+1)}$ are the $n \in N$ initially provided points to be fit, $A \in R^{(n+1)x(n+1)}$ is the coefficient matrix which is computed according to Algorithm 9.1 of [8].

The validation of the smoothness, mathematical correctness and continuity of the fitting task was done primarily through unit-testing, although later some hardware tests were also conducted.

2.8 Avoiding Obstacles Through Path Optimization

Due to the rather simple, and as of yet not kinematically aware, path planning algorithm, resulting paths can sometimes get too close to obstacles. This is especially the case when a plan already is close to a wall and the planned path is inflated to the width of the AMR. This can be observed in the left image of Figure 5, where the original planned path does not touch the corner, but the inflated one around it does. To tackle this issue, an iterative optimization algorithm is employed [9]. It samples the current path, which is in NURBS format, in small, incremental steps. For each of these sampled points the distance to the nearest wall is then computed. If this distance falls below a certain threshold, e.g. the robot width, the closest control point is moved away from the wall using a potential field approach until it reaches the minimal allowed distance to the wall. This process is repeated until no more control points have to be moved or the maximum amount of iterations is reached.

The validation of the implementations correctness was verified with unit-tests.

3 Results and Discussion

3.1 Path Downsampling

The resulting implementation of the modified *Ramer-Douglas-Peuker* algorithm brought the amount of waypoints (and therefore by extension the network usage) down to an average of less than 10%. For the path displayed in Fig. 3, the number of waypoints was brought down from 479 to 31, along with the corresponding network usage decreasing from 270.89 to 18.27 KB/s.

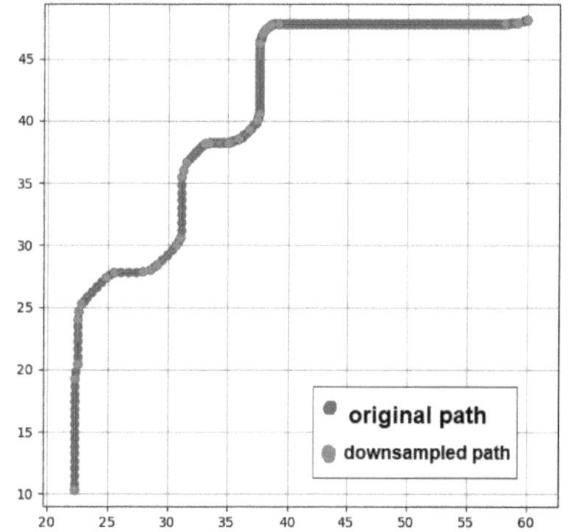

Figure 3: The original path and the downsampled one layed on top of each other.

3.2 NURBS Fitting

An exemplary result of a NURBS being fit to a set of waypoints in a maze is displayed in Figure 4. Along with a multitude of unit-tests, the computed NURBS where visually inspected for their smoothness. Later, the new algorithm was tested also on real robots, who showed no more of the typical 'drive straight to waypoint, rotate towards next and drive straight towards it' -behavior, but rather smooth, continuous motion along the entirety of the path.

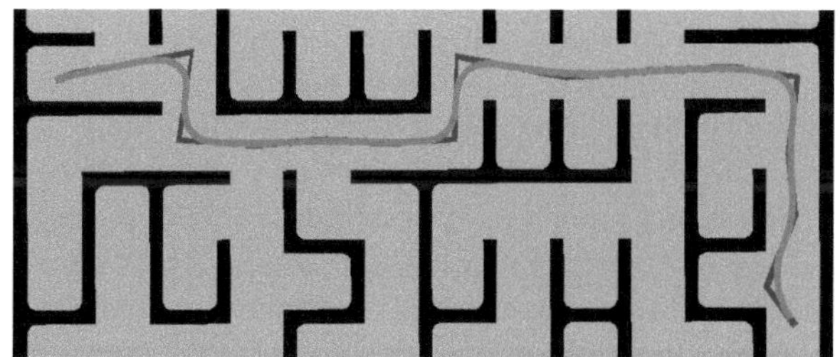

Figure 4: Original set of waypoints connected by straight lines (dark line in the background) and the interpolated NURBS curve (smooth line on top).

3.3 Zone Leasing

The rewrite of the zone leasing code, to incorporate a look-ahead-controller like behavior, was successful. Due to the new predictive nature of the lease requests, the operating of the whole robot fleet (seen in Fig. 1) was much smoother compared to the old version. Robots requested and released the leases correctly and, after finding the correct parameters, hardly ever showed the previous behavior of stopping and immediately starting again. Also, the issue of repeatedly requesting and returning leases when driving right on the edge of a zone was solved.

3.4 Avoiding Obstacles Through Path Optimization

Running the iterative algorithm multiple times yields very satisfactory results. A before and after image can be seen in Figure 5. The effectiveness of the algorithm was proven by unit-testing, where a path was deliberately planned too close to a wall and then iteratively optimized. Throughout these tests all of the paths were optimized successfully.

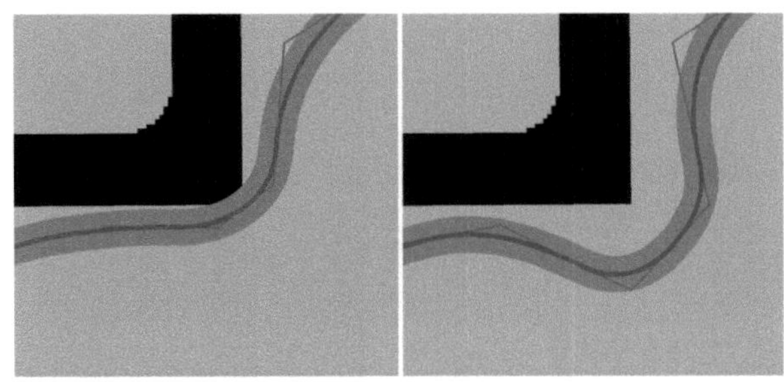

Figure 5: Before (left) and after (right) optimizing the path to not go through walls.

The application of this algorithm is a first step towards kinematically aware global path planning. Since robot kinematics differ from vendor to vendor, the optimized paths for each of them will also differ, e.g. a bigger robot might require a larger distance to the wall as it drives around a corner. To make this possible, the algorithm is made parametrizable.

4 Conclusion

The above work shows that there is still some improvements to be made regarding autonomous mobile robots for warehouse operation. However, it also shows that these challenges are often very much solvable and mostly come down to engineering problems. Research into AMRs of the past decades, at the Fraunhofer IPA and around the world, was fundamental for the upcoming automation and robotization of factory floors. As of now, there are already many companies looking into automating their intra-logistics. To be able to keep up with the growing demand and scaling requirements of these companies, further research and development, for example in the area of cloud-computing, will be necessary.

Acknowledgement

The work has been carried out at NODE Robotics GmbH, Stuttgart and supervised by the Institute for Electrical Engineering in Medicine, Universität zu Lübeck. Research was funded by the Fraunhofer Institute for Manufacturing Engineering and Automation through the European Union's *EXIST* program.

Author's Statement

Conflict of interest: Authors state no conflict of interest.

5 References

[1] Stanford Artificial Intelligence Laboratory et al., *Robotic Operating System.* Available: `https://www.ros.org` [last accessed on 2021-12-06].

[2] F. Zehra, M. Javed, D. Khan and M. Pasha, *Comparative Analysis of C++ and Python in Terms of Memory and Time.* 2020.

[3] NODE Robotics GmbH, *NODE Robotics' Website.* Available: `https://www.node-robotics.com` [last accessed on 2022-01-20].

[4] F. Duchon, et al., *Path Planning with Modified A Star Algorithm for a Mobile Robot.* In: In: Procedia Engineering, vol. 96, pp. 59–69, 2014.

[5] F. Duchon, et al., *Transactional Mutex Locks.* European Conference on Parallel Processing, Springer, pp. 2–13, 2010.

[6] L. Dalessandro, et al., *A Method for Simplifying Ship Trajectory based on Improved Douglas–Peucker Algorithm.* In: European Conference on Parallel Processing, vol. 166, pp. 37–46, 2018.

[7] D.F. Rogers, *An Introduction to NURBS: with Historical Perspective.* Morgan Kaufmann, 2001.

[8] L. Piegl and W. Tiller, *The NURBS Book.* 2nd Edition, Springer, pp. 364–373, 1995.

[9] B. Lau, C. Sprunk and W. Burgard, *Kinodynamic Motion Planning for Mobile Robots using Splines.* In: IEEE/RSJ International Conference on Intelligent Robots and Systems, pp. 2427–2433, 2009.

Analysis of the impact of super-resolution on medical CT-Scans for classification and segmentation

Niklas Christoph Koser [1], Hellena Hempe [2] and Mattias Paul Heinrich [3]

[1] Medical Informatics, University of Luebeck, niklas.koser@student.uni-luebeck.de
[2] Institute of Medical Informatics, University of Luebeck, hellena.hempe@uni-luebeck.de
[3] Institute of Medical Informatics, University of Luebeck , mattias.heinrich@uni-luebeck.de

Abstract

Recently, super-resolution techniques use deep neural networks and have shown remarkable results. However, most super-resolution techniques focus on creating photo-realistic images, which is not always suitable for medical applications. It is important that relevant structures are enhanced, and no artifacts are added. In order to generate high-resolution CT-scans, we use a modified Generative Adversarial Network (GAN) architecture and a ResUNet. We then evaluated the generated CT-Scans using a classification and segmentation task. The classification tasks involve extracting patches surrounding a candidate for a lung nodule, which are then classified. As part of the segmentation task, we segment the pulmonary airways into their individual sections. We could show that the CT-Scans generated with super-resolution for both tasks lead to an improvement of the results. CT-Scans generated from the GAN allowed segmentation of previously undetectable bronchi.

1 Introduction

Computerized tomography (CT) and magnetic resonance imaging (MRI) are among the most important diagnostic techniques in medicine today. These techniques allow us to obtain detailed information about the structure of the body and its pathology. With a higher resolution, texture information, pathologies, and anatomical structures become more visible. The high-resolution of the image facilitates easier diagnosis by experts and leads to better classification and localization of pathologies in medical images [1].

The image resolution is dependent on the device selected. Improved image quality can be achieved by using new sensors, higher radiation doses, or contrast agents. Nevertheless, new sensors are expensive, contrast agents can only be administered during imaging, and higher radiation doses negatively impact human health [2]. Another approach to increase the resolution of an image is super-resolution (SR). This is an ill-posed problem in which a corresponding high-resolution (HR) image is reconstructed from a low-resolution (LR) image.

We can currently differentiate between approaches based on direct reconstruction (resolution recovery) of LR images by using differential errors and those based on artificial neural networks (ANN). By comparison with direct methods, ANNs offer a powerful strategy for generating plausible-looking, high-quality images of anatomical structures [3].

In previous SR work, the mean square error (MSE) was used to evaluate how similar a generated SR image and a ground truth image are. MSE has the disadvantage that relevant perceptual differences can only be measured to a limited extent due to its pixel-based approach. In recent SR applications, peak signal-to-noise ratios (PSNR) and structural similarity indices (SSIM) are used to evaluate photo-realistic images [4]. Nevertheless, it was shown that these metrics are not fully reliable. The reason is that the PSNR is calculated from the MSE, and there is a relationship between the PSNR and SSIM [5]. Therefore, a high PSNR or SSIM cannot guarantee realistic imaging [4]. For medical applications, however, obtaining a realistic image is not necessarily desired. It is far more important to artificially improve existing features than to add new ones. In order to evaluate the potential benefits of SR in medical imaging, we compared the outcomes of CT-Scans generated with SR techniques on a classification and segmentation task.

In this study, we investigate whether we can use current SR techniques based on Super-Resolution Generative Adversarial Network (SRGAN)[4] and Residual UNet (ResUNet) [6] [7] to reconstruct HR CT-Scans based on appropriately downscaled CT-Scans, since high and LR image pairs are usually not available in medicine. As part of the evaluation, we will perform segmentation and classification tasks to evaluate the generated SR CT-Scans.

2 Material and Methods

Data and Preprocessing

For the classification task we use the CT Thorax volumes provided by the Lung Nodule Analysis (LUNA) Grand

Challenge[1]. It contains 888 CT-Scans with additional annotations for the position of lesions and lung nodules. In order to use only CT-Scans with an appropriate resolution for our objective, we exlude all CT-Scans with an axial slice thickness larger than 1.5 mm.

For the subsequent segmentation task, 60 CT-Scans of the lung are used (40 CT-Scans come from the Lung Image Database Consortium image collection (LIDR-IDRC)[2] and the other 20 CT-Scans comes from the EXACT'09[3] dataset) with a provided segmentation mask of the lung airways[4]. The lung airway was divided into 32 anatomical regions considering the anatomical topology.

For both tasks, we create a downsampled set. Using the PyTorch area mode, we sampled the CT-scans down based on the scale factor. Because the size of the CT-Scans changes when downsampled, we apply trilinear interpolation to resize the downsampled CT-Scans back to their original size. The Pytorch area mode uses adaptive average pooling to assign to each output pixel the average of all pixels within a given region of the input.

Super-Resolution Architectures

For our experiments, we use two different SR architectures. The first architecture is the ResUNet, composed of an encoder and decoder part. The encoder part of our ResUNet consists of three residual convolutions (ResConv), which always halve the input size. For the decoder part, we use upsample layers because of unsightly checkerboard patterns when using transposed convolution. After each upsampling step, there is another ResConv. The ResConv consists mainly of a batch normalization, a ReLU, and a convolution layer. We use padding mode reflect on the convolution due to black stripe edges for all SR architectures.

In order to design our SRGAN, we utilize a discriminator network that is alternately optimized with a generator network according to the adversarial min-max problem. Our objective is to train a generator to produce SR CT-Scans from LR CT-Scans that the discriminator cannot differentiate from ground truth images. A discriminator is trained to assess between the generated HR images and the ground truth images. Based on the implementation of Lornateng[5], our generator is quite similar to the one proposed by Ledig et al. [4]. Due to the fact that the SRGAN generator is designed to handle two-dimensional data, we adapted its architecture for three-dimensional data. Additionally, we removed the upsampling component of the generator because the downscaled CT-Scan has the same size compared to the original resolved CT-Scan. In this work, we completely replace the discriminator of the SRGAN with a 3D ResNet-34

to determine whether the input is fake or not [7].

The SRGAN's core strength is the definition of the loss function, which has a significant impact on the quality of the resulting image. As well as using the MSE, they included a Perceptual Loss in their generator. This is calculated by squaring the distance between two pre-trained feature maps of the VGG-19. It is more comparable to a perceptual similarity. In our Perceptual Loss, we use a pre-trained ResNet on 3D medical data [8] and calculate the Perceptual Loss from the extracted feature maps instead of the VGG net. Additionally, we replaced the binary-cross-entropy of the min-max loss with a Wasserstein loss to become more stable and less sensitive to changes in hyperparameters [9].

Classification Task

We extracted 64x64x64 patches around the position of the lung nodules from the LR CT-Scan and the HR CT-Scan to train the SR task on the LUNA dataset. As output, we receive a reconstructed SR CT Patch based on the LR CT-Scans. The patches were normalized to the interval [0,1]. The dataset was then split into 60% training, 20% validation, 20% test data for training of the SR task. For Comparing the quality of the generated CT-Patches of the SRGAN and the ResUNet we train a 3D ResNet-34 with the original resolved 3D CT-Patches from the validation and test data to determine whether the candidate is a lung nodule. Becuase of the imbalance dataset, we chose to take all positive candidates and add seven times as many negative candidates from the test and validation data. Due to the imbalanced nature of the data, we apply a five-fold cross-validation and use class weights. To determine the quality of the SR CT patches, we test the ResNet-34 on the original resolved CT patches, the downsampled CT patches, and the SR CT patches after training.

Segmentation Task

In order to segment the lung airways properly, a full CT-Scan of the lung is required. Because training on a full CT-Scan is memory intensive, we extract $64x64x64$ random patches and train the SR task on them. In order to ensure that the patches not only contain background information, we only use patches that do not consist of 75% background voxels. The test CT-Scans are divided into 64x64x64 patches, which are then reassembled after determining the individual patches for segmentation. We use 32 CT-Scans for training, 8 for validation, and 20 for testing.

In order to segment the pulmonary airways, we use a nnUNet [10] that has been previously trained on the HR CT-Scans of the Airways Dataset. To evaluate the quality of the generated SR CT-Scans of the ResUNet and the SRGAN, we perform a segmentation analysis with the nnUNet on the SR, the HR and the LR CT-Scans. As a result, we obtain a segmentation mask for each CT-Scan, which we evaluate using the Dice score based on the given ground truth segmentation masks.

---

[1]A. A. A. Setio, et al., "Automatic detection of large pulmonary solid nodules in thoracic CT-Scans,"

[2]S. G. Armato III, et al., "The lung image database consortium (lidc) and image database resource initiative (idri): a completed reference database of lung nodules on CT-Scans,"

[3]P. Lo, et al., "Extraction of airways from ct (exact'09),"

[4]Z. Tan, et al., "Sgnet: structure-aware graph-based network for airway semantic segmentation,"

[5]GitHub repository: https://github.com/Lornatang/SRGAN-PyTorch

LUNA	AUROC	PR-AUC	Precision	Recall	F1-Score
HR (ORG)	$0.963 \pm .01$	$0.942 \pm .03$	$0.950 \pm .02$	$0.907 \pm .02$	$0.926 \pm .02$
SRGAN (4x)	$\mathbf{0.968 \pm .02}$	$\mathbf{0.951 \pm .02}$	$\mathbf{0.965 \pm .01}$	$\mathbf{0.915 \pm .02}$	$\mathbf{0.936 \pm .02}$
ResUNet (4x)	$0.958 \pm .02$	$0.935 \pm .03$	$0.967 \pm .01$	$0.891 \pm .03$	$0.924 \pm .02$
LR (4x)	$0.953 \pm .02$	$0.928 \pm .04$	$0.955 \pm .02$	$0.895 \pm .03$	$0.920 \pm .02$
SRGAN (8x)	$0.938 \pm .01$	$0.914 \pm .02$	$0.956 \pm .02$	$0.875 \pm .03$	$0.907 \pm .03$
ResUNet (8x)	$0.920 \pm .02$	$0.895 \pm .03$	$0.953 \pm .02$	$0.877 \pm .04$	$0.907 \pm .03$
LR (8x)	$0.929 \pm .02$	$0.902 \pm .04$	$0.955 \pm .02$	$0.872 \pm .03$	$0.904 \pm .03$

Table 1: Comparison between LR CT-Scan (LR), HR CT-Scan and the reconstructed SR CT-Scans by a SRGAN and a ResUNet with different resolutions on the LUNA dataset.

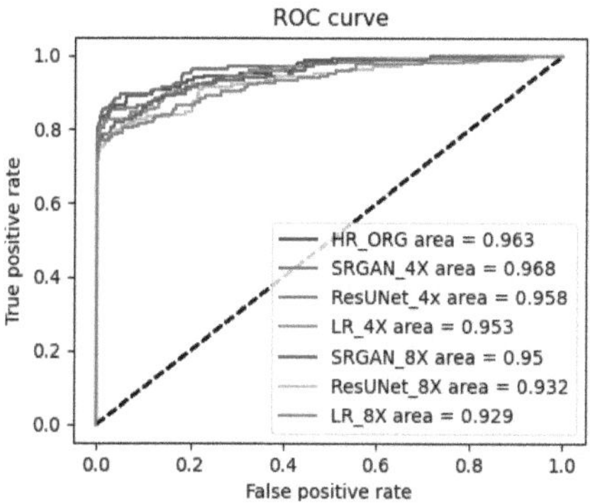

Figure 1: Representation of the ROC curves and corresponding areas of the experiments in table 1

3 Results and Discussion

Classification Task

For comparision our results, we use the area under the receiver operating characteristic (AUROC), the area under the precision-recall curve (PR-AUC), precision, recall, and the F1 score. In table 1 we present the metrics average over the individual folds for the comparison of the different Patches. The table shows that the metrics measured on the patches with different resolutions are all close to each other. However, we can observe that when we start from the HR CT-patches, the values of the metrics decrease with an increased downsample factor. This can be seen by comparing the results on the HR CT-patches and the CT-patches downsampled by a factor of 8. The PR-AUC and the recall have decreased. This is probably because the information was lost during downsampling, leading to several lung nodes on the LR patches being classified as false positives. To ensure that we have a significantly better result on the HR CT-patches than on the LR CT-patches, we calculate the L1 and L2 norm between the ResNet-34 output and the ground truth values. We use these for a Wilcoxon test to show that there is a weak significance ($p = 0.045 < 0.05$). The improvement in results between LR CT-Patches and those generated by SRGAN or ResUNet does not warrant much improvement. The table as well as the plot of the ROC curves in Fig. 1 confirm this. Thus, we can see that by using an SRGAN or ResUNet, we could slightly improve the CT-Scans downsampled by a factor of 8 and still fall below the results on the CT-Scans downsampled by a factor of 4. On the LR 4x patches, we find that there is no significant difference between the HR and the LR CT-Scans. However,

the computed patches of the SRGAN perform so well that we are even better than on the HR patches. Comparing the computed patches on a ResUNet and SRGAN, we discover that the SRGAN gets higher results than the ResUnet. By looking at the images, we can conclude that the generated ResUNet patches appear more realistic than the SRGAN patches. The PSNR and SSIM values are greater than those obtained by the SRGAN by at least a factor of 0.1. Overall, we achieved good results even with the LR patches, which implies that the classification task was too easy. This is why we decided to add a second task, where downsampling will have a greater impact.

Segmentation Task

The labels are divided into four anatomical levels to compare their results. Layer one describes the trachea, which can still be segmented well with the LR CT-Scans. The second layer comprises all primary bronchi, the third layer all lobar bronchi, and the final layer combined the tertiary bronchi. The tertiary airways, in particular, require a high level of resolution and detail density for segmentation. Since we are interested in recovering fine structures and HR details based on the LR CT-Scan by SRGAN and ResUNet, respectively, the segmentation of the tertiary labels gives us a reliable measure of the quality of the generated SR CT-Scans. Using a boxplot, we compare the Dice values of the tertiary labels between the different resolved CT-Scans (see Fig. 2). We see that the median on the LR CT-Scan is zero. This means that a tertiary label could not be segmented for many of the samples. Also, the mean of the Dice values on the tertiary labels on the LR images is significant lower at around 0.15 compared to the mean on the HR CT-Scans. This shows us that downsampling has a significant impact on the segmentation of the tertiary bronchi, and thus we lose relevant structures necessary for segmentation. With the ResUNet and the SRGAN, we increased the mean and the median of the Dice Scores significantly compared to the LR CT-Scans. The results show that we can recover the fine structures from the LR CT-Scans so that this has a positive effect on the segmentation. With the help of a paired t-test, we could show that we got even significantly higher Dice values by the superresolution ($p = 0.0012 < 0.05$). However, compared to the HR CT-Scans, the CT-Scans reconstructed with SRGAN and ResUNet do not reach the Dice values obtained with the HR CT-Scans. Based on the box plot, the SRGAN and the ResUNet provide similar results. The average and median Dice values on the tertiary labels are a bit higher on the SRGAN computed CT-Scans. When we examine the obtained Dice values for the lobar bronchi, we can observe a higher significant difference between the SRGAN and ResUNet values. Therefore, we decided to conduct a paired T-test on the whole set of labels. The Dice values obtained on the SRGAN are significantly higher than those obtained on the ResUNet ($p = 0.001 < 0.05$). Similar to the results on the LUNA dataset, the ResUNet computed CT-Scans have a higher PSNR and SSIM value than the SRGAN computed CT-Scans. One explanation can be provided by the fact that we selected the MSE as an opti-

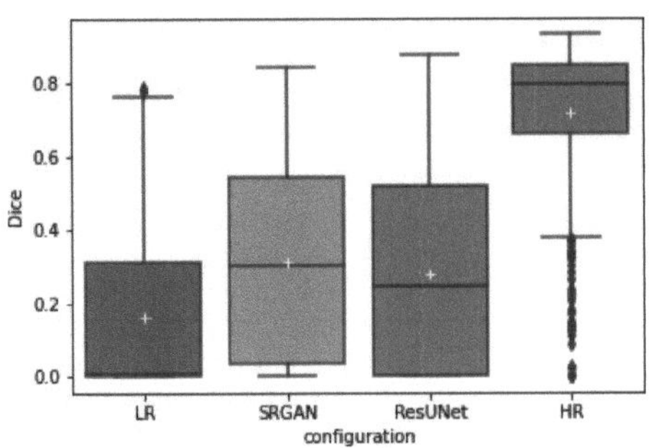

Figure 2: Boxplot of the reached dice scores of the tertiary labels on the LR, ResUNet, SRGAN and HR

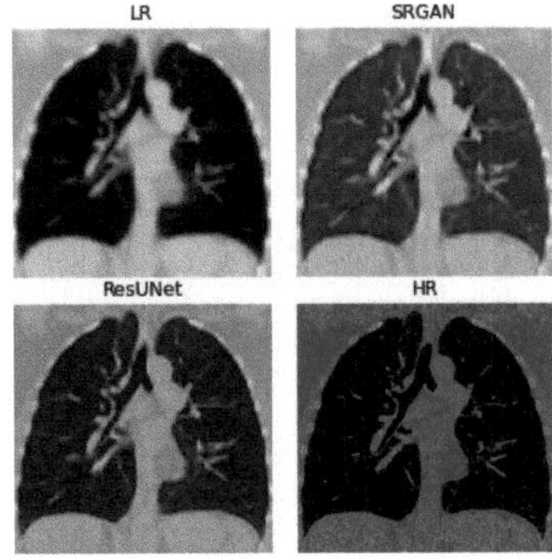

Figure 3: Viszualisation from one same axial slice of the LR (4x) CT-Scan in comparison to the generated SR CT-Scans of the ResUNet and the SRGAN, as well as the HR CT-Scan.

mization function for the ResUNet. In our introduction, we noted that a low MSE results in a large PSNR value. Considering the dependence between the SSIM and the PSNR, it is not unlikely that the SSIM value gets high. Upon viewing the generated CT-Scans in Fig.3, it is apparent that the ResUNet CT-Scan are blurrier and do not possess the same level of detail depth as the SRGAN CT-Scans.

4   Conclusion

Our research showed that by using the SRGAN and ResUNet, it is possible to increase the resolution of LR CT-Scans to improve results for segmentation and classification tasks. The segmentation task, in particular, demonstrated our ability to recover small structures, because we were able to segment tertiary bronchi on the SR CT-Scans that had been difficult to segment on the LR CT-Scan. When using the commonly used metrics of PSNR and SSIM for SR, we could show that the generated CT-Scans with an lower PSNR and SSIM value on the SRGAN than on the generated CT scans of the ResUNet led to better classification and segmentation results. This shows that Medical data does not always need to appear photo-realistic, but the data must still remain interpretable for an expert. Due to this, we would like to emphasize that PSNR and SSIM values for medical purposes should not be taken too seriously.

Acknowledgement

The work has been carried out at the University of Luebeck and supervised by the Institute of Medical Informatics, University of Luebeck.

Author's Statement

Conflict of interest: Authors state no conflict of interest. Informed consent: Only publicly available data were used for this study; no patient data were acquired.

5   References

[1] J. S. Isaac and R. Kulkarni, "Super resolution techniques for medical image processing," in *2015 ICTSD*, pp. 1–6, 2015.

[2] Z. Chen, X. Guo, P. Y. M. Woo, and Y. Yuan, "Super-Resolution Enhanced Medical Image Diagnosis With Sample Affinity Interaction," *IEEE Trans Med Imaging*, vol. 40, pp. 1377–1389, May 2021.

[3] O. Oktay, E. Ferrante, K. Kamnitsas, M. Heinrich, W. Bai, J. Caballero, *et al.*, "Anatomically constrained neural networks (acnns): Application to cardiac image enhancement and segmentation," *IEEE Trans Med Imaging*, vol. 37, no. 2, pp. 384–395, 2018.

[4] C. Ledig, L. Theis, F. Huszar, J. Caballero, Cunningham, *et al.*, "Photo-Realistic Single Image Super-Resolution Using a Generative Adversarial Network," *arXiv:1609.04802*, May 2017.

[5] D. R. I. M. Setiadi, "PSNR vs SSIM: imperceptibility quality assessment for image steganography," *Multimedia Tools and Applications*, vol. 80, pp. 8423–8444, Mar. 2021.

[6] O. Ronneberger, P. Fischer, and T. Brox, "U-Net: Convolutional Networks for Biomedical Image Segmentation," *arXiv:1505.04597*, May 2015.

[7] K. He, X. Zhang, S. Ren, and J. Sun, "Deep Residual Learning for Image Recognition," *arXiv:1512.03385*, Dec. 2015.

[8] S. Chen, K. Ma, and Y. Zheng, "Med3d: Transfer learning for 3d medical image analysis," *arXiv preprint arXiv:1904.00625*, 2019.

[9] M. Arjovsky, S. Chintala, and L. Bottou, "Wasserstein GAN," *arXiv:1701.07875*, Dec. 2017.

[10] F. Isensee, P. F. Jaeger, S. A. A. Kohl, J. Petersen, and K. H. Maier-Hein, "nnU-Net: a self-configuring method for deep learning-based biomedical image segmentation," *Nature Methods*, vol. 18, pp. 203–211, Feb. 2021.

Uncertainty Estimation and Calibration of Neural Networks

Lakshith Nagarur Lakshminarayana Reddy [1], Audrey Galametz [2], Philipp Gruening[3]

[1]Robotics and Autonomous Systems, Universität zu Lübeck [lakshith.nagarurlakshminarayanareddy@student.uni-luebeck.de]

[2] Airbus AI Research, Airbus Defense Space GmbH, Ottobrunn, [audrey.galametz@airbus.com]

[3] Institute for Neuro- and Bioinformatics, Universität zu Lübeck [gruening@inb.uni-luebeck.de]

Abstract

Deep neural networks have gained a lot of popularity in the field of machine-learning and are now being used for a wide range of tasks. Unfortunately, these neural networks suffer from an overconfidence in their predictions in particular when the prediction is wrong or when there is a perturbation in the data. Accurately modeling predictive uncertainty and calibrating these neural networks are essential. Although Bayesian neural networks have been shown to be a promising approach for reliably quantifying uncertainty, they are still computationally expensive. We therefore focus our work on non-Bayesian techniques and explore e.g., ensemble techniques (deep ensembles) and post-processing calibration methods (temperature scaling) to estimate predictive uncertainty and to calibrate neural networks. We also explore the possibility to combine ensemble techniques and temperature scaling to further optimise uncertainty estimates.

1 Introduction

Deep neural networks (NN) will one day be used to make critical decisions in domains such as transportation (e.g. autonomous driving, flying) or medical image diagnosis. Airbus for example is working towards including more and more artificial intelligence including NN in its products and will need to ensure that the NN models they use are robust and the uncertainties of their predictions under control. The Airbus AI research team in which this work was performed is currently exploring techniques to make these uncertainty estimates the most trustworthy possible.

Despite often reaching high accuracy, neural networks suffer from a miscalibration of the estimated prediction uncertainty, e.g., when they are exposed to new input data never seen in training. Unfortunately, these overconfident predictions can cost a lot in real world applications and could at worst be seriously harmful [1]. A reliable estimate of uncertainty is therefore necessary in AI-based systems used in real life applications. There are two sources of uncertainties: aleatoric and epistemic uncertainty. The aleatoric uncertainty occurs when the training data is corrupted or noisy or when a new model input differs from the data the model was trained on; this uncertainty in the real world is unavoidable and cannot be avoided even by training with large datasets. Epistemic uncertainty occurs when the parameters are poorly learnt; this uncertainty can be diminished with good machine-learning engineering e.g., by increasing the data size and finding the most suitable ML architecture [2].

There has been a lot of research carried out in the field of uncertainty quantification in recent years. A large major-ity of the work has concentrated on Bayesian Neural Networks [3] and Bayesian variational inference [4]. However, training and optimizing Bayesian neural networks parameters are difficult and are computationally expensive in practical applications. Complementary non-Bayesian techniques include ensemble techniques such as deep ensembles (i.e., combining multiple independently-trained models [5] and Monte-Carlo dropouts (MC dropouts, i.e., enabling dropout during inference/test time and averaging over multiple approximations) [6]. These techniques may be favored because of their ease of implementation comparing to Bayesian neural networks. They are also robust at exhibiting higher predictive uncertainty especially for out-of-distribution data (OOD).

In this work, we explore 'simple' and robust uncertainty calibration techniques: *(i)* temperature scaling (designed to re-calibrate over-confident neural networks and optimal for independent and identically distributed (IID) dataset) and deep ensembles and other ensemble techniques (e.g., snapshot ensembles). We also explore the option to calibrate deep ensemble recipes following published works that show that combining ensembles with data augmentation could worsen the model uncertainty [7, 8].

2 Problem definition and metrics

In this work, we mainly focus on multi-class classification problem with input $x \in \mathcal{X}$, labels with $\in \mathcal{K}$ classes with output $y \in \{1, ..., K\}$. Given input sample x, the probabilistic predictive distribution of NN is $p_\theta(y|x)$, the softmax output. The predicted class is $\hat{y} = \mathrm{argmax}(p_\theta(y|x))$ and the confidence score $\hat{q} = \mathrm{max}(p_\theta(y|x))$, the maximum of the softmax probability. This section will introduce a few metrics used in uncertainty quantification.

2.1   Reliability diagram

Reliability diagram are used to visualize model (mis)calibrations. The validation data are first sampled in bins of confidence score. The reliability diagram shows the expected accuracy as a function of these binned confidence scores [9]. If the model is calibrated, the accuracy and confidence score should be matched (identity function) i.e., the correct number of predicted class is equal to the probability of this class being correctly predicted. Any deviation from the identity function represents miscalibration (see Fig 1).

2.2   Expected calibration Error

The Expected Calibration Error (ECE) is a calibration metric, where the predicted confidence score is first divided into M equally-spaced bins and one can then take the weighted average of the difference between accuracy and confidence score of each bin [9]:

$$ECE = \sum_{m=1}^{M} \frac{|B_m|}{n} \left| \mathbf{acc}(B_m) - conf(\mathbf{B}_m) \right|. \quad (1)$$

2.3   Negative Log Likelihood

The Negative Log Likelihood (NLL), often known as cross entropy loss, is a metric used to evaluate model prediction. The goal during training is often to minimize the NLL given the predictions and targeted labels:

$$NLL = -\sum_{i=1}^{n} \log(\mathbf{p}(\mathbf{y_i}|\mathbf{x_i})) \quad (2)$$

2.4   Entropy

A measure of uncertainty is often estimated using the entropy function, often known as Shannon's Entropy and denoted by $\mathcal{H}(.)$. It is also known as total uncertainty as it is a combination of both model and data uncertainties:

$$\mathcal{H}(\mathbf{x}) = -\sum_{i=1}^{n} \mathbf{P}(\mathbf{x_i})\log(\mathbf{P}(\mathbf{x_i}) \quad (3)$$

3   Methods

3.1   Temperature scaling

Temperature scaling is a simple post-processing calibration technique and an extension to the platt-scaling technique (commonly used for the calibration of binary classification models) [10]. Temperature scaling is used to calibrate neural networks which are overconfident in their predictions

(see Fig 1). The temperature (T) is a single scalar parameter (> 0). The optimal temperature T is obtained by minimizing the NLL on a held-out validation dataset [9]. Given logits (z_i), the scaled confidence score is

$$\hat{q}_i = \max_{\mathbf{k}}(\sigma_{SM}(z_i/T)^k). \quad (4)$$

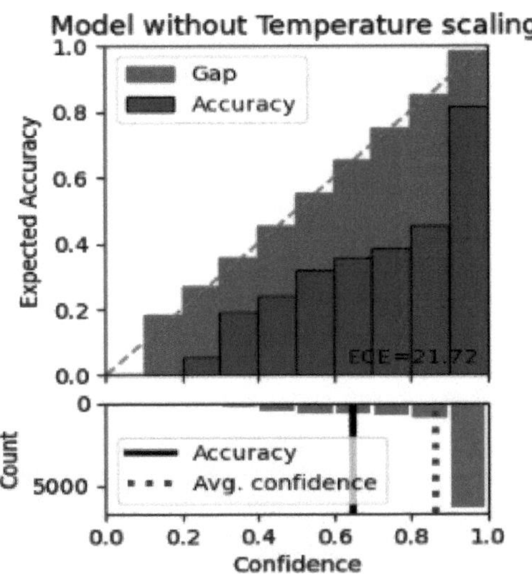

Figure 1: Example of a reliability diagram (blue). Strong divergence from the diagonal (purple histogram or 'Gap') shows a miscalibration of model uncertainty estimates.

When $T > 1$, the temperature scales the softmax output and increases the output entropy. The temperature does not affect the maximum of the softmax prediction p_i i.e., the predicted class y_i remains the same after the temperature scaling so the procedure therefore does not affect (i.e., improve) the model accuracy.

3.2   Deep Ensembles

While ensembles of models are now commonly used to improve the accuracy of classification models, ensemble techniques were first used to estimate uncertainty [5]. In ensemble techniques, the neural networks are initialized with random weights and converge/explore different modes in function space. During inference time, the predictions of all individually-trained model are then combined/averaged which enables an improved uncertainty estimation [11].

$$P_E(y|x) = M^{-1} \sum_{m=1}^{M} \mathbf{P}_{\theta_\mathbf{m}}(\mathbf{y}|\mathbf{x}) \quad (5)$$

3.3   Re-calibrating ensembles

We see that deep ensembles are good at estimating uncertainties. They may need to be further calibrated however for some particular regimes or use cases (limited training data

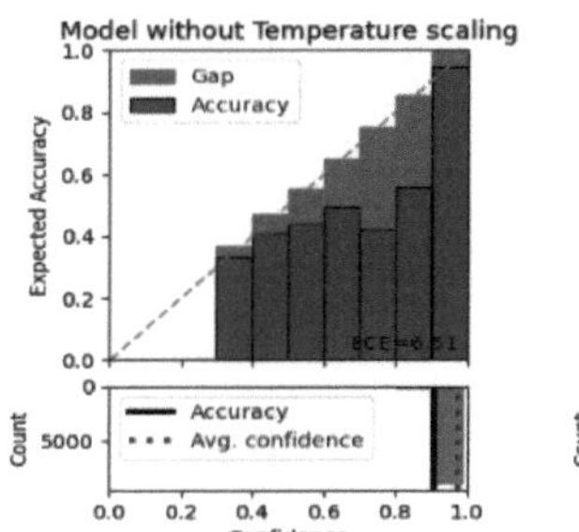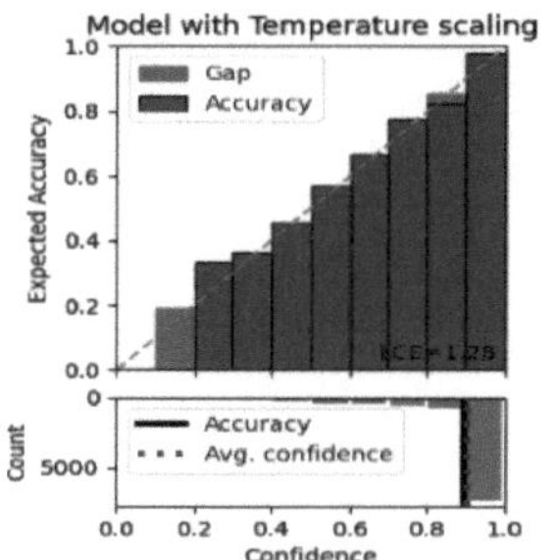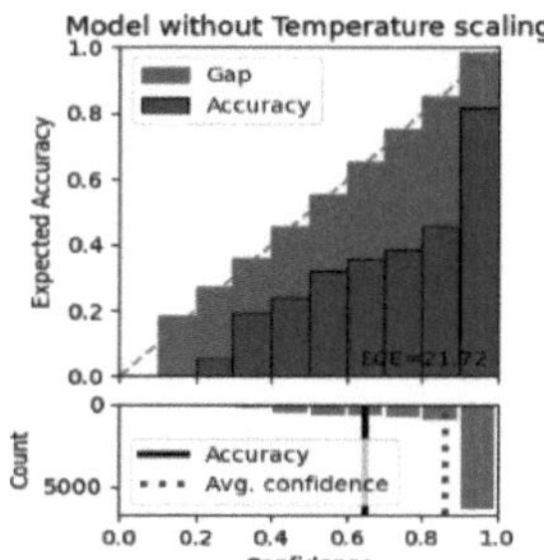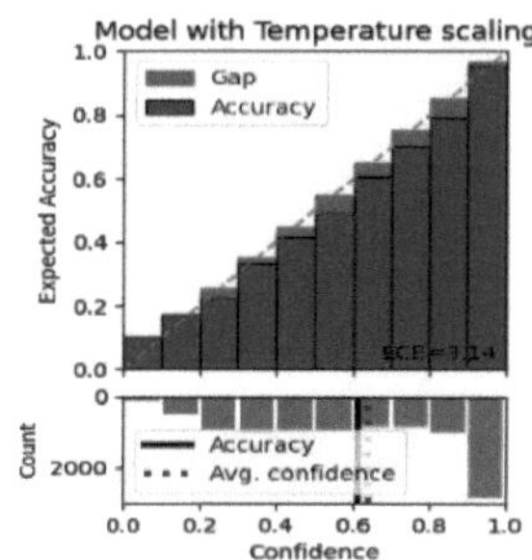

Figure 2: Reliability diagram for CIFAR10 test dataset before (left) and after (right) calibration

Figure 3: Reliability diagram for CIFAR100 test dataset before (left) and after (right) calibration

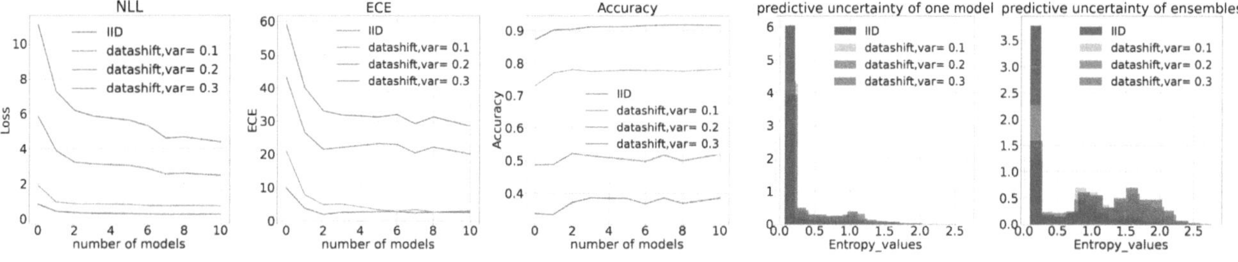

Figure 4: NLL, ECE, accuracy and predictive entropy of individual versus ensemble models for IID data as well as input data with noise of different intensity.

samples and/or when trained with complex data augmentation). When models are particularly overconfident (e.g., for OOD data), we can further increase the predictive uncertainty by calibrating individually trained models and ensembling them or ensemble the trained models and then calibrate [8]. Also wen2020combining stated that combining deep ensembles and data augmentation might worsen the uncertainty estimates; calibration could also be essential in this case.

4   Experimentation Setup

In this work[1], we use CIFAR10 and CIFAR100 datasets to test the efficiency of temperature scaling at calibrating uncertainty. We train Densenet40 and Resnet50 architectures similarly to the original article (200 epochs with learning rate 0.001). For the deep ensemble work, we use the publicly available EUROSAT dataset which contains 10 different classes of satellite images with total of 27000 images [12]. This dataset is first split into 70:15:15 (train:valid:test) then trained with a small custom model built of 4 convolutional layers followed by ReLU activation functions and batch normalization and one linear layer as a final layer. We also train a Resnet50-like network using its pre-trained weights. For tuning hyperparameters, we use Raytune gridsearch (publicly outsourced for Pytorch and Tensorflow). In our experiments, the learning rate is 0.001 with batch size 128. The number of models to be ensembled is chosen to

be 10. The complete implementation is done using Pytorch framework.

5   Results and Discussion

Temperature scaling: We have applied this post-processing calibration method on an image classification problem based on the CIFAR10 and CIFAR100 dataset. (see Fig 2, 3). We measure an ECE (with M = 10 bins) before temperature scaling of 6.51% and 21.72% for CIFAR10 and CIFAR100 respectively. After calibration, the ECE drops to 1.28% and 3.14% respectively showing the neural network uncertainties have been efficiently re-calibrated.

Ensembles: We have ensembled 10 models. The accuracy of the ensemble model is significantly increased compared to individual models performances. We are also able to show an improvement in uncertainty estimates with e.g., a higher uncertainty for misclassification in the IID data. To further our analysis, we added Gaussian noise (with mean = 0 and var =0.1, 0.2, 0.3) to the test inputs to introduce the shift in data to analyse how ensemble methods performs with regards to OOD/corrupted test inputs, the ensembles outperformed in all the metrics(NLL, ECE and accuracy) comparing to individual models (see Fig 4). We also show that calibration of ensembles is necessary especially for OOD data or extremely corrupted test inputs where there is a further improvement in ECE and NLL, however, vanilla or standard ensemble method would be enough for IID data(see Fig 5).

---

[1] We also apply similar experiments to an Airbus-driven use case (object classification on runway images). We will however focus this report on results obtained on public dataset for the sake of IP-protection.

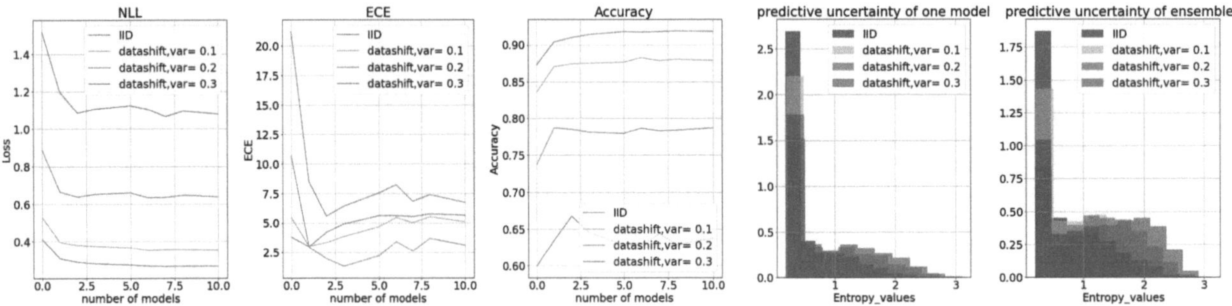

Figure 5: NLL, ECE, accuracy and predictive entropy of individual calibrated and ensembled models for IID data as well as input data with noise of different intensity.

6 Conclusion

In this work, we have explored a simple and easy post-processing calibration technique to calibrate the overconfident neural network models, but they are limited to IID datasets, they perform poorly when there is a small shift in the data. We have also explored deep ensembles technique and they exhibit higher uncertainty especially when there is a shift in the data or OOD data,at the core we have also shown why sometimes even calibration of deep ensembles is necessary.

Acknowledgement

The work has been carried out at Airbus AI Research, Airbus Defense and Space GmbH, Ottobrunn under the supervision of Dr. Audrey Galametz (Airbus), and Philipp Gruening, Institute for Neuro- and Bioinformatics, Universität zu Lübeck.

Author's Statement

Conflict of interest: Authors state no conflict of interest.

7 References

[1] D. Amodei, C. Olah, J. Steinhardt, P. Christiano, J. Schulman, and D. Mané, "Concrete problems in ai safety," *arXiv preprint arXiv:1606.06565*, 2016.

[2] A. Kendall and Y. Gal, "What uncertainties do we need in bayesian deep learning for computer vision?" *arXiv preprint arXiv:1703.04977*, 2017.

[3] W. J. Maddox, P. Izmailov, T. Garipov, D. P. Vetrov, and A. G. Wilson, "A simple baseline for bayesian uncertainty in deep learning," *Advances in Neural Information Processing Systems*, vol. 32, pp. 13 153–13 164, 2019.

[4] A. Graves, "Practical variational inference for neural networks," *Advances in neural information processing systems*, vol. 24, 2011.

[5] B. Lakshminarayanan, A. Pritzel, and C. Blundell, "Simple and scalable predictive uncertainty estimation using deep ensembles," *arXiv preprint arXiv:1612.01474*, 2016.

[6] Y. Gal and Z. Ghahramani, "Dropout as a bayesian approximation: Representing model uncertainty in deep learning," in *international conference on machine learning*. PMLR, 2016, pp. 1050–1059.

[7] Y. Wen, G. Jerfel, R. Muller, M. W. Dusenberry, J. Snoek, B. Lakshminarayanan, and D. Tran, "Combining ensembles and data augmentation can harm your calibration," *arXiv preprint arXiv:2010.09875*, 2020.

[8] R. Rahaman and A. H. Thiery, "Uncertainty quantification and deep ensembles," *arXiv preprint arXiv:2007.08792*, 2020.

[9] C. Guo, G. Pleiss, Y. Sun, and K. Q. Weinberger, "On calibration of modern neural networks," in *International Conference on Machine Learning*. PMLR, 2017, pp. 1321–1330.

[10] J. Platt *et al.*, "Probabilistic outputs for support vector machines and comparisons to regularized likelihood methods," *Advances in large margin classifiers*, vol. 10, no. 3, pp. 61–74, 1999.

[11] S. Fort, H. Hu, and B. Lakshminarayanan, "Deep ensembles: A loss landscape perspective," *arXiv preprint arXiv:1912.02757*, 2019.

[12] "Eurosat dataset," https://github.com/phelber/EuroSAT#.

Feature Identifier for Underwater Ultrasonic Multipath Diffraction Model

Garrett Mulkerin [1], Fabian John [1], Horst Hellbrück [1]

[1] Technische Hochschule Lübeck - University of Applied Sciences, Germany, Department of Electrical Engineering and Computer Science, Center of Excellence CoSA garrett.mulkerin@stud.th-luebeck.de, fabian.john@th-luebeck.de, horst.hellbrueck@th-luebeck.de

Abstract

Ultrasonic pulses are widely used in underwater localization applications. This paper applies a spectral ultrasonic multipath propagation model [1] to train a neural network for precise underwater object localization with the auto-machine learning platform Modulos. Modulos auto-machine learning allows us to run unsupervised calculations. We observed The mean absolute error of a generated prediction to analyze the effectiveness of the new model. To do this, we put three datasets into Modulos. We created the datasets using the spectral ultrasonic multipath propagation model, with two being modified to explore different applications. These two datasets used segments of the original data vector to explore different Modulos generated solutions and computation times. All datasets produced an absolute mean error of 0.00708 m or less.

1 Introduction

One of the most used abilities of Machine Learning is data prediction. This ability allows high complexity mathematics to be simplified. The Multipath Diffraction model, developed within the funded research project EXTENSE, is validated for free-water object localization tasks and outputs a spectrum for a given transducer object geometry. The model utilizes distances, frequencies, and reflection coefficients to generate the spectral output. In measurements, we know these factors before getting the spectrum. However, for real-world applications, the factors vary. Therefore it would be beneficial to have a model that does the opposite. Using the spectrum, we calculate features such as the line of sight to the object. Extracting these features by hand would have a high level of complexity. Therefore, we use the Modulos Auto Machine Learning (autoML) tool. This tool allows us to generate solutions without the use of manual calculation. Using Modulos also allows us to analyze variations of the problem.

EXTENSE must prove the possible usefulness of machine learning spectral analysis within underwater localization purposes. By developing a method to create a model for a single input identification, we expand the EXTENSE localization system by applying this method to multiple inputs.

The development of the machine learning results provides us with the following benefits:

- Expanding current EXTENSE applications

- Insight on autoML outcomes with spectral inputs.

The paper is structured as follows: Section 2 discusses the background material needed for the project and the methods used in data generation and solution computation. Section 3 discusses the results of the generation solutions and addresses possible solution modification. Finally, Section 4 is a conclusion that addresses the use of the solutions and discusses future work with the proposed results.

2 Material and Methods

2.1 Related Work

An underwater ultrasonic multipath diffraction model was developed to create data points similar to those measured in real-world applications [1]. This model calculates the effect that various geometries have on a pulse-echo. They created Mathematical models for diffraction, line of sight transmission, and reflection. The model takes in several geometrical and physical parameters shown in Figure 1.

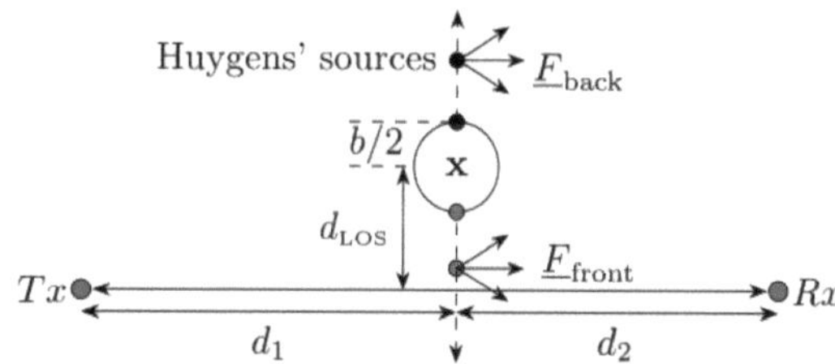

Figure 1: Visualization of diffraction model with components used for data generation ©2019 IEEE, adapted from [2].

These inputs vary depending on the object and where a

measurement is taken. The model calculates a spectrum output similar to one made using real-world measurements using these inputs. This model only works for calculating the spectrum but not any of the input components. The primary goal of this paper was to take this model and invert it using the same parameters as shown in Figure 2.

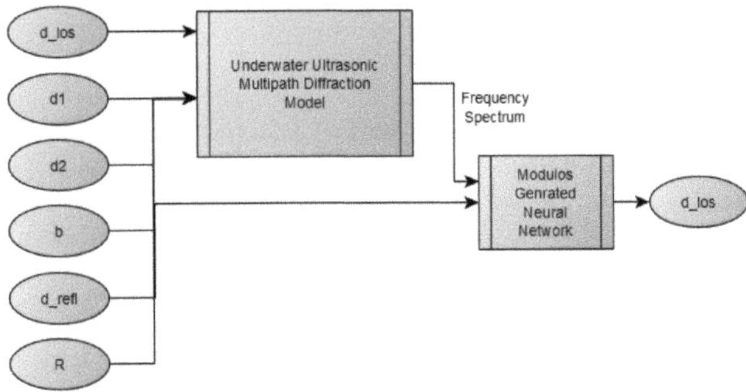

Figure 2: Block diagram showing input and output as well as the use of underwater multipath propagation model.

2.2 Viability of Spectrum Inputs

As spectrum vectors are quite large, we must explore the viability of using a neural network with spectrum input. Previous studies have used spectral components and machine learning to test their viability [3, 4, 5]. As stated in [3], many spectrum analyses are still done manually. To improve this, they show the viability in spectrum analysis using 5G spectrum within a machine learning program. This report shows that a spectrum is a valuable component in automated calculations but is complicated enough that machine learning is a valuable tool for this purpose.

In the field of medical research, a study was done on magnetocardiography [6]. This research attempted to detect and locate heart disease. To do this, they used feature extraction and machine learning to help identify heart disease characteristics quickly. The features extracted were time domain, frequency domain, and information theory features. The study shows the possibility of using a frequency domain input and several other factors to achieve machine learning results.

Another project utilized Audio spectrum to identify mechanical vibrations[7], focusing more on identification using spectral analysis. This study uses a similar regression method to the one we wanted to create, with the difference being that our model allows more variation with the network inputs.

2.3 Modulos Tool

As seen in Figure 2, we generate our neural network with the tool Modulos. Modulos is an Auto Machine Learning tool that creates various neural network solutions for a single dataset. To do this, a user chooses desired network output, minimized factor, autopause criteria, feature extraction, and whether it is a categorical network or regression-based. Once we choose these criteria for a workflow, the tool searches for the best neural network solution. The final result is

chosen based on your minimized factor criteria, and the neural network solutions could be very different between workflows even if the data is similar. The discovered solution is available for download and applicable for any desired implementation. These solutions are in the form of clients within python terminals and jupyter notebook.

2.4 Dataset Generation

To test the Modulos results, we explore possible solutions that find the line of sight distance known as d_los. The spectral ultrasonic multipath model is the primary data generation method of the Modulos workflow inputs. To create a single dataset, we take the model's output and add the original inputs of the same model to create a single string of 297. We repeat this process until there are enough strings for neural network training. Variation within a single generation was needed to make the network training useful. We must modify the spectral ultrasonic multipath Matlab generation slightly to do this. First, we set d_los to a vector that increments evenly for a set amount of instances, as seen in the following code block.

```
d_los = 0:4e-4:5e-2;
rng_vals = zeros(125,5);
[~, y] = us_mp_mdl.get_rx_db(0);
y_store = zeros(length(d_los), length(y));
```

For each instance, before spectrum generation, the other inputs, such as distance one and two, are randomized to create different spectrum outputs for similar d_los. We set this random data to ranges for each variable we want to give variety to in each iteration. The following block shows the random generation loop, including the spectral ultrasonic multipath model call line.

```
for idx = 1:length(d_los)

us_mp_mdl.d1 = d1min+rand(1)*(d1max-d1min);
rng_vals(idx,1) = us_mp_mdl.d1;
us_mp_mdl.d2 = d2min+rand(1)*(d2max-d2min);
rng_vals(idx,2) = us_mp_mdl.d2;
us_mp_mdl.b = bmin+rand(1)*(bmax-bmin);
rng_vals(idx,3) = us_mp_mdl.b;
us_mp_mdl.d_refl = d_reflmin+rand(1)
    *(d_reflmax-d_reflmin);
rng_vals(idx,4) = us_mp_mdl.d_refl;
us_mp_mdl.R  = Rmin+rand(1)*(Rmax-Rmin);
rng_vals(idx,5) = us_mp_mdl.R;
[f, y_store(idx,:)] =
    us_mp_mdl.get_rx_db(d_los(idx));

end
```

We then place the variables in the model, generating a spectrum. Once generated, we extract a usable portion of the spectrum and place the data within a comma-separated values file.

To explore possible results, we created more datasets using slightly different criteria. The two sets created utilized the same approach as the original dataset but kept different components. One utilizes half of the initially used spectrum, resulting in fewer inputs for solution generation. This variation will show if calculation time reduction is possible since it uses fewer parts. The other utilizes only the usable spectrum without the other inputs. This variation shows the possibility to predict inputs without external information besides the spectrum itself.

2.5 Modulos Workflow Generation

Modulos doesn't accept only the Matlab-generated code. Modulos needs a compressed tape archive file that contains both the Matlab data and a dataset structure file to make the dataset. This structure file allows Modulos to correctly extract the data with their labels, shown in the following code block.

```
[{"name": "dlos",
"path": data_dLos.csv",
"type": "table"},
{"_version": "0.2"}]
```

We created a batch script to simplify the generation of multiple datasets for testing purposes. This script took the output file created by Matlab and moved it to a separate folder containing a structure file made using the name of the generated Matlab file. Once in this new location, it generated the needed tape archive file for Modulos upload. After creating this file, it is copied and renamed to the appropriate title of which dataset was used. We then upload the dataset to the Modulos tool.

Once uploading a dataset, we create the workflow. First, we chose the dataset to investigate and select the desired variable search, in this case, d_los. We chose to minimize absolute mean error (AME) in prediction for our objective. We also allowed the calculations to run until finding no improvements after 200 solutions using a Bayesian optimizer.

2.6 Solution Evaluation

To view the solutions, we need to extract the Modulos result and use Python terminals to view the data. Modulos recommends using a Linux operating system to run the clients for optimal functionality. Once the result is downloaded and extracted, we use the batch client provided to calculate individual results. This client requires data to be structured in the format detailed in the following example code in the correct order of the input data of the simulation. Note that not all data points for a set were listed.

```
sample_dict =
{'1000': -2.3086,
'1001':-2.3226,
'd1': 0.050113,
'd2': 0.019438,
'sample_ids_generated': '106'}
```

The Modulos solution provides the training dataset to run a validation check yourself. However, we use the same structure to calculate the results of external data. This external

data is applicable from any source, such as real-world measurements or generated data if put into the correct structure. This structure allows us to take the solution and use it in other applications.

3 Results and Discussion

All three variations received similar results for their simulated solutions. Each variation improved from its worst result by at least a factor of ten.

Table 1: Computation Results of Training.

Workflow	Time	Solutions
Original	2:05:20:06	532
Less Spectrum	0:17:11:21	224
Spectrum Only	1:07:33:16	217

Table 2: Score Results of Training.

Workflow	Best AME	Worst AME
Original	0.00536	0.0628
Less Spectrum	0.00632	0.0368
Spectrum Only	0.00708	0.169

While there seems to be a disparity within the length of time for each version, we must acknowledge the inherent randomness in neural networks. Rerunning the simulations could produce different solution computation times. However, it takes about one to two days for Modulos to find the neural network of the simulation using autopause criteria regardless of which dataset style you select.

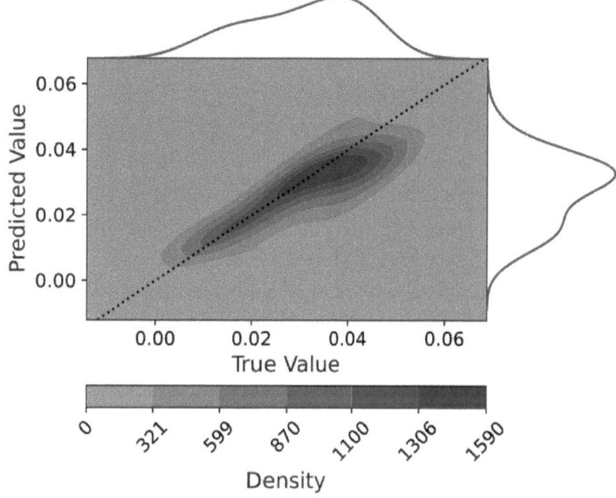

Figure 3: Kernel density plot of Original Data.

Figure 3 shows that the results of the density kernel follow the regression line quite well for values between zero and approximately 0.03 m. After this point, the prediction trends toward predicting values slightly less than the actual value. While this trend is still close to desired for this set of data, there is a possibility that using this network for larger d_los applications would have undesirable predictions.

While all three datasets create solutions with acceptable absolute mean error values, each set has some advantages. The original provides the most accurate results for use in further applications. However, the solution created using less spectrum shows similar results and requires fewer inputs than the original. The spectrum only set provides a result similar to the original, but it allows the advantage of not knowing the other input factors.

Note that there are two ways we could change the solution computation to receive better results possibly. However, neither of these methods guarantees a better approach. Increasing the pause criteria to 300 or higher instead of 200 could, in the end, provide better results due to requiring more simulations. Similarly, setting a goal mean absolute error would also create more simulation results. The problem with both methods is that these increase the possibility of the programs running indefinitely without guaranteeing results. The top solutions of all three sets resulted in very little difference, sometimes only a difference of one ten-thousandth. Allowing more simulations to occur could drastically increase simulation time while providing only marginal benefits.

4   Conclusion

The models made using Modulos show a viable solution creation method for this particular spectrum analysis. The models each have advantages but individually also have very acceptable error rates. We generate models for any component desired within a few days using this method. The abilities of Modulos also allow multiple solution generations to happen simultaneously. These abilities provide a helpful and relatively quick method of future model generation. By creating this input model, we move to explore real-world measurements.

To further develop this model, it would be helpful to create a similar model for all of the different input variables used in the original Diffraction model. Making these similar models creates a possibility that a single spectrum input extract all input information. By doing this, we expand the possible solutions that we will implement in future measuring equipment. If possible, this would allow the calculation of all inputs using a measured spectrum reading from a pulse-echo system. This expanded model would create a method of underwater location to develop the EXTENSE project further.

To show the complete utility of the model, we need to perform real-world measurement validation. This validation is needed to show that the models created here do not have an issue with the overfitting of the data used for generation.

Acknowledgement

This publication results from the research of the Center of Excellence CoSA at the Technische Hochschule Lübeck and is funded by the Federal Ministry of Economic Affairs and Energy of the Federal Republic of Germany (Id 03SX467B, Project EXTENSE, Project Management Agency: Jülich PTJ). It is the Product of the Module "Scientific Project", which is part of the Applied Information Technology Curriculum. Horst Hellbrück is an adjunct professor at the Institute of Telematics of the University of Lübeck.

Author's Statement

Conflict of interest: Authors state no conflict of interest.

5   References

[1] F. John, M. Cimdins, and H. Hellbrück, "Underwater ultrasonic multipath diffraction model for short range communication and sensing applications," *IEEE Sensors Journal*, vol. 21, no. 20, pp. 22 934–22 943, 2021.

[2] M. Cimdins, S. O. Schmidt, and H. Hellbrück, "Modeling the magnitude and phase of multipath uwb signals for the use in passive localization," in *2019 16th Workshop on Positioning, Navigation and Communications (WPNC)*, 2019, pp. 1–6.

[3] V. Nagpure, S. Vaccaro, and C. Hood, "A case study of spectrum analysis using unsupervised machine learning," in *2019 IEEE International Symposium on Dynamic Spectrum Access Networks (DySPAN)*, 2019, pp. 1–2.

[4] M. Kulin, T. Kazaz, I. Moerman, and E. De Poorter, "End-to-end learning from spectrum data: A deep learning approach for wireless signal identification in spectrum monitoring applications," *IEEE Access*, vol. 6, pp. 18 484–18 501, 2018.

[5] X. Ma, S. Ning, X. Liu, H. Kuang, and Y. Hong, "Cooperative spectrum sensing using extreme learning machine for cognitive radio networks with multiple primary users," in *2018 IEEE 3rd Advanced Information Technology, Electronic and Automation Control Conference (IAEAC)*, 2018, pp. 536–540.

[6] R. Tao, S. Zhang, X. Huang, M. Tao, J. Ma, S. Ma, C. Zhang, T. Zhang, F. Tang, J. Lu, C. Shen, and X. Xie, "Magnetocardiography-based ischemic heart disease detection and localization using machine learning methods," *IEEE Transactions on Biomedical Engineering*, vol. 66, no. 6, pp. 1658–1667, 2019.

[7] J.-S. Liang and K. Wang, "Vibration feature extraction using audio spectrum analyzer based machine learning," in *2017 International Conference on Information, Communication and Engineering (ICICE)*, 2017, pp. 381–384.

Multi-omics data analysis of patients with colorectal carcinoma

Tim Lenfers [1], Axel Künstner [2], Michael Kohl [3], Thorben Sauer [3], Timo Gemoll [3], Hauke Busch [2]
[1] Medical Informatics, Universität zu Lübeck, tim.lenfers@student.uni-luebeck.de
[2] Lübeck Institute of Experimental Dermatology, University of Lübeck,
axel.kuenstner@uni-luebeck.de, hauke.busch@uni-luebeck.de
[3] Section for Translational Surgical Oncology & Biobanking Department of Sugery , University of Lübeck,
Michael.Kohl@uksh.de, thorben.sauer@student.uni-luebeck.de, Timo.Gemoll@uksh.de

Abstract

Due to the development of new high-throughput methods for the analysis of biological processes, more and more data are generated which have to be analyzed by bioinformatic processes. The resulting so-called omics data provide a perspective of the biological processes. The integration of different omics data, in other words, the integration of different biological perspectives, is a promising approach for the diagnosis of complex diseases, as well as the prognosis of their disease courses. However, the analysis of these data brings its own challenges. The aim of this work is to provide an insight ino the data processing pipeline and challenges working with multi-omics data using the example of biomarker discovery and creation of predictive machine learning models on two target variables.

1 Introduction

More than five percent of all Germans are developing colorectal cancer during their lifetime. It is the third most common cancer in men, and the second most common in women [1]. Colorectal cancer is a subgroup of malignant tumors of the intestine, of which colorectal carcinomas, located in the colon and rectum, account for more than 95% of these tumors. Depending on the stage of the disease detected, the five-year survival rate with treatment by surgery and chemotherapy reduces dramatically. If cancer only has been found in the colon or rectum (Union for International Cancer Control (UICC) Stage I) the five-year relative survival rate is 89.9%. However, if cancer could already metastasize to other organs (UICC Stage IV) the five-year relative survival rate reduces to 14.2% [2]. It is possible that recurrences will occur after successful therapy and that the recurrence is located in the same place as the primary tumor or a different place. The average five-year recurrence rate for people with successful surgery is 5% for UICC Stage I disease, 12% for UICC Stage II disease, and 33% for UICC Stage III disease. Depending on various risk factors, these values can fluctuate. In order to ensure a cure of colorectal carcinomas, it is therefore important to recognize possible risks for recurrences at an early stage. For early detection, multi-omics data from patients can be analyzed to detect biomarkers. The discovered biomarkers can then be used to predict recurrences [3]. In this work, the relation between metastatic illnesses and recurrence of primary tumor of colorectal carcinoma based on multi-omics data will be analyzed.

2 Material and Methods

2.1 Multi-omics

The central dogma in molecular biology describes the flow of genetic information in the cell from DNA (deoxyribonucleic acid) to mRNA (messenger ribonucleic acid) to protein in living beings [4]. To understand complex biological diseases such as colorectal carcinoma, one can study measured omics data which can be used as biomarkers to investigate the flow of this genetic information and find malfunctions within the different steps of the pathway which are leading to the disease [5].

The study of DNA, mRNA, and proteins is broadly denoted as genomics, transcriptomics, and proteomics respectively. Genomics describes the exploration of the genetic building plan of a cell. It is used to investigate specific genes. In transcriptomics analysis, the transcribed genetic material is used to examine genes that are actively expressed. Therefore, it can be analyzed what is happening at the cellular levels. Proteomics helps to understand information flow within the cell and organism by identifying the expression or abundance of proteins. All omics are part of the same biological information pipeline, and each of these biological perspectives can be measured using specialized high-throughput technologies [6].

Biomarkers are measurable indicators of biological processes that have e.g. diagnostic, prognostic, and therapeutic significance. They can indicate the risk or progression of a disease and are therefore important for early targeted intervention [7].

This work focuses on the integration of multiple omic datatypes as the usage of single omics can deteriorate performance due to inherent characteristics [6].

2.2 Dataset

For this analysis, a demo-dataset from the University of Lübeck Section for Translational Surgical Oncology & Biobanking Department of Surgery was used, which was modeled on patient data of the hospital. The provided dataset consists of related clinical and multi-omics data of patients with colorectal cancer. Information that might lead to the identification of real clinical data was anonymized in advance. Written and informed consent is obtained by the Translational Surgical Oncology & Biobanking Department of Surgery prior to collection of the data and preserving the biological specimens. The collection and processing of the data have been cleared by the ethics committee of the University of Lübeck. The clinical part of the dataset consists of several parameters regarding the patient's personal and therapeutic data. Copy number variations, oncogene mutations (as single nucleotide variants), proteomics and transcriptomics data is available as omics data. The oncogene mutations were selected based on the set of known mutated genes in cancer.

2.3 Machine Learning

In health data analysis, statistical modeling has been the de-facto standard. With the increase of affordable computational power and high-throughput omics data, the success of machine learning models and artificial intelligence technologies has become more popular in various fields.

Machine Learning models can be used to mine information based on the given data without introducing statistical assumptions, which is the case with the use of conventional statistics-based models.

The objective of machine learning models is to use the acquired knowledge from the given data to understand its interrelationships and use it to make forecasts for future data measures [6].

In this work, conventional machine learning methods were used because the use of complex deep learning methods requires larger data sets to avoid overfitting, i.e. avoiding memorization of the data set.

2.4 Challenges in multi-omics analysis

Using machine learning on multi-omics data through high-throughput methods brings its own challenges.

At first, the omics data of different high-throughput sources are usually heterogeneous. Different normalization and scaling techniques are used in the analysis of the different omics. This leads to different dynamic ranges and distributions. Furthermore, some omics analysis are prone to generate more sparse data or outliers which have to be taken into account. Imputation and outlier detection should thus be considered for each omic data before integration. Imputation is a statistical process in which missing values in a data set are substituted by an estimate of the distribution of this variable. [8]

When using clinical patient data for disease classification, some classes occur less frequently than others, which leads to an imbalance in the dataset. If a machine learning model is trained with an imbalanced dataset, the model may be overfitted. The model has i.e. a high accuracy on the training data but underperforms in the generalization step, the usage on novel data.

In addition, multi-omics datasets have typically much fewer observed samples than multi-omics features and suffer therefore from the classical *curse of dimensionality* problem. The *curse of dimensionality* describes that the resulting high-dimensional space of the trained models often contains correlated and redundant features which can mislead the trained algorithm. Therefore, the selection of important features is an important step in this data processing pipeline. When using machine learning methods, it should be noted that each method brings its own strengths and weaknesses. Hence, it is essential to apply different models to the issue and to compare them. [9]

Last but not least, the explainability of the models should be considered. Since the issue of this work relates to the discovery of biomarkers, a transformation of the features should be avoided, where inference to the original features is no longer possible.

2.5 Data integration and analysis

Various workflow pipelines are possible for integrating the multi-omics data, as can be seen in figure 1. The workflow pipeline consists of four steps. Stage 1 contains the raw data. In this stage, individual preprocessing steps take place for each omic data type. Depending on the integration method, preprocessing steps for integration are performed in stage 2. Thereafter, the omics data will be integrated into a joined model. In the concatenation-based integration approach, the omics data will be joined into a single matrix, which will be used to train one single model.

In model-based integration, a separate learning procedure is used for each omic data type. Subsequently (Stage 3), the models are connected with each other.

When using the transformation-based approach, each omic is transformed into a new representation using graphs or kernel matrices. Afterwards, all new representations are combined into one before constructing a model. The transformation process causes the original features to be lost, which is why the transformation-based approach is not suitable for biomarker discovery and is therefore not examined in this work. [6]

2.6 Metrics

To evaluate the models performance the following metrics were used.

Root mean squared error (RMSE) is the square root of the mean of the square of all errors. The larger the RMSE, the worse the fit of the model.

$$RMSE = \sqrt{\frac{1}{n}\sum_{i=1}^{n}(S_i - O_i)^2} \qquad (1)$$

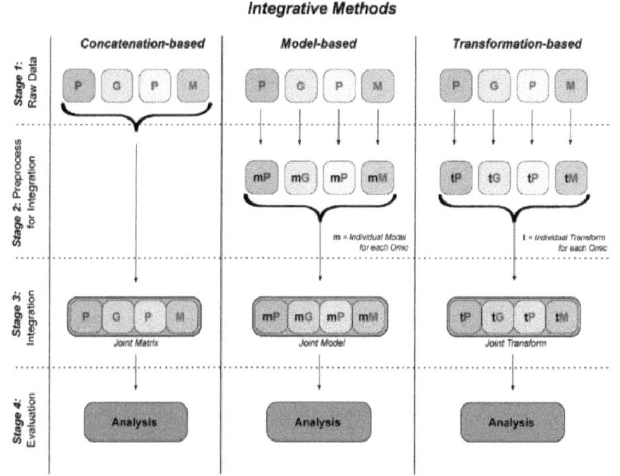

Figure 1: workflow pipelines for different types of integration methods for multi-omics analysis.
P: Phenotype, G: Genomics, P: Proteomics, M: Metabolomics [6]

where O_i are the observations, S_i predicted values of a variable, and n the number of observations available for analysis.

The mean absolute error (MAE) measures the error between paired observations expressing the same phenomenon.

$$MAE = \frac{\sum_{i=1}^{n} |y_i - x_i|}{n} \qquad (2)$$

where y_i is the prediction and x_i is the true value.

2.7 Data processing pipeline

In this work, the concatenation integration approach was used for the data processing pipeline as this allows the multi-omics data types to be considered together. At first, the individual data was preprocessed. The single nucleotide variants were filtered based on a variant allele frequency of at least 5 %, coverage of 100x, and only frameshift, missense, nonsense, and splice-site mutations were considered. The transcriptomics and proteomics data were batch corrected using combat [10]. In addition, the proteomics data had to be imputated using random forest imputation due to the sparsity of the data.

Subsequently, the continuous omics data was normalized and scaled using a z-score normalization before the integration step.

For integration, different models have been trained for comparison using 10-fold cross validation to avoid overfitting. The models used were Naïve Bayes, K-Nearest Neighbors (KNN), Support Vector Machines (SVM) with a linear kernel, Random Forest (RF), and a basic 3-layer Multilayer perceptron (MLP).

The training of the models was separated into training with and without previous feature selection. To select the most important features the wrapper method *recursive feature elimination* (RFE) was used and the top 40% most important features were selected. [11] The recursive feature elimi-

nation method repeatedly applies a predictive model on different subsets to select the most prominent features in each round. Since this procedure selects the most informative variables for the target variable, this procedure can already be used as biomarker discovery. In addition, an embedded feature selection method was used within the random forest classification method, since the selection of key features is part of the learning procedure. A principal component analysis could not be used to remove redundant data as this would result in a loss of biomarkers.

The models were trained on the target variables survival time (in months) as a regression task and secondary tumor as classification. The secondary tumor variable contained the following classes: none, recurrence, metastasis, secondary carcinoma, and the combination of these classes. To reduce the risk of overfitting introduced by class imbalances, all classes with less than four occurrences were removed. This led to the fact that only the classes *none* and *metastasis* remained, thus the classification task became a binary classification problem.

The data processing pipeline was implemented using the R programming language and the R package *caret* for the machine models and feature selection [12].

3 Results and Discussion

The following section shows the results for the biomarker discovery and the models' performance. The results were only obtained for the target variable survival time as the models did not converge for the target variable secondary tumor.

Table 1 shows the ranked features of the feature selection method "RFE" based on the scoring given by the corresponding algorithm. The score should be considered as a comparative value between the features and not as an absolute one. It can be seen that there are no clear important features, as none of the scores stand out unambiguously. The features p1, p2, and p3 in table 1 are proteins, the others are clinical features.

Table 1: Found biomarkers from RFE

variable	score
progress	3.093407
p1	3.092895
age	2.761403
p2	2.682206
tumor1	2.606259
status	2.592535
p3	2.222712
karnof	2.191458

In table 2 the performance evaluation of the tested models is shown for the regression task target survival time in months. The best results were obtained using a random forest model without the selection of features in a previous step, which is marked as bold in the table 2. It should nevertheless be taken into account that the random forest model has a feature selection method build in. The next best model is the model "KNN" without a feature selection step, followed by the random forest with the rfe feature selection step.

Table 2: Model performances for survival time

Feature selector	Model	RMSE	MAE
-	knn	48.68	40.17
-	mlp	105.84	93.55
-	nnet	99.75	80.37
-	**rf**	**41.44**	**26.07**
-	svmLinear	492.45	369.68
rfe	kknn	65.19	57.33
rfe	mlp	69.76	49.53
rfe	nnet	99.75	80.37
rfe	rf	59.03	48.29
rfe	svmLinear	68.43	56.16

4   Conclusion and Outlook

In this work, the data processing pipeline and integration of multi-omics data could be demonstrated using the example of biomarker discovery and machine learning. No clearly important features could be identified during the biomarker discovery task. Therefore the initial task to identify novel biomarkers could not be achieved. It was nevertheless possible to create a decent model for survival time prediction. However, it was not possible to train a model to predict metastasis using the target variable *secondary tumor* as the classes could not be separated.

In this work, the difficulty in dealing with biological, especially multi-omics data, became clear. On the one hand, special attention must be paid to the privacy of patients when obtaining data, since biological data always uniquely identify individuals. On the other hand, the processing of multi-omics data is challenging due to its characteristics such as heterogeneity and sparsity. This was observed, for example, in the processing of proteomics data, where missing values had to be filled by imputation before further processing. The usage of imputation can also introduce an algorithm-related bias.

In future work other feature selection methods should be taken into account to get another perspective on the most significant features. Also, different integrative methods described in 2.5 should be tested in order to find out correlations between the omics data types which could lead to better understanding and better results. Finally, other clinical parameters should be evaluated as possible outcome variables to gain further insight from the data set.

Acknowledgement

The work has been carried out at the Group for Medical Systems Biology of the Lübeck Institute of Experimental Dermatology in cooperation with the University of Lübeck Section for Translational Surgical Oncology & Biobanking Department of Sugery.

Author's Statement

Because the data used were modeled on patient data, clinical information had to be anonymized in this work for privacy reasons.

5   References

[1] Ferlay J, Ervik M, Lam F, Colombet M, Mery L, Piñeros M, Znaor A, Soerjomataram I, and Bray F, "Cancer today." https://gco.iarc.fr/today, 2020.

[2] G. Herold, *Innere Medizin 2020*. De Gruyter, 2020.

[3] E. Osterman and B. Glimelius, "Recurrence Risk After Up-to-Date Colon Cancer Staging, Surgery, and Pathology: Analysis of the Entire Swedish Population," *Diseases of the Colon and Rectum*, vol. 61, pp. 1016–1025, Sept. 2018.

[4] F. H. Crick, "On protein synthesis," *Symposia of the Society for Experimental Biology*, vol. 12, pp. 138–163, 1958.

[5] B. Alberts, "Molecular biology of the cell 5E," *Garland science*, pp. 906–911, 2008.

[6] P. S. Reel, S. Reel, E. Pearson, E. Trucco, and E. Jefferson, "Using machine learning approaches for multiomics data analysis: A review," *Biotechnology Advances*, vol. 49, p. 107739, July 2021.

[7] K. Strimbu and J. A. Tavel, "What are biomarkers?:," *Current Opinion in HIV and AIDS*, vol. 5, pp. 463–466, Nov. 2010.

[8] A. R. T. Donders, G. J. M. G. van der Heijden, T. Stijnen, and K. G. M. Moons, "Review: A gentle introduction to imputation of missing values," *Journal of Clinical Epidemiology*, vol. 59, pp. 1087–1091, Oct. 2006.

[9] B. B. Misra, C. Langefeld, M. Olivier, and L. A. Cox, "Integrated omics: Tools, advances and future approaches," *Journal of Molecular Endocrinology*, vol. 62, pp. R21–R45, Jan. 2019.

[10] Y. Zhang, G. Parmigiani, and W. E. Johnson, "ComBat-seq: Batch effect adjustment for RNAseq count data," *NAR Genomics and Bioinformatics*, vol. 2, p. lqaa078, Sept. 2020.

[11] B. F. Darst, K. C. Malecki, and C. D. Engelman, "Using recursive feature elimination in random forest to account for correlated variables in high dimensional data," *BMC genetics*, vol. 19, no. 1, pp. 1–6, 2018.

[12] M. Kuhn, J. Wing, S. Weston, A. Williams, C. Keefer, A. Engelhardt, T. Cooper, Z. Mayer, B. Kenkel, R. C. Team, M. Benesty, R. Lescarbeau, A. Ziem, L. Scrucca, Y. Tang, C. Candan, and T. Hunt, "Caret: Classification and Regression Training," Oct. 2021.

Measurement data analysis of pulse waves using machine learning on a microcontroller system

Felix Wätge [1], Sebastian Hauschild [2] and Horst Hellbrück [2]

[1] Applied Information Technology, Luebeck University of Applied Sciences, felix.waetge@stud.th-luebeck.de
[2] Luebeck University of Applied Sciences, Department of Electrical Engineering and Computer Science, Center of Excellence CoSA, {sebastian.hauschild,horst.hellbrueck}@th-luebeck.de

Abstract

To save computing capacity on integrated circuits, the use of machine learning in measurement data analysis is investigated. It is tested whether it is possible to reliably distinguish the measuring subjects from each other by using machine learning on integrated circuits with limited resources based on recorded raw data of a photoplethysmography (PPG) measurement. For processing the data, different network structures, based on fully connected layers or convolutional layers, are trained on the recorded data and compared against each other for their test accuracy. The evaluation of the comparison shows that both chosen network structures have a still insufficient accuracy when applied to test data to be used for person distinction.

1 Introduction

The use of highly integrated microcontrollers with small form factors is necessary to make systems resource-efficient, cost-effective, space-saving and scalable. In particular, wireless distributed systems in medical technology benefit from the efficient design due to lower energy consumption and portability. At the same time, data processing requirements are becoming more complex, so the more rudimentary but energy-efficient system architecture is complementary to the necessary system performance. Current developments show that the integration of artificial intelligence (AI) into resource-efficient systems enables significantly more complex evaluations than conventional machine learning methods.

Using a microcontroller-based system and machine learning methods, it is to be analyzed whether persons can be reliably distinguished on the basis of their pulse shape. The required measurement data are to be collected for several test persons with all the limitations that could arise from the use of integrated circuits.

The paper is organized as follows: First, related work is discussed in section 2. In section 3, the approach to the measurement idea is explained. Section 4 gives an overview of the hardware and software used. The configuration of the measurement sensors and the procedure for acquiring measurement data are also described here. Finally, the parameters used to train the neural network are defined. The results of the training processes are then discussed in section 5. Finally, a summary of the work results and an outlook on future work approaches is given in section 6.

2 Related Work

The measurement data acquisition of this work is based on the measurement method of pulse oximetry. In this optical measurement, the skin and the vascular structure near the surface are transilluminated with the aid of two LEDs. At the same time, the reflected light is determined with a photodetector. A nearly static or slowly changing fraction (DC) of the reflected light is thereby determined by respiratory movements, sympathetic nervous system activity, and thermoregulation, in addition to previously unknown factors. This component is superimposed by "a pulsatile (AC) physiological waveform attributed to cardiac synchronous changes in the blood volume with each heart beat" [1].

In addition to blood volume, the degree of oxygen saturation also determines how much of the light is reflected by the blood. Oxygen is transported through the body with the help of hemoglobin as a component of the blood. The proportion of light absorbed differs between oxygenated and non-oxygenated hemoglobin. Additionally, the named difference in light absorption varies when considering different wavelengths. Fig. 1 shows the relative absorption coefficient of oxygenated and non-oxygenated hemoglobin for different wavelengths of light.

By using two LEDs of different wavelengths, the absorption coefficients at two different working points can be used to determine oxygenation [4].

Continued analysis of the underlying measurement data using machine learning could already be used to detect peripheral aterial disease in a proof-of-concept study [5]. The investigations of this work differ from this, apart from the deviating objective of the classification, by an alternative

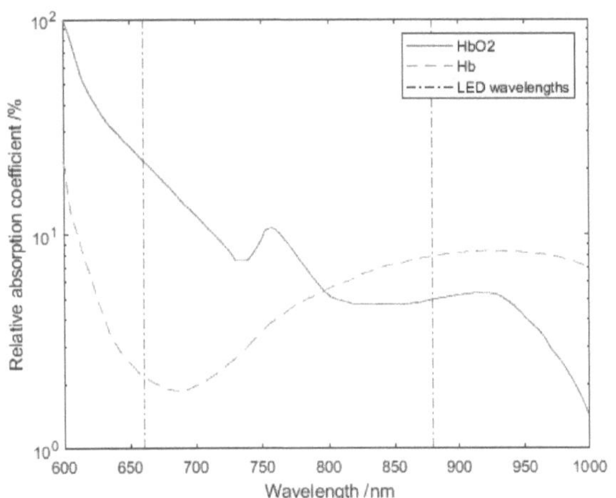

Figure 1: Relative absorption coefficients of oxygenated and non-oxygenated hemoglobin [2], [3].

processing of the raw signals as well as a broader frequency spectrum to be examined.

3   Idea

This section explains the basic approach that will be used for person distinction. As a basis for this work it is assumed that in the measured raw signals of the pulse oximetry measurement beside the static signal components also higher frequency signal components are contained. Furthermore, it is assumed that the named signal components are determined in their occurrence and intensity of occurrence by individual characteristics of the test subjects. Fig. 2 sketches two exemplary time courses of a measuring channel of the pulse oximetry measurement with two different test subjects, whereby the forms of the signal curves differ from each other. The individual signal shapes result in different

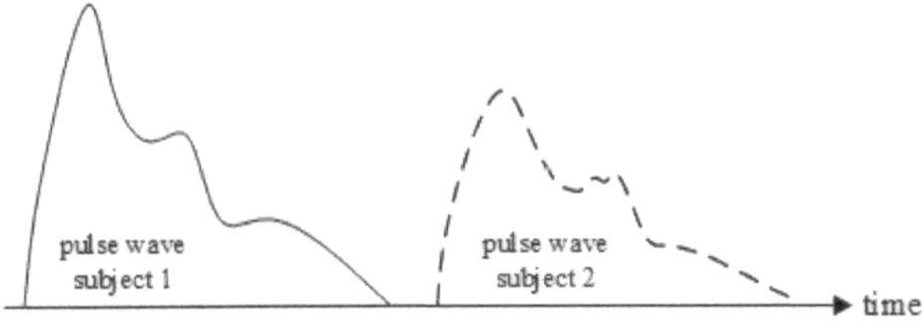

Figure 2: Schematic representation of a measurement curve of two different subjects.

frequency spectra, which can be calculated from the measurement signals using a fast fourier transformation (FFT). Provided that the frequency spectra differ sufficiently from each other, these could be used to differentiate the subjects. In order to be able to limit the complexity of the calculations required for this, the evaluation of the signal spectra is to be carried out with the aid of a neural network.

The exact application and implementation of this approach will be explained in the following section.

4   Implementation

This section describes the hardware components used. Also presented is the software used to implement the signal processing on the microcontroller. Subsequently, the configuration of the hardware for recording the measurement data and the selection of the measured subjects are discussed.

4.1   Hardware

The development board "NUCLEO-H7A3ZI-Q" from STMicroelectronics is used for the investigations of this work. The development board contains a STM32H7A3 SoC, an Arm Cortex-M7 CPU with 280Mhz clock frequency. Besides the 32-bit FPU, the advantages of the controller are the large memory areas of Flash (2048 kBytes) and RAM (1184 kBytes) [6].

For the absorption measurement described in section 3, the MAX30102 sensor from Maxim Integrated is used. This sensor has been specially developed for the use of portable measurement technology for the determination of oxygen saturation and heart rate. As a complete package with LED's (660nm, red light and 880nm, infrared light), photo detectors and an internal AD conversion with up to 18 bit signal resolution, the sensor is available ready for use as a breakout module. The internal measurement electronics can be configured for up to 3200 measurement cycles per second, whereby the available resolution of the AD conversion decreases with increasing measurement rate [7].

4.2   Software

The online tool "Edge Impulse" is used to create and train the neural network. Live data from the target hardware can be imported using the CLI provided. Basic mechanisms for signal processing are additionally available as easily configurable building blocks. The features generated in this way are used to train a neural network. All components used can subsequently be further modified, tested or downloaded as a library for the most common development interfaces.

4.3   Sensor configuration

To record the measurement data, the sensor is configured in SPO2 mode with a measurement frequency of 1 kHz. This means 1000 measurements for both LEDs. The comparison current for the AD conversion is defined to the lowest value of 2048 nA. The maximum possible LED pulse duration to be configured at the measurement frequency of 1 kHz is about 188 μs per sampling period. For the power supply of the two LEDs a value of 6.2 mA (corresponding to 0x1F as register value) is set.

4.4   Recording of measurement data

The selection of the test persons is done randomly and includes five different persons, which differ from each other regarding age, gender as well as athletic fitness. For data acquisition the sensor is positioned at the upper phalanx of the index finger (opposite to the fingernail) of the test subject. To ensure an even contact pressure during the measurement, the sensor is fixed to the finger phalanx with medical tape. Subjects are instructed to place their hand quietly on the table during the measurement data acquisition without applying any additional pressure to the finger or the sensor. In order to obtain a sufficient number of valid pulse shapes for the training process, the measurement data are recorded for ten minutes. The selected measurement duration also allows the measurement data to be generously divided between the training and test data sets.

4.5 Training process

Following the preparation of the raw data, the selected configurations of the processing blocks are explained. Window sizes of one second (1000 samples) and two seconds (2000 samples) are compared for processing the measurement data. Since the pulse shape should repeat at least every two seconds under physically normal conditions, these two variants appear to be sufficient. To facilitate the structure of the software implementation, the window offset is defined to the same value. Using the Spectral Analysis DSP block, an eighth order high pass filter with a cutoff frequency of 10 Hz is applied to the data. The FFT length is defined to 1024 values, the number of spectral power peaks considered is increased to 100 values. The remaining settings remain unaffected.

Table 1: Layer overview of the configured neuronal network, version 1 with dense (fully connected) layers

#	Layer	# neurons	dropout factor
1	dense layer	200	-
2	dropout layer	-	0.1
3	dense layer	100	-
4	dropout layer	-	0.1
5	dense layer	50	-
6	dropout layer	-	0.1
7	dense layer	25	-

For the structure of the neural network, the first approach uses a conventional variant of fully connected layers (FC) in series, each connected by a dropout layer. The parameters used are listed in table 1.

Table 2: Layer overview of the configured neuronal network, version 2 with convolutional layers

#	Layer	Parameter description
1	reshape layer	columns: 10
2	2D conv. layer	filters: 8, kernel size & layers: 3
3	flatten layer	-
4	dense layer	neurons: 16

As a second approach, a more complex convolutional network (CN) is configured and trained. The parameters used for configuration are listed in table 2. Finally, the training process is run for 100 epochs with a learning rate of 0.001.

The training and testing results are explained in the evaluation section.

5 Evaluation

In the following section, the results of the training processes with the named parameters are presented and compared with each other. All obtained accuracies are summarized in table 3 in addition to the following explanation. For the

Table 3: Summary of classification accuracies when applied to training, validation or test dataset

Network	window size	Accuracies (train/ validation/ test)
FC	1000	38.0 %/ 34.6 %/ 28.5 %
FC	2000	23.5 %/ 38.9 %/ 26.0 %
CN	1000	61.3 %/ 34.2 %/ 27.2 %
CN	2000	66.3 %/ 34.2 %/ 25.8 %

training process of the neural network based on fully connected layers (FC), a training accuracy of 23.5% is obtained with a window size of 2000 samples. Even though the obtained accuracies of the validation (38.9%) and test (26.0%) datasets are slightly higher, the overall accuracy appears to be significantly too low for real-world applicability for live classification. With training, validation and test accuracies of 38%, 34.6% and 28.5%, respectively, adjusting the window size to 1000 samples also does not result in sufficient improvement. In fig. 3, the classification results obtained by applying the trained FC network with a window width of 1000 samples to the test dataset are shown in the form of a confusion matrix. It is noticeable that one of the data labels was almost completely eliminated during training.

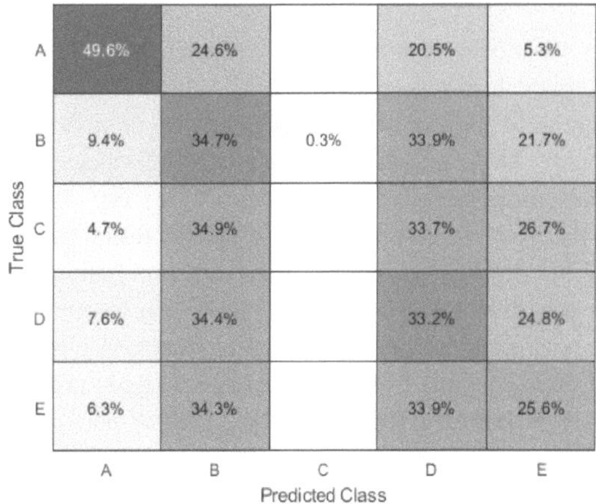

Figure 3: Confusion matrix for FC network test classification with subject A - E and 1000 samples per window.

For the second neural network considered with 2D convolutional layers (CN), the training process yields a training accuracy of 61.3% for a window width of 2000 samples. However, the significant gap with the accuracies obtained when applied to the validation dataset (34.2%) and the test dataset (25.8%) suggests that this must be an over-fitting instead of an improvement in overall accuracy. Reducing the window width again results in no significant change. However, the use of the second network configuration (CN) shows a more balanced probability distribution when applied to a test data set. The determined classification probabilities with an application to the test data set are shown in fig. 4.

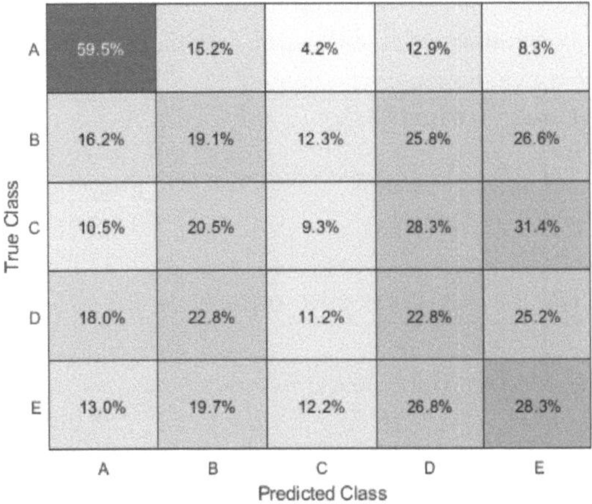

Figure 4: Confusion matrix for CN network test classification with subject A - E and 1000 samples per window.

The test accuracy of 27.2% determined here is still not suitable for applying the overall system for a live classification. Nevertheless, the results obtained using the convolutional network appear more suitable as a basis for future investigations.

6   Conclusion and Future Work

In this work it was investigated whether it is possible to differentiate test subjects from each other on the basis of measurement data collected in the course of a PPG measurement. The evaluation of the measured values and the spectral components contained was carried out using machine learning. In summary, it can be stated that the investigated system configurations of the data processing as well as the neural networks could not lead to a training result which is applicable for a live classification.

Based on the results of this work, deeper signal processing seems to be necessary to achieve the goal of person differentiation nevertheless. One possible adaptation would be to perform the high-pass filtering already before transferring the data samples into Edge Impulse. Since the signal processing blocks in Edge Impulse can only be processed in parallel and not serially, the input signals would have to be cleaned of heart and respiration rate components. Alternative analysis options such as the creation of spectrograms would not be possible without prior filtering. The use of spectrograms also enables a change in the data basis, from raw data classification to image classification. Since some proven neural networks are already available for this purpose, applying transfer learning as in [5] would be a way to bypass the search for a suitable neural network. The use of existing networks is also possible in Edge Impulse using Keras source code, so that the possibilities for rapid source code generation could still be used.

Acknowledgement

This publication is a result of the research of the Center of Excellence CoSA and funded by the Joachim Hertz Foundation (Project "PASBADIA"). Horst Hellbrück is adjunct professor at the Institute of Telematics of University of Lübeck.

Author's Statement

Conflict of interest: Authors state no conflict of interest.

7   References

[1] J. Allen, "*Photoplethysmography and its application in clinical physiological measurement,*" *Physiological Measurement*, vol. 28, no. 3, pp. R1–R39, feb 2007. [Online]. Available: https://doi.org/10.1088/0967-3334/28/3/r01

[2] Natural Phenomena Simulation Group. *Human Skin Data*. University of Waterloo. Last accessed 2022-01-24. [Online]. Available: http://www.npsg.uwaterloo.ca/data/skin.php

[3] S. Prahl. *Tabulated Molar Extinction Coefficient for Hemoglobin in Water*. Last accessed 2022-01-24. [Online]. Available: https://omlc.org/spectra/hemoglobin/summary.html

[4] Maxim Integrated. *GUIDELINES FOR SPO2 MEASUREMENT*. Last accessed 2022-01-24. [Online]. Available: https://www.maximintegrated.com/en/design/technical-documents/app-notes/6/6845.html

[5] J. Allen, H. Liu, S. Iqbal, D. Zheng, and G. Stansby, "*Deep learning-based photoplethysmography classification for peripheral arterial disease detection: a proof-of-concept study,*" *Physiological Measurement*, vol. 42, no. 5, p. 054002, may 2021. [Online]. Available: https://doi.org/10.1088/1361-6579/abf9f3

[6] STMicroelectronics. *STM32H7A3ZI Product Page*. Last accessed 2022-01-24. [Online]. Available: https://www.st.com/en/microcontrollers-microprocessors/stm32h7a3zi.html?ecmp=tt9470_gl_link_feb2019&rt=ds&id=DS13195

[7] Maxim Integrated, *MAX30102*, ref. 1, 10/18 ed. [Online]. Available: https://datasheets.maximintegrated.com/en/ds/MAX30102.pdf

Extending ROS2SWARM to UAVs: a Proof of Concept

Christian Charles[1], Tanja Katharina Kaiser[2], and Heiko Hamann[2]

[1] Robotics and Autonomous Systems, Universität zu Lübeck, christian.charles@student.uni-luebeck.de
[2] Institute of Computer Engineering, Universität zu Lübeck, {kaiser, hamann}@iti.uni-luebeck.de

Abstract

Programming reusable swarm robot controllers is challenging due to the lack of common software architectures and the variety of available swarm robotic platforms. The ROS 2 package ROS2SWARM addresses this problem by providing modular and platform-independent swarm behaviors that can be combined to create more complex behaviors. ROS2SWARM currently supports mobile ground robots with a differential drive, such as the TurtleBot3 Waffle Pi. As the use of Unoccupied Areal Vehicles (UAVs) is becoming increasingly popular, we present a concept and first results for extending the package to natively support multi-rotor UAVs. We discuss the hardware chosen for development and testing, and address challenges that arise during implementation. Based on our first results and the presented concept, we are confident that the extension of ROS2SWARM to UAVs is possible.

1 Introduction

Robotic swarms [1] are decentralized systems consisting of many simple robots that solve tasks together. Swarms have no single point of failure due to their de-centrality resulting in robustness against failure of individuals. But the available variety of swarm robot platforms makes the development of reusable software challenging. Hardware components can differ greatly from one another and different software architectures may be required for control, which makes adaptation and reuse of robot controllers difficult.

With the introduction of the Robot Operating System 2 (ROS 2), a middleware for robots is available that is suitable for use in robotic swarms. In contrast to the first ROS version, that relied on a central managing master, ROS 2 is decentral utilizing the Data Distribution Service (DDS) to implement a Publish-Subscribe model allowing robots to exchange messages. Consequently, ROS 2 is more suitable for swarms than its predecessor and research starts to leverage its potential [2, 3].

ROS2SWARM [4] is a ROS 2 package that provides swarm behaviors out of the box in a modular architecture allowing to easily combine or extend behaviors to create more complex behaviors. The package supports all ROS and ROS 2 compatible mobile ground robots and currently supports the TurtleBot3 Burger, the TurtleBot3 Waffle Pi,[1] and the Jackal unmanned ground vehicle[2] out of the box. Thus, swarm behaviors are limited to 2D space so far. To make the package even more universally applicable, we aim to extend it to support Unoccupied Areal Vehicles (UAV). In this paper, we present a concept and first results for such an extension of ROS2SWARM to multi-rotor UAVs. This comes with several additional challenges such as the added dimension the robots move in, and that flying objects pose a considerable safety threat to people and the environment. We address these challenges in our concept for the extension of the ROS2SWARM presented in this paper.

First, we introduce the background in Sec. 2, including the fundamentals of ROS2SWARM, control of UAVs and especially UAV swarms. In Sec. 3, we present our concept to extend the existing ROS2SWARM package for the use with UAVs including first results. Sec. 4 summarizes our results and gives an outlook on the further development and adaptation of the ROS2SWARM package.

2 Background

ROS2SWARM is a modular and easily extendable ROS 2 package for swarm robotics. It provides a variety of swarm behaviors for mobile ground robots and is designed to be easily extendable both regarding swarm behaviors and robot platforms. It can be used with any ROS 2 compatible robot platform that can be controlled via a motor command containing a linear velocity along the x-axis and an angular velocity around the z-axis. Additionally, the robot platform requires a 2D LIDAR providing sensor data in a LaserScan message to execute the implemented swarm behaviors. Robot platforms supporting only ROS can be integrated using a network bridge to exchange messages between ROS and ROS 2.[3] ROS2SWARM differentiates two types of swarm behaviors or patterns: movement patterns and voting patterns. Movement patterns, such as flocking or aggregation, determine the movement of the robots. By contrast, voting patterns provide collective decision-making mechanisms, such as the majority rule or the voter model.

---

[1] http://turtlebot3.robotis.com
[2] https://clearpathrobotics.com/
jackal-small-unmanned-ground-vehicle

[3] https://github.com/ros2/ros1_bridge

Figure 1: 3DR Iris in the Gazebo simulator. The quadrotor UAV has an external front facing GPS module that includes a magnetometer. The internal Pixhawk flight controller provides accelerometer, gyroscope, barometer and a second magnetometer. We extended the basic UAV model with a LIDAR on top.

Both pattern types can be used on their own or combined with other patterns to generate more complex behaviors. Patterns can be adapted to new hardware or new scenarios via parameter files. To avoid collisions between robots or between robots and the environment, ROS2SWARM includes a hardware protection layer. It runs separately from the executed swarm behavior and prevents robots from getting closer to an obstacle than a predefined threshold distance using a repulsive potential field approach.

So far, ROS2SWARM has been successfully tested on three mobile ground robot platforms with differential drives: the TurtleBot3 Burger, the TurtleBot3 Waffle Pi, and the Jackal unmanned ground vehicle. To increase flexibility in the choice of robot hardware, we show the first steps to extend the package to include multi-rotor UAVs.

UAVs are cost-effective and versatile tools that can, for example, perform precise environmental monitoring tasks, such as mapping and monitoring crops [5, 6]. Albani et al. [7] demonstrated crop observation by a swarm of UAVs as a collaborative swarm task using ROS. When dealing with autonomous air vehicles, safety and collision avoidance are essential to protect people and the environment. In addition, reliability is key during extensive missions or missions in difficult terrain that allow only limited access to the robot hardware. Yasin et al. [8] discussed a variety of different active and passive sensors for collision avoidance. When compared, active sensors, such as LIDARs or sonars, have a higher power consumption as they emit signals in order to capture information. Passive sensors, such as cameras, rely on external transmission sources (e.g., sunlight). However, they often require more power for data processing as the gathered data may need to be filtered and processed to find the points of interest or to calculate distances.

Especially when it comes to safety issues, it is good practice

not to rely on just one sensor type but to combine different sensor types to compensate for weaknesses (e.g., sensitivity to light or materials of certain sensors) or to enhance the quality of noisy data [9]. While LIDARs, for example, are capable of reliably capturing information at long distances, unlike ultrasonic sensors, they tend to have difficulties with transparent materials (e.g., glass). But also inconsistencies in a LIDAR's point cloud may lead to over-segmentation of thin objects (e.g., power lines) and can be handled by using an additional camera sensor and an algorithm fusing sensor data [10]. Yasin et al. [8] also discuss multiple methods to execute collision avoidance with the acquired sensor data. For example, geometric approaches use geometric attributes like position and velocity to ensure a minimal distance between objects. Force field methods, by contrast, make use of potential fields, attracting or repulsing robots towards target points, or away from other robots. The current hardware protection layer in ROS2SWARM is based on such a force field method for movement in 2D space and has to be extended to movement in 3D space, to be suitable for use with UAVs.

3   Concept

First, we discuss our choice of the UAV platform for the first integration of UAVs into ROS2SWARM, and present required modifications to the package afterwards. In the current form of the ROS2SWARM, the modifications only affect the movement patterns, since voting patterns rely solely on the exchange of ROS 2 messages that are independent from the robot hardware.

3.1   UAV and Simulation Environment

In order to be able to test the adaptations to the ROS2SWARM package, we extend the package to UAVs by using one specific platform first. The integration of support for more UAV platforms requires only minor modifications as soon as this first platform is included. Our main criterion to choose the UAV hardware is that it supports ROS 2. Furthermore, we want to implement the extension to UAVs in simulation before testing it on real robots to avoid possible endangerment of people and the environment as well as the destruction of the real hardware. As ROS2SWARM already provides simulations for the supported mobile ground robots in the Gazebo simulator, a second criterion for the choice of the UAV hardware is that it can be simulated in Gazebo.

Several off-the-shelf UAVs, including some of DJI's platforms and the crazyfly 2.1,[4] come with a ROS SDK. ROS 2 support is not offered by the manufacturer, but sometimes possible using packages developed by the ROS or UAV community. Alternatively, these UAVs could be used with the rosbridge between ROS and ROS 2. But to keep the setup simple and to leverage the full potential of ROS 2,

---

[4] https://www.bitcraze.io/products/crazyflie-2-1

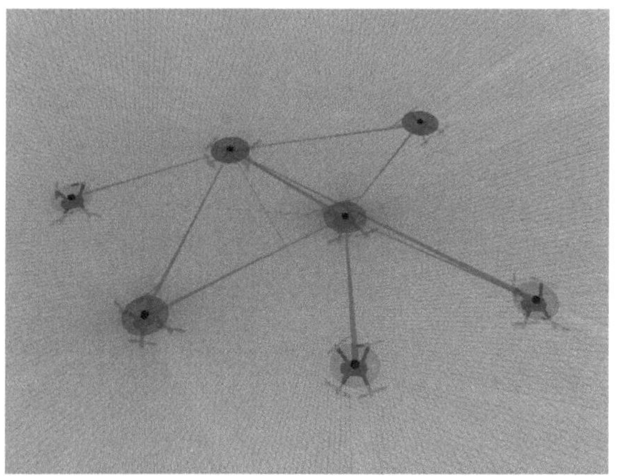

Figure 2: Simulated swarm of seven 3DR Iris UAVs in Gazebo.

we restrict us to UAV hardware natively supporting ROS 2 here.

A different option are custom built UAVs. In this case, the parts requiring ROS 2 support are the flight controller, a companion computer, and the sensors. Hardware following the Pixhawk standard[5] for flight controllers and supporting the PX4 autopilot software stack fulfill our needs. The PX4 autopilot controls UAVs motors by accepting high level commands published to ROS 2 topics via a companion computer and a software bridge provided by PX4[6].

When adapting ROS2SWARM to UAVs, relying on such a common flight controller with open hardware not only allows us to reduce costs and minimize the complexity of the swarm members, but also simplifies porting to UAVs with PX4 compatible controllers. Those range from educational UAV kits like CLOVER[7] and PX4 Devkits[8] to in self-built UAVs compliant with the listed airframes in the PX4 developer guidelines. [9]

However, the Gazebo[10] simulator does not yet offer many multi-rotor UAV models with hardware following the Pixhawk standard and supporting the PX4 autopilot. To date, only one UAV model with ROS 2 support, the 3DR Iris Quadrotor[11], is available as a simulated model that also supports the simulation of multiple UAVs in Gazebo. The 3DR Iris comes with a Pixhawk autopilot system and a GPS module which includes a magnetometer. Since the Pixhawk flight controller takes over low level control for steering and stabilization, we can use such a typical UAV model for the high level programming of the ROS 2 control expecting similar performance on other UAVs with the same flight controller. However, the 3DR Iris has no LIDAR, but the current implementation of ROS2SWARM is heavily based on LIDAR data.

---

[5] https://pixhawk.org
[6] https://docs.px4.io/master/en/ros/ros2_comm.html
[7] https://coex.tech/clover
[8] https://px4.io/devkits
[9] https://docs.px4.io/master/en/airframes
[10] http://gazebosim.org
[11] http://www.arducopter.co.uk/iris-quadcopter-uav.html

Therefore, we extend the default 3DR Iris model with a simulated generic 2D LIDAR mounted on top of the UAV, as visualized in Fig. 1. We chose a 2D LIDAR here, as the current implementation of ROS2SWARM makes use of 2D sensor data. In future work, we will exchange it with a 3D LIDAR. Running the simulation with several UAVs (see Fig. 2) requires a dedicated launch file in order to prevent duplicate name spaces of communication ports and ROS 2 sensor topics.We implement a launch script that automatically creates unique name spaces for each simulated UAV. This separation of UAV name spaces also enforces their computational separation as if decentralized hardware was used on a swarm of real UAVs.

3.2 Adaptation of the ROS2SWARM Package

UAVs have six controllable degrees of freedom (DOFs) in contrast to the two controllable DOFs of mobile ground robots with differential drives, for which ROS2SWARM is currently designed. Consequently, requirements for sensors (e.g., coverage above or below the UAV) differ and additional control signals are required for the movement pattern provided by ROS2SWARM. We address those challenges regarding movement and perception separately.

3.2.1 Movement

A fundamental difference of UAVs to mobile ground robots with a differential drive is the ability of UAVs to maneuver in all directions of a 3D space without prior rotation around their yaw axis. We choose an incremental process in adapting the package to multi-rotor UAVs to include successively the control of more degrees of freedom.

For the first modifications of the ROS2SWARM package to support UAVs, we simplify sensing, by assuming that the UAV is always aligned in the direction of flight, which also us to also simplify movement. For this purpose, we also fix the flight altitude targeted by all UAVs. Based on these two assumptions, we can directly transfer the control for locomotion in 2D space based on rotation around the yaw axis and translation along the x-axis to UAVs (i.e., 2 controllable DOFs, 3 total DOFs). The ROS2SWARM package can therefore be retained in its current form and only a few minor modifications are required to adapt the 2D drive commands to the required 3D drive commands of the UAVs. This allows us to first get the system up and running with the UAVs and to ensure basic functionality. Based on this first modification, the package can be gradually modified until it supports movement in 3-dimensional space (i.e., 6 controllable and total DOFs).

To date, we implemented a first ROS 2 controller in Python 3, that initiates takeoff to a targeted altitude where it navigates the UAV via flight commands based on rotation and translation. This behavior equals the simple drive pattern provided by ROS2SWARM to let a robot drive straight forward. Starting from this simple UAV support, we plan to increase the complexity of the implementation by extending control to six DOFs and by calculating sensor orientation based on the UAV's orientation. In the process, we will adjust the movement pattern included in ROS2SWARM package gradually to support those additional DOFs. In particular, we will also adjust the hardware protection layer to ensure a safe execution of the ROS2SWARM package on UAVs.

3.2.2 Perception

When moving freely in three-dimensional space, other robots or obstacles can also be above or below an autonomous agent. A 3D LIDAR mounted on top or beneath the UAV fuselage can provide sensor coverage on all sides. Since ROS2swarm is currently designed for 2D scan data, potential field calculations for the movement patterns must be extended to support 3D sensor data.

As discussed before, additional sensors can be used to implement reliable obstacle avoidance, ensuring safety of persons, environment, and robot hardware. Additionally, collisions between swarm members can be prevented by regularly exchanging positions and flight directions with their neighbors [3]. Positions and flight direction can then be used to determine motor commands to avoid collisions.

4   Conclusion

We presented a concept to extend ROS2SWARM to UAV swarms and showed first results in the implementation. Special attention was paid to achieve compatibility to different multi-rotor UAV models and good transferability into the real world, which is possible using Pixhawk flight controllers with the ROS 2 interface provided by the PX4 autopilot. For further porting of the swarm behaviors implemented by ROS2SWARM, start scripts and a first UAV controller that initiates takeoff and ensures proper control via execution of flight commands for rotation and translation were implemented in Python 3. To be able to execute the movement patterns included in ROS2SWARM, we extended the sensor system of the 3DR Iris with a 2D LIDAR sensor. Based on these first works on the extension of ROS2SWARM to UAVs, we will test the swarm behaviors included in the ROS 2 package in their current implementation for 2D use by restricting the UAVs to a fixed flight altitude next. This allows us to identify any difficulties that may arise. We will then incrementally extend included behaviors for movement in 3D space. This will probably be challenging due to the required detection of other swarm members and obstacles in three dimensions. Furthermore, hardware protection must be further improved and extensively tested to ensure the safety of the swarm system.

In total, our first steps and plans to extend ROS2SWARM are promising. Although UAV swarms bring additional challenges, such as an increased requirement for safety, we are confident to integrate full support for UAVs to the existing ROS 2 package in future and thus including not only out of the box support for more robot platforms, but also for movement of robots in 2D and 3D.

Acknowledgement

The internship was carried out at the Institute of Computer Engineering.

Author's Statement

Conflict of interest: Authors state no conflict of interest.

5   References

[1] H. Hamann, *Swarm Robotics: A Formal Approach.* New York, USA: Springer US, 2018.

[2] A. Barciś, M. Barciś, and C. Bettstetter, "Robots that sync and swarm: A proof of concept in ROS 2," in *2019 Int. Symp. on Multi-Robot and Multi-Agent Systems (MRS)*, 2019, pp. 98–104.

[3] J. P. Queralta, Y. Xianjia, L. Qingqing, and T. Westerlund, "Towards large-scale scalable MAV swarms with ROS2 and UWB-based situated communication," *arXiv preprint arXiv:2104.09274*, 2021.

[4] T. K. Kaiser, M. J. Bergmann, T. Plattenteich, L. Schilling, G. Schildbach, and H. Hamann, "ROS2swarm - a ROS 2 package for swarm robot behaviors," [in press].

[5] H. Hoffmann, R. Jensen, A. Thomsen, H. Nieto, J. Rasmussen, and T. Friborg, "Crop water stress maps for an entire growing season from visible and thermal UAV imagery," *Biogeosciences*, vol. 13, no. 24, pp. 6545–6563, 2016.

[6] J. M. Peña, J. Torres-Sánchez, A. I. de Castro, M. Kelly, and F. López-Granados, "Weed mapping in early-season maize fields using object-based analysis of unmanned aerial vehicle (UAV) images," *PloS one*, vol. 8, no. 10, p. e77151, 2013.

[7] D. Albani, T. Manoni, D. Nardi, and V. Trianni, "Dynamic UAV swarm deployment for non-uniform coverage," in *Proc. of the 17th Int. Conf. on autonomous agents and multiagent systems*, 2018, pp. 523–531.

[8] J. N. Yasin, S. A. Mohamed, M.-H. Haghbayan, J. Heikkonen, H. Tenhunen, and J. Plosila, "Unmanned aerial vehicles (UAVs): Collision avoidance systems and approaches," *IEEE Access*, vol. 8, pp. 105 139–105 155, 2020.

[9] N. Gageik, P. Benz, and S. Montenegro, "Obstacle detection and collision avoidance for a UAV with complementary low-cost sensors," *IEEE Access*, vol. 3, pp. 599–609, 2015.

[10] Z. Ma, W. Yao, Y. Niu, B. Lin, and T. Liu, "UAV low-altitude obstacle detection based on the fusion of LiDAR and camera," *Autonomous Intelligent Systems*, vol. 1, no. 1, pp. 1–10, 2021.

Bluetooth Fingerprinting Localization

Mohammad Javaad Zamani [1], Oliver Harnack [2], and Georg Schildbach [3]

[1] Robotics and Autonomous Systems, Universität zu Lübeck, mohammad.zamani@student.uni-luebeck.de
[2] Technology Management Research & Development Dräger Safety AG & Co. KGaA, oliver.harnack@draeger.com
[3] Institute for Electrical Engineering in Medicine, Universität zu Lübeck, georg.schildbach@uni-luebeck.de

Abstract

Indoor localization estimation is an essential part of cloud asset management services nowadays. This paper demonstrates the proof of concept for a low energy demanding localization solution based the received signal strength indicator (RSSI) for a Bluetooth Low Energy (BLE) device using a deep recurrent neural network. The inherent problem of using RSSI is its susceptibility to noise and fluctuation over time. In order to remove the noise and estimate the position of the target location, a deep neural network architecture was proposed. Moreover, an average moving filter reduces the outliers' effects and leads to more accurate results. A real experiment has been conducted in an indoor environment to evaluate the deep neural network architecture. The preliminary results of the trained network show great potential and could maintain the error within 1.5m for approximately 88% of the samples.

1 Introduction

Cloud base technologies are playing a crucial role in businesses nowadays. Cloud asset management is an optimized and secure way of managing physical devices. This technology helps companies to monitor resources, record and track assets throughout their life cycle. A device itself is required to have some smart features to be used in such a system. An essential feature for such a device is to send its location information. However, many of these devices are going to be used in an indoor environment where GPS signal is very weak or nonexistent. Imagine a device that collects a gas concentration in an indoor environment and sends the measurements to the cloud. In this paper, we demonstrate a proof of concept for estimating the location of such a device based on the Bluetooth signal strength indicator (RSSI). BLE sends packets, useful for other devices, at a fixed interval in the so-called advertising state. BLE v4.0 protocol uses 40 channels separated by 2 MHz distances and only advertises on the channels 37, 38, and 40, as shown in Fig.1. The idea is to estimate the target location based on captured

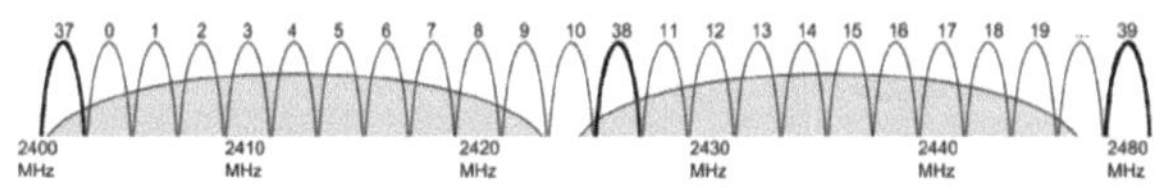

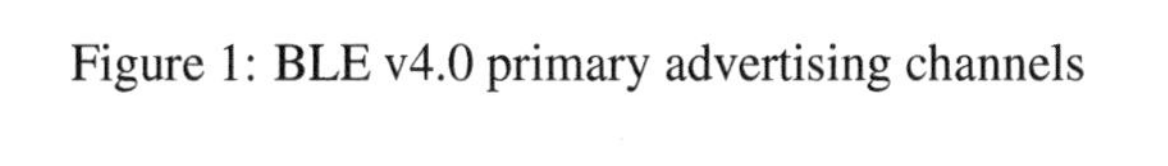

Figure 1: BLE v4.0 primary advertising channels

advertising signals from known fixed access location points (APs). There are also similar advertising channels for WIFI signals. However, the primary benefits of using BLE are its low cost and low power consumption. The most crucial challenge, on the other hand, is the fluctuation of the

RSSI due to the complex indoor environment and noise [1]. In the literature, there are two major algorithms which are used for localization based on RSSI of both BLE and WIFI signals [2]. Tirlateration method uses the theoretical signal propagation model to calculate the distances between the unknown location and known access points (APs). Because the RSSI values do not always follow the theoretical propagation model, this method cannot estimate the unknown location with great accuracy. In contrast, the fingerprinting method measures the advertising RSSIs of the APs from some known locations called reference points (RFs) in an offline phase. This data can be then used to estimate the target location with different algorithms in an online phase. Among such algorithms, there are RSSI fingerprinting approaches, which use parametric probabilistic models such as Gaussian [3] or Gaussian mixture model [4] to estimate the target location based on fingerprinting data. The noneparametric methods also attempted for the problem, which use no assumption on RSSI's PDF and achieve better results in more general situations but they also need more data to be able to produce the accurate results [5]. Recent success of deep learning enables even higher accuracy of location estimation based on architectures such as Recurrent Neural Network (RNN). A RNN is a class of Artificial Neural Networks (ANN) where the output is not only dependent on the current input value but also the historical data [6]. In order to get the best estimation of the target location, this paper focused on a special version of RNN called long short-term memory (LSTM) architecture which helps exploiting the sequential correlation between the RSSI measurements in time and along the trajectory of the movement.

2   Material and Methods

In this section, we first introduce the proposed network architecture for BLE fingerprinting localization, then we present the data collection process and the details of the training phase. Finally, we introduce the software and hardware used in the experiment.

2.1   Network Architecture

The proposed architecture is shown in Fig.2. The input to the network can be represented with the matrix X defined as:

$$X = \begin{bmatrix} x_1, & x_2, \dots & x_m \end{bmatrix}, \tag{1}$$

where each vector $x_i \in \mathbb{R}^n$ contains the RSSI values form

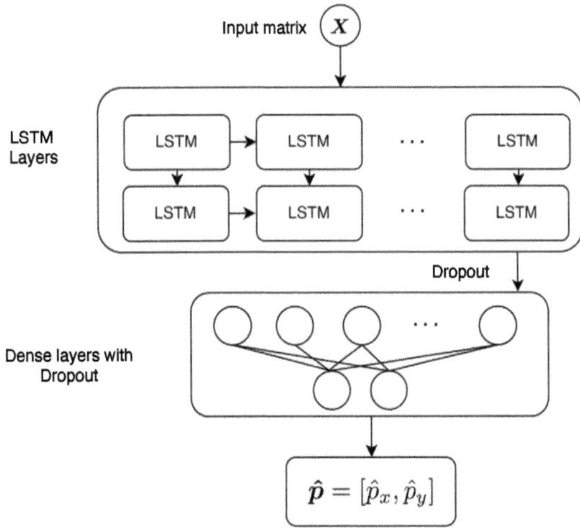

Figure 2: Proposed architecture

all APs. The matrix X has, therefore the dimension of $n \times m$, where the n is the number of APs and m is the number of consequent samples in a scan interval. A preprocessing will be performed before feeding the network with the raw RSSI values. Firstly, a moving average filter should apply to the existing values. Fig.3 depicts an example of such a filter with the window size of 3. Secondly, the missing values are replaced with -100. The preprocessed data is then ready to be altered by a series of LSTM Layers.

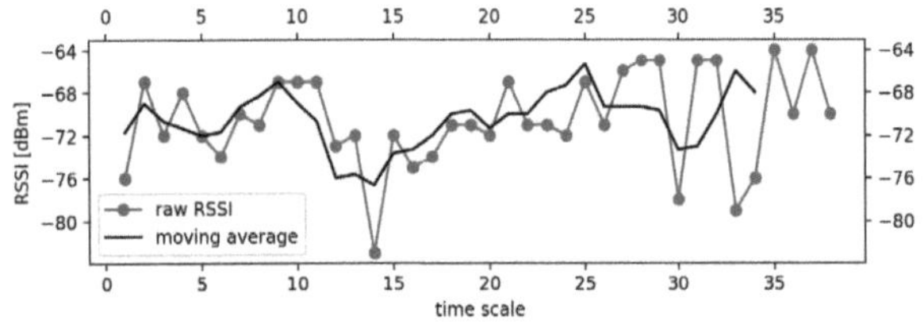

Figure 3: Moving average filter

These LSTM layers are supposed to extract the high-level features of the RSSI time series. Each LSTM cell takes the current input and the information from the previous time step. Fig.4 illustrates the internal mechanism of a typical LSTM cell, where x_t is the current input and C_{t-1}, h_{t-1} are the memory cell and the hidden state in the time step $t-1$, respectively. Throughout the paper, we use the symbol W to denote weights, symbol b for biases and σ for the sigmoid activation function.

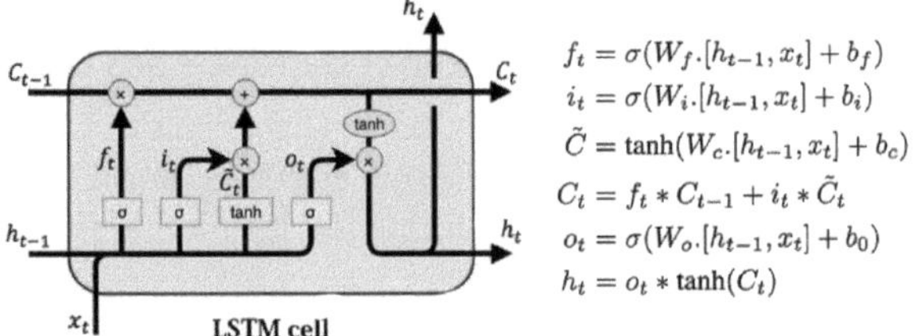

Figure 4: Structure of an LSTM cell

The first task of the cell is to decide which information should be thrown away from the cell state. This is done by a sigmoid layer called "forget gate":

$$f_t = \sigma(W_f.[h_{t-1}, x_t] + b_f) \tag{2}$$

The next task is to decide which information to store in the cell state. This includes two parts: First, a sigmoid layer called "input gate layer"decides which values will be updated. A tanh layer then creates the new candidate values for the next state $\tilde{C}_t$. Combining these two steps, we get the update state C_t as follows:

$$i_t = \sigma(W_i.[h_{t-1}, x_t] + b_i) \tag{3a}$$

$$\tilde{C} = \tanh(W_c.[h_{t-1}, x_t] + b_c) \tag{3b}$$

$$C_t = f_t * C_{t-1} + i_t * \tilde{C}_t \tag{3c}$$

Finally the output of the cell is just the filtered version of the cell state and includes two steps, which can be formulated as:

$$o_t = \sigma(W_o.[h_{t-1}, x_t] + b_0) \tag{4a}$$

$$h_t = o_t * \tanh(C_t) \tag{4b}$$

In this particular architecture, we used two series of such LSTM cells. The output of the final layer then goes through a dropout layer to prevent overfitting. A fully connected (or dense layer) with a ReLU activation function and a subsequent dropout layer are used to estimate the final location. This final layer should act as a regression layer and output the final estimated position. The optimum number of layers is determined empirically by monitoring the loss during several training experiments.

2.2   Collecting data and training

Based on the network discussed in the previous section, the development of our BLE localization model is divided in two phases: 1) Collecting data: The experiment done by installing four BLE V4.0 beacons at the corners of the office building of Dräger Safety. In order to send the commands to the scanner device, a custom BLE service is implemented.

The process can begin by positioning the device at the desired reference point then connecting to the BLE service using an Android or IOS device. After setting the position in the application, the scanner will collect 60 RSSI samples from each nearby beacon and send them to the server. This process should be repeated at every reference point. 2) Creating dataset and training: We first apply the moving average filter to the RSSI values. The RSSI values are then divided into training (80 %) and test parts (20%). In the next step, we create 100 random trajectories, each of which beginning with a random node as the starting point (each reference point can be seen as a node). We randomly choose each next node from all nodes within a two-node distance until we reach the desired path length. An example of such a path is shown in Fig.5. Our dataset generator uses these trajectories and outputs an array containing all the trajectory nodes plus an RSSI value assigned to each node. These RSSI values are chosen randomly from their respective training or test split. In this way, even the same trajectory could be slightly different due to assigning different RSSI values to their nodes. Although the path length of 8 had the best performance in our experiment, due to the specific scan interval constrain that we have during the online phase, we chose a path length of 3. That means the network is trained to see the RSSI of the last three steps. However, in the software implementation, we found it slightly easier to generate the path length of 24 and divide each to 8 paths before feeding them to the network.

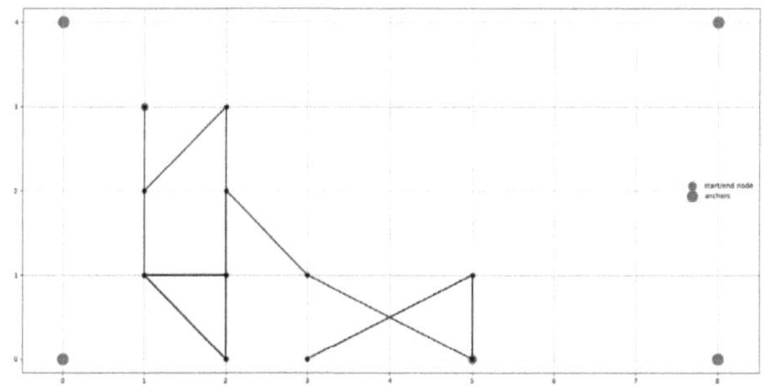

Figure 5: Random trajectory inside the grid area contains all the RFs. Nodes are located at the position of RFs

Given the estimated location outputted by the network and actual locations of the training nodes, the mean squared error can be calculated as the loss and can be back-propagated to obtain the new weights. As shown in Fig 6, the validation loss increased after it reached its minimum at the 43rd epoch. Accordingly, we chose the weights corresponding to epoch 43 as our final weights.

2.3   Software and Hardware

For the Scanner, we used a Dräger custom BLE board. These Boards are equipped with BLE v5.0 capable nRF52840 chip and an NB-IoT/LTE-M RF Transceiver nRF9160 chip with additional GPS Receiver capabilities. We used an open-source real-time operating system called Zephyr OS for BLE-Scanning, BLE-Services, as well as

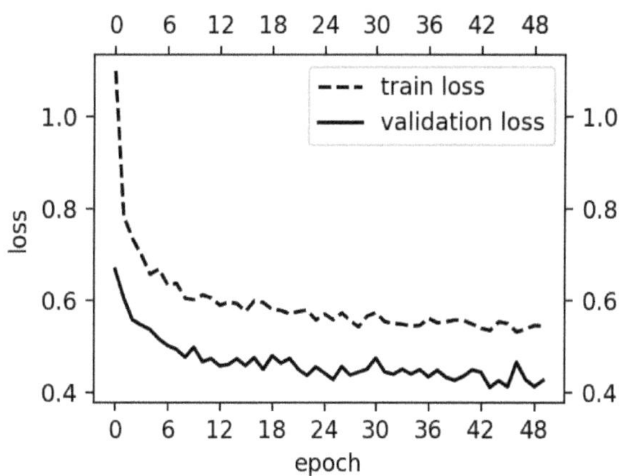

Figure 6: training and validation losses

sending the data to the server. We then used Tensorflow to implement the network architecture.

3   Results and Discussion

In order to estimate the performance of the network, we measured RSSI values along random walk trajectories similar to the training trajectories. The samples are collected from the entire tested area to ensure the results remain valid for all possible nodes. Figs.8-10 illustrate the qualitative network's performance along different paths, where actual trajectories are shown in gray vs. the predicted trajectories in black. The Cumulative Density Function (CDF) of the Errors is shown in Fig.7. Approximately 88% of the errors are within the 1.5 meters distance and the network could estimate the true locations with the mean error of 0.7 meters. Due to the constrains typical for an asset management system, such as minimizing the energy consumption and using minimum processor's resource, we used only four APs in this experiment, sent advertising signals sparingly to minimize energy consumption, and used just three recent advertised RSSIs. Considering all constrains this method is highly suitable for asset management systems.

4   Conclusion

This paper investigated the performance of a deep recurrent neural network architecture for a BLE indoor localization. The LSTM layers are shown to be an effective tool to reduce the noise effect of the raw RSSI and extract the high level representation from the RSSI time series. A real experiment has been conducted in an indoor environment to measure the performance of the method. We used only four beacons to send the location advertising signals. Increasing the number of beacons and collecting the RSSI samples in a 3D fashion are expected to increase the accuracy of the results further. The optimum number of beacons and their optimum distances are subjected to further investigations in future works. Implementing the network made it evi-

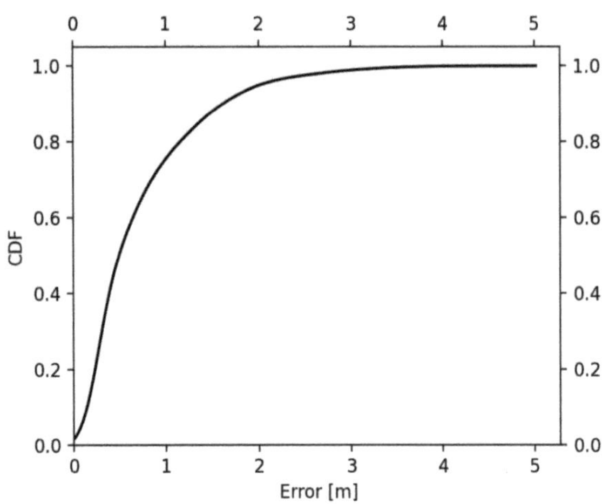

Figure 7: Cumulative Density Function of the Errors

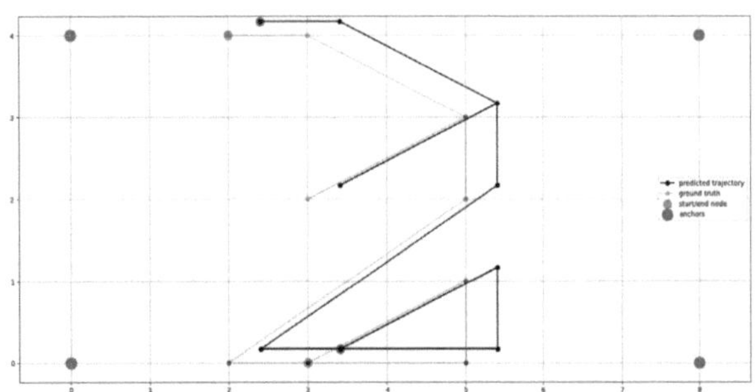

Figure 8: predicted trajectory from the random walk 1

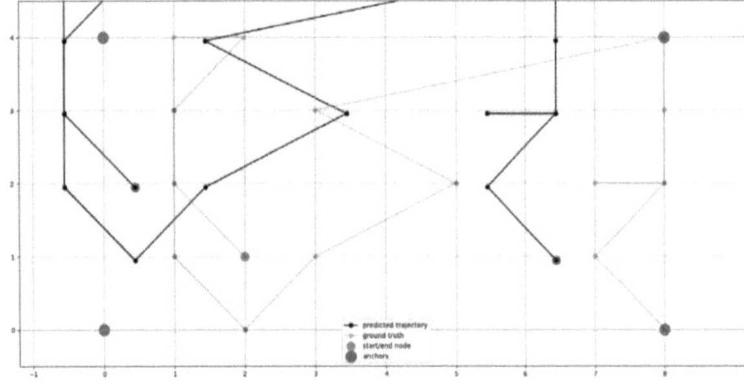

Figure 9: predicted trajectory from the random walk 2

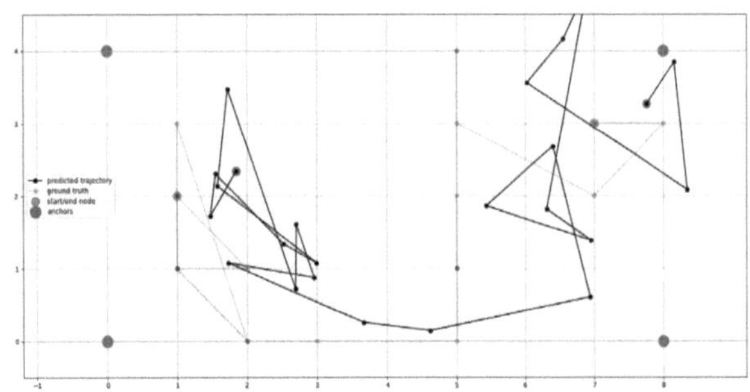

Figure 10: predicted trajectory from the random walk 3

dent that collecting RSSI samples accounts for most of the working loads and is very time-consuming. Also, a poten-

tial issue is a change in the environment, caused by weather changing the furniture or during the crowded hours. Therefore, an investigation for adapting the networks to the new environment with transfer learning methods [7] could be a good starting point to solve the above issues.

Acknowledgement

The work has been carried out at Dräger Safety AG & Co. KGaA. Technology Management Department.

Author's Statement

Authors state no conflict of interest.

5   References

[1] T. Tan, L. Zhang, and Q. Li, "An efficient fingerprint database construction approach based on matrix completion for indoor localization," *IEEE Access*, vol. 8, pp. 130 708–130 718, 2020.

[2] L. Y.-j. XI Rui and H. Meng-shu., "Survey on indoor localization," *Computer Science*, vol. 43(4), pp. 1–6, 2016.

[3] S. Kumar, R. M. Hegde, and N. Trigoni, "Gaussian process regression for fingerprinting based localization," *Ad Hoc Networks*, vol. 51, pp. 1–10, 2016. [Online]. Available: https://www.sciencedirect.com/science/article/pii/S1570870516301834

[4] P. Malekzadeh, M. Salimibeni, M. Atashi, M. Barbulescu, K. N. Plataniotis, and A. Mohammadi, "Gaussian mixture-based indoor localization via bluetooth low energy sensors," in *2019 IEEE SENSORS*, 2019, pp. 1–4.

[5] C. Pozuelo, I. Jiménez, A. Guerrero-Curieses, J. L. Rojo-Álvarez, E. Everss, M. Wilby, and J. Ramos, "Nonparametric model comparison and uncertainty evaluation for signal strength indoor location," *IEEE Transactions on Mobile Computing*, vol. 8, pp. 1250–1264, 09 2009.

[6] A. Sherstinsky, "Fundamentals of recurrent neural network (RNN) and long short-term memory (LSTM) network," *CoRR*, vol. abs/1808.03314, 2018. [Online]. Available: http://arxiv.org/abs/1808.03314

[7] F. Zhuang, Z. Qi, K. Duan, D. Xi, Y. Zhu, H. Zhu, H. Xiong, and Q. He, "A comprehensive survey on transfer learning," *CoRR*, vol. abs/1911.02685, 2019. [Online]. Available: http://arxiv.org/abs/1911.02685

6

E-Health

Towards Clinical Workflow Support for Robotic-Assisted Interventions

Anne Tjorven Büßen [1,2], Christian Kaethner [2], Elisabeth Preuhs [2], Marcus Pfister [2], Markus Kowarschik [2] and Mandy Ahlborg [3,4]

[1] Medical Engineering Science, Universität zu Lübeck, anne.buessen@student.uni-luebeck.de

[2] Advanced Therapies, Siemens Healthcare GmbH, Forchheim, Germany, {christian.kaethner,elisabeth.preuhs, marcus.pfister,markus.kowarschik}@siemens-healthineers.com

[3] Institute of Medical Engineering, Universität zu Lübeck, ahlborg@imt.uni-luebeck.de

[4] Fraunhofer Research Institution for Individualized and Cell-Based Medical Engineering, Lübeck, Germany, mandy.ahlborg@imte.fraunhofer.de

Abstract

Mechanical thrombectomy is an effective treatment for patients suffering from acute ischemic stroke. However, this form of treatment can only be performed by trained physicians in facilities equipped for such procedures, so not every patient has access to this kind of treatment within an appropriate time frame. Thus, it is an intriguing field of research to perform such a treatment remotely with the help of a dedicated robotic system. In this work, first thoughts towards a support system have been developed that is meant to guide the communication between the remote and local staff and to support the intervention itself. Clinical relevance and specifications have been defined with clinical workflow experts and were turned into conceptual prototypes that were refined in several iterations. The final prototypical concept lays the groundwork for an initial functional implementation and practical test-runs of this system.

1 Introduction

Each year, 15 million people suffer from a stroke worldwide. About 5 million survive without consequences, 5 million stay permanently disabled and 5 million of them pass away. This makes stroke the second leading cause of death and the high percentage of patients in permanent need of care causes a burden on society as well as healthcare systems [1]. With a fast diagnosis and appropriate treatment, the chances for a positive outcome are increased drastically. A stroke can either be caused by a ruptured vessel in the brain, leading to a hemorrhagic stroke, or by an occluded vessel, causing an ischemic stroke. Both lead to an insufficient supply of blood in the affected areas of the brain and neurons begin to die off, which causes a wide range of neurological symptoms. Since treatments of hemorrhagic and ischemic strokes differ significantly, a distinction between the two types is essential [2]. In 76.3% to 80.5% of the cases, the strokes are caused ischemic. Because their incidence is substantially higher than for hemorrhagic ones, this work focuses ischemic strokes [3].

With the onset of symptoms related to a stroke and the emergency call, the stroke pathway is set in motion. The first steps are partially performed during transportation to a hospital and include the aquisition of patient data and symptoms, blood tests and standard diagnostic procedures, such as measurement of vital parameters and monitoring of the electrocardiogram.

Brain imaging, usually performed in-hospital, is crucial for the diagnosis of an ischemic stroke. Computed tomography (CT) and magnetic resonance imaging (MRI) are the standard techniques to rule out hemorrhages, detect occlusions, define their precise location and evaluate the severity of the stroke [2].

Ischemic strokes can either be treated by a chemical dissolving of the thrombus or by an endovascular mechanical removal. The goal is the reperfusion of brain tissue to save hypoxic areas. Depending on the stroke, both options can be combined in support of each other [4].

Tissue plasminogen activator was approved for the treatment of stroke in 1996 and initiates fibrinolysis. During this natural process, the protein fibrin is broken down into smaller degradation products and the thrombus dissolves [5]. Thrombolysis does not decrease the mortality of ischemic stroke patients, but improves the outcome of surviving patients. A critical aspect of thrombolysis therapy is the swift initiation of treatment. It was shown that the timespan between when the patient was last seen well and the drug administration should not be longer than 3 to 6 hours [6]. Since 2015, multiple trials have shown that an intervention via mechanical thrombectomy is a promising alternative to thrombolysis. This treatment decreases the mortality rate, increases the chance for a positive outcome and enlarges the time window for an impactful intervention to up to 24 hours. Also for patients with large-vessel occlusions, who barely benefit from thrombolysis, treatment becomes sensible [7]. During a mechanical thrombectomy, medical devices are inserted into the patient's vascular system via a small incision, usually in the groin, and are used to remove the thrombus mechanically. To perform angiograms and secure a safe and fast access to the thrombus for all devices, a guide wire and a guide catheter are maneuvered towards the occlusion in the beginning of the intervention. The insertion and exchange of further devices may follow in preparation of the final clot-removal, for which multiple techniques have been proposed and investigated [8]. The application of an aspiration system and a stent retriever have shown the best results. Aspiration can suck the thrombus completely or partially into a catheter which is then pulled out. Suction additionally prevents potential fragments of the thrombus from moving further distally. For the use of a stent retriever, first a micro-wire and -catheter are moved through the thrombus. After the removal of the micro wire, the stent retriever is introduced and set free for expansion by removing the micro-catheter. During a waiting period of a few minutes, it fuses with the thrombus and is pulled back into a catheter. This method can be combined with aspiration to increase the safety of the intervention [8].

For navigation during this endovascular intervention, live imaging of the vascular system and the instruments is realized by fluoroscopy. To enhance the contrast in the vascular structures, the acquisition of the X-ray images is combined with the injection of iodinated contrast agent. Technologically, this is realized by a C-arm system, that can be moved flexibly around the patient table to depict the desired angulation relative to the patient. Especially in neurological interventions, a biplane imaging system can be used to view two perpendicular planes simultaneously. This eases navigation and reduces the duration of the intervention.

During the procedure, the physicians need to stand next to the patient to navigate the devices, so they are exposed to ionizing radiation. Studies with interventional cardiologists have shown that they suffer from an increased risk for cataract. A high lifetime dose also implies an increased risk for stochastic effects of ionizing radiation. Due to this, the operator needs to be shielded in the most effective way. Therefore, operating teams wear protective gear during the interventions. Unfortunately, those garments are heavy and can lead to orthopedic problems [9].

To provide a more ergonomic protection of the operator from radiation during endovascular procedures, a robotic system could be used. It was shown that the use of such a system can reduce the radiation exposure of the main physician by 95% [9]. It consists of a bedside unit that controls the devices and a dedicated lead shielded workstation, usually in the same or a neighbouring room. Current systems have been shown to support cardiovascular interventions [9] and the treatment of cerebral aneurysms [10], but the introduction of a similar system for ischemic stroke interventions is conceivable and will be assumed in this work.

With the introduction of robotic systems to the interventional suite, a remote control of such systems becomes a feasible concept. Especially time-crucial interventions, such as a mechanical thrombectomy may benefit from this concept, because they are only performed in designated stroke centers. In many regions, the distance to those specialized clinics is long and affects the patient's chance for recovery. Thus, the possibility of a remote-controlled robotic system would increase the proportion of population with access to the right care in a shorter time frame. In such a setup, the local staff is responsible for aspects such as patient monitoring and the handling of the robotic system, whereas the actual procedure is performed by an experienced physician who can work remotely.

In 2019, a feasibility study reported the successful performance of five remote robotic-assisted interventions in cardiology. In those cases, the communication between both parties was ensured by an audio- and video-connection. The operator was provided with live medical imaging and the patients' vital parameters and a trained team, well known to the operator, was locally on stand by at all times [11].

To establish this way of working, it needs to become feasible for a remote operator and a local team who aren't used to work in this pairing. The goal of this work is to design an application to support and guide the procedure towards a safe and smooth routine.

2   Material and Methods

The work is methodically structured into three task areas. A theoretical workflow for remotely performed mechanical thrombectomies needs to be developed, initial prototypes for a tool to support this workflow in practical use need to be designed and evaluated and finally implemented to build a working prototype application. Based on the understanding gained in one of the areas, the others may get adapted throughout the project. The goal for this developed tool is to improve the remote work experience with the robotic system and to increase confidence for all involved parties.

In the first phase, conventional mechanical thrombectomy procedures have been reviewed, focusing on their workflows and the exchange of information during the procedure. Therefore, papers describing different methods of thrombectomies, videos and live-cases have been studied.

On this basis, an abstract workflow has been designed, covering the essential aspects of the mechanical thrombectomy techniques. This has been transferred into a general workflow for robotic-assisted thrombectomies with the goal for the robotic system to not dictate the workflow, but to assist and adapt to the existing ones.

With the extension to a remote scenario, it becomes essential to distinguish between the local team's and the remote operator's site. Their tasks and their involvement differ significantly throughout the procedure, so for every step in the abstract workflow, the actual actions of both sites, the displayed information they need and the information that should be transferred to the other team have been defined.

Those considerations according to a remote robotic-assisted workflow were supplemented with the results of a meeting with an experienced interventional neuroradiologist, who presented a clinical view on remote operations and expounded features relevant in daily use. Challenging sit-

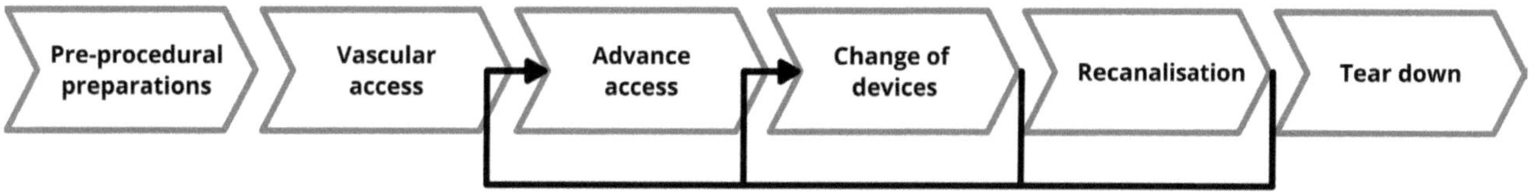

Figure 1: Abstract workflow of a robotic-assisted mechanical thrombectomy. The black arrows represent the opportunity to advance newly inserted devices or to repeat the recanalisation.

uations during remote interventions that require dedicated support were identified and strategies for associated assistance were developed.

Based on this first phase of research and the worked out specifications, a conceptual non-functional prototype of a support tool was developed. This developed model will not be part of the final functional prototype, but enables the presentation of a potential result designed according to predefined specifications within a short period of time. In multiple internal iterations of reviews, as well as with the interventional neuroradiologist, substantial changes of the prototype may be requested. Those can be realized faster and with less effort in the conceptual protoype than in the implemented version of this tool.

Once the evaluation results converge, an implementation can be realized and tested with two different locations for the local and remote tool.

3   Results and Discussion

In the first phase of the project, the theoretical workflow shown in Fig. 1 was created. Its six main steps include pre-procedural preparations, vascular access, advancement of the medical devices towards the thrombus, change of devices, recanalization and tear down. Depending on the chosen technique, it may be necessary to continue the advancement of the access after the exchange of devices. Based on the success of the recanalization, additional iterations of the procedure may be necessary. Those opportunities are implied by the black arrows underneath the boxes in Fig. 1.

With a focus on remote interventions, especially in the beginnig of the procedure and during the change of devices, the handover of responsibility between remote and local site needs to be smooth and safe. Therefore, it has to be considered that the local team does not necessarily work with the robotic system on a regular basis. Both parties may barely know and trust each other and in some scenarios may not even speak the same native language or dialect.

Based on this groundwork, different specifications for a support tool for distributed workflows have been defined, on which basis the prototype has been designed. In this section, the main requirements are presented.

A detailed workflow of each procedure should be automatically tracked and adapted towards the operator's decisions. This is less important to the operator themselves, but the local team can keep track of the procedure, gets early information about potential complications and feels more involved throughout the intervention. The current step in the workflow could be tracked by manual input, the analysis of currently loaded devices or the evaluation of device positions in the live-imaging. This knowledge may also lead to a more intelligent support system that offers certain options or images right then when they are needed.

At all times, audio- and video-connection between the teams should be maintained. Different local camera settings are possible and need to be further investigated.

Also, a display of the currently loaded devices in the different cassettes of the robotic system could be helpful to compensate for the missing ability of the remote operateur to reassure visually or haptically on that matter.

In terms of communication between both teams, some instructions from the operator to the local team need to be unmistakable, for example the change of devices. Those should not only be requested orally, but confirmed with the developed tool. This way, the responsibility is clearly defined and could even contribute to a documentation of the intervention. To improve the workflow, the most likely used devices for the current step should be suggested when a change of devices is requested. Additionally, a choice from all devices that are locally available is necessary. To improve safety, the robotic system could get locked after this instruction. This way, the local team can open the robotic system without worrying about movements of the patient table, C-arm or an unannounced application of radiation.

In the beginning of the procedure, the local team should update on the status of loaded devices, vascular access, the connection of the robotic system to the sheath and finally confirm the preparedness for the start of procedure. The operator confirms the knowledge of the patient data and the intervention, including the planned use of devices, and declares the final start of the intervention. Those specifications resulted in the design of an interactive monitor for each site. Potential screens from the prototype phase are shown in Fig. 2. There are multiple tiles to toggle between displays and camera angles, view and change four loaded devices, manage the audio connection and a timer and to locally report issues. Additionally, the local site can see some workflow steps ahead with specially marked warnings when a change of devices becomes necessary.

A personal adaption of the system is desirable. The workflow and the preferred pre-defined devices may vary between operators and should be known to the system. The user interface could be modified according to personal preferences and presented informations and assistances may vary depending on the level of familiarity between the teams and their experience with the robotic system.

In the development of the prototype, a balance between a

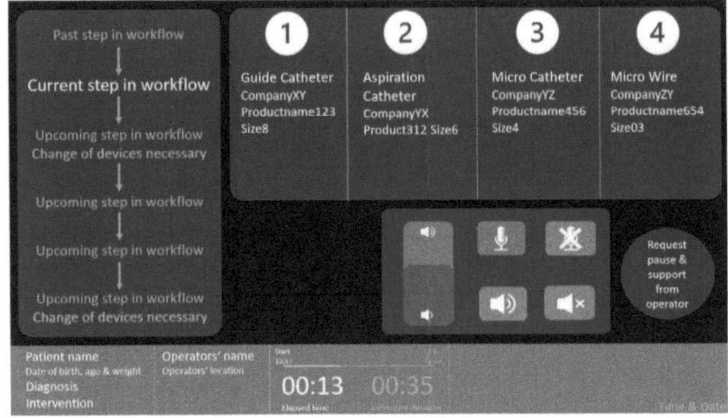

Figure 2: Screenshots of a prototype for the workflow support system. Local version (top) and remote version (bottom).

strict guidance through manual confirmations and the minimization of additional tasks had to be found. Discussions with further clinical workflow experts need to be held to validate the usability of the tool. Once the conceptual prototype settles to one version after some iterations of adjustment, a functional prototype will be implemented. First test runs with that version in a setting of two separated workplaces will show if it can be used pleasantly.

4 Conclusion

The introduction of remote robotic-assisted interventions into clinical routines has a long way to go. However, especially the treatment of acute ischemic stroke via mechanical thrombectomy is an attractive application field because many patients might benefit from an improved accessibility. This work focussed not only on making this technique possible, but also practicable and trustworthy in clinical use. Potential sources of miscommunication between a local team and the remote operator were identified and an application to prevent them was developed. Practical test runs with a functional prototype need to validate whether this tool actually improves the communication.

Acknowledgement

The work has been carried out at Advanced Therapies, Siemens Healthcare GmbH, Forchheim, Germany and supervised by the Institute of Medical Engineering, Universität zu Lübeck.

Authors' Statement

Conflict of interest: The authors state no conflict of interest. Disclaimer: The concepts and results presented in this paper are based on research and are not commercially available.

5 References

[1] A. D. Lopez and C. C. J. L. Murray, *The global burden of disease, 1990–2020.* Nature medicine, vol. 4, no. 11, pp. 1241–1243, 1998.

[2] B. C. V. Campbell and P. Khatri, *Stroke.* The Lancet, vol. 396, no. 10244, pp. 129–142, 2020.

[3] V. L. Feigin, C. M. Lawes, D. A. Bennett and C. S. Anderson, *Stroke epidemiology: a review of population-based studies of incidence, prevalence, and case-fatality in the late 20th century.* The Lancet Neurology, vol. 2, no. 1, pp. 43–53, 2003.

[4] A. Ciccone et al., *Endovascular treatment for acute ischemic stroke.* New England Journal of Medicine, vol. 368, no. 10, pp. 904–913, 2013.

[5] T. Brott and J. Bogousslavsky, *Treatment of acute ischemic stroke.* New England Journal of Medicine, vol. 343, no. 10, pp. 710–722, 2000.

[6] J. M. Wardlaw et al., *Recombinant tissue plasminogen activator for acute ischaemic stroke: an updated systematic review and meta-analysis.* The Lancet, vol. 379, no. 9834, pp. 2364–2372, 2012.

[7] R. G. Nogueira et al., *Thrombectomy 6 to 24 hours after stroke with a mismatch between deficit and infarct.* New England Journal of Medicine, vol. 378, no. 1, pp. 11–21, 2018.

[8] A. M. Spiotta et al., *Evolution of thrombectomy approaches and devices for acute stroke: a technical review.* Journal of neurointerventional surgery, vol. 7, no. 1, pp. 2–7, 2015.

[9] G. Weisz et al., *Safety and feasibility of robotic percutaneous coronary intervention: PRECISE (Percutaneous Robotically-Enhanced Coronary Intervention) Study.* Journal of the American College of Cardiology, vol. 61, no. 15, pp. 1596–1600, 2013.

[10] V. M . Pereira et al., *First-in-human, robotic-assisted neuroendovascular intervention.* Journal of neurointerventional surgery, vol. 12, no. 4, pp. 338–340, 2020.

[11] T. M. Patel, S. C. Shah and S. B. Pancholy, *Long distance tele-robotic-assisted percutaneous coronary intervention: a report of first-in-human experience.* EClinicalMedicine, vol. 14, pp. 53–58, 2019.

Development and comparison of PPGs for the acquisition of vital data with intelligent evaluation

Jannik Dann [1], Sebastian Hauschild[2], Horst Hellbrück [3]
[1] Applied Information Technology, Luebeck University of Applied Sciences, jannik.dann@stud.th-luebeck.de
[2] Luebeck University of Applied Sciences, sebastian.hauschild@th-luebeck.de
[3] Luebeck University of Applied Sciences, horst.hellbrueck@th-luebeck.de

Abstract

A photoplethysmograph (PPG) is a medical device for monitoring blood flow and transportation of substances in the blood. It consists of a light source and a photo detector for measuring transmitted and reflected light signals. The problem of today's PPG measurements is the ambient light interfering the measurement. To overcome this problem extensive techniques are required for compensation. Thus, the a new method is tested using modulation techniques. High frequency modulations of the LED´s create enough space to the interference to filter it by high pass filter components. This paper focuses on the hardware implementation of a PPG device and achieves a pulse measurement using modulation techniques.

1 Introduction

A photoplethysmograph (PPG) is a low-cost optical technique that can be used to detect blood volume changes in the microvascular bed of tissue. These medical devices are already available as wearable devices. The PPG device uses a light source and a light detector in order to perform these measurements. Typically light emitting diodes (LED) are used in combination with a photo-diode. The sensor placed directly on the skin, typically a fingertip or wrist. Depending on the current change of blood volume, more or less light gets absorbed by the skin. As a result, the received light intensity at the photo diode deviates depending on the change of blood volume. Further signal processing leads to a visible pulse. Due to the device being available as a wearable, long term measurements are interfered by motion artifacts. These motion artifacts are the result of ambient light, creating a superposition with the LED light. In addition the position of the LED's is changing, which impacts the light intensity of the wavelength entering the skin and thus the reflected light. Overall only in certain condition a reliable PPG measurement can be performed. Existing solution for these problem are using an extensive workaround by using motion sensors for compensation. In this paper, the development of a new technique for signal processing using signal modulations is developed and tested. This approach will only take use hardware components and will perform a heart rate measurement. The following sections will provide information about related work, the implementation and challenges in setting up the sensor, the hardware design and a final evaluation of resulting signals during a real pulse measurement.

2 Material and Methods

2.1 Related Methods

The compensation of motion artifacts is a well discussed topic in the field of wearable devices. Thus, many approaches are developed tackling the problem of motion artifacts. These approaches are using extensive techniques for compensation such as independent component analysis (ICA), adaptive noise cancellation (ANC) and Fourier series analysis (FSA). These methods focus on a digital pre- and post processing of the signal to regenerate the origin signal without noise interference by motion artifacts [2]. All of these methods are using complex algorithms, which requires fast micro controllers performing these algorithms and calculation in a real time application. In addition to this, a two channel PPG is tested using different light intensities and a subtraction method for compensating [4]. Also learning based approaches are tested in order to overcome the problem by using AI-technology. The AI-Algorithm searches for implausible features and corrects them [2]. All of these methods are providing a feasible solution and compensate motion artifacts to satisfaction. Nevertheless, all of these methods require extensive techniques and go ahead with complex algorithms and a lot calculation power. The approach of this work only focuses on a hardware implementation. For later evaluation, the hardware is supposed to provide a reference level, which is also going to be processed by a micro controllers for compensation. Nevertheless, the approach brings a reduction in complexity, expense and costs to overcome the problem of motion artifacts. The following section will cover the main idea, and will provide detailed information about the used modulation technique.

2.2 Concept of signal modulation

In this section, the general idea of how to overcome the problem of motion artifacts only by using hardware components is going to be explained. Thus, the problem is explained on a technical basis in a first step and the corresponding hardware solution in a second step.

Whenever the patient moves, ambient light creates a superposition with the LED's light at the photo diode being the light detector, which reflections are supposed to be evaluated after a signal processing to determine certain vital parameters. In addition, a position change of the wearable device also causes the reflection to fluctuate in its intensity causing the input signal to be incorrect. To overcome this problem, the implementation aims for decoupling the LED light and the interfering ambient light. This decoupling method is a well known technique, since frequency selective filters can separate higher and lower frequencies very effectively. Thus, the LED is driven by current source, which provides a sinusoidal current flow with a frequency being way higher than the frequency of the interfering ambient light. This leads to a sinusoidal light intensity and a sinusoidal output voltage of the photo diode. By only processing the amplitude of the desired frequencies of the LED the ambient light is not longer part of the signal processing. A filter and demodulation process of the signal converts the amplitude into a DC-voltage, which is later used as a reference voltage. The blood-pulse of the patient will enter the the input signal as an amplitude modulation, which passes filters and will be recognizable as a periodical fluctuation on the reference voltage. A schematic of this procedure is shown in figure 1. The reference voltage (DC) and the amplitude of the pulse (AC) are supposed to have a relation. Thus, the reference level can be used for stabilizing the pulse amplitude in a post processing calculating in order to overcome motion artifacts. The following section will provide a detailed description of how the idea is implemented in a hardware design, containing the sensor and a hardware solution for signal processing.

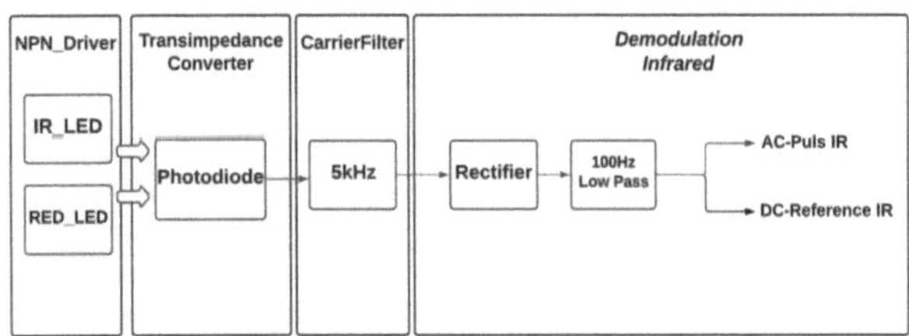

Figure 1: Overview Schematic.

2.3 Implementation

The whole circuit is designed to run at a unipolar voltage of 5V. To achieve a bipolar behavior for AC-signal processing, a virtual ground at $\frac{V_{CC}}{2}$ is created, which is the reference voltage for a major part of the signal processing.

The LED-driver is build by a class A transistor amplifier with adjustable amplitude and offset, which is used as a voltage controlled current source. The oscillating voltage is provided by signal generator for test purposes. The frequency being tested is 5kHz. Its important to consider, that the biasing point for the LED-driver is supposed to provide a signal, that can be fully visible at the output of the photo diode. In case the signal intensity leads to the saturation of the photo diode, a clipping effect will occur. The clipping effect has its impact on the positive half-wave, which can not exceed the supply voltage. However, this limitation could cause the demodulated DC-level to be wrong. Which overall leads to a trade off between signal strength of the sender LED and the ability of eliminating external light intensity. To overcome this problem partly, the bias current needs to be as low as possible. This leads to the requirement of a small modulation amplitude, because otherwise the bottom half of the amplitude gets clipped. In addition the LED needs to be operated in bias-point being linear.

The first signal processing step is performed by a transimpedance converter. This type of amplifier uses an operational amplifier to turn the photo current into an output voltage. With an adjustable resistor in the negative feedback branch of the amplifier, the gain of the output voltage is adjustable [7]. Its important to consider, that an high amplification easier leads to clipping effects in the output voltage.

A multiple feedback filter (MFB) second order is implemented with a center frequency of 5kHz in order to filter the signal [7]. The band-pass filter provides a signal without any DC-level at its output. For demodulation, a full rectification is done by a operational amplifier circuit. Providing a precise rectification without any losses due to diodes forward voltages.

Further LF-low pass filters with a corner at frequency at 100Hz are used to turn the rectified signal into a DC-voltage [7]. Although a low pass filter with a lower corner frequency would be able of providing a better DC-level with a higher attenuation on the ripple, a higher corner frequency is required due to the spectrum of the pulse itself. The pulse itself is elaborated to have spectral components with reach frequencies up to 100Hz. Thus, to allow all pulse-determining frequencies to pass, a corner frequency at 100Hz is required [6]. Already after the low pass filter, the pulse is going to be visible.

A hardware design was made in Target3001 in order to test the performance of the described technique.

3 Results and Discussion

This section focuses on the evaluation of the described method and hardware implementation in section 2.3. Therefore the signal processing is tested by measurements, which were applied to the fingertip. The signal processing procedure follows the schematic in figure 1. The used biasing signal as well as sensor components are described in Section 2.3. Figure 2 shows the detection of the infrared LED at the photo diode with a frequency of 5kHz.

As noticeable the output voltage is measured with $V_{off} = 2.15\,V$ and $V_{ampl} = 0.875\,V$ amplitude. The feedback

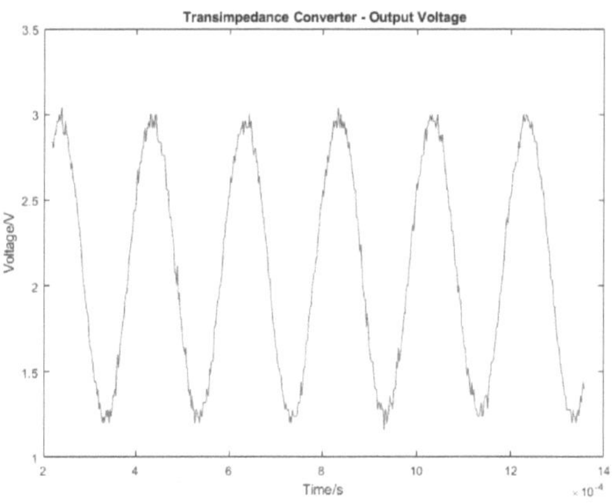

Figure 2: Output Signal Transimpedance.

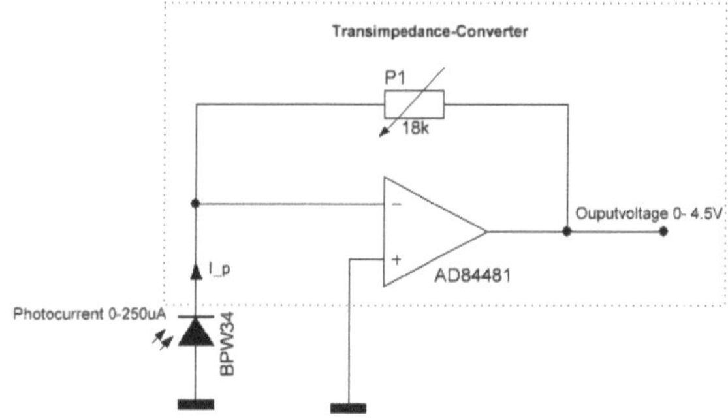

Figure 3: Circuit of Transimpendance - Converter

resistor P1 of the trans-impedance converter is determined in order to not previously limit the saturation of the photo diode. According to the data-sheet, the photo diode has a maximum output current of $250\mu A$ [1]. This results in $\frac{4.5V}{250\mu A} = 18k\Omega$. Increasing the resistor P1 even further, would directly decrease the ability of compensating ambient light, by resulting in clipping effects even faster. The circuit, containing trans-impedance converter as well as the photodiode, is shown in figure 3. The Equation 2 shows the Leftover Light Intensity LLI [lx] in dependence of the light sensitivity of the photo diode, the measured path attenuation and the LED current. According to the data-sheet ,the photo diode provides a sensitivity of $0{,}0833\frac{\mu A}{lx}$ and a saturation thresh hold at 3000 lx [1]. In a first step, the relation between driving current of the LED's and the received light intensity are calculated in Equation 1.

$$\frac{lx}{I_{LED}} = \frac{\Delta I_p * \frac{1}{0.0833}\frac{lx}{\mu A}}{\Delta I_{LED}[mA]]} = 60\,\frac{lx}{mA} \qquad (1)$$

The ratio derived in Equation 1 is supposed to stay constant. However, changing positions, different skin types and further influences can change this value, leading to a deviating factor in the evaluation of LLI. This test measurement provided a sensitivity of $60\,\frac{lx}{mA}$, leading to a LLI value according to Equation 2.

$$LLI[lx] = 3000lx - 30mA * 60\,\frac{lx}{mA} = 1200lx \qquad (2)$$

By having the gain of the trans-impedance converter adapted to the maximum photo current of the BPW34, the ratio between LLI and maximum light intensity should match the ratio of maximum signal voltage compared to the supply voltage. Equation 3 provides the comparison of both ratios as an evidence of matching the theoretical approach with the hardware solution.

$$\frac{1200lx}{3000lx} = \frac{2}{5} = \frac{5V - 3V}{5V} \qquad (3)$$

After having received an expected signal at the trans-impedance converter, further processing is evaluated in-cluding the carrier filter as well as the demodulation process. The filter attenuates the carrier signal by 250mV, which equals -0,66dB. This attenuation is expected and does not have any effect on further signal processing. Due to the band pass characteristic, the DC components in the signal are eliminated as well. The filter has a virtual ground reference at 2.5V. Thus, the sinusoidal signal is centered around the 2.5V reference.

The filter provides the signal for the rectifier circuit, which achieves a full rectification of the input signal by using two operational amplifiers. The operational amplifiers being used are the AD4841 [5], which have sufficient characteristics such as slew rate and Gain-Bandwidth Product (GBP) to process a 5kHz input signal. The fully rectified signal can be viewed in figure 4. Due to the required 2.5V reference, the rectified signal has a 2.5V offset.

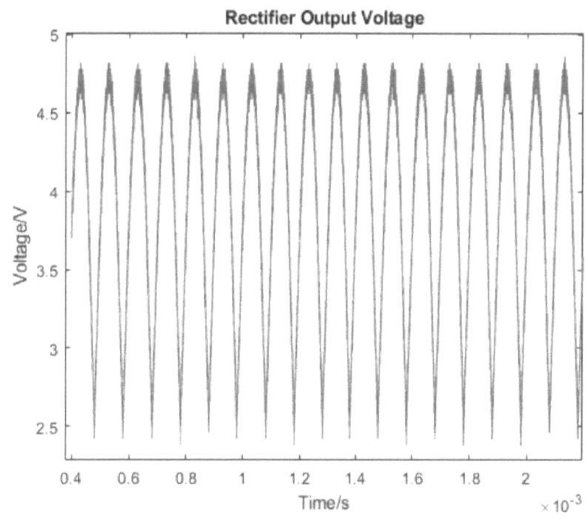

Figure 4: Output Signal Rectifier.

In the last step, the signal is filtered by a 100Hz low-pass filter. This signal contains two information's, which are the reference level and the pulse itself. As described in Section 2.2, the reference level can provide information about the current path attenuation. Taking the DC-level of the reference voltage into account by comparing it with the amplitude of the pulse, a relation can be set up, providing the change of a stabilized pulse amplitude depending on the current reference level. A test measurement shows clear relation between two signals. Figure 5 shows the detection of a pulse during a test measurement on the fingertip. The period of the pulse is measured with $T_{Period} = 0.8s$, which equals a heart rate of 75BPM.

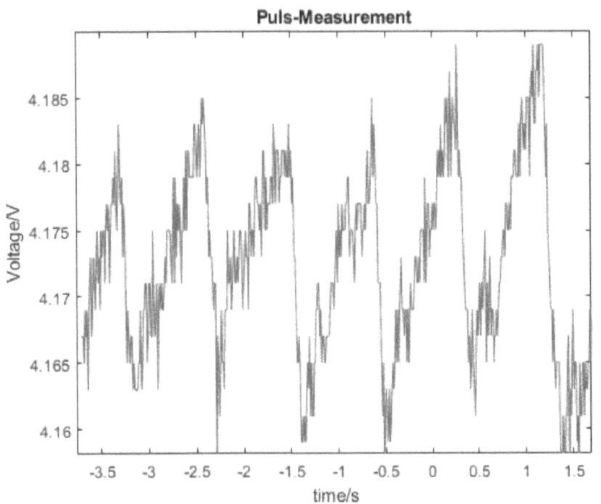

Figure 5: Pulse Measurement.

The plus gets visible on the reference voltage after the 100Hz low pass filter. A comparison between DC-level and pulse amplitude shows, that dependencies exist. A first test measurement showed, that depending on the pressure the offset of the reference level changes. Simultaneously the pulse amplitude changed as well. Overall a higher pressure on the sensor leads to a higher reference voltage and a lower pulse amplitude.

The circuit board provides the expected characteristics concerning the hardware implementation. The relation between the reference voltage and the pulse- amplitude can be confirmed, but in order to take advantage of it more analyses is required. Furthermore, the work showed that the advantage of modulation techniques require a lot attention to the biasing of LED and photo diode, which depends on the sensor being used.

4 Conclusion

The work showed, that its possible to perform a pulse measurement by using modulation techniques. To make the pulse visible, its necessary to implement filters, rectifiers and amplifiers for demodulation. Furthermore, different obstacles need to be taken into account by such implementation. Its important to put enough effort into a good sensor being responsible for the input signal from the photo diode. The sensor should have permanently mounted components providing a good repeatability. Otherwise biasing issues will occur at the input, making an evaluation and data acquisition more difficult by resulting in clipping effects. The biasing of the current sources as well as the amplification of the trans-impedance converter require attention to achieve accurate and processable measuring results. Test measurements have shown that reference level and pulse amplitude have a relation, which comes out to be anti proportional. For further developments, an additional signal processing of the pulse would lead to a more accurate evaluation. Therefore, additional filter and amplifiers have to be implemented. Furthermore, the implementation of a second channel could enable enable an oxygen measurement according to Section 1. In addition, the modulation frequency for the second channel should be chosen, which has to make a compromise between the 100Hz and the corner frequency determined by the LED inertia. Moreover, its still to be tested, how satisfactory the motion artifacts can be eliminated using the DC - reference level provided by the hardware. Thus, additional elaborations are required for testing a motion artifact compensation using the reference voltage.

Acknowledgement

This publication is a result of the research of the Center of Excellence CoSA. Horst Hellbruck is adjunct professor at ¨ the Institute of Telematics of University of Lubeck.

Author's Statement

Conflict of interest: Authors state no conflict of interest.

5 References

[1] OSRAM 2022, *BPW34*. Available: https://dammedia.osram.info/media/resource/hires/osram-dam-5488305/BPW%2034_EN.pdf.

[2] Dongyeol Seok, Sanghyun Lee, Minjae Kim, Jaeouk Cho, Chul Kim, *Motion Artifact Removal Techniques for Wearable EEG and PPG Sensor Systems*. Available: https://www.frontiersin.org/articles/10.3389/felec.2021.685513/full

[3] David Pollreisz, Nima TaheriNejad, *Detection and Removal of Motion Artifacts in PPG Signals*. Available: https://www.researchgate.net/publication/335056984_Detection _and_Removal_of_Motion_Artifacts_in_PPG_Signals

[4] Jongshill Lee , Minseong Kim, Hoon-Ki Park,In Young Kim , *Motion Artifact Reduction in Wearable Photoplethysmography Based on Multi-Channel Sensors with Multiple Wavelengths*. Available: https://www.mdpi.com/1424-8220/20/5/1493/html

[5] Analog Elektronics, *ADA4807*. Available: https://www.analog.com/media/en/technical-documentation/data-sheets/ADA4807-1_4807-2_4807-4.pdf

[6] Kai Li, Heinz Rüdiger, Tjalf Ziemssen, *Spectral Analysis of Heart Rate Variability: Time Window Matters*. https://www.frontiersin.org/articles/10.3389/fneur.2019.00545/full

[7] U. Tietze, Ch. Schenk, *Halbleiter-Schaltungstechnik*, Springer 1999

Mapping, Modeling and Standardization of Clinical Data into the HL7 FHIR Specification for the EU Project PanCareSurPass

Anke Neumann [1, 2], Ann-Kristin Kock-Schoppenhauer [2], Desiree Grabow [3], Diana Walz [3], Anna-Liesa Filbert [3], Thorsten Langer [4], Josef Ingenerf [2, 5]

[1] Medical Informatics, Universität zu Lübeck, anke.neumann@student.uni-luebeck.de

[2] IT Center for Clinical Research, Lübeck, Lübeck, Universität zu Lübeck, {annkristin.kockschoppenhauer, josef.ingenerf}@uni-luebeck.de

[3] Division of Childhood Cancer Epidemiology, German Childhood Cancer Registry, Institute of Medical Biostatistics, Epidemiology and Informatics, Universitätsmedizin der Johannes Gutenberg-Universität Mainz, {desiree.grabow, dianwalz, anfilber}@uni-mainz.de

[4] Pediatric Oncology and Hematology, Department of Pediatrics and Adolescent Medicine, Universitätsklinikum Schleswig-Holstein, Campus Lübeck, thorsten.langer@uksh.de

[5] Institute of Medical Informatics, Universität zu Lübeck, josef.ingenerf@imi.uni-luebeck.de

Abstract

Childhood cancer patients have an increased risk of developing health problems known as late effects, which lead to increased morbidity and mortality. In response to these challenges, the Survivorship Passport was developed providing a Treatment Summary and an individualized care plan. The aim of this work was to introduce and pilot the Survivorship Passport in Germany. Therefore, a practical implementation strategy was developed to use the SurPass for person-centered care in scenarios in which electronic health data come from national cancer registries and hospital-based late effects clinics. The challenge was to match the existing datasets and achieve maximum harmonization. A mapping was performed to identify data elements from the minimal dataset compared to a long-term observation study. These data elements had to be matched with the register data and made available in FHIR interoperability standard. A program for mapping and transferring the data FHIR profiles was implemented for this project.

1 Introduction

In Germany, more than 80% of children and adolescents survive five years after cancer therapy [1]. The number of survivors is growing every year and exceeds 500,000 in Europe [2], [3]. People with cancer treatments like chemotherapy or radiotherapy, particular if they have to be applied in childhood, can have late effects and therefore need to be closely monitored. For high-quality and appropriate care, named long-term follow-up, of (adult) survivors of childhood cancer, it is crucial, among other things, that survivors are informed about their personal risk for long-term consequences. With this knowledge, survivors can manage their own care and the needs for support with their healthcare professionals. The major challenge is that only a minority of adult survivors of childhood cancer receive high-quality, long-term follow-up care that is tailored to the individual.

To manage this challenge, the Survivorship Passport (SurPass) [4] was developed. On the one hand, it provides survivors with an overview of their treatment, and on the other hand, it gives personalized recommendations for follow-up care, which is based on a combination of the internationally approved International Guideline Harmonization Group (for Late Effects of Childhood Cancer) guidelines and the PanCareFollowUp guidelines. The SurPass prototype has already been tested in a local study at a single center with more than 300 survivors [4].

The EU-funded PanCareSurPass (PCSP) project is investigating how to implement the SurPass on a larger scale [5]. To develop a strategy for Europe-wide implementation, the consortium of 17 partners from 7 European countries includes experts in interoperability, e.g. Health Level Seven International (HL7), epidemiologists, clinicians and other stakeholders. Fig. 1 shows the different stakeholders and the structure of the SurPass. In addition, the SurPass is being further developed. This will require data from electronic patient records and registries.

1.1 PanCareSurPass in Germany

The German representations for the PanCareSurPass project are the Department of Pediatrics and Adolescent Medicine of the Universitätsklinikum Schleswig-Holstein (UKSH), Campus Lübeck and the German Childhood Cancer Registry (GCCR) hosted by the University Medicinal Center Mainz. The GCCR records all cancer diagnoses of children

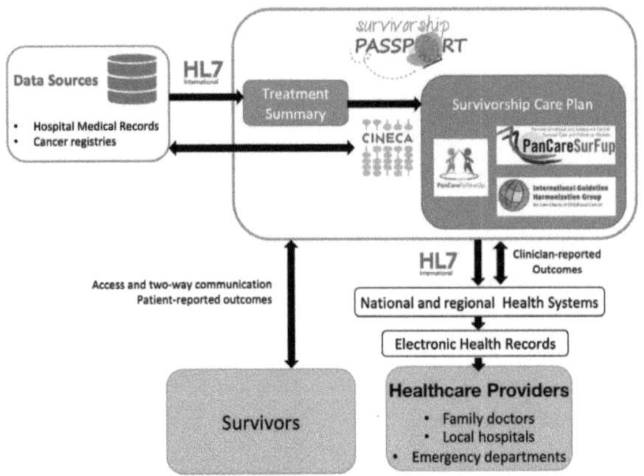

Figure 1: Data transfer to and from the SurPass between existing data sources and integration within the health system. CINECA is the Italian project partner responsible for creating the digital SurPass.

and adolescents in Germany. Therefore, the GCCR will host the PanCareSurPass platform throughout Germany. In order to create an individual care plan with the help of the SurPass, treatment and clinical data are required in addition to general patient and diagnosis data. The necessary data needs will be different data sources and transformed in a HL7 FHIR (Fast Healthcare Interoperability Resources) format. Part of the data needed for the PanCareSurPass project is already collected in a long-term late-effects follow-up study at the UKSH, other sources are the clinical data from the electronic health informations system and registry data from the GCCR. The IT Center for Clinical Research Lübeck (ITCR-L) provides technical support for a long-term follow-up study in a clinical trial management system, e.g. using electronic Case Report Form (eCRF). The ITCR-L will enable the data transfer of additional clinical data to the PanCareSurPass platform in a standardized format.

2   Material and Methods

2.1   Health Level Seven International

The mission of HL7 is to provide standards that empower global health data interoperability. HL7 provides a comprehensive framework and related standards for the exchange, integration, sharing, and retrieval of electronic health information that supports clinical practice and the management, delivery and evaluation of health services [6]. The most recent standard Fast Healthcare Interoperability Resources (FHIR) and their profiles facilitate seamless data exchanges in order to improve the interpretability and, therefore the quality and safety of care.
In PanCareSurPass a mapping of given proprietary data into standardized FHIR resources have to be created. In this context new profiles for the FHIR resources used are developed and refined. Profiles of resources allow contextual adaptation of resources. Especially the Logical Model of the Min-

imalTreatmentSummary (MTS) with the prototype profiles of resources were used for this work.

2.2   ISO/TR 12300:2014

ISO/TR 12300:2014 "Health informatics - Principles of mapping between terminological systems" provides a methodology for measuring the degree of equivalence between a source and a target [7]. ISO/TR12300 distinguishes five ratings of equivalence:

1 : Equivalence of meaning; lexical, as well as conceptual.

2 : Equivalence of meaning, but with synonymy.

3 : Source concept is broader and has a less specific meaning than the target concept/term.

4 : Source concept is narrower and has a more specific meaning than the target concept/term.

5 : No map is possible. No concept was found in the target with some degree of equivalence (as measured by any of the other four ratings).

The degree of equivalence is calculated by summing all data items with rank 1-4 divided by the number of all data items. ISO/TR 12300:2014 deals with terminology mapping in general. However, the methods presented are also applicable to semantic mapping of data elements or concepts, such as those applied by Rinaldi et al [8].

2.3   Evaluation

Following the steps described in section 3, an evaluation is carried out with the help of suitable experts. The experts include medical staff from UKSH at the clinical level and qualified data-steward professionals from the GCCR who have evaluated the interim results. Six experts were involved in each step of the evaluation.

3   Results and Discussion

For the progress of a project it is necessary to know the tasks and goals of the stakeholders. These were discussed and developed in the first meetings. The Fig. 2 shows a mapping diagram with the stakeholders, their data sources for the data to be mapped, and possible output formats.

3.1   Early mapping

Using a standard template and internationally agreed Coding Systems (e.g., ICD-O-3 for oncological diagnoses, ATC codes for medication data), the PanCare treatment summary is personalized for each survivor. It contains all relevant information about the survivor's diagnosis, cancer treatment, and any other relevant health data with potential long-term impact in a standardized format that can be used by healthcare professionals to quickly understand the medical history

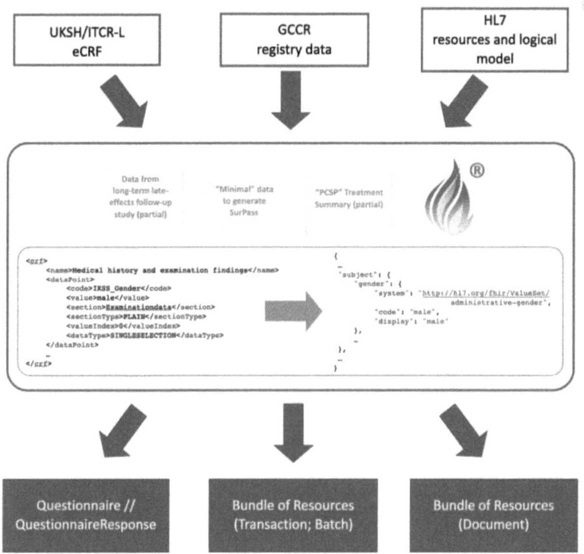

Figure 2: In the first level, the mapping diagram shows the different data sources. In the second level, the mapping between the relevant sources and bindings is shown and as an example of converting the gender from XML to JSON. In the third level shows the possible output formats for the German sub-project.

of each survivor [9]. A primary task is to gain interoperability between clinical records transfer of information to the SurPass and provision of information from the SurPass back to national electronic health records (EHR) and information systems. The comparison of data element definitions was based on the methodology of ISO/TR 12300 as first used by Brammen et al [10]. The minimum required record mentioned earlier, MTS, is required to create the SurPass. It contains personal data (gender and date of birth), initial treatment data (such as chemotherapy, stem cell transplant, radiotherapy, and surgical therapy), other information and relevant events, and information on progression and relapse after the end of therapy. The MTS includes 42 data elements with different encodings and data formats. Whereas the long-term follow-up already contains 14 different eCRFs with at least 13 data elements. For matching purposes, the eCRFs related to the patients' mental life situation were excluded, as such aspects will be considered much later in the further development of the SurPass. There remained four eCRFs from long-term follow-up that were considered. These contained 192 data elements. Because the goal was to cover the required MTS data elements (source for ISO-rating) in long-term follow-up (target for ISO-rating), only these 42 data elements were used for record equivalence matching.

Of the 42 data elements, only 16 could be successfully matched. Thereby 8 data elements corresponded to ISO-rating equal, two data elements were synonyms (ISO-rating 2), one data element of the source was more comprehensive than the target and five data elements of the source were more limited than the target. 26 data elements could not be successfully matched (ISO-rating 5). This corresponds to

an equivalence of only 38% between MTS and long-term follow-up.

Data that are not available in the source systems and cannot be mapped to standard templates according to MTS must be collected in other ways (possibly twice) by adding them as study characteristics. Therefore, the 26 unmatched data elements were added to the appropriate eCRF. In doing so, the elements were not only introduced lexically correct but also the data types and possibly necessary codings were taken into account. Of the 26 new data items introduced, 21 were treatment-specific questions. Four data items required special coding and one data field required a date.

This was followed by an evaluation by the experts. The experts did not see any errors in the adjustments made and welcomed the consideration of data types and encodings. In the future, it is very likely that more than the MTS data will need to be collected and transmitted. A further mapping with additional data elements to a complete TreatmentSummary is necessary for a more exact determination of the individual treatment plan of a patient and the further plan in PanCareSurPass. Based on the procedure presented here, this should not pose any further problems. However, an evaluation by experts should never be disregarded.

3.2   Modeling in output format

The relevant eCRFs are exported from the clinical trial management system as an XML file. The data export was created in the software specifically for this project and contains only the necessary data elements from the respective eCRFs. JSON was chosen as the transmission format of the MTS to the SurPass.

For the output in JSON format, which is the result of the mapping of the data according to the template of the MTS, a Java program was written. The program reads the exported eCRFs of all selected patients. Based on the GermanSurPassID the eCRFs of a patient are selected together with the enclosed necessary nodes. The output XML has a structure within the eCRFs in which the data elements are mapped into their own "data points". Unfortunately, the data points do not have attributes, but only child elements. In these child elements the name of a datapoint is marked with the element "code" and the respective entry with "value". The structure is shown in Fig. 2 using the gender data element. Therefore, the code of a data point must be matched with the data element to be determined from the eCRF.

The resources used in the MTS were represented using Java classes. The classes contain the necessary data elements of the resources to be mapped, as seen in the Differential Table of FHIR resources. For each patient, an instance of the MTS class is created to replicate the MTS. This contains the other classes (resources) to represent a separate MTS for each patient. In total, an MTS contains 8 different classes. Within the classes, terminologies and ontologies are used for the comparability of medical data. These are represented in FHIR with the data type "CodeableConcept". A CodeableConcept contains a coding that is built from three main components:

- system: Uniform Resource Identifier of the terminology system

- code: Symbol in syntax defined by the system

- display: Representation defined by the system

This structure (example in Fig. 2) was taken into account in the Java classes to allow comparability of the registered codes and to enable a later import into FHIR. The attribute "system" was defined as a constant according to HL7 specifications. The attribute "code" can be taken from the value element of the corresponding data point, since the mapping already takes the correct terminologies into account. Display is introduced automatically by means of the code in the SurPass platform at the latest. This saves work, because otherwise the complete terminologies would have to be stored in the program.

Using the Jackson Application Programming Interface, the Java classes were converted into JSON objects by means of an object mapper and then written to a JSON file.

With the publication of the FHIR profiles it would be possible to create or validate the resources directly via FHIR. Then no classes would have to be rebuilt.

4   Conclusion

Mapping of necessary data for an MTS and the already ongoing long-term follow-up according to ISO/TR 12300:2014 allowed to identify missing data elements. These data elements were completed in the appropriate eCRFs for a complete mapping to the MTS. Subsequently, the data could be transformed into a JSON file using a custom-written program. This is based on HL7 FHIR resources and is similar to an export from FHIR. In this way, sample data could already be transferred to the GCCR. With the help of this sample data the profiles of the resources will be tested and further developed. This approach moves the entire PanCareSurPass forward and gradually allows for the inclusion of other necessary data elements. Moreover, it is possible to transfer more data faster according to the required format for the further development of SurPass.

Acknowledgement

The work has been carried out at IT Center for Clinical Research, Lübeck (ITCR-L), Lübeck and supervised by the Institute of Medical Informatics, Universität zu Lübeck. It was realized in cooperation with the Department of Pediatrics and Adolescent Medicine and the German Childhood Cancer Registry and is part of the project "PanCareSurPass", the fourth EU-funded PanCare project (grant agreement No. 899999).

Author's Statement

Conflict of interest: Authors state no conflict of interest.

5   References

[1] F. Erdmann, P. Kaatsch, D. Grabow, C. Spix, *German Childhood Cancer Registry - Annual Report 2019 (1980-2018)*, Institute of Medical Biostatistics, Epidemiology and Informatics (IMBEI) at the University Medical Center of the Johannes Gutenberg University Mainz, 2020.

[2] S. Francisci, S. Guzzinati, L. Dal Maso, C. Sacerdote, C. Buzzoni, A. Gigli, & AIRTUM Working Group, *An estimate of the number of people in Italy living after a childhood cancer.* International journal of cancer, vol. 140, no. 11, pp. 2444–2450, 2017.

[3] SIOP Europe, *SIOP Europe Strategic Plan Update (2021-2026).* Available: https://siope.eu/media/documents/siop-europes-strategic-plan-update-2021-2026.pdf [last accessed on 2022-01-14].

[4] R. Haupt, S. Essiaf, C. Dellacasa, C. M. Ronckers, S. Caruso, E. Sugden, L. Zadravec Zaletel, M. Muraca, V. Morsellino, A. Kienesberger, A. Blondeel, D. Saraceno, M. Ortali, L. C. M. Kremer, R. Skinner, J. Roganovic, F. Bagnasco, G. A. Levitt, M. De Rosa, M. Schrappe, L. Hjorth, R. Ladenstein, PanCareSurFup, ENCCA Working Group; ExPo-r-Net Working Group, *The 'Survivorship Passport' for childhood cancer survivors.* European journal of cancer, Oxford, vol. 102, pp. 69–81, 1028.

[5] PanCareSurPass, *What is PanCareSurPass?.* Available: https://www.pancaresurpass.eu/what-is-pancaresurpass/ [last accessed on 2022-01-10].

[6] Health Level Seven International, *About HL7.* Available: http://www.hl7.org/about/index.cfm?ref=nav [last accesssed on 2022-01-21]

[7] ISO/TR 12300:2014, *Health informatics – Principles of mapping between terminological systems.* 2014.

[8] E. Rinaldi, S. Thun, *Mapping from openEHR to FHIR and OMOP CDM to support interoperability for infection control.* GMS Med Inform Biom Epidemiol., 2021.

[9] PanCareSurPass FHIR Implementation Guide, *PanCareSurPass FHIR Implementation Guide.* Available: https://build.fhir.org/ig/hl7-eu/pcsp/toc.html [last accessed on 2022-01-21].

[10] D. Brammen, P. Eggert, B. Lucas, L. Herrmann-Langford, J. C. McClay, *Comparing the German Emergency Department Medical Record with the US HL7 Data Elements for Emergency Department Systems,* Studies in health technology and informatics, vol. 247, pp. 216-220, 2018.

Implementation Guide Rare Disease Documentation - Enabling of the FAIR Principles for Rare Diseases

Miriam Hübner [1], Josef Schepers [2] and Josef Ingenerf [3]

[1] Medical Informatics, Universität zu Lübeck, miriam.huebner@student.uni-luebeck.de
[2] Berlin Institute of Health, Berlin, josef.schepers@bih-charite.de
[3] Institute of Medical Informatics, Universität zu Lübeck, ingenerf@imi.uni-luebeck.de

Abstract

There are approximately 300 million people worldwide suffering from rare diseases. Before they will receive the correct diagnosis, they typically undergo a tough process of numerous examinations, misdiagnoses and mistreatments. CORD-MI (Collaboration on Rare Diseases) aims to increase the detection, documentation and recognition of rare diseases. This paper discusses the required steps to adapt the MII (Medical Informatics Initiative) information model and its implementation to requirements in the context of rare diseases. An Implementation Guide has been developed that contains constraints for structured processing of health data to enable secondary use. The ambition is to initiate an impetus for more accurate and detailed documentation. The work done to this extent must be regarded as a compromise between data requirements and data availability, setting the direction of the way forward and serving as a guidepost. Adequate documentation is imperative for a sustainable improvement of treatment for patients suffering from rare diseases.

1 Introduction

According to estimates, four million people in Germany suffer from one of the approximately 8.000 known rare diseases - worldwide it affects about 300 million people. Many of the rare diseases manifest in early childhood, as a large proportion of these diseases are genetic [1]. Often a lot of time passes before a correct diagnosis is made. This time is characterized by numerous examinations, consultations, misdiagnoses and incorrect treatments. In addition, there is a lack of treatment methods, expertise, knowledge and research in the field of rare diseases. Furthermore, existing knowledge is difficult to access - which prolongs the diagnosis. However, a concrete diagnosis is of enormous importance for the patient and for research - even if there is no therapy for this disease yet.

The Medical Informatics Initiative (MII) seeks to improve patient care and research opportunities. A main approach is the development and implementation of an information model that makes the content, structures and coding of the data generated in the healthcare system interoperable [2]. The MII Use Case "Collaboration on Rare Diseases" (CORD-MI) is specifically designed for the area of rare diseases. One approach is to implement the FAIR-Principles (Findability, Accessibility, Interoperability, Reuse) to the documentation, use and reuse of data from patients with rare diseases. For this purpose, the information model of the MII will be adapted to the needs of rare diseases [3]. Another measure is to strengthen accurate coding of rare diseases in order to make rare diseases more visible.

Since individual rare diseases are very infrequent locally, it is important to collaborate both nationally and internationally. Findability and accessibility are prerequisites for the approach to collaboration. Reusability is the prerequisite for the execution. The use of appropriate standards that ensure both structural and semantic interoperability is essential to enable this collaboration [4]. In this Paper, it will be discussed which steps are necessary to enable the FAIR principles for rare diseases, which challenges arise and how these can be handled. This is demonstrated by showing the necessary measures that are required to adapt the MII information model and its implementation to rare diseases. The result of this work is an Implementation Guide (IG) that contains the previous achievements.

2 Material and Methods

2.1 Fast Healthcare Interoperability Resources (FHIR)

Generally, autonomously developed software systems based on proprietary data models are used in healthcare sector. This leads to the problem that different IT systems cannot communicate well with each other or cannot even communicate at all. A software and hardware-independent communication standard is indispensable for the exchange of data in healthcare. FHIR is a freely available standard that supports data exchange and communication between different healthcare software systems and enables persistent storage [5]. It is composed of three main elements: Resources, references and profiles. The predefined resources, delimited according to different areas of healthcare, provide a basis for achieving interoperability. They are well-defined logically discrete entities and cover around 80% of the usual usage scenarios. A FHIR profile is the definite description of a specific data object for the respective context. It is important to select the right resources for the specific context and then adapt them to the requirements of the particular use case. Therefore, it is necessary to evaluate what additional information is required that is not included in the core. Extensions can be used to extend the core specification with the required elements. A value set can be used to define the codes required for a specific use case [5].

2.2 Required steps for reuse of patient data

FHIR enabled interoperability is a remarkable approach to the other three components of the FAIR principles: Findability, Accessibility and Reusability. The main focus of the MII is to ensure that the healthcare data generated in university hospitals is made reusable, i.e. available for secondary use, especially for clinical research. Thus, the goal is to transform the various types of collected patient data into knowledge. Methods such as machine learning rely on a certain documentation quality and structure of data. Therefore widely applicable classifications and terminologies (structured documentation) are used in addition to free text description (unstructured documentation). Structured documentation makes the input more processable for the computer, is subsequently more comparable, and improves the feasibility of research. In order to create the required basis for an adequate analysis many preliminary steps need to be completed. These steps can be understood as inspired by the KDD (Knowledge Discovery in Databases) process [6]. At the beginning is the locally created documentation of many types of patient data. The second step is to select the needed documentations and patient data for analytical reuse. The following step is to preprocess the data in a standardized and structured manner. Data Integration Centers (DICs) are needed for the locally generated data of the university hospitals to ensure that the data stored decentralized is available in a structured form for data exchange between patient care and research across different facilities [7]. This step itself depends on the constraints that are specified in the developed IG. After successful standardized data processing, the patient data can be used to analyze and evaluate various questions.

2.3 Implementation Guide

A proper primary documentation of patient data is enormously helpful for later analyses and research. In the case of reuse at least the selection, preprocessing and transformation has to be executed carefully. Therefore, the planning and development of a good specification is indispensable. The IG for the field of rare diseases describes how the healthcare data should be documented where new documentation is possible, or should be preprocessed in ETL-chains (Extract, Transform, Load) where reuse is aimed for. The goal of the IG is to combine structured clinical care documentation with science-oriented registry documentation. The german health care system differs from those of the other countries. Therefore, the preliminary work of the MII that is adjusted to the german healthcare system will be used as a basis and will be extended accordingly. To foster the necessary international collaboration, the information model is aligned to other international data models for rare diseases that have already been developed like the Rare Disease Registration Infrastructure Common Data Set (ERDRI CDS) and the French National Rare Diseases Minimum Data Set (RD MDS). For this purpose, the different data sets are compared and significant attributes for rare diseases not yet included in the MII model are added to the rare disease model. The new attributes are distributed among the matching existing modules of the MII, and in addition their descriptions and cardinalities are defined. Furthermore, specific modules for different disease groups will be added step by step – such as the registries of the specific European Reference Network (ERN). The IG is created using Simplifier's Markdown-based IG editor. Simplifier is a digital archive for all FHIR resources and provides hosting of IGs for international visibility. All required and created files can be uploaded to a Simplifier project and included in the IG using Markdown commands [8].

2.4 Challenges of coding rare diseases

According to § 301 and § 295 SGB V (*Sozialgesetzbuch* V), the encoding of diagnoses with the International Statistical Classification Of Diseases And Related Health Problems, 10th revision, German Modification (ICD-10-GM) is prescribed by law for billing purposes. As a result, the encoding of diagnoses in Germany has so far been dominated by ICD-10-GM. However, it is not sufficient for coding rare diseases – only 500 rare diseases can be specifically coded. Some other of the rare diseases are found hidden in subcategories of ICD-10-GM but are not usable for specific coding. An example is the code E75.2 (*Sonstige Sphingolipidosen*) – only one code is used for at least seven different rare diseases. However, the accurate diagnosis is required for many purposes, such as patient recruitment for therapy studies or for analyses in general. This problem can be remedied by the use of ORPHAcodes (specifically geared to rare diseases) that currently enable the encoding of more than 5000 different rare diseases [9]. The usage is expected to be made mandatory in 2023 by the *Digitale Versorgung und Pflege-Modernisierungs-Gesetz* (DVPMG) [10].

3 Results and Discussion

3.1 Components of the IG

The developed IG is accessible at Simplifier.net [11]. The IG will be expanded on an ongoing basis and has to be regarded as a work in progress.

In the following, an overview of the different components (see Fig. 1) is given with the help of excerpts from the IG.

Figure 1: Structure of the Implementation Guide.

UML (Unified Modeling Language) diagrams (see Fig. 2) serve as an abstract version of the information model and help to clarify the interconnections between concepts. The information model has so far been extended especially by various attributes for the capture of genetic examinations, specific coding of rare diseases and family-related data.

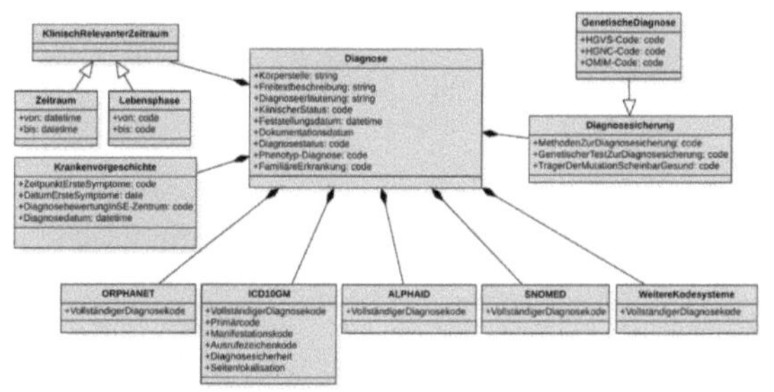

Figure 2: UML diagram for a diagnosis in CORD-MI.

FHIR Logical Models (see Fig. 3) serve as an abstract data set description to bridge different modeling worlds and is used to map data elements and their descriptions.

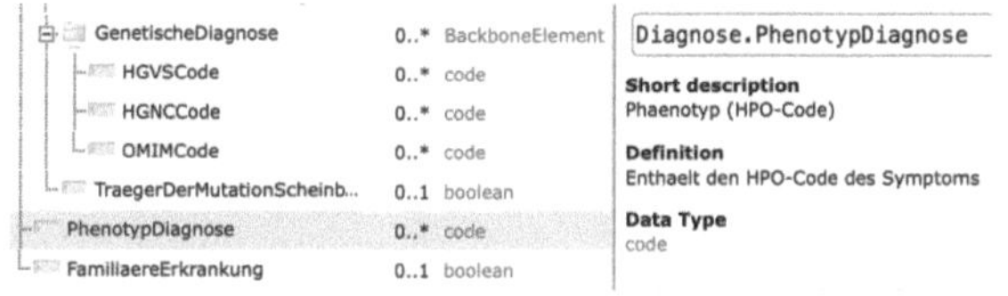

Figure 3: Extract of a logical model for a diagnosis.

In addition to profiles that are currently being developed (such as the adjustment of the MII profiles Condition (see Fig. 4) and Patient), the IG also contains tables that map the contents of the FHIR Logical Model to the profile components.

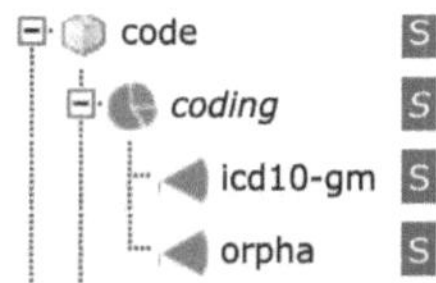

Figure 4: Small snippet of the profile Condition. Adjustment of the *Todesursache* (Cause of death) – encoding with ORPHAcodes in order to make rare diseases visible.

To ensure that the defined properties of FHIR profiles are actually fulfilled, it is useful to construct example patient data (see Fig. 5). This example illustrates once again the importance of coding a rare disease by using an ORPHA-code, as this allows the coding of a specific disease and not just a summary of a variety of rare diseases.

```
"code": {
    "coding": [
        {
            "system": "http://fhir.de/CodeSystem/bfarm/icd-10-gm",
            "version": "2021",
            "code": "E75.2",
            "display": "Sonstige Sphingolipidosen"
        },
        {
            "system": "http://fhir.de/CodeSystem/bfarm/alpha-id",
            "code": "I118641",
            "version": "2021"
        },
        {
            "system": "http://www.orpha.net",
            "code": "333",
            "display": "Farber-Lipogranulomatose"
        }
```

Figure 5: Code snippet of a created example diagnosis.

The created examples can be validated with Simplifier. Feedback and error messages can be used to validate the profile. The file format of the test patients can be JSON (JavaScript Object Notation) or XML (Extensible Markup Language). Another important component are the use case specific ValueSets which will be used. For genetic diagnostics which will be enabled in CORD-MI due to the many gene-related rare diseases, the use of HGNC (HUGO Gene Nomenclature Committee) and HGVS (Human Genome Variation Society) codes should be mentioned.

3.2 Statistical analysis

With FHIR Search, which e.g. is integrated in the R-script fhircrackr, patient data can be transformed into a statistically analyzable format and the analysis in R can begin. For the analysis, various exercises for test evaluation were developed within CORD-MI, such as age and gender distributions (see Fig. 6).

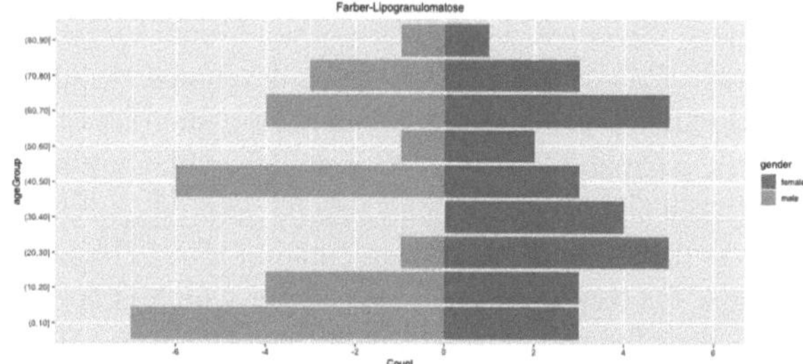

Figure 6: Age and gender distribution for the ORPHAcode 333 (*Farber-Lipogranulomatose*).

Performing CORD-MI-defined test analyses with synthetic test data provided by CORD-MI demonstrates that analysis of prepared, standardized data can work. The conducted analyses serve as exercises for later analyses with real patient data which then can be interpreted in the future and possibly lead to new insights in the field of rare diseases.

4   Conclusion

An optimization of research and therapy for patients with rare diseases can be advanced through meaningful, specific documentation and proper encoding. To improve reuse of medical data, structured documentation is enormously important. In contrast to other medical areas - such as medical image processing - patient data (such as previous illnesses, symptoms, etc.) cannot yet be retrieved in a completely structured form. However, even parts of the structured documentation (such as coding with ICD-10-GM) are not yet designed for people with rare diseases. Specific coding with ORPHAcodes is essential to make rare diseases visible. The IG developed during the internship, which follows FAIR principles and provides instructions for structured documentation, ensures that patient data can be improved for primary and secondary use in the future and thus help people with rare diseases in various areas. The information model in the IG must be seen as a compromise between data requirements and data availability. The IG will be expanded in the context of further requirements of specific rare diseases and should be considered as a work in progress. Analyses with structured test data have shown the usefulness of these efforts and serve as practice for the later analyses with real patient data. In addition to the creation of example patient data for the validation of profiles, an HL7 balloting must be carried out – as soon as all profiles have been finalized. This is an official review of the profiles, followed by comments on deficiencies in the specification. After this process, the specification will be revised.

The milestones achieved so far can be considered as preliminary work and signposts for national and international collaboration. The collaboration in the EU research project Screen4Care (Shortening the path to rare disease diagnosis by using newborn genetic screening and digital technologies) should be emphasized. One goal is the development of AI-based algorithms, which should shorten the diagnostic process for patients [12]. However, for these algorithms to provide meaningful results, the creation of the non-proprietary "Screen4Care Common Data Model for RD" is necessary. This data model has to be developed by adapting and extending international standards, such as FHIR. However, compatibility with other models which are more common outside Germany must also be ensured. A second project is the work on the documentation tool "TBase" of the Clinic for Nephrology at the Charité which is already used in the clinical routine in nephrology and neurology. Our team is working on the implementation of additional special masks for rare diseases, which may encourage the creation of more precise documentation in the future.

Acknowledgement

The work has been carried out at Berlin Institute of Health, Berlin and supervised by the Institute of Medical Informatics, Universität zu Lübeck.

Author's Statement

Conflict of interest: Authors state no conflict of interest.

5   References

[1] Achse. *Seltene Erkrankungen.* Available: https://www.achse-online.de/de/die_achse/Seltene-Erkrankungen.php [last accessed on 2022-01-15].

[2] T Ganslandt et al.. *Der Kerndatensatz der Medizininformatik-Initiative: Ein Schritt zur Sekundärnutzung von Versorgungsdaten auf nationaler Ebene.* In: Forum der Medizin-Dokumentation und Medizin-Informatik. vol. 20, no. 1, pp. 17–21, 2018

[3] Medizininformatik-Initiative. *Use Case CORD-MI.* Available: https://www.medizininformatik-initiative.de/en/CORD [last accessed on 2022-01-16].

[4] Lehne, M., Sass, J., Essenwanger, A. Schepers, J, Thun, S. *Why digital medicine depends on interoperability.* npj Digit. Med. 2, 79 (2019). https://doi.org/10.1038/s41746-019-0158-1

[5] Health Level 7. *FHIR Overview - Architects.* Available: http://www.hl7.org/fhir/overview-arch.htmlprinciples [last accessed on 2022-01-13].

[6] U. Fayyad, G. Piatetsky-Shapiro, P. Smyth. *From Data Mining to Knowledge Discovery in Databases.* AI Magazine Volume 17 Number 3, pp 37-54, 1996

[7] Medizininformatik-Initiative. *Datenintegrationszentren.* Available: https://www.medizininformatik-initiative.de/de/konsortien/datenintegrationszentren [last accessed on 2022-01-13].

[8] Simplifier.net. *Features.* Available: https://simplifier.net/features [last accessed on 2022-02-07].

[9] BfArM. *Alpha-ID-SE.* Available: https://www.bfarm .de/EN/Code-systems/Terminologies/Alpha-ID-SE/_node.html [last accessed on 2022-02-04].

[10] BfArM. *Alpha-ID-SE.* Available: https://www.bfarm .de/DE/Kodiersysteme/Terminologien/Alpha-ID-SE/_node.html [last accessed on 2022-02-04].

[11] MII. *Medizininformatik Initiative - CORD - Implementierungsleitfaden.* Available: https://simplifier.net/guide/MedizininformatikInitiative-CORD-ImplementationGuide/Beschreibung2 [last accessed on 2022-01-15].

[12] Screen4Care. *Press Release.* Available: https://www .screen4care.eu [last accessed on 2022-01-14].

7

Image Processing

k-Space Modification and Visualization Tool
– An application to explore the k-space –

Ricardo Sarau [1]

[1] Robotics and Autonomous Systems, Universität zu Lübeck, ricardo.sarau@student.uni-luebeck.de

Abstract

This scientific paper presents the results of a project assigned by the Institute for Robotic and Cognitive Systems. The goal was to create a tool that helps researchers understand the k-Space and how modifications are affecting it. Solutions that are currently available could not be used for this project, because they do not fulfill the necessary requirements and are not fit for further development. As a result, a new application was deemed necessary and the development was completely executed by the author. At the start, the necessary materials like the Python OpenCV package were determined and the graphical user interface was planned. After that, a Model-View-Controller software architecture was implemented with all the required functions. These allow users to freely move and change images in image and k-space. A final comparison with the requirements showed that the finished tool fulfilled them sufficiently.

1 Introduction

A previous study by the Institute of Medical Engineering and the at the Universität zu Lübeck showed how Convolutional Neural Networks (CNNs) can help with image-guided radiotherapy for cancer patients with head tumors. An important part of this study is the k-space, which was used to withdraw the information relevant to the CNN from magnetic resonance (MR) images [1].

An application was deemed necessary by the Institute for Robotics and Cognitive Systems to help researchers in finding new use-cases for k-space visualization like in the mentioned study and to increase understanding of the k-space. The requirements for a sufficient k-space exploration were set as a result and current solutions like k-Space Tutorial [2], K-space Explorer [4] or Journey through k-space [5] were analyzed based on these requirements. The analyzed applications could not fulfill them and were not suitable for further development. Hence a new application, the Modification and Visualization Tool, was fully developed by the author. This tool allows users to freely modify images in image space and k-space by applying the implemented functions. The general structure of the application as well as some chosen functions are presented in the following. Class and function names specific to the application are written in italic, to distinguish them from the other technical terms.

1.1 Objects in k-space

The k-space is an extension of the Fourier space that is used to represent the 2D or 3D spatial frequency information of objects. The k-space in itself is raw data acquired during MR imaging. Depicted in the data are transversal components created by exciting protons in the imaging object with magnetization. One transverse slice then consists of a frequency encoding direction along horizontal and vertical axis [2]. The result can be seen in Fig. 1, where frequency differences along the respective axis due to the square in the image space are visualized in the principal axes of the k-space.

1.2 Image space and k-space transformation

The phase and frequency encoded data in k-space are related to the spatial data in the image space through the Fourier transformation. It is not possible to take one point from the k-space and find one point in the image space that corresponds to it because every point in the k-space contains spatial frequency and phase information about every pixel in the image space. As a general rule, the outer data points of the k-space matrix, which are the high spatial frequencies, contain information regarding edges and details. Furthermore, the inner data points, which are the low spatial frequencies, contain information for the contrast and general shape of the image [3]. This concept is further elaborated in section Results and Discussion.

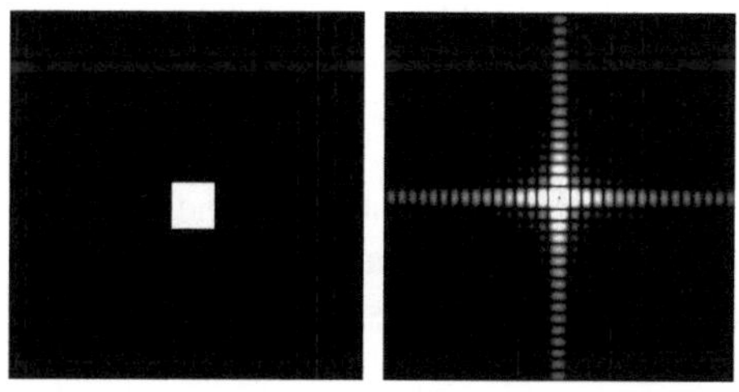

Figure 1: square in image space (left) and k-space (right)[3]

2 Material and Methods

The development of the k-space Modification Visualization Tool was done with the programming language Python. Three Python packages should be pointed out since they were critical for development. First, the Tkinter package that was used to create the GUI. Next, the OpenCV package, which provided most of the functions used for modifications. Lastly, the NumPy package was primarily used for the FFT and inverse FFT, but also for some other necessary calculations.

Testing was done with an MR transversal slice image of a human head. Though it should be mentioned that every image in a commonly used format could be experimented with inside the tool, by uploading the image.

2.1 Start of the project

The general requirements for the project are that it should be a PC application, have modifications in image space and k-space with simultaneous visualization of the changes as well as general functions like loading and saving of images. Additionally, it is required that the tool is structured in a way that allows it to be used for further development and it should be user-friendly by current industry standards. Available applications like k-Space Tutorial [2], K-space Explorer [4] or Journey through k-space [5] were evaluated based on these requirements. Since none fulfilled the requirements and most necessary modifications were missing in all of them, the development of a new tool was started.

In the first development step, the graphical user interface (GUI) was planned. Planning of the GUI consisted of sketching the general layout and determining the actions the user can take to interact with the tool. This included defining how each function, which was part of the requirements, would be activated and visualized.

2.2 Software Architecture

The software architecture, Fig. 2, was defined by using the Model-View-Controller (MVC) architectural pattern. If this pattern is applied in an application, then it is divided into three parts Model, View and Controller to separate the presented information from its internal representations during processing and storing[6]. As a result, it is easier to maintain and implement new functions, which will help in further development as required.

So the view displays information to the user, which in the case of this project is done through the previously mentioned GUI. Image matrices are visualized inside the GUI, which are stored and processed before output by two controller classes.

Classes that belong to the controller handle user input, interact with the model and return output to the view. In this application one controller class, the *Image Handler*, manages the image matrices for the image space and another, the *K-Space Handler*, manages the k-space image matrices. The processing done inside these classes is mainly the respective FFT and inverse FFT functions, but also prepara-

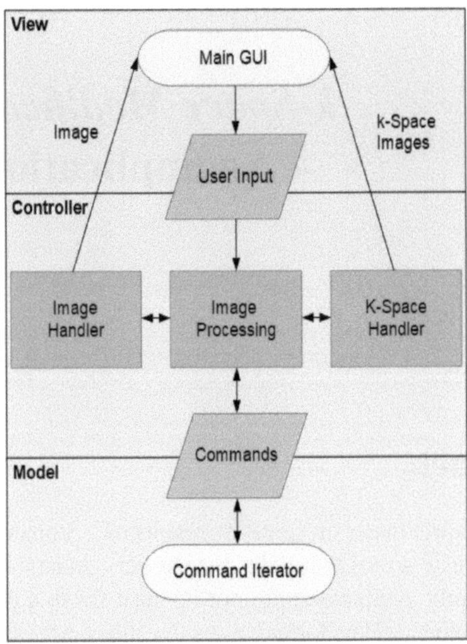

Figure 2: Architecture of the k-space Tool

tions before visualization. All the modifications, for which examples are shown later in section Results and Discussion are computed in the controller class *Image Processing*.

The last part of MVC, the model, is used for storing and retrieving data usually by connecting it to a database. Although some data needs to be stored, a database is not necessary for this project. All the necessary information instead is stored inside a collection that is managed by the model class *Command Iterator*.

Factory, Abstract Factory and some other design patterns were examined, in order to find a structure for the interaction between the different parts of the MVC architecture. In the end no design pattern was directly applied, but the iterator and observer design patterns were used as inspiration. The observer design pattern addresses the issue that many objects are depending on the state of one object by defining a subscription mechanism. The iterator design pattern is helpful for defining a structure in which a collection of complex data structures needs to be traversed without exposing the complexity.

The resulting observer-like structure compiles user input into the complex *Commands* data structure, to classify its current state. Each modification inside the *Image Processing* class observes this state and compares it to its activation conditions. If they are fulfilled, the function becomes active and the modification it contains is processed. This has two main benefits: First, the *Commands* data structure contains all necessary information to process modifications so it can be used to store information inside the collection for the model. *Commands* can be reactivated in order by traversing through the collection, which makes it possible to reverse or to reproduce user actions and was inspired by the iterator design pattern. Secondly, it is easier to maintain and add new modifications, because they only have to be added to the list of subscribers instead of being implemented into a more complex decision tree.

3 Results and Discussion

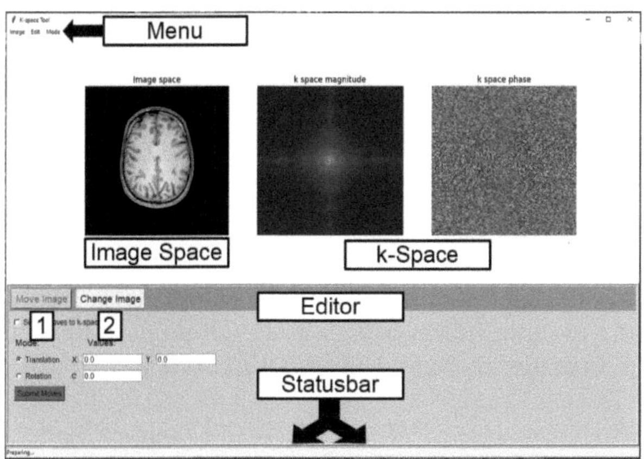

Figure 3: GUI of the k-space tool visualizing image space and k-space as well as the editor, menu and statusbar. Editor with two tabs: (1) Move Image and (2) Change Image

The GUI of the k-Space Modification and Visualization Tool is shown in Fig. 3. To interact with the presented images of image space and k-space, the user can utilize the editor below that. The editor has two tabs, which consist of functions that can be used to apply modifications. Tab 1 in Fig. 3 has functions for moving images by translation or rotation. Tab 2 allows the user to change images by adding shapes or removing parts of the image. Additionally, there are modes, which change how modifications are applied. These can be chosen inside the menu, which also allows the user to open and save images as well as reset and repeat steps. A status bar on the bottom additionally informs the user about the last changes done.

3.1 Translation and Rotation

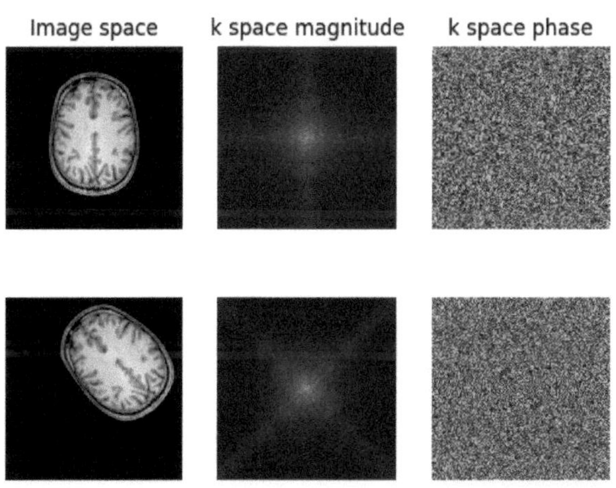

Figure 4: Translation and rotation in the image space

Fig.4 shows the MR transversal slice image before changes in the upper half with its respective representations. The lower half of the figure depicts the same MR transversal slice after a slight translation to the right and a counter-clockwise rotation of the head by 45 degrees in the image space. The results of those changes in the k-space are also visualized in the magnitude and phase representations next to that. There are small differences in the k-space phase due to translation and rotation in the image space, but since these are hard to make out, is the main focus on the magnitude representation.

The k-space magnitude corresponding to the MR transversal slice image shown before the changes are depicted in a "+"-form since the dominant spatial frequencies are in the horizontal and vertical direction. If these edges of the head now rotate by a certain degree, then the corresponding dominant spatial frequencies are at a different angle and the k-space magnitude shows this by depicting the "+"-form in the rotated angle. Since a translation has no effect on this relation, there are no changes due to the translation. Mathematically a rotation by θ is computed by changing each image pixels position (x_1, x_2) to get the rotated image matrix f. This applied to the k-space matrix F and each point position (k_1, k_2) is shown in (1) [7].

$$F(k_1 cos\theta - k_2 sin\theta, k_1 sin\theta - k_2 cos\theta) \\ \leftrightarrow f(x_1 cos\theta - x_2 sin\theta, x_1 sin\theta - x_2 cos\theta) \tag{1}$$

For the translation each image pixels position (x_1, x_2) is shifted by the shifting parameters a and b in the image matrix m. This is then applied to the k-space matrix M with its pixel position vectors (k_1, k_2), shown in (2) [7].

$$F(k_1, k_2) exp[-j(ax_1 + bx_2)] \leftrightarrow f(x_1 + a, x_2 + b) \tag{2}$$

3.2 Adding shapes to images

Figure 5: Replacing higher frequencies with a square shape

The user can freely choose form, size, color and position for adding a form. By applying one black square over the image center in the k-space, a high pass filter can be emulated. After that modification is processed, only the edges remain visible in the image domain, which is shown in Fig. 5. The reason for this emulation of a high pass filter is that the lower frequencies lie in the center of the k-space matrix. If those values are set to zero by replacing the central values with a square-shaped zero matrix, like in Fig. 5, all the lower frequencies are essentially filtered out without a specific filter algorithm.

3.3 Remove parts of images

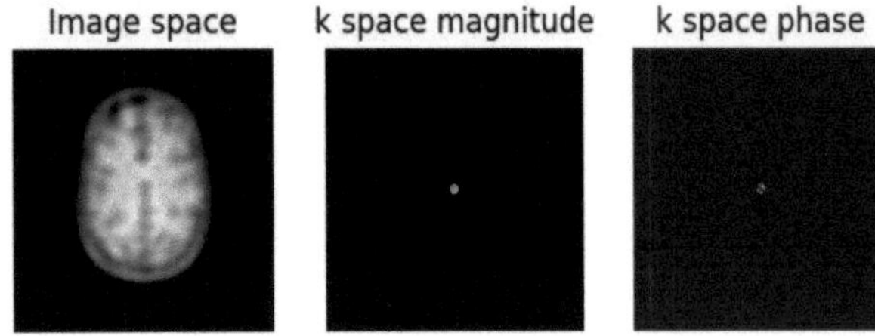

Figure 6: Removal of higher frequencies with a circle form

A low pass filter can be emulated by the same principle. The resulting image in the image domain has no contours and loses some details, shown in Fig. 6. Only small circles of the original k-space images remain. The higher frequencies, as they are in the periphery, were removed from the k-space. This is done in the figure by computing a zero matrix with the same shape as the original image and then computing a circle containing only ones inside the zero matricx. Calculating the Hadamard product of the k-space matrix and the changed zero matrix will result in a matrix with only the central values of the k-space and everything else set to zero. Essentially filtering out the higher frequencies.

3.4 Real and Imaginary Mode

The tool also has, in addition to the magnitude and phase k-space representations, the *Real/Imaginary Mode*. In this mode the real and imaginary values are extracted from the k-space, to get the real and imaginary part representation. Furthermore is it possible to use the *Single Edit Mode* of the tool, which causes changes to no longer be applied to the whole k-space, but instead to only be applied to one k-space representation.
Modifications inside the k-space with real and imaginary part representation will lead to a different result inside the image space compared to the magnitude/phase representation, where the same modification has been applied. These differences can be attributed to the amount and type of information encoded in the k-space representation for a specific image.

4 Conclusion

This paper presented the capabilities of the k-Space Modification and Visualization Tool through some applications of the implemented functions and how they affect the image space or the k-space respectively. All the necessary requirements regarding modifications have been fulfilled within this project and the structure should allow for further development without any problems.
An area for improvement though is the GUI, which is design-wise not up to current software standards and lacks some quality of life functions. Furthermore, despite the tool's capability of emulating certain results through specific available modifications or combinations, it would be beneficial if some modifications could be applied directly through respective functions as a low pass filter for example. The application of CNNs on the k-space showed that the k-space should still be explored for new use cases. Researchers can freely apply changes to different parts of image and k-space with this tool and observe the results, which can help find those new use cases.

Acknowledgement

The work has been carried out at the Universität zu Lübeck and was supervised by Marius Krusen, Institute of Robotics and Cognitive Systems, Universität zu Lübeck.

Author's Statement

Conflict of interest: Authors state no conflict of interest.

5 References

[1] M. Wattenberg and J. Hagenah and C. Schareck and F. Ernst and M. Koch, *Head Movement Detection from Radial k-Space Lines using Convolutional Neural Networks – A Digital Phantom Study*. Institute of Medical Engineering, Universität zu Lübeck, Institute for Robotics and Cognitive Systems, Universität zu Lübeck.

[2] D. Moratal and A. Vallés-Luch and L. Martí-Bonmatí and M.E. Brummer, *k-Space tutorial: an MRI educational tool for a better understanding of k-space.* Biomed. Imaging Interv. J. 4., 2008.

[3] A. D. Elster, *Location of Spatial Frequencies.* Available: https://mriquestions.com/locations-in-k-space.html [last accessed on 2022-01-23]. Mallinckrodt Institute of Radiology, Washington University School of Medicine.

[4] Gergely Biró, *K-space Explorer.* Available:https://github.com/birogeri/kspace-explorer[last accessed on 2022-01-23].

[5] Ali Raza Shahid and Mehmood Qureshi and Hammad Omer, *Journey through k-space: an interactive educational tool.* Biomedical Research.V. 28., Nr.4.,2017.

[6] Zanfina Svirca, *Everything you need to know about MVC architecture.* Available:https://towardsdatascience.com/everything-you-need-to-know-about-mvc-architecture-3c827930b4c1[last accessed on 2022-01-23].

[7] J.J.K. O'Ruanaidh and T. Pun, *Rotation, scale and translation invariant digital image watermarking.* p. 536-539, IEEE International Conference on Image Processing (ICIP 97), 1997.

Photogrammetry Pipelines under Varying Illumination Conditions

Zachary Krouse [1], Ralph Hänsel [2],

[1] Electrical Engineering and Computer Science, Lübeck University of Applied Sciences, zachary.krouse@stud.th-luebeck.de

[2] Electrical Engineering and Computer Science, Lübeck University of Applied Sciences, ralph.haensel@th-luebeck.de

Abstract

We investigate how photos taken from different illumination levels can be translated into three-dimensional models. Photogrammetry pipelines struggle with variances in the illumination of the subject matter. These variances in lighting exist as the process of gathering photos may take some time, especially if the subject matter is rather large such as a university campus. In order to achieve the best possible 3D render, the most robust against changes in illumination structure from motion pipeline needs to be found. Using three photosets from different times of day, several reconstruction setups are tested and compared. The results are evaluated based on the images used, feature points created, and quality of the render. After investigating it would seem that for the most points MeshRoom using a mixture of SIFT and A-KAZE descriptors is the best while for using the most images COLMAP utilized the highest amount

1 Introduction

The three-dimensional reconstruction of a scene can be performed by structure from motion (SfM) algorithms. These algorithms incorporate a set of 2-dimensional images from a monocular camera. First, the poses of the cameras need to be estimated based on some characteristic image points. Then the 3D reconstruction for most valid points is performed. The reconstructed 3D points form the visual map.

In order to create a 3D model, there need to be plenty of photos and what would be optimal is that all these photos were taken under the same conditions. As it can take some time to fully capture the point of interest, certain aspects may have changed such as the illumination. Varying illumination poses a problem as colors will differ, see figures 1 and 2. All of this will lead to issues with matching up the images to each other when trying to create a model during the feature extraction and description sub-processes.

In this paper, we focus on the investigation of photogrammetry pipelines under varying illumination conditions. We use SfM pipelines to reconstruct a 3D model of a scene using images of a said scene with varying lighting between the images. The image sets were taken using the same camera but at different times. Then we input all the images into the pipeline setups and compare the render results.

The paper is structured as follows. Section II starts with the discussion of related work into the idea and how it gets implemented. In section III we evaluate the results. Section IV ends with summarizing our conclusions and hinting at the future outlook.

Figure 1: Daytime Image for Comparison

Figure 2: Evening Image for Comparison

2 Material and Methods

2.1 Related Work

There are a wide variety of options when putting together an SfM pipeline, from different pipelines [1] to different pieces such as feature descriptors.[2] Comparisons have been made between them, though have yet to be fully fleshed out. The specific problem of illumination invariance has been an issue for photogrammetry for a while and so of course there have been attempts to solve it, such as an algorithm based on a cost function [3], or using deep neural networks. [4]

Simone Bianco et. al. [1] investigated several state of the art incremental SfM pipeline options for how well they could rebuild various scenes. Within their investigation, it was found that COLMAP had the best average results. This outcome shows that when choosing what pipeline to use, COLMAP is a solid choice for a wide variety of scenarios. Though this result only works with scenes created inside their blender setup without any variance in illumination amongst the images, nor did it compare with another popular state of the art software called MeshRoom. Both of these will be expanded upon later in the paper.

Tareen and Saleem [2] compared different feature detector descriptor algorithms. After their investigation, they found SIFT to be the most overall accurate algorithm, and determined a benchmark between several descriptors. While they did compare the descriptors' ability to handle a variety of circumstances, they did not compare the robustness of the descriptors to changes in illumination.

David Lowe [5] created the SIFT method for matching images of an object or scene back in 2004 and it is still one of if not the most accurate algorithms there are today. SIFT stands for scale-invariant feature transform, and it was designed to be invariant to image scale and rotation with robustness against affine distortion, the addition of noise, change in 3D viewpoint, and change in illumination.

Pablo Alcantarilla et. al. [6] developed an alternative method that is also available to use in the MeshRoom pipeline, that being A-KAZE. A-KAZE is an accelerated version of KAZE which exploits non-linear scale-space through non-linear diffusion filtering developed in 2013 and is named after the Japanese word for wind. These features are invariant to scale rotation and limited affine, and due to using non-linear scale-space have more distinctiveness at varying scales.

Max Mehltretter & Christian Heipke [3] proposed a way to overcome the issue of illumination variance using a cost function to estimate dense depth. Their results point towards the concept achieving state of the art levels of robustness against changing illumination. Though this new algorithm is still under testing and as such not available for use.

M. Meshry et. al. [4] proposed an alternative involving training a deep neural network the learn the mapping of initial renderings to the actual photos. This rerendering network also takes notes of not only scene changes between illumination or weather but also transient objects like pedestrians. Though the computational power in order to accomplish this investigation took 8 GPUs, a massive amount that the average consumer would not have available but does bode well for future possibilities of improvement within the field.

2.2 Idea

The essential idea for this investigation is to look for a feasible method to create a 3D map of campus buildings, and an issue with doing such a large-scale project is that photo collection can take time and as such there will be lighting differences amongst the photo set. Another way the problem appears is for when construction occurs and then a map would need to be updated. Now, these new images would be from a completely different time than the original set it's being updated into. In order to figure out the best method for creating a model, possibilities need to be compared.

Looking into different pipelines lead me to investigate what was shown to be the best, COLMAP, and the newer state of the art pipeline, MeshRoom. Then within MeshRoom, there are different descriptors available to choose from. As illumination variance mainly affects feature identification and matching the investigation continued, comparing several of the descriptor options. Namely the two big name descriptors, SIFT and A-KAZE, along with modified versions such as DSP_SIFT (Domain-Size Pooling [7]). Within MeshRoom there is also the ability to run both SIFT and A-KAZE at the same time which will also be tested.

2.3 Implementation

In order to implement first had to collect photos at different times of day that have noticeably different levels of light. Then take said data sets and input them into the different SfM pipelines in order to compare, also test different feature extracting types within the pipelines. Then draw conclusions from the feature points used, images used, and quality of the 3D results.

The photos were all taken using the same camera with the same settings enabled. The camera's metadata was as follows, its make: Canon EOS M50 Mark II, model: EF-M15-45mm f/3.5-6.3 IS STM, focal length: 15mm, and sensor width: 22.3mm.

These photos consisted of three sets of data, one of a bright sunny day in the early afternoon, the second was a cloudy afternoon, and the third was a cloudy evening near sunset. The three data sets consisted of the following number of images: 102, 88, and 133, for a total of 323 photos.

The focus of the images was an outdoor area where two buildings connected, a parking lot next to them, and numerous trees spread around. This area gives a combination of both man-made and natural structures to compare the data amongst. Though as the images were taken on different days certain characteristics of the scene changed over time. The primary representative of this change were the leaves on the trees, as for the sunny data set, they had yet to fall,

Figure 3: Daytime Image for Descriptor Comparison

Figure 4: Afternoon Image for Descriptor Comparison

but for the other two, they were mostly all on the ground by this point.

As for the SfM pipelines, both COLMAP and MeshRoom were left at default settings for each's first attempt. Then as finding and matching features between data sets was the area of interest alternative descriptors were tested within the MeshRoom pipeline. For the SIFT descriptor attempts the default vocabulary tree was used, but as it was designed for SIFT it could not be used for the A-KAZE algorithm. As for that, the exhaustive image matching method was used.

3   Results and Discussion

In this section, we evaluate the different pipelines and descriptors used in section IV. First, we analyze the impact from the time difference between the three photosets. For comparison, a similar image with varying lighting was found between the three sets of data and the features found and used will be compared. Second, the investigation between the reconstruction methods begins, looking into features detected, images used, and quality of the 3D model.

3.1   Detection

For the first area of interest, a similar set of angled pictures of the primary building was chosen. All three images were run through MeshRoom using the SIFT descriptor and so the dots on the images are the SIFT features identified, matched, and used, and in that order, the legend would be blue, orange, and red. MeshRoom reported that for all the

Figure 5: Evening Image for Descriptor Comparison

images 20000 features points were identified as is according to the default settings. Then for the matched features both the day and evening photos had over 4100 features, but the afternoon only had 1300 features matched, a significant decline from the other two results. Though as for what features resulted in a point being made for the 3D render the day image had 2596 features used, and the afternoon image had only 635. Meanwhile the evening image did not get used for the render despite having more matched features than the afternoon image. This may have been due to its features only matching with one other image when the requirement for a point is three matches.

With images in figures 3, 4, and 5 a few things can be noted. A lot of the points that were matched in the evening image actually came from the ground in front of the building which was not the case for the other two. Another difference between where features were found was that the day photo did not have any points in the sky as there were no clouds to give edges and it did find features further away on the wall whereas the other two lacked down there. Despite the difference in the presence of leaves on the trees between the photos all three appear to see similar features in that region of the images.

3.2   Reconstruction

| | | | Images used | | | |
Pipeline	Descriptor	Total	Day	After-noon	Evening	Feature Points
COLMAP	SIFT	202	100	57	45	34539
MeshRoom	SIFT	124	99	15	20	41940
MeshRoom	DSP_SIFT	32	32	0	0	8647
MeshRoom	A-KAZE	38	38	0	0	7812
MeshRoom	SIFT + A-KAZE	196	97	47	52	92309

Table 1: Results across Different Pipeline Setups

As for the different pipelines and descriptors used table 1 displays the statistics found. The best two results are achieved by COLMAP and MeshRoom when using both SIFT and A-KAZE. COLMAP used the most images, day images, and afternoon images. While MeshRoom using two

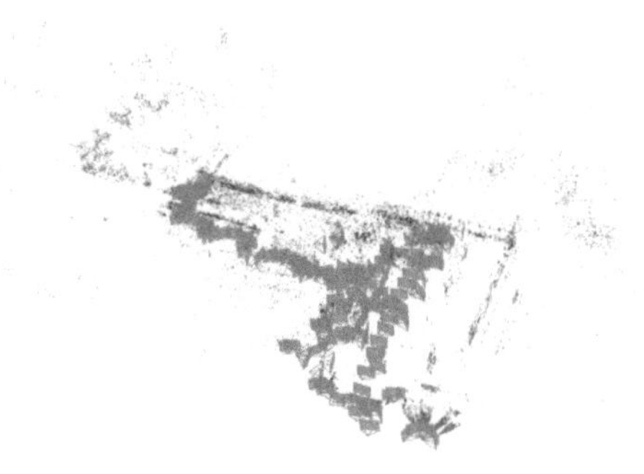

Figure 6: COLMAP 3D SIFT Image

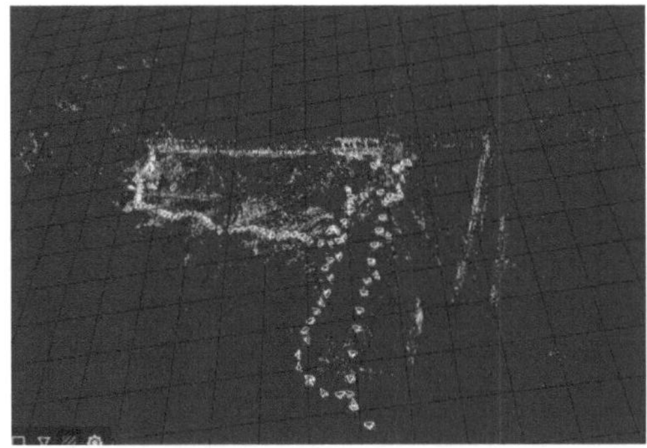

Figure 7: MeshRoom 3D SIFT Image

descriptors used the most evening photos, and had the most feature points. Neither DSP_SIFT nor A-KAZE by itself within the MeshRoom pipeline worked very well and neither used any images outside of the sunny set. Its interesting that adding A-KAZE improved the SIFT MeshRoom pipeline to such a degree when the A-KAZE trial by itself consisted of so few images and feature points.

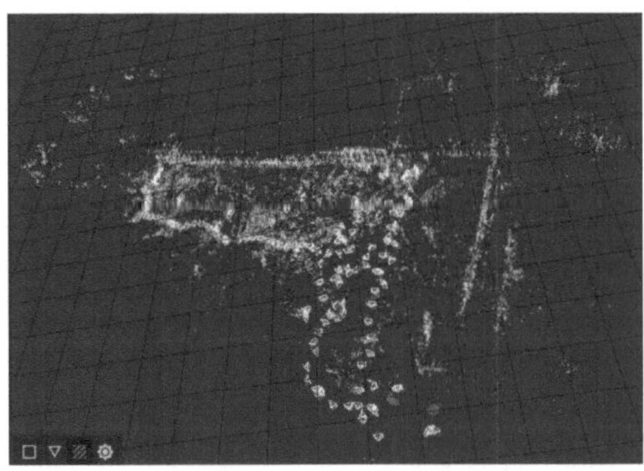

Figure 8: MeshRoom SIFT + A-KAZE 3D Image

The images in figures 6, 7, and 8 were the result of the reconstructions and there is a noticeable difference between the MeshRoom SIFT reconstruction and the other two, it appears to have fewer points despite actually having more than the COLMAP reconstruction. Though a lot of the points it seems to be lacking are actually from the environment around the buildings.

4   Conclusion

In summary, two structure from motion pipelines and several feature descriptors within were compared for their robustness to changes in illumination. Using three photosets of the same area at different times of day it was found that for the most points MeshRoom using a mixture of SIFT and A-KAZE descriptors is the best while for using the most images COLMAP utilized the highest amount. As for what the future holds, as neural networks become more prominent, promise is held with them to beat out or enhance SIFT based algorithms.

Acknowledgement

The work has been carried out during the Scientific Project module at Technische Hochschule Lübeck.

Author's Statement

Conflict of interest: Authors state no conflict of interest.

5   References

[1] S. Bianco, G. Ciocca, and D. Marelli, "Evaluating the performance of structure from motion pipelines," *Journal of Imaging*, vol. 4, no. 8, 2018. [Online]. Available: https://www.mdpi.com/2313-433X/4/8/98

[2] S. A. K. Tareen and Z. Saleem, "A comparative analysis of sift, surf, kaze, akaze, orb, and brisk," in *2018 International Conference on Computing, Mathematics and Engineering Technologies (iCoMET)*, 2018, pp. 1–10.

[3] M. Mehltretter and C. Heipke, "Illumination invariant dense image matching based on sparse features," in *Conference: 38. Wissenschaftlich-Technische Jahrestagung der DGPF und PFGK18 Tagung in München*, 03 2018.

[4] M. Meshry, D. B. Goldman, S. Khamis, H. Hoppe, R. Pandey, N. Snavely, and R. Martin-Brualla, "Neural rerendering in the wild," in *2019 IEEE/CVF Conference on Computer Vision and Pattern Recognition (CVPR)*, 2019, pp. 6871–6880.

[5] D. G. Lowe, "Distinctive image features from scale-invariant keypoints," *International Journal of Computer Vision*, 2004.

[6] P. Fernández Alcantarilla, "Fast explicit diffusion for accelerated features in nonlinear scale spaces," in *British Machine Vision Conference (BMVC)*, 09 2013.

[7] J. Dong and S. Soatto, "Domain-size pooling in local descriptors: Dsp-sift," in *2015 IEEE Conference on Computer Vision and Pattern Recognition (CVPR)*, 2015, pp. 5097–5106.

Building a sparse visual map of the THL campus including multiple buildings

Oke Petersen [1], Ralph Hänsel [2],
[1] Electrical Engineering and Computer Science, Luebeck University of Applied Sciences, oke.petersen@stud.th-luebeck.de
[2] Electrical Engineering and Computer Science, Luebeck University of Applied Sciences, ralph.haensel@th-luebeck.de

Abstract

Fields like augmented reality or autonomous navigation/guidance often require a 3-dimensional sparse map as foundation. Starting point is the acquisition of photographic images from the campus. These 2-dimensional images are than used to create the 3-dimensional visual map. The urban landscape of the Technische Hochschule Lübeck (THL)-campus with many connected buildings or open spaces brings certain challenges to the image acquisition. The success of a sparse visual map depends on the quality of the image dataset. In this paper methods and strategies are developed to acquire a meaningful image dataset for the visual mapping of the THL-campus. The encountered error sources and measures preventing their appearance are presented. The reconstruction results of multiple parts of the campus showed that the developed acquisition strategies can be used to create a feasible sparse reconstruction.

1 Introduction

The aim of this paper is to describe the process of generating a sparse visual map of the THL-campus. This sparse reconstruction [1] can be afterwards used to perform a variety of more detailed modeling techniques like dense reconstruction to further increase the level of detail. The target group are particularly beginners in the fields of photogrammetry and 3-dimensional reconstruction, who look for more successful reconstruction results with the state-of-the-Art Structure from Motion (SfM) software tools. Especially strategies and setup recommendations are given for a more structured image acquisition and subsequent image processing.

At first the paper gives introduction to the general idea of photogrammetry. Later on methods and strategies to improve the reconstruction results are presented. Thereafter the process of reconstruction the THL-campus with the defined strategies and guidelines is described from the image acquisition to the resulting sparse visual map.

2 Photogrammetry

The term Photogrammetry describes the science and technologies used to obtain information about the physical environment with the help of photographic cameras [2] [3]. Its the foundation of most SfM-algorithms to extract 3-dimensional information of an object from 2 dimensional perspective images [4]. This chapter explains the theoretical fundamentals used to create a visual map of the THL-campus.

Starting point for the reconstruction are the information provided by the acquired images. Each point in the image can be described with a x and y-coordinate (called image coordinates), which depicts a pixel value on the 2-dimensional image sensor. In the first step all images of the set are examined to determine distinctive keypoints in the images. If a keypoint is detected, the point and the local structure around the point (summarized in a feature descriptor) is saved [5]. In the second step two images are paired together to match common features. If the same feature appears on both images of the pair, the corresponding image coordinates can be used to create a geometrical model. The model can than be used to determine the corresponding camera pose between the image pair (Fig.1). This step is repeated for every possible image pair of the set.

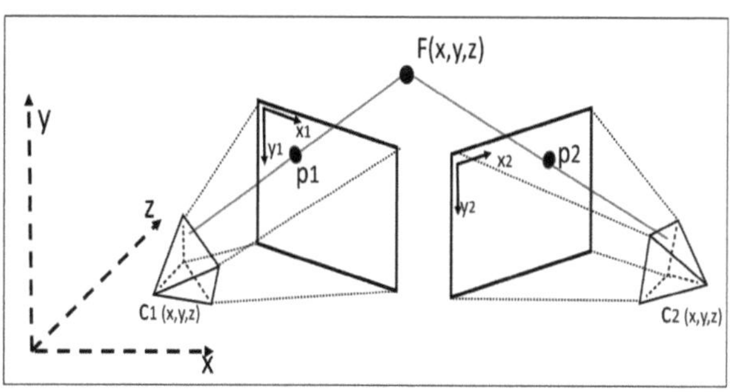

Figure 1: Fundamental model of Photogrammetry: The camera coordinates and the images coordinates create a geometrical model to determine the position of the feature

Given is a feature F which is captured from at least two different viewpoints. Depending on the perspective, the point appears on different locations on the acquired images and therefore has different image coordinates (p1 and p2). This can be used to derive the position and orientation of the camera in 3-dimensions (camera pose c1 and c2) [6] [1]. With the camera coordinates and the image coordinates of the feature, a straight line can be created in the individual image of the pair. Finally, the intersection of both lines depicts the 3-dimensional position of the feature (feature coordinates)

3   Material and Methods

Key for reproducible and sophisticated reconstruction results is a viable image database of the target object. The predominant share of possible sparse reconstruction problems and errors have their roots in insufficient image and meta data. This chapter describes the used Software tool and the developed methods to create a sufficient database for the visual map. The used camera setup and the subsequent acquisition strategies heavily depend on the characteristics of the object, that have to be reconstructed. In the case of the THL-campus a more urban landscape with larger open areas in form of parks and parking lots, plain surfaces, window fronts of buildings, larger connected building complexes and narrow alleys between objects have to be considered. In addition, its very likely that on some images moving objects in form of people or cars appear and may disturb the reconstruction.

Acquisition strategies

Universal Practices

The following recommendations should be considered in every image acquisition regardless of the object, surrounding terrain, size and camera setup [4]:

- Avoid moving objects like cars or people in the scenes to decrease image distortions

- Even lighting on every scene wherever possible. Direct sunlight may create strong shadows or highlights specific features more than normal.

- Shoot scene with at least 60 % overlapping pictures to increase the number of identical pixels per image. 3D reconstruction algorithms look for similar pixel values between each image of a set. A high overlap between the pictures the probability, that the algorithm detects dependencies more reliable.

Dome strategy

[7] The dome strategy is in theory the best way to acquire images of an object (Fig. 2). The placement of the acquired images results in a dome like structure around the object. This creates a strong parallax with many different views on the object. Although this can only be done for objects up to a certain size without the use of drones (width and height smaller than approximately 5m).

Circle Strategy

Used on large open spaces like parks, meadows or parking lots. In predefined distances 360-degree shots were taken with approximately 60 % overlapping share. The used distances depend on the size of the area and the surrounding objects. Over the course of the paper and with the experience of multiple reconstruction iterations, a distance of

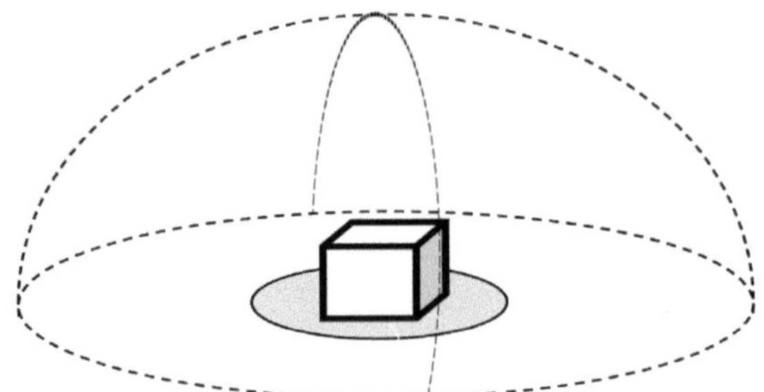

Figure 2: Dome Strategy: The images of the object are acquired in a dome shape around the object

3-5m turned out to provide sufficient data sets. Figure 3 shows two circle shots with an overlapping share between the individual images:

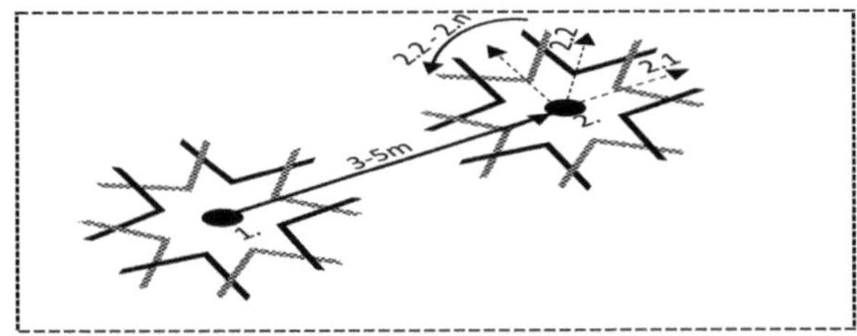

Figure 3: Circle Strategy: 360° image acquisition with an overlapping share between each image.

Left-to-Right Strategy

This strategy is advantageous in narrow and corridor shaped landscapes The image sets are taken from the left site to the right site with a minimum of 60 % overlapping share between each image (Fig. 4)

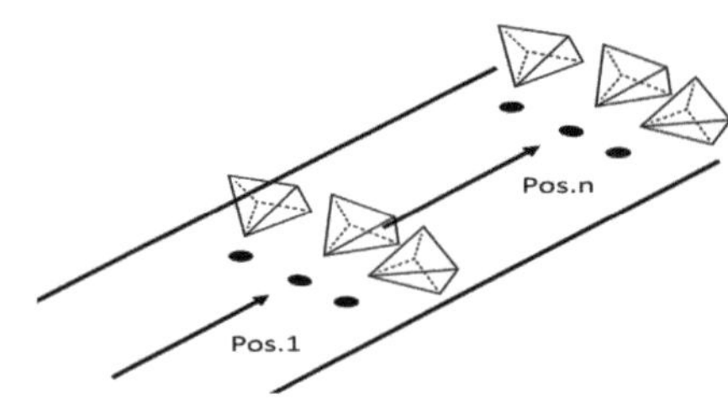

Figure 4: Left-to-Right Strategy: Alternating shots for capturing corridor shaped structures

The Acquisition Plan

Based on the introduced acquisition strategies and a location map of the campus, an acquisition plan was created. The plan contains which parts of the campus are already covered and which parts are still have to be shot. Furthermore the plan proposes an advantageous acquisition strategy based on the location. If prepared thoroughly, both aspects can simplify the actual image acquisition significantly and prevents possible errors on larger spaces or longer sessions. To ease the subsequent reconstruction process steps like segmentation and to increase the probability of a successful reconstruction, a further breakdown of the location into different sectors is recommended.

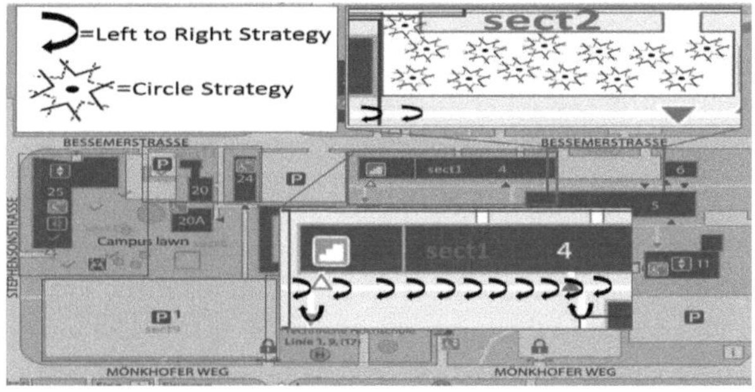

Figure 5: The Acquisition Plan: Used to acquire the images for the reconstruction in a more structured way

Segmentation of the image database

Experience has demonstrated that the usage of a large image database right from the beginning leads often to poor reconstruction result. Instead, the database should be split into multiple segments. These segments are easier to handle regarding error detection in the image pool and are later on faster to reconstruct. If neighboring segments both are successfully reconstructed with a satisfactory degree of detail, both image pools can be combined in a uniform database. This process can then be applied for every segment to create a large database with sophisticated results. This comes with a certain risk. If the space between two segments is not sufficiently captured on each of the parts of the database, the reconstruction of the combined segments might not deliver the result from each individual reconstruction. The algorithm simply can't detect the link between both segments and therefore often neglects one part of the database. If this occurs, additional images between two neighboring segments are required. This can be done with additional acquisition session on the object.

4   Implementation

For the reconstruction of the THL-campus the SfM-algorithm of the software Meshroom was used. Meshroom offers a customizable image processing pipeline from camera parameter extraction up to the actual SfM reconstruction part with a convenient user interface. Each step of the reconstruction is depicted as individual block connected with wires to the subsequent block. Additional blocks can be added and connected with low effort to the existing structure. The setup provided by Meshroom can already be used with just minor changes to the block architecture. All blocks behind the SfM-part are not essential for the reconstruction and can be deleted. By adding a "ConvertSfmFormat"-Block to the pipeline the resulting reconstruction model can be exported into different data for-

mats depending on the individual needs. Most applications are compliant with the .ply-format, to visualize the created pointcloud. Meshroom offers own visualizing functionalities, although not on a statisfactory level. Softwares like Blender or Meshlab can be recommended to depict and edit the pointcloud in a user friendly way. Fig. 6 depicts an exemplary scene on the campus with the corresponding reconstruction result in and Fig. 7.

Figure 6: Exemplary input image of the acquired image data set

Figure 7: Reconstruction result of the section in figure 6 depicted as pointcloud in Blender

5   Results

As already mentioned in chapter 3, the key for successful reconstructions is viable image data of the scene. For the creation of a visual map of the THL-campus, a first image database was created. The database contains all acquired images of the campus according to the created acquisition plan. The database forms the foundation for all upcoming reconstruction efforts. Table 1 summarizes the most significant metrics of the achieved results so far.

Encountered errors in the reconstruction

During multiple reconstruction runs of image data from the campus, some errors appeared with a higher frequency. The section below lists the most common errors and suggests actions to decrease their probability of appearing in the resulting reconstruction.

Table 1: Summary of the achieved reconstruction results

Metric	value
Total number of images acquired	≈ 5000
Covered area of the campus	$11/ \approx 20$ sections
Number of reconstruction iterations	< 20
Average percentage of matched images per reconstruction	$\approx 40\,\%$
Number of sections combined to a coherent model	7 sections
Size of created point cloud	≈ 200.000 points

Ghost artifacts

Due to errors in the loop closing, some reconstruction iterations lead to implausible reconstruction results. Parts or objects from the acquired images appeared multiple times in the reconstruction model (Fig. 8). These duplicates are often placed roughly around the original object but with a slight shift, tilt or angle.

Figure 8: Pillar of the pavilion (right side) appears as undesired ghost artifact on the resulting reconstruction (left side)

The analysis of the affected data sets showed multiple sources for this kind of error:
- High number of redundant images in the data set
- Environmental influences(snow, uneven lighting)
- Inconsistent distance to the object
As a result the whole scene around the Ghost Artifact has to be shoot again or the image data has to be corrected.

Smooth or flat surfaces

Many algorithms use edge detection algorithms to determine the significant features of an image. On flat or smooth surfaces like building facades, larger Windows fronts or even concrete floors are the results often unsatisfactory. Fig. 9 below shows the detected features in an image of the THL-campus. Rougher surfaces like the facade provide a lot of remarkable features (indicated by the orange dots) that can be extracted. The windows however don't provide a large number of distinctive features for the reconstruction.

6   Conclusion & Future Work

Over the duration of this paper a first image database over larger connected sections of the THL-campus were acquired. The subsequent reconstruction efforts with the software tool Meshroom showed, that the developed methods

Figure 9: Untextured surfaces like windows provide not enough remarkable features to extract

and strategies for the image acquisition can be used to create meaningful reconstruction results. Common error sources, as mentioned in the paper, occurred far less likely and the overall number of successfully matched images per section during the reconstruction was further enhanced. With the results of this paper many directions are hereafter possible. The current image database could be expanded to the whole THL-Campus with a resulting spatial reconstruction. In the next phase a dense reconstruction and subsequent meshing based on the current model could further improve the degree of detail. The use of technologies like LIDAR (Light detection and ranging) could also lead to promising results for a specific sections of the campus

Acknowledgement

The work has been carried out with the Luebeck University of Applied Sciences in the department of Electrical Engineering and Computer Science

Author's Statement

Conflict of interest: Authors state no conflict of interest.

7   References

[1] J. Iglhaut, C. Cabo, S. Puliti, L. Piermattei, J. O'Connor, and J. Rosette, "Structure from motion photogrammetry in forestry: a review," *Current Forestry Reports*, Sep. 2019.

[2] K. Schindler, "Mathematical foundations of photogrammetry," ETH Zurich, Switzerland, Tech. Rep., 2015.

[3] P. R. Fergus, "Multi-view stereo & structure from motion," in *Vision-Conferenz*, 2012.

[4] A. Eltnera and G. Sofia, "Structure from motion photogrammetric technique," *Developments in Earth Surface Processes*, Jan. 2020.

[5] C. Stachniss, "Visual features: Descriptors(sift, brief, and orb)," Universität Bonn, Tech. Rep., 2014.

[6] D.-I. M. Gandyra, "Prinzipien und verfahren der 3d-messtechnik," Tech. Rep., 2020. [Online]. Available: https://www.dr-gandyra.com/fachliches/3d-messtechnik/s32/

[7] S. Lachambre, S. Lagarde, and C. Jover, *Unity Photogrammetry Workflow*, Unity, Jun. 2017.

Prospects of Quantum Computing for Medical Imaging

Henrike Goetze [1]

[1] Medical Engineering Science, Universität zu Lübeck, henrike.goetze@student.uni-luebeck.de

Abstract

Quantum computing and especially quantum image processing are trending topics for researchers now a day. By exploiting quantum mechanics, quantum computers could achieve computing performance that dwarfs all previous computing possibilities. In the field of healthcare, where the amount of data is increasing more and more, quantum computers could make a great contribution in various applications such as image processing. Yet the physical realisation of quantum computers has been very challenging, making it difficult to say whether they could ever be used in everyday clinical practice. This paper gives a brief introduction into quantum computing fundamentals and healthcare applications. Aim of this paper is to investigate how quantum computing could be integrated in image processing and to discuss the application feasibility on the basis of current challenges and limitations.

1   Introduction

Quantum computing is a revolutionary computing method based on quantum-mechanical phenomena, first proposed in the 1980s. The introduction of Shor's algorithm for factoring large numbers in 1994, which, if implemented on a quantum computer, would exponentially speed up the calculation, then demonstrated the unique potential of quantum computers and catalyzed interest among researchers [1]. At this time, the results could only be of theoretical interest, but today, nearly 25 years later, efforts to create a quantum computer in practice have made huge progress. In theory, quantum computing has the potential to address computational problems that no classical computer could realistically master, also called "quantum supremacy", which motivates commercial interest in different application domains, such as chemistry, medicine, and financial services.

The general long-term goal would be to develop quantum computers that can reliably perform tasks that are difficult to solve with conventional computers and therefore reach quantum superiority. Quantum computing in the medical field is an area that has been rapidly attracting more attention over the last years. According to *Web of Science* there has been a steep increase in publications and citations of papers on quantum computing in the healthcare sector since 2018 as shown in Fig. 1. In 2021, publications in this sector were cited over 800 times in total and the trend for 2022 shows there will most likely be even more. Based on this increasing interest, this paper aims to investigate potentials of deploying quantum computing to improve medical image processing but must also critically question the feasibility of functioning quantum computers in everyday clinical practice due to significant limitations.

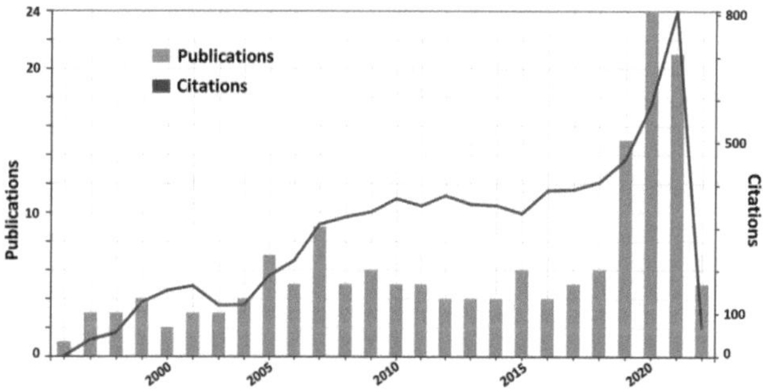

Figure 1: Published and cited journal articles of quantum computing in the medical sector over time.

1.1   Fundamentals

A traditional computer uses long strings of "bits" as a basis. Bits have only two possible values, either zero or one. Quantum computers, in turn, use quantum bits, also called qubits, as basic units. Unlike Boolean variables, they can also exist in superposition of their possible observable states before measurement. The principle of quantum **superposition** states that quantum particles can be overlaid and thus can exist in multiple states or locations at the exact same time [2]. The quantum particle is calculated as a probability distribution in which it can exist simultaneously at all locations at all times with varying probabilities. The other main property of qubits is **entanglement**. It stands for the ability of quantum particles to correlate their measurement results with each other. The correlation is so strong that two or more quantum particles can be inextricably linked and influence each other [2]. Due to entanglement large amounts of data can be processed simultaneously, because the operation can be carried out on only a few quantum particles in a massively parallel manner. In addition to simulating quantum computers on conventional computers, several approaches are being pursued by researchers to create a practical quantum computer, for example using free electrons in superconductors, magnetic resonance with nuclear spin or entangled photons as qubits. It still remains unclear which approach is ahead of the others [3].

2 Quantum Computing in Healthcare

Quantum computing offers great potential in a wide range of applications, such as search and optimization, simulation of quantum systems and cryptography, which are fundamental tasks for the healthcare sector [2]. An enormous load of data has to be processed in the healthcare sector, this includes clinical trials, registries of disease, Electronic Health records and data acquired by medical imaging devices. This data has been increasing with an annual growth rate of 36% [2]. This poses a great challenge to data processing that can be solved with the access to large-scale, parallel computing bringing this huge amount of data into clinical use. In the following some of the potential application areas of quantum computing in the healthcare sector are presented.

Molecular Simulations: Molecular structures with many interacting electrons are very complex and simulation would take a classical computer a long time to process. This could be exponentially sped up with the possibility of many parallel calculations in quantum computing [2].

Radiotherapy: Radiography treats cancer through radiation beams. Computing the exact direction and energy of these radiation beams is a complex optimization problem with thousands of variables. Extensive computer simulations are required to find an optimal solution of destroying cancerous cells while sparing healthy tissue. The capacity of quantum computers could allow advanced precision and multiple simulations simultaneously, to allow comparison between all possible approaches and to reach the optimal radiation planning faster and with less side effects [2].

Data Security: Healthcare data requires a high level of data privacy. Quantum computing with its novel characteristics could open up new paths for security of healthcare. Moreover, previous classical encryption methods could become obsolete, if quantum computers become commercially available [9]. So once quantum computers are operational, medical data would be at risk and the quantum encryption algorithms could be the only way to adequately protect this sensitive data.

Medical Imaging: Diagnosing a disease as early as possible is crucial for accurate treatment and lessening the stress on the patient and the healthcare system. Medical Imaging is a crucial diagnosis tool that could potentially be enhanced by the use of quantum computing. Data collected by medical devices e.g. CT and MRI scanners need to be reconstructed into an image. However, most of the used image processing techniques are considered time consuming for a real time application that is sensitive to incorporated

delays [7]. A variety of scientific papers deal with the approach of Quantum computing offering a solution that could eventually lead to exponential speed up in execution time compared to classical image reconstruction and image understanding. This will be discussed here in more detail.

2.1 Quantum Image Reconstruction

In general, quantum image reconstruction can be divided into three steps as shown in Fig. 2.

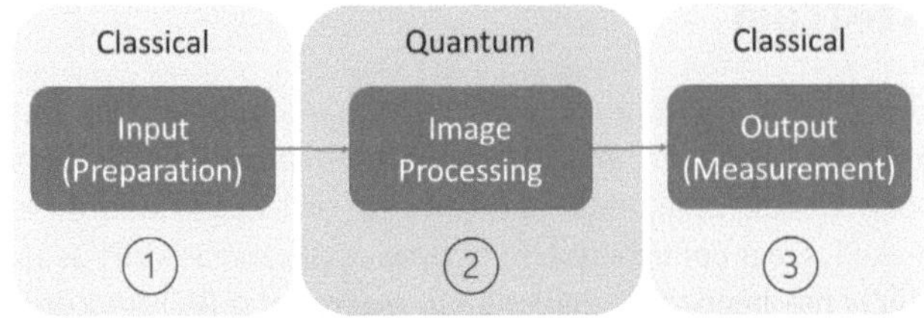

Figure 2: Flowchart of quantum image reconstruction. Modified from [7] and [8].

First, as an input, quantum medical imaging algorithms take a quantum state representing the raw data output from a medical imaging device. This is also known as image preparation. Second, the quantum image is being processed, which is the main step of the quantum image processing. Finally, the result is obtained. Because the algorithm's outputs are also quantum states, reading out all the pixels in a quantum image is generally not efficient. Directly reading out the pixel values of an $N \times N$ quantum reconstructed image requires the time complexity $\mathcal{O}(N^2)$ [6].

Image reconstruction algorithms often make use of relations between a target image and its representation in the frequency (Fourier) domain, making the Fourier transform one of the most important algorithms in image processing [6]. Thus, its quantum version is of great importance to quantum image processing. The Quantum Fourier transform (QFT) differs from the classical Fourier transform in that it takes a quantum state in which the initial data has been encoded into probability amplitudes and can so exploit the advantages of quantum mechanics which could be applied in MRI and CT image reconstruction [7].

MRI: The classical method for MRI image reconstruction uses algorithms that reconstruct the image from input data collected in the frequency domain, called k-space. The reconstruction algorithm mostly used is an inverse Fast Fourier transform, which reconstructs the density of nuclear spins $\rho(x, y)$ of hydrogen nuclei. The classical reconstruction requires $\mathcal{O}(N^2 \log N)$ time for an $N \times N$ image. Since the input data is collected in the Fourier space, the quantum algorithm for MRI image reconstruction is straightforward. The reconstruction of the quantum state $\rho(x, y) \, |x\rangle \, |y\rangle$ is also a simple 2D inverse Fourier transform of the data in k-space. The reconstruction time of a $N \times N$ image here is reduced to $\mathcal{O}(\log^2 N)$ [6].

CT: CT scanners use projections of line integrals, called Radon transform for image reconstruction. Classical reconstruction via Fourier slice theorem takes the data from the

Radon transform applied on a function $F(x, y)$ as an input and reconstructs another function $G(x, y)$ that is close to the original function $F(x, y)$. For a given set of parallel projections at N discrete angles, the reconstruction of a $N \times N$ image needs an execution time of $\mathcal{O}(N^2 \log N)$. For the quantum approach of the Fourier slice theorem we need to assume that the input data is provided as a quantum state in two registers. Then the Quantum Fourier transform is applied. Next, a linear interpolation from polar coordinates to Cartesian coordinates is needed. The final step is a 2D inverse Fourier transform to reconstruct the image. The run time for this quantum image reconstruction algorithm takes $\mathcal{O}(s^2 \log N + \log^2 N)$ where s is a constant describing the number of points used to interpolate data from polar to Cartesian coordinates [6].

2.2 Quantum Image Understanding

In addition to image reconstruction, it is equally important for diagnosis, prognosis, and treatment planning that the captured image is understood. Machine Learning is an effective tool that is used for image understanding tasks such as image classification, segmentation, and localization for example for the detection and classification of tumors. Quantum computers can have the potential to outperform classical computers on machine learning tasks [4]. The idea behind leveraging quantum computing for machine learning tasks is the parallelism of calculations that a quantum computer can conduct.

Kinshuk Sengupta et al. [5] have experimentally implemented a new image classification and segmentation prototype method conducted over images of COVID-19 patients, introducing the use of Quantum Machine Learning in medical practice. They compared COVID-19 CT pneumonial images with non-COVID images. The investigated novel algorithmic implementation using quantum neural network (QNN) outperformed the conventional deep learning models for this specific classification task. The results suggest that quantum neural networks would perform better in classifying the image features compared to deep learning in terms of model effectiveness and also training time.

Convolutional neural networks (CNNs) consist of convolutional layers which extract features and pooling layers. These sample down the convolved features. It has become a successful model for image understanding which automatically learns the needed features and extracts them for medical image understanding. However, CNNs reach their limits if the given dimension of data or model becomes too large [10]. Quantum Convolutional Neural Networks (QCNNs) that extend the convolution layer and the pooling layer to quantum systems, offer a new solution to this problem. The advantages that superposition and parallel computation of quantum computers offer, are reduced learning time and evaluation time [10].

3 Discussion

3.1 Opportunities

The previous sections of this paper have outlined the potential of quantum computers in medical image processing and other medical applications. Quantum computers offer new possibilities with their quantum properties, attracting a lot of attention from most industrial sectors and research fields. In their paper Kinshuk Sengupta et al. have experimentally shown the superiority of quantum machine learning over conventional machine learning algorithms. Both a higher accuracy and a shorter execution time could be shown [5]. To incorporate these simulations into hardware and run them on a real biomedical device for seamless analysis would now be the next step.

There are various examples showing the potential of quantum machine learning to enable and innovate future artificial intelligence applications that could deliver results with great time savings. Therefore, quantum-based machine learning algorithms could serve as building blocks for quantum image classification and recognition. Medical image understanding might benefit from this potential. With their capacity to handle large amounts of data, quantum computers could not only add medical images but also background information of the patient to calculations. If both the image and the patient history could be processed by the machine learning model it could result in a new level of prediction precision.

The quantum Fourier transform could also offer great time savings over its conventional counterpart in image reconstruction if input data is provided as a quantum state.

3.2 Weaknesses

As promising as these applications sound, the real benefit of quantum computing is still unclear. Quantum computers have to overcome significant hardware challenges to revolutionize computing performance. Quantum computers face problems in the following areas.

Noise Sensitivity: One of the major differences between a classical and a quantum computer is their processing of unwanted disturbances. Qubits and quantum gates are very sensitive to noise and small errors can eventually lead to wrong outputs of the computation. In mid-2018, the error rate for 2-qubit operations was more than a few percent [1]. For a quantum system to reliably process information, the qubits are required to be in isolation and to be operated at temperatures close to the absolute zero, which constrains application in comparison with the high accuracy of classical computation methods and limits the size of quantum circuits that are reliable [9]. In these extreme conditions qubits can stay in coherence for about 0.1 ms [4]. To correct the noise errors, quantum error correction algorithms (QEC) would have to be run [1]. However, QEC are very resource intensive and require additional physical qubits, since qubits' states cannot simply be copied because it would cause decoherence. So the information the qubits contain must be "spread out" among many other qubits.

This again leads to an increase of the system's vulnerability with added qubits [9]. Another point to consider is that because of their imperfect performance, it can be difficult to verify the correctness of quantum algorithms. Since we are talking about medical images being processed, it is most important to achieve reliable results.

Data Loads: Converting large amounts of data to quantum states dominates the computation time and thus slows down the process so that the exponential acceleration of quantum image operations may not exist [1].

Immediate State: In classical computing systems one can debug by reading the immediate state store in memory, which is infeasible in quantum computing systems, since quantum states cannot be copied for later examination and measuring a quantum state leads to its collapse [1]. Therefore, new methods for debugging must be developed for the design of large-scale quantum computing systems.

Given the current state of quantum computing and the significant challenges that still must be overcome, some quantum computing researchers have expressed scepticism about the physical feasibility of scalable quantum computers. Without a viable solution for the correction of errors in a quantum system, it is unlikely that highly complex quantum programs would ever run correctly. It is also not certain that quantum computers can really be used universally for a wide range of tasks or whether they might be best suited to only a specifically tailored narrow range of applications, as simulating systems dominated by quantum effects and thus could not bring any advantage over conventional computers. Moreover, the hardware and software of conventional computers are constantly being improved. If classical computers become powerful enough to simulate the presented quantum computer tasks without drawbacks before their reliability is sufficiently developed, they could quickly lose their usefulness.

4 Conclusion

Quantum computers have opened up completely new possibilities with their quantum properties. They have the potential to enhance speed and efficiency of computations compared to conventional computers which is arousing increasing interest from researchers. In the healthcare sector, where processing more and more data is becoming an increasingly big challenge, this could bring great benefits to the goal of providing better healthcare to the patients. This research explored quantum computing solutions from the perspective of healthcare image processing, which is one of the most promising areas in the field of research for the improvement of diagnosis and treatment. It is shown that, although quantum computers have a theoretical advantage compared to conventional computing, their practical benefit is still not there and we are still a long way from quantum supremacy. Actual Quantum computing hardware must first become reliable and needs to overcome major challenges and limitations, most applications yet are only tested by simulating quantum states on classical computer. It is not clear yet, if and when quantum computers will be accessible for everyday use in medicine and medical image processing. It must be noted, however, that only a selection of sources were examined for this work. There are many more papers on this topic that deserve attention.

Acknowledgement

The work has been carried out at Siemens Healthineers, Erlangen supervised by Sebastian Schroth. I would like to thank Prof. Rafecas from the Institute of Medical Engineering, Universität zu Lübeck for her insightful comments.

Author's Statement

Conflict of interest: Authors state no conflict of interest.

5 References

[1] National Academics of Sciences, Engineering and Medicine, *Quantum Computing: Progress and Prospects*. Washington, DC, The National Academies Press, 2019.

[2] R. U. Rasool, H. F. Ahmad, W. Rafique, A. Qayyum and J. Qadir, *Quantum Computing for Healthcare: A Review*. TechRxiv, 2021.

[3] D. F. Parsons, *Possible medical and biomedical uses of quantum computing*. Neuroquantology 9.3, 2011.

[4] D. Solenov, J. Brieler and J. F. Scherrer, *The potential of quantum computing and machine learning to advance clinical research and change the practice of medicine*. Missouri Medicine, 115(5), p. 463, 2018.

[5] K. Sengupta and P. R. Srivastava, *Quantum algorithm for quicker clinical prognostic analysis: an application and experimental study using CT scan images of COVID-19 patients*. BMC Medical Informatics and Decision Making, 21(1), pp. 1-14, 2021.

[6] B. T. Kiani, A. Villanyi and S. Lloyd, *Quantum medical imaging algorithms*. arXiv e-prints, arXiv-2004, 2020.

[7] O. Al-Taani1a, A. M. Alqudah and M. Al-Bzoor, *Implementation and Analysis of Quantum Fourier transform in Image Processing*. Jordan Journal of Electrical Engineering, 5(1), pp. 11-26, 2019.

[8] S. Chakraborty, S.B. Mandal and S. H. Shaikh, *Quantum image processing: challenges and future research issues*. International Journal of Information Technology, pp. 1-15, 2018.

[9] J. Preskill, *Quantum computing in the NISQ era and beyond*. Quantum 2, p. 79, 2018.

[10] S. Oh, J. Choi and J. Kim, *A tutorial on quantum convolutional neural networks (QCNN)*. 2020 International Conference on Information and Communication Technology Convergence (ICTC), IEEE, pp. 236-239, 2020.

Optimization-based Correction of Image Distortion in OCT for Analysis of Corneal Curvatures

Maron Dolling [1], Hinnerk Schulz-Hildebrandt [2] and Reginald Birngruber [2,3]

[1] Medical Engineering Science, Universität zu Lübeck, maron.dolling@student.uni-luebeck.de
[2] Institute of Biomedical Optics, Universität zu Lübeck, {hinnerk.schulzhildebrandt, reginald.birngruber}@uni-luebeck.de
[3] Wellman Center for Photomedicine, Massachusetts General Hospital, Harvard Medical School, Boston, MA, USA

Abstract

The corneal filler implantation is a novel, minimally invasive therapy approach to correct presbyopia and astigmatism by injecting a biocompatible viscous implant into a fs-laser-cut stromal pocket. For this method, an in-process control is necessary to determine the changes of the refractive power by quantification of the corneal curvature. Optical Coherence Tomography (OCT) is a well established modality capable of imaging corneal tissue. However, the design of most OCT suffer from optical fan distortion and a spherical cornea appears as a toric surface equivalent to an astigmatism of up to 3.5 ± 1 dpt. This work proposes an optimization-based calibration with a sphere phantom that reduces the fan distortion created error in radius measurement to < 0.03 mm with a standard deviation < 0.015 mm (equivalent to ca. 0.2 ± 0.07 dpt corneal refractive power), enabling precise measurement of astigmatism angle and strength of samples without sacrificing advantages of OCT.

1 Introduction

The age-related elasticity decrease of the ocular lens and resulting loss of accomodative power is called presbyopia. It affects people in their mid-forties and older. Based on the demographics of the world's population, it is estimated that currently nearly 25% are affected [1]. Presbyopia leads to difficulties focusing near objects and therefore has a significant impact on reading and computer work. This results in a high interest in treatment.

Although there are several non-invasive therapy methods like reading glasses, they are often perceived as uncomfortable or are not applicable in some situations. Hence, many surgical treatment options have been developed (e.g. intraocular lens implantation, laser refractive surgery, etc.). However, presbyopia usually does not justify highly invasive and irreversible surgical interventions. Moreover, the development of presbyopia over time requires an adjustable visual correction procedure and current procedures may be not eligible at all due to thin corneal tissue.

The novel corneal filler approach uses a femtosecond laser-cut pocket inside the stroma, filled with a viscous filler material to modify the corneal profile, creating a bifocal corneal refractive power [2], [3]. In contrast to several other refractive surgeries, the corneal filler is less invasive, does not ablate corneal tissue and may offer easy adjustment.

To accurately measure the eye's refractive power, including astigmatism, before and after filler injection, the curvatures of the different surfaces must be precisely measured. Three-dimensional optical coherence tomography (OCT) poses an appropriate modality, if the OCT related optical distortion can be compensated. A spherical object typically appears to have an astigmatism, which results in an angle-dependent difference in refractive power of up to 3.5 dpt, assuming the examined object has the refractive index of corneal tissue. This is due to the geometry of the OCT scanner: In order to acquire a 3D-scan, most OCT systems have two separated mirrors in the scanner, which cannot both be placed in the focal point of the collimating lens, resulting in different depth relations for the two scanning directions x and y. As a result, the incoming beam is scanned differently in x- and y-direction, causing an optical distortion of the OCT image [4]. Several approaches which require phantoms were investigated to address distortion in OCT [5], [6]. However, eye phantoms or lens constructions require precise production. Distortion correction with a grid target [4] for topographic is insufficient because the fan distortion is smaller than the OCT resolution.

This work introduces an optimization-based approach for image field correction that overcomes these limitations by using a rather simple spherical phantom as a standard which offers three-dimensional structural information. The correction enables measurements of refractive power and angle of an astigmatism with an accuracy of at least 0.2 ± 0.07 dpt, which satisfy the precision requirements in ophthalmology.

2 Material and Methods

2.1 OCT System

In this work, the Telesto Spectral-Domain-OCT (Thorlabs, Newton, NJ, USA) with a central wavelength of 1300 nm was used. It achieves a lateral resolution of 15.63 µm and an axial resolution of 6.15 µm, and with $1024 \times 512 \times 512$ voxels a field of view (FOV) of approximately $6.3 \times 8 \times 8$ mm ($z \times x \times y$).

2.2 Calibration Standards

In order to measure the fan distortion caused astigmatism, ball bearing spheres with a radius of 8, 8.5 and 9 mm were used. They provide surface dimensions similar to the corneal curvature and comply with the quality grade G10 (class 2), which, according to DIN 5401 : 2002-08, guarantees a limit in radius deviation of $\pm$ 9.75 µm. This may result in a deviation in refractive power of a corneal surface by 0.044 dpt maximum, which is acceptable in terms of ophthalmology. Hence, the sphere was considered ideal and the deviation between its measured and extracted surface was the basis for the quantification of the fan distortion. This work exploits not only the sphere's constant curvature characteristics but its expansion in depth and therefore three-dimensional displacement, compared to a planar phantom (e.g. OCT grid target).

2.3 Curvature Measurement

In order to quantify the fan distortion caused astigmatism, the spherical standard was scanned in 3D with the OCT device in different depths. The surface was extracted using a max-pooling in every a-scan. Outliers were removed and only the biggest connected structure accepted as surface data to minimize impact of noise. The surface was regarded as a point cloud with cartesian coordinates (x, y, z) in mm to avoid interpolation errors of filled data in an array.

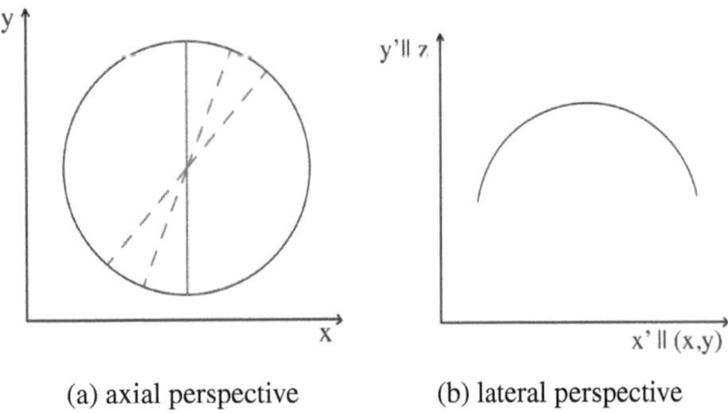

(a) axial perspective   (b) lateral perspective

Figure 1: Schematic representation of a sphere's dome curvature measurement method. Semicircles are fitted into the dome in rotational steps (a) and transferred into a new coordinate system (b).

As a robust method for astigmatism quantification, there were, stepwise, semicircles inside the dome of the sphere regarded (s. Fig. 1a), spanned by all points that have a certain angle to the x-axis within a tolerance of 0.05 rad. These semicircles were transformed into a two dimensional rotating coordinate system (s. Fig. 1b) with axes (x', y') and approximated with a curve fit with following condition:

$$y' = z = y_0 + \sqrt{r^2 + x'^2}, \quad x' = \sqrt{x^2 + y^2}, \quad (1)$$

where y_0 denotes an offset on the y-axis and r being the fitted radius of the semicircle and therefore representing the radius of the spherical phantom at the regarded angle. r does uniquely correspond to a refractive power P of the cornea as follows:

$$P = \frac{n_1 - n_2}{r}, \quad (2)$$

with refractive indices $n_1 \approx 1$ (air) and $n_2 \approx 1.376$ (cornea) [7]. Different rotation angles at measurement were investigated to rule out any possible astigmatism of the spherical standard itself, although this is, due to the high precision of the phantoms, not to be expected. The residual sum of squares (RSS) per datapoint was considered for goodness-of-fit estimation.

2.4 Parameterized Distortion Model

A nonlinear scaling factor for the x- and y-coordinates is proposed. This yields

$$x_c = q_{10} + q_{11}x + q_{12}x^2, \quad y_c = q_{20} + q_{21}y + q_{22}y^2 \quad (3)$$

with scalar coefficients q_{ij} and corrected coordinates (x_c, y_c). A two-dimensional polynom of 4th degree was used to approximate the distorted surface in the z-dimension, correcting the z-coordinates to

$$\begin{aligned} z_c = z + p_{00} + p_{10}x_c + p_{01}y_c + p_{11}x_cy_c + p_{20}x_c^2 + \\ p_{02}y_c^2 + p_{21}x_c^2y_c + p_{12}x_cy_c^2 + p_{22}x_c^2y_c^2, \end{aligned} \quad (4)$$

with scalar coefficients p_{kl}. Since this parameterization does neglect the depth-dependency of the distortion, there also was a multi-depth-method with further coefficients investigated and compared to the proposed single-depth-method. The additional term yields

$$\begin{aligned} x_c' = x_c(s_1 + s_2z + s_3z^2), \\ y_c' = y_c(s_1 + s_2z + s_3z^2), \end{aligned} \quad (5)$$

with scalar coefficients s_m. These parameters may then be applied onto coordinates of points of an object of examination to correct their displacement caused by the fan distortion and other systematic influences.

2.5 Loss Function

Within an optimization process, the loss metric denotes the function which output scalar value is to minimize. An optimization routine exploiting the Downhill-Simplex-Algorithm [8] with termination criteria of at most 1200 iterations or change in value of the loss function of less than

10^{-5} during one step ('TolFun') was chosen. Using the algorithm described in section 2.3, the coefficient-dependent loss function $L(p_{ij}, q_{kl})$ $(L(p_{ij}, q_{kl}, s_m)$ including depth-dependency) uses the mean deviation of the measured radii r_i in the corrected structure at n different angles from the phantoms real radius $\hat{r}$ and therefore describes the strength of the fan distortion. In addition, the difference of the initially measured apex depth $\hat{z}$ to the corrected apex depth z_{min} was added as a correction term to avoid the structure being shifted as a whole. In conclusion, the optimization process is described as:

$$L = \frac{1}{n}(\sum_{i=1}^{n} |\hat{r} - r_i|) + |\hat{z} - z_{min}| \longrightarrow \min. \qquad (6)$$

3 Results and Discussion

3.1 Single-Depth Correction

Fig. 2 shows the semicircle fit based astigmatism measurement results of an 8 mm radius sphere phantom, with and without correction. The solid line shows the astigmatism mainly caused by the fan distortion. The dashed line shows the measured astigmatism after applying the correction. It appears that the method is capable of correcting the astigmatism almost completely. Analyses of spheres of different

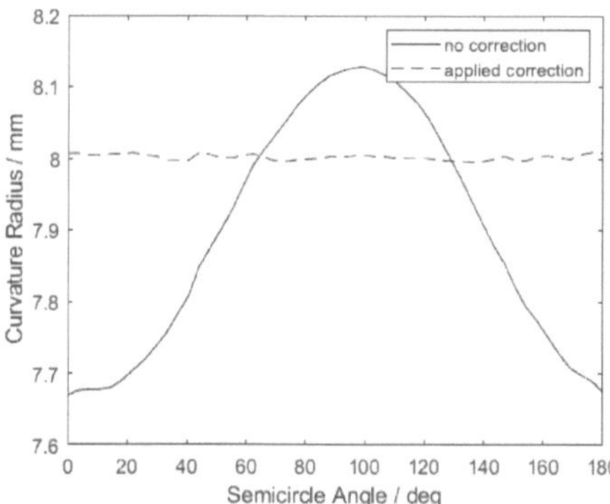

Figure 2: Measured radii of semicircles fitted to the phantom's surface in rotational steps, with and without correction. To expect is a horizontal line at 8 mm. The correction coefficients were calculated using a spherical phantom with its apex in 1 mm depth.

sizes rule out an overfitting on a certain object size. Table 1 shows that all phantoms, without any correction, are imaged with a difference in mean radius of their curvature from the ground truth of more than 0.1 mm, whereas, with applied correction, the difference stays below 0.03 mm. Furthermore, the correction reduces the standard deviation σ_r of the refractive powers below 0.015 mm, whereas it is more than 0.15 mm for imaging without correction. This is crucial also for precise determination of the astigmatism angle. The mean RSS for the circle fit was 2.4×10^{-3} without and 9.8×10^{-4} with correction. However, this method does not hold for the whole 3D-FOV, as can be seen in Fig. 3. While the mean of the measured refractive power deviation

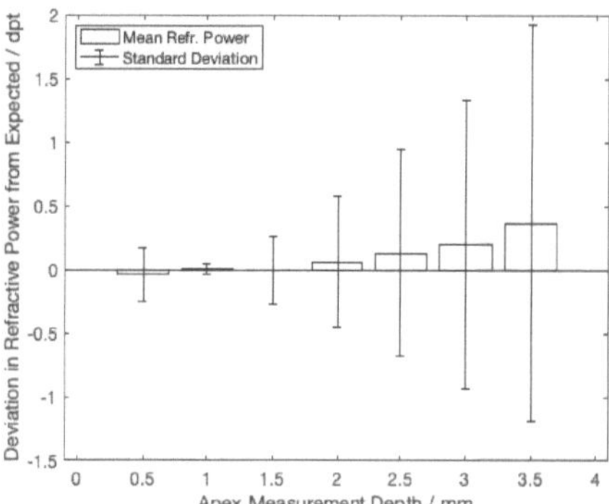

Figure 3: Mean deviation of the refractive power of the semicircle curvatures of spherical phantom (radius 8 mm) from expected (47.00 dpt) in different measurement depths of a corrected image. Calibration was performed with phantom measurements in 1.0 mm apex depth.

expected	$\hat{r}/mm$	8.000	8.500	9.000
no	$\bar{r}/mm$	7.895	8.394	8.891
correction	σ_r/mm	0.165	0.193	0.218
applied	$\bar{r_c}/mm$	8.002	8.519	9.026
correction	σ_{r_c}/mm	0.004	0.011	0.014

Table 1: Expected radii $\hat{r}$, mean measurement of the radii $\bar{r}, \bar{r_c}$ of the spherical phantoms and their standard deviation σ_r, σ_{r_c} of image without correction and with correction, calculated on the 8 mm radius phantom, in 1 mm depth.

stays well below 0.5 dpt for apex depths up to 3.5 mm of the spherical surface, the standard deviation increases, indicating a strong residual astigmatism. Nevertheless the correction is sufficient in an axial FOV of 0.5 mm around the calibration depth. This margin is of adequate size since it is possible to arrange the object of examination in a certain measurement depth with an uncertainty of less than 0.1 mm, due to the fine suspension mechanic of the OCT. The calibration took 10-20 minutes in MATLAB, depending on hardware capability and terminated due to TolFun-criteria. The sphere was removed and replaced in the setup five times to ensure reproducibility, which showed a mean deviation in radius measurement of 0.0049 mm after one calibration.

3.2 Multi-Depth Correction

Multi-depth calibration results in a trade-off between corrected FOV size and precision. Fig. 4 shows the mean deviation of corneal refractive power of the phantom at different angles, in different measurement depths. It is possible to achieve deviations of less than 0.5 dpt in several depths, however it is much less precise for every depth on its own compared to single-depth-correction. Regarding its standard deviation, the residual astigmatism can lead to errors in the angle estimation. The measurement results can be seen in Table 2. Due to the necessity of using multiple images of a phantom in different depths for one optimization iteration, an increased number of parameters and consequently computational complexity, the computation time was about 2.2 hours and terminated due to the iteration limit. Since the physical distortion model poses many degrees of freedom with its quadratic components, it serves the theoretical distortion possibility. Therefore, the residual error is either a consequence of short computation time or a lack of convergence capability of the loss metric.

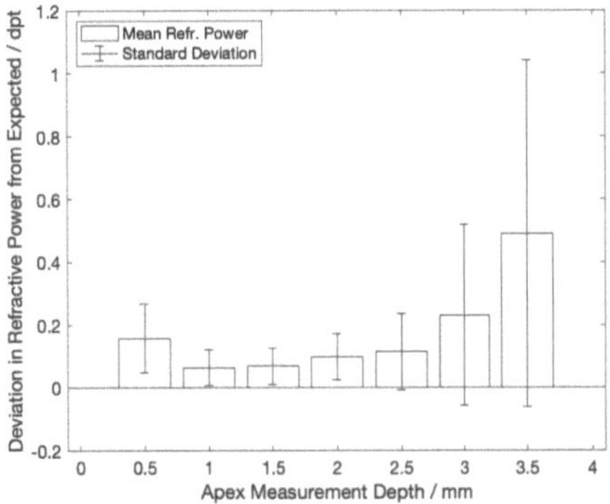

Figure 4: Mean deviation of the refractive power of the semicircle curvatures of spherical phantom (radius 8 mm) from expected (47.00 dpt) in different measurement depths of a corrected image. Calibration was performed with phantom measurements in 1.0, 1.5, 2.0 and 2.5 mm apex depth.

Depth / mm	no correction $\bar{r}/mm$	σ_r	applied correction $\bar{r_c}/mm$	σ_{r_c}
0.5	7.901	0.157	7.977	0.019
1.0	7.903	0.176	7.991	0.010
1.5	7.912	0.204	8.011	0.010
2.0	7.907	0.230	8.015	0.013
2.5	7.905	0.259	8.011	0.021
3.0	7.900	0.300	7.990	0.049
3.5	7.874	0.360	7.947	0.092

Table 2: Mean measurement of the radii $\bar{r}, \bar{r_c}$ of the spherical phantom and their standard deviation σ_r, σ_{r_c} in different depths of image without correction and with correction, calculated in 1.0, 1.5, 2.0 and 2.5 mm depth on sphere phantom with radius 8 mm.

4 Conclusion

The optimization-based calibration enables quantification of a corneal radius with at least 0.03 ± 0.015 mm and an astigmatism with at least 0.2 ± 0.07 dpt accuracy. This complies with the precision requirements in ophthalmology and is calculated in under 20 minutes. An approach to correct the whole 3D-FOV results in a trade-off between FOV size and correction accuracy. Although it was possible to achieve sufficient precision, the computation time of at least 2 hours exceeded the limits of practicability by far. Because it is easy to stick to a certain measurement depth experimentally, a single-depth calibration is enough to realize the required precision of 0.5 dpt. In consequence, it is the method of choice for introduced experimental setups. Although it is, in theory, possible to use one calibration measurement for multiple sample examination series, the aim should be to perform calibration prior to every series whenever one can not ensure that there was no change on the optical setup or adjustment of the OCT internal reference length. However, whenever samples with multiple layers of media with different refractive indices need to be measured, it may be beneficial to apply a multi-depth correction with a sufficient FOV if the different surfaces exceed the area of proper correction.

Acknowledgement

This work has been carried out at the Schulz-Hildebrandt group in the Institute for Biomedical Optics, University of Lübeck. It was supervised by Hinnerk Schulz-Hildebrandt, PhD and Prof. Reginald Birngruber, MD, PhD. Additional support was given by the members of the Schulz-Hildebrandt group Tim Eixmann, Martin Ahrens, PhD and Tabea Kohlfärber, cand. PhD.

Author's Statement

Conflict of interest: Authors state no conflict of interest.

5 References

[1] Holden, B. et al. (2008). *Global Vision Impairment due to Uncorrected Presbyopia.* Arch Ophthalmol. 26:1731–1739.

[2] Kassumeh, S. et al. (2020). *Corneal Stromal Filler Injection as a Novel Approach to Correct Presbyopia—An ex vivo Pilot Study.* In: Trans Vis Sci Tech. 9(7):30.

[3] Wertheimer, C. et al. (2020). *Refractive Changes after Corneal Stromal Filler Injection for the Correction of Hyperopia.* In: J Refract Surg. 36(6):406-414.

[4] Ortiz, S. et al. (2009). *Optical Coherence Tomography for Quantitative Surface Topography.* In: Applied Optics, 48(35):6708-6715.

[5] Ortiz, S. et al. (2010). *Optical Distortion Correction in Optical Coherence Tomography for Quantitative Ocular Anterior Segment by Three-Dimensional Imaging.* Opt. Express 18, 2782-2796.

[6] Tan, B. et al. (2021). *Ultrawide Field, Distortion-Corrected Ocular Shape estimation with MHz Optical Coherence Tomopgraphy.* In: Biom. Opt. Express 12, 5770-5781.

[7] Bergua, A. (2017). *Das menschliche Auge in Zahlen.* Chapter 11, Tab. 11.1. Springer-Verlag GmbH Deutschland. ISBN: 978-3-662-47283-5.

[8] Nelder, J. and Mead, R. (1965). *A Simplex Method for Function Minimization.* In: Comput. J., 7:308-313.

Setup for 3D Robot Model Detection and Simulation in Mixed Reality using HoloLens 2

Gesche Held [1] and Jan Krieglstein [2]

[1] Robotics and Autonomous Systems, Universität zu Lübeck, gesche.held@student.uni-luebeck.de

[2] Fraunhofer Institute for Manufacturing Engineering and Automation IPA, jan.krieglstein@ipa.fraunhofer.de

Abstract

The programming of industrial robots classically requires expert knowledge and can therefore only be performed by specially trained professionals. To facilitate and accelerate robot programming, the idea is to use a 3D interface as provided by the Hololens Mixed Reality (MR) glasses from Microsoft. In order to program and simulate a robot in MR, a 3D robot hologram must first be superimposed onto the real robot, which requires knowledge of its pose. To calculate this pose, a 3D object detection pipeline is implemented based on point clouds of the scene and the known robot model. The focus of this paper is to provide an overview of the required components and the basic setup to obtain and process point clouds of both the scene and the robot model. To conclude, the detection pipeline is presented and possible approaches for improvement are identified.

1 Introduction

1.1 Mixed Reality and HoloLens 2

Milgram and Kishino [1] introduced Mixed Reality (MR) in the mid-nineties as the merging of real and virtual worlds by combining both Augmented Reality (AR) and Virtual Reality (VR). The categorization of VR, MR and AR with respect to the physical and the digital world is shown in Fig. 1.

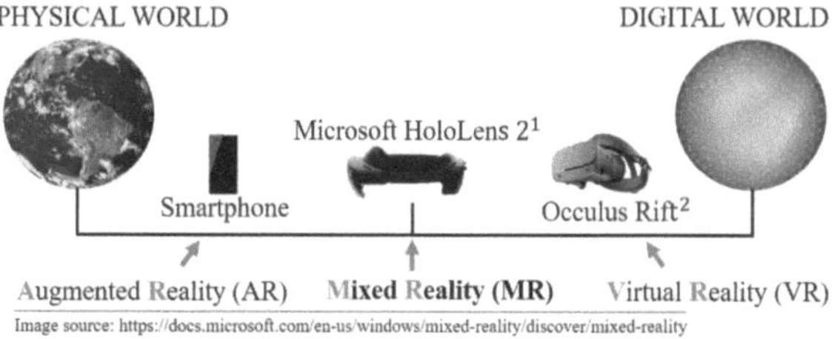

Figure 1: Categorization of AR, MR and VR with respect to the physical world (left) and the digital world (right). Possible devices are illustrated above for each category: a classical smartphone for AR tasks, Microsoft HoloLens 2[1] or Occulus Rift[2] as MR / VR devices.

AR is closer to the physical world where our reality is augmented by non-interactable holograms. Closer to the digital world is VR with a fully artificial digital environment. Interaction is only possible with controllers. MR lies in between as a transition between AR and VR, where interaction between physical and virtual elements is possible [2]. This

---

[1] https://www.microsoft.com/en-us/hololens/hardware

[2] https://www.oculus.com/rift/

concept can be used to display a hologram of a physical robot and superimpose it directly onto the real robot [3].

Microsoft's HoloLens 2 glasses, released in 2019, are a MR Head Mounted Display. It has sensors for hand- and eye-tracking, inertial measurement unit (IMU), RGB camera and a time-of-flight depth sensor. With its built-in sensors, HoloLens 2 can record a spatial map of the environment to create spatial awareness. This spatial map is of particular interest for this project, since it can be extracted as a point cloud and used for 3D object detection.

1.2 Related Work

For 3D object detection, there are two main approaches: marker based and model based approach. For the marker based approach, a physical marker must be placed visibly in advance. This marker is localized by the MR Device. With a previous calibration of robot and marker, the transformation device-to-robot can be obtained [4]. On the other hand, model based approaches require a 3D model with knowledge of the current robot configuration, such as joint states. Ostanin et. al [3] successfully developed a model based approach to detect a robot and superimposed it with a hologram. For this purpose, 3D point clouds of both the scene where the real robot is located and the robot model were used. The model cloud was preprocessed into a convex hull in order to eliminate the intersections of the individual joint clouds. They also deleted the ground points from the scene using *random sample consensus algorithm* (RANSAC), in order to remove object connections across the ground. In addition, they performed a density-based spatial clustering of the scene to group related points. Finally, they calculated the volume of each cluster found in the scene and compared

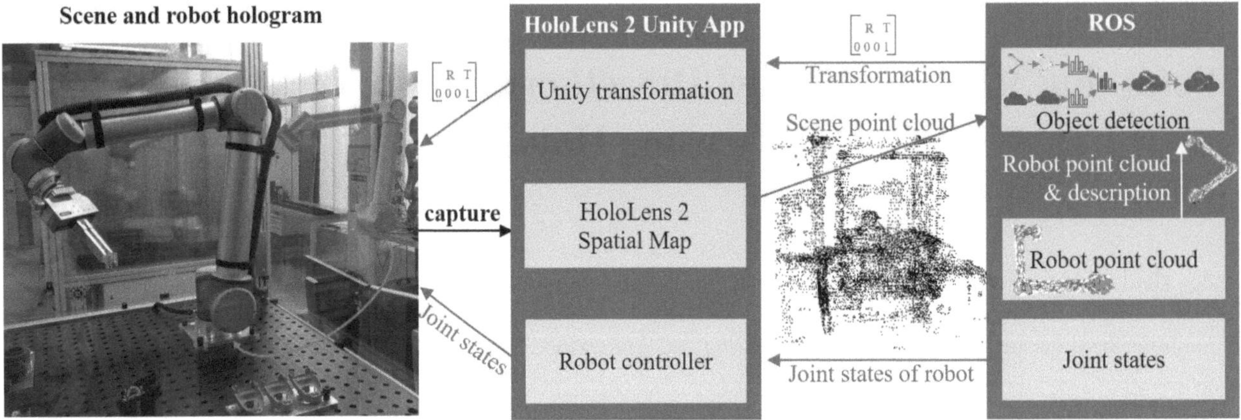

Figure 2: Project overview. The process starts by extracting the spatial map from the environment (left) in form of a point cloud (indicated by *capture*). This scene cloud is published by the Unity application running on HoloLens 2 (left block) and forms one input to the ROS node *Object detection* (right block). The model input for the object detection pipeline is processed in an other ROS node *Robot point cloud*. In this node, the point clouds of the individual robot joints are combined to a complete robot point cloud. Based on those two inputs, the object detection node calculates the pose of the real robot and outputs a transformation matrix for the robot hologram. The image of the scene and the robot hologram on the left is taken by the HoloLens, with the real robot on the left and the robot hologram on the right. The illustrated scene point cloud in the middle is the original cloud of the scene.

them with the volume of the model. Thus, the most similar object to the model in terms of volume was used to calculate the pose of the robot through an *Iterative Closest Point* (ICP). Furthermore, they have used the hologram to program the robot and simulate movements in MR.

As in [3], in this approach, 3D point clouds of both the scene and model were used to perform reliable model based detection. Point clouds can represent not only xyz coordinates, but also RGB values, intensity and normals of individual points. A common tool for point cloud processing is the *Point Cloud Library*[3] (PCL). Helpful tools such as modules for cloud processing itself, keypoint detection and description as well es filtering and segmentation are only an insight into the capabilities of PCL. To decrease the computation time, keypoints were first detected and then described. Hereby, a distinction was made between local and global descriptors [5], [6]. Local descriptors characterize a keypoint based on its local neighborhood, while global descriptors analyse the input cloud as a whole.

Stancelova et. al [7] compared different feature detector and descriptor pairs available on PCL. Relevant results were achieved by three detectors: *Intrinsic Shape Signatures 3D* (ISS3D), *Normal aligned Radial Feature* (NARF), *Harris3D* and by three descriptors: *Point Feature Histogram* (PFH), *Fast Point Feature Histogram* (FPFH) and *Signature of Histogram of Orientation* (SHOT). In terms of recall and precision, the following four pairs formed a pareto front.

ISS3D and FPFH achieving the highest precision, followed by NARF and PFH. But Harris3D and FPFH achieved the best recall, followed by NARF and SHOT.

2 Hardware, Software and Methods

2.1 Hardware and Software

A Microsoft HoloLens 2 was used as the MR device. The application for HoloLens 2 was developed with Unity3D[4] 2020.3.16f1. To obtain the spatial map, the Microsoft Mixed Reality Toolkit[5] (MRTK) for Unity was included. This toolkit provides useful features for HoloLens applications - such as a spatial awareness feature to access the spatial map captured by the HoloLens (scene cloud in Fig. 2). To add MRTK features to Unity applications, the Mixed Reality Feature Tool[6] v1.0.02104.4 Beta was used. The robot control, model cloud generation and detection pipeline were performed on a Ubuntu computer in ROS Melodic[7] (ROS block in Fig. 2). Unity itself offers with *Unity-Robotics-Hub*[8] a powerful tool to load a robot hologram into the application based on its description and mesh files (*URDF Importer*), as well as to establish a connection between Unity and ROS (*ROS-TCP-Endpoint/-Connector*). An overview of the entire project process is illustrated in Fig. 2.

2.2 Robot Hologram and Control

To control the robot hologram corresponding to the physical robot, a custom controller script for Unity was implemented which subscribes to the *joint_states* topic, published by the robot simulator or the robot driver (robot controller in Fig. 2). For the final transformation obtained by the detection pipeline in ROS, a second script is needed to convert and

---

[3] https://pointclouds.org/

[4] https://unity.com/

[5] https://docs.microsoft.com/en-us/windows/mixed-reality/mrtk-unity/?view=mrtkunity-2021-05

[6] https://docs.microsoft.com/en-us/windows/mixed-reality/develop/unity/welcome-to-mr-feature-tool

[7] https://www.ros.org/

[8] https://github.com/Unity-Technologies/Unity-Robotics-Hub

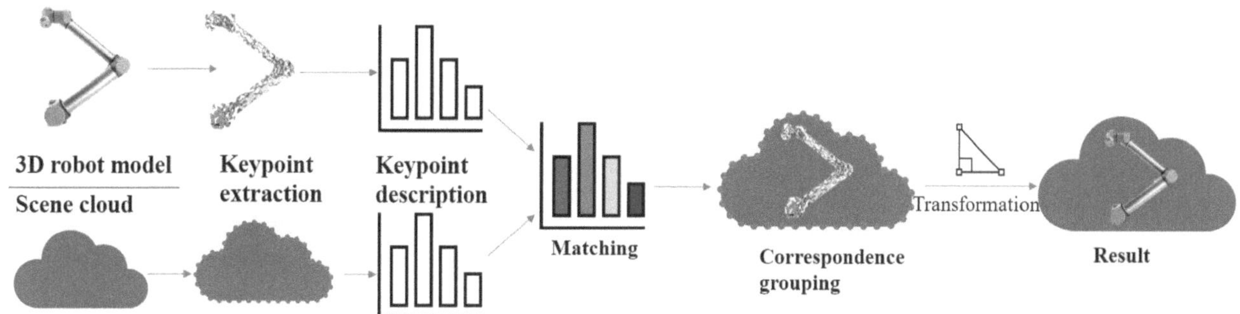

Figure 3: Overview of the object detection pipeline. It starts by computing keypoints of both the point clouds of the scene and the model cloud with an ISS3D detector. These keypoints are then described by a local descriptor called FPFH. This is followed by a matching process to find point-to-point correspondences between model descriptors and scene descriptors. The last step is a correspondence grouping where the found correspondences are group together and possible model transformations are calculated. The resulting transformation is then applied to the robot hologram.

apply the ROS transformation to the Unity hologram (Unity transformation in Fig. 2).

2.3 Point Cloud Processing and Detection Pipeline

The point cloud processing can be separated into the following three parts: (i) scene extraction and preprocessing, (ii) point cloud generation of the model cloud and (iii) the detection pipeline itself. The latter is illustrated in Fig. 3.

First, within the Unity application, the scene point cloud is extracted by the MRTK feature *SpatialAwareness*. After the scene cloud is published to ROS, RANSAC is used, as in [3], to eliminate the ground points and thus remove object connections through the ground.

The second part, the generation of a robot model cloud, requires a manual steps, namely converting the mesh files of the individual links into point clouds. Then, the point clouds of the links are merged in a ROS node (Robot point cloud in Fig. 2) to form a combined model cloud based on the robot description.

An overview of the detection pipeline of the third step can be found in Fig. 3. The detection part itself is performed within the ROS node (Object detection in Fig. 2). Inputs are the extracted and preprocessed scene cloud from the first step and the model cloud form the second step mentioned before. Then, keypoints are first extracted from both point clouds using ISS3D, which are then described with the FPFH descriptor. Both algorithms are suitable for object detection [5] and also in combination together they can achieve good results [7]. The ISS3D keypoint detector takes a point cloud with xyz information as input and computes viewpoint independent local features based on an eigenvalue analysis. The FPFH descriptor works based on xyz-keypoint data and their corresponding surface normals. The difference between two neighboring points is computed and the k-nearest neighbors are compared and weighted. Then, the keypoint descriptors are matched together based on a *Fast Library for Approximate Nearest Neighbors* (FLANN),

which is a search algorithm used to establish point-to-point correspondences between model descriptors and scene descriptors. Point pairs are added to a correspondence vector if they are similar enough, i.e. if the euclidean distance is less than a certain threshold. A *geometric correspondence consistency grouping* (GCCG) is applied at the end to find clustered correspondences between scene and model, group them correctly and compute possible transformations.

The chosen algorithms in the pipeline result in 15 parameters comprising the search radii for keypoint detection and descriptions, thresholds for matching and grouping, and parameters for normal computation and correspondences.

3 Results and Discussion

So far, an initial pipeline is implemented with the following process: extract and generate the clouds and preprocess them. For the scene cloud, the ground points are removed with RANSAC and the model cloud is downsampled to reduce the number of points. Afterwards, the keypoints are detected for both clouds using ISS3D and described using FPFH. Finally, the descriptors are matched and clustered by GCCG.

The scene point cloud of a UR10[9] robot cell was recorded at the Fraunhofer Institute. This scene cloud has a total of 52,678 points. After RANSAC and manual cropping, the scene data set is reduced to 16,638 points. The UR10 model cloud from the manual conversion of the meshes provided by *Universal Robots*[10] (UR) has 428,682 points. To align the density to the measured robot, the model cloud is downsampled to 1,353 points. However, out of the 959 detected scene keypoints, with 34 on the robot, and the 95 model keypoints, no corresponding points are found during matching with manual parameter tuning. Possible reasons are that the model cloud and the measured cloud are too different, as it can be seen in Fig. 4, or that the parameters

---

[9]https://www.universal-robots.com/products/ur10-robot/
[10]https://www.universal-robots.com/

need more comprehensive tuning. Furthermore, the quality of the measured point cloud could be too low to result in matching keypoints. The physical robot was placed in a cell with acrylic glas, which could have increased the point cloud uncertainty. Also, the gripper mounted on the robot was not part of the robot model cloud, as well as the cable drag, which also influenced the measured point cloud.

Figure 4: Comparison of the model cloud (left) and the real robot cloud (right) with an additional gripper attached. The highlighted, thicker dots are the detected keypoints.

Possible approaches for the optimization of the detection algorithm are identified below. To improve the quality of the measured point cloud, the cloud could be recorded in a scene without acrylic glas around the robot and possibly with a longer recording time, giving the Hololens' *Simultaneous Localization And Mapping* (SLAM) algorithm more data to refine the measured point cloud. In addition, it was pointed out that the detection pipeline in its current structure does not find good matches. It needs further processing and testing of several parameters in combination. Up to this point, only one combination of detector and descriptor has been tested. In order to achieve improvements, it is advisable to also apply other pairs tested in [7], as for example NARF and SHOT or HARRIS3D and FPFH. If the pipeline results in a good estimation, an ICP algorithm can be performed at the end to refine the resulting transformation.

4 Conclusion

It was shown which requirements are necessary to connect the HoloLens with ROS for message transmission and control and to extract or generate the required point clouds. This includes Unity3D, ROS, Unity Robotics Hub, MRTK, PCL and manual processing steps.

In order to be able to program and simulate robots in MR, the robot must first be detected in the environment and superimposed with a hologram. To accomplish this task, some preparation is required. Since there is not yet a single program that can be used to detect and control a robot in the scene based on model detection, the setup had to be created first. For this setup, the used softwares and libraries were presented in this paper. Using a self implemented robot controller script, the robot hologram can be placed in the same starting position as the physical robot and movements can be simulated in MR. However, the detection pipeline requires optimization. But as already mentioned, possible optimization approaches exists and will be further investigated in the following steps.

Acknowledgement

The work has been carried out at the Fraunhofer Institute for Manufacturing Engineering and Automation IPA in Stuttgart and supervised by Prof. Dr. Floris Ernst, the Institute for Robotics, Universität zu Lübeck. This work received funding by the State Ministry of Baden-Wuerttemberg for Economic Affairs, Labour and Housing Construction under the grant Zentrum Kognitive Robotik.

Author's Statement

Conflict of interest: Authors state no conflict of interest. All authors contributed to the final manuscript.

5 References

[1] P. Milgram and F. Kishino, *A Taxonomy of Mixed Reality Visual Displays*. In: IEICE TRANSACTIONS Information Systems, vol. 77, no. 12, pp. 1321-1329, 1994.

[2] Microsoft, *What is Mixed Reality?*. Available: https://docs.microsoft.com/en-us/windows/mixed-reality/discover/mixed-reality [last accessed on 2022-01-12], 2021.

[3] M. Ostanin, S. Mikhel, A. Evlampiev, V. Skvortsova, A. Klimchik, *Human-Robot Interaction for Robotic Manipulator Programming in Mixed Reality*. In: 2020 IEEE International Conference on Robotics and Automation (ICRA), pp. 2805–2811, 2020.

[4] L. Kästner, V. Frasineanu and J. Lambrecht, *A 3d-Deep-Learning-based Augmented Reality Calibration Method for Robotic Environments using Depth Sensor Data*. In: 2020 IEEE International Conference on Robotics and Automation (ICRA), pp. 1135–1141, 2020.

[5] X. Han, J. Sin, J. Xie, N. Wang, W. Jiang, *A comprehensive Review of 3d Point Cloud Descriptors*. arXiv preprint arXiv:1802.02297, vol. 2, 2018.

[6] C. Mateo, P. Gil, F. Torres, *A Performance Evaluation of Surface normals-based Descriptors for Recognition of Objects using CAD-Models*. In: 2014 11th international conference on informatics in control, automation and robotics (ICINCO), vol. 2, pp. 428–435, 2014.

[7] P. Stancelova, E. Sikudova, Z. Cernekova, *3D Feature Detector-Descriptor Pair Evaluation on Point Clouds*. In: 2020 28th European Signal Processing Conference (EUSIPCO), pp. 590–594, 2021.

Segmentation of Infants in Depth Data:
Evaluation of Thresholding Approaches

Anne Stein [1,2], Karen Otte [2,4], Hanna Marie Röhling [2,4], Frédéric Li [3] Adeel Nisar [3], and Marcin Grzegorzek [3].

[1] Medical Informatics, Universität zu Lübeck, anne.stein@student.uni-luebeck.de

[2] Motognosis GmbH, {as, karen.otte, hanna.roehling}@motognosis.com

[3] Institute of Medical Informatics, Universität zu Lübeck, {fr.li, m.nisar, marcin.grzegorzek}@uni-luebeck.de

[4] Experimental and Clinical Research Center Max Delbrück Center for Molecular Medicine and Charité, Universitätsmedizin Berlin, corporate member of Freie Universität Berlin, Humboldt-Universität zu Berlin and Berlin Institute of Health

Abstract

The neurological development of infants can be assessed through the analysis of their Fidgety Movements. Those are small, spontaneous, repetitive movements that occur in infants from 3 to 5 months of age. Currently, they have to be assessed by a trained professional which limits their use for screening purposes. The ScreenFM project aims to explore different machine-learning-based approaches to screen babies and thus allow for an automated, instrumentalized recognition of Fidgety Movements. Therefore image and depth data of infants are collected and processed. For a clean analysis of movements, a robust segmentation of the infant is necessary. First experiments using thresholding approaches for the segmentation were evaluated. However, the depth data of lower elevations were not always clearly identifiable and thus cannot be clearly segmented.

1 Introduction

The automatic screening of newborns for early detection of neurological dysfunctions is the goal of the ScreenFM project. Infants are screened for the presence and quality of *Fidgety Movements* (FM), which occur in infants from 3 to 5 months of age. FM are spontaneous movement patterns of moderate speed with variable acceleration. They appear in neck, trunk, and limbs and take place in all directions. To observe these patterns well, the babies should lie on their backs and should be awake [1]. The presence of FM is highly associated with a child's development. If the FM are present and normal, infants usually develop as expected. In contrast, infants who do not have those movements are at high risk of neurological deficits, particularly cerebral palsy or developmental retardation [1][2]. Abnormal or sporadic FM are nonspecific and have a less predictive value. They often occur in infants later diagnosed with autism. The early screening of babies can lead to early intervention. Currently, few infants are screened for FM since specialists are needed and screening is only done in suspected cases like premature babies [1].

In order to screen more children for these movements, the ScreenFM project explores different technical approaches for the recording and detection of FM with the aim of narrowing down the most precise method.

A robust segmentation of the infant is needed to perform a clear motion analysis. One of the most established methods for segmentation is thresholding [3].

In this paper, the recording setup and different infant-segmentation approaches based on depth data are explored.

2 Material and Methods

This section describes the recording setup and the methods used for normalizing the data and segmenting the baby.

2.1 Recording Setup

During the recording sessions, a fixed setup with different sensors was used (see Fig. 1).

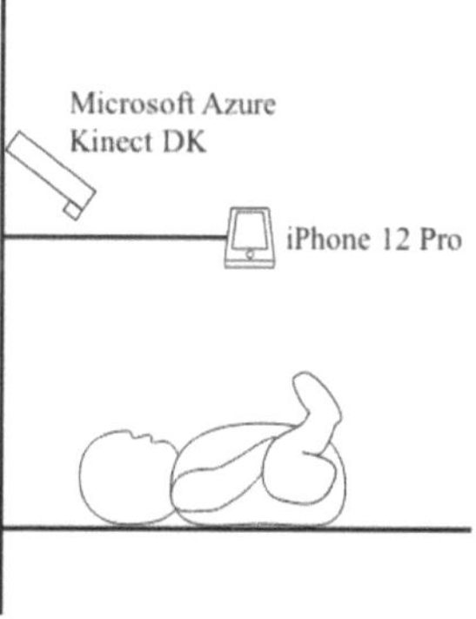

Figure 1: Recording setup with a Microsoft Azure Kinect DK and an iPhone 12 Pro.

For visual recording of the movements, a Microsoft Azure Kinect DK (hereinafter "Kinect") and an iPhone 12 Pro (hereinafter "iPhone") were utilized. Both devices were mounted above a stretcher, where an infant was laid on its back and centered within both camera frames. The Kinect recorded RGB, infrared and depth data, as well as the camera orientation as *Inertial Measurement Unit* (IMU) data. Only depth data and RGB video were recorded on the iPhone. The pixels in the depth data contain the value of the distance from the camera to its respective point in the real world measured in metric units. Depth data was captured on both the Kinect and iPhone with a dimension of 576x640 and 192x256, respectively.

Since actual data acquisition has not yet started, all methods have been applied to a single frame from a test recording with a baby puppet.

2.2 Infant Segmentation in Depth Data

In what follows, each sequential step applied to segment the depth data is presented. All methods were implemented using Python 3.8. For some methods additional packages and libraries were used, specifically numpy 1.20.3, openCV 4.5.4.58, scikit-image 0.18.3, Kinect SDK 1.4.1 and the pyk4a Kinect wrapper 1.3.0 [4].

2.2.1 Preprocessing Steps

1. **Depth data to 3D point cloud conversion** To level the sensor tilt in the image, a 3D point cloud was generated from the depth image. The Kinect SDK used with python wrapper pyk4a [4] already provided a function (depth_image_to_point_cloud()). To calculate the real world coordinates for the iPhone point cloud, the x- and y-coordinates of the image were substracted by the principal point of the camera. The result was then multiplied by the corresponding depth-value and divided by the focal length of the camera [5].

2. **Leveling sensor tilt** To normalize the input for later image processing methods, the image was aligned perpendicular to the sensor plane. For the Kinect, the tilt angles (pitch and roll) were calculated using the internal IMU data. In pitch direction (around the x-axis) the point cloud was rotated $-90°$ + pitch-angle of the sensor to be perpendicular to the sensor. To be parallel to the sensor plane in roll direction, the point cloud was rotated by the roll-angle (around the y-axis) of the sensor (see Fig. 2).

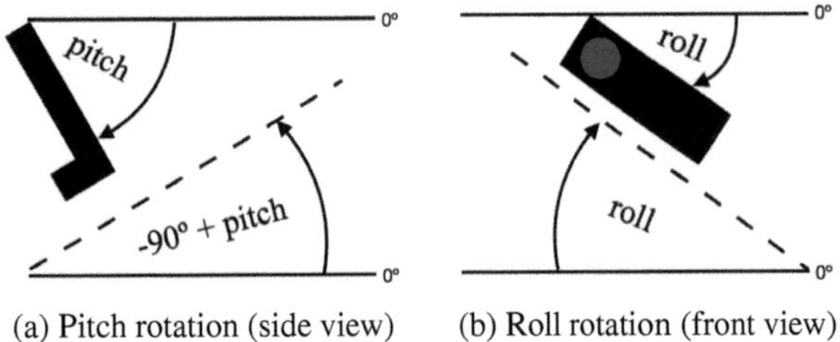

(a) Pitch rotation (side view)   (b) Roll rotation (front view)

Figure 2: Schematic rotation in pitch and roll direction.

3. **3D point cloud to depth image** Segmentation using image processing techniques requires a reverse transformation of the 3D point cloud to a depth image. The Kinect SDK already provided the corresponding function (convert_3d_to_2d()) [4]. The inverse of the operation described in step one was applied to the iPhone data. If negative coordinates were present, the coordinates were shifted to positive values to arrange it in an array. For coordinates mapped multiple times to the same pixel, the median of the depth values was calculated.

2.2.2 Infant Segmentation

A threshold approach and subsequently object detection were used to detect an infant. Different approaches are presented in the following and discussed afterwards.

1. **Thresholding**

   - **Global binary threshold** This is a static self-implemented threshold of two values (t_{low} and t_{high}). The image was divided in two classes. Pixels of the source image *src* that had a depth value between t_{low} and t_{high} got the value one in the result image *dst*, all other pixels got the value zero as shown in (1).

   $$\text{dst}(x, y) = \begin{cases} 1 & t_{low} < \text{src}(x, y) < t_{high} \\ 0 & \text{otherwise} \end{cases} \quad (1)$$

   - **Adaptive mean threshold** The threshold value was calculated by computing the mean of the local neighborhood of a pixel called *blocksize*, subtracted by an *offset* [6]. This method is implemented in the scikit-image-library ("local threshold", method="mean") [6] and in the openCV library ("adaptive mean threshold") [7]. The openCV implementation, however, could only be applied to arrays with a range of values from 0 to 255 which required a depth-value-scaling first.

   - **Adaptive gaussian threshold** The threshold value was calculated by the weighted sum (cross correlation with a Gaussian window) of a local neighbourhood, the *blocksize*, subtracted by an *offset* [7]. This method is also part of the scikit-image-library ("local threshold", method="gaussian") [6] and the openCV library ("adaptive gaussian threshold") [7]. Depth values had to be scaled for the openCV implementation as mentioned in the previous method.

2. **Object detection** To segment the infant the correct area had to be detected and cropped. Considering that, the contours of the segmented areas were first determined using the openCV method findContours() [8]. Then, the center of each contour was determined and the Euclidean distance to the center of the image was calculated. The contour closest to the center of the image which is bigger than 1500 px was assumed to be

the contour of the baby and cropped. The size condition was added to avoid accidentally segmenting tiny blobs which might be closer to the image center than the infant. This method can only be used with the recording setup described in Section 2.1 or similar and is not universal.

3 Results and Discussion

The results of the methods described in the previous section are presented and discussed in this section.

3.1 Preprocessing

Since the camera intrinsics of the iPhone were not yet available, the preprocessing could only be applied to the Kinect data. It was noted that even after leveling the sensor tilt, not all points had the same distance to the sensor. The difference between the closest and furthest point were still about 70 mm, although 0 mm would have been expected in an ideal world. In the future, it could be investigated whether this can be further improved by using the distortion information from the camera intrinsics.

In order to apply image processing techniques, the point cloud was converted back to a depth image. To verify results of the method presented in Section 2.2.1, the unprocessed 3D point cloud was converted back to a depth image. Calculating the difference between original depth image and generated depth image was zero at each point of the image. Accordingly, the conversion back to depth images can be applied without quality degradation.

When applying a rotation of the point cloud, however, the transformation caused some gaps. Those could be closed by a dilation with a 3x3 kernel (see Fig. 3(a) and 3(b)).

(a) Before dilation   (b) After dilation

Figure 3: Section of depth image after point cloud rotation and reverse transformation to a depth image.

3.2 Infant Segmentation

The focus of the segmentation was to keep the outer shape of the infant and to not cut any body parts off. The threshold values were determined manually by incremental adjustments. For this purpose, coarser parameter combinations (t_{low}, t_{high} within a range of 500–900, *blocksize* within a range of 10–200, both in steps of 20 and *offset* within a range of -30–0 in steps of 1) were first applied. The best combination was then adjusted in smaller steps. In the end, selection was based on visual impression. In the future, however, automated methods for determining the parameters (e.g., by comparing with ground truth segmentation)

have to be evaluated due to the amount of data.

The segmentation of the iPhone data was performed directly on the recorded depth data. Additionally those depth values were first scaled to millimeters for data normalization. The results of the global binary threshold (Fig. 4(b) and 5(b)) have achieved the least accurate segmentation. Nevertheless, this method was the only one that keeps both arms and hands at least implied for Kinect data.

Since the puppet had smaller depth values than the stretcher, the adaptive thresholds used negative offset values to increase the threshold value and cut less parts of the puppet off.

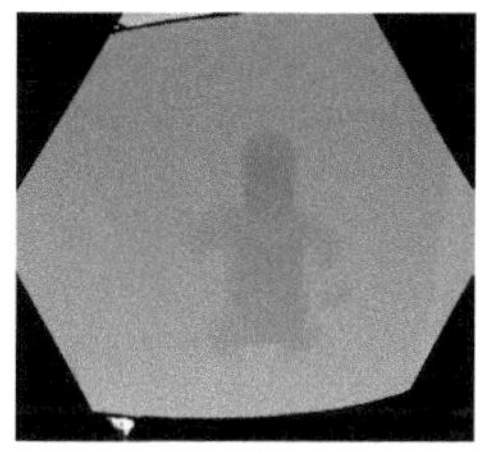
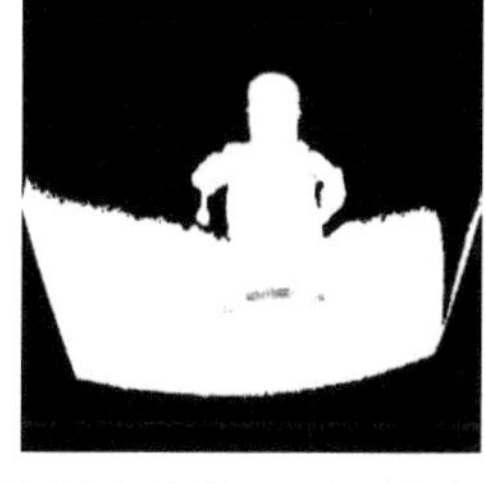

(a) Kinect depth recording   (b) Global binary threshold
$t_{low} = 580, t_{high} = 690$

(c) Local threshold (mean)   (d) Local threshold (gaussian)
blocksize=101, offset=-1   blocksize=171, offset=-23

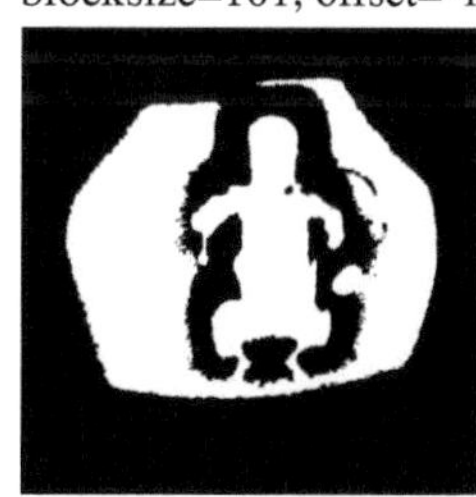
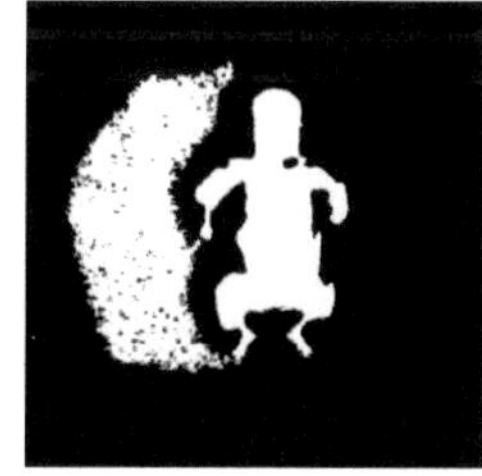

(e) Adaptive mean threshold   (f) Adaptive gaussian threshold
blocksize=125, offset=-4   blocksize=175, offset=-1

Figure 4: Different threshold approaches and object detection applied to Kinect data after preprocessing steps.

The adaptive mean threshold (Fig. 4(e) and 5(e)) as well as the adaptive gaussian threshold (Fig. 4(f) and 5(f)) of the openCV implementation produced rather inaccurate results with a higher segmentation of the background. When selecting the parameters, it was noticed that especially the segmentation of the limbs caused problems. If these were included, more background was added to the segmentation. The implementations of the scikit-image-library ("local threshold") for the adaptive mean threshold (Fig. 4(c) and 5(c)) and the adaptive gaussian threshold (Fig. 4(d) and 5(d)), on the other hand, segmented the puppet most accurately. Nevertheless, even with this approach, one arm was not visible in each case. This arm could not be segmented successfully with any parameter combination. Missing limbs may have to be completed by adding segmented RGB data or by using other object detection techniques in the future. The gap in the neck area was probably caused by a medical tube on the puppet and should not occur in real data.

(a) iPhone depth recording

(b) Global binary threshold
$t_{low} = 535, t_{high} = 805$

(c) Local threshold (mean)
blocksize=31, offset=-0.03

(d) Local threshold (gaussian)
blocksize=45, offset=-0.01

(e) Adaptive mean threshold
blocksize=29, offset=-0.01

(f) Adaptive gaussian threshold
blocksize=55, offset=-0.01

Figure 5: Different threshold approaches and object detection applied to raw iPhone depth data.

Although the scikit-image implementations performed the same operations as the ones of openCV, their results differed in terms of accuracy. This could be related to the preceding scaling needed for the openCV implementations. Overall, the resolution of the Kinect depth data was substantially more accurate than the one of the iPhone. This could be seen, for example, on the puppet's left arm, which was not mapped in the depth data (see Fig. 5(a)) and therefore could not be segmented correctly without adding RGB data. Nevertheless, the Kinect and iPhone data should be usable for a segmentation leading to a clean feature extraction.

4   Conclusion

Overall, various methods for segmenting infants in depth data have been explored and promising results have been achieved by the presented approaches. However, improvements are needed, especially in the segmentation of limbs. The next steps for the ScreenFM project will be the optimization of the segmentation and testing with real data followed by the feature extraction for movement analysis. In addition, it is necessary to evaluate which combination of sensors is most suitable for recording and detecting FM.

Acknowledgement

This work has been carried out at Motognosis GmbH and supervised by the Institute of Medical Informatics, Universität zu Lübeck. The ScreenFM project is funded by the German Federal Ministry of Education and Research.

Author's Statement

Anne Stein and Hanna Marie Röhling are paid employees at Motognosis GmbH. Karen Otte is stakeholder of the company. All other authors state no conflict of interest.

5   References

[1] C. Einspieler, R. Peharz, P. B.Marschik, *Fidgety movements – tiny in appearance, but huge in impact.* In: Jornal de Pediatria (Versão em Português), Volume 92, pp. S64–S70, 2016.

[2] H. FR Prechtl, C. Einspieler, G. Cioni, A. F. Bos, F. Ferrari, D. Sontheimer, *An early marker for neurological deficits after perinatal brain lesions.* In: The Lancet, Volume 349, Issue 9062, pp. 1361–1363, 1997.

[3] N. Singh, S. Goyal, *Determination and Segmentation of Brain Tumor Using Threshold Segmentation with Morphological Operations.* In: Pant M., Ray K., Sharma T., Rawat S., Bandyopadhyay A. (eds) Soft Computing: Theories and Applications. Advances in Intelligent Systems and Computing, Volume 584. Springer, Singapore, 2018.

[4] pyk4a, 08/2021, *pyk4a source code.* Available: https://github.com/etiennedub/pyk4a [last accessed on 2022-01-15]

[5] Apple Documentation ARKit, n.d., *Displaying a Point Cloud Using Scene Depth.* Available: https://developer.apple.com/documentation/arkit/ environmental_analysis/displaying_a_point_cloud_using_ scene_depth [last accessed on 2022-01-02].

[6] scikit-image development team, n.d., *Module: filters.* Available: https://scikit-image.org/docs/stable/ api/skimage.filters.html [last accessed on 2021-12-30].

[7] openCV documentation, 12/2021, *Miscellaneous Image Transformations.* Available: https://docs.opencv.org/4.x/d7/d1b/group__imgproc__ misc.html#gaa42a3e6ef26247da787bf34030ed772c [last accessed on 2021-12-30].

[8] openCV documentation, 01/2022, *Structural Analysis and Shape Descriptors.* Available: https://docs.opencv.org/4.x/d3/dc0/group__imgproc__ shape.html#gadf1ad6a0b82947fa1fe3c3d497f260e0 [last accessed on 2022-01-04].

Implant positioning with an automated landmark regression in 3D CBCT scans using CNNs

Ziad Al-Haj Hemidi [1], Hrvoje Starcevic [3], Mladen Berekovic [2], and Javad Ghofrani [2],

[1] Medical Informatics, Universität zu Lübeck, ziad.alhajhemidi@student.uni-luebeck.de
[2] Institute of Computer Engineering, Universität zu Lübeck, firstname.lastname@uni-luebeck.de
[3] Starcevic Dental Practice, Benkovačka 2a 10000 Zagreb (Croatia), hrvoje@drstarcevic.hr

Abstract

Planning implant treatments requires special expertise which is met differently by each dentist. This procedure is time consuming and prone to human error. The purpose of this work is to design, develop, and evaluate a method to automatically annotate important landmarks using Convolutional Neural Networks (CNNs) to aid the process implant treatment planing. The method is based on 3D heatmap regression with the use of a fully-convolutional neural network. It takes 3D patches of the Cone Beam Computed Tomography (CBCT) volumes and results in 3D heatmaps. The landmark coordinates are determined by the heatmaps. Quantitative evaluation was performed using the point-to-point error for each landmark. With a mean point-to-point error of 88.89 ± 11.92 mm the results presented a high detection error, however, the work proposed a promising technique making the process of automated implant treatment planning feasible.

1 Introduction

Dental implants, also known as endosseous implants, are surgical components that affiliate with bone tissue of the jaw. They are used to support dental prostheses e.g., crowns and bridges. Planning the treatment procedure takes the dentist to manually determine vertical and horizontal dimensions from static landmarks of the skull. This is prone to human error, quiet difficult and time consuming. Such work could be accelerated and automated with a combination of a low dose imaging technique and sophisticated computer algorithms. Cone Beam Computer Tomography (CBCT) is a widely used modality in the field of orthodontics due to low radiation exposure and low time consumption [1]. It enables fast and precise measurements of skull dimensions making it suitable for this kind of task. Deep Learning (DL), on the other hand, showed promising results in the field of medical image processing in the last decade [2] making it a fitting technique to automate this work. The jaw dimensions are determined from static points, hence a technique for automatic landmark localization could be researched where dimensions correspond to distances between the landmarks. This paper proposes a method for automated landmark localization in 3D CBCT volumes using 3D heatmap regression with Convolutional Neural Networks (CNNs). The used materials and methods will be presented in section 2. Conducted experiments and obtained results will be documented in section 3. The work will be briefly discussed in section 4 and finally, section 5 will conclude this project and give a future outlook.

2 Material and Methods

In this work twenty-four 3D CBCT volumes with a full skull view are acquired, annotated and preprocessed. The acquisition is deployed by the dentists involved in this work and saved as $DICOM^1$ files. Due to space limitations no CBCT scan is illustrated, but can be observed in [1]. In total, nineteen landmarks are manually annotated using the medical imaging software $Slicer^2$ for supervised learning. The landmarks are positioned as follows: 1. Anterior nasal spine, 2. Point A, 3. Point B, 4. Pogoinion, 5. Gnathion, 6. Menton, 7. Foramen mentale right side, 8. Foramen mentale left side, 9. Gonion right, 10. Gonion left, 11. Condylus lateral right, 12. Condylus lateral left, 13. Proccessus Coronoideus right, 14. Proccessus Coronoideus left, 15. Lateral Zygomatic right, 16. lateral Zygomatic left, 17. Posterior nasal spine, 18. Orbitale inferius right, and 19. Orbitale inferius left.

As a preprocessing step the volumes are resized to a unit size of $250 \times 250 \times 250$ with a voxel spacing of $0.8 \times 0.8 \times 0.8$ and bone masked using Otsu's [3] thresholding method.

The proposed method is a 3D approach to an exercise of the lecture of Advanced Methods of Medical Image Analysis [4]. Starting the training with randomly drawing an initial shape (a shape is the coordinates of all landmarks in one volume) in each iteration, a landmark offset is determined by subtracting the initial shape from the other ones. With the landmark offset, target heatmaps are then generated using a Gaussian distribution with a mean in the target land-

---

¹https://www.dicomstandard.org
²https://www.slicer.org

mark. A D-dimensional heatmap image $G_i(x) : \mathbb{R}^D \to \mathbb{R}$ of a target landmark L_i, where $i = \{1, \ldots, N\}$ and N is the total number of annotated landmarks with coordinate $\overset{*}{\mathbf{x}}_i \in \mathbb{R}$, is defined as the following Gaussian distribution:

$$G_i(\mathbf{x}; \sigma_i) = \frac{\gamma}{(2\pi)^{D/2}\sigma_i^D} exp(-\frac{||\mathbf{x} - \overset{*}{\mathbf{x}}_i||_2^2}{2\sigma_i^2}). \quad (1)$$

Outputs of (1) result in high value heatmap pixels that are near to the target coordinate $\overset{*}{\mathbf{x}}_i \in \mathbb{R}$, and smoothly, but rapidly decrease as further the heatmap pixels get from the coordinate $\overset{*}{\mathbf{x}}_i$. The parameter γ acts as a scaling factor to avoid numerical instabilities during training due to otherwise very small values. The standard deviation σ_i defines the peak width of the Gaussian function in the heatmap image for a landmark L_i.

The CNN in Fig. 1 resembles a Fully Convolutional Neural Network (FCNN) [5] that enables image to image (heatmap) mapping. CNNs that directly regress landmark coordinates e.g., [6] require dense layers with many network parameters to model image to coordinate mapping, hence the proposed method is based on regressing heatmap images [7], which result in the pseudo-probability of a landmark located at a certain pixel position.

A stack of N 3D patches centred at the landmarks L_i is given as input to the CNN, resulting in N heatmaps H_i. Narrower heatmap peaks are predicted where the CNN is confident that the prediction is correct, and wider peaks for more uncertain ones. A regression loss between H_i and G_i is then backpropagated through the CNN and optimized. The final coordinate $\hat{x}_i \in \mathbb{R}^D$ of each landmark L_i is then obtained by taking the coordinate where the predicted heatmap takes the highest value. The regression losses used, are the mean squared errors (MSE), and the AWing loss proposed by [8]. For optimization, two different optimizers are chosen, the first one is Adam [9] and the second is Lamb [10].

Lastly, the predictions are evaluated using conventional metrics from the literature. The point-to-point error for each landmark L_i in a volume v is defined by the Euclidean distance between the predicted coordinate $\hat{x} \in \mathbb{R}^D$ and the target coordinate $\overset{*}{\mathbf{x}} \in \mathbb{R}^D$.

For a landmark L_i in a volume v, the point-to-point error is defined as:

$$ppE_i^{(v)} = ||\overset{*}{\mathbf{x}}_i^{(v)} - \hat{\mathbf{x}}_i^{(v)}||_2. \quad (2)$$

The number of predicted landmarks L_i which fall outside a certain error radius called r for all volumes is defined as $numOut_r$ the number of outliers that is defined as follows:

$$numOut_r = |\{(v, i)|ppE_i^{(v)} > r\}|. \quad (3)$$

For all landmarks of a single volume v, a volume specific point-to-point error $vppE^{(v)}$ is be defined as:

$$vppE^{(v)} = \frac{1}{N} \sum_{i=1}^{N} ppE_i^{(v)} \quad (4)$$

where N is the total number of all annotated landmarks in a volume. This is different from the $numOut_r$, since the

number of outliers gives a general insights on the total number of outliers in all volumes, and the $vppE$ shows the number of volumes in which outliers of a certain error are.

3 Experiments and Results

For training the data was prepossessed and saved as tensors with shape $18 \times 1 \times 250 \times 250 \times 250$. Six images were discarded due to missing information in the DICOM headers which led to errors while reading the volumes.

The data was split in train and test sets using k-fold with a k of five and shuffled data, that resulted in a train set with size $15 \times 1 \times 250 \times 250 \times 250$ with the corresponding fifteen landmark sets and a test set with size $3 \times 1 \times 250 \times 250 \times 250$ with the corresponding three landmark sets.

The training set was used as input to the proposed method where in each iteration 3D-Patches with size $batchsize \times 19 \times 23 \times 23 \times 23$ were extracted and the landmark offset was obtained by selecting a random initial shape and subtracting it from the other available shapes. Using the Gaussian distribution (1) and the landmark offset, the target heatmaps were generated, where in the case of MSE, γ was set empirically to be twenty to avoid small errors, and in the case of the AWing γ was set to one. The heatmap dimensions of the targets and the networks prediction were $batchsize \times 19 \times 17 \times 17 \times 17$.

Random rotations in a range of $[-45°, 45°]$ were applied on the available data on the fly during training, but also a small random translation was added to the landmark offset in each iteration, making it more robust against rotations and translations.

For experiments five trials were set up using random configurations drawn from a predefined set, in which the learning rates, dropout probabilities, batchsizes, number of epochs, loss functions, and optimizers are set with different possible initialization. For the learning rate, a linear distribution between 10^{-4} and 10^{-2} was configured. For the batchsize a random choice between $3, 5$, and 15. For the number of epochs a choice between $50, 100$, and 150, the loss function was set to be chosen as either MSE loss or AWing loss and the optimizer was set with a choice between the Adam or the Lamb optimizers. This gave an overview of the best parameter combination possible and discarded the ones which are counterproductive.

The framework *ray-tune*[3], which included a integration for the used DL framework *PyTorch*[4], was used for experiments. Non-converging trials were stopped with a scheduler not to waste resources. Trial two, four, and five were stopped early in the first few iterations. Trial three was stopped after 64 out of 150 epochs due to the non coverage of the loss. The execution time for the whole experiment was 75 minutes.

The trials were compared by the ability to converge the loss to a minimum. The five trials were drawn randomly from the set as shown in Table 2. Trial one was chosen, since it

---

[3] https://docs.ray.io/en/latest/tune/index.html
[4] https://pytorch.org/docs/stable/index.html

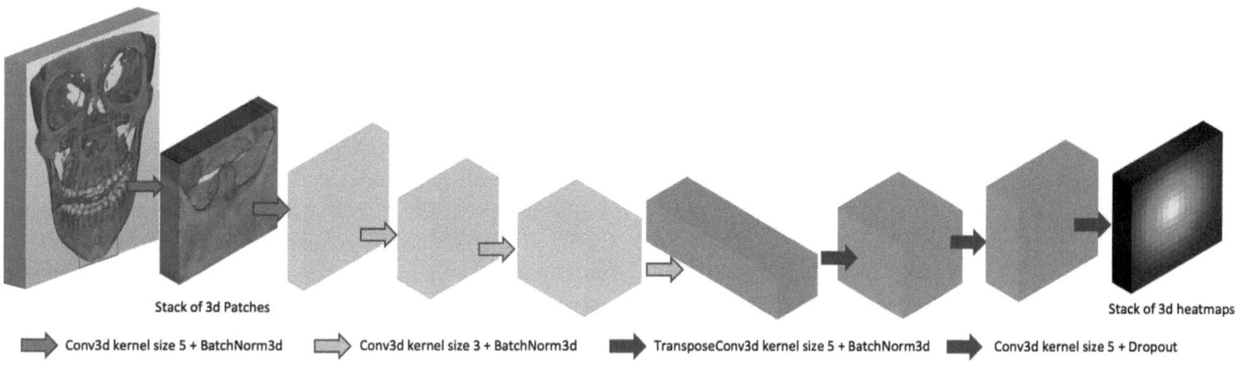

Figure 1: The architecture of the used FCNN in this work is illustrated here. The input of the network, as shown on the left side, is a stack of 3D patches centered at each landmark. In the contracting path the 3D patches are first convolved with 3D kernels of size five and then convolved three times with kernels of size three. The bottleneck is upsampled using two 3D transposed convolutions with kernels of size five. All the mentioned convolutions and transposed convolutions are followed by 3D batch normalizations and activation functions. The last convolution comprises a kernel of size five and a dropout layer outputting a stack of 3D heatmaps for each landmark.

Table 1: A specific volume point-to-point errors was calculated for each volume of the test dataset over all the landmarks and measured in millimeters.

Volume	vppE
Volume I	28.4256
Volume II	28.6197
Volume III	31.8401

converged to a minimum loss of 1.16194. The CNN was applied on the test data and the evaluations were performed on the outputs. Table 4 lists the 3D point-to-point errors (2) for

Table 2: Here the different randomly drawn configuration sets for each of the five trials are presented. The number of epochs is defined by #epochs.

Trial	dropout rate	learning rate	optimizer	loss function	batchsize	#epochs
Trial one	0.34	0.005	Adam	AWing	5	150
Trial two	0.40	0.007	Adam	AWing	15	100
Trial three	0.41	0.005	Adam	MSE	5	150
Trial four	0.25	0.006	Adam	AWing	15	100
Trial five	0.21	0.001	Lamb	AWing	15	50

each landmark. Resulted predictions example in comparison to the manual annotations is presented in Fig. 2. Landmark outliers (3) which lied outside certain radii that were set empirically to be $80, 100, 120$, are documented in Table 3. The volume specific point-to-point errors $vppE$ (4),

Table 3: The number of outlier landmarks are listed for three empirically set radii.

$numOutliers$		
$r = 80$	$r = 100$	$r = 120$
13(68%)	7(37%)	2(10%)

were obtained from the three test volumes for all the landmarks and are presented in Table 1.

Table 4: The 3D ppE for each landmark over the training volumes are shown here. With R/L, meaning the right an left side of the skull.

Landmark	3D		
	mean	median	std
Anterior Nasal Spine (ANS)	86.9175	87.4760	1.6323
Point A	77.3507	74.2458	5.5424
Point B	88.2892	88.1603	3.5914
Pogonion (Pg)	107.6562	98.3552	17.9674
Menton (Me)	110.7312	113.1427	10.5960
Gnathion (Gn)	102.5895	103.9153	4.2894
Foramen mentale (Fm) R	64.1969	61.0713	12.1732
Foramen mentale (Fm) L	118.9408	118.4374	14.3572
Gonion (Go) R	59.5152	67.8498	17.5949
Gonion (Go) L	128.0890	126.5877	3.3382
Condylus laterals (Co) R	82.1980	85.7260	14.6827
Condylus laterals (Co) L	77.6545	78.4632	2.0841
Processus Coronoideus (PrCo) R	59.6544	51.8158	14.4527
Processus Coronoideus (PrCo) L	87.9949	91.2897	5.9654
Lateral Zygomatic (Zyg) R	83.8939	88.3786	8.0392
Lateral Zygomatic (Zyg) L	103.6899	95.5512	35.4361
Posterior Nasal Spine (PNS)	37.0022	33.7897	17.6829
Orbitale inferius (Or) R	86.1056	86.4382	5.0507
Orbitale inferius (Or) L	126.3541	115.7003	31.9516
Average	88.8855	87.7050	11.9173

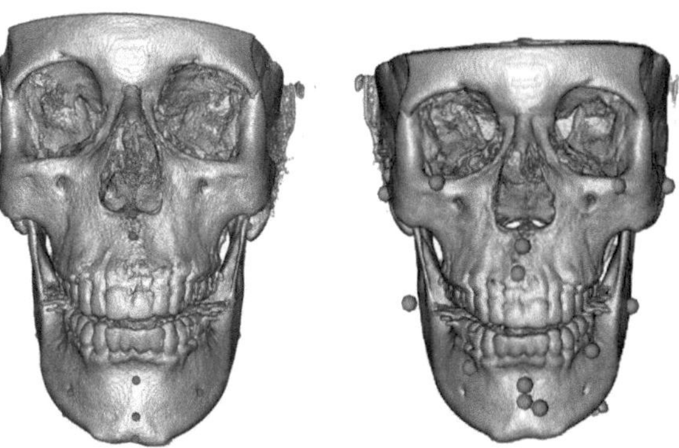

Figure 2: On the left side, the manually annotated target landmarks on the test data are shown. The right side shows the automatic annotated landmarks of the proposed method in the anterior view.

4 Discussion

For this work the acquisition of enough data was laborious because of limited availability and ethical constraints, but also because later unusable scans were removed resulting in using 18 final volumes. A situation occurred while reading in the images where the landmark positions did not correspond to the volume positions due to the different coordinate systems. DICOM data is usually presented in the LPS (Left-Posterior-Superior) coordinate system and the used software *Slicer* uses RAS (Right-Anterior-Superior). Header tags in the DICOM data should contain information about voxel spacing, slice depth, voxel origins, and the affine transformation matrix to transform the data into the volume coordinate system with the first voxel centered in position $(0, 0)$. Unfortunately, in this case, the affine transformation matrix was not suitable to convert the data correctly into the voxel coordinate system. During testing, a problem caused by the conversion between the aforementioned coordinate systems arose, where the predictions did not correspond spatially to the test volumes and two 90 degrees rotations had to be applied in the RL dimension and the PA dimension for spatial alignment. Time limitations led to the incapacity of solving aforementioned problem, but it will be corrected in future work. The visual results in Fig. 2 showed that landmark correlations were learned and the shape regression performed well in learning the spatial relations and alignments between the landmarks. The visualizations also demonstrated that the method still lacks the performance of finding the corresponding positions on the volumes as demonstrated by the quantitative evaluation with a mean ppE of 88.89 ± 11.92 mm in 3D. The number of outliers showed how many landmarks lie outside a certain error which gave insights on the tendency to correctly localize the landmarks. Employing the volume specific point-to-point error to observe possible volumes that were insufficiently annotated was obtained showing that the performance for all tested volumes did not vary a lot, where the best volume had a $vppE$ of 28.4 mm and the worst performing volume had a $vppE$ of 31.84 mm.

5 Conclusion

A technique for automated regression of nineteen anatomical landmarks in 3D CBCT scans was proposed and evaluated. On one hand, the results demonstrated that this method was able to learn landmark correlations, but on the other hand, the evaluation showed that the technique performed inadequately on localizing the landmarks in the volumes due to a small dataset and the mentioned coordinate system issues.

However, better results could be achieved, if more volumes are obtained, annotated, and transformed properly. Whilst, the usage of some techniques like elastic deformations or the employment of synthetic data generation methods could enrich the dataset. At the same time, having more data will enable splitting the dataset into training, validation, and test sets. Introducing such a validation set and defining a proper

metric could enable better convergence and avoid possible overfitting during the training process. Further experiments in this work will be conducted by deploying more research and obtaining more real or synthetic data, and fine-tuning the proposed method. With the mentioned improvements, this work will automatically localize important landmarks with high precision. This enables the automatic measurement of vertical and horizontal jaw dimensions, which is still obtained manually by dentists, aiding the laborious process of implant treatment planning through shortening the time consumption and lowering the expenses.

Acknowledgement

The work has been carried out at the Institute of Computer Engineering, Universität zu Lübeck.

Author's Statement

Conflict of interest: Authors state no conflict of interest.

6 References

[1] M. Kumar, M. Shanavas, A. Sidappa, and M. Kiran, "Cone beam computed tomography - know its secrets," *Journal of international oral health : JIOH*, vol. 7, pp. 64–8, 02 2015.

[2] A. Esteva, K. Chou, S. Yeung, N. Naik, A. Madani, A. Mottaghi, Y. Liu, E. J. Topol, J. Dean, and R. Socher, "Deep learning-enabled medical computer vision," *NPJ Digital Medicine*, vol. 4, 2021.

[3] N. Otsu, "A threshold selection method from gray-level histograms," *IEEE Transactions on Systems, Man, and Cybernetics*, vol. 9, no. 1, pp. 62–66, 1979.

[4] M. Heinrich, "Fvmbv-advanced methods of medical image analysis exercise sheet on heatmap cnns," Accessed: February 2022. [Online]. Available: https://moodle.uni-luebeck.de/pluginfile.php/371864/mod_resource/content/0/Exercise6_Sheet.pdf

[5] J. Long, E. Shelhamer, and T. Darrell, "Fully convolutional networks for semantic segmentation," 2015.

[6] Q. Ma, E. Kobayashi, B. Fan, K. Nakagawa, I. Sakuma, K. Masamune, and H. Suenaga, "Automatic 3d landmarking model using patch-based deep neural networks for ct image of oral and maxillofacial surgery," *The International Journal of Medical Robotics and Computer Assisted Surgery*, vol. 16, no. 3, p. e2093, 2020. [Online]. Available: https://onlinelibrary.wiley.com/doi/abs/10.1002/rcs.2093

[7] J. J. Tompson, A. Jain, Y. LeCun, and C. Bregler, "Joint training of a convolutional network and a graphical model for human pose estimation," in *Advances in Neural Information Processing Systems*, Z. Ghahramani, M. Welling, C. Cortes, N. Lawrence, and K. Q. Weinberger, Eds., vol. 27. Curran Associates, Inc., 2014. [Online]. Available: https://proceedings.neurips.cc/paper/2014/file/e744f91c29ec99f0e662c9177946c627-Paper.pdf

[8] X. Wang, L. Bo, and L. Fuxin, "Adaptive wing loss for robust face alignment via heatmap regression," 2020.

[9] D. P. Kingma and J. Ba, "Adam: A method for stochastic optimization," 2017.

[10] Y. You, J. Li, S. Reddi, J. Hseu, S. Kumar, S. Bhojanapalli, X. Song, J. Demmel, K. Keutzer, and C.-J. Hsieh, "Large batch optimization for deep learning: Training bert in 76 minutes," 2020.

Pose-Based Control in Real-World Robotic Application

Ron Keuth [1], Alexander Bigalke [2], and Mattias Heinrich [2]

[1] Medical Informatics, Universität zu Lübeck, ron.keuth@student.uni-luebeck.de,
[2] Institute of Medical Informatics, Universität zu Lübeck, {alexander.bigalke, mattias.heinrich}@uni-luebeck.de

Abstract

Deep neural networks set a new state-of-the-art in a variety of computer vision tasks. However, their performance is often only examined on benchmark datasets with the lack of an implementation in real-world application. In this paper, we investigate therefore the use of deep learning in an image- and point cloud-based pose control on a Jackal robot. Without any suitable annotated data for the robot's domain, we separate the pose estimation from its classification to use a pretrained keypoint R-CNN for the image-based pose estimation. We design a set of handcrafted features, which allow the use of a simple ML estimator for the pose classification, yielding a remarkable weighted average F1-score of 0.9419. We train a DGCNN for the point cloud-based pose estimation examining projected 2D keypoints and a synthetic dataset as possible sources for supervision.

1   Introduction

There are huge application fields, including the medical domain, where robotics could be used to support humans at various monotonous or physically demanding tasks. However, to do this, the robot must be able to interpret its environment and to take commands in an intuitive way, e.g., by voice or via gestures. This can be achieved due to the recent development in the field of computer vision especially of artificial neural networks and the gain in computational power of mobile robots. However, the most proposed new methods in the literature only achieve impressing results on benchmark datasets with a lack of an evaluation in a real-world application.

In this work, we implement a pose-based robot navigation on a small robot using deep learning-based computer vision on images and point clouds. Because of the lack of appropriate annotated data for an end-to-end training approach, we present an alternative pose classification pipeline consisting of a pose estimation and a joint-based classifier. This separation allows us to use a pretrained keypoint Region-Based Convolutional Neural Network (R-CNN) [1] for the pose estimation, that shows excellent off-the-shelf performance in our robotic domain. We use the pose estimation's joints in handcrafted features followed by a Linear Discriminant Analysis (LDA) for the pose classification. Each detected pose triggers its associated driving command for the robot. We train and evaluate our pipeline using a self recorded dataset and achieve good performance, even with limited training data available.

Because of the drawbacks of our image-based pose classification like the lack of geometric scene information, distorted features due to body alignment, and possible conflicts with the strict data policies in the medical domain [2], we want to expand our approach to point clouds. For a point cloud-based pose estimation, we train a Dynamic Graph CNN (DGCNN) [3]. However, contrary to the R-CNN, a pretrained model is not available and, thus, we have to deal with the lack of annotations. We experiment with two approaches for inferring the ground truth from other data to overcome this issue. The first approach maps the 2D joint predictions to the nearest valid point of the point cloud. The second approach uses the SURREAL dataset [4] for training and validates its performance on the robot's point cloud. While SURREAL introduces a domain gap to our robotic domain, training on the mapped 2D joints yields qualitatively satisfying results and presents a first step towards a point cloud-based pose estimation on the robot.

2   Material and Methods

2.1   The Jackal Robot

The robot we chose is the Jackal model from Clearpath Robotics, which is a small unmanned ground vehicle. It consists of a metal chassis, four independent motors for the wheels, and a computer with Ubuntu 16 and Robot Operating System (ROS) melodic preinstalled. Furthermore, the installed stereo camera provides both RGB images and point clouds of the environment. The NVIDIA GTX 1050 4GB allows fast inference with deep neural networks.

2.2   Image-Based Pose Classification

Our pose-based control pipeline takes an image and maps it on one of five body poses (see Fig. 1). Every pose is associated with a simple driving command for the robot, e.g. drive for-/backwards, turn right/left. Our general pipeline for pose-based control consists of two elements: the pose

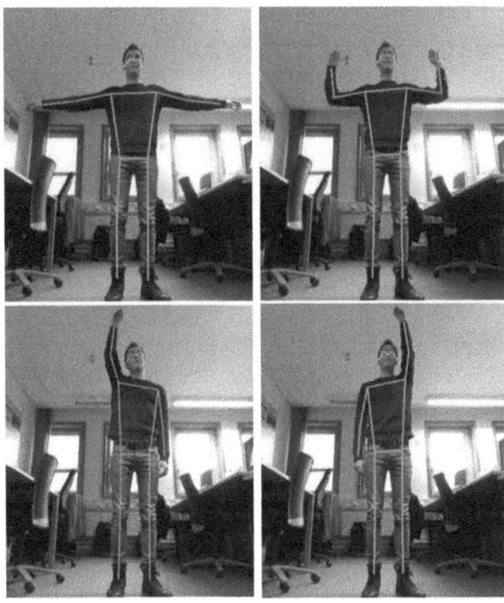

Figure 1: Visualization of the four poses: T-, double L-, and right/left I-pose. The keypoint estimation by the R-CNN is shown as skeleton.

estimation and pose classification. We use a pretrained keypoint R-CNN, that was trained on the COCO [5] dataset, which performs excellently on the robot's images. We then use the predicted keypoints to extract two types of handcrafted features, which are used by a Maximum Likelihood (ML)-estimator for the pose classification.

2.2.1 Handcrafted Features

Given the predicted keypoints $X \in \mathbb{R}^{N \times M}$ from the R-CNN, where N determines the number of keypoints with M dimensions (here $M = 2$), we extract two types of handcrafted features.
The first feature type is the Euclidean distance between all N keypoints in the distance matrix $D \in \mathbb{R}^{N \times N}$ with entries

$$d_{ij} = ||X[i] - X[j]||_2 \cdot \frac{1}{2.3 \cdot d_{\text{torso}}} \quad (1)$$

We want to normalize the values of D by the body size of the person and therefore calculate the torso size d_{torso} as the norm of the vector between the center points of the two shoulder and hips keypoints. Multiplying the torso size with a factor of 2.3 is a reasonable approximation of the actual body size in our data. Normalizing with the body height has the advantage to be independent of the person's current pose.
As the second feature type, we use the angle of the two shoulder and elbow joints. A joint is modeled via two vectors $\vec{u}$ and $\vec{v}$. We use their dot product for the calculation of its angle φ and normalize it by dividing with π to obtain roughly the same value space like our normalized Euclidean distance.

$$\varphi = \arccos\left(\frac{\vec{u} \cdot \vec{v}}{||\vec{u}|| \cdot ||\vec{v}||}\right) \cdot \frac{1}{\pi} \quad (2)$$

For example, the left shoulder joint is represented by two vectors, pointing from the left shoulder's to the left elbow's and to the right shoulder's keypoint, respectively.

2.2.2 Pose Classification

All features are concatenated to a vector $\vec{f} \in \mathbb{R}^F$ (here $F = 49$). We use the four joint angles of shoulders and elbows and only the lower triangle matrix of the Euclidean distance matrix D because of its symmetry and its zero main diagonal. Based on $\vec{f}$, we classify a pose using a Maximum Likelihood (ML) estimation. Although an ML estimation is a naive classification approach estimating a parametric density p (here Gaussian with the same covariance matrix) for each class over the training data, we argue that our pose class ω should be sufficiently separable from each other due to our features. During inference, the classification is

$$\text{class}(\vec{f}) = \arg\max_j p(\omega_j|\vec{f}) \quad (3)$$

We do not use a maximum a-posterior estimation because the estimated priors over our dataset would mislead due to its class imbalanced and in addition, we want to consider every pose class with the same prior. For preprocessing, we use an LDA to reduce the dimensionality of the feature space by projecting it to its most discriminative directions to increase its linear separability.

2.2.3 Dataset and Evaluation

We fit the ML estimator using 154 images from four different people, captured ourselves with the Jackal robot. The model is validated via five fold cross-validation using the F1-score per class. The normalization and high abstraction of our features allow us to consider this relatively small amount of data as sufficient. We use the same argument for the possible lack of disjointness (one person could be in more than one of the five folds).

2.3 Point Cloud-Based Pose Estimation

We use the DGCNN for a pose estimation on point clouds to extend our pose-based control pipeline to 3D. The DGCNN's architecture is based on edge convolutions, which capture local geometric structures in point clouds using a knn-graph and combining in this way the local and global features of point clouds. For an estimated pose $Y \in \mathbb{R}^{K \times 3}$ on a point cloud $P \in \mathbb{R}^{N \times 3}$, all K joints are modeled via the learned softmax-normalized weight map $W^{N \times K}$. So in the end, every k-th joint y is just a weighted linear combination of all points in the point cloud.

$$y_k = \sum_{i=1}^{N} \vec{p_i} \cdot w_{ik} \quad (4)$$

2.3.1 Dataset

Because there exists no suitable dataset for the Jackal robot's point clouds, we train the DGCNN from scratch. Therefore, we collected data ourselves by using the Jackal's stereo camera. The associated software kit provides a point cloud, that is registered to the left camera frame. In the end, our dataset contains three sequences of two people; two used for training with 4066 frames in total and one with 1027 frames for testing.

Because our data lack of ground truth and due to the time consuming of manually annotating them ourselves, we generate pseudo annotations using a pretrained keypoint and mask R-CNN. For each frame, the predicted segmentation mask is used to cut out the human part of the overall scene point cloud, and the predicted keypoints are then mapped to the nearest valid points of the point cloud. We use a subset of the SURREAL [4] for a second source of supervision for training and validates its performance on the Jackal's point cloud.

3 Results and Discussion

The Jackal's hardware turned out to be capable of running our pose based control pipeline at around three frames per second. We consider this processing speed as fast enough to react in real-time, still with room for improvement in the efficiency of our implementation. We are happily surprised how robust the pose estimation of the pretrained R-CNN works, even with the domain gap due to the frog perspective of the robot to the COCO dataset. Our pose-based control has no issues navigating the Jackal robot between two persons.

3.1 Image-Based Pose Classification

The confusion matrix of our proposed image-based pose classification is shown in Fig. 2 and its F1-score in Table 1. Overall, our pipeline achieves a remarkable weighted average F1-score of 0.9419. However, considering the confusion matrix, some rare misclassified poses point out the limitation of our image-based approach. Most of these errors (except the false-positives from the normal pose) are based on people standing at a high angle to the camera, which leads to a distorted joint angle and keypoint distances. Solving this problem is our fundamental motivation for a point cloud-based pose estimation, taking depth information into account. Finally, we want to point out the closed world assumption of the classifier as its biggest drawback. Thus, we have to model every pose that should not be interpreted as a command, even if its similar to another pose (like "taking a picture" to the double L-pose) into the rest class *neutral*. This rest class is becoming very heterogeneous, rendering it unsuited for a parametric density estimation in most circumstances.

Table 1: F1-Score from the five-fold cross validated pose classification. The weighted average is 0.9419 with a standard deviation of 0.0224 over all folds.

Pose	T	double I	double L	left I	right I	neutral
F1-Score	0.95	1.	0.86	1.	0.91	0.94

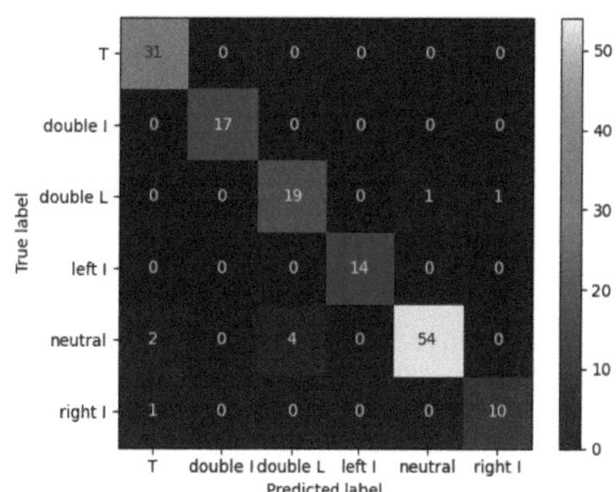

Figure 2: Confusion matrix of the five-fold cross validated pose classification.

Table 2: Average $L1$-norm d between predicted joints and pseudo ground truth.

ground truth used for training	d in cm
pseudo ground truth	10.92 ± 12.43
SURREAL	27.86 ± 23.15

3.2 Point Cloud-Based Pose Estimation

We show qualitative results of our point cloud-based pose estimation in Fig. 3. On the one hand, our two trained DGCNNs both learned to predict a plausible human pose. On the other hand, however, the domain gap from the SURREAL to our data is too large for a reliable and robust prediction.

For a quantitative evaluation, we compute the distance between the prediction of the models and the pseudo ground truth using the $L1$-norm (see Table 2). However, as a result of the normalization of the point cloud to $[-1, 1]$ and consequently the loss of the correct axis ratio, the results can not easily be converted back into metric values. One rough approximation for the distance in centimeter is the body size of the person in our validation sequence; $d_{\mathrm{cm}} = d_{L1} \cdot \frac{176}{2}$. While the model trained on the pseudo ground truth achieves an acceptable mean error of $10\,\mathrm{cm}$, training on the SURREAL dataset almost increases the error by factor of 3. This quantitative results go hand-in-hand with the qualitative results shown in Fig. 3. Besides the domain gap, the underwhelming performance of the model trained on the SURREAL data is also probably based on the different annotation keypoint positions; the 2D joints are just mapped on the surface. However, the SURREAL represents their anatomically correct internal positions (see Fig. 4).

4 Conclusion and Outlook

In this work, we proposed a novel pipeline for pose-based control of a Jackal robot. Our pipeline consists of a pretrained keypoint R-CNN for the image-based pose estimation and an ML-estimator for the pose classification using

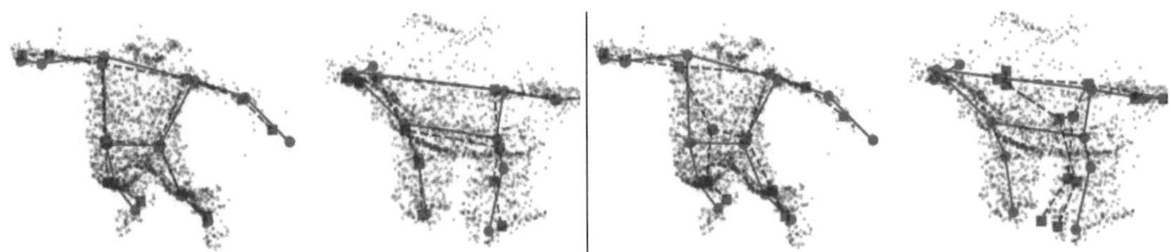

Figure 3: Qualitative evaluation on Jackal's point clouds. The red, round markers are the pseudo ground truth and the green, squared ones with dashed lines are the prediction of the DGCNN. Left: the DGCNN trained on the pseudo ground truth, right: the one trained on SURREAL data.

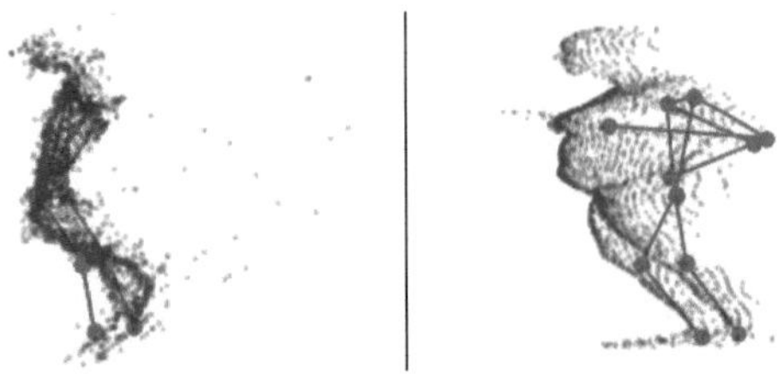

Figure 4: Visualization of the domain gap between the Jackal's and SURREAL point cloud due to the Jackal's poor depth estimation. Skeleton represents the pseudo ground truth for the Jackal's (left) and the ground truth on the SURREAL's point cloud (right).

handcrafted features. Having implemented our pipeline on the Jackal robot, we consider its hardware as suitable for using out-of-the-box computer vision algorithms for real-time pose estimation and classification. The pretrained keypoint R-CNN generalizes very well to the robot's frog perspective. We showed that our handcrafted features are suitable for a pose classification task and allow us to use a simple LDA and ML estimator as a classifier, fitted with a relatively small dataset, yielding a remarkable weighted F1 score of 0.9419. This pose classification accuracy is in line with our successfully real-world test of navigating the Jackal between two people only using poses. In addition, we believe that our classification approach with our type of handcrafted features is applicable out-of-the-box to the point cloud-based pose estimation.

However, the current approach is frame-by-frame based, leading to some flickering in the prediction between two consecutive frames. Thus, a solution should be considered that takes the temporal context into account, like using a first order IIR lowpass filter on the probability value of the different pose densities. Another interesting new approach could use a hand gesture to control the Jackal. This type of control would be more intuitive and might look a little more majestic than using the whole body. A big problem that we are not able to solve is the closed world assumption of the pose classifier, resulting in a very heterogeneous rest class *neutral* for all poses without a command purpose.

As our second contribution, we made the first step towards pose estimation on point clouds. To deal with the lack of annotated data in this situation, we examined two different sources of supervision to train a DGCNN. In the first approach, we mapped the image-based R-CNN joint predic-

tions to the nearest valid point of the corresponding point cloud. Secondly, we used the synthetic SURREAL dataset with its annotated data for training. While our first approach gives acceptable results, the domain gap between the synthetic and the Jackal's data turned out to be too huge to get a robust pose estimation on our point clouds. The gap is mainly based on the poor depth estimation by the robot's stereo camera, which is fine for navigating in a scene but has a too bad resolution for a human body pose which yields strong artifacts and a skewed body pose.

Acknowledgement

The work has been carried out at the Institute of Medical Informatics, Universität zu Lübeck.

Author's Statement

Conflict of interest: Authors state no conflict of interest.

5   References

[1] K. He, G. Gkioxari, P. Dollár, and R. Girshick, "Mask R-CNN," *IEEE Transactions on Pattern Analysis and Machine Intelligence*, vol. 42, no. 2, pp. 386–397, 3 2020.

[2] M. R. Silas, P. Grassia, and A. Langerman, "Video recording of the operating room—is anonymity possible?" *Journal of Surgical Research*, vol. 197, no. 2, pp. 272–276, 8 2015.

[3] Y. Wang, Y. Sun, Z. Liu, S. E. Sarma, M. M. Bronstein, and J. M. Solomon, "Dynamic graph Cnn for learning on point clouds," *ACM Transactions on Graphics*, vol. 38, no. 5, p. 13, 1 2019.

[4] G. Varol, J. Romero, X. Martin, N. Mahmood, M. J. Black, I. Laptev, and C. Schmid, "Learning from Synthetic Humans," in *CVPR*, 2017.

[5] T. Y. Lin, M. Maire, S. Belongie, J. Hays, P. Perona, D. Ramanan, P. Dollár, and C. L. Zitnick, "Microsoft COCO: Common objects in context," in *Lecture Notes in Computer Science*, vol. 8693 LNCS, no. PART 5. Springer Verlag, 5 2014, pp. 740–755.

Towards a Horizontal Gaze Nystagmus Detector on a Mobile Platform

Paul Kaftan [1], Katharina Gercke [2], Andreas Trabhardt [3], and Jan Ehrhardt [4]

[1] Medical Informatics, Universität zu Lübeck, paul.kaftan@student.uni-luebeck.de
[2] Medical Engineering Science, Universität zu Lübeck, katharina.gercke@student.uni-luebeck.de
[3] Drägerwerk AG & Co. KGaA, Lübeck, andreas.trabhardt.contractor@draeger.com
[4] Institute of Medical Informatics, Universität zu Lübeck, ehrhardt@imi.uni-luebeck.de

Abstract

Horizontal gaze nystagmus (HGN) is an indicator for high blood alcohol concentration and central nervous system depressants, so it is part of the standard procedure in roadside impairment testing. In this work, we develop and evaluate a pipeline to track horizontal gaze direction as the basis for HGN detection. First, we employ a convolutional autoencoder network to predict eye landmarks from which we then compute the horizontal gaze index (HGI), a novel robust proxy for gaze direction. We implemented our approach in Android and process the camera feed of a modern smartphone in real-time. Using training from synthetic images and a small fine-tuning data set, we manage to close the domain gap to the real-world application via heavy data augmentation. Future research will develop HGN detection based on the HGI tracking proposed here.

1 Introduction

Horizontal gaze nystagmus (HGN) evaluation is a component of the Standardized Field Sobriety Test (SFST) defined by the American National Highway Traffic Safety Administration (NHTSA). It has been shown that HGN is a reliable indicator of a high blood alcohol concentration as well as central nervous system depressants [1].

In an effort to make HGN testing more objective, we propose and evaluate a mobile pipeline to quantify the horizontal gaze. For HGN detection, we need to monitor horizontal gaze over time. In this work, we focus on the prerequisite that is predicting gaze from eye images. In the context of roadside testing, our system needs to work in uncontrolled environments. With this in mind, we train a convolutional neural network (CNN) that locates eye landmarks. On that basis, future work will develop the temporal evaluation.

Our first contribution is the definition of three robustly recognizable eye landmarks for horizontal gaze estimation. Based on these three landmarks, we propose a novel horizontal gaze index (HGI) for quantification, which is not sensitive to the movement of a handheld camera, nor the distance between camera and eye. We also outline a training scheme for CNN-based eye landmark detection from a large synthetic and a small real-world data set, which we evaluate in terms of real-world performance. Finally, we implemented the horizontal gaze tracking in a library that is easily imported into Android or iOS apps and can process a real-time camera feed with 16 frames per second (FPS) on modern smartphones.

2 Material and Methods

2.1 Horizontal Gaze Nystagmus Detection

As a nystagmus is a rapid movement of the eye, detecting HGN can be solved by performing gaze estimation with high temporal resolution. Normally, gaze estimation aims to predict the three-dimensional gaze vector. However, given that it is sufficient to detect horizontal change in gaze for HGN, we chose to simplify the problem. We track three well discernible landmarks in images of the eye. The pupil center c defines the origin of the gaze vector and we use the inner eye corner i (lacrimal caruncle) and the outer eye corner o as reference points. The distance between center and eye corners acts as an easy-to-compute proxy variable for horizontal gaze direction. For combining the distances into one parameter, we propose the horizontal gaze index (HGI):

$$\text{HGI}(i, c, o) = \frac{d(c, i) - d(c, o)}{d(c, i) + d(c, o)} \tag{1}$$

The index is calculated from the three detected landmark points $i, c, o \in \mathbb{R}^2$ (see Fig. 1). We measure the Euclidean distance d from the center to both eye corners and use their sum for normalization. The HGI being in range $[-1, 1]$ leads to intuitive values, as Fig. 1 demonstrates.

Because our system is supposed to work on smartphones and in real-time, we do not try to model the eyeball in 3D, which could yield more accurate gaze estimations. Instead, we simplify the problem and only use the 2D projection in images.

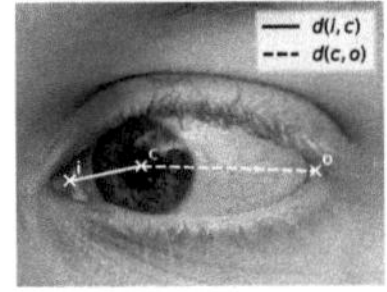 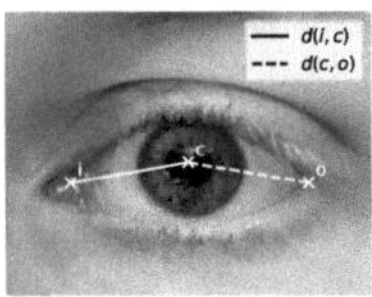 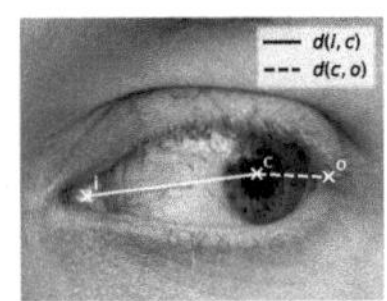

(a) HGI = −0.41   (b) HGI = −0.01   (c) HGI = 0.41

Figure 1: Visualization of our horizontal gaze index (HGI) to quantify gaze direction from three eye landmarks.

When performing HGN detection with a mobile phone, we cannot assume the camera is always a fixed distance from the eye. Also, the examiner and the subject will not be perfectly stationary during the recording and therefore the eye will not have a fixed position in the frame. These disturbances lead to projective transformations of the image. The HGI is invariant to similarity transforms because of the normalization and the usage of relative distances between the pupil center and the eye corners. We argue that this invariance provides enough robustness for our task and extending it would require too much additional processing.

2.2 Eye Landmark Localization

We apply a CNN to detect the three eye landmarks because it can achieve fast inference with accurate results in many different scenarios, if trained accordingly. Conventional approaches would need a lot of manual parameter tuning to work in various lighting conditions, whereas this can be automatically taught to a sufficiently complex CNN.

Predicting landmark positions is a regression task of the 2D coordinates. Past research on regression with neural networks showed it to be difficult to train due to the high influence of outliers. Instead, we model landmark regression as a per-pixel likelihood prediction, adopting the state-of-the-art approach from human pose estimation (e.g. [2]). Thus, the network output is a likelihood heat-map for each landmark (see Fig. 2). For training, we generate the target heat-map for each landmark by computing a 2D Gaussian distribution with its mean at the landmark point and a very small variance of 10^{-4}.

We adapt the CNN architecture from U-Net [3] and use it to infer the landmark heat-maps from an input eye image. For our task, we use three contracting convolutional blocks with max-pooling in the encoder and three expanding blocks with bilinear upsampling in the decoder. The skip-connections help restoring spatial information when decoding hidden representations.

Finally, the landmark prediction is taken at the highest likelihood pixel in its respective heat-map. Before this step, the predicted heat-maps are smoothed in order to reduce noise and the influence of local maxima.

For a more robust HGI evaluation over time, we perform a plausibility check on the predicted landmarks. The pupil center c has to lie in between the eye corners i and o horizontally, the angle between the vectors $(c - i)$ and $(c - o)$ needs to be in range $[130°, 230°]$, and the height difference between i and o has to be less than 20% of the image height.

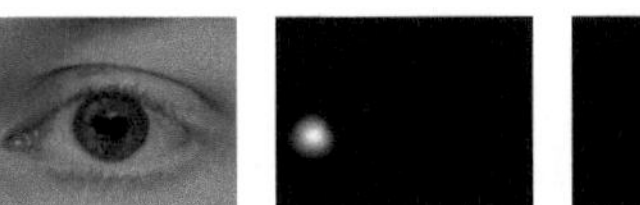  

Figure 2: An example input image for our CNN and its output per-pixel likelihood (heat-maps) of the inner eye corner i, pupil center c, and outer eye corner o (from left to right).

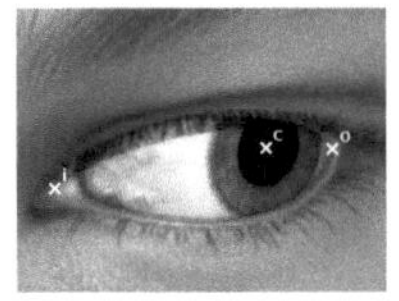 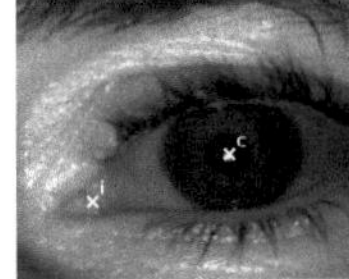 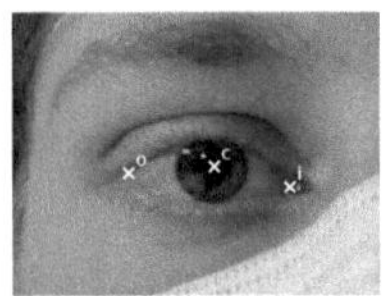

(a) Training image.   (b) Fine-tuning image.   (c) Test image.

Figure 3: Example images from our three data sets with ground truth annotations. (a) shows a synthetic image generated by UnityEyes [4]. (b) shows an image from the MMU Iris Database 2 [5] we use for fine-tuning. (c) is an example from our own test set.

Landmarks are considered plausible if and only if all three conditions are met. Parameters were chosen empirically to discard impossible landmark predictions while not being restrictive to the application.

2.3 Training and Test Data

This section describes the CNN input data. An example for each data set can be found in Fig. 3.

To gain access to a large quantity of annotated eye images for training, we use UnityEyes (UE) [4], a tool that generates eye region images with random anatomy, textures, lighting and gaze direction. The provided vertices of the generated 3D model are suitable as ground truth landmarks. Each image is converted to grayscale, cropped to a bounding box resulting from Haar cascade eye detection from OpenCV, and then resized to 128×128 pixels. The eye detection on synthetic data helps excluding images that contain artifacts from the random generation.

For fine-tuning the models, we chose the Multimedia University Iris Database 2 (MMU2) [5], which includes ten grayscale images for each of the 100 subjects. The images were captured in a light-controlled environment. We selected and annotated 262 images by hand. Images excluded were either overexposed or cut off at one of the eye corners. We resized the images to 192×256 pixels, keeping their original aspect ratio.

For evaluation, we created our own test data set called StudentData (SD) by taking photos of fellow students in different offices and laboratories with a modern smartphone. The camera was handheld and we only used room lighting. We collected a total of 94 images of left and right eyes from 16 subjects. 16 of the images depict an eye with glasses. Images were converted to grayscale and resized to 192×256 pixels.

2.3.1 Data Augmentation

Especially the synthetically generated UE images are not representative of images taken in real-world scenarios. Similarly, due to the highly controlled nature of the MMU2 data set, these images also appear vastly different from images in the real-world application. Therefore, we employ a suite of spatial and pixel-level transformations for data augmentation (DA). We randomly apply affine and perspective transformations as well as change the contrast, resolution, and noise.

2.4 Experimental Setup

We train our models for 100 epochs with a learning rate of 10^{-3} and choose the specific epoch where the model performs the best on the validation split. All input images are z-standardized with the gray value mean and standard deviation calculated over the training data set. Performance is measured by comparing predicted landmarks $p \in \mathbb{R}^2$ with the ground truth $\hat{p} \in \mathbb{R}^2$.

We propose the percentage of eye (PoE) distance as a scale invariant metric. It normalizes the Euclidean landmark distance d with the eye diameter defined as the distance between ground truth landmarks i and o.

$$\mathrm{PoE}(p, \hat{p}, i, o) = \frac{d(p, \hat{p})}{d(i, o)} \cdot 100\%$$

For fine-tuning, we load the pre-trained weights and freeze the first encoder stage. All other parameters are trained with a learning rate of 10^{-4}.

For every experiment, we test model performance on all three data sets in a five-fold cross-validation manner. Reported metrics are averaged over the five folds.

2.5 Mobile Implementation

Our CNNs are trained in PyTorch. In order to run inference on a mobile device, we convert them into the Open Neural Network Exchange (ONNX) format, from which an NCNN [6] model is created. NCNN in C++ allows for fast CNN inference on smartphone processors. We implemented the gaze estimation pipeline as a native C++ library, which we then imported into a Java Android prototype app via the Java Native Interface (JNI). Importantly, this implementation allows us to use the library in any Android and iOS app.

We run our pipeline on a real-time camera feed, from which we extract images of size 192×256. We report the measured frame rate on a *One Plus 8 Pro*.

3 Results and Discussion

As can be seen in Fig. 4 in the top row, landmark predictions under good lighting conditions are accurate, even when wearing glasses. When light falls in at an unusual angle, however, the CNN can produce implausible results

Table 1: Landmark prediction error of our U-Net with different training data sets: UnityEyes (UE) and Multimedia University Iris Database 2 (MMU2) with or without data augmentation (DA). Our main benchmark is the StudentData (SD) set. All values given in mean percentage of eye (PoE) distance in %.

Train (+ fine-tune) data	Test data set		
	UE	MMU2	SD
UE	1.44	34.28	22.91
UE (w/ DA)	1.58	27.90	13.45
MMU2	38.42	2.00	36.76
MMU2 (w/ DA)	42.00	2.00	35.85
UE + MMU2	3.08	2.64	14.92
UE + MMU2 (w/ DA)	3.00	2.25	5.32

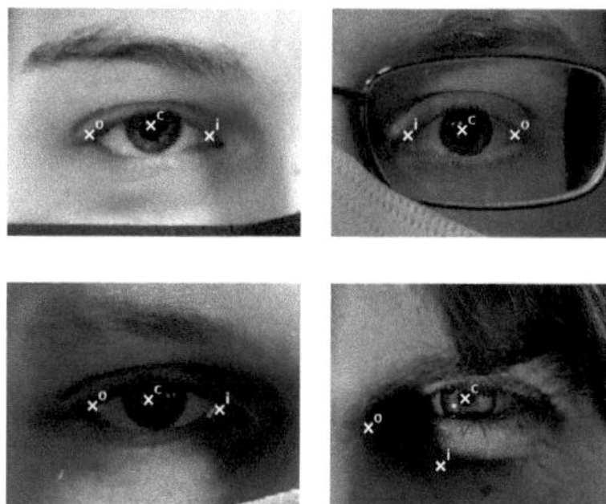

Figure 4: Visual results of the landmark localization for images from our test data set.

(bottom right). However, a frame with this kind of prediction will be discarded due to failing the plausibility check, as the center landmark does not lie in between the two corners.

Note that in the following paragraphs, we report landmark localization error of our CNN before discarding implausible results, unless it is stated otherwise.

3.1 Effect of Training Data Set

Table 1 shows that when only training our models on UE, we get a high test error of 13.45% PoE on SD. Therefore, we argue that synthetic images alone are not sufficient for the application to real-world images. Training on MMU2 yields worse test results with 35.85% PoE. This means that a highly uniform data set taken in a controlled environment generalizes worse than varied synthetic data. These results validate our decision to use UE as a starting point before fine-tuning with MMU2.

3.2 Effect of Fine-Tuning

After training with UE, the mean PoE reached 13.45% (see Table 1). Fine-tuning with MMU2 reduced the error by more than a half to 5.32%. Therefore, the combination

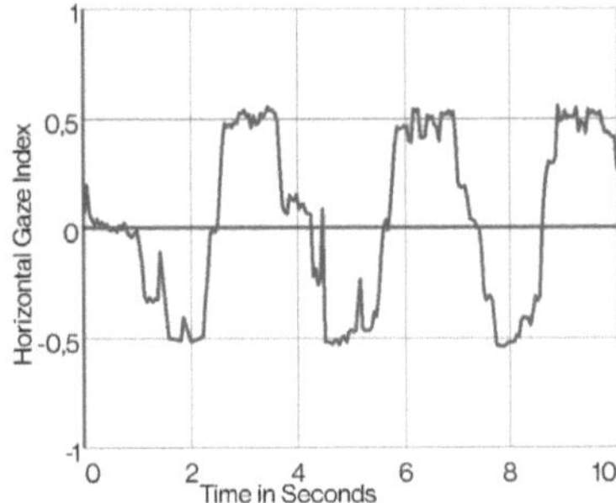

Figure 5: HGI measured over 10 seconds. The subject was asked to alternate between looking inwards and outwards.

of data sets is much more generalizable to SD than each of them alone. Assuming, we were able to bridge the domain gap to SD perfectly, the best possible results would be 1.44% or 2.00% PoE, coming from UE or MMU2, respectively.

3.3 Effect of Data Augmentation

Comparing the results with DA, as described in the previous paragraphs, to results without DA, we can see the importance of it. Table 1 shows, that omitting DA during training with UE leads to almost double the error on SD (22.91% vs. 13.45%) and after fine-tuning without DA we measured almost triple the PoE (14.92% vs. 5.32%).

Therefore, a heavy DA suite is necessary to train generalizable networks from the synthetic UE and the low-variety MMU2 data set.

3.4 Gaze Evaluation Over Time

While we saw substantial improvements in the generalization ability of our models with fine-tuning, our best error of 5.32% PoE is still considerably higher than the best possible baseline (1.44% or 2.00% PoE). When excluding results due to a failed plausibility check, we reached 4.03% ± 7.23% PoE. The high standard deviation results in the recorded HGI over time being noisy (see Fig. 5).

It remains to be tested if a reliable HGN detection can be made from that data. If this is not the case, we suggest capturing and annotating a specialized data set containing smartphone videos of subjects' eyes in different positions and environmental conditions for training. As we have shown, pre-training with the synthetic UE images has a positive impact on the results and should still be employed.

With our current implementation, we already achieved a real-time capable frame rate, processing approximately 16 FPS. Further improvements can easily be made by replacing the standard convolutional blocks in the U-Net with ones specifically designed for mobile applications.

4 Conclusion and Outlook

In this work, we developed and evaluated a system to track a subject's horizontal gaze for the detection of HGN on a mobile device. Our approach uses the novel robust HGI for quantifying gaze direction based on CNN-detected landmarks of the eye. We showed the importance of finding a suitable training data set and augmenting training data to bridge the domain gap between synthetic or low-variety data sets and real-world images.

Further research needs to be done to determine if an HGN can be accurately detected on the basis of our application. In theory, the HGI is a well-suited parameter to track horizontal gaze in smartphone videos because of its invariance to similarity transforms. However, noisy landmark localization might lead to false positive nystagmus predictions.

Fine-tuning the models with MMU2 was highly beneficial to the landmark accuracy, so we suspect that a large data set comprising images from highly varied environments would boost the performance much further. Additionally, generating annotated eye videos could lead to even better predictions by taking the information of multiple successive frames into account.

Acknowledgement

The work has been carried out at Drägerwerk AG & Co. KGaA, Lübeck and supervised by the Institute of Medical Informatics, Universität zu Lübeck.

Author's Statement

Conflict of interest: Authors state no conflict of interest.

5 References

[1] A. J. Porath and D. J. Beirness, "Predicting categories of drugs used by suspected drug-impaired drivers using the drug evaluation and classification program tests," *Traffic Injury Prevention*, vol. 20, no. 3, pp. 255–263, 2019.

[2] T. Pfister, J. Charles, and A. Zisserman, "Flowing convnets for human pose estimation in videos," *CoRR*, vol. abs/1506.02897, 2015.

[3] O. Ronneberger, P. Fischer, and T. Brox, "U-net: Convolutional networks for biomedical image segmentation," 2015.

[4] E. Wood, T. Baltrušaitis, L.-P. Morency, P. Robinson, and A. Bulling, "Learning an appearance-based gaze estimator from one million synthesised images," in *Proceedings of the Ninth Biennial ACM Symposium on Eye Tracking Research & Applications*, 2016, pp. 131–138.

[5] Multimedia University. (2006) Multimedia university iris database 2. Accessed: 2021-11-15. [Online]. Available: https://andyzeng.github.io/irisrecognition

[6] Tencent. (2017) Ncnn. Accessed: 2021-11-3. [Online]. Available: https://github.com/Tencent/ncnn/tree/20210720

8

Biomedical Engineering

Feasibility Study of the SonoBox: Effect of Water Turbulence on Ultrasound Images and Importance of Initial Gain for Final Reconstruction

David Sindermann [1], Jonas Osburg [2], Sven Böttger [2], Floris Ernst [2]

[1] Biomedical Engineering, Lübeck University of Applied Sciences, david.sindermann@stud.th-luebeck.de
[2] Institute for Robotics and Cognitive Systems, Universität zu Lübeck, {osburg, boettger, ernst}@rob.uni-luebeck.de

Abstract

To avoid harmful X-ray application on children, a sonographic robotic system (SonoBox) which facilitates contactless diagnosis of forearm fractures is developed. In this study two emerging issues in the development of the SonoBox are explored: The influence of water turbulence on ultrasound images as well as the influence of the initial gain of the ultrasound station on the subsequent reconstruction of the image in the PLUS tool. An experimental setup was designed to simulate varying water turbulence while taking ultrasound images of a fixed forearm phantom in a bucket. Only small deviations up to 2.3% were measured between predefined landmarks. From this it can be concluded that water motion doesn't severely influence sonographic images of the bone. After reconstructing several 3D images with the image processing software 3D Slicer and PLUS, it is recommended to use a low gain around 59 dB to avoid a loss of detail and artifacts. Overall, these studies are the basis for further development and implementation of the SonoBox.

1   Introduction

Distal forearm fractures are the most occurring fractures in childhood [1]. Either ultrasound imaging or X-ray is used to perform a diagnosis. Yet it is known, that children show a five times higher sensitivity to ionizing radiation than adults [2] and 81% of all performed X-ray at children with suspected distal forearm fractures from 0 - 12 years can be avoided [3]. This represents a considerable potential for the avoidance of X-ray radiation exposure and thus the protection of children. Ultrasound imaging is widely available in non-hospital settings, since it is a cost effective and non-invasive technology [4]. Real-time acquisition is possible, but the diagnosis quality is so far dependent on the operator. Ackermann et al. [5] states that ultrasound should be the primary imaging method in children and pediatric forensic medicine. It should only be supplemented by additional radiological diagnostics if necessary [5].

To facilitate the diagnosis of forearm fractures the SonoBox is developed. A prototype of it can be seen in Fig. 1. It is a robotic system which facilitates quick contactless ultrasound imaging and reconstructs a 3D image of the alleged fracture to enable an accurate diagnosis by the operator. As ultrasound is diffracted by air, the SonoBox operates in a bucket filled with water. The arm of the patient will be placed in the bucket and the robotic system moves along defined trajectories while the ultrasound probe is taking ultrasound images. Hence, the SonoBox won't get in contact with injuries and it is a pain-free procedure.

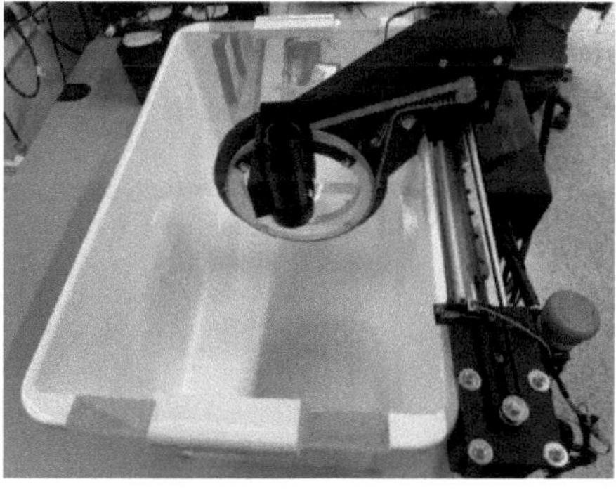

Figure 1: Prototype of the SonoBox is fixed to a bucket. The two trajectories enable a quick contactless ultrasound imaging of the forearm which has to be placed in the loop.

An ultrasound transducer emits high frequency sound pulses into the medium and receives the corresponding echoes. The echoes are subsequently processed to create the ultrasound image. Most ultrasound systems assume a sound travel speed of 1,540 m/s in soft tissue to calculate the covered distance of the waves. If the density of the medium changes, distortion is created [6].

Additionally, most ultrasound stations enable a modification of the gain. Gain is a uniform amplification of the ultrasonic signal that is returning to the transducer after it has travelled through the tissue [7]. The returning signal is amplified to make objects easier to see.

The aim of our study was to refine the development of the

SonoBox. During the operation of the SonoBox, rotational motion of the ultrasound transducer leads to water turbulence and eventually to air bubbles. Therefore it was necessary to investigate the effect of water turbulence on the ultrasound image quality. Furthermore, the effect of initial gain on the final reconstruction was investigated using the open-source software PLUS and 3D Slicer.

2 Material and Methods

2.1 Water Turbulence Investigation

A biomechanical pediatric forearm phantom was used for functional testing of the system. The phantom consisted of a cast silicone-based phantom with two embedded 3D-printed bone models [8]. The whole setup can be seen in Fig. 2.

In order to simulate water turbulence a 15 Watt aquarium water pump with 13 and 16 mm nozzles was used. The pump and the forearm phantom were submerged at the same depth of water. The water had room temperature of 23 °C. The phantom was clamped between the ends of a bucket. The distance between the nozzle and the forearm phantom was approximately 12 cm. The ultrasound transducer was mounted vertically in the water with transverse orientation over the forearm phantom. A L12-3 broadband linear array ultrasound transducer of PHILIPS was used, because of its high frequency range from 3 – 12 MHz. In order to acquire doppler images, the ultrasound probe X6-1 of PHILIPS was used, which has a bigger field of view. Both probes were connected to the ultrasound system EPIQ 7 of PHILIPS. The bucket was inlayed with acoustic foams to prevent the ultrasound waves to be reflected from the inner walls of the bucket.

Figure 2: Experimental setup to evaluate the effects of water turbulence on the ultrasound image: The phantom was fixed in a bucket and above an ultrasound transducer is placed.

In total, three setups were conducted with different velocities of flow of the water pump. Following specifications were set for the three trials:

- No water turbulence: An ultrasound image of the forearm phantom was acquired while the water wasn't in motion.

- Water turbulence: An ultrasound image of the forearm phantom was acquired with a velocity of flow of 163.74 cm/s set by the pump.

- Stronger water turbulence: An ultrasound image of the forearm phantom was acquired with a velocity of flow of 313.92 cm/s set by the pump.

For each setup ultrasound images of the forearm phantom were acquired. The ultrasound station was set to the following settings: intensity gain 52, angle 30, depth 16 cm and frequency (volume acquisition) 30 Hz.

Afterwards the measurement tool of the PHILIPS ultrasound software was used to measure predefined landmarks in the ultrasound image, which can be seen in Fig. 3. These distances were used to evaluate the effect of water turbulence on the visualisation of the bones in the image.

In order to show the varying movement of the water depending on the setup, doppler images were acquired additionally with the same procedure. Therefore, the ultrasound probe X6-1 of PHILIPS was used because of its 100° field of view.

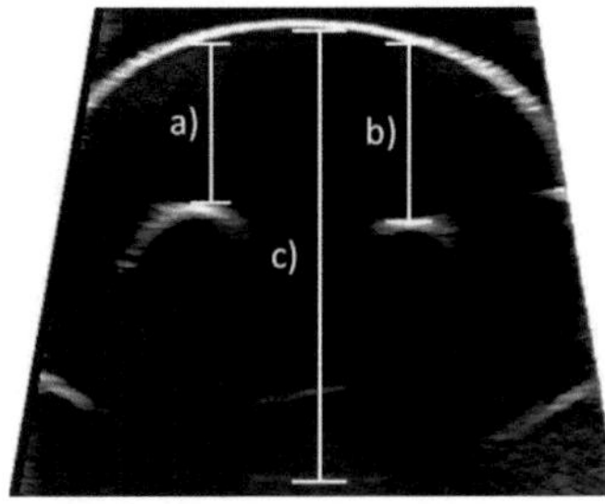

Figure 3: Predefined landmarks a), b) and c) were used to compare the influence of water turbulence on the ultrasound image.

2.2 Reconstruction Data Acquisition

The same pediatric forearm phantom [8] was used for the investigation of the effect of initial gain on the final reconstruction. The phantom was fixed in a bucket with sponges. The ultrasound system Toshiba Aplio 50 (SSA-770A) was used. The ultrasound transducer L14-5 was attached to a robot arm. The whole setup can be seen in Fig. 4. The robot arm was programmed to follow a trajectory along the phantom with constant velocity. The robot arm moved along the phantom, holding the transducer under water surface while recording ultrasound images.

Before recording images, a spatial and temporal calibration of the ultrasound probe was performed. For this purpose a freehand tracked ultrasound calibration application (fCal) was used. In order to reconstruct the single ultrasound images to a 3D object the data was transferred from the ultrasound station to the PLUS toolkit (`http://www.plustoolkit.org/`). By means of the medical image processing software 3D Slicer (`https://www.slicer.org/`) post-processing of the images was done.

Figure 4: Experimental setup to evaluate the effects of initial gain on final reconstruction: A forearm phantom is fixed within a bucket and an ultrasound transducer is mounted to a robot arm above.

The gain in the preamplifier of the ultrasound system is fixed to 14 dB and the variable gain amplifier sets the additional gain from 0 to 75 dB [9]. In total, three 3D reconstructions of the forearm phantom were acquired with varying gains: 59 dB (60%), 74 dB (80%) and 89 dB (100%). This data was used to investigate the effect of the amplification of the echo-signals returning on the final reconstruction and post-processing with 3D Slicer.

3 Results and Discussion

3.1 Landmark Analysis

In Fig. 5 the recordings of the ultrasound images with varying water turbulence can be seen. In all images the predefined landmarks a) and b) can be clearly identified. The only differences are small artifacts at the bottom of the images, which make it difficult to identify landmark c). Otherwise, no clear artifacts are visible due to water movement.
Table 1 shows the measured distances of the landmarks for the different setups.

Table 1: Distances of the landmarks depending on water turbulence.

flow velocity	distance (a)	distance (b)	distance (c)
without	17.3 mm	18.3 mm	48.7 mm
163.74 cm/s	16.9 mm	18.2 mm	48.2 mm
deviation	-2.3%	-0.5%	-1%
313.92 cm/s	17.1 mm	18.6 mm	48.4 mm
deviation	-1.2%	0.16%	-0.6%

The setup without external influence of water turbulence is regarded as ground truth. Only small deviations up to 2.3% can be seen between the different setups. In total the deviations vary downwards and upwards. No correlation can be identified.
The deviations can be explained due to measurement inaccuracies of the operation. The predefined landmarks were not easily measured with the tool of the ultrasound system. Another source of error might be that the forearm phantom was mounted insufficiently and thus was able to move which led to a change of angle and the distance.

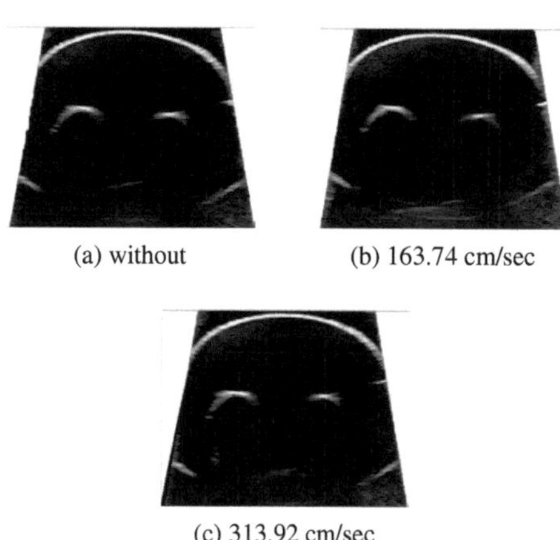

(a) without   (b) 163.74 cm/sec

(c) 313.92 cm/sec

Figure 5: Ultrasound images based on varying water currents.

Another experiment with the X6-1 probe was performed to prove the presence of water turbulence. Therefore, Doppler sonography was done. While little doppler effects can be seen in Fig. 6 (b), it increases steadily from (d) to (f). It elucidates that water motion is increasing, but it has no influence on the ultrasound images.
Hence, there is no correlation discernable. It can be concluded that water turbulence has no influence on the ultrasound images. The reason for the deviations at the distances can be due to measurement inaccuracies.

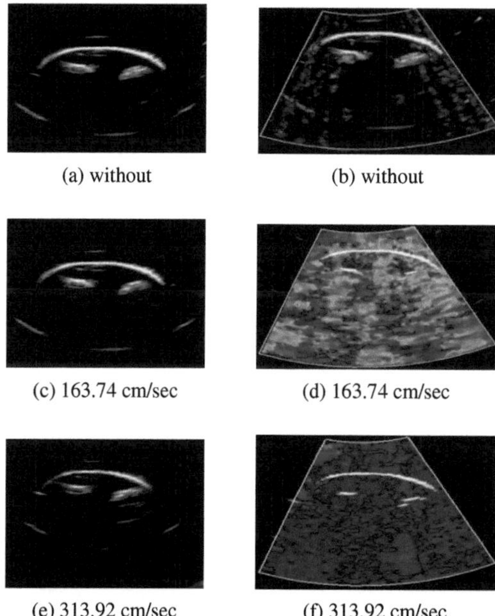

(a) without   (b) without

(c) 163.74 cm/sec   (d) 163.74 cm/sec

(e) 313.92 cm/sec   (f) 313.92 cm/sec

Figure 6: Ultrasound images based on varying water currents in comparison with their doppler images.

3.2 Analysis of Gain Modification

In order to make the images comparable, a region of interest (ROI) needed to be defined in 3D Slicer. All overlapping layers were removed so that only the bone is visible. The defined ROI for the three gains can be seen in Fig. 7. It becomes clear that with higher gain the reconstruction quality decreases.

As bone, in general, has a high density, the ultrasound signal is reflected strongly and thus there is a clear signal independently from the gain. Through the gain, noise gets amplified as well which leads to a greater influence of artifacts and a loss of detail. This can be seen especially in Fig. 7(c).

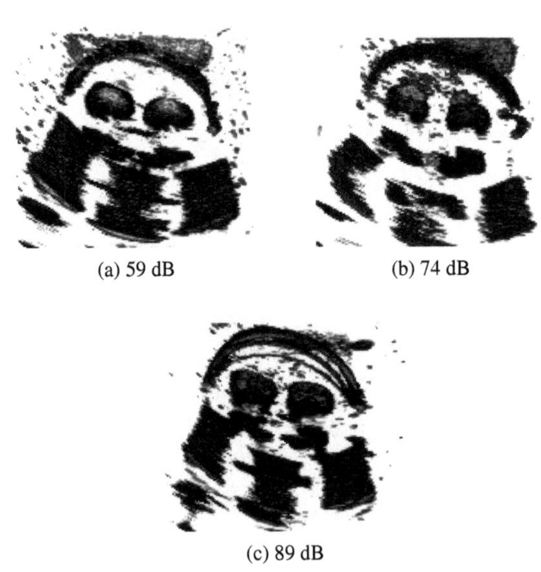

Figure 7: 3D reconstructed ultrasound volumes based on varying gains.

Overall, the impact of the initial gain could be demonstrated as well as how a reconstruction with PLUS toolkit and post-processing with 3D slicer could be done. Yet it became clear that post-processing cannot compensate for poor data acquisition. Therefore, a low gain is recommended to provide a good foundation for a successful reconstruction and to avoid blurriness and artifacts.

4 Conclusion

This study compared the effects of varying water turbulence on the ultrasound image. It became clear that the investigated water turbulence has no effect on the ultrasound image. Thus, no consideration needs to be given to excessively rapid rotational movements of the robot during development of the SonoBox. Still it needs to be considered that the creation of air bubbles could lead to artifacts.

Furthermore, it needs to be emphasized that the initial set gain at the ultrasound station influences the final reconstruction quality of the PLUS toolkit. To reduce artifacts, it is recommended to choose a gain around 59 dB for the reconstruction. These studies gave important insights for the further development and implementation of the SonoBox.

Acknowledgement

This work has been carried out at the Institute of Robotics and Cognitive Systems, Universität zu Lübeck and supervised by Jonas Osburg.

Author's Statement

Authors state no conflict of interest.

5 References

[1] R. Kraus, D. Schneidmüller, C. Röder, *Häufigkeiten von Frakturen der langen Röhrenknochen im Wachstumsalter*. Deutsches Ärzteblattt, vol. 102, pp. 838-842, 2005.

[2] ICRP, *The 2007 Recommendations of the International Commission on Radiological Protection*. Ann ICPR 2007, vol. 37 p. 123, 2007.

[3] O. Ackermann, P. Wojciechowski, M. Dzierzega, K Grosser, A. Schmitz-Franken, H. Rudolf, K Eckert, *Sokrat II – An international, prospective, multicenter, phase IV diagnostic trial to evaluate the efficacy of the wrist SAFE algorithm in fracture sonography of distal forearm fractures in children*. Ultraschall in der Medizin, vol. 40, no. 3, pp. 349-358, 2019.

[4] G. Schmid, B. Kühnast, M. Heise, T. Deutsch, T. Frese, *Ultrasonography in assessing suspected bone fractures: a cross-sectional survey amongst German general practitioners*. BMC Family Practice, vol. 21, no. 9, p. 2, 2020.

[5] O. Ackermann, J. Simanowski, K. Eckert, *Fracture Ultrasound of the Extremities*. Ultraschall in Med., vol. 12, no. 28, p. 20, 2020.

[6] H. Shin, R. Prager, H. Gomersall, N. Kingsbury, G. Treece, A. Gee, *Estimation of Average Speed of Sound Using Deconvolution of Medical Ultrasound Data*. Ultrasound in Medicine Biology, vol. 36, no. 4, pp. 623-636, 2010.

[7] V. Gibbs, D. Cole, A. Sassano, *Ultrasound Physics and Technology*. Elsevier Health Sciences, p. 9, 2011.

[8] H. El-Shaffey, S. Klein, F. Hainer, *Development of a Novel Pediatric Forearm Fracture Treatment: Simulation, Prototype and Evaluation*. Current Directions in Biomedical Engineering, vol. 3, no. 2, pp. 325-329, 2017.

[9] Toshiba Medical Systems Corporation *Service Manual for Diagnostic Ultrasound System aplio Models SSA-700/SSA770A*. ch. 4 p. 9, 2001.

Fingerprinting via the Human Proteome as a Localization Scheme for Nanobots

Lena Felicitas Unger [1], Regine Wendt [2], Florian-Lennert Lau [2], and Stefan Fischer [2]

[1] Biomedical Engineering, Lübeck University of Applied Sciences, l.unger@student.uni-luebeck.de
[2] Institute of Telematics, Universität zu Lübeck, {wendt, lau, fischer}@itm.uni-luebeck.de

Abstract

The use of nanoscale devices in medical scenarios could pave the way to precise treatment and true continuous surveillance. The localization of nanobots in the human body is a crucial element to enable diagnostic ability. So far, localization schemes mainly rely on mathematical principles to delineate the positions of nanobots in the human body. A different approach to determining the position is local pattern recognition, so-called fingerprinting. In this paper, we analyze and evaluate different substances in the environment of the nanobots to find fingerprints for areas of the human body. The human proteome, resembling the entirety of human proteins, grants an inventory of many different genes. A plethora of genes marks different proteins with certain tissue specificities. We present fingerprints using tissue-specific genes for the most significant organs and tissues. With unique gene combinations for each tissue, the human proteome poses a promising localization method for nanobots in the human body.

1 Introduction to Nanodevices in Medical Applications

To explore the cause of physical discomfort, people often seek the assistance of a physician, consult (online) information sources, or talk to friends and family. While successful treatment at a later stage is feasible for some diseases, certain disorders need to be detected as soon as possible. This early diagnosis could be challenging as the patient might not have any symptoms until later. Persons with chronic illnesses are in even greater need of constant monitoring due to possible immunodeficiency. Regular examinations at the physician can reduce the risk, but even daily visits do not ensure continuous surveillance. Not to mention that daily medical appointments are difficult to reconcile with everyday life. The vision of nanonetworks is that *nanoscale devices* (nanodevices), for example, patrol the body, take measurements wherever necessary, and send collected data to the outside. Therefore, e.g., a tumor can be detected before it starts metastasizing. Even better, these machines may immediately work on problems they detect within the body, such as cancer cells, arteriosclerosis, or *human immunodeficiency viruses* (HIV). To correctly detect abnormalities, it is essential to connect the measurements of the nanobots to the corresponding organs or tissues in the human body. That requires a reliable localization of the nanobots in question.

Current localization schemes make use of mathematical algorithms like *Function Centric Nano-Networking* (FCNN) [1]. Using biological or chemical values is an entirely new approach outlining a promising possibility to differentiate particular body regions. Fingerprinting aims to determine the position of nanobots via local pattern recognition. An individual fingerprint is assigned to each body region by the properties of the environment in question. By identifying these fingerprints, nanobots can determine their position in the human body and communicate it to the outside.

2 Fingerprinting Approaches

To the best of our knowledge, no method of assigning histological fingerprints to different body regions exists. Therefore, this research study aims to analyze various histological values in the human body regarding their suitability for fingerprinting and find a reliable method. Possible candidates are systems comprising the whole body, like the lymphatic, endocrine, and cardiovascular system.

The lymphatic system must be excluded because nanobots are envisioned to run in the blood circuit. Hence, the values used for fingerprinting have to be present in the cardiovascular system.

The endocrine system and the hormonal balance in the human body include several different parameters which could serve as fingerprinting characteristics. Certain hormones are produced in different glands of the body and then distributed throughout the endocrine system. However, this distribution poses a problem for the clear differentiation of body regions, as there are no single hormones only present in certain regions.

Three methods show favorable characteristics for fingerprinting and are worthy of further analysis, including blood gases, trace elements, and the human proteome.

2.1 Requirements

To be a reliable source for the localization of nanobots, the parameters used for fingerprinting have to fulfill specific requirements:

1. For a distinct differentiation between body regions, either a singular compound only existing in specific regions or a compound with significantly differentiable concentrations in different regions could be chosen.

2. The existence and the concentration of the compound in question should be stable and not influenced by other factors, e.g., diseases or physical activity, for reliable and continuous monitoring of patients.

3. To make reasonable differentiations inside the human body, research for a sufficient number of organs or tissues has to be available. Explicitly critical organs like the heart, lungs, liver, stomach, and kidneys should be covered.

2.2 Suitability of Fingerprinting Approaches

The first compounds under consideration are the gases in the cardiovascular system. A general blood gas analysis measures O_2, CO_2, and pH levels. Arteries, veins, and smaller vessels can be distinguished by those characteristics. Oxygen enters the body by breathing and subsequently arrives in the cardiovascular system through the alveolar walls of the lung. It travels to the pulmonary capillaries, binds hemoglobin, and finally reaches tissues, organs, and cells through diffusion. This path occurs inside the arteries, transporting oxygenated blood towards tissues and organs. The veins then transport deoxygenated blood back to the heart. Exceptions are the arteries and veins of the lung, in which case the arteries transport deoxygenated blood and the veins oxygenated blood. Another localization differentiation can be made inside the arteries and veins if one follows the decrease in oxygen concentration towards the capillaries and organs. For a general localization of regions in the human body, this method is not suitable as, e.g., the difference in O_2 concentration could provide information about the distance from the aorta but no clear differentiation of body regions or organs. In addition, physical activity influences O_2 levels in the blood [2] and can therefore not ensure continuous monitoring. The fluctuations in concentration could lead to uncertainties or even wrong diagnostics. Hence, blood gases are excluded as a fingerprinting compound, as the general differentiation of body regions is the primary goal.

Trace elements that occur in minute concentrations in many environments pose another possible fingerprinting characteristic. Essential trace elements can evoke a deficiency syndrome if absent or too low in concentration. The number of essential elements varies slightly throughout literature, as some are presumably essential but not proven yet. Essential trace elements coinciding throughout literature include Cobalt, Copper, Iodine, Iron, Manganese, Molybdenum, Selenium, and Zinc [3], [4], [5]. Due to the same

trace elements being present in different tissues, the concentration of the elements is the focus of interest for fingerprinting. Four of the eight previously mentioned essential trace elements occur in all examined tissues: Copper, Manganese, Selenium, and Zinc. The concentration gradients in different tissues of those four elements can provide insight into the position in the human body. The presence in the respective tissues is the relevant factor for the remaining elements. Reported concentrations for trace elements in different human tissues vary widely [5]. Apart from the variability, the concentrations are minuscule, constituting a problem for the use in fingerprinting due to inadequate reliability. The compounds blood, packed blood cells, and muscle tissue, as reported by J. Versieck [5], cannot be used to identify different body regions. That only leaves urinary excretion, lung tissue, liver tissue, and kidney tissue. These tissues are too few to provide a valuable differentiation between different human body parts. Furthermore, it is impossible to differentiate the concentrations for different ages and sexes, again posing a problem for fingerprinting. Due to those reasons, we omit trace elements as a compound for fingerprinting.

The human proteome represents the entirety of proteins in the human body. It is based on the human genome, the genetic material. Genetic material contains all the information needed to define how an organism is built and is crucial to ensuring the functionality and maintenance of a living organism in the way it is intended. The genome describes the entirety of biological information for one organism. In order to obtain the human proteome from the human genome, individual protein-coding genes (specific sequences of bases) are copied into RNA molecules, called the transcriptome [6]. In the next step, the translation, a polypeptide chain made of amino acids is assembled. Each amino acid is coded by three bases from the RNA molecule. The polypeptide chain is then folded in several steps, acquiring a three-dimensional shape – a protein. The assembly of proteins (gene expression) differs for human tissues and cells, leading to a specific proteome in each cell line and tissue [7]. By measuring the occurrence of particular genes in different tissues, the proteins can be determined. A plethora of research follows up on classifying all human genes according to specificity and occurrence throughout the human body [7], [8]. This information makes it feasible to extract individual fingerprints for different regions via the human proteome.

3 The Human Proteome as a Fingerprinting Scheme

The distribution and tissue specificity are taken into account for the choice of protein-coding genes for the fingerprints. Uhlén et al. published a database covering more than 90% of the putative protein-coding genes available at the *Human Protein Atlas* (HPA) [8]. Performing immunohistochemistry on 44 human tissues complemented by RNA-sequencing on 37 of them, protein and mRNA expression data were derived [9].

From the approximately 20,000 human protein-coding genes, around 9,000 genes show low tissue specificity and represent the housekeeping proteome [8]. The remaining genes show an elevated expression in particular tissues, subdivided into (i) tissue enriched genes with mRNA levels in one tissue at least four-fold higher than the maximum of any other tissue, (ii) group enriched genes with four-fold higher mRNA levels in a small group of tissues, and (iii) tissue enhanced genes with mRNA levels in a particular tissue at least four-fold higher than the average in all other tissues [8]. As for the distribution, genes can be (i) detected in a single tissue, (ii) detected in some tissues (more than one but less than one-third of all), (iii) detected in many tissues (at least one-third but not all), or (iv) detected in all tissues [8]. The distribution visualizes the number of genes showing detectable mRNA levels in different tissues. Gene detection is done by the normalization of transcriptomics data at hand from several different research groups [8]. A gene has detectable levels of mRNA molecules if the *normalized expression* (NX value) is above one [8].

3.1 Creating Fingerprints

For the sample space from which the genes for the fingerprints are chosen, genes not occurring in the respective tissues and genes with low tissue specificity are excluded. Table 1 gives an overview of the occurrence of genes according to the HPA in the major organs.

Table 1: Occurrence of Genes in Major Organs.

Tissue	Detected	Not Detected	Elevated Genes
Brain	16,227	3,443	2,587
Heart	14,409	5,261	387
Intestine	15,609	4,061	764
Kidney	14,823	4,847	413
Liver	14,110	5,560	936
Lung	15,021	4,649	239
Pancreas	14,490	5,180	422
Stomach	14,707	4,963	159

Furthermore, the elevated genes can be subdivided into four groups regarding the distribution, as mentioned earlier. Genes only detected in a single tissue are advantageous to ensure as unique fingerprints as possible. Since not all organs feature more than one gene from this category, genes detected in some tissues are also considered.

A comparison of the gene data of all relevant tissues showed that none of the tissues have a complete overlap of elevated genes. For a reliable localization, the number of genes should be as high as necessary but as low as possible to be still dependable. For our studies, we chose an amount of five genes for each fingerprint. Upon further research on the prevalence of genes in the examined tissues, the number of genes might differ from five.

The key element for choosing genes for the fingerprints is the NX value. The gene expression is estimated by measuring the average *transcripts per million* (TPM) for different

datasets of the human tissues in question [8]. For comparability of the different datasets, the numbers are normalized in several steps. The higher the NX value, the higher the mRNA level of a gene in the respective tissue. A *tissue specificity score* compares the tissue with the highest mRNA level of a gene to the tissue with the second-highest mRNA level. Therefore, the remaining genes suitable for the fingerprints are sorted for their tissue specificity score. The genes with the highest scores are subsequently chosen for the fingerprints. Table 2 shows the gene combination for the fingerprint of the heart with the corresponding distributions and tissue specificity scores.

Table 2: Fingerprint of the Heart.

Gene	Tissue Distribution	Tissue Specificity Score
MYL7	Detected in single	822
NPPA	Detected in some	418
TNNI3	Detected in some	367
NPPB	Detected in some	296
TNNT2	Detected in some	241

To guarantee unique fingerprints, we compared the genes chosen for the fingerprints. There is a maximum overlap of one gene between two fingerprints, ensuring unique gene combinations.

4 Conclusion and Outlook

The human body provides a plethora of possibilities to find unique combinations of substances for different body regions. In order to detect fingerprints via nanobots, the substance in question must be allocated in the cardiovascular system, show significant contrasts in occurrence or concentration in different body regions, and cannot be affected by diseases or physical activity. The human proteome, which resembles the entirety of human proteins, grants an inventory with many different genes which code different proteins with certain tissue specificities. By using the gene specificity and distribution throughout different tissues, unique combinations can be found, accounting for only one body region at a time. Each fingerprint identified in this work consists of five protein-coding genes, showing no intersections in the most crucial genes, ensuring an exclusive localization for each tissue.

The localization of nanobots utilizing histological values is a novel method for which no research is available. In the scope of this work, possibilities, difficulties, and a promising fingerprinting method were pointed out. Fingerprinting via the human proteome is a suitable method to determine positions in the human body. The proteins appearing in different areas can give a definite localization and constitute a promising new method. Further research on the human proteome and the possible detection methods by nanobots has to be carried out to integrate the localization via fingerprinting into existing simulations.

4.1 Future Research

It is still unclear how nanobots will detect the fingerprints. There are different designs of nanobots which could be electrical or biological. A way of recognizing genes in the human body could be through DNA-tile-based nanobots [10]. Those nanobots consist of assembled DNA molecules while an arbitrary number of DNA strands have sticky ends, meaning that the end of the DNA strand is open and can bind another DNA molecule. With the different designs of open strands, all logic operators can be applied using DNA-tile-based nanobots. For 2-gene fingerprints, detection could be realized by a 2-bit-AND tile-based nanobot. Two open strands form counterparts for the fingerprint genes and send a message of localization to the outer environment only if both of the genes bind to the nanobot.

Furthermore, it is worth investigating if only one gene could be sufficient for localization in the human body. The total mRNA concentration and the relative concentration compared to other tissues are provided by the HPA [8]. With this information, concentration gradients can be examined, and conclusions for the localization can be drawn. If the communication of nanobots among each other allows it, the concentrations of different genes can be compared, therefore determining the position of each nanobot.

Depending on the type of nanobot and detection method, the fingerprint genes require more research regarding the biological structure and binding affinity. It has to be ensured that the nanobot binds to the correct genes or measures the correct concentrations, which highly depends on the chemical pull between the chosen nanobot and the gene in question.

A simulation framework for the detection via fingerprints has to be developed to test different scenarios. It would be advantageous to integrate the localization into existing simulations for nanobots in the human body, e.g., BloodVoyagerS which is a body simulator module for the network simulator ns-3 [11]. BloodVoyagerS simulates the blood flow in the human body through the major vessels and integrates nanobots into the flow. It then depicts the distribution of nanobots throughout the human body.

Acknowledgement

Parts of this work have been supported by the German Research Foundation (DFG) in the context of the project NaBoCom: Connecting in-body communication with body area networks.

It has been carried out at the Institute of Telematics, Universität zu Lübeck.

Author's Statement

Conflict of interest: Authors state no conflict of interest.

5 References

[1] M. Stelzner, F. Dressler, and S. Fischer, "Function centric nano-networking: Addressing nano machines in a medical application scenario," *Nano Communication Networks*, vol. 14, pp. 29–39, Dec. 2017. [Online]. Available: https://doi.org/10.1016/j.nancom.2017.09.001

[2] H. Mairbäurl, "Anpassung des erythrozytären Sauerstofftransports an Belastung und Höhe," *Jahrbuch 2014 Österreichischer Gesellschaft für Alpin- und Höhenmedizin*, pp. 161–185, Jan. 2014.

[3] N. Hassanin, M. M. Elkishki, and L. H. Fawzy, "Trace elements and their relation to diabetes mellitus and obesity," *Journal of Recent Advances in Medicine*, vol. 2, pp. 128–132, Jan. 2021. [Online]. Available: https://jram.journals.ekb.eg/article_123247.html

[4] R. Brigelius-Flohé and P. E. Petrides, *Löffler/Petrides Biochemie und Pathobiochemie*. Springer-Verlag, 2014, ch. Essentielle Spurenelemente, pp. 736–744. [Online]. Available: https://doi.org/10.1007/978-3-642-17972-3_60

[5] J. Versieck, "Trace elements in human body fluids and tissues," *Critical reviews in clinical laboratory sciences*, vol. 22, pp. 97–184, 1985. [Online]. Available: https://pubmed.ncbi.nlm.nih.gov/3891229/

[6] T. A. Brown, *Genomes*. Wiley-Liss, 2002, ch. 3: Transcriptomes and Proteomes. [Online]. Available: http://www.ncbi.nlm.nih.gov/books/NBK21128/

[7] P. J. Thul and C. Lindskog, "The human protein atlas: A spatial map of the human proteome," *Protein Science: A Publication of the Protein Society*, vol. 27, pp. 233–244, Jan. 2018.

[8] "The human protein atlas." [Online]. Available: https://www.proteinatlas.org/

[9] M. Uhlén et al., "Tissue-based map of the human proteome," *Science*, vol. 347, p. 1260419, Jan. 2015.

[10] F.-L. Lau, F. Büther, R. Geyer, and S. Fischer, "Computation of decision problems within messages in DNA-tile-based molecular nanonetworks," *Nano Communication Networks*, vol. 21, Sep. 2019.

[11] R. Geyer, M. Stelzner, F. Büther, and S. Ebers, "Bloodvoyagers: simulation of the work environment of medical nanobots," in *Proceedings of the 5th ACM International Conference on Nanoscale Computing and Communication*. ACM, 2018, pp. 1–6.

Economic Evaluation of Rotational Atherectomy in Coronary Heart Disease Performed in a Tertiary Hospital

Julia Bandura [1], Joachim Weil [2,3]

[1] Biomedical Engineering, Lübeck University of Applied Sciences, julia.bandura@stud.th-luebeck.de
[2] Sana Klinik Lübeck, Medizinische Klinik II, Lübeck, joachim.weil@sana.de
[3] Universität zu Lübeck

Abstract

Coronary Artery Disease (CAD) causes around 40 % of the cardiovascular deaths worldwide. Percutaneous Coronary Interventions (PCI) and Rotational Atherectomy (ROTA) are non surgical methods to treat CAD. ROTA is especially useful in calcified coronary lesions were the use of PCI alone is limited. However, it is often assumed that ROTA is an expenisve procedure compared to PCI. The economic outcome of 906 consecutive ROTA and PCI procedures in a tertiary hospital in Germany was analyzed in a retrospective analysis. Both groups were compared in regard to their baseline patient characteristics. Clinical outcome was evaluated with the material costs given by the *WaveMark™ Supply Management* system and the documented Diagnosis Related Group (DRG) flatrate. The groups differed in a few baseline characteristics and ROTA had a higher risk profile. Although the material costs were higher for ROTA, considering the reimbursement through the DRG flatrate, ROTA is not more expensive than PCI for the hospital.

1 Introduction

Diseases of the cardiovascular system are the most common causes of deaths worldwide [1]. Around 40 % of those deaths are caused by coronary artery disease (CAD). CAD is an atherosclerotic disease that leads to stenosis of the coronary arteries. If the conservative medical treatment is not sufficient an invasive revascularization is the treatment of choice. Percutaneous Coronary Interventions (PCI) are one of the standard revascularization techniques used today [2]. It is a non-surgical catheter based method where the inflation of a balloon at the catheter tip widens the diameter of the narrowed vessel by compressing the plaque. Ablative methods and atherectomy are added to standard balloon PCI where it is not sufficient to compress the plaque but it is necessary to remove it. In Rotational Atherectomy (ROTA) a catheter with an elliptically shaped burr at the tip rotates with high speed. This rotating burr removes the stenosis by fragmenting the plaque into particles that are small enough to get washed away with the blood flow. ROTA is used for removing inelastic plaque like calcified or resistant fibrotic tissue. The healthy elastic tissue of the vessel wall bends away and only the inelastic plaque is removed. However, ROTA is not suitable for the ablation of soft thrombotic plaques [3]. PCI or ROTA can be combined with Drug Eluting Stents (DES).

It is often assumed that ROTA is a more expensive intervention compared to PCI because of the high material costs. But it is not proven that higher material costs also result in a higher total hospital stay cost since many factors have to be considered. In Germany the *Statutory Health Insurance Reform Act of 2000* a DRG-based hospital payment method was introduced. It is based on an algorithm assigning patients to a specific DRG decided on factors like major diagnosis, procedures, secondary diagnoses, and patient characteristics. The DRG code together with the length of patient stay determines the reimbursement (DRG flatrate) of the case [4]. Automated inventory management systems are used by hospitals to reduce the expense of inventory management. The cardiac catheter laboratory of the *Sana Kliniken Lübeck GmbH* works with the *WaveMark™ Supply Management* system to document the material cost for each procedure in the catheter laboratory. With the documented DRG flatrates for each case and the material costs given by the *WaveMark™ Supply Management* system the economic outcome for ROTA and PCI was retrospectively analyzed to gain a better insight of the true in-hospital costs at a tertiary hospital without onsite cardiac surgery.

2 Material and Methods

The data of 906 consecutive PCI and ROTA procedures, performed in the time from April 2019 to December 2020, was retrospectively analyzed. Material costs and product quantity were given by the Wavemark™database and exported to an Excel (*Microsoft Office 365*) spreadsheet. With the procedure numbers from this table the corresponding clinical data was found in the clinical database in the form of procedural reports, laboratory reports, as well as physician's letters. The anonymized clinical data was stored in an Excel spreadsheet.

To facilitate the comparison of the PCI- and ROTA-group only elective and not Impella-supported procedures were included into the final analysis. Since the DRG flatrate for in-hospital patients is paid per stay and not per procedure, only patients with one procedure per stay were selected.

The *PCI complication - Mayo score* was calculated via the *QXMD* app/wep page [5]. The *CathPCI Bleeding Risk* score was calculated at the *Tools.Acc.org* wep page and app [6].

Descriptive analyses were used for this project. To describe the baseline patient characteristics and economic outcome continuous variables were written as mean with standard deviation and categorical variables were reported as frequencies with percentage. Comparison of means for the baseline patient characteristics was done using the T-test and comparison of proportions for the baseline patient characteristics with chi-squared. $P \leq 0.05$ was considered statistically significant. A part of the economic outcome (costs and DRG flatrate) is presented as percentages because of the clinical data security policy. All calculations were done using Excel and/or Matlab 2019b (*MathWorks®*). The graphs were created with Matlab 2019b.

3 Results and Discussion

3.1 Baseline Patient Characteristics

From originally 906 procedures 530 (314 PCI and 216 ROTA) met the inclusion criteria and were chosen for the evaluation of the economic outcome. The baseline patient characteristics of the 530 procedures are shown in Table 1. ROTA has a higher risk profile in some baseline characteristics than PCI. ROTA patients were older than the PCI patients with a mean age of 70.56 ±9.73 years and 68.54 ±11.06 years (p = 0.03), respectively. There was no statistical difference in proportions of sexes among both groups (p = 0.07). The mean BMI was 28.20 in both groups, with a standard deviation of 4.82 for ROTA and 4.61 for PCI (p = 0.98).

89,35 % had arterial hypertension in the ROTA-group and 81.53 % in the PCI-group (p = 0.01) indicating a higher baseline risk for ROTA. There was no statistical difference for the number of patients with a history of nicotine abuse or who are still active smokers among groups (p = 0.18), as well as for the number of patients with genetic predisposition for cardiac diseases (p = 0.17). However, the ROTA-group had a higher proportion of patients with diabetes mellitus type II (p = 0.04). In both groups a type of lipometabolic disorder was documented in most cases (p = 0.59). Arterial hypertension, nicotine abuse, genetic predisposition, diabetes mellitus type II, lipometabolic disorder, and the male sex were considered as cardiovascular risk factors. The number of cardiovascular risk factors per patient was higher in the ROTA-group (p < 0.01).

The amount of patients with renal insufficiency stage 3 or higher was greater in the ROTA-group (p < 0.01). Single vessel disease was almost twice as frequent in the PCI-group as in the ROTA-group (p < 0.01) while double vessel disease was practically of the same frequency in both groups (29.63 % for the ROTA-group and 29.94 % in the PCI-group, p = 0.94). Triple vessel disease was more common in the ROTA-group with 58.80 % and 46.82 % in the PCI-group (p < 0.01). The occurrence of a ST-Segment-Elevation Myocardial Infarction (STEMI) prior to the intervention (anytime prior to the intervention, no acute STEMI) did not differ among groups (p = 0.95). The mean baseline hemoglobin was higher in the PCI-group (p < 0.01). More patients in the ROTA-group than in the PCI-group already had a PCI anytime before the considered procedure (p = 0.01). No cardiogenic shock within 24 hours prior to the intervention was observed in the ROTA-group but one case (0.32 %) was reported in the PCI-group (p = 0.41). No patients of the ROTA-group but three patients (0.96 %) of the PCI-group were on dialysis (p = 0.15). There was no statistical difference in the mean glomerular filtration rate among groups (p = 0.21), as well as in the mean serum creatinine value (p = 0.51) and in the number of patients with left main coronary disease (LMCAD, p = 0.74). For both groups class I and II were the most common NYHA classifications with 83.33 % in the ROTA-group and 95.86 % in the PCI-group (p < 0.01). A thrombus on angiography was observed more often in the PCI-group (p = 0.03).

The mean CathPCI Bleeding risk was low in both groups (p = 0.28). The most common Mayo risk category for the ROTA-group was *low risk* and *very low risk* for the PCI-group. Only a few patients of both groups (less than 5 % of cases) were assigned to the Mayo risk categories *high risk* and *very high risk*. Only in the *very low risk* and *moderate risk* categories a statistical difference was observed among the groups (p < 0.01 and p = 0.02, respectively).

A statistical difference in the baseline patient characteristics were observed in 15 of 33 categories: age, the number of patients with arterial hypertension, diabetes mellitus type II, number of cardiovascular risk factors, the number of patients with renal insufficiency stage 3 or higher, the number of pathological coronary vessels, the number of patients with previous PCI, hemoglobin, New York Heart Association (NYHA) classes, thrombus on angiography, and Mayo risk score categories *very low risk* and *moderate risk*. This is inevitable since ROTA is a treatment for calcified lesions which should result in a different patient group.

3.2 Economic Outcome

The economic outcome was described with the overall DRG flatrate, mean DRG flatrate, mean material cost, and product quantity. During PCI an average of 8 ±3.49 products was used while during ROTA an average of 13.45 ±4.06 products was used. This was as expected because ROTA is more complex than standard PCI and the number of pathological coronary vessels to treat was higher in the ROTA-group. This is also evident in the distribution of material cost. The mean material cost of ROTA was almost four times as high as PCI material cost. As shown in Fig. 1 the mean material cost of ROTA is 78.72 % of the com-

Table 1: Baseline patient characteristics, p $\leq$ 0.05 is represented by *

	ROTA	PCI
Age (yrs)*	70.56 ±9.73	68.54 ±11.06
Males, *n (%)*	160 (74.07)	210 (66.88)
Females, *n (%)*	56 (25.93)	104 (33.12)
BMI	28.20 ±4.82	28.20 ±4.61
BMI >30, *n (%)*	68 (31.48)	99 (31.53)
Arterial hypertension, *n (%)**	193 (89.35)	256 (81.53)
Nicotine abuse, *n (%)*	102 (47.22)	130 (41.40)
Genetic predisposition, *n (%)*	58 (26.85)	68 (21.66)
Diabetes mellitus Type II, *n (%)**	75 (34.72)	83 (26.43)
Lipometabolic disorder, *n (%)*	152 (70.37)	214 (68.15)
Cardiovascular risk factors*	3.43 ±1.13	3.06 ±1.25
Renal insufficiency >3, *n (%)**	62 (28.70)	54 (17.20)
Single vessel disease, *n (%)**	25 (11.57)	73 (23.25)
Double vessel disease, *n (%)*	64 (29.63)	94 (29.94)
Triple vessel disease, *n (%)**	127 (58.80)	147 (46.82)
Prior STEMI, *n (%)*	21 (9.72)	30 (9.55)
Hemoglobin (g/dL)*	13.47 ±1.55	13.88 ±1.57
Prior PCI, *n (%)**	139 (64.35)	167 (53.18)
Cardiogenic shock within 24 h, *n (%)*	0 (0.00)	1 (0.32)
Currently on dialysis, *n (%)*	0 (0.00)	3 (0.96)
Glomerular filtration rate (mL/min/1.m²)	70.01 ±19.47	72.25 ±21.71
Serum creatinine (mg/dL)	1.07 ±0.31	1.05 ±0.44
LMCAD, *n (%)*	9 (4.17)	15 (4.78)
NYHA 1-2, *n (%)**	180 (83.33)	301 (95.86)
NYHA 3-4, *n (%)**	36 (16.67)	13 (4.14)
Thrombus, *n (%)**	1 (0.46)	10 (3.18)
CathPCI Bleeding Risk	1.82 ±1.32	1.97 ±1.76
Mayo - very low risk, *n (%)**	80 (37.04)	162 (51.59)
Mayo - low risk, *n (%)*	81 (37.50)	101 (32.17)
Mayo - moderate risk, *n (%)**	45 (20.83)	41 (13.06)
Mayo - high risk, *n (%)*	9 (4.17)	8 (2.55)
Mayo - very high risk, *n (%)*	1 (0.46)	2 (0.64)

bined mean material cost of PCI and ROTA. The share of mean PCI material cost is 21.28 %. Levin et al. also reported a higher material cost for ROTA than for PCI. In their study the ROTA catheterization laboratory cost was

approximately 1.5 times higher than for PCI [7].

The overall DRG flatrate shown in Fig. 1 is the combined DRG flatrate of both groups. Although the ROTA-group included 98 procedures less than the PCI-group, it holds a bigger share on the overall combined DRG flatrate (56.41 % ROTA-group and 43.59 % PCI-group).

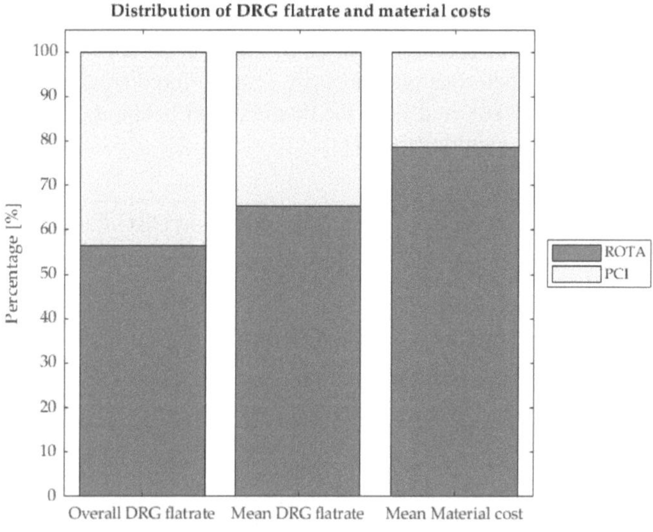

Figure 1: Distribution of overall and mean DRG flatrate and material costs of both groups. The Overall DRG flatrate is the sum of the total DRG flatrate of both groups. The Mean DRG flatrate is the sum of the mean DRG flatrate of both groups and the Mean Material cost is the sum of the mean material costs of both groups.

In Fig. 1 the combined mean DRG flatrate of both groups can be seen, too. The share of the mean DRG flatrate of the ROTA-group in this case is 65.29 % and the share of the PCI-group is 34.71 %. This means that in average for every patient undergoing ROTA the DRG flatrate is almost twice as high as for a patient undergoing standard PCI. The patients in the ROTA-group are assigned to a higher DRG flatrate because the factors like major diagnosis, procedures, secondary diagnoses, and patient characteristics must be considered as more costly to treat.

Table 2 shows the relation of the material costs and the DRG flatrate for both groups. For the considered time period from April 2019 to December 2020 the share of the material costs from the DRG flatrate for the ROTA-group was 34.82 %. More than one third of the DRG flatrate has to be used to pay the material costs of the ROTA procedures. Whereas the share of the material costs from the DRG flatrate for the PCI-group was 17.71 %, which is less than a fifth of the DRG flatrate. This seems like a huge difference but if the overall DRG flatrate is subtracted by all material costs the difference for each group is not that much apparent. So with material costs subtracted from the overall DRG flatrate the ROTA-group holds a proportion of 50.62 % and the PCI-group holds 49.38 % (see Table 2) for the whole time period of April 2019 to December 2020. The proportion of the ROTA-group is higher even though 98 procedures less were considered for this group.

As mentioned before the ROTA procedure is way more complex and different devices have to be used which results in a higher material cost. Although the material cost is quite high in regard to the DRG flatrate of the ROTA-goup it still holds a bigger share of the overall DRG flatrate that the hospital has received for this time period although less ROTA procedures were done. While ROTA procedures are expensive in material costs they are still economical.

A similar effect was observed by Levin et al. Although the mean catheterization laboratory costs differed significantly among ROTA and PCI, the median total hospital cost did not differ significantly [7].

Table 2: Relation of material costs and DRG flatrate

	ROTA	PCI
Ratio of material costs to DRG flatrate, %	34.82	17.71
Distribution of overall DRG flatrate without material costs, %	50.62	49.38

4 Conclusion

Although this study was performed diligent, it still has some limitations. The DRG flatrate is not only used to settle the material costs of a procedure but all costs caused during the patients stay at the hospital which involves medicine, personnel cost, meals, and more. In this study only the material costs were considered. To get a better and fuller view on the economics of the procedures it is necessary to investigate other costs, too. Another limitation is that this study is retrospective. The data was not specifically gathered for this purpose.

In conclusion, both procedures PCI and ROTA have application and are necessary to treat different kind of lesions. Although the material costs for ROTA are expensive, it is still cost effective because of the higher DRG flatrate for ROTA patients. Thus ROTA is not more expensive for the hospital than PCI. And as shown by Levin et al. the median total hospital costs did not differ among ROTA and PCI [7].

Acknowledgement

The work has been carried out at the Sana Kliniken Lübeck GmbH and was supervised by Prof. Dr. med. Joachim Weil, Universität zu Lübeck and Sana Kliniken Lübeck GmbH.

Author's Statement

Conflict of interest: Authors state no conflict of interest.

5 References

[1] Amini, M., Zayeri, F., & Salehi, M. (2021). *Trend analysis of cardiovascular disease mortality, incidence, and mortality-to-incidence ratio: results from global burden of disease study 2017.* BMC Public Health, doi: 10.1186/s12889-021-10429-0

[2] Hartmann, F. (2020). *Koronare Herzkrankheit.* In U. Stierle & J. Weil (Eds.), Klinikleitfaden Kardiologie (Siebte Ausgabe) (Siebte Aus, pp. 95–180). Munich: Urban & Fischer. doi: https://doi.org/10.1016/B978-3-437-22285-6.00004-X

[3] De Belder, M., & Kunadian, B. (2008). *Rotational and directional atherectomy.* Essential Interventional Cardiology, 1, 165–176. doi: 10.1016/B978-0-7020-2981-3.50017-9

[4] Quentin, W., Geissler, A., Scheller-kreinsen, D., & Busse, R. (2010). *DRG-type hospital payment in Germany : The G-DRG system.* Euro Observer, 12(September 2014), 4–6. Retrieved from https://eurodrg.projects.tu-berlin.de/publications/DRG-type hospital payment in Germany.pdf

[5] QXMD. (2015). *Calculator PCI Complication - Mayo.* Retrieved from https://qxmd.com/calculate/calculator_139/pci-complication-mayo

[6] American College of Cardiology. (2017). *Cath-PCI Bleeding Risk Calculator.* Retrieved from https://tools.acc.org/CathPCIBleedRisk/?redirect=f#!/content/calculator/

[7] Levin, T. N., Holloway, S., & Feldman†, T. (1998). *Acute and late clinical outcome after rotational atherectomy for complex coronary disease.* Catheterization and Cardiovascular Diagnosis, 45(2), 122–130. doi: 10.1002/(SICI)1097-0304(199810)45:2<122::AID-CCD5>3.0.CO;2-E

Test System for Qualification of a Mass Flow Meter for Medical Ventilation

Wiebke Heyer [1]
[1] Medical Engineering Science, Universität zu Lübeck, wiebke.heyer@student.uni-luebeck.de

Abstract

Mass flow meters are important components of a medical ventilator and are necessary for monitoring and controlling the gas volumes supplied to patients. To qualify a mass flow meter for future use in a ventilator, it must first be integrated into a development environment to test its functionality. For this purpose, a test setup including a SpeedGoat Target Machine, which operates the I2C (Inter-Integrated Circuit) interface of the sensor, is used. In a Simulink model, the functionality for the communication via I2C is programmed. While using multiple sensors with the setup, limitations of the system were identified and attempts were made to improve performance. Two of the desired three mass flow meters can be successfully read out in the required sampling time simultaneously. For the communication with the desired three sensors further measures must be taken to maximize the processing speed via the I2C interface.

1 Introduction

In a medical ventilator flow meters play an important role in addition to differential pressure sensors to ensure that the vital exchange of gases in the patient's lungs can take place to a sufficient extent. In order to be able to monitor and control the gas volumes supplied to the patient as precisely as possible, a variety of requirements are placed on the sensor technology. Not only the volume of the supplied gases, but also their mixing ratio and the timing of the supply must be monitored in order to then design a control system based either on pressure measurement or on flow measurement.

Additionally, ventilators are used in different environments causing the exact requirements to vary. In emergency medicine a smaller dynamic range is needed; instead, a low pressure drop across the flow meter is important to avoid straining the turbine and thus extending battery life. In clinical devices, on the other hand, more comprehensive functions are desired, such as differentiation between adult and pediatric patients, which requires high measurement accuracy and fast response times [1].

In the wake of the Covid-19 pandemic, further requirements arose; flow meters had to be mass produced quickly and cheaply, drawing on existing resources to meet the increasing demand for ventilators. In this respect, not only the hardware of the sensors had to be optimized, but also the software had to be designed in such a way that interfaces could be programmed quickly and easily. For this purpose, Sensirion AG developed the mass flow meter SFM3019, which was optimized based on the already existing SFM3000. This flow meter is to be qualified for use in a new ventilator and will be integrated into a test setup in the course of this. First, the communication via the I2C interface of the sensor will be established and subsequently the setup will be extended so that up to three sensors can be controlled simultaneously. This extension is important because typically up to three flow meters are used in a ventilator, two in the inspiratory branch, distinguishing between oxygen and air supply, and one in the expiratory branch.

2 Material and Methods

2.1 Test Setup

To qualify a sensor, various components of the sensor must be tested, such as measurement accuracy, pressure drop, frequency response, and stability of communication. However, before actual measurements can be made, a test system must be created to operate the sensor's interface and provide a development environment in which the necessary functionality can be programmed. Building on this, the setup can then be extended for qualification. The requirement for the test system is that three mass flow meters can be read out simultaneously with a maximum sampling time of 1 ms.

The setups main component is the SFM3019, a digital mass flow meter for high-volume applications by Sensirion AG. It measures unidirectional in a flow range from -10 slm to 240 slm and is calibrated for air, oxygen and mixtures thereof with an accuracy of ±3 % and a step response time < 3 ms. The output signal of the sensor is a 16-bit digital signal with an update rate of maximal 2.2 kHz and is internally linearized and temperature compensated. The design of the flow channel results in a low pressure drop of < 0.8 mbar at 60 slm and < 5 mbar at 200 slm. Additionally, the sensor measures temperature in a range from - 20 °C to

Figure 1: Overview of the test setup: development computer running Simulink connected via Ethernet to the SpeedGoat target machine communicating via I/O module with the mass flow meter [2, 7].

85 °C with an accuracy of ±3 °C [2]. The flow meter is operated with Sensirions patented *CMOSens* Technology, which combines the sensor components for measuring flow and temperature with the signal-processing circuitry on a single CMOS (Complementary Metal-Oxide Semiconductor) chip. Flow is measured with a thermal measurement principle using a heating element located in the center of a pressure-stabilized membrane surrounded by two temperature sensors. A flow of gas over the membrane results in a thermal transfer of heat and thus in a difference of temperature between the upstream and downstream sensor. The membrane is integrated on the CMOS chip as well as a signal amplifier and an analog-to-digital converter [3].

Furthermore, the chip includes the digital signal-processing circuitry with an I2C communication interface. I2C stands for Inter-Integrated Circuit and describes a serial data bus consisting of two 8-bit orientated bidirectional wires, which is why the interface is also referred to as Two-Wire-Interface. Developed by Philips Semiconductors (now NXP Semiconductors) I2C is widely used to connect device-internal components over short distances. It consists of a hierarchic bus system using the master-slave-principle where each slave has a unique address and the number of slaves is only limited by the bus capacity and the number of available addresses. Two wires, SDA (Serial Data Line) and SCL (Serial Clock Line), are needed for communication between master and slave, where both can act as either transmitter or receiver. The I2C bus speed depends on the clock frequency generated by the master. In standard mode the data transmission rate is 100 kbit/s, in faster modes 400 kbit/s, 1 Mbit/s or 3.4 Mbit/s. The ultra-fast mode with 5 Mbit/s is a unidirectional mode [4].

The master initiates the data transfer by generating the start condition followed by the 7-bit slave address with the most significant bit first. Bit 0 of the address byte is either 0 indicating that the master will write a byte to the slave or 1 indicating that the master will read from the slave. Each written or read byte must be acknowledged by the receiver and after all bytes are written or read the master sends a stop condition. The communication is entirely controlled by the master [4].

The physical interface consists of the two I2C wires connected via pull-up resistors to an I/O module of a SpeedGoat Real-Time Target Machine. Target machines are typically used in rapid control prototyping systems for real-time applications of control algorithms created with Simulink, for example. In this test setup a SpeedGoat Performance Real-Time Target Machine with a configurable I/O-module

IO323 is used, which combines analog and digital input and output channels with a selection of standard interfaces as SPI, PWM and I2C [5]. Custom configurations can be achieved via *bitstream* files according to the requirements of the application. The target machine is connected to a computer running MATLAB and Simulink in the version 2021b on Windows 10. To create real-time applications from Simulink models the extension Simulink Real-Time is required. Simulink Real-Time allows the user to expand their Simulink models with SpeedGoat I/O driver blocks to run the model as a real-time application on their target machine. Both, the SpeedGoat system and Simulink Real-Time are expressively designed to be used together [6].

Fig. 1 shows the complete testing environment, beginning with the physical system, in this case the SFM3019 sensor, connected via an I/O module to the SpeedGoat target computer, where a multicore CPU runs a Simulink Real-Time kernel, FPGAs, I/O and protocol interfaces. Furthermore, the development computer running MATLAB and Simulink including MATLAB Coder, Simulink Coder and Simulink Real-Time is connected to the target computer via Ethernet. This environment lets the user design a model in Simulink, automatically builds and downloads the application from Simulink to the target computer using automatic code generation and during the real-time execution offers the possibility to tune and log signals in Simulink [7].

2.2 Simulink Model

The Simulink model created for the real-time application must include a driver block dedicated to the setup of the I/O module in use, in this case the IO323. In the setup the desired configuration file (bitstream) for the module must be selected. Further SpeedGoat driver blocks used for the communication with the flow meter are an *analog output* block to control the voltage supply of the sensor, an *I2C master write* block to send commands to the slave and an *I2C master read* block to read data send by the slave. In the properties of the two I2C driver blocks the I2C channel in use depending on the selected *bitstream* file must be specified, as well as the desired sampling time which must be an integer multiple of the fixed step size specified for the entire model. Lastly the I2C clock frequency generated by the master depends on the FPGA frequency and is defined by a clock divider.

Required inputs for the *I2C master write* block are the slave ID, a 7-bit identifier which is specified by the manufacturer of the sensor, the number of bytes to write to the slave and the data to be written. The data has to be presented in a vector of the size 64x1 limited by the hardware buffer size, where each component of the vector represents one byte of data. Via the *I2C write block* commands can be send to the sensor. Following the I2C protocol the master initiates the communication with the start condition followed by the slave address and the write bit. After this first byte is acknowledged by the slave the master sends either two bytes each acknowledged by the sensor or two bytes for the

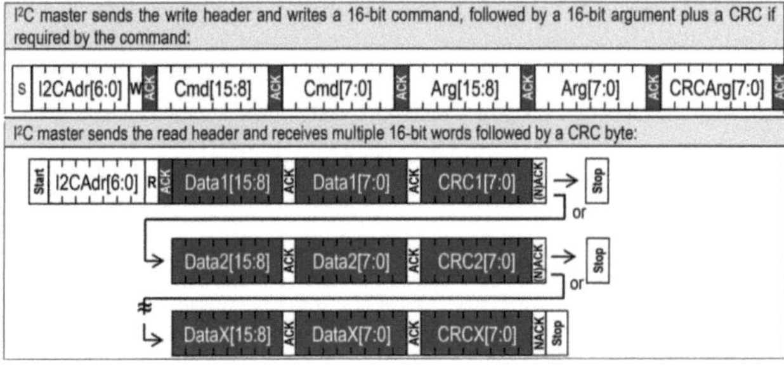

Figure 2: I2C sequences to write commands to the sensor (top) and read data from the sensor (bottom) [2].

command and additionally an argument consisting of two bytes as well, followed by a checksum byte. The *number of bytes to write* parameter must be adjusted accordingly. The sequence is shown in Fig. 2.

For the SFM3019 there are three commands to start a continuous measurement of flow and temperature depending on the type of gas. Possible gas types are oxygen, air and a mixture thereof. While the continuous measurement is running no other command can be send to the sensor apart from the stop command and a command to update the oxygen concentration in the gas mixture in case the measuring mode is set to mixtures of air and oxygen. The new volume fraction of oxygen in per mille must be converted according to the hexadecimal system and then send in two bytes as the argument of the command. To ensure communication reliability a checksum must be calculated additionally, for that the cyclic redundancy check (CRC) with an 0x31 polynomial is used. The checksum covers the two bytes of the argument. Further commands include the configuration of the averaging mode, a soft reset, entering the sleep mode, reading the product identifier and the scale factor, offset and flow unit [2].

The generation of commands in the Simulink model can be controlled via a drop-down menu and edit boxes in case a command needs an argument to be sent along with it. If a checksum is necessary, it is calculated automatically via an integrated MATLAB function and written to the sensor along the command and argument. The driver block input parameter describing the number of bytes to write is adjusted to either two bytes or five bytes automatically.

Required inputs for the *I2C master read* driver block are the slave ID and the number of bytes the master should read from the slave. The output consists of the data retrieved from the slave stored in a vector of the size 64x1 and an additional optional output to display the state of the acknowledge error flag. According to the I2C communication protocol, shown in Fig. 2, the master sends the address followed by a read bit. After that first byte is acknowledged by the slave, the sensor controls the data line and sends a total of nine bytes to the master, regardless of the command send prior. The sequence consists of multiple 16-bit words each followed by an 8-bit checksum. After every checksum byte further transmission can be aborted by a stop condition sent by the master. If one of the "start continuous measurement"

commands has been sent the measured flow value is updated with a maximum rate of 2.2 kHz and the temperature value with an update rate of 112.5 Hz. Whenever the master sends the read header the slave answers with the latest flow value as a signed 16-bit integer and the associated checksum as the first three bytes followed by the temperature and associated checksum in bytes four to six. The flow and temperature values are represented by a two's complement and have to be converted to physical values by dividing by a scale factor and subtracting an offset. The last three bytes of the nine-byte sequence are used to return a status word and again a checksum. The status contains the currently running measurement command, information on the averaging mode and the defined gas concentration in case of a measurement in gas mixtures. Flow and temperature can be displayed in real-time using a scope, the status and the three checksums are analyzed in dedicated MATLAB functions and displayed on the dashboard. Users can therewith check if the retrieved data is valid and if the sensor accepted a sent command.

2.3 Extension to Multiple Flow Meters

After one sensor has been successfully integrated into the test setup, the setup is expanded to independently control and read the desired three sensors. However, since all three sensors have the same I2C address, which is fixed by the manufacturer, they cannot be controlled via a single I2C interface. The number of slaves with different addresses would theoretically only be limited by the bus capacity, but due to this circumstance three I2C interfaces are needed. Since the SpeedGoat I/O module used is a configurable module, appropriate interfaces can be retrofitted by adapting the configuration of the module via a *bitstream* file provided by SpeedGoat.

The adaptation of the Simulink model can essentially be done by copying the existing structure. It should be paid attention that the configuration file is edited in the setup driver block and that the power supply for the additional sensors is ensured. There must be a separate pair of read/write blocks for each of the three channels. However, using multiple I2C driver blocks significantly increases the Task Execution Time (TET) of the application. This is because the *I2C master read* block waits for the I2C to be terminated also physically, it first initiates the I2C frame, then waits until the data is written to the frame and the frame is terminated. After the frame is terminated, the data has to be written from the FPGA to the CPU and only after this is completed the *I2C master read* block terminates its execution. Additionally, the time the transmission to the CPU takes adds up accordingly. The described effects lead to the fact that with the use of three I2C channels the TET exceeds the sampling time of the model during the readout of the sensors and thus leads to a CPU overload. A solution to the problem would be to reduce the sampling rate, but this would result in the sampling rate of the flow measurement no longer meeting the requirements of the ventilation system.

Another approach is to increase the I2C clock frequency

that is set by the clock divider in the *I2C master* driver blocks. The clock divider defines for how many FPGA ticks the I2C clock signal remains LOW, respectively HIGH. As for one full clock cycle one bit can be transmitted in I2C, this value defines the overall speed of the transmission. The clock divider accepts values in the range from 4 to 65535 based on the fixed FPGA frequency of 75 MHz for the IO323. This leads to an I2C frequency range of about 574 Hz to 9.4 MHz. In practice this frequency is not achievable for I2C communication, as the physical voltage lines cannot switch from LOW to HIGH with this frequency. Additionally, the frequency can be limited by the slaves. The SFM3019 flow meter allows an maximum I2C transmission speed of 1 MHz, which corresponds to the *Fast Mode* of the I2C bus. With increasing cable length, a further limitation of the bus speed can be caused, resulting in a maximum clock frequency of 750 kHz while preserving a stable connection.

3 Results and Discussion

With the presented test setup consisting of the Sensirion SFM3019 mass flow meter, the SpeedGoat target machine and the development computer including Simulink Real-Time, first a single sensor was successfully integrated. With a maximum I2C clock frequency of 750 kHz, a TET of 0.276 ms and thus a maximum sampling rate of 2.2 kHz could be achieved. With the developed Simulink model, all required commands can be sent to the sensor, including three commands to start and stop a continuous flow and temperature measurement in different media, the command to update the oxygen concentration during the measurement and the output of different status messages. The measured values for flow and temperature are correctly converted into physical values according to the provisions of the data sheet and their validity is determined by means of a checksum calculation.The functionality of the mass flow meter can be successfully tested with the described setup.

When connecting multiple sensors, a limitation of the system is reached. The use of multiple *I2C master* driver blocks leads to a significant increase in task execution time with the result that the required sampling time of 1ms cannot be met when connecting more than two sensors. However, by maximizing the I2C clock frequency and shortening the cables used, it was possible to successfully maintain communication with two of the three mass flow meters. With a resulting TET of approximately 0.786 ms during communication with two sensors, the application's 1 ms sampling time requirement is met.

4 Conclusion

With the test setup presented and the Simulink model created, two mass flow meters were successfully integrated and their functionality tested. In order to achieve the desired number of three sensors, further measures must be taken: the Task Execution Time must be further reduced so that the required sampling time can be met. An approach to be followed in the future is to apply explicit partitioning in order to distribute the calculation for each *I2C master read* block to another core on the CPU. Partitions can be created by using referenced models in the root level of the main model.

Acknowledgement

The work has been carried out at Drägerwerk AG & Co. KGaA, Lübeck and supervised by Prof. Dr. Philipp Rostalski, Institute for Electrical Engineering in Medicine, Universität zu Lübeck.

Author's Statement

Conflict of interest: Authors state no conflict of interest.

5 References

[1] D. Träutlein, *Gas Flow Sensing Technology in Medical Ventilation Devices*. Sensirion AG. Available: https://www.sensirion.com/fileadmin/user_upload/cus tomers/sensirion/Dokumente/0_Corporate/Specialist_ Articles/Sensirion_Specialist_Article_Ventilation_EN .pdf [last accessed on 2022-01-07].

[2] Sensirion AG, *Datasheet SFM3019 series*. Sensirion AG. 2020. Available: https://www.sensirion.com/fileadmin/user_upload/cus tomers/sensirion/Dokumente/5_Mass_Flow_Meters/ Datasheets/Sensirion_Mass_Flow_Meters_SFM3019_ Datasheet.pdf [last accessed on 2022-01-07].

[3] Sensirion AG, *CMOSens® Technology for Gas Flow and Differential Pressure*. Sensirion AG. Available: https://www.sensirion.com/en/about-us/company/technology/cmosens-technology-for-gas-flow/ [last accessed on 2022-01-07].

[4] NXP B.V., *I2C-bus specification and user manual*. 2021. Available: https://www.nxp.com/docs/en/user-guide/UM10204.pdf [last accessed on 2022-01-07].

[5] Speedgoat GmbH, *Performance real-time target machine*. Speedgoat GmbH. Available: https://www.speedgoat.com/products-services/real-time-target-machines/performance [last accessed on 2022-01-07].

[6] The MathWorks, Inc., *Simulink Real-Time*. The MathWorks, Inc.. Available: https://www.mathworks.com/products/simulink-real-time.html [last accessed on 2022-01-07].

[7] Speedgoat GmbH, *Simulink Real-Time Workflow*. Speedgoat GmbH. Available: https://www.speedgoat.com/learn-support/simulink-real-time-workflow [last accessed on 2022-01-07].

Development of an infrared radiation sensor system for non-contact temperature measurement during breathing gas conditioning

Celina Sophie Lemke [1], Norbert Linz [2], Ludger Tappehorn [3]

[1] Medical Engineering Science, Universität zu Lübeck, celina.lemke@student.uni-luebeck.de
[2] Institute of Biomedical Optics, Universität zu Lübeck, norbert.linz@uni-luebeck.de
[3] Drägerwerk AG & Co. KGaA, ludger.tappehorn@draeger.com

Abstract

During ventilation the breathing gas has to be warmed and humidified for patient safety. For regulating the temperature it is necessary to use sensors in the system. Currently Pt100 temperature sensors are used which can cause hygiene and/or user problems. Therefore, a method for contactless temperature measurement with infrared (IR) radiation shall be evaluated. We investigated different cuvettes/materials for contactless temperature measurement. Temperatures with an IR sensor and two reference sensors were determined to evaluate the measurement system. Three series of tests were performed – 1. dry gas and constant flow, 2. humidified gas and constant flow, 3. humidified gas and ventilation. It could be shown that an aluminium cone and a thin synthetic measuring window achieved the best results for contactless temperature measurement. We were able to show that the measurement with IR radiation is a good alternative to the current technology and hygiene and user problems are circumvented. It also offers a gain in terms of sustainability.

1 Introduction

Respiration is responsible for the supply of oxygen in the human organism. The conductive bronchial system provides the mucociliary clearance during inspiration and expiration. Mucociliary clearance describes the clearing function of the conductive system with producing mucus inside the respiratory system. The mucus transports dust and pathogens to the air tube (trachea) to avoid infections and other contaminations [1].

During ventilation, the patient is supplied with breathing gas, which is initially dry and cold. For this reason, the gas is humidified and heated to guarantee the clearing function. The humidification process can be carried out via *Heated Humidifiers* (HH) (active humidification). A water chamber is heated by a heating plate, which produces steam. The inspiratory air is passed through the water chamber where it is both heated and moistened (pass-over humidifier). Heated breathing hoses are mostly used to avoid a temperature drop on the way to the patient. In addition, they prevent the formation of condensate because of the temperature differences from inside and outside the hose. 37 °C is conditioned at the water chamber and 39 °C shortly before the patient to achieve an optimal gas temperature in case of invasive ventilation [1]. Because of the heating process it is necessary to measure the temperature right before the patient to ensure that the temperature of the breathing gas is not too high and the control circuit of the humidifier is correctly working. This measuring is normally carried out with

Pt100 resistance temperature sensors [2]. There are systems with sensors for multiple use that are inserted into the ventilation hose and sensors for single use that are integrated into the hose. The disadvantage of insertable temperature sensors is that user errors, such as incorrect insertion of the sensors, can occur. This leads to incorrect measurement und leakages in the breathing system. Another problem is the aspect of hygiene. Reusable sensors are only disinfectable and can not be sterilised. For this reason the risk of cross-contamination is high. To solve the issue of cross-contamination some manufacturers use integrated sensors. The disadvantage of these sensors is that they are thrown away with the tube. In addition, there are higher manufacturer costs due to directly installed electronics, and disposable systems offer a major disadvantage in terms of sustainability. The entire electronics are disposed with the plastic tube, which does not conserve resources and also creates a large and poorly separable waste production [4].

For this reason an infrared (IR) radiation sensor system for non-contact temperature measurement is to be developed, with which contamination can be prevented and the sensors can be easily attached onto the hose from the outside. IR radiation starts at the visible boarder at the wavelength of 780 nm up to 1 mm. The wavelength is related to the spectral radiance, which is moreover temperature-dependent. A body temperature of 37 °C corresponds to 310.15 K and emits a wavelength in the range of $3.5 - 60\,\mu\text{m}$. The maximum radiance is at approx. $10\,\mu\text{m}$ (*Wien's law of displacement*).

The relationship between the temperature of an ideal radiator (*Black body*) and its emitted spectral radiance $L_{\lambda,S}$ is described by Planck's law of radiation with [5]:

$$L_{\lambda,S} = \frac{c_1}{\pi \cdot \Omega_0} \cdot \frac{1}{\lambda^5 \cdot \left[e^{c_2/(\lambda \cdot T)} - 1\right]}. \tag{1}$$

$L_{\lambda,S}$: Spectral radiance $/\mathrm{W}/(\mathrm{m}^3 \cdot \mathrm{sr})$
Ω_0: Solid angle $/°$
λ: Wavelength $/\mathrm{m}$
T: Temperature $/\mathrm{K}$
$c_1 = 2 \cdot \pi \cdot c_0^2 \cdot h = 3.741832 \cdot 10^{-16}\,\mathrm{W} \cdot \mathrm{m}^2$
$c_2 = \frac{h \cdot c_0}{k} = 1.4388 \cdot 10^{-2}\,\mathrm{m} \cdot \mathrm{K}$
c_0: Speed of light
h: Planck's quantum of action
k: Boltzmann-constant

In the measuring system with IR sensors, the emitted IR radiation of a radiating body is detected and the temperature is calculated with (1). A contactless measurement of surfaces is possible with this technology [3]. Since 2009, Drägerwerk AG & Co. KGaA holds a patent in Germany and the US under the disclosure document DE102007037955, which describes the measurement of temperature with an infrared sensor but has not yet been technically implemented [6]. The patented technology is based on a hollow body that protrudes into the flow channel of the breathing tube, whereupon the latter takes on the heat of the surrounding breathing gas and emits infrared radiation. The sensor detects the radiation from the breathing gas, or measuring window, and converts it into an electrical signal to calculate the temperature. In this work, a technical implementation of the measuring system is to be carried out. The system is to be evaluated with various materials for non-contact temperature measurement. The sensor requires a measuring accuracy of $\pm\,2\,°\mathrm{C}$ due to the ISO standard for an active humidification system (*ISO 80601-2-74*).

2 Material and Methods

For our study, we used an IR sensor, the MLX90614ESF-AAA-000-SP (Melexis N.V., Belgium). The sensor has an accuracy of $\pm\,0.5\,°\mathrm{C}$ and a measurement resolution of $0.2\,°\mathrm{C}$. As part of a bachelor's thesis [7] an electrical circuit was previously designed that implements the data readout of the IR sensor with a PIC10F206 microcontroller (Microchip Technology Inc., USA). The DB9 connector allows serial communication via USB port with a computer. The circuit is seen in Fig. 1.

To run the sensor and read out the data on a computer we have established a programme based on a diploma thesis [4] with LabVIEW (V. 17.0, National Instruments Corp., USA). The programme enables the communication via serial port (DB9) with help of the VISA (Virtual Instrument Software Architecture) add-on of LabVIEW. The sensor supplies two temperatures - the measured object temperature and the temperature of the sensor housing. To integrate the sensor into the ventilation system it was connected between the inspiration hose and a y-piece or other connector

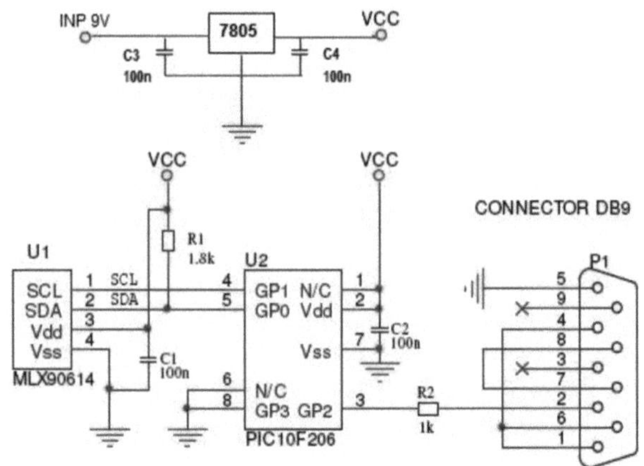

Figure 1: An electrical circuit that implements the data readout of the MLX90614 IR sensor with a PIC10F206 microcontroller.

using a cuvette. Dräger sells a CO_2 sensor (6871950, [8]) that is placed in front of the patient's endotracheal tube with the help of a cuvette. Since the sensor housing and cuvette fit together perfectly, the IR sensor was integrated into this sensor housing. This enabled to use the cuvettes already available.

2.1 Cuvettes with different measuring windows

To determine the temperature of the breathing gas it was necessary to obtain an optimal measuring window. Table 1 shows the investigated measurement windows with their geometry and layer thickness.

Table 1: Cuvettes with different measuring windows.

Cuvette acronym	Material	Geometry	Layer thickness
M1	Synthetic	Hollow body (flat)	0.4 mm
M2	Aluminium (varnished black)	Hollow body (conical)	0.4 mm
M3	Aluminium (varnished black)	Flat window	0.4 mm
M4	Synthetic	Transparent flat window	0.1 mm
M5	Sapphire glass	Transparent flat window	1.01 mm

Two different approaches were investigated. On the one hand we wanted to measure the gas temperature directly through the window. In this case a high IR transmission is necessary and refractions and reflections should be avoided. On the other hand, we also tested an indirect measurement. Here, it was assumed that the measuring window has the temperature of the gas and the temperature of the window was determined. In this case it is important to achieve a high thermal conductivity so that the window adopts the temperature of the gas as accurately as possible.

2.2 Test cases

To test the system first we carried out three different series of measurements:

1. Measurement with dry gas and constant flow

2. Measurement with active humidification and constant flow

3. Measurement with active humidification and ventilation

2.2.1 Measurement with dry gas and constant flow

For the first test case we used a pressure regulator to set the correct pressure and flow out of the central air supply. Behind the regulator we connected a TSI mass flow meter (4045, TSI Incorporated, USA) followed by a heated inspiration breathing hose. Normally, the hose is controlled by a humidifier. But since we wanted to use dry gas we connected the hose to a power supply monitored by a multimeter. This enables to adjust the power specifically and thus test different target temperatures. Behind the breathing hose we connected the cuvette with two reference temperature sensors (NTC sensor FN 0001 K, Ahlborn Mess- und Regelungstechnik GmbH, Germany) before and behind it. The reference sensors have an accuracy of $\pm$ 0.5 °C. Because the part of the cuvette is not heated we assumed that the IR sensor should measure the mean value between the two reference sensors because of the cooling path. The values of the reference sensors were recorded with an Almemo data logger (2890-9, Ahlborn Mess- und Regelungstechnik GmbH, Germany). In addition we logged the ambient temperature with a separate sensor to be able to assess possible temperature gradients. We have set the Almemo to measure the temperature every 1 s, the IR sensor approx. every 300 ms. The IR sensor was plugged to the circuit board which was supplied with a 9 V block battery and connected to a computer via USB. A scheme of the test setup is shown in Fig. 2. We performed measurements for different flows and temperatures. The temperatures of the breathing hose were set to a range of 30 - 33 °C and 39 - 42 °C for 10 l/min, 20 l/min and 30 l/min flow. Every measurement was performed with every cuvette for a time of approx. 30 min.

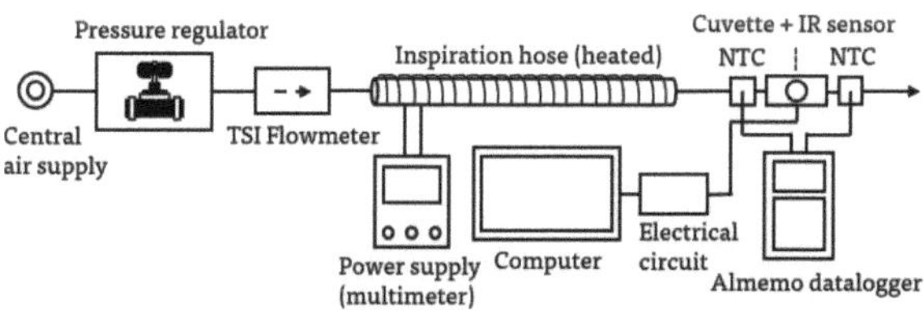

Figure 2: A scheme of the first test setup.

2.2.2 Measurement with active humidification and constant flow

The second test case was very similar to the first setup (cf. 2.2.1). Here a Aquapor H300 humidifier (Drägerwerk AG & Co. KGaA, Germany) was added between the flow meter and the inspiration hose. This allowed the heated hose to be connected directly to the humidifier whereupon the multimeter was no longer necessary. The humidifier was set to the maximum humidity level to produce the greatest amount of humidity. Since the humidifier's instructions for use (IFU) requires a starting phase of 30 min and reaches a constant level after approx. 20 - 25 min, the tests were carried out for 35 min. These were performed with 10, l/min, 20 l/min and 30 l/min flow and the humidifier was operated in invasive mode, which corresponds to a chamber temperature of 37 °C and a patient temperature of 39 °C.

2.2.3 Measurement with active humidification and ventilation

As the measuring system proved to work reliably in the previous test cases a ventilation setting could be carried out. The evaluation with a ventilator corresponded to the real use case. For the third test case we used the Savina 300 Select ventilator (Drägerwerk AG & Co. KGaA, Germany). The humidifier was connected behind the inspiration valve, with the same settings as in section 2.2.2. The inspiration hose was connected to the humidifier and the sensors. These were followed by y-piece which allows the connection of a test lung (SelfTestLung, MP02400 [8]) and the expiration hose. The expiration hose led to the expiration valve of the ventilator. The tests were performed for 35 min.

2.3 Evaluation method

To compare the values of the IR sensor with the reference sensors we first determined the mean value and standard deviation of each sensor when the system was in steady state with Microsoft Excel (V. 2008). For the test setup no. 1 this was the case after approx. 20 min, for test case no. 2 - 3 after approx. 25 min. After determining the mean value and standard deviation we determined the mean value of the two reference sensors. This value was defined as the expected value/target-value for the IR sensor. The target-value could be compared to the actual-value.

3 Results and Discussion

As described in section 2 we evaluated the measurements with comparing the mean values of the reference sensors with the values of the IR sensor. We analysed the values after the system was in steady state. For dry gas experiments the steady state was reached after approx. 10 min. Because of the equality to the other measurement we compared the values from 20 to 30 min. For active humidification the values from 25 to 35 min were used. The results of test case no. 1 showed that except from M5 all cuvettes for all temperatures and flows measured temperatures inside the tolerance of $\pm$ 2 °C. M5 measured too low temperatures with a difference of $\Delta = -3.83$ °C averaged for the experiment with 10 l/min flow, the other results were inside the required range, too.

Test case no. 2 revealed a very similar behaviour. For 10 l/min flow M5 produced a lower temperature measurement with $\Delta = -2.85$ °C and for other flows not. Here we obtained a measurement below the limit value for M3 with $\Delta = -2.19$ °C. In some graphs we could recognize overshoots for M3 during the heating phase. A leakage could be figured out at the location of the measuring window of M3. So condensate could pass through and aggregated in front of the IR sensor which led to an unprecise measurement. An example measurement for the overshoot is seen in Fig. 3 with the comparison to the other four cuvettes. We would like to emphasise that the measurements were carried out consecutively and that slight temperature deviations may have occurred in the system. However, M1 mostly caused too high temperatures and M5 mostly too low temperatures.

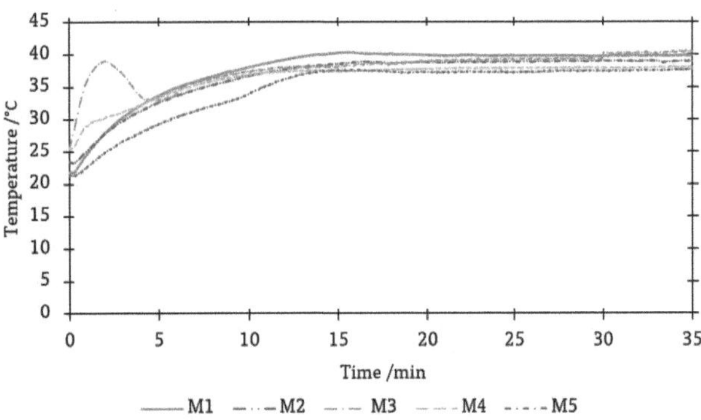

Figure 3: An example measurement of the five cuvettes. Cuvette M3 led to an overshoot in comparison to M1, M2, M4 and M5.

Test case no. 3 showed different results. Here the cuvette M1 indicated a higher measurement with $\Delta = +2.84$ °C. It could be observed that the leakage of M3 resulted in an overshoot at the beginning again. Obviously leakage must be prevented to avoid influence of condensation to the measuring system.

In addition to the five cuvettes we performed tests with a neonatal disposable cuvette from Dräger. Neonatal breathing hoses have a smaller flow channel. Here we obtained lower temperatures than during the measurements with the adult disposable cuvette. This could be an indication that the smaller size of the flow channel leads to a lower temperature measurement. It may be because not enough IR radiation is emitted through the size and volume to pass through the measurement window.

Another point we want to mention is the influence of flow. All over we could measure lower temperatures for the same cuvettes with lower flows. This underlines our assumption that the volume could have an influence on the emission. The flow resistance could be another impact factor as it also has an effect on the flow. Overall, the heating phase of the system should still be evaluated for further tests.

4   Conclusion

We were able to show that temperature measurement with IR radiation during breathing gas conditioning is a good alternative to Pt100 temperature sensors. The IR measurement method offers a great advantage due to hygiene and sustainability. Overall, it was observed across all tests that the aluminium cone (M2) and the transparent synthetic measuring window (M4) produced the most consistent results, which is why these materials and geometries should be further investigated. In addition, a measurement with two IR sensors is conceivable, which could carry out a reference measurement from both sides of the flow channel.

Acknowledgement

The work has been carried out at Drägerwerk AG & Co. KGaA and supervised by Dr. rer. nat. N. Linz, Institute of Biomedical Optics, Universität zu Lübeck.

Author's Statement

Conflict of interest: Authors state no conflict of interest.

5   References

[1] H. Lang, *Beatmung für Einsteiger.* vol. 3, Springer, pp. 7–88, 2020.

[2] H.-R. Tränkler and L. M. Reindl, *Sensortechnik: Handbuch für Praxis und Wissenschaft.* vol. 2, Springer, pp. 904–909, 2015.

[3] M. Piasecka, D. Michalski and K. Strak, *Comparison of two methods for contactless surface temperature measurement.* In: EPJ Web of Conferences (EDP Sciences), vol., p. 02094, 2016.

[4] T. Kutschina, *Untersuchung einer berührungslosen Atemgastemperaturmessung mittels Infrarot als Eingangsgröße für die Regelung der Temperatur im Zusammenhang mit der Atemgaskonditionierung für die Langzeitbeatmung.* Diploma thesis, 2012.

[5] F. Berhnhard, *Handbuch der Technischen Temperaturmessung.* vol. 2, Springer, pp. 1165–1193, 2014.

[6] J. Koch, *Temperature-measuring device for a respiration humidifier,* DE102007037955. 2009.

[7] D. Hoepfner, *Untersuchungen zur Eignung der Gastemperaturmessung mittels Infrarotthermometer bei der Atemgaskonditionierung.* Bachelor's thesis, 2011.

[8] Drägerwerk AG & Co. KGaA *Accessory catalogue 2022.* Available: https://www.draeger.com/Library/Content/accessory-ca-9066485-en.pdf [last accessed on 2021-12-01], 2021.

Development of a sustainable concept of a biodegradable and climate-neutrally burning single-use patient hose system

Celine Hügel [1], Christian Neuhaus [2]
[1] Biomedical Engineering, Luebeck University of Applied Sciences, celine.huegel@stud.th-luebeck.de
[2] Weinmann Emergency Medical Technology GmbH + Co. KG, C.Neuhaus@weinmann-emt.de

Abstract

Single-use patient hose systems are mainly produced from synthetic polymers. The goal of this project is to investigate the concept of a sustainable patient hose system by using biobased and biodegradable materials. Substitution materials were searched in the literature and results applied to the product life cycle. PLA, natural rubber, and cellulose were found suitable substitutes. Resource extraction and processing of PLA components would result in 2.93 times lower CO_2 emissions compared to respective synthetic polymer components. In order to move towards a sustainable concept, energy required throughout the entire product life cycle must be obtained from renewable sources.

1 Introduction

In recent years, the trend away from reusable products towards the use of single-use consumables in the field of medicine has increased globally [1]. The risk of contamination and infection along with the requirement for costly hygienic reprocessing, among other reasons, may be keen drivers of this shift. At Weinmann Emergency Medical Technology GmbH Co. KG (WM EMT), about 98 % of sold patient hose systems for the emergency ventilator MEDUMAT Standard[2] are single-use consumables [2]. The patient hose system delivers breathing gas from the ventilator to the patient. It consists of several components (see Fig. 1), all of which are made from synthetic polymers such as polypropylene (PP), (low-density) polyethylene ((LD-) PE), polyvinyl chloride (PVC), or polycarbonate (PC). Since the single-use patient hose system is classified as infectious medical waste, it is disposed by burning in medical waste incinerator releasing greenhouse gases and toxic pollutants into the environment [3]. If such waste disposal infrastructure is not in place, it takes several decades for synthetic polymers to decompose.

Alternative materials, so called bioplastics that can function as substitutes have been increasingly researched. While synthetic plastics are fossil-based and non-biodegradable, bioplastics are distinguished by their raw material (biobased or fossil-based) and the ability to biodegrade. *Biobased* refers to a product made from renewable raw material, so called biomass. This biomass can be derived from agricultural feedstock (e.g corn, sugarcane, rice, soy), agricultural waste and discarded biomass, or from fast-growing microorganisms such as bacteria, fungi, and algae [4]. *Biodegradability* refers to those polymers biodegrading to CO_2, methane, water, and biomass through the natural action of microorganisms. This process is not time-bound.

Three different groups of bioplastics are distinguished: 1) fossil-based biodegradable plastics, 2) biobased non-biodegradable plastics, and 3) biobased biodegradable plastics (BBPs). [3].

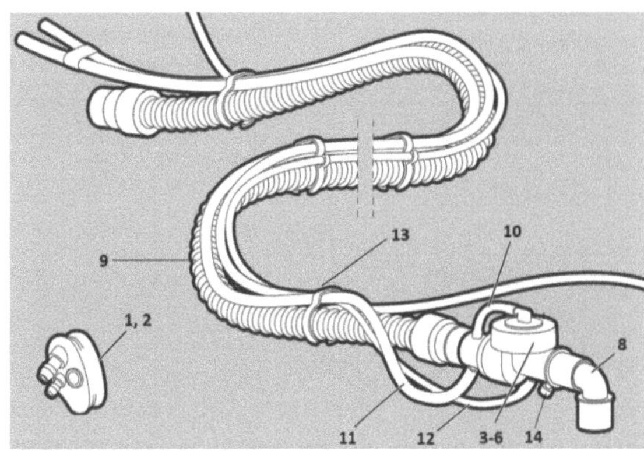

Figure 1: WM EMT single-use patient hose system. Numbers are linked to Tab. 1. Adapter and flat pouch are not displayed.

In this work, a climate-neutral concept of a biodegradable single-use patient hose system for a WM EMT ventilator was investigated. The patient hose system shall be biodegradable and its packaging shall further be compostable in a natural environment. Since sustainability requires independence of fossil resources, only BBPs are evaluated as substitutes for synthetic plastic components. Once suitable substitutes were found, aspects of the products' life-cycle, from resource extraction, product manufacture, distribution and application to the final disposal of patient hose system components were considered. Finally, a visual demonstrator was developed to showcase differences in the appearance of BBP components.

2   Material and Methods

Baseline data   Data on material and manufacturing method components were acquired from internal documentation for each component and listed in Table 1. Information on weight and distribution of components was further collected. In order to find suitable substitutes, currently used synthetic plastics were distinguished based on thermoplastic/elastomer behaviour as well as biodegradability/compostability requirements of substitutes. Three groups evolved: 1) biodegradable material to substitute thermoplastics (PP, PE, PVC, PC, Nylon), 2) biodegradable material to substitute elastomers (silicone, TPE), and 3) compostable material to substitute thermoplastics (LDPE).

Table 1: Current materials used for the single-use patient hose system components (1-14) and packaging (15). Information on manufacturing method is further displayed: CP-Chemical Processing, IM-Injection Moulding, E-Extrusion.

	Component	Material	Manufacture
1	Silicone block	Silicone	CP
2	Insert	PC	IM
3	Main body, Patient valve	PP	IM
4	Cover, Patient valve	PP	IM
5	Membrane, Control valve	Silicone	CP
6	Diaphragm, Check valve	Silicone	CP
7	Adapter	PP	IM
8	Elbow adapter	PP	IM
9	Creased tube	Tube: PE	E
		Cuff: TPE	E
		Helix: PP	IM
10	PEEP control hose	Silicone	CP
11	Oxygen control hose	PVC	E
12	Pressure measurement hose	Silicone	CP
13	Clip	PP	IM
14	Luer-lock plug	Nylon	-
15	Flat pouch	LDPE	E

Literature research   Characteristics and substitution potential of BBPs were mainly extracted from the book chapter *The Role of Biodegradable Plastic in Solving Plastic Solid Waste Accumulation* by Dilkes-Hoffmann et al. Cellulose was considered for packaging due to its establishment in food packaging. Natural rubber was evaluated as potential substitute for elastomer components. For thermoplastic components, TPS, PHA, and PLA were investigated since they hold the largest market share within the BBP sector and are most often represented in life cycle assessment (LCA) studies [3], [5]. Substitutes were deemed suitable if:

- replacement with substitute recommended by experts,
- compatibility with conventional processing methods,
- raw material is available at large scale, and
- substitutes can cope with moisture

Other mechanical and physical properties as well as biocompatibility were not considered yet at this stage of product development. For PLA, an exemplary literature research on LCA publications was conducted using PubMed database. The search strategy included the following terms: (PLA OR Polylactic Acid) AND (Life Cycle Assessment OR LCA). Papers were reviewed by headline and abstract screening, followed by full text acquisition if suitable. Paper references were also included in the review process. LCA data were applied to aspects of resource extraction, product manufacture, distribution, application, and disposal of PLA components. For resource extraction and processing, CO_2 equivalents were calculated for synthetic components and PLA components by using weight and global warming potential over 100 years (GWP_{100}): CO_2 eq. = weight x GWP_{100}. The GWP_{100} represents the heat trapping ability of different gases using CO_2 as benchmark [2]. CO_2 equivalents represent the total warming effect with respect to quantity.

Development of a visual demonstrator   A visual demonstrator was developed using: 1) prefabricated components made from substitute material, 2) components constructed in Creo Parametric 4.0 (PTC Inc.) and three-dimensional (3D) printed with substitute material, or 3) original components that were coloured according to the expected level of transparency when fabricated with a substitute material.

3   Results and Discussion

3.1   Substitution materials

A literature research was conducted on characteristics of natural rubber, cellulose, TPS, PHA, and PLA. Cellulose was found suitable to replace the LDPE packaging. Natural rubber is considered a suitable substitute for components made from silicone and TPE. PLA is considered suitable to replace PP, PE, PC, PVC and nylon [3]. TPS and PHA are both deemed unsuitable due to the inability of TPS to cope with humidity, and the yet unreliable availability of biomass and thus costly production of PHA at commercial scale. The final concept of substitution is shown in Tab. 2.

Table 2: Substitution categories based on biodegradability/compostability and thermoplastic/elastomer behaviour.

Component	Original	Substitute
1-14	PP, PE, PC, PVC, Nylon	PLA
	Silicone, TPE	Natural rubber
15	LDPE	Cellulose

3.2   Product life cycle

The product life cycle was divided in resource extraction and product manufacture, distribution, and disposal of patient hose system components. No GWP is associated with the application phase.

Resource extraction and product manufacture PLA can be processed similarly to synthetic plastic components via injection moulding and extrusion. Thus, only raw material extraction and PLA resin production would change in the production chain. The environmental impact of PLA is discussed in literature by means of LCA. Two suitable studies were found for PLA, quantifying the GWP_{100} in a cradle-to-gate LCA setting (see Table 3) for PLA resin obtained from corn [6] and sugarcane [7]). Cradle-to-gate considers all resources (diesel, fertilizers, water, electricity, etc.) necessary to produce 1 kg of PLA resin.

Table 3: Contribution of steps to cradle-to-gate GWP.

Author	Processing step	GWP / CO_2 eq./kg PLA
Vink et al.	Corn production	0.25
	CO_2 uptake	-1.83
	Dextrose production	0.29
	Lactic acid production	1.16
	Lactide production	0.54
	PLA production	0.20
Morão et al.	Sugarcane production	0.41
	CO_2 uptake	-1.83
	Sugarmill	-0.21
	Lactic acid production	1.45
	Lactide production	0.52
	PLA production	0.17

The cradle-to-gate GWP for PLA resin produced from corn is 0.61 CO_2 eq./kg PLA resin. For PLA resin made from sugarcane, a cradle-to-gate GWP of 0.51 CO_2 eq./kg PLA resin was obtained. Differences occur due to variations in the agricultural system and production processes. During plant growth, CO_2 is captured and later incorporated into the PLA resin, specifically 1.83 CO_2 eq./kg for both sugarcane and corn. With sugarcane, an additional negative GWP was achieved in the sugar mill by using its by-products (bagasse) to generate electricity and steam [7]. For both studies, lactic acid production, particularly the energy consumption during the fermentation process, contributed most to the GWP of PLA [7]. A comparison of CO_2 equivalents for PLA and synthetic polymers was drawn. The latter were calculated with $GWP_{100PVC} = 1.9$, $GWP_{100PP} = 1.6$, $GWP_{100PE} = 1.8$, and $GWP_{100PC} = 4.1$ [6]. Resource extraction and processing resulted in 2.93 times higher CO_2 emissions for thermoplastic components made from synthetic plastic compared to thermoplastic PLA components (see Figure 2, overall). The manufacturing of components by extrusion or injection moulding remains unchanged. Thus, only insignificant reduction in GWP is expected in this step. In order to move towards a carbon-neutral resource extraction and manufacturing phase, the required energy must be obtained from renewable resources.

Distribution Components are manufactured externally (North America, Asia, Europe) and distributed for final as-

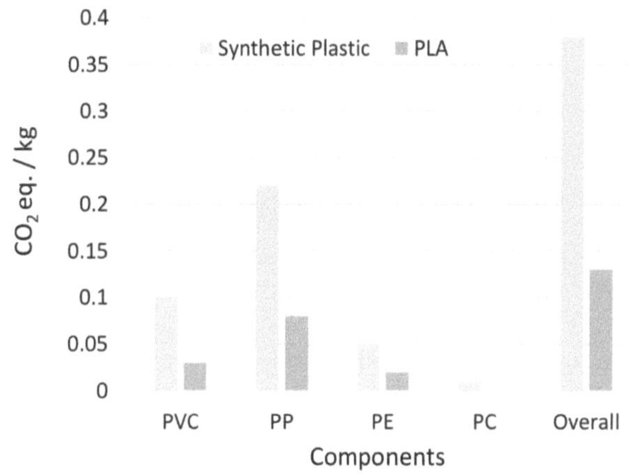

Figure 2: Comparison of CO_2 eq. for resource extraction and processing of synthetic polymer and PLA components.

sembly to the production line of the company. Data on distances travelled by all components individually throughout the supply chain were obtained from internal sources and summarized by transportation method. In total, components (1-15) travel 11,614 km by truck, 102,915 km by ship, and 8,911 km by plane. Previous work has found a CO_2 eq. of 1.39 kg for all components of a WM EMT single-use patient hose system. To reach a carbon-neutral supply chain, transportation would have to be switched exclusively to electricity or carbon-neutral fuels obtained from renewable energy. The required global infrastructure may not yet be in place for all routes of transportation. This also applies to the distribution of manufactured goods to customers. The centralization of manufacture to one location would further minimize the number of distribution paths. Ideally, overall manufacturing of components would take place at the production line in the company, which would supersede the need for distribution within the manufacturing process.

Disposal Biobased plastics generally account for zero emission when disposed in most end-of-life (EOL) scenarios (except landfills, where methane production is assumed): Different to synthetic plastics, the amount of CO_2 released in the atmosphere upon disposal of BBPs equals the amount of CO_2 bound during plant growth [3], [8]. Incineration is considered a closed-loop process with the transformation to CO_2 and ashes (biomass) and the recovery of thermal energy during the incineration process [3]. This further applies to natural rubber. In contrast, the CO_2 released into the atmosphere upon incineration of synthetic plastic cannot be compensated by the life-cycle process itself [3], [4], [8]. Mechanical recycling is considered the most sustainable EOL scenario for PLA. This process involves separation of plastic waste by its type, followed by shredding, cleaning and reprocessing of the small plastic flacks to form a new product. Mechanical recycling thus prolongs service life and reduces the requirement for raw materials [3]. However, hygienic reprocessing and insufficient recycling infrastructure may pose barriers to this concept. The cellulose packaging would both compost and incinerate with a net zero emission.

Figure 3: Original components (top) and components made from biobased and biodegradable materials (bottom) for (from left to right): creased breathing tube, adapter, clip, oxygen supply hose, and pressure control hose.

3.3 Demonstrator

A visual demonstrator of a biodegradable single-use patient hose system was developed. Adapters and clips were 3D printed using the 100 % biodegradable PLA filament. A premanufactured biodegradable natural rubber hose was used to substitute for the positive end-expiratory pressure (PEEP) control hose. The compostable cellulose packaging was obtained from a conventional food item. For the ceased tube, an optical simulation was achieved by painting a standard tube with latex milk. Other components remained unchanged. Original components and substitute material components can be seen in Fig. 3. Most synthetic plastic components and silicone hoses have a transparent to milky appearance. This will change when using components made from BBP and natural rubber, which appear milky to opaque (when using a standard process). While the appearance of natural rubber components cannot be altered, a low level of transparency might be achieved for PLA components, depending on material composition. Customer acceptance of this appearance remains unknown and is subject of further investigation. Transparency of tubes and hoses might be considered crucial in the context of ventilation to ensure these components remain free from occlusion.

4 Conclusion

Biobased and biodegradable bioplastics have been investigated to replace synthetic plastics of a WM EMT single-use patient hose system. Three key findings were acquired: 1) Biodegradable PLA, natural rubber, and cellulose were found eligible substitutes for synthetic polymers of the single-use patient hose system, 2) biobased and biodegradable plastics account for a net zero emission upon incineration, and 3) energy required for all phases of the life cycle, especially product manufacture and distribution, must be obtained from renewable sources to facilitate a sustainable concept. Using biobased biodegradable raw material is a step towards a sustainable concept of a biodegradable carbon-neutral single-use patient hose system.

Acknowledgement

The work has been carried out at Weinmann Emergency Medical Technology GmbH + Co. KG and supervised by Prof. Dr. Philipp Rostalski, Universität zu Lübeck, whom the authors wish to thank for his review.

Author's Statement

Conflict of interest: Authors state no conflict of interest.

References

[1] E. S. Windfeld and M. S. L. Brooks, "Medical waste management - A review", *Journal of Environmental Management*, vol. 163, pp. 98–108, 2015, ISSN: 10958630.

[2] K. Fredriksdotter, "Ökonomischer und ökologischer Vergleich von Einweg- und Mehrweglösungen in der Medizintechnik am Beispiel von Patientenschlauchsystemen", 2019.

[3] L. S. Dilkes-Hoffman, S. Pratt, P. A. Lant, and B. Laycock, *The role of biodegradable plastic in solving plastic solid waste accumulation*. Elsevier Inc., 2018, pp. 469–505, ISBN: 9780128131404.

[4] E. Zanchetta, E. Damergi, B. Patel, T. Borgmeyer, H. Pick, A. Pulgarin, and C. Ludwig, "Algal cellulose, production and potential use in plastics: Challenges and opportunities", *Algal Research*, vol. 56, no. July 2020, p. 102 288, 2021, ISSN: 22119264.

[5] T. A. Hottle, M. M. Bilec, and A. E. Landis, "Sustainability assessments of bio-based polymers", *Polymer Degradation and Stability*, vol. 98, no. 9, pp. 1898–1907, 2013, ISSN: 01413910.

[6] E. T. Vink and S. Davies, "Life Cycle Inventory and Impact Assessment Data for 2014 Ingeo® Polylactide Production", *Industrial Biotechnology*, vol. 11, no. 3, pp. 167–180, 2015, ISSN: 15509087.

[7] A. Morão and F. de Bie, "Life Cycle Impact Assessment of Polylactic Acid (PLA) Produced from Sugarcane in Thailand", *Journal of Polymers and the Environment*, vol. 27, no. 11, pp. 2523–2539, 2019, ISSN: 15728919.

[8] W. J. Groot and T. Borén, "Life cycle assessment of the manufacture of lactide and PLA biopolymers from sugarcane in Thailand", *International Journal of Life Cycle Assessment*, vol. 15, no. 9, pp. 970–984, 2010, ISSN: 09483349.

Comparison of ad- and desorption of sevoflurane at charcoal filters in a passive and active scavenging set-up at an anesthesia device

Lisa Marie Nacke [1], Ralf Heesch [2], and Max Christoph Urban [3]

[1] Biomedical Engineering, Luebeck University of Applied Sciences, lisa.marie.nacke@stud.th-luebeck.de
[2] Drägerwerk AG & Co. KGaA, Lübeck, ralf.heesch@draeger.com
[3] Institute of Applied Sciences, Luebeck University of Applied Sciences, max.urban@th-luebeck.de

Abstract

The health care sector contributes 5-10 % of global greenhouse gas emissions and anesthetic gases account for a major part of the CO_2-emissions of an operating theatre [1] - [3]. Recent studies research into the reduction of volatile agent usage focusing on the reduction of freshgas flow [4]. Besides this the adsorption and recycling or destruction of volatile agents is a way to decrease the environmental impact [5]. To compare the adsorption capacity of charcoal filters used with an active or passive scavenging system the adsorption capacity of ten filters from ZeoSys Medical GmbH and Dräger Safety AG & Co. KGaA with different freshgas flows was evaluated. It could be shown that the adsorption and desorption of volatiles depends on the freshgas flow and the scavenging system, whereas the latter has a greater impact. Additionally, it turned out, that volatile agent gets washed out of the charcoal when flushing it directly after saturation.

1 Introduction

Today hospitals strive for less environmental impact under the slogan "Green Hospital". One part of "Green Hospital" is "Green Anesthesia". Under this title measures to reduce the CO_2 equivalent emissions caused by anesthesia are discussed. General anesthesia is either based on intravenous medication, on volatile agents or on a combination of both. Total intravenous anesthesia works with hypnotics, opioids and muscle relaxants that are administered intravenously and does not make use of volatile agents. For an inhalation anesthesia volatile agents such as Isoflurane, Sevoflurane or Desflurane are used instead of intravenous hypnotics [6]. Depending on the set freshgas flow anesthetic agent from the anesthesia machine gets into the anesthetic gas scavenging system (AGSS). In recent studies volatile agents have been found to have a higher Global Warming Potential (GWP_{100}) than CO_2. Meaning that sevoflurane (SEV) contributes 130, isoflurane 510 and desflurane 2,540 times more to the greenhouse effect than the same amount of CO_2 would in 100 years [7]. The scavenging systems of operating theaters evacuate the anesthetic agent from the operating theater but mostly the exhaust air from all parts of the hospital is guided to the roof-top and then released into the surrounding without prior filtering. The standard ISO 80601-2-13 requires a flow rate of either 25 l/min – 50 l/min (low-flow transfer and receiving system) or 50 l/min – 80 l/min (high-flow transfer and receiving system) to evacuate the surplus gas from the anesthesia machine (active setting)

[8]. However, there is the possibility to use a passive AGSS with a charcoal filter to adsorb volatile agent. The adsorption capacity of charcoal is dependent on the grain size, evenness of packing, concentration, humidity, the characteristics of the charcoal and the flow velocity [9]. The experiments in this paper analyze the adsorption capacity of charcoal filters with a passive and active AGSS setting. Until the completion of the paper, the authors were not aware of any standard-compliant combination of a charcoal filter with an anaesthesia machine.

2 Material and Methods

The set-up can be seen in Fig. 1. As a source of volatile agent the anesthesia device Atlan A350 (Drägerwerk AG & Co. KGaA) with a sevoflurane vaporizer (Vapor 3000, Drägerwerk AG & Co. KGaA) was used. The charcoal filter housing (CONTRAfluran, ZeoSys Medical GmbH) with a filling volume of 1.9 L was placed in a SENSOfluran^{PLUS} (ZeoSys Medical GmbH) adapter. To introduce pure air the Luftikus (Drägerwerk AG & Co. KGaA) was used. The flow sensors TSI 4000 Series, Model 4040 monitored the flow rate in the Luftikus and exhaust limb. Both flow sensors have been calibrated on measuring air. Their accuracy is $\pm$ 2 % of the measured value [10]. The flow rate was set according to the TSI sensor. The concentration of volatile agent in the exhaust gas was measured with the 1312 Photoacoustic Multi-gas Monitor (Innova AirTech Instruments A/S). The consumption of volatile anesthetic

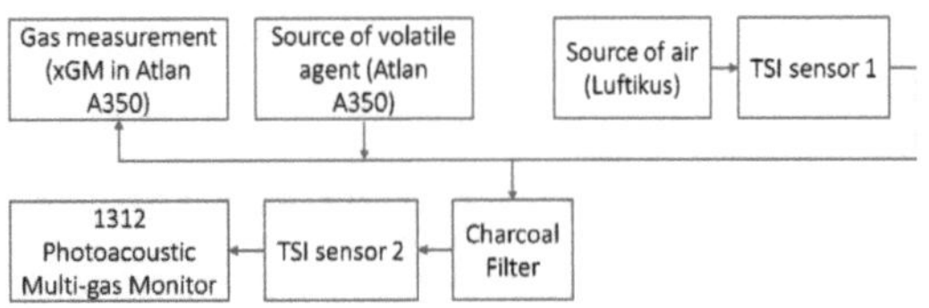

Figure 1: Set-up of the experiment. Luftikus and Atlan A350 as flow sources, TSI sensors to measure the flow rate and xGM and photoacoustic multi-gas monitor to measure agent concentrations.

Table 1: Characteristic Charcoal

	Supplier	Expiry Date	Condition
C1F1 / F2	ZeoSys	22-12-31	pre-saturated
C1F3 / F4	ZeoSys	22-12-31	new
C1.1F1 - F4	ZeoSys	22-05-31	new
C2F1 / F2	Dräger	-	new

agent was measured with the integrated gas module (xGM) of the Atlan A350. The anesthetic agent consumption was calculated with an accuracy of $\pm$ 25 % or $\pm$ 1 ml by the xGM, the higher value is applicable. The canisters were weighted with the scale NewClassic MS (Mettler Toledo) with an accuracy of 1 % of the measured value. For the experiments two different charcoals in ten different canisters were used (see Table 1). When using Dräger charcoal it was filled into the CONTRAfluran filter housing. The work reported in this paper was conducted with sevoflurane and dry air from the central gas supply in the Dräger laboratory exclusively. All experiments were conducted under controlled conditions. A patient was not involved at any time.

To compare the behaviour in a passive scavenging set-up and active set-up saturation and flushing was conducted with different flows and concentrations but always the same experimental set-up. The hypothesis is, that independent from freshgas flow and concentration no sevoflurane can be washed out of the charcoal once adsorbed.

3 Results and Discussion

3.1 Saturation process

For full saturation the filters C1F3, C1F4, C2F1 and C2F2 have been saturated with sevoflurane. Full saturation has been determined to be the point at which the measured concentration suddenly rises exponentially (breakthrough of the filter). To accelerate the experiment, the maximum concentration of 8 Vol.% was set on the vaporizer with a fresh-gas flow of 15 l/min on the Atlan A350, which allowed for quick saturation of the charcoal (clinical range of volatile agent approx. 1.4 - 3.3 Vol.%). To check whether the actual flow rate has an impact on the measured concentration another 35 l/min from the Luftikus was added to the flow from the anesthesia device for 1/3 of the saturation time. With these changes no direct impact of the disposal flow

could be observed. Over the whole course of the saturation phase the charcoal adsorbed the sevoflurane well for the filter C1F4, C2F1 and C2F2. The measured sevoflurane concentration in the exhaust limb was under 20 ppm for the whole time until breakthrough. Filter C1F3 showed a different concentration pattern. This filter showed a constant rise in concentration than an abrupt breakthrough. The adsorption capacity for the concentration of 8 Vol.% on the vaporizer and the high flow rates was measured to be 225.4 ml for C1F3, 311.5 ml for C1F4, 318.4 ml for C2F1 and 309.4 ml for C2F2. Thus both charcoals have approximately the sames adsorption capacity when used in the same set-up.

3.2 Flushing - alternating flow settings

It has been assumed that the behaviour during flushing can be seen immediately when exposed to different air flow rates. The first experiment was conducted with a step-wise flow pattern sweeping from 1 l/min to 5, 25 and 50 l/min and back to 1 l/min. Each setting was held for three minutes. This experiment was done with six filters (C1F1, C1F2, C1F3, C1F4, C2F1 and C2F2). The flushing started at different times after full saturation. As C1F1 and C1F2 have been used in a hospital before, the exact date of full saturation cannot be verified. The assumption is, that those filters were stored at least weeks after saturation. Filter C1F3 was stored five days after saturation and all other filters were flushed directly after saturation.

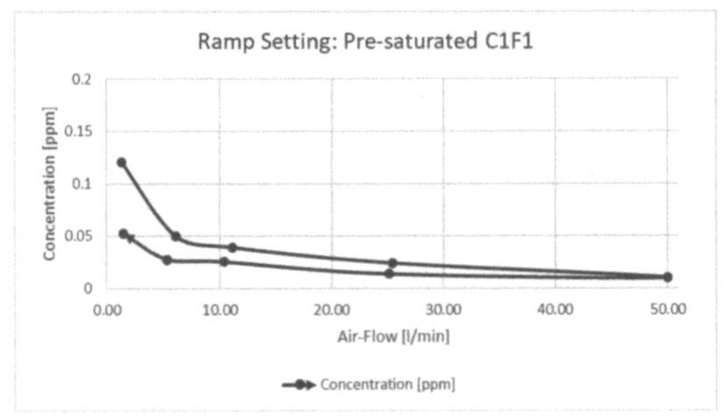

Figure 2: SEV concentration in ppm during the step-wise flow pattern. Flushing weeks after full saturation. (C1F1)

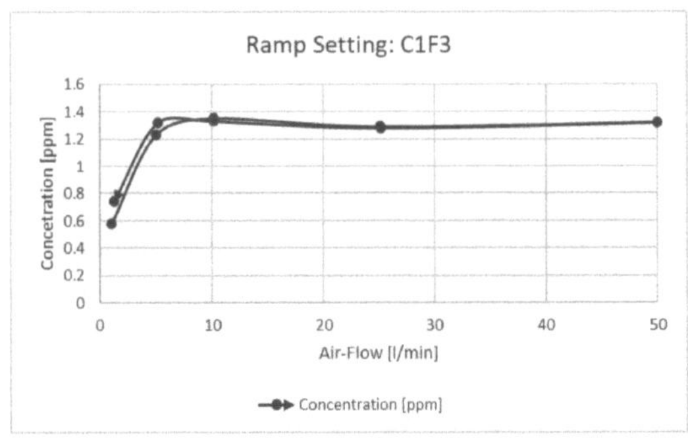

Figure 3: SEV concentration in ppm during the step-wise flow pattern. Flushing five days after full saturation. (C1F3)

When looking at Fig. 2 - Fig. 5 it can be seen that the duration between full saturation and flushing has an im-

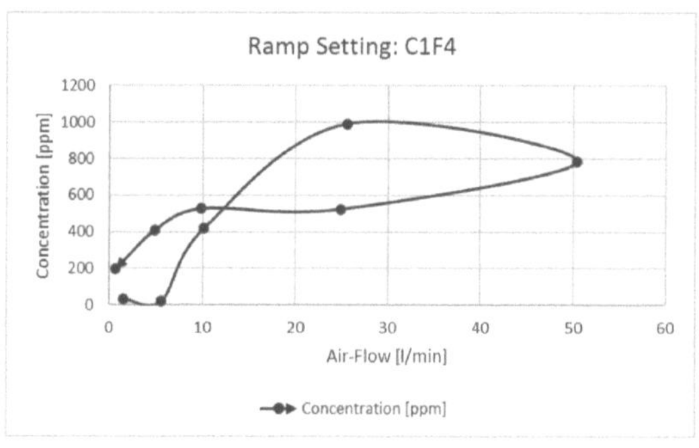

Figure 4: SEV concentration in ppm during the step-wise flow pattern. Flushing directly after full saturation. (C1F4)

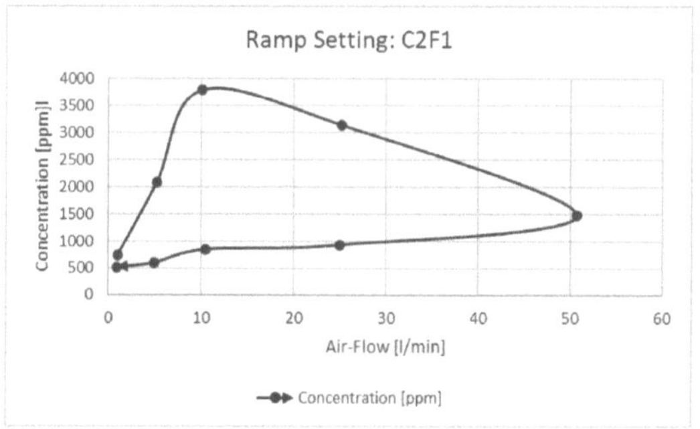

Figure 5: SEV concentration in ppm during the step-wise flow pattern. Flushing directly after full saturation. (C2F1)

pact on the results. The measured concentrations are higher when flushing directly after fully saturating the filter. For C1F1 and C1F2 the measured concentrations lied between 0.00977 ppm and 0.121 ppm. For the filter C1F3 the concentrations measured were between 0.578 ppm and 1.35 ppm. Thus, they were 11- to 60-times higher than for C1F1 and C1F2. C1F4 showed concentrations between 21 ppm and 993 ppm and C2F1 and C2F2 showed concentrations between 519 ppm and 3,791 ppm, and 338 ppm and 3,078 ppm. It could be shown that flushing the filters directly after saturation washed out prior adsorbed sevoflurane.

3.3 Flushing - continuous flows

When considering the clinical set-up the filters are exposed to a constant flow of 25 l/min – 80 l/min, when placed between active AGSS and wall inlet [8]. Thus, the initial short flushing times gave a first impression of the behaviour of the charcoal but did not represent the actual clinical set-up. Therefore, filter C1F4, C2F1 and C2F2 meant to be flushed with 45 l/min for 180 min. C1F4 and C2F2 could be flushed for 180 min. The test had to be stopped for C2F1 after ten minutes because too much sevoflurane was washed out (possible contamination of the laboratory). The reason for that could not be identified.

$$m\_ml = \frac{m\_g}{\rho}. \tag{1}$$

The weight loss of the charcoal filters m_g can be converted into ml sevoflurane m_ml by considering the density

of sevoflurane with $\rho=1.52$ g/cm^3 [11]. During 180 min flushing C1F4 lost 49.04 g and C2F2 64.32 g. Referring to (1) C1F4 lost 32.26 ml and C2F2 42.32 ml of sevoflurane in 180 min. The amount of sevoflurane dosed by the vaporizer but not adsorbed by the charcoal is neglected.

3.4 Continuous flow with low concentrations

The next four filters were tested with low concentrations and low (1 l/min and 4 l/min) as well as high flow rates (35 l/min). To simulate a day with an active AGSS in an OR the filters C1.1F1 and C1.1F3 were saturated for 120 min and flushed 60 min three times (simulation of not disconnected AGSS between surgeries). Disregarding this sequence, the filters were further saturated until the breakthrough. The filters C1.1F2 and C1.1F4 were continuously saturated with a flow of 4 l/min and 1 l/min.

Table 2: Weight gain in 120 min saturation phase. (C1.1F1)

Phase	δ Weight [g]	SEV [ml]	δ to calculated [ml]
Sat. 1	+ 95.28	89.7	27.02
Sat. 2	+ 107.38	87.8	17.16
Sat. 3	+ 103.16	87.6	19.73

Table 3: Weight loss in 60 min flushing phase. (C1.1F1)

Phase	δ Weight [g]	SEV [ppm]
Flush. 1	- 8.27	0.00482 - 0.0223
Flush. 2	- 1.00	44.2 - 53.9
Flush. 3	-13.39	550 - 877

Table 4: Weight gain in 120 min saturation phase. (C1.1F3)

Phase	δ Weight [g]	SEV [ml]	δ to calculated [ml]
Sat. 1	+ 11.7	28.5	20.8
Sat. 2	+ 19.54	28.6	15.74
Sat. 3	+ 30.73	28.5	8.22

Table 2 shows a difference between the calculated and the weighted amount of adsorbed sevoflurane. To compare the consumption displayed by the Atlan A350 with the actual weight increase, the weight difference is converted in ml sevoflurane (see (1)). Table 3 shows that during a flushing phase of 60 min the weight of the filter decreases. A similar behaviour can be observed for the filter C1.1F3 which was saturated with the similar flow pattern but a vaporizer setting of 4 Vol.% (see Table 4 and Table 5). Filter C1.1F1 (active) adsorbed 265.1 ml, C1.1F2 (passive) 243.1 ml, C1.1F3 (active) 208.8 ml and C1.1F4 (passive) 121.9 ml. No major difference in adsorption capacity in a passive or active set-up can be reported.

Table 5: Weight loss in 60 min flushing phase. (C1.1F3)

Phase	δ Weight [g]	SEV [ppb]
Flush. 1	- 8.63	3.97 - 22.7
Flush. 2	- 1.61	15.0 - 31.7
Flush. 3	- 0.01	26.1 - 62.9

4 Conclusion

These experiments are the first to compare the use of charcoal filters with an active and passive AGSS. It could be shown, that both the charcoal from ZeoSys Medical GmbH and Dräger Safety AG & Co KGaA were able to adsorb sevoflurane with various flow rates. However, it could be shown that over the course of time sevoflurane can be washed out of the charcoal especially when flushing it directly after saturation (mostly active set-up). To verify a standard-compliant set-up for charcoal filters at anesthesia devices further research on its behaviour is necessary. Furthermore, it is important to note the limitations of the experiments:
1. Dry gas from the central gas supply was applied to the filters. In a clinical set-up the gas from the re-breathing system is more humid due to moisture from the patient and soda lime especially during phases of low fresh gas flow, thus the co-adsorption of water in the charcoal should be analyzed further. 2. The weight in- and decrease of the filters has to be interpreted carefully as it is possible that the dry gas initially reduces the humidity in the filter and thus also the weight. 3. The unevenness of packing or channel formation in the charcoal can lead to different adsorption capacities and might influence the results. 4. The displayed consumption on the Atlan A350 shows the amount of sevoflurane that left the vaporizer. Due to leakages and absorbed sevoflurane in components of the system, the actual amount of sevoflurane reaching the filter might have been less. 5. As the experiments had to be monitored by a person continuously the goal was to quickly saturate the filters. To achieve this, the applied concentrations were higher than in a clinical set-up. 6. As the experiments have been conducted with a small amount of filters, the experiments should be conducted again with more filters.

Acknowledgement

The work has been carried out at Drägerwerk AG & Co. KGaA, Lübeck and supervised by the Institute of Applies Sciences, Luebeck University of Applied Sciences.

Author's Statement

Conflict of interest: Ralf Heesch is employee of Drägerwerk AG & Co. KGaA. Lisa Marie Nacke is employee of Dräger Medical Deutschland GmbH.

5 References

[1] M. J. Eckelman and J. Sherman, *Environmental Impacts of the U.S. Health Care System and Effects on Public Health*, PLOS ONE, vol. 11, no. 6, p. e0157014, 2016.

[2] M. J. Eckelman, J. D. Sherman and A. J. MacNeill, *Life cycle environmental emissions and health damages from the Canadian healthcare system: An economic-environmental-epidemiological analysis*, PLOS Medicine, vol. 15, no. 7, p. e1002623, 2018.

[3] A. J. MacNeill, R. Lillywhite and C. J. Brown, *The impact of surgery on global climate: a carbon footprinting study of operating theatres in three health systems*. The Lancet Planetary Health, vol. 1, no. 9, pp. e381–e388, 2017.

[4] K. L. Zuegge et al., *Provider Education and Vaporizer Labeling Lead to Reduced Anesthetic Agent Purchasing With Cost Savings and Reduced Greenhouse Gas Emissions*. Anesthesia & Analgesia, vol. 128, no. 6, pp. e97–e99, 2019.

[5] V. Rauchenwald et al., *New Method of Destroying Waste Anesthetic Gases Using Gas-Phase Photochemistry*. Anesthesia & Analgesia, vol. 131, no. 1, pp. 288–297, 2020.

[6] J. Schulte am Esch, H. Bause and E. Kochs, *Anästhesie: Intensivmedizin, Notfallmedizin, Schmerztherapie*. Thieme, Stuttgart, 2007.

[7] S. M. Ryan and C. J. Nielsen, *Global Warming Potential of Inhaled Anesthetics: Application to Clinical Use*. Anesthesia & Analgesia, vol. 111, no. 1, pp. 92–98, 2010.

[8] ISO copyright office, *ISO 80601-2-13: Medical electrical equipment - Part 2-13: Particular requirements for basic safety and essential performance of an anaesthetic workstation*. Geneva, 2011.

[9] F. a. P. Maggs and M. F. Smith, *Adsorption of anaesthetic vapours on charcoal beds*, Anaesthesia, vol. 31, no. 1, pp. 30–40, 1976.

[10] TSI GmbH, *Mass Flowmeters for Gases*. Available: https://tsi.com/getmedia/2e8285c3-6ca8-47f3-b395-36d90cf04ad2/Flowmeters-Brochure_A4_2980399_RevP_Web?ext=.pdf [last accessed on 2022-01-02].

[11] R. Sun et al., *A cost comparison of methohexital and propofol for ambulatory anesthesia*, Anesth Analg, vol. 89, no. 2, pp. 311–316, 1999.

The Effect of Hygienic Processing and Short Wire Lengths on the Performance of Hot-wire Flow Sensors

Joanna Burbat [1], Alexander Struth [2],

[1] Biomedical Engineering, TH Luebeck, joanna.burbat@stud.th-luebeck.de
[2] Mechatronic Engineer, Drägerwerk AG Co. KGaA, alexander.struth@dräger.de

Abstract

Continous improvement and research of ventilators are necessary to ensure the safety of the patient and to keep high quality standards in the medical market. The hygiene of reusable flow sensors and better and faster verification conditions in manufacturing are research topics in Dräger's flow sensor department. The sensors operate by the principle of hot-wire anemometry with thin hot-wires, which lengths and functionality are significant for their technical performance. In this work, testing strategies for an improved manual and machinery hygienic processing procedure were evaluated by analysing the sensors functionality after processing. Additionally, the impact on sensor performance by using shorter hot-wire lengths was examined. The work showed, that for the latter test, a statistical study with more sensors is necessary in order to give more valid relationships between the parameters. The hygienic test revealed, that some sensors showed ruptured wires after the processing and others were not cleaned insufficiently.

1 Introduction

Flow sensors are an essential accessory in ventilation, since they cover a variety of monitoring functions such as measurement of volume parameters, flow curves and alarm triggering. Their performance is highly dependent on ultra thin hot-wires, which in case of contamination or damage can cause measurement errors. Reusable sensors can be cleaned, disinfected and sterilized. Their correct processing is very important to prevent cross-contaminations or nosocomial infections, which can be life-threatening for patients [1]. Current processing of Dräger sensors in hospitals is not always performed correctly, which may lead to contamination residuals inside the sensor or rupturing of the hot-wires. The sensors which are examined here are the reusable SpiroLife and the disposable Spirolog sensor.

The thin platin hot-wire (d= 13 μm, l= 4-5 mm) represents the core element of the hot-wire anemometry principle, which is used to measure the flow rate indirectly [2], [3]. The wire is heated up to around 180 °C, significantly higher than the ambient temperature [4]. The heat dissipation from the wire to the surrounding area is dependent on the flow velocity. The higher the velocity, the more the wire is cooled down which results in a lower electrical resistance of the wire (see Fig. 1) [3], [4]. The cooler the wire, the more current is required to keep the constant wire temperature of around and a constant resistance. From the required heating current, conclusions are drawn regarding the flow rate [3],[2]. Moreover, during the sterilization processes, which apply high temperatures of around 134 °C, the thin hot-wires of the flow sensors expand [4]. To prevent ruptur-

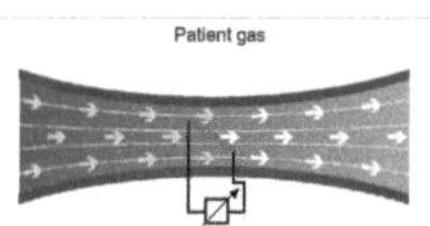

Figure 1: Principle of the hot-wire anemometry [4]

ing by tension, all sensors are manufactured with intended longer length. A disadvantage of long wires lengths is the higher statistical deviation of the sensor performance properties [5].

In this work, an improved manual and machinery hygienic processing procedure for reusable flow sensors was tested. The procedure was developed in a way to prevent a wrong application by hospital personal and thus to ensure the function of the hot-wires. The focus lied on the evaluation of the used test strategies and the comparison of the sensor performance before and after the processing. Additionally, this work also analysed the impact on sensor performance when shorter wire lengths are used. A comparison was made between the current long wire sensors and sensors with shorter wires. Under the assumption, that the deviation improves, less measuring points could be used in verification processes in the flow sensor's production, which may save time and costs.

2 Material and Methods

Since not all hospitals have the opportunity to process medical products in a disinfection machine, both the manual and the machinery processing was tested. For each processing method, only one processing cycle as well as a load capacity test was carried out, which will be explained more detailed in the following sections. Before the tests, the sensors were contaminated with a defined test contamination, being a mixture of mucin and BSA (bovine serum albumin), marked with an orange dye, hereinafter referred to as Orange-G (compare Fig. 2). After soiling the sensors, the Orange-G dried for at least 24 hours. For each test (one processing cycle and load capacity test), the quality of cleaning was examined microscopically and the sensor performance was compared before and after the procedure.

Figure 2: Sensors marked with the Orange-G

For the processing and the wire length tests, standard flow curves based on Dräger's production data for new manufactured sensors, were measured and evaluated. The measurement set-up consist of a flow generating machine, an incident flow element which connects the sensor and the machine, the sensor and a cable which is attached to the electronic board (compare Fig. 3).

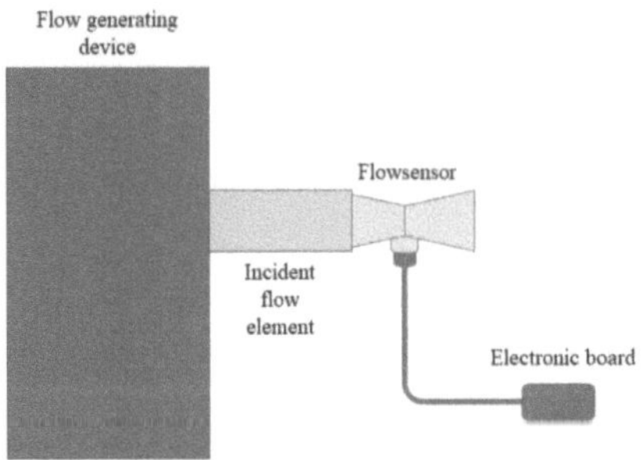

Figure 3: Measurement setup for the sensor performance after the processing tests

The flow generating device produces different flow rates and guides it through the incident flow element to create a laminar flow profile. The element is connected with the flow sensor on one side and open on the other side. The plug of the sensor is attached to the cable of the platine with the electronic algorithm. A software, called D-Space, records the sensor voltages which leads to the flow rates by more complex calculations, subsequently evaluated with Excel. The calculations are comparable with the sensor algorithm on the platine. Before each measurement, the sensors were calibrated. The interesting parameter examined during the measurements, is the deviation between the measured flow rate and the real flow rate provided by the flow generator. For the hygienic processing tests, two unprocessed reference sensors (1 Spirolog and 1 SpiroLife sensor) were additionally examined within every performance measurement.

2.1 One processing cycle

In the manual cleaning test, five flow sensors were positioned in a container and attached inside via an adapter (3 Spirolog and 2 SpiroLife sensors). The container was filled with 500 ml of cleaning agent and swivelled gently for three times before and after the exposure time of 15 minutes, based on the cleaning part of the current instruction for use [6]. In a next step, the adapter with the sensors was placed in a second container filled with 500 ml water and again swivelled gently three times. After this, the sensors were removed from the container and dried for 30 minutes in a dust-free environment. For the machinery cleaning test, two SpiroLife sensors were fixated by using the machinery adapters for the disinfection device (Miele PG8536). Within one cycle, the sensors ran through defined rinse and drying programs, which are similar to the programs in the Central Sterile Supply Department (CSSD) of hospitals.

2.2 Load capacity test

The load capacity test represents the worst case and was structured in a different way. Here for the manual method, five clean sensors (3 SpiroLife and 2 Spirolog sensors) were placed with the adapter inside the container, which was filled with 500 ml water. The container was shook roughly for one minute in a randomized movement pattern. After the shaking process, the sensors were removed for drying for 30 minutes. The three sensors for the machinery procedure ran through 25 cycles inside the disinfection machine, where they were dried inside afterwards.

2.3 Different lengths of hot-wires

For this test, two short and two current long hot-wire SpiroLife sensors were compared under different measurement conditions. It was examined, which impact the wire lengths showed when the sensor position was turned by 90 degrees and when the gas type was changed from air to oxygen.

3 Results and Discussion

3.1 One processing cycle

The evaluation of the microscopic photographs showed, that only one of five sensors (Spirolog) could be cleaned properly by using the manual cleaning adapter. For the other four sensors, droplets marked with Orange-G were found on the sensor walls and residuals near the plug insert. The wires appeared clean for all sensors (compare Fig.4).
The performance test showed, that all sensors were located in the area of the allowed deviation for production limits. Differences among the curves cannot be interpreted clearly,

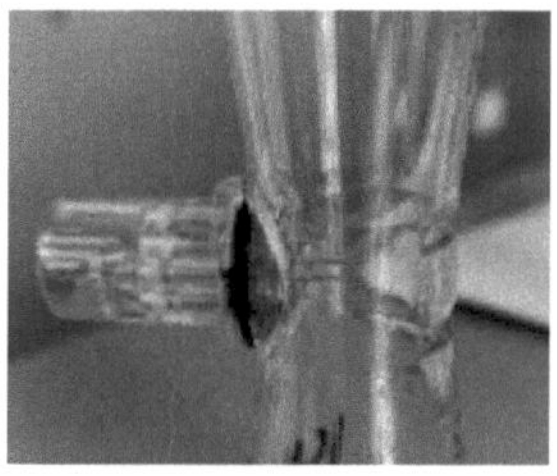

(a) Spirolog sensor with Orange-G residuals

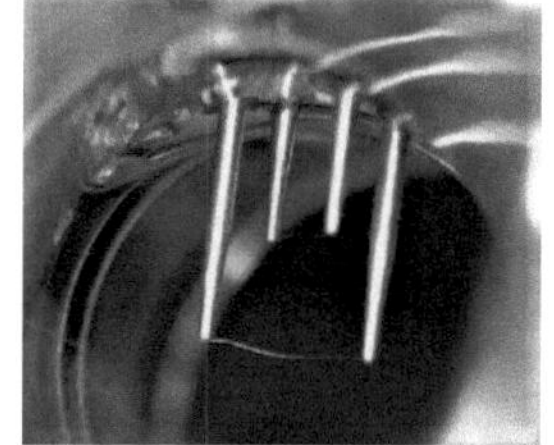

(b) Clean wire of the Spirolog

Figure 4: Exemplary sensor after one manual processing cycle

since many factors need to be included. Reasons for the difference between the curves before the contamination could be a different wire position, since the plug insert could not be positioned equally for each sensor, temperature differences or different electronic parameters during the test procedure. The shift of the curves after the processing may have appeared due to water residuals as well as the reason mentioned above. Additionally, different ambient conditions may have changed the ambient temperature and pressure, which is considered in the flow calculations.

Fig. 5 shows the deviation difference for all sensors between the state after processing and before contamination (compare equation 1).

$$\Delta deviation[\%] = deviation_2[\%] - deviation_1[\%] \quad (1)$$

with the parameters being the deviation between the generated and the measured flow for before the processing ($deviation_1$) and after the processing ($deviation_2$). All sensors except sensor S2 (dark dashed dotted) showed clear differences. Since the two reference sensors R1 (dashed) and R2 (dotted) showed less than 0.5 % of deviation difference, it can be assumed that the processing with the manual adapter altered the conditions of the flow sensors.

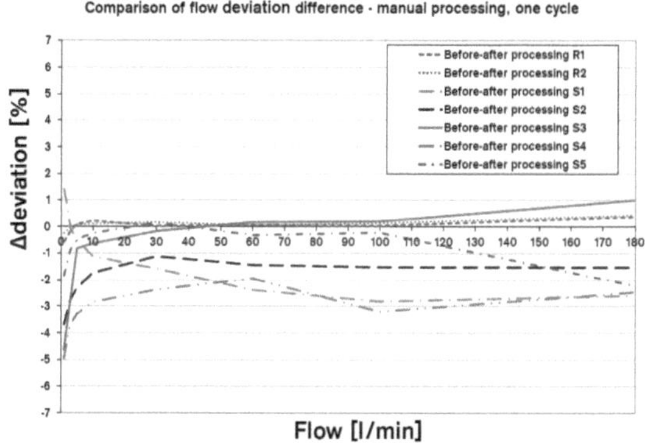

Figure 5: Deviation difference of sensor curves: Before contamination and after one manual processing cycle

The machinery processing showed succesful results regarding the quality of cleaning. No orange residuals were left on the sensors. After one cycle, the wires were undamaged. Here, all sensors were inside the allowed deviation

after processing. The deviation difference was equally distributed for all flow rates regarding sensor S2 (dashed dotted) and the reference sensor R1 (dashed). In contrary for sensor S1 (solid curve) and reference sensor R2 (dotted) the deviation difference increased for low flow rates. Since the same effect occurs for the reference sensor, which was not part of the processing test, it is not clear if the deviation difference for S1 and S2 occurred totally from the processing step.

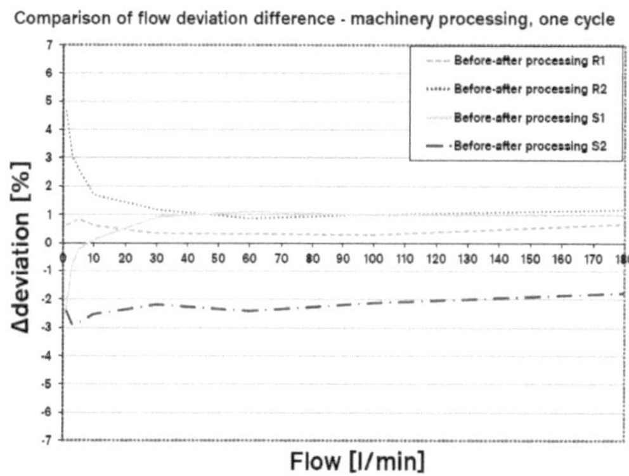

Figure 6: Deviation difference of sensor curves: Before contamination and after one machinery processing cycle (compare equation 1)

3.2 Load capacity test

After 15 cycles of manual processing, only one of five sensors was damaged. For sensor S4, the temperature wire was ruptured after cycle 13. The performance measurement examined, that this sensor already measured strong deviations compared to the state before processing after one cycle. All other sensors lied in the allowed deviation after 15 cycles of randomized strong shaking. A few of the sensors did not show any difference between the cycle 1,5,10 and 15, which may indicate, that the processing did not impact their performance. The sensors which were processed by the disinfection automate attached to an adapter, had more critical results. The first sensor lost its temperature wire after ten cycles and the second sensor lost the same wire after 19 cycles. Due to this fact, the third sensor was removed from the automate after cycle 9 in order to record a standard curve before potential wire rupturing. After nine cycles, the sen-

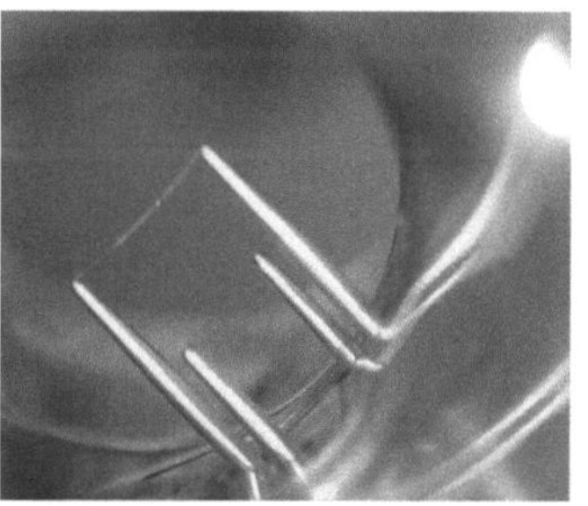

Figure 7: Missing temperature wire (marked with arrow) after ten cycles in the disinfection automat

sor was still in the allowed flow deviation. The deviation was <1 % for flows of 100 and 180 L/min and around 1% for flows smaller than 100 L/min. There was also no significant difference between the deviation before and after the processing. A conclusion for the lower wire rupturing could be the targeting of the water jet in exactly this area, which may deliver too much pressure for the thin wire. Another assumption could be, that the wire was still contaminated with Orange-G or other residuals and the physical properties of the wire changed and may led to a rupturing. An explanation for this is, that a contaminated wire has more molecules in the region of contamination. These molecules result in a higher resistance of the wire, which is unlike the state of calibration, and make the wire either more conductive or work as an isolator. The changing properties may be the cause for a rupturing wire but cannot be proven for this test.

3.3 Different lengths of hot-wires

All sensors lied in the allowed deviation but differed among each other, because of the different physical properties of the sensors and due to the wire position. Short wires are located differently inside the sensor and see the highest flow velocity in the middle of the sensor at high flows, because of the rectangular flow profile inside the sensor. The deviation decreases for higher flows for all short wire sensors. A reason could be the physical effect of convection. The hot wire emits thermal energy, which has higher impact for low flow rates than for high flow rates. Due to this reason, the working area of the sensor starts for higher flows, which leads to decreased deviation in this region. By turning the plug insert from down to behind with air as a gas type, there was a visible change in deviation for one of the longer hot-wires. The short wires showed almost no deviation difference. The reason is, that the wire is located in a different position inside the convective cloud, which is emitted by the hot wire and influenced by gravity. Changing the gas from air to oxygen shows opposite effects for long and for short hot wires while using the same plug insert position. For the latter ones, the deviation increases by using air as a gas, especially for low flow rates. In this case, it is not clear if the differences occured from calculation or from physical errors. A certain shift in the curves is a normal behavior when changing the gas, since the thermal conductivity changes [3]. For fluctuations less than 0.5 %,

it is not clear to say if this fluctuation is caused by physical effects, measurement impacts or just the production tolerances. Generally, the curves of the short wires may lie less close to each other for most of the measurements, since the longer the wire, the higher the deviation due to a higher measurement signal (see Fig. 8).

4 Conclusion

This work gave an insight into two very important life cycle engineering topics regarding flow sensors in ventilation devices. It showed,that hygienic processing requires a defined procedure to avoid failure in the medical device. The tests demonstrated that the current procedure still leads to rupturing of the wires. Additionally, more sensors must be tested to give a proper statistical distribution of the sensor performance after processing. The examination of the second test set the starting point for an interesting research regarding the question, how different wire lengths affect the sensor performance. Nevertheless, more sensors need to be investigated under different measurement conditions to give a more precise statement about the impacts on performance changes.

Acknowledgement

The work has been carried out at Drägerwerk AG & Co. KGaA and supervised by Prof. Dr.-Ing. Stefan Müller, Technische Hochschule Lübeck.

Author's Statement

Authors state no conflict of interest.

5 References

[1] S.K. Fridkin & S.F. Welbel & R.A. Weinstei, *Magnitude and Prevention of Nosocomial Infections in the Intensive Care Unit.*.Infectious Disease Clinics of North America, 1997

[2] G. Wiegbleb, *Gasmesstechnik in Theorie und Praxis: Messgeräte, Sensoren, Anwendungen.*.Springer Vieweg, Wiesbaden, 2016

[3] Dr. R. Best-Timmann, *Dokumentation Flowmessung mit Spirolog/SpiroLife*. Dräger intern, Lübeck, 2015.

[4] R. Kramme, *Medizintechnik: Verfahren - Systeme - Informationsverarbeitung.*.Springer Reference Technik, Berlin, 2016

[5] J. Loeser, *FMS1 Flowkennlinien ASPF*. Dräger intern, Lübeck, 2021.

[6] Drägerwerk, *IfU SpiroLife Flow Sensor 9038622 me*. Dräger, Lübeck, 2017

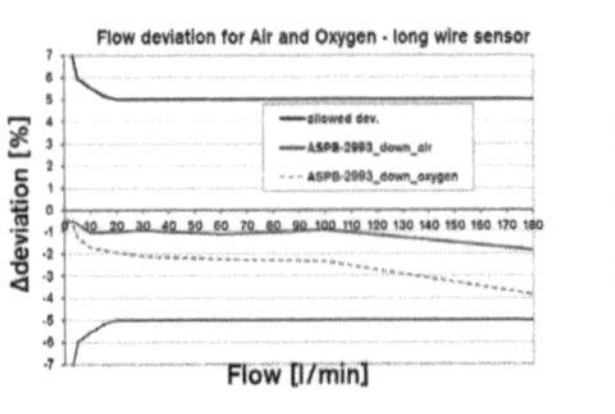
(a) Long wire sensor

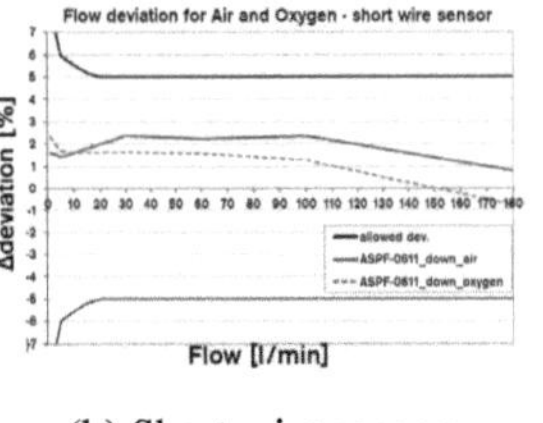
(b) Short wire sensor

Figure 8: Gas type dependency on flow deviation for the long wire sensor (a) and the short wire sensor (b);solid line = air, dotted line = oxygen)

Development and validation of a test method using laser Doppler vibrometry for ultrasound-based air detectors

Rebecca Barth [1], Matthias Bäcker [2]

[1] Biomedical Engineering, Lübeck University of Applied Science, rebecca.barth@stud.th-luebeck.de
[2] Fresenius Medical Care Deutschland GmbH, Schweinfurt, matthias.baecker@fmc-ag.com

Abstract

In a dialysis machine, air detectors are used for sensing air infusion of purified blood returned into the patient's body. The functionality of these air detectors is checked by testing blocks. The blocks themselves are regularly calibrated by the drip chamber test device. Based on laser Doppler vibrometry, the functionality of the drip chamber test device is to be calibrated using a new test method. For this purpose, investigations concerning the experimental setup and the measurement parameters were carried out and optimized iteratively. Finally a test method validation was performed to validate the developed method. The findings of this work result in a well-functioning performance of the test method proving the calibration of the drip chamber test device.

1 Introduction

A treatment opportunity for patients with kidney diseases is hemodialysis, which is an important process to purify blood. In this process, blood is drawn from a blood vessel and then passed through an artificial kidney, called dialyzer. This dialyzer is used to filter and purify the blood outside the body. The purified blood is then returned to the patient. All steps from pumping the blood to regulating the purification process are controlled by a dialysis machine [1].

For every medical device, it must be ensured that the patient is not exposed to any danger or even harm. In case of a dialysis machine, a protective system against air infusion into the patient when blood is returned to the body is needed [2]. An air detector to sense dissolved air based on the usage of ultrasound is built into the dialysis machine. It is based on the transmission of ultrasonic waves. Ultrasonic waves are emitted, pass through the venous drip chamber, which contains the purified blood of the patient and are detected. The quantity of passing air is then monitored by the strength of the transmitted signal, which is characterized by the coupling voltage.

The calibration of such air detector is based on the usage of blocks made of polyvinyl chloride. These blocks are equivalent to a blood-filled drip chamber without air and a blood-filled drip chamber including air in terms of their properties to propagate an ultrasonic signal. A specific test device, the so-called drip chamber test device is used to check these blocks. The verification of the functionality of the drip chamber test device has to be ensured by a new test method, which is based on the principle of laser Doppler vibrometry. This work deals with the development and validation of this test method to establish the metrological traceability.

2 Material and Methods

2.1 Experimental Setup

Laser Doppler vibrometry is an optical and contactless test method to measure vibrations of surfaces or objects. The measurement principle is shown in Fig. 1. A laser beam is divided into measuring and reference beam by beam splitter 1. The reference beam passes the Bragg cell and is then directed onto the photodetector. The Bragg cell is used to detect the direction of the measurement object. The measuring beam passes directly the lens and hits the measurement object. The beam is scattered at the vibrating surface of the object and undergoes a change of frequency and phase depending on the velocity and deflection of the target. The backscattered beam and the reference beam are superimposed on the photodetector. It results in a motion-dependent modulation of the detector signal [3].

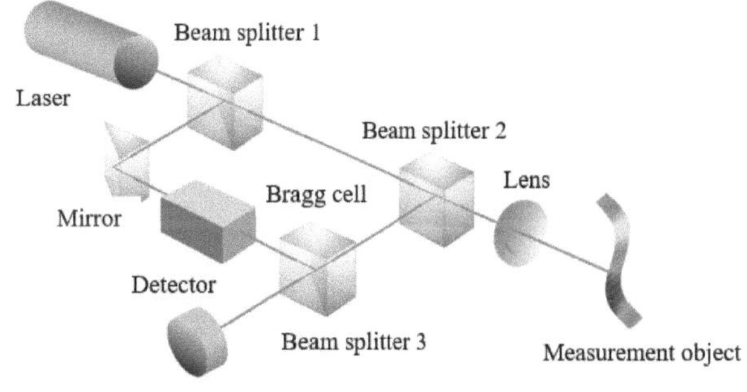

Figure 1: Schematic depiction of the measurement principle of laser Doppler vibrometry [3]

The experimental setup used in this work is shown in Fig. 2.

The LD4008 unit contains a circuit board and provides an ultrasonic transmitter clocking with an amplitude. Furthermore, it generates a TTL triggering signal for the vibrometer. The ultrasonic transmitter converts the electrical signal coming from the LD4008 unit in a mechanical ultrasonic signal using piezo effect. The holding unit directly next to the ultrasonic transmitter shall ensure a reproducible positioning of the testing blocks. The blocks are inserted into the holding unit and pressed against the transmitter with a defined force. For this, the holding unit is constantly supplied by compressed air coming from compressed air equipment and controlled by a lever. The vibrometer generates the laser beam hitting the receiver side of the testing block, is backscattered at the oscillating surface of the block due to piezo effect and experiences a frequency and phase change. This information collected by the vibrometer is send to a computer.

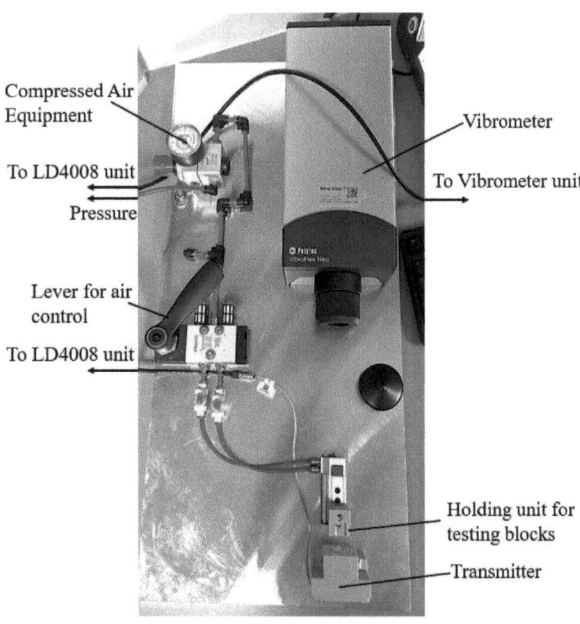

Figure 2: Photo of experimental set-up of the measurement system

2.2 Development of a Test Method

Various examinations were done to identify factors influencing the performance of the test method. These factors can be assigned to the experimental setup or measurement parameters of the vibrometer. Influencing factors of the setup were pressure, operating temperature, excitation voltage and focus of the laser on the block.
Factors influencing the measurement parameters are cursor positions and time domain averaging. Cursor positions indicate the limitation of the signal to a certain range. Time domain averaging means measuring the oscillations several times and then be averaged over the course of the measurement.
A further investigation was examining the noise component of the signal to investigate whether there was enough light or not due to scratched surface of testing blocks.

2.3 Preparation of a Test Method Validation

Subsequently a test method validation for determining the measurement uncertainty was performed. It serves to validate the developed test method. Among other things, the reproducibility and repeatability is assessed. Repeatability is the examination of the measurements when the same operator measures the same part with the same measuring equipment. Reproducibility describes the variation when different operators measure the same part with the same measuring equipment [4].
The measurements can be influenced by various uncertainty components. These uncertainty components are assigned either to the measurement system or the measurement process. Measurement system is part of the measurement process, which describes uncertainty influences without operator, measurement object influences or environmental influences shown in Table 1 [4]. Measurement process includes all relevant uncertainty influences, among others repeatability or comparability due to operator influence.

Table 1: Measurement uncertainty components assigned to measurement process or system [4]

Variable	Uncertainty influence	Assignment
u_{RE}	Resolution of display	System
u_{EVR}	Repeatability (on reference)	System
u_{EVO}	Equipment variation - Repeatability (on test object)	Process
u_{AV}	Appraiser variation - Comparability (operator influence)	Process

The test method validation was performed with six selected testing blocks, which were chosen from 30 blocks based on preliminary tests. Of these 30 testing blocks, the six selected blocks provided a minimum value twice, a typical value twice and a maximum value twice in terms of their properties to propagate an ultrasonic signal.
The six selected blocks were measured three times by three operators. These measurements were done in randomized order. Each measurement consisted of ten individual measurements, from which the calibration value was formed.
The number of total repetitions, which must be greater than 30 to obtain a valid result according to Fresenius' SOP-000810, is calculated from the number of operators, testing blocks, implementations and factor levels according to (1) [5].

$$n_{\text{repetitions}} = 3 \cdot 6 \cdot 3 \cdot 1 - 1 = 53 \geq 30 \qquad (1)$$

In the test method validation, the final determination is calculation of the uncertainty components to get the expanded uncertainty u_{MP} for the measurement process.

3 Results and Discussion

3.1 Test Method

The investigations showed different findings, which are briefly explained in the following. The top of Fig. 3 shows the oscillations of a transmitter measurement (without an inserted block) detected by the vibrometer. The transmitter clocking is clearly visible in the bottom of Fig. 3.

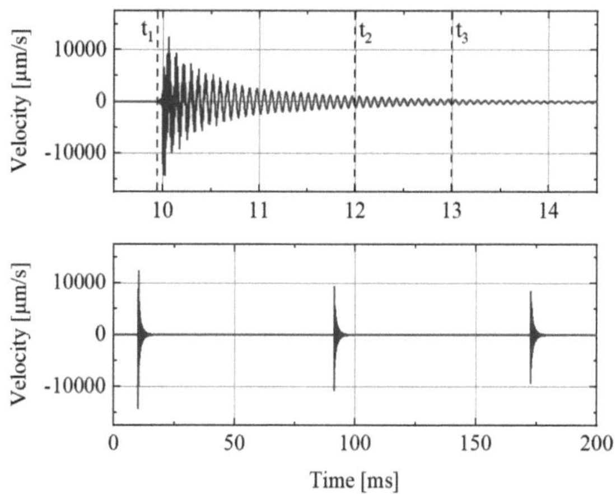

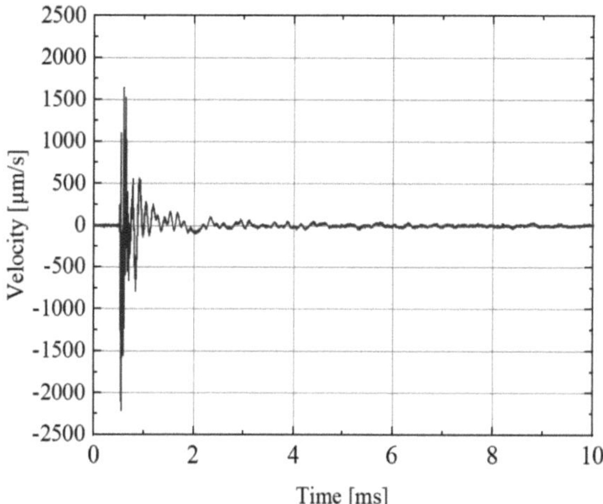

Figure 4: Vibration of the surface of a block in response to a single excitation pulse of the ultrasonic transmitter

Figure 3: The top indicates one period of an ultrasonic transmitter signal without inserted block and in the bottom three periods of an ultrasonic transmitter signal are shown. Both measured by the vibrometer.

It was found out that the oscillation amplitude depends on the excitation voltage. Therefore, the excitation voltage is adjusted depending on the value of the amplitude. The variable to be determined is then defined as effective value of the vibration velocities, referred as root mean square (RMS value) in a time-defined range. This is calculated by the device software according to (2) with the measured values y_n and the number of samples $n_{samples}$ in a time domain.

$$RMS = \sqrt{\frac{\sum y_n^2}{n_{samples}}} \qquad (2)$$

The time domain is limited by cursor positions. Throughout the tests different positions were investigated shown in the top of Fig. 3 with an ultrasonic signal. During oscillation, the signal becomes smaller and the samples contain less useful information. Therefore, the cursor positions should limit the signal to a minimum of samples with useful information. In the test method, the cursor positions $t_1 = 0.47$ ms and $t_2 = 1.5$ ms are applied.

As mentioned further, in Fig. 3 are three periods of an ultrasonic transmitter signal shown. The amplitude of these three signals changes slightly over the course of time. That means, the lower the time domain averaging, the less accurate the measurement result. Therefore, a time domain averaging of 600 of these periods is set.

Fig. 4 shows a clear ultrasonic signal with an inserted block. This is much weaker than the transmitter signal, since the signal passed through the block is attenuated and thus only a fraction of the signal is detected by the laser. Furthermore, less information is contained in the samples more quickly, as the samples with useful information ends faster.

During the measurement it was observed that the measured signal could be affected by a poor focus on the block

and laser signal level due to low coupling. Therefore, the blocks calibrating the drip chamber must be prepared for measurement. For this purpose, the block is greased with lithium soap grease on the transmitter side before each measurement to get a good coupling.

Furthermore, the investigation of the pressure of the holding unit leads to no appreciable influence on the RMS value, but a pressure is needed to keep the blocks in reproducible positioning and support a good coupling. Therefore, the pressure is set to 2.5 bar.

Moreover, as experimental results revealed, the self-heating of the measuring instrument must be taken into account. Therefore, the warm-up phase of at least 30 minutes must be considered before measurements can be started.

3.2 Test Method Validation

According to SOP-000810 the determination of uncertainty components can be done by statistical evaluation due to test method validation or use of prior information. To calculate the standard uncertainties mentioned in Table 1 both approaches were used.

The resolution for the used measuring head of the vibrometer is calculated using the prior information of the data sheet. Therefore, the used velocity measuring range is 10 mm/s, the sampling frequency is 500 kHz and a typical resolution in this range is $3 \frac{\frac{nm}{s}}{\sqrt{Hz}}$. This results in a resolution of the measuring system of 2.12 μm/s [6].

The tolerance resolution of the measuring system is recommended by SOP-000810 to not exceed 5% of the tolerance to be validated. Thus, the minimum detectable tolerance was calculated and results in 42.24 μm/s [5].

The standard uncertainty u_{RE} is determined using the minimum detectable tolerance as prior information and leads to a resolution u_{RE} of 12.19 μm/s [7]. The determination of the uncertainty components u_{EVR}, u_{EVO} and u_{AV} are done by test method validation with the software Minitab.

Table 2: Results of measurement uncertainty components

Variable	Uncertainty influence	Result
u_{RE}	Resolution	12.19 µm/s
u_{EVR}	Repeatability (to reference/ to testing block)	13.49 µm/s
u_{EVO}	Repeatability	13.49 µm/s
u_{AV}	Comparability (operator influence)	7.16 µm/s

The results are summarized in Table 2. All uncertainties are now merged to form the combined standard uncertainty according to (3). In this case the uncertainties u_{AV} and u_{EVO} are used for calculation. According to SOP-000810 uncertainty u_{RE} must not be considered, because it is smaller than uncertainty u_{EVO} [5].

$$u(y) = \sqrt{\sum u_{AV}^2 + u_{EVO}^2} = 15.27\,\mu m/s \qquad (3)$$

The expanded uncertainty u_{MP} for the measurement process is calculated using the combined standard uncertainty $u(y)$ and the appropriate k factor according to (4). Coverage factor k leads to a probability of 95 % the measurement variable is located within the assigned interval [5].

$$u_{MP} = k \cdot 15.27\,\mu m/s = 30.54\,\mu m/s \qquad (4)$$

In order to establish metrological traceability, measurements were done with the drip chamber test device and the vibrometer taking into account this expanded uncertainty u_{MP}. The measured values from the drip chamber test device were then correlated with measured values from the vibrometer. It resulted in a positive increasing straight line shown in Fig. 5. This means that there is a direct relation between coupling voltage and vibration velocity. The calibra-

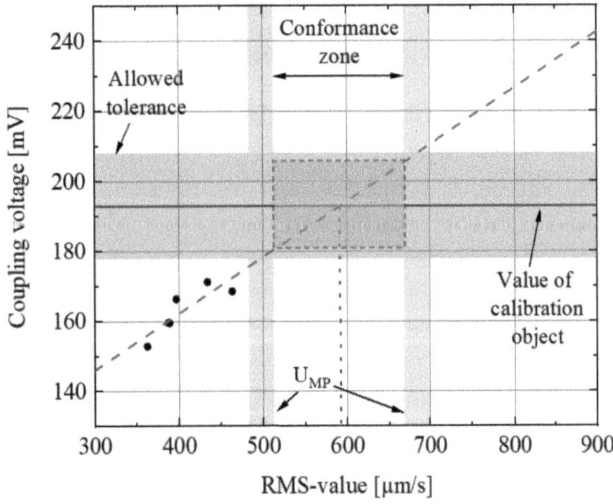

Figure 5: Correlation between coupling voltage and root mean square

tion object of the drip chamber test device is characterized by a coupling voltage of 193 mV with an allowed tolerance of ±15 mV. With the determined measurement uncertainty u_{MP} of the vibrometer, this results in the indicated conformance zone. When measuring the calibration object with the vibrometer, it must lie within this zone in order to prove the functionality of the drip chamber test device.

4 Conclusion

This paper discussed the development of a new test method based on the principle of laser Doppler vibrometry and the validation by test method validation.

It resulted in a well-functioning performance of the test method, which could prove the calibration of the drip chamber test device and establish metrological traceability in the future. For this purpose, the extrapolation must be substantiated with further measurements of modified solid blocks. In addition the measurement of the calibration object itself based on the established test method needs to be done. Finally, a measurement device can be developed from this. The measurement of vibrations by laser Doppler vibrometry for the application of calibrating the drip chamber test device is a new field, which need to be tested further.

Acknowledgement

The work has been carried out at Fresenius Medical Care Deutschland GmbH, Schweinfurt and supervised by Prof. Dr.-Ing. Dipl.-Ing. Stefan Müller, Department of Applied Natural Sciences, Technische Hochschule Lübeck.

Author's Statement

Conflict of interest: Authors state no conflict of interest.

5 References

[1] Hämodialyse verstehen. Available: www.fresenius.de/media/Haemodialyse_verstehen.pdf [last accessed on 2021-12-20].

[2] International Electrotechnical Commission, *General standard IEC 60601-1*. VDE, pp.26-27, 2012.

[3] Laser-Doppler-Vibrometrie. Available. www.polytec.com/de/vibrometrie/technologie/laser-doppler-vibrometrie [last accessed on 2022-01-17].

[4] Gage Repeatability and Reproducibility (R&R). Available: https://sixsigmastudyguide.com/repeatability-and-reproducibility-rr/ [last accessed on 2022-01-10].

[5] Christoph Brückner, *Validierung von Testmethoden für physikalische Messgrößen*. Standard Operating Procedure für Fresenius Medical Care Schweinfurt, Schweinfurt, 2019.

[6] Polytec GmbH, *VibroFlex Connect Datenblatt*. Waldbronn, pp. 6-7, 2019.

[7] Verband der Automobilindustrie, *Prüfprozesseignung*. Berlin, 2011.

Update of an existing usability engineering process to IEC 62366-1:2015/A1:2020 and ISO 13485:2016 compliant software tool validation for usability process

Mitakshya Karki [1],
[1] Biomedical Engineering, Luebeck University of Applied Sciences, mitakshya.karki@stud.th-luebeck.de

Abstract

During the development process of optoacoustic imaging devices, manufacturers at iThera Medical Gmbh are obliged to carry out the usability engineering process following the standard EN 62366-1:2015/A1:2020 which relates to user safety, effectiveness and efficiency of medical devices. The manufacturers shall also follow the standard ISO 13485:2016 for an efficient quality management system during the product development lifecycle. Research and review of the standards, including an integrative approach were performed as usability is linked with quality risk management and product development. Also, the software tool (Polarion Application Lifecycle Management 21 (Polarion ALM 21)) validation for the usability process included the creation, review and approval of the tool's user needs and requirements, test specifications, and test plans. The mentioned methods led to a successful update of usability process into the upgraded standard edition to overcome the gap from old standard version and to an ISO 13485:2016 compliant software tool validation fulfilling the intended purpose for usability process.

1 Introduction

The current usability standard, IEC 62366-1:2015/A1:2020 has been updated from IEC 62366-1:2015 in June 2020 with few, but significant changes. The upgraded standard strictly focuses on the application of the usability engineering process to advance medical device usability in terms of safety, effectiveness and efficiency of the device. The Usability Engineering is defined as "application of knowledge about human behaviour, abilities, limitations, and other characteristics to the design of medical devices (including software), systems and tasks to achieve adequate usability" [1]. One of the salient changes in the new version of the standard is the definition of the correct use for each hazard related use scenarios in summative evaluation. This is considered as an important addition to define success and failure for each evaluated task . The manufacturers shall now select hazard-related use scenarios based on the severity of the potential harm and based on other circumstances specific to the medical device and the manufacturer [1]. Another considerate change is the interface with risk management. In this context, manufacturers have to follow the revised risk management standard ISO 14971:2019. ISO 13485 is the International Standard which points the requirements for a quality management system which should be used by an organization involved in one or more stages of the medical device life-cycle. The different stages include design and development, production, storage and distribution, installation, servicing, final decommissioning, and disposal of medical devices [2]. During the design and devel-

opment phase, the manufacturers are obliged to carry out the usability engineering process following the IEC 62366-1:2015/A1:2020, and also the proper validation of the application of computer software used in quality management system referencing ISO 13485:2016, chapter 4.1.6, 4.2.5 and 7.5.6 [2]. The software tool used in the project is Polarion ALM 21 which provides a single solution for the entire development processes of a medical device combining requirements management, quality assurance, and application lifecycle management (ALM). The iThera Medical Gmbh is involved in the development and manufacturing of the optoacoustic imaging systems for preclinical and clinical macroscopy (MSOT inVision, MSOT Acuity Echo) and mesoscopy (RSOM Explorer P50, RSOM Explorer C50 research system), however the software tool only provides the structure and guidelines for the creation of a complete usability file in general and not for any specific devices as mentioned above. This software tool for usability is then validated which also enables the complete traceability throughout the device's usability lifecycle. When combined with compliant electronic signatures, the traceability soars in providing the complete proof of validation evidence.

2 Material and Methods

- The existing usability standard operating procedure (SOP) of the manufacturer dealing with the optoacoustic imaging devices and IEC 62366-1:2015/A1:2020 was briefly reviewed. The overall update in the SOP for the usability process was done in Polarion. The definitions and terms relating to usability, regulatory requirements, overall usability process flowchart were updated according to the new standard in the SOP. The interfaces of the usability process with the risk management and product development were identified in the usability steps. The formsheets templates for the Usability File Summary, Use Specifications, Use Scenario Specifications, Use Scenario and Task Analysis, Formative Usability Evaluation Plan, Formative Usability Evaluation Report, Summative Usability Evaluation Plan and the Summative Usability Evaluation Report were finalized according to templates provided by external consultants and approved. These templates describe the major requirements and informations required for a complete usability file summary. In addition, the certain important definitions and parts of the formsheets according to the requirements in IEC62366-1:2015/A1:2020 were divided into different work items and created as shown in Table.1. Finally, the updated SOP for usability was sent for review and approval.

Table 1: Formsheets and their work item types

Formsheets	Work item types
Use Specifications	Intended purpose, User Group, User Profile, Use Environment
Use Scenario Specification	Use Scenario, Usability Task
Use Scenario and Task Analysis	Usability Finding

- For the Software tool validation for Usability process in Polarion, ISO 13485:2016 was briefly reviewed and started as follows:

  - The work instruction was created which described the structure and data model in Polarion for the documentation of the usability engineering process for the general optoacoustic device that manufacturers are involved for development. This explained the procedure for creating formsheets, creation of links between the work item types within the formsheets in Polarion to show the linkage between usability processes for a complete Usability File.

  - In the document "Assessment and Validation Plan", the software tool assessment (risk assessment, regulatory assessment) and validation plan (tool description, tool's intended use and method) was described.

  - The document "User Needs and Requirements" was written. The user needs and requirements included that user shall be able to create formsheets and work items in Polarion that described the usability engineering processs. The user shall also be able to create a complete usability file which comprised of basic structure of intended purpose, user groups, user profile, use environment, use scenario, usability task, usability findings, summative and formative evaluation reports which the manufacturers need to complete with informations from the usability process of the new device developed. The user shall also be able to link work items as shown Fig.2. Additionally, all the needs and requirements also included the structure and data model from work instruction.

  - In the document "Test specifications", the test cases and their expected result were written according to the user needs and requirements and linked with one another. The test cases are written to check if the user needs and requirements are fulfilled or not. The test cases were also linked to the test setup (Required Test Equipment and/or Software Tools).

  - In the document "Plan of Test Runs", the test runs were created and they were linked to the respective test cases. The validation was completed by running the tests individually, and checked if the actual results from the test runs matched with the expected results. To perform the test runs, one of the company's device project was also considered and the contents from the Usability file of the project were migrated to the new documents created from the formsheets templates. Finally, a short summary was written in the document "Test Run Summary Report" for the complete validation of the usability software tool.

3   Results and Discussion

3.1   Usability process with interface to risk management and product development according to 62366-1:2015/A1:2020

Fig.1 reflects the overall usability process in the version of the revised SOP including the interfaces with risk management and product development. The existing version of SOP was successfully updated and approved. According to the new amendment in the standard 62366-1:2015/A1:2020 , the manufacturer shall create, implement and maintain a usability engineering process as defined in Clause 5, for safety of the patient and other users [1]. This also increases the efficiency of the device. Following the standard, the usability process starts with the preparation of use specifications of the device (input from product development file) which includes the intended medical indication, use environment and patient population intended for the device. The manufacturers are obliged to carry out the risk management process which determines the known or foreseeable hazards and hazardous situation associated with the medical device,

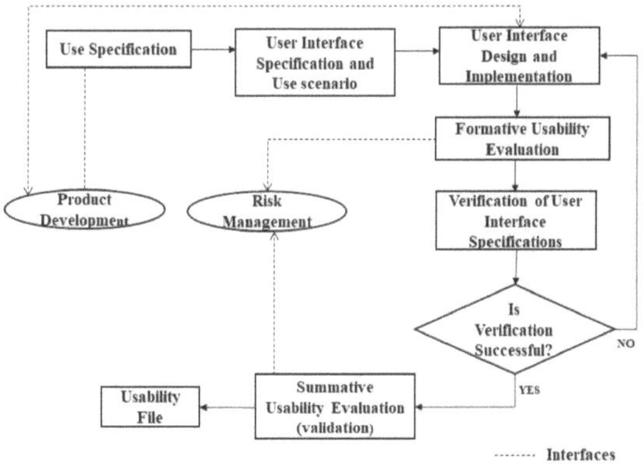

Figure 1: Overall Usability process including interfaces with risk management and product development

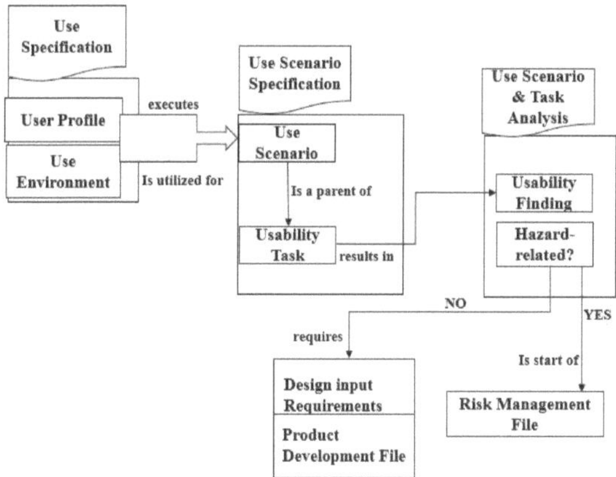

Figure 2: Links between the formsheets in usability process (the arrows define the link roles)

and their probability of occurrence. During the identification of hazards and hazardous situations, the following shall be considered [1]:

- Use Specifications, including User Profile

- Information on hazards and hazardous situations known for similar types of devices in case of user interfaces

- Identified Use errors. These errors are the errors described in 62366-1:2015/A1:2020, subclause 3.21.

The description of each identified use related hazard shall describe usability tasks performed in a specific use environment and their sequences as well as the severity of the associated harm [3]. The identification of user interface specifications which identifies user interface characteristics with testable technical requirements relevant to the user interface, includes the requirements related to safety as described in 62366-1:2015/A1:2020, subclause 5.2 [1]. The formative usability evaluation describes if the user interface design has been met according to the design specifications (specified in use specifications and product development file), with the changes in the design in case of failed verification. The summative evaluation is an important factor conducted in the final product which considers: hazard related use scenario with the intent to obtain objective evidence of safe use of user interface, procedure of involving the participants who are the representatives of the intended user profile during evaluation.

3.2 Gap analysis of the existing Usability process SOP with regard to 62366-1:2015/A1:2020

The section of regulatory requirements in the new version of the SOP describing the regulations to be followed by the manufacturer during the usability process was updated with regard to IEC 62366-1:2015/A1:2020 and IEC TR (Technical Report) 62366-2:2016. The major revisions [1] in the standard considered for the update of existing SOP are :

- Inclusion of use difficulties in summative evaluation: The use difficulty is generally due to use error done by user during operation where the use difficulty arises when the user almost commits an error but prevents it. These use difficulties are used to establish the interface of summative evaluation with risk management.

- The primary operating functions which are the main functions relating to user interfaces related to safety does not need to be determined. Instead, hazard related use scenarios shall be described.

- All kinds of hazardous situations are considered (62366-1:2015/A1:2020, Table B.1 in Annex B). The hazardous situations are no longer limited to only physical hazards.

- The user profile describing the knowledge, skills, and abilities that may have an impact on design decisions must be considered, hence included in the Use Scenario Specification template.

3.3 Successful creation of links between work items

Following the updated version of SOP and the work instruction for software tool validation process, the relation between different work items described in the usability formsheets were successfully determined and programmed using different link roles as shown in Fig.2. These linkages provide for a better understanding of the inter connection within the overall usability engineering process in Polarion.

3.4 ISO 13485 compliant validation

The standard ISO 13485 focuses on the specific requirements like customer and regulatory requirements for a quality management system to be used by an organization. The organization shall identify the regulatory requirements and thereby incorporate them into their quality management system. According to ISO 13485:2016, chapter 4.1.6 [2], "the organization shall document the validation procedures for the validation of the application of computer software used in the quality management system". The methods required for software validation shall be comparative to risks related to the software use. In the assessment and validation plan created during the software tool validation for usability, the electronic records are assessed under 21 CFR (Code of Federal Regulations) Part 11 which describes criteria acceptance by FDA (Food and Drug Administration) for the electronic records, electronic signatures, and handwritten signatures executed to electronic records as equivalent to paper records and handwritten signatures executed on paper [4]. The FDA was also considered since the optoacoustic imaging devices of the manufacturer also has the US market. The risk has been assessed as indirect, but even moderate since the failure of the software tool leads to reduction/impairment of a quality process which indirectly affects product safety, performance and effectiveness. The validation is further required for moderate risks. The user needs and requirements were validated by carrying out successful test runs for each test specifications. The proof of validation were stored in electronic format in Polarion since the manufacturers shall maintain the records (see ISO 13485:2016, chapter 4.2.5) providing evidence of validation for an effective quality management system [2]. The process of the validation approach should also be documented by the manufacturer including the re-validation criteria and approval to changes. The documented procedure shall establish the controls required to [2]:

- review and approve documents before issuing

- update and re-approve documents

- determination of revisions and current versions of the required documents

- easily identifiable and legible documents

All the documents required for validation of the usability software tool were reviewed and approved as described according to the ISO 13485:2016, chapter 7.5.6 [2].

4 Conclusion

In the study, we demonstrated that the use of IEC and ISO standards for the update of usability engineering processes has the potential to increase the safety, efficiency and effectiveness of the medical device. The ISO 13485:2016 provides requirements for manufacturers for software tool validations, documentation procedures, and records for effectively implementing an improved quality management system. The updated SOP and software tool validation for usability is sufficient to perform the usability processes effectively and also to store the usability reports electronically. However, further additions regarding the ergonomic features according to the European Union Medical Device Regulation 2017/745 (EU MDR 2017/745 Annex I Chapter 1 part 5) [5] should be implemented in the Usability SOP

since various risks of injury related to physical properties should either be excluded or minimized for the operation of safe product.

Acknowledgement

I sincerely would like to thank Mr.Ingmar Thiemann, iThera Medical GmbH, München, Germany for conducting and supervising the internship and the management team of iThera Medical GmbH, München, Germany for providing a work place. I would also like to thank Prof.Dr.Folker Spitzenberger, Center for Regulatory Affairs in Biomedical Sciences (CRABS), Technische Hochschule Lübeck, Lübeck for his continous guidance and supervision during the project.

Author's Statement

The general overview for the Usability SOP and validation procedure belongs to the company affiliated with the internship. Informed consent has been given for the publication of the paper.

5 References

[1] *Medical devices - Part 1: Application of usability engineering to medical devices (IEC 62366-1:2015 + IEC 62366-1:2015/A1:2020).* In: Official Journal of CENELEC, 2020

[2] *International Organization for Standardization. Technical Committee Quality management and corresponding general aspects for medical devices, ISO 13485: Medical Devices-Quality Management Systems-Requirements for Regulatory Purposes. ISO, 2003*

[3] Jorien van der Pejil, Jan Klein, Christian Grass and Adidnda Freudenthal , *Design for risk control: The role of usability engineering in the management of use-related risks.*Journal of Biomedical Informatics, 2012

[4] *Part 11, Electronic Records; Electronic Signatures - Scope and Application. Guidance document,* 2003 Available: https://www.fda.gov/regulatory-information/search-fda-guidance-documents/part-11-electronic-records-electronic-signatures-scope-and-application [last accessed on 2021-12-14]

[5] *European Union, Regulation (EU) 2017/745 of the European Parliament and of the Council of 5 April 2017 on medical devices, amending Directive 2001/83/EC,Regulation (EC) No 178/2002 and Regulation (EC) No 1223/2009 and repealing Council Directives 90/385/EEC and 93/42/EEC. Official Journal of the European Union, 2017.*

Signal processing of gas sensor measurements for system identification and modelling of an anesthetic dosage unit

Carlotta Tebbe [1], Jana Tomat [2]

[1] Medical Engineering Science, Universität zu Lübeck, carlotta.tebbe@student.uni-luebeck.de
[2] Drägerwerk AG & Co. KGaA, jana.tomat@draeger.com

Abstract

Considering the increasing number of prolonged and more complex operations, automation of routine manual control tasks during surgeries, like setting an anesthetic gas concentration, would reduce the anesthetist's workload and offers the opportunity to achieve the preset target concentrations more efficiently. In order to automate such functions, an automatically controllable dosage unit is needed.

The considered anesthetic dosage unit delivers a pulsed gas concentration. These pulses can be characterized by their known width and frequency. To evaluate the deployed concentration, a suitable averaging method (recursive average estimation) was selected to smoothen the measurement signal. The smoothing process needs to be dynamically optimizable to adapt to the pulse parameters in runtime. Further experimental investigations indicate that the system can be identified as a first order system with delay characterized by dead time and time constant depending on flow settings.

1 Introduction

A prerequisite for the performance of surgical procedures is the transfer of the patient into a painless, unconscious and relaxed state (anesthesia) [1]. Regulating the anesthetic gas concentration in the patient's inspiratory air to an individual set point with an anesthetic dosage unit, is one of the central anesthesiologist's tasks. To relieve the anesthetist it would be appropriate to automate the control of the anesthetic gas concentration delivery.

The dosage unit considered in this article delivers a pulsed gas concentration that needs to be filtered, to accurately determine the average concentration getting administered to the patient's ventilation system. By measuring step responses, the system order can be identified and a mathematical model of the system can be established. The intended controller can be derived using that model.

1.1 System Modelling

System modelling describes the process of developing abstract models of a system to help understanding its functionality. A model is a mathematical representation of a physical, biological or information system. By describing the relation between inputs and outputs of the system, it allows to make predictions about the behaviour. For dynamical systems the effects of actions do not occur immediately, but usually evolve over time. In this article the anesthetic dosage unit at hand will be treated as a so-called Black-Box model, that disregards the internal workings and only considers the input and output relationship [2].

1.2 Dosage Control

To design a model-based controller for the output concentrations of the anesthetic dosage unit of an anesthesia machine, a mathematical model of the corresponding system is required. Derighetti [3] has shown that gas dosage and ventilation systems can be modeled mathematically as first order systems with dead time (T_{delay}) (see also [4], [5]). It is to be experimentally determined whether the first order system is applicable in the considered dosage unit.

In Fig. 1 the relevant inputs and outputs for the device at hand are illustrated. The anesthetic dosage unit takes a target concentration value as user input. To fulfill the request, the gas mixture accumulates with anaesthetic agent by passing through the dosage unit at a certain flowrate [L/min].

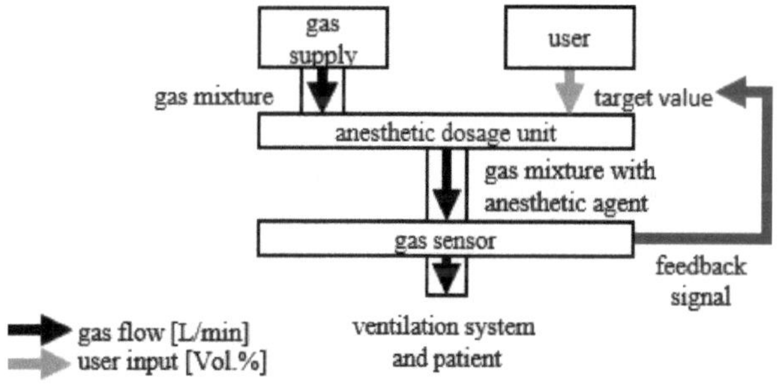

Figure 1: Control-loop for anaesthetic agent in gas mixture with user input.

The output gas concentration of the device is measured by a gas sensor before being delivered into the ventilation system and administered to the patient.

The pulsed output concentration delivered by the device is illustrated in Fig. 2. The pulses can be described by their pulse width and frequency. Due to the spatial distance between patient and device, changes in user input will not take effect at the patient immediately. Furthermore the applied pulsed concentration gets mixed into the gas volume of the breathing circuit for anesthesia and therefore reaches the patient smoothly. Therefore, to accurately detect if the average delivered concentration into the system corresponds to the target concentration, the pulses have to be low pass filtered over time.

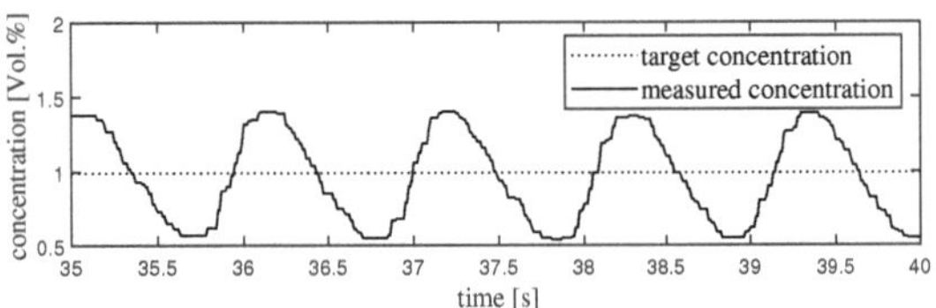

Figure 2: Pulsation of the deployed gas concentration around the target value. [flow: 3 L/min, target concentration: 1 Vol.%]

To achieve the desired filtering result, requirements have been defined. The filtered signal should follow the system dynamics and show the applied average concentration over time. The smoothing method should be easy to implement and ideally have low computational effort. The filter parameters should also be changeable during runtime to adapt to the varying system dynamics. Lastly the method should not introduce large additional delays into the system.

2   Material and Methods

2.1   Averaging Methods

There are several different approaches to average the input signal during runtime. MathWorks® provides the DSP System Toolbox™ for MATLAB® with a MovingAverage function. This function computes the moving average of the input signal over time. There are different possibilities to implement the filter, for example sliding window method or exponential weighting [6]. Furthermore a Drägerwerk internal Toolbox in MATLAB® provides a "Recursive Average Estimation" method.

2.1.1   Sliding Window

The sliding window approach considers a range of the input data to calculate the average. The range is called 'window' and is specified by its length. This window moves over the data and calculates the statistics over the current sample and the (window length - 1) previous samples. Therefore, the approach remembers previous data [6]. Choosing the window length influences the output significantly. Larger windows lead to an output closer to the stationary statistic of the data, while smaller windows can keep up and smooth rapidly changing data [6].

2.1.2   Exponential Weighting

The exponential weighting method computes a set of weights and applies these weights to the input data recursively. The weights for the data samples decrease exponentially over time, but never reach zero. Therefore, recent data has higher weights and more influence on the statistics at the current sample, than previous data. A forgetting factor $\lambda \in [0,1]$ can be defined, that determines the rate of change of the weighting factor. A higher forgetting factor takes previous data longer into account. To weigh recent data more, a value close to zero has to be chosen. A forgetting factor of 1 indicates infinite memory [6].

2.1.3   Recursive Average Estimation

An alternate approach to smoothen the pulsed signal without the DSP System Toolbox™ for MATLAB® is the recursive average estimation. This method follows (1) to calculate the average comparable to the exponential weighting method. $x(t)$ represents the input value, which is used to calculate the evolving average, the output values $y(t)$. The $\alpha \in [0,1]$ influences the tracking speed and variation of the mean similar to the forgetting factor (section 2.1.2). A small α allows fast signal tracking, while large α values are better for slow signal tracking.

$$y(t) = (1 - \alpha) \cdot x(t) + \alpha \cdot y(t-1). \tag{1}$$

To find an α value that represents a given time constant τ, (2) can be used.

$$\alpha = e^{-T_{sample}/\tau} \tag{2}$$

The variable T_{sample} describes the sampling time interval of the discrete MATLAB® model while τ must be calculated during runtime in relation to the pulse frequency.

2.2   Step Response

In order to determine the system order, experimentally obtained step responses can be used. In this article, the step response describes the dynamic of the pulsed gas concentration after changes in the user input.

In general, the step response describes how information represented in the time domain is being modified by the system [7] after a step in the input function. The input function, also called step function, is characterized by its sudden change from an initial value, often 0, to a final value. For an ideal step function, the time required for this value change approaches zero. A first order delay element, for example a PT1 element, is a transfer element whose output variable grows exponentially, with a certain initial increase, asymptotically towards a final value after a step change of the input variable [8]. For first order systems, the step response starts to increase immediately after the step function is applied without dead time. For many practical purposes, the result of an analysis of a system under the assumption of first order behavior is sufficiently accurate even with dead time, but this is to be verified for the specific case [9].

2.3 Test Setup

The experiments to identify the system order were performed on a hardware-in-the-loop (HiL) experimental setup. A HiL-setup includes the real components to be tested embedded in a simulated application environment. With the help of a host PC, the components can be controlled with appropriate software and can therefore be tested according to the application. The concluding simulation shows, how the systems will respond to different stimulation during runtime [10], [11]. The HiL-setup used consists of an anaesthetic dosage unit, gas sensor and in contrast to the final product a second validation gas sensor. The real-time computing and control unit of the anaesthetic dosage unit is taken over by a dSPACE-System [12].

The basis for the control of the hardware is a Simulink model, that includes the product software and enables the implementation of all operating functions. The model is delivered to the dSPACE-System in C-Code. With dSPACE ControlDesk® it is possible to change setting parameters, as well as readout, visualize and save measured values during runtime. This enables the implementation of functions in the product, that are not part of the serial product.

3 Results and Discussion

3.1 Selection of the averaging method

To accurately determine the average delivered concentration over time, the pulses need to be filtered. To select an appropriate averaging method, several different dosing settings will be evaluated. The dosing settings vary in gas-mixture-flow and target-concentration and are based on real operational scenarios. The considered flow set points are 0.2, 0.5, 1, 2, 5 and 10 L/min while the target concentrations are 0.2, 0.5, 1, 2, 4, 8 and 16 Vol.%.

The comparison of the available averaging settings for the different dosing points show that the averaging method is depending on the set target concentration which influences the pulse width and frequency. Therefore, the averaging of the pulses needs to adapt during runtime with the pulse frequency as well. For higher frequencies, the window size (section 2.1.1) must be smaller compared to lower frequencies, to preserve the dynamic of the system. The forgetting factor used in section 2.1.2 and section 2.1.3 needs to be closer to 1 for lower frequencies than for higher frequencies. It needs to be acknowledged, that the sliding window method provided by Simulink cannot be modified during runtime and is therefore not appropriately applicable in this case. Only one previously determined window size can be applied, but for the different pulse settings a dynamically changing sliding window would be needed to accurately describe the system.

As seen in Fig.3, all three applied averaging methods achieve an appropriate filtering result and minimize the pulse amplitude to emphasize the system dynamic, but show significant differences. At first the exponential weighting method follows the pulses closer and shows a stronger ex-

ponential increase than the other methods, while the sliding window is initially nearly linear. The recursive average estimation method has a slower increase, but averages the pulses stronger in the beginning and therefore better represents the dynamic.

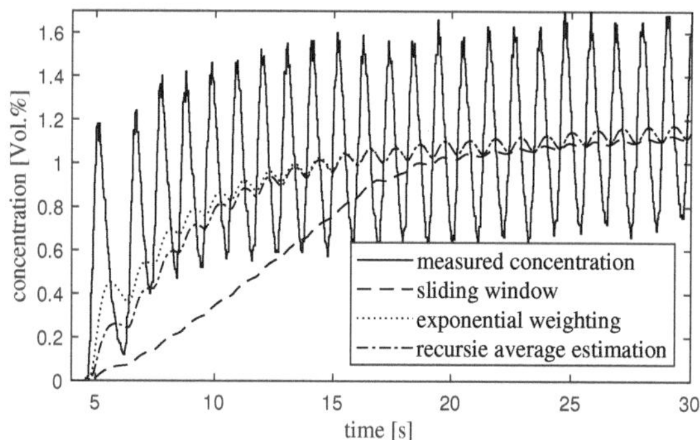

Figure 3: Comparison of the different averaging methods. The averaging parameters are chosen in relation to the pulse parameters. [flow: 3 L/min, target concentration: 1 Vol.%]

The recursive average estimation method (section 2.1.3) fulfills the previously defined requirements best and is therefore selected to conduct further investigations.

3.2 System Identification

The method selected in section 3.1 is used as an approximation of the step response. With help of the MATLAB®-function *'tfest'* a transfer function is estimated based on a data set consisting of the moving average, the step-function and the sample time. To use the *'tfest'*-function, the system order must be specified initially. The function was tested for different system orders and it was found, that a PT1 system with dead-time best fits the dynamic. As seen in Fig. 4, the PT2 system does not follow the actual system dynamic as closely as the PT1 system and is therefore not the best representation of the system.

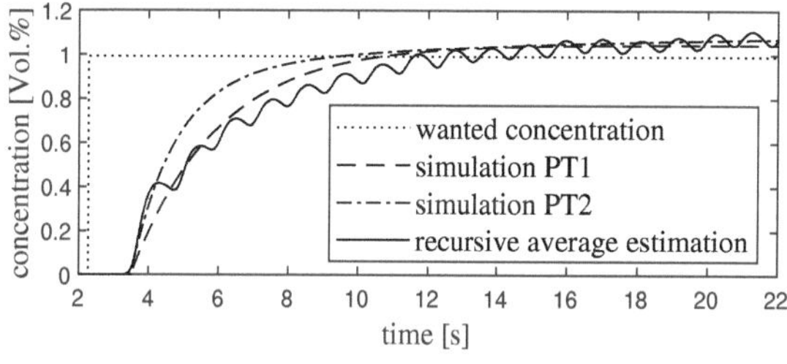

Figure 4: Comparison of the system orders PT1 and PT2 that are suitable for simple modelling. [flow: 2 L/min, target concentration: 1 Vol.%]

Due to the varying dosing settings, the parameters of the PT1 system change with the operational conditions. The conducted experiments show a dependency between the flow [L/min] as a variable system input and the PT1 system parameters τ and T_{delay}. Using the dosing settings defined in section 3.1 the parameters τ and T_{delay} are determined and averaged over five different measurements at the respective dosing set points. Between the parameters and the flow a two-term power dependency (3) can be found, where t is τ or T_{delay} respectively. Fig. 5 shows the dependency between set flow [L/min] and system parameter τ by visualizing the actually measured points and the estimated function.

$$t[s] = a * \dot{V}[L/min]^b + c \qquad (3)$$

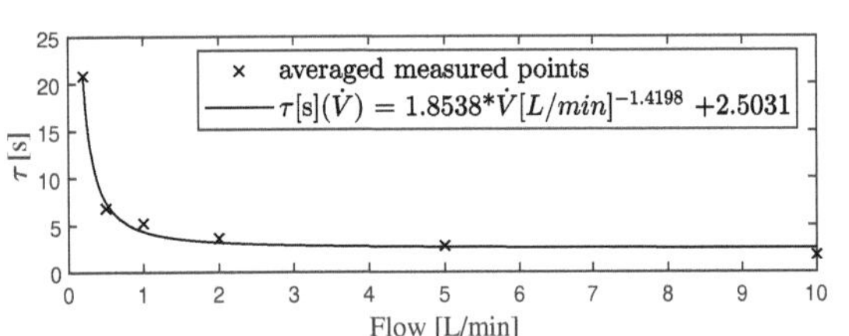

Figure 5: Dependency between set flow [L/min] and PT1 system parameter τ.

For T_{delay} another two-term power dependency (3) can be established as shown in Fig. 6.

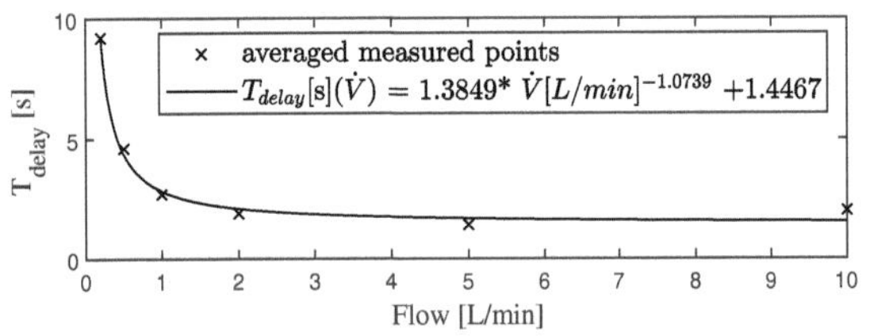

Figure 6: Dependency between set flow [L/min] and PT1 system parameter T_{delay}.

The system is therefore sufficiently accurate identified as an adaptable PT1 system with dead time.

4 Conclusion

Due to the anesthetic dosage unit delivering a pulsed gas concentration, it is necessary to average the signal to determine the dosage accuracy sufficiently. In consideration of the requirements defined in section 1.2 for the averaging method and the in section 3 experimentally obtained results the recursive average estimation is chosen to represent the dynamic of the system. With the averaged system dynamics, the system order is determined to be an adaptive PT1 system where τ and T_{delay} are depending on the set flow. Furthermore, validation of the model will have to be carried out by comparing the model output to validation measurements to ensure that the model can accurately represent the device response and can be used to simulate different dosing points.
Based on the PT1 system a controller could then be implemented that regulates and supervises the delivered gas concentration and is able to correct or provide feedback ensuring the safety of the patient.

Acknowledgement

The work has been carried out at Drägerwerk AG & Co. KGaA, Lübeck and supervised by P. Rostalski, Institute for Electrical Engineering in Medicine, Universität zu Lübeck.

Author's Statement

Conflict of interest: Authors state no conflict of interest.

5 References

[1] H.–W. Striebel, *Anästhesie, Intensivmedizin, Notfallmedizin*, Schattauer Verlag, 2016

[2] K. J. Astrom and R. M. Murray, *Feedback Systems: An Introduction for Scientists and Engineers*, Princeton University Press, 2008

[3] M. P. Derighetti, *Feedback control in anaesthesia*, ETH Zürich, PhD thesis, Zürich 1999

[4] K. S. Käther, *Development of a Multivariable Controller for Simultaneous Oxygen and Anesthetic Gas Concentration Control of a Semi-Closed Anesthesia Machine*, TH Lübeck, Masterthesis, Lübeck 2016

[5] D. Schmidt-Riese Nunes, *Modellbildung und robuster Reglerentwurf zur Anästhesiegasregelung im Atemkreis*, TU Darmstadt, Masterthesis, Darmstadt 2014

[6] MATLAB®, 9.9.0.1467703 (R2020b), *Moving Average*.
Available: https://de.mathworks.com/help/dsp/ug/sliding-window-method-and-exponential-weighting-method.html [last accessed on 2021-12-22].

[7] S. Smith, *Digital Signal Processing: A Practical Guide for Engineers and Scientists*. Elsevier Science, Germany 2003.

[8] H. Unbehauen, *Regelungstechnick 1. Klassische Verfahren zur Analyse und Synthese linearer kontinuierlicher Regelsysteme, Fuzzy-Regelsysteme*, Vieweg & Sohn Verlag, Wiesbaden 2007

[9] R. Parthier, *Messtechnik. Grundlagen und Anwendung in der elektrischen Messtechnik*, 8.Auflage, Springer Vieweg, Wiesbaden 2016

[10] J. Schäuffele and T. Zurwka, *Automotive Software Engineering.*, Wiesbaden, Vieweg + Teubner, 2010

[11] K. Borgeest, *Elektronik in der Fahrzeugtechnik. ATZ / MTZ-Fachbuch*. Wiesbaden: Springer Fachmedien, 2010

[12] dSPACE GmbH, *dSPACE Rapid Prototyping Systems*.
Available: https://www.dspace.com/de/gmb/home/products/systems/functp.cfm [last accessed on 2022-02-04].

Analysis of the measurement process for the focal spot size of an X-ray tube assembly

Laura Stehr [1], Ronald Tuchen [2]
[1] Medical Engineering Science, Universität zu Lübeck, laura.stehr@student.uni-luebeck.de
[2] Philips Medical Systems DMC GmbH, Hamburg, ronald.tuchen@philips.com

Abstract

To achieve the best image quality combined with as less radiation dose as possible, a small, defined focal spot size is needed for X-ray tube assemblies. For the determination of this spot, special focal spot cameras are used. These cameras have to fulfill requirements set by national and international standards. To use the most practical and efficient method for the measurement of focal spot sizes, a medical device manufacturer decided to analyze the determination of the enlargement factor of a focal spot camera. Therefore, the relevant requirements were collected and analyzed. Theoretical considerations about an alternative way to determine the enlargement factor have been done. These consist of the use of the theorem of intersecting lines with an additional pinhole to minimize the amount of influencing parameters. In next steps, a mathematical model and practical tests have to prove whether the considerations are correct, and the alternative provides an improvement.

1 Introduction

The best quality in X-ray images with as less radiation exposure as possible is a goal every manufacturer of X-ray systems has. As the X-ray tube assemblies are medical devices, there are many requirements from national and international standards that have to be met during the production. Manufacturers have to ensure, to always be compliant with the latest edition of the standards.

An important characteristic of an X-ray tube assembly is the focal spot, i.e. the spot on the anode from where the radiation exposures. To determine the size of this spot with a focal spot camera in the most accurate way and still in compliance with the requirements, the current way of working was analyzed and reviewed for potential improvement.

1.1 X-Ray radiation

To generate X-ray radiation, fast electrons need to be decelerated. Therefore, electrons are boiled out of the cathode filament and are accelerated through an electric field to the anode. To achieve that, an acceleration voltage U_a from 10 kV up to 500 kV [1], depending on the purpose of the use of the radiation, is applied between the anode and the cathode. To prevent collisions between air molecules and the electrons, anode and cathode are located in a vacuum. A schematic example of an X-ray tube is shown in Fig.1. The dotted line shows the way of the electrons. When these hit the anode, X-ray radiation is generated due to interactions between the electrons and the atoms in the material of the anode [2]. The interrupted line shows the way of the radia-

tion leaving the tube housing through the ray exit window.

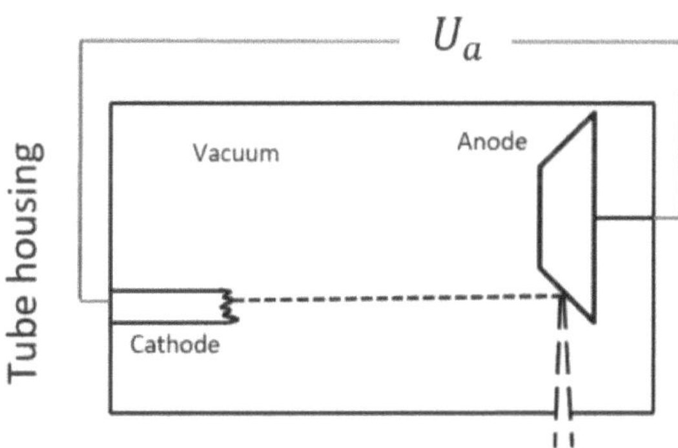

Figure 1: Example schema of an X-ray tube assembly. The dotted line shows the way of the electrons. The interrupted line shows the way of the radiation leaving the tube housing through the ray exit window.

1.2 Focus of the electron beam

The electrons, which are boiled out of the material of the filament, would accelerate through the electric field in a divergent beam [2]. The result would be a large *effective focal spot* on the anode. For best results in imaging diagnosis, this focal spot should be as small as possible with a defined size. Otherwise, the edges of the objects in the image would be blurred. Therefore, the electrons must be focused.

A *Wehnelt cylinder* can be used around the filament [2]. This cylinder has a negative charged potential and pushes off the electrons that diverge too far on the way to the anode. The result is a focused beam through the vacuum.

To be able to have two different sized focal spots, two filaments with different sizes can be used in the cathode.

In Fig. 2 two different sized focal spots as the result of such a construction are shown.

These pictures were recorded with a focal spot camera, which will be further described in 2.2.

Figure 2: Example of different sizes of focal spots. A small focus on the left and a large focus on the right.

2 Material and Methods

2.1 Requirements for X-ray tube assemblies

Manufacturers of X-ray tube assemblies have to fulfill several requirements during the production to ensure a safe product for users and patients. Some of these requirements are set by the International Electrotechnical Commission (IEC). This organization has the goal to standardize all interests related to electrical and electronical topics on a global level [3]. The IEC publishes international standards for the production of medical electrical equipment. The most important one regarding focal spot cameras is the international standard IEC 60336:2020 Ed. 5.0 *"Medical electrical equipment - X-ray tube assemblies for medical diagnosis - Focal spot dimensions and related characteristics"*. It describes test methods with digital detectors for determining the following tasks:

- focal spot dimensions [. . .];

- line spread functions;

- one-dimensional modulation transfer function;

- focal spot pinhole radiograms [3]

A *focal spot camera* is used to determine all of the above mentioned. Utilizing this equipment during the production of X-ray tubes assemblies ensures to meet the needed requirements. For the equipment itself, requirements are set as well. Among others the following ones are listed by the IEC 60336:

- determinations for the evaluation of the focal spot characteristics

- focal spot camera set-up

- determination of focal spot dimensions and nominal focal spot values

- alternative measurement methods for determining nominal focal spot values [3]

In edition 5 of the standard, analog detectors are replaced with digital detectors because they are the standard nowadays.

2.2 Focal spot camera

A focal spot camera is a measuring equipment to determine the width and length of the focal spot of an X-ray tube assembly. Therefore, it is directly mounted on the ray exit window of the tube housing. The X-ray beam passes a slit or pinhole diaphragm and is detected with a digital detector. Fig. 3 shows the geometrical setup of the camera. According to the IEC standard 60336, important distances are the ones between the effective focal spot on the anode and the diaphragm, i.e., m and the diaphragm and the digital detector, i.e., n.

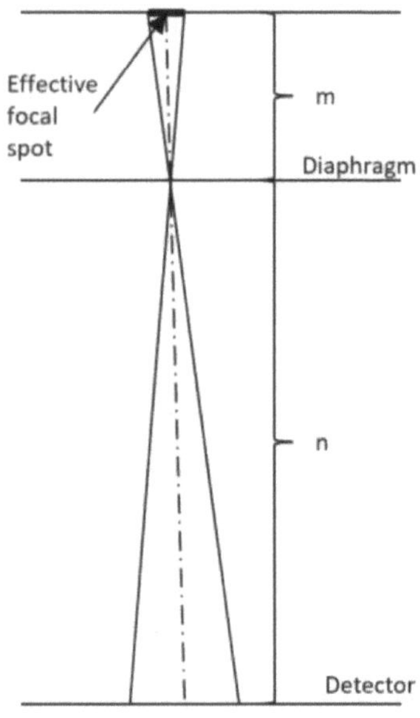

Figure 3: Geometrical setup of parameters acc. to IEC 60336:2020 [3].

2.3 Detector

The X-ray beam can be detected with different kinds of detectors: analog detectors or digital detectors.

Radiographic films are used as *analog detectors*. These films consist of a base material that is surrounded with a photographic emulsion, which is protected by a gelatin coating [4]. Silver halide microcrystals can be used as photosensitive elements [4]. The further processing of a radiographic film is equal to photographic films. So after exposure with radiation the film is processed and the analog X-ray image can be used for diagnosis.

A first step towards digital detectors is the digital processing of radiographic films, i.e., the picture on the film is converted to a digital image and can be further processed with a computer.

One kind of digital detectors are *energy-integrating detectors* (EID). They consist of a scintillator and a light sensor. The X-ray photons that hit the scintillator are converted into photons of visible light [5]. After that, e.g. a photodiode absorbs these photons and, by measuring the amount of light, an electrical signal proportional to the incoming energy is generated [5].

Another kind of digital detectors are *photon-counting detectors* (PCD). These do not convert X-ray photons to light photons. Instead, they consist of a large voltage applied semiconductor layer. When an X-ray photon hits the semiconductor diode it generates charges which are positive or negative [5]. These are divided fast and through the moving charges an electrical pulse is generated. This pulse is read out via wires that are attached to the electrodes [5].

The further processing of the signals is done with a software.

2.4 Determination of the enlargement factor

To determine the size of the effective focal spot, the enlargement factor E and the size of the focal spot on the detector need to be defined. With this two parameters the effective focal spot size on the anode can be calculated by dividing the measured size through the enlargement factor.

$$\text{effective focal spot size} = \frac{measured\ focal\ spot\ size}{enlargement\ factor} \tag{1}$$

The IEC 60336:2020 defines the calculation of the enlargement factor E as

$$E = \frac{n}{m}. \tag{2}$$

The result of this calculation shall be defined with an accuracy of $\pm 3\%$ [3].

The schematic drawing of a focal spot camera mounted on an X-ray tube assembly in Fig. 4 details the parameters. It shows, that m is composed of the distance between the anode and the housing (t) and the distance between the housing of the tube where the camera is mounted and the diaphragm (c). The parameter n is equal to the distance between the diaphragm and the active layer on the detector (a).

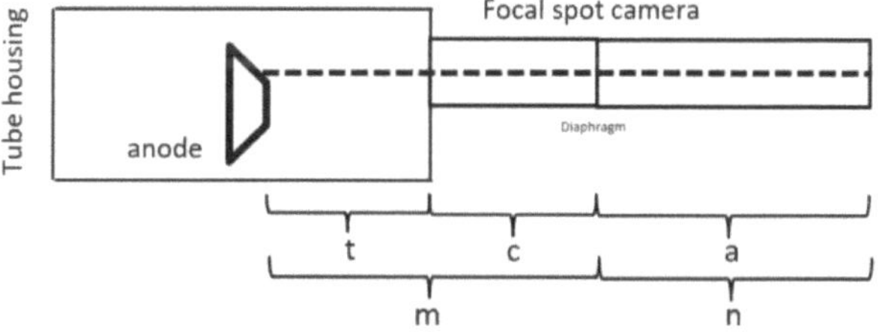

Figure 4: Schematic drawing of a camera mounted on an X-ray tube assembly.

Using these parameters, the calculation of E would be as follows:

$$E = \frac{a}{t + c}. \tag{3}$$

To determine the value of c, the length of the part of the camera, that is mounted on the tube housing, to the diaphragm is measured with measuring tools.

The value of the parameter a is composed of the measured distance from the diaphragm to the plane on which the detector is attached and the active layer of the detector. As the detector is included in a housing, the distance from the bearing area to the active layer has to be added. This has to be specified by the manufacturer of the detector and has certain uncertainties due to manufacturing tolerances.

Besides, the value for the parameter t, i.e., the distance from the effective focal spot on the anode to the camera, cannot be determined by measuring, but has to be calculated from different parameters.

As the three parameters are independent and the factor E is not directly measured but calculated indirectly, the rules from the *Gaussian error propagation* need to be followed. All individual uncertainties incorporate in the uncertainty of the enlargement factor ΔE as follows:

$$\Delta E = \sqrt{\left(\frac{\partial E}{\partial a} \cdot \Delta a\right)^2 + \left(\frac{\partial E}{\partial c} \cdot \Delta c\right)^2 + \left(\frac{\partial E}{\partial t} \cdot \Delta t\right)^2} \tag{4}$$

The insertion of the partial derivatives leads to:

$$\Delta E = \sqrt{\frac{a^2(\Delta c^2 + \Delta t^2) + \Delta a(c + t)^2}{(c + t)^4}} \tag{5}$$

Equation 5 shows, that the parameter t influences the uncertainty of the enlargement factor immensely. As this parameter cannot be measured directly, this is a potential source of inaccuracy [6].

The IEC gives manufacturers the possibility to determine the required parameters in other ways as described in the standard, as long as the used method does not provide smaller focal spots [3]. Another way to determine the enlargement factor can be used, as long as this method provides results with an accuracy of $\pm 3\%$ or better.

3 Results and Discussion

3.1 Alternative determination of the enlargement factor

Instead of using a focal spot camera with one pinhole, a camera with two pinholes could be used to detect two optical focal spots on the detector at the same time.

Due to the *theorem of intersecting lines* it would be possible to calculate the enlargement factor with the distance between the two focal spots on the detector, b, and the distance between the two pinholes, p, as shown in Fig. 5.

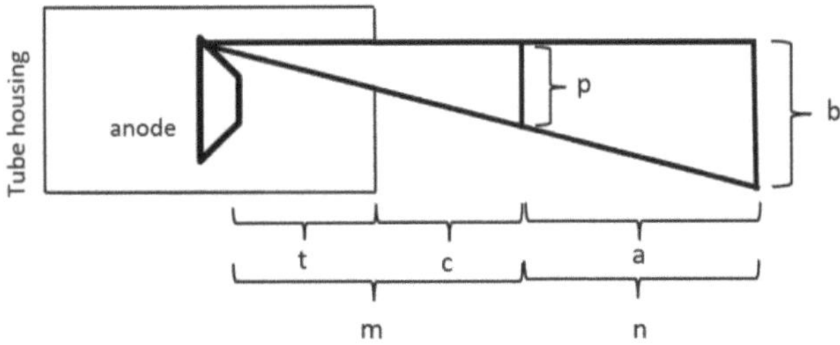

Figure 5: Schematic drawing of a different setup to use the theorem of intersecting lines.

This would minimize the amount of parameters and with that the amount of uncertainties.

Using the theorem of intersecting lines, the equation of the parameters would be:

$$\frac{n+m}{m} = \frac{b}{p}. \tag{6}$$

Combining (6) and (2) leads to the following equation

$$E = \frac{b}{p} - 1. \tag{7}$$

To verify that this determination of the enlargement factor is more, or at least just as, accurate and within the required accuracy of $\pm$ 3%, the relative error of the enlargement factor is calculated as:

$$\frac{\Delta E}{E} \tag{8}$$

Again, due to the independence of the parameters, the *Gaussian error propagation* needs to be considered for the determination of ΔE:

$$\Delta E = \sqrt{\left(\frac{\partial E}{\partial b} \cdot \Delta b\right)^2 + \left(\frac{\partial E}{\partial p} \cdot \Delta p\right)^2} \tag{9}$$

Equation (9) consists of the partial derivatives of E with respect to b and p, as well as the particular uncertainties Δb and Δp of these parameters.

The partial derivatives are determined as the following:

$$\frac{\partial E}{\partial b} = \frac{1}{p} \tag{10}$$

$$\frac{\partial E}{\partial p} = -\frac{b}{p^2} \tag{11}$$

Inserting (10) and (11) in (9) leads to:

$$\Delta E = \sqrt{\frac{\Delta b^2 p^2 + \Delta p^2 b^2}{p^4}} \tag{12}$$

This shows, that for the alternative determination of the focal spot size the parameter p has the greater influence on the accuracy of the enlargement factor.

4 Conclusion

Considering the theoretical calculations above, the alternative procedure to determine the enlargement factor E, that has been described in this article, could decrease the uncertainties of E significantly. The advantage of this procedure is based on the fact, that the pinhole distance p is known exactly. In a next step, the calculations set out above have to be implemented in a protocol that includes all factors of influence and the exact way of measuring or calculating for these factors. Finally, practical tests should be performed, which comprise a series of measurements in order to determine optimal parameters with minimal uncertainties. The crucial test for this alternative procedure would be to obtain a relative error for the enlargement factor E that is equal to or below the required accuracy of $\pm$ 3%.

Acknowledgement

The work has been carried out at a Philips Medical Systems DMC GmbH and supervised by the Institute for Physics, Universität zu Lübeck. I sincerely thank the involved people from this company for providing workspace, material and knowledge.

Author's Statement

Conflict of interest: Author states no conflict of interest.

5 References

[1] T. M. Buzug, *Computed Tomography - From Photon Statistics to Modern Cone-Beam CT*. Springer Verlag, Berlin Heidelberg, 2008.

[2] W. Schlegel, C. P. Karger, O. Jaekel *Medizinische Physik*. Springer Spektrum, Berlin, 2018.

[3] International Electrotechnical Commission (IEC), *IEC 60336:2020 Medical electrical equipment - X-ray tube assemblies for medical diagnosis - Focal spot dimensions and related characteristics*. IEC, Genf, 2020.

[4] D. Oborska-Kumaszynska, S. Wisniewska-Kubka, *Analog and digital systems of imaging in roentgenodiagnostics*. In: Polish Journal of Radiology: Volume 75: Number 2 - March 2010, pp. 73-81, 2010.

[5] M. J. Willemink, M. Persson, A. Pourmorteza, N. J. Pelc and D. Fleischmann, *Photon-counting CT: Technical Principles and Clinical Prospects*. In: Radiology: Volume 289: Number 2 - November 2018, pp. 59–62, 2018.

[6] M. Nishiki, S. Yanagite, n. Nishikawa, H. Sugimura, *Measurement of the x-ray effective focal spot size with edge response analysis using digital detectors*. In: Journal of Medical Imaging: Vol. 7(2) - Mar/Apr 2020, 2020

Correlation between cochlear implant MAPping and protocol data from patients based on Support Vector Machines

Daseul Jeong [1], Tim Jürgens [2], Jonnas Park [3]

[1] Hörakustik und Audiologische Technik, Universität zu Lübeck, daseul.jeong@student.uni-luebeck.de
[2] Institut für Akustik, Technische Hochschule Lübeck, tim.juergens@th-luebeck.de
[3] Lehrstuhl der Hals-Nasen-Ohren-Heilkunde der Universität Witten/Herdecke, jonas.park@uni-wh.de

Abstract

In recent years growing attention is being paid toward machine learning to analyze a large-scale datasets and this technology is already applied in medical areas successfully. Cochlear implant (CI) fitting is the adjustment of individual settings according to patients' needs. The MAPping process is done at first activation and adjustments of the MAPs are done whenever the patients experience problems with their CI. It takes constant effort to obtain proper settings, because the settings depend strongly on the individual. It may be very potentially helpful to analyze the correlations between different MAPpings and the problems of patients in order to understand the structures behind them. The goal of this study is to use the supervised learning method "support vector machines" (SVM) for this task. To achieve the desired accuracy of label prediction, feature extraction, and feature selection are conducted. The accuracy of the trained SVM is 92.3% in the given dataset.

1 Introduction

It is important to find optimal settings or MAPs in individual Cochlear implant (CI) users for good rehabilitation. After first activation of a CI, these MAPs are usually adjusted several times within a short period to find the better MAPs and then in regular intervals over several years because of physiological changes at the electrode-nerve interface. The change of these setting as a function of time is rather individual and adjustments are time-consuming. Furthermore, every patient explains their problems differently, or patients have no idea what the problems are and they are just anxious, so it is quite tough to figure out the real problems.

The goal of this study is to find relationships between the MAPping datasets and the patients' problems using a machine learning method. There are already several applications of machine learning to the research of cochlear implants [1]. This paper uses a supervised learning methods, support vector machine (SVM) to find relationships between patients' datasets. In this work, the MAPping datasets are analyzed to find correlations to the individual patients' CI MAPping and their problems by using CI in every-day life. Based on these correlations, it will be possible to find underlying structures and insight based on the change of MAPping.

2 Material and Methods

2.1 Patients and Datasets

Datasets of twenty-five patients with unilateral or bilateral cochlear implants were investigated (12 male and 13 female) with mean age 60 years. The datasets at Katholisches Krankenhaus (KKH) in Hagen, Germany, are the primary datasets used in this work. All patients were implanted with Nucleus CI512 cochlear implant. The Nucleus CI512 cochlear implant has 22 half-banded platinum electrodes, molded in a perimodiolar shape. The length of the electrode array is 19 mm. Electrode diameter at apical end and basal end is 0,5 mm and 0.8 mm, respectively. Input frequency range is between 118 Hz and 7789 Hz. When the patients visit the hospital because of problems or annual tests, they fill out protocols.

Table 1: Brief information on symptom classes and labels. A label covers more than one problem and noise means noise while riding a bicycle or driving a car, etc.

Symptom classes		Labels
Sound		
	Beep Sound	1
	Whistle Sound	1
	Hollow Sound	1
Echo		2
Noise		3

These patient protocols, their explanation about their problems, and newly obtained MAPping datasets will be used as the main dataset for this study. The datasets consist of changes of MAPping over time and classes. Each MAPping dataset has a class, in this case, one of three different symtopms, as shown in Table 1. There are more than three classes, however, the other cases are rarely found. So, these datasets are excluded in the process of data preprocessing. A MAPping dataset consists of threshold levels(T-levels) and comfort levels(C-levels or sometimes called M-levels). T-levels are the minimal amount of electrical stimulation required for the auditory system to perceive sound and C-levels are the upper limits of electrical stimulation judged to be most comfortable, or loud but comfortable. So, the MAPping datasets report on differences between the previous and the newly obtained MAPping dataset from T-levels and C-levels, respectively. Lastly, the MAPping datasets and the symtopms are combined. The size of the datasets is 37518 * 4 and because of different observation times, the amount of information from each patient is different.

2.2 Machine learning algorithm

The purpose of pattern classification is to get a model which maximizes its performance or the accuracy based on the training dataset. Normally, training of a model is done by dividing the full dataset into ground-truth and test dataset. However, if the classifier is over-fit to the training dataset, the model begins to memorize training data and fails to predict labels, when new datasets are used. The main motivation of a support vector machine (SVM) is to separate several classes in the training set with a surface, such as a line or a curve, that maximizes distances between them.

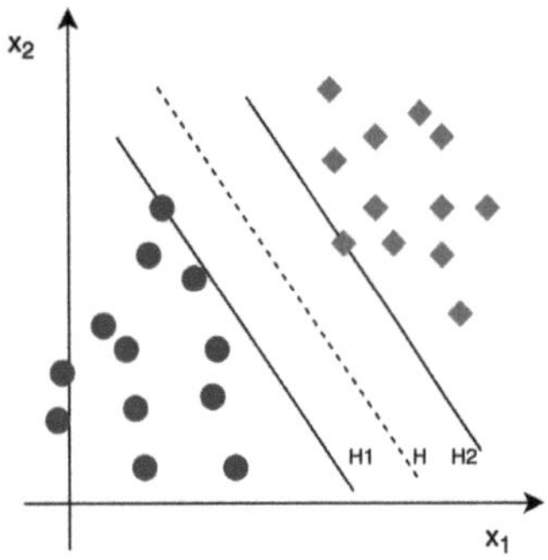

Figure 1: The maximum margin hyperplane. The margin is the distance from the hyperplane - H to the closest points in either class, which touch the parallel dotted lines, H1 and H2.

Fig 1 shows the basic idea of SVM. Two types are separated by hyperplanes (H). The distance between the hyperplane and the training data closest to the hyperplane is called margin. The generalization ability is maximized if the optimal separation hyperplane is selected as the separation hyperplane. There are two parallel hyperplanes called H1 and H2. The distance between H1 and H2 is maximized and there is optimally no data between the two hyperplanes [2], [4].

2.3 Model development

The workflow of this study, including data collection and model development, is shown in Fig 2. In the following sections, each step is described in detail.

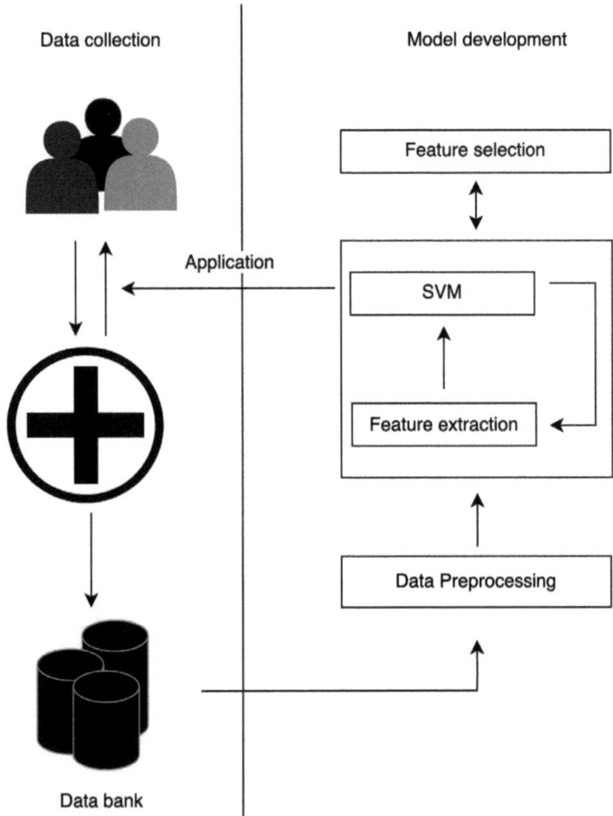

Figure 2: Sketch of workflow of this study. It consists of two different parts. The first part is the data-collection part, where patients' information is collected in the hospital and digitalized. In the second part, the model is developed, where data-preprocessing, feature extraction, and feature selection are implemented [4].

2.3.1 Data preprocessing

In the whole dataset, a few patient MAPs were missing T-level and C-level values in frequency ranges. In the part of data preprocessing, the datasets with missing values were dropped and the differences of T-level and C-level values were saved in combination with their labels.

2.3.2 Feature extraction

In Section 2.1 it was mentioned that the final size of the whole dataset is 37518 * 4. Each row dataset has forty-four values from T-level and C-level and one label. However, it might be difficult to find important information on the correlation between T-level and C-level values and labels without feature extraction. Feature extraction is a useful approach to analyze large-scale datasets. This approach is used not only to reduce the dimension of high-dimensional datasets but also to find more relevant information than raw datasets for developing a model. A lot of feature extraction methods and applications are available [3], but still the performance of the model will strongly be affected by the type of datasets. So, in this work, basic features are used. At first, the differences between previous fitting values and the newly found fitting values are calculated. According to the frequency ranges, these differences are divided into three different groups, in other words, into low-frequency, middle-frequency, and high-frequency. To make sure that the three groups have different importance and effect on the results, respectively, different features are calculated, for example, in the low-frequency ranges, mean value and summation and in middle-frequency ranges, only summation of values is calculated. Through this process, different weights were assigned to each feature group.

2.3.3   Model Train and Test

In the Model Development part, the preprocessed datasets with labels are divided into training datasets and test datasets. 70% of the whole datasets, which is 25137*4, will be used to train the model. There are several cross-validations methods, for example, simple K-folds cross-validation or Leave One Out cross-validation. However, in this work, K-fold shuffle cross-validation is applied to avoid some tendencies of specific patients and to ensure that the trained datasets are randomly chosen [5]. As Section 2.2 mentioned, SVM is used not only as a classifier but also to measure if appropriate features are selected. This feature selection is used to maximize the performance of classifiers in supervised learning. With the help of the given labels or ground truth, it is possible to find more relevant features with the given labels. So, a better feature combination results in better performance. To find the better feature combination, several features are calculated, changed, and then calculated manually. There are no specific guidelines to find feature combinations. That is why a feature selection process is necessary and the process is manually conducted [3], [6]. After manual tests of feature selection by comparing the accuracies between the previous and the newly calculated features, when the desired accuracy (higher than 90 % in this study is achieved, it is time to conduct feature importance. Feature importance is a technique for giving scores to the used features that indicate the relative importance of each feature to divide the given dataset into the symptom classes. The finally found optimal features are weighted superpositions of the initial MAP-based values. Therefore, it is difficult to interpret what they are. However, it might be possible to get insight into the features. Furthermore, dimensionality reduction of data is an important further goal of feature extraction. In this study, there are three different features, it is not necessary to implement feature reduction but still feature selection is helpful to find which electrodes should be adjusted more carefully than others.

Table 2:   The accuracies of the trained model with 5 times cross-validation. The mean value of the 5 tests is 93.9%.

Cross-validation	Accuracy
# 1	0.935
# 2	0.918
# 3	0.944
# 4	0.90
# 5	0.918
mean	0.923

3   Results and Discussion

Table 2 shows the accuracies of the SVM classifier with 5 cross-validations and their mean value. The high accuracies of more than 0.93 indicate that calculated features are properly chosen and the developed model is trained well. The next step is to investigate if accuracy can be further improved by evaluating the importance of the features and if it is necessary to apply feature selection.

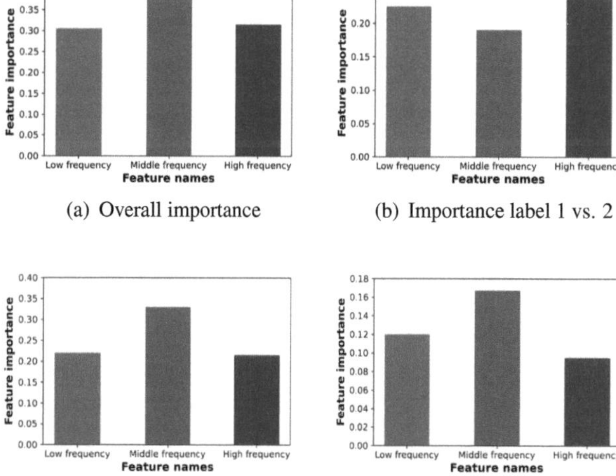

Figure 3:   Barplots of the feature importance from three different features. Fig 3(a) shows that three features have different importance and the middle frequency feature is the most important to divide the datasets into three different groups. The remaining three barplots explain which features are more important than others distinguishing label 1 from label 2, label 2 from label 3, and label 1 from label 3, respectively.

Fig 3(a) shows the feature importances between the features and the labels. Although only three different features are used to divide the datasets into three different groups, it still appears that the features are properly selected. In addition, Fig 3(b), Fig 3(c), and Fig 3(d) indicate which features have more effects on finding specific labels. For example, the low-frequency features and the high-frequency features have a stronger effect on distinguishing label 1 from label 2 in Fig 3(b). Because of the limitation of the number of datasets and patient's protocols, each label has more than one symptom class in this study. However, these insights of understanding the relationships between CI MAPping values and problems would be valuable to build a more statistically appropriate and more systematic method in future work. However, it needs to be pointed out how feature selections are conducted. As mentioned in Section 2.3.2, several mathematical values are calculated, at first. In addition, it should be decided how many T-level values and C-level values are used for one feature, in other words, the number of the used T-level values and C-level values are the same or different, and if the number of the used values are different from each feature, how to decide the number and how to ensure that it is the best choice for the better model, etc. That is why manual tests are necessary to find the most appropriate features. In this study, a lot of combinations of features and different numbers of T-level values and C-level values were tested. At the end of the model development part, the trained model was able to predict labels quite well with high accuracies in Table 2. Consequently, when new patients with the same symptoms come for their MAPping procedure, it would be possible to find better MAPping values not only based on experts' experiences but also based on analytical reasons by using the trained model.

4 Conclusion

The objective of this work was to find the relationships between CI MAPping datasets and questionnaires of patients. To reach this objective, a data-driven model was developed. With the help of the developed model, the dataset was classified according to the desired labels with an accuracy of 93.9%. Furthermore, it is possible to find which frequency areas should be adjusted based on the problems. However, although the accuracy is quite high and it is now known which features are more crucial than others during the fitting process, it is hard to say that better results occur always in the fitting process. There are several reasons for that. First of all, the number of patients is relatively small and it results in a lack of diversity of symptoms and information. Under the condition of this work, it was easily possible to divide the datasets into the correct smptom classes because there are only three different classes. Furthermore, these datasets were collected during several years by different people. As a result, there may be some variability due to protocols being written from different points of view. Consequently, it was very difficult to digitize the protocols. As a solution, it will be better to make a kind of standard protocol formula to check what problems are and how they are treated in the fitting process. Besides, this approach for data collection would be stable and more clear than just protocols. This is because it occurs often that patients use their own words. In some protocols, some strange words are found, which are not used in real life, but the solution was similar to a common problem. If there are well-categorized multiple choices, it is clear that the quality of datasets will be higher, more advanced models could be developed due to cleaner data. This approach may have massive positive effects on the fitting process.

Acknowledgement

This work was carried out at Katholisches Krankenhaus Hagen and was supported by the Institute of Acoustics, Technische Hochschule Lübeck.

Author's Statement

Conflict of interest: Authors state no conflict of interest.

5 References

[1] Crowson, Matthew G and Lin, Vincent and Chen, Joseph M and Chan, Timothy CY, *Machine Learning and Cochlear Implantation—A Structured Review of Opportunities and Challenges*. Otology & Neurotology: pp.e36-e45, 2020.

[2] Cortes, Corinna, and Vladimir Vapnik, *Support-vector networks*. Machine learning 20.3: pp.273-297, 1995.

[3] Ghojogh, Benyamin and Samad, Maria N and Mashhadi, Sayema Asif and Kapoor, Tania and Ali, Wahab and Karray, Fakhri and Crowley, Mark, *Feature selection and feature extraction in pattern analysis: A literature review*. arXiv preprint arXiv, 2019.

[4] Cervantes, Jair and Garcia-Lamont, Farid and Rodríguez-Mazahua, Lisbeth and Lopez, Asdrubal, *A comprehensive survey on support vector machine classification: Applications, challenges and trends*. Neurocomputing: pp.189-215, 2020.

[5] Berrar, Daniel, *Cross-Validation*. : pp.542-545, 2019.

[6] Guyon, Isabelle and Elisseeff, André, *An introduction to variable and feature selection*. Journal of machine learning research: pp.1157-1182, 2003.

The Effect of Mechanical Stimulation using Focused Low-Intensity Pulsed Ultrasound (FLIPUS) on Osteocytes

Aseel Khaffaf [1], Regina Puts [2], and Kay Raum [3]

[1] Biomedical Engineering, Luebeck University of Applied Sciences, aseel.khaffaf@stud.th-luebeck.de
[2] Charité – Universitätsmedizin Berlin, BIH Center for Regenerative Therapies, regina.puts@bih-charite.de
[3] Charité – Universitätsmedizin Berlin, BIH Center for Regenerative Therapies, kay.raum@charite.de

Abstract

Mechanically stimulated cells receive physical stimuli and convert them into biological responses. The mechanical forces can be supplied to a biological system in the form of ultrasound. Low-Intensity Pulsed Ultrasound (LIPUS) is a therapeutic technique used for bone healing. However, the biological mechanisms of this method are still poorly understood. Focused LIPUS (FLIPUS) *in-vitro* set-up was previously developed at the BIH Center for Regenerative Therapies. It provides acoustic intensity controlled and artefact-free stimulation conditions. In this work, the effect of FLIPUS stimulation of MLO-Y4 osteocytes on expression of genes associated with bone homeostasis was assessed 3 hours after the 10-min stimulation. As a result enhanced expression of Cox-2, Cx43, Cyr61 and Birc5 genes, which are associated with cell survival, differentiation, communication and bone formation, was observed. RANKL, OCN, SOST and E11 genes did not show significant change between stimulated and unstimulated starved MLO-Y4 cells at the selected time point.

1 Introduction

Bone tissue is categorized as a specialized connective tissue, whose functionality is highly reliant on the activity of four cell types, i.e., osteoblasts, osteocytes, osteoclasts and osteogenic precursors [1]. Osteocytes account for more than 90% of the total bone cells, making them the most commonly found cell-type in mature bone tissue.

During the preceding decade, research has been focusing on the responsiveness of bone tissue to mechanical stimulation [2]. This is due to the latter's role in the enhancement of fracture healing. In bone tissue, it has been proposed that osteocytes are the pivotal cell type, which is receptive to mechanical stimulation. Additionally, they serve as the primary regulators of bone mechanosensation and mechanotransduction.

Mechanical stimuli, sensed by cells, are converted into a biological response, which in turn can influence and control cellular functions including migration, cell survival, cell-cycle progression, and differentiation [3]. Various techniques for mechanical stimulation have been classified, among these is Low-Intensity Pulsed Ultrasound (LIPUS). Ultrasound is a longitudinal pressure wave propagated in the given medium. LIPUS is a therapeutic technique with an intensity value below $100 \, \text{mW/cm}^2$. It is acknowledged as a non-invasive and non-ionizing therapy approach for fracture healing. However, certain inconsistencies in the success of this therapeutic method in clinic have been reported, showing both positive and lack of effects [4]. This is related to the fact, that LIPUS-triggered biological mech-

anisms are not fully understood. Most of the *in-vitro* LIPUS cell-stimulation set-ups, described by the scientific community, have no control over the acoustic dose delivered to the cells. Additionally, several unwanted artifacts are associated with the aforementioned set-ups, such as temperature elevation and inhomogenous intensity distribution. Hence, a modified *in-vitro* set-up with the use of Focused LIPUS (FLIPUS) was introduced by Puts et al. [5]. The FLIPUS set-up has been developed to guarantee mechanical stimulation of cells in a homogenous far-field of focused transducers with a specified acoustic intensity in addition to the elimination of undesirable artefacts. The established set-up allows stimulation of adherent cells cultured in 24-well plates under sterile conditions at $37°C$, supplied by 95% air and 5% CO_2 mixture in quadruplicate.

In order to get a further insight into the ultrasound-mediated enhanced bone regeneration, the effect of FLIPUS on expression of genes associated with cell survival and bone tissue homeostasis was investigated 3 hours after 10 minutes of mechanical stimulation of osteocytes, using quantitative real-time reverse transcriptase (qRT-PCR) technique.

2 Material and Methods

2.1 Cell culture

Murine osteocyte-like cell line MLO-Y4 was generously provided by Prof. Martina Rauner from Dresden University of Technology. MLO-Y4 cells were cultured in alpha-MEM media (Corning, Manasses, VA, USA) supplemented with

2.5% fetal bovine serum (FBS; Biochrom GmbH, Berlin, Germany), 2.5% calf serum (CS; Sigma Aldrich, St.Loius, MI, USA), 1% penicillin-streptomycin (P/S; Corning, Manasses, VA, USA) and 200mM Glutamax (Glux; Gibco, Life Technology Corperation, Grand Island, NY, USA). For each experiment, MLO-Y4 cells were seeded at 10^5 cells per mL in a 24-well plate coated with 0.15 mg/ml type I rat tail collagen (Corning, Manasses, VA, USA). Afterwards, the cells were starved in the same media, but supplemented with 0.75% fetal bovine serum, 0.75% calf serum, 1% penicillin-streptomycin and 200mM Glutamax. After 3 hours of starvation, the cells were transferred to FLIPUS set-up for mechanical stimulation.

2.2 FLIPUS for mechanical stimulation

FLIPUS set-up, shown in Fig.1, was used for stimulation of MLO-Y4 cells cultured in 24-well plates. Details on the set-up are provided by Puts et al. [5]. The water tank was filled with sterilized, deionized, degassed water and maintained at 37°C.

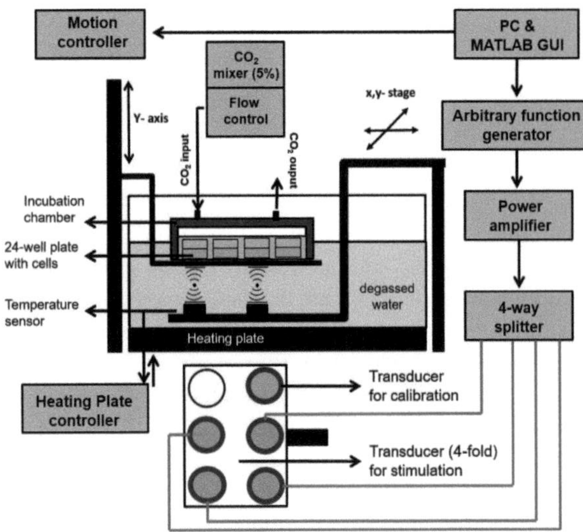

Figure 1: Focused Low-Intensity Pulsed Ultrasound (FLIPUS) set-up for in-vitro mechanical stimulation of cells cultured in a 24-well plate [5].

The well plate was placed inside the incubation chamber and attached to the set-up via magnet holders. The incubation chamber inlet was connected to a CO_2/O_2 mixer to mimic the incubator-like environment. Air bubbles created during the insertion of the incubation chamber with the well plate into the water tank were removed from the bottom of the well plate prior to stimulation. Using the automated protocols programmed with MATLAB 2009a, the cells were stimulated with FLIPUS for 10 minutes at a frequency of 3.6 MHz, pulse repetition frequency of 100 Hz, duty cycle of 27.8% and spatial average temporal average intensity (I_{SATA}) of 44.5 mW/cm^2.

2.3 qRT-PCR analysis

In order to investigate the genes expressed after FLIPUS stimulation, qRT-PCR was performed. MLO-Y4 cells were stimulated for 10 minutes with FLIPUS and incubated for 3 hours. Control samples were treated identically, only without the FLIPUS signal. Then mRNA from the stimulated and control samples was extracted using NucleoSpin RNA II kit (Machery-Nagel GmbH, Düren, Germany) and mRNA concentration for each sample was measured by NanoDrop spectrophotometer. The complementary DNA was reverse transcribed from $1\mu g$ RNA in qScript cDNA superMix (Quanta Bio-Sciences Inc., Gaithersburg, MD, USA) using Mastercycler ep gradient S. Subsequently, qRT-PCR was performed using Roche Light-Cycler 480 II. The primers used for qRT-PCR analysis are osteoclacin (OCN), receptor activator of nuclear factor kappa- ligand (RANKL), sclerostin (SOST), cysteine-rich angiogenic inducer 61 (Cyr61), connexin-43 (Cx43), cyclooxygenase-2 (Cox-2), baculoviral inhibitor of apoptosis repeat-containing 5 (Birc5), and podoplanin (E11). The primer sequences are shown in Table1. Each sample was quantified in duplicate, then normalized to the corresponding C_t value of hypoxanthine-guanine phosphoribosyltransferase (HPRT) gene.

Gene	Table 1: Primers list Primer sequence (5´–3´) Forward (F), Reverse (R)
OCN	F: CTTTAGGGCAGCACAGGT
	R: CTTTAGGGCAGCACAGGT
RANKL	F: AGGAGGGAGCACGAAAAACT
	R: AAGGGTTGGACACCTGAATG
SOST	F: AGCCTTCAGGAATGATGCCAC
	R: CTTTGGCGTCATAGGGATGGT
Cyr61	F: AGGGTCTGCCTTCTGACTGA
	R: AGGGTCTGCCTTCTGACTGA
Cx43	F: CGGTTGTGAAAATGTCTGCTATG
	R: GGCACAGACACGAATATGATCTG
Cox-2	F: ATCCTGAGTGGGGTGATGAG
	R: GGAACTGCTGGTTGAAAAGG
Birc5	F: CATCGCCACCTTCAAGAACT
	R: AAAACACTGGGCCAAATCAG
E11	F: TTGGAATCATAGTTGGCGTCT
	R: TTGGAATCATAGTTGGCGTCT
HPRT	F: TGTTGTTGGATATGCCCTTG
	R: ACTGGCAACATCAACAGGACT

2.4 Statistical analysis

All data were statistically analyzed using one-way ANOVA followed by a post-hoc multi-comparison Tukey-Kramer test. The data was considered statistically significant if $p \leq 0.05$ and presented as mean value ± standard deviation. All statistical analyses were performed using the MATLAB statistics toolbox (The Math-Works, Natick, MA, USA).

3   Results and Discussion

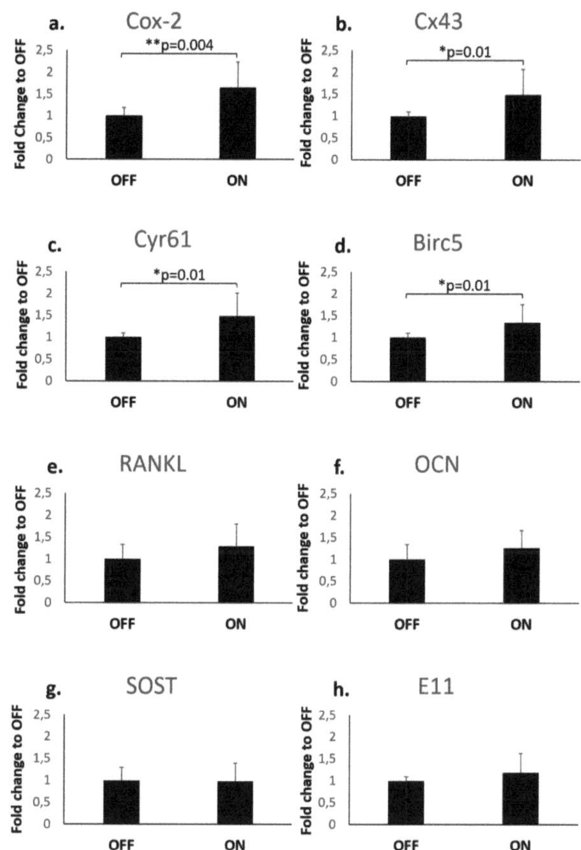

Figure 2: Gene expression quantified via qRT-PCR 3 hours after the mechanical stimulation of the starved MLO-Y4 cells with FLIPUS for 10 minutes: n=5 (number of trials), duplicate samples for ON and OFF conditions per trial. OFF and ON stand for unstimulated and stimulated starved cells, respectively. (a-h) statistical analysis of Cox-2, Cx43, Cyr61, Birc5, RANKL, OCN, SOST and E11, respectively. Values are expressed as mean ± standard deviation. The relative gene expression was normalized to HPRT. * significant difference p<0.05. ** significant difference p<0.01.

Osteocytes are the main bone cell type controlling mechanosensation and mechanotransduction due to their morphological structure and paracrine functions. The dendritic processes of osteocytes allow them to be interconnected, which enables cellular communication and substance exchange. Several studies have indicated that mechanical stimulation enhances osteocytes communication and survival [6], [7]. Current research have investigated the effect of *in-vitro* mechanical stimulation with ultrasound on expression of genes related to bone tissue homeostasis in MLO-Y4 osteocytes.

In this manuscript, the expression of OCN, RANKL, Cx43, Cox-2, SOST, E11, Cyr61 and Birc5 genes in response to FLPIUS were evaluated by qRT-PCR 3 hours after the 10-min-long stimulation. Fig.2 shows expression profiles of the aforementioned genes after statistical treatment of the data obtained by qRT-PCR. OFF and ON stand for unstim-

ulated and stimulated starved cells, respectively.

Cox-2 (Fig.2.a), Cx43 (Fig.2.b), Cyr61 (Fig.2.c) and Birc5 (Fig.2.d), which are linked to cell communication, bone formation, osteogenic differentiation and cell survival, respectively, were up-regulated in the starved cells 3 hours after the FLIPUS stimulation in comparison with the unstimulated starved cells and the increase was statistically significant.

An enhancement of Cox-2 expression was detected in the stimulated MLO-Y4 cells (Fig.2.a). Cox-2 is a rate-limiting enzyme in the production of prostaglandin E2 (PGE2), a crucial signaling molecule for bone formation in mature skeleton [2]. These results are in agreement with previously published data, where increased release of PGE2 was observed after a mechanical stimulation with shear stress [8]. Thus, the enhancement in Cox-2 expression after the FLIPUS stimulation, might suggest that the generated mechanical stimuli augments bone formation.

Osteocytes in lacuno-canalicular network are interconnected by dendrites. These dendrites allow the communication and signals transmission between osteocytes through hemichannels and gap junctions generated by connexin proteins [2]. Hence, the increase of Cx43 observed in this experiment (Fig.2.b) implies an enhancement in the communication, signal transmission and substances exchange between osteocytes under mechanical stimulation. This effect was also previously observed in the shear stress-treated osteocytes [8].

Cyr61 protein, which is linked to osteogenic differentiation, resides in the extracellular matrix and is situated in mineralized tissues [9]. In addition, it modulates Wnt signaling by suppressing SOST, leading to an increase in bone formation. The stated enhancement in this gene after FLIPUS stimulation (Fig.2.c) implies an improvement in bone mass regulation by controlling osteocyte function.

Birc5 gene is crucial for cell division in addition to its role in the inhibition of cell death through apoptotic and autophagic pathways [10]. The noted improvement of Birc5 gene in stimulated cells (Fig.2.d) is an indication of a decrease in cell apoptosis and an increase in cell proliferation. An improved survival of MLO-Y4 has been previously established by Puts et al. (data not shown).

In contrast, other genes including RANKL (Fig.2.e), OCN (Fig.2.f), SOST (Fig.2.g) and E11 (Fig.2.h) were not drastically altered by FLIPUS stimulation and the data did not indicate statistical significance. The genes RANKL, OCN, SOST and E11 are associated with osteoclastogenesis, bone mineralization, inhibition of bone formation and osteocytes' dendrite elongation, respectively [11], [12]. Thus, FLIPUS might not have an effect on these pathways, at least 3 hours after the stimulation of the osteocytes.

4   Conclusion

Although LIPUS has a great potential for bone fracture regeneration, the biological mechanisms of the technique are not fully understood and the clinical data reports controversial findings. Hence, Focused Low-Intensity Pulsed

Ultrasound (FLIPUS) *in-vitro* set-up was developed to enable artefact-free and acoustic intensity-controlled mechanical cell stimulation in order to investigate the underlining mechanisms. The effect of FLIPUS stimulation of osteocytes on expression of several genes, having impact on bone homeostasis, have been evaluated using qRT-PCR. The enhanced expression of Cox-2, Cx43, Cyr61 and Birc5 compared to unstimulated cells was observed, implying pro-regenerative nature of the FLIPUS technique. Other genes, i.e., RANKL, OCN, SOST and E11 have not been affected by the FLIPUS stimulation at 3 hours after the 10-min-long stimulation.

The conducted experiment is the first step for verifying FLIPUS role in the enhancement of osteocytes regenerative properties. Further investigations are needed to study FLIPUS-targeted signaling pathways such as TGF-β, β-catenin/Wnt, ATP and Cox-2/PGE2 signaling pathways. Additionally, further experiments and analysis are required in order to study the morphological changes in response to mechanical stimulation and resulting intercellular communication.

Acknowledgement

The work has been carried out at Charité-Universitätsmedizin Berlin, BIH Center for Regenerative Therapies and supervised by Prof. Henrik Botterweck, Institute of Biomedical engineering, Luebeck University of Applied Sciences. The authors would like to thank Nirina Beilfuß and Hui Zhang for their technical assistance.

Author's Statement

Authors state no conflict of interest.

5   References

[1] Rinaldo Florencio-Silva, Gisela Rodrigues da Silva Sasso, Estela Sasso-Cerri, Manuel Jesus Simões, and Paulo Sérgio Cerri, *Biology of Bone Tissue: Structure, Function, and Factors that Influence Bone Cells.* BioMed Research International, 2015.

[2] Meng Chen Michelle Li, Simon Kwoon Ho Chow, Ronald Man Yeung Wong, Ling Qin, and Wing Hoi Cheung, *The Role of Osteocytes-Specific Molecular Mechanism in Regulation of Mechanotransduction – A systematic review.* Journal of Orthopaedic Translation, vol. 29 ,pp. 1–9, 2021.

[3] Frederic Padilla, Reigna Puts, Laurence Vico, Alain Guignandon, and Kay Raum, *Stimulation of Bone Repair with Ultrasound.* Advances in Experimental Medicine and Biology, pp.385–427, 2016.

[4] R. Puts1, L. Vico, N. Beilfuß, M. Shaka, F. Padilla,4 and K. Raum, *Pulsed Ultrasound for Bone Regeneration - Outcomes and Hurdles in the Clinical Application: A Systematic Review.* Cells and Materials, vol. 42, pp.281–311, 2021.

[5] Regina Puts, Karen Ruschke, Thomas H. Ambrosi, Anke Kadow-Romacker, Petra Knaus, Klaus-Vitold Jenderka and Kay Raum, *A Focused Low-Intensity Pulsed Ultrasound (FLIPUS) System for Cell Stimulation: Physical and Biological Proof of Principle.* IEEE transactions on ultrasonics, ferroelectrics and frequency control, vol. 63, no. 1, 2016.

[6] Benxu Cheng, Shujie Zhao, Jian Luo, Eugen Sprague, Lynda Bonewald, and Jean X. Jiang, *Expression of Functional Gap Junctions and Regulation by Fluid Flow in Osteocyte-Like MLO-Y4 Cells.* Journal of bone and mineral research, vol. 16, no. 2, 2001.

[7] Tatsuya Shimizu, Naomasa Fujita, Kiyomi Tsuji-Tamura, Yoshimasa Kitagawa, Toshiaki Fujisawa, Masato Tamura, and Mari Sato, *Osteocytes as Main Responders to Low-Intensity Pulsed Ultrasound Treatment during Fracture Healing.* Scientific reports, vol. 11, 2021.

[8] Priscilla P. Cherian, Arlene J. Siller-Jackson, Sumin Gu, Xin Wang, Lynda F. Bonewald, Eugene Sprague, and Jean X. Jiang, *Mechanical Strain Opens Connexin 43 Hemichannels in Osteocytes: A Novel Mechanism for the Release of Prostaglandin.* Molecular Biology of the Cell, vol. 16, pp.3100–3106, 2005.

[9] Gexin Zhao, Bau-Lin Huang, Diana Rigueur, Weiguang Wang, Chimay Bhoot, Kemberly R Charles, Jongseung Baek, Subburaman Mohan, Jie Jiang, and Karen M Lyons, *CYR61/CCN1 Regulates Sclerostin Levels and Bone Maintenance.* The official journal of the American Society for Bone and Mineral Research vol. 33, no. 6, pp. 1076–1089, 2018.

[10] Sally P. Wheatley, and Dario C. Altieri, *Survivin at a Glance.* Journal of Cell Science, vol. 132, 2019.

[11] Emily G. Atkinson, Alejandro Marcial, Zuleima Sánchez,Christian Porter, and Lilian I. Plotkin, *MLO-Y4 Osteocytic Cell Clones Express Distinct Gene Expression Patterns Characteristic of Different Stages of Osteocyte Differentiation.* Actualizaciones en Osteología, vol. 19, no. 7, 2018.

[12] Katharina Jähn, Deborah J. Mason, Jim R. Ralphs, Bronwen A.J. Evans ,Charles W. Archer, R. Geoff Richards, and Martin J. Stoddart, *Phenotype and Viability of MLO-Y4 Cells is Maintained by TGFβ3 in a Serum-Dependent Manner within a 3D-Co-Culture with MG-63 Cells.* International Journal of Molecular Science, vol. 13, no. 3, 2017.

Modelling of a Positive End-Expiratory Pressure Valve with a Voice-coil Actuator

Robin Brütt [1] Georg Männel [2]

[1] Biomedical Engineering, Luebeck University of Applied Sciences, robin.vinzenz.bruett@stud.th-luebeck.de
[2] Fraunhofer IMTE, georg.maennel@imte.fraunhofer.de

Abstract

In artificial ventilation devices the positive end-expiratory pressure (PEEP) valve is important in the regulation of the pressure at the patient airways. The goal of this work is to derive a model that sufficiently captures the dynamics of the PEEP valve and can be used for model-based control. This paper describes the development of a model for common PEEP valve designs. As this is a system with one degree of freedom, a simplified dynamic model is used as a basis, which, due to the membrane, partly consists of a mass-spring-damper system. A valve test rig was developed to record various series of measurements and to identify unknown parameters with the help of the acquired data and the grey-box model method. It was possible to determine a suitable flow factor ($23.77 \cdot 10^{-3}$) which describes the correlation between flow, valve opening and upstream pressure.

1 Introduction

The positiv end-expiratory pressure (PEEP) valve is one major component in anaestesia devices and ventilators. Such devices have an inspiratory side through which air is actively delivered into the patients airways and an expiratory side through which the air is released from the patients lungs. The PEEP valve is located on the expiratory side of these devices. During ventilation, a PEEP valve ensures that the pressure in the lungs does not drop to the ambient pressure during expiration. The functional residual capacity and thus the surface area for gas exchange is increased, which prevents the alveoli from collapsing. The increased airway pressure also reduces the work the patient has to do in case of spontaneous breathing [1]. Until now, PEEP valves have been controlled in such a way that a constant upstream pressure and a steady flow or zero flow can be set [2]. The goal of this work is to develop a model that describes the PEEP valve which can be used for controller design. A grey-box model approach was chosen as the basis for modelling. The modelling process consists of setting up a simplified dynamic model that implies a mass-spring-damper behaviour and the derivation of the underlying differential equation of this system, combined with measurement data, which is used to determine unknown parameters. A valve test rig was developed to record the measurement data.

2 Material and Methods

The PEEP valve has an inlet and an outlet. When installed, the patient breathes into the valve's inlet. The air exhaled by the patient flows into the inlet until it meets the membrane and the direction of flow is reversed by 180 degrees. In some literature this design is also called U-turn design [3]. After the flow hits the membrane, the fluid exits the valve through a side outlet. The balance of forces on the membrane can be altered by a piston pushing against the membrane from the outside. This influences the size of the opening between the valve inlet and the membrane. The force of the piston is provided either pneumatically or electromechanically. Due to the high requirements on the control speed, electromechanical solutions such as voice-coil actuators are preferred [2].

2.1 Simplified Dynamic Model

As basis for the modelling of the PEEP valve a simplified dynamic model is chosen as it is a system with one degree of freedom and this modelling methodology was already used by Song et al. [4] and Darby et al. [5] [6] for the mathematical description of pressure relief valves. Figure 1 shows a schematic view of the PEEP valve with different forces that act on the valve, especially the valves membrane.

Due to the fact that the valve membrane has a mass and is moving in the direction of the degree of freedom a mass-spring-damper model is chosen to derive the equations of motion. Since the flow exits the inlet against the membrane and it is therefore displaced in the direction of the flow, the damping force $F_{damping}$ is defined against the direction of the motion of the membrane with the displacement x (see figure 1). The velocity of the displacement $\dot{x}$ is multiplied with the damping coefficient d:

$$F_{damping} = \dot{x}d \qquad (1)$$

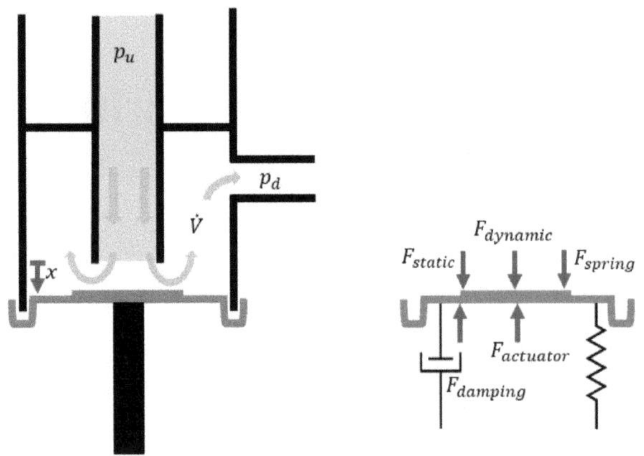

Figure 1: Schematic illustration of the PEEP valve (left) and forces acting on the membrane (right).

In previous experiments, it was found that the spring properties of the membrane cannot be considered linear. The equation for the recoil force F_{spring} of the membrane is therefore described as a nonlinear function:

$$F_{spring} = p_1(x_{max} - x)^{p_2} \qquad (2)$$

As can be seen from the right-hand illustration in figure 1, the static pressure force and the dynamic pressure force apply from the inside of the membrane. The static pressure force F_{static} is defined as the product of the static upstream pressure p_u and the area of the valve inlet A.

$$F_{static} = p_u A \qquad (3)$$

If a flow $\dot{V}$ through to the valve exists, a so-called dynamic pressure is also acting on the membrane in addition to the static pressure. The general equation for this dynamic pressure can be found in [7]. If this is divided by the area of the valve inlet, it results in:

$$F_{dynamic} = \frac{\dot{V}^2 \rho}{2A} \qquad (4)$$

Inside of the voice-coil actuator is a proportional solenoid which applies a force that is proportional to the input signal [8]. In this case the input signal is a square wave signal with duty cycle (DC). The equation of the force applied by the voice-coil actuator $F_{magnetic}$ is therefore:

$$F_{magnetic} = q_1 DC + q_2 \qquad (5)$$

To combine these forces, Newtown's second law is applied, which states that the mass m times the acceleration $\ddot{x}$ of a body is equal to the sum of all the forces acting on it. From this follows:

$$m\ddot{x} = -F_{magnetic} - F_{damping} + F_{spring} + F_{static} + F_{dynamic} \qquad (6)$$

By rearranging (6) the following differential equation is derived which gives information about the internal states of the valve:

$$\ddot{x} = -\frac{1}{m}\left(q_1 DC + q_2 + \dot{x}d - p_1(x_{max} - x)^{p_2} - p_u A - \frac{\dot{V}^2 \rho}{2A}\right) \qquad (7)$$

Table 1: Identified Parameters

Parameter	Symbol	Value	Unit
Inlet area	A	$3.8 \cdot 10^{-4}$	m^2
Inlet radius	r_i	0.011	m
Maximum valve opening	x_{max}	0.0035	m
Membrane mass	m	0.005	kg
Density of air	ρ	1.225 [7]	kg/m^3
Restoring force coefficient	p_1	0.02146	N/m
Restoring force exponent	p_2	6.361	
Actuator's sensitivity	q_1	0.05706	N/DC
Acuator's offset	q_2	-0.5161	N

As equation for the output flow, the general equation for globe valves [7] is used, since the geometry of these valves is very similar to that of the PEEP valve.

$$\dot{V} = A_o \frac{1}{\zeta} \sqrt{\frac{2\Delta p}{\rho}} \qquad (8)$$

The drag coefficient ζ is needed to describe the pressure losses resulting from the complex geometry. The area of the opening A_o can be described as the circumference of the valves inlet times the opening of the valve:

$$A_o = x 2\pi r_i \qquad (9)$$

By inserting (9) into (8) and the combining of all constant values into one flow factor S the equation can be simplified. Additionally Δp is assumed to be equal to the upstream pressure p_u due to the fact that the valve outlet is to ambient air. This leads to the following expression:

$$\dot{V} = xS\sqrt{p_u} \qquad (10)$$

2.2 Parameter Identification

Some of the parameters mentioned in equations (1) to (10) are geometrical (A, r_i, x_{max}) and can therefore be determined by simple measurement. Since the membrane is detachable from the valve, its weight (m) could also be determined. The parameters describing the restoring force (p_1, p_2) of the membrane could therefore also be identified with the help of a force-displacement measurement. By dismounting the voice-coil actuator and measuring the force exerted by it at certain duty cycles, the parameters for equation (5) could also be identified (q_1, q_2). The identified parameters are given in table 1.

Since the damping is a dynamic parameter which depends, in part, on the force applied by the actuator, the damping coefficient in (1) can only be identified by using series of measurements in which the membrane is stimulated by different pressures and actuator forces. Similar to the damping coefficient, the flow factor in equation (10) can only be determined by means of experimental data. It cannot be determined with sufficient accuracy with the help of the literature, as the resistance coefficient ζ is not known for the complex geometry of the valve. In order to identify this missing parameters, a test rig was developed that allows the controlled adjustment of the upstream pressure, the flow

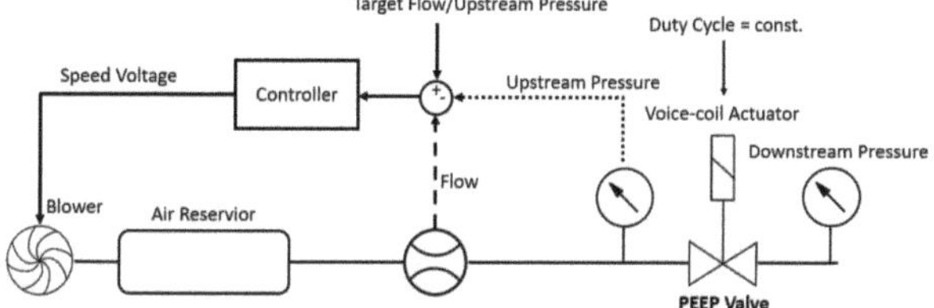

Figure 2: Schematic illustration of the test rig with pressure (dotted line) and flow (dashed line) feedback

and the duty cycle of the actuator. Additionally the actuator can be replaced by an adjusting screw that allows the adjustment of defined valve openings.

2.3 Valve Test Rig

The test rig used for running measurement series on the PEEP valve consists of a blower, an air reservoir, a flowmeter and the PEEP valve connected in series (see figure 2). Right before and after the valve the static upstream and downstream pressure is measured. The accuracy of the flow sensor is ±0.05 l/min and of the pressure sensors ±350 Pa. For the measurements the PEEP valve is controlled by a square wave signal with different duty cycles but alternatively the voice-coil actuator can be replaced by an adjusting screw. Due to a magnet between the bolt of the screw and the membrane, a coupling is ensured. With the help of this screw, the valve opening can be adjusted precisely.

3 Results and Discussion

Two broad series of measurements were recorded with the test rig. In the first measurement series the adjusting screw was used to get information about the position of the membrane as function of the static upstream pressure and the flow. By means of this measurement information about the flow-pressure-position characteristic gets available. In the second measurement series the flow through the valve was controlled by the blower, such that the flow is constant at a desired target flow (see dashed line in figure 2). With the measured upstream pressure and flow during this measurement the flow factor in (10) and the damping coefficient in (1) can be identified by system identification using MATLAB System Identification Toolbox.

3.1 Flow-Pressure-Position Characteristic

The membrane was adjusted in steps of 0.25 mm from 0 mm to 3.5 mm opening and a pressure-flow curve was recorded for each position by ramping up the blower voltage from the minimum to the maximum value. Figure 3 shows the result of this measurement. For every valve position another flow-pressure curve were measured which are indicated as black lines in the figure. The larger the opening, the greater the flow that can be measured at the same pressure. However, as the opening of the valve increases, this effect decreases, as can be seen from the increasing closeness of the flow-pressure curves in the upper range. At a certain opening

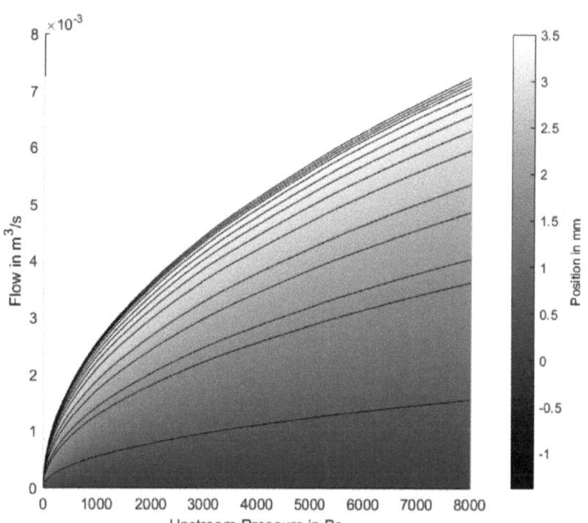

Figure 3: Flow-Pressure-Position Characteristic Map. Darker areas indicate a small valve opening. The brighter the area, the more the valve is open. The maximum opening was 3.5 mm.

degree, the flow is no longer dependent on the opening and depends only on the prevailing upstream pressure. The values between these curves were filled by interpolation which gives the completed characteristic map in figure 3. The current valve opening can be calculated by means of this map and the measurement of the actual upstream pressure and flow.

3.2 Parameter Estimation

For the parameter identification a measurement was recorded where the flow was stepped up to create a stair function. To generate this profile the flow increased from 0 to 290 l/min by 10 l/min every 10 seconds. The static upstream pressure was no controlled variable and was measured as it was resulting from the flow and the input of the voice-coil actuator. Figure 4 shows the measured pressure curves for the different duty cycles. For all duty cycles the same behaviour can be observed. First, the pressure rises steeply to open the valve initially. After a short oscillation process, the static upstream pressure then starts to decrease from a higher value, while the flow continues to increase as desired. This can be explained by the flow-pressure-position characteristic (see figure 3) and the basic equation for pressure loss $\dot{V} = \sqrt{\frac{\Delta p}{R}}$ [7]. If a higher flow is to be set and the static pressure is to decrease at the same time, the valve opening must continue to enlarge, which results in a decrease in the flow resistance R. However, this process is limited by the maximum possible opening of the valve. If the maximum valve opening is reached, a further increase in flow can only be achieved by an increase in pressure. This is the reason why after a certain time all pressure curves in figure 4 are lying one over the other. At this point the upstream pressure is completely dependent on the flow.

With the known flow-pressure-position characteristic the valve opening can be determined at any time from the prevailing static upstream pressure and flow. With the help

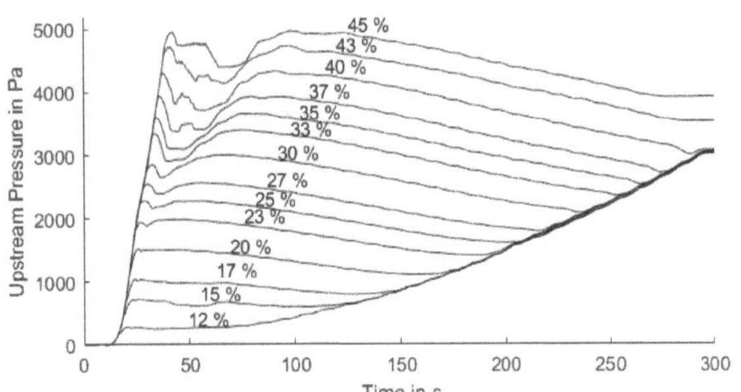

Figure 4: Measured upstream pressures for different duty cycles. The percentages on the lines indicate the respective duty cycle.

of the recorded characteristic map, $x = f(p_u, \dot{V})$ can be solved for each flow-pressure pair and inserted into the output equation (10). This avoids the calculation of the valve position via the differential equation (7) and the flow factor can be estimated independently of the damping coefficient. Using the information about the actual flow, the flow factor is now adjusted using the MATLAB System Identification Toolbox so that the calculated flow matches the measured flow as good as possible. This is done for all 14 measurements from the flow controlled measurement. SI-units were used for this simulation (valve position: m; pressure: Pa; flow: m^3/s). A mean value is calculated out of all calculated flow factors. The achieved fit between measured flow values and simulated flow values with estimated flow factor lies between 93.88 and 99.97% for all measurements. The averaged flow factor S was determined to be $23.77 \cdot 10^{-3}$ (SD = $0.315 \cdot 10^{-3}$). With the identified flow factor and the recorded measurements, another parameter estimation was carried out, whereby the valve opening was determined using equation (7) and the damping coefficient could be optimised using the MATLAB System Identification Toolbox. With the available measurements, an average damping coefficient of 3805.1 kg/s (SD = 748.9 kg/s) was calculated. The fit of the flow simulated with this damping coefficient with the real flow was in the range of 57.14 and 76.53 %. Due to the high standard deviation of the estimated damping coefficients and the low fit with the actual flow, the estimation of the damping coefficient cannot be classified as successful so far. Since damping is a highly dynamic parameter, it can be assumed that the estimation was insufficient due to the lack of dynamics in the measured data. Measured values with greater dynamics in flow and pressure are currently being recorded in order to carry out further parameter estimations for the damping coefficient.

4 Conclusion

The underlying mathematical model of the valve was established and almost every parameter could be identified. Most of the parameters were identified by measuring the geometry and masses and isolated tests on the membrane and actuator. Only the damping coefficient and flow factor had to be identified from the combination of the underlying differ-

ential equation and the measured data. A valve test rig was developed to record this data. Through the finding about the flow-pressure-position characteristics the valve opening, a previously unknown internal state, was made estimable and an identification of the flow factor was possible. An identification of the damping coefficient was not successful so far, due to the low dynamics in the recorded measurement data. The next steps are to record new data sets with more dynamics in the flow and pressure in order to provide an appropriate foundation for the estimation. As soon as a complete model of the PEEP valve is available it should be used for the implementation of a model-based control.

Acknowledgement

This work was done in collaboration with the Fraunhofer Research Institution for Individualized and Cell-Based Medical Engineering (IMTE) and the Institute for Electrical Engineering in Medicine (IME).

Author's Statement

Conflict of interest: Authors state no conflict of interest.

5 References

[1] R. F. Schmidt, F. Lang, M. Heckmann, *Physiologie des Menschen. mit Pathophysiologie*. SpringerMedizin Verlag, Heidelberg, Germany, 2010

[2] M. Borrello, *The Application of Controls in Critical Care Ventilation*. Phillips Respironics, Carlsbad, Canada, 2021

[3] P. Bergvist and L. Kemmler, *Voice Coil Controlled Inspiration and Expiration Valves*. KTH Industrial Engineering and Management Machine Design, Stockholm, Sweden, 2012.

[4] X. G. Song, Y. C. Park, J. H. Park, *Blowdown prediction of a conventional pressure relief valve with a simplified dynamic model*. Dong-A University, Busan, Republic of Korea, 2011.

[5] R. Darby, *The dynamic response of pressure relief valves in vapor or gas service, part I: Mathematical model*. Texas AM University, College Station, Texas, USA

[6] A. A. Aldeeb, R. Darby, S. Arndt, *The dynamic response of pressure relief valves in vapor or gas service. Part II: Experimental investigation*. Texas AM University, College Station, Texas, USA

[7] S. Bschorer, *Technische Strömungslehre 11. Auflage*. Fakultät Maschinenbau, Technische Hochschule Ingolstadt, Ingolstadt, Germany, 2018.

[8] A. Büngers, *Elektromagnetische Aktoren 3*. Technische Hochschule Mittelhessen, Gießen, Germany, 2012.

Feedback Linearization of a Blower in an Anesthesia Workstation

Jannik Prüßmann [1], Georg Männel [2], and Philipp Rostalski [2,3]

[1] Medical Engineering Science, Universität zu Lübeck, jannik.pruessmann@student.uni-luebeck.de
[2] Fraunhofer Research Institution for Individualized and Cell-Based Medical Engineering, Lübeck,
georg.maennel@imte.fraunhofer.de
[3] Institute for Electrical Engineering in Medicine, Universität zu Lübeck, philipp.rostalski@uni-luebeck.de

Abstract

Blowers, which are often used in ventilation units of anesthesia workstations, are known to have non-linear dynamics. To make them accessible for the large field of linear control theory their model needs to be linearized. This paper uses a gray box model of a blower driven by a brushless direct current (BLDC) motor to design a controller based on feedback linearization. The controller is designed to achieve approximate linear closed loop behavior of the system. The controller is then evaluated concerning reference tracking and disturbance rejection in simulation and on a demonstrator system. The evaluation showed that an approximate linear closed loop behavior was achieved. Nevertheless, the dynamics of the system using the derived controller on the demonstrator lag behind the results obtained in simulation.

1 Introduction

Blowers are widely used in ventilation units of anesthesia workstations. They are known to have a non-linear behavior regarding the pressure generated at the outlet.In control theory linear systems are usually advantageous, thus a Jacobian linearization of the system at the operating point is usually implemented. Such a linearization is only valid close to an equilibrium point, which is impractical for applications, where the system is to follow a reference and not to operate around a fixed operating point, e.g. in lowest control level in the hierarchical control structure such as described by [2] for a ventilation unit. Therefore, a controller to control the blower's pressure output designed by feedback linearization [1] is derived in this paper. The controller is intended to ensure a linear closed loop behavior and reference tracking to target pressure values. The controller shall also reject disturbances occurring from the flow through the system. To design the feedback linearization, a non-linear model of the blower is to be established, which is done using a gray box modeling approach. The feedback linearization is then implemented and evaluated regarding it's capability of reference tracking and disturbance rejection in simulation and on a demonstrator system.

2 Material and Methods

2.1 Blower Model

The examined blower is actuated by a brushless direct current (BLDC) Motor that drives a fan wheel. By rotation of this impeller, air is aspired from the blower's pneumatic in-

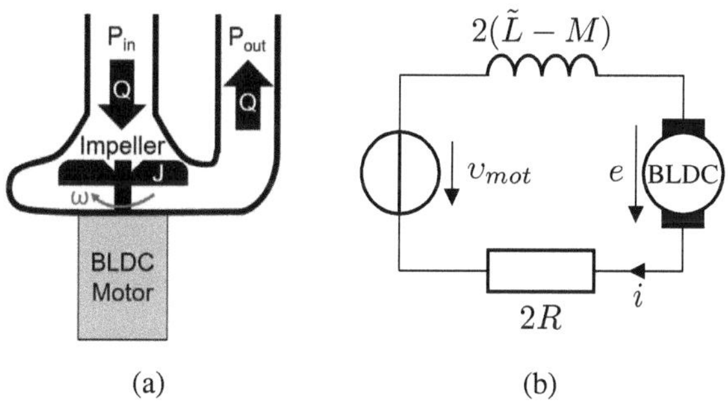

Figure 1: (a) Schematic illustration of the blower. The impeller is actuated by the BLDC motor, resulting in the indicated directions of rotation and air flow. (b) Equivalent circuit of the dynamics of the BLDC motor. Slightly altered from [6].

put and pressurized. Due to the pressure gradient, the air is then emitted to the blower's pneumatic output as shown schematically in Fig. 1a. In BLDC motors the rotor is a permanent magnet and the rotating magnetic field is induced by current through the stator windings [3]. To establish and maintain a consistent rotation of the rotor, the currents through the stator windings are electrical commuted and thus energized in a sequence, that needs to match to the current rotor position and rotation speed [4].

In accordance to [5], the electrical dynamic equation as well as the mechanical dynamics can be assumed to follow the structure of a regular direct current (DC) motor. Therefore, it is assumed that phase resistances and inductances are identical, the mutual inductance M is covered by $L = \tilde{L} - M$, and the commutation is not modeled. The electrical dynamics are illustrated by the electric equivalent circuit in Fig. 1b and can be described by

$$v_{mot} = 2L \cdot \frac{\mathrm{d}i}{\mathrm{d}t} + 2R \cdot i + 2e,$$

where v_{mot} is the motor supply voltage, L the phase inductance, R the phase resistance, i the stator winding current and e the resulting voltage from back electromotive force (back-EMF). This back-EMF results from magnetic induction and can be calculated as $e = k_e \cdot \omega$, which shows it's proportionality to the rotational speed ω and the motor's characteristic back-EMF constant k_e [5]. The mechanical torque T_{mech} of the motor is proportional to the current through the motor. Therefore, it can be expressed as

$$T_{mech} = \eta_{mot} \cdot k_m \cdot i,$$

which is derived using the motor's mechanical power with η_{mot} being the motor's efficiency coefficient and k_m being the characteristic torque constant of the motor. The dynamic of the mechanical system can be described with respect to Newton's second law as

$$T_{mech} = J \cdot \dot{\omega} + b \cdot \omega,$$

with inertia J and damping coefficient b. This yields the following linear state-space representation of the motor dynamics state transition

$$\begin{bmatrix} \frac{\mathrm{d}i}{\mathrm{d}t} \\ \frac{\mathrm{d}\omega}{\mathrm{d}t} \end{bmatrix} = \begin{bmatrix} -\frac{R}{L} & -\frac{k_e}{L} \\ \frac{\eta_{mot}k_m}{J} & -\frac{b}{J} \end{bmatrix} \cdot \begin{bmatrix} i \\ \omega \end{bmatrix} + \begin{bmatrix} \frac{1}{2L} \\ 0 \end{bmatrix} \cdot v_{mot}. \quad (1)$$

The dynamic behavior of the motor is disturbed by the occurring air flow Q since the flow impacts a torque on the rotor and thereby has an effect on $\dot{\omega}$ and consequently on the revolutions per minute n. According to [6], the blower's static characteristics can be approximated assuming the blower to be an ideal pressure source with an additional blower inherent resistance e.g., an additional throttle. With the Bernoulli equation [7] and the turbulent flow caused on openings and vents [8] the pressure change across the blower is thus,

$$\Delta p_{blower} = \Delta p_{blower}^{ideal} - \Delta p_{blower}^{thr},$$

with the non-linear terms

$$\Delta p_{blower}^{ideal} = \frac{1}{2}\rho_{out}\left(\frac{n}{r_b}\right)^2$$

and

$$\Delta p_{blower}^{thr} = \frac{1}{2}\rho_{out}\left(\frac{C_b^{thr}}{A_b^{thr}}\right)Q^2,$$

with ρ as the density of the conveyed gas, r_b the fan area radius, C_b^{thr} as the throttle's flow coefficient and A_b^{thr} as characteristic throttle area, which can all be determined experimentally. These variables are summarized in the matrices A and B, which were identified on experimental data to obtain an approximate gray box model. The resulting non-linear state-space model yields

$$\dot{x} = f(x) + g(x)u = A\begin{bmatrix} x_1 \\ x_2 \end{bmatrix} + B\begin{bmatrix} u_1 \\ u_2 \end{bmatrix} \quad (2)$$

$$y = h(x) = k_{blower} \cdot x_1^2,$$

where x_1 corresponds to the revolutions per minute n, x_2 to the motor current i in (1), $u_1 = Q$, $u_2 = v_{mot}$, the system output is $y = \Delta p_{blower}$ and k_{blower} is the system gain. Thus, by estimation from data, an approximate, non-linear expression for the blower's pressure change depending on revolutions per minute n and volume flow Q is given. Note that the disturbing effect of the flow on the output is not considered in the gray box model.

2.2 Closed Loop Control using Feedback Linearization

The common Jacobian linearization is established on the derivatives in an equilibrium point, thus the obtained linearization is only locally valid and can be arbitrarily deviating from the non-linear function when the operating point differs. Feedback linearization overcomes this limitation by a non-linear change of coordinates and non-linear state feedback [9]. The general idea of feedback linearization is, to find a control input u that cancels all non-linearities in the system dynamics. Applying a state-feedback in a closed loop control using such control input can lead to a linear input-output model and can thus make the system accessible for control strategies that require linear systems [1].

In order to linearize the input-output behavior, a coordinate change $z = T(x)$, transforming the system into the normal form is to be found. This transformation is required to be a diffeomorphism, thus the function and its inverse are both smooth and differentiable [1]. Furthermore, the Lie derivative is introduced as

$$L_f h(x) = \frac{\mathrm{d}h(x)}{\mathrm{d}x} f(x) = 2k_{blower}x_1 \cdot A\begin{bmatrix} x_1 \\ x_2 \end{bmatrix}$$

$$L_g h(x) = \frac{\mathrm{d}h(x)}{\mathrm{d}x} g(x) = 2k_{blower}x_1 \cdot B\begin{bmatrix} u_1 \\ u_2 \end{bmatrix}.$$

Note that u_1 can not be manipulated directly as it is corresponding to the flow through the system. Thus the system is to be controlled via u_2. According to [1] a transformed system needs to be produced, whose states are the output y and the first $(r - 1)$ Lie derivatives of the output, where r is the relative degree of the system. As $r = 2$ in the blower model, the transformation is

$$z = T(x) = \begin{bmatrix} z_1 \\ z_2 \end{bmatrix} = \begin{bmatrix} y \\ \dot{y} \end{bmatrix} = \begin{bmatrix} h(x) \\ L_f h(x) \end{bmatrix}.$$

Since [1] showed that z is a diffeomorphism, it is differentiable and thus, we can find a new transformed system

$$\dot{z}_1 = z_2$$
$$\dot{z}_2 = z_3 \quad (3)$$
$$\dot{z}_3 = L_f h(x) + L_g h(x)u,$$

which is derived using $L_g L_f^{r-1} h(x) \neq 0$ and $L_g L_f^{r-2} h(x) = 0 \ \forall x$ in a neighborhood of defined operating point [1]. If the input u is now chosen to satisfy

$$u = \frac{1}{L_g h(x)}(v - L_f h(x))$$

the last line of the transformed system in (3) becomes $\dot{z}_3 = v$ and thus, the transformation yields a linear system. The transformed input v can now be used for state-feedback using a standard state-feedback control law of $v = -Kz$ for pole placement to achieve control goals and system stability in the same manner as if the system was linear. For the blower control, the target closed loop dynamics are that of a linear first order system with the transfer function $H(s) = 1/(0.3s + 1)$. The necessary control gains to achieve the targeted dynamics were determined by pole placement [1]. The obtained controller's output is the supply voltage that is to be applied to the BLDC motor.

3 Results and Discussion

In order to evaluate the feedback linearization based controller of the ventilation unit's blower, a model was created using MATLAB Simulink [10]. The model contains the developed blower model from subsection 2.1 including the linear state-space model of the motor dynamics from (1) and the non-linear relation of the pressure change produced by the blower as in (2). Furthermore the Simulink model contains a controller consisting of a Kalman filter for state estimation of the motor speed and the feedback linearization approach as illustrated in subsection 2.2.

In addition to the simulation the controller was implemented on a demonstrator system. Therefore, C Code was generated from the MATLAB Simulink model and then incorporated in the modularized system described in [2] by replacing the previously used controller for the blower with the controller designed by feedback linearization. The reference inputs to determine pressure and positive end-expiratory pressure (PEEP) set values were given from a test-sequence for the evaluation. While running the test-sequence on the system, the system's outputs were measured, including the controlled variable, which is the pressure output of the blower P_{out}. The experimental setup is shown in Fig. 2. Note that the pneumatic system remains closed as long as the PEEP valve is closed. The PEEP valve is intended to hold a positive pressure during expiration in mechanical ventilation and can be described as a pressure relief valve. Therefore, a force is applied to the PEEP valve depending on a pulse width modulation (PWM) signal applied to the corresponding actuator. If the pressure in the system exceeds the thereby set PEEP value, the PEEP valve opens and a disturbance air flow is released. The system further contains two check valves, which allows air flow only from the blower's pneumatic output to the input.

3.1 Reference Tracking

To evaluate the reference tracking achieved using the controller, the PEEP is hold fixed whilst different steps in the pressure reference value are performed. This allows to observe the closed loop system's response to different steps. In total 14 different positive and negative steps are applied to the reference pressure signal and each step is hold for three seconds. This procedure is repeated entirely for 14 different

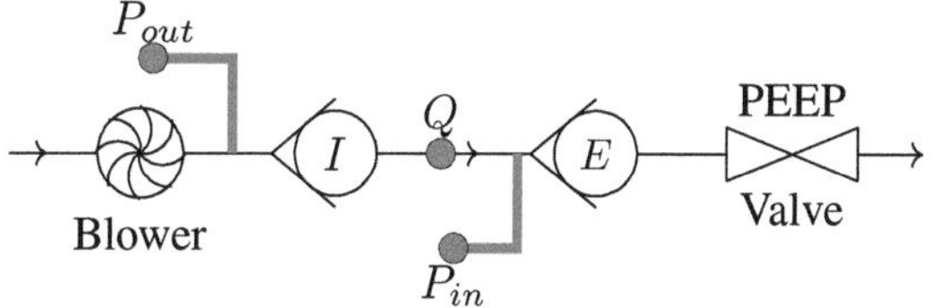

Figure 2: Experimental setup with sensors Q for flow, P_{out} and P_{in} for the pressure out- and input to the system, check valves I and E, and blower and PEEP valve.

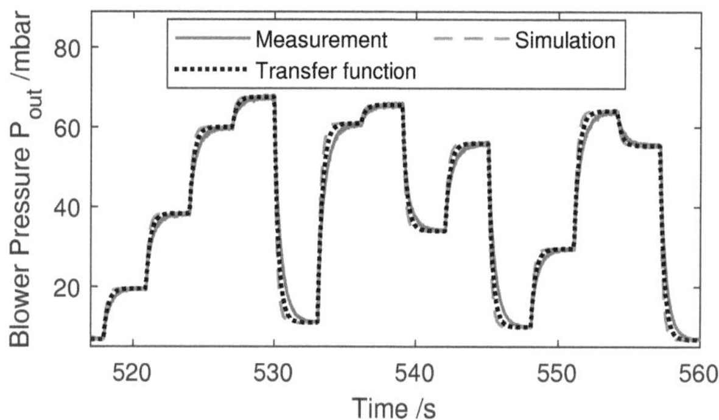

Figure 3: Comparison of the course of pressure P_{out} produced by the blower in simulation and measured on the demonstrator system from one repetition of the reference tracking evaluation. Additionally the control goal, given by the transfer function, is depicted.

PEEP values to observe the reference tracking under different situations of disturbance. The reference pressure value is thereby bound by the maximum pressure that the blower can produce for all PEEP values to a maximum of 66 mbar. The results for reference tracking of the closed loop system using the feedback linearization approach as controller are illustrated in Fig. 3, where the pressure curve for one repetition of the test-sequence is depicted. The average T90 time targeted in the transfer function is 0.6591 sec. For the implementation on the demonstrator an average T90 time of 0.8508 sec with a standard deviation of 0.0954 sec was measured. Thus, the data measured on the demonstrator system indicates slower dynamics than formulated as control goal in the transfer function. Nevertheless, the controller is capable of achieving reference tracking. In the simulation an average T90 time of 0.4149 sec with a standard deviation of 0.001 sec was achieved. Therefore, the simulation shows faster dynamics than targeted. These observations may occur due the unconsidered flow in the output model in (2).

3.2 Disturbance Rejection

The air flow through the PEEP valve has an effect on the pressure in the system and is thus a disturbance. To evaluate the controllers capability of rejecting the pressure change produced by this disturbance, the disturbance response of the system is observed by fixing the reference pressure, whilst changing the PEEP valve position every three sec-

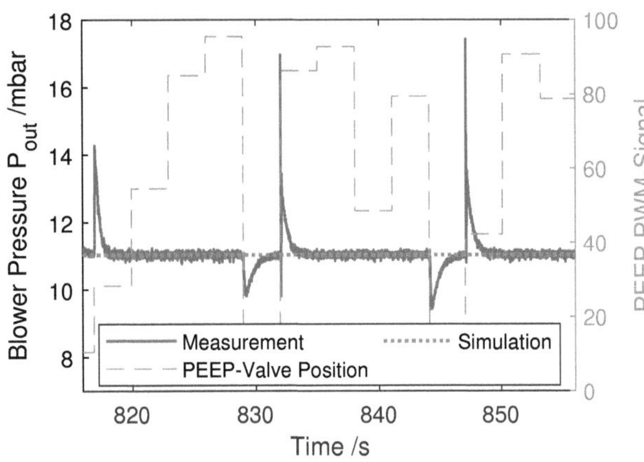

Figure 4: Comparison of the course of pressure P_{out} produced by the blower in simulation and measured on the demonstrator system in the disturbance rejection evaluation with reference pressure value of 11.03 mbar. Changes of the PEEP valve PWM signal are indicated.

onds. Therefore, 14 positive and negative steps are applied directly to the PEEP valve's actuator PWM signal throughout the blower's controller is set to hold the fixed reference pressure. This procedure is repeated with 14 different reference pressure values to take into account, that the disturbance response may vary with the applied pressure.

To illustrate the disturbance rejection of the closed loop system, Fig. 4 shows an exemplarily pressure course for one repetition. It is observable, that in simulation the controller was capable to mostly reject the disturbance, showing no or only minimal change in the pressure after the indicated PEEP valve position changes. In the measurement data the disturbance rejection works significantly worse, partially deviating massively from the set value of 11.03 mbar. Especially, changing from a far open to a mostly closed PEEP valve poses a problem to the controller. When the PEEP valve allows for high air flow, the motor speed has to be strongly increased in order to achieve the reference pressure. If the PEEP valve is then abruptly closed, the pressure in the system rises quickly until the rotational speed of the fan is reduced. Particularly, the system lacks options to reduce the pressure when the PEEP valve is closed, apart from unintended leakage. Thereby, also the check valves may influence the disturbance rejection. Besides the check valves, neglecting the flow in the output model can be a reason for the significantly better performance in simulation. Subsequently, the controller, which is designed with respect to the model, does not perfectly match the plant dynamics.

4   Conclusion

Since the evaluation showed that feedback linearization is generally capable to fulfill the desired reference tracking, it is a promising approach for the control of the ventilation unit's blower in an anesthesia workstation. Although the control goals in terms of rise time were not yet achieved, the closed loop system behaved like a first order linear system as desired. Moreover, the evaluation revealed several entry points for further research and controller improvements. Therefore, modifications to the model might be necessary to include the flow disturbance in the gray box model. Further experiments concerning the effect of the check valves could be of interest.

Acknowledgement

The work has been carried out at Fraunhofer Research Institution for Individualized and Cell-Based Medical Engineering, Lübeck and supervised by the Institute for Electrical Engineering in Medicine, Universität zu Lübeck.

Author's Statement

Conflict of interest: Authors state no conflict of interest.

5   References

[1] M. A. Henson, *Nonlinear process control*. Prentice Hall, Englewood Cliffs, New Jersey, 1997.

[2] G. Männel, M. Siebert, C.Brendle and P. Rostalski, *Robust model predictive control of an anaesthesia workstation ventilation unit*. at - Automatisierungstechnik, vol. 68, no. 11, pp. 941–952, 2020.

[3] G. Henneberger, *Electrical machines I: basics, design, function, operation*. Lecture Notes, Aachen University, 2003.

[4] S. V. Gadewar, and A. M. Jain, *Modelling and simulation of three phase BLDC motor for electric braking using MATLAB/Simulink*. International Journal of Electrical, Electronics and Data Communication, vol. 5 no.7, pp. 48–53, 2017.

[5] A. Mertens, *Elektrische Antriebstechnik II: Lecture Notes* . Leibniz Universität Hannover, 2008.

[6] C. Hoffmann, *Design and control of a novel portable mechanical ventilator*. Master Thesis, Technische Universität Hamburg-Harburg, 2011.

[7] H. Watter, *Hydraulik und Pneumatik: Grundlagen und Übungen-Anwendungen und Simulation..* Friedr. Vieweg & Sohn Verlag | GWV Fachverlage GmbH, Wies- baden, 2007.

[8] L. Böswirth, *Technische Strömungslehre: Lehr- und Übungsbuch*. Friedr. Vieweg& Sohn Verlag | GWV Fachverlage GmbH, Wiesbaden, 7. Edition, 2007.

[9] B. Jakubczyk, *Feedback linearization of discrete-time systems*. Systems & Control Letters, vol. 9, pp. 411–416, 1987.

[10] MATLAB, *R2021b*. The MathWorks Inc., Natick, Massachusetts, 2021.

Identification of crosstalk in respiratory sEMG signals using K-means clustering in frequency domain

Sanaz Delavari [1], Julia Sauer [2], and Thomas Handzsuj [3]
[1] Biomedical Engineering, Luebeck University of Applied Sciences, sanaz.delavari@stud.th-luebeck.de
[2] Institute for Electrical Engineering in Medicine, Universität zu Lübeck, j.sauer@uni-luebeck.de
[3] Drägerwerk AG & Co. KGaA, Luebeck, Deutschland, Thomas.Handzsuj@draeger.com

Abstract

Monitoring breathing effort is crucial in respiratory care, and can be accomplished with several practical methods. Surface electromyography (sEMG) provides a non-invasive way to measure muscle activity for patients undergoing assisted mechanical ventilation. However, during measurements of the diaphragm and intercostal muscles, crosstalk can be detected from nearby muscles, causing interference with the signal of interest. The objective of this work is to investigate the presence of crosstalk in sEMG signals by means of spectral analysis. To this aim, K-means clustering is applied to the normalized cumulative power spectral density (CPSD) to detect meaningful clusters. Besides, the visualization of power spectral density (PSD) is presented using normalized CPSD. Performance evaluation is carried out by comparing the clustering results with esophageal pressure (p_{es}) in a visual way. Our approach could distinguish inspiratory activity from noise in intercostal muscle and detect crosstalk in costal margin exemplary clinical dataset.

1 Introduction

Mechanical ventilation plays an essential role in daily clinical practice, providing common treatments for patients suffering from respiratory insufficiency. The purpose of a mechanical ventilation system is to restore the gas exchange function and reduce the respiratory muscle effort so that the patient can recover from acute illness. Currently, assisted mechanical ventilation is used in care unit [1].

Generally, selecting the right mode and level of ventilation for the patients without any knowledge of their breathing activity is challenging [3]. Besides, a number of muscles are involved in respiration, among which the diaphragm and intercostal muscles are the primary ones. During inspiration diaphragm contracts, and the volume of the chest cavity increases. Consequently, the lungs expand. Furthermore, intercostal muscles, located between the ribs, contribute to inspiration by pulling the rib cage upward and outward. Whereas, the act of exhalation is primarily passive [2]. As a consequence, monitoring respiratory muscle activity has vital importance in respiratory care, and can be accomplished with several practical methods [3], [4]. Currently, clinicians identify inspiratory efforts by invasively measuring the esophageal pressure (p_{es}). However, p_{es} measurement requires the placement of a balloon catheter with sufficient volume in the esophagus. Therefore, the use of p_{es} is relatively rare in many clinics, despite its many advantages [5]. Alternatively, a non-invasive technique referred to as surface electromyography (sEMG) has been developed to measure the effort of muscles. The recording is conducted using a single electrode or an array of electrodes placed on the skin surface over the muscles [5]. Nevertheless, sEMG measures the surrounding muscle activity which is far away from the location of the electrode. Crosstalk refers to EMG signals from muscles that are adjacent to the one of interest. Having a global view of muscle activity may be useful but is a problem when selective information is required. Thus, the identification and removing such additional activity is vital [6].

Studies indicate that several physiological and anatomical factors influence the properties of the recorded sEMG signals. For instance, muscle anatomy, the active parts of muscle during recording, the level of muscle activity, and the firing rate are the factors, affecting the characteristics of sEMG signal. Another significant aspect is the condition in which the sEMG signal is recorded. The position of the electrodes and volume conductors have a profound impact on the properties of sEMG signals [7]. Therefore, we expect varying spectral properties of the sEMG signal. The present work examines if crosstalk can be identified in the frequency domain. In particular, a clustering method is applied to the frequency domain data to investigate the hypothesis of the presence of crosstalk or noise. Moreover, the time-varying spectral properties of the sEMG signal are illustrated using normalized cumulative power spectral density (CPSD).

The following section provides basic information about the recorded clinical dataset. Additionally, it explains the procedure for investigating the hypothesis of crosstalk. Section three demonstrates the results and discusses the detection of existing extra breathing effort. As a final point, the conclusion summarizes the proposed method and results.

2   Material and Methods

2.1   Clinical data

The dataset utilized in this work was provided by a clinical partner, the Charité Universitätsmedizin Berlin, and recorded in a clinical trial. The test subject was ventilated using a pressure support mode. The sEMG data was measured using two electrode pairs placed bilaterally in the second intercostal space on the parasternal line and bilaterally at the lower costal margin on the midclavicular line and one reference electrode placed above the sternum in which the sampling frequency (f_s) was 1000 Hz [5]. Based on the visual inspection of the costal margin data, there was extra activity (crosstalk) in the time range between 1346 sec. and 1358 sec but not in the signal from Intercostal muscles in the same time range.

2.2   Data analysis

The thoracic sEMG signals are contaminated strongly by cardiac electrical activity and this must be suppressed before analyzing the sEMG spectrum. In this work, We have employed an algorithm proposed by Peterson et. al. [4], which suppresses cardiac artifacts in the wavelet domain. Short-term Fourier transform (STFT) is employed for spectral to retain time information by applying the traditional Fourier transform through a fixed-size time window. STFT is applied to the sEMG signals $s_1(n)$ and $s_2(n)$, where $s_1(n)$ refers to the costal margin and $s_2(n)$ to intercostal signals. Both signals are assumed to contain no cardiac artifacts or other disturbances after denoising in the wavelet domain. An important tradeoff in the short-time spectral analysis is time-frequency resolution. We chose window widths long enough to gain detailed frequency features. Data in blocks of 512 samples were multiplied by Hamming windows with 90 percent overlap and transformed by the FFT algorithm to determine the PSD. The PSD of the received signals $s_1(n)$ and $s_2(n)$ can be estimated by its STFT as follows:

$$P(f,t) = |X(f,t)|^2 \qquad (1)$$

where $X(f,t)$ is the STFT of the signal with f in $[0, f_s/2]$. Fluctuations in the PSD are a consequence of the nature of the data and the FFT, inevitably making it difficult to have well-structured features for further study. To overcome the barrier set by the fluctuations in the PSD, an extended approach is proposed for frequency domain analyses, namely the cumulative power spectral density (CPSD). The CPSD of the received signals $s_1(n)$ and $s_2(n)$ is defined as follows:

$$P_{\mathrm{cum}}(f,t) = \sum_{f'=0}^{f} P(f',t) \qquad (2)$$

where CPSD is related to the PSD using (2) [9]. The influence of fluctuations seen in real PSD, obtained from the STFT, can be reduced using integrated PSD while preserving the essential information.

In order to enable the clustering to focus only on the spectral characteristic, independent of the broad variety of power itself, the measured CPSD is normalized using (3).

$$\hat{P}_{\mathrm{cum}}(t,f) = \frac{P_{\mathrm{cum}}(t,f)}{P_{\mathrm{cum}}(t,f_s/2)} \qquad (3)$$

where $\hat{P}_{\mathrm{cum}}(t,f)$ is the normalized CPSD.

In the next step, we investigate the hypothesis of the presence of crosstalk using the K-means clustering. The K-means algorithm clusters the data by minimizing a criterion known as the inertia or within-cluster sum-of-squares [10]. From a physiological point of view, two categories were considered as the number of clusters. because we hope to distinguish between activity and non-activity. Then, the algorithm is applied to the normalized CPSD data in which the frequency range is the number of features and the time segments are the number of samples.

3   Results

Fig. 1 shows the recorded sEMG signals where all electrocardiography parts have been removed and the corresponding envelope for (a) the costal margin muscle and (b) the intercostal muscle. Fig. 2 (a) and (b) depict the PSD of each time slice and the equivalent normalized CPSD for costal margin muscles, respectively. Whereas, Fig. 3 (a) illustrates the PSD and (b) normalized CPSD of intercostal muscles. Each line in the PSD plot indicates the power in specific time duration. Each power line is subject to fluctuations, whereas the normalized CPSD is smoother and less noisy. The result of clustering of intercostal muscle using all frequencies has been illustrated for two arbitrary features in Fig. 4, in which two clusters are separated from each other. Fig. 5 (a) shows the p_{musRef}, derived from p_{es} [5] and represent the p_{es} variations where p_{es} is the ground truth and is conducted to follow the muscle activity. The projection of the clusters to the sEMG envelopes results in Fig. 5 (b) and (c) with two clusters. The start point of boost in each p_{musRef} peak indicates the breathing effort and subsequently the muscle activity. Two clusters are detected in Fig. 5 (b) and alternately assigned to the activity boosts. An exception is the activity boost in the middle, which is assigned to cluster2 instead of cluster1. The assignment of the minima and the transition between the activity boosts is also not clear. When comparing to p_{musRef} in the plot (a), cluster1 rather corresponds to the positive swings in p_{musRef}. Whereas, the cluster2 in Fig. 5 (c) correlates with p_{musRef} plot. So, cluster2 is subject to muscle activity.

4   Discussion

The PSD plot is associated with several fluctuations, making the further steps in data analysis much more challenging. The normalized CPSD visualization, on the other hand, results in smooth plots such that the fluctuations are reduced due to the PSD integration and the removal of absolute power values. Therefore, normalized CPSD is consid-

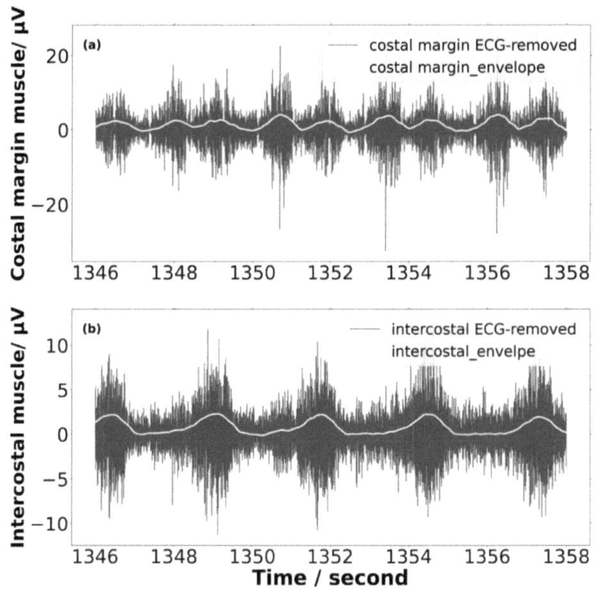

Figure 1: Small sequence of clinical denoised sEMG signals and the corresponding envelope for (a) the costal margin muscle and (b) the intercostal muscle

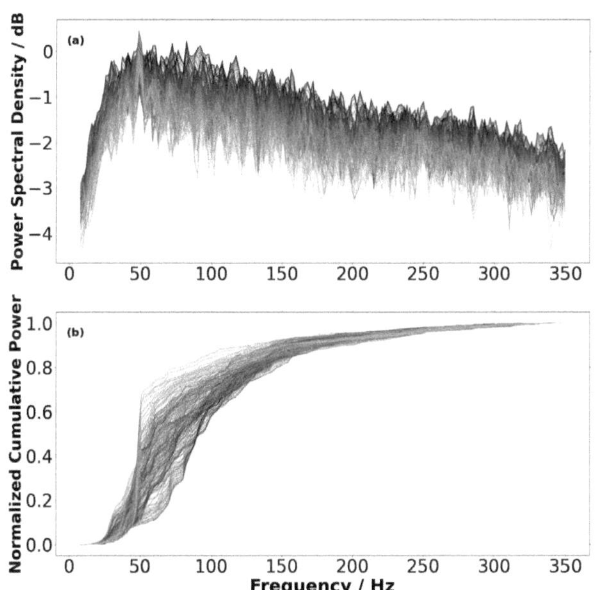

Figure 2: (a) The Power Spectral Density (PSD) (b) and the coresponding normalized cumulative Power Spectral Density (CPSD) of each time segment for the ECG-removed costal margin muscle

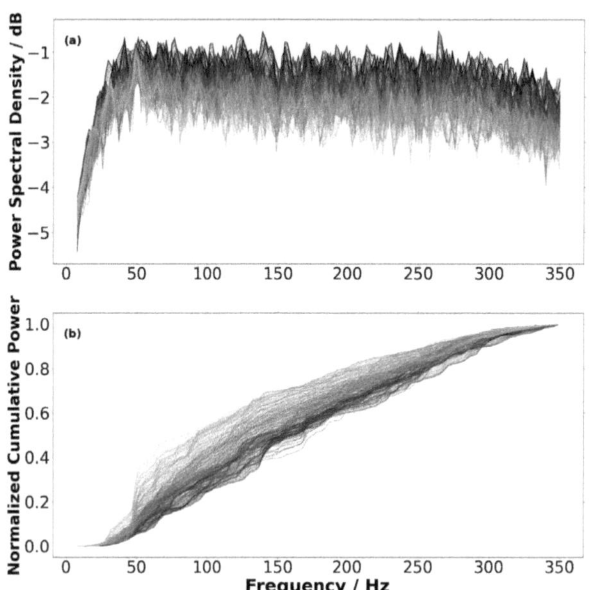

Figure 3: (a) The Power Spectral Density (PSD) (b) and the coresponding normalized cumulative Power Spectral Density (CPSD) of each time segment for the ECG-removed intercostal muscle

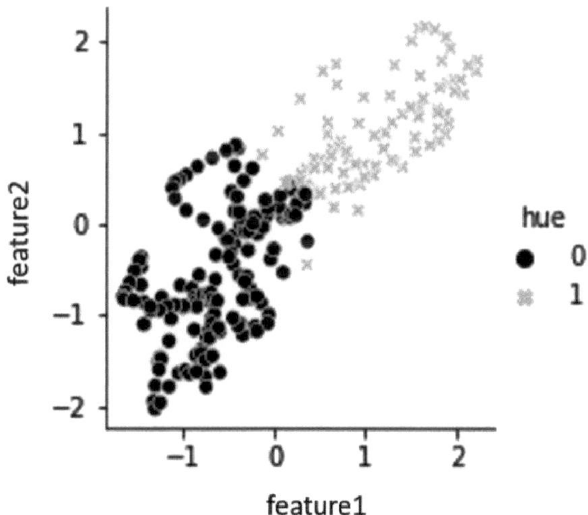

Figure 4: Scatter plot of K-means clustering for two features in the normalized CPSD of the denoised sEMG signal of the intercostal muscle

ered the efficient method to reduce the effect of the power fluctuations [8]. Visualization of K-means clustering with many features is not possible. Therefore, two arbitrary features were selected to illustrate the results. p_{es} shows the actual muscle activity. The bursts in the costal margin envelope, the start point of increase in each p_{musRef} peak, are inspiratory activities. However, the presence of additional bursts where the p_{musRef} is of positive value, points out the possibility of extra activity. Since the K-means clustering is not able to classify all the extra bursts in one cluster, we hypothesize that in the costal margin signal, the measured activity assigned to cluster 2 was not caused by inspiratory muscles. So it might be crosstalk, with different spectral properties than the inspiratory activity. On the other hand, based on p_{musRef} plot, there is no additional activity in the intercostal signal. In this case, K-means clustering is able to separate all bursts from the baseline which means the clustering method can recognize the muscle activity from the background noise in the intercostal muscle.

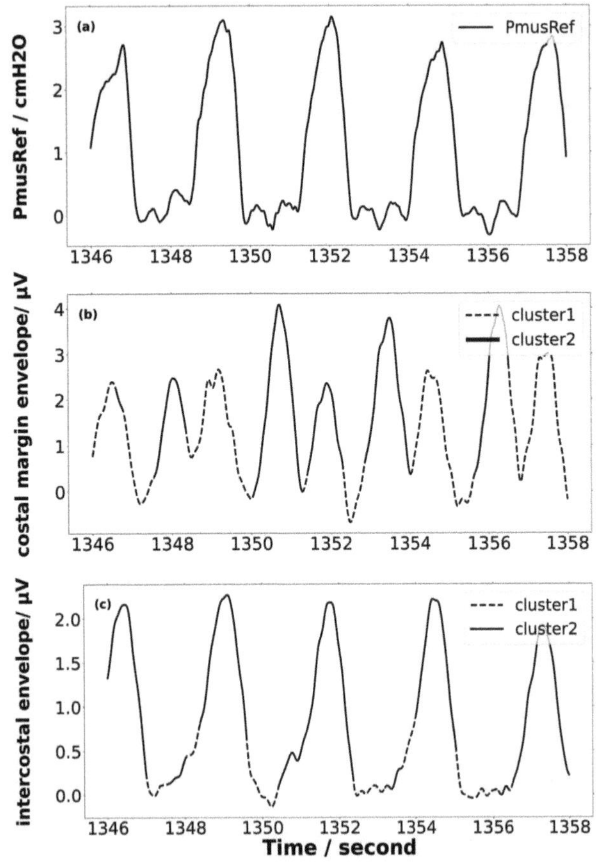

Figure 5: (a) Small section of p_{musRef} derived from esophegeal pressure and represent the p_{es} (b) Visualization of two clusters resulting from K-means clustering on costal margin envelope signal (c) Visualization of two clusters resulting from K-means clustering on intercostal envelope signal

5   Conclusion

Monitoring a patient's breathing effort is important in acute care. Accurate monitoring and interpretation of equivalent data lead to early identification and treatment of impending respiratory failure. The sEMG provides a noninvasive method to follow muscle activity. The presented work shows the new visualization which makes the acquired data be more understandable. Moreover, it has the ability to present spectral data with less influence of spectral fluctuations. The proposed clustering method indicates that crosstalk might be detected by the failing ability to decide between activity and non-activity. The method might be useful for selecting the channel with less crosstalk. This work needs to be validated on a larger dataset from different patients and the detection of crosstalk needs more formalization.

Acknowledgement

The work has been carried out at Drägerwerk AG & Co. KGaA, Luebeck, Germany and supervised by Prof. Dr. Philipp Rostalski, Institute for Electrical Engineering in Medicine, Universität zu Lübeck.

Author's Statement

Conflict of interest: Thomas Handzsuj is an employee of Drägerwerk AG & Co. KGaA and holds several patents in the area of sEMG.

6   References

[1] W. Verbrugghe and P. G. Jorens, *Neurally adjusted ventilatory assist: a ventilation tool or a ventilation toy?.* Respiratory Care, vol. 56, no. 3, pp. 327–335, 2011.

[2] A. Biel, *Trail Guide to the body: A Hands-on Guide to Locating Muscles, Bones, and More.* United States: Books of discovery, 2014.

[3] E. Petersen, J. Graßhoff, M. Eger, and P. Rostalski, *Surface EMG-based Estimation of Breathing Effort for Neurally Adjusted Ventilation Control.* IFAC-PapersOnLine, vol. 53, no. 2, pp. 16323–16328, 2020.

[4] E. Petersen, J. Sauer, J. Graßhoff and P. Rostalski, *Removing cardiac artifacts from single-channel respiratory electromyograms.* IEEE Access, pp. 30905–30917, 2020.

[5] J. Graßhoff, E. Petersen, F. Farquharson, M. Kustermann, H. J. Kabitz, P. Rostalski and S. Walterspacher, *Surface EMG-based quantification of inspiratory effort: a quantitative comparison with Pes.* Critical Care, vol. 25, no. 1, pp. 1–12, 2021.

[6] D. Farina, R. Merletti, B. Indino and T. Graven-Nielsen, *Surface EMG crosstalk evaluated from experimental recordings and simulated signals.* Methods of information in medicine, vol.43, no.01, pp. 30–35, 2004.

[7] D. Farina, C. Cescon, and R. Merletti, *Influence of anatomical, physical, and detection-system parameters on surface EMG.* Biological cybernetics, vol. 86, no. 6, pp. 445–456, 2002.

[8] S. C. Lee and R. D. Peters, *A New Look at an Old Tool-the Cumulative Spectral Power of Fast-Fourier Transform Analysis.* arXiv preprint arXiv, pp. 0901–3708, 2009.

[9] A. Nasser, A. Mansour, K. C. Yao, H. Abdallah and H. Charara,*Spectrum sensing based on cumulative power spectral density.* EURASIP Journal on Advances in Signal Processing, vol.2017, no.1, pp. 1–19, 2017.

[10] A. Likas, N. Vlassis, and J. J. Verbeek,*The global k-means clustering algorithm.* Pattern recognition, vol.36, no.2, pp. 451–461, 2003.

Optical Characterization of Microfluidic System for Multi-Infusion Systems

Rabia Cece[1], Saif Abdul-Karim[2], Jörg Schroeter [2], and Stephan Klein [2]

[1] Biomedical Engineering, Lübeck University of Applied Sciences, rabia.cece@stud.th-luebeck.de

[2] Medical Sensors and Devices Laboratory, Lübeck University of Applied Sciences, saif.abdul-karim@th-luebeck.de

[2] Medical Sensors and Devices Laboratory, Lübeck University of Applied Sciences, joerg.schroeter@th-luebeck.de

[2] Medical Sensors and Devices Laboratory, Lübeck University of Applied Sciences, stephan.klein@th-luebeck.de

Abstract

Syringe pumps are widely used devices in intensive care units. Patients who are on the intensive care generally require more than one medication simultaneously. The combination of more than one infusion line into a single central line is called multi-infusion. Multi-infusion might be a cause of flow variability and medication dosing error. The aim of this study is to identify the effect of multi-infusion setups on the received medication dose expressed as mass flow rate and to observe the differences between two multi-infusion setups. For this purpose, light absorbance measurements were carried out. Two different paediatric infusion setups from the University Hospital Campus Luebeck (UKSH) in Germany and the Medical University Utrecht (UMC) in the Netherlands were compared. An infusion set with lower dead volume shows considerably higher flow rate variation compared to an infusion set with higher dead volume. Results showed that varying one flow rate of mediation affects the measured mass flow rate of other medications. The measured mass flow rates deviated from the set mass flow rates.

1 Introduction

Infusion therapy with multi-infusion systems is a difficult treatment method since the flow rate of one medication affects the flow of other medications in the central line and results in dosing errors due to flow rate fluctuations[1]. The amount of medication required may vary depending on the patient's condition. Therefore, it is sometimes necessary to change the flow rate of the medication during treatment. However, these changes lead to dosing errors due to flow rate fluctuations. The particular reason of flow rate variability is pressure changes. These pressure changes are especially observed when flow rates are changed or after the pumps are started[2]. Dosing errors because of pressure changes and consequent flow rate variability can sometimes result in death. Dosing error may also be caused by dead volume which is a significant phenomenon related to infusion systems. Dead volume is the total volume from the mixing point to the catheter tip[3]. Low dead volume is an important phenomenon that must be taken into account in order to administer the medication as accurately as possible to the patient. Spectrometric method is commonly used for the evaluation of dead volume and flow rate variability[3]. By measuring the light absorbance of the liquid passing through the central line, mass flow rate can be calculated with (1) and (2). The relation between concentration and light absorbance is determined with Beer-Lambert Law. Beer's Law states that absorbance is proportional to

the concentrations of a sample. A solution with higher concentration absorbs more light and transmits less light than solution of a lower concentration. The concentration can be estimated with (1) by means of the performed calibrations.

$$c = k * A + b \qquad (1)$$

A is the absorbance of dye, c is the concentration, k is the coefficient and b is the intercept. The values a and b are acquired by finding the linear relationship between the concentration and corresponding absorbance measurement during calibration [modelling]. If the concentration of a solution is calculated in accordance with Beer's law, then the mass flow rate can be calculated according to the following formula:

$$m = c * V * M \qquad (2)$$

m stands for mass flow rate (mg/h). c states concentration (mmol/L). V is total volume flow (ml/h) and M is the molar mass of the dye powder which is depicted in Table 2. There is a significant point need to be noticed for light absorbance measurements. Light source must be heated at least for an hour because the black body radiation (thermal electromagnetic radiation)[4] curve vary with the increasing temperature of light source. Therefore, light source should be heated for a while to obtain more stable spectrum.

There are some studies about the investigation of the effect of varying flow rates[5] in the literature, but in general, only the effect of one flow rate change is observed. For this reason, the main focus of this study is to observe the impact of simultaneously varying flow rates of the three syringe pumps on the mass flow rate. Additionally, the effect of dead space volume in two setups is analyzed in this study.

2   Material and Methods

2.1   Experimental Setup

In two in vitro set-up, flow rate variability and effect of different dead volumes were evaluated. The illustration of the setups is depicted in Fig.1. Three syringe pumps with 50 ml BBraun syringes from UKSH were connected to a infusion line. Three pressure sensors and flow sensors were inserted in infusion line and these lines were connected to a five valves manifold. A single infusion line comes from manifold was placed into a flow cell which is connected to a light source (AvaLight-HAL) and a spectrometer (AvaSpec-2048L-USB2). For an effective light absorbance measurement, the incident light should be as straight as possible and not scattered. For this purpose, rigid connectors and focusing lenses were used to connect the light source and the spectrometer to the flow cell. Additionally, a material prepared from PTFE and foam was used in accordance with the shape of the light source's output in order to prevent scattering and to take light reference. To attenuate the incoming light, PTFE was used because the electromagnetic spectrum of the of PTFE is similar with the spectrum of white light. The single line through the flow cell is connected to a catheter (vygon, Paediatric Multicath 2, inner diameter 1.5 mm, outer diameter 4.5 mm) attached to the collecting tank. In Setup Lübeck, this single line is connected to a filter (Sterifix, 0.2μ) and then to a catheter (Arrow, DE-12854-UKSH, CVC set: 4 Lumen 8.5FR,2.9mmx16cm) with larger diameter, unlike Setup Utrecht. The dead space volume for Setup Lübeck is 4.7 ml and for Setup Utrecht it is 3.15 ml. The block diagram of the setups is depicted in Fig.1. Before starting the experiment, the light source must

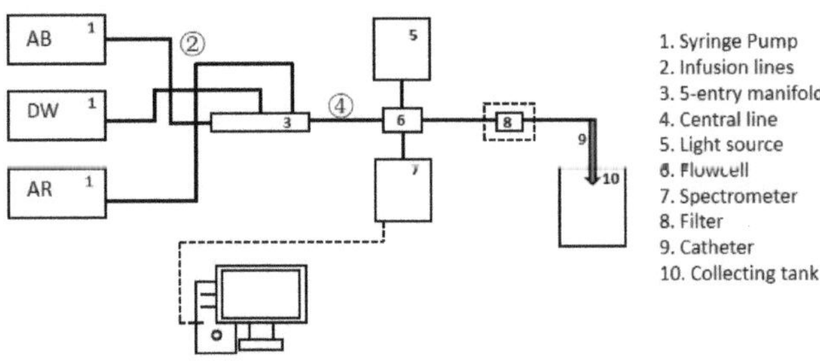

Figure 1: The illustration of experimental setups.

be run for at least one hour to obtain a stable light absorption spectrum. Thereafter, calibration should be performed for red solution (AR) and blue solution (AB) of different concentrations (see concentrations in Table 1). Distilled water was used as the reference solution in each calibration. After calibration, absorbance can be carried out. Light absorbance of the solutions were measured with AvaSoft 8.11 software. Experiments were carried out with flow rate 1 (FR1) values for approximately one hour or one and half

hour, and then flow rate 2 (FR2) values for the same duration. Different fluids were used for each syringe pump. In all experiments, syringe pump 1 (S. pump1) was used for the injection of the AB, syringe pump 2 (S. pump2) for the distilled water, and Syringe pump 3 (S. pump3) for the AB. The three liquids used, the molar mass of the dye powders used, the concentration of the prepared solutions, the first set flow rate values, and the second set flow rate values of the three syringe pumps are shown in the Table 2. The red and blue solution used were prepared using Allura Red AC (dye content 80%, SIGMA-ALDRICH) and Acid Blue 129 (dye content 25%, SIGMA-ALDRICH), respectively. An important point to be noticed when preparing the

Table 1: Concentration of solutions

Red solution (mmol/l)	Blue solution (mmol/l)
0.05	0.01
0.1	0.03
0.2	0.07

Table 2: Parameters of the Multi-Infusion System

Parameters	S. Pump1	S. Pump2	S. Pump3
Dye powder	AcidBlue129	Dw	AlluraRedAC
Molar mass	0.115g/mmol	-	0.397g/mmol
Conc.	0.07mmol/L	-	0.1mmol/L
FR1(ml/h)	10	7	5
FR2(ml/h)	6	2	7
Height	0	0	0

AB is to leave the solution to stir for a day at 90 degrees Celsius. Afterwards, the solution must then be filtered to remove big particles in solution so that precipitation of AB can be avoided.

3   Results

3.1   Absorbance Rectification

The absorption spectrum of AB and AR are different from each other. The absorption peak of AR is around 500 nm (see Fig. 2) and the absorption peak of AB is around 630 nm (see Fig. 3). There is absorption at both 630 nm and 500 nm for the AB, whereas there is no absorption around 630 nm for the AR. Since the absorption value obtained at 500 nm is not purely red dye absorption, but the sum of red and blue dye absorption, this results in an incorrect concentration of the AR. Hereby, the absorption of AB at 500 nm must be subtracted from the total absorption at 500 nm to find the correct concentration of the red solution. To find the absorbance of the AB at 500 nm, absorbance linearity curve was formed by measuring absorbance of AB with 0.01, 0.03, 0.07, 0.1 mmol/L concentrations, respectively (see Fig.4). To find the absorbance of the AB at 500 nm, absorbance linearity curve was formed by measuring absorbance of AB with 0.01, 0.03, 0.07, 0.1 mmol/L concentrations.

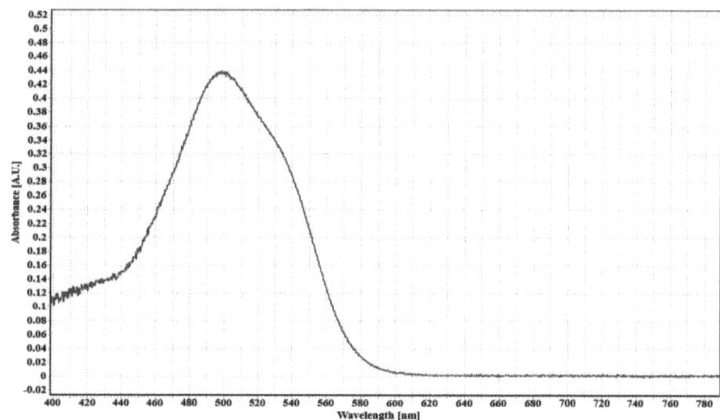

Figure 2: Absorbance spectrum of red solution with 0.1 mmol/L concentration.

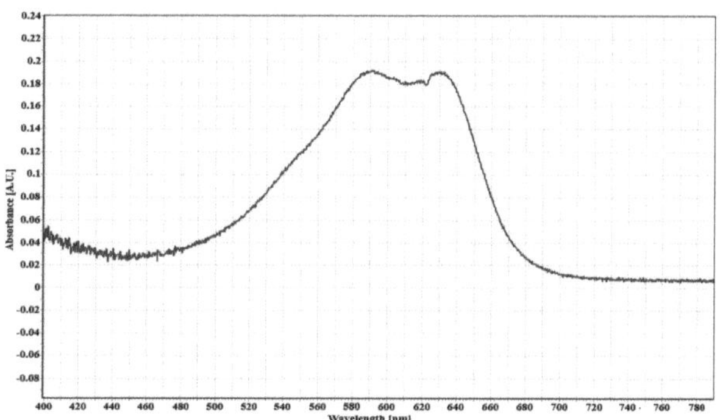

Figure 3: Absorbance spectrum of blue solution with 0.07 mmol/L concentration.

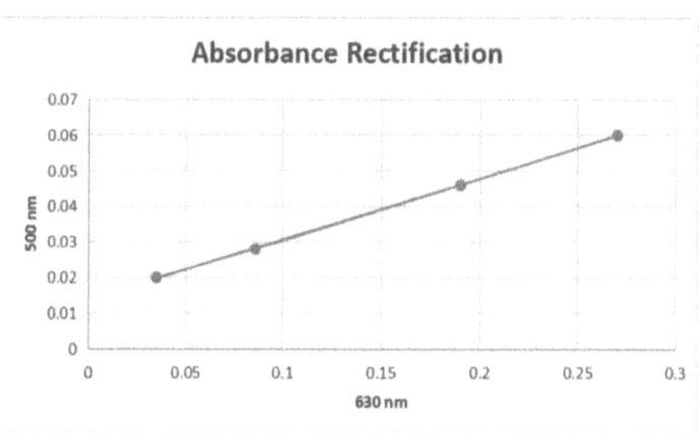

Figure 4: Absorbance rectification line.

3.2   Analysing of mass flow rate

To understand the effect of flow rate variation of three different solutions and the effect of different setups on mass flow rate, 63 light absorbance measurements were conducted. It takes approximately 14 minutes for mixed solution to reach the flow cell in Setup Lübeck, and only 2-3 minutes in Setup Utrecht (Fig.5 and Fig.6). In Fig.5, the mass flow rate of the AB is slightly below the set value before and after the flow rate change. Whereas the mass flow rate is below the set value at the beginning of the measurement, it is too high for the AR after the flow rate change. In Fig.6, the mass flow rate of AR is even lower compared to Fig.5 before changing of flow rates. There is a mistake at the flow rate change point in Fig.5 and Fig.7. Also, the mass flow values are not correct in all figures except Fig.9 due to miscalculations. A peak at the beginning of the measurements can be clearly seen in the figures apart from Fig.5

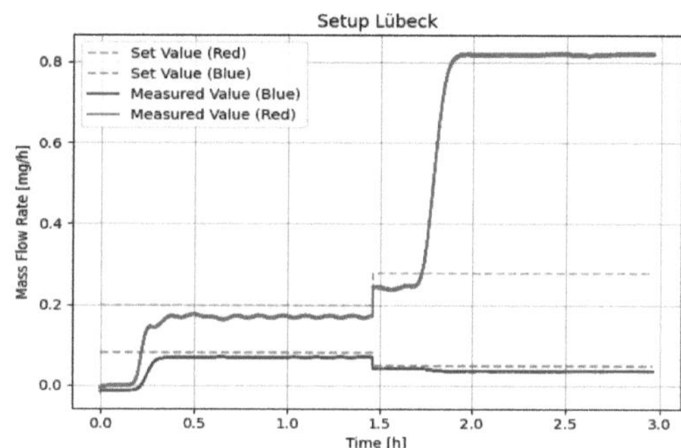

Figure 5: Mass flow rate of AR and AB in the central line. Flow rate of AB was set from 10 ml/h to 6 ml/h. Flow rate of dw was set from 7 ml/h to 2 ml/h. Flow rate of AB was set from 5 ml/h to 7 ml/h.

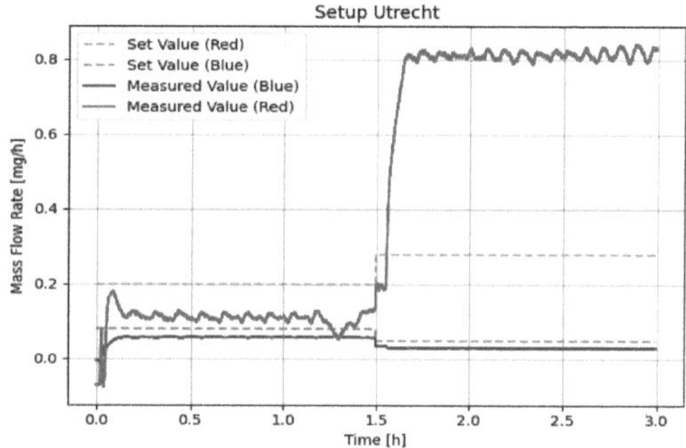

Figure 6: Mass flow rate of AR and AB in the central line. Flow rate of AB was set from 10 ml/h to 6 ml/h. Flow rate of dw was set from 7 ml/h to 2 ml/h. Flow rate of AR was set from 5 ml/h to 7 ml/h.

and Fig.9. According to the graphs of Setup Utrecht, strong fluctuations of the mass flow rate of AR can be observed. With respect to Fig.9, the mass flow rate of AR is around

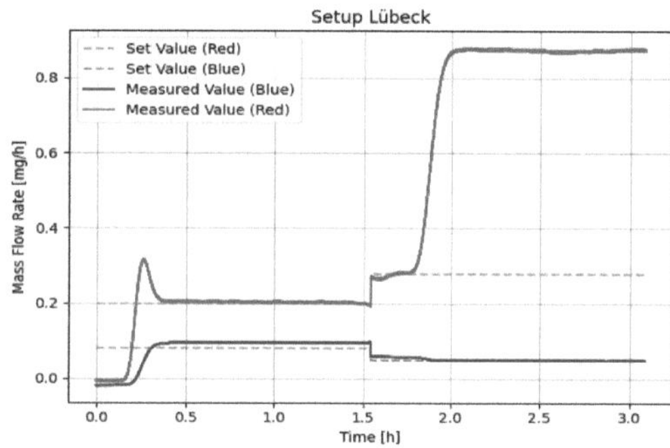

Figure 7: Mass flow rate of AR and AB in the central line. Flow rate of AB was set from 10 ml/h to 6 ml/h. Flow rate of dw was set from 7 ml/h to 2 ml/h. Flow rate of AR was set from 5 ml/h to 7 ml/h.

0.24 mg/h before flow rate change and 0.35 mg/h after flow rate change. There is approximately 0.04 mg/h dosing error for AR. The mass flow rate of AR is almost stable around 0.08 mg/h. While the mass flow rate of AB fits the set line, the mass flow rate of AR is higher than the set value. For

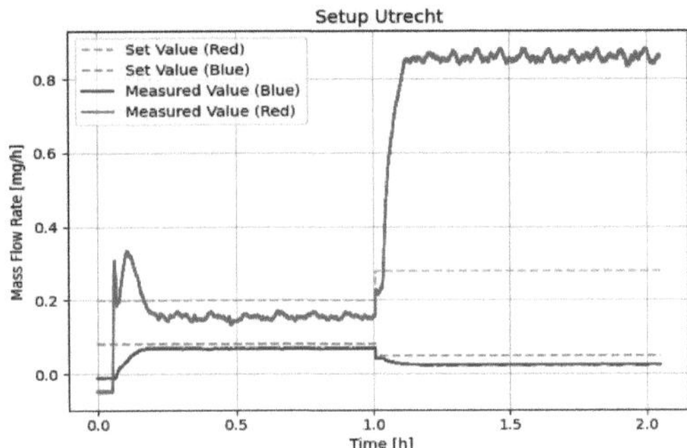

Figure 8: Mass flow rate of AR and AB in the central line. Flow rate of AB was set from 10 ml/h to 6 ml/h. Flow rate of dw was set from 7 ml/h to 2 ml/h. Flow rate of AR was set from 5 ml/h to 7 ml/h.

a better understanding the effect of multiple varying flow rates, one measurement was carried out by changing just the flow rate of AR.

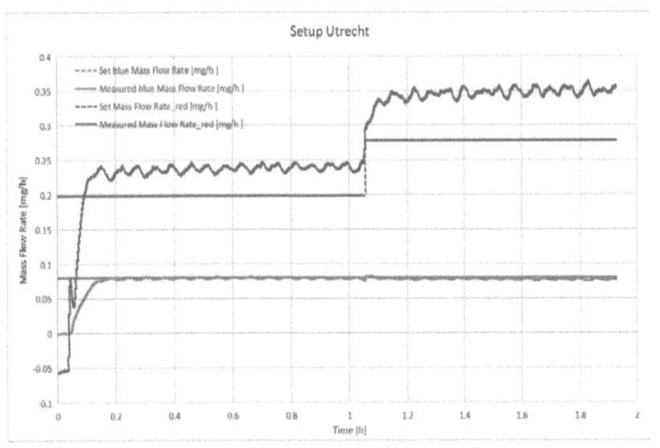

Figure 9: Mass flow rate of AR and AB in the central line. First flow rates are as indicated in Table 2. Only the flow rate of AR was changed from 5 ml/h to 7 ml/h.

4   Discussion

The difference between Setup Lübeck and Setup Utrecht in terms of dead space volume can be observed with the shorter and longer duration until the mass flow rates increase. In Setup Lübeck, it takes longer to reach the set value after starting the syringe pumps and changing the flow rates compared to Setup Utrecht. In addition, Setup Utrecht shows considerably higher flow rate change efficiency compared to Setup Lübeck. The main purpose of this study is to investigate the effect of multiple flow rates simultaneously. However, changing three flow rates at the same time instead of changing just one flow rate complicates the situation more because the flow rate of solutions affects each other's flow rate and this case must be considered in the calculations but this case was not considered in the calculations of Fig.5, Fig.6, Fig.7 and Fig.8. It was only considered in the calculations of Fig.9. Even though this case was considered and the absorbance of the AB at 500 nm was subtracted

from the total absorbance at 500 nm, the mass flow rate of AR is still higher than the set value. This will be further examined in the future work. The cause of the peak in Fig.6, Fig.7, and Fig.8 is due to push-out effect. If the valves of the manifold are turned before the measurement starts, the graphs show a push-out effect.

5   Conclusion

Evaluation of mass flow rate is significant to prevent dosing errors in multi-infusion systems. Different setups with different features can impact the measured mass flow rate differently. Therefore, it is important to characterize the used system to prevent the potential risks. The effect of varying flow rates at the same time were investigated in this study. However, this goal could not be fully accomplished due to limited time. In the future, the algorithms for evaluating the concentration measurements and calculating the mass flow rates are being consistently further developed.

Acknowledgement

The work has been carried out at Medical Sensor and Devices Laboratory, Luebeck University of Applied Sciences. The authors thank UMC Utrecht for providing materials for test.

Author's Statement

The course of the project was planned by Rabia Cece and Saif Abdul-Karim. The experiments were carried out by Rabia Cece. Saif Abdul-Karim and Stephan Klein supervised the project. Jörg Schroeter helped building up the setups.

6   References

[1] Snijder, Roland A. *Impact of physical parameters on dosing errors due to a syringe exchange in multi-infusion therapy.* Pharmaceutical Technology in Hospital Pharmacy 2.2 (2017): 85-96.

[2] Décaudin, Bertrand, *Impact of multiaccess infusion devices on in vitro drug delivery during multi-infusion therapy.* Anesthesia Analgesia 109.4 (2009): 1147-1155.

[3] Snijder, R. A., Konings, *Flow variability and its physical causes in infusion technology: a systematic review of in vitro measurement and modeling studies.* Biomedical Engineering 60.4 (2015): 277-300.

[4] Ranganath, G. S. *Black-body radiation.* Resonance 13.2 (2008): 115-133.

[5] Haaijer, Kelly. "Modelling multi-infusion: the effects of flow rate changes and air on drug administration.

9

Biomechanics

Mechanical characterisation of carbon fiber reinforced orthoses for the lower extremity

Jannes Brüling [1], Jörg Schroeter [2], and Robert Wendlandt [2]

[1] Biomedical Engineering, Luebeck University of Applied Sciences, jannes.brueling@stud.th-luebeck.de

[2] Biomechanics, Universität zu Lübeck, {joerg.schroeter, robert.wendlandt}@uksh.de

Abstract

Ankle-foot orthoses (AFOs) enable natural and safe walking for people with calf muscle weakness. To maximize the effect of an ankle foot orthosis on walking, the rotational stiffness of the AFO must be matched to the patient's body weight, activity level and clinical appearance. In practice, the rotational stiffness usually depends on the experience of the prosthetist, as the ankle foot orthosis is individually produced from carbon fiber prepregs. To improve adjustment of the rotational stiffness and thus the care of the patient, a test method using a built-in lower limb phantom was developed to determine rotational stiffness of AFOs. The feasibility of the test method for AFOs was successfully demonstrated. In addition, the influence of different AFO-designs on rotational stiffness was investigated.

1 Introduction

Ankle foot orthoses (AFOs) are traditionally prescribed for patients suffering from calf muscle weakness. Typical diseases that can result in weakness of the calf muscles are neuromuscular diseases, like poliomyelitis or cerebral palsy [1][2]. Calf muscles weakness has a strong impact on the gait pattern of the patient, which is characterised by an increased dorsiflexion in the stance phase and reduced push-off power in the push off-phase.[1] Ankle foot orthoses help to correct the gait pattern and enable fluid and safe walking [3]. Advanced ankle foot orthoses can withstand and even support high biomechanical loads like jumping and running. During dorsiflexion in the stance phase the ankle foot orthosis stores energy and releases it during the push-off phase in form of an external plantarflexion moment. Therefore, dorsiflexion of the foot ankle is restricted, the probability of injury is reduced, and the overall energy efficiency of walking increases significantly [4]. Since the plantar flexion moment provided by the orthosis is proportional to the rotational stiffness of the AFO, the rotational stiffness must be adapted to the applied load of the patient. Traditionally, the stiffness is selected depending on the body weight, size, clinical appearance, and activity level of the patient. A high stiffness offers on the one hand a high restriction of dorsiflexion but decreases the capacity of releasing energy in the push off phase on the other hand. To provide the best care for the patient, a good balance between restriction of dorsiflexion and the capacity to release push off energy must be made. Carbon fiber composites have shown superior functional performance for springs of ankle foot orthoses and are nowadays the primarily used material [5]. In addition to standardized carbon fiber springs, such as the

Carbon Ankle Seven (Otto Bock HealthCare Deutschland GmbH, Duderstadt), carbon fiber springs made of prepreg mats are commonly used. They are produced by the prosthetist themselves and are fitted to the anatomy of the lower limb of the patient. Thus, they offer patients a comfortable wearing and are adjustable to the individual needs. Therefore, it is the standard procedure in patient care. Since mechanical parameters of ankle foot orthoses made of carbon fiber prepregs are rarely quantified, the stiffness relies on the experience of the prosthetist. To make the fabrication process more standardizable, a testing method has been developed to assess the rotational stiffness of ankle foot orthoses. Furthermore, the influence of different AFO-prepreg-designs on the overall rotational stiffness has been evaluated. This study improves the understanding of different AFO-prepreg-designs on the overall rotational stiffness and enables the prosthetist to match AFOs to the patient-specific needs.

2 Material and Methods

The rotational stiffness of seven different carbon fiber ankle foot orthoses for children with calf muscle weakness have been investigated (Fig. 1). Six of the seven models are AFOs made of prepreg mats, and one model is an AFO with a build-in Carbon Ankle Seven spring, which serves as a reference. For the ankle foot orthoses, made of prepreg, the influence of two different design factors on the rotational stiffness are analyzed. As shown in Table 1, one factor is the variation of the approach of the spring, which can be lateral, dorsal distal, dorsal proximal or dorsal normal, and the other factor is the spring shape, which is either flat or anatomic. Anatomic in this case means that the spring shape is adapted to the anatomical shape of the heel and lower leg.

Table 1: AFO-Designs

		Spring shape	
		anatomic	flat
	Lateral	–	type 2
Spring	Dorsal normal	type 3	type 4
approach	Dorsal distal	type 5	type 6
	Dorsal proximal	type 7	–

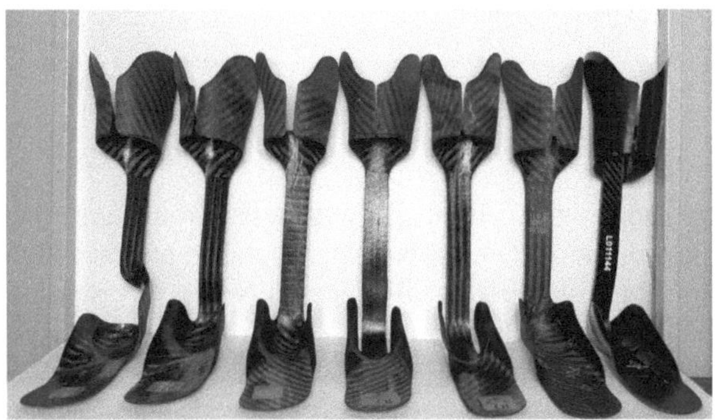

Figure 1: AFO-Designs: type 2 to 7 and the AFO with a build in Carbon Ankle Seven (from left to right)

Two possible design-variants were excluded from the testing series since they are not used in patient care. The orthoses were provided by Technische Orthopädie Lübeck Jens Becker GmbH, headquartered in Lübeck, Germany.

To simulate realistic loading and load distribution on the orthoses, the AFOs are attached to a phantom lower limb in the testing setup. The phantom lower limb and the AFOs were based on the plaster cast of the lower limb of a five-year-old child (weight: 10-20 kg; shoe size: 26 (EU)) that was already made during the fitting of an AFO. Based on the plaster cast, a positive model was created. After smoothening the surface of the limb positive model, a plastic mold was made of the positive. This plastic mold served as a casting mold for the final lower limb phantom. The inner structure of the phantom consists of a steel threaded rod (M12), which is additionally stiffened with a matching steel tube. This structure can be considered as completely rigid and serves as the skeleton of the lower limb. The ankle joint is realized by a fork joint (Misumi; 751.1-12-48-M12-KL) and a pedestal bearing (Misumi; C-BGHKA6901ZZ-20). This uniaxial joint only enables movement in the sagittal plane, so that the rotational stiffness along this plane can be analyzed. The inner foot skeleton consists of a 3D printed PLA component. The inner structure is embedded into a silicone matrix with a shore hardness of 10 A. (Dragon Skin 10 Slow, Smooth-On, Inc., USA + 10 % of mixture weight Silicone Thinner, Smooth-On, Inc., USA) The silicone matrix mimics soft tissue bulk and provides a realistic support for the orthoses. For the testing procedure a nylon stocking is pulled over the lower limb phantom to ensure realistic friction behaviour at the interface.

To simulate dorsiflexion of the lower limb phantom, it is pushed against a steel bar at the forefoot, which is elevated

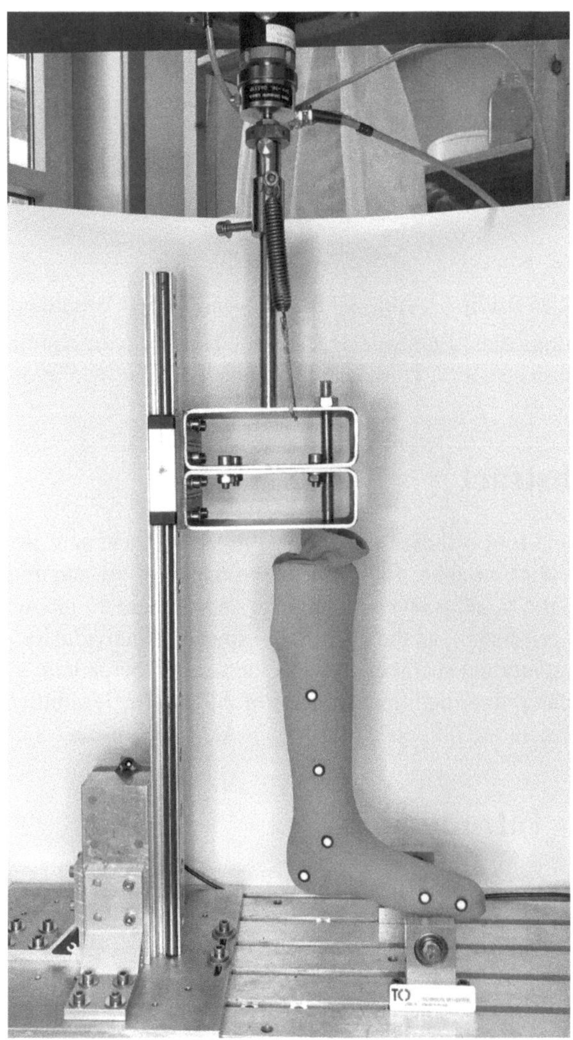

Figure 2: Test setup - rotational stiffness acquisition of the lower limb phantom

and fixed by ball bearings. This setup minimizes static friction to the plantar surface of the phantom-AFO complex during testing. The line of contact is located 80 mm horizontal from the ankle joint of the lower limb phantom in this test setup. As shown in Fig. 2, the lower limb phantom is attached rigidly to a linear rail to ensure a pure vertical movement. The linear rail is realized by a fixation of the lower limb phantom to two rigid steel profiles, which are mounted on top of a slide, that glides with low friction on a rail. Furthermore, the rail is rigidly fixed via aluminium components in the material testing machine (Zwick Roell, 1456). The material testing machine applies a force via a pressure stamp with a constant velocity of 5 mm/min on the steel profile. During the testing, the force is measured by a ±2kN force sensor, which is located above. As soon as a preload of 5 N is reached, the testing machine starts to record the applied force. The measurement ends as soon as a displacement of 40 mm is reached, or the applied force exceeds a value of 180 N, to prevent the ankle to touch the setup and to limit the load within the body weight of the child. To compensate for the weight of the lower limb phantom, the square steel profile is connected via a spring to the testing machine. Due to the fixed horizontal distance between the ankle joint of the lower limb phantom and the contact with the elevated steel bar, a constant lever arm is used to calculate the torque during testing of the limb-AFO complex from the recorded force. The rotation of the phantom limb-AFO construct around the ankle is tracked via markers, which are placed along the phantom limb, and a camera system (framerate: 2 fps; resolution: 2 MP) placed in front of the test setup. To synchronise the acquired force data and the rotation around the phantom ankle joint, a LED light turns of at the beginning of the measurement. Since the lower limb phantom has a certain inherent rotational stiffness, the rotational stiffness of the lower limb phantom without an AFO has been determined prior.

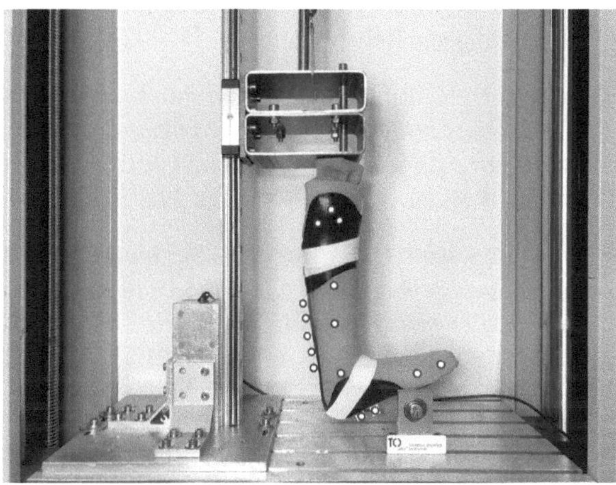

Figure 3: Testing procedure - rotational stiffness acquisition of AFOs

The testing procedure for the different AFO-designs follows a common test protocol. After attaching and fixing the ankle foot orthosis to the lower limb phantom and placing the markers (according to Fig. 3) six measurements are executed. The first measurement from each test series is only used for settling and preconditioning effects and is eliminated. The five latter measurements are used for further evaluation. For video evaluation the GOM correlate software (GOM GmbH, GOM correlate 2021 free version) was used. The software determined the rotation around the ankle joint by tracking the relative motion of the markers. After synchronising the rotation data and the acquired force data, the rotational stiffness as a function of the rotation angle is calculated from the ratio of the change of torque and change of angle via a R-script. To obtain the stiffness of the orthoses alone, the stiffness of the phantom is deducted from the result. Additionally, the mean value of rotational stiffness of each AFO-design was calculated.

3 Results and Discussion

3.1 Assessment of the rotational stiffness

The rotational stiffness of the AFO and the phantom over the dorsiflexion angle is shown in Fig. 4. During the measurements, it was observed, that the force increased non-linearly. Possible reasons for this behaviour are non-linear mechanical properties of the silicone rubber, static friction in the ankle joint and static friction to the plantar surface of the lower limb phantom. The influence of the latter was minimized by the described test set-up. The tracking and evaluation of the rotation around the ankle joint via markers worked very well and showed a high reliability.

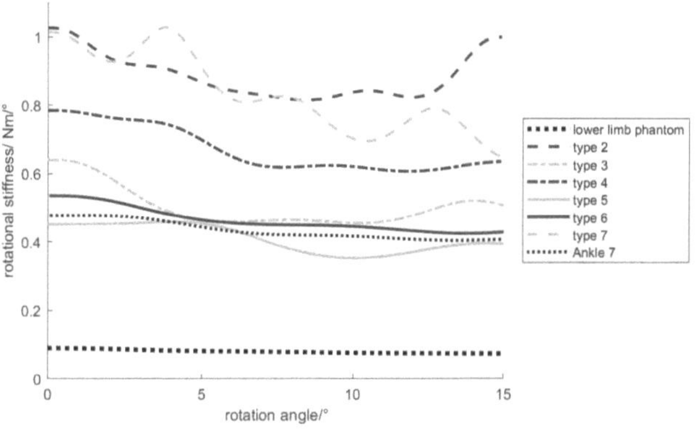

Figure 4: Rotational stiffness of each AFO over the ration angle (0°-15°)

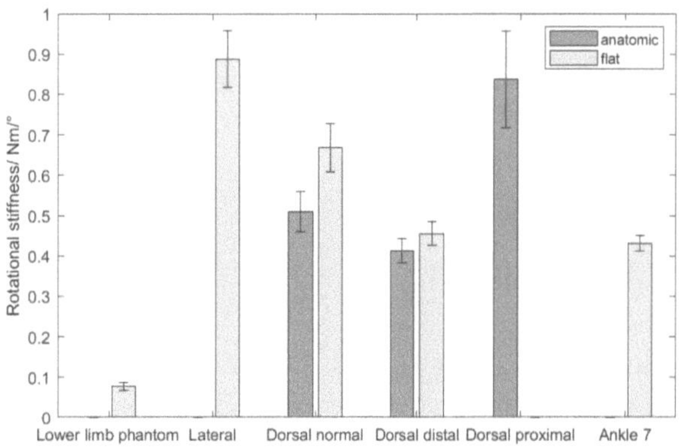

Figure 5: Mean values and standard deviations of the rotational stiffness of each AFO

In Fig. 4 it can be observed, that the acquired rotational stiffness of the AFO depends on the rotation angle. At the beginning the rotational stiffness decreases with an increasing rotation angle. This effect is caused by a compression of the silicon of the lower limb phantom, especially on the calf attachment (see Fig. 3).

Under real conditions, compression of soft tissue occurs as well at the beginning of AFO loading. However, the AFO is closely attached to the lower leg, so that the effect of tissue compression might be smaller. Still, it needs to be considered that the effect of the AFO is not influenced by the mechanical properties alone but the fixation to the lower leg needs to be considered as well. In Fig. 5 it can be seen that the standard deviation increases with the mean rotational stiffness of the AFO, due to greater compression of the silicon at the calf part. The non-linear course of the rotational stiffness of the AFO, shown in Fig. 4, could have been caused by the mechanical properties of the carbon springs, like alignment of carbon fibres, and sudden release of energy due to static friction at the AFO plantar surface. In summary, the developed test method is a promising tool for the characterisation of the rotational stiffness of ankle foot orthoses in the sagittal plane.

3.2 Influence of different AFO-designs on the rotational stiffness

As it was expected, the attachment of an AFO to the lower limb phantom increased the rotational stiffness of the lower limb phantom significantly. As shown in Fig. 5, the AFOs with a lateral and dorsal proximal spring approach exhibited the greatest rotational stiffness. The rotational stiffness of the AFO with a dorsal distal spring approach was lower than the rotational stiffness of the AFO with a dorsal normal spring approach, due to the larger spring. Furthermore, it can be observed, that the anatomic shape of the AFO strut reduced the rotational stiffness.The Carbon Ankle Seven increased the rotational stiffness of the lower limb phantom by a factor of approximately five and showed a smoother stiffness behaviour over rotation compared to the AFO-designs made of prepreg mats. This may be caused by the production process of the AFOs. The AFOs made by prepreg are likely to exhibit an uneven distribution of resin along the spring and the Carbon Ankle Seven was produced by standardized production procedure.

4 Conclusion

In this study a testing method with an incorporated lower limb phantom has been developed to assess the rotational stiffness of ankle foot orthoses in dorsiflexion. Feasibility of the testing method for AFOs was demonstrated successfully. Nevertheless, the elimination of some drawbacks, like high compression of the lower limb phantom, might further improve the accuracy of the test method.

Acknowledgement

The support of the biomechanics institute of the university of Lübeck is gratefully acknowledged. A special thank go to Dr.-Ing Robert Wendlandt and Dipl.-Ing. Jörg Schroeter for their assistance and support throughout the project and Orthopädie Lübeck Jens Becker GmbH (Lübeck, Germany) for providing the ankle foot orthoses.

Author's Statement

Conflict of interest: The samples have been provided free of charge by TO Becker. TO Becker didn't participate in the evaluation of the results nor in the writing of the paper.

5 References

[1] Hilde E. Ploeger, Sicco A. Bus, Frans Nollet, Merel-Anne Brehm, *Gait patterns in association with underlying impairments in polio survivors with calf muscle weakness*,In: Gait & Posture, Volume 58, pp. 146-153, 2017.

[2] Waterval, Niels and Brehm, Merel-Anne Harlaar, Jaap and Nollet, Frans, *Individual stiffness optimization of dorsal leaf spring ankle–foot orthoses in people with calf muscle weakness is superior to standard bodyweight-based recommendations*, Journal of Neuro-Engineering and Rehabilitation, p. 18, 2021.

[3] Ploeger, H. E., Bus, S. A., Brehm, M.-A., and Nollet, F., *Ankle-foot orthoses that restrict dorsiflexion improve walking in polio survivors with calf muscle weakness*. In: Gait Posture, Volume 40(3), pp. 391–398, 2014.

[4] D.J.J. Bregman, J. Harlaar, C.G.M. Meskers, V. de Groot, *Spring-like Ankle Foot Orthoses reduce the energy cost of walking by taking over ankle work*, In: Gait & Posture, Volume 35, pp. 148-153, 2012.

[5] Lusardi, M., Nielsen, C. C., and Milagros, J., *Orthoses in Rehabilitation, Orthotics and Prosthetics in Rehabilitation: A Multidisciplinary Approach*, Elsevier, St. Louis, MO, pp. 181–455, 2013.

Stenosis Simulation of the Superficial Femoral Artery Using an Adaptive 3D-printed Actuator: Development and Verification

Annika Dell [1,2], Franz Wegner [4], Thorsten Buzug [2,3], and Thomas Friedrich [2]

[1] Medical Engineering Science, Universität zu Lübeck, Annika.Dell@student.uni-luebeck.de
[2] Fraunhofer Research Institution for Individualized and Cell-Based Medical Engineering IMTE
[3] Institute of Medical Engineering, University of Lübeck
[4] Department of Radiology and Nuclear Medicine, University of Lübeck

Abstract

A novel hydraulic actuator was conceived and developed using multi-material additive manufacturing, motivated by the need for a simulator that can be adjusted to create varying degrees of vascular stenosis and several stenosis geometries. This adaptive actuator allows the simulation of varying degrees of vascular stenosis, provides a proof-of-concept for future vascular simulators, and can be used in various types of medical imaging modalities. The actuator was tested to determine how to achieve the desired degree of stenosis: a linear correlation between the stenosis-inducing mechanism and the degree of stenosis, as well as consistently reproducible stenoses of the SFA were achieved.

1 Introduction

The aim of this paper is to design and construct a 3D-printed adaptive stenosis actuator in order to simulate different degrees of superficial femoral artery (SFA) stenoses, as well as verify its functionality [2]. Lower extremity peripheral arterial disease (PAD) is the third leading cause of atherosclerotic cardiovascular morbidity, following coronary arterial disease and stroke [2]. An estimated 200 million people suffer from peripheral arterial disease, making PAD a global issue [1], [2]. The SFA is the vessel most frequently afflicted by PAD, which is why it has been chosen to be emulated with this actuator [3]. Due to the prevalence and clinical impact of vascular disease, vascular phantoms are an essential part of medical research [2]. They allow the simulation of complex pathologies and can often closely emulate vascular structures physiologically and geometrically. Specifically stenosis phantoms present a practical and modifiable way to test interventional devices and procedures without the need for animal testing [4]. Currently, simple 3D-printed vessel phantoms with non-variable geometries are often used in emulating real vasculature. It would be beneficial to have an *adaptive* stenosis simulator, which would allow for varying degrees of stenosis without requiring several models. This work presents such an adaptive stenosis simulator, and is the continuation of an initial stenosis actuator design presented by Dell et al. [2]. Here, fundamental changes to the first design, as well as verification of the functionality of this actuator, are described. The necessary characteristics of this adaptive stenosis actuator include: small size, uncomplicated construction, simple working mechanism to allow for varying degrees of stenosis, swift means of altering the geometry, and omission of metal components.

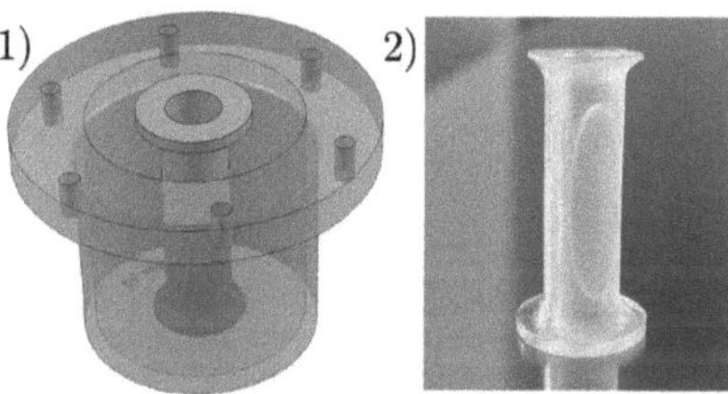

Figure 1: **1)** CAD rendering of the assembled inner and outer actuator components. The inner actuator component is opaque; the outer actuator component is transparent [2]. **2)** 3D printed inner actuator component with a flexible one-sided patch, shown in white. Adapted from Dell et al. [2].

2 Material and Methods

2.1 Construction

The adaptive hydraulic actuator is comprised of three fundamental parts: the lid, the hard outer actuator component, and the partially soft, compressible inner actuator component [2]. The vessel phantom is threaded through the inner component. The CAD modeling program SolidWorks 2020 (Dassault Systemes, France) was used to construct all components. The Stratasys J850 (Stratasys Ltd., Minnesota, USA), a multi-material polyjet printer, is used to print the vessel phantom and the actuator [2]. Stratasys Agilus30 White, a soft, flexible 3D-printing material with a Shore hardness of 30A, is used to construct the vessel phantoms. The vessel phantoms measure 10 cm in length, and have a lumen diameter of 5 mm and a wall thickness of 0.5 mm [2]. The dimensions of the vessel phantom aim to emulate that of a superficial femoral artery [5], [6]. Stratasys

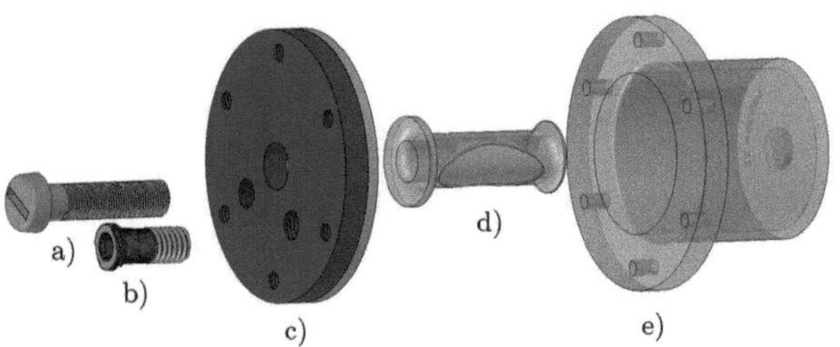

Figure 2: CAD rendering of the adaptive actuator show-ing the exploded view of the parts. **a)** M6 plastic bolt, **b)** Luer-M6 hollow adapter, **c)** lid with openings for a) and b), respectively, an opening for the vessel phantom, and six holes for M3 screws, **d)** inner actuator component, **e)** outer actuator component. M3 screws for the lid are not shown.

VeroClear, a firm 3D-printing material, is used to construct the outer actuator component. The CAD rendering of the outer component is shown in Fig. 1.1 without the lid, al-lowing the view of the water chamber and the nested inner component [2]. Fig. 1.2 shows the 3D-printed inner actua-tor component, printed with hard VeroClear material (clear, firm), featuring a flexible Agilus30 White patch (white, flexible). Though several variations are possible, the variant described and tested in the scope of this work has only one flexible patch that allows for a one-sided stenosis. Other possibilities include variants that have two patches, allow-ing for two oppositely positioned stenoses [2]. It is sig-nificant to note that several other stenosis geometries and degrees of stenosis severity can easily be created by alter-ing the patch and inner component design. To make con-nections to Luer systems, such as stopcocks and syringes, possible, a Luer adapter is printed using Formlabs Form 3B (Formlabs, Somerville, MA, USA) using Model V2 resin. The Luer to M6 adapter is screwed into the actuator lid, and allows pressure measurement in the chamber. The actuator is shown in Fig. 2 b). The lid c) in Fig. 2, is constructed with a mixture of different Stratasys Vero materials.

2.2 Mechanism

The design of this actuator was inspired by the pneumatic simulator by Maréchal et al. [8], though this actuator uses a hydraulic mechanism rather than a pneumatic one, mak-ing use of the incompressible nature of water. The explo-sion view of the CAD rendering of the adaptive actuator in Fig. 2 shows all parts of the actuator and helps visualize its function. The lid (c) is attached securely to the actuator (e) using plastic M3 bolts (not shown) wrapped in Teflon tape to insure watertightness. The Agilus30 White coating on the lid seals the chamber once the bolts are tightened. Wa-ter pressure can be raised using a bolt (a), after the chamber created between (d) and (e) is filled with room-temperature deionized water. For each turn of the bolt, a small amount of water is displaced in the chamber, causing the soft patch of the inner actuator component (d) to yield to the hydraulic pressure and be displaced inwards. This causes pressure on

the water-filled vessel phantom that is filled with water in-side the actuator, creating an artificial stenosis in the vessel. In Fig. 3, inside the actuator, the bolt to induce water pres-sure (II), as well as the sensor used for pressure measure-ment (III) are shown with the 3D-printed actuator.

2.3 Measurements and Testing

The overarching goal of this work is to develop a 3D-printed actuator that can be adjusted to simulate varying degrees of vascular stenosis, specifically in this vessel model of a su-perficial femoral artery. The following measurements were performed to determine if a) the results of this actuator are reproducible, i.e., does it yield the same degree of steno-sis each time if adjusted the same way, and b) what is the relationship between water pressure in the actuator cham-ber and the resulting diameter within the stenosis in the vessel. To determine reproducibility of the results, as well

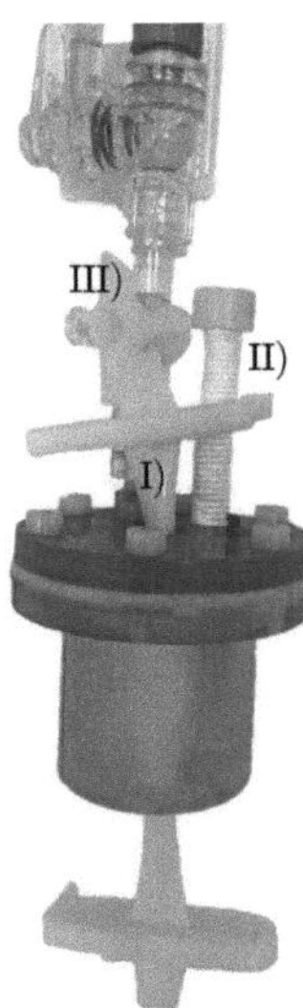

Figure 3: The entire experimental setup, showing the actua-tor and **I)** vessel phantom containing water, clipped on each end, **II)** M6 plastic bolt, **III)** Invasive blood pressure (IBP) sensor, connected via luer lock adapter.

as check the watertightness of the actuator, an intra-arterial blood pressure (IBP) monitor is used. The IBP monitor used is a Siemens Sirecust 1260 in combination with a Siemens Sirem Module (Siemens, Erlangen, Germany). To prepare for pressure measurement, the water chamber is filled with deionized water and all bubbles in the system are removed, the sensor is connected via Luer lock, and the chamber sealed by inserting an M6 plastic bolt wrapped in Teflon tape into the lid. The vessel phantom is filled with 0.75 ml of deionized room-temperature water, is inserted into the actuator, and is clamped with 3D printed clips, leaving 1 cm of vessel on each end. Fig. 3 shows the described assembly. To alter the chamber pressure, and thus achieve the stenosis of the vessel phantom, the bolt in the lid is turned. To test reproducibility, the water pressure is measured with the IBP sensor after each turn, waiting one minute before measuring for the pressure to stabilize. The pressure is noted and the bolt is turned again, repeating for a total of nine turns after an initial measurement before turning the bolt. The measuring process is repeated three times. This way, there are three comparable sets of pressure data that were recorded under the same conditions. If the variability of the data can be determined to be negligibly small, it can be assumed that the actuator produces reliable results.

In order to quantify the degree of stenosis, x-ray images are taken in an AngioSuite. The imaging system used is the Philips Azurion Clarity IQ (Philips, Netherlands). The vessel model is filled with 0.75 ml of a 1:1 ratio of Imeron 300 (Bracco, Buckinghamshire, UK) contrast agent and deionized water and placed inside the actuator. An x-ray is taken of the actuator at the same position after each turn at a tube voltage of 60 kV. This way, there is an image for each bolt turn, so the stenosis diameter in the vessel phantom can be measured. Fig. 4 shows an example of an image resulting after nine turns, the maximal degree of stenosis is thereby simulated. To determine the relationship between the num-

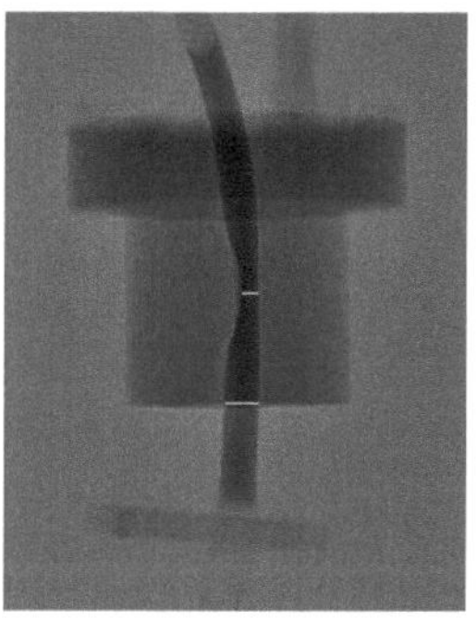

Figure 4: X-Ray image of the actuator and vessel phantom after nine bolt turns. The white lines mark the measurement lines for the prestenotic and stenotic diameters. Values are noted in columns two and three in Table 5, respectively.

ber of turns and the resulting stenosis, the stenoses are measured. The stenotic diameter is the lumen diameter at the most narrow point of stenosis. The prestenotic diameter is a reference value consistently taken at the same location using the Agfa IMPAX EE PACS client (Agfa-Gevaert N.V., Belgium). Measurements were performed manually by two separate observers and the results were averaged. The difference between between observers is within a tenth of a millimeter for each measurement, allowing averaging of the data without notable skewing. The stenotic and prestenotic diameter measurement lines can be seen in Fig. 4, which shows resulting x-ray image after nine bolt turns. Table 5 shows the averaged measured diameters.

3   Results and Discussion

The hydraulic actuator simulates varying degrees of vascular occlusion reliably and functions accurately: upon turning the bolt, the rising hydraulic pressure in the chamber causes a stenosis in the vessel phantom inside the actuator. As described above, the reproducibility of the actuator results is tested using an IBP monitor. The statistical

analysis of the variability of the data is determined with a repeated measures ANOVA, which compares the means of three or more matched groups – in this case, the three measurements. Using GraphPad Prism (Graphpad Software Inc., California, USA), a repeated measures ANOVA is calculated, yielding F(2,18) = 0.7552, p = 0.4842. A resulting $p > 0.05$ allows the assumption that no significant statistical difference of the means can be assumed between the three sets of data, and supports the reliability claim of the actuator's performance and supports the statement that the actuator is fully watertight. In Table 5, the averaged measured diameters of the stenotic and prestenotic parts of the vessel phantom are displayed, along with the percent of occlusion within the vessel. In order to find the relationship between the number of bolt turns and the diameter of the lumen in the stenosis at the point of maximal occlusion, a linear regression is performed.   As expected, the stenosis

Vessel Diameters			
Bolt Turns	Prestenotic ⌀ (mm)	Stenosis ⌀ (mm)	Percent Stenosis
0	6.7	6.4	4%
1	6.6	6.3	5%
2	6.6	6.2	6%
3	6.4	5.3	17%
4	6.3	4.5	29%
5	6.4	4.2	35%
6	6.3	3.8	40%
7	6.4	3.3	48%
8	6.3	3.0	53%
9	6.3	2.7	58%

Table 5: Vessel diameters and percent occlusion of the vessel phantom for each bolt turn after averaging the measurements of two trained users.

diameter falls linearly with the rising number of bolt turns. More specifically, the linear equation for the linear model correlating stenosis diameter (d_{sten}), and number of turns (b_{rot}), is d_{sten}= -0.4582b_{rot} + 6.592. The visually fitting linear regression model in Fig. 6 is supported by an R^2 = 0.9735, indicating that the model fits the data very well. In addition, an inspection of the residuals plot yields no unwanted pattern in residuals, meaning that the R^2 value can be trusted. This result is calculated using GraphPad Prism (Graphpad Software Inc., California, USA). The stenosis diameters from Fig. 6 are shown in Table 5, column three. As can be seen in Table 5, a variety of degrees of occlusion can be simulated in a single device, from very mild to severe stenosis. For example, four turns deliver a mild stenosis of 29%, six turns deliver 40%, and nine turns produce an occlusion of 58%.

Several applications are possible for an adaptive stenosis actuator: Vaalma et al. describe Magnetic Particle Imaging (MPI) as very accurate tool for stenosis quantification [2], [9], [10]. Furthermore, MPI can quantify the lumina of endovascular stents without the influence of material induced artifacts [7]. A hydraulic actuator, such as the one described in this work, would be very beneficial for a wide range of

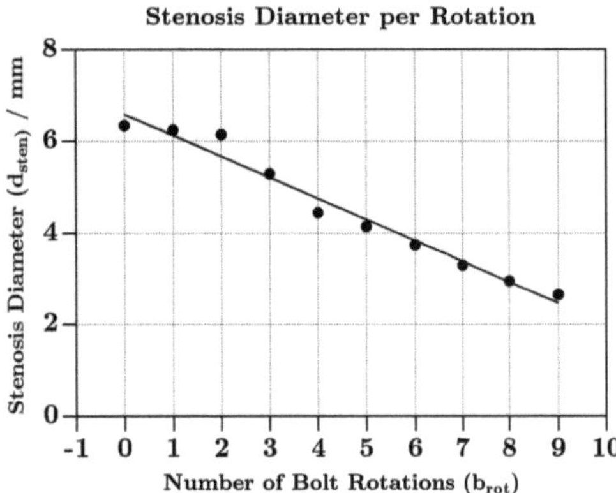

Figure 6: This graph displays the linear fit performed on the plotted stenosis diameter (mm) within the vessel phantom, in relation to the number of bolt rotations of the bolt in the actuator.

in vitro experiments regarding cardiovascular applications of MPI, as well as studies using MRI or CT [2]. The actuator poses several benefits: small size, simple construction, adjustability, and lack of metal artifacts, characteristics that are useful in a variety of studies and applications. Due to the progressive characteristic of vascular stenoses, the actuator offers the possibility for instrument testing, as well as hemodynamic studies regarding different degrees of stenoses [2]. Additive manufacturing allows for quick alteration of such an actuator – different vessel sizes and various water chamber dimensions, for example, could quickly be achieved [2]. In addition, the results of this work serve as a proof of concept that the Stratasys J850 PolyJet printer is capable of producing membrane-like functional structures in otherwise firm printed components for either pneumatic or hydraulic applications, or other mechanisms entirely.

4   Conclusion

An adaptive 3D-printed actuator using a hydraulic mechanism to simulate different degrees of stenosis can be used in a wide variety of clinical research applications, with the benefit of omitting metal artifacts in medical images and avoiding animal testing [2]. 3D-printed constructs allow realistic simulation of several vascular pathologies, have the benefits of swift construction, design freedom, and a wide variety of materials, which make adaptive mechanisms like the one shown here possible. The integration of an elastic 3D-printed material, used here in both vessel phantom and actuator patch, paves the way for simulation studies with highly adjustable models and many applications in medical imaging.

Acknowledgements

The work has been carried out at the Institute of Medical Engineering, University of Lübeck, in cooperation with the Fraunhofer Research Institution for Individualized and Cell-Based Medical Engineering IMTE, and was supervised by Dr. Thomas Friedrich.

Author's Statement

Conflict of interest: This work is closely related to a preliminary work published previously in Transactions on Additive Manufacturing Meets Medicine (Vol. 3, No. 1) titled "Stenosis simulation of femoral arteries using an adaptive 3D-printed actuator" by Dell et al.

5   References

[1] Fowkes et al., *Comparison of Global Estimates of Prevalence and Risk Factors for Peripheral Artery Disease in 2000 and 2010: a Systematic Review and Analysis*. Lancet, vol. 382, no. 9901, pp. 1329–1340, 2013.

[2] Dell et al., *Stenosis Simulation of Femoral Arteries Using an Adaptive 3D-Printed Actuator* Transactions on Additive Manufacturing Meets Medicine, vol. 3, no. 1, 2021.

[3] R. Walden, R. Adar, Z. J. Rubinstein and A. Bass., *Distribution and Symmetry of Arteriosclerotic Lesions of the Lower Extremities: an Arteriographic Study of 200 Limbs*. Cardiovascular and Interventional Radiology, vol. 8, no. 4, pp. 180–182, 1985.

[4] Th. Friedrich, F. Wegner, T. M. Buzug *3D-printing of Elastic Stenosis Phantoms*. Transactions on Additive Manufacturing Meets Medicine, vol. 2, no. 1, 2020.

[5] W. Schäberle, *Ultrasound in Vascular Disagnostics - Ultraschall in der Gefäßdiagnostik*. Springer-Verlag Heidelberg, 3. Edition, 2010

[6] Temelkova-Kurktschiev et al. *Intima-Media-Dicke bei Gesunden ohne Risikofaktoren für Arteriosklerose*. Deutsche Medizinische Wochenschrift, vol. 126, no. 8, pp. 193-197, 2001.

[7] W. Schäberle, *Sonographic Classification of Carotid Stenoses*. Gefässchirurgie, vol. 25, pp. 91–104, 2020.

[8] Maréchal et al., *Modelling of Anal Sphincter Tone based on Pneumatic and Cabledriven Mechanisms*, 2017 IEEE World Haptics Conference (WHC), IEEE, Munich, Germany, 2017.

[9] Wegner et al., *Magnetic Particle Imaging: In Vitro Signal Analysis and Lumen Quantification of 21 Endovascular Stents*. International Journal of Nanomedicine, vol. 16, pp. 213–221, 2021.

[10] Vaalma et al., *Magnetic Particle Imaging (MPI): Experimental Quantification of Vascular Stenosis Using Stationary Stenosis Phantoms* PLOS ONE vol. 12, no. 1, 2017.

Development of a prototype for a material pumping system for additive manufacturing with biological materials

Sophie Westmeyer [1] and Andre Behrends [2]

[1] Medical Engineering Science, Universität zu Lübeck, sophie.westmeyer@student.uni-luebeck.de

[2] Fraunhofer Research Institution for Individualized and Cell-Based Medical Engineering, Lübeck, andre.behrends@imte.fraunhofer.de

Abstract

The development of micropumps with different geometries and operating mechanisms that change for different applications has become an important area of research. Approaches to geometric structures that exhibit diodicity were investigated in a material conveying system for bioprinting. First, research was carried out on systems that had already been analysed. From the information gained, new concepts were designed, and prototypes were manufactured from a mixture of different structures to further improve the functionality. By testing these concepts, it was found that a parallel connection of the system with built-in structures can increase the performance in terms of flowrate by 64 %. The concept of a series connection and the one with a diffuser-nozzle integration, in contrast, decreases the performance. The other concepts in this work also show no functional improvements, but the principles could be improved in further research, as they probably still have potential.

1 Introduction

The aim is to achieve an effective and controlled material transport through fluidic passages to the printhead of a modified FDM printer of cell suspension to be used as material for bioprinting. To fulfill this effect, the geometry of the material transport system has the task of providing a preferential flow direction. That is, a forward motion that is subjected to little to ideally no resistance in order to achieve the lowest possible energy loss during reverse flow. Passive check valves, which open in only one direction due to pressure from the fluid and close automatically in response to back pressure, are usually used for this implementation. To avoid the mechanical influence of these check valves that cause cell damage, no moving parts are used. Flow obstacles, such as those present in the Tesla valve, create laminar flow in the preferred direction and turbulence in the opposite direction.[1] Diodicity D_i, the ratio of the volume flow in the pass direction to the volume flow in the blocking direction, expresses the efficiency of the valve, and can be calculated as follows; the volume flow in passing direction $\dot{V}_d$ divided by the volume flow in blocking direction $\dot{V}_s$ [2]:

$$D_i = \frac{\dot{V}_d}{\dot{V}_s} \qquad (1)$$

Today there are many variants of micropumps for different applications. Centrifugal pumps (flow pumps) and peristaltic-, rotary lobe- and diaphragm pumps (positive displacement pumps) are pumps that make the liquid flow through mechanically moving components.[3] Looking at the requirements for the application studied here, the diaphragm pump principle is the most promising. Unlike centrifugal and rotary lobe pumps, there are no moving individual parts that come into contact with the pumped medium, which reduces the risk of clogging cells in the pumped medium. The pump chamber is absolutely sealed by the diaphragm to protect the drive, so this pump is suitable for pumping harmless and dangerous suspensions. In addition, the diaphragm pump cannot cause wear or incrustation of the components. They are insensitive to continuous loads and easy to maintain. [3] [5] Cell damage would be difficult to avoid by crushing the cells when the peristaltic pump tubing is squeezed, or by the rapidly rotating propeller blades of the centrifugal pump. However, there has been research on a medium pressure ring pump that works similarly to a peristaltic pump, which was used for the same application and caused little cell damage.[7] In addition, the diaphragm pump as a microtransport system has already been researched in several scientific publications and proved to be a functional micropump with a high volume flow in the direction of passage that comes very close to the requirements.[8] [9] [10] [11]

2 Material and Methods

The pumping concepts elaborated and tested in this work consist of a piezoelectric drive and several structural elements, which have already been discussed separately in some scientific publications and led to positive results.[6] [9] [11] The prototypes were modeled using the AutoCAD program and then 3D printed with Stratasys J850. The material of the prototypes is a Photopolymer (VeroClear).

2.1   Diaphragm Pump

The diaphragm pump operates as an oscillating displacement pump, with the diaphragm completely sealing the pumping chamber, thus keeping the drive protected. Due to the low and controllable pressure, the fluids are handled gently [4]. The inlet valve is opened due to the pressure difference in the suction phase, allowing the fluid to be sucked into the larger pumping chamber. During the reversal point, the maximum pumping chamber volume, no fluid delivery takes place. When the diaphragm is moved downward, the fluid flows against the inlet valve, which is closed as a result. At the same time, the outlet valve is opened due to the pressure difference between the pumping chamber and the discharge chamber. The fluid is compressed by the reduction in volume and expelled. The pump is driven piezoelectrically or electromagnetically. [8]

2.2   Structural Design

Figures 1 and 2 illustrate the cross-sections of the different concepts as exploded views and their dimension. The individual structural elements can be seen, which provide a better volume flow in the direction of flow. All cross-sections show a pump body with crescent-shaped structures. In 2019, it was found by Da Zhao et al. [9] that the use of a quadruple crescent-shaped structure with $\delta = 0.6\,\mathrm{mm}$ (see Fig.1) produces an improved flow rate because high diodicty develops through the structures. The dimensions of the structures and pump bodies are mostly taken from the publication and replicated (Fig. 1). This principle represents concept 1 here.

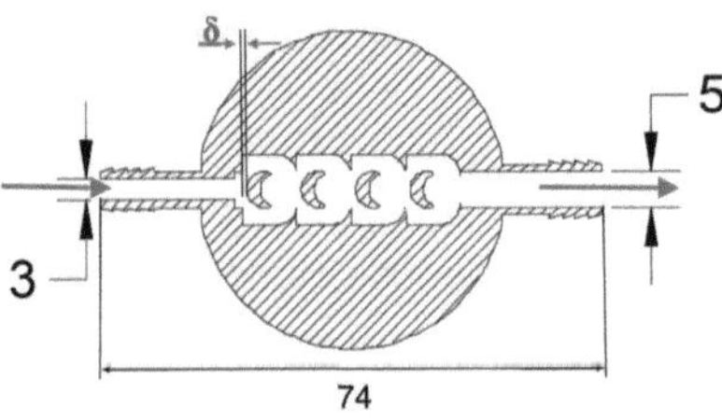

Figure 1: Concept 1: Pump body with crescent shaped elements as Built-In structures, its entrance on the left (diameter: 3 mm) and outlet (diameter: 5 mm) on the right and δ as the gap from the intermediate crescent to the inlet of the crescent-shaped structure, flow direction indicated by arrows.

In order to test whether the function of the pump improves even further, additional elements are also used which, among other things, take over the function of a weak valve. The elements are nozzle-diffusor elements and elastic fins integrated at a certain angle in front of and behind of the structures. The following concepts are also being investigated: a parallel connection, which has a Y-piece to allow the tubes and thus the fluid to merge again after parallel

pumping (Fig. 2 d)), a series connection (Fig. 2 e)) and the concept shown in Figure 2 c). The piezo diaphragm is located directly above the structures in all concepts except concept 4, where two synchronously running piezo actuators are directly opposite of each other and the structures are implemented in front of and behind the pumping chamber.

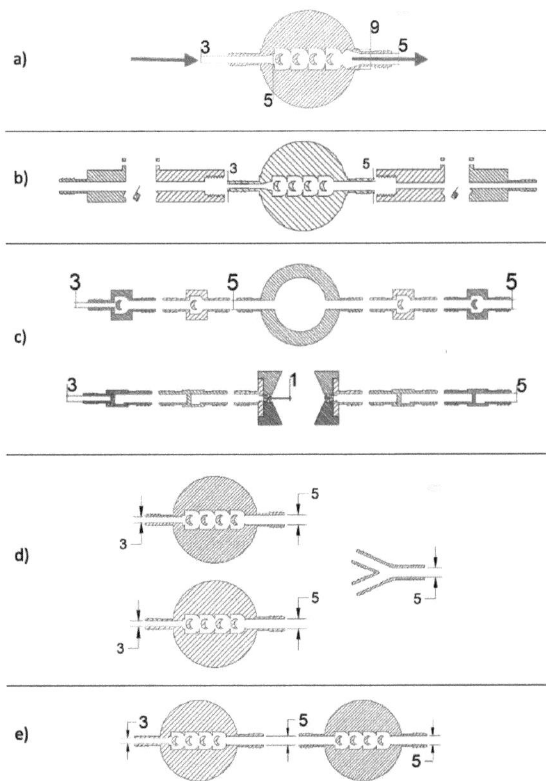

Figure 2: Cross-sections of concepts 2-6 (a-e) with flow direction from left to right, from inlet (3 mm) to outlet (5 mm).
a) Concept 2: crescent-shaped elements and nozzle-diffusor elements at the inlet and outlet, flow direction indicated by arrows
b) Concept 3: crescent-shaped elements, elastic valves in additional tubes respectively in front of and behind the structures
c) Concept 4: opposite piezo diaphragms between four crescent-shaped elements
d) Concept 5: Parallel connection of two pump bodies with crescent-shaped elements with inlet and outlet respectively left and right of the structures
e) Concept 6: series connection of two pump bodies with crescent-shaped elements

2.3   Measurements

Two preparation tests are performed at the beginning. In these preparation tests, the water cups are below the pump level. In test A, there is still no water in the pump system, only the hose is in the water. Now the frequency is changed during the pumping process and the process in the system is observed. In test B the entire system is filled with water before the pump is switched on.The experimental setup is further adapted and test C is performed, which is explained in Section 3. Accordingly, in test C the water cups are finally placed at the height of the pump so that the water level is above the system. This test setup (Fig. 3) is used for further tests.

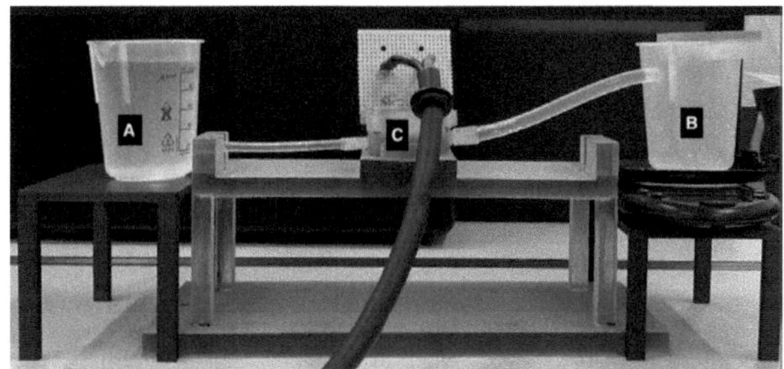

Figure 3: Test setup C, where the water level is above the system, including two water cups as inlet (A) and outlet reservoir (B) connected with tubes to the pump body (C), voltage connection via a circuit board.

First, concept 1 (Fig. 1) with the Built-In elements, based on [9], is tested again with the conditions present here in order to clarify the sole effect of the structure once again and to be able to compare it with the other concepts. All tests are performed with a voltage of 180 volts, the highest voltage that can be set on the available device. Through the data from [9], the information is taken that a high voltage performs well. The frequency is varied and the resulting volume flows are compared at the end (Fig.4). A scale and a stopwatch are used to calculate the volume flow. The prototypes of the different concepts are tested, evaluated and compared one after the other. A total of three repetitions of the 15 - 55 Hz measurement series are performed in steps of five with a square wave function. Additional care must be taken to ensure that the system is vented using a syringe before starting a test. The mean of the three measurement repetitions is then calculated and plotted.

3   Results and Discussion

In test A and B, where the water level is below the pump level, no positive results are obtained. Test A shows that the force of the pump at the optimum frequency here is only able to transport the water upwards along the hose by a few millimeters. This could be due to the increasing weight of the water in the hose. After some time a critical weight is reached and the water level does not change anymore, but oscillates at one point at the same frequency. Test B already shows a very slight pumping action in the desired direction. However, after a short time or changing the here optimal frequency, the pumping process is carried out in the other direction and the pump empties again with water. Here a sufficiently large force of the pump is missing, in order to come against gravitational forces. Test C shows positive results. Shortly after the pump is switched on, the outflow

of water droplets is observed. Here, the force of the pump does not have to work against gravity. There is equilibrium at the inlet and outlet. The water level in the inlet cup drops and the liquid is transported to the collecting cup. From this point on, a change in weight can be seen on the display of the scale under the collection cup. The different volume flow rates (ml/min.) that are shown at different set frequencies are summarized by Figure 4.

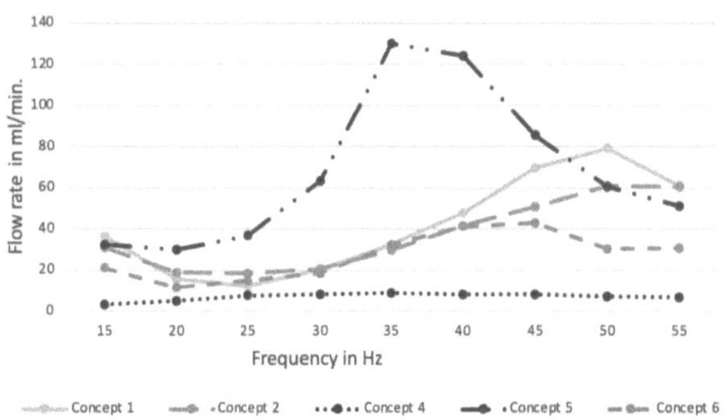

Figure 4: Comparison of the performance of concept 1 (built-ins only), 2 (built-ins+nozzle-diffusor elements), 4 (opposite piezos), 5 (concept 1 parallel) and 6 (concept 1 in series) by the curves shown in relation to the maximum volume flows and frequencies.

No results of concept 3, which integrates the elastic fins, could emerge because the construction turned out to be leaky and no improvement could be worked out due to time constraints. A strategy could be worked out here in further research. Concept 2 is similar to concept 1 in terms of the curve. An improvement of the volume flow can be seen at low frequencies 20 Hz, 25 Hz and 30 Hz, otherwise the optimum values are lower than those of concept 1, the volume flows have deteriorated. Presumably, the structures are somewhat more usable for this frequency range. The lower volume flow can probably be explained by the different geometry at the inlet and outlet. This suggests that the abrupt change in the flow path, as in Concept 1, rather than the gradual change by nozzle-diffuser elements, as here, results in better performance. If the curve progression of concept 4 is considered, significantly lower volumetric flow values for all frequencies can be seen. Here, there is only a maximum of 8.8 g/min. at 35 Hz. Since two piezo diaphragms operate synchronously, it would be expected that the system would exert more pressure and thus transport more volume at once, resulting in a larger volume flow. With this design, it is not possible to tell if there are still air bubbles in the pumping chamber, which would explain the much lower output. A second assumption is that the gap at the inlet and outlet is too small and the power of the pump is not sufficient to pump as much volume per time unit as possible. An improvement of the system could be created in further investigations. In the test with concept 5, the first pass was aborted because the water volume in the reservoir was not sufficient. Due to the transport speed created here, the complete amount of water was pumped into the collection cup before the end of the time (1 min.). In order not to have to change anything in the setup, the time was reduced to 30 seconds and the mean values were doubled. The curve visible here has its global maximum 130.1 ml/min at 35 Hz. If this value is compared with the highest value of concept 1, it shows that the system has resulted in a 64.3 % increase in performance. To obtain an accurate result, the test should be repeated again with sufficient liquid and beaker volume. However, an increase in performance due to concept 5 can be clearly seen. The values of the pump chambers connected in series (concept 6) also show no visible increase in performance compared to concept 1.

When switching from a square wave function to a sine wave function, a significant change in flow velocity can be observed at the same frequencies. In some cases, the liquid comes out of the output hose only drop by drop. Also, the liquid oscillates in a much smaller range, which does not cause such high turbulence. The transport would be gentler, but much slower. This behavior is generally also true at higher frequencies, in the range of 100 Hz, and when the square wave function is set. Due to time constraints, this phenomenon was only observed and not documented.

4 Conclusion

In this work, several concepts for valveless material transport of cell suspensions for bioprinting were developed. Because of the absence of valves, attention was paid to high diodicity. Designs, some from previously published scientific papers, were drawn upon and merged. The resulting concepts were then prototyped, tested and evaluated. Volumetric flow rate was calculated for output power comparison. A piezoelectric diaphragm was used as the actuator. To increase the diodicity, elements that had shown a positive effect on the output power in previous work were used. The elements used were built-in structures, diffuser-nozzle elements, and elastic fins. Finally, the documented volume flow rates of the different concepts were compared. In summary, it is clear from the tests that the optimal frequency to be set in order to achieve a high volume flow rate is highly dependent on the system geometry. Furthermore, the performance of concept 1 can be significantly improved by a parallel circuit, but with different frequencies. The other concepts did not achieve better values, for example the series connection (concept 6). However, concept 4 can be improved in further research, since potential for improvement can be assumed for this system, by increasing the inlet and outlet diameter sufficiently. Furthermore, it would be recommended to test the design of concept 3 again in a sealed condition and observe if the diodicity and performance increases. It should also be noted that in the tests conducted here, there was a voltage limit from the device and thus further research could include increasing the voltage values that drive the piezo diaphragm.

Acknowledgement

The work has been carried out at Fraunhofer Research Institution for Individualized and Cell-Based Medical Engineering (IMTE), Lübeck and supervised by the Institute of Medical Engineering (IMT), Universität zu Lübeck.

Author's Statement

Conflict of interest: Authors state no conflict of interest.

5 References

[1] B.C. Bobusch (2015), *"Fluidic Devices for Realizing the Shockless Explosion Combustion Process"*. (Dissertation), Technischen Universität Berlin.

[2] I. Klammer (2010), *"Bionische Mikrofluidventile aus Polydimethylsiloxan nach dem Vorbild der Venenklappen"* (Dissertation), RWTH Aachen.

[3] Die Chemie Schule, *Vakuumpumpe.* Available: `https://www.chemie-schule.de/ KnowHow/Pump` [last accessed on 2022-01-12].

[4] Techniker Schule Butzbach, *Membranpumpen.* Available: `https://projektwiki.zum. de/wiki/Techniker_Schule_Butzbach/ Verfahrenstechnik/Fördertechnik/ Membranpumpen,` [last accessed on 2022-01-12].

[5] Star Pump Alliance, *Kolbenmembranpumpe*, Available:`https://de.starpumpalliance.com/ kolbenmembranpumpe` [last accessed on 2022-01-12].

[6] A. Aboubakri, V. Ebrahimpour Ahmadi and A. Kosar, *"Modeling of a Passive-Valve Piezoelectric Micro-Pump: A Parametric Study"*. Micromachines, 2020, 11, 752.

[7] K. Uesugi et al. , *"Survival Rate of Cells Sent by a Low Mechanical Load Tube Pump: The "Ring Pump""*. Micromachines, 2020, 11, 447.

[8] A., Bußmann et al. *"Microfluidic Cell Transport with Piezoelectric Micro Diaphragm Pumps"*. Micromachines, 2021, 12, 1459.

[9] D. Zhao, L. He, Wei Li, Y. Huang and G. Cheng., *"Experimental analysis of a valveless piezoelectric micropump with crescent-shaped structure"*. J. Micromech. Microeng. 29 (2019) 105004 (9pp).

[10] C. Yamahata et al., *"Pumping of mammalian cells with a nozzle-diffuser micropump"*. Lab Chip, 2005, 5, 1083–1088.

[11] Q. Yan, Y. Yin, W. Sun and J. Fu. *"Advances in Valveless Piezoelectric Pumps"*. Appl. Sci. 2021, 11, 7061.

Finite Element Analysis to determine Mechanical Properties of Materials from Bending Test

Muhammed Musa [1], Robert Wendlandt [2],
[1] Biomedical Engineering, Luebeck University of Applied Sciences, muhammed.musa@stud.th-luebeck.de
[2] Biomechanics Laboratory, Clinic for Orthopedic and Trauma Surgery, University Medical Center, robert.wendlandt@th-luebeck.de

Abstract

The utilization of finite element analysis in biomechanical engineering is on the rise, with the highly efficient computational capabilities allowing for more complex models to be generated and analyzed. In order for fracture plate implants to comply with all regulatory requirements and obtain a CE-Certificate, a destructive material testing such as a mechanical four-point bending test as described by the ASTM F382 is required. Material properties can then be determined from these tests. In orthopedic surgeries, fracture plates are implanted to link broken bones together, however cases of detachment or failure under heavy loads occurs. It is therefore important to carry out simulations to determine properties such as load carrying capacity and stability of fracture plates implants. Five different materials were used in this study to determine which one will perform better under certain conditions, this report aims to create a standard testing scenario for the fracture plate implants.

1 Introduction

Osteosynthesis is an operative procedure for the treatment of bone fracture. It is applied to join bone fragments in the anatomical shape, while providing mechanical stability and ultimately to reinstate the function of the bone. Different osteosynthesis techniques are available in the following forms such as screw fixation, plate fixation, wire fixation, external fixation, dynamic hip screw and intramedullary nail osteosynthesis. In a four-point bending test, a piece of material is put through uni-axial stress, namely tension and compression. The aim is to determine properties such as flexural strength, deflection and modulus of elasticity. Aside from experimental testing, finite element analysis (FEA) has evolved to be a powerful tool in the analysis of stresses and strains within structures during static and dynamic load situations. It also offers detailed information, which cannot be determined with experimental methods [1]. FEA has the capability to analyze the impact of various parameters on implant components during pre-clinical testing without prototype production [1]. To conduct FEA-simulations, the stress-strain characteristics of the materials is required. It is usually available in tabular form, after tensile testing of the material batch has been conducted. In 1943 an equation was first proposed by Ramberg and Osgood [2], which analytically describes the stress-strain relationships in materials. This mostly applies to metals that harden with plastic deformation. The Ramberg-Osgood equation usually describes the non-linear relation of the one-dimensional elastoplastic behavior of materials [3]. In 2014 Kato et al [4] demonstrated a method to determine

both uni-axial tension and compression stress-strain curves from the result of a single four-point bending test as shown in figure 1.

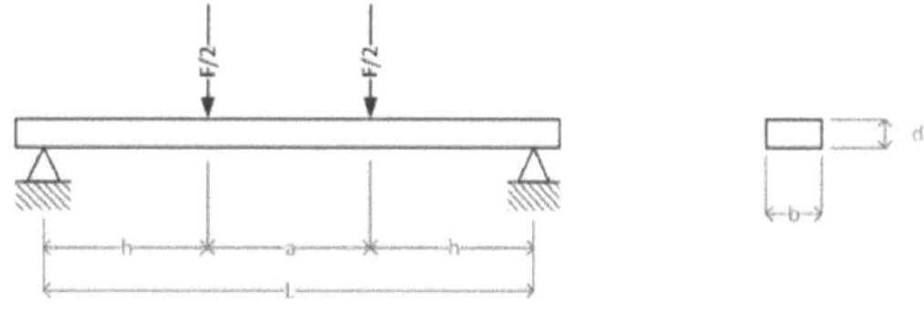

Figure 1: Description of a four-point bending test.

In this paper, the approach of Ramberg and Osgood and Kato is presented and pursued. The aim is to calculate characteristic material values from a bending test, which would conventionally be taken from a tensile test, and to set up a material model. First a model is created in the FEA-simulation program Ansys (Student Version 2019 R1) to simulate a four-point bending test resulting in force-deformation data with known materials. The estimated material models are improved iteratively and in the end, a material model of the respective material batch should be available, which allows to predict the material behavior in future simulations.

2 Material and Methods

Material data were got from the experimental data overview of the engineering virtual organization for cyber design [5].

These data are from the physical measurements of the materials. Five materials were used for this study. They are FC0205 – Iron Carbon Copper Steel, 6061651 Aluminum, TI6AL4V - Titanium alloy, TI6AL6V2SN – Titanium alloy and Ti8al1mo1v – Titanium alloy.

These material data were prepared by carrying out the following procedures using the software R (Version 4.0.3). Sparse data point interpolation and estimating the elastic region using a kernel mean shift clustering procedure [6]. Furthermore, the data are idealized, with an extrapolation to let the data start at 0 strain and 0 MPa stress. In a further step, the yield point is determined in the idealized data using a criterion of 0.2 % plastic deformation. A Ramberg-Osgood fit (RO-fit) was carried out for the idealized data, and an Ansys Model was created from the RO-fit. The created Ansys model was entered in Multi-linear kinetic/isotropic hardening.

Ansys was used for the design and simulation of a four-point bending test of the osteosynthesis plate. The space claim feature of Ansys was used to design the plate of dimensions 150 x 15 x 3 mm as shown in figure 2.

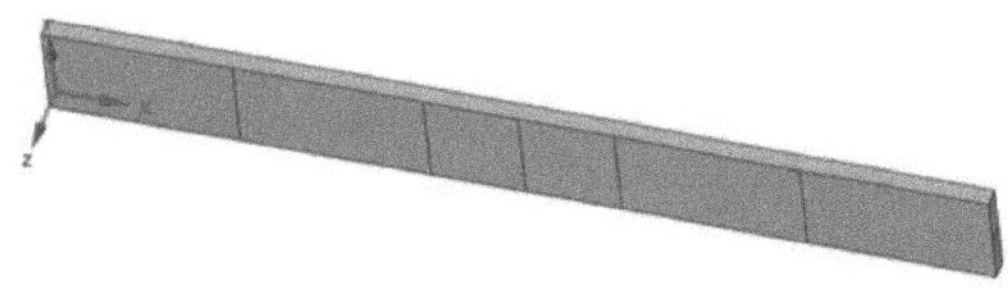

Figure 2: Model of the osteosynthesis plate (150 x 15 x 3 mm) designed by Ansys.

The points of application of the force are set at 50 mm in x-direction (25 mm in each direction from the center) as shown by point B in figure3 and the supports are 50 mm further away as shown in point A in figure 3.

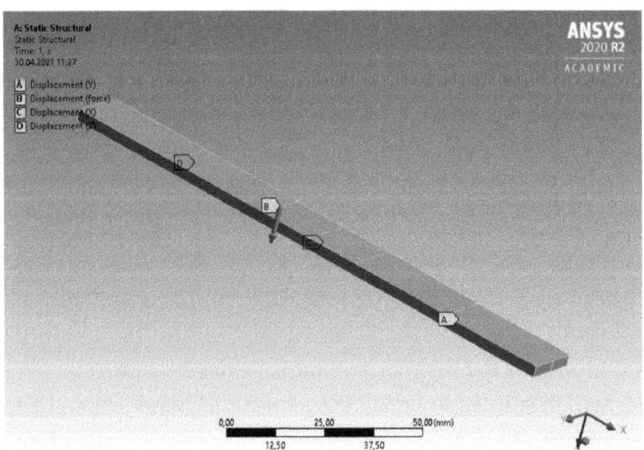

Figure 3: Model showing boundary conditions.

After the simulation with Ansys, the force displacement characteristics were evaluated. The deformation at the point of force application, the deformation at the center and the

reaction at the supports are evaluated using R studio.

The force-displacement data are converted to the stress-strain domain and interpolated to data of 1000 data points using generalized additive models with integrated smoothness estimation. The strain is estimated using the measured deformation at the points of force-application, the displacement at the center and the difference between both displacements (further denoted as delta displacement) by means of a circle fit algorithm. Furthermore, the elastic region was estimated, this was done to each of the three displace data sets. In the next step, the data sets are idealized by an extrapolation to allow the data start from 0 MPa stress and 0 strain, this data is corrected by Kato's equation with a yield point offset of 0.2 %. The yield point and high strain behaviors was then obtained and further analyzed. To further understand the yield point and strain behaviors of our materials, another simulation was carried out on Ansys with estimations from the first results. A sensitivity analysis of the yield point using different thresholds – 0.1 %, 0.01 %, 0.05 % respectively, enabling the comparison of the yield point behaviors under different conditions. Consequently, a Ramberg-Osgood fit was used to create new Ansys models in order to carry out iterative simulations, these are simulations of the new Ansys models scaled with different factors (0.5, 1, 2, 3, 4). The results from this iterations were evaluated and analyzed further.

3   Results and Discussion

The whole study was set up to determine a suitable simulation tool for the determination of material properties. Five material were used in this project. After the simulation on Ansys, the deformation at the point of force application, the deformation at the center and the reaction at the supports are exported. These exported data are evaluated using R studio. In R studio, the strain is estimated using the measured deformation at the points of force-application, the displacement at the center and by the difference between both displacements using a circle fit algorithm as shown in figure 4.

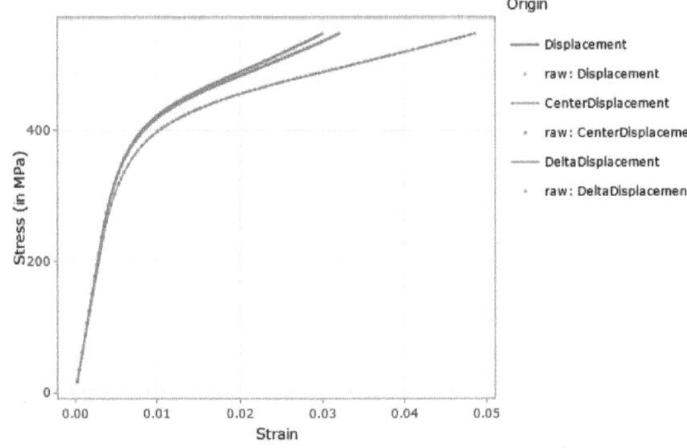

Figure 4: Raw displacement data from the simulated bending test (6061t65al).

The elastic region is estimated by means of linear regression. The elastic region is estimated for the displacement at the point of force application, the displacement at the center and the displace difference called delta displacement as shown in figure 5.

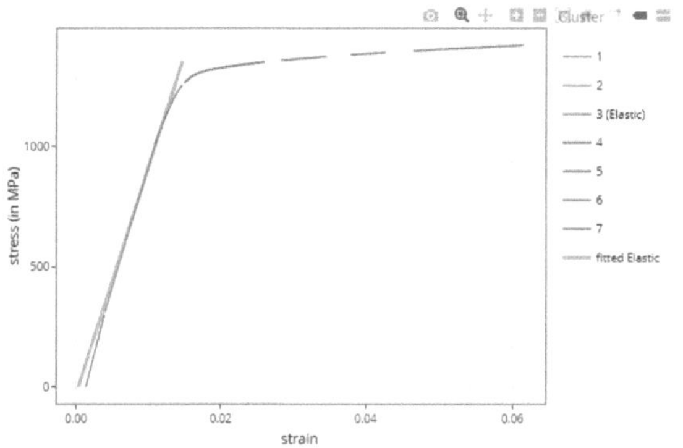

Figure 5: Estmating elastic region by linear regression (6061t65al).

After estimating the elastic region, the data are idealized with an extrapolation to let the data start at 0 strain and 0 MPa stress as shown in figure 6.

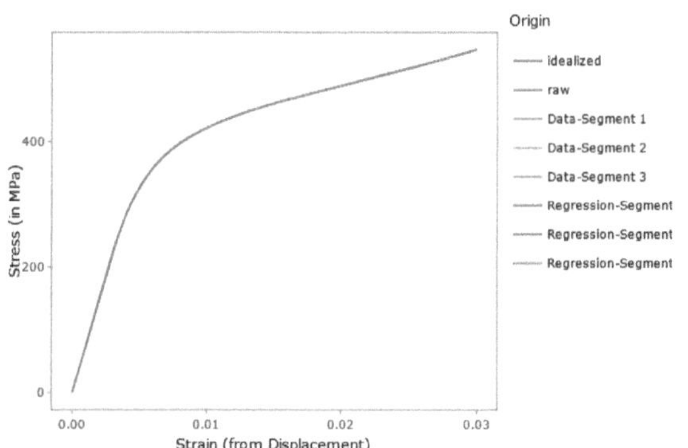

Figure 6: idealized displacement data.

The data are corrected according to Kato's equation. During a four-point bending test, pure bending occurs between the two points of the force application and there is a constant bending moment subjected. In the elastic region, stress and strain can be calculated from the linear equations of beam theory. With the onset of plastic deformation, the distribution of the stress in the beam is becoming non-linear as the out-most part of the beam (furthest from the neutral axis) is already subjected to plastic deformation while the inner parts are still in the elastic domain. To compensate for this effect, the relationship between stress and strain from Kato [4] is additionally taking the slope of the load-strain curve into account. The effect of this compensation is visible in figure 7, the stress-strain curves („naive" from the linear beam equations and corrected with the compensation for non-linear effects) are equal in the elastic range and start to differ with the onset of plastic deformation showing the expected outcome of smaller stress after the compensation. This process was done according to the yield-point threshold of 0.2 %, (see figure 7), the difference between the yield

point for the naive stress and the yield point for the corrected stress can easily be deduced. This process was then repeated for the other three yield point thresholds – 0.01 %, 0.05 % and 0.1 % respectively.

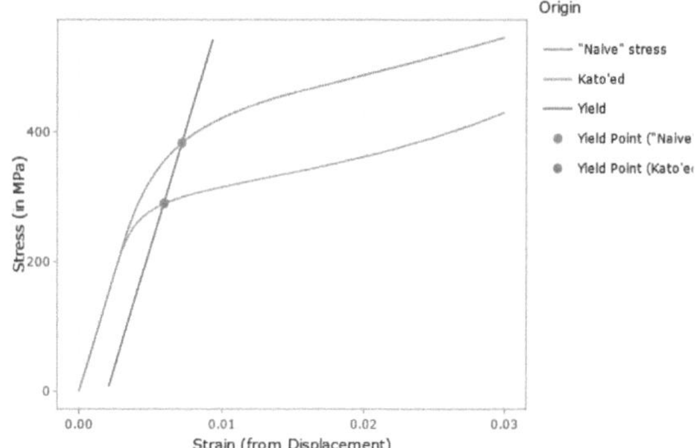

Figure 7: corrected data with Kato's equation.

The yield points from this threshold were compared and the relative error calculated. The thresholds of 0.1 % and 0.2 % shows the least relative error for all five materials. These two yield point thresholds were used further in the study.

At this point material models were then created with different factors and imported back into Ansys for further simulations. figure 8 and figure 9 show the new material models created.

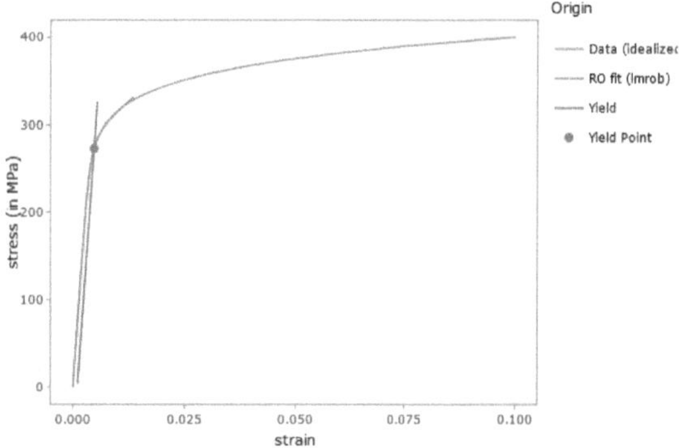

Figure 8: RO- fit of material data.

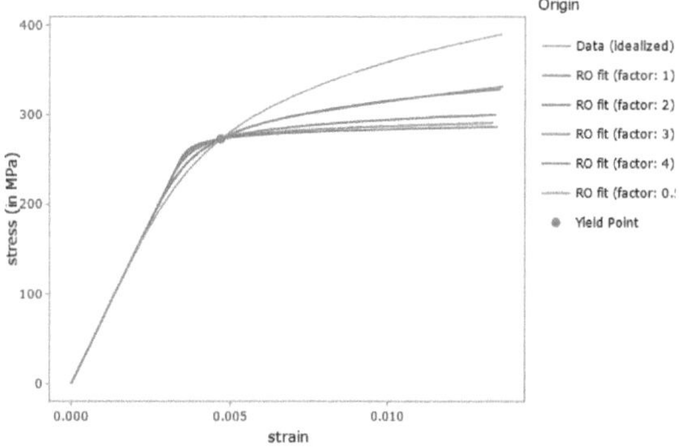

Figure 9: Scaled RO- fit of material data.

Following the new simulated bending test with the sets of 5

different material models, the sparse data points are interpolated. Finally the difference integral between the iterations and measured data are estimated as shown in the figure 10.

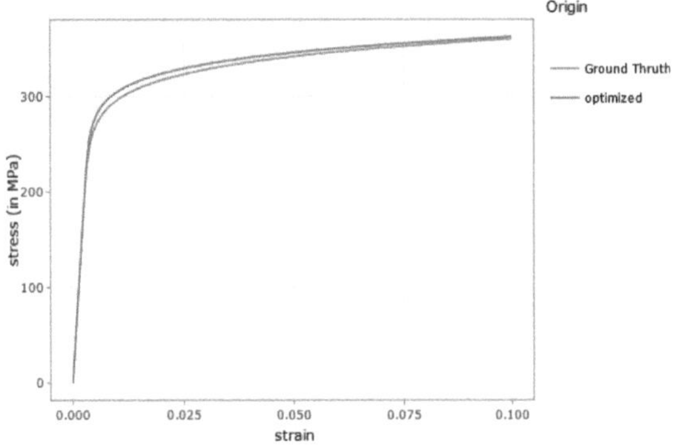

Figure 10: difference between iterations and measured data.

4 Conclusion

This research gives a rundown for the derivation of material properties for the numerical testing of osteosynthesis plates. To start with, osteosynthesis plates were considered before the tensile and bending test for the determination of material properties. The procedure for reconstructing the stress-strain curve of a rectangular sample typical for the material was explained and the results evaluated. It can be noted that the combination of simulation, Kato equation and Ramberg-Osgood equation provides a good approximation of the stress-strain curve. The simulation model in Ansys simulates the real test setup adaptable, if a different material or a different specimen is used. For example, instead of the simulated rectangular body made of solid material, a more realistic plate with drilled holes could be tested or the plate is screwed onto bone-like material. Furthermore, after correction of the force-deformation curve, the Kato equation can make a first approximation of the yield strength, which otherwise can only be seen in the stress-strain diagram of a tensile test and not from a bending test due to non-linear strain distributions in the material. Furthermore, the calculation of the strain from the deformation leads to a wrong assumption in the plastic range, but for the calculation of the yield strength this is to a certain extent irrelevant. With the help of the Ramberg-Osgood equation a complete approximation of the stress-strain curve typical for the material is achieved. This has been validated by the addition of reference values from a tensile test and provides a plausible material model for the considered material after the optimization process.

Acknowledgement

This work has been carried out as Biomechanics Laboratory, Clinic for Orthopedic and Trauma Surgery, university Medical Center, University of Lübeck

Author's Statement

Authors state no conflict of interest. Informed consent.

5 References

[1] Wieding, J., Souffrant, R., Fritsche, A., Mittelmeier, W. and Bader, R., *Finite element analysis of osteosynthesis screw fixation in the bone stock: an appropriate method for automatic screw modelling*. PloS one, 7(3), p.e33776. 2012.

[2] RAMBERG, W., OSGOOD, W. R., *Description of Stress-Strain Curves by Three Parameters*, National Advisory Committee on Aeronautics, Technical Note, 902, 1943.

[3] Sireteanu, T., Mitu, A.M., Giuclea, M., Solomon, O. and Stefanov, D., 2014. *Analytical method for fitting the Ramberg-Osgood model to given hysteresis loops*. Proceedings of the Romanian Academy-A, 15(1), pp.35-42.

[4] Kato, H., Tottori, Y. and Sasaki, K., 2014. *Four-point bending test of determining stress-strain curves asymmetric between tension and compression*. Experimental Mechanics, 54(3), pp.489-492.

[5] EVOCD, Engineering Virtual Organization for Cyber Design https://icme.hpc.msstate.edu 03 Dec. 2020.

[6] Chacon, José E., and Tarn Duong. *Multivariate kernel smoothing and its applications*. CRC Press, 2018.

10

Biomedical Optics

Development and Evaluation of an Arduino based EOM Regulation

Stefan Meyer [1], Tonio Kutscher [3], Philipp Lamminger [3], Moritz Wiggert [2] and Sebastian Karpf [3]

[1] Medical Engineering Science, Universität zu Lübeck, stefan.meyer@student.uni-luebeck.de

[2] Biophysics, Universität zu Lübeck, moritz.wiggert@student.uni-luebeck.de

[3] Institute of Biomedical Optics, Universität zu Lübeck, (t.kutscher, phi.lamminger, sebastian.karpf)@uni-luebeck.de

Abstract

In biomedical applications electro-optical modulators (EOMs) help generating short laser pulses from continuous wave laser systems. A big disadvantage of those EOMs is their heat sensitivity. With difference in temperature the suppression of the EOM decreases due to a drift in destructive interference condition. Until now, the bias voltage of the EOM had to be adjusted with a potentiometer to reachieve a high suppression. To optimize the regulation of this EOM voltage we have implemented an Arduino based circuit which provides a continuous operation mode of the EOM. To evaluate the circuit it was executed in a long-term testing and compared to the manual regulation with a potentiometer. It was shown that the suppression with active regulation is even better than the manual operation. Also the testing shows that the maximum suppression could be guaranteed over a long period of time compared to a big drifting without active regulation.

1 Introduction

The generation of short laser pulses is one of the biggest milestones in the evolution of laser technologies especially for the use in medical applications [1]. Short and ultra-short laser pulses can be generated by using Q-switched laser systems or mode coupling. Q-Switched laser systems need optical elements to modulate the light of continuous wave (cw) or pre-modulated lasers [1]. EOMs are often used for this purpose. The main task of an EOM is to suppress the incoming laser light as good as possible and only letting the light out within a very short moment so the outgoing pulse can pass through possible amplification stages. But if the suppression provided by the EOM decreases, the modulation of the light gets worse until there are no pulses anymore but cw light. A big disadvantage of EOMs is that they are heat sensitive [2], [4] but through lasing and ambient upheating the temperature of those modulators increases with time until it gets saturated. With increasing heat the efficiency of suppression decreases and the operating bias voltage of the EOM has to be changed to optimize suppression. At the Institute for Biomedcial Optics in Luebeck a potentiometer is often used to change the voltage of the EOM but due to high sensitivity of those potentiometers it is difficult to set the perfect bias voltage manually and it is cumbersome to change the bias voltage frequently. In this internship an Arduino based circuit was implemented that measures the laser light behind the EOM and sets a bias voltage at the minimum of the sensor values which are decreasing due to increasing suppression. To provide a good operation over a long time the circuit should set a new voltage if the value is drifting. Such regulation would optimize the persistent operation of an EOM and eliminate the need for operator intervention, allowing the user to focus on other tasks.

2 Material and Methods

2.1 Electro-optical modulators

The EOMs used in the experiments are based on a Mach-Zehnder modulator (MZM). In this principle, the incident light beam is split into two beams. One beam has to pass a crystal, which shows a strong electro-optical effect. By applying a voltage to this crystal, a change of the refractive index occurs, resulting in a phase modulation. After the beam has passed the crystal, both beams are reunited [3]. Due to the phase modulation, interference phenomena occur at this point, which in turn lead to a pulsed beam. However, since the refractive index is temperature-dependent, temperature fluctuations have a large influence on the effectiveness of an EOM at constant operating voltage. Therefore, there are several ways to maintain the efficiency of an EOM. For example, to put the EOM in an environment where there are no temperature fluctuations or to change the operation voltage. This developed regulation utilizes the latter mechanism, since this is easier to implement. The EOM regulation was tested with three different laser systems. Fig. 1 illustrates the principle construction, in which the EOM is located between the seed diode and the amplifier stages.

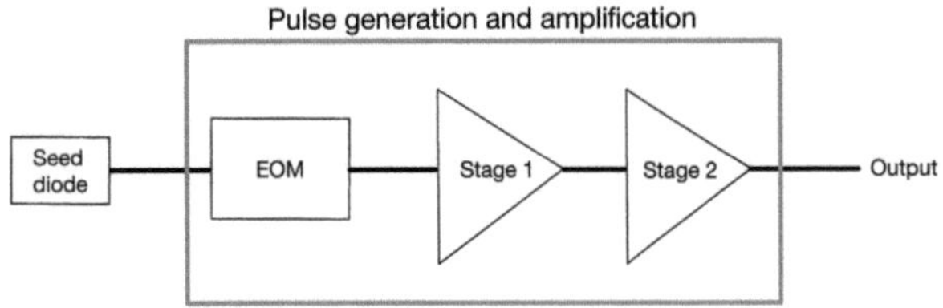

Figure 1: Schematic representation of the used laser setups.

2.2   Regulator components

In order to design a regulation which is simple to use and easy to replicate, an Arduino based development was chosen. Therefore the main component of the regulator is an Arduino Uno Revision 3, which can be easily programmed with the company owned Arduino IDE. Further needed components are a LCD display with 16 characters on 2 lines, two buttons for a reset and save, a switch to activate or deactivate the regulation, a rotary encoder for manual operation, some resistors and a photodiode to measure the light behind the EOM. To provide an EOM bias voltage between 0 V and 10 V a pulse width modulation (PWM) module is used. This module converts the PWM signal into a direct current voltage in the needed range. Since the Arduino Uno can only provide a maximum output voltage of 5 V, the PWM module must be operated with at least 10 V to amplify the Arduino voltage. As the PWM module itself also needs a certain voltage, it is operated with 12 V and thus the voltage range from 0 V to 10 V can be guaranteed. The module can supply a voltage between 0 V and 10 V or between 0 V and 24 V depending on the position of a jumper. Since a voltage above 10 V can destroy certain EOMs, the maximum voltage is limited to this value with the help of the jumper. The final circuit diagram of the regulation is shown schematically in Fig. 2.

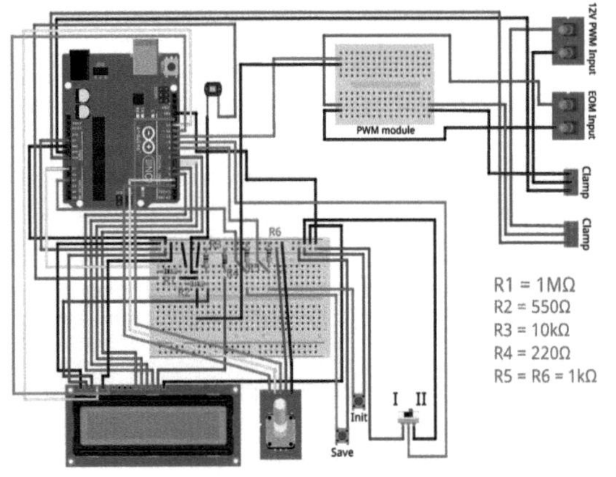

Figure 2: Schematic representation of the regulation circuit which provides a stable operating of EOMs by adjustment of the EOM bias voltage.

Since the circuit should be easy to rebuild, it is represented by a plug-in diagram. To install the circuit in a portable box, a 3D design template was used [6]. The size of the box was set to 15x15x5 cm, which makes the regulator compact, but with enough space inside for the electronic components. After the front and back of the box were adapted to the circuit, the components could be printed using an original Prusa i3 MK3S+ printer from Prusa Research a.s. and then assembled.

2.3   Regulation algorithm

To find a first minimum, a push button was implemented to start the ramp from 0 V to 10 V. While the voltage is continuously increased, the photodiode checks the applied light intensity at the 1% EOM tap output in serial to that, where 1000 sensor values are averaged each time. Depending on the EOM used, the measured intensity decreases at a certain voltage until it increases again at some point. The resulting minimum corresponds to a maximum suppression of the EOM. After the large ramp is finished, the voltage at which the minimum light intensity was measured is set. Fig. 3 shows a ramp from 0 V to 10 V and the measured light intensity.

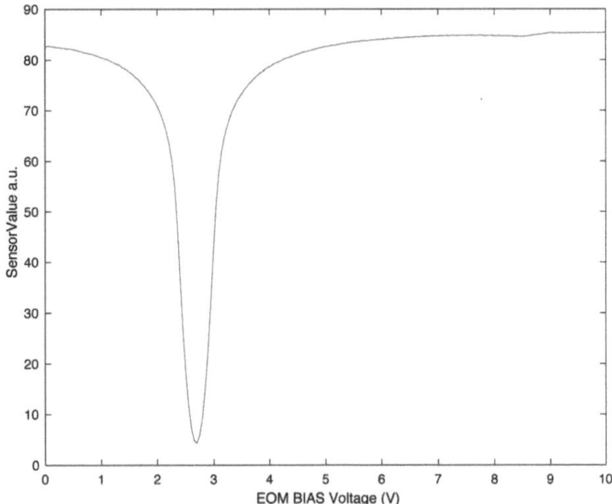

Figure 3: Measured light intensity levels behind the EOM while running a 10 V EOM-Voltage ramp on a 1064 nm laser system. As can be seen here the maximum suppression of the EOM was achieved with an operation voltage of approx. 2.7 V.

An algorithm was developed to ensure automatic operation of the control system. When the control mode is activated and the ramp was driven already or an old minimum was found in the EEPROM, the regulator checks 10 averaged sensor values every 1.5 seconds. For the averaging, 1000 sensor values are used to minimize the influence of a possible electronic noise of the photodiode. If one of the averaged sensor values deviates from the previous minimum by more than 3 percent, a mini-ramp of 0.05 V in total is to be performed. The midpoint of the mini-ramp is the voltage value at which the previous minimum was located. If a new minimum of an averaged sensor value is detected in this small ramp, the algorithm checks whether it was measured at an edge of the ramp. If this is the case, a mini-ramp starting at that point will be run again until the minimum is no longer measured at an edge. Once this has occurred, a new minimum has been found and the EOM is in a good operation mode again. Fig. 4 shows the complete flowchart of the regulation algorithm.

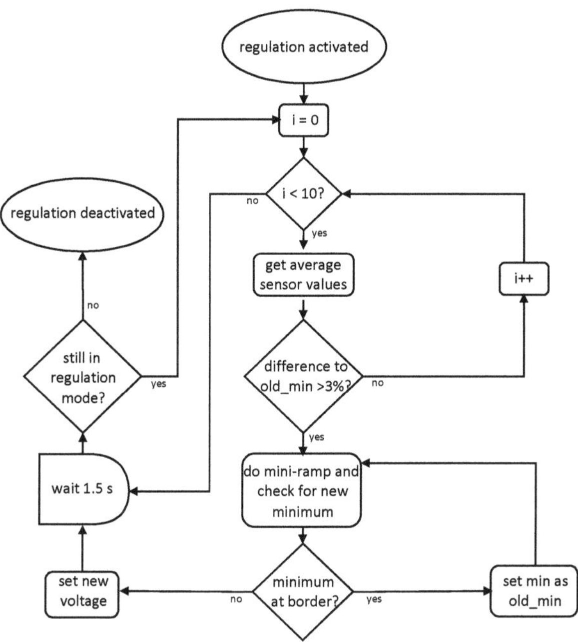

Figure 4: This flowchart shows the algorithm which is used to check the sensor values if there is a difference of more than 3% in respect to the last minimum. If this is the case the algorithm searches a new minimum with a mini voltage ramp of 0.05 V.

3 Results and Discussion

3.1 Comparison to manual operation

To be able to determine the suppression of the laser light, the optical spectrum analyzer AQ6374 from YOKOGAWA was used. The following results were measured with a 1064 nm laser setup using the EOM NIR-MX-LN-40 from iXblue. To determine the maximum suppression, the 10 V ramp was run in each case without subsequently activating the regulation. Table 1 lists the results.

Table 1: Maximal suppression of manual operation and maximal suppression found by the regulation at different power levels.

Power	Regulation	Manual
60 µW	-34.35 dBm	-39.07 dBm
200 µW	-38.88 dBm	-39.66 dBm
445 µW	-42.78 dBm	-43.70 dBm
720 µW	-43.26 dBm	-43.70 dBm

Several measurements were carried out with increasing the power applied to the measuring diode. As the results show the difference between the manually found intensity minimum and the minimum found by the regulation becomes smaller the more power is available for the measurement. Thus, at 60 µW power, the values still differ significantly with a difference of 4.72 dB, but at a power of 720 µW, the

difference decreases to 0.44 dB. To determine the attenuation by the EOM, the minimum suppression was estimated as well. At 60 µW, this was -9.35 dBm, resulting in a difference of 29.72 dB which corresponds to an thousandfold attenuation. It should be noted here that the values for one determination method, e.g. manual operation, cannot be directly compared since these were recorded at different powers. Thus, only the differences of the two methods within the same power class shall be considered.

3.2 Long-term testing

In order to assess the influence of a temperature difference due to an increasing ambient temperature and due to lasing within the laser system, two tests were performed over one and a half hours. The tests were accomplished using a pre-modulated 1550 nm system with a 10 Gb/sec "Bias Ready" Intensity Modulator from JDS Uniphase. Since the EOM had to cool down again for the second test, this test was performed two days later. The laser system was reworked by another intern, resulting in different power values being measured. However, this had no influence on the use of regulation, since only the amplifier stages behind the EOM were revised. Therefore, the results are reproduced in normalized form at the end. To determine the values, a power meter (PM100D, Thorlabs Inc.) was used, which was connected to a 1 percent output after the first amplifier stage. Every 5 minutes the present value was averaged over 100 values and stored. For the first test, the 10 V ramp was run, the minimum set and the regulation activated. The results of the test are shown in Fig. 5.

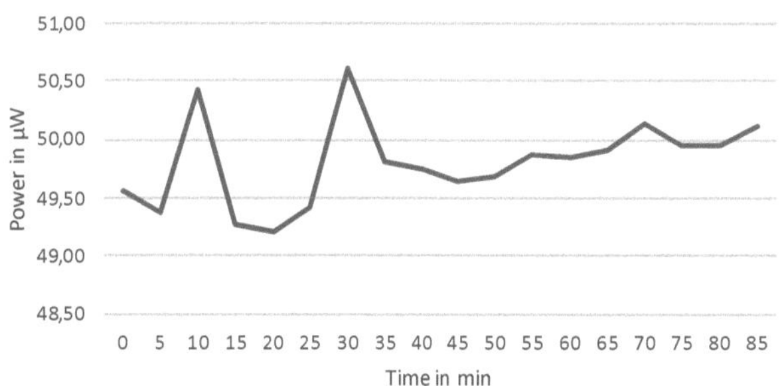

Figure 5: With a power meter measured power levels every 5 minutes over a time period of one and a half hour. The spikes could cause of the mini-ramp of 0.05 V which maybe was performed at the same time as the value has been measured.

The curve shows two spikes at 10 and 30 minutes, probably due to the fact that the mini ramp of the regulation was active and therefore the power fluctuates a little. The maximum variation during one regulation was less than 2 µW. Overall, the maximum difference in the graph is only 1.4 µW which corresponds to about 3%. In addition, it can be observed that the power of the laser increases with time, but this is a characteristic of the laser system and cannot be explained by a reduced efficiency of the regulation. In total, the voltage had to be adjusted by 0.85 V during the test run. In the second test, the 10 V ramp was performed again to determine the minimum. However, the control was now deactivated in order to record the behavior of the EOM. As already mentioned, the amplifier stages have been revised in the meantime, which is why the starting values of the two curves are different. This is why the curves have been normalized which can be found in Fig. 6, where the reduced efficiency of the EOM without regulation is clearly seen as the averaged power at the power meter triples and then saturates over time. When the regulation is enabled, the power is nearly constant at the minimum, ensuring efficient operation.

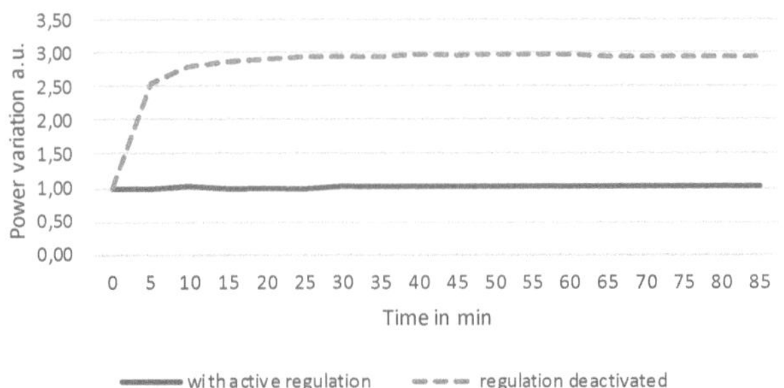

Figure 6: Normalized long-term testing power levels with and without active regulation measured with a power meter (PM100D, Thorlabs Inc.).

To illustrate the influence on the maximum pulse power, the pulses behind the second stage were measured with an oscilloscope (SDA830Zi-B, Teledyne Lecroy). The results can be found in Fig. 7. As it can be seen here, decreasing laser light suppression due to the temperature dependence of the EOM corresponds to decreasing maximum pulse power. Overall, the voltage difference between the start and the new minimum was 0.97 V after the test.

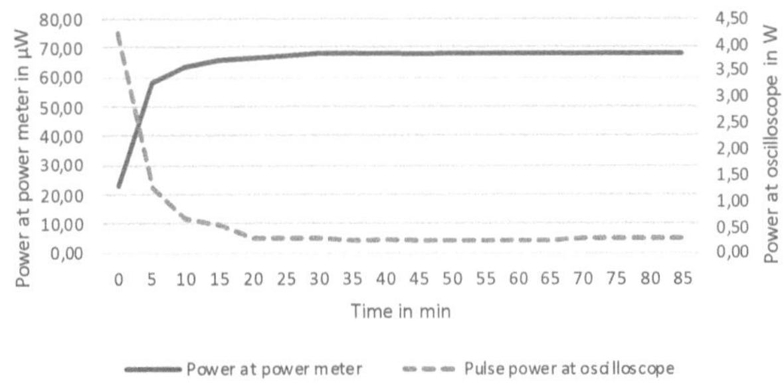

Figure 7: Comparison of middled power which was measured by a power meter with maximum pulse power of the laser system at an oscilloscope (SDA830Zi-B, Teledyne Lecroy).

3.3 Discussion

The results are generally in line with expectations and the regulation guarantees a well operating of the EOMs, but there are some points that can be criticized here. For example, the results would be even better if all experiments had been performed with the same and unchanged laser system. However, this was not possible because several participants were working on the setups. It must also be considered whether the algorithm used is optimal or whether a "perturb and observe" [5] algorithm would not be the better choice to achieve a faster voltage adaptation and less fluctuation of the voltage during a regulation since the step size in that algorithm decreases continuously.

4 Conclusion

In conclusion, the developed regulation unit can keep up with the manual operation and even ensure more efficient operation of EOMs in long term. Moreover, it could be shown how high the loss of the magnitude of the pulse power is after some time if the operating voltage of the EOM is not adjusted. As expected, saturation occurs after about half an hour and the EOM was fully warmed up. This means that especially at the beginning of a measurement, when the laser has just been switched on, an automated control of the operating voltage makes sense, since the drift is greatest here. But further experiments with different laser systems are needed to confirm the benefit of the regulation.

Acknowledgement

The work has been carried out and was supervised by the Institute of Biomedical Optics, Universität zu Lübeck.

Author's Statement

Conflict of interest: Authors state no conflict of interest.

5 References

[1] H. J. Eichler and J. Eichler, *Laser: Bauformen, Strahlführung, Anwendungen*. Springer Verlag, vol. 8, 2015.

[2] F. Lehner, *Aufbau und Analyse eines elektrooptischen Modulators zur schnellen Polarisationsschaltung*. Fakultät für Physik, LMU München, 2011.

[3] R. Mehra, H. Shahani and A. Khan, *Mach Zehnder Interferometer and its Applications*. International Journal of Computer Applications, 0975 – 8887, 2014.

[4] H. G. Heinol, *Untersuchung und Entwicklung von modulationslaufzeitbasierten 3D-Sichtsystemen*. Fachbereich Elektrotechnik und Informatik, Universität Siegen, 2001.

[5] S. Lotz, C. Grill, M. Göb, W. Draxinger, J. P. Kolb and R. Huber, *Cavity length control for Fourier domain mode locked (FDML) lasers with µm precision*. Biomedical Optics Express, vol. 12, no. 5, pp. 2604-2616, 2021.

[6] Makerbot Thingiverse 2016, *The Ultimate box maker*. Available: https://www.thingiverse.com/thing:1264391 [last accessed on 2022-01-05].

Developing a catadioptric Imaging System for multispectral Segmentation of the Retina on a single Camerachip

Fabian Henk [1], Malte Gienow-Broers [2] and Mathias Beyerlein [3]

[1] Medical Engineering Science, Universität zu Lübeck, fabian.henk@student.uni-luebeck.de
[2] Medical Optics, Technische Hochschule Lübeck, malte.gienow-broers@th-luebeck.de
[3] Medical Optics, Technische Hochschule Lübeck, mathias.beyerlein@th-luebeck.de

Abstract

In this project, an innovative optical system suitable for multispectral retinal imaging has been developed and theoretically analyzed. The working principle of the proposed system is snapshot imaging technology. This is advantageous for ophthalmic imaging, since the system is fast and robust. The spectral filtering component is composed of a dichroic tribandfilter and mirror combination. A catadioptric imaging setup has been developed to minimize chromatic effects over a wide spectral range. This cost-effective multispectral setup is modular, based on stock optics and could offer additional diagnostic value compared to standard white-light imaging. The system is diffraction limited in the desired wavelength range from 400 nm to 700 nm and over the full field of view of $\pm 5°$ in x-direction and $\pm 2°$ in y-direction, respectively. This work focuses on a theoretical and practical proof of concept of the optical design.

1 Introduction

Snapshot multispectral imaging is a technique to acquire spectral and spatial object information in a single image. Using images with different spectral information is helpful for ophthalmic diagnostics, as it enhances white-light imaging [1]. Multispectral imaging helps to make a wide range of relevant structures visible, such as the retinal pigment epithelium (RPE) or the choroid layer which is located behind the RPE. Imaging various tissue depths require different illumination wavelengths as the absorption coefficients for melanin and hemoglobin are wavelength-dependent [1]. In addition, this would also allow characteristic fluorescence spectra to be used for the detection of biomarkers that are important for the diagnosis of certain diseases such as age-related macular degeneration and diabetic retinopathy [1], [2]. Therefore, multispectral systems are suitable and necessary for modern fundus imaging.

There are several ways to achieve multispectral snapshot images [3], some are using dichroic filter stacks to spectrally separate the image information on a single camera sensor. Dichroic filters are beamsplitters based on the interference of light waves. Specific wavelengths are transmitted while others are reflected. Another method uses multiband-filters which differentiate between several bands depending on their design. Tribandfilters for example reflect and transmit three bands resulting in six channels overall. Instead of using a monochrome sensor for each channel, it is possible to detect the spectral information by using two separate color cameras (RGB or CYMG Bayer matrix), one for the reflected and one for the transmitted light [4]. This project

can be understood as a synthesis of both spectral band splitting concepts. The innovative approach is to double pass a triband filter in such a way that the reflected and transmitted information is mapped onto a single RGB camera chip. This work focuses on a proof of concept of the optical design. Considering physiological requirements is not part of this work.

2 Material and Methods

The project is divided into a theoretical simulation of the optical setup and a practical realization of the optical design. Generally, a fundus camera consists of an illumination beam path and an imaging beam path. The illumination optics need to image the light source into the eye's pupil to homogeneously illuminate the retina. The imaging optics lead the collimated light coming from the pupil of the eye through a complex optical system and focus it on a camera sensor. This pupil may have a diameter of about 4 mm for the reason of pupil segmentation. A proper imaging ray path is substantial as it determines the field of view and image resolution.

Table 1: Utilized imaging optics by Edmund Optics (EO)

	Part	EO no.
a	Off-Axis Parabolic Mirror f'=54.4 mm	35490
b	Convex Mirror f'=-200 mm	87660
c	Multiband Dichroic Filter	87286
d	Plain Mirror Protected Aluminium	30286
e	Aspheric Achromate f'=50 mm	49665

The imaging simulation has been carried out in Zemax OpticStudio®. The simulation starts at the entrance pupil and ends at the image surface. It does not include the illumination part. Table 1 shows the utilized stock optics for imaging. Fig. 1 shows the final imaging setup in OpticStudio. This modular setup consists of three exchangeable sections. The first section is an afocal mirror telescope which, on principle, has no chromatic aberration. This particular one is derived from a symmetric Offner system which has advantageous imaging properties, but other telescope types are generally possible. An Offner telescope has a magnification of -1, can be corrected for all primary aberrations and is designed for finite conjugates [5]. However, this derived Offner variation has significant differences, as it is afocal and uses two off-axis parabolic mirrors for focusing and collimation. The secondary mirror is convex, its curvature mainly influences the telescope's exit pupil position. Additionally, a rectangular field stop is located in the intermediate image plane on the secondary mirror surface to restrict the field of view.

Transmitted and reflected rays each carry three specific spectral bands and are imaged on different sensor areas by the third section, a stock aspherized achromate. Inside both of the two areas three spectral bands coincide, but the separation of the two areas from each other is large enough to not overlap. This separation is ensured by the secondary mirror's field stop. The simulated field angle is not the same for each direction. It is $\pm5°$ in x-direction and $\pm2°$ in y-direction, respectively. These values were chosen as a trade-off between image size and tilt angle of the filtering components. To increase the tilt angle would mean deviating more from the designed incidence angle of the dichroic filter which induces a spectral shift [6]. Although the field of view is quite narrow, it is sufficient to either image the optical nerve disc or macula.

The practical setup includes both illumination and imaging optics. The imaging setup is tested by illuminating an artificial model eye with a LED light source and shown in Fig. 2. The illumination components are listed in Table 3.

Table 3: Utilized illumination optics

	Part	Supplier
a	VIS/NIR LED OCL-480 GIR	OSA Opto Light
b	Lens AC254-030-A f'=30 mm	Thorlabs
c	Plain Mirror G340-086-000	Qioptiq Photonics
d	Beamsplitter BSW26R	Thorlabs
e	Retinoscopy Trainer	Heine Optotechnik

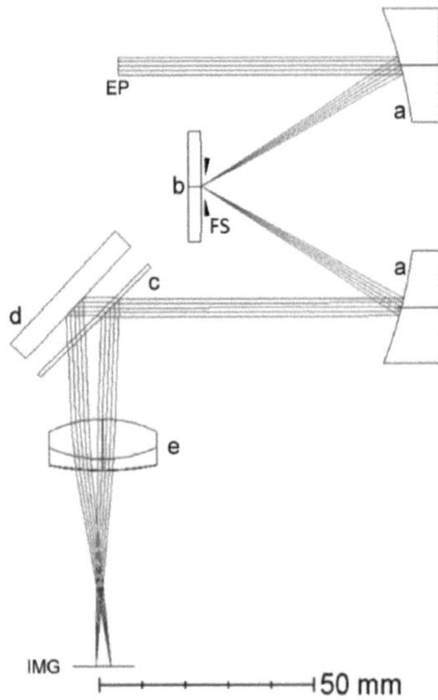

Figure 1: The imaging setup consists of an afocal mirror telescope (a, b), a multispectral filtering unit (c, d) and a focusing lens (e). The field stop (FS) is located on the mirror surface b. The entrance pupil (EP) is the first simulated surface located in front of the primary mirror. The image plane (IMG) is the sensor area and the last simulated surface behind the focusing lens. The by the filter c reflected and transmitted on-axis rays are shown simultaneously.

The second section is the multispectral filtering unit. A dichroic filter and a mirror are tilted at 45° and $\pm1°$ with respect to each other. The filter characteristics are shown in Table 2.

Table 2: The six bands of the tribandfilter 87286

Reflection	Transmission
1 400-440 nm	2 440-490 nm
3 490-520 nm	4 520-570 nm
5 570-610 nm	6 610-720 nm

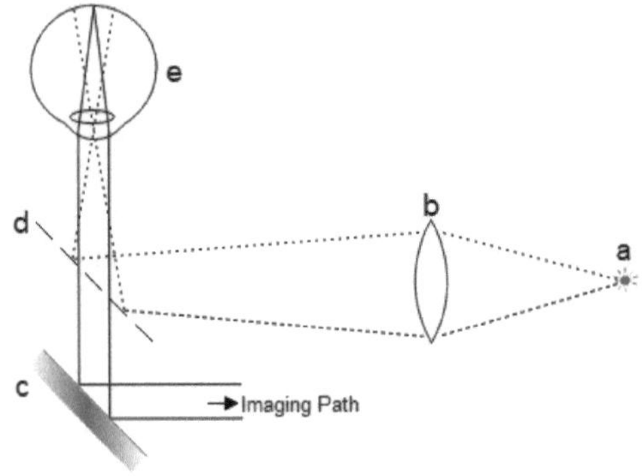

Figure 2: The illumination setup contains a LED (a) that is imaged into the pupil of the model eye (e) which has an adjustable diameter. To divide illumination and imaging light a plate beamsplitter (d) is used. A flat mirror (c) reflects the light in the direction of the imaging ray path.

As described, the proposed recovering of spectral information works in combination with a RGB or CYMG sensor. For now, it is sufficient to assess the image quality with a monochrome camera. The utilized model is the EVO8051MFLGEC by SVS-Vistek. The 4/3" sensor has 3296 x 2472 pixels with a size of 5.5 μm.

Most optomechanical components such as the parabolic mirror mounting plates, the flat mirror and filter holders or the LED, camera and model eye stages had to be customized and were realized by 3D-printing using PLA material.

3 Results and Discussion

3.1 Theoretical Analysis

The imaging setup of Fig. 1 has been optimized and analyzed in OpticStudio. The optimization criterion was the root-mean-square (RMS) spotradius. The use of stock optics mainly limited the optimization to the air spaces between the optics and the y-decentration of the asphere. All parameters of the reflected and transmitted parts were weighted equally. Three wavelengths have been chosen for both analyses. They match the center wavelengths of the six spectral bands described in Table 2.

3.1.1 Reflected Rays

The three simulated wavelengths are 420 nm, 505 nm and 590 nm. They are included in Fig. 3 which shows the spot diagrams for selected field angles on the image surface. The diffraction limit is indicated by the Airy disc, a circle around the spots.

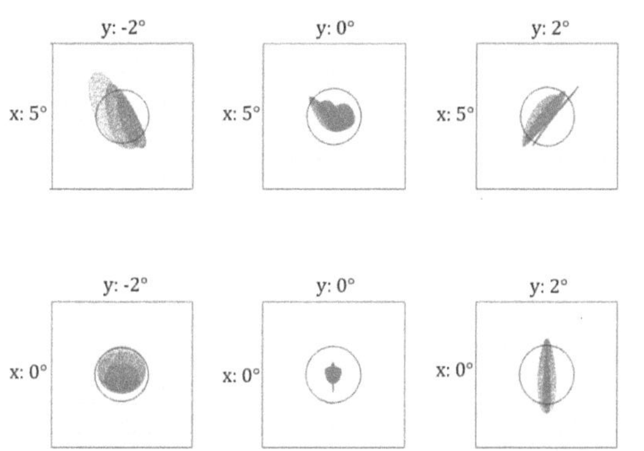

Figure 3: The simulated spot sizes are shown. The maximum field angles are 5° in x- and ±2° in y-direction, respectively. There are no negative x-field angles depicted due to mirror symmetry with respect to the y-axis. The wavelength-averaged Airy disc radius is 7.703 μm.

As the spot sizes are either smaller or about the size of the Airy disc, the imaging setup is of diffraction limited performance. The chromatic aberration in the image plane is negligible because the diffraction limit is reached for the whole simulated wavelength range.

Table 4: Zernike Fringe Coefficients (reflected)

Z	Name	Wave contribution
4	Defocus	-0.1152
5	Astigmatism x	0.0804
6	Astigmatism y	0.0973
7	Coma x	-0.0017
8	Coma y	-0.0028
9	Spherical Aberration	0.0012
10	Trefoil x	0.0030
11	Trefoil y	-0.0054
12	Sec. Astigmatism x	0.0013
13	Sec. Astigmatism y	-0.0010

A specific aberration analysis to identify the remaining wave aberrations in the field corners has been carried out, too. Table 4 shows the most important wave aberrations in Zernike fringe notation for the 5° x- and -2° y-direction field angle and for the center wavelength of the second reflected band (505 nm). This analysis can be done for one field angle and wavelength at a time. The unit *waves* refers to an optical path difference relative to an ideal spherical reference wave. The main wave contributions are made by the third order aberrations, in this case defocus and astigmatism, leaving the wavefront properties piston, tilt x and tilt y (Z1- Z3) aside. Higher order aberrations (HOA) are present but of minor significance. Out of those trefoil aberration is the most pronounced.

3.1.2 Transmitted Rays

The simulated wavelengths are 460 nm, 550 nm and 660 nm. Fig. 4 shows the transmitted spot sizes including all three wavelengths.

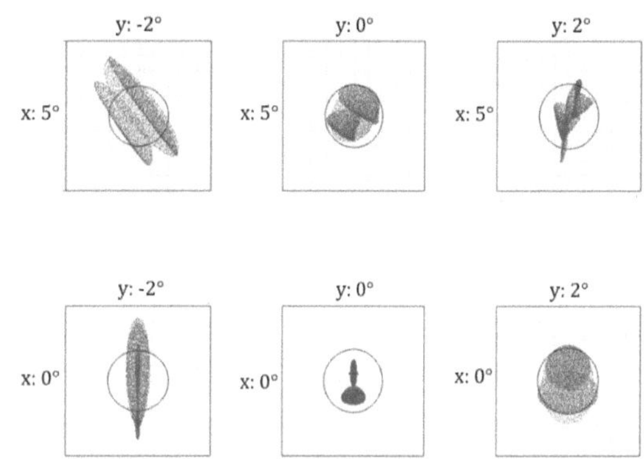

Figure 4: The simulated spot sizes are shown, the field angles remain the same. The averaged Airy disc radius increased to 8.412 μm due to the longer wavelengths in the reflected bands.

The spot sizes of the transmitted rays are at least in one direction diffraction limited. Table 5 shows the corresponding wave aberration coefficients for the 5° x- and -2° y-direction field angle at 550 nm.

Table 5: Zernike Fringe Coefficients (transmitted)

Z	Name	Wave contribution
4	Defocus	0.1410
5	Astigmatism x	-0.1284
6	Astigmatism y	-0.3256
7	Coma x	0.0006
8	Coma y	0.0121
9	Spherical Aberration	0.0007
10	Trefoil x	0.0134
11	Trefoil y	0.0059
12	Sec. Astigmatism x	0.0026
13	Sec. Astigmatism y	-0.0004

The transmitted rays have a similar aberration contribution.

The leading coefficients are third order astigmatism and defocus but there is also an influence of coma in y-direction present. Out of the HOA there is trefoil aberration again, especially in x-direction.

3.2 Practical Demonstration

A real version of the whole setup has been built on an optical bench. For qualitative assessment, images of the optical nerve disc of the model eye have been captured. These are shown in Fig. 5.

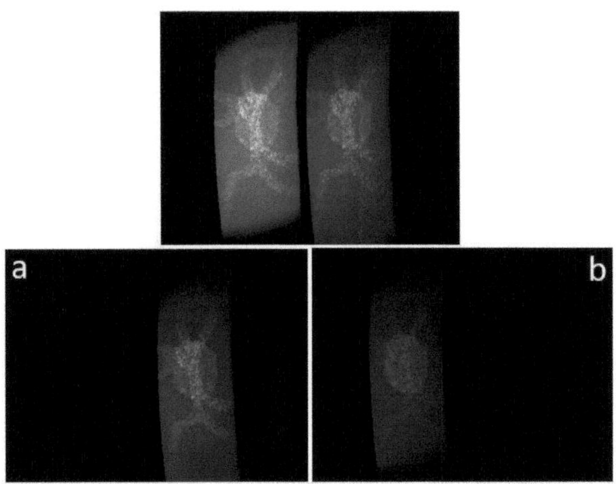

Figure 5: The model eye's retina can be seen. The multispectral filtering unit produces two versions of the same object in one image. Two of the six color channels (a, b) are isolated by narrow bandpass filters placed in front of the LED to visualize the filtering process. Image a: 500 nm, image b: 550 nm.

The size of the two segments can be adjusted by the rectangular aperture on the secondary mirror. The straight aperture edges appear slightly bent due to the distortion introduced by the mirror telescope. This could be corrected digitally. The image quality seems to be sufficient, but a proper adjustment of the mirror telescope with a wavefront sensor and resolution measurements are part of the ongoing project. Another point that has to be evaluated is the dichroic mirror's filtering characteristics depending on the rays' incident angle. The real incident angles on the filter greatly differ from the designed angle of $\pm 1°$ and this theoretically induces a shift to shorter wavelengths. To what extent this is applicable is a question of further investigation. A customized coating of the dichroic filter could be a suitable solution for this problem.

4 Conclusion

A catadioptric system for multispectral fundus imaging has been developed. The chosen components are cost-effective and generally modular. The multispectral filtering unit is based on a tilted dichroic tribandfilter and a mirror. There are six spectral bands overall. The simulated field of view is $\pm 5°$ in x- and $\pm 2°$ in y-direction. The theoretical performance is diffraction limited for both reflected and transmitted on-axis rays. Third order astigmatism and coma, defocus and also trefoil error dominate the remaining wave aberrations for the outmost field angles.

The practical setup includes optics for retina illumination of the model eye. First monochrome images were taken as proof of concept, a real system would take color images. Their quality is briefly assessed by visual inspection. The spectral channels, lightsource and filter characteristics are not yet optimized as physiological requirements will be defined later on. The concept is working, but an exact adjustment including resolution measurements needs to be done in the future. Also, the spectral information needs to be recovered.

Acknowledgement

The work has been carried out at the Medical Optics Laboratory of the Technische Hochschule Lübeck. It has been additionally supervised by Prof. Dr. Maik Rahlves, Institute of Biomedical Optics, Universität zu Lübeck.

Author's Statement

Conflict of interest: Authors state no conflict of interest.

5 References

[1] C. Zimmer, D. Kahn, R. Clayton, P. Dugel and K. Bailey Freund, *Innovation in Diagnostic Retinal Imaging: Multispectral Imaging.* Retina Today, Issue October 2014, pp. 94–99.

[2] M. Hammer et al., *Ocular fundus auto-fluorescence observations at different wavelengths in patients with age-related macular degeneration and diabetic retinopathy.* Graefes Arch Clin Exp Ophthalmol. 246, 105-114, 2008.

[3] N. A. Hagen and M. W. Kudenov, *Review of snapshot spectral imaging technologies.* Optical Engineering, 52(9), 090901, 2013.

[4] G. Themelis, Jung S. Y. and V. Ntziachristos, *Multispectral imaging using multiple-bandpass filters.* Optics Letters, 33(9), pp. 1023–1025, 2008.

[5] H. Gross, F. Blechinger and F. Achtner, *Handbook of Optical Systems Volume 4: Survey of Optical Instruments.* Wiley-VCH, Weinheim, 2008.

[6] P. Favreau et al., *Thin-film tunable filters for hyperspectral fluorescence microscopy.* Journal of Biomedical Optics 19(1), 011017, 2014.

Supporting mechanical and electrical engineering for an optical coherence elastography setup towards clinical deployment

Abdulmelik Mohammed [1] Nicolas Detrez [2] Ralf Brinkmann[3] Fred Reinholz [4]

[1] Biomedical Engineering, Luebeck University of Applied Sciences, abdulmelik.adilu.mohammed@stud.th-luebeck.de
[2] Medical Laser Center Luebeck, Luebeck, nicolas.detrez@uni-luebeck.de
[3] Medical Laser Center Luebeck, Luebeck, ralf.brinkmann@uni-luebeck.de
[4] Institut für Biomedizinische Optik ,Universität zu Lübeck, fred.reinholz@uni-luebeck.de

Abstract

The application of optical coherence elastography (OCE) requires a stable sample holder system and a mechanical adjustable clip-on system to supply pressurized air on the surface of the sample from the air puff system. This project seeks to design and produce the fore-mentioned systems. Additionally, the force from the air puff unit needs calibration before using it in the current state. The intended mechanical parts are manufactured at workshop in UzL using aluminium. Force measurements of the air from the air puff unit for variable vertical distances are realized using an external reference force sensor. The data is logged into an excel sheet and fitted using MATLAB to find an adequate polynomial coefficients. The polynomial coefficients are used to correct the calculated value from the air puff to the actual practical value given by the force sensor. This is necessary because the math behind the calculation does not cover the non ideal properties of the nozzles and hoses. The availability of a stable sample holder, clip-on system and the precise knowledge of the force of air on the sample being probed lead to good conditions for OCE imaging.

1   Introduction

Among the leading causes of death, cancer takes second place. Tumor margin evaluation plays a crucial role during the diagnostic and treatment phase of cancer[1]. During breast resection surgery for breast cancer removal, it is a common problem that the patient loses healthy tissue due to poor segmentation of margins. Twenty percent of patients who went through BCS (breast conserving surgery) need re-excision due to weak margin assessment between healthy and cancerous tissue. In addition, inadequate margins in BCS are the cause of local recurrence of breast cancer[2]. Recently the OCE (optical coherence elastography) has gained more attention in its ability to determine tissue stiffness. That would allow performing morphological segmentation of tumor tissue characteristics similar to the conventional histopathology examination[3], considered as the gold standard in this field[4]. Biomechanical properties exhibited by cells within biological tissue can provide a piece of vital information like shear modulus G, which is responsible for the ability of tissues to maintain their form. In the case of compressional elastography, the Young modulus E, which is the measure of elasticity, is used to characterize the stiffness of a tissue [5]. The Youngs modulus E is different from one tissue to another and used as a differential marker for the clinical diagnostics of tumor cells [6]. As Zaitsev et al [5] reported, there are several ways to cause tissue deformation for elastography measurement, including manual C-OCE(compression optical coherence to-

mography) and piezo actuator. In this paper, a dedicated air puff mechanism is adopted, for inducing tissue deformation. Coupled with the OCT system, that allows assessing tumor margins during brain surgery. According to the literature review, this method is docile and the first of its kind concerning direct and online measurement of force. However, a precise design to integrate the conventional OCT system with the air puff mechanism is required. In this paper, a customized stable sample holder and clip-on system are presented to support the development of the OCE device. Additionally, the calibration process of the air from the air puff generator is presented. The sample holder and the clip-on system are mechanical designs. The calibration process guarantees a predefined amount of force of air to be applied to the tissue phantom during a routine clinical usage of the OCE unit. The three tasks presented here are mainly playing a supporting role.

2   Material and Methods

2.1   Selection of Design Software

The CAD programm used for this internship was SOLID-WORKS, version 2019/20, from Dassault systems. For the purpose of data processing MATLAB from MathWorks Inc was used.

2.2   Sample holder

A custom sample holder with a dimension of $142 \times 140 \times 120\ mm^3$ is designed using SOLIDWORKS. The main aim of this mechanical design is to make the reference glass slide as stable as possible to avoid measurement error. As shown in Fig.1 the design uses a massive cube-like structure that serves as the base and holds the sample for OCE imaging. On the surface of the cube-like shape, there are two rods mounted vertically to fix the rectangular-shaped stage in place. The horizontal distance between two rods is 24 mm. The stage has a slit compartment on its front side to hold the reference glass slide in place for optical phase stability test. This compartment has a depth of 12.5 mm. The stage glides up and down along with the reference glass slide with the help of a micrometer screw. One end of the ball-tipped micrometer screw touches the bottom surface of the stage. A brass-bushing is inserted in the hole, where the micrometer screw intersects with the body of the sample holder. To move the stage up and down, the operator rotates the knob at the end of the micrometer screw. Two identical one newton compression springs are mounted onto the two rods above the surface of the vertically sliding stage to keep the stage stable during the course of usage.

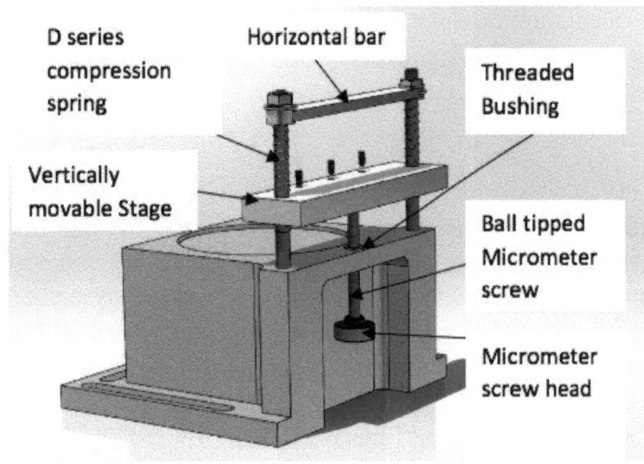

Figure 1: CAD model of the Sample-holder looking from the back.

2.3   Clip-on system

An adjustable clip-on system with a dimension of $50 \times 50 \times 16\ mm^3$ is designed to enable delivery of pressurized air on the surface of the sample being imaged. The total weight of the clip-on should be low-weight to be carried effectively. As shown in Fig.2 the clip-on system is composed of a hexagonal adapter, that holds the nozzle and the luer lock adapter in place. Each pair of 4-40 threaded cage assembly rods create vertical adjustability. These are firmly attached to the square-shaped adapter that is not shown on Fig.2. The clip-on assembly can move in horizontal and vertical directions. The nozzle of the air puff will be inclined at a given angle to have a precise alignment with the focal point of the OCT. Unfortunately, the result from this section can not be presented because of NDA.

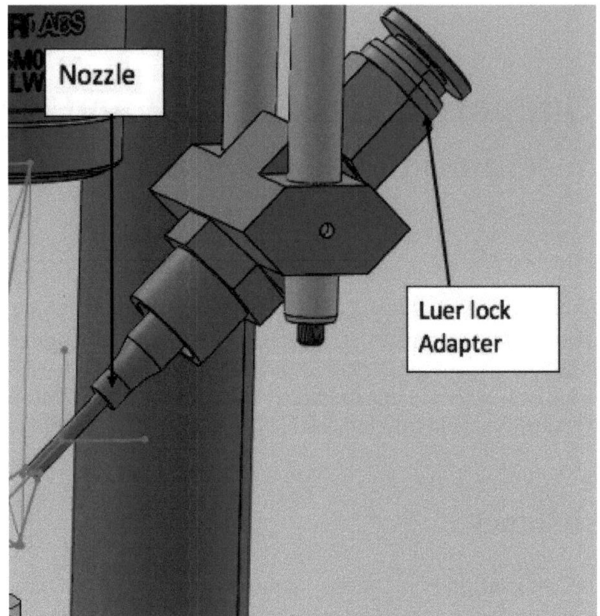

Figure 2: Design concept of Clip-on system

2.4   Optimization of Force of Air-puff

For the proper function of the OCE device, it is necessary that the force of air from the air puff generator is equal in value with what is effectively delivered to the sample from the nozzle output. But, due to practical inaccuracies of the real setup which are not covered by the theoretical equations, there is a difference between those forces. This means, in other words, there is a need for a calibration process. This is accomplished with the help of an external force sensor. The force F from the air puff is calculated from the mass flow measured with a flow sensor inside the air puff system. For the calibration, the internal force F from the air puff will be compared with the effective output force measured with the reference force sensor at the nozzle output. The difference delta F between these two forces is fitted and compensated with the calculated polynomial. Ideally, the delta F should be zero. The external reference force sensor already comes calibrated from the supplier using an absolute gauge sensor. This sensor measures the force F at vertical heights of 2 μm and, from 2 to 8 mm, with a 2 mm steps. As shown in Fig.3, the vertical height represents the vertical distance between the nozzle tip and the sensor head. The flow value spans from 0.5 l/min to 18 l/min with 0.5 l/min increment intervals, that corresponds to a theoretical force between 0 mN and 100 mN. At each vertical height, the measurement of the effective force takes place at 36 distinct flow values. The height difference affects the force distribution and because of that an averaging process is necessary to ideally average out the effects of the height. After taking the mean from all heights, the data is processed using MATLAB to find a fitting polynomial coefficients. The resulting polynomial coefficients are inserted into the LabVIEW program, which is an interface software, that the air puff system is operated with, to correct the internal force of air from the air puff.

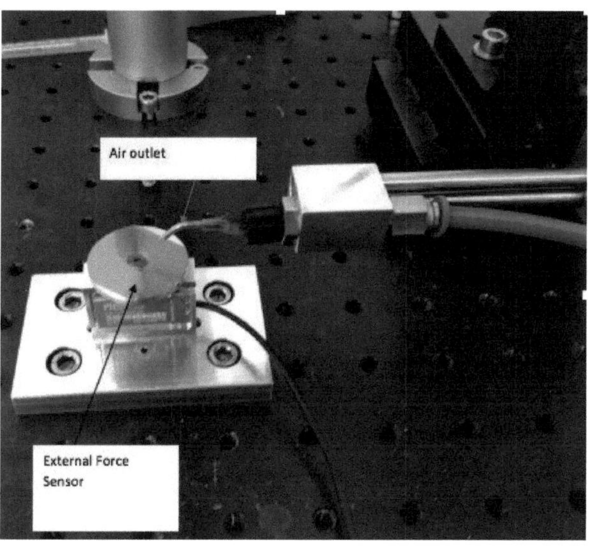

Figure 3: Measurement setup of air puff force from the air puff unit using KD34s external force sensor from ME-measurement systems at distinct heights.

3 Results and Discussion

The manufactured sample holder and the clip-on system assembly are shown in Fig.4. Both systems are manufactured at a workshop in UzL. The demand for lightweight, reasonable cost, ease of machining and form made aluminium the material of choice to manufacture both systems. Only the two mountable rods of the sample holder are made from brass material. Brass is also relatively easy to work on a lathe and stable enough to cut M6 thread onto it. The sample holder and the clip-on system weigh 2 kg and 300 g, respectively. The heavyweight of the sample holder increases its stability during the measurement procedure. The lightweight of the clip-on system makes it easy to mount it without compromising the load-carrying capacity.

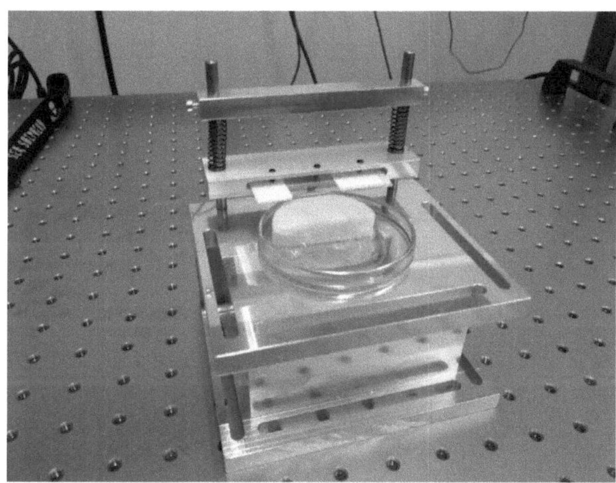

Figure 4: Sample holder and tissue phantom shown from the front side during a trial measurement.

After putting together all the components of the sample holder and the clip-on, a kinematic test was done to check if

Table 1: List of materials for the sample holder and clip-on system

Item	Type	Manufacturer
Threaded bushing	Brass	Thorlabs
Removable adjuster knob	Plastic	Thorlabs
Set screw	Alloy steel nylon tipped	Thorlabs
Compression spring	D-series	Misumi
Cage assembly rod	4-40	Thorlabs
Luer lock adapter	Threaded	Thorlabs
Nozzle	Conic	Vieweg
Fine hex adjuster	Threaded	Thorlabs
Micrometer screw	Adjustment screws	Thorlabs

a smooth and functional movement between the interlinked components can be initiated. The result showed a relatively good outcome for the current application. The results from the air puff force calibration are presented in graphs Fig.5, Fig.6 and Table.2. For comparison Fig.5 shows the force value from the external force sensor for flow values between 0.5 l/min and 18 l/min before and after polynomial correction of the internal air puff force. The curve before polynomial correction represents the theoretically calculated internal air puff force before the calibration process. The curve after polynomial correction represents the force of the air puff after a calibration is done. The now corrected curve has lower values after the flow point 11 l/min. This is because of the main unknown factors like the practical flow resistance of the hose system and the nozzles are not covered with the theoretical equations. This factors are usually higher than assumed by the equations. The quality of fit after the polynomial correction of the force value has shown a large improvement as presented in Table.2.

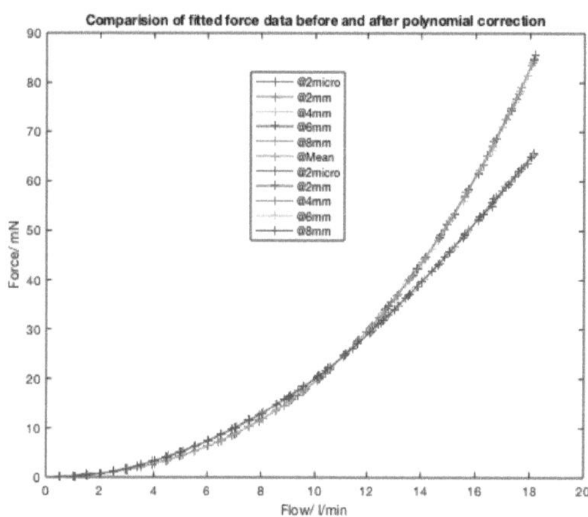

Figure 5: Plot of Force(mN) Vs Flow(l/min) of the internal air puff force before and after polynomial correction at different heights. The now corrected internal air puff force has lower values of force after the flow 11 l/min.

The relative error and quality of fit after polynomial correction are presented in Table.2. The relative error is calculated from the absolute error. The absolute error represents the difference between the internal air puff force and the ef-

Table 2: Summary of error, E.B: Mean relative error before polynomial correction, E.A: Mean relative error after polynomial correction, RSB: Quality of fit before polynomial correction, RSA: Quality of fit after polynomial correction

Height	E.B %	E.A %	RSB %	RSA %
2μm	- 0.0468	0.0236	99.8	99.93
2mm	0.9928	0.185	99.64	99.93
4mm	-0.286	0.79	99.58	99.9
6mm	0.805	-0.481	99.78	99.92
8mm	-0.836	-0.821	99.64	99.91
Mean	0.755	-0.373	99.92	99.92

fective force measured using an external force sensor. As expected, the errors decrease after the polynomial correction. The quality of the fit also improved. Using the mean data ensures the best quality fit. Final testing of the Force calibration is done at 8 mm, that is the maximum vertical height, as shown in Fig.6. As sought, the internal force from the air puff unit now agrees very well with the data found from the external sensor.

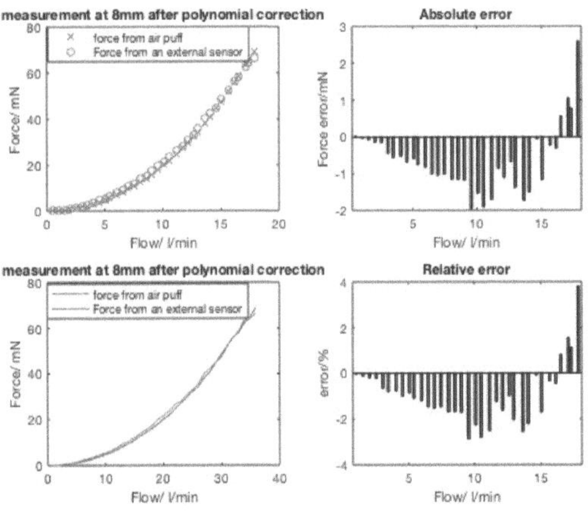

Figure 6: The result of test run after polynomial correction at 8 mm height. On the left side it shows how the Internal force and the measured force from the external sensor fit together because of the calibration. On the right side the absolute and relative error are shown depicting the difference between those two forces.

4   Conclusion

In this paper, a mechanical and electrical engineering approach to support the development of OCE towards clinical deployment is presented. A stable and robust sample holder system, which helps to firmly hold in place the sample, is designed and manufactured. A Clip-on system to deliver a specific amount of pressurized air to the sample surface is also presented. The sample holder and the Clip-on system constitute important parts of the whole OCE system. The force F of the air from the air puff system was calibrated and surveyed. All three completed tasks have proven essential to the ongoing research and development of the desired OCE device. Finally, the tasks presented in this paper are only part of a big research project currently going on in MLL and due to NDA agreement with MLL some results are not explicitly presented.

Acknowledgement

The work has been carried out at Medizinische Laserzentrum Lübeck(MLL), Lübeck and supervised by Dr. Fred Reinholz, Institut für Biomedizinische Optik, Universität zu Lübeck.

Author's Statement

Conflict of interest: Authors state no conflict of interest.

5   References

[1] Zhu et al. , *Differentiation of breast tissue types for surgical margin assessment using machine learning and polarization-sensitive optical coherence tomography.* Biomedical Optics Express , Vol. 12, No. 5 Optical Society of America, p. 3021-3036,2021.

[2] Kennedy et al. , *Diagnostic accuracy of quantitative micro-elastography for margin assessment in breast-conserving surgery* . Cancer research , Vol. 80, No. 8 ,AACR, p. 1773-1783, 2020.

[3] Gubarkova et al., *Diagnostic accuracy of cross-polarization OCT and OCT-elastography for differentiation of breast cancer subtypes: Comparative study* . Diagnostics , Vol. 10, No. 12 ,Multidisciplinary Digital Publishing Institute ,p. 994, 2020.

[4] Plekhanov et al., *Histological validation of in vivo assessment of cancer tissue inhomogeneity and automated morphological segmentation enabled by Optical Coherence Elastography.* Scientific reports , Vol. 10, No. 1 ,Nature Publishing Group ,p. 1-16, 2020.

[5] Zaitsev et al. , *Practical obstacles and their mitigation strategies in compressional optical coherence elastography of biological tissues* . Journal of Innovative Optical Health Sciences ,Vol. 10, No. 06 ,World Scientific, p. 1742006, 2017.

[6] Parmar, Asha Sharma, Gargi Sharma, Shivani Singh, Kanwarpal , *Portable Optical Coherence Elastography System With Flexible and Phase Stable Common Path Optical Fiber Probe.* IEEE Access , Vol. 9, IEEE, p. 56041-56048, 2021.

Wavefront measurement with swept source OCT for the determination of astigmatism in the eye

Michel Wunderlich [1], Michael Stender [2]
[1] Medical Engineering Science, Universität zu Lübeck, michel.wunderlich@student.uni-luebeck.de
[2] Heidelberg Engineering GmbH, michael.stender@heidelbergengineering.com

Abstract

Measuring refractive error in the human eye is an essential part of ophthalmology. This examination is usually performed with an autorefractometer. In this work, a method based on swept source OCT is used, which measures the phase and magnitude information of the backscattered light to fit the Zernike polynomials of the underlying wavefront. Here only the two Zernike terms describing the astigmatism are considered. To test the method, several series of measurements were performed with an artificial eye and rotatable cylindrical lenses. In addition, a simulation was created which can generate ideal and artificially manipulated measurements. For an artificial eye, an accuracy of the astigmatism angle of 3.2 degrees could be achieved. The precision for this angle is 1.3 degrees. The measurement results for human eyes are not yet satisfactory, but it could be shown that a determination of astigmatism is possible.

1   Introduction

Optical coherence tomography is already an important part of modern ophthalmology. It enables high-resolution imaging of the anterior and posterior segments of the eye [1]. It is based on the interference of a broadband light source. Taking a simple Michelson interferometer as an example, interference only occurs when the optical path length of the sample arm is approximately the same as the optical path length of the reference arm. For coherence to occur, the path length difference must be smaller than the coherence length of the light source. Thus, the depth-dependent scattering and reflectivity in the tissue can be measured and a depth section of the tissue can be reconstructed. Here, however, no image of the eye is to be generated, but the aberration of the eye is to be measured on the basis of the complex signal.

1.1   Fourier Domain OCT

There are different methods to perform optical coherence tomography, only Fourier Domain OCT is considered here [2]. One possible implementation is spectral domain OCT, in which a broadband light source is used. The interference between the sample and reference arms is broadened so that the individual spectral components can be measured with local resolution, for example with a line scan camera. Another implementation is swept source OCT; instead of a broadband light source, a narrowband light source is used here, which can sweep through its wavelength range very quickly. Thus, the individual spectral parts are not measured locally, but differentiated in time. A Fourier transformation of the wavenumber-dependent intensity yields a complex signal that contains information about the phase and amplitude of the incident light from the respective depth. This information is used in the following to reconstruct the wavefronts aberrated by the eye [4]. In this work a swept source OCT is used, which acquires A-scans with a frequency of 100kHz.

1.2   Zernike Polynomials

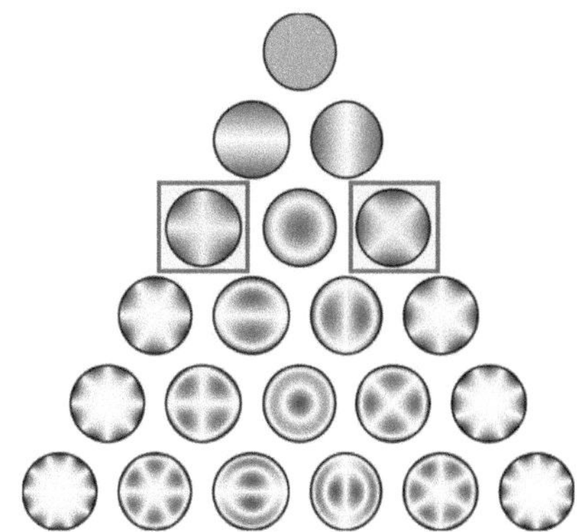

Figure 1: Shown are the first 21 Zernike polynomials. For the aberration measurements in this work, the oblique and vertical astigmatism is evaluated (marked) [3].

The aberrated wavefronts can be described by the Zernike polynomials [5], which are shown in Figure 1. The individual Zernike polynomials are orthogonal to each other and defined on the unit circle. These are used here to fit the measured phase information to the Zernike polynomials using the least squares method, where only the two first order astigmatism terms are considered. In the following they are named as:

$$A_{oblique} = \sqrt{6}\rho^2 sin(2\varphi) \qquad (1)$$

$$A_{vertical} = \sqrt{6}\rho^2 cos(2\varphi) \qquad (2)$$

Where $0 \leq \rho \leq 1$ is the radial distance and $0 \leq \varphi < 2\pi$ the angle on the unit circle. The aim is to measure coefficients $c_{1,2}$ of these two terms to determine angle α and magnitude M of astigmatism of the wavefront aberrated by the optics of the eye:

$$\phi_{oblique} = c_1 \cdot A_{oblique} \qquad (3)$$

$$\phi_{vertical} = c_2 \cdot A_{vertical} \qquad (4)$$

$$\alpha = \frac{1}{2} arctan(\frac{c_2}{c_1}) \qquad (5)$$

$$M = \sqrt{(c_1^2 + c_2^2)} \qquad (6)$$

In the following, α is given in degrees and M in diopter.

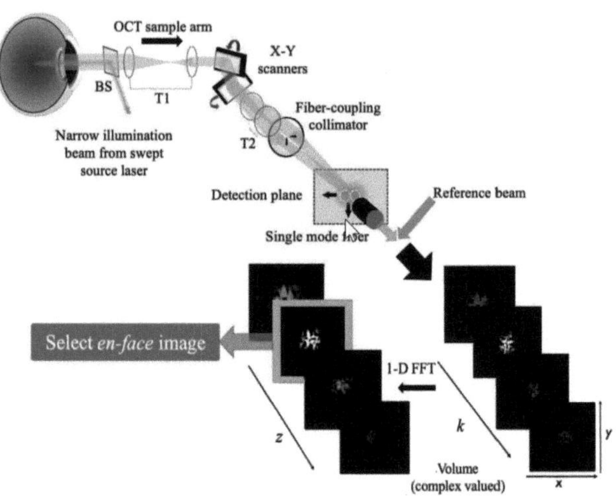

Figure 2: Simplified representation of the experimental setup [4].

2 Material and Methods

The setup used is similar to the setup described in [4]. A thin needle beam, whose origin is the light source of the OCT, is focused on the retina of the eye. This beam is backscattered and expanded. Due to the aberrations of the cornea (or the additional cylindrical lenses used here), the backscattered wavefront is aberrated. Using two galvanometer scanners, this wavefront is scanned in a spiral at a sampling frequency of 100kHz. 20000 sample points are recorded. Each sampled signal is individually interfered with the reference beam. The resulting intensity signal $I(k)$ is Fourier transformed. Thus, the signal is obtained, which now consists of magnitudes and phase information of the respective axial layers. By interpolation, the point spread function of the optical system can be reconstructed. An overview of the setup is shown in Figure 2. The en-face layer, where the signal is maximum, is used for further processing. A 2d discrete Fourier transform is used to propagate the wavefront into the plane of the pupil [6]. By using k-means clustering [7], the signal terms are separated from the noise terms and circularly masked. The derived phase data are used to fit the astigmatism terms from (1) and (2) using the least squares method.

2.1 Simulation

In order to generate ideal measurement results, a simulation was created in Matlab (MATLAB R20121b, The MathWorks Inc., USA), which generates artificial recordings. These recordings can be used to generate a lower limit of the possible measurement accuracy and repeatability. In addition, this allowed the influence of various disturbance factors to be analyzed. The thin needle beam is backscattered on the retina at a simulated scatterer. Propagation of the backscattered light is realized by a Fourier transform of the magnitudes and phase data of the backscattered light. By adding the phase error, which is caused by the simulated refractive power of the eye, the aberrated wavefront is obtained. Propagation of the wavefront back to the retinal plane and subsequent spiral scanning yields the complex signal that would also be measured by the real measurement setup.

3 Results and Discussion

The following measurement results for an artificial eye and the human eye were recorded and processed with the presented measurement setup. MATLAB was used for the simulated data.

3.1 Simulation

The simulated data sets were used to investigate how high the accuracy can be for ideal recordings. To determine the accuracy of the astigmatism angle, a simulated cylindrical lens of 4 diopters was rotated 180 degrees in 6 degree increments. To determine the accuracy of magnitude, simulated cylindrical lenses between 2 diopters and 7 diopters in 0.5 diopter steps were used for the measurements. The residual was calculated in which the distance between measured and adjusted value was calculated. The results are shown in Figure 3 and 4. For the measurement of the angle, the root mean square error (RMSE) is 0.75 degrees. The regression with the comparison line gives an R^2 of 0.9998. The RMSE for the measurement of the magnitude is 0.14 diopters, R^2 is 0.99.

The results with the simulated data show that a precise and accurate result can be achieved with this method. Both the angle and the magnitude of the astigmatism can be determined to create lenses that could subjectively correct the astigmatism of a patient sufficiently well [8]. Using astigmatism, it was shown that a reconstruction of the wavefront can be implemented with the recorded phase and amplitude data. In the future, other zernike terms will be implemented and evaluated.

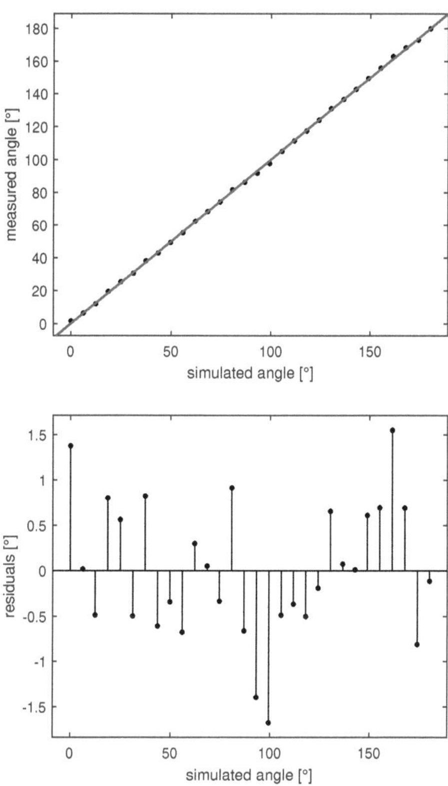

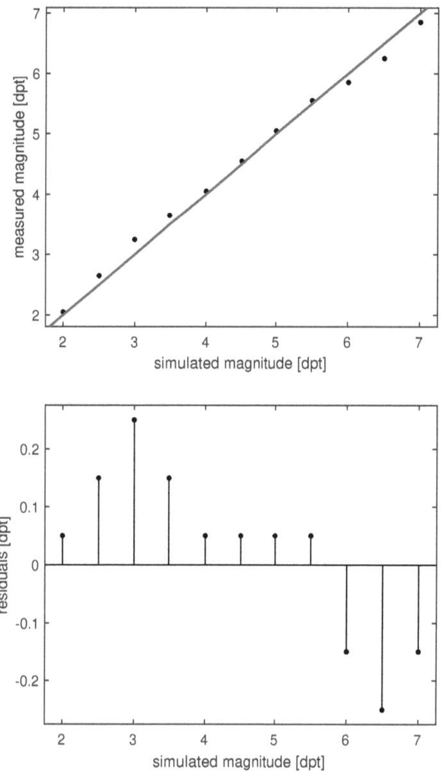

Figure 3: (top): The measured astigmatism angle is plotted against the simulated angle. The ideal result is shown as a line. (bottom): Calculated residual of the measurement shown above.

Figure 4: (top): The measured astigmatism magnitude is plotted against the simulated magnitude. The ideal result is shown as a line. (bottom): Calculated residual of the measurement shown above.

3.2 Artificial Eye

An artificial eye and a rotatable cylindrical lens were used to determine the accuracy of the angle and magnitude of astigmatism. For the determination of the angular accuracy, a cylindrical lens with 3.3 diopters was used and 5 measurements were performed in 20 degree steps. For the determination of the accuracy of the magnitude, several measurements were performed in each case with cylindrical lenses with focal lengths of 200mm, 300mm, 400mm and 500mm. The RMSE for the angular accuracy is 3.2 degrees. The results are shown in Figure 5 and 6. In addition, an identical measurement was performed 45 times. The magnitude has a precision of 0.42 diopters and the angle has a precision of 1.3 degrees.

If an artificial eye is used instead of a simulation, the result is not sufficient to correct the astigmatism. This is mainly due to the fact that the determination of the magnitude is not accurate. One reason may be that the optical properties of the artificial eye could not be adjusted well, the distance between the defocus lens and the retina could not be measured with sufficient accuracy. The effect of the strong defocus made it difficult to determine the derived wavefronts. However, with more work it will be possible to produce better results. This approach will be taken in the future.

3.3 Human Eye

Instead of the artificial eye, a human eye was now used in which a test person looks into the beam of the sample arm. The rotatable cylindrical lens was set to 0 degrees in the first series of measurements and to 90 degrees in the second series. The cylindrical lens had a power of 5 diopters. The result is shown in Figure 7. It can be seen from the measurements with the real eyes that the procedure can work. The two measurement positions can be clearly distinguished. One problem is the recording time of the measurement, with 20000 sample points and a recording frequency of 100 kHz a recording time of 20 ms results. This is too long to prevent the occurrence of saccades [9]. The occurrence of saccades leads to short changes of the wave field, which makes the evaluation difficult. One solution is to increase the switching frequency or to filter out the motion by additional post-processing. Both are to be implemented in the future.
The results are not yet satisfactory, but it could be shown that a determination of astigmatism is definitely possible.

4 Conclusion

In this paper, a method was presented to measure the refractive error of the eye using astigmatism as an example. It was shown with simulated data that a sufficient accuracy tends to be possible [8]. However, these results could not yet be measured with real recordings. In the future, the aim is not only to measure astigmatism, but also to determine other Zernike terms with sufficient accuracy.

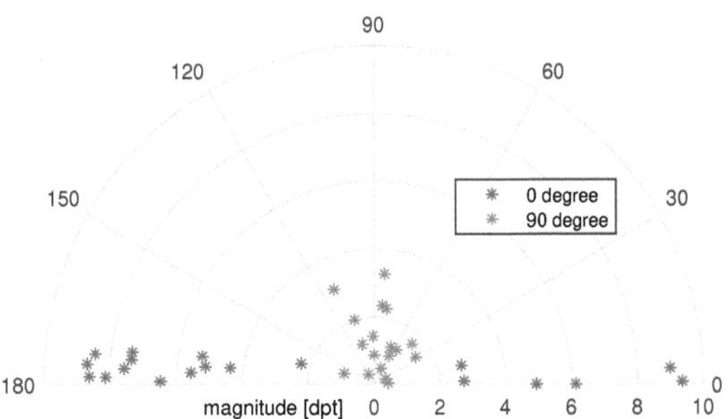

Figure 7: A real eye and a rotatable 5 diopter cylinder lens were used and several measurements were made for 0 and 90 degrees rotated cylinder lens.

Acknowledgement

The work has been carried out at Heidelberg Engineering and supervised by Michael Stender (Heidelberg Engineering) and Prof. Huber, Institute of Biomedical Optics, Universität zu Lübeck.

Author's Statement

Conflict of interest: Authors state no conflict of interest.

5   References

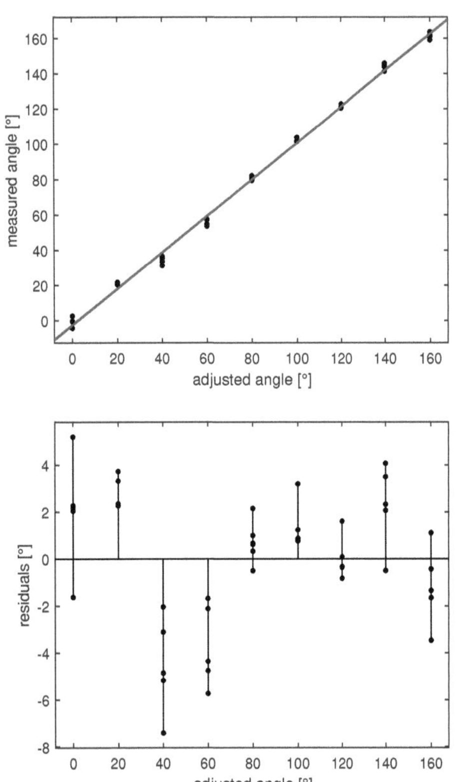

Figure 5: (top): The measured astigmatism angle is plotted against the adjusted angle of the artificial eye. The ideal result is shown as a line. (bottom): Calculated residual of the measurement shown above.

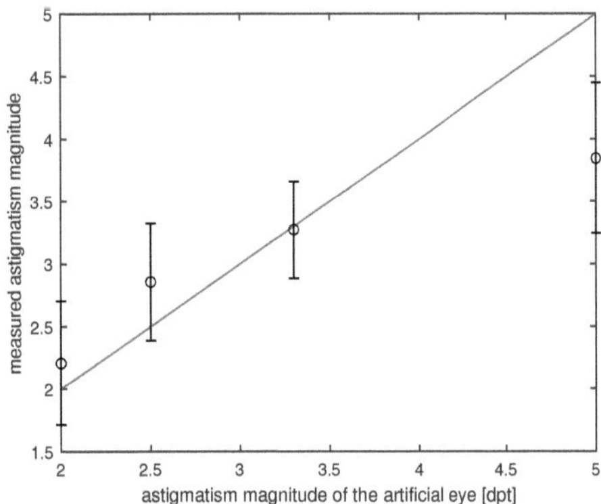

Figure 6: The measured astigmatism magnitude is plotted against the adjusted magnitude of the artificial eye. The ideal result is shown as a line. The RMSE is displayed as an error bar.

[1] N. Yoshimura and M. Hangai, *OCT Atlas*. chap. 1, (Springer, 2014).

[2] W. Drexler and J. G. Fujimoto, *Optical Coherence Tomography: Technology and Applications*. chap. 1, (Springer, 2008).

[3] Nschloe, <https://creativecommons.org/licenses/by-sa/4.0>, via Wikimedia Commons

[4] S. Georgiev et al., *Digital ocular swept source optical coherence aberrometry*. In: Biomedical Optics Express, Vol. 12, no. 11, 2021.

[5] M. Born, E. Wolf, *Principles of Optics. Electromagnetic Theory of Propagation, Interference and Diffraction of Light*. chap 9, (Elsevier, 1980).

[6] J. W. Goodman, *Introduction to Fourier optics. 3rd edition*. chap. 4, (Roberts & Co., 2005).

[7] J. MacQueen, *Some methods for classification and analysis of multivariate observations*. In: Berkeley Symposium on Mathematical Statistics and Probability, Vol. 5.1, 281-297, 1967.

[8] U. v. Pape, *Wavefront Sensing in the Human Eye*. Dissertation, 2002.

[9] O. Bayer, *Sakkaden zu bewegten und stationaren Zielen*, Dissertation, 2005.

Instrumental Development and Analysis of Meniscus Method as a Contact Angle Measurement Technique for a Drop Shape Analyzer

Sirajum Muneera Progga [1,2], Niels Delbos [2], and Stefan Müller [3]

[1] Biomedical Engineering, Lübeck University of Applied Sciences, sirajum.muneera.progga@stud.th-luebeck.de

[2] Krüss GmbH, {sirajum.progga, n.delbos}@kruss.de

[3] Medical Sensors and Devices Laboratory, Department of Applied Natural Sciences, Lübeck University of Applied Sciences, stefan.mueller@th-luebeck.de

Abstract

Wettability is a significant parameter for surface investigation, and contact angle (CA) is the measurement index that reflects the wettability of a surface. There is a diverse range of applications of wettability and CA measurement in the field of biomaterials. This paper describes the pre-development of meniscus method as a CA measurement technique in a drop-shape analyzer. It includes the instrumental implementation as well as the algorithmic development of this technique. The immersion and withdrawal of a fiber attached to a drop-shape analyzer via special accessories forming a meniscus with a sample liquid was captured with a high-resolution camera. Afterwards, the recordings were processed to develop an algorithm which calculated the receding CA (RCA) occurring between the fiber and the liquid interfaces. A cross-validation with alternate methods and devices evaluated the performance of the said implementation, which provided consistent values of CAs as with the developed algorithm.

1   Introduction

Biomaterials can be described as natural or artificial materials which facilitate replacement, enhancement or restoration of biological tissues while interacting with a biological environment [1]. Applications of biomaterials include: medical implants, blood-contacting medical devices, tissue engineering substrates, contact lenses [2]. Since biomaterials come into contact with biological environments, biocompatibility must be ensured in order to avoid adverse effects during an interaction between biomaterials and biological tissues [3]. Physiochemical properties such as wettability and surface tension (SFT) play a significant role in biocompatibilty. Wettability of a solid surface is quantified by the contact angle (CA) among solid/liquid-vapor interfaces. Some of the major techniques for CA measurement include: sessile drop method, pendant drop method, Washburn method, Wilhelmy balance method. The meniscus method described in this paper is a CA measurement technique for solid/liquid interfaces based on the single fiber Wilhelmy method. Traditionally the Wilhelmy method involves a thin plate of a solid substrate attached to a microbalance which is then submerged into a non-penetrating liquid. CA is measured from the force acting on the solid due to wetting and dewetting. In case of single fiber Wilhelmy method with a force tensiometer, the CA is measured

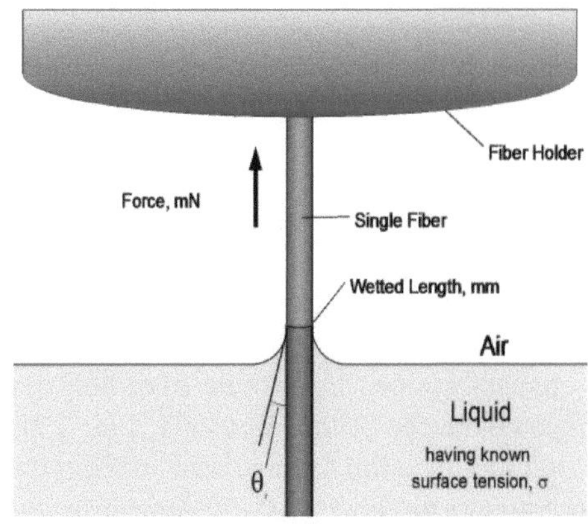

Figure 1: Receding contact angle θ between fiber and liquid interfaces during single fiber Wilhelmy method [4]

according to the following formula:

$$\cos\theta = \frac{F - F_b}{l\gamma}.\tag{1}$$

where l is the wetted circumference of the fiber, F is the total force acting on the fiber at any submerged position and F_b is the buyoant component of the force which is a result of the displacement of the liquid during wetting and dewetting of the fiber, γ is the surface tension and θ is the receding CA occurring between the liquid and the fiber during dewetting of the fiber from the liquid [5][6].

The meniscus method can be considered as an optical variation of the single fiber Wilhelmy balance method discussed above. While the CA in the single fiber Wilhelmy method is determined with the help of a very precise microbalance on a force tensiometer [2], the meniscus method measures the CA using an optical tensiometer by analysing a shadow image of the resulting meniscus between solid fiber and liquid. As opposed to the single fiber Wilhelmy method, meniscus method can be used to determine CA without having previous knowledge about the wetted length of the sample fiber or the surface tension of the sample liquid. Fig. 1 shows the receding CA occurring in the respective meniscus formed during dewetting of fiber from a liquid.

This paper describes the pre-development of implementing the meniscus method as a CA measurement technique on a drop shape analyzer DSA100S manufactured by KRÜSS GmbH. It includes the steps and achievements leading towards this development starting from the instrumental implementation, image acquisition with the setup, algorthmic development to calculate the CA as well as the validation of the obtained outcomes.

2 Material and Methods

2.1 Instrumental Implementation and Image Acquisition

The meniscus method was implemented to a standard KRÜSS drop-shape analyzer DSA100S. Typically a drop-shape analyzer (DSA) is used for obtaining CA between a liquid and a solid surface by analysing an image of a drop of the liquid placed on the solid surface which is recorded using a high resolution camera. The meniscus method is far from the regular CA measurement techniques that are included in a DSA such as: sessile drop or pendant drop method. Therefore, a modification to the instrument was required to integrate meniscus method to it. This was achieved by using special accessories including a fiber pick-up element clipped to a single fiber holder attached to a syringe connected to the DSA100S. Fig. 2 shows the instrumental modification for meniscus measurement. The fiber was fixed to the fiber pick-up element using a light-curated glue. In this case, human hair was used as a sample fiber and distilled water was used as a sample liquid. The used portion of the hair was cut as short as possible in order to avoid any curling of the surface. The diameter of the sample hair was 73.5 µm which was measured by using a calibration tool CP35 of diameter 2 mm. This tool was used to determine the pixel scale per mm and afterwards ImageJ software was used to obtain the actual diameter. Once the accessories were attached to the DSA100S, the setup was illuminated with a high powered monochromatic blue LED of wavelength 470 nm. The immsersion and withdrawal of the fiber from the liquid in a glass vessel was recorded using a high-resolution camera. The entire procedure was controlled by an automation program with defined dosing position and speed created on the ADVANCE software (ver-

sion 1.14) developed by KRÜSS GmbH, which ensured a reproducable measurement. The withdrawal speed of the fiber must be slower than the immersion speed, since the point of interest here is the meniscus formation during the fiber being pulled out from the liquid. Recordings were obtained as '.AVI' files for varying illumination settings on the DSA100S, exposure setting on the ADVANCE software, z-axis position, zoom, focus at maximum zoom and rotations of the single fiber holder.

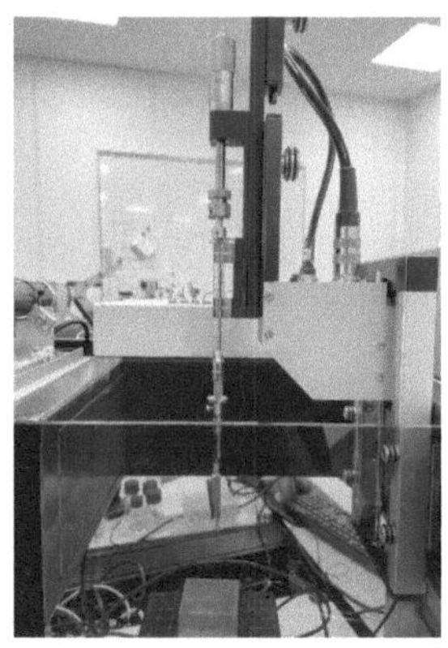

Figure 2: **Instrumental implementation** (*Left*) Accessories attached to the syringe of the DSA100S for meniscus measurement. (*Right*) A closer look to the instrumental modification showing a fiber pick-up element clipped to a single fiber holder attached to the syringe. A glass vessel is placed on the stage containing the sample liquid

2.2 The Meniscus Contact Angle Algorithm

2.2.1 Edge Detection

Once the recordings were obtained for various experimental settings, individual frames were extracted from all the recordings. In order to develop the meniscus CA algorithm, firstly, the edges of the formed meniscus during fiber-liquid interaction must be detected for every measurement condition. Edge of a digital image can be defined as a sharp change from a minimum pixel intensity to a maximum pixêl intensity. The minimum pixel intensity is considered as threshold, which means a pixel with lower intensity than the threshold is defined as black and with higher intensity than the threshold is defined as a white pixel [7]. Due to varying measurement conditions, frames from different recordings had different thesholds. Therefore, the threshold needed to be optimized in a way which could ensure edge detction regardless of the measuremnt conditions. To achieve that the frames were converted to grayscale images and mean pixel intensities were calculated for all cases. Using the calculated mean values, an inverse binary thresholding was applied to the grayscale images which converted any pixel above the threshold to black and pixel below the threshold to white [7]. Afterwards, a Canny edge detection was applied for this threshold. The Canny edge detection algorithm contains noise filtering and double thresholding and is able to detect a broad range of edges [8]. Later on, a morphological closing operation was applied on the detected contours to enhance the detected edges while reducing remaining imperfections [9]. Fig. 3 shows an individual frame and its detected edges as described.

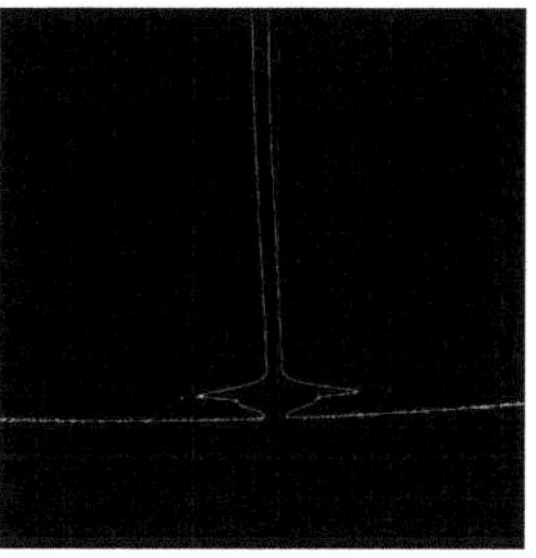

Figure 3: **Contour detection** (*Left*) Individual frame extracted from a recording with optimal illumination, maximum focus and zoom. (*Right*) Detected contours using Canny edge detection and morphological closing operation

2.2.2 Numerical Optimization

Since the desired CA is a result of interaction between the fiber and liquid surfaces, the fiber contour must be separated from the meniscus contour to differentiate these two surfaces from each other. This was achieved by selecting four different regions of interest (ROIs) containing either only the fiber contour or the meniscus contour. For a successful numerical optimization of the receding CAs at the formed meniscus, a linear fitting function was applied on the fiber contour and a theoretical non-linear curve containing a combination of an exponential and a linear equation was applied on the meniscus contour. The intersection point

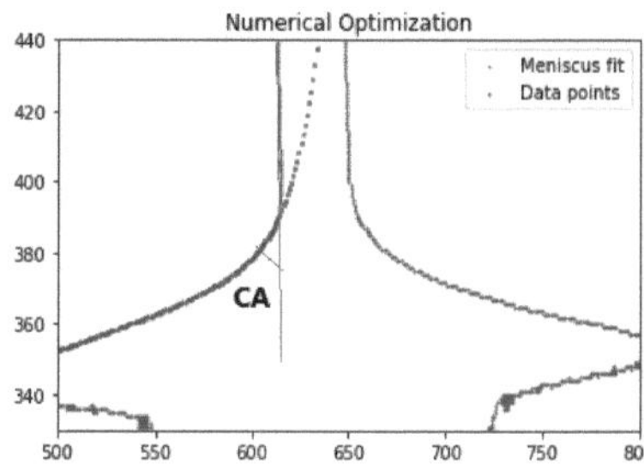

Figure 4: **Numerical optimization of fiber and meniscus contour** Fitting functions (*in red*) applied to the ROIs of the left side of the fiber-liquid mensicus. CA represents the contact angle formed at the intersection of the two fittings.

of both contour fittings was found by solving the two fitting equations. Slopes for both fitting functions were calculated at the intersection points. Finally, CAs for both sides of the meniscus were determined using the calculated slopes earlier and the mean value of the two CAs was obtained. Fig. 4 illustrates example fittings for the fiber ROI and and the mensicus ROI and the CA at their intersection on the left side.

2.3 Cross Validation with Alternative Methods and Devices

In order to validate the outcome of the developed CA algorithm a set of measurements were performed using force and optical tensiometry. A fishing line made of polyamide with a diameter of 330 µm was used as a sample fiber and double-distilled water was used as a sample liquid for these measurements. At first, the CA was measured using the DSA100S and the CA algorithm follwoing the optical method described earlier.

Afterwards, a standard KRÜSS force tensiometer K100 was used to determine the CAs occuring between fiber and liquid interface. In order to do that at first the SFT of double-distilled water must be known. Therefore, it was measured with the Wilhelmy SFT technique, which included immersion of a thin platinum plate into the vessel containing double distilled water. The resulting SFT was 72.5 mN/m. Using this value, a dynamic CA method based on Wilhelmy technique was used for CA measurements. The single fiber holder containing the fiber in a fiber pick-up element was attached to the microbalance of the instrument. The fiber was dipped in a glass vessel containing the liquid at a 3mm/min speed and CAs were measured during the wetting and dewetting of the fiber. The entire procedure was automated using the ADVANCE software. Three different portions of the sample fiber with the same fiber diameter were used to obtain comparable measurements.

Lastly, another KRÜSS drop shape analyzer DSA30M was used to determine the static CA based on sessile drop method. Microliter drops of the sample liquid were placed on the fiber and the drops were illuminated from behind with a monochromatic blue LED of wavelength 460 nm and captured by a high-resolution camera with the support of the ADVANCE software. The static CA defined by an elliptical fitting on the drop was obtained from this measurement.

3 Results and Discussion

The meniscus images processed by the CA algorithm developed earlier provided a mean CA of $37.29 \pm 2.96°$ for a polyamide/double-distilled water interface. The K100 used a dynamic CA method based on the Wilhelmy technique which measured advancing and receding CAs as well as CA hysteresis for the immersed part of the sample fiber. For three different sample fiber portions, the obtained mean ACAs between the fiber/liquid interfaces are $99.03 \pm 1.37°$, $92.96 \pm 1.75°$ and $89.89 \pm 2.37°$. And mean RCAs are $41.60 \pm 1.38°$, $41.61 \pm 1.11°$ and $40.26 \pm 1.51°$. For validation purpose only RCAs are relevant as mentioned earlier. Fig. 5 shows the CA values for K100 and DSA100S measurements. The CA obtained using the meniscus algorithm seem to slightly deviate from the RCAs obtained with dynamic CA method. This deviation indicates further requirement for numerical optimization of the algorithm e.g. applying a spline fitting on the meniscus contour.

Fig. 5 also includes the static CA measured with the DSA30M using the sessile drop method, which seems to

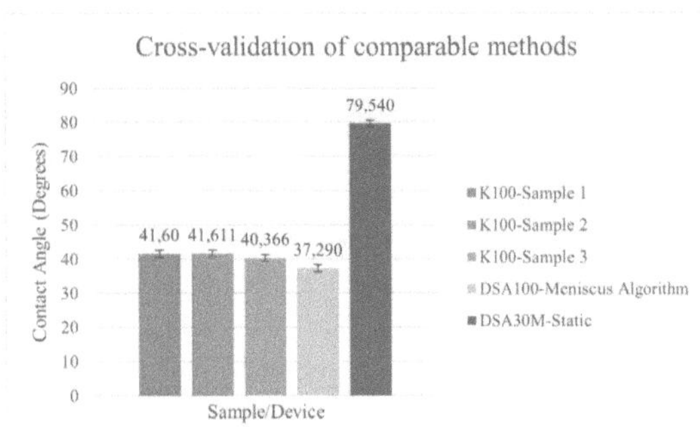

Figure 5: CAs obtained using comparable CA measurement techniques and devices for a polyamide/double-distilled water interface. The error bars indicate standard deviations from the mean CA values obtained duirng each set of measurements

be inconsistent with the remaining CA values. This inconsistency occures due to the difference in measurement techniques. Both meniscus method and dynamic CA method are based on vertical motion of the solid sample. The surface heterogenisity throughout the wetted portion of the solid sample gives rise to dynamic values of CAs for the same sample resulting to a hysteresis of the CAs. On the other hand, static CA is measured by placing a tiny drop of sample liquid onto the sample solid while everything remains in the exact same position throughout the measurement. However, for the same samples the static CA is typically within the range of advancing and receding CAs, which has been validated here by the obtained value of $79.54 \pm 5.94°$.

The instrumental setup of the meniscus method on the DSA100S was tested for different fibers such as human hair, carbon fiber (CF), nylon and fishing line. However, fibers with smaller diameters such as CF is out of scope for this report due to higher experimental efforts e.g. complicated fiber preparation, upgrading instrument optics with a microscope objective. Another point of consideration for this measurement was the influence of vibration. Use of anti-vibration tables and optimal environmental conditions are recommended during such measuremnts. Lastly, fibers prone to absorbing the sample liquid can get swollen after several measurements which can lead to extension of the wetted perimeter, thus change in CA values. Therefore, thorough environmental control must be maintained during measurements.

4   Conclusion

Wettability of surfaces plays a vital role in the biomedical industry. Therefore, various static and dynamic CA measurement techniques have been developed to precisely determine this parameter ensuring safe interactions between biomaterials and their surrounding environments. The meniscus method described in this paper offers an optical adaptation for the single fiber Wilhelmy method on a DSA.

The instrumental setup and the developed CA algorithm for this method provided successful determination of CA. CAs for the same fiber measured with comparable methods and devices validated the outcome. The CA algorithm could be further optimized numerically before its integration to the ADVANCE software as a measurement technique.

Acknowledgement

The work has been carried out at KRÜSS GmbH Wissenschaftliche Laborgeräte (Borsteler Chaussee 85, 22453 Hamburg) and supervised by the Medical Sensors and Devices Laboratory, Lübeck University of Applied Sciences.

Author's Statement

Conflict of interest: Authors state no conflict of interest.

5   References

[1] National Institute of Biomedical Imaging and Bioengineering, *Biomaterials*. Available: https://www.nibib.nih.gov/science-education/science-topics/biomaterials [last accessed on 2022-01-06].

[2] Kara L. Menzies, and Yoshito Ikada, *The Impact of Contact Angle on the Biocompatibility of Biomaterials*. Optometry and Vision Science, vol. 87, no. 1040-5488, pp. 387–399, 2010.

[3] Lotfi M'Hamdi, Nejib M'Hamdi, and Naceur M'Hamdi, *Cell Adhesion to Biomaterials: Concept of Biocompatibility*. 2013.

[4] FK,KO,TW, *Wettability of carbon fibers using single-fiber contact angle measurements – a feasibility study*. Krüss GmbH, Appl. Note AR271e

[5] Oleg N. Tretinnikov, and Lyndon Jones, *Dynamic wetting and contact angle hysteresis of polymer surfaces studied with the modified Wilhelmy balance method*. American Chemical Society, *Langmuir*, Vol. 10, No. 5, pp. 1606–1614, 1994.

[6] C. Rulison, *Contact Angle Determinations by the "Straw" Method and Packed Cell Method*. Krüss GmbH, Appl. Note AR206e.

[7] Prathima Guruprasad, *Overview of Different Threshold Methods in Image Processing*. 2020.

[8] Weibin Rong, Zhanjing Li, Wei Zhang, and Lining Sun, *An Improved Canny Edge Detection Algorithm. IEEE International Conference on Mechatronics and Automation*, pp. 577–582, 2014.

[9] Megha Goyal, *Morphological Image Processing. International Journal of Computer Science Technology*, Vol. 2, No. 4, pp. 161–165, 2011.

Simulation and optimization of blood flow in a rectangular microchannel

Karan Grewal [1,2], Reza Behroozian [2], Benjamin Kern [2] and Stefan Müller [2]

[1] Biomedical Engineering, Luebeck University of Applied Sciences, karan.grewal@student.uni-luebeck.de
[2] Medical Sensors and Devices Laboratory, Luebeck University of Applied Sciences

Abstract

Fast and portable blood testing devices are important as ever before. In this paper a test chip for optical blood measurement was analyzed and optimized. A rectangular microchannel having a dynamically changing geometry must fill with blood evenly and fast. First, a mathematical analysis assuming blood as a Newtonian Poiseuille flow within Ansys Fluent using a two-phase VOF-model and a calculation based on the electric circuit analogy were carried out. Secondly, the microfluidic device was manufactured and tested, comparing mathematical with experimental data. Three different designs were analyzed, where the filling time from the first design to the third design improved by $16\,\mathrm{s}$. Experimental data for design one and two yielded similar results compared to the mathematical data, while two out of three designs were manufactured. Flow separation in the first and second design was predominant, whereas it was greatly reduced in the third design.

1   Introduction

Microfluidic devices transport very small amounts of liquid, resulting in cost efficient and portable blood testing devices. In traditional fluidics the effect of inertia, applied pressure and potential energy are dominant, but in microfluidics the liquid within the device moves due to capillary motion. Capillary motion depends on the interplay between liquid surface tension, geometry and material surface chemistry. In this paper we present a framework for the geometry optimization of a microchannel with rectangular cross-section used in optical blood analysis. The optical measurement device is used in emergency situations where measurement time decides over patient life. Consequently, the goal is to improve the blood filling time of the test chip while ensuring an evenly filled microchannel aiming fast and reliable measurements.

2   Fundamentals

Understanding blood flow in microchannels requires basic understanding of hydrodynamic fluid behavior. Blood is a non-Newtonian fluid resulting in a non-constant viscosity. The viscosity μ for blood depends on the shear-rate $\dot{\gamma}$, hence blood is a shear-thinning fluid [1]. Viscosity also has an Arrhenius dependence on temperature T, meaning that fluid viscosity increases exponentially with decreasing temperature. Generally, problem statements involving blood assume a constant temperature at $37\,°\mathrm{C}$. Viscosity models like the Power-law, Carreau and Cross model represent the shear-rate dependant blood viscosity $\mu_{\dot{\gamma}}$.

2.1   Flow assumptions

Flow in this work is approximated as incompressible ($\rho = const.$) and the system as adiabatic ($T = const$) because no major flow pressure and temperature variations are present.

For the mathematical calculation and CFD simulation the blood is assumed to be Newtonian. This results in the incompressible Navier-Stokes equation for uniform-viscous Newtonian fluids neglecting gravity and assuming conservation of mass [2]:

$$\rho\frac{\partial \vec{v}}{\partial t} = -\rho\,\vec{v}\,\nabla\vec{v} - \nabla\rho + \mu\nabla^2\vec{v} \qquad (1)$$

The momentum change on the left side of the equation is described by the sum of convective-, pressure- and viscous forces on the right side. Assuming a long rectangular channel and a unidirectional flow velocity, the unsteady and convection terms are assumed to be zero, resulting in:

$$\rho\frac{\partial \vec{v}}{\partial t} = \mu\nabla^2\vec{v} \qquad (2)$$

This pressure-driven motion is termed Poiseuille flow, having a parabolic velocity profile and is used to describe the volume flow rate Q. The Reynolds number Re defines the ratio of inertial to viscous forces, describing whether the flow is laminar or turbulent. The fluid flow in a microchannel is predominantly laminar with low Reynolds number ($Re << 1$, height and width dependent), where viscous forces are dominant and the no-slip condition valid.

2.2   Flow model

Capillary pressure difference depends on the wettability of the microchannel material. A surface is considered as wettable if the contact angle θ of a liquid on the surface is $< 90\,°$. The wettable microchannel generates a concave liquid-air interface with the wetting liquid (liquid in contact with the wettable microchannel surface), moving the wetting liquid into the microchannel. Capillary pressure is defined as the difference in pressure across the interface between two immiscible fluids, resulting in the following equation:

$$\Delta p = p_c = p_{air} - p_{blood} \qquad (3)$$

A microfluidic device with a capillary action driven flow is filled with a wetting fluid ($\theta < 90°$) and a non-wetting fluid ($\theta > 90°$). The pressure difference Δp is defined with the non-wetting fluid air p_{air} and the wetting liquid pressure p_{blood}, thus the capillary pressure for a rectangular microchannel with four contact angles θ_t, θ_b, θ_l and θ_r representing the four fluid contact surfaces in the microchannel, width w and height h can be determined with [3]:

$$\Delta p = -\sigma_{ba} \left[\frac{cos(\theta_t)+cos(\theta_b)}{h} + \frac{cos(\theta_l)+cos(\theta_r)}{w} \right] \qquad (4)$$

The surface tension σ_{ba} in $\mathrm{N\,m^{-1}}$ is the blood-air surface tension dependent on the temperature. A temperature T dependent surface tension equation derived from experimental data for blood in the range of 20 °C to 40 °C is expressed as [4]:

$$\sigma_{ba}(T) = (-0.473 \cdot T + 70.105) \cdot 10^{-3} \qquad (5)$$

The volumetric flow rate Q for pressure-driven Poiseuille flow in a rectangular microchannel with high aspect ratio ($\frac{h}{w} << 1$) can be described with [5]:

$$Q = \frac{h^3 w \Delta p}{12 \mu L} \left[1 - 0.63 \frac{h}{w} \right] \qquad (6)$$

Then the flow velocity v is derived from the volumetric flow rate Q and the cross-sectionional area A:

$$v = \frac{Q}{hw} = \frac{h^2 \Delta p}{12 \mu L} \left[1 - 0.63 \frac{h}{w} \right] \qquad (7)$$

The time t needed to move the advancing blood-air meniscus through the microchannel can be calculated with the volumetric flow rate Q and the volume V at position L:

$$t = \frac{Lhw}{Q} = \frac{L^3 6 \mu}{\Delta p h^5} \left[1 - 0.63 \frac{h}{w} \right]^{-1} \qquad (8)$$

Losses in laminar flow are mostly minor losses due to bends or valves. In a microchannel with $Re << 1$ the hydraulic resistance R_h dependent on the geometry is predominantly decreasing flow rate Q. For the aspect ratio $\frac{h}{w} << 1$ the hydraulic resistance simplifies to:

$$R_h = \frac{12 \mu L}{h^3 w} \left[1 - 0.63 \frac{h}{w} \right]^{-1} \qquad (9)$$

Combining equation 4, 6 and 9 based on the simplifications established in equation 2 the Poiseuille flow in a microchannel has the following relationship:

$$\Delta p = Q R_h \qquad (10)$$

Equation 10 concludes the following: (a) pressure difference Δp is proportional to volumetric flow rate Q and hydraulic resistance R_h. (b) hydraulic resistance R_h is constant if fluidic and geometric conditions are constant; then

$R_h \propto L$ is valid. This implies that the relationship in equation 10 can be used to derive flow rate, velocity, flow time, and pressure for a rectangular microchannel.

2.3 Electric circuit analogy

The equations established in section 2.2 are valid in the context of a straight microchannel with fixed length, but microchannel networks used in microfluidic devices consist of canals, valves and mixing chambers. In order to apply the established fluid model onto a variety of microchannel geometries the electric circuit analogy is used. Some physical similarities are the volume flow rate, the hydraulic resistance and the pressure drop opposing to the electric current, the electric resistance and the voltage drop [6]. Analog to linear electric resistors the laminar fluidic resistor is a two-terminal (one input and output terminal) passive fluidic element where the resistance is dependent on the channel length. If n hydraulic resistors in a microfluidic network are arranged in series, the total resistance is the sum of n hydraulic resistances:

$$R_{h,eq} = R_{h,1} + R_{h,2} + ... + R_{h,n} \qquad (11)$$

For resistors connected in parallel the reciprocal equivalent fluidic resistance for n fluidic resistors is the sum of reciprocal single laminar fluidic resistors:

$$\frac{1}{R_{h,eq}} = \frac{1}{R_{h,1}} + \frac{1}{R_{h,2}} + ... + \frac{1}{R_{h,n}} \qquad (12)$$

For parallel microfluidic channels Kirchhoff's current law is valid, assuming that the sums of flows into a node should be equal to the flows leaving the node. The total pressure drop $\Delta p_{c,eq}$ in a serial fluidic network is the sum of the individual pressure drops in the circuit $\Delta p_{c,eq} = \Delta p_{c,1} + \Delta p_{c,2} + ... + \Delta p_{c,n}$ based on Kirchhoff's voltage law.

3 Material and Methods

Electric circuit analogy and the established fluid model were used to improve the geometry with the goal to decrease filling time. The CFD simulation had the goal to describe the filling behavior of the test chip.

3.1 Experimental Procedure

The test chip was placed on graph paper and the blood was tempered to 37 °C, while the reference temperature was 25 °C. The experiments were filmed with a high resolution camera. The blood samples were treated with sodium citrate to prevent coagulation. The measured hematocrit (HCT) was 0.42.

3.2 Design and manufacturing

Fig. 1 shows the design iterations D1 at the top , D2 in the middle and D3 at the bottom having a total length $L = 60$ mm. Due to the use in an optical sensor, the height h and width w in the measurement windows of the test chip can not be changed. In the three designs the main measurement window (dotted rectangle) has a width of 6 mm and the sensor spots (solid rectangle) are 3 mm wide.

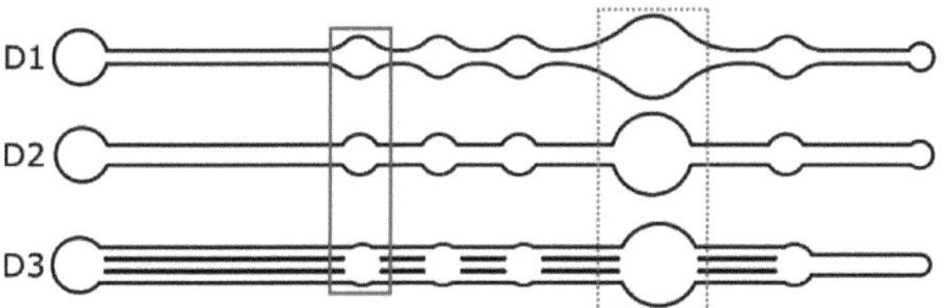

Figure 1: Design D1 (top), D2 (middle) and D3 (bottom)

In D1 the channel width between sensor spots and measurement window is 1 mm. For D2 the measurement window and sensor spots are not streamlined and the channel width is increased to 1.5 mm, with the goal to increase the volumetric flow rate and decrease the hydrodynamic resistance. In D3 the channel width is 0.8 mm, with a three channel design functioning as parallel resistors, increasing the volumetric flow rate by decreasing hydrodynamic resistance significantly while the cappillary pressure is constant. Capillary pumps function similar to geometry D3, where the parallel microchannels are increased while increasing flow rate. A channel width smaller than 0.8 mm could not be manufactured with the available manufacturing technique, even though small channel sizes would benefit the viscosity behavior of blood. The microchannel geometry was cut into a double sided adhesive spacer tape with a thickness of 100 µm. A top and bottom plate enclose the spacer tape, manufactured with a laser cutter. A coating with polyvinyl alcohol for increasing the wettability is printed onto the top and bottom plates, while the microchannel walls were not coated. Lastly, the spacer tape is bonded with the top and bottom plates.

3.3   Simulation and calculation

The mathematical input parameters were determined through experimental data. Blood velocity in the input region prior to the first optode with a length of 20 mm is determined through a high speed camera, then the contact angles for calculating the capillary pressure are changed until the experimentally determined input velocity matches the calculated input velocity. The simulation was carried out in Ansys Fluent using the Volume of fluid model, while multi zone meshing with unstructured hexa- and tetrahedral elements was used [7]. The channel is filled with air at P_{amb}. Blood entered the microchannel at Δp creating a negative outlet pressure difference. The viscosity was set to 0.005 Pa s, which is the upper limit for blood reaching Newtonian behavior for a HCT at 0.45 and $T = 37\,°C$. The channel surface wettability was realized using the wall-adhesion function with contact angles for the top θ_t and bottom surface θ_b are set to 78 °, whereas the contact angles for left θ_l and right surface θ_r are set to 103 °. The system is assumed to be adiabatic and the flow behavior was set to laminar. For the calculation the microchannel geometry was divided into small resistors $R_{h,n}$ arranged in series or in parallel with a fixed length ΔL, allowing the width w to change dynamically along channel length L.

4   Results and Discussion

Table 1 shows the microchannel filling times. The experiment and simulation data were averaged over ten runs. The calculation, CFD simulation and the experiments yield similar results. Filling time from D1 to D2 improved marginally by 3.7 % in the simulation and 3.3 % in the calculation. For D3 it improved by 43.2 % in the simulation and 46.1 % in the calculation compared to D1. As for the experiments the filling time from D1 to D2 improved by 4.4 %.

Table 1: Filling time of three microchannel in seconds

Design	Calculation	Simulation	Experiments
D1	38.09	38.00	37.67
D2	36.85	37.00	36
D3	20.06	21	-

Table 2 shows the absolute error between the calculation methods. The error between the calculation and the simulation is below 5 %. Between simulation and experiment the absolute error lies below 3 % and between calculation and experiment below 2.5 %.

Table 2: Absolute error E between methods in percent

Design	Calculation-Sim.	Sim.-Ex	Ex.-Calculation
D1	0.23	0.88	1.10
D2	0.42	2.78	2.3
D3	4.47	-	-

In Fig. 2 the volumetric flow rate Q of the blood-air meniscus along the microchannel length L is shown. In D1, D2 and D3 the flow rate in the sensor spots peaks at $0.515\,\mu l/s$ and in the measurement window at $1.04\,\mu l/s$. However, in the straight channel region the flow rate increased from $0.15\,\mu l/s$ in D1 to $0.24\,\mu l/s$ in D2 resulting in an improvement of 60 %. For D3 the flow rate in the straight channel increased significantly to $0.3\,\mu l/s$ resulting in a 126 % increase compared to D1.

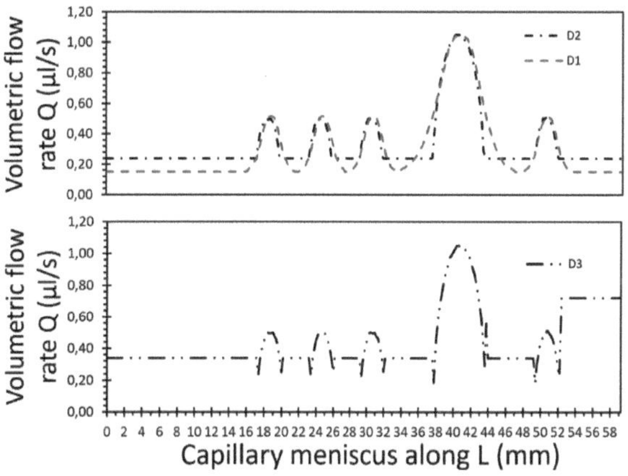

Figure 2: Flow rate Q along microchannel in D1 (top), D2 (top) and D3 (bottom)

In Fig. 3 the flow velocity v of the blood-air meniscus along the microchannel length L is shown. In D1, D2 and D3 the velocity in the sensor spots is peaking at $1.7\,mm\,s^{-1}$ and in the measurement window at $1.8\,mm\,s^{-1}$. However, in the straight channel region the velocity increased from $1.5\,\mathrm{mm\,s^{-1}}$ in D1 to $1.6\,\mathrm{mm\,s^{-1}}$ in D2 resulting in an improvement of $6\,\%$. For D3 the velocity in the straight channel increased significantly to $4.3\,\mathrm{mm\,s^{-1}}$ resulting in an $185\,\%$ increase compared to D1.

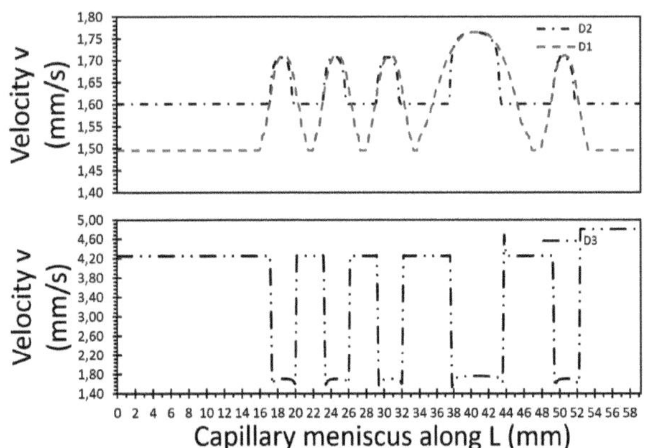

Figure 3: Velocity v along microchannel in D1 (top), D2 (top) and D3 (bottom)

According to the calculation results from design D1 and D2, the volume flow rate and the velocity can not be improved purely through changing the width w of the microchannel, which results in an higher aspect ratio. D3 shows that decreasing the aspect ratio, with multiple microchannels in parallel leads to a noticeable speed improvement. Another benefit of D3 is the fact that blood is a shear-thinning fluid, so that the fluid viscosity will decrease in a channel with smaller aspect ratio further, if the fluid is assumed non-Newtonian using a viscosity model. In the CFD simulation the flow profile in the sensor spots and the optical measurement window can be seen in Fig. 4.

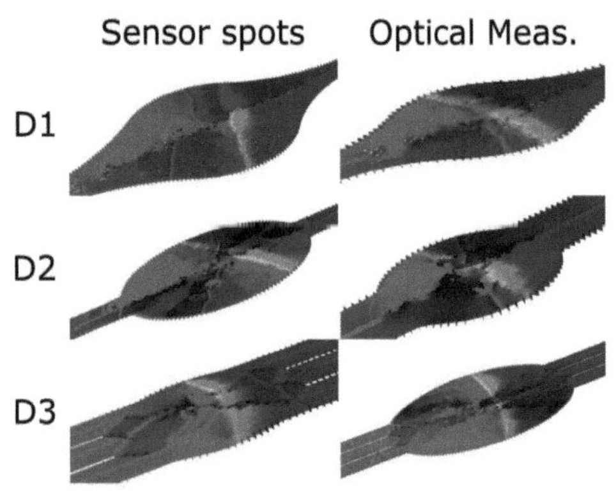

Figure 4: Flow profile in the sensor spot and optical measurement window from design iteration D1 to D3

For D1 and D2 the velocity at the left and right boundary is not the same leading to flow separation. This was observed in both the experiments and simulation. In D3 the sensor spot and the optical measurement window were filled evenly, preventing flow separation between boundary and main flow.

5   Conclusion

A framework for analyzing microchannels with dynamically changing geometry size was developed. The calculation method based on the electric circuit analogy showed how blood flow is influenced by different microchannel geometries. CFD simulation and experimental data matched with the calculations. Assumptions like a constant blood viscosity did not have a significant impact on filling time. Design D3 was the best design according to the simulation and calculation but has to be proven in experiments. A different manufacturing method gives the opportunity of manufacturing D3 and have more stable experimental results. The goal of analyzing and designing a fast and evenly filling test chip was reached ensuring fast and reliable optical blood measurements. However, the CFD simulation and the calculations are only valid if blood is assumed to be Newtonian and the geometry stays simple. For more complicated geometries with mixing chambers or valves, the usage of a velocity model assuming blood as a non-Newtonian fluid is crucial. Even though this would result in increased computation time and calculation complexity.

Acknowledgement

The work has been carried out in the Medical Sensors and Devices Laboratory at the University of Applied sciences.

Author's Statement

Conflict of interest: The corresponding author states that there is no conflict of interest.

6   References

[1] P. C. Barman, R. R. Kairi, A. Das, and R. Islam, *An Overview of Non-Newtonian Fluid*, Intern. Jour. Appl. Scie. and Engin., vol. 4, no. 2, p. 97, 2016.

[2] A. L. Gerhart, T. H. Okiishi, J. I. Hochstein, and P. M. Gerhart, *Munson, Young, and Okiishi's: fundamentals of fluid mechanics*, SI version. 2021.

[3] A. Olanrewaju, M. Beaugrand, and D. Juncker, *Capillary microfluidics in microchannels: From microfluidic networks to capillaric circuits*, Lab on a Chip, vol. 18, Jul. 2018.

[4] J. Rosina, E. Kvašňák, D. Šuta, H. Kolářová, J. Málek, and L. Krajčí, *Temperature dependence of blood surface tension,*" Physiol Res, pp. S93–S98, 2007.

[5] H. Bruus, *Theoretical microfluidics*, Oxford; New York: Oxford University Press, 2008.

[6] K. W. Oh, K. Lee, B. Ahn, and E. P. Furlani, *Design of pressure-driven microfluidic networks using electric circuit analogy*, Lab on a Chip, vol. 12, no. 3, pp. 515–545, 2012.

[7] Ansys Confidential, *Fluent Multiphase*, presented in Lecture 3, 2015. Accessed: Jan. 16, 2022. [Online].

Laboratory setup for wavefront measurement with Swept-Source OCT with defocus correction

Anika Reinecke[1,2], Michael Stender [2] and David Klinger [2]

[1] Medical Engineering Science, Universität zu Lübeck, anika.reinecke@student.uni-luebeck.de
[2] Heidelberg Engineering GmbH, {anika.reinecke,michael.stender,david.klinger}@heidelbergengineering.com

Abstract

Today wavefront analysis is important in ophthalmology for measuring the wavfront errors of the human eye. One method for determination of the wavefront aberrations is an autorefractometer. However, this method usually only provides information about the defocus and astigmatism of the eye. In some cases, e.g. before refractive laser surgery, more detailed information about the wavefront error is needed. Therefore measurements have to be made with a wavefront sensor. This paper deals with a new approach to calculate the wavefront error using the phase information of Swept-Source optical coherence tomography (SS-OCT) images. For preventing sensitivity loss in the OCT configuration for ametropic eyes, a labratory setup with a defocus correction was developed. The optics design is developed in Zemax (Zemax LLC.) and proves that the defocus correction can be done with focus tunable liquid lenses. The simulation of the hardware setup is done in SolidWorks (Dassault Systèmes SolidWorks Corp.) and shows that the optical setup is integratable into a 30 mm cage system from Thorlabs, Inc.

1 Introduction

1.1 Wavefront measurement

The first wavefront sensor was the Hartmann-Shack sensor, which exists in its current form since 1970. The sensor consists of a micro lens array which focuses the light onto an optical detector, usually a CCD or CMOS camera. Depending on the slope of the wavefront, the focus points on the camera chip are shifted relative to a reference point. By determining the difference between the ideal focus position and the measured one, the local slope of the wavefront can be calculated. With the help of a feedback system and adaptive optics, for example deflectable micro mirror arrays, the wavefront aberrations can be corrected [1], [2]. In 1977, this technique was used for the first time in astronomy [3], and in 1994, aberrations in the human eye were measured in this way for the first time [4].

A method to correct wavefront errors without adaptive optics is to use digital adaptive optics (DAO). In this case, algorithms correct the optical aberrations in a post-processing step. An advantage of DOAs is that no additional hardware modifications are necessary to calculate the wavefront error [5].

1.2 Swept-Source OCT

Optical coherence tomography (OCT) is an optical imaging technique that was invented in the 1980s. Since its invention, the technique has been continuously improved and is now an important feature in ophthalmology.

The base of this technique is a Michelson interferometer with a short coherent infrared light source. The light is divided by a beam splitter and is directed into a reference arm and a sample arm. The reference arm beam is back reflected by a mirror, the sample arm beam enters the eye and is reflected by the different layers of the retina. The OCT signal can be determined by the interference of light from the reference arm and the sample arm.

Due to the short coherence length of the light source, interference only occurs when the optical path length of the two arms is within the range of the coherence length.

There are two main techniques in OCT: Time-Domain and Fourier-Domain OCT.

A variant of Fourier-Domain OCT is Swept-Source OCT. A narrow-band light source with a tuneable wavelength range is used. During one sweep, each wavelength component of the interference signal is detected sequentially by a high-speed photodetector. Afterwards, the signal, that already contains depth information, is Fourier transformed [6], [7]. In order to obtain high lateral resolution in addition to high axial resolution, hardware-based adaptive optics have been successfully combined with OCT in the past. [8].

1.3 OCT and wavefront measurement

Since OCT is an interferometric technique, information about the phase of the light is included. With the help of DAO, this phase information can be used to correct aberrations in post-processing and to achieve diffraction-limited lateral resolution [3], [5].

Most DAO algorithms use a methode for optimise the image quality based on a sharpness metric. The coefficients of a phase correction function are iteratively changed until a predefined value of image sharpness is reached [5].

The digital lateral shearing based digital adaptive optics (DLS-DAO) algorithm, which is used here, was developed by Abhishek Kumar and is based on a volumetric point spread function (PSF) scanning. The DLS-DAO does not require any iteration steps. Besides the improved performance by avoiding the iteration, this DAO algorithm has the advantage that no a priori knowledge about the system is required, for example wavelength, focal length, numerical aperture (NA) or information about the detector pixel size [3], [9].

The main difference to a scanning OCT configuration is that the illumination of the retina is provided by a separate, narrow illumination beam which is stationary (Fig. 1). The illumination creates an almost diffraction-limited guide star on the retina. The PSF created by the guide star is de-scanned in the detection fiber plane using XY-galvoscanners. In that way a 3D-OCT dataset is generated. Afterwards, the slope of the wavefront is calculated using the DLS-DAO algorithm [10].

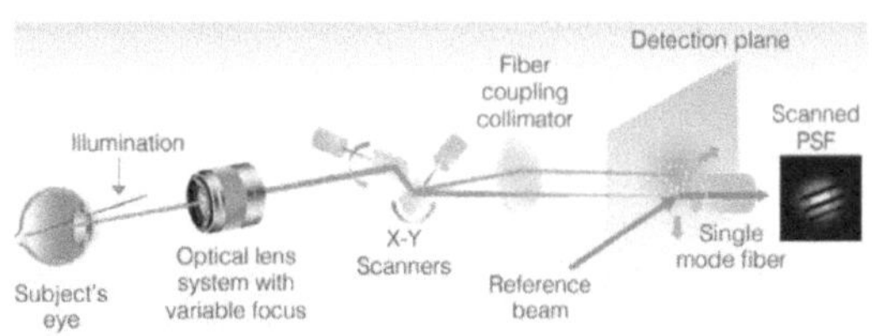

Figure 1: PSF scanning principle (by Abhishek Kumar) OCT configuration with a narrow fixes illumination beam and a de-scanned PSF in the detection fiber plane.

Due to its small beam diameter, the illumination has a low NA and therefore ensures only minor aberrations in the beam while passing the eye. The detected beam, which was reflected back from the retina, has a NA of approx. 6 mm (diameter of the pupil of the eye). This means the NA of the detected beam is 10 times larger than the one of the illumination beam. This leads to a detection of many aberrations.

The necessary phase stability is given by a fast sweep rate of the light source. The B-scan rate must be larger than 1.5 kHz [10].

If the defocus in the eye is higher than 2 diopters, the spot size of the PSF is blurred and more than 10 times larger than the diffraction limited spot size. Scanning the PSF already leads to a loss of sensitivity. If aberrations such as a high defocus are also present, the loss of sensitivity is even high and the collection efficiency decreases fast. If the sensitivity is too low, the phase noise increases and the quality of the aberration calculation decreases [9].

The aim of this work is to show that it is possible to design and realise a laboratory setup with a defocus correction by using tunable liquid lenses with variable focal length. This should improve the collection efficiency for ametropic eyes.

2   Material and Methods

The concept of this SS-OCT setup is a de-scanning configuration with a fixed illumination spot on the retina. A swept-source laser (SSL) with a central wavelength of 1060 nm is used as light source. A fiber coupler is used to split the light from the SSL into a reference arm and a sample arm. The setup of the sample arm is shown in Fig. 2.

In the sample arm, the light from the SSL is injected via a collimator and it passes through an achromat into a 90:10 (90 % transmission; 10 % reflection) beam splitter. The light that is reflected by the beam splitter reaches the eye via an objective lens (OL). This represents the injection arm. The lens behind the collimator and the objective lens form a telescope. The focal lengths and distances of the two lenses must be selected so that a narrow, collimated beam with a diameter of approx. 0.6 mm reaches the cornea. The beam is focused by the optical system of the eye on the retina and produces a guide star which has a diffraction limited diameter of approx. 60 μm.

The back reflected light from the retina passes through the optical media of the eye and the pupil of the eye through the objective lens and the beam splitter into the detection arm. In the detection arm, the light passes through two tunable liquid lenses (TL) with adjustable focal lengths, which form a telescope. The focal lengths of the two lenses must be chosen in such a way that the beam is collimated behind the second TL and the optical system has a total de-magnification of 2.5. This ensures that the Nyquist sampling criterion is fulfilled and that no clipping of the beam occurs on the following x- and y-scanning mirrors. After the light has passed the scanning mirrors, it is focused on the tip of a single mode fiber by a collimator.

Afterwards, the interference signal from the sample arm and the reference arm is detected by a dual balance detector.

The optics design program Zemax (Zemax LLC.) was used to plan the laboratory setup. Optics design programs can be used for the design and analysis of imaging and illumination systems. In this work, the program was used to select suitable lenses and optical elements (e.g. collimators, beam splitter). In simulating the optical setup, it is possible to verify that the beam in the injection arm, with the selected lenses and at a certain distance between the lenses, is collimated and meets the cornea with a diameter of approx. 0.6 mm. In the same manner, properties of the optical system of the detection arm can be simulated at different focal lengths of the TLs. It can be verified whether the beam is collimated after the second TL and whether the optical system in the detection arm leads to an total de-magnification of 2.5. In this way, it can be simulated if a defocus in the eye can be compensated by using tunable lenses.

Afterwards, the computer-aided design program (CAD) Solidworks (Dassault Systèmes SolidWorks Corp.) was used to plan the laboratory setup. In Solidworks, the integration of the laboratory setup into the 30 mm cage system from Thorlabs, Inc. was simulated. To hold the x- and y-galvoscanning mirrors a component was designed and was manufactured using a 3D printing procedure.

3   Results

3.1   Optical Setup

The schematic of the reference arm setup with defocus correction is shown in Fig. 2. In the injection beam path, the light is coupled out with the collimator C1 (F230APC-1064, Thorlabs, Inc.). The collimated light beam passes through an achromat (AC254-250-C-ML, Thorlabs, Inc.) with a focal length of 250 mm. In combination with the objective lens, the achromat creates a telescope. The magnification has to be created in such a way that the spot size on the cornea is approximately 0.6 mm. A beam splitter cube splits the beam so that a part of the light is directed in the eye. An achromat (AC254-150-C-ML, Thorlabs, Inc.) with a focal length of 150 mm is used as OL.

The back scattered light passes through the OL and through two tunable liquid lenses (Optotune Switzerland AG) with an adjustable focal length.

For an emmetropic eye, the beam exits the eye as a collimated beam. The focal length of the first TL is set to 150 mm and the focal length of the second TL is set to 32.9 mm. If an ametropia, e.g. hyperopia, exists in the measured eye, the beam exits the eye as a divergent beam. The focal length of the tunable lenses has to be adjusted. The focal length of the first TL is set to 72.2 mm and the focal length of the second TL is set to 14.9 mm. The lenses are arranged with a distance of 70 mm.

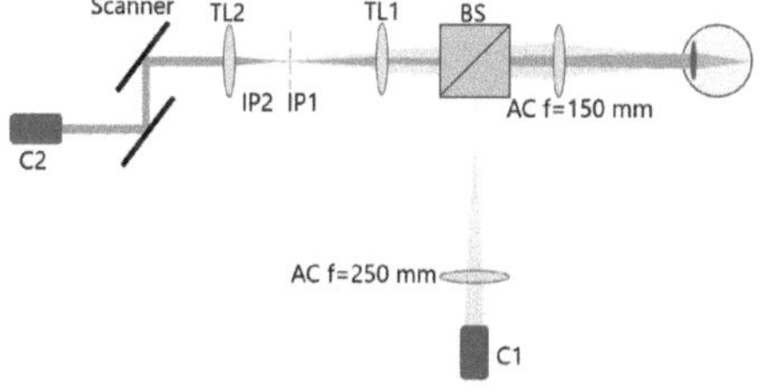

Figure 2: Schematic of the reference arm setup. The collimator C1 injects the beam of the SSL, narrow illumination beam creates a guide star on the retina. Tunable lenses (TL1&TL2) create a telescope with adjustable image plane (IP) in between. The red beam shows the beampath of the detection beam for an emmetropic eye. The position of the IP is at IP1. The yellow beam shows the beampath of the detection beam for a hyperopic eye with a defokus of >20 dpt. The position of the IP2 is shifted towards TL2. Behind TL2 the beam is collimated and passes the scanning mirrors. The collimator C2 couples the light into the detection fiber.

In comparison to the emmetropic eye, the image plane which is located between the two tubeable lenses is shifted in direction of the second tuneable lens.

The galvo-scanner are located behind the TLs in the detection beam path. At the end of the detection beam path the collimator C2 (F260APC-1064, Thorlabs, Inc.) couples the collimated beam into a detection fibre.

3.2   CAD-Design

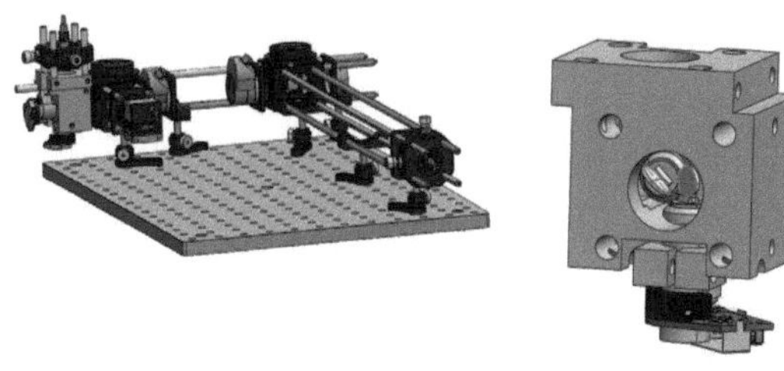

Figure 3: CAD simulation of the labrabory setup (left) with the new cube for the galvo scanner (right), integrable into a 30 mm cage system.

The optical design was integrated into the 30 mm cage system from Thorlabs and realised with a breadboard set-up. The component that holds the galvo scanners and the detection collimator unit was designed and is also compatible with the cage system. The galvo scanning mirrors which are insertable into the component, have a distance of 8 mm to each other. A small distance between the mirrors is important to keep the field curvature as small as possible.

4   Discussion

An available tool in Zemax (Zemax LLC.) is the spot diagram. Spot diagrams are maps that show where rays with preselected wavelengths intersect a selected surface. Therefore, it is a suitable tool to determine the beam diameter on surfaces. The beam radius is given as RMS (root mean square) radius.

An analysis with the standard spot diagram (Fig. 4) shows a RMS radius of 255 µm on the cornea. This represents an illumination beam diameter of 0.51 mm on the cornea. The required narrow beam for the illumination of 0.6 mm is therefore achieved.

The simulation of the defocus correction shows the following results:

For an emmetropic eye an analysis with the standard spot diagram (Fig. 5, a) in Zemax shows a RMS radius of 99 µm behind the second TL. This equals a de-magnification of 2.57 and matches the required de-magnification of approximately 2.5.

The analysis with the standard spot diagram for a hyperopic eye (Fig. 5, b) shows a RMS radius of 104 µm behind the second TL. This equals a de-magnification of 2.44 and the required de-magnification of approximately 2.5 is achieved. The spot diagrams were generated on the surfaces of the scanner mirrors. Since the mirrors have a tilt angle of 45°, the spot on the mirror is not round but oval.

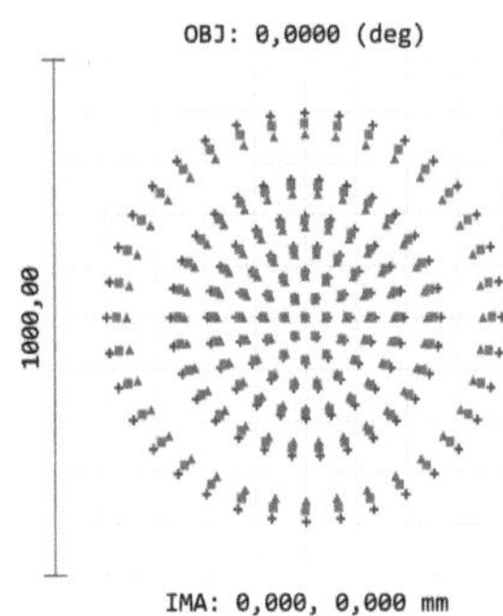

Figure 4: Spotdiagram of the beam on the cornea:
An analysis of the injection beam shows a RMS radius of 255 µm on the cornea. The blue crosses represent a wavelength of 1050 nm, the green squares a wavelength of 1100 nm and the red triangles a wavelength of 1150 nm.

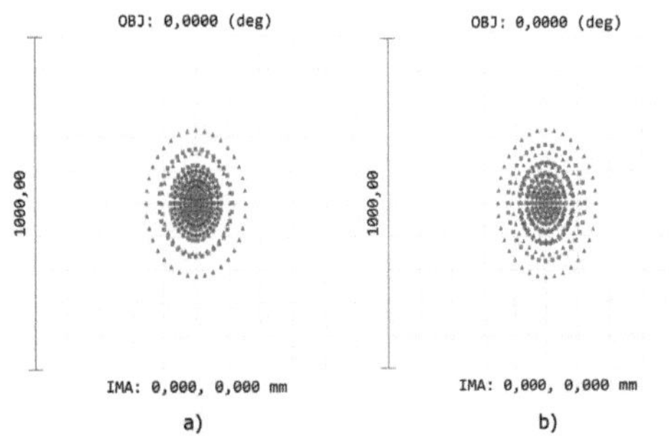

Figure 5: Spotdiagram after second TL for an emmetropic (a) and a hyperopic eye (b):
The spot diagram for an emmetropic eye shows a RMS radius of 99 µm. The spot diagram for an hyperopic eye shows a RMS radius of 104 µm. The blue crosses represent a wavelength of 1050 nm, the green squares a wavelength of 1100 nm and the red triangles a wavelength of 1150 nm.

5   Conclusion

In this paper, a concept for the optics design and CAD simulation for a new laboratory setup for wavefront measurement with swept-source OCT including defocus correction was shown.
The simulation and analysis in Zemax shows that it is possible to use tunable liquid lenses for defocus correction in the labratory setup in order to prevent sensitivity loss for ametropic eyes. By using two TLs, it is possible to collimate the detection beam and maintain a total demagnification of 2.5 even though the refractive error of the measured eyes varies.
Currently we are working on proving that the laboratory setup functions in practice.

Acknowledgement

The work has been carried out at Heidelberg Engineering GmbH, and supervised by Prof. Dr. rer. nat. Robert Huber, Institute of Biomedical Optics, Universität zu Lübeck.

Author's Statement

Conflict of interest: Authors state no conflict of interest.

6   References

[1] J. Hartmann, *Bermerkungen ueber den Bau und die Justierung von Spektrographen.* In: Zeitschrift fuer Instrumentenkunde, Vol. 20, pp. 47, 1900.

[2] R. Shak, *Production and use of a lecticular hartmann screen.* In: J. Opt. Soc. Am. A 11,pp. 656-661, 1971.

[3] A. Kurma, *Digital adaptive optics for achieving space-invariant lateral resolution in optical coherence tomography.* Center for Medical Physics and Biomedical Engineering, Vienna, 2015.

[4] J. Liang, B. Grimm, S. Goelz and J.F. Bille, *Objective measurement of wave aberrations of the human eye with the use of a Hartmann–Shack wave-front sensor.* In: JOSA A, Vol. 11, pp. 1949-1957, 1994.

[5] J.R.. Fienup and J.J.. Miller *Aberration correction by maximizing generalized sharpness metrics.* In: J. Opt. Soc. Am. A 20, pp 609-620, 2003.

[6] W. Drexler and J.G. Fujimoto *Optical coherence tomography: technology and applications.* Springer: Berlin New York, 2008.

[7] M. Kaschke, K.-H. Donnerhacke and M.S. Rill, *High resolution imaging in microscopy and ophthalmology : new frontiers in biomedical optics.* Springer: Cham, Switzerland, 2019.

[8] M. Ruggeri et al. *Combined anterior segment OCT and wavefront-based autorefractor using a shared beam.* In:Biomed. Opt. Express, Vol. 12, pp. 6746-6761, 2021.

[9] A. Kurma, S. Georgiev, M. Salas and R.A. Leitgeb, *Digital adaptive optics based on digital lateral shearing of the computed pupil field for point scanning retinal swept source OCT.* In:Biomed. Opt. Express, Vol. 12, pp. 1577-1592, 2021.

[10] S. Georgiev et al. *Digital ocular swept source optical coherence aberrometry.* In:Biomed. Opt. Express, Vol. 12, pp. 6762-6779, 2021.

Development of a Measuring Chamber for a Fluorescent Opto-chemical Sensor for In-Vitro Blood Analysis

Daniela Trujillo Cabrera [1,2], Reza Behroozian [2], Benjamin Kern [2], Stefan Müller [2]

[1] Biomedical Engineering, Luebeck University of Applied Sciences, daniela.trujillo.cabrera@stud.th-luebeck.de

[2] Medical Sensors and Devices Laboratory, Luebeck University of Applied Sciences, {reza.behroozian, benjamin.kern, stefan.mueller}@th-luebeck.de

Abstract

Opto-chemical sensors enable the determination of pH using a dye, whose fluorescence strongly depends on the pH. In this study, the dye HPTS is used. To develop a measuring chamber for an opto-chemical sensor, it was necessary to characterize the HPTS first. Afterwards, CAD prototypes of a chamber were tested until it was enabled to perform reproducible measurements. HPTS solutions of 500 µmol/L with pH from 6.0 to 8.0 were used. A mirror of 7 mm diameter with a centered hole of 3 mm was located around the opto-chemical sensor in order to get a stable reference of the excitation. A linear relationship between pH and emission intensity of HPTS was demonstrated using the measurement chamber. While the emission intensity at an excitation wavelength of 405 nm decreases with increasing pH, the opposite is true at an excitation wavelength of 456 nm. The replication error of the measurements was 0.35%.

1 Introduction

Knowing the pH of various body fluids such as blood or urine is of great importance in medicine, as it enables the diagnosis of diseases and pathologies. The physiological pH range of the human blood is between 7.35 and 7.45 [1]. A lowered value is called acidosis while an elevated value is called alkalosis. These conditions are associated with severe symptoms and usually require immediate medical treatment as various body functions depend on the physiological pH.

The pH can be determined, among other techniques, with so-called opto-chemical sensors, in which a dye is used whose fluorescence strongly depends on the pH. Opto-chemical sensors have several advantages over other methods. They are insensitive to electromagnetic fields, can be excellently miniaturized, and are inexpensive to manufacture in large quantities. One well-suited dye for optical pH sensing is 8-Hydroxypyrene-1,3,6-trisulfonic acid (HPTS). It has two excitation peaks at 405 nm and 450 nm and an emission peak at 512 nm[2],[3]. The use of opto-chemical sensors requires a suitable light source for excitation, a cavity to hold the sample and a detector to measure the fluorescence.

For use in a mobile blood analyzer, a measuring chamber is to be developed that enables reproducible measurements of the pH using HPTS. Therefore, the optical behavior of HPTS in buffer solutions of different pH was characterized first. Since the sensor concept requires the simultaneous acquisition of excitation and emission to compensate for disturbances, the measurement chamber was optimized in particular so that the detection of excitation was stable.

2 Material and Methods

Different materials and devices were involved in the development and testing of the measuring chamber for the fluorescence opto-chemical sensor. The buffer solutions and the HPTS used to prepare the samples are listed in Table 1.

Table 1: Chemical materials

Description	Article number	Company
HPTS	102411760	Aldrich
Buffer – pH 6.0	1.99036.0500	Supelco
Buffer – pH 6.86	1.99068.0500	Supelco
Buffer – pH 7.0	1.09407.0500	Supelco
Buffer – pH 7.4	11237.02500	Morphisto
Buffer – pH 8.0	1.99038.0500	Supelco

The devices and equipment used in the process are shown in Table 2.

Table 2: Devices and equipment

Description	Article number	Company
Fluorescence spectrometer	108525	PerkingElmer
Spectrometer	ULS2048	Avantes
Voltage source	E3631A	Hewlett-Packard
3D Printer	CR-10S	Shenzhen creality 3D
Optical fiber	15-01-01-04	Modellbau Schoenwitz
Pipette	K18843H	Eppendorf

2.1 HPTS Characterization Process

The optical properties of HPTS were verified using a laboratory fluorescence spectrometer first. Therefore, samples with different pH from 6.0 to 8.0 were measured. This allowed to inspect the optimal excitation wavelengths that should be used by measuring the excitation spectrum while the emission wavelength remained stable at 512 nm. The behavior of the fluorescence depending of the pH was also inspected.

The optimal concentration of HPTS had to be obtained, to ensure the optimal scale according to the detector range. This was done by preparing a dilution series from 1 to 20 µmol/L in the pH range from 6.0 to 8.0. Afterwards, the stability of HPTS to the incidence of light, the so-called photobleaching, was investigated, by taking measurements every 5 minutes for a period of 1.5 hours. A long-time test was performed to assess the stability of the sample over time. This test consisted in daily measurements over 20 days. The measurements were carried out only for a pH of 7.4 at a temperature of 25 °C for the two excitation wavelengths of 405 nm and 450 nm.

Additionally, the effect of temperature on HPTS emission was assessed from 20 ° C to 50 ° C in 5 °C steps. Finally, the relationship between excitation and emission was measured, using neutral density filters to decrease the intensity of the light source.

2.2 Measuring Chamber Development

The measuring chamber was developed in different stages. At first, a custom fluorometer was built using the rail system by LINOS for optical test setups. The spectrometer ULS2048L from Avantes was used as detector. The used optical fiber had a diameter of 1.5 mm and an acceptance angle of 30°. The volume of the sample in this stage was 3 ml. As stated in the literature, the angle used between the light source and the detector was 90° [4].

Subsequently, the miniaturization of the measuring chamber was developed. To achieve that, CAD prototypes were designed using the software SolidWorks and printed with a black filament. In this stage the sample volume was 100 µL and it was deposited in a reusable minicuvette.

After that, the prototype was adapted to measure the sample using a test-chip with 100 µm sample layer. This was in order to measure optodes of 3 mm of diameter. In this stage the test-chip was completely inside the chamber and the sample was exchanged manually via syringes.

Once a stable model using the completely covered prototype was verified, a final version of the prototype was built. The main motivation for this was to be able to have a slot for inserting and removing the test-chip as it would be implemented in a mobile device. That can be seen in Fig. 7.

To include the excitation and emission wavelengths of the system in the same graph, the measurements were performed using wavelengths from 380 to 700 nm. Integration time was adjusted for detector sensitivity. The spectra were normalized to counts per second and recombined afterwards. This procedure avoided the use of additional ele-ments such as a reference photodiode to verify the intensity of the LED.

Controlling the intensity of the light emitted by the LED guaranteed that the emission from the sample was not disturbed, ensuring the reliability of the system. Changes in the LED light intensity result in a proportional change of the fluorescence of the sample. This would introduce an error in the pH determined by the device.

The measurement of the LED intensity must be performed without interfering with the absorption effect of the different pH of the HPTS dilutions. To achieve that, different tests were carried out by using mirrors of different diameters with a central hole around the measured opto-chemical sensor.

The aim was to maximize the LED intensity in direct reflection, thus minimizing the effect of absorption from the HPTS itself on the detector signal. This way, the LED intensity is expected to be received relatively unchanged, which would allow to evaluate the correct operation of the LED over time. Five different situations were tested from a fully matte black foil to a fully reflective foil (mirror). Additionally to these two cases, the other three conditions were mirrors of 5mm, 7mm and 10mm diameter. All the five conditions had a 3mm centered hole for the optode. These five different cases were performed for five different distances between the optode and the fiber tip. These distances correspond to 1.2 mm, 3.2 mm, 5.2 mm, 7.2 mm and 9.2 mm.

3 Results and Discussion

3.1 HPTS Characterization Tests

The optimal concentration of HPTS for characterizing its optical properties using the fluorescence spectrometer was found to be 15 µmol/L.

The results of excitation spectrum of HPTS at emisison wavelength of 512 nm for the different pH values can be seen in Fig. 1.

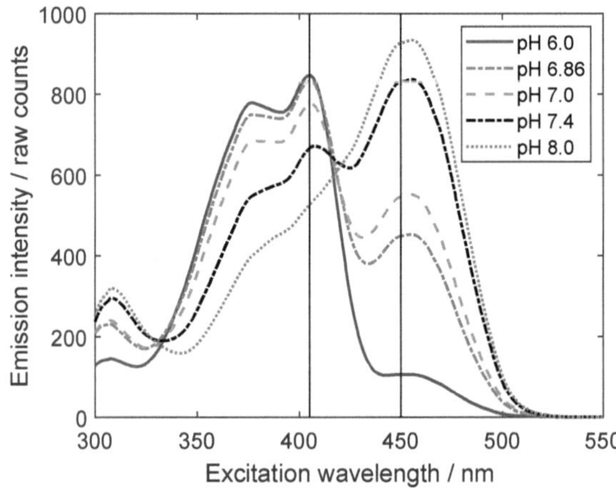

Figure 1: Excitation spectrum of HPTS at various pH-values with a set emission wavelength of 512 nm.

As shown in Fig. 1, the excitation spectrum of HPTS has

two noticeable peaks. That is the main reason to choose 405 nm and 450 nm as excitation wavelengths for the characterization tests. The emission of HPTS with different pH values at a constant temperature of 25 °C and an excitation wavelength of 405 nm can be seen in Fig. 2.

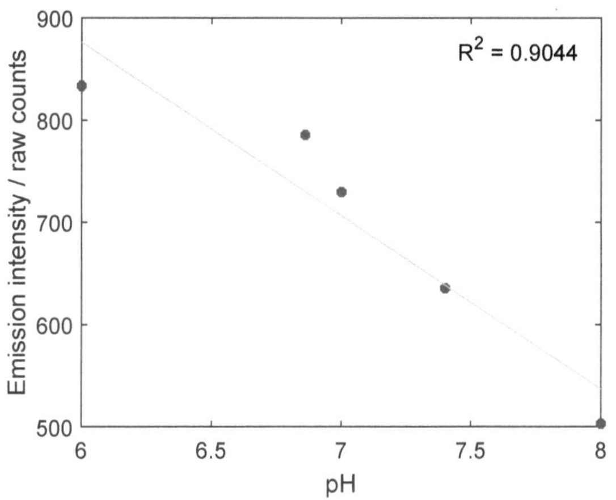

Figure 2: pH-dependence of the emission intensity of HPTS (Conc. 15 µmol/L) at excitation wavelength of 405 nm.

Fig. 2 shows that the intensity of the emission decreases when the pH increases, following a linear behavior. Linked to this, Fig. 3 shows the emission of the HPTS with pH values from 6.0 to 8.0 at a constant temperature of 25 °C and an excitation wavelength of 450 nm.

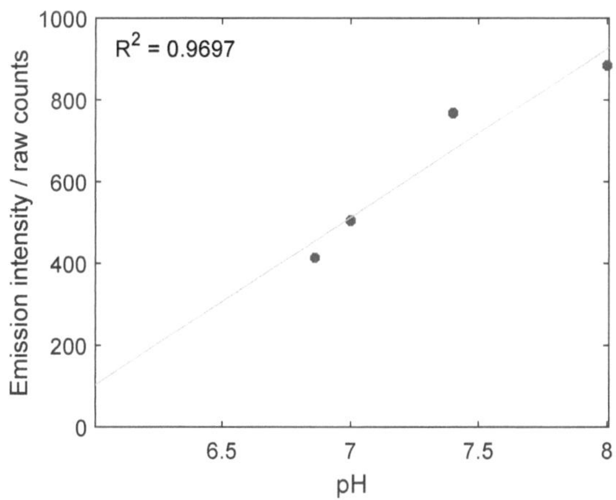

Figure 3: pH-dependence of the emission intensity of HPTS (Conc. 15 µmol/L) at excitation wavelength of 450 nm.

In Fig. 3, in contrast to the results in Fig. 2, the intensity of the emission increases with increasing pH. This is in agreement with results observed in past studies and literature [5], [6].
The absolute error of the photobleaching test was 1.23 counts and the relative error was 0.19%. This means that the exposure of the sample to the light does not affect the result obtained and therefore, several measurements can be carried out using the same sample without alterations. The absolute error of the long-time test was 8.75 counts and the

relative error was 1.25%. The main factor of this error is the loss of liquid in the sample over the time, which causes the increase of the concentration. It must be pointed out that the temperature of the sample must be constant for every measurement, to avoid errors due to temperature effect.

The effect of the temperature was assessed both for 405 nm and for 450 nm excitation. In the first case, the emission decreased approximately 1% for each 5 °C increase. In contrast, for the second case, a 0.5% increase was observed in the emission every time the temperature increased by 5 °C. These results correspond to the sample of pH 7.4.

Besides, the ratio of the intensity of light incident on the sample and the intensity of emission was also verified using neutral density filters. It was confirmed that both the emission intensity and the intensity of the light source decrease proportionally.

3.2 Measuring Chamber Tests

According to the two optimal excitation wavelengths of HPTS, the LEDs VLMU3100 with a central wavelength of 405 nm and 150224BS73100 with central wavelength of 456 nm were selected.

To have a stable light source, a regulated current driver was implemented. It guarantees a constant current of 20 mA through the LED.

Since the volume of the optode is 0.71 nL, it was necessary to increase the concentration of HPTS to detect an appropriate amount of emission. The new concentration used was 500 µmol/L.

The emission results obtained with the final prototype for the two-excitation wavelengths are shown in Fig. 4 and Fig. 5. The emission intensity of HPTS for the different pH values follow the same behavior of the reference values of the characterization stage.

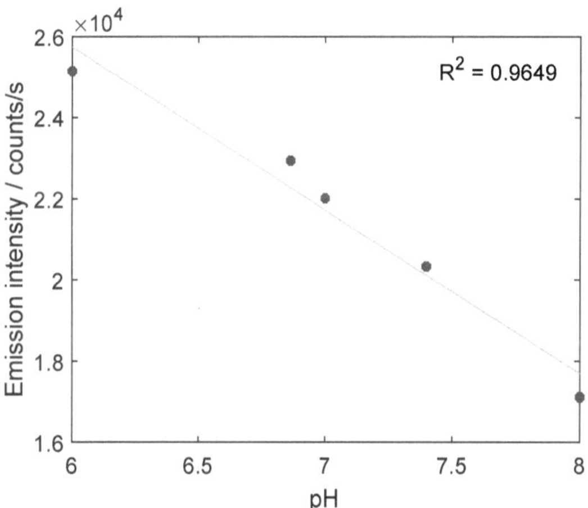

Figure 4: pH-dependence of the emission intensity of HPTS at 500 µmol/L with the final prototype at excitation wavelength of 405 nm.

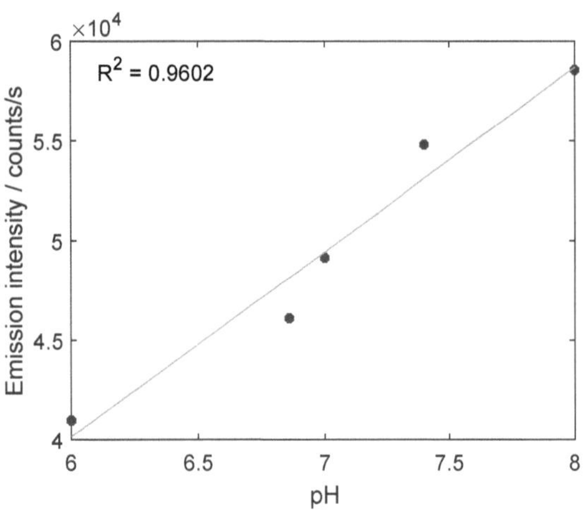

Figure 5: pH-dependence of the emission intensity of HPTS at 500 μmol/L with the final prototype at excitation wavelength of 450 nm.

Fig. 6 shows the error between the excitation intensity of pH 6.0 and 8.0 at 405 nm. This was performed with the five-mirroring conditions and the five distances between the optode and the fiber tip.

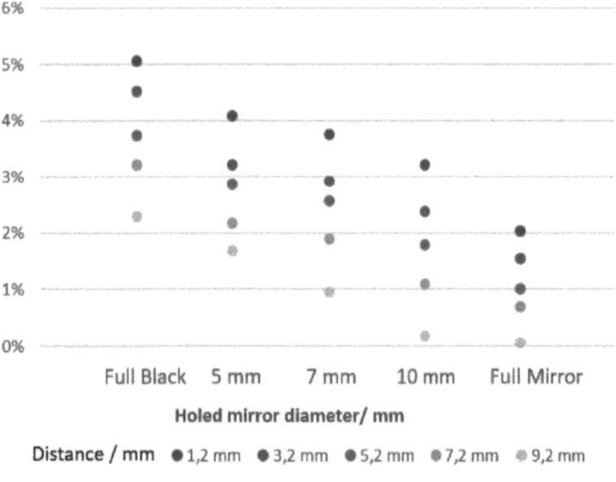

Figure 6: Mirroring conditions versus distance between optode and fiber tip.

According to the results shown in Fig. 6, the mirror with 7 mm diameter and the distance of 5.2 mm between the optode and the fiber tip were chosen. Since these reduced the effect of absorption to an error of 2.58% maintaining an adequate level of emission intensity of the sample.

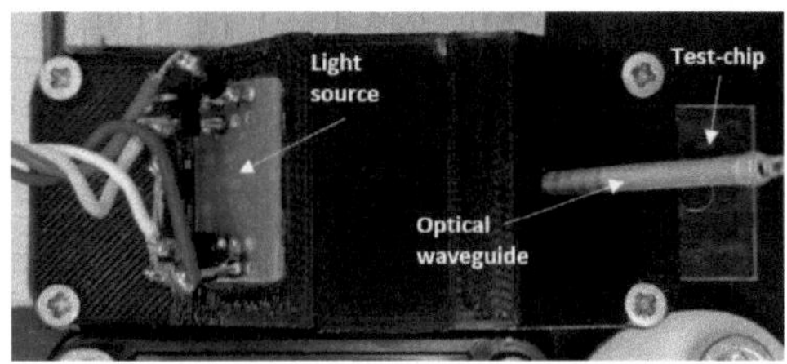

Figure 7: Final measuring chamber prototype.

The final prototype shows in Fig. 7 has a measurement relative replication error of 0.35%. This means that, the effect of the mechanical action of inserting the test-chip has no significant influence on the results.

4 Conclusion

The data obtained made it possible to characterize HPTS regarding different aspects such as optimal concentration, photobleaching, long-time behavior, effect of the temperature and emission behavior at different pH. All that information can be used in the design of algorithms that enable a reliable determination of blood pH.

In addition, it was possible to build and test a stable and reproducible measuring chamber to assess the fluorescence of HPTS at different pH. The system used an HPTS solution with a concentration of 500 μmol/L as fluorescent compound. Moreover, the final measuring chamber enabled a repeatability error of 0.35% at an optode of 3 mm diameter. The measurements therefore had a high precision.

Acknowledgement

The work has been carried out and supervised by the Medical Sensors and Devices Laboratory at Luebeck University of applied sciences.

Author's Statement

Conflict of interest: Authors state no conflict of interest.

5 References

[1] J. K. Tusa and H. He, *Critical care analyzer with fluorescent optical chemosensors for blood analytes*, Royal society of chemistry, pp 2643, 2005.

[2] AAT Bioquest, *HPTS spectrum*. Available: https://www.aatbio.com/fluorescence-excitation-emission-spectrum-graph-viewer/HPTS [last accessed on 2022-01-05].

[3] O. S. Wolfbeis, E. Fürlinger, H. Kroneis and H. Marsoner, *A study on fluorescent indicators for measuring near neutral ("Physiological") pH – values*, Springer, pp 120 - 123, 1983.

[4] P. So and C. Dong, *Fluorescence spectrophotometry*, Macmillan Publishers Ltd, pp 3, 2002.

[5] A. Steinegger, O. S. Wolfbeis and S. M. Borisov, *Optical sensing and imaging of pH values: spectroscopies, materials, and applications*,Chem. Rev, pp 12364, 2020.

[6] M. J.P. Leiner, *Optical sensors for in vitro blood gas analysis*, Elsevier Science S.A, 1995.

Measurement of the optical power distribution in the core and inner cladding of a double clad fiber

Merle Loop [1], Stefan M. Jensen [2], Madhu Veettikazhy[3], and Peter E. Andersen [3]

[1] Medical Engineering Science, Universität zu Lübeck, merle.loop@student.uni-luebeck.de
[2] Department of Photonics Engineering, Technical University of Denmark, s193649@student.dtu.dk
[3] Department of Health Technology, Technical University of Denmark, {madve,peta}@dtu.dk

Abstract

Double clad fibers with a singlemode beam output from the core and a multimode supporting inner cladding are a subject of current research in multiphoton endoscopy. Since only the fundamental mode in the core excites fluorescence, a fast and reliable method to distinguish the power in the core from that in the inner cladding is needed to measure the intensity that excites fluorescence and to avoid damage of samples and future patients. In this work, a portable setup was used to acquire images of the end facet of the double clad fiber using two different cameras, analyze the power distribution in the fiber and compare the data based on the different camera specifications. The data give a first indication that a camera with a high dynamic range is required for the most accurate calculations.

1 Introduction

The two photon laser scanning microscopy (TPLSM) [1] is a valuable tool for studying morphological, physiological and pathological features on a cellular and subcellular level [2], [3]. Therefore, the potential of TPLSM for the diagnosis of various cancers such as bladder [4], breast [5] and epithelial tissue [3], is the subject of current research. In the future, TPLSM may provide a faster, non-invasive and thus more patient-friendly, alternative to surgical biopsies [2], [3]. In order to access internal organs with this imaging technique, an endoscopic solution is required.

The conditions for two-photon (TP) excitation were described by Goeppert-Mayer in her doctoral dissertation [6]. To enable this non-linear effect, a very high photon density is required, which is achieved today by spatially and temporally focused laser pulses. In endoscopic set-ups, optical fibers are usually used to guide these laser pulses from the laser outside the patient into the tissue of interest inside the patient and to collect the fluorescence signal of the excited molecules. Double clad fibers (DCFs) are used in many multiphoton endoscope projects because they combine the advantages of singlemode (SMFs) and multimode fibers. The core of a DCF provides a well-focusable gaussian singlemode pulse to excite the fluorescent molecules, while the multimodal inner cladding with its higher numerical aperture (NA) and larger area efficiently collects the isotropically emitted fluorescence signal and guides it back through the fiber [2].

However, if the excitation beam is not perfectly coupled into the core, the cladding also delivers a part of the excitation light to the distal end. This leads to the problem of estimating the applied excitation intensity for the gener-

ation of fluorescence. Therefore, it would be of great relevance to have a method available to distinguish between the light in the core that excites fluorescence and the unwanted multimodal cladding light that carries power but cannot excite fluorescence. In reference [7] the authors explain a method (constant-intensity-extrapolation, CIE2) with reliable results that uses an iris with an adjustable aperture to measure the power as a function of the area of the aperture. Interpolation and extrapolation of these values provides a linear dependency of the power on the area and allows to extract the power in the core. However, accurate positioning an adjustment of an iris in the micrometer range is a challenge. To overcome these problems, I propose a faster and simpler method of analyzing the camera images of the end facet of the fiber. A comparison was drawn between two cameras with different digital outputs to evaluate if a specific type of camera is required to accurately calculate the power distribution in the core and inner cladding. A simulation software was used to demonstrate the sensitivity of the power content in the core of the fibre and of the intensity profile in the DCF on small changes in the coupling process.

2 Material and Methods

2.1 Double Clad fibers

The DCF used in the experiments is the SMM 900 (Fibercore House) with a core, an inner cladding and outer cladding with diameters of 3.6 µm, 100-104 µm, 125 µm and 245 µm, respectively. The structure of the fiber is visualized in Fig. 1 and illustrates the significant difference between core and inner cladding area. The NA of the core is 0.18-0.20 and the NA of the first cladding is given as 0.24-0.28 in the data sheet of the fiber. In contrast to SMFs, light can propagate in both areas, the core and the inner cladding of the DCF. The refractive index decreasing from the core to the outer cladding allows light to be guided by total internal reflection.

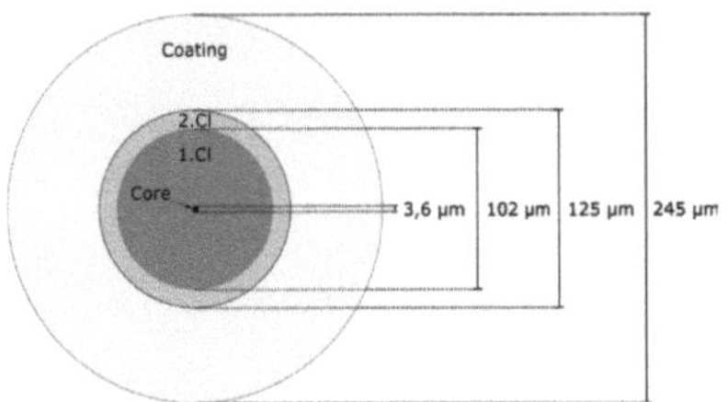

Figure 1: Scheme of the SMM 900 as a typical DCF. Coupled light can propagate in the core and the inner cladding (1. CL). The outer cladding (2. CL) enables the total internal reflection in the inner cladding.

2.2 Simulations

When using DCFs in endoscopes, it might be necessary to monitor the power distribution of the fibre, as the coupling into the core of a DCF is very sensitive and can change due to slight movements or bending. Then not only the power in the core decreases, but higher modes gain power and the intensity profile changes. To simulate the coupling of a gaussian beam into the DCF and to obtain an indication of whether singlemode coupling is possible with the given components, the open-source beam propagation software called BPM-Matlab [8] can be used. The program calculates the propagation of the electric field in optical fibers with arbitrary refractive indices for the specified fibers and beam properties. BPM-Matlab is described in detail in [8]. To study the potential effects of imperfect alignment, various offsets and angles of the gaussian input beam relative to the fiber's end facet were analyzed.

2.3 Experiments

To investigate the power content in the cladding and the core after coupling a laser beam into a DCF, the setup shown in Fig. 2 was used. The light source was a Superluminescent Diode (SLD830S-A20W, Thorlabs) with a center wavelength of 830 nm and a 55 nm 3 dB bandwidth to simulate the properties of a pulsed laser at 800 nm. The output was fiberbased and coupled via a mating sleeve to another SMF, whose output is collimated by a fiber coupler (PAF2P-A10B, Thorlabs) and further coupled into the DCF using the same fiber coupler. On this proximal end of the fiber, the fiber itself was plugged in a bare fiber adapter attached to the FiberPort, whose mechanical features allow the position of the fiber to be adjusted in x, y and z axes relative to the fixed lens. On the distal end it is fixed to a 3-axis stage. A beam expander with a lense (L1) and an objective (NA 0.66, 40x, Leica) magnified the fiber's

end facet onto the camera chip. Two cameras were used, the Thorlabs (CS165MU/M, Thorlabs, Peak Quantum efficiency (PQE): 69 % at 575 nm, Number of active pixels: 1440 x 1080) with 10-bit digital output and the air-cooled ORCA-Fusion (C14440-20UP, Hamamatsu, Number of active pixels: 2304 x 2304, PQE: 80 % at 550 nm) with 16-bit. Neutral density filters (NE50A-B, NE20A-B, Thorlabs) were placed in front of the camera to attenuate the optical power reaching the detector elements to avoid saturation. Based on different sensitivities, the transmission on the ORCA-Fusion is reduced to 10^{-9} of the initial optical power, on the Thorlabs to 10^{-7}. To provide a fast and flexible method, most components of the setup (from the first fiber coupler to the filters and even the Thorlabs camera) were mounted on a breadboard (MB3030/M, Thorlabs) to enable a portable solution. The cameras were connected to a computer via USB and the ORCA-Fusion is controlled by the software Micro-Manager 2.0 [9], the Thorlabs by Thor-Cam (Version 3.6.0.6, Thorlabs). To vary the used pixel depth, hence the maximum pixel value, the exposure time could be changed. In each case, a stack of ten images was taken with the same exposure time, without saturating the pixels and without touching the setup.

A self-written code in MATLAB (MathWorks) subtracts

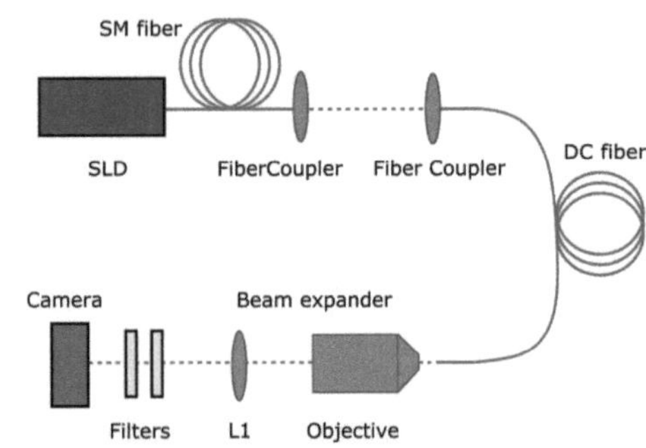

Figure 2: Scheme of setup. The SLD's output is coupled to a SMF, a Fiber coupler collimates the beam and the same Fiber coupler focuses the beam onto the DCF which is fixed in a bare fiber adapter. A beam expander with an objective and L1 magnifies the fiber's end facet. Neutral density filters attenuate the power that reaches the camera.

the background and creates an averaged image of the acquired image stack. The software ImageJ [10] was used to calculate the power by the multiplication of intensity and area in the two parts of the fiber from the averaged image. The intensity is given in pixel values and is specified for the cladding by the mean value of the average intensities of four large sections. The average intensity of the entire core area was defined as the core intensity. In order to determine the core area reproducibly, it was defined by an intensity threshold. Therefore, the core area includes the pixels with an intensity between the maximum pixel value and $1/e^2$ of the maximum pixel value (mode field diameter). The cladding area is calculated by its radius, the pixel size and the magnification of the beam expander.

3 Results and Discussion

3.1 Results

3.1.1 Simulations

Table 1 shows the dependence of the coupling efficiency into the core and thus of the power fraction in the core compared to the total power in the fiber on different offsets and angles. The proportion in the core decreases with increasing angle and displacement. An offset of a few micrometers has a large effect on the coupling into the core with a diameter of only 3.6 µm. As the power fraction in the

Table 1: Results of BPM-Matlab simulations. Proportion of power in the core of the total power guided in the fiber. Depending on angle and offset in one dimension of the incoming beam relatively to the fiber's end facet.

Angle / °	Offset / µm			
	0	1	2	5
0	0.814	0.745	0.473	0.041
2	0.793	0.735	0.462	0.039
5	0.726	0.653	0.405	0.032
10	0.512	0.458	0.261	0.016

core decreases, more power is gained by various modes in the inner cladding, changing the intensity distribution in the fiber. Examples of imperfect coupling simulations are represented in Fig. 3, where the left side shows the intensity of a non-gaussian core mode at an angle of 5° and 2 µm displacement. The right side displays the intensity profile when the input beam is displaced by 5 µm, with the multimodes visible in the cladding. The circumstance that a fraction of 1.0 is not reached even without angle and displacement is due to other specified input parameters such as the beam waist or the wavelength of the input beam.

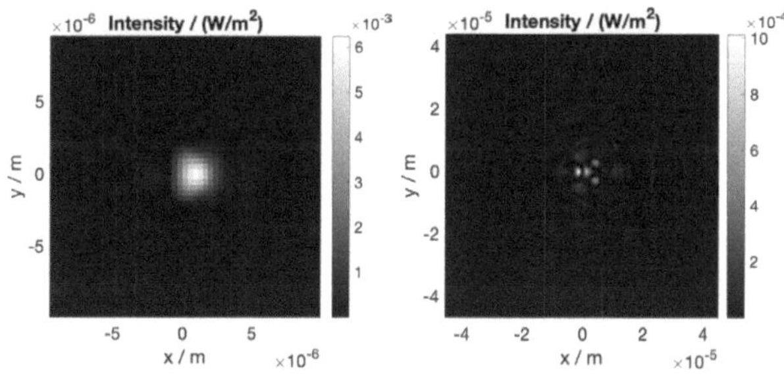

Figure 3: Intensity profiles simulated by BPM-Matlab. On the left side the intensity distribution is shown for an input beam at an angle of 5° and 2 µm offset in one dimension, on the right at an offset of 5 µm.

3.1.2 Experiments

The analysis of the intensity profiles of the acquired images indicates a gaussian beam profile. Therefore, a sufficient coupling efficiency can be assumed without large misalignment. As an example, in Fig. 4, the core and a plot profile

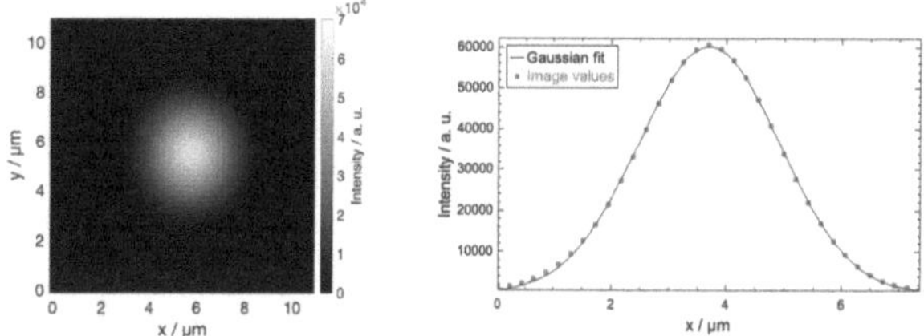

Figure 4: The core of the DCF is shown in a ten times averaged ORCA-Fusion image (left). A gaussian function is fitted to the data corresponding to a line drawn through the center of the spot (right).

with a gaussian fit of a 10x averaged image taken with the ORCA-Fusion are shown.

Fig. 5 visualizes the averaged images of the ORCA-Fusion (left) and the Thorlabs (right) on a logarithmic scale. In these representations, the cladding can be distinguished from the background even on images of the Thorlabs. Examining the intensity values in Fig. 5, it can be observed when the background is set to near zero (here it is set to 10^{-9} to enable a logarithmic plotting), most of the cladding area has a value of one.

The calculated core power fraction of the total power based

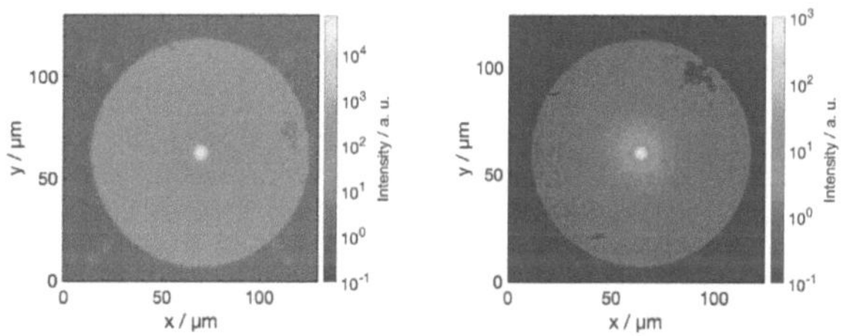

Figure 5: Ten times averaged images are displayed on a logarithmic scale. On the left side the images have been acquired with the ORCA-Fusion, on the right side with the Thorlabs.

on ORCA-Fusion images for selected exposure times and used pixel depths is listed in Table 2. Higher exposure time results in higher maximum pixel values and an increase in the calculated power fraction in the core. At higher exposure times, above 0.5 ms, the calculated power proportion in the core stabilizes at a fraction of approximately 0.6. The power proportion calculated with the Thorlabs images provides results with small variation. The average of five measurements yields 0.473, with a standard deviation of 0.013.

Table 2: Calculated power content in the core for selected exposure times and used pixel depths.

Exposure time / ms	Maximum pixel value	Proportion of core power to total power
0.08	60809	0.598
0.05	37920	0.597
0.03	22013	0.581
0.01	6823	0.520

3.2 Discussion

The coupling of light into the core of a DCF is very similar to the coupling into a SMF. The comparison between the simulation results in Table 1 and the alignment sensitivities of SMFs investigated in [11] shows a comparable and expected trend of a strong dependence of the lateral misalignment on the coupling efficiency. Since the impact of the angle depends highly on the NA of the core, it is significantly lower for the simulated SMM 900 than for the fiber used in [11] with a NA of only 0.063.

The calculations in Table 2 show an increase in the core power by raising the maximum pixel value. A higher maximum pixel value corresponds to a wider dynamic range that was used and provides a higher resolution by converting the intensity into pixel values. Therefore, with over 60 times more available pixel values than the Thorlabs, the ORCA-Fusion can more accurately represent the true intensity. The very low resolution in the lower intensity range of the Thorlabs, which can be seen in Fig. 5, leads to the assumption that the intensities in this range (of the cladding and background) cannot be represented well enough for a reliable calculation. This could then explain why the power fraction in the core calculated on the basis of the experimentally obtained images with the Thorlabs deviates by approximately 11 percentage points from that calculated with the ORCA-Fusion.

Since the results are based on a single data set, further confirming data are required. Other camera properties such as quantum efficiency and readout noise or a light source with a different wavelength should also be considered and investigated.

4 Conclusion

The paper addresses the relevance and problem of determining the power in the core of a DCF. To provide a method images of the fiber's end facet were acquired and analyzed to calculate the power in the core and inner cladding. To give an indication of what specifications the camera should fulfill for correct results, two cameras with different dynamic ranges were used. A difference in the calculated results was found and the indication was made that the camera with the higher bit depth, thus the better intensity resolution, provides the more accurate results. Further data needs to be collected from various cameras to confirm this indication and provide a guidance on a reliable method.

Acknowledgement

The work has been carried out at Department of Health Technology, Technical University of Denmark (DTU) and supervised by R. Huber, Institute of Biomedical Optics, Universität zu Lübeck. Thanks to all members of the Biophotonic Imaging Group, especially to A. L. Borre, who always tried to answer all my questions, and to P. E. Andersen and R. Huber, who made my internship at DTU possible.

Author's Statement

Conflict of interest: Authors state no conflict of interest.

5 References

[1] W. Denk, J. H. Strickler and W. W. Webb, *Two-Photon Laser Scanning Fluorescence Microscopy*. Science, vol. 248, p. 73-76, 1990.

[2] V. Kučikas, M. P. Werner, T. Schmitz-Rode, F. Louradour and M. A. M. J. van Zandvoort, *Two-Photon Endoscopy: State of the Art and Perspectives*. Molecular Imaging and Biology, Springer Nature, 2021.

[3] M. C. Skala, J. M. Squirrell, K. M. Vrotsos, J. C. Eickhoff, A. Gendron-Fitzpatrick, K. W. Eliceiri et al., *Multiphoton Microscopy of Endogenous Fluorescence Differentiates Normal, Precancerous, and Cancerous Squamous Epithelial Tissues*, Cancer Res, 65, p. 1180-1186, 2005.

[4] S. Mukherjee, J. S. Wysock, C. K. Ng, M. Akhtar, S. Perner, M.-M. Lee, et al., *Human bladder cancer diagnosis using Multiphoton microscopy*, Proc SPIE Int Soc Opt Eng., 2009.

[5] Y. K. Taoa, D. Shenb, Y. Sheikineb, O. O. Ahsena, H. H. Wang, D. B. Schmolze, et al., *Assessment of breast pathologies using nonlinear microscopy*, PNAS, vol. 111, no. 43, p. 15304-15309, 2014.

[6] M. Göppert-Mayer, *Über Elementarakte mit zwei Quantensprüngen*, Annalen der Physik, vol. 401, issue 3, p. 273-294, 1931.

[7] T. Walbaum, A. Liem, T. Schreiber, R. Eberhardt and A.. Tünnermann, *Measurement and removal of cladding light in high power fiber systems*, of SPIE, vol. 10513, 105130U, 2018.

[8] M. Veettikazhy, A. K. Hansen, D. Marti, S. M. Jensen, A. L. Borre, E. R. Andresen, et al., *BPM-Matlab: an open-source optical propagation simulation tool in MATLAB*. Optics Express, vol. 29, no. 8, p. 11819-11832, 2021.

[9] A. D. Edelstein, M. A. Tsuchida, N. Amodaj, H. Pinkard, R. D. Vale, and N. Stuurman, *Advanced methods of microscope control using µManager software*, Journal of Biological Methods, vol. 1, no. 2, 2014.

[10] W. S. Rasband, *ImageJ*, U. S. National Institutes of Health, Bethesda, Maryland, USA, https://imagej.nih.gov/ij/, 1997-2018.

[11] J. M. Martin, *Coupling Efficiency and Alignment Sensitivity of Single Mode Optical Fibers*, University of Central Florida, STARS, Retrospective Theses and Dissertations, 434, 1979.

Using a multiphoton system for structure generation by photopolymerization

Melanie Wacker [1], Markus Böttcher [2], and Martin Schütte [2]

[1] Medical Engineering Science, Universität zu Lübeck, melanie.wacker@student.uni-luebeck.de
[2] LaVision BioTec GmbH - A Miltenyi Biotec Company, Bielefeld, {boettcher, schuette}@miltenyi.com

Abstract

Two-photon (2P) lithography is a powerful tool to generate arbitrary 3D structures. This technique is used, among other applications, for the creation of organ-on-a-chip-systems, allowing to study the human physiology and its dysfunctions and pathogenesis. A 2P setup was used to generate structures within the photopolymer Ultracur3D® RG 50 by photopolymerization. Small parameter studies were carried out to find appropriate settings for the initiation of polymerization. After structuring, the 2P setup was used in its original purpose as a microscope and imaging of the polymerized structures was performed. Burns created within the photopolymer during the structuring process caused difficulties in the imaging. A first 3D structure with an approximate size of $200 \times 200 \times 95 \,\mu m$ was generated and imaged successfully. Regarding the goal of creating an organ-on-a-chip-system, the photopolymer will be replaced with a material that is suitable for cell cultivation.

1 Introduction

With two-photon (2P) lithography, arbitrary 3D structures with a resolution below the diffraction limit can be generated [1], [2]. This is one of the many advantages of this technique over conventional lithography. The 2P lithography is based on the nonlinear effect of multiphoton absorption. To induce the simultaneous absorption of two photons, a very high irradiance is required, achievable with a focused femtosecond pulsed laser. When the focus is inside a photosensitive polymer material, also known as a photoresist, the formation of radicals, propagation reactions and eventually crosslinking occur. The material can be polymerized this way. By moving the focused laser beam laterally and axially within the photoresist, any 3D structure can be generated. This offers another advantage over conventional lithography, which is based on the principle of one-photon (1P) absorption. Using a UV photoresist, light in the UV range is applied in the case of conventional lithography. Along the entire laser beam located inside the photoresist, 1P excitation occurs, resulting in polymerization. In the case of 2P lithography, light twice the wavelength is required. The photoresist is transparent to the 2P beam outside the focus and only absorbed within. Therefore, in conventional lithography, 3D structures can only be built up by applying the photoresist layer-by-layer, whereas in 2P lithography, structures can also be generated within the resist. The non-irradiated and thus non-polymerized parts of the material can then be developed with an appropriate solvent, leaving only the 3D structure at the end. A liquid photoresist solidified by photopolymerization is called a negative photoresist. In contrast, there are also positive photoresists. In that case, the 2P absorption causes the breaking of the polymer chains of the photoresist, thus increasing the solubility of the irradiated parts. In the development step, the irradiated parts are removed and the non-irradiated parts form the 3D structure.

The structuring technique of 2P lithography has various applications. One of these involves the field of tissue engineering and regenerative medicine [3]. In recent decades, the gap between the demand for organ transplantation and the availability of organ donors has been widening. With the development of new biomaterials that better mimic natural tissue and organ structures, 3D scaffolds can be generated, followed by seeding with the patient's own cells. In the future, this technique should enable the possibility of generating individual parts or entire organs that can then be implanted into the patient, replacing the pathological structure. Another field of application are organ-on-a-chip-systems [4]. The aim is to reproduce the most important structural and functional features of a physiological organ or tissue as a small biomimetic system with microfluidics on a chip. Compared to the usual widely used 2D cell cultures, this offers the advantage that interactions between cells as well as between cells and their environment can be simulated and observed not only in two, but in three dimensions. By means of microfluidic channels, small amounts of fluid can be controlled, nutrients and other substances can be added to the system and concentration gradients can be generated. Because of these features, organ-on-a-chip-systems help in the study of human physiology and its dysfunctions and pathogenesis. They are used in different research fields such as medicine, drug development and toxicology. Various organs have already been developed as organ-on-a-chip-systems, e.g. livers, lungs, hearts, kidneys and intestines.

A multiphoton system was used to create structures in a polymer by photopolymerization. Small parameter studies were carried out to find suitable settings for the initiation of polymerization. First attempts to image the generated structures with the same 2P setup were started.

2   Material and Methods

An experimental 2P setup based on the commercially available multiphoton microscope "TriM Scope II" (LaVision BioTec GmbH) was used. A tunable femtosecond laser source (Light Conversion, Cronus-2P) with a repetition rate of 76.8 MHz was used at the wavelengths of 780 nm and 800 nm with an output power of 1.083 W and 0.780 W, respectively. The emitted fs-pulses are passed through an assembly of electro-optic modulator, half wave plate and polarizing filter for the variable adjustment of the power. A pair of galvanometric mirrors is used to move the laser beam laterally. A 4x air objective (Olympus, NA = 0.28) focuses the light onto the sample. The lateral focus spot size is approximately 1 µm as the full width at half maximum (FWHM) of the point spread function, depending on the numerical aperture (NA) of the objective and the wavelength (λ) as described in (1) [5].

$$FWHM_{x,y} = \frac{0.32\lambda}{\sqrt{2}NA}2\sqrt{\ln 2} \qquad (1)$$

Axial scanning is performed by the movement of the objective. The sample consisting of photoresist and object slide is placed on a xy-stage. The fluorescent light emitted by the sample as a consequence of the 2P absorption is collected by the objective, separated from the excitation light via a dichroic mirror and directed to the detector. Then, a photomultiplier tube (PMT, Hamamatsu) detects the fluorescence. In addition, the dichroic mirror can be moved out of the beam path to view the sample in transmitted light with a scientific complementary metal-oxide-semiconductor (sCMOS) camera (pco.panda 4.2). The ImSpector software (LaVision BioTec GmbH) is used to set the measurement parameters and to acquire 2P images. Furthermore, the software has a Python interface. Using Python, arbitrary scan patterns can be generated. As a photoresist the photopolymer Ultracur3D® RG 50 (BASF 3D Printing Solutions GmbH) was used, which is based on (meth)acrylate resin. The working wavelengths specified by the manufacturer are in the UV range at 355 nm, 385 nm and 405 nm. Due to the 2P excitation, the structuring was performed at wavelengths of 780 nm and 800 nm.

3   Results and Discussion

The basic idea was to apply a small drop of the photopolymer to an object slide and to create a structure within it us-

ing the 2P setup. To ensure that the structure adheres to the object slide and thus to prevent the structure from floating away, the focus was always initially placed on the surface of the slide. This requires sufficient contrast between slide and photoresist. The structure was then generated layer-by-layer upwards.

3.1   Determining polymerization parameters in 2D

To determine the parameters for polymerization, first tests were performed in 2D and evaluated in the transmitted light image acquired with the sCMOS camera. The structures had a size of 200 x 200 µm squares, which were generated line by line. The following measurement was implemented in Python. For a fixed laser power, the exposure time per plane was increased in 10 s steps in x-direction, starting with 10 s. In y-direction, the laser power was varied in 10 mW steps, starting with a power of 35 mW for a fixed exposure time. The distance between two squares was 50 µm in both x- and y-direction. Depending on the exposure time, the time per pixel and frame was different, resulting in a different writing speed. For a 200 x 200 µm structure consisting of a 64 x 64 pixel array generated at a line frequency of 200 Hz, an exposure time of 10 s means a 30-fold repetition of the scan. This results in an effective line writing speed of $85 \frac{mm}{s}$. If all other parameters remain unchanged, increasing the exposure time means slowing down the effective line writing speed.

Various measurements were performed at 780 nm and 800 nm. A result for 780 nm is shown in Fig. 1 as a transmitted light image. In this case, no structures were visible below a power of 65 mW, regardless of the exposure time. Moreover, 10 s exposure time was not sufficient to induce polymerization. Regardless of the wavelength, burns could always be seen in the structures, especially at higher powers and longer exposure times. Since they did not occur regularly, they could be due to impurities in the polymer. Another reason for the burns could be the use of an air ob-

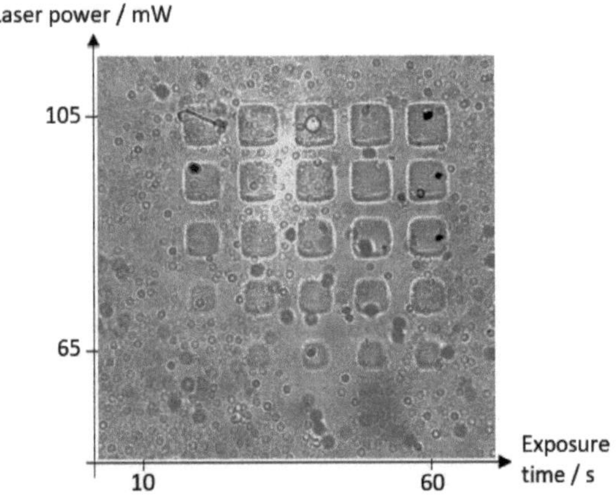

Figure 1: Transmitted light image: Structure generation of 200 x 200 µm squares with 780 nm at different laser powers and exposure times.

jective which can lead to strong aberrations due to the mismatch of the refractive indices, resulting in intensity peaks in the beam path [5].

3.2   2P Imaging of 2D structures

After structuring, 2P images of the polymerized structures were acquired using the same 2P setup. The result of a measurement is shown in Fig. 2, where the imaging was done at 800 nm with 97 mW laser power. The image has a size of 447 x 447 μm with 563 x 563 pixels. It corresponds to one plane from a 88 μm z-stack with a step size of 2 μm, each plane acquired at a line frequency of 800 Hz. Imaging was performed as a z-stack to observe the extent of the structure in depth. It also demonstrated that no excitation and polymerization occurred out-of-focus. Major challenges arose during imaging. The burns within the photoresist, described in the previous section, resulted in greatly increased signals causing a saturation and automatic shutdown of the PMT. Moreover, if the power was too high and the imaging speed was too slow, the power density was sufficient to polymerize the surrounding photoresist, which was still liquid. Therefore, direct imaging after structuring should be avoided. The polymer drop should be rinsed with a developer solution, such as distilled water in the case of Ultracur3D, removing the unpolymerized parts and leaving only the cured structure. Thus, the only risk while imaging is to further cure the polymerized structure, which could even be beneficial and provide more stability.

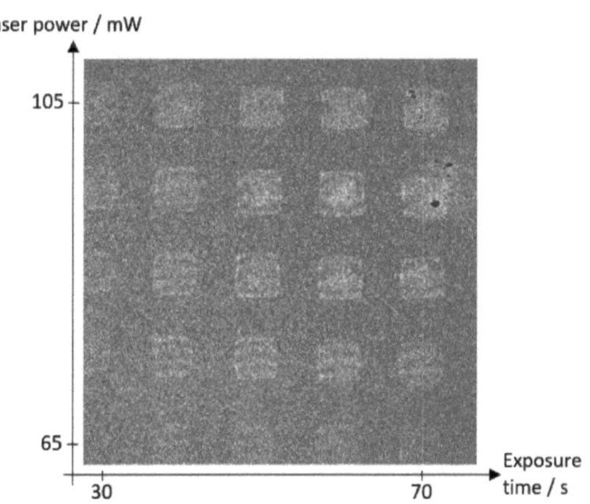

Figure 2: 2P Image of 2D polymerized 200 x 200 μm structures acquired with 800 nm.

3.3   Generation of 3D polymerized structures

The previously described structuring procedure has been extended by a dimension that enables the generation of 3D structures. Different depths in z-direction with different step sizes were tested. A large z-range was selected to guarantee that the first structural layer was really on the surface. The 3D structuring was performed at 780 nm on the one hand with 95 mW laser power and 40 s exposure time per plane and on the other hand with 105 mW laser power and

Table 1: Settings for the 200 x 200 μm structures shown in Fig. 3. The structuring was done at 780 nm.

No.	laser power [mW]	exposure time per z-plane [s]	z-range [μm]	step size in z [μm]
1	95	40	442	2
2	95	40	72	3
3	95	40	54	3
4	95	40	54	3
5	105	30	123	3

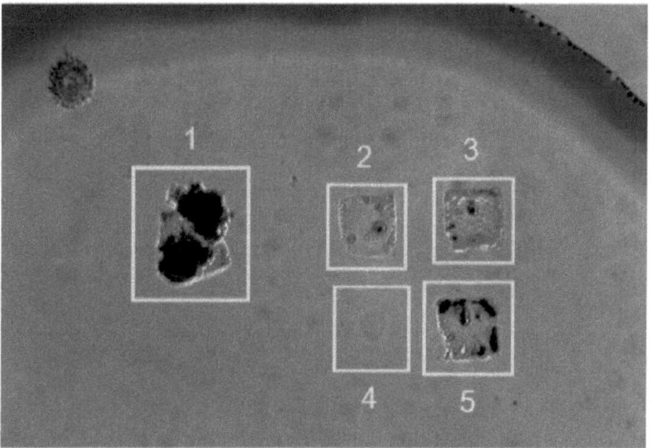

Figure 3: Transmitted light image of 3D polymerized 200 x 200 μm structures generated with 780 nm. Structuring parameters were set according to the values listed in Table 1.

30 s exposure time per plane. These parameter sets were chosen for two reasons. First, for these settings no burns inside the polymer could be seen in Fig. 1. Secondly, good adhesion to the object slide should be obtained. Particularly structures that were generated with lower laser power and shorter exposure time, were washed away during the development step with the solution. Thus, the choice of parameters is always a compromise between good adhesion and the prevention of burns. The distance between two z-images was set quite small to ensure crosslinking between the z-planes. In Table 1, the different structuring parameters are listed. The corresponding results were imaged in transmitted light with the sCMOS camera (Fig. 3). Although there were no problems with the selected parameters in the 2D case, burns could be seen in the 3D case. In the test no. 1, they occurred very strong, especially in the upper layers. Due to the large z-range, polymerization was induced close to or even beyond the surface of the photoresist. Local overheating may have occurred below the surface, which might explain this result. That is the reason why smaller z-ranges and a larger step size were tested. Then, burns occurred to a much smaller extend.

3.4   2P Imaging of 3D polymerized structures

Finally, the imaging of the 3D polymerized structures from Fig. 3 was done. For this purpose, the polymer drop was developed with double distilled water. Because the structure from test no. 5 was still in good condition after the development step, this one was selected for imaging. 800 nm at 14 mW laser power was used. The scan field had a size of 331 x 342 µm with 92 x 95 pixels, which was slightly larger than the structure. The imaging was done as a z-stack with a step size of 3 µm, covering a range of 225 µm and making it about twice the z-range of the structuring. The scan was performed with a line frequency of 200 Hz. After imaging, the maximum intensity projection (MIP) was determined in each of the three directions, which are here referred to as top (x-y-plane), front (x-z-plane) and side (y-z-plane) view. As already described in section 3.2, burns caused problems while imaging as a shutdown of the PMT. To prevent this, a lower laser power must be used. However, this had the consequence that the contrast decreased for weaker structures, especially in the edge areas. Thus, the contours of the object were often difficult to identify. Nevertheless, an attempt was made to determine the possible size of the structure, using the MIP representation. The measured extent in x-direction was 184 µm for the top view and 202 µm for the front view. The size in y-direction was 188 µm for the top view and 166 µm for the side view. The difference of 20 µm in both directions can be explained by unclearly recognizable object contour, which was defined by hand. The extent in the z-direction was determined as 95 µm in both the front and side view and corresponded to the height of the structure. This was smaller than the selected 123 µm z-range (see Table 1) due to the fact, that the structuring was always started clearly below the surface of the object slide.

4 Conclusion

A 2P setup can be used for structure generation by photopolymerization. With 780 nm, a first 3D structure with an approximate size of 200 x 200 x 95 µm was generated from the photopolymer Ultracur3D® RG 50. In addition, it was successfully shown that it is also possible to acquire 2P images of the polymerized structures with the same 2P setup, using 800 nm. However, many improvements are still needed. For a successful structuring, a clear representation of the object slide surface in the 2P image is necessary. This could be achieved by a fluorescent coating on the slide or by adding a fluorescent dye to the photoresist. In addition, there must be sufficient adhesion between the slide and the polymerized structure. For this purpose, slides made out of other materials than glass could be tested. Furthermore, the burns within the polymerized structure must be reduced, as they cause problems during imaging. To achieve this, an immersion objective will be used to reduce aberrations and thus intensity peaks in the beam path. Despite the mentioned challenges, the results shown are very promising, so that two major modifications will be made in the next step. Especially with the regard to the goal of creating an organ-on-a-chip-system, other photoresists need to be used. These should have properties such as biocompatibility and cell viability, so that cultivation with cells will be possible. Moreover, the writing speed should be increased. For this purpose, a spatial light modulator (SLM) will be integrated into the present 2P setup. A SLM consists of many small liquid crystals, which can be used to modulate the phase of the light [6]. This leads to a change of the intensity distribution within the photoresist. With a SLM, arbitrary 3D structures can be generated directly, which results in an increase in writing speed. Due to the characteristics of the test setup, the laser power in the focus is only in the lower mW-range. For the use of a SLM, an increase of the laser power by a factor of 10 is required.

Acknowledgement

The work has been carried out at LaVision BioTec GmbH - A Miltenyi Biotec Company and supervised by Prof. Dr. Robert Huber, Institute of Biomedical Optics, Universität zu Lübeck.

Author's Statement

Conflict of interest: Authors state no conflict of interest.

5 References

[1] C. LaFratta, J. Fourkas, T. Baldacchini, and R. Farrer, "Multiphoton fabrication," *Angewandte Chemie International Edition*, vol. 46, no. 33, pp. 6238–6258, 2007. [Online]. Available: https://onlinelibrary.wiley.com/doi/abs/10.1002/anie.200603995

[2] E. D. Lemma, B. Spagnolo, M. De Vittorio, and F. Pisanello, "Studying cell mechanobiology in 3D: the two-photon lithography approach," *Trends Biotechnol.*, vol. 37, no. 4, pp. 358–372, 2019.

[3] P. Bajaj, R. M. Schweller, A. Khademhosseini, J. L. West, and R. Bashir, "3D biofabrication strategies for tissue engineering and regenerative medicine," *Annu. Rev. Biomed. Eng.*, vol. 16, no. 1, pp. 247–276, 2014.

[4] Q. Wu, J. Liu, X. Wang, L. Feng, J. Wu, X. Zhu, W. Wen, and X. Gong, "Organ-on-a-chip: recent breakthroughs and future prospects," *BioMedical Engineering OnLine*, vol. 19, no. 1, p. 9, 2020.

[5] W. R. Zipfel, R. M. Williams, and W. W. Webb, "Nonlinear magic: multiphoton microscopy in the biosciences," *Nature Biotechnology*, vol. 21, no. 11, pp. 1369–1377, 2003.

[6] L. Yang et al., "Parallel direct laser writing of micro-optical and photonic structures using spatial light modulator," *Optics and Lasers in Engineering*, vol. 70, pp. 26–32, 2015. [Online]. Available: https://www.sciencedirect.com/science/article/pii/S0143816615000263

Concept for OCT optimized laser therapy of basal carcinoma

Isabelle Skambath [1], Tim Eixmann [2], Hinnerk Schulz-Hildebrandt [3], Gereon Hüttmann [2,3]

[1] Medical Engineering Science, Universität zu Lübeck, isabelle.skambath@student.uni-luebeck.de
[2] Medizinisches Laserzentrum Lübeck Gmbh, tim1.eixmann@uni-luebeck.de
[3] Institute of Biomedical Optics, Universität zu Lübeck, {hinnerk.schulzhildebrandt, gereon.huettmann}@uni-luebeck.de

Abstract

Basal cell carcinoma (BCC) is the most common form of cancer around the world. An early diagnosis is critical to prevent extensive tissue damage and frequent recurring. In recent years, optical coherence tomography (OCT), a non-invasive imaging technique, is increasingly used in the diagnosis of BCC as it allows visualization not only of the surface but also of the dermis and epidermis. It has also been shown that laser therapy allows for a minimally invasive removal of BCC. The aim of this work is to develop a holistic concept of a theragnostic system for the diagnosis and treatment of BCC using AI-based OCT data imaging in combination with laser therapy. Current OCT systems do not provide the right requirements, which lead to the design of a new combined system. A prototype of the handpiece for the OCT device was designed, built and tested.

1   Introduction

Basal cell carcinoma (BCC), a non-melanotic form of skin cancer, is the most common form of cancer around the world [1]. In Germany, about 230,000 people developed non-melanotic skin cancer for the first time in 2016 of which about three quarters are BCC [2]. It occurs mostly in sun exposed areas of the skin. People who suntan tend to have a higher risk to develop this type of skin cancer. Superficial tumor nests occur in the epidermis and can grow down to the top layers of the dermis and can reach up to two millimeter depth [3], [4]. Even though BCC has a very low risk of metastasizing or becoming fatal, it can grow and cause severe tissue damage, for example in bones or cartilage. Hence, an early diagnosis and treatment is necessary to prevent damage and recurring appearance [1], [5]. The current diagnosis includes the observation of the clinical appearance along with dermatoscopy, because the clinical results alone do not provide an explicit characterization into the histological subtypes [6]. The diagnosis is often confirmed with a biopsy of the lesion to examine the tumor thickness and resection borders. Optical coherence tomography (OCT) can be used to detect skin changes caused by BCC. It provides a noninvasive possibility to visualize tumor borders and to measure the thickness of the pathology. This offers an advantage compared to the standard diagnosis with histopathology, which takes more time and causes iatrogenic trauma [3], [6]. Currently, the most common form of treatment is, depending on the risk of recurrence, the surgical removal or excision. For tumors with a lower risk and a thickness up to two millimeters there are also different therapy options, for example photodynamic therapy or laser treatment [6], [8]. The last-mentioned alternative

is currently recommended only for tumor growth, with contraindications against surgical removal or topical treatments [6]. Since histological verification is cumbersome and involves discomfort and iatrogenic trauma for the patient, a non-invasive system for combined diagnosis and treatment for early detection of BCC is desirable.

Our approach is using an OCT-system for AI (artificial intelligence)-based diagnosis. OCT is a noninvasive imaging technique that uses low-coherent light to produce cross-sectional images. It is a well-established diagnostical tool in Ophthalmology and widely used in Dermatology. Current OCT-systems, for example VivoSight^{TM} by Michaelson Diagnostic or LC-OCT, DAMAE Medical France are not suitable for this application because of the weight of their handpiece, price and field of view. Furthermore, it is not possible to combine these systems with laser therapy because they are not working near real-time, a requirement for AI-based diagnostic.

As part of this project, the Medical Laser Center Lübeck is developing a new OCT device especially for the AI-based theragnostic. The important requirements for the imaging system are soft real-time acquisition, a lateral resolution below 25 µm, a penetration depth of 1-2 mm and a field of view of 10 mm × 10 mm to resolve pivotal structures in the skin to detect BCC. In addition, the handpiece should not be heavier than half a kilogram, so the combined system including the laser is lighter than a kilogram.

The objective of the present paper is the concept design and evaluation of an OCT system for laser combined BCC theragnostic. This includes choosing the wavelength for the OCT and design a handpiece for the correct arrangement of all optical components, which was then tested to validate the alignment and measure the FOV and resolution.

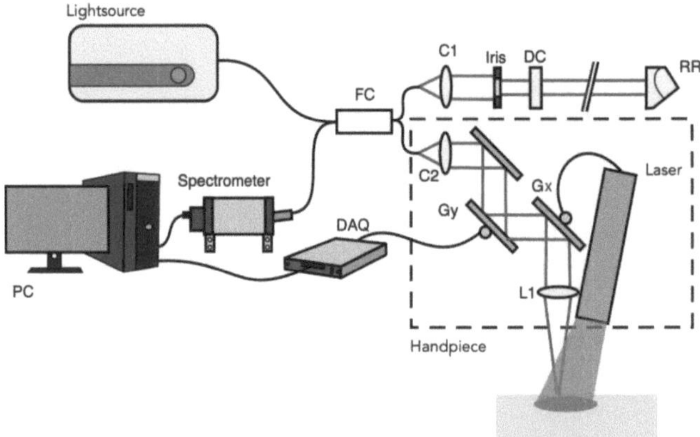

Figure 1: Schematic illustration of the OCT setup including the laser for treatment. FC: fiber coupler; C1, C2: collimators; DC: dispersion compensation; RR: retroreflector; Gx, Gy: galvanometric mirrors; L1: scanning lens; DAQ: digital-analog-converter.

2   Material and Methods

2.1   OCT

Fig.1 schematically shows the entire setup of the OCT system. The light source by Thorlabs GmbH emits low-coherent light with a broad bandwidth that is split up by the fiber coupler into a reference and a sample arm. The designed handpiece contains the sample arm of the system, it includes a collimator (C2), a prism to direct the collimated beam onto the galvanometric mirrors (Gx, Gy) and the scanning lens (L1) to focus the beam. The spectrometer measures the spectral interference intensity dependent on the frequency of the recombined reference and the sample beam. For the acquisition, processing, and visualization of the OCT images we are using the ThorImage software (Thorlabs GmbH).

2.2   Requirements for OCT Device

2.2.1   Wavelength

The right wavelength is essential for the right balance of penetration depth and axial resolution in OCT. Two different OCT systems of the company Thorlabs with different center wavelengths were compared, the Ganymede with $\lambda = 930$ nm and the Telesto with $\lambda = 1310$ nm. To determine the depth of penetration, the loss of signal-to-noise ratio (SNR) was measured over depth and averaged for different measurement points. An apparent attenuation coefficient α was calculated as

$$I(z) = I_0 \cdot e^{-\alpha \cdot z}. \tag{1}$$

by a fit to the measured SNR loss. The functions for the Ganymede and Telesto can be viewed in Fig.2. The measurements were performed on healthy volunteers, and the same skin area was measured for both systems. For the Ganymede, the coefficient was 6.3 mm^{-1}, for the Telesto 3.0 mm^{-1}, so the Telesto can be expected to have twice the penetration depth. Therefore, in the project we use a 1325 nm OCT system with a bandwidth of 150 nm, since

the increased measuring depth allows to visualize the layers of dermis and epidermis.

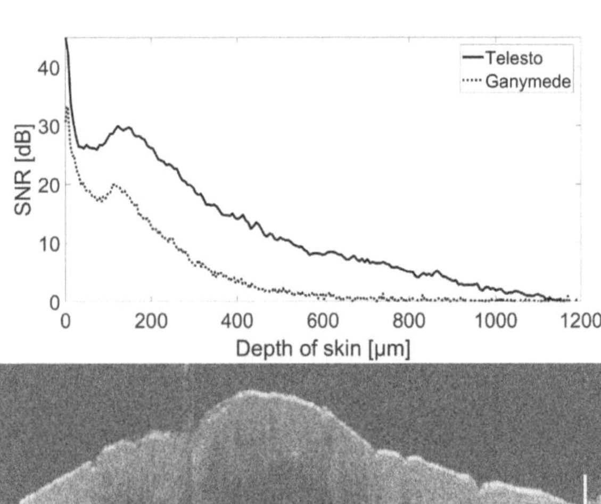

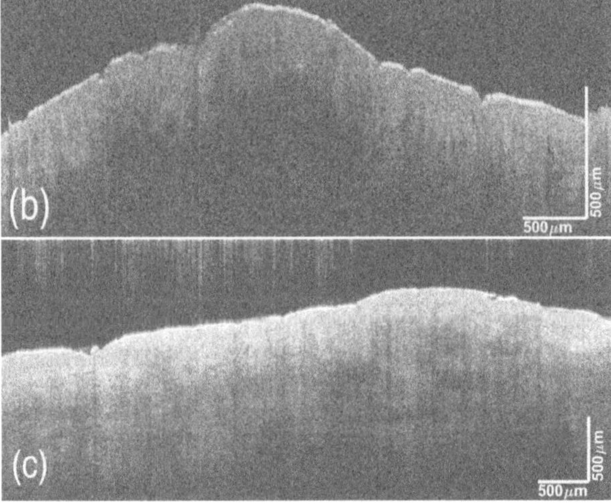

Figure 2: (a) Signal-to-noise over depth for the Telesto and the Ganymede, measured on healthy volunteers. Picture (b) shows the skin images with the Ganymede and (c) with the Telesto.

The lateral resolution and the field-of-view are determined by the focusing objective and the alignment of the optical components [7]. High axial resolution below 25 μm is requested in order to obtain the important skin structures for the AI-based diagnostic to work. The size of the desired field of view is 10 mm × 10 mm, above this size the BCC is often too deep to use laser therapy as possible treatment. For the combination with laser therapy, it is also necessary to have soft real-time acquisition. The advantage of the new combined system should be, that the therapy starts right after the diagnostic without long waiting time for the patient.

2.2.2 Lab Prototype

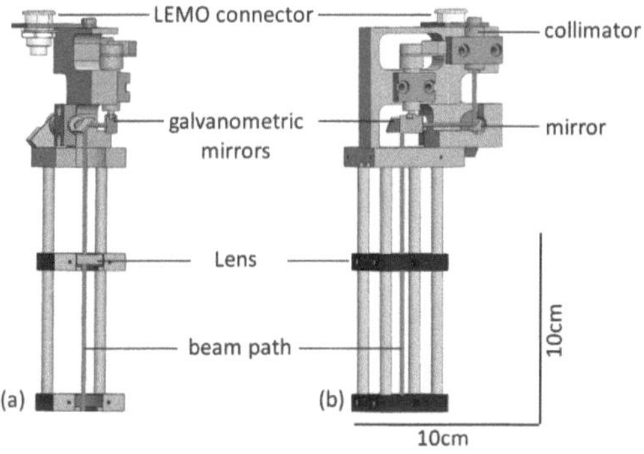

Figure 3: The design of the handpiece with all optical components. Both pictures show the beam path. In the left picture the sectional view of the front can be seen.

The OCT handpiece was designed using SolidWorks regarding all requirements. Additionally, to the given requirements, a mirror (89625, Edmund Optics GmbH) to redirect the light beam was included. It is mounted on a turnable part to align the collimated beam with the galvanometer. To prevent aberrations, the galvanometric mirrors should have the minimal distance possible, without restricting their movement. Therefore, the distance was set to 1.5 mm. We used two galvanometric mirrors with silver coating Cti 6210 KB (MediaLas Electronics GmbH). The construction was build out of aluminum by scientific workshops (WW) of the Universität zu Lübeck. The design measures 57 mm (width) × 71.5 mm (height) × 75 mm (length) and weighs 167 grams. The height depends on the lens that is used, we used an achromatic lens with 75 mm focal length (45805, Edmund Optics GmbH) so the height results in 216 mm. There are different options to insert the laser and its components into the OCT-system. If the galvanometric mirrors are shared for both systems, the distance can be short and the chosen focal length small. Otherwise, the working distance has to be longer in order to include the laser beam from the side below the lens. Thus, this is a prototype to test and measure the image quality. The lens is mounted on a cage system with no cover for easy adjustment. This leads to an adjustable operating distance to test different possibilities for the laser. The whole construction with all optical components can be seen in Fig. 3

Summarized, the necessary optical components included into the handpiece are: two galvanometer, including their electrical ports, a collimator to connect the light source, an adjustable lens mounting to test different optics and a port for all electrical interfaces. Additionally, most of the electrical and light ports are on the same side. We used

the F280APC-C collimator (Thorlabs GmbH) for coupling light into and out of fibers into our system, the resulting beam width is 3.4 mm.

2.3 Measurements

To test the designed handpiece and measure the field of view and the resolution, it was connected to the Telesto (Thorlabs GmbH).

Zemax was used to simulate the image of a collimated beam in order to determine the distance of the optimal focus and the focus width. Then the focus width was recorded and analyzed with BeamGage, a system for Camera-based beam profile measurement, and the FWHM beam diameter was calculated in MATLAB®. For determining the lateral resolution, the USAF resolution test chart was used. The field of view (FOV) was determined as the unshaded pixel, while the pixel spacing was measured by the width of the measured lines of the USAF target. The FOV was then calculated as

$$FOV[mm] = FOV[px] \times pixelspacing[mm/px]. \quad (2)$$

The chart has a size of 15 mm × 15 mm, which is bigger than the desired field of view.

3 Results and Discussion

To define the resolution, the device was simulated with Zemax and tested with a lens with 75 mm focal length. The expected lateral resolution calculated by simulation was FWHM = 24 µm. After adjusting the galvanometer and the prism, the beam diameter was measured with BeamGage, the measurement can be seen in Fig.4. The resulting diameter is FWHM = 27 µm. So the result is a few micrometer larger than the expectation, and also larger than the previously described and necessary lateral resolution.

It was shown, that the FOV is restricted through the lens mounting and the voltage restriction for the galvanometer.

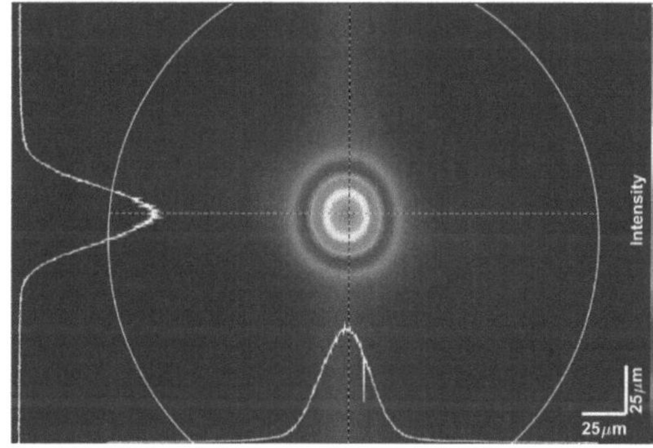

Figure 4: Beam diameter acquisition with BeamGage.

The desired FOV of 10×10 mm was surpassed. Nevertheless, it should be considered that the FOV will decrease for lenses that are smaller than an inch in diameter, which could be considered to for a narrower handpiece.

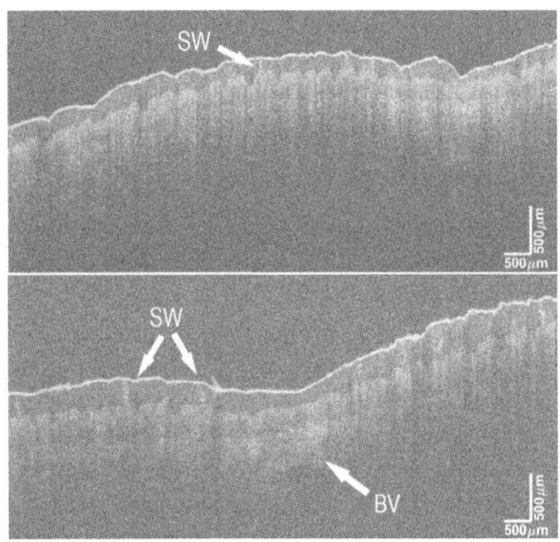

Figure 5: Sectional view of the finger of a young, healthy male subject. Upper image: acquisition with our optics, lower image shows the same skin obtained with the Telesto system in combination with a commercially available handpiece (OCT-G-1300, Thorlabs). SW marks sweat glands, BV marks blood vessels.

In Fig.5 a comparison between our system and the Telesto by Thorlabs can be seen. For this, the healthy skin of a patient was recorded with both systems. It can be seen that the penetration depth and the resolution of the new system is as good as the image with Telesto in combination with a commercially available handpiece (OCT-G-1300, Thorlabs), which is to be expected because we are using the same system. The layered structure and features like sweat glands can be recognized. All in all, it has been shown that the new system meets the expected requirements in terms of FOV and lateral resolution.

4 Conclusion

A new OCT device for BCC diagnostic was successfully designed and tested. The results show that our handpiece was lighter than the required 500 grams, and it surpassed the requested FOV of 10 mm × 10 mm. When comparing the images of our system with the current clinical systems, it shows that we achieved almost the same resolution and penetration depths. More lenses with shorter focal length should be tested to offer the possibility for a shorter operating distance while providing a narrower focus. To obtain a high resolution in all skin layers, different options for focal extension, for example Bessel beam or multifocals should be considered.

The next step for the project is the implementation of the laser into the OCT-system. For this, it needs to be discussed where the laser should be included. Regarding the performance power of the laser and the sensitivity of the galvanometric mirrors, it needs to be determined if it is possible to guide the laser beam through the OCT beam path. This could lead to a shorter operating distance, which could offer an improvement for the resolution. The other possibility would be to include the laser beam below the lens, which necessitates that the laser has its own control device to direct the laser beam to the desired skin area.

Acknowledgement

The work has been carried out at the Institute of Biomedical Optics, Universität zu Lübeck. In addition, we want to thank Prof Dr. med. Julia Welzel and Dr. med. Sandra Schuh for their support with the measurements in Augsburg.

Author's Statement

Conflict of interest: Authors state no conflict of interest.

5 References

[1] Skin Cancer Foundation, *Basal Cell Carcinoma Overview*. Available: https://www.skincancer.org/skin-cancer-information/basal-cell-carcinoma/ [last accessed on 202-01-03].

[2] Robert Koch-Institut, *Krebs in Deutschland für 2015/2016*, Gesellschaft der epidemiologischen Krebsregister in Deutschland e. V.: vol. 12, pp.68-69, 2020.

[3] E. Sattler, R. Kästle, J. Welzel, *Optical coherence tomography in dermatology*, Journal of Biomedical Optics, 2013.

[4] M. Mackiewicz-Wysocka, M. Bowszyc-Dmochowska, D. Strzelecka-Węklar, A. Dańczak-Pazdrowska, Z. Adamski, *Basal cell carcinoma – diagnosis*, Contemporary Oncology/Współczesna Onkologia, vol. 17(4), pp. 337-342, 2013

[5] American Academy of Dermatology Association, *Skin Cancer Types: Basal Cell Carcinoma Overview*. Available: https://www.aad.org/public/diseases/skin-cancer/types/common/bcc [last accessed on 202-01-04].

[6] AWMF 2017/2018, *S2k-Leitlinie Basalzellkarzinom der Haut*. Available: https://www.awmf.org [last accessed on 202-01-04].

[7] W. Drexler, J.G. Fujimoto, *Optical Coherence Tomography, Technology and Applications*, Springer, 2015

[8] T. Soleymani, M. Abrouk, K.M. Kelly, *An Analysis of Laser Therapy for the Treatment of Nonmelanoma Skin Cancer*, Dermatologic Surgery: vol. 43(5), pp 615-624, 2017

Optimization of a Master Oscillator Power Amplifier system at a wavelength of 1550 nm

Moritz Wiggert [1], Tonio Franz Kutscher [3], Philipp Lamminger [3], Stefan Meyer [2] and Sebastian Karpf [3],

[1] Biophysics, Universität zu Lübeck, wiggert@physik.uni-luebeck.de

[2] Medical Engineering Science, Universität zu Lübeck, stefan.meyer@student.uni-luebeck.de

[3] Institute of Biomedical Optics, Universität zu Lübeck, (t.kutscher, phi.lamminger, sebastian.karpf)@uni-luebeck.de

Abstract

We are optimizing a pulsed high-power fiber laser light source in the 900 nm to 940 nm regime for a two-photon, spectro-temporal Imaging by diffractive excitation (SLIDE) microscope. This technique allows for ultra fast in tissue scanning microscopy. Due to a lack of gain media for fast tunable lasers in this wavelength range we want to utilize Raman Scattering in a two-level cascade Raman shift from 1550 nm to 1800 nm. Later this light will be frequency doubled to reach the aimed for wavelength of 900 nm. For an efficient energy transfer of the first Raman shift to 1665 nm via Stimulated Raman scattering (SRS) a tunable ring laser is built to function as a Raman probe laser. This laser can be tuned from 1624 nm to 1747 nm and provides a narrow band output power of $-9\,\mathrm{dB\,m}$ at a wavelength of 1665 nm.

1   Introduction

In biomedical in-tissue imaging the available laser excitation wavelengths are an important factor. The photons need to have a sufficient penetration depth to reach the area of interest. There also has to be a range of wavelengths available when working with different fluorophores or auto fluorescent molecules. In scanning microscopy the imaging speed is often limited by the scan speed of the galvanometric scanner. A new technic called spectro-temporal Imaging by diffractive excitation (SLIDE) can help boost imaging speed by orders of magnitudes. In SLIDE a diffractive element in combination with a fast swept laser is used for the line scans [1], [2]. This, in combination with two-photon microscopy allows for deep optical penetration and fast data acquisition in tissue. For biological tissue the optical window for sufficient penetration depth reaches from 650 nm to 900 nm. The final aim of this work is to built a laser with a wavelength of 900 nm. For this part of the spectrum there are no feasible gain media to realize fast swept lasers with sufficient photon density to achieve two-photon absorption.

Fiber lasers are a good choice when building biomedical optical instruments. They are rather robust and stable after initial setup and can also be modified to later be used endoscopically. As a bonus, fiber laser components around the 1550 nm wavelength region are readily available and inexpensive as this is close to the center wavelength of the telecommunication c-band used in large quantities worldwide. To convert the 1550 nm laser to a 900 nm laser two effects can be utilized. First, two Raman shifts via Stimulated Raman Emission to 1665 nm and from there to 1800 nm are required [3]. Afterwords this light can be fre-

quency doubled to 900 nm via Second Harmonics Generation by using a periodically poled lithium niobate (PPLN) crystal. For spontaneous Raman emission at infrared wavelength a high photon density is needed. Therefore the initial 1550 nm laser has to be of sufficient power. We estimate a required output power of at least 10 kW. To achieve this we want to use a three stage Master Oscillator Power Amplifier (MOPA).

2   Material and Methods

2.1   Master Oscillator Power Amplifier

A Master Oscillator Power Amplifier (MOPA) can be used to achieve high power pulses in a fiber laser system. A master oscillator or seed pulse with the desired laser parameters e.g. pulsewidth and wavelength is amplified by a number of consecutive amplification stages. In the stages the pulse passes through an optically pumped active fiber. The active fiber is doped with rare earths such as Erbium or Ytterbium. The dopant atoms are put into an excited state by absorbing the pump energy. The photons of the seed pulse can then stimulate the excited atoms to emit an identical photon.

The great advantages of these systems are the power decoupled beam parameters. The output power can be changed to a large extent without fear of beam quality deterioration. Michalska and Swiderski report a total amplification of 51 dB in three amplification stages [4]. A setup similar to the one proposed by Michalska and Swiderski was built at the institute by Annika Hunold. Hunold reports on an amplification of 39.64 dB after the first two stages. Beginning with a seed pulse power of 2.57 mW this results in an output power of 23.6 W [5]. In comparison with similar systems at the institute this output power fell short. To avoid possible damage to components caused by amplified spontaneous emission (ASE) pulses a pulse power of at least 130 W was aimed for as seed for the stronger pumped third amplification stage.

The first two amplification stages where reworked to reliably provide at least 130 W of peak pulse power. The schematics of the reworked system can be seen in Fig. 1. From a 1550 nm continuous wave laser diode, pulses are modulated by an electro-optical modulator (EOM). For protection of previous optical components a tap isolator (TAPI) with a 1 % port (TAP) for beam observation is added. The amplification occurs when the pulses are passing the erbium doped fiber (EDF) that is pumped by a 976 nm pump diode. This diode is protected by a pump protector (PP). A wavelength division multiplexer (WDM) separates the signal- and pump light. At the end of the first two stages a band pass (BP) is used to clean up the signal. The third stage is pumped in the forward direction. It utilizes a Erbium-Ytterbium co doped double cladding fiber (EYDF) . The first cladding, where the multi mode pump light is guided can handle more power than the core in single cladding fibers. A more power tolerant pump signal combiner (PSC) is used to combine pump- and signal light. The 9 W 976 nm multi mode pump diode is protected by a multi mode pump protector (MM-PP). The pump power is manually adjusted via the current flow through this diode. At the output the amplified pulse can exit the system.

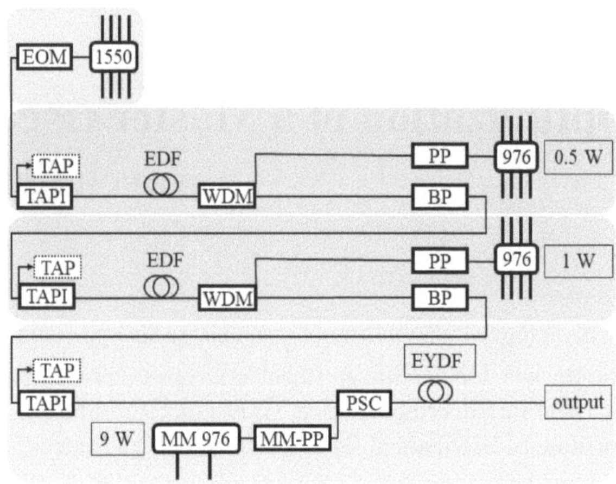

Figure 1: Conceptualized depiction of the MOPA setup. The red box (top) represents the creation of the seed pulse. The blue boxes (middle) are core pumped amplification stages one and two. Stage two is a copy of stage one but utilizes a stronger pump diode and a longer EDF. The green box (bottom) contains the third, cladding pumped amplification stage. (*1550*) narrow band 1550 nm seed diode, (*EOM*) eloctro-optical modulator, (*TAPI*) tap isolator, (*TAP*) 1 % port, (*EDF*) Erbium doped fiber, (*WDM*) wavelength division multiplexer, (*PP*) pump protector, (*976*) single mode pump diode with optical pump power, (*BP*) band pass filter, (*EYDF*) Erbium-Ytterbium co doped fiber, (*PSC*) pump signal combiner, (*MM-PP*) multi mode pump protector, (*MM 976*) 976 nm multi mode pump diode

2.2 Ring Laser

For the generation of the SRS probe a second laser system is required. A tunable ring laser was built for this purpose. A booster optical amplifier (BOA) with an ASE central wavelength of 1692 nm provides the gain. This component is attached on each side to the polarization maintaining fiber PM1550. All the other components utilize the single mode fiber SMF-28. The BOA is protected by two optical isolators (TAPI), one on each side. They also ensure that light can only travel in the ring in forward direction. The Polarization Controller is required to reliably fit the polarization of the light inside the SM fiber to the PM fiber. With the Fabry-Pérot-Interferometer (FFP-TF) the resonant wavelength can be tuned by applying and changing a voltage. A 80:20 coupler allows 80 % of the light to leave the ring. This light will be used as Raman seed. The other 20 % stay inside the ring for continuous lasing. The schematic setup of this system can be seen in Fig. 2.

2.3 Stimulated Raman Emission

Raman scattering occurs as a nonlinear effect when high photon densities propagate through a medium such as fused silica fibers in optical wave-guides. The photons are scattered inelastically and thus lose some of their energy. This causes a wavelength shift towards longer wavelengths. In fused silica the scattered photons are shifted by 13 THz to 15 THz [6]. With an initial seed wavelength of 1550 nm or a frequency of 194 THz the first Raman shift will lead to a range of frequencies from 179 THz to 181 THz or range of wavelength from 1657 nm to 1675 nm, respectively. For stimulated Raman emission a probe-laser is added into the system. The wavelength of the probe is chosen to fall within the Raman spectrum. This probe acts as a new seed laser. The power of the source is then efficiently drained for up to 90 % towards the wavelength of the probe [3]. With a chosen probe wavelength of 1665 nm the second Raman shift will fall into a range of 1796 nm to 1818 nm.

3 Results and Discussion

3.1 Optimization of amplification stage one and two

Several modifications where made to reliable achieve the aimed for 130 W of peak pulse power after the second amplification stage. The measurements for the power after the EOM where taken by connecting the output of the EOM directly to a photo diode (Finsar 50 GHz dual-window Photodetector). This photo diode converts the incoming optical power to a voltage with a factor of $16.25\,\mathrm{V\,W^{-1}}$. The diode was in turn connected to an oscilloscope (Teledyne Lecroy SDA 830Zi-B) where the average of the resulting peak voltage was read to calculate the pulse peak power.

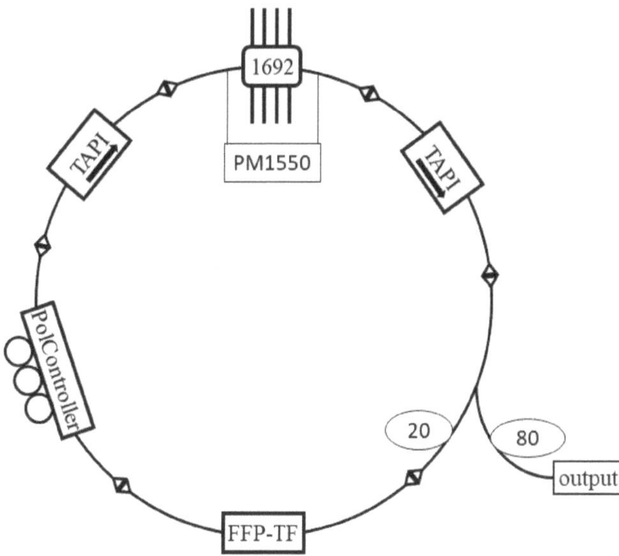

Figure 2: Schematic depiction of the Raman probe laser. All connectors of the individual components where simply plugged into each other using a mating sleeve. This is symbolized with two adjacent triangles. The booster optical amplifier with an ASE central wavelength of 1692 nm was the only component attached to a *PM1550* fiber. The other components utilized *SMF-28* fiber. (*TAPI*) tap isolator with transmission direction, (80:20) coupler, (*FFP-TF*) Fabry-Pérot-Interferometer, (*POLController*) polarization controller

The power measurements for the output of the second stage where taken with the output of the second stage connected first to two consecutive 99:1 couplers. The resulting attenuation by 40 dB was necessary to avoid destruction of the photo diode. The electrical pulses needed by the EOM to generate optical pulses where provided by an arbitrary waveform generator (Tektronix AWG 7122B). They had a pulse length of 100 ps and a repetition rate of 5.33 MHz. Hunold reports on a peak pulse power of 2.57 mW after the EOM [5]. The seed diode was turned up to its full 30 mW output power and two damaged APC connectors where exchanged. Now 7 mW peak pulse power was measured after the EOM and 50.6 W after the second stage. The amplification stages were then changed to a backwards pumped approach. This was done to facilitate a more efficient pump to signal conversion. The pulse at the beginning of the active fiber might not have enough energy to stimulate all exited dopant atoms. In a forward pumped system the most energy is stored at the beginning of the active fiber and might be wasted. Additionally the pump diode in the second stage was swapped from a 0.5 W diode to a 1 W diode. Also, the length of the active fiber of the second stage was changed from 2.5 m to 5 m to allow for more absorption of the light from the stronger pump diode. These changes resulted in a significant change in peak pulse power from 26 W to 115 W.

During measurements the peak pulse power fluctuated noticeably with several pulse trains collapsing to less then half of the maximum peak power. After some investigation

the fault was found at the fusion splice of the 500 mW pump diode to the pump protector. After optimizing this splice the pulses stabilized. As a final measure to optimize power output the seed diode was premodulated. The electro-optical modulator (EOM) that is used to generate pulses from the continuous wave emitted from the seed diode is not perfectly blocking the seed light in its off state. This causes the actively pumped gain fiber to be partially depleted by light leaking through the EOM. To avoid this the seed diode is turned on and off synchronous to the EOM. The seed diode needs some time to start lasing at full power so the initial seed pulse is weaker. It does however receive more gain in the active fiber. The peak power of the premodulated pulse after the EOM was 6.3 mW. Compared to the 7 mW of the unmodulated pulse this is a loss of 10 %. The peak power of the premodulated pulse after the second stage was 131 W. Compared to the 112 W of the unmodulated pulse this is a gain of 17 %. The measurement of the pulse peak power were repeated on several days. In the described state of the setup the lowest measured peak output power was 131 W. On other days a peak power of up to 170 W was measured.

For the third stage only a single measurement was taken with a thermally sensitive probe (Thorlabs S470C 0.25 μm to 10.6 μm up to 5 W with the powermeter Thorlabs PM100D). As the deposited energy on the probe is integrated over time this measurement can be viewed like that of a continuous wave. Multiplying the measured power by the inverse of the duty cycle gives an estimate of the maximum possible pulse peak power. With a measured cw power of 2 W and a duty cycle of 0.000533 the resulting pulse peak power is 3.7 kW. The described changes of the setup result in a total gain of 43.5 dB in the first two stages. With the calculated value for the pulse power of the third stage the whole system provides a gain of 57.69 dB.

3.2 Raman Probe Laser

The tunable wavelengths of the ring laser where analyzed with a optical spectrum analyzer (OSA)(Yokogawa AQ6374 350 nm to 1750 nm). The resonant wavelength could be selected by changing the voltage across the Fabry-Pérot filter. At a voltage of 3.7 V a first small peak at 1620 nm was visible. Shortly after that at 1624 nm a clear peak 30 dB above ASE was visible. The aimed for 1665 nm peak had a signal strength of 40 dB above ASE. At the end of the available range of the (OSA) at 1747 nm and a voltage of 17.4 V across the Fabry-Pérot filter the signal was at 48 dB above ASE. The spectra of each measurement can be seen in Fig. 3. Each spectrum was represents a wavelength sweep form 1610 nm to 1750 nm.

3.3 Discussion

The output power generated by the described three stage MOPA system at 1550 nm is still not sufficient to generate spontaneous Raman Scattering. The calculated pulse amplification for the third stage is only an estimate of the highest possible pulse power, with the actual pulse power probably

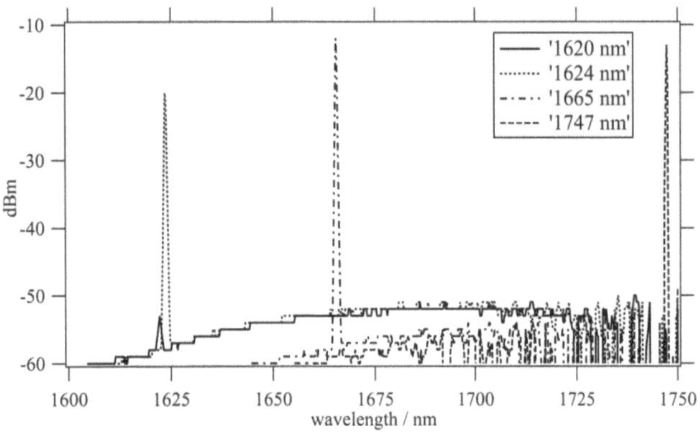

Figure 3: Lasing capabilities of the ring laser at different wavelengths.

being much lower. The stability of the output power on a pulse to pulse time scale was increased dramatically by optimizing the fusion splice from the pump diode of the first stage to its pump protector.

The stability of the system on a day to day timescale is much harder to measure. Ambient and system temperature seem to be a factor in pulse output power with the temperature changing by more than $10\,^{\circ}\mathrm{C}$ on a single day, the pulse power after the second stage also changed by up to $40\,\mathrm{W}$ on a day to day basis. A prolonged tracking of ambient temperature in combination with peak pulse power could help to clarify if a relation between these parameters exists.

After pumping the third stage for roughly one minute the tap isolator at the end of the second stage heated up and the signal from the $1\,\%$ port of the isolator vanished. The reason for this could not be identified. For a reliable operation of the system further investigation and fixing of this problem is needed.

Reaching the aim of $10\,\mathrm{kW}$ of peak pulse power with the three amplification stages with their current components seems improbable. A stronger multi mode pump diode or a fourth amplification stage are probably required.

4   Conclusion

We conclude that further improvements and additional components are required to achieve sufficient amplification for a cascade Raman shift at infrared wavelength. The Raman probe laser is working as expected and can probably be used for SRS in this setup once the spontaneous Raman emission threshold is reached.

Acknowledgement

The work has been carried out and was supervisedby the Institute of Biomedical Optics, University of Luebeck.

Author's Statement

Conflict of interest: Authors state no conflict of interest.

5   References

[1] S. Karpf, C.T. Riche, D. Di Carlo et al. *Spectro-temporal encoded multiphoton microscopy and fluorescence lifetime imaging at kilohertz frame-rates* Nat Commun 11, 2062 (2020)

[2] S. Karpf, B. Jalali, *Fourier-domain mode-locked laser combined with a master-oscillator power amplifier architecture* Opt. Lett., 44(8), 1952 (2019)

[3] M. Eibl, S. Karpf, H. Hakert, T. Blömker, J. P. Kolb, C. Jirauschek, R. Huber, *Pulse-to-pulse wavelength switching of a nanosecond fiber laser by four-wave mixing seeded stimulated Raman amplification* Opt. Lett. 42, 4406-4409 (2017)

[4] M. Michalska, J. Swiderski, *Highly efficient, kW peak power, 1.55 μm all-fiber MOPA system with a diffraction-limited laser output beam.* Appl. Phys. B 117, 841–846 (2014)

[5] A. Hunold *Development of a two-photon microscope for biomedical autofluorescence imaging,* masters thesis, Institut für Biomedizinische Optik, Universität zu Lübeck (2020)

[6] R. H. Stolen and E. P. Ippen, *Appl. Phys. Lett. 22, 276* (1973)

11

Medical Imaging

Revisiting the Iterative Image Registration Technique *deeds* in terms of Speed Performance

Sreejith Kannath Kalam [1], Christian Weihsbach [2], Mattias Heinrich[2],
[1] Robotics and Autonomous Systems, Universität zu Lübeck, sreejith.kannathkalam@student.uni-luebeck.de
[2] Institute of Medical Informatics, Universität zu Lübeck, {christian.weihsbach, mattias.heinrich}@uni-luebeck.de

Abstract

The advancement in automated image registration, particularly in the rapid evolution of deep learning over the past decade has improved the state of the art performance and speed. Apart from deep learning-based image registration techniques, traditional iterative methods are also widely used due to their alignment quality. This paper attempts to analyze the computation improvement of deformable discrete image registration technique *deeds*, by testing a ported version in both Graphics Processing Unit(GPU) and Central Processing Unit(CPU) cores, using a highly optimized deep learning framework. The experiment was carried out with four sets of input 3D data of cubic sizes on four versions of *deeds* algorithm. The Average computation time for processing input cubic tensors of different sizes was estimated and compared. An increase in computation speed is observed.

1   Introduction

Medical images acquired in Magnetic Resonance Imaging(MRI) and Computed Tomography(CT) play a significant role in diagnosing diseases of internal organs. The alteration of abnormalities in internal organs requires periodical monitoring and progress quantification. Therefore medical image registration is a highly relevant preprocessing step in medical image analysis. Since manual image registration methods demand high expertise, it is more vulnerable to errors. Before the recent growth of deep learning-based methods, traditional methods were widely used in the industry. There is considerable research interest in proposing improved registration techniques, each aiming for faster and more accurate registrations between the source and target images. We can decompose any image registration techniques into three components: a transformation, a similarity measurement, and an optimization technique (for finding optimal transformation parameters) between source and target image [1]. Each component increases the computational complexity of the overall algorithm. Thus, improving each element's speed and efficiency is an important research objective. The traditional image registration process usually involves iterative steps for these components [2]. Whereas in learning based methods the similarity quantification can be done faster by extracting features using a convolution neural network [3]. Usage of iterative-based image registration algorithms take more time compared to the deep learning techniques [4]. Although several deeplearning methods have a speed benefit, some still cannot reach the alignment quality of iterative methods [5]. The comprehensive medical image registration challenge Learn2Reg [5] demonstrates that the conventional image registration tech-

niques are more robust and accurate than learning-based techniques, especially for large internal motion and multimodal fusion tasks [5]. After the release of Compute Unified Device Architecture(CUDA), an Application Programming Interface(API), in 2007, more experiments on GPU accelerated image registration became more common, Especially in CT modality [6]. Also, the cost efficient GPU cores motivated a widespread use of medical imaging tools outside research facilities, from small-scale clinical facilities to more expensive operating theaters for diagnosing purposes [7]. This motivated us to revisit a conventional image registration technique to improve speed performance by using GPU power. The work examines whether the speed of the traditional image registration algorithm *deeds* [8] can be improved.

2   Material and Methods

As part of the experiment, a widely used medical image registration algorithm *deeds* was chosen for conducting experiment and analysis. The algorithm employs a discrete dense displacement sampling for the deformable registration of high-resolution CT volumes [8]. The work involves testing the computation speed on four different implementations of the algorithm as shown in Table 1. The source algorithm was programmed in C++. The program is ported to Python language to take advantage of easier libraries and frameworks. Then the algorithm is integrated with Pytorch for processing the variables as Tensors, both in CPU and GPU. 3D Tensors of different sizes were used for testing the computation time as shown in Table 3.

2.1   The deeds

deeds is a deformable non-rigid-based algorithm. It involves a discrete optimization technique on a non-convex cost function. The core of the algorithm involves in creating a minimum spanning tree on the input image grid, both in source and target images. This is established to create independence between distant voxels during transformation. This indeed increases the sparsity and enhances the alignment quality and speed, especially on 3D volumetric images [8]. *deeds* also showed a good performance compared to other competitive image registration algorithms[9].

The algorithm involves the following steps. Initially, both the source and target image grids are deduced into patches, such that a single node represents each patch. Assuming that the neighboring voxels have high similarity, a neighborhood of six nodes was taken and a minimum spanning tree is established on the image grid based on this and more sparsity is achieved for reducing computation complexity. A non-parametric optimization is done on the graph with an energy cost function using dynamic programming. The energy cost function has two components. A unary data cost function D and a pair-wise regularisation cost function R. The unary data cost function measures the similarity between the voxel in the source image and the displaced voxel in target image. The energy cost function is defined as follows. Let a random voxel or node of the patch be $p \epsilon P$ and f_p be its corresponding label. So in a neighborhood of (ϵN), where p and q are directly connected, The energy cost function can be calculated as [8]

$$E(f) = \sum_{p \epsilon P} D(f_p) + \alpha \sum_{(p,q)\epsilon N} R(f_p, f_q). \quad (1)$$

The term α can be tuned to adjust the influence of regularisation. As an extension, the 'Self-Similarity Context'(SSC) descriptor can be integrated into *deeds* to improve multimodal registration [10].

2.2   Computation Speed Estimation and Analysis

The work was carried out on four versions of the algorithm as shown in Table 1. The initial version of source code written in C++ was implemented in 2014 by the author of *deeds*. Three enhanced version of the source is created as shown in Table 1. An equivalent translation of the code to python is created. This ported version is integrated with Pytorch, aiming an optimized speed performance. All the three versions described above are created for processing in CPU. The Pytorch integrated version is further updated for GPU processing. Four relevant functions in the algorithm from each version are tested for computation time estimation and comparison. The functions and their short description is given in Table 2.

To check whether each version outputs the same as the source code, the functions were then unit tested with multiple random tensors with increasing cubic sizes. Tensor sizes are shown in Table 3. The input parameters speci-

Table 1: Tested code versions

S.No	Version	Description
1	deeds[8] C++	Source Code
2	deeds Python equivalent	Naive python Code
3	deeds Pytorch Optimized	CPU Processed
4	deeds Pytorch Optimized	GPU Processed

Table 2: Functions of interest

Name	Description
`WarpAffine()`	Warping Segmentation
`ConsistentMapping()`	Inverse mapping
`Volfilter()`	Gaussian filter
`UpsampleDeformations()`	Upsampling

fied for testing the individual functions are as follows. The `WarpAffine()` function is tested by an affine matrix of random values. For `UpsampleDeformations()`, the input deformation field is up-sampled to double of input size with trilinear interpolation. The function `Volfilter()` performs gaussian filtering on input image. This was created using mesh grid and convolution of gaussian kernel on the grid. A kernel size 5x5x5 is used for testing. The `ConsistentMapping()` is tested on a deformation field. These functions in each version are tested and average computation speed is calculated. The average computation time of the functions is estimated out of 100 repetitions on CPU (Apple M1, 8GB RAM, 8-core CPU) and GPU (NVIDIA GeForce RTX 2080 Ti, 11018 MiB, CUDA 11.5 (495.29.05)).

3   Results and Discussion

The average computation time obtained after running the experiments are shown in Table 4 and Table 5.

The attempt to process the equivalent python version of deeds algorithm resulted in significantly high computation time especially in certain functions like `UpsampleDeformations()` and `ConsistentMapping()` as shown in Table 5.

By analyzing the source code it is observed that the custom made trilinear interpolation function used inside both `UpsampleDeformations()` and `ConsistentMapping()` requires more execution time. When optimized with Pytorch interpolation function a considerable hike in computation speed is obtained. This can be observed in Table 4. The `WarpAffine()` and `Volfilter()` functions in both versions are similar. But

Table 3: Test Data

Test Case	Size
1	32x32x32
2	64x64x64
3	128x128x128
4	256x256x256

Table 4: Mean computation time in seconds of functions of interest on test data:-Torch Optimized Versions. *C/M(CUDA out of memory)

Function	Size	CPU	GPU
UpsampleDeformations()	32	0.83	0.0069
	64	6.17	0.0075
	128	48.05	0.0076
	256	399.69	C/M
WarpAffine()	32	0.133	0.0179
	64	0.947	0.0187
	128	5.465	0.0317
	256	57.790	0.7057
ConsistentMapping()	32	1.723	0.388
	64	5.945	0.7811
	128	34.433	3.159
	256	239.67	C/M
Volfilter()	32	0.05407	0.0253
	64	0.1198	0.0264
	128	0.1928	0.131
	256	4.2979	0.746

Table 5: Mean computation time in seconds of functions of interest on test data :- C++ V/S Python equivalent, *C/M(CUDA out of memory)

Index	Size	C++	Python
UpsampleDeformations()	32	0.2556	217.063
	64	1.8345	C/M
	128	14.756	C/M
	256	130.258	C/M
WarpAffine()	32	0.2264	0.2162
	64	1.6965	1.5965
	128	1.6835	1.6347
	256	13.212	11.7885
ConsistentMapping()	32	59.42	514.725
	64	C/M	C/M
	128	C/M	C/M
	256	C/M	C/M
Volfilter()	32	0.7122	0.6862
	64	5.6902	5.721
	128	45.495	46.032
	256	C/M	C/M

the optimized versions indicate a considerable improvement can be achieved using the Pytorch framework. The powerful convolution functions, especially in 3D data offered by these frameworks might be the reason. The results obtained from processing the Pytorch optimized version on GPU demonstrates room for development of the computation speed. Especially estimated computation time in GPU indicates around 100 fold improvement in speed in some functions. Especially the `UpsampleDeformation()` and `ConsistentMapping()` function which took significant amount of time in CPU appeared to be faster, even with increase in input size as shown in Figure 1. Further optimized functions like `WarpAffine()` processed with

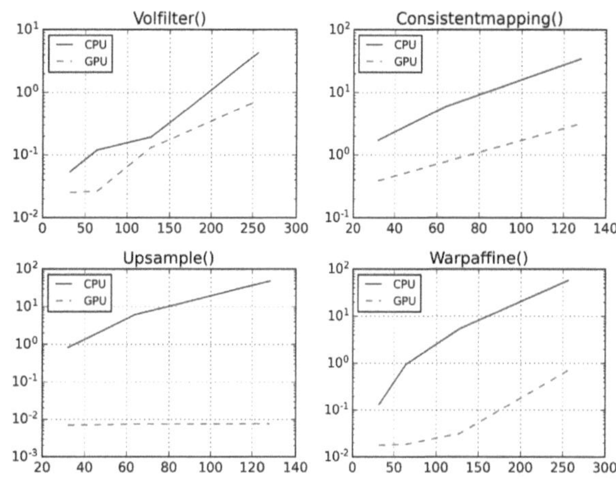

Figure 1: Average computation time(logarithmic scale) V/S Voxel size

considerably less change in computation time with an increase in the input size.

At the same time, with an increase in the size of input, the GPU demands more CUDA memory. For the input size of 256x256x256 the GPU memory was overloaded and could not process the tensor. Therefore, it is observed that some functions demand more memory allocation with the increase in the input size. The cost of more advanced hardware may hinder the usage of GPU cores for processing algorithms on small scales. Moreover in this work computation speed analysis was done only with custom-made inputs. Testing with real medical image data is required for further analysis. This requires porting complete version of the source code into python and Pytorch versions.

The experiment indicates a significant improvement in computation speed, when optimized with deep learning framework.

4 Conclusion

We have attempted to optimize a traditional iterative image registration method *deeds* using a popular deep learning framework PyTorch, And performed a speed comparison study. The experiment could find an improvised computation performance on the optimized method when processed in GPU. The error in translating to python equivalent implementation of source code must be put for analysis in future work. The experiment was carried out only for a few functions and some important functions in the algorithm such as functions for creating minimum spanning tree, regularisation cost, data cost need to be ported, and an integration test of the whole deeds algorithm is needed to be carried out for further analysis. Comparing the original C++, Pytorch(CPU), and Pytorch(GPU) execution times we show that the execution speed of deeds can be improved while outputting identical registered image quality.

Acknowledgement

The work has been carried out at the Institute of Medical Informatics as part of an internship, under the supervision of Prof. Dr. Mattias Heinrich and M.Sc. Christian Weihsbach, Universität zu Lübeck.

Author's Statement

Conflict of interest: Authors state no conflict of interest.

5 References

[1] V. Mani and S. Arivazhagan, "Survey of medical image registration," *Journal of Biomedical Engineering and Technology*, vol. 1, no. 2, pp. 8–25, 2013.

[2] F. P. Oliveira and J. M. R. Tavares, "Medical image registration: a review," *Computer Methods in Biomechanics and Biomedical Engineering*, vol. 17, no. 2, pp. 73–93, 2014. PMID: 22435355.

[3] Y. Fu, Y. Lei, T. Wang, W. J. Curran, T. Liu, and X. Yang, "Deep learning in medical image registration: a review," *Physics in Medicine & Biology*, vol. 65, no. 20, p. 20TR01, 2020.

[4] H. R. Boveiri, R. Khayami, R. Javidan, and A. Mehdizadeh, "Medical image registration using deep neural networks: A comprehensive review," *Computers & Electrical Engineering*, vol. 87, p. 106767, 2020.

[5] L. Hansen and M. P. Heinrich, "Revisiting iterative highly efficient optimisation schemes in medical image registration," in *International Conference on Medical Image Computing and Computer-Assisted Intervention*, pp. 203–212, Springer, 2021.

[6] A. Eklund, P. Dufort, D. Forsberg, and S. M. LaConte, "Medical image processing on the gpu–past, present and future," *Medical image analysis*, vol. 17, no. 8, pp. 1073–1094, 2013.

[7] R. Shams, P. Sadeghi, R. A. Kennedy, and R. I. Hartley, "A survey of medical image registration on multicore and the gpu," *IEEE signal processing magazine*, vol. 27, no. 2, pp. 50–60, 2010.

[8] M. P. Heinrich, M. Jenkinson, S. M. Brady, and J. A. Schnabel, "Globally optimal deformable registration on a minimum spanning tree using dense displacement sampling," in *Med Image Comput Comput Assist Interv.*, pp. 115–122, Springer, 2012.

[9] Z. Xu, C. P. Lee, M. P. Heinrich, M. Modat, D. Rueckert, S. Ourselin, R. G. Abramson, and B. A. Landman, "Evaluation of six registration methods for the human abdomen on clinically acquired ct," *IEEE Transactions on Biomedical Engineering*, vol. 63, no. 8, pp. 1563–1572, 2016.

[10] M. P. Heinrich, M. Jenkinson, B. W. Papież, M. Brady, and J. A. Schnabel, "Towards realtime multimodal fusion for image-guided interventions using self-similarities," in *International conference on medical image computing and computer-assisted intervention*, pp. 187–194, Springer, 2013.

Monocular Depth Estimation for Bronchoscopic Navigation

Rakesh Reddy Kondeti [1] Marian Himstedt [2] Mattias Heinrich [2]
[1] Robotics and Autonomous Systems, University of Lübeck, rakesh.kondeti@student.uni-luebeck.de
[2] Institute of Medical Informatics, University of Lübeck, {marian.himstedt, mattias.heinrich}@uni-luebeck.de

Abstract

Depth estimation from monocular coloured images is an important task in localization and 3D reconstruction pipelines for bronchoscopic navigation [6]. Several methods are proposed to address this task on the natural scenes. However, lack of labeled data and presence of scaled texture makes it difficult to apply these methods for bronchoscopy navigation. We implemented a fully convolutional U-net architecture trained on a simple phantom dataset involving 16 sequences , a two-step approach for estimating depth from the RGB images. Our network projects the input RGB images onto the domain of BRDF images (CT renderings) and then the depth is estimated from the predicted BRDF images to cope up with domain adaptation. The results show that the depth images obtained through this network are very much similar (quantitatively and qualitatively) to the ground truth labels. Additionally, the second part of the network can be trained with the CT renderings of the specific patient in pre-operative conditions to not to loose patient specificity [1].

1 Introduction

World Health Organization has reported that *Lung Cancer* is the second most leading causes of cancer related deaths worldwide, accounting for nearly 2.21 million cases in 2020. With the widespread use of computed tomography(CT), pulmonary lesions have become common findings inside the lungs [5]. However, most of the lesions found are benign and it is important to distinguish between benign/cancer lesions. Nodule biopsy is needed for the diagnosis confirmation. Bronchoscopic navigation is a technique that assists the bronchoscope to safely reach the pulmonary lesions by avoiding unnecessary surgical iterventions [5]. Hence, the ease of navigation to the biopsy sites or bronchoscopic navigation is therefore paramount [1].

Because the convolutional neural network (CNN) preserves spatial information and uses weight sharing, CNNs are known to perform better with less computational costs. In this paper, a fully convolutional U-net (adapted from cycleGAN) is implemented to yield the depth images from coloured images. In encoder-decoder architecture of the U-net, the meaningful features are learnt in the contracting path by projecting the higher dimensional image space to lower dimensional feature space. In the expansion path, the image is reconstructed using the features learnt in the contracting path.

2 Material and Methods

2.1 Dataset

Lack of labeled data or ground truth depth images is itself a significant bottleneck. However, this paper makes use of the first publicly available simple phantom bronchoscopy

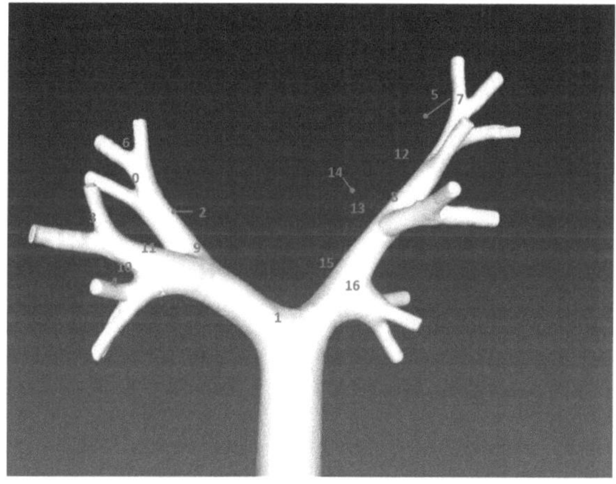
Figure 1: Phantom bronchial tree
This is a simple model which is made of plastic and closely resembles the bronchial tree. There are 17 nodes and are numbered correspondingly.

dataset with aligned 3D ground truth depth images, provided by Visentini-Scarzanella et al. [1]. The phantom dataset is a simple plastic model which consists of 17 terminal and intermediate nodes (Fig. 1), each consisting of $1k-4k$ paired images. The RGB image, corresponding BRDF image and depth image sampled from a random sequence are shown in Fig. 2

In the Fig. 2, the BRDF image can be assumed as a textureless image. Given the light source and the viewing direction, the Bidirectional Reflectance Distribution Function (BRDF) fully defines the reflectance property (the ability to take light from one direction and reflect it into another direction) of the material. The Z images are the depth heatmaps.

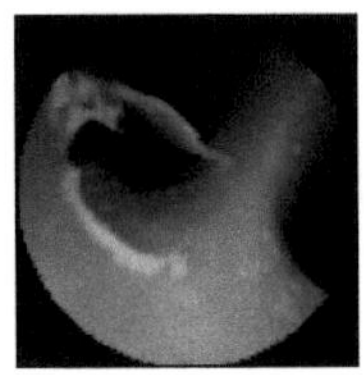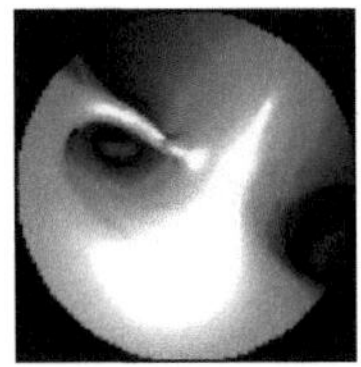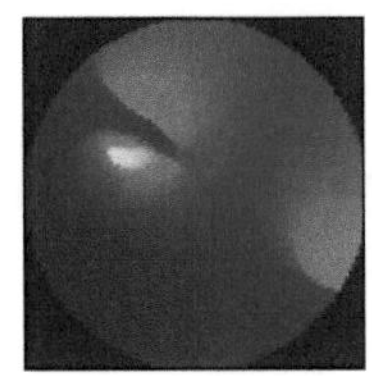

Figure 2: Left: RGB image; Middle: BRDF (Textureless) image; Right: Z (Depth)image
A camera is passed into the phantom model to fetch RGB images. Thus obtained RGB images are later projected on to the domain of CT renderings to obtain the BRDF images

2.2   Related Work

Visentini-Scarzanella et al. presented a fully convolutional encoder-decoder architecture and used the same phantom dataset for training the model. This network consists of one fundamental block called β block. The β block consists of 3×3 convolutional layer followed by batch normalization and rectified linear unit (ReLu). The β blocks are concatenated thrice to give a α block. Each α block is followed by a pooling operation to downsample the input [1].

As the loss function, Marco used MSE loss combined with 0.1 times of total variation(TV) penalty. Given x is the input to network and $H(x)$ as the output, the cost function (J) over all the samples Ω is shown as:

$$J = \frac{1}{\Omega} \sum_{i \in \Omega} (x_i - H(x_i))^2 + 0.1 \times |\nabla H(x_i)| \quad (1)$$

2.3   Our Network

Our approach also includes the intermediary representation of BRDF images. Our network is rather a complex architecture when compared to Visentini-Scarzanella et al. network [1]. We are using U-net architecture with skip connections as shown in Fig. 3

The image I is downsampled by performing 4×4 convolutional operation with a stride 2. In the encoder path, each downsampling is followed by batch normalization and leaky rectified unit (Leaky ReLu) with a slope of 0.2. In the bottleneck layer, the feature maps is downsampled to resolution of 1×1. To upsample the image in the decoder path, the image is first resized and then the convolution operation is performed. This implementation of resize and then convolution avoids the checkerboard artifacts which are appeared when used the traditional transposed convolution operation. The upsampled image is then passed through batch normalization layer before concatenating with the skip connections. Then the concatenated feature map is passed through the rectified linear unit (ReLu) and the image is upsampled again. The final layer in this network is tanh activation layer. This Unet is initialized with *kaiming normal* weights.

2.4   Loss Function

The loss function used for training the above mentioned network consists of three terms: 1) L1Loss 2) SSIM loss and 3) Mean gradient loss.

2.4.1   L1 Loss

The loss function is very important to train the network. It says our model how far is the estimated value away from the true value. The L1 loss, otherwise called Mean Absolute value calculates the absolute difference between the actual ground truth label and predicted label. Unlike the MSE loss, L1 loss reduces the average error and makes it a decent fit for the image reconstruction.

2.4.2   SSIM Loss

Structural Similarity Index Measure (SSIM) is a comparison metric between two images, a reference image and the predicted image. Most of the loss functions calculate the difference in the values of the corresponding pixels in the reference and predicted image (MSE Loss, L1 Loss,..). However, all the natural images are considered to be structured: the pixels are dependent on neighbouring pixels, and this dependency exhibhit strong information regarding the structure of all the objects in the scene [3]. SSIM is a perceptual metric which tries replicate this behaviour by separating similarity measurement into three comparisions: Luminance, Contrast and Structure [3].

1. Luminance: Luminance is a photometric measure of the light emitted per unit area in a given direction. For a given image, it is measured as the average of all the pixel values.

$$\mu_x = \frac{1}{N} \sum_{i=1}^{N} x_i, \quad (2)$$

   where μ_x is the luminance value for a given image x and N is the number of the pixels.

2. Contrast: We use the standard deviation (square root of variance) as an estimate of the signal contrast. The unbiased estimate of the contrast is given as

$$\sigma_x = \left(\frac{1}{N-1} \sum_{i=1}^{N} (x_i - \mu_{x_i}) \right)^2 \quad (3)$$

3. Structure: The input signal is normalized (divided) by its own standard deviation, so that the two signals being compared have unit standard deviation. This allows for a more robust comparison.

$$\frac{x - \mu_x}{\sigma_x} \quad (4)$$

After obtaining the luminance, contrast and structure for individual reference signal x and predicted signal y, the luminance comparision function $l(x, y)$, contrast comparison function $c(x, y)$ and the structure comparison function $s(x, y)$ are calculated as follows:

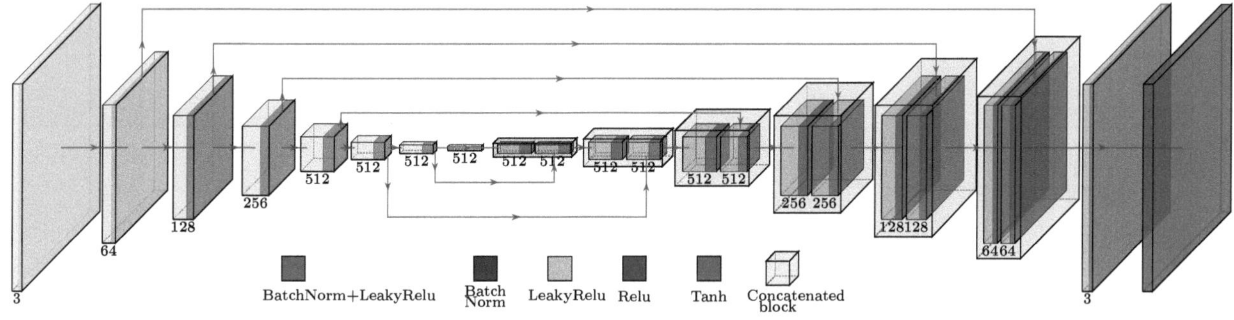

Figure 3: Unet with skip connections

$$l(x,y) = \frac{2\mu_x\mu_y + C_1}{\mu_x{}^2 + \mu_y{}^2 + C_1}, \tag{5}$$

$$c(x,y) = \frac{2\sigma_x\sigma_y + C_2}{\sigma_x{}^2 + \sigma_y{}^2 + C_2}, \tag{6}$$

$$s(x,y) = \frac{\sigma_{xy} + C_3}{\sigma_x + \sigma_y + C_3}, \tag{7}$$

where C_1, C_2 and C_3 are the small constants to avoid instability when denominator is close to zero and $\sigma_{xy} = \frac{1}{N-1}\sum_{i=1}^{N}(x_i - \mu_{x_i})(y_i - \mu_{y_i})$
Now, SSIM value of two images x and y is given by,

$$SSIM(x,y) = \frac{(2\mu_x\mu_y + C_1)(2\sigma_{xy} + C_2)}{(\mu_x^2 + \mu_y^2 + C_1)(\sigma_x^2 + \sigma_y^2 + C_2)} \tag{8}$$

2.4.3 Mean Gradient Error

The above used L1 loss does pixel wise comparison and SSIM loss acts as perceptual loss to better reconstruct an image with respect to the human visual system. However, edges in the images are important features as they contribute to significant features of the objects. The *mean gradient error* (MGE) is used as edge wise comparison in this loss function [2].

MGE is specifically designed to capture the difference in edges of the two given objects. The edges in the direction of X-axis are captured using the horizontal sobel kernel and edges in the direction of Y-axis are captured using the vertical sobel operator. These kernels when convolved over the image, gives the corresponding gradients.

$$G_x = \begin{bmatrix} -1 & 0 & 1 \\ -2 & 0 & 2 \\ -1 & 0 & 1 \end{bmatrix} * A \tag{9}$$

$$G_y = \begin{bmatrix} -1 & -2 & -1 \\ 0 & 0 & 0 \\ 1 & 2 & 1 \end{bmatrix} * A \tag{10}$$

$$G = \sqrt{G_x^2 + G_y^2} \tag{11}$$

where G_x and G_y are the gradient images in the direction of X and Y- axis of the image A. G is the gradient image of the image A. The symbol $*$ represents the convolution operation according to the signal processing.

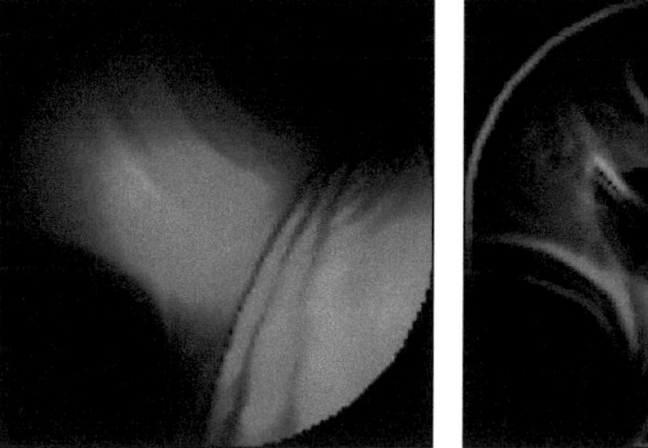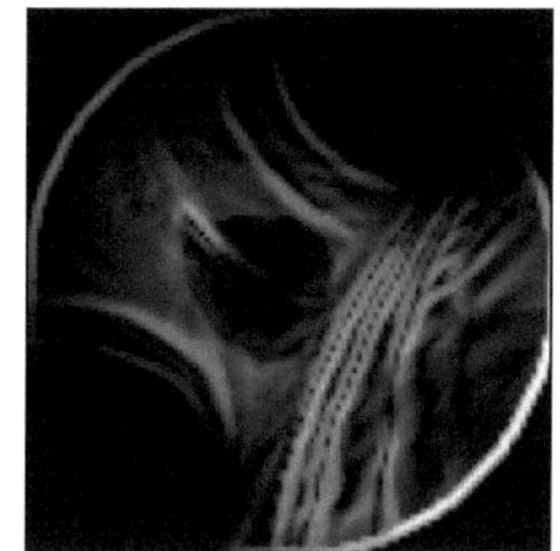

Figure 4: Image and corresponding gradient image

After obtaining the edges or gradients of the individual images (reference and predicted), the mean squared error is calculated between these two gradient images

$$MGE = \frac{1}{n}\frac{1}{m}\sum_{i=1}^{N}\sum_{j=1}^{M}\left(G(i,j) - \hat{G}(i,j)\right), \tag{12}$$

where $G(i,j)$ and $\hat{G}(i,j)$ are the corresponding pixel values of the gradient images of refernce image and predicted image. Similar to L1 loss or MSE loss, MGE loss computes the pixel wise gradients, which helps in reconstructing sharp images.

2.4.4 Mixed Loss

Mixed loss is the combination of all the above mentioned loss functions. The final loss function which is used in project is 70% of L1 Loss, 20% of SSIM loss and 10% of MGE loss.

$$Loss = (0.7 \times L1Loss) + (0.2 \times SSIMLoss)$$
$$+ (0.1 \times MGELoss) \tag{13}$$

3 Results and Discussion

3.1 Experiments and Results

In this project, the first network is trained for translating the RGB image to BRDF image and the second network is for predicting depth map from the BRDF images. Both the networks are trained for 100 epochs and are similar to

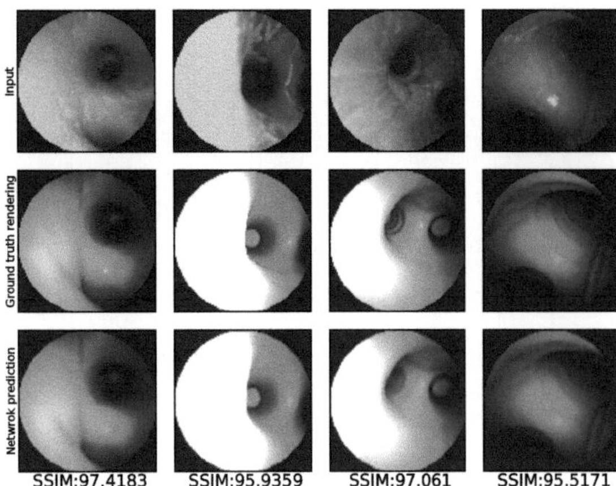

Figure 5: Sample results for the RGB to BRDF network
Top row: input frame; Middle row: ground truth rendering;
Bottom row: network prediction

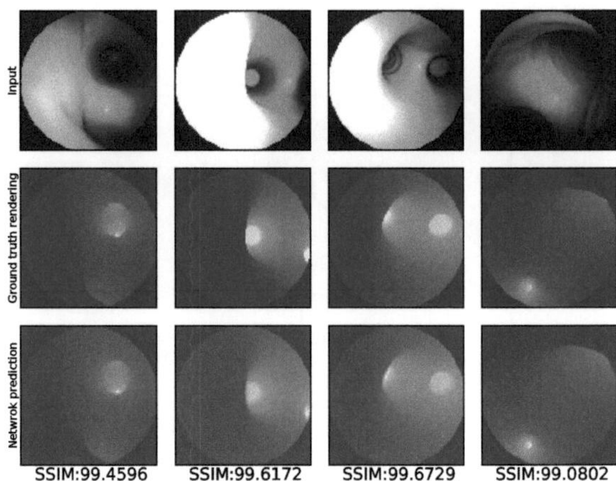

Figure 6: Sample results for depth estimator network
Top row: input rendering; Middle row: ground truth;
Bottom row: network estimation

network architecture shown in Fig. 3. All the sequences of the simple phantom dataset are considered and divided into 70% for the training, 20% for network validation and 10% for testing purposes. For the data augmentation, random rotation and Gaussian blur are applied with a probability of 50%. All the experiments are conducted on Nvidia Tesla T4 GPU.

It is important to note that the images shown in Fig. 5 and Fig. 6 are from the testing sequence. Meaning: the network has not seen these frames before. Regarding the RGB to BRDF network, the network is trained for 100 epochs and the training time is 7.02 hours. Despite the huge training time, the time taken during prediction is only 35.26 ms per frame. The network performs well on the simple phantom dataset as seen from the results in Fig. 5. It can be seen that the 10% of the MGE loss helped in edge reconstruction, but are smoothened as the images are slightly blurred.

Concerning the depth estimation network, the results can be seen from the Fig. 6. The time taken for training the network for 100 epochs is 7.24 hours, while it is only 30.68 ms per frame during prediction. Qualitatively and quantitatively (SSIM metric value) , the network yields similar results when compared to ground truth labels.

4 Conclusion

This paper investigates a U-net architecture for depth estimation from a single monocular RGB image for bronchoscopy navigation. It is also observed that the implemented SSIM loss and MGE loss helped in better reconstruction of a sharper image. The future scope of this work is to further investigate the loss function and fine tune the hyperparameters to make the network more robust.

Acknowledgement

The work has been carried out at the Institute of Medical Informatics, and supervised by Dr. Marian Himstedt, Institute of Medical Informatics, Universität zu Lübeck

Author's Statement

Conflict of interest: Authors state that they have no conflict of interest.
Informed consent: This article does not contain patient data

5 References

[1] Visentini-Scarzanella, Marco, et al. "Deep monocular 3D reconstruction for assisted navigation in bronchoscopy." International journal of computer assisted radiology and surgery 12.7 (2017): 1089-1099.

[2] Lu, Zhengyang, and Ying Chen. "Single image super resolution based on a modified u-net with mixed gradient loss." arXiv preprint arXiv:1911.09428 (2019).

[3] Wang, Zhou, et al. "Image quality assessment: from error visibility to structural similarity." IEEE transactions on image processing 13.4 (2004): 600-612.

[4] Cheng, Kai, et al. "Depth Estimation for Colonoscopy Images with Self-supervised Learning from Videos." International Conference on Medical Image Computing and Computer-Assisted Intervention. Springer, Cham, 2021.

[5] Asano, Fumihiro, Ralf Eberhardt, and Felix JF Herth. "Virtual bronchoscopic navigation for peripheral pulmonary lesions." Respiration 88.5 (2014): 430-440.

[6] Karaoglu, Mert Asim, Nikolas Brasch, Marijn Stollenga, Wolfgang Wein, Nassir Navab, Federico Tombari, and Alexander Ladikos. "Adversarial Domain Feature Adaptation for Bronchoscopic Depth Estimation." In International Conference on Medical Image Computing and Computer-Assisted Intervention, pp. 300-310. Springer, Cham, 2021.

Towards PET-CT Imaging of Zebrafish: Testing and Characterization of a X-Ray Tube and Flat Panel Detector

Leoni de Graaf [1], Steven Seeger [2], Ezzat Elmoujarkach [2], Andreas Bolke [2] and Magdalena Rafecas [2]

[1] Medical Engineering Science, Universität zu Lübeck, leoni.degraaf@student.uni-luebeck.de

[2] Institute of Medical Engineering,Universität zu Lübeck, {seeger, elmoujarkach, bolke, rafecas}@imt.uni-luebeck.de

Abstract

Within the framework of the MERMAID project, an innovative Positron Emission Tomography (PET) system for adult zebrafish has been developed. The aim of this study was to extend the PET system to include a Computed Tomography (CT) device to obtain both, metabolic and anatomical information (PET-CT). To this aim, we set up and characterized a non optimized x-ray tube and x-ray detector which were already available at the Institute of Medical Engineering (IMT), and investigated its suitability for zebrafish imaging by processing planar radiographs. We have verified that the detector behaves linearly over the emission current range. The best Contrast-to-Noise ratio (CNR) in the radiographs of two different phantoms was obtained with the highest tube voltage of 40 kV. This study lays the groundwork for the long-term goal of developing an integrated PET-CT device.

1   Introduction

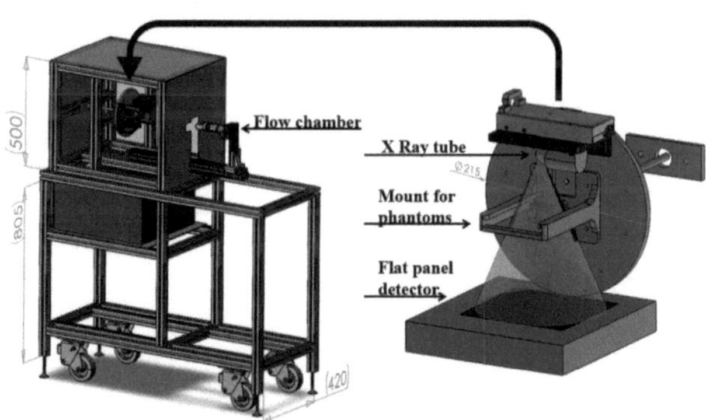

Figure 1: Left: PET prototype. Right: Setup for testing and evaluating the x-ray tube and detector. Dimensions are in mm.

Zebrafish have become a meaningful laboratory animal in biomedical research due to their small body size, their fully sequenced genome, excellent reproducibility and the transparency of their embryos [1]. In contrast to embryos adult zebrafish are not transparent. To make metabolic processes visible, the nuclear medical imaging technique Positron Emission Tomography (PET) is one of the most suitable methods. In biomedical research rodents are often the animals of choice, for which appropriate PET systems are commercially available. These scanners, however, are inappropriate for aquatic animals like zebrafish, partly because of insufficient spatial resolution and because the aquatic environment poses a major challenge. The development of a PET device dedicated to in-vivo adult zebrafish

imaging is one of the main goals of the MERMAID *Multi-Emission Radioisotopes - Marine Animal Imaging Device* project. The immobilization and water supply of the fish is ensured by a specially designed flow chamber [2]. PET is often combined with Computed Tomography (CT) in one device. This offers the advantage of applying both imaging modalities without large time delay and without changing the position of the fish. The CT information is commonly used to provide a morphological framework and for attenuation correction of the PET image. The main goal of this work was to characterize, set up an x-ray source and an independent x-ray detector, available at the IMT, and to test their functionality and suitability for imaging zebrafish.

2   Material and Methods

Specifications and functionalities mentioned in this section were extracted from the associated data sheets.

2.1   Setup

The x-ray tube and detector were placed inside the PET prototype to assess how much space is required. We mounted the x-ray tube on an aluminium disc, on which the PET detectors will be placed as well, and the detector below the tube (see Fig.1). The source to image distance is 178 mm. With this setup, the radiation protection regulations are complied with. At present, it only allows 2D imaging.

2.2   X-Ray Tube

The x-ray source is within a 40 kV tube from the company Moxtek (1001, Moxtek, Orem/USA) with a weight of approximately 450 g. The x-ray tube contains the electron source, a tungsten filament cathode, and a stationary tungsten anode within an evacuated ceramic-metal envelope and aluminum housing. The tube window is directed sideways to the anode and consists of a 0.127 mm thick beryllium layer. The cone beam angle is 46° and the focal spot is 300 μm small. The tube voltage and the emission current can be adjusted manually or from an external control computer. The voltage can be set from 0 to 40 kVp (peak voltage) and the current from 0 to 100 μA with 1/10 scaling factor. A detailed x-ray spectrum of the source is not available, but we can make the following considerations: typical filtered bremsstrahlung spectra show a distribution with no x-ray contribution below about 10 keV. The average x-ray energy is typically about one third to one half of the highest x-ray energy in the spectrum [3]. In our case, this means that the energy of the photons generated with the x-ray tube is in the range from about 10 to 20 keV. The most prevalent characteristic x-rays in the diagnostic energy range result from K-shell vacancies. For tungsten, the three prominent lines on the bremsstrahlung spectrum lie above 40 kV. Characteristic x-rays other than generated by K-shell transitions are unimportant in diagnostic imaging because they are almost entirely attenuated by the x-ray tube window [3]. Therefore, the characteristic radiation can be neglected for the x-ray tube used here.

2.3 X-Ray Detector

We use a CMOS area detector from the company Hamamatsu Photonics (C7942SK-05, Hamamatsu Photonics, Hamamtsu/Japan) where photons in the visible light spectrum generate electrons in photodiodes. To convert the incoming x-ray photons into visible photons, gadolinium oxysulfide scintillators covered with 1 mm carbon fiber are used. The photons are converted into electric charge in the junction capacitance of each photodiode. The detector operates in charge integration mode. The output signal voltage $V(t)$ can be expressed by:

$$V(t) = G \cdot Q(t) = G \cdot I(t) \cdot t_1 \qquad (1)$$

where G is the amplifier gain, $Q(t)$ the integrated charge, $I(t)$ the photodiode photocurrent and t_1 is the integration time. According to the manufacturer the number of electrons generated is proportional to the energy of the incoming photons. The control unit includes A/D converters, memory, an interface circuit and a programmable logic device that controls these components. A vertical shift register reads out the accumulated charge row by row [4]. Therefore, the displayed gray values are proportional to the sum of the energies of the incoming photons. The detector is designed for an energy bandwidth of 20 to 150 keV. The pixel size is 50 x 50 μm^2 and the digital video output is 12-bit in size, for a total of 4096 shades of grey. In total there are 2316 x 2316 active pixels, which equates 5.3 MP. The detector comes with a digital frame grabber card from National Instruments (IMAQ-PCI-1424, National Instruments, Austin/USA). To test the detector, we used the demonstration software provided by the manufacturer. It consists of

a simple viewer to acquire and save an image. The images are saved as TIFF files. Note that a typical exposing time for a CT slice is between 0.5 to 2 seconds to reduce motion artifacts [5]. We used MATLAB to view, postprocess and evaluate the images. The radiation exposure time affects the lifetime of the detector and the radiation dose absorbed by the zebrafish; therefore, the exposure time should be kept as short as possible. For example, the lifetime of C7942CA-22, a comparable Flat panel detector from Hamamatsu, is 152 days when it is irradiated with a tube voltage of 80 kV over 4 hours per day [4]. To spare the electronics in the housing of the detector, we collimated the beam with lead plates to allow for a field of view (FOV) of 29 mm in diameter.

2.4 Phantoms

We 3D printed two different phantoms mimicking the step wedge phantom used in radiology, thus allowing us to investigate two different materials and thicknesses. The first phantom (Phantom P) was printed with a FDM (Fused Deposition Modeling) printer (S3, Ultimaker, Utrecht/Netherlands) (see Fig. 2). It has six steps consisting of solid polyactide (PLA) and a seventh step filled with air. PLA has a density of 1.24 g/cm^3 [6], which is comparable to the density of human bones [7] and is part of the fish chamber as well. Air has a density of 1.29 kg/m^3 [8]. The phantom has a total height of 30 mm, and each step is 29 mm wide with a height difference of 5 mm. The second phantom (Phantom W) was printed with a formlabs SLA (stereolithography) printer (Form 3+, formlabs, Somerville/USA). It has six hollow steps made of clear resin with 1 mm wall thickness and a seventh step filled with air. It has also a width of 29 mm, a total height of 34 mm and 5 mm height difference. This phantom can be filled with water, which density is comparable to soft tissue. To place the phantoms between tube and detector, we printed a mount as well (see Fig.1). The phantom height was chosen to roughly match the current diameter of the fish chamber. The distance of the tube to the highest step is 42 mm.

Figure 2: 3D printed phantoms made of solid PLA (left) and waterfillable made of resin (right).

2.5 Image Acquisition and Postprocessing

First we implemented the control of the x-ray tube using MATLAB®, so that we can start the exposure and adjust the tube voltage, emission current and exposure time via an external computer. To have an impression of the attenuation of the phantoms, we have calculated the negative logarithm of the ratio between the image values with the phantom ($I_{P/W}$) and without the phantom (I_0). This is given by lambert-beers law [5]:

$$\mu \cdot x = -ln\left(\frac{I_{P/W}(x)}{I_0}\right) \qquad (2)$$

where μ is the linear attenuation coefficient and x the thickness of the material. This is an approximation for monochromatic rays and uniform material.

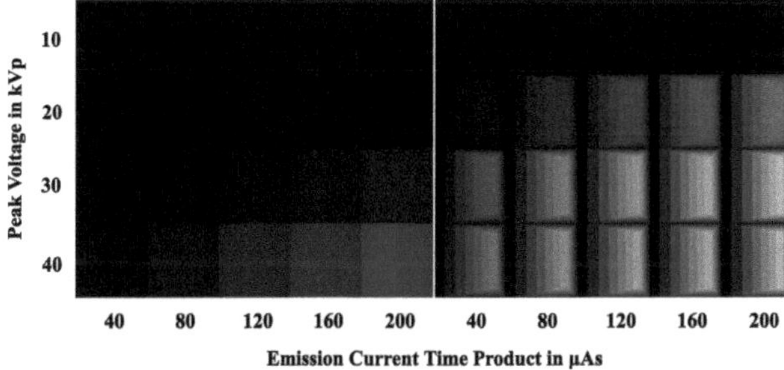

Figure 3: Image Montage. Left: Raw images without phantom. Right: Postprocessed Images of Phantom P for all considered tube settings.

2.6 Determination of the Contrast

To distinguish different anatomical structures, a good contrast is essential. It is also useful to evaluate the suitability of detector and tube. For the contrast both, the subjective contrast of the object itself and the detector contrast play a role. The latter is determined by how the detector maps the detected energy into an output signal. Both contrasts depend on the tube energy spectrum. Once a digital image is acquired, a series of processing steps are performed as part of the acquisition software or post processing. Therefore, a more meaningful and frequently used measure in assessing digital images related to contrast is the Contrast-to-Noise ratio (CNR) [3]:

$$CNR = \left(\frac{A - B}{\sigma}\right) \qquad (3)$$

where A is the mean pixel value in the Region of Interest (ROI_1), B is the mean pixel value in the background ROI (ROI_2) and σ is the standard deviation of the pixel values in ROI_2. To distinguish two different object patterns from two different ROIs, the CNR must exceed 3 to 5. This factor is known as the Rose Criterion, which takes into account the perception capabilities of the human eye. [5]. For further investigations we have to consider that the product of tube current and time (short mAs or μAs) directly influences the number of detected x-ray photons, thereby affecting the Signal-to-Noise Ratio (SNR) and the CNR. Doubling of the mAs increases the SNR by $\sqrt{2}$, and the contrast consequently improves [3]. We increased the exposure time

product (μAs) linearly from 0 to 200 μAs and the tube voltage from 10 to 40 kV.

3 Results and Discussion

3.1 Characteristic Detector Curve

Fig. 4 shows the mean values of the 12-bit gray scale image versus 40 to 200 μAs for four different tube energies. No object was placed between the x-ray tube and the detector. The first thing to observe is the offset caused by dark noise. The incline for the 10 kV line is under 1 %. This lack of a well-defined response is consistent with the information about the detector, that it is not sensitive for energies below 20 kV. For this reason we will only use tube voltages higher than 20 kV in the following steps. In summary, it can be seen that the detector behaves linearly over the adjustable range and no saturation occurs. With a suitable software we could increase the frame time in the future and thus the time over which the charge is integrated. Since the change in the slope of each line is not linear there seems to be a non-linearity of the detector with respect to the energies. To be able to make an exact statement about the reasons for this, we need further investigations in the future.

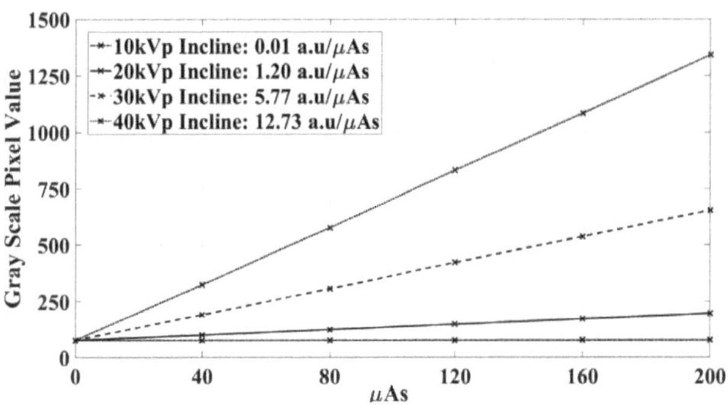

Figure 4: Characteristic detector curve. Pixel value versus µAs for four different tube voltages.

3.2 CNR

Fig. 3 shows the raw image intensities I_0 and the postprocessed images I_P exemplary for all used tube settings. Fig. 5 shows the CNR for both phantoms between two successive stair steps and the ROIs used. Each ROI includes 40000 pixels. For Phantom P only the last three settings fulfill the Rose Criterion; for the Phantom W, the last two. The CNR with the first six settings (up to 30 kV and 40 μAs) are not sufficient for distinguish the steps according to the Rose Criterion. Only the CNR to air meet the Rose Criterion. The final settings should depend on the dimensions and composition of the final fish chamber. In general, the decision should be made according to the ALARA ("as low as reasonably achievable" exposure to x-ray) principle [5]. The exposure is approximately proportional to the square of the kVp value in the diagnostic energy range [3]. Therefore the relative radiation exposure of a beam generated with a tube voltage of 30 kVp compared to a 40 kVp with the same tube current and exposure time is $(40/30)^2 = 1.78$. The output radiation exposure therefore increases by about 78%, which is why the lower voltage should be chosen, if possible. To further increase the contrast between different materials, filters can be used in the future to remove low energy photons, which scatter more. For example, molybdenum could be a possible filter because it effectively attenuates photons with energies under 10 keV [3].

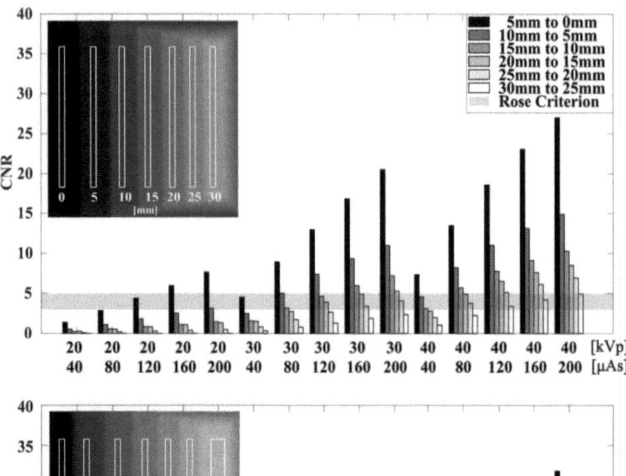

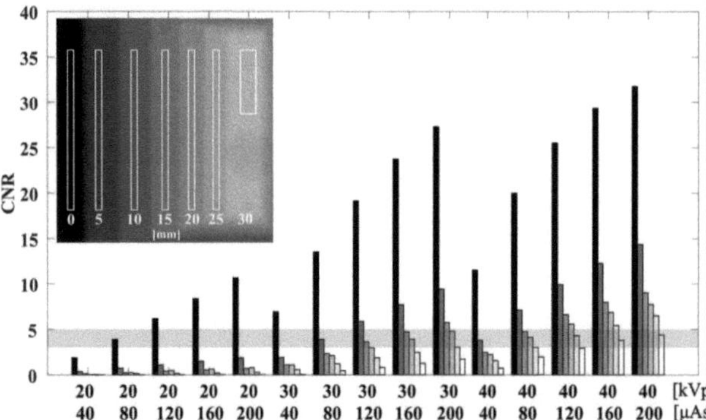

Figure 5: CNR and Rose Criterion. Top: PLA phantom. Bottom: waterfilled phantom. In the left corner: Images with 40 kVp and 200 μAs and the ROIs. To avoid a trapped air bubble the 7th ROI differs from the others. Rose Criterion represented by the grey horizontal band.

4 Conclusion

In this work we characterized the x-ray tube and detector and has shown that they are capable of producing good and uniform planar x-ray images. In future work, we will assess which settings are best appropriate with respect to CNR and which to avoid. Nevertheless we have to see how the results translate from the planar radiographs into CT images. In particular, the tube performance needs to be further investigated. We need to evaluate if the limited emission current is sufficient to achieve good contrast on CT images because therefore the mAs should be as high as possible, taking the radiation dose into consideration. But the exposure time should be kept as well as short as possible to avoid motion artifacts and heat damage of the tube. To determine the appropriate dosage for a zebrafish, fundamental research stud-

ies are necessary in the future. To improve the contrast, additional filters will be considered. Furthermore, we still have to implement the joint rotation of the tube and the detector to make CT possible. Additionally we plan to develop phantoms that mimic the zebrafish inside the fish chamber to better evaluate the suitability of the setup for imaging zebrafish. In summary, this work provides the fundamental experimental results of the x-ray image needed to integrate the CT-component into the PET for use with marine life.

Acknowledgement

This work was carried out in cooperation with the Institute of Medical Engineering, Universität zu Lübeck. Special thanks to Dirk Steinhagen for technical support, as well as to Maja Frerkes and Nadine Möller, who helped with the setup and the 3D printing procedures.

Author's Statement

Authors state no conflict of interest.

5 References

[1] T. Y. and T.I. Choi, Y. R. Lee, S. K. Choe, C. H. Kim, *Zebrafish as an animal model for biomedical research.* Experimental and Molecular Medicine, vol. 53, no. 3, pp. 310-317, 2021

[2] M. Zvolsky, S. S. Seeger, M. Schaar, C. Schmidt, M. Rafecas, *MERMAID - A PET Prototype for Small Aquatic Animal Imaging.* IEEE Nuclear Science Symposium and Medical Imaging Conference. IEEE, pp. 1-2, 2019.

[3] J. T. Bushberg, J.M. Boone, *The essential physics of medical imaging.* Lippincott Williams & Wilkins, vol. 2, pp. 98-102, pp. 261-263, p. 369, p. 135, p. 114, 2002.

[4] *Technical Note: x-ray detectors.* HAMAMATSU PHOTONICS K.K., Solid State Division, 2021, Available: https://www.hamamatsu.com/jp/en/support/resources/ technical-notes.html, [last accessed on 2022-02-04]

[5] S. R. Cherry, J. A. Sorenson, M. E. Phelps, *Physics in nuclear medicine.* Elsevier Health Sciences, vol. 4, p. 347, p. 81, pp. 244-246, pp. 433-434, 2012.

[6] Simplify3D, 2022, *Filament Properties Table.* Available: https://www.simplify3d.com/support/materials-guide/properties-table/, [last accessed on 2022-02-04]

[7] H. E. Meema, S. Meema, *Compact Bone Mineral Density of the Normal Human Radius.* Taylor and Francis, vol. 17, no. 4, pp. 342-352, 1978.

[8] *NIST: X-Ray Mass Attenuation Coefficients.* Available: https://physics.nist.gov/PhysRefData/XrayMassCoef/ tab2.html, [last accessed on 2022-02-04]

Analysis and implementation of semi-supervised segmentation methods using autoencoders

Friederike Schneider[1], and Jan Ehrhardt [2]

[1] Medical Informatics, Universität zu Lübeck, friederike.schneider@student.uni-luebeck.de

[2] Institute for Medical Informatics, Universität zu Lübeck, jan.ehrhardt@uni-luebeck.de

Abstract

The lack of annotated image data is an important issue in medical segmentation. The idea of semi-supervised learning is to use both annotated and unannotated images for the training of a segmentation network to improve the outcome even if the number of annotated data is small. This paper analyses the effect of semi-supervised approaches for brain lesion segmentation in comparison to a classical supervised training of a U-Net. Therefore, two different semi-supervised approaches, both based on an autoencoder architecture are presented. The results show that there are slight improvements of the segmentation results for datasets with a small amount of labeled data. For more than 150 labeled data there is no difference between the semi-supervised approaches and the comparative approach of a simple U-Net.

1 Introduction

The training of networks for lesion segmentation requires pixel-level annotations based on domain experts. To produce manual annotations of medical data is a time consuming and tedious task. Thus annotated data are often only available for a small subset of the data. A common way to generate additional annotated data and prevent over fitting is data augmentation. The high correlation between the augmented data and the original data causes that this method addresses the problem of a limited amount of annotated data only partly [1]. A further common technique is transfer learning. The knowledge gained from an already trained network is stored and applied to a different but related problem. Due to the diversity between the different medical imaging modalities and the different body tissue structures this technique also solves the problem of limited annotated data only partly [2]. The idea of semi-supervised learning (SSL) is to use weak labeled or unlabeled data additionally to the few strong labeled ones to improve the performance of a model. In this paper two different approaches for semi-supervised learning are presented. The first approach is a multi-class attention-based framework which uses generated pseudo annotations for unlabeled data. The second approach is a shared encoder double task network which uses reconstruction knowledge for the segmentation.

2 Material and Methods

In this paper two semi-supervised learning (SSL) approaches are implemented and compared regarding their impact on the segmentation results for datasets with a low number of labeled data.

2.1 Dataset

The BRATS2020 challenge data [3] covering 369 multi-modal MR image sequences of brain tumor patients were used. Each sequence consists of one Flair image, an T1 image, an T1ce image, a T2 image and a segmentation mask per patient. For a first experiment, 2D slices through the tumor center were extracted from the 3D volumes. In the second experiment 3D volumes were used. The segmentation mask is a manual annotation of the gadolinium (GD) enhancing tumor, the peritumoral edema (ED), and the necrotic and non-enhancing (NCR/NET) tumor core. For this approaches ED and NCR/NET are summerized into "tumor". The segmentation mask consists of a background class (label 0), a tumor class (label 1) and a edema class (label 2).

2.2 Architecture

The encoder of both approaches as well as all decoders contain of 4 resolution levels. Starting with an image input size of 4x64x96x96 (CxDxHxW) for 3D data and 4x96x96 (CxHxW) for 2D slices, respectively, down to 128x4x6x6 (CxDxHxW) and 128x6x6 (CxHxW)for the latent space. The number of feature maps is doubled with each time of downsampling and halved with each upsampling step.

Multi-Class Attention-based Framework
The architecture of the model is derived from the idea of Chen et al. [4]. They introduce an attention-based semi-supervised approach. But in difference to our multi-modal and multi-class frameworks they only use the flair data and sum edema and tumor to one class to have a binary foreground background classification problem.

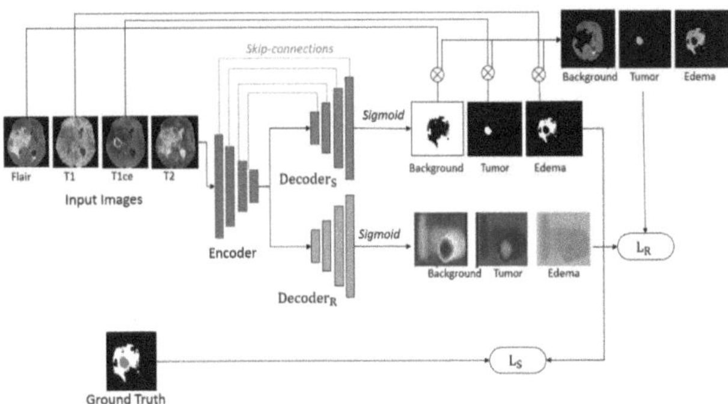

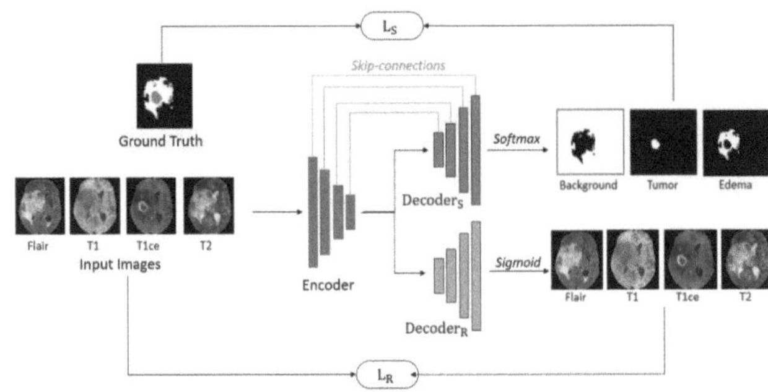

Figure 1: **Architecture of Multi-Class Attention-based Framework.** Segmentation loss L_S is computed between the segmentation predictions for background, tumor and edema and the ground truth. Reconstruction loss L_R is computed between the output of the reconstruction decoder and the class separated images calculated by the attention mechanism.

Figure 2: **Architecture of Shared Encoder Double Task Framework.** Segmentation loss L_S is computed between the segmentation predictions for background, tumor and edema and the ground truth. Reconstruction loss L_R is computed between the reconstructed images and the input images.

The architecture of the segmentation path is similar to the architecture of a U-Net [5] and is summarized in Figure 1. It consists of one encoder and two decoders for image segmentation ($Decoder_S$, Figure 1) and image reconstruction ($Decoder_R$, Figure 1). The idea of sharing one encoder is that the reconstruction and segmentation mechanisms both rely on the same latent space and therefore the calculation for the segmentation is based on the features learned through the reconstruction. It has skip connections between the decoder and segmentation encoder to allow the transfer of detail pixel location information. Additionally short skip connections are used to improve the network accuracy even further [6]. The sequence of each resolution level follows the construction of the following double convolution block.

$$ConvolutionalLayer \rightarrow InstanceNormalization$$
$$\rightarrow ReLUactivation \rightarrow ConvolutionalLayer$$
$$\rightarrow InstanceNormalization \rightarrow ReLUactivation$$

For the output activation of the segmentation decoder a sigmoid activation is chosen. One hot encoding is applied for the pixel classification. The output of the segmentation decoder is one segmentation mask per class.

The reconstruction path is similar to the segmentation path, except that it has no skip connections. The aim of the reconstruction decoder is to reproduce the intensities of the input image for the background, tumor and edema regions. An attention mechanism performs a pixelwise multiplication between the predicted segmentations of $Decoder_S$ (Figure 1) and the input image, which is compared to the reconstructed regions ($Decoder_R$ output, Figure 1).

Shared Encoder Double Task Framework

The architecture of the shared encoder double task framework is based on an autoencoder path for reconstruction and an autoencoder like path for segmentation. Both paths are sharing the same encoder (see Figure 2). The idea of reusing reconstruction knowledge is also used for that approach but instead of generating pseudo labels which

are used for the segmentation, the network learns to fulfill two tasks, reconstruction of the image and segmentation. The encoder and decoder of the segmentation path are connected with skip connections. For the reconstruction path there are no skip connections.

The architecture of the segmentation path of the shared encoder double task framework and the multi-class attention-based framework only differ in their output activation. While for the multi-class attention-based framework the sigmoid activation is used, the softmax function is chosen for the shared encoder double task framework. The structure of the reconstruction decoder is also similar to the one of the multi-class attention-based framework, but it lacks the attention mechanism. Instead the reconstructed images are compared directly to the input images. It has a four channel output. One for each image type ($Decoder_R$ output, Figure 2).

2.3 Preprocessing

In a preprocessing step 3D data are cropped to a size of 128x192x192 (DxHxW) and 2D slices to a size of 192x192 (HxW) to reduce the amount of background pixels. Reasoned by limited storage capacity the 3D data need further size reduction. They are interpolated by a trilinear interpolation to a size of 64x96x96 (DxHxW). A flip, an intensity and an affine transformation are applied for the data augmentation.

The whole data are separated into training data (70%) validation data (15%) and test data (15%). The data affiliation is set in the beginning and remains the same for each run.

2.4 Training

For both architectures, the reconstruction path is carried out for unlabeled data and the segmentation for labeled data. The amount of labeled data starts with 50 samples and is increased each run by 50 up to 250 samples. The labeled samples are randomly selected from the training data. The remaining data are treated as unlabeled data. In each of the

100 epochs of the training the reconstruction path is trained before the segmentation path.

For the shared encoder double task architecture the network is pretrained in 100 epochs only to reconstruct the patient data. During the training of the segmentation path the encoder stays fixed. The encoder gets optimized only by the reconstruction to ensure that the segmentation results rely on the features learned during the training of the reconstruction path.

Two Adam optimizer, one for segmentation and one for reconstruction, are used for the training. The learning rate is set to 0.001. Additionally, a learning rate scheduler is used (StepLR with step size = 90 and gamma = 0.5).

2.4.1 Loss Functions

Multi-Class Attention-based Framework

For the segmentation path, the ground truth mask is compared with the model output of the segmentation decoder using a generalized dice loss [7] (1).

$$L_S = 1 - 2\frac{\sum_{l=1}^{3}\omega_l \sum_n r_{ln} * p_{ln}}{\sum_{l=1}^{3}\omega_l \sum_n r_{ln} + p_{ln}} \quad (1)$$

For the attention mechanism, the reconstruction decoder is trained to reconstruct the masks for the different classes multiplied with the original data. For the ground truth tumor image, the predicted tumor mask is multiplied with the T1ce image input data. For the ground truth edema image, the predicted edema mask is multiplied with the T1 image input data. For the ground truth background image the predicted background mask is multiplied with the Flair image. The L_R loss is calculated by summing up the MSE-Loss between new created ground truth images and the output of the reconstruction decoder multiplied by the weight how often the class occurs in the ground truth data (2).

$$L_R = \frac{\sum_i \tilde{y}_b^{(i)}}{n} MSE[\hat{y}_b, x_{Flair} \odot \tilde{y}_b]$$
$$+ \frac{\sum_i \tilde{y}_t^{(i)}}{n} MSE[\hat{y}_t, x_{T1ce} \odot \tilde{y}_t] \quad (2)$$
$$+ \frac{\sum_i \tilde{y}_e^{(i)}}{n} MSE[\hat{y}_e, x_{T1} \odot \tilde{y}_e]$$

For the visualization of the loss curves the L_R loss is not summed up but calculated separately for each class.

Shared Encoder Double Task Framework

Same as for the segmentation path of the multi-class attention-based framework (MCAbF) the generalized dice loss [7] (1) is used for the training and evaluation of the output of the segmentation decoder.

For the reconstruction path a simple L_1 Loss is calculated between the output of the reconstruction decoder and the input patient data.

3 Experiments and Results

The results of all experiments are listed in the following tables. To ensure the validity of the results a 5-fold cross validation has been carried out.

Architecture	50	100	150	200	250
MCAbF					
Background	**0.994**	**0.994**	**0.994**	0.994	0.994
Tumor	0.447	0.468	0.509	0.497	0.513
Edema	0.364	0.379	0.412	0.425	0.440
SEDTF					
Background	0.993	**0.994**	**0.994**	**0.995**	0.995
Tumor	**0.493**	**0.558**	**0.588**	0.619	0.632
Edema	**0.430**	0.449	**0.518**	0.503	0.521
U-Net					
Background	0.990	0.992	**0.994**	**0.995**	**0.996**
Tumor	0.443	0.492	0.570	**0.633**	**0.696**
Edema	0.392	**0.455**	0.473	**0.518**	**0.569**

Table 1: 2D slices. Dice score per run (number of labeled samples: 50, 100, 150, 200, 250) for the three different architectures (Multi-Class Attention-based Framework: MCAbF; Shared Encoder Double Task Framework: SEDTF; Simple segmentation network: U-Net).

Architecture	50	100	150	200	250
MCAbF					
Background	0.989	0.989	0.997	**0.998**	0.997
Tumor	0.537	0.697	0.742	**0.845**	0.707
Edema	0.564	0.630	0.0639	0.757	0.649
SEDTF					
Background	**0.997**	**0.998**	**0.998**	**0.998**	**0.998**
Tumor	**0.659**	**0.738**	**0.787**	0.842	0.891
Edema	**0.660**	**0.703**	**0.758**	0.779	0.817
U-Net					
Background	**0.997**	0.997	**0.998**	**0.998**	**0.998**
Tumor	0.633	0.693	0.774	0.841	**0.898**
Edema	0.636	0.675	0.744	**0.787**	**0.831**

Table 2: 3D Data. Dice score per run (number of labeled samples: 50, 100, 150, 200, 250) for the three different architectures (Multi-Class Attention-based Framework: MCAbF; Shared Encoder Double Task Framework: SEDTF; Simple segmentation network: U-Net).

The architectures and training setups explained in the previous chapter have been used for the experiments of this paper. In first experiments the multi-class attention-based framework has been carried out for the 2D slices. Its outcome is worse than the outcome of the U-Net. The approach from Chen et al. [4], on which the MCAbF is based on, deals with a binary classification. The MCAbF faces a three class segmentation problem. In a further experiment the data have been adjusted to turn it into a binary classification problem. Therefore edema and tumor lesions were summarized into one class (Table 3). This leads to an improvement of the results.

Due to the finding that the MCAbF does not work properly for multi-class segmentation, the architecture of the shared encoder double task framework was developed. In comparison to a simple U-Net segmentation architecture, this architecture shows an improvement of the segmentation results (Table 1).

Architecture	50	100	150	200	250
MCAbF					
Background	0.976	0.984	0.981	0.978	0.980
Foreground	0.599	0.610	0.621	0.618	0.623

Table 3: 2D Slices. Dice score per run (number of labeled samples: 50, 100, 150, 200, 250) for the adjusted (tumor and edema summarized) data using multi-class attention-based framework (MCAbF).

In the last experiments the network architectures have been adjusted for 3D data. For all architectures the results for the 3D data (Table 2) are better than the results of the 2D slices (Table 1). Same as for the 2D slices, the shared double task architecture achieves an improvement of the outcomes when the amount of labeled data is small. If the amount of the labeled data is 150 or more, there is nearly no difference between the shared encoder double task framework and the U-Net.

4   Discussion and Conclusion

The results of the experiments show that there is an improvement of the segmentation results using the shared encoder double task architecture compared to a simple U-Net if the number of annotated data is small. Especially for the runs with 150 and less labeled data, the dice score for edema and tumor lesions of the shared encoder double task architecture is higher than the dice score of the U-Net for those runs.

The multi-class attention-based framework extends the binary segmentation framework presented by Chen et al. [4] to a multi-class segmentation problem. The results of this paper show that this kind of attention mechanism seems to work for binary classification, but is less successful for multi-class segmentation. Comparing the MCAbF results listed in Table 1 and the results for the 2D slices, two class variation in Table 3, the same effects show up for the 2D slices. A possible reason could be the basic idea of this approach is to learn the structure of the classes. If the foreground subject is split into edema and tumor, the class imbalance between the different classes is bigger. Additionally there is more variance of the appearance of the classes between the data of the patients.

All presented architectures, semi-supervised as well as the U-Net, perform better for the 3D data than for the 2D slices. This effect can be explained by the variance of the data in the 2D slices. In contrast to the 3D data the 2D slices only cover one layer of the transverse plane. Depending on the location of the tumor and edema lesion in the brain, the surrounding brain structure is different. Therefore, there is considerably more variance in the data. The network has to deal with different shapes of the tumor and additionally distinguish between the different brain regions. For the 3D data the variance of the brain regions is lower since they capture the full brain.

Acknowledgement

The work has been carried out and supervised by the Institute for Medical Informatics, Universität zu Lübeck.

Author's Statement

Conflict of interest: Authors state no conflict of interest. Informed consent: Only publicly available data were used for this study; no patient data were acquired.

5   References

[1] Tajbakhsh, N., Jeyaseelan, L., Li, Q., Chiang, J. N., Wu, Z., & Ding, X., *Embracing imperfect datasets: A review of deep learning solutions for medical image segmentation*, Medical Image Analysis(2020), 63, 101693.

[2] Nikolas Adaloglouon 2020-11-26 *Transfer learning in medical imaging: classification and segmentation*, Available: https://theaisummer.com/medical-imaging-transfer-learning/, AI Summer [last accessed on 2021-12-30]

[3] Center for Biomedical Image Computing & Analytics, *Brain Tumor Segmentation (BraTS) Challenge 2020: Scope*, Available: https://www.med.upenn.edu/cbica/brats2020/ [last accessed on 2021-11-22]

[4] Chen, Shuai and Bortsova, Gerda and Juárez, Antonio García-Uceda and van Tulder, Gijs and de Bruijne, Marleen, *Multi-task attention-based semi-supervised learning for medical image segmentation*. In: International Conference on Medical Image Computing and Computer-Assisted Intervention, Springer, Cham, pp.457–465, 2019.

[5] Ronneberger, O., Fischer, P., Brox, T., *U-Net: Convolutional networks for biomedical image segmentation*. In: MICCAI, Springer, Cham, pp. 234–241, 2015

[6] Bittner, K., Liebel, L., Körner, M., & Reinartz, P., *Long-Short Skip Connections in Deep Neural Networks for DSM Refinement*. The International Archives of the Photogrammetry, Remote Sensing and Spatial Information Sciences, 43(B2), 383-390, 2020.

[7] Sudre, C. H., Li, W., Vercauteren, T., Ourselin, S., & Cardoso, M. J. , *Generalised dice overlap as a deep learning loss function for highly unbalanced segmentations*. In: Deep learning in medical image analysis and multimodal learning for clinical decision support, Springer, Cham, pp. 240–248, 2017.

Evaluation on the allocation of planar detector blocks for a rotational partial-ring positron emission tomography scanner

Hong Phuc Vo [1], Ezzat Elmoujarkach [2], Andreas Bolke [2], Steven Seeger [2], and Magdalena Rafecas [2]

[1] Biomedical Engineering, Lübeck University of Applied Sciences, hong.phuc.vo@stud.th-luebeck.de

[2] Institute of Medical Engineering, Universität zu Lübeck, {elmoujarkach, bolke, seeger, rafecas}@imt.uni-luebeck.de

Abstract

To construct a prototype for small-fish positron emission tomography based on a partial ring of four rotating detector blocks, we have studied in silico three possible configurations covering different fields-of-view: a) Symmetric, b) Asymmetric-1, and c) Asymmetric-2. Their imaging performance was evaluated using Monte-Carlo simulations and image-quality phantoms; the simulated data were reconstructed using the *Maximum Likelihood Expectation Maximization* algorithm. The results show that the performance of Symmetric and Asymmetric-1 is similar, while the spatial resolution of Asymmetric-2 is severely affected by parallax errors. Therefore, detector blocks should be placed in the symmetric configuration for better image quality.

1   Introduction

Positron Emission Tomography (PET) is a nuclear imaging method widely used in the clinics and biomedical research. Whereas PET scanners dedicated to rodents exist, there is no available device for fish, although zebrafish and similar species are also relevant animal models in medicine. The first system for PET imaging of small aquatic animals, MERMAID (Multi-Emission Radioisotopes - Marine Animal Imaging Device), is currently under development at the Institute of Medical Engineering, University of Lübeck [1]. Its goal is to support biomedical research in small fish to study fish anatomy and physiology, as well as to trace the path of nutrients in sustainable aquaculture environments.

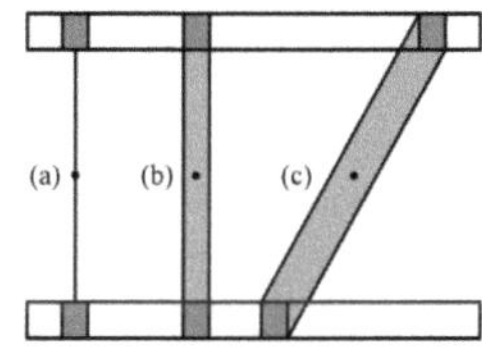

Figure 1: (a) LOR; (b) direct VOR, and (c) oblique VOR with parallax error.

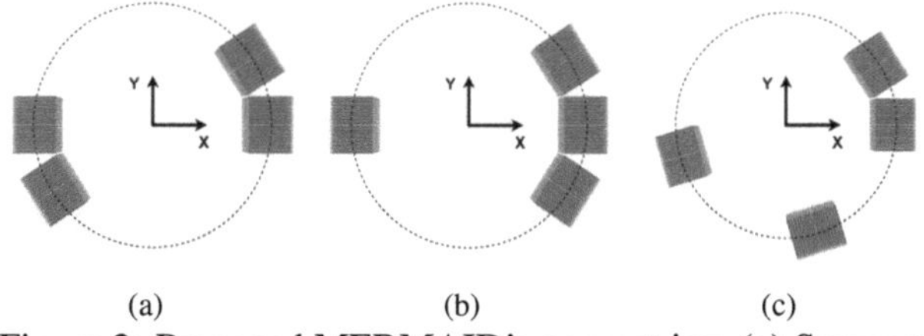

Figure 2: Proposed MERMAID's geometries: (a) Symmetric; (b) Asymmetric-1; and (c) Asymmetric-2. (The dashed circle indicates the rotation path.)

PET is based on β^+-decay of radionuclides. Each positron annihilation results in two collinear 511 keV photons. The goal of PET scanners is thus to detect as many annihilation photon pairs as possible. Therefore, modern PET scanners have a full-ring geometry. When two PET detectors units register quasi-simultaneously two 511 keV photons, this pair is considered a coincidence and assigned a *line-of-response* (LOR), a connection between the two involved PET detector units (Fig. 1a). The origin of the annihilation photons, on the other hand, is assumed to be within a volume-of-response (VOR) specified by the volume of the two detector units (Fig. 1b). Compared to detector units directly facing each other, when the units are oblique, the VOR is enlarged as a result of crystal penetration (Fig. 1c). Consequently, the unknown location of the radiation origin is affected by a larger uncertainty (*parallax error*) [2].

Currently, only four detector blocks are available, which are insufficient to build a static full-ring scanner with a diameter suitable for small aquatic animals, but might be enough if detector rotation is considered. This study thus aimed to determine the best allocation of these four detector blocks to construct a partial-ring scanner, including rotation to compensate for the missing detectors. Three different detector allocations were considered (see Fig. 2): a) Symmetric, b) Asymmetric-1, and c) Asymmetric-2.

2   Material and Methods

2.1   Simulations and Phantoms

Independent Monte-Carlo simulations for all geometries were executed with radioactive sources to generate data for later reconstruction and performance evaluation. To this aim, we have used the simulation software *Geant4 Application for Tomographic Emission* (GATE) version 9.1 [3]. We have simplified positron emission and subsequent annihilation by the emission of two collinear 511 keV photons. According to the National Electrical Manufacturers Association (NEMA) protocol for performance measurements of small animal PET [4], point sources and a NEMA phantom are required to characterize the performance of a scanner.

Point sources are adopted to characterize the spatial resolution. In the transversal plane, two lines of point sources, placed along the X-axis and at an angle of $110°$ to the X-axis, respectively, are used to see the effects of the geometries inside and outside of the symmetry axes. An additional line of points sources was simulated in the axial direction. Each point source is a sphere with a radius of $0.3\,mm$ and activity of $1\,MBq$. The total acquisition time, $180\,s$, is kept constant for all configurations to ensure the same number of expected emissions. To guarantee uniform coverage of the field-of-view (FOV), the symmetric and asymmetric scanner need to be rotated $180°$ and $360°$, respectively.

The NEMA phantom's aim is to determine the image quality; its detailed construction is described in the NEMA NU 4 - 2008 protocol [4]. In this study, we have down-scaled the phantom by a factor of 0.6 to better match the size of small aquatic animals. In addition, we have modified the phantom to resemble a situation with an active lesion ("hot") inside a less radioactive background ("warm"). The phantom consists of three parts: I) rods region: 5 rods of different diameters filled with radioactivity inside a "cold" background (without activity); II) uniform region: a cylindrical chamber uniformly filled with radioactivity, and III) cylinders region: chamber like II) but including two cylinder inserts, one filled with nonradioactive air ("cold cylinder") and a smaller one ("hot cylinder") with a total radioactivity five times higher than the background. The activity concentration in the rods and uniform regions is $18\,kBq\,mm^{-3}$, the acquisition time is $300\,s$, and each system performs one full acquisition as in the point sources case.

Each planar detector block consists of two 8x8 LYSO crystal modules. Each crystal has a size of $1.12\,mm$ x $1.12\,mm$ and a length of $15\,mm$. As in the real hardware, there is a small gap of $0.28\,mm$ between the detector modules. These detector blocks can rotate around the scanner center with a radius of $33\,mm$; the angle between two adjacent blocks is $33°$. During this study, scanners with a radius of $24\,mm$ and an angle between two adjacent blocks of $45°$ were also examined but are not shown in this work. Systems with a radius of $33\,mm$ returned a slightly larger FOV and would be easier to construct in reality. The point sources and each part of the NEMA phantom were simulated once; for the cylinders region, five independent simulations were run to better take into account the effects of statistical fluctuations.

2.2 Image Reconstruction

We have implemented the *Maximum Likelihood Expectation Maximization* (MLEM) algorithm [5], which is the gold standard for image reconstruction in nuclear imaging. To calculate the system matrix required by MLEM, the multi-ray method described in [6] was implemented: For each pair of detectors, several lines are traced; their end-points lie within the crystals, and are obtained from random sampling of a 2D uniform distribution for the cross section and an exponential distribution for the depth, in accordance with Lambert-Beer law. As a result, the depth of interaction (DOI) is better modeled as when using the conventional Jacobs' algorithm. Finally, the intersection lengths between lines and image voxels are computed. Despite the simplifying assumptions, our system model is sufficient for the purpose of comparison. Each simulated data set is reconstructed until convergence. The stopping criterion is based on the similarity of reconstructed images, such that the mean squared error between two subsequent images is less than 10^{-4} for three continuous iterations. The reconstructed images have a voxel size of $0.25\,mm$ in each dimension.

2.3 Metrics

The metrics were obtained as described in the NEMA protocol [4].

Full width at half maximum (FWHM): The spatial resolution is calculated as the FWHM of the reconstructed point spread functions (PSF) along the radial, tangential, and axial directions of each point source. The FWHM is computed by linear interpolation between neighboring pixels at half the maximum value of the PSF.

Recovery Coefficient (RC): The purpose of RC is to estimate the system's ability to reconstruct small lesions and is thus a measure of the spatial resolution. The RC of each rod is determined using the rods region of the NEMA phantom. The RC (ideally 1) corresponds to the mean value of each rod profile divided by the mean value of the uniform region.

Uniformity: This measurement aims to quantify possible truncation or sensitivity issues. The central part of the uniform region is assigned a volume-of-interest with a diameter of 75% of the physical diameter. Its mean, maximum, minimum, and percentage standard deviation are reported.

Spill-Over Ratio (SOR): This metric is the relative activity in the hot and cold cylinders to the mean activity in the background (warm). The SOR quantifies possible truncation and parallax effects, and their compensation by the reconstruction model. Ideally, the SOR value is 0 for the cold cylinder as it does not contain any activity; similarly, the SOR of the hot cylinder should be 5 since the hot cylinder has five times more activity than the background.

3 Results and Discussion

The motivation behind each configuration are: a) *Symmetric*: most of the direct LORs will be detected with the cost of a smaller FOV (radius: $9\,mm$). b) *Asymmetric-1*: it can detect more oblique LORs and cover a larger FOV (radius: $9.5\,mm$); however, the number of detected events at direct LORs will decrease. c) *Asymmetric-2*: the idea is to enlarge the FOV by capturing more oblique LORs. Therefore, a block is placed at 90° to the stand-alone block. However, the 90-degree block cannot detect any direct LORs. As a result, its FOV is the largest of all configurations (radius: 10 mm), but the images might become distorted because of larger parallax errors. For a better interpretation of the results, the number of events detected by each geometry for all phantoms is shown in Fig. 3. In all cases, the symmetric configuration provides the highest detection efficiency.

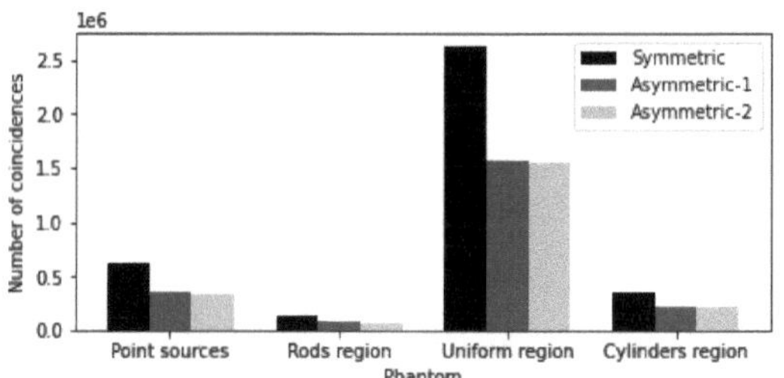

Figure 3: Number of coincidences from each simulation.

3.1 Spatial resolution

FWHM of point sources: The reconstructed sections are shown in Fig. 4. Qualitatively, there is no significant difference between Symmetric and Asymmetric-1, while Asymmetric-2 clearly blurs point sources. Quantitatively, Symmetric and Asymmetric-1 yield similar FWHM in axial and tangential directions while Symmetric performs slightly better in the radial direction (Table 1). The FWHM of Asymmetric-2 is worse in the radial and tangential directions compared to the other geometries but performs similar in the axial direction. The inferior performance of Asymmetric-2 might be attributed to the fact that it features considerably more oblique LORs than Symmetric and Asymmetric-1. In addition, the parallax error is more pronounced in the oblique pair of detectors, which is not fully compensated by the system response model.

Table 1: Mean FWHM (in mm).

Geometry	axial	radial	tangential
Symmetric	0.92 ± 0.05	1.14 ± 0.05	1.25 ± 0.07
Asymmetric-1	0.92 ± 0.06	1.24 ± 0.06	1.25 ± 0.09
Asymmetric-2	0.92 ± 0.06	1.86 ± 0.11	1.81 ± 0.03

RC for the NEMA phantom: Considering the rods region, all three geometries are able to reconstruct 4 out of 5 rods (radius of 1.5 mm, 1.2 mm, 0.9 mm, 0.6 mm, and 0.3 mm, respectively), while the smallest rod is slightly visible in Symmetric's image as shown in Fig. 4. Moreover, Symmetric's rods appear to be brighter, more homogeneous and have well-defined borders as a consequence of more direct LORs and higher statistics (Fig. 3). On the contrary, although Asymmetric-2 and Asymmetric-1 had similar detection efficiency (Fig. 3), a higher blurring affects the former's rods as a result of the higher number of oblique LORs and subsequent increase in parallax error. In Fig. 5, the RCs of each system are reported. It is clear that Symmetric has the

best RC at all radii, followed by Asymmetric-1 with a small margin, and Asymmetric-2 is outperformed by the others.

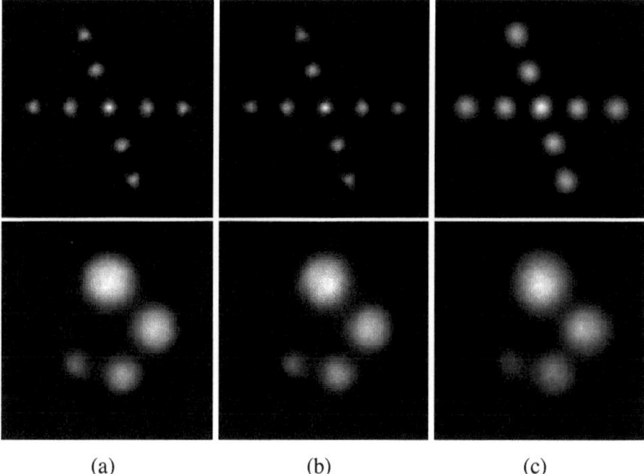

(a) (b) (c)

Figure 4: Reconstructed sections (sum of 20 central transversal slices) for (a) Symmetric, (b) Asymmetric-1, and (c) Asymmetric-2. Top row: The two lines of point sources; bottom row: Rods region of the NEMA-Phantom.

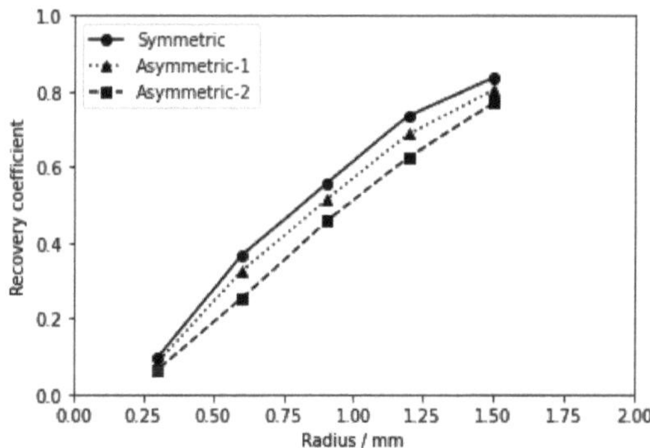

Figure 5: Recovery Coefficients of geometries.

3.2 Uniformity and Spill-Over Ratio

In the uniformity test, all geometries performed similarly in terms of percentage standard deviation (%STD) although the values for the mean, maximum (Max), and minimum (Min) of Symmetric are twice as high as the values of the others (Table 2). The difference could be explained by a pair of direct detector blocks less in asymmetric geometries, which accounts for most of the events detected in our simulated PET system (Fig. 3). Besides, both Symmetric and Asymmetric-1 yield a well-defined homogeneous region with some spill activity outside of the physical region, as shown in Fig. 6. However, the edge of the uniform region is better defined with Asymmetric-2 than for the other geometries, and the region appears much more uniform. The latter effect could be explained by the worse spatial resolution which characterizes this geometry (i.e. broader PSF) and acts by blurring every source inside the FOV. The broader PSF is probably caused by the parallax error, which is more severe in pairs of oblique detectors and not fully compensated for by our simplified system model. On the other hand, the sharper contour could be the result of better sampling the outer region of the FOV using Asymmetric-2. In summary, Symmetric and Asymmetric-1 perform better at the FOV's center, whereas Asymmetric-2 has higher image quality at the edge of the FOV.

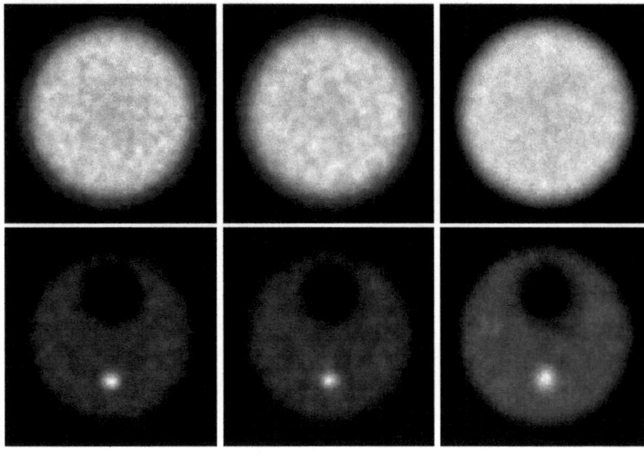

(a) (b) (c)

Figure 6: Reconstructed sections (sum of 20 central transverse slices) for (a) Symmetric, (b) Asymmetric-1, and (c) Asymmetric-2. Top row: Uniform region; bottom row: cylinders region.

In the cylinders region, the reconstructed sections (Fig. 6) again confirm the observation of the spatial resolution and uniformity test. The mean values from five independent simulations are reported in Table 3. All systems have comparable SOR in the cold region, which indicates similar image quality in the cold region as well as equal performance in reconstruction compensation for parallax effects, data truncation, etc. The SOR in the hot region indicates the ability to reconstruct a lesion with a radioactive background; Symmetric and Asymmetric-1 have similar performance while the value of Asymmetric-2 is significantly lower, resulting in poorer contrast.

Overall, the bad performance of Asymmetric-2 allocation in all metrics could be explained by the missing direct LORs of the 90-degree detector block, which contains the most important information for the reconstruction. Besides, the parallax error also increases when the detectors are oblique. This effect could be reduced with a more accurate system model for the reconstruction. Hence, as long as the FOV of the system is acceptable for small-fish PET, a rotational partial-ring PET scanner should be designed to capture as many coincidences in the direct LORs as possible.

Table 2: Uniformity test.

Geometry	Mean	Max	Min	%STD
Symmetric	16.79	31.23	7.60	15.41
Asymmetric-1	8.41	15.17	3.78	14.36
Asymmetric-2	8.13	14.46	3.55	15.48

Table 3: Mean SOR.

Geometry	cold cylinder	hot cylinder
Symmetric	0.11 ± 0.05	2.2 ± 0.5
Asymmetric-1	0.14 ± 0.07	2.1 ± 0.5
Asymmetric-2	0.11 ± 0.06	1.7 ± 0.5

4 Conclusion

The performance of simulated rotational PET scanners with different allocations of four detector blocks was measured to determine the optimal configuration for small-fish PET. From the given data, we conclude that there is no significant difference in terms of spatial resolution, uniformity, and SOR between Symmetric and Asymmetric-1. In contrast, Asymmetric-2 enlarges and blurs the phantom, which demonstrates the importance of direct LORs over oblique LORs. Therefore, as long as the FOV is large enough for the required studies, Symmetric or Asymmetric-1 should be chosen over Asymmetric-2. Considering the requirements of the project, our team has decided to adopt the symmetric solution due to its lower mechanical complexity.

Acknowledgement

The work has been carried out and supervised by the Institute of Medical Engineering, Universität zu Lübeck.

Author's Statement

Conflict of interest: Authors state no conflict of interest.

5 References

[1] M. Zvolský, S. Seeger, M. Schaar, C. Schmidt and M. Rafecas, "MERMAID - A PET prototype for small aquatic animal imaging," *2019 IEEE NSS/MIC*, pp. 1–2, 2019.

[2] S. R. Cherry, J. A. Sorenson, M. E. Phelps, *Physics in Nuclear Medicine*, ch. 18, pp. 397–343. Philadelphia: Saunders, 2012.

[3] D. Sarrut *et al.*, "Advanced Monte-Carlo simulations of emission tomography imaging systems with GATE," *Phys. Med. Biol.*, vol. 66, no. 10, pp. 881—901, 2021.

[4] *NEMA standard publication NU4-2008: performance measurements of small animal positron emission tomographs.* National Electrical Manufacturers Association, Rosslyn, VA, 2001.

[5] L. A. Shepp and Y. Vardi, "Maximum likelihood reconstruction for emission tomography," *IEEE Trans. Med. Imag.*, vol. 1, pp. 113—122, 1982.

[6] S. Moehrs *et al.*, "Multi-ray-based system matrix generation for 3D PET reconstruction," *Phys. Med. Biol.*, vol. 53, no. 23, 2008.

[7] F. Jacobs, E. Sundermann, B. De Sutter, M. Christiaens, I. Lemahieu, "A fast algorithm to calculate the exact radiological path through a pixel or voxel space," *J. Comput. Inf. Technol.*, vol. 6, no. 1, 1998.

12

Medical Electronics

Design of an actuator control circuit with crowbar protection for a neonatal incubator

Luca Belloi [1]

[1] Biomedical Engineering, Luebeck University of Applied Sciences, luca.belloi@stud.th-luebeck.de

Abstract

Power MOSFETs are commonly used in medical devices as semiconductor actuators due to their robustness and simplicity of control characteristics. However, the drain-source short-circuit (DSS) represents a critical failure mode for a power MOSFET because it leads to the inability of deactivating the actuator. This study presents the design of a heater control circuit for a neonatal incubator that implements the working principle of crowbar circuits to achieve fail-safe protection in case of DSS failure. Even though crowbar circuits are mostly used as protection against over-voltage events, the results show the main advantages brought by this implementation: great effectiveness of the circuit in recognizing the DSS failure and protecting the patient, and high reliability of the control circuit expressed as MTBF (Mean Time Between Failures).

1   Introduction

A key function of a neonatal incubator is to control the temperature inside the patient cabin. A control circuit uses a Power MOSFET as an actuator to activate and deactivate a heater element to generate the requested temperature.

However, the control circuit must also be able to assess circuit faults such as the DSS, which represents one the most common electrical failure mode for a power MOSFET [1]. When a DSS failure occurs, the power MOSFET is always conductive and the heater element cannot be deactivated. The temperature inside the patient cabin increases above the maximum temperature allowed and the patient is harmed.

To mitigate the risk for the patient, a redundancy design is the most common approach but it results in a more expensive and complex system.

In this study, a fail-safe approach is implemented because is the most protective and effective method for the patient. Fail safe solutions only requires to terminate the risk situation, and not to mitigate the effect of it. This translates to a simpler design, which means less components into play and then to a more reliable control circuit.

The control circuit shown in this paper uses a power MOSFET to command a heater element inside the patient cabin of a neonatal incubator. In case of DSS, the control circuit activates a custom crowbar circuit to mitigate the risk.

Crowbar circuits are highly recognized to be a reliable fail-safe solution due to their intrinsic simplicity [2] and they are mainly used as over-voltage detectors.

The crowbar concept is revisited and adapted to detect a DSS failure mode by realization of a dedicated logic network. The crowbar principle brings to the control circuit the main advantages of effectiveness, simplicity and reliability against DSS failures.

2   Material and Methods

In this study, the common assumption that the failure rate is constant in electronics components is applied [3],[4]. The determination of the failure rate (therefore the reliability) of the control circuit is obtained through the indications reported in the MIL-HDBK-217 (Military Handbook: Reliability Prediction of Electronic Equipment). The military book assigns different failure rate based on the nature of the electronic component. To achieve a reliable control circuit, one should use the lowest amount of electrical components and, possibly, not programmable integrated circuits (ICs) [5]. Based on this consideration, the block diagram of the control circuit is presented in (Fig.1), while in (Fig.2), the blocks 1. and 2. are shown as schematic circuit.

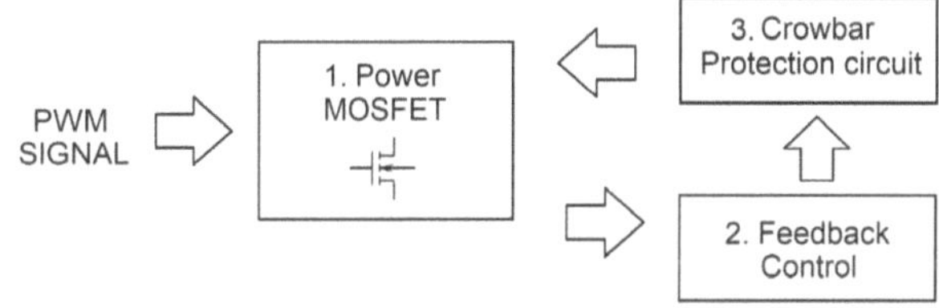

Figure 1: Control circuit block diagram

The pulse width modulation (PWM) signal represents the control signal sent by a micro-controller. A 32bit ARM processor manufactured by ST-Microelectronics was used and programmed in C++ to generate a PWM signal. The "PWM Control" signal of (Fig.2) is used to activate/deactivate a power MOSFET accordingly.

The "PWM Control" is a 3.3V logic signal that commands a PMOS power MOSFET in a high-side switch configuration. A heating element capable of 200W power and sup-

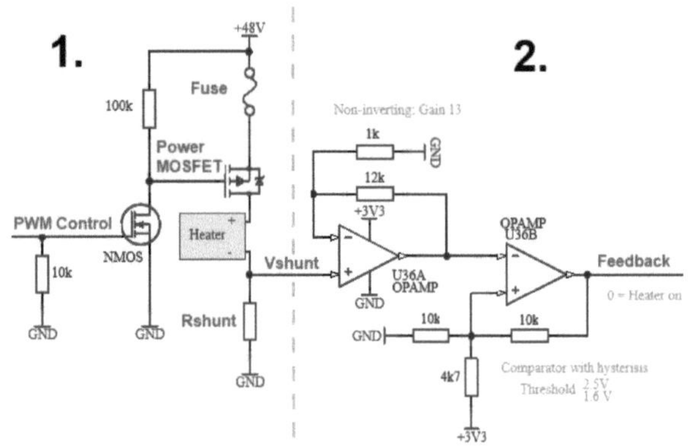

Figure 2: Schematic circuit of block 1. and 2.

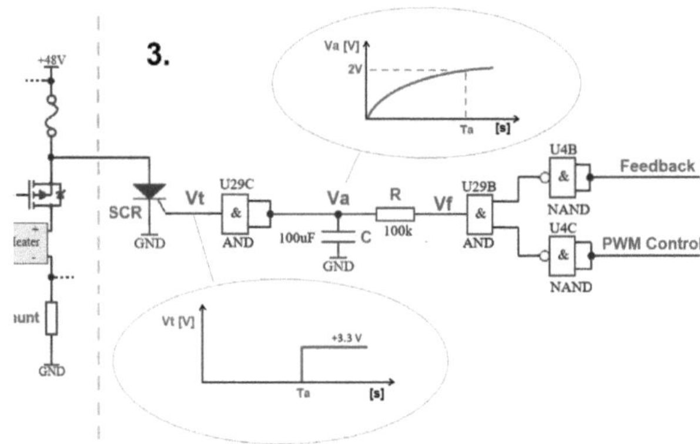

Figure 3: Schematic circuit of block 3

plied by 48VDC is mounted underneath a metal plate to create a heater and positioned inside the patient cabin.

A shunt resistor, R_{shunt}, of $62m\Omega$ is used to assess the status of the actuators. When the "PWM Control" signal assumes a high logic level, the power MOSFET is activated and the shunt resistor produces a voltage of around 250mV across its terminal, V_{shunt}. The shunt resistance should be as low as possible, so that no perturbation of the system is introduced. However, if the resistance value is too low, the drop voltage produced might not be high enough to be distinguishable from the noise level. During testing, a value of 250mV was considered strong enough to be immune to the background noise. The signal V_{shunt} is then amplified, (Fig.2) right side, by a non-inverting operational amplifier with a total gain of 13 to have a maximum output of 3.3V for a maximum input signal of 250mV.

Then, a dual threshold inverting Schmitt-Trigger is introduced to achieve good noise immunity and correct reconstruction of the signal generated by the shunt resistor. The higher the hysteresis of the Schmitt-Trigger, the higher is the immunity to noise. However, when the hysteresis is too large, the circuit is not able to distinguish between normal signal variations and changes due to noise (signal insensitivity) [6].

In this study, a 25% V_{shunt} voltage drop is considered as the maximum allowed due to noise or unwanted variation, leading to the following thresholds:

$$V_{upper.thr} = 2.418V \qquad (1)$$

$$V_{lower.thr} = 1.616V \qquad (2)$$

The output of the Schmitt-Trigger is used as feedback signal ,"Feedback" of (Fig.2), to determine the status of the actuators.

The Schmitt-Trigger is designed as an inverting type to obtain a negative logic, which is preferred due to its intrinsic robustness and noise immunity [7].

The last block of (Fig.1), Crowbar Protection Circuit, is responsible for the DSS failure detection, and the related schematic is shown in (Fig.3).

When a deactivation of the heater is requested, "PWM Control = 0", but the heater is still activated, "Feedback" = 0 (negative logic is applied), a DSS failure is detected and the

crowbar protection is triggered.

In (Fig.2), when a short-circuit between the Drain-Source is present, the power MOSFET is always conductive, no matter the value assumed by the "PWM Control" signal.

This situation is translated in form of truth table (Table 1) and implemented by the combination logic circuit of (Fig.3) with the 3.3V logic ports U4B,U4C and U29B.

The signal V_f, is used to determine if a failure has been detected ($V_f = 1$ failure detected).

Table 1: Truth table for the combination logic circuit of U4B,U4C and U29B

PWM Control	Feedback	V_f
0	1	0
0	0	1
1	1	0
1	0	0

Finally, V_f is filtered by a low-pass filter formed by R and C, to make sure that the noise and unwanted spikes are rejected. The low pass filter is then followed by a buffer gate (U29C) to allow the decoupling between the RC filter and the SCR. The buffer gate needs a minimum input voltage, V_{IH}, to recognize the input voltage as a high logic level (around 2V).

The voltage created by the RC filter, V_a, must reach at least 2V before the buffer transmits the logic level to the next stage. It is possible to determine the value T_a as the minimum time needed to the filter RC to reach 2V by:

$$2V = 3.3V \cdot (1 - e^{\frac{-T_a}{100K\Omega \cdot 100uF}}) \qquad (3)$$

and $T_a = 9.130$ s.

That is, the DSS failure must last for about 9 seconds to be detected by the system. A failure mode and effect analysis (FMEA) study was performed to understand the characteristics required by the control circuit, such as the intervention time in case of heater failure. An intervention time of around 9 seconds is fully accepted and provide great immunity from noise.

Finally, when V_a reaches 2V a failure trigger signal, V_t, is used to activate an SCR (Silicon Controlled Rectifier) in crowbar configuration. The SCR connects the fuse of (Fig.2) directly to ground, so, when the SCR is activated, $V_t = 3.3V$, a high current is drawn directly from the power supply +48V to ground.

The power supply provides under short circuit a maximum current of 25A, which is 5 times higher than the one for which the fuse is rated (FF=fast fuse, 5A) and so the fuse breaks in about about 2ms (manufacturer datasheet).

In conclusion, when the power MOSFET shows a DSS failure, the main fuse is blown by short-circuiting the power supply to ground and the heater deactivated. Various simulations were carried out with the simulation software LT-Spice to confirm the correct behaviour of the circuit, and a PCB prototype was designed and developed with the software Altium Designer.

3 Results and Discussion

The simulations and the theoretical assumptions were verified by measuring the real signals in the prototype. The PWM signal is generated by a micro-controller with a maximum frequency of 0.5Hz to simulate a PID controller whose parameters are calculated every 2 seconds.

In (Fig.4) the "PWM Control" signal is shown (3.3V logic with a duty cycle of around 60%) and it is used as input of the circuit to activate and deactivate the power MOSFET.

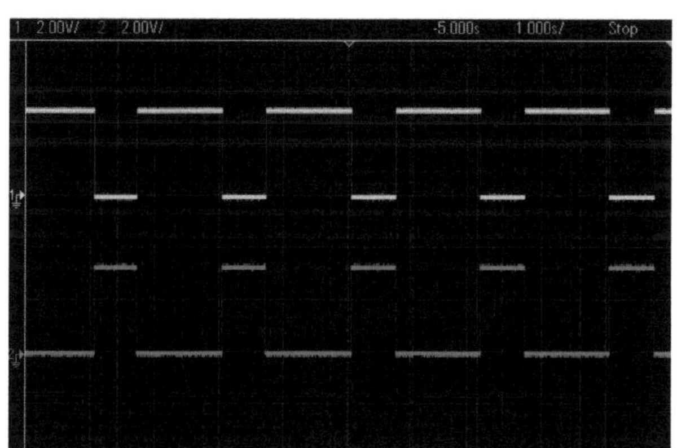

Figure 4: Trace 1 (above): PWM Control signal, trace 2 (below): Feedback signal

The "Feedback" is generated at the output of the Schmitt-trigger and it is visible in (Fig.4) as trace 2. Since a negative logic is applied, the "Feedback" signal has a duty cycle which is the complementary of the "PWM Control" signal, around 40%. This result confirms the correct generation of the feedback signal.

The DSS failure leads to a power MOSFET which is always conductive and the heater cannot be switched off. On one hand, it is important that the crowbar protection circuit correctly detects a failure, on the other, that no false failure is triggered.

For this reason block 3 was at first tested without any DSS, as to verify the normal operation of the block. When no DSS is present, the failure signal V_f should be a clean low level signal. However, during the measurements, spikes

were visible with an average duration of around 1.8ms and random period. When the CPU generates the PWM Control signal, some time is required before the Feedback signal is fully generated due to the inertia required to fully switch off and on the heating element. The RC filter fully absorbed this unwanted spikes and the measurements of V_a showed a clean and noiseless low level signal.

Then, a DSS failure was created by soldering together the drain and source of the power MOSFET. By keeping the "PWM Control" constantly to a low logic level, the values of V_a and V_t were measured.

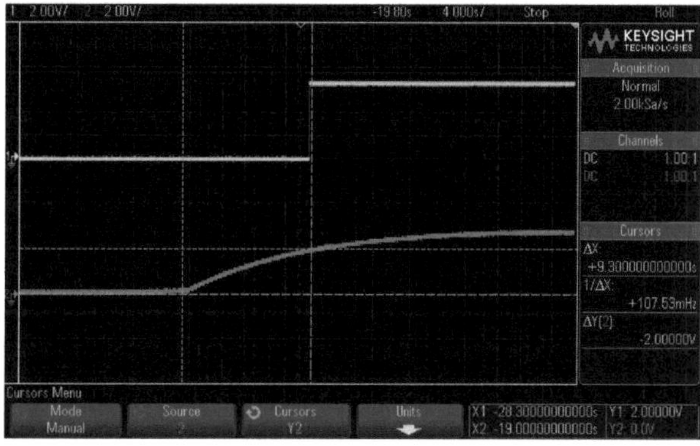

Figure 5: Trace 1 (above): $V_t(t)$, trace 2 (below): $V_a(t)$

The failure signal V_f sharply assumes a high logical level, inducing an exponential charge of V_a, as visible from the oscilloscope measurements of (Fig.5). As soon as V_a reaches the value of 2.13V, V_t, activates the SCR and the fuse is blown. A total of 5 different prototypes were used to obtain an average value for the activation time, T_a and these are reported in the (Table 2).

Table 2: Measured values of T_a and T_{Fuse} in 5 different prototypes

T_a [s]	T_{Fuse} [ms]
9.36	1.52
9.41	1.34
9.55	1.67
9.27	1.13
9.42	2.12

As illustrated in (Table 2), the values are very similar and they differ due to the slightly different threshold voltage of the SCR used in each prototype. The measured average T_a is equal to 9.41s which is coherent with the theoretical one. The time to fully break the fuse, T_{fuse}, was measured as the rising time from 0V to +48V of the voltage across the fuse itself. The fuse clearly breaks in a time span coherent with the one reported in the manufacturer data-sheet (a maximum of 2ms). Finally, the failure rate of the circuit is calculated, based on the guidelines given by the MIL-HDBK-217. In (Table 3), the components of the control circuit are classified based on the different section of the military book.

The manual provides for each component a mathematical description to obtain the failure rate. For resistors the fail-

Table 3: Classification of the control circuit components based on the classification groups of MIL-HDBK-217.

Component group	Handbook section	Total quantity
Resistors	9.1	11
Capacitors	10.10	1
Integrated Circuits	5.1	5
SCRs	6.1	1
Fuse	22.1	1

ure rate of a single resistor, λ_p, (hence the reliability) is calculated as follows:

$$\lambda_p = \lambda_b \cdot \pi_R \pi_Q \pi_E \tag{4}$$

The unit of λ_p is: $\frac{failures}{10^6 hours}$ and the base failure rate is:

$$\lambda_b = 4.5 \cdot 10^{-9} e^{12(\frac{T+273}{343})} e^{\frac{S}{6}(\frac{T+273}{373})} \tag{5}$$

T is the ambient temperature [°C] and S is the ratio between the operating power to rated power; S=1 and T = 35°C for all resistors while λ_b = 0.00085, π_R = 1.0, π_Q = 15 and π_E = 1.0.
The failure rate for 11 resistors, λ_{Res}, is:

$$\lambda_{Res} = 11 \cdot \pi_R \pi_Q \pi_E = 11 \cdot 0.0127 = 0.1406 \tag{6}$$

A similar approach is followed for the other components group and the total failure rate of the control circuit, λ_{TOT}, is then:

$$\lambda_{TOT} = \lambda_{Res} + \lambda_{Cap} + \lambda_{ICs} + \lambda_{SCR} + \lambda_{Fuse} \tag{7}$$

$$= 0.1406 + 0.034 + 0.513 + 0.3770 + 0.010 \tag{8}$$

$$= 1.046 \tag{9}$$

To calculate the MTBF, (7) is reversed, leading to a failure every 930319 hours of usage. If 10 years of continuous usage is considered (about 87600 hours), the probability of finding the circuit functional is defined by:

$$P_t = e^{\frac{-t}{MTBF}} \tag{10}$$

and by inserting the values in (10), the result is:

$$P_{87600} = e^{\frac{-87600}{930319}} = 0.91 = 91\% \tag{11}$$

4 Conclusion

The proposed control circuit is able to operate the heater of a neonatal incubator and to protect the safety of the patient in case of a DSS failure of the power MOSFET. This study does not present a solution to prevent the DSS of a MOSFET, but a solution to put the system in a fail-safe condition. The control circuit is effective and able to detect a DSS failure. To achieve this result, the working principle of a crowbar circuit is adapted and implemented to create a reliable fail-safe mechanism, which mitigates the risk for the patient in about 9s. The control circuit is also highly reliable over the time as resulting from the MTBF and probability calculation.

Acknowledgement

The work has been carried out at LMT Medical Systems, Lübeck, Germany and supervised by Prof. Dr.-Ing. Dipl.-Ing. Stefan Müller, Department of Applied Sciences, Technische Hochschule Lübeck.

Author's Statement

Conflict of interest: Author state no conflict of interest.

5 References

[1] Jayanth Bhargav, *Average ppm for Various MOSFET Failure Modes - Development of Design Tools and Application of IoT for Condition Monitoring of PV Systems* . Springer, 2020.

[2] Mohamed Ebeed, Omar NourEldeen, A.A.Ebrahim3, *Assessing Performance of Crowbar Protections with The DFIG under Grid Fault*. International Journal on Power Engineering and Energy (IJPEE), April 2013.

[3] Joseph B. Bernstein, Moshe Gurfinkel, *Electronic circuit reliability modeling*. University of Maryland, ScienceDirect, Dec 2006.

[4] Omid Alavi, Abbas Hooshmand Viki and Sadegh Shamlou, *A Comparative Reliability Study of Three Fundamental Multilevel Inverters Using Two Different Approaches*, Alavi2016ACR, 2016

[5] Vaishali Hegde, *Reliability in the Medical Device Industry*, Springer, Jan 2008

[6] Marco Donato, Fabio Cremona, *A noise-immune subthreshold circuit design based on selective use of Schmitt-trigger logic*, Conference: GLSI, ACM VLSI, 2012

[7] Li Ding and Pinaki Mazumder, *On Circuit Techniques to Improve Noise Immunity of CMOS Dynamic Logic*, citeseerx.ist.psu.edu, 2008

Design and Development of Field Generator for Continuous Flow Magnetic Spectroscopy

Keana Josephine Ridder [1], Ankit Malhotra [2,3], Dirk Steinhagen [3], Arthur-Vincent Lindenberg [2,3], and Kerstin Lüdtke-Buzug [3]

[1] Biomedical Engineering, Luebeck University of Applied Sciences, keana.josephine.ridder@stud.th-luebeck.de

[2] Fraunhofer Research Institution for Individualized and Cell-Based Medical Engineering IMTE, Lübeck, {ankit.malhotra, arthur-vincent.lindenberg}@imte.fraunhofer.de

[3] Institute of Medical Engineering, Universität zu Lübeck, {steinhagen, luedtke-buzug}@imt.uni-luebeck.de

Abstract

The novel continuous flow magnetic particle spectroscopy (MPS) device offers the opportunity of steering and manipulating nucleation and growth of magnetic nanoparticles (MNPs) during synthesis process. Due to this a quality control of the MNPs can be realized which will be in particular beneficial regarding mass production for industrial use. Therefor, a field generator attachable to a microfluidic system is designed and developed. This field generator was manually produced and analyzed regarding electric characteristics, magnetic field strength and mutual inductance. Finally, a frequency spectrum analysis was applied for comparing the signal received by inserting MNPs with the reference measurement. The developed field generator is able to record the signal of the MNPs. This can be used to characterize the magnetic properties of the inserted MNPs. Prospective continuous flow MPS devices could apply multiple field generators for monitoring magnetic characteristics of synthesized MNPs during nucleation and growth phase in real time.

1   Introduction

In 2005, Gleich and Weizenecker proposed magnetic particle imaging (MPI) which is a tomographic imaging modality [1] [2] [3]. This method uses magnetic nanoparticles (MNPs) as signal source and visualizes the related spatial distribution [2] [4]. A few possible applications of MPI are e.g. cardiovascular interventions, sentinel lymph node biopsy in breast cancer staging and molecular imaging [4] [5].

However, magnetic particle spectrometers (MPS) are used to characterize magnetic nanoparticles (MNPs) regarding their magnetic properties. Therefor, a homogeneous time varying magnetic field is applied and the change in the non-linear magnetization is measured [6].

The signal chain of the state of the art MPS is shown in Fig. 1. First, the computer is used to steer the data acquisition process. Secondly, the sinusoidal signal is amplified by a power amplifier [6]. Afterwards, the signal passes the impedance matching unit for suppressing higher harmonics produced by the power amplifier and for matching the impedance of power amplifier and send coil [6]. Furthermore, the signal produced by send coil and MNPs is received by the receive coil [6]. Finally, the signal needs to be band stop filtered and amplified for further signal processing [6].

This work deals with the design and the development of field generators, consisting of send and receive coils, for a novel continuous flow MPS device. In contrast to the commonly used MPS, no impedance matching unit and no band stop filter are needed at all. Here, a compensation unit is used instead. The compensation unit can be externally placed and subsequently adjusted.

In the following sections the fabrication process of the field generator will be presented. Additionally, testing procedures regarding electrical characteristics, magnetic field, coupling factor and spectrum generation will be described and interpreted.

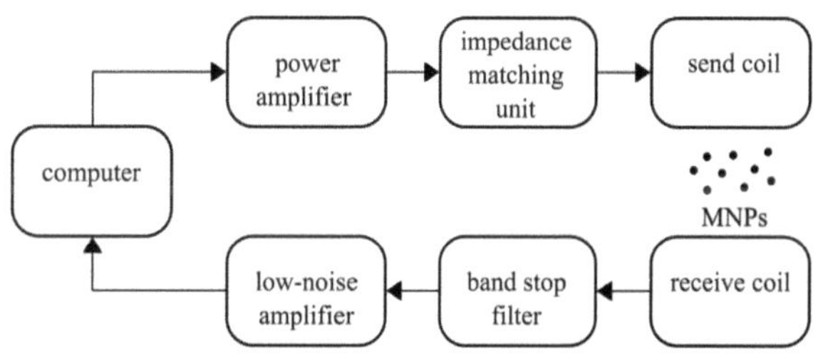

Figure 1: Signal chain of conventional MPS [6].

2 Material and Methods

2.1 Fabrication of Field Generator

To receive the signal of the superparamagnetic iron-oxide nanoparticles a magnetic particle spectrometer is built. The magnetic particle spectrometer consist of two concentric solenoids, which are known as send coil and receive coil. Moreover, the magnetic field strength generated by the send coil should be minimum 20 mT as commonly used [2].

The send coil is manufactured by winding a copper wire of 0.4 mm in diameter around a supporting structure. Thereby every single layer, including the final layer, of the solenoid is fixed with epoxy adhesive.

The receive coil is also manufactured by winding a copper wire of 0.4 mm in diameter around a supporting structure. Additionally, a cast consisting of Polyoxymethylen (POM) is constructed for manufacturing the receive coil. This cast is placed around the receive coil and is filled with epoxy adhesive.

Send and receive coil are then placed into the drying oven at 100 °C for 20 minutes. Finally, after finishing the drying process the coils and the supporting structure are separated. In the case of the receive coil the cast is removed prior to that. To assemble the field generator the receive coil is placed inside the send coil. The manufactured solenoids are shown in Fig. 2.

The send coil has an inner diameter of 8.07 mm and consists of three layers with 22 windings per layer. Whereas the receive coil consists of two layers with 20 windings per layer and has an inner diameter of 3.91 mm.

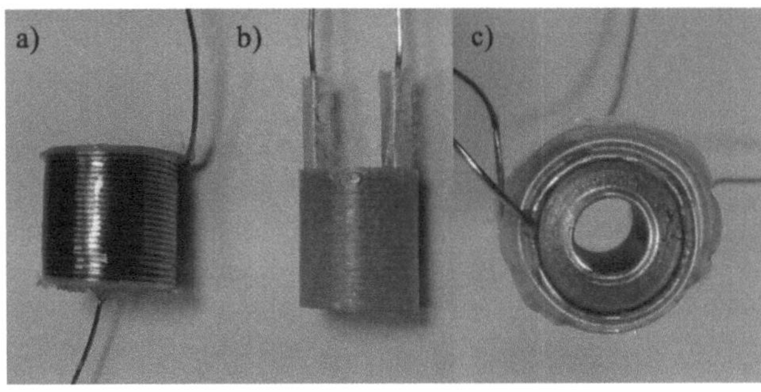

Figure 2: a) Send coil, b) Receive coil, and c) Field generator.

2.2 Electric Characteristics

To measure the electric characteristics such as inductance and impedance of send and receive coil an impedance analyzer (Keysight Technologies, Inc., E4990A Impedance Analyzer) is used. Therefor, the start frequency is determined as 10 kHz and the stop frequency as 100 kHz to visualize the inductance change depending on the frequency. Additionally, a marker is set to 25 kHz to receive the impedance as a function of this particular frequency. Thus, the marker frequency corresponds to the drive frequency of MPI [3]. Measuring the electric characteristics at the drive frequency of MPI is reasonable because the electrical behavior of solenoids is frequency dependent.

2.3 Measurement of Magnetic Field

Furthermore, the voltage inside the field generator and the send coil is measured by using a calibration coil as flux meter.

For test purpose the field generator is used, as described in 2.1. Furthermore, the field generator is fixed by a hinged bracket while the calibration coil (one layer, five windings, inner diameter of 2.98 mm) is fixed at a sliding bed, which slides in millimeter range. Due to this measurement setup an approximately concentric position of all three solenoids is received. However, the receive coil is open looped. Due to this the mutual inductance is taken into account which will be present in the intended setting. Additionally, the same setup, without the inserted receive coil, is used to measure the voltage inside the send coil to determine the difference.

Therefor, the input signal of the send coil is generated by a frequency generator (Rigol DG1022, Batronix GmbH & Co. KG) and is linear amplified by a power amplifier (2105, AE Techron Inc.). The resulting voltage signal of the calibration coil and the send coil current are detected by an oscilloscope (HD06104-MS, Teledyne Technologies Inc.). Here, the sliding bed is moved with a step size of 0.5 mm to detect the voltage at defined measuring points. With the function generator a 25 kHz sinusoidal with an amplitude of 1 V_{PP} is applied. The measurement is performed at a constant send coil current of 3 A which results in a send coil voltage of 12 V for the field generator and 12.3 V for the send coil, respectively.

2.4 Identification of Coupling Factor

The coupling factor is experimentally determined to find out the dimension of mutual inductance between the send and the receive coil.

Therefor, two different conditions of the field generator are under investigation. On the one hand the receive coil is open looped and on the other hand the receive coil is short circuited.

The input signal of the send coil is generated by a frequency generator (Rigol DG1022, Batronix GmbH & Co. KG) and is linear amplified by a power amplifier (2105, AE Techron Inc.). The resulting voltage signal at the receive coil is detected by an oscilloscope (HD06104-MS, Teledyne Technologies Inc.).

Here, the voltage of the receive coil is measured at a send coil current of 1 A. The function generator produces a 25 kHz sinusoidal signal with an amplitude of 1 V_{PP}.

2.5 Frequency Spectrum Analysis

For doing a frequency spectrum analysis of the MNPs signal the signal chain is arranged, as shown in Fig. 3. Here, the signal applied to the send coil is generated by a function generator (Rigol DG1022, Batronix GmbH & Co. KG) and is transmitted towards a power amplifier (2105, AE Techron Inc.). Afterwards, the received signal is coupled into the receive coil and is amplified (SR560 low-noise preamplifier, Standford Research Systems). Furthermore, the data are measured and stored by the oscilloscope (HD06104-MS, Teledyne Technologies Inc.). To receive the frequency spectrum a Fast Fourier Transformation (FFT) is performed to the data.

For reducing the noise produced by interfering signals a compensation unit is integrated. The compensation unit consists of a second field generator which is connected in series. When connecting the send coils in series the direction of the winding needs to be opposite. For the receive coils the same arrangement is valid.

For performing a frequency spectrum analysis a commercial sample of Resovist® (Schering AG) coming with a concentration of 0.5 mmol Fe/ml is used [7]. Therefor, a piece of Tygon tube with 3.0 mm in outer diameter is prepared with 10 µL Resovist® as a sample. Both endings of the tube are sealed by hot glue to prevent a leakage.

Here, the input signal is a 25 kHz sinusoidal signal with an amplitude of 2 V. The resulting send coil voltage is 25.4 V. However, the low noise amplifier is set to a gain of 100.

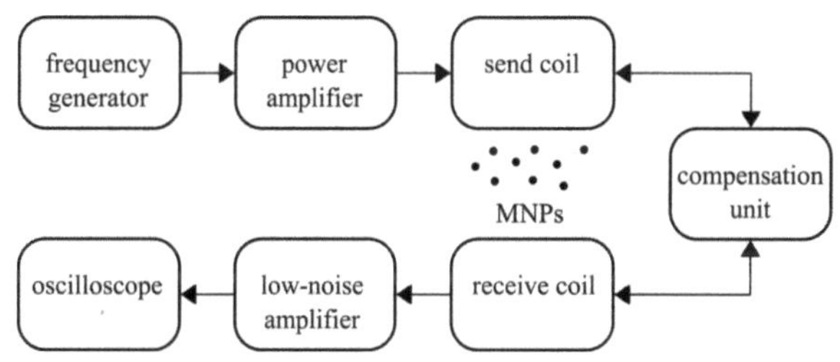

Figure 3: Setup to measure the signal produced by MNPs.

3   Results and Discussion

As described in 2.3 the voltage signal in z-direction is measured. The induced voltage of the field generator and the induced voltage of the send coil is highest at the center and approximately symmetrically decreasing in both directions, as shown in Fig. 4. The maximal achieved voltage signal in z-direction amounts 131 mV with the receive coil inside of the send coil and 133 mV without the receive coil inside, as shown in Fig. 4. By applying Faraday's law of electromagnetic induction the maximum achieved magnetic field is calculated. With the receive coil inside the send coil the maximum magnetic field is 37.56 mT. Considering the case that no receive coil is inside the send coil maximum performance is 38.14 mT. Hence the required excitation field of minimum 20 mT is achieved.

For calculating the coupling factor k between send and receive coil the inductance is needed. The measurement is carried out as described before in 2.4. The resulting voltage with receive coil in short circuit condition is 3.4 V while open loop condition produces a voltage of 4.28 V. With

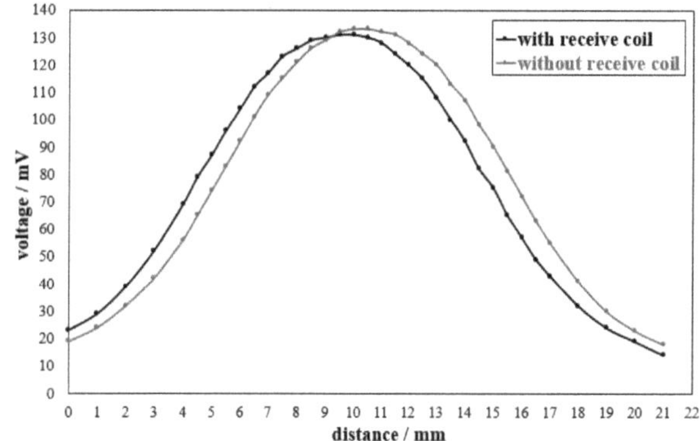

Figure 4: Voltage signal in z-direction as function of distance measured with calibration coil (5 windings, wire diameter of 0.28 mm and inner diameter of 2.98 mm).

given voltage, frequency and current the inductance is calculated. Applying this calculated values to (1) the coupling factor is calculated as 0.45. The coupling factor in general describes how much of the voltage from send coil is induced to receive coil. Assuming that a coupling factor of 0 means no mutual induction whereas a coupling factor of 1 indicates maximal mutual induction. Based on this assumption the calculated coupling factor indicated an acceptable coupling between send and receive coil for the application.

$$k = \sqrt{1 - \frac{L_s}{L_o}} = \sqrt{1 - \frac{2.16 \cdot 10^{-5} H}{2.72 \cdot 10^{-5} H}} = 0.45 \ . \qquad (1)$$

In Table 1 the inductance and the impedance of all manufactured send and receive coils are visualized. These values are determined as described in 2.2. The resulting variation of inductance and impedance among the coils is due to the manual manufacturing process. To achieve enhanced conformity the manufacturing process of send and receive coil could be improved further. For improving the signal quality of the frequency spectrum the coils used for the field generator and the compensation unit should have as similar values as possible.

Taking this into account the resulting frequency spectra are shown in Fig. 5. This frequency spectra are produced by applying FFT at a sampling rate of 5 MS/s. The Resovist® sample produces even and odd harmonics of the fundamental frequency, which is 25 kHz. Odd harmonics are repre-

Table 1: Inductance and impedance of manufactured send and receive coils.

Coil Type and No.	Inductance	Impedance
Send No. 1	24.978 µH	356.85 m
Send No. 2	24.791 µH	358.20 m
Send No. 3	25.379 µH	346.88 m
Send No. 4	24.461 µH	351.89 m
Receive No. 1	3.273 µH	128.75 m
Receive No. 2	3.250 µH	119.30 m
Receive No. 3	3.343 µH	124.41 m
Receive No. 4	3.302 µH	129.93 m

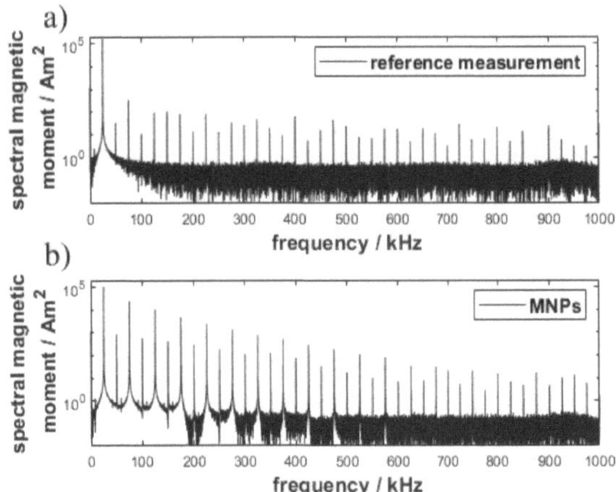

Figure 5: a) Spectrum of reference measurement, and b) Spectrum of MNPs in the frequency range of 0 kHz to 1000 kHz with compensation unit.

senting the MNPs signal whereas even harmonics are due to interfering signals for instance from earth magnetic field [3]. In contrast to the reference measurement, the amplitudes of the harmonics are increased when applying the MNPs inside the field generator. Due to this harmonics up to a frequency of 625 kHz are clearly detectable.

To calculate the compensation received by applying the compensation unit the voltage change of having connected the compensation unit in series (U_2) or not (U_1) is inserted in (2). The maximum achievable compensation a is calculated as 33.58 dB which indicates differences between the applied coils.

$$a = 20 \cdot log(\frac{U_1}{U_2}) = 20 \cdot log(\frac{2.1\,V}{0.044\,V}) = 33.58\,dB\ . \quad (2)$$

4   Conclusion

The developed field generator for continuous flow magnetic spectroscopy is able to detect a clear signal of the tested magnetic nanoparticles up to a frequency of 625 kHz. Moreover, the maximal produced excitation field of the field generator therefor is 37.56 mT. The analyzed sample consists of 10 µL Resovist® (Schering AG) coming with a concentration of 0.5 mmol Fe/ml [7].

Due to the application of a compensation unit in series with the field generator improving the signal quality of the frequency spectrum is possible. Therefor, the electrically connected send coils and accordingly receive coils need to show similar electrical characteristics to compensate one another. The achieved compression is 33.58 dB which means that the manually fabrication process has a large impact on the producible excitation signal of each coil in particular.

In future a novel continuous flow MPS device can benefit from this developed field generator. Thereby multiple field generators could be applied to the device to monitor magnetic characteristics of synthesized magnetic nanoparticles

during nucleation and growth phase in real time.

Acknowledgement

The work has been carried out at the Institute of Medical Engineering, Universität zu Lübeck and supervised by the Fraunhofer Research Institution for Individualized and Cell-Based Medical Engineering IMTE, Lübeck.

Author's Statement

Conflict of interest: Authors state no conflict of interest.

5   References

[1] B. Gleich and J. Weizenecker, *Tomographic imaging using the nonlinear response of magnetic particles.* Nature, vol. 435, pp. 1214–1217, 2005, DOI:10.1038/nature03808.

[2] N. Panagiotopoulos et al., *Magnetic particle imaging: current developments and future directions.* International Journal of Nanomedicine, vol. 10, pp. 3097–3114, 2015, DOI:10.2147/IJN.S70488.

[3] S. Biederer et al., *Magnetization response spectroscopy of superparamagnetic nanoparticles for magnetic particle imaging.* Journal of Physics D: Applied Physics, vol. 42, no. 20, 2009, DOI:10.1088/0022-3727/42/20/205007.

[4] T. M. Buzug et al., *Magnetic particle imaging: Introduction to imaging and hardware realization.* Zeitschrift für Medizinische Physik, vol. 22, no. 4, pp. 323-334, 2012, DOI:10.1016/j.zemedi.2012.07.004.

[5] N. Garraud, R. Dhavalikar, L. Maldonado-Camargo, D. P. Arnold and C. Rinaldi, *Design and validation of magnetic particle spectrometer for characterization of magnetic nanoparticle relaxation dynamics.* AIP ADVANCES, vol. 7, 2017, DOI:10.1063/1.4978003.

[6] S. Biederer et al., *A Spectrometer for Magnetic Particle Imaging.* In: IFMBE Proceedings, Springer-Verlag Berlin Heidelberg, pp. 2313–2316, 2008, DOI:10.1007/978-3-540-89208-3_555.

[7] Schering AG, *Zusammenfassung der Merkmale des Arzneimittels (SPC) Resovist® 0.5 mmol Fe/ml Injektionslösung, Fertigspritze.* Available: https://myhealthbox.eu/de/view/589292/d94093af0ab11e07354502bad4aeb40a/leaflet [last accessed on 2022-01-15].

Analysis for the design of a Motion Detection and Orientation Device

Milton David Pujos [1], and Jörg Schroeter [2]

[1] Biomedical Engineering, Luebeck University of Applied Sciences, milton.david.pujos.yanzapanta@stud.th-luebeck.de

[2] Medizinische Sensor- und Gerätetechnik, Luebeck University of Applied Sciences, joerg.schroeter@th-luebeck.de

Abstract

Seizure detection devices are supposed to improve safety and provide assurance to patients with epilepsy. Motion detection sensitivity, security of collected data, or operating time are some of the features to consider in an orientation and motion detection device. First, it was determined the most suitable components for a device that detects movement and orientation. Three inertial measurement units (IMU) and three microcontrollers (MCU) were chosen for testing to decide the most appropriate ones for the device to be developed. The performed tests were focused on power consumption, proper operation, and sensor outputs. Finally, it was concluded that the device to be developed should have an IMU that incorporates an accelerometer, a gyroscope, and a magnetometer, and also a microcontroller which has power saving modes, and communication protocols.

1 Introduction

Patients with epilepsy put a particular emphasis on the necessity for seizure detection devices to be comfortable, usable, reliable, and nonintrusive. These devices should be designed with good motion detection sensitivity to avoid false alarms. In addition, its physical design and a long battery life are relevant features [1]. Therefore, as a first approach, it has been decided to design a device with the minimum requirements, depicted in Table 1, that are focused on the development of a prototype that allows later adding functionalities to the existing ones or implementing changes without the need of a new design. It is estimated that the device will operate continuously for several days; a minimum time cannot be established until the corresponding clinical studies with patients have been conducted.

Table 1: Minimum requirements of the device

Parameter	Description
Type	Wearable device
Battery life	Several days
Outputs/Features	Orientation and motion detection

The orientation or attitude of a body is calculated based on angles measured from a local reference frame (n-frame) to another reference frame on the body (b-frame). The navigation angles Pitch, Roll and Yaw are commonly used to determine the orientation of a body. Another way to represent attitude is by quaternions which are an extension of complex numbers [2]. A quaternion has 4 components: a real one and 3 imaginary ones represented by "i", "j" and "k". The advantage of using quaternions is that they decrease the

processing time or computational load compared to navigation angles since they do not use trigonometric functions; additionally, they avoid the presence of singularities when Pitch is $\pm90°$.

In order to achieve the requirements, the performance of the main components of the device will be analyzed, and then choose the ones that provide the results that fulfill the needs of the project. The project is in its initial stage, after choosing the components for the device to be developed, a prototype will be produced to conduct trials with patients.

2 Material and Methods

After carrying out an online search in the market of electronics based on the requirements, it was identified that the device to be developed will mainly consist of an inertial measurement unit (IMU) and a microcontroller (MCU).

An Inertial Measurement Unit (IMU) is a sensor that measures acceleration and angular velocity [3]. The IMU consists of an accelerometer and a gyroscope embedded in a single device. It can also be found with a magnetometer integrated into the chip. IMU outputs can be combined by sensor fusion algorithms and thus determine the orientation of a body. Accelerometer signals are also useful for setting a threshold to allow motion detection.

A microcontroller unit (MCU) is a computer on a chip which has a microprocessor and peripherals such as memory, inputs/outputs (GPIO), timers, communication protocols, interrupts, or clock circuits integrated together on a single chip. Furthermore, power modes allow the microcontroller to optimize power consumption [4].

It is proposed that the measurements of power consumption and variation of the sensor outputs will allow the selection of the suitable IMU and UCM for the project. In addition, characteristics of the components that might facilitate the implementation of the device should be analyzed.

2.1 MCU tests

MCUs were selected for their power consumption, processing power, CPU, and communication protocols. Considering that it is intended to develop a wearable device, it is recommended that it should have Bluetooth Low Energy (BLE) because it will facilitate the transmission of data acquired by the sensors or will allow to check continuously its operation status, moreover that BLE operates with low current consumption. Three MCUs were chosen for testing their performance: PSoC 6 (Cypress), PSoC 4 (Cypress) and STM32WB (STMicroelectronics). Development boards were used to easily evaluate the functionalities of each MCU.

Tests were performed by alternating the operating mode between a power saving and active mode. A Cassy Module was used to measure power consumption, it is a device for recording measurement data through analog inputs [5]. Power consumption was determined by measuring the current since the supply voltage was constant (DC value). Each development board was connected to a USB port and to the Cassy Module, without any other device to avoid noisy measurements. A periodic signal of 1 second for the PSoC MCUs and 0.5 seconds for the STM32WB was generated in order to distinguish the change between power modes. The Cassy Module was configured to acquire data for 4 seconds with a sample rate of 1 ms.

For Cypress MCUs, the firmware was developed with the software PsoC Creator 4.4. It was necessary to use the Watch Dog Timer (WDT) in order to generate a periodic signal every fixed time that triggers an interrupt that was executed even when the MCU is in a low power mode such as Sleep or Deep Sleep Mode [6]. The WDT requires a low-frequency clock to operate. PSoC development boards have an integrated Watch Crystal Oscillator (WCO) of 32.768 kHz with an accuracy of 150 PPM (0.015%). The crystal accuracy is a parameter to consider in the design because it can lead to a time deviation, for instance an accuracy of 150 PPM can produce a time deviation of 12.96 seconds after one day of operation. Deep Sleep Mode was used in these MCUs since it could wake up using WDT [7].

The STMicroelectronics microcontroller was programmed using Mbed OS which automatically manages power modes without any configuration or command in the firmware. Therefore, no WDT configuration or interruption was required. Additionally, several nested loops in the firmware were necessary in order to show the changes between Deep Sleep Mode and Active Mode because power optimization is managed by the MCU itself.

2.2 IMU tests

IMUs were selected based on their current consumption, measurement range, data output frequency, available filters, and signal processing capabilities. Three IMUs were used for testing their performance: LSM6DSL (STMicroelectronics), BMX160 (Bosch) and BNO055 (Bosch).

Firmware was developed on PSoc 4 for data acquisition from IMUs every 500 ms. Two types of tests were performed:

- One for simultaneous acquisition of IMUs outputs which were connected to the MCU through the I2C communication pins as it is shown in Fig. 1. The I2C protocol only requires two pull up resistors connected to its two lines (SDA and SCL) [8]. Measurements were taken with the IMUs in a steady state for at least 5 minutes, without any kind of external movement. The standard deviation of the measurements was calculated to be able to compare them.

- Another one to determine the power consumption of each IMU by using the Cassy Module configured for an acquisition of 4 seconds with a sample rate of 0.5 ms. In this case, only the IMU to be measured was connected to the I2C bus to avoid noise in the measurements.

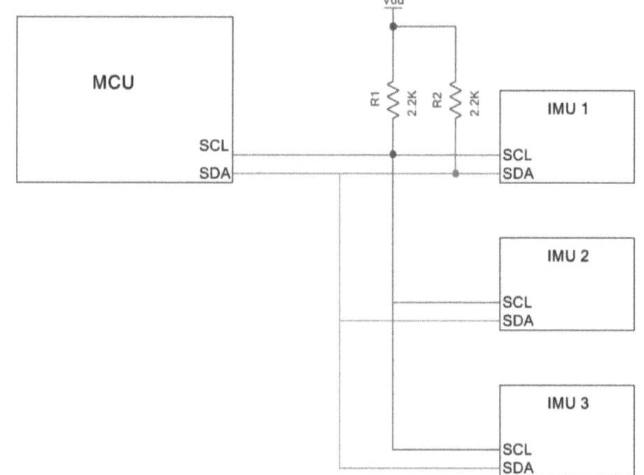

Figure 1: Scheme of the built circuit for communications over I2C.

IMU sensors were configured with similar parameters, as it can be seen in Table 2, 3 and 4, in order to be able to take measurements and compare between them. BNO055 was configured to operate in Fusion Mode as NDOF (9 degrees of freedom) which calculates absolute orientation from sensor fusion of the accelerometer, gyroscope, and magnetometer. The orientation calculated by the BNO055 is given in terms of quaternions and navigation angles. Additionally, it has a magnetometer calibration function that runs in the background automatically. BNO055 uses its default settings in NDOF, they cannot be changed by the user. BMX160 and LSM6DSL were configured with similar operating characteristics to those of the BNO055.

Table 2: Accelerometer configuration

Parameter	BNO055	BMX160	ST LSM6DSL
Range	±4g	±4g	±4g
Bandwidth	62.5 Hz	40.5 Hz	52 Hz

Table 3: Gyroscope configuration

Parameter	BNO055	BMX160	ST LSM6DSL
Range	±2000 dps	±2000 dps	±2000 dps
ODR	100 Hz	100 Hz	104 Hz

Table 4: Magnetometer configuration

Parameter	BNO055	BMX160	ST LSM6DSL
ODR	20 Hz	25 Hz	not applicable

3   Results and Discussion

3.1   Power consumption measurements

BNO055 worked in Suspend and Normal mode. However, it was necessary to establish a wait time of at least 45 ms to change from Suspend to Normal mode, otherwise the outputs were wrong. That time delay limits the maximum output data rate of the sensor, therefore, BNO055 should operate in low power mode, but this mode is automatically managed by the sensor itself and depends on its motion and a sensitivity level defined in the firmware. BNO055 average current consumption in Suspend mode was 2.1 mA and 12.4 mA in Normal mode. BMX160 current consumption was 1.1 mA in Sleep mode and 3.2 mA in Normal mode. The BMX160 IMU had lower current consumption than the BNO055 in both modes of operation (saving and active mode) as can be seen in Table 5.

Table 5: IMU current consumption

IMU	Power saving mode	Active mode
BNO055	2.1 mA	12.4 mA
BMX160	1.1 mA	3.2 mA

Regarding the microcontrollers, PSoC 4 had the highest current consumption, which is almost twice that of the other MCUs in Deep Sleep mode (Fig. 2), which is relevant to consider because the device is intended to work most of the time in a power saving mode.

PSoC 6 has two embedded processors. The measurements in Fig. 3 were taken with one of the processors (CM0) in Deep Sleep Mode permanently and the other processor (CM4) alternating between Deep Sleep Mode and Active Mode. PsoC 6 measurements were affected by interference that could be due to electronic elements on the testing board.

STM32WB MCU presented the lowest current consumption (Fig. 4), but it cannot be assured that using peripherals such as I2C or BLE the results will remain just as efficient. Table 6 shows that PsoC 6 is a good option to choose because its power consumption values are similar to

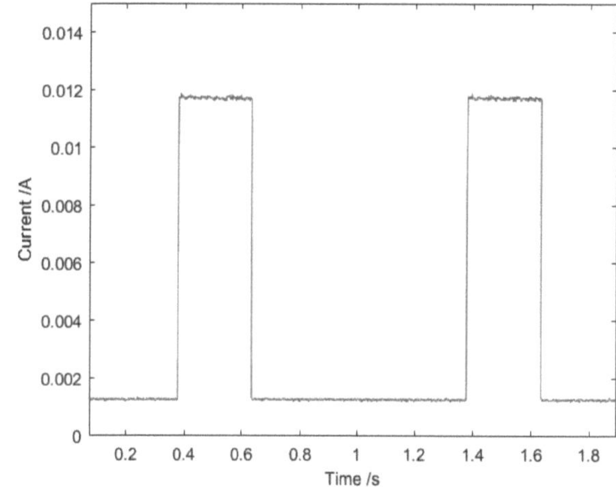

Figure 2: Current consumption on PsoC 4, Deep Sleep Mode (bottom of signal) and Active Mode (top of signal).

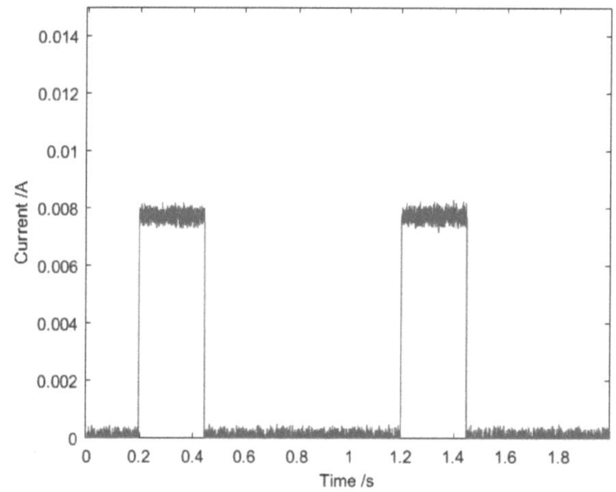

Figure 3: Current consumption on PsoC 6, Deep Sleep Mode (bottom of signal) and Active Mode (top of signal).

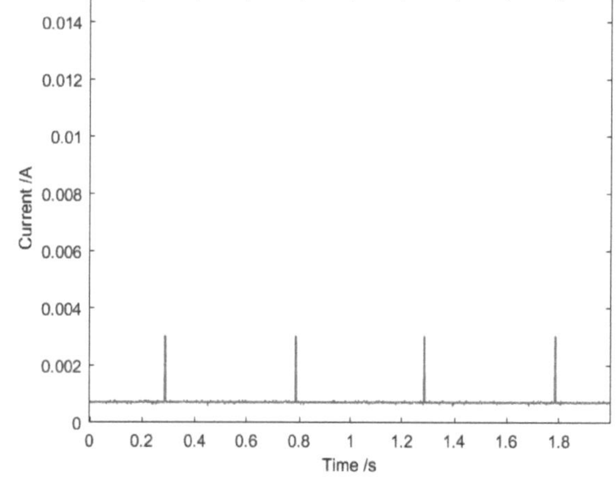

Figure 4: Current consumption on STM32WB, Deep Sleep Mode (bottom of signal) and Active Mode (peaks).

STM32WB in power saving mode.

Table 6: MCU current consumption

MCU	Power saving mode	Active mode
PSoC 4	1.37 mA	11.71 mA
PSoC 6	0.7 mA*	7.73 mA
STM32WB	0.72 mA	3.04 mA

*from datasheet

3.2 Steady state measurements

Results varied depending on the sensor of each IMU. Considering only the accelerometer, ST LSM6DSL sensor was the best with a standard deviation (SD) of 4.47×10^{-3} g, followed by the BMX160 (SD=4.86×10^{-3} g) and finally the BNO055 (SD=6.71×10^{-3} g). For the gyroscope, BNO055 sensor presented the best results (SD=5.66×10^{-2} dps), then there was the BMX160 (SD=8.59×10^{-2} dps) and finally the ST LSM6DSL (SD=11.8×10^{-2} dps). The LSM6DSL was intended to analyze the outputs from the accelerometer and gyroscope, it does not have a magnetometer, and considering its poor response in the case of the gyroscope, this sensor was discarded. The lack of a magnetometer does not allow to determine relative orientation between two or more devices. Regarding the magnetometer, BNO055 had better results (SD=1.25 uT) than the BMX160 (SD=1.6 uT).

3.3 General overview

The most suitable MCU for the project is the PSoC 6, it has a good power consumption performance that can be controlled and configured in the firmware as opposed to the STM32WB MCU which is recommended as a second option.

BNO055 IMU is the best option based on its embedded functionalities such as sensor fusion, continuous calibration of the magnetometer and determination of orientation by quaternions and navigation angles. However, BNO055 has a higher power consumption, but this can be optimized to achieve an extended autonomy. BMX160 IMU is a good option, it has a lower power consumption, but lacks BNO055 functionalities that must be implemented in the microcontroller (MCU), thus increasing the firmware complexity and computational workload.

It was not possible to compare the obtained current consumption with the values given in the datasheets because manufactures do not clearly describe the information according to the configuration and employed peripherals.

4 Conclusion

The motion detection and orientation device should mainly consist of a microcontroller (MCU) and an inertial measurement unit (IMU), based on the conducted online market search. The MCU should incorporate at least serial communication protocols, Bluetooth Low Energy (BLE), power optimization modes and Watch Dog Timer (WDT). The IMU should consist of an accelerometer, a gyroscope, and a magnetometer. According to power consumption and variation of the sensor outputs, BNO055 IMU is recommended for the project because of its embedded functionalities. Moreover, PSoC 6 MCU is also suitable due to its current consumption and because it allows power management configuration in the firmware.

Acknowledgement

The work has been carried out at GAIA AG and supervised by the Medizinische Sensor- und Gerätetechnik, Luebeck University of Applied Sciences.

Author's Statement

Conflict of interest: Authors state no conflict of interest.

5 References

[1] E. Bruno, P. Viana, M. Sperling and M. Richardson, *Seizure detection at home: Do devices on the market match the needs of people living with epilepsy and their caregivers?*. Epilepsia, vol. 61, no. S1, 2020.

[2] P. Corke, *Robotics, vision and control: fundamental algorithms in MATLAB*, Springer, 2013.

[3] H. Lui, *Robot Systems for Rail Transit Applications*, Elsevier, 2015.

[4] X. Fan, *Real-time embedded systems: design principles and engineering practices*, Elsevier, 2015.

[5] LD DIDACTIC Group, *Sensor-CASSY*. Available: https://www.ld-didactic.de/software/524221en/Content/CASSYs/Sensor-CASSY/Introduction.htm [last accessed on 2022-01-15].

[6] Cypress Semiconductor, *PSoC 4 BLE Architecture Technical Reference Manual*, Rev. *D, 2017.

[7] K. Sadasivam and M. Kingsbury, *PSoC 4 Low-Power Modes and Power Reduction Techniques*. Infineon Technologies AG, 2021.

[8] NXP Semiconductors, *I2C-bus specification and user manual*, Rev. 6, 2014.

Design and Construction of Surface EMG Simulator

Aleksandra Filippova [1]

[1] Biomedical Engineering, Lübeck University of Applied Sciences, aleksandra.filippova@stud.th-luebeck.de

Abstract

Physiological signals from the human body may tell a lot about health condition of the organism. The aim of this project is the creation of a universal device for simulation of electromyography signals. The main purpose of the device is simulation of analog signals for medical electronics and software testing. Since electrical signals and muscle pressure are closely related, it is possible to produce a signal, imitating a breath muscle pressure, which can be supplied to an ASL500 lung simulator. To achieve the aim of the project, the STM32 microcontroller based on a single board, requiring no additional modules was used. It produces analog surface electromyogram signal patterns from real patients as well as simulated one. It can be also used as an external source of muscle pressure analog signal for the ASL 5000 lung simulator. The device showed a satisfactory results, but still requires further improvement to use with lung simulator.

1 Introduction

The project is related to *electromyography (EMG)*, a technique to measure and record electrical activity produced by the muscles of human body. Monitoring the activity of the respiratory muscles is of critical importance in respiratory care and can be achieved continuously and noninvasively using *surface electromyography (sEMG)*. The electric potential generated by muscle cells is detected by a device, called *electromyograph*, when they are electrically or neurologically activated. The signals can be analyzed to detect abnormalities, activation level, recruitment order, or to analyze the biomechanics of human or animal movement. In peaceful breathing, diaphragm and intercostal muscles are involved. However, in case of respiratory distress the breathing effort is changed. That is why collecting a signal, produced by muscles, involved in breathing, is highly important for patients, suffering from respiratory diseases. Generally, a respiratory surface EMG requires five electrodes, which are attached to the skin of patient's chest. The scheme is shown in Fig. 1. Each electrode measures the voltage with respect to the jointly connected reference. Since one electrode is used as a reference, the result contains four signals: two from intercostal muscles and two from the diaphragm. Examples of signals collected from surface respiratory EMG are shown on the Fig. 2. It is clearly seen from the Fig. 2, that electrical signals from respiratory muscles are extremely small; moreover, often they look similar to electrical noise, because of their chaotic shape and very small amplitude. Therefore, collecting and processing of such a tiny signal is not a simple task. Good amplifiers and sensitive electronics are required to prepare such signals for the further processing with PC or microcontroller. In addition to this, when medical software or hardware is being tested, it is not always possible to have a

real patient to make a measurement.

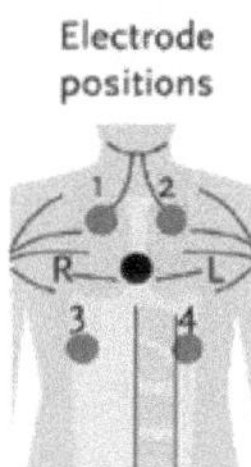

Figure 1: scheme of electrode attachment for measuring respiratory EMG [6]. 1,2 – for intercostal muscles, 3,4 – for diaphragm. Central point – reference.

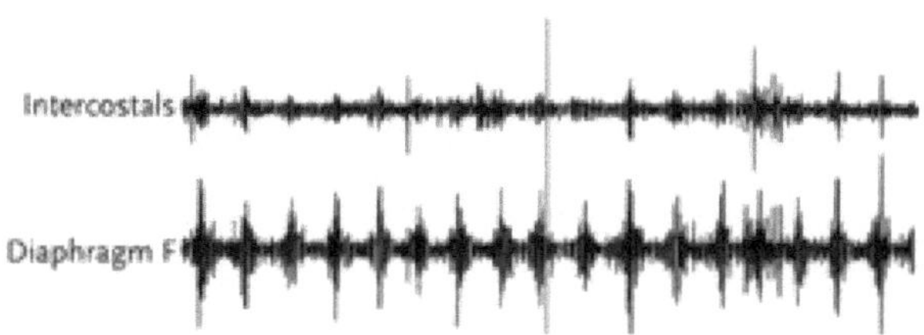

Figure 2: example of the EMG signals from intercostals and diaphragm [6]. The amplitude of these signals is in the range of hundreds of microvolts (uV), time of measurement – approx. 2 minutes.

For these cases, a more convenient solution is to have a simulated analog signal, resembling real data, already collected and pre-processed by an amplifier. From the shapes of surface EMG signals, it is seen, that electrical signal produced by contraction of the *diaphragm* have more accurate waveform and larger amplitude, than those from the intercostal muscles. That is why diaphragmatic signals were chosen for simulation.

2   Materials and Methods

The developed device has four operating modes, listed in Table 1.

Table 1: operating modes of the device

Mode	Function	Description
1	sEMG	Based on real patient's data; filtering algorithm: HP200
2	sEMG	Based on real patient's data; filtering algorithm: SWT
3	sEMG	Simulated signal
4	Muscle pressure	Represents muscular pressure, for working with ASL 5000 lung simulator

Change of operational modes is indicated by LED. For the first operational mode LED blinks one time at the beginning of simulation, for the second mode – two times and so on. First two modes are used to simulate EMG signal using patterns, based on raw data, taken from real patients. These signal patterns were obtained after pre-processing of a raw signal, collected directly from the human body. The data was filtered from the inevitably occurring cardiac artifacts using different filtration algorithms [3]. Therefore, operation in the first mode represents data, filtered by HP200 (fourth-order Butterworth high-pass filter with a cutoff frequency of 200 Hz) algorithm, in the second mode – by SWT (stationary wavelet transform) algorithm. These algorithms were suggested by authors [3] as providing the most accurate and precise filtering from cardiac artifacts. In third mode, a fully artificial EMG signal is generated. For that, a model of random noise, modulated by a sinusoidal wave was used, which is quite similar to real EMG signal (see Fig. 2).

Muscle activity is directly related to the pressure produced for breathing. Therefore, it was decided also to add a possibility to simulate a signal resembling a muscle pressure. Fourth mode represents a respiratory muscle activity, produced by human during the breathing process. In this mode of operation, the device can be used with ASL 5000 lung simulator (Fig. 3) as an external source.

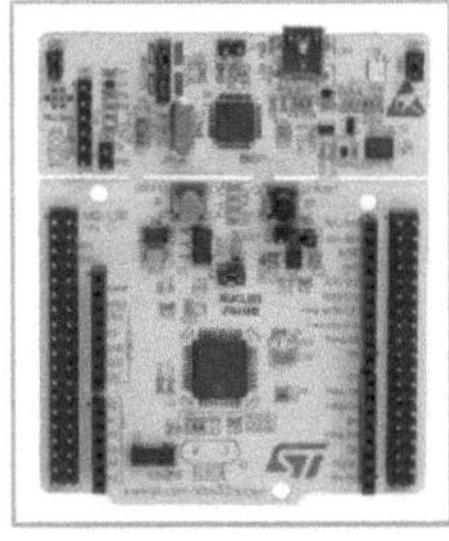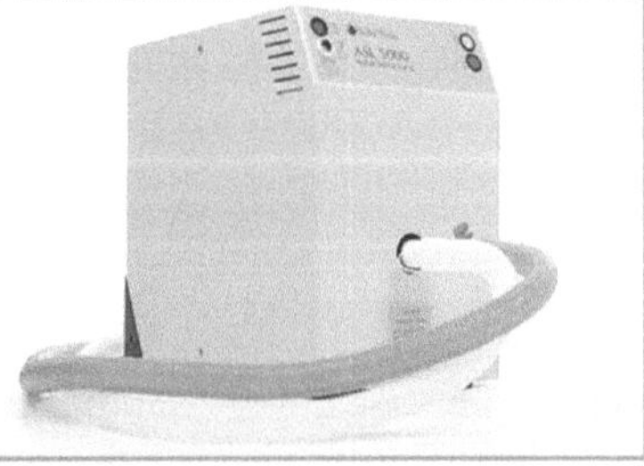

Figure 3: STM32 NUCLEO-F44RE development board (left), ASL 5000 lung simulator (right) [1].

ASL 5000 software suggests using simple sinusoidal half-wave to simulate muscle pressure [1]. However, for more precise result, a flexible dynamical model was implemented. This model, suggested by [2], has an electrical analogy of RC- circuit, shown on Fig. 4.

The input of the system is the voltage U produced by the central nervous system. Output is the voltage U_{rm}, which corresponds to the voltage delivered by the respiratory muscles. The contraction corresponds to the charge and the relaxation to the discharge of a capacitance C_m through a resistance R_m. The capacitance C_m characterizes the distension property of respiratory muscles and the resistance R_m corresponds to the resistance produced via tendons and intramuscular structures such as myofibrils and connective tissue.

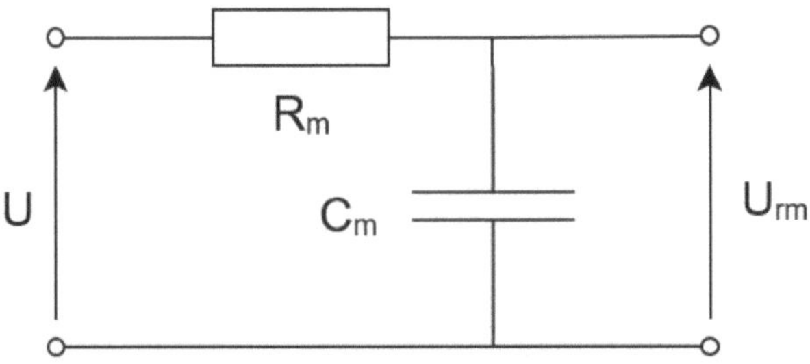

Figure 4: electrical analogy for muscular activity during breathing cycle.

Therefore, activity of muscles P_{mus} according to chosen model can be described by following:

$$P_{mus} = \begin{cases} 1 - exp(-\frac{f_v + 4P_{0.1}}{10}t), & 0 < t \leq T_I \\ -exp(-\frac{f_v + \frac{P_{0.1}}{2}}{10}t), & T_I < t \leq T_{tot} \end{cases} \quad (1)$$

The ventilatory dynamics is clinically characterized by the ventilatory frequency f_v (in cycles per minute, cpm) and the mouth occlusion pressure $P_0.1$ (cmH_2O per 0.1s). T_I is the time of inspiration, T_{tot} is a total time of a single breath. These parameters can be changed in the source code to create different scenarios of ventilatory dynamics. Some examples are shown in Fig. 5.

To complete the task, a development board STM32NUCLEO-F446RE (Fig.3) was used. There are other constructions which are based on single-board computer (Raspberry PI) and additional modules [6]. However, often it is not necessary to have the power of a single-board computer for this kind of simulation. STM32 board provides embedded ST-LINK debugger and programmer, Arm Cortex M4 core operating at 180MHz, 512 Kbytes of Flash memory and 128 Kbytes SRAM. The development board includes also an embedded digital-to-analog converter (DAC), which was used to generate analog signals with a certain shape. A button is used to switch between four possible modes and an LED indicates which mode was chosen. Power supply and programming is performed via single connection to PC/laptop by USB Type-A to Mini-B cable. For programming and debugging a STM32 CubeIDE was used.

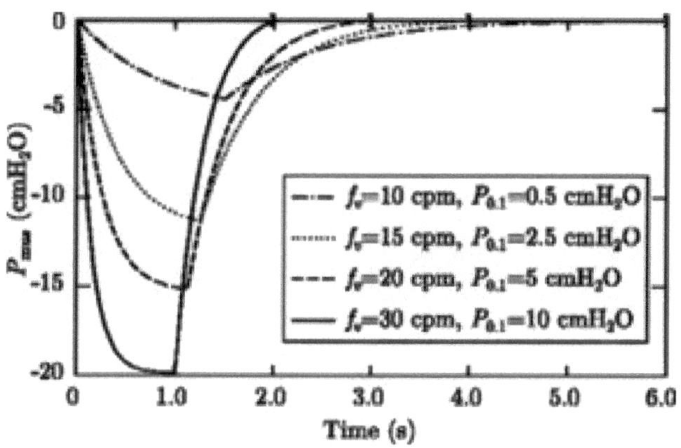

Figure 5: time series of the muscle pressure simulated for four different combinations of parameters [2].

3   Results and Discussion

The algorithm used in the work is briefly described below. The variable *data-number* indicates the mode of operation. The peripherals have been configured in such a way that an interrupt occurs when the button is pressed. Further, in the interrupt handler function data, which is an output to the DAC, is cyclically switched. Two data sets are stored in 8-bit format as *data1.h* and *data2.h* (DAC was configured to have 8-bit resolution as well). These data sets are used in modes 1 and 2 to produce analog signal. For modes 3 and 4 signal is generated in real time. In mode 3, a pseudo-random noise is modulated by a sinusoidal wave. In mode 4, expression (1) is used for signal generation. .

Signals, produced by the device are shown below on Fig. 6 – 9.

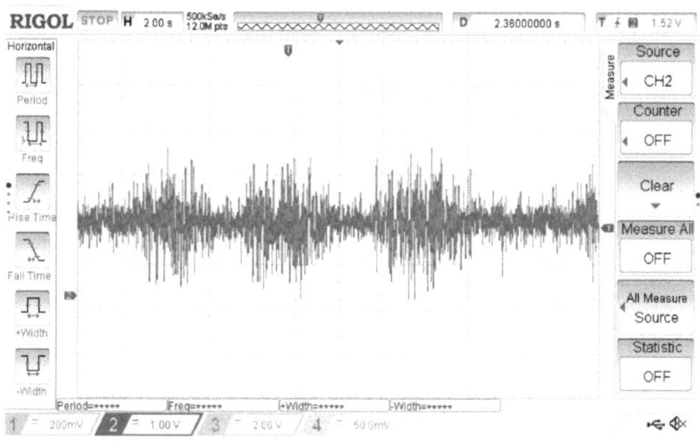

Figure 6: first operation mode, real patient's data, HP200 filtering algorithm.

Signals, produced by a device are quite similar to real patient data, as seen from the comparison of simulated signal (Fig. 6-8) to real data from the patient, shown in Fig. 10. On this figure data filtered by HP200 and SWT algorithms from [3] is plotted. In addition, noise amplitude and period of simulated EMG signal can be changed to achieve more realistic waveform. However, generated signals have max. peak-to-peak amplitude of 3.3V (which is a limitation of board operating voltage), imitating already amplified EMG

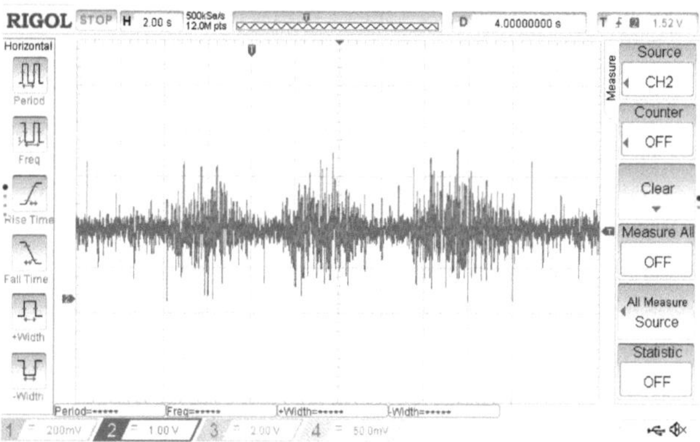

Figure 7: second operation mode, real patient's data, SWT filtering algorithm.

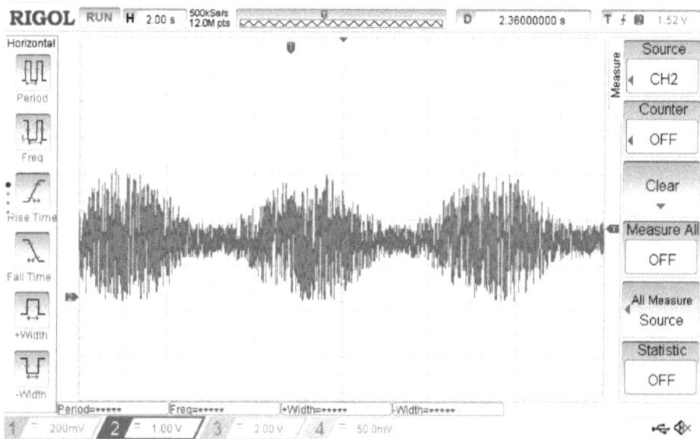

Figure 8: third operation mode, simulated signal.

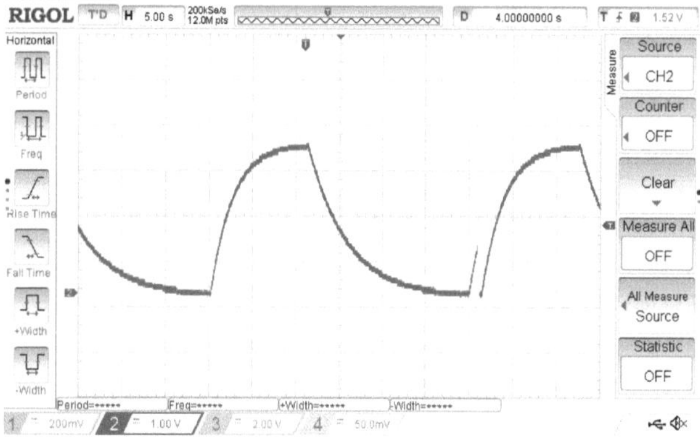

Figure 9: fourth operation mode, simulated muscular activity.

signal. This makes further processing of such a signal far easier in comparison with raw EMG data in the range of microvolts.

As seen from the Fig. 11, ASL 5000 lung simulator software received and plotted signal from external analog input. The lung simulator performed some kind of simulation with external analog signal, however, it seems to be not fully correct. Moreover, "Large volume error" occurs. This may be connected with input voltage. Analog input of the ASL 5000 operates in the range from 0 to 10V [1], with a bias of 5V, representing a zero effort. The STM32 develop-

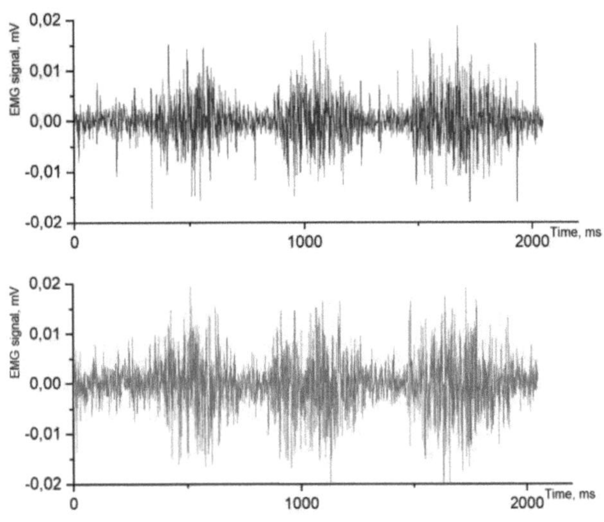

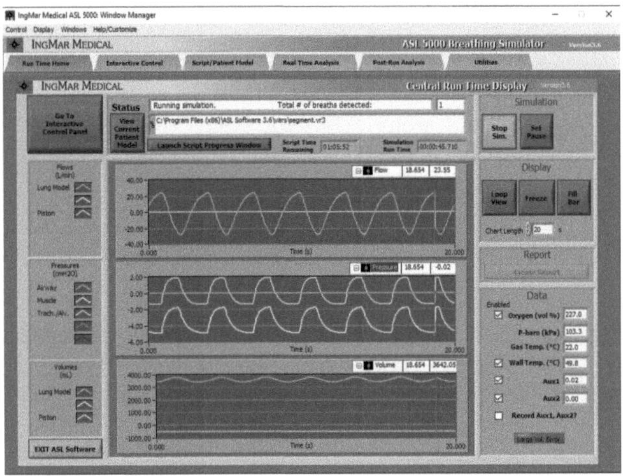

Figure 10: raw surface electromyogram data. Cardiac artifacts removed by HP200 algorithm (top) and SWT algorithm (bottom). Original data taken from [3].

Figure 11: input signal is plotted as "Muscle pressure" in ASL 5000 software.

ment board, in contrast, produces max. 3,3V output voltage. Therefore, this incongruity may lead to incorrect simulation results. The solution here is to add a constant voltage component of 5V.

4   Conclusion

In this work, a universal device for simulating signals produced by breathing muscles was developed. Usually, during the testing procedures of medical devices or software, it is quite hard and inconvenient to collect physiological parameters from a real patient. Moreover, this kind of data is rare and often not provided for public access, due to complicated procedure for measurements and for ethical and data protection issues. In this situation, it is good to have a compact device, which can produce various analog signals similar to what can be measured from a human body. The device can

generate four different signals – surface electromyogram from real patient raw data, fully simulated electromyogram signal and muscle pressure pattern, which is based on electrical analogy and depends only on two parameters – ventilatory frequency and mouth occlusion pressure. This mode can be used as an external input to lung simulator. However, for the last option a device requires further improvement. Surface EMG simulation seems to be quite reliable, when the original and simulated signals are visually compared. The board is supplied from usual USB cable and can be used with a personal computer or a laptop, to avoid interferences from power network. The board is compact enough to make a portable simulation device. Possibilities of development board (core, memory) give the opportunity for further development and generation of more precise and complicated data (for example, apply more complicated algorithms for signal generation, add user's input to manually change parameters of signals, LCD for displaying the parameters, etc.).

Acknowledgement

The work was carried out at Institute for Electrical Engineering in Medicine, University of Lübeck and supervised by Carlos Castelar Wembers, Jan Graßhoff and Prof. Dr. Georg Schildbach, Institute for Electrical Engineering in Medicine, University of Lübeck.

Author's Statement

Author state no conflict of interest.

5   References

[1] *User's Manual ASL 5000*, SW 3.6, Rev. 2, ©IngMar Medical, LLC. 2020

[2] Fresnel, E., Muir, JF., Letellier, C., *Realistic human muscle pressure for driving a mechanical lung*. EPJ Nonlinear Biomed Phys 2, 7 (2014). https://doi.org/10.1140/epjnbp/s40366-014-0007-8

[3] E. Petersen, J. Sauer, J. Graßhoff and P. Rostalski, *Removing Cardiac Artifacts from Single-Channel Respiratory Electromyograms*. in IEEE Access, vol. 8, pp. 30905-30917, 2020, doi: 10.1109/ACCESS.2020.2972731.

[4] *Description of STM32F4xx HAL drivers*. ©2014 STMicroelectronics

[5] Jason H.T. Bates, *Lung Mechanics. An inverse modelling approach*. Cambridge University Press, 2009

[6] P. Stagge, *Testmöglichkeiten eines sEMG-Verstärkers*. Projektpraktikum, Universität zu Lübeck, 2019

13

Auditory Technology

Method of Expansion Conv-TasNet to Multiple Input to handle Realistic Room Reverberation

Ann-Katrin Griedelbach [1], Ole Thomsen [2], and Alfred Mertins [3]

[1] Auditory Technology, Universität zu Lübeck, annkatrin.griedelbach@student.uni-luebeck.de
[2] Institute for Signal Processing, Universität zu Lübeck, o.thomsen@uni-luebeck.de
[3] Institute for Signal Processing, Universität zu Lübeck, alfred.mertins@uni-luebeck.de

Abstract

Speech separation refers to the extraction of different speakers from one ore more mixtures. This could improve speech intelligibility of hearing aid users in cocktail party situations. In this work a two channel fully-convolutional time-domain audio separation network (Conv-TasNet) with a multi-phase gammatone filterbank (MP-GTF) as a static encoder is presented. It will be tested on the open source dataset for speech separation LibriMix. To obtain more realistic scenarios a method is introduced to expand LibriMix to two channels with binaural room impulse responses (BRIR). A database of measured BRIRs with everyday background noise is used for this purpose. First results for single-channel Conv-TasNet and anechoic LibriMix dataset are a scale invariant signal to noise ratio improvement (SI-SNR) of 13.7 dB. Future experiments will show whether there is an advantage from the multiple input for the performance of Conv-TasNet.

1 Introduction

Even at a loud cocktail party, the human auditory system gives us the ability to focus on a speaker. This challenge is much higher for the hearing impaired. To facilitate speech understanding in noisy environments with different simultaneous speakers for hearing aid users, it is necessary to focus on different sound sources. The anechoic scenario thus describes a mixture waveform $x(t)$ consisting of the waveforms s_i in the sum of C speakers:

$$x(t) = \sum_{i=1}^{C} s_i(t) \,. \tag{1}$$

However, this is not a realistic scenario because the transmission channels h_i from the sources to the microphones are not included:

$$x(t) = s_{r1}(t) + s_{r2}(t) = s_1(t) * h_1(t) + s_2(t) * h_2(t) \,. \tag{2}$$

These are unknown and cannot be measured in realistic scenes, making the problem of speech separation far more challenging. Blind source separation methods need to perform the separation based on the measurement signals alone. For better speech intelligibility in cocktail party situations, it is necessary to separate different speakers from each other, while removing disturbing reverberation and background noise. In this paper, the Conv-TasNet [1] is first applied to the LibriMix dataset [2] for the anechoic case. Then a dataset of binaural room impulse responses [3] is used to extend the dataset with them. The intention in prospect is to create a two-channel dataset that contains more realistic scenarios with unknown transmission channels and daily life background noise. Experiments for expanding Conv-TasNet to multiple input show a good performance [4]. So Conv-TasNet is extended to stereo input to handle the extended LibriMix dataset. At least the encoder of Conv-TasNet is replaced by multi-phase gammatone filterbank described in [5].

2 Material and Methods

2.1 LibriMix

LibriMix is an open source dataset for speech separation. It is based on the data set LibriSpeech for automatic speech recognition. The dataset includes different cases for two and three speakers, with and without background noise. The used training dataset contains 101 hours of LibriSpeech with no noise, where the longer audiodata is shortened to the length of the shorter one. The data was downsampled to 8kHz sampling rate. Each utterance occurs only once. The level of the speakers was randomized between -25 dB and -33 dB loudness units relative to full scale. There are 45 hours of training data and 5 hours of evaluation data. The SNR of the final dataset is normal distributed with a mean of 0 dB and 4.1 dB standard deviation.

2.2 Conv-TasNet

Most of neuronal networks for source separation use a time-frequency representation of the speech mixture. Due to the uncertainty relation, long time windows are necessary for high frequency resolution. This leads to high latency which is not suitable for real-time applications as in hearing devices. Time-domain audio separation networks (TasNet) use only the mixture waveform in time domain to avoid this problem. Conv-TasNet is a fully-convolutional network. It consists of three modules: an encoder, which transforms the input waveform into feature space, then a separator, which estimates a multiplicative mask for each source, at last the decoder, which uses the masked feature representation and reconstructs a waveform for each source [1].

2.2.1 Encoder/Decoder

The encoder consist of a 1-D-convolution which transforms overlapping time segments $x \in \mathbb{R}^{1 \times L}$ of the input mixture into N-dimensional space. The convolution can be presented as a matrix multiplication:

$$w = \mathcal{H}(xU) \tag{3}$$

U presents N encoder basis functions with length L. To be sure w is non-negative $\mathcal{H}$ is a rectified linear unit (ReLU). The reason for non-negativity is the intention to find masks representing the proportion of sources in the mixture. Equivalent to the encoder, the decoder is also a 1-D convolution, which can be written as matrix multiplication:

$$\hat{s}_i = d_i V \, , \tag{4}$$

where $\hat{s}_i$ is the reconstructed waveform for each source. V are the decoder basis functions with length L and d_i is the representation of the i-th source, calculated by applying the determined masks of the separator on the output of the decoder w [1].

2.2.2 Separation

At the beginning the data is layerwise normalized and a bottleneck layer reduces the number of channels. Then the separation module consists of repeats of stacks of 1-D-convolutional blocks with exponentially increasing dilation in each stack. This has the advantage to consider a large temporal range with a relatively small number of weights. A flowchart of the separation can be seen in Fig. 1. In Fig. 2 the structure of a convolutional block can be seen. A block consists of a separated convolution. The convolution is divided into a depthwise convolution over the time domain with the increasing dilation factor and a pointwise convolution, i.e. a convolution with kernel size of one, which transforms the feature space. The separated convolution has the advantage of a smaller model size instead of the standard convolution due to the reduced kernel dimension. After both convolutions a parametric retified linear unit (PReLu) as activation function is added. Also a normalization process takes place. Two different normalizations can be chosen depending on the causal system. The global normalization is calculated over time as well as over the feature space, which has the disadvantage that the case is not causal and thus not suitable for the real-time application. In the other

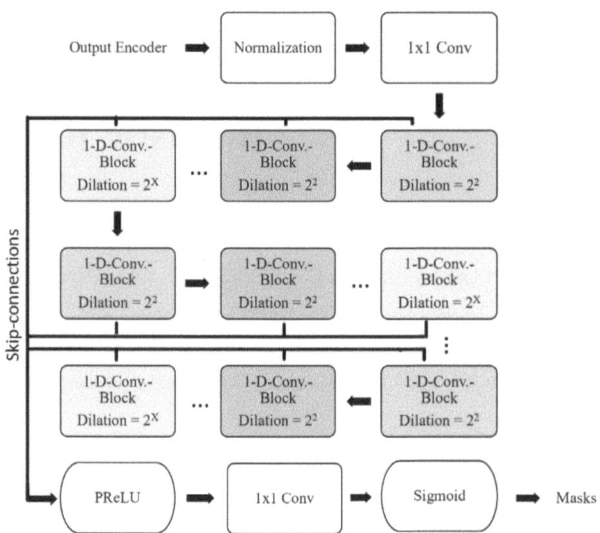

Figure 1: The design of the seperation module: each row represents a repetition of the 1-D convolutional blocks with increasing dilation factor.

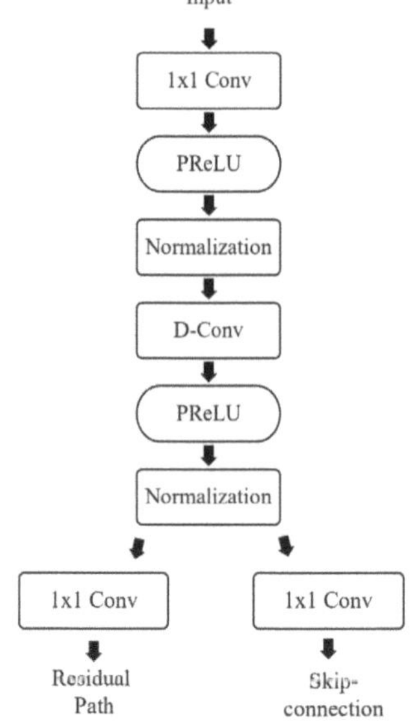

Figure 2: The design of the convolutional blocks.

case, cumulative normalization is always performed only up to the current time, which provides applicability for the causal case. A 1-D convolutional block ends in splitting into a skip-connection path and a residual path. The residual path leads into the following convolution block. The sum of all skip connection paths results in the output of the temporary convolutional net. To create the masks for the source estimation a PRelu followed by a point-wise convolution with a sigmoid function as activation function is used [1].

2.3 Training

For training, the audio data was divided into windows of 4 seconds length with 1.5 seconds overlap. The optimizer used is the Adam optimization algorithm with a starting learning rate of 10^{-3}. If the accuracy of the validation did not increase for 3 consecutive epochs, the learning rate was halved. As accuracy and loss the scale invariant SNR was used:

$$\begin{cases} s_{target} := \frac{\langle \hat{s}, s \rangle s}{\|s\|} \\ e_{noise} := \hat{s} - s_{target} \\ SI - SNR := 10 \log_{10} \frac{\|s_{target}\|^2}{\|e_{noise}\|^2} \, , \end{cases} \tag{5}$$

where s_{target} is the estimated source normalized with the original source and e_{noise} are all signal parts not included in the original source. The dataset was trained for 100 epochs. In each epoch, the entire training data set was used. Gradient clipping with maximum L_2-norm of 5 was used. Parameters applied for Conv-TasNet are presented in Table 1. Parameters and non-causal normalization were chosen because they gave the best results in [1]. The network was implemented in Tensorflow.

Table 1: Parameters Conv-TasNet

Parameter	Values
Number of filters in encoder	512
Length of filters	16
Number of channels in bottleneck and residual paths	128
Number of channels in skip-connection paths	512
Number of channels in convolutional blocks	128
Kernel size of convolutional blocks	3
Number of convolutional blocks in each repeat	8
Number of Repeats	3

For determination of the loss during training the SI-SNR between the estimated waveform of the source $\hat{s}_i$ and the clean waveform s_i needs to be calculated. The problem is that the permutation of the output is unknown. Since Conv-TasNet predicts both sources based on the same mixture at the same time, the output can be $[\hat{s}_1, \ \hat{s}_1]$ or $[\hat{s}_2, \ \hat{s}_1]$. To avoid the label permutation problem, utterance-level Permutation Invariant Training (uPit) technique was used [6]. A pairwise SI-SNR score for all possible permutations was calculated and the permutation of the sources was switched for loss calculation to the permutation with the highest score.

2.4 Extending LibriMix to stereo

The room impulse response (RIR) describes the transmission path from a sound source to the receiver in time domain. It represents the direct sound and the reflections arriving. The RIR depends on a variety of factors, such as the room dimensions, the positions of the source and receiver, furniture, etc.. The head-related impulse response (HRIR) describes the transmission path to the right and left ear channel, includes the influencing properties of head,

torso and pinna. The combination of RIR and HRIR is the BRIR. It describes the transmission path to the ear channels in a reverberant environment [7]. The selected database from BRIRs consists of one of a series of recordings in everyday environments with natural background noise [3]. The BRIRs were measured with a head-and-torso simulator (HATS). The measurements with the in-ear channels are taken from the database. The total of 32 BRIR measurements took place in three different environments with different head and speaker positions. Environments were an office room, a countyard and a fully occupied cafeteria [3]. For application of the BRIR database to LibriMix the mixtures had to be new generated. The level of the speaker is already randomized in LibriMix dataset [2]. To avoid a double level change through the convolution with BRIR, which are caused by the different distances to the receiver, the audio data was normalized. For a mixture, an environment was selected randomly. Then both sources were randomly placed in this environment on different positions with the same head orientation of the receiver. The right and left transmission path for the allocated positions were convolved with the sound sources and summed up for new stereo mixtures, representing the right and the left ear of the receiver.

2.5 Extension of Conv-TasNet

For adaption of the Conv-TasNet to the stereo mixture, the input x of the encoder had to be changed in dimension from $x \in \mathbb{R}^{1 \times L}$ to a concatenated representation of the stereo mixture waveform $x \in \mathbb{R}^{2 \times L}$. The encoder was replaced by the multi-phase gammatone filterbank (MP-GTF) introduced by [5]. The MP-GTF has the advantage that it is based on human auditory perception. It also requires fewer filters and still improves the SI-SNR of Conv-TasNet over the regular encoder. Another advantage is that, the encoder no longer needs to be learned, because the filters are already fixed. For MP-GTF the auditory gammatone filterbank was modified. For keeping short latency for causal application the impulse response of the filters had a length of 2 ms. The frequency range was adjusted to the sampling rate of 8000 Hz to a maximum frequency of 4000 Hz. As adaption to the non-negative input of the separation module, each filter includes a negative version to ensure there is at least one filter containing energy. Also, for each centre frequency, several filters were generated for different phase shifts [5]. Output of the Conv-TasNet should be the clean single channel sources. So Conv-TasNet should perform dereverberation of the signal in addition to separation. To consider the time delay caused by the BRIR during training, sources were convolved with the first direct sound peak.

3 Results and Discussion

Conv-TasNet has achieved an SI-SNR improvement from a mean of 0 dB to 13.7 dB after 100 epochs of training. The SI-SNR improvement of training and test data in each epoch is shown in Fig. 3. The course shows no signs of overfitting.

This would mean that the network is over-specified on the training data and would result in a poorer performance on the test data. Also, both curves still show a slight incline, which shows that further training would still be possible. Anyway, the following comparable data were also determined after 100 epochs. The SI-SNR improvement is lower than the performance with the common wsj0-2mix dataset, based on read speech from Wall Street Journal [8], of 15.4 dB [1]. wsj-2mix shows very good results with the Conv-TasNet, however, the application of neural networks trained with wsj-2mix dataset showed low generalization with other datasets [2] [9]. A better generalizability of LibriMix was shown by [2], which explains why the Conv-TasNet performs poorer than with the wsj0-2mix. Conv-TasNet trained by [2] with a more than three times larger trainings dataset of LibriMix yielded a scale invariant signal to distortion ratio of 14.7 dB. The larger dataset requires significantly more time capacity to train Conv-TasNet. To prepare for the expansion of the network and the data set, the smaller one is adequate.

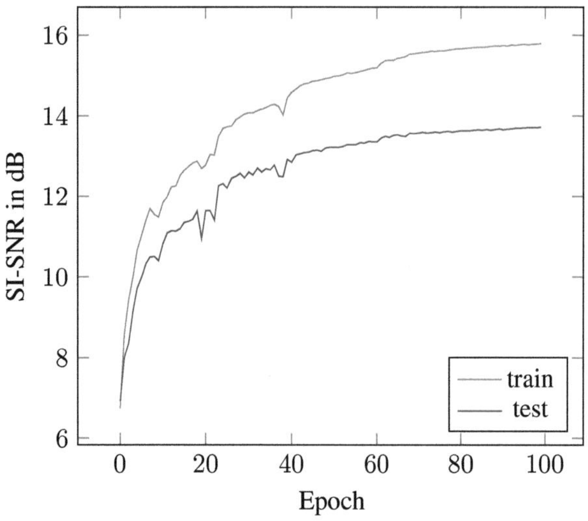

Figure 3: Development of accuracy over the epochs.

4 Conclusion

Conv-TasNet showed a high SI-SNR improvent of 13.7 dB for single-channel speech seperation with the LibriMix data set. A method was presented to improve Conv-TasNet for more realistic scenarios. The encoder of Conv-TasNet was replaced by a static multi-phase gammatone filterbank with a two-channel input. The adjusted Conv-TasNet was applied to the open source dataset LibriMix extended for more realistic scenarios by connecting it with a database of BRIR measured in everyday environments. Future experiments will show whether the adjustment of Conv-TasNet will provide benefits for more realistic scenarios.

Acknowledgement

The work has been carried out and supervised by the Institute for Signal Processing, Universität zu Lübeck.

Author's Statement

Conflict of interest: Authors state no conflict of interest.

5 References

[1] Y. Luo and N. Mesgarani, "Conv-tasnet: Surpassing ideal time-frequency magnitude masking for speech separation," *IEEE/ACM Transactions on Audio, Speech, and Language Processing*, vol. 27, no. 8, pp. 1256–1266, 2019.

[2] J. Cosentino, M. Pariente, S. Cornell, A. Deleforge, and E. Vincent, *LibriMix: An Open-Source Dataset for Generalizable Speech Separation*, 2020.

[3] H. Kayser, S. D. Ewert, J. Anemüller, T. Rohdenburg, V. Hohmann, and B. Kollmeier, "Database of multichannel in-ear and behind-the-ear head-related and binaural room impulse responses," *EURASIP Journal on Advances in Signal Processing*, vol. 2009, no. 1, pp. 1–10, 2009.

[4] C. Han, Y. Luo, and N. Mesgarani, "Real-time binaural speech separation with preserved spatial cues," in *ICASSP 2020 - 2020 IEEE International Conference on Acoustics, Speech and Signal Processing (ICASSP)*. IEEE, 04.05.2020 - 08.05.2020, pp. 6404–6408.

[5] D. Ditter and T. Gerkmann, "A multi-phase gammatone filterbank for speech separation via tasnet," in *ICASSP 2020 - 2020 IEEE International Conference on Acoustics, Speech and Signal Processing (ICASSP)*. IEEE, 04.05.2020 - 08.05.2020, pp. 36–40.

[6] M. Kolbaek, D. Yu, Z.-H. Tan, and J. Jensen, "Multitalker speech separation with utterance-level permutation invariant training of deep recurrent neural networks," *IEEE/ACM Transactions on Audio, Speech, and Language Processing*, vol. 25, no. 10, pp. 1901–1913, 2017.

[7] K. Iida, *Head-Related Transfer Function and Acoustic Virtual Reality*. Singapore: Springer Singapore Pte. Limited, 2019.

[8] J. R. Hershey, Z. Chen, J. Le Roux, and S. Watanabe, "Deep clustering: Discriminative embeddings for segmentation and separation," in *2016 IEEE International Conference on Acoustics, Speech and Signal Processing (ICASSP)*. IEEE, 20.03.2016 - 25.03.2016, pp. 31–35.

[9] B. Kadıoğlu, M. Horgan, X. Liu, J. Pons, D. Darcy, and V. Kumar, "An empirical study of conv-tasnet," in *ICASSP 2020 - 2020 IEEE International Conference on Acoustics, Speech and Signal Processing (ICASSP)*, 2020, pp. 7264–7268.

Development of a systematic procedure for creating new lists for the Freiburg monosyllabic speech test

Thomas Schwarz [1], Marlitt Frenz [2], and Hendrik Husstedt [2]

[1] Auditory Technology, Universität zu Lübeck, thomas.schwarz@student.uni-luebeck.de
[2] German Institute of Hearing Aids, Lübeck, {m.frenz, h.husstedt} @dhi-online.de

Abstract

The Freiburg speech test is a commonly used speech test in German-speaking countries. The test material of the Freiburg monosyllabic speech test partly contains words which are currently not used frequently in everyday life. In this paper, a possible systematic procedure for compiling new test lists for the Freiburg monosyllabic speech test is presented. Different, current language corpora were automatically searched for monosyllabic nouns. The words were sorted according to criteria for a speech intelligibility test and target values for a list composition corresponding to the specifications of DIN EN ISO 8253-3:2012-08 were formulated. According to the presented procedure, new lists for the Freiburg monosyllabic speech test can be generated. To this day, the project is still in process and the performing acuracy and speed as well as the amount of output is still to be discussed.

1 Introduction

Speech audiometry is an important criterion for the fitting of hearing aids and is used for differential diagnostics in the field of hearing disorders. In German-speaking countries the Freiburg monosyllabic speech test (FMST), developed by Hahlbrock, is the most used test for this purpose [1]. The result of the FMST and the result of the pure tone audiometry are decisive for the indication of hearing aids and thus for getting reimburstment by the National Association of Statutory Health Insurance Funds, according to the Heil- und Hilfsmittelrichtlinie §21/22 [2]. The test inventory is standardized in DIN 45621-1:1995-08 and consists of 20 lists including 20 monosyllabic nouns, which are presented via headphones or via loudspeaker in a free field according to the definition of test [3]. Speech comprehension is measured as the number of correctly repeated test stimuli to the number of presented words in a test list.

The voice recordings for the test material of the FMST as it is used today date back to 1969 [1]. In the past, the FMST has been criticized on various points. A central aspect of the critics is the lack of topicality of the words of the test in their usage, but also psychological inhibitions to repeat the words directed at the examiner, such as "Schwein" (pig), regional language peculiarities, such as "Grog" (grog) and contextual reference were criticized, e.g., [4], [5]. Since language changes over time and there are new, easier ways to generate speech via text-to-speech-synthesis, the production of new test material is necessary and feasible with little effort. In this paper, a structured procedure for creating new lists of monosyllabic nouns for the speech test is presented with the relevant aspects to be considered. Starting with the selection of possible speech corpora and the extraction of monosyllabic nouns. Then unsuitable words are sorted out and the words are grouped into lists taking into account the number of phonemes and phonemic balance.

2 Material and Methods

2.1 Requirements for Freiburg monosyllabic speech test

For FMST, the requirements according to DIN EN ISO 8253-3:2012-08 apply. Here it is defined that the test lists of the testmaterial must be phonemically balanced. This means that the distribution of the phonemes of the speech material should be as close as possible to the phoneme distribution of the German language. Furthermore, the standard requires the perceptual balance of the test lists. It means that the result of the test must not depend on the selection of the test list. [6]

Other requirements that apply to the FMST in particular, apart from those relevant to voice recording, were formulated by Hahlbrock when the language test was created. Optimally, the monosyllabic nouns should be equally known throughout the German-speaking area. Each list has to include the same number of phonemes and shall use a similarly structure in terms of word length. In addition, Hahlbrock also paid attention to an uniform distribution of initial phonemes in the lists, since he stated this is important for speech understanding and the perceptual balance of the test lists. [7]

2.2 Choice of corpora

To meet the requirement for a speech intelligibility test to contain only words from speech actually used in everyday life, different speech corpora should be searched for potential test material. The speech corpora used to compile the lists of monosyllables were required to contain current language from the last two decades, to consist of free everyday speech, to be standard German and to include a frequency indication for the occurrence of the words in everyday speech. Optimally, the speech corpus should be spoken German.

2.3 Extracting monosyllabic nouns

The selection of monosyllabic nouns from the speech corpora and subsequent processing was performed using an individual Matlab script with version Matlab R2020a.

To extract monosyllabic nouns from a speech corpus, the respective corpus was first searched for capitalized words in order to filter out words containing only one vowel letter (including "y"), one diphthong, or two consecutive monophthongs. In order to sort out abbreviations and meaningless words from the monosyllable collection, as they occur depending on the composition of the corpora, the collected words were checked again for contained special characters and inner majuscules.

The extracted monosyllables were matched, ensuring that each monosyllable appears only once in the collection.

2.4 Sorting of the monosyllabic nouns

Even though it is impossible to make the word material of the speech test equally difficult for all regions, social classes, educational levels and age groups, this goal should be pursued [7]. For this reason, the collected monosyllables were checked one by one according to the following criteria.

As it is known, anglicism are common in the German language nowadays. However, particular words add new phonemes to the speech corpora, e.g. the phoneme "th" as in "thrill". Additionally, anglicism are more popular in the native tongue of younger people, than elder ones. By that, it was decided to sort them exclude them of the collection.

In the case of corpora with spoken language, filler words such as "Ähm" are occasionally transcribed as well. These tokens sometimes meet the criteria for automatic extraction of monosyllabic nouns and must therefore be sorted out.

The same holds for unambiguous expletives, regionalisms, technical language, fashionable language, lyricism and fantasy words. Numerals were also excluded, since a related test, the Freiburg multi-syllabic speech test, consists of numbers only, which could lead to confusions. The information for this revision was taken from the Duden (dictionary of the German language). Care was taken to ensure that the selected monosyllables were in the basic, colloquial, or elevated vocabulary of the Duden.

2.5 Grouping of the collected monosyllabic nouns into lists

The composition of the lists, in terms of number and length of the monosyllables, is based on the structure of the original lists of the FMST defined by Hahlbrock, as shown in Fig. 1 [7]. In order to be able to determine the length of the words in phoneme number, the collected monosyllables had to be transformed in phoneme transcription. In the compilation of the words into lists, attention was also paid to the phonemic balance of the test words according to DIN EN ISO 8253-3:2012-08 [6].

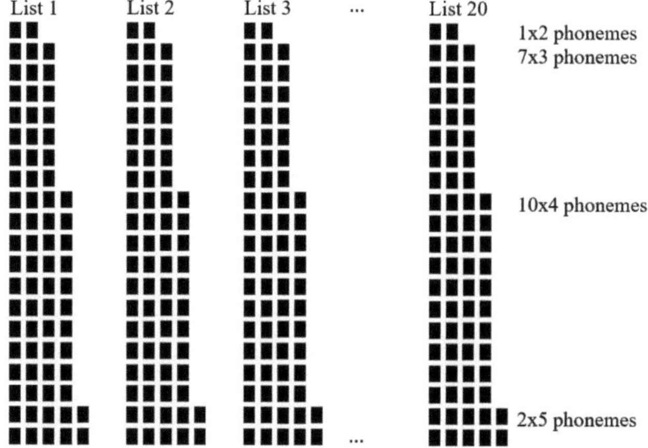

Figure 1: Basic structure of the 20 lists of the Freiburg monosyllable test with representation of the word length. Each rectangle corresponds to a phoneme. Lists 5 and 15 differ by 1 phoneme at the shortest word.

3 Results and Discussion

3.1 Choice of corpora and extraction monosyllabic nouns

As corpora with spoken language, the speech corpora of the Database for Spoken German (DGD) were accessed. The largest amount of data here was provided by the "Forschungs- und Lehrkorpus Gesprochenes Deutsch (FOLK)-corpus". As a conversation corpus consisting of private, institutional, and public environments transcribed conversations, it meets all the conditions imposed on the speech corpus in subsection 2.2 for the extraction of monosyllables. Nevertheless the word inventory of the FOLK corpus alone was not sufficient as the sole source to generate an adequate number of new test lists. Other corpora used from this database for the extraction of the monosyllables all include free speech from the last two decades. Individual subject areas that are likely to be clustered in some of the speech corpora differ between the corpora (For example, care was taken in the selection of corpora not to take only corpora from war narratives.) [8].

By adding other written corpora in the further course, a heterogeneous language corpus can finally be assumed. A total of 1329 monosyllables were extracted automatically from the DGD corpora.

The largest total number of monosyllables was obtained from the written "German mixed typical speech corpus" of the Leipzig Corpora Collection (LCC) from 2011. For this corpus, the material is taken from various written sources, such as Wikipedia, public web, random webpages, newscrawl (can also include older material than 2011) and news [9]. A total of 7210 monosyllables were extracted automatically from the LCC corpora.

Additional words were added to the monosyllable collection from the word rank lists of the "Deutsches Referenz Korpus" (DeReKo). This word rank list is a collection of different word form and basic form lists with a special consideration of the frequency of use [10]. A total of 2326 monosyllables were extracted automatically from the DeReKo corpora.

After matching the total extracted monosyllables from the different speech corpora, 8133 different tokens resulted.

3.2 Sorting of the monosyllabic nouns

The monosyllables were filtered again by a person according to the criteria mentioned in subsection 2.4, so that ultimately 805 different monosyllables are available as potential test material.

Despite clearly formulated criteria for sorting out unsuitable words, there were uncertainties in dealing with individual words. In the case of city and river names such as "Köln", "Bern" or "Rhein" there were considerations about using the words as test material, since it can be assumed that they are generally known. However, they also represent a regionalism in the respective and surrounding area, so that they were finally sorted out. Other regionalisms that were excluded are for example "Schmarrn" or "Gracht"(dutch canal). As short forms, for example, the words "Lok" as an abbreviation for "Lokomotive" (Locomotive) and "Klar" as a short form for "Eiklar" (Egg white) were taken out. In addition to the words that were clearly defined in advance to be sorted out, names of fantasy creatures were also taken out of the word collection, since they can only be recognized depending on prior knowledge, such as "Sphinx" (Sphinx), "Schlumpf" (smurf), "Ork" (orc), "Troll" (troll) or "Greif" (griffin).

The dominant number of tokens that were sorted out from the collection were meaningless strings of letters that met the criteria for automated recognition of monosyllabic nouns as described in subsection 2.4. At this point, if there is another trial with more language corpora, it would probably make sense to collect, for example, only words as monosyllables that are present in at least one-third of the language corpora.

In principle, an attempt should be made to make the entire step independent of any person and to automate it. One possible approach would be an application programming interface (API) that can be used to access an online dictionary in order to automatically read out the existence and usage of a word. Such a process would save time and be more objective.

3.3 Grouping of the collected monosyllabic nouns into lists

In order to determine the words according to their length in phoneme number, they first had to be transcribed into their phonemes. For this purpose, the online tool "G2P" (Grapheme to Phoneme) of the Bavarian Archive for Speech Signals was used [11]. For processing with Matlab, the monosyllables were converted into a Sampa script, a computer-readable phonetic alphabet. In Sampa, each phoneme is represented by one or two strings, for example, the German word "Zahn" would be formulated "t s a: n" in Sampa. The transcribed words were randomly sampled to check the accuracy of the transcription. No incorrect transcription could be detected.

Divided by length in phoneme number, one gets a good overview of the number of available monosyllables per length, shown in Table 1. The proportions of words of different length correspond well to the proportions targeted by Hahlbrock in the list composition, as shown in Fig. 1. With 25 monosyllables consisting of 2 phonemes, these are a limiting factor in the list composition. More than 25 lists cannot be compiled from the collected material under the given specifications. Therefore, it can be a reasonable approach to neglect this requirement, e.g., by using a one-phoneme word ("Ei" [egg]) or three-phonemes word instead.

Table 1: resulting number of monosyllables sorted by their number of phonemes

Number of phonemes	Monosyllable count
2 phonemes	25 words
3 phonemes	254 words
4 phonemes	356 words
5 phonemes	141 words

In order to be able to fulfill the phonemic balance in the test lists required in DIN EN ISO 8253-3:2012-08 for speech tests, a target distribution of the phonemes was first required. The values should indicate the proportion of each phoneme in speech spoken in everyday life. These values were taken from the phoneme statistics of the Verbmobil project [12]. The project ranged from 01/01/1993 to 08/31/2000 and includes 4387118 words. There were 47 phonemes distinguished in this observation. The "glottal stop" (consonantal sound produced by obstructing airflow in the glottis) was included in the transcription to Sampa by the G2P online-tool, but was not considered in the evaluation of the monosyllable length, nor in the calculation of the percentage frequency. Since it is unrealistic to achieve a balanced distribution of 47 different phonemes in lists containing 73 phonemes each, seven phoneme groups were formed as suggested in DIN EN ISO 8253-3:2012-08: voiced plosives, voiceless plosives, voiced fricatives, voiceless fricatives, nasal consonants, long vowels and short vowels. The phoneme groups with associated target percentage per list is shown in Table 2.

Table 2: Phoneme groups with their share of the total speech corpus in percent and target values for phonemically balanced lists of monosyllabic nouns formulated on this basis.

Phoneme group	Percentage	Target interval
voiced plosives	7.8 %	6 ± 4
voiceless plosives	12.3 %	9 ± 4
voiced fricatives	10.9 %	8 ± 4
voiceless fricatives	14.6 %	11 ± 4
nasal consonants	15.6 %	11 ± 4
long vowels	11.2 %	8 ± 4
short vowels	27.6 %	20 ± 4

The target interval specified in Table 2 refers to the absolute number of phonemes in the respective phoneme class that may occur in a list of the FMST in order to obtain a phoneme distribution similar to the German language. For example, to produce a distribution of voiced plosives in a list similar to the German language, exactly 6 of the total 73 phonemes in the list would have to be voiced plosive phonemes. However, a deviation of ± 4 phonemes is accepted. The interval size is basically variable and should be discussed after attempts with different sizes.

The goal in the following is to distribute the monosyllables among the lists following the structure of Fig. 1 and to obtain a phoneme group distribution within the list, where all target intervals from Table 2 are met. In the course of this search, various interval limits are tried out in order to allow as little deviation as possible and necessary in the end. In a first try, already 16 lists were found that matched the criteria.

4   Conclusion

The procedure described in this paper for the creation of new lists for the FMST, to extract monosyllables from speech corpora, to check them for certain criteria for use in a speech test, and then to arrange them into lists according to the construction scheme already used by Hahlbrock, leads to good results.

It has not yet been finally clarified, how many lists can be created taking into account the requirements of DIN EN ISO 8253-3:2012-08. Future goals are comparing the monosyllables automatically with a dictionary in favor of objectivity and optimize the time of work.

Acknowledgement

The work has been carried out at German Institute of Hearing Aids, Lübeck. I would like to thank my supervisors, who accompanied me with great commitment during my work.

Author's Statement

Conflict of interest: Authors state no conflict of interest.

5   References

[1] A. Winkler and I. Holube, *What do we know about the Freiburg speech test?*. Oldenburg, Audiological Acoustics, vol. 53 pp. 146–154, 2014.

[2] G-BA, *Richtlinie des Gemeinsamen Bundesausschusses über die Verordnung von Hilfsmitteln in der vertragsärztlichen Versorgung*, Bundesanzeiger (BAnz AT 15.04.2021 B3), 2021.

[3] German Institute for Standardization e.V. *DIN 45621-1:1995-08, Sprache für Gehörprüfung - Teil 1: Ein- und mehrsibige Wörter*, Beuth Verlag GmbH, 1995.

[4] T. Steffens, *Verwendungshäufigkeit der Freiburger Einsilber in der Gegenwartssprache*, HNO, vol 64 pp. 549-556, 2016.

[5] H. Bangert, *Probleme bei der Ermittlung des Diskriminationsverlustes nach dem Freiburger Sprachtest*, Audiologische Akustik, vol 19 pp. 166-170, 1980.

[6] German Institute for Standardization e.V., *DIN EN ISO 8253-3:2012-08, Akustik - Audiometrische Prüfverfahren - Teil 3: Sprachaudiometrie*, Beuth Verlag GmbH, 2012.

[7] K.-H. Hahlbrock, *Grundlagen und praktische Anwendungen einer Sprachaudiometrie für das deutsche Sprachgebiet*, Georg Thieme Verlag Stuttgart, 1957.

[8] T. Schmidt, *DGD - die Datenbank für Gesprochenes Deutsch*, Zeitschrift für germanistische Linguistik, Walter de Gruyter GmbH, vol. 45 pp. 451-463, 2017.

[9] D. Golhahn, T. Eckart and U. Quasthoff, *Building Large Monolingual Dictionaries at the Leipzig Corpora Collection: From 100 to 200 Languages*, Proceedings of the 8th International Language Resources and Evaluation (LREC'12), 2012.

[10] Leibniz-Institut für Deutsche Sprache, *DeReWo - Korpusbasierte Wortlisten Stand:2012*, http://www.ids-mannheim.de/kl/projekte/korpora/, accessed 17.01.2022 23:13.

[11] U.D. Reichel and T. Kisler, *Language-independent grapheme-phoneme conversion and word stress assignment as a web service*, Elektronische Sprachverarbeitung. Studientexte zur Sprachkommunikation, TUDpress, Dresden, vol 71, pp 42-49, 2014.

[12] Bayerisches Archiv für Sprachsignale,*BAStat - Statistics of Conversational German*, Ludwig-Maximilians-Universität Müchen, https://www.phonetik.uni-muenchen.de/forschung/Bas/BasPHONSTATeng.html #PhonBi, accessed 17.01.2022 23:20

Using Monte-Carlo simulations to optimize an adaptive Bayesian algorithm for a spatial hearing study

Stefanie Goicke [1, 2], Florian Denk [3] and Tim Jürgens [2]

[1] Student of Auditory Technology, Universität zu Lübeck, stefanie.goicke@student.uni-luebeck.de
[2] Institute of Acoustics, University of Applied Sciences Lübeck, tim.juergens@th-luebeck.de
[3] German Institute of Hearing Aids, Lübeck, f.denk@dhi-online.de

Abstract

The QUEST algorithm is a Bayesian adaptive algorithm for measuring perception thresholds in psychoacoustic experiments. Here it is used in a spatial hearing study to estimate the distance between a sound source relative to two other sound sources. The evaluation of the QUEST algorithm is the preparation for spatial hearing tests with subjects to find out if individual acoustic directional cues have an effect on allocentric localisation ability. To this end, the different parameters of the algorithm were evaluated in this paper to find the best conditions for the measurements by using Monte-Carlo simulations with 500 iterations for each combination of parameters. The best was achieved for a psychometric function slope of $15\%/°$, a starting point of $8°$ away from the real threshold and an estimated standard deviation of $10°$ are the best parameters for all simulated psychometric functions, because of the smallest standard deviation and amount of outliers that did not reach the real threshold.

1 Introduction

In psychoacoustic measurements, there are various different methods to find reliable thresholds. One method is to use adaptive algorithms to get a successive approximation of the threshold. While doing so, the task is made more difficult, as long as the subject responds correctly, whereas the task is made easier, when the subject's response is wrong. One of these adaptive algorithms is the QUEST algorithm [1]. It is a Bayesian adaptive algorithm which applies the maximum likelihood estimation to optimally vary the next stimulus and which considers prior knowledge about the underlying psychometric function or measured data to estimate the threshold.

The QUEST algorithm has already been used for several psychoacoustic measurements, including spatial hearing experiments, e.g., [2] tested spatial hearing of blindfolded subjects in two conditions. The first task was a minimum audible angle task, where subjects heard two consecutive stimuli and rated which one was perceived more to the right. In this task no internal spatial representation is required, subjects use their egocentric frame, which means that they referred to their body as the center of environment [3]. The second task was a bisection task in which subjects heard three consecutive stimuli and rated if the second stimulus was perceived closer to the left or to the right reference stimulus. This kind of spatial hearing tasks requires an internal representation and the usage of their allocentric frame, which means that sound sources are localised using an external object as reference [3]. Previous studies found that individuals mostly rely on their egocentric frames when carrying out spatial hearing tasks [3]. Visually handicapped

people show deficits in spatial hearing when using their allocentric frame, but they perform similarly to sighted people in egocentric tasks [3]. It is still not clear, if or how much visual input is required to calibrate spatial hearing. This cross-sensory calibration seems to depend on the complexity of the spatial hearing task [3], [4], [5].

Aspects of spatial hearing can also be simulated for headphone presentation using virtual acoustics, i.e., convolving stimuli with the transfer functions between a source and the subjects' ear to create the impression of a sound source at a desired position in space. Usually, non-individual Head-Related Transfer Functions (HRTF) are used for many daily life applications such as virtual reality headsets. The microphone position during the recordings of the HRTF is important for the preservation of spectral cues [6] and thereby for the localization ability when using non-individual HRTF [7]. Spatial hearing experiments can also be done using virtual acoustics. For example, [9] investigated spatial hearing of visually handicapped and sighted subjects using an experiment in virtual acoustics where subjects discriminated whether two successive stimuli, which were convolved with dummy head HRTF, were perceived from the same or from different directions. Subjects used their egocentric frame, so that no significant differences between visually handicapped and sighted people was found.

To the authors´ knowledge, no research on the influence of individual acoustic directional cues on allocentric localization ability was done so far. The goal of this study is to evaluate and optimize the QUEST algorithm using Monte-Carlo methods for a prospective spatial hearing study to be done as a research project. The research question of the project will be if individual acoustic directional cues have an effect on the allocentric localization ability, which will be tested with blindfolded and visually handicapped people.

The experiment is based on [2] and will be performed in real and in virtual acoustics. Therefore the best parameters to assess the real threshold and a similar psychometric function as in [2] have to be found. Thus, the QUEST algorithm and the adjustable parameters have to be evaluated.

2 Material and Methods

The spatial hearing experiment will be placed in the anechoic chamber of the University of Applied Sciences in Lübeck. A sketch of the setup for the real acoustic measurements can be seen in Fig. 1, which is adapted from [2]. Subjects are sitting in a circular arc of 13 loudspeakers at a distance of three meters, spanning $\pm24°$ with a spacing of $4°$. Blindfolded subjects will hear three successive noise bursts as stimuli, the first coming from one of the first three loudspeakers on the left serving as the first reference. The second is the probe stimulus which comes from a random loudspeaker. The third stimulus serves as a second reference and is always 10 loudspeakers apart of reference 1. For the virtual acoustic measurements the setup will be simulated using a HRTF-database [10] and stimuli will be presented through headphones.

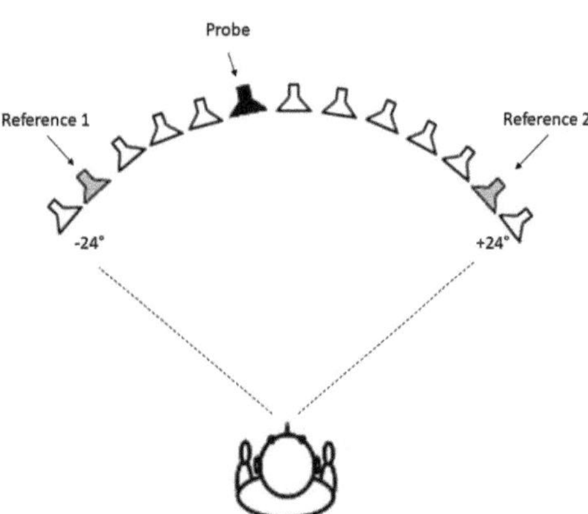

Figure 1: Experimental setup for spatial hearing study, adapted from [2]: subjects sit in circular arc of 13 loudspeakers ($\pm24°$ in steps of $4°$) and rate which distance between stimuli is smaller (reference 1 - probe or probe - reference 2).

In this bisection task, subjects rate whether the probe stimulus has a smaller perceived distance to reference 1 or 2. Therefore they have to use their allocentric frame. The position of the probe stimulus is determined by the Bayesian adaptive QUEST algorithm [1] that finds the threshold for a reliable detection of relative differences between the sound sources. This algorithm considers prior knowledge of the psychometric function (see Fig. 3) and already tested conditions. For each trial the QUEST algorithm estimates the threshold by calculating the maximum likelihood estimation and sets this estimate as the next measured threshold.

As an adjustment to the hardware setup which will be used in the research project with loudspeaker angles at $4°$ separation, this threshold is rounded to the nearest angle of a loudspeaker position. This procedure is applied for 40 times, to reach the real threshold most accurately. The probability density function of the threshold used inside the QUEST algorithm is visualized in Fig. 2. Dashed lines represent the maximum likelihood estimation of all 40 trials, the solid line is the maximum likelihood estimation after trial 40 and represents the final estimated threshold.

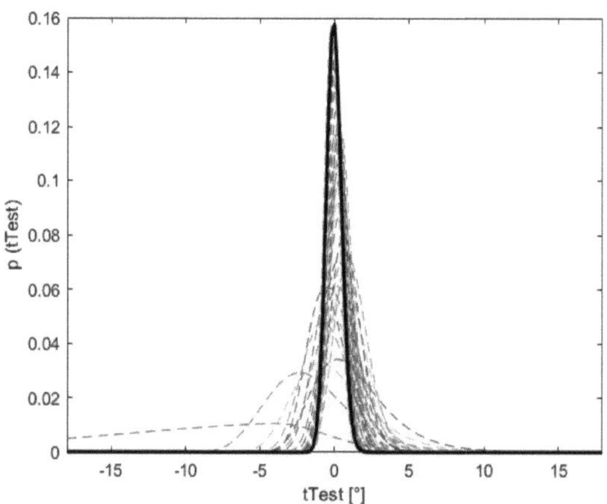

Figure 2: Development of probability density function (PDF) for 40 trials using the QUEST algorithm to calculate the maximum likelihood estimation; the position of the maximum of the PDF of the solid line represents the estimated threshold after 40 trials.

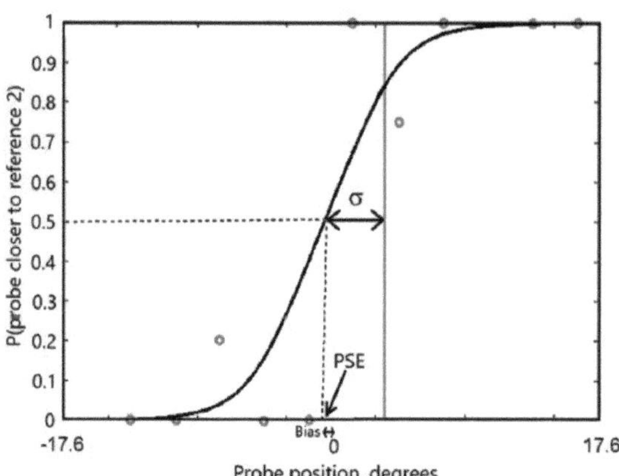

Figure 3: Determined psychometric function of [2] for the spatial bisection task with a slope of approximately $10\%/°$ and the standard deviation σ (licensed under Creative Commons CC BY).

The QUEST algorithm is implemented in the publicly available psychophysic toolbox for MATLAB (The MathWorks) [11]. Here it was evaluated for the spatial hearing study by using 500 Monte-Carlo simulations for each combination of parameters. The real threshold $tActual$ is defined as a positive value and it is the relative distance to reference 2 and is here tested for distance from $1°$ to $10°$ because the real threshold is not yet known. An initial threshold *tGuess* ($0° : 4° : 12°$), the estimation of thresholds' standard deviation *tGuessSD* ($10°, 14° or 18°$) over all simulations as well as the slope of the underlying psychometric function ($5\%/°, 10\%/°, 15\%/° or 20\%/°$) are estimated. The response of the virtual subject is simulated on the basis of the underlying psychometric function. The next threshold is estimated by the QUEST algorithm using the maximum likelihood estimation with the latest threshold and the subject's response. A comparable psychometric function from [2] is shown in Fig. 3. Marked points (o) are the tested values for one exemplary subject and the measured standard deviation σ. The determined threshold is marked by the vertical line at a probability of 80% and leads to a threshold of approximately $7°$.

3   Results and Discussion

Fig. 5 shows boxplots of the *tTest* thresholds as an output of the QUEST algorithm for different slopes of the psychometric function (different panel columns), different standard deviations (SDs), and different real underlying thresholds (tActual, abscissa of each panel). The closer the initial guess of threshold *tGuess* is to the real threshold, the better matches the real threshold. Based on the results of [2] and their measured threshold of approximately $7°$, best results were achieved when using *tGuess* $= 8°$.

The median *tTest* values (red lines in each panel of Fig. 5) match well with the real thresholds *tActual* across all conditions. However, the spread of different *tActual* outcomes varies considerably across panels. A small slope of $5\%/°$ leads to a large variance of reached thresholds *tTest* and as a result to a high root-mean-square-error (RMSE) of approximately $2°$ depending on the standard deviation. Steeper slopes lead to smaller variances, but there are more outliers which means there are more *tTest* values outside of the 1.5 times interquartile range. The RMSE is the smallest for a slope of $20\%/°$ and a SD of $14°$.

Fewer outliers, approximately four, and a little higher RMSE of $0.67°$, are reached for a slope of $15\%/°$ and a SD of $10°$. However the RMSE is not the smallest of the simulations, this condition seems to receive the best results, because it is a compromise of a small RMSE and a small amount of outliers. These chosen parameters are now used to visualize the psychometric function of the QUEST algorithm. The result is shown in Fig. 4. All tested thresholds and their simulated answers (x: wrong; o: right) are marked. Compared to the psychometric function of [2] in Fig. 3 the slope is steeper. For a similar psychometric function as in [2] a slope of $10\%/°$ would be necessary, which would lead to a higher RMSE according to Fig. 5.

The evaluation of the QUEST algorithm is based on simulations which can explain the variability of the measured data from [2]. The response of real subjects is based on their perception and leads to an individual probability distribution for each subject. The underlying probability density function for the simulated answer of a fictional subject in the QUEST algorithm is a simulation for an average subject and is not necessarily comparable to real subjects answers which may lead to differences in psychometric functions. Since there is a resolution of $4°$ between two loudspeakers in the experiment and suggested thresholds have to be rounded to this resolution, an increased RMSE may be possible in the actual experiment, because of a smaller amount of outliers.

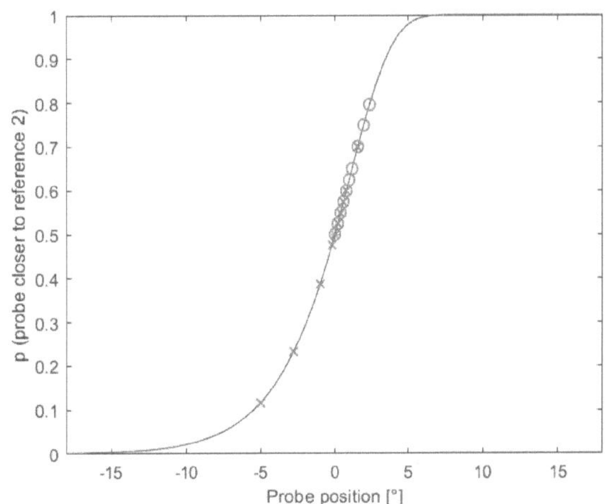

Figure 4: Simulation of psychometric function in QUEST algorithm with slope $15\%/°$; tested thresholds are marked (o: right; x: wrong).

4   Conclusion

In the QUEST algorithm, there are many parameters to adjust its performance to an experiment. In this paper the QUEST algorithm was evaluated for a spatial hearing experiment using several loudspeakers. The best condition of parameters found here is a standard deviation of $10°$ and a starting point for measurements of $8°$ to reach the real threshold after 40 trials. The best fit is reached for a slope of the underlying psychometric function of $15\%/°$. These settings of parameters are promising to approach the real threshold reliably with a small root-mean-square error.

Acknowledgement

The work has been carried out at the University of Applied Sciences in Lübeck and is supported by the institute of acoustics. We also thank the reviewers for helpful comments on an earlier version of this paper.

Author's Statement

Conflict of interest: Authors report no conflict of interest.

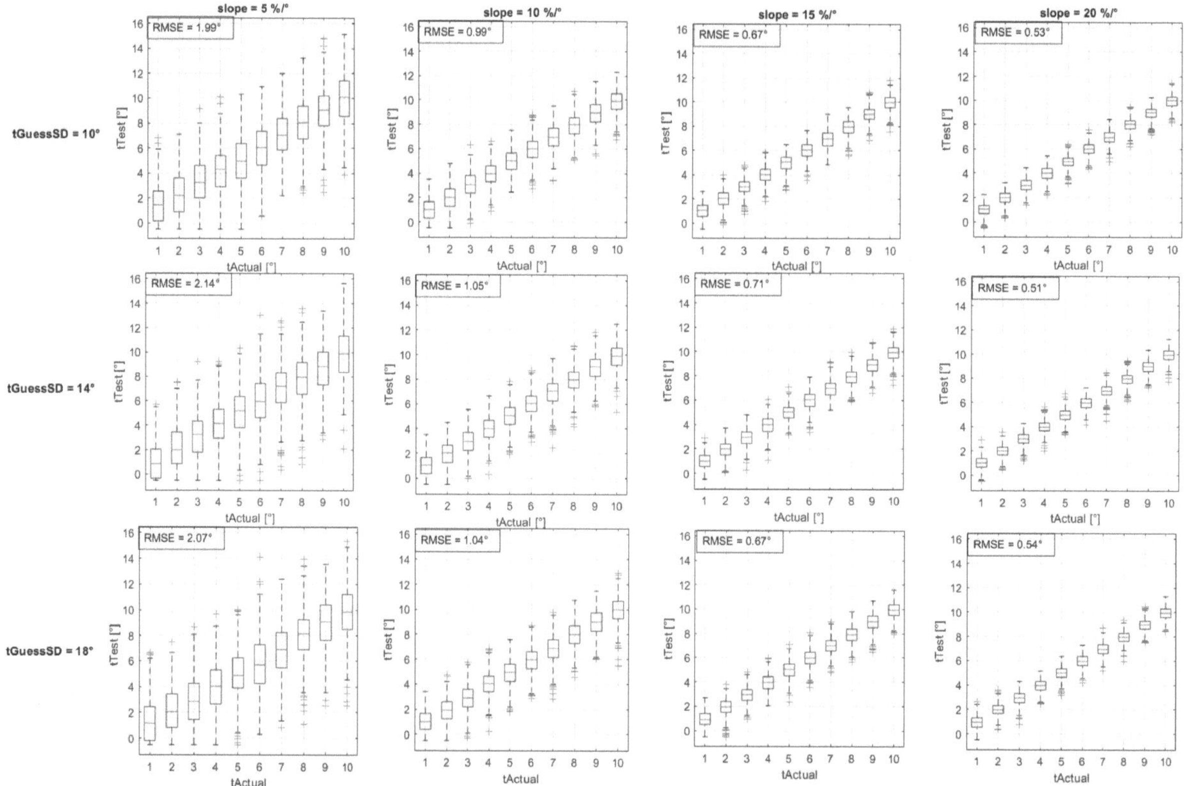

Figure 5: Boxplots for different values of parameters in the QUEST algorithm: determined thresholds *tTest* for given real thresholds *tActual* for several slopes of psychometric functions (panel columns) and standard deviations (rows of panels); The root-mean-square-error (RMSE) is given in the top left corner of each panel as a value for accuracy of QUEST algorithm.

5   References

[1] A. B. Watson and D. G. Pelli, *QUEST: A Bayesian adaptive psychometric method*, Perception psychophysics 33.2 (1983): 113-120.

[2] E. Aggius-Vella, A. J. Kolarik, M. Gori, S. Cirstea, C. Campus, B. C. J. Moore and S. Pardhan, *Comparison of auditory spatial bisection and minimum audible angle in front, lateral, and back space*, Scientific reports 10.1 (2020): 1 9.

[3] P. Voss, *Auditory spatial perception without vision*, Frontiers in psychology 7 (2016): 1960.

[4] M. Gori, G. Sandini, C. Martinoli and D. C. Burr, *Impairment of auditory spatial localization in congenitally blind human subjects*, Brain 137.1 (2014): 288-293.

[5] E. Aggius-Vella, C. Campus and M. Gori, *Different audio spatial metric representation around the body*, Scientific reports 8.1 (2018): 1-9.

[6] F. Denk, S. D. Ewert and Birger Kollmeier, *Spectral directional cues captured by hearing device microphones in individual human ears*, The Journal of the Acoustical Society of America 144.4 (2018): 2072-2087.

[7] F. Denk, S. D. Ewert and B. Kollmeier, *On the limitations of sound localization with hearing devices*, The Journal of the Acoustical Society of America 146.3 (2019): 1732-1744.

[8] F. L. Wightman and D. J. Kistler, *Headphone simulation of free-field listening. I: stimulus synthesis*, The Journal of the Acoustical Society of America 85.2 (1989): 858-867.

[9] S. Goicke, J. Tchorz and T. Jürgens, Examination of the relative localization ability in the horizontal plane in blind and visually impaired people using headphones and virtual acoustics ;*Untersuchung der relativen Lokalisationsfähigkeit in der Horizontalebene bei blinden und sehbehinderten Personen über Kopfhörer mittels virtueller Akustik*, Fortschritte der Akustik-DAGA (2021).

[10] H. Wierstorf, M. Geier and S. Spors, *A free database of head related impulse response measurements in the horizontal plane with multiple distances*, Audio Engineering Society Convention 130. Audio Engineering Society, 201, e-Brief 6

[11] D. H. Brainard, *The psychophysics toolbox*, Spatial vision 10.4 (1997): 433-436.

Infinite Science